PHYSICAL CONSTANTS

Quantity	Symbol	Value
Universal gravitational constant	G	6.674×10^{-11} m^3/(kg·s^2)
Speed of light in vacuum	c	2.998×10^{8} m/s
Elementary charge	e	1.602×10^{-19} C
Planck's constant	h	6.626×10^{-34} J·s
		4.136×10^{-15} eV·s
	$\hbar = h/(2\pi)$	1.055×10^{-34} J·s
		6.582×10^{-16} eV·s
Universal gas constant	$R = N_A k_B$	8.314 J/(mol·K)
Avogadro's number	N_A	6.022×10^{23} mol^{-1}
Boltzmann constant	k_B	1.381×10^{-23} J/K
		8.617×10^{-5} eV/K
Coulomb force constant	$k = 1/(4\pi\epsilon_0)$	8.988×10^{9} N·m^2/C^2
Permittivity of vacuum (electric constant)	$\epsilon_0 = 1/(\mu_0 c^2)$	8.854×10^{-12} C^2/(N·m^2)
Permeability of vacuum (magnetic constant)	μ_0	$4\pi \times 10^{-7}$ T·m/A
Electron mass	m_e	9.109×10^{-31} kg
		0.000 548 580 u
Electron rest energy	$m_e c^2$	0.5110 MeV
Proton mass	m_p	1.673×10^{-27} kg
		1.007 276 5 u
Proton rest energy	$m_p c^2$	938.272 MeV
Neutron mass	m_n	1.675×10^{-27} kg
		1.008 664 9 u
Neutron rest energy	$m_n c^2$	939.565 MeV
Compton wavelength of electron	$\lambda_C = h/(m_e c)$	2.426×10^{-12} m
Stefan-Boltzmann constant	σ	5.670×10^{-8} W/(m^2·K^4)
Rydberg constant	$R = -E_1/(hc)$	1.097×10^{7} m^{-1}
Bohr radius of hydrogen atom	$a_0 = \hbar^2/(m_e k e^2)$	5.292×10^{-11} m
Ionization energy of hydrogen atom	$-E_1 = m_e k^2 e^4/(2\hbar^2)$	13.61 eV

Volume 1

THIRD EDITION

Physics

Alan Giambattista

Betty McCarthy Richardson

Robert C. Richardson

Mc Graw Hill Education

PHYSICS, VOLUME ONE, THIRD EDITION

Published by McGraw-Hill Education, 2 Penn Plaza, New York, NY 10121.

This book is printed on acid-free paper.

1 2 3 4 5 6 7 8 9 0 DOW/DOW 1 0 9 8 7 6 5

ISBN 978-0-07-734052-0
MHID 0-07-734052-3

Senior Vice President, Products & Markets: *Kurt L. Strand*
Vice President, General Manager, Products & Markets: *Marty Lange*
Vice President, Content Design & Delivery: *Kimberly Meriwether David*
Managing Director: *Thomas Timp*
Brand Manager: *Thomas Scaife*
Director, Product Development: *Rose Koos*
Product Developer: *Eve Lipton*
Marketing Manager: *Nick McFadden*
Director, Content Design & Delivery: *Linda Avenarius*
Program Manager: *Lora Neyens*
Content Project Managers: *Dana M. Pauley, Tammy Juran*
Buyer: *Laura M. Fuller*
Design: *David W. Hash*
Content Licensing Specialists: *Carrie Burger (Photo), DeAnna Dausener (Text)*
Cover Image: *©Mike Kemp/Getty Images/RF*
Compositor: *Laserwords Private Limited*
Printer: *R. R. Donnelley*

Library of Congress has cataloged the main title as follows:

Giambattista, Alan
Physics / Alan Giambattista (Cornell University), Betty McCarthy Richardson (Cornell University), Bob Richardson (Cornell University).—Third edition.
pages cm
ISBN 978-0-07-351215-0 (alk. paper)
1. Physics—Textbooks. I. Richardson, Betty McCarthy, author. II. Richardson, Robert C. (Robert Coleman), 1937–2013, author. III. Title.
QC21.3.G537 2015
530—dc23

2014026758

www.mhhe.com

About the Authors

Alan Giambattista grew up in Nutley, New Jersey. Although he started college as a piano performance major, by his junior year at Brigham Young University he decided to pursue a career in physics. He did his graduate studies at Cornell University and has taught introductory college physics ever since. When not found in the classroom or at the computer keyboard working on *Physics*, he can often be found at the keyboard of a harpsichord or piano. He has been a soloist with the Cayuga Chamber Orchestra and has given performances of the Bach harpsichord concerti at several regional Bach festivals. When the long upstate New York winter is finally over, he loves to sail on Cayuga Lake. Alan met his wife Marion in a singing group. They live in an 1824 parsonage built for an abolitionist minister, which is now surrounded by an organic farm. Besides making music and taking care of the house, cats, and gardens, they love to travel together. They also love to spend time with their charming granddaughter, Ivy.

Betty McCarthy Richardson was born and grew up in Marblehead, Massachusetts, and tried to avoid taking any science classes after eighth grade but managed to avoid only ninth grade science. After discovering that physics explains how things work, she decided to become a physicist. She attended Wellesley College and did graduate work at Duke University. While at Duke, Betty met and married fellow graduate student Bob Richardson and had two daughters, Jennifer and Pamela. Betty began teaching physics at Cornell in 1977 with Physics 101/102, an algebra-based course with all teaching done one-on-one in a learning center. From her own early experience of math and science avoidance, Betty has empathy with students who are apprehensive about learning physics. Betty's hobbies include collecting old children's books, reading, enjoying music, travel, and dining with royalty. A highlight for Betty during the Nobel Prize festivities in 1996 was being escorted to dinner on the arm of King Carl XVI Gustav of Sweden. Currently she is spending time enjoying grandsons Jasper (once the 1 m child in Chapter 1), Dashiell and Oliver (the twins of Chapter 12), and Quintin, the later arrival.

Robert C. Richardson was born in Washington, D.C., attended Virginia Polytechnic Institute, spent time in the United States Army, and then returned to graduate school in physics at Duke University where his thesis work involved NMR studies of solid helium-3. In the fall of 1966 Bob began work at Cornell University in the laboratory of David M. Lee. Their research goal was to observe the nuclear magnetic phase transition in solid helium-3 that could be predicted from Richardson's thesis work with Professor Horst Meyer at Duke. In collaboration with graduate student Douglas D. Osheroff, they worked on cooling techniques and NMR instrumentation for studying low-temperature helium liquids and solids. In the fall of 1971, they made the accidental discovery that liquid helium-3 undergoes a pairing transition similar to that of superconductors. The three were awarded the Nobel Prize for that work in 1996. Before his death in 2013, Bob was the Vice Provost for Research, emeritus, and the F. R. Newman Professor of Physics at Cornell. He also enjoyed gardening, photography, and spending time with his grandsons.

Dedication

For Charlotte, Denisha, Ivy, Julia, and Katie

Alan

In memory of daughter Pamela, and husband of 50 years Bob, and for Quintin, Oliver, Dashiell, Jasper, Jennifer, and Jim Merlis

Betty

Brief Contents

Contents

List of Selected Applications

Biology/Life Science

Bone density and osteoporosis, Ex. 1.1, p. 4
Surface area of alveoli in the lung, Ex. 1.7, p. 10
Number of cells in the body, Ex. 1.10, p. 14
Speed and acceleration of animals, Sec. 2.3; P: 7, 27, 75, 76
Can the lion catch the buffalo?, Sec. 2.3, p. 35
Doppler echocardiography, Ex. 2.6, pp. 41–42
Tensile and contact forces in the body, Sec. 4.1, pp. 93–94
Traction apparatus, Ex. 4.1, p. 95
Traction on a foot, Ex 4.1, p. 95
Newton's third law: swimming, walking, skiing, Sec. 4.4, p. 104
Tensile forces in the body, Sec. 4.7, p. 118
Effects of acceleration on the human body, Sec. 4.10, pp. 129–130
Speed in centrifuge, Ex. 5.2, p. 158
Centrifuges, Ex. 5.2, p. 158; Ex. 5.4, p. 162
Effects of acceleration on organisms, Sec. 5.2; Ex. 5.4, p. 162
Energy conversion in jumping animals, Sec. 6.7, p. 214; Ex. 6.12, pp. 215–216; PP 6.12, p. 216
Molecular motors in bacteria and in muscles, Ex. 6.13, p. 217; PP 6.13, p. 217
Protecting the body from injury, Sec. 7.3, pp. 236–237; Ex. 7.2, 238; PP 7.2, p. 238
Ballistocardiography, Sec. 7.4, p. 243
Jet propulsion in squid, Ex. 7.5, pp. 243–244
Exercise is good for you, Sec. 8.4, p. 279
Posture and center of gravity of animals, athletes, Sec. 8.4, p. 281; CPP 8.9, p. 288
Conditions for equilibrium in the human body, Sec. 8.5, pp. 289–290
Forces on human spine during heavy lifting, Sec. 8.5, pp. 292–293
Torque and equilibrium in the human body, Sec. 8.5, p. 289; Ex. 8.10, p. 290; PP 8.10, p. 290
Flexor versus extensor muscles, Sec. 8.5, p. 289
Force to hold arm horizontal, Ex. 8.10, p. 290
Conservation of angular momentum in figure skaters, divers, Sec. 8.8, pp. 298–299
Pressure on divers and animals underwater, Ex. 9.3, p. 329
Sphygmomanometer and blood pressure, Sec. 9.5, p. 333
Specific gravity measurements in medicine, Sec. 9.6, p. 335
Animals manipulating their densities to float or sink, Sec. 9.6, pp. 333–335; Ex. 9.8, p. 337
Specific-gravity measurements of blood and urine, Sec. 9.6, p. 335
Speed of blood flow, Ex. 9.9, p. 340
Plaque buildup and narrowed arteries, Ex. 9.9, pp. 340–341
Arterial flutter and aneurisms, Sec. 9.8, p. 344
Narrowing arteries and high blood pressure, Sec. 9.9, p. 347
Arterial blockage, Ex. 9.12, p. 348
How insects can walk on the surface of a pond, Sec. 9.11, p. 350
Surfactant in the lungs, Sec. 9.11, p. 350
Lung pressure, Ex. 9.14, p. 351
Elastic properties of bone, tendons, ligaments, and hair, Secs. 10.2–10.4, pp. 365–374
Compression of the femur, Ex. 10.2, p. 367
Osteoporosis, Sec. 10.3, p. 368
Bone structure, Sec. 10.3, p. 370
Size limitations on organisms, Sec. 10.3, p. 370
How walking speed depends on leg length, Ex. 10.10, p. 386
Sensitivity of the human ear, Sec. 11.1, p. 403
Seismic waves used by animals, Sec. 11.2, p. 406
Ultrasonography, Ex. 11.5, p. 415
Frequency ranges of animal hearing, Sec. 12.1, pp. 434–435
Sound waves from a songbird, Ex. 12.2, p. 438
The human ear, Sec. 12.6, pp. 446–448
Echolocation by bats and dolphins, Sec. 12.9, pp. 455–456
Ultrasound and ultrasonic imaging, Sec. 12.9, pp. 456–457
Temperature conversion, Sec. 13.2, 469; Ex. 13.1, p. 470
Regulation of body temperature, Ex. 13.1, p. 470; Sec. 13.7, p. 487
Breathing of divers, Ex. 13.6, pp. 480–481
Temperature dependence of biological processes, Sec. 13.7, pp. 485–487
Diffusion of O_2, water, platelets, Sec. 13.8, pp. 488–489; Ex. 13.9, p. 489
Why ponds freeze from the top down, Sec. 14.5, p. 516
Using ice to protect buds from freezing, Sec. 14.5, pp. 511–512
Temperature regulation in the human body, Sec. 14.7, p. 520
Forced convection in the human body, Sec. 14.7, p. 520
Convection and radiation in global climate change, Sec. 14.7, pp. 520–521; Sec. 14.8, p. 521
Thermal radiation, Sec. 14.8, p. 521
Thermography, Sec. 14.8, p. 525
Heat loss and gain by plants and animals, Ex. 14.12, p. 522; Ex. 14.14, p. 525; PPs 14.13, p. 524, 14.14, p. 525
Changes in internal energy for biological processes, Ex. 15.1, p. 541
Entropy and evolution, Sec. 15.8, p. 559
Hydrogen bonding in water and in DNA, Sec. 16.1, pp. 575–576
Electrolocation in fish, Sec. 16.4, p. 592
Gel electrophoresis, Sec. 16.5, pp. 596–597
Transmission of nerve impulses, Sec. 17.2, p. 627
Electrocardiographs, electroencephalographs, and electroretinographs, Sec. 17.2, p. 628
Potential differences across cell membranes, Sec. 17.2; Ex. 17.11, p. 640; PP 17.11, p. 640
Neuron capacitance, Ex. 17.11, p. 640
Defibrillator, Ex. 17.12, p. 643
Propagation of nerve impulses, Sec. 18.10, pp. 689–690
Effects of current on the human body, Sec. 18.11, p. 690
Defibrillator, Sec. 18.11, p. 690
Magnetotactic bacteria, Sec. 19.1, pp. 707, 711
Medical uses of cyclotrons, Sec. 19.3, p. 721
Mass spectrometry, Sec. 19.3, pp. 718–719
Electromagnetic blood flowmeter, Sec. 19.5, pp. 725–726
Magnetic resonance imaging, Sec. 19.8, pp. 737–738
Magnetoencephalography, Sec. 20.3, p. 770
Infrared detection by snakes, beetles, and bed bugs, Sec. 22.3, p. 827
Thermograms of the human body, Sec. 22.3, pp. 827, 828
Fluorescence, Sec. 22.3, p. 828
Biological effects of UV exposure, Sec. 22.3, p. 829
X-rays in medicine and dentistry, CAT scans, Sec. 22.3, p. 830
Navigation of bees, Sec. 22.7, p. 848

Chemistry

Geology/Earth Science

Astronomy/Space Science

Architecture

Technology/Machines

Transportation

Sports

Everyday Life

Preface

Physics is intended for a two-semester college course in introductory physics using algebra and trigonometry. Our main goals in writing this book are

- to present the basic concepts of physics that students need to know for later courses and future careers,
- to emphasize that physics is a tool for understanding the real world, and
- to teach transferable problem-solving skills that students can use throughout their lives.

We have kept these goals in mind while developing the main themes of the book.

NEW TO THE THIRD EDITION

Although the fundamental philosophy of the book has not changed, detailed feedback from instructors and students using the previous editions has enabled us to continually fine-tune our approach. Some of the most important enhancements in the third edition include:

- Based on a review of the content lists for physics in the *Preview Guide for the MCAT[2015] Exam*, coverage of the following topics has been added or expanded: mechanical advantage, turbulence, surface tension, attenuation of sound waves, paramagnetism and diamagnetism, circular polarization, and lens aberrations.
- MCAT review questions have been moved online so actual questions from the 2015 MCAT exams can be made available to students.
- Starting with Chapter 4, Review & Synthesis problems appear at the end of every chapter instead of after related groups of chapters.
- To help students see that the physics they are learning is relevant to their careers, the third edition includes 116 new **biomedical applications** in the end-of-chapter Problems, 12 new biomedical Examples, and 10 new text discussions of biomedical applications.
- A **list of selected biomedical applications** appears on the first page of each chapter.
- Ninety-five new **Ranking Tasks** have been included in the Checkpoints, Practice Problems, and end-of-chapter Problems.
- New **Checkpoints** have been added to the text to give students more frequent opportunities to pause and test their understanding of a new concept.
- Every chapter includes a set of **Collaborative Problems** that can be used in cooperative group problem solving.
- The **Connections** have been enhanced and expanded to help students see the bigger picture—that what may seem like a new concept may really be an extension, application, or specialized form of a concept previously introduced. The goal is for students to view physics as a small set of fundamental concepts that can be applied in many different situations, rather than as a collection of loosely related facts or equations.
- Many of the legends have been expanded to help students learn more from the illustrations.

- Most marginal notes from the previous edition have been incorporated into the text for better flow of ideas and a less cluttered presentation.
- Multiple-Choice Questions that are well-suited to use with **student response systems** are identified with a "clicker" icon. Answers to even-numbered questions are not given, for instructors who track student performance using "clickers."

Some chapter-specific revisions to the text include:

- In **Chapter 1**, the general guidelines for problem solving have been expanded.
- **Chapter 2** introduces motion diagrams earlier and uses them extensively. Students are asked to construct or to interpret motion diagrams in Checkpoints, Examples, Practice Problems, and end-of-chapter Problems.
- **Chapters 3** continues the increased emphasis on motion diagrams. Motion with constant acceleration is now introduced first with motion diagrams, before other representations (graphs and equations).
- In **Chapter 4** the introduction of forces as interaction partners in Section 4.1 now includes an explicit reference to Newton's third law. More prominence is given to the specific identification of forces; the student is asked to state *on* what object and *by* what other object a force is exerted. A Connection has been added to reinforce a central theme in Newton's laws: no matter what *kinds* of forces are acting on an object, we always add them the same way (as vectors) to find the net force.
- **Chapter 6** is enhanced with a new problem-solving strategy box on how to choose between alternative problem-solving approaches (energy vs. Newton's second law). The explanation of why the change in gravitational potential energy is the *negative* of the work done by gravity is simpler and more intuitive. Chapter 6 also uses energy graphs more frequently.
- **Chapter 7** now includes a text discussion of ballistocardiography.
- **Chapter 11** discusses the use of seismic waves by animals to communicate and to sense their environment. The presentation of interference and phase difference has been simplified.
- **Chapter 12** contains an expanded discussion of audible frequency ranges for various animals. The presentation of the (nonrelativistic) Doppler effect is more straightforward, with emphasis on the relative velocities of the wave with respect to source and observer. A new problem-solving strategy box for the Doppler effect has been added.
- **Sections 15.5–15.7** contain improved explanations of heat engines and heat pumps.

COMPREHENSIVE COVERAGE

Students should be able to get the whole story from the book. The previous editions have been tested in our nontraditional course, where students must rely on the textbook as their primary learning resource because there are no lectures. Nonetheless, completeness and clarity are equally advantageous when the book is used in a more traditional classroom setting. *Physics* frees the instructor from having to try to "cover" everything. The instructor can then tailor class time to more important student needs—reinforcing difficult concepts, working through Example problems, engaging the students in peer instruction and cooperative learning activities, describing applications, or presenting demonstrations.

A CONCEPTS-FIRST APPROACH

Some students approach introductory physics with the idea that physics is just the memorization of a long list of equations and the ability to plug numbers into those equations. We want to help students see that a relatively small number of basic physics concepts are applied to a wide variety of situations. Physics education research has shown that students do not automatically acquire conceptual understanding; the concepts must be explained and the students given a chance to grapple with them. Our presentation, based on years of teaching this course, blends conceptual understanding with analytical skills. The "concepts-first" approach helps students develop intuition about how physics works; the "formulas" and problem-solving techniques serve as *tools for applying the concepts.* The **Conceptual Examples** and **Conceptual Practice Problems** in the text and a variety of ranking tasks and Conceptual and Multiple-Choice Questions at the end of each chapter give students a chance to check and to enhance their conceptual understanding.

INTRODUCING CONCEPTS INTUITIVELY

We introduce key concepts and quantities in an informal way by establishing why the quantity is needed, why it is useful, and why it needs a precise definition. Then we make a transition from the informal, intuitive idea to a formal definition and name. Concepts motivated in this way are easier for students to grasp and remember than are concepts introduced by seemingly arbitrary, formal definitions.

For example, in Chapter 8, the idea of rotational inertia emerges in a natural way from the concept of rotational kinetic energy. Students can understand that a rotating rigid body has kinetic energy due to the motion of its particles. We discuss why it is useful to be able to write this kinetic energy in terms of a single quantity common to all the particles (the angular speed), rather than as a sum involving particles with many different speeds. When students understand why rotational inertia is defined the way it is, they are better prepared to move on to the more difficult concepts of torque and angular momentum.

We avoid presenting definitions or formulas without any motivation. When an equation is not derived in the text, we at least describe where the equation comes from or give a plausibility argument. For example, Section 9.9 introduces Poiseuille's law with two identical pipes in series to show why the volume flow rate must be proportional to the pressure drop per unit length. Then we discuss why $\Delta V/\Delta t$ is proportional to the fourth power of the radius (rather than to r^2, as it would be for an ideal fluid).

Similarly, we have found that the definitions of the displacement and velocity vectors seem arbitrary and counterintuitive to students if introduced without any motivation. Therefore, we precede any discussion of kinematic quantities with an introduction to Newton's laws, so students know that forces determine how the state of motion of an object changes. Then, when we define the kinematic quantities to give a precise definition of acceleration, we can apply Newton's second law quantitatively to see how forces affect the motion. We give particular attention to laying the conceptual groundwork for a concept when its name is a common English word such as *velocity* or *work.*

WRITTEN IN A CLEAR AND FRIENDLY STYLE

We have kept the writing down-to-earth and conversational in tone—the kind of language an experienced teacher uses when sitting at a table working one-on-one with a student. We hope students will find the book pleasant to read, informative, and accurate without seeming threatening, and filled with analogies that make abstract concepts easier to grasp. We want students to feel confident that they can learn by studying the textbook.

Although we agree that learning correct physics terminology is essential, we chose to avoid all *unnecessary* jargon—terminology that just gets in the way of the student's understanding. For example, we never use the term *centripetal force*, since its use sometimes leads students to add a spurious "centripetal force" to their free-body diagrams. Likewise, we use *radial component of acceleration* because it is less likely to introduce or reinforce misconceptions than *centripetal acceleration.*

ACCURACY ASSURANCE

The authors and the publisher acknowledge that inaccuracies can be a source of frustration for both the instructor and students. Therefore, throughout the writing and production of this edition, we have worked diligently to eliminate errors and inaccuracies. Maureen Ross and her team at diacriTech conducted an independent accuracy check of all new and revised material in the final draft of the manuscript. They then coordinated the resolution of discrepancies between the accuracy check and the end-of-book answers. The page proofs of the text were proofread against the manuscript to ensure the correction of any errors introduced when the manuscript was typeset. The end-of-book answers were then re-checked by Ralph McGrew.

PROVIDING STUDENTS WITH THE TOOLS THEY NEED

Problem-Solving Approach

Problem-solving skills are central to an introductory physics course. We illustrate these skills in the Example problems. Lists of problem-solving strategies are sometimes useful; we provide such strategies when appropriate. However, the most elusive skills—perhaps the most important ones—are subtle points that defy being put into a neat list. To develop real problem-solving expertise, students must learn how to think critically and analytically. Problem solving is a multidimensional, complex process; an algorithmic approach is not adequate to instill real problem-solving skills.

Strategy We begin each Example with a discussion—in language that the students can understand—of the *strategy* to be used in solving the problem. The strategy illustrates the kind of analytical thinking students must do when attacking a problem: How do I decide what approach to use? What laws of physics apply to the problem and which of them are *useful* in this solution? What clues are given in the statement of the question? What information is implied rather than stated outright? If there are several valid approaches, how do I determine which is the most efficient? What assumptions can I make? What kind of sketch or graph might help me solve the problem? Is a simplification or approximation called for? If so, how can I tell if the simplification is valid? Can I make a preliminary estimate of the answer? Only after considering these questions can the student effectively solve the problem.

Solution Next comes the detailed *solution* to the problem. Explanations are intermingled with equations and step-by-step calculations to help the student understand the approach used to solve the problem. We want the student to be able to follow the mathematics without wondering, "Where did that come from?"

Discussion The numerical or algebraic answer is not the end of the problem; our Examples end with a *discussion.* Students must learn how to determine whether their answer is consistent and reasonable by checking the order of magnitude of the answer, comparing the answer with a preliminary estimate, verifying the units, and doing an independent calculation when more than one approach is feasible. When several different approaches are possible, the discussion looks at the advantages and disadvantages of each approach. We also discuss the implications of the answer—what can we learn from it? We look at special cases and look at "what if" scenarios. The discussion sometimes generalizes the problem-solving techniques used in the solution.

Practice Problem After each Example, a Practice Problem gives students a chance to gain experience using the same physics principles and problem-solving tools. By comparing their answers with those provided at the end of each chapter, students can gauge their understanding and decide whether to move on to the next section.

Our many years of experience in teaching the college physics course in a one-on-one setting has enabled us to anticipate where we can expect students to have difficulty. In addition to the consistent problem-solving approach, we offer several other means of assistance to the student throughout the text. A boxed problem-solving strategy gives detailed information on solving a particular type of problem, and an icon for problem-solving tips draws attention to techniques that can be used in a variety of contexts. A hint in a worked Example or end-of-chapter problem provides a clue on what approach to use or what simplification to make. A warning icon emphasizes an explanation that clarifies a possible point of confusion or a common student misconception.

An important problem-solving skill that many students lack is the ability to extract information from a graph or to sketch a graph without plotting individual data points. Graphs often help students visualize physical relationships more clearly than they can with algebra alone. We emphasize the use of graphs and sketches in the text, in worked examples, and in the problems.

Using Approximation, Estimation, and Proportional Reasoning

Physics is forthright about the constant use of simplified models and approximations in solving physics problems. One of the most difficult aspects of problem solving that students need to learn is that some kind of simplified model or approximation is usually required. We discuss how to know when it is reasonable to ignore friction, treat g as constant, ignore viscosity, treat a charged object as a point charge, or ignore diffraction.

Some Examples and Problems require the student to make an estimate—a useful skill both in physics problem solving and in many other fields. Similarly, we teach proportional reasoning as not only an elegant shortcut but also as a means to understanding patterns. We frequently use percentages and ratios to give students practice in using and understanding them.

Showcasing an Innovative Art Program

In every chapter we have developed a system of illustrations, ranging from simpler diagrams to elaborate and beautiful illustrations, that brings to life the connections between physics concepts and the complex ways in which they are applied. We believe these illustrations, with subjects ranging from three-dimensional views of electric field lines to the biomechanics of the human body and from representations of waves to the distribution of electricity in the home, will help students see the power and beauty of physics.

Helping Students See the Relevance of Physics in Their Lives

Students in an introductory college physics course have a wide range of backgrounds and interests. We stimulate interest in physics by relating its principles to applications relevant to students' lives and in line with their interests. The text, Examples, and end-of-chapter problems draw from the everyday world; from familiar technological applications; and from other fields such as biology, medicine, archaeology, astronomy, sports, environmental science, and geophysics. (Applications in the text are identified with a text heading or marginal note. An icon ( ) identifies applications in the biological or medical sciences.)

The **Everyday Physics Demos** give students an opportunity to explore and see physics principles operate in their everyday lives. These activities are chosen for their simplicity and for their effectiveness in demonstrating physics principles.

Each **Chapter Opener** includes a photo and vignette, designed to capture student interest and maintain it throughout the chapter. The vignette describes the situation shown in the photo and asks the student to consider the relevant physics. A reduced version of the chapter opener photo and question indicate where the vignette topic is addressed within the chapter.

Focusing on the Concepts

By identifying areas where important concepts are revisited, the **Connections** allow us to focus on the basic, core concepts of physics and reinforce for students that all of physics is based on a few, fundamental ideas. A marginal Connections heading and summary adjacent to the coverage in the main text help students easily recognize that a previously introduced concept is being applied to the current discussion.

The exercises in the **Review & Synthesis** sections help students see how the concepts in the previously covered group of chapters are interrelated. These exercises are also intended to help students prepare for tests, in which they must solve problems without having the section or chapter title given as a clue.

Checkpoint questions encourage students to pause and test their understanding of the concept explored within the current section. The answers to the Checkpoints are found at the end of the chapter so that students can confirm their knowledge without jumping too quickly to the provided answer.

Applications are clearly identified as such in the text with a complete listing in the front matter. With Applications, students have the opportunity to see how physics concepts are experienced through their everyday lives.

connect icons identify opportunities for students to access additional information or explanation of topics of interest online. This will help students to focus even further on just the very fundamental, core concepts in their reading of the text.

ADDITIONAL RESOURCES FOR INSTRUCTORS AND STUDENTS

McGraw-Hill SmartBook™

Powered by the intelligent and adaptive LearnSmart engine, SmartBook is the first and only continuously adaptive reading experience available today. Distinguishing what students know from what they don't, and honing in on concepts they are most likely to forget, SmartBook personalizes content for each student. Reading is no longer a passive and linear experience but an engaging and dynamic one, in which students are more likely to master and retain important concepts, coming to class better prepared. SmartBook includes powerful reports that identify specific topics and learning objectives students need to study. These valuable reports also provide instructors insight into how students are progressing through textbook content and are useful for identifying class trends, focusing precious class time, providing personalized feedback to students, and tailoring assessment.

How does SmartBook work? Each SmartBook contains four components: Preview, Read, Practice, and Recharge. Starting with an initial preview of each chapter and key learning objectives, students read the material and are guided to topics for which they need the most practice based on their responses to a continuously adapting diagnostic. Read and practice continue until SmartBook directs students to recharge important material they are most likely to forget so as to ensure concept mastery and retention.

ALEKS® Math Prep for *Physics*

ALEKS Math Prep for *Physics* is a web-based program that provides targeted coverage of critical mathematics material necessary for student success in *Physics.* ALEKS uses artificial intelligence and adaptive questioning to assess precisely a student's preparedness and deliver personalized instruction on the exact topics the student is most ready to learn. Through comprehensive explanations, practice, and feedback, ALEKS enables students to quickly fill individual knowledge gaps in order to build a strong foundation of critical math skills.

Use ALEKS Math Prep for *Physics* during the first six weeks of the term to see improved student confidence and performance, as well as fewer dropouts.

ALEKS Math Prep for *Physics* Features:

- **Artificial Intelligence:** Targets gaps in student knowledge
- **Individualized Assessment and Learning:** Ensure student mastery
- **Adaptive, Open-Response Environment:** Avoids multiple-choice questions
- **Dynamic, Automated Reports:** Monitor student and class progress

Please visit www.aleks.com/highered/math for more information about ALEKS. ALEKS is a registered trademark of ALEKS Corporation.

McGraw-Hill Connect® Physics

McGraw-Hill Connect® Physics to accompany *Physics* offers online electronic homework, an eBook, and a myriad of resources for both instructors and students. Instructors can create homework with easy-to-assign, algorithmically generated problems from the text. This feature also offers the simplicity of automatic grading and reporting.

- MCAT review materials are available online. These include links to practice tests. After the revised MCATs have been administered in 2015, actual questions from those past tests will be made available online for student practice.
- The end-of-chapter problems and Review & Synthesis exercises appear in the online homework system in diverse formats and with various tools.
- The online homework system incorporates new and exciting interactive tools and problem types: ranking problems, a graphing tool, a free-body diagram drawing tool, symbolic entry, a math palette, and multipart problems.
- Mimicking the interaction with a tutor or professor by providing students with detailed explanations and probing questions, several comprehensive tutorial problems cover the main topics of the course. These give students a way to help learn the concepts in a careful, thoughtful way and guide them to a deeper understanding of the material.

Instructors also have access to PowerPoint lecture outlines, an Instructor's Resource Guide with solutions, suggested demonstrations, electronic images from the text, clicker questions, quizzes, tutorials, interactive simulations, and many other resources directly tied to text-specific materials in *Physics.* Students have access to self-quizzing, interactive simulations, tutorials, selected answers for the text's problems, and more.

See www.mhhe.com/grr to learn more and to register.

Online Physics Education Research Workbook

To help professors integrate new research on how students learn, Drs. Athula Herat and Ben Shaevitz of Slippery Rock University have written a workbook to accompany *Physics.* This workbook contains questions and ideas for classroom exercises that will get students thinking about physics in new and comprehensive ways. Students are led to discover physics for themselves, leading to a deeper intuitive understanding of the material. A group of professors who use new ideas from Physics Education Research in the classroom reviewed the workbook and suggested changes and new problems. By providing the workbook in an online format, professors are free to use as much or little of the material as they choose.

Electronic Book Images and Assets for Instructors

Build instructional materials wherever, whenever, and however you want!

Accessed from the Connect Physics website to accompany *Physics*, an online digital library containing photos, artwork, interactives, and other media types can be used to create customized lectures, visually enhanced tests and quizzes, compelling course websites, or attractive printed support materials. Assets are copyrighted by McGraw-Hill Higher Education, but can be used by instructors for classroom purposes. The visual resources in this collection include

- **Art** Full-color digital files of all illustrations in the book can be readily incorporated into lecture presentations, exams, or custom-made classroom materials.
- **Photos** The photos collection contains digital files of photographs from the text, which can be reproduced for multiple classroom uses.
- **Worked Example Library, Table Library, and Numbered Equations Library** Access the worked Examples, tables, and equations from the text in electronic format for inclusion in your classroom resources.

Also residing on the Connect Physics website are PowerPoint Lecture Outlines, ready-made presentations that combine art and lecture notes for each chapter of the text.

Computerized Test Bank Online

A comprehensive bank of test questions in multiple-choice format at a variety of difficulty levels is provided within a computerized test bank powered by McGraw-Hill's flexible electronic testing program—EZ Test Online (www.eztestonline.com). EZ Test Online allows you to create paper and online tests or quizzes in this easy-to-use program!

Imagine being able to create and access your test or quiz anywhere, at any time without installing the testing software. Now, with EZ Test Online, instructors can select questions from multiple McGraw-Hill test banks or create their own, and then either print the test for paper distribution or give it online. See www.mhhe.com/grr for more information.

Electronic Books

If you or your students are ready for an alternative version of the traditional textbook, McGraw-Hill brings you innovative and inexpensive electronic textbooks. By purchasing E-books from McGraw-Hill, students can save as much as 50% on selected titles delivered on the most advanced E-book platforms available.

E-books from McGraw-Hill are smart, interactive, searchable, and portable, with such powerful built-in tools as detailed searching, highlighting, note taking, and student-to-student or instructor-to-student note sharing. E-books from McGraw-Hill will help students to study smarter and quickly find the information they need. E-books also save students money. Contact your McGraw-Hill sales representative to discuss E-book packaging options.

Personal Response Systems

Personal response systems, or "clickers," bring interactivity into the classroom or lecture hall. Wireless response systems give the instructor and students immediate feedback from the entire class. The wireless response pads are essentially remotes that are easy to use and engaging, allowing instructors to motivate student preparation, interactivity, and active learning. Instructors receive immediate feedback to gauge which concepts students understand. Questions covering the content of the *Physics* text (formatted in PowerPoint) are available on the Connect Physics website for *Physics.*

Instructor's Resource Guide

The *Instructor's Resource Guide* includes many unique assets for instructors, such as demonstrations, suggested reform ideas from physics education research, and ideas for incorporating just-in-time teaching techniques. The accompanying Instructor's Solutions Manual includes answers to the end-of-chapter Conceptual Questions and complete, worked-out solutions for all the end-of-chapter Problems from the text. The Instructors Resource Guide is available in the Instructor Resources on the Connect Physics website to accompany *Physics.*

Student Solutions Manual

The *Student Solutions Manual* contains complete worked-out solutions to selected end-of-chapter problems and questions, and to selected Review & Synthesis problems. The solutions in this manual follow the problem-solving strategy outlined in the text's Examples and also guide students in creating diagrams for their own solutions.

For more information, contact a McGraw-Hill customer service representative at (800) 338–3987, or by email at www.mhhe.com. To locate your sales representative, go to www.mhhe.com for Find My Sales Rep.

Acknowledgments

We are grateful to the faculty, staff, and students at Cornell University, who helped us in a myriad of ways. We especially thank our friend and colleague Bob Lieberman who shepherded us through the process as our literary agent and who inspired us as an exemplary physics teacher. Donald F. Holcomb, Persis Drell, Peter Lepage, and Phil Krasicky read portions of the manuscript and provided us with many helpful suggestions. Raphael Littauer contributed many innovative ideas and served as a model of a highly creative, energetic teacher.

We are indebted to Tomás Arias, David G. Cassel, Edith Cassel, Glenn Fletcher, Glenn Case, Chris Henley, Dorothy Holland-Minkley, and Leaf Turner for many helpful discussions while they taught Physics 1101–1102 using the previous edition. We thank our enthusiastic and capable teaching assistants and, above all, the students in Physics 1101–1102, who patiently taught us how to teach physics.

We are deeply thankful for the generosity of Ralph McGrew, who gave us many invaluable suggestions to improve the accuracy and clarity of this edition. We also wish to thank Zhaolong Liu of the Beijing Institute of Technology, who found a few errors as she worked on the Chinese translation of *College Physics*.

We are grateful for the guidance and enthusiasm of Thomas Timp, Tomm Scaife, and Eve Lipton, our editors at McGraw-Hill, whose tireless efforts were invaluable in bringing this project to fruition. *Grazie mille* to Linda Davoli for meticulous copyediting enlivened by a great sense of humor. We also thank Julie De Adder for helping us find so many excellent photos for the book. We are grateful to Jolynn Kilburg and Dana Pauley, our production managers; their steady hands at the tiller helped ensure the high quality of this publication. We would like to thank the entire team of talented professionals assembled by McGraw-Hill to publish this book, including Dan Wallace, Nick McFadden, Lora Neyens, Tammy Juran, Sandra Schnee, David W. Hash, Carrie Burger, DeAnna Dausener, and many others whose hard work has contributed to making the book a reality.

We are grateful to Ralph McGrew as well as Maureen Ross and her team at diacriTech for accuracy-checking the manuscript, writing solutions to the end-of-chapter problems, and for many helpful suggestions.

Our thanks to Michael Famiano, Todd Pedlar, John Vasut, Janet Scheel, Warren Zipfel, Rebecca Williams, and Mike Nichols for contributing some of the medical and biological applications; to Nick Taylor and Mike Strauss for contributing to the end-of-chapter and Review & Synthesis problems; and to Nick Taylor for writing answers to the Conceptual Questions.

From Alan: Above all, I am deeply grateful to my family. Marion, Katie, Charlotte, Julia, and Denisha, without your love, support, encouragement, and patience, this book could never have been written.

From Betty: I thank my daughter Pamela's classmates and friends at Cornell and in the Vanderbilt Master's in Nursing program who were an early inspiration for the book, and Dr. Philip Massey who was very special to Pamela. I thank Alex, Damon, Dave, and Graham, of the English rock band *blur*, who love physics and are inspiring young people of Europe to explore the wonders of physics through their work with the European Space Agency. Finally I thank my daughter Jennifer, grandsons Jasper, Dashiell, Oliver, and Quintin, and son-in-law Jim who endured protracted hours of distraction while this book was being written.

REVIEWERS, CLASS TESTERS, AND ADVISORS

This text reflects an extensive effort to evaluate the needs of college physics instructors and students, to learn how well we met those needs, and to make improvements where we fell short. We gathered information from numerous reviews, class tests, and focus groups.

The primary stage of our research began with commissioning reviews from instructors across the United States and Canada. We asked them to submit suggestions for improvement on areas such as content, organization, illustrations, and ancillaries. The detailed comments of these reviewers constituted the basis for the revision plan.

We then recruited three groups of professors to help guide the updated content. A group of professors who use electronic media and online homework in their classes advised us about updates to the Connect website. Professors who use the latest research in physics education in their courses helped us develop the online workbook and other supplemental materials. Finally, Professors Michael Famiano of Western Michigan University, Todd Pedlar of Luther College, and John Vasut of Baylor University suggested new ways to incorporate applications to biology and medicine throughout the text.

Considering the sum of these opinions, this text now embodies the collective knowledge, insight, and experience of hundreds of college physics instructors. Their influence can be seen in everything from the content, accuracy, and organization of the text to the quality of the illustrations.

We are grateful to the following instructors for their thoughtful comments and advice:

REVIEWERS AND CONTRIBUTORS FOR PHYSICS AND COLLEGE PHYSICS

David Aaron *South Dakota State University*
Bruce Ackerson *Oklahoma State University*
Iftikhar Ahmad *Louisiana State University–Baton Rouge*
Rhett Allain *Southeastern Louisiana University*
Deepthi Amarasuriya *Northwest College*
Peter Anderson *Oakland Community College*
Vasudeva Rao Aravind *Clarion University*
Martina Arndt *Bridgewater State University*
Bijaya Aryal *Lake Superior State University*
Karamjeet Arya *San Jose State University*
Charles Bacon *Ferris State University*
Yiyan Bai *Houston Community College*
Becky Baker *Missouri State University*
Arlette Baljon *San Diego State University*
David Bannon *Oregon State University*
Natalie Batalha *San Jose State University*
David Baxter *Indiana University*
Raymond Benge *Tarrant County College*
Philip Best *University of Connecticut–Storrs*
George Bissinger *East Carolina University*
Julio Blanco *California State University, Northridge*
Werner Boeglin *Florida International University–Miami*
Thomas K. Bolland *The Ohio State University*
Richard Bone *Florida International University*
Catalina Boudreaux *The University of Texas–San Antonio*
Arthur Braundmeier, Jr. *Southern Illinois University–Edwardsville*
Mike Broyles *Collin College*
Hauke Busch *Augusta State University*
David Carleton *Missouri State University*
Paul Champion *Northeastern University*
Soumitra Chattopadhyay *Georgia Highlands College*
Lee Chow *University of Central Florida*
Rambis Chu *Texas Southern University*
Francis Cobbina *Columbus State Community College*
John Cockman *Appalachian State University*
Stuart Cohen *Horry-Georgetown Technical College*
David Cole *Northern Arizona University*
Joshua Colwell *University of Central Florida*
Teman Cooke *Georgia Perimeter College*
Andrew Cornelius *University of Nevada–Las Vegas*
Carl Covatto *Arizona State University*
Michael Crescimanno *Youngstown State University*
Michael Crivello *San Diego Mesa College*
Nathaniel Cunningham *Nebraska Wesleyan University*
Jack Cuthbert *Holmes Community College*
Orville Day *East Carolina University*
Keith Dienes *University of Arizona*
Russell Doescher *Texas State University–San Marcos*
Gregory Dolise *Harrisburg Area Community College–Harrisburg*
Aaron Dominguez *University of Nebraska–Lincoln*
Donald Driscoll *Kent State University–Ashtabula*
James Eickemeyer *Cuesta College*
Steven Ellis *University of Kentucky–Lexington*
John Farley *The University of Nevada–Las Vegas*
Abu Fasihuddin *University of Connecticut–Storrs*
Gerald Feldman *George Washington University*
Jerry Feldman *The George Washington University*
Sharmanthie Fernando *Northern Kentucky University*
Frank Ferrone *Drexel University*
John Fons *University of Wisconsin–Rock County*
Lyle Ford *University of Wisconsin–Eau Claire*
Gregory Francis *Montana State University*
Carl Frederickson *University of Central Arkansas*
Delena Bell Gatch *Georgia Southern University*

Margaret Geppert *Harper College*
David Gerdes *University of Michigan*
Igor Glozman *Highline Community College*
Jim Goff *Pima Community College–West*
Greg Gowens *University of West Georgia*
Omar Guerrero *University of Delaware*
Gemunu Gunaratne *University of Houston*
Robert Hagood *Washtenaw Community College*
Ajawad Haija *Indiana University of Pennsylvania*
Hussein Hamdeh *Wichita State University*
James Heath *Austin Community College*
Paul Heckert *Western Carolina University*
Thomas Hemmick *Stony Brook University*
Athula Herat *Slippery Rock University*
Joseph Hernandez *Bridgewater State University*
Lynn Higgs *University of Utah*
Derrick Hilger *Duquesne University*
Gerald Hite *Texas A&M University–Galveston*
Laurent Hodges *Iowa State University*
William Hollerman *University of Louisiana–Lafayette*
Klaus Honscheild *The Ohio State University*
James Ho *Wichita State University*
Chuck Hughes *University of Central Oklahoma*
Yong Suk Joe *Ball State University*
Linda Jones *College of Charleston*
Nikolaos Kalogeropoulos *Borough of Manhattan, Community College/CUNY*
Daniel Kennefick *University of Arkansas*
Robert Klie *The University of Illinois–Chicago*
Raman Kolluri *Camden County College*
Dorina Kosztin *University of Missouri–Columbia*
Liubov Kreminska *Truman State University*
Allen Landers *Auburn University*
Eric Lane *University of Tennessee at Chattanooga*
Mary Lu Larsen *Towson University*
Kwong Lau *University of Houston*
Paul Lee *California State University–Northridge*
Geoff Lenters *Grand Valley State University*
Goh Hock Leong *National Junior College–Singapore*
Alfred Leung *California State University–Long Beach*
Pui-Tak Leung *Portland State University*
Jon Levin *University of Tennessee, Knoxville*
Mark Lucas *Ohio University*
Hong Luo *University at Buffalo*
Lisa Madewell *University of Wisconsin–Superior*
Ponn Maheswaranathan *Winthrop University*
Rizwan Mahmood *Slippery Rock University*
Eric Mandell *Bowling Green State University*
George Marion *Texas State University–San Marcos*
Pete Markowitz *Florida International University*
Perry Mason *Lubbock Christian University*
David Mast *University of Cincinnati*
Lorin Swint Matthews *Baylor University*
Mark Mattson *James Madison University*
Richard Matzner *University of Texas*
Shannon Mayer *University of Portland*
Dan Mazilu *Virginia Polytechnic Institute & State University*
Joseph McCullough *Cabrillo College*
Rahul Mehta *University of Central Arkansas*
Nathan Miller *University of Wisconsin–Eau Claire*
John Milsom *University of Arizona*
Rabindra Mohapatra *The University of Maryland–College Park*
Kin-Keung Mon *University of Georgia*
Ted Morishige *University of Central Oklahoma*
Krishna Mukherjee *Slippery Rock University*
Mohammed Saber Musazay *King Fahd University of Petroleum and Minerals*
Hermann Nann *Indiana University*
Meredith Newby *Clemson University*
Bruce Palmquist *Central Washington University OR Case Western Reserve University*
Russell Patrick *Southern Polytechnic State University*
Galen Pickett *California State University–Long Beach*
Christopher Pilot *Maine Maritime Academy*
Amy Pope *Clemson University*
Scott Pratt *Michigan State University*
Michael Pravica *University of Nevada–Las Vegas*
Roger Pynn *Indiana University*
Oren Quist *South Dakota State University*
W. Steve Quon *Ventura College*
David Raffaelle *Glendale Community College*
Gordon Ramsey *Loyola University–Chicago*
Natarajan Ravi *Spelman College*
Steven Rehse *Wayne State University*
Robert Balogh-Robinson *Marist College*
Michael Roth *University of Northern Iowa*
Alberto Sadun *University of Colorado–Denver*
G. Mackay Salley *Wofford College*
Phyllis Salmons *Embry Riddle Aeronautical University*
Alvin Saperstein *Wayne State University*
Jyotsna Sau *Delaware Technical & Community College*
Ben Shaevitz *Slippery Rock University*
Douglas Sherman *San Jose State University*
Natalia Sidorovskaia *University of Louisiana–Lafayette*
Bjoern Siepel *Portland State University*
Chandralekha Singh *University of Pittsburgh*
Joseph Slawny *Virginia Polytechnic Institute & State University*
Henry Smith *River Parishes Community College*
Clark Snelgrove *Virginia Polytechnic Institute & State University*
John Stanford *Georgia Perimeter College*
Elizabeth Stoddard *University of Missouri–Kansas City*
Donna Stokes *University of Houston*
Michael Strauss *University of Oklahoma*
Cynthia Swift *Santiago Canyon College*
Colin Terry *Ventura College*
Michael Thackston *Southern Polytechnic State University*
Cheng Ting *Houston Community College–Southeast*
William Tireman *Northern Michigan University*
Bruno Ullrich *Bowling Green State University*
Gautam Vemuri *IUPUI*
Melissa Vigil *Marquette University*
Judy Vondruska *South Dakota State University*
David Weaver *Chandler-Gilbert Community College*
Carlos Wexler *University of Missouri–Columbia*
Joe Whitehead *University of Southern Mississippi*
Daniel Whitmire *University of Louisiana–Lafayette*

Donald Whitney *Hampton University*
Craig Wiegert *University of Georgia*
Arthur Wiggins *Oakland Community College*
Luc Wille *Florida Atlantic University*
Jonathan E. Williams *Bowling Green State University–Firelands College*
Suzanne Willis *Northern Illinois University*
Weldon Wilson *University of Central Oklahoma*
Scott Wissink *Indiana University*
Yumei Wu *Baylor University*
Sanichiro Yoshida *Southeastern Louisiana University*
Andrew Young *Casper College*
David Young *Louisiana State University*
Richard Zajac *Kansas State University–Salina*
Zhixian Zhou *Wayne State University*
Yifu Zhu *Florida International University*
Raymond Zich *Raymond Zich*
Steven Zides *Wofford College*

Volume 1

THIRD EDITION

Physics

CHAPTER

1

Introduction

NASA's Mars rover *Curiosity* landed on the surface of Mars in August 2012. One of the mission's primary objectives is to determine whether Mars ever had an environment capable of supporting microbial life. This photo taken by *Curiosity* shows a rock outcrop that contains rounded pieces of gravel. The size, shape, and composition of the gravel led scientists to conclude that a stream once flowed here.

NASA's many successful missions to Mars have sent back a wealth of geologic data. However, in 1998, a simple mistake caused the loss of the *Mars Climate Orbiter* as it entered orbit around Mars. In this chapter, you will learn how to avoid making this same mistake. (See p. 9.)

BIOMEDICAL APPLICATIONS

- Bone density and osteoporosis (Ex. 1.1)
- Surface area of alveoli in the lung (Ex. 1.7)
- Blood flow rates (Probs. 37, 42, 75)
- Mass dependence of metabolic rate (Prob. 5)
- Smallest and largest organisms (Probs. 72, 73)

Concepts & Skills to Review

- Algebra, geometry, and trigonometry (Appendix A)
- How to Succeed in Your Physics Class (see the text website connect)

1.1 WHY STUDY PHYSICS?

Physics is the branch of science that describes matter, energy, space, and time at the most fundamental level. Whether you are planning to study biology, architecture, medicine, music, chemistry, or art, some principles of physics are relevant to your field.

Physicists look for patterns in the physical phenomena that occur in the universe. They try to explain what is happening, and they perform experiments to see if the proposed explanation is valid. The goal is to find the most basic laws that describe the universe and to formulate those laws in the most precise way possible.

The study of physics is valuable for several reasons:

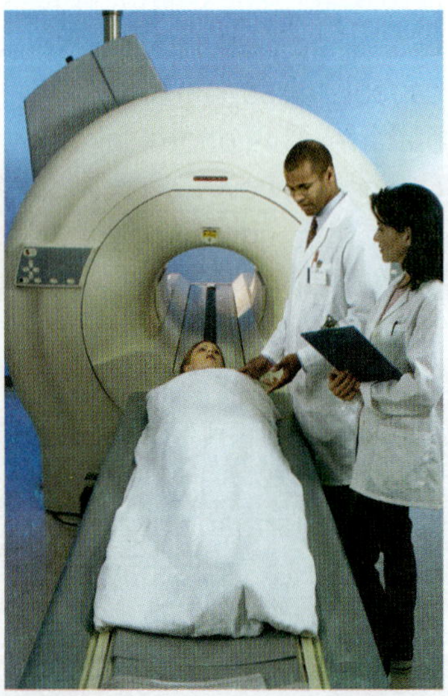

A patient being prepared for magnetic resonance imaging (MRI). MRI provides a detailed image of the internal structures of the patient's body.

- Since physics describes matter and its basic interactions, all natural sciences are built on a foundation of the laws of physics. A full understanding of chemistry requires a knowledge of the physics of atoms. A full understanding of biological processes in turn is based on the underlying principles of physics and chemistry. Centuries ago, the study of *natural philosophy* encompassed what later became the separate fields of biology, chemistry, geology, astronomy, and physics. Today there are scientists who call themselves biophysicists, chemical physicists, astrophysicists, and geophysicists, demonstrating how thoroughly the sciences are intertwined.
- In today's technological world, many important devices can be understood correctly only with a knowledge of the underlying physics. Just in the medical world, think of laser surgery, magnetic resonance imaging, instant-read thermometers, x-ray imaging, radioactive tracers, heart catheterizations, sonograms, pacemakers, microsurgery guided by optical fibers, ultrasonic dental drills, and radiation therapy.
- By studying physics, you acquire skills that are useful in other disciplines. These include thinking logically and analytically; solving problems; making simplifying assumptions; constructing mathematical models; using valid approximations; and making precise definitions.
- Society's resources are limited, so it is important to use them in beneficial ways and not squander them on scientifically impossible projects. Political leaders and the voting public are too often led astray by a lack of understanding of scientific principles. Can a nuclear power plant supply energy safely to a community? What is the truth about global climate change, the ozone hole, and the danger of radon in the home? By studying physics, you learn some of the basic scientific principles and acquire some of the intellectual skills necessary to ask probing questions and to formulate informed opinions on these important matters.
- Finally, by studying physics, we hope that you develop a sense of the beauty of the fundamental laws that describe the universe.

1.2 TALKING PHYSICS

Some of the words used in physics are familiar from everyday speech. This familiarity can be misleading, however, since the scientific definition of a word may differ considerably from its common meaning. In physics, words must be precisely defined so that anyone reading a scientific paper or listening to a science lecture understands exactly what is meant. Some of the basic defined quantities, whose names are also words used in everyday speech, include time, length, force, velocity, acceleration, mass, energy, momentum, and temperature.

In everyday language, *speed* and *velocity* are synonyms. In physics, there is an important distinction between the two. In physics, *velocity* includes the *direction* of motion as well as the distance traveled per unit time. When a moving object changes direction, its velocity changes even though its speed may not have changed. Confusing the scientific definition of *velocity* with its everyday meaning will prevent a correct understanding of some of the basic laws of physics and will lead to incorrect answers.

Mass, as used in everyday language, has several different meanings. Sometimes *mass* and *weight* are used interchangeably. In physics, mass and weight are *not* interchangeable. Mass is a measure of inertia—the tendency of an object at rest to remain at rest or, if moving, to continue moving with the same velocity. Weight, on the other hand, is a measure of the gravitational pull on an object. (Mass and weight are discussed in more detail in Chapter 4.)

There are two important reasons for the way in which we define physical quantities. First, physics is an experimental science. The results of an experiment must be stated unambiguously so that other scientists can perform similar experiments and compare their results. Quantities must be defined precisely to enable experimental measurements to be uniform no matter where they are made. Second, physics is a mathematical science. We use mathematics to quantify the relationships among physical quantities. These relationships can be expressed mathematically only if the quantities being investigated have precise definitions.

1.3 THE USE OF MATHEMATICS

A working knowledge of algebra, trigonometry, and geometry is essential to the study of introductory physics. Some of the more important mathematical tools are reviewed in Appendix A. If you know that your mathematics background is shaky, you might want to test your mastery by doing some problems from a math textbook. You may find it useful to visit www.mhhe.com to explore the Schaum's Outline series, especially the Schaum's Outlines of *Precalculus, College Physics,* or *Physics for Pre-Med, Biology, and Allied Health Students*.

Mathematical equations are shortcuts for expressing concisely in symbols relationships that are cumbersome to describe in words. Algebraic symbols in the equations stand for quantities that consist of numbers *and units*. The number represents a measurement and the measurement is made in terms of some standard; the unit indicates what standard is used. In physics, a number to specify a quantity is useless unless we know the unit attached to the number. When buying silk to make a sari, do we need a length of 5 millimeters, 5 meters, or 5 kilometers? Is the term paper due in 3 minutes, 3 days, or 3 weeks? Systems of units are discussed in Section 1.5.

There are not enough letters in the alphabet to assign a unique letter to each quantity. The same letter V can represent volume in one context and voltage in another. Avoid attempting to solve problems by picking equations that seem to have the correct letters. A skilled problem-solver understands *specifically* what quantity each symbol in a particular equation represents, can specify correct units for each quantity, and understands the situations to which the equation applies.

Ratios and Proportions In the language of physics, the word **factor** is used frequently, often in a rather idiosyncratic way. If the power emitted by a radio transmitter has doubled, we might say that the power has "increased by a factor of 2." If the concentration of sodium ions in the bloodstream is half of what it was previously, we might say that the concentration has "decreased by a factor of 2," or, in a blatantly inconsistent way, someone else might say that it has "decreased by a factor of $\frac{1}{2}$." The *factor* is the number by which a quantity is multiplied or divided when it is changed from one value to another. In other words, the factor is really a ratio. In the case of the radio transmitter, if P_0 represents the initial power and P represents the power after new equipment is installed, we write

$$\frac{P}{P_0} = 2$$

It is also common to talk about "increasing 5%" or "decreasing 20%." If a quantity increases n%, that is the same as saying that it is multiplied by a factor of $1 + (n/100)$. If a quantity decreases n%, then it is multiplied by a factor of $1 - (n/100)$. For example, an increase of 5% means something is 1.05 times its original value, and a decrease of 4% means it is 0.96 times the original value.

Physicists talk about increasing "by some factor" because it often simplifies a problem to think in terms of **proportions**. When we say that A is proportional to B (written $A \propto B$), we mean that if B increases by some factor, then A must increase by the same factor. In other words, the ratio of two values of B is equal to the ratio of the corresponding values of A, expressed as $B_2/B_1 = A_2/A_1$. For instance, the circumference of a circle equals 2π times the radius: $C = 2\pi r$. Therefore $C \propto r$. If the radius doubles, the circumference also doubles. The area of a circle is proportional to the *square* of the radius ($A = \pi r^2$, so $A \propto r^2$). The area must increase by the same factor as the radius *squared*, so if the radius doubles, the area increases by a factor of $2^2 = 4$. Written as a proportion, $A_2/A_1 = (r_2/r_1)^2 = 2^2 = 4$.

Example 1.1

Osteoporosis

Severe osteoporosis can cause the density of bone to decrease as much as 40%. What is the bone density of this degraded bone if the density of healthy bone is 1.5 g/cm^3?

Strategy A decrease of n% means the quantity is multiplied by $1 - (n/100)$.

Solution $1.5\ \text{g/cm}^3 \times [1 - (40/100)] = 1.5\ \text{g/cm}^3 \times 0.60 = 0.90\ \text{g/cm}^3$

Discussion Quick check: The final density is a bit more than half the original density, as expected for a 40% decrease.

Practice Problem 1.1 Red Blood Cell Count

A hospital patient's red blood count (RBC) is 5.0×10^6 cells per microliter ($5.0 \times 10^6\ \mu\text{L}^{-1}$) on Tuesday; on Wednesday it is $4.8 \times 10^6\ \mu\text{L}^{-1}$. What is the percentage change in the RBC?

Example 1.2

Effect of Increasing Radius on the Volume of a Sphere

The volume of a sphere is given by the equation

$$V = \frac{4}{3}\pi r^3$$

where V is the volume and r is the radius of the sphere. If a basketball has a radius of 12.4 cm and a tennis ball has a radius of 3.20 cm, by what factor is the volume of the basketball larger than the volume of the tennis ball?

Strategy The problem gives the values of the radii for the two balls. To keep track of which ball's radius and volume we mean, we use subscripts "b" for basketball and "t" for tennis ball. The radius of the basketball is r_b and the radius of the tennis ball is r_t. Since $\frac{4}{3}$ and π are constants, we can work in terms of proportions.

Solution The ratio of the basketball radius to that of the tennis ball is

$$\frac{r_b}{r_t} = \frac{12.4\ \text{cm}}{3.20\ \text{cm}} = 3.875$$

The volume of a sphere is proportional to the cube of its radius:

$$V \propto r^3$$

Since the basketball radius is larger by a factor of 3.875, and volume is proportional to the cube of the radius, the new volume should be bigger by a factor of $3.875^3 \approx 58.2$.

Discussion A slight variation on the solution is to write out the proportionality in terms of ratios of the corresponding sides of the two equations:

$$\frac{V_b}{V_t} = \frac{\frac{4}{3}\pi r_b^3}{\frac{4}{3}\pi r_t^3} = \left(\frac{r_b}{r_t}\right)^3$$

Substituting the ratio of r_b to r_t yields

$$\frac{V_b}{V_t} = 3.1875^3 \approx 58.2$$

which says that V_b is approximately 58.2 times V_t.

continued on next page

Example 1.2 continued

Practice Problem 1.2 Power Dissipated by a Lightbulb

The electric power P dissipated by a lightbulb of resistance R is $P = V^2/R$, where V represents the line voltage. During a brownout, the line voltage is 10.0% less than its normal value. How much power is drawn by a lightbulb during the brownout if it normally draws 60.0 W (watts)? Assume that the resistance does not change.

CHECKPOINT 1.3

If the radius of the sphere is increased by a factor of 3, by what factor does the volume of the sphere change?

1.4 SCIENTIFIC NOTATION AND SIGNIFICANT FIGURES

In physics, we deal with some numbers that are very small and others that are very large. It can get cumbersome to write numbers in conventional decimal notation. In **scientific notation**, any number is written as a number between 1 and 10 times an integer power of ten. Thus the radius of Earth, approximately 6 380 000 m at the equator, can be written 6.38×10^6 m; the radius of a hydrogen atom, 0.000 000 000 053 m, can be written 5.3×10^{-11} m. Scientific notation eliminates the need to write zeros to locate the decimal point correctly.

Learn how to use the button on your calculator (usually labeled EE) to enter a number in scientific notation. To enter 1.2×10^8, press 1.2, EE, 8.

In science, a measurement or the result of a calculation must indicate the **precision** to which the number is known. The precision of a device used to measure something is limited by the finest division on the scale. Using a meterstick with millimeter divisions as the smallest separations, we can measure a length to a precise number of millimeters and we can estimate a fraction of a millimeter between two divisions. If the meterstick has centimeter divisions as the smallest separations, we measure a precise number of centimeters and estimate the fraction of a centimeter that remains.

Significant Figures The most basic way to indicate the precision of a quantity is to write it with the correct number of **significant figures**. The significant figures are all the digits that are known accurately plus the one estimated digit. If we say that the distance from here to the state line is 12 km, that does not mean we know the distance to be *exactly* 12 kilometers. Rather, the distance is 12 km *to the nearest kilometer*. If instead we said that the distance is 12.0 km, that would indicate that we know the distance to the nearest *tenth* of a kilometer. More significant figures indicate a greater degree of precision.

Rules for Identifying Significant Figures

1. Nonzero digits are always significant.
2. Final or ending zeros written to the right of the decimal point are significant.
3. Zeros written to the right of the decimal point for the purpose of spacing the decimal point are not significant.
4. Zeros written to the left of the decimal point may be significant, or they may only be there to space the decimal point. For example, 200 cm could have one, two, or three significant figures; it's not clear whether the distance was measured to the nearest 1 cm, to the nearest 10 cm, or to the nearest 100 cm. On the other hand, 200.0 cm has four significant figures (see rule 5). Rewriting the number in scientific notation is one way to remove the ambiguity. In this book, when a number has zeros to the left of the decimal point, you may *assume a minimum of two significant figures*.
5. Zeros written between significant figures are significant.

Example 1.3

Identifying the Number of Significant Figures

For each of these values, identify the number of significant figures and rewrite it in standard scientific notation.

(a) 409.8 s
(b) 0.058 700 cm
(c) 9500 g
(d) 950.0×10^1 mL

Strategy We follow the rules for identifying significant figures as given. To rewrite a number in scientific notation, we move the decimal point so that the number to the left of the decimal point is between 1 and 10 and compensate by multiplying by the appropriate power of ten.

Solution (a) All four digits in 409.8 s are significant. The zero is between two significant figures, so it is significant. To write the number in scientific notation, we move the decimal point two places to the left and compensate by multiplying by 10^2: 4.098×10^2 s.

(b) The first two zeros in 0.058 700 cm are not significant; they are used to place the decimal point. The digits 5, 8, and 7 are significant, as are the two final zeros. The answer has five significant figures: 5.8700×10^{-2} cm.

(c) The 9 and 5 in 9500 g are significant, but the zeros are ambiguous. This number could have two, three, or four significant figures. If we take the most cautious approach and assume the zeros are not significant, then the number in scientific notation is 9.5×10^3 g.

(d) The final zero in 950.0×10^1 mL is significant since it comes after the decimal point. The zero to its left is also significant since it comes between two other significant digits. The result has four significant figures. The number is not in *standard* scientific notation since 950.0 is not between 1 and 10; in scientific notation we write 9.500×10^3 mL.

Discussion Scientific notation clearly indicates the number of significant figures since all zeros are significant; none are used only to place the decimal point. In (c), if the measurement was made to the nearest gram, we would write 9.500×10^3 g to show that the zeroes are significant.

Practice Problem 1.3 Identifying Significant Figures

State the number of significant figures in each of these measurements and rewrite them in standard scientific notation.

(a) 0.000 105 44 kg (b) 0.005 800 cm (c) 602 000 s

Significant Figures in Calculations

1. When two or more quantities are added or subtracted, the result is as precise as the *least precise* of the quantities (Example 1.4). If the quantities are written in scientific notation with different powers of ten, first rewrite them with the same power of ten. After adding or subtracting, round the result, keeping only as many decimal places as are significant in *all* of the quantities that were added or subtracted.
2. When quantities are multiplied or divided, the result has the same number of significant figures as the quantity with the *smallest number of significant figures* (see Example 1.5).
3. In a series of calculations, rounding to the correct number of significant figures should be done only at the end, *not at each step*. Rounding at each step would increase the chance that roundoff error could snowball and adversely affect the accuracy of the final answer. It's a good idea to keep *at least two* extra significant figures in calculations, then round at the end.

Example 1.4

Significant Figures in Addition

Calculate the sum 44.560 05 s + 0.0698 s + 1103.2 s.

Strategy The sum cannot be more precise than the least precise of the three quantities. The quantity 44.560 05 s is known to the nearest 0.000 01 s, 0.0698 s is known to the nearest 0.0001 s, and 1103.2 s is known to the nearest 0.1 s. Therefore the least precise is 1103.2 s. The sum has the same precision; it is known to the nearest tenth of a second.

continued on next page

Example 1.4 continued

Solution According to the calculator,

$$44.560\,05 + 0.0698 + 1103.2 = 1147.829\,85$$

We do *not* want to write all of those digits in the answer. That would imply greater precision than we actually have. Rounding to the nearest tenth of a second, the sum is written

$$= 1147.8 \text{ s}$$

which has five significant figures.

Discussion Note that the least precise measurement is not necessarily the one with the fewest number of significant figures. The least precise is the one whose rightmost significant figure represents the largest unit: the "2" in 1103.2 s represents 2 tenths of a second. In addition or subtraction, we are concerned with the precision rather than the number of significant figures. The three quantities to be added have seven, three, and five significant figures, respectively, but the sum has five significant figures.

Practice Problem 1.4 Significant Figures in Subtraction

Calculate the difference 568.42 m − 3.924 m and write the result in scientific notation. How many significant figures are in the result?

Example 1.5

Significant Figures in Multiplication

Find the product of 45.26 m/s and 2.41 s. How many significant figures does the product have?

Strategy The product should have the same number of significant figures as the factor with the least number of significant figures.

Solution A calculator gives

$$45.26 \times 2.41 = 109.0766$$

Since the answer should have only three significant figures, we round the answer to

$$45.26 \text{ m/s} \times 2.41 \text{ s} = 109 \text{ m}$$

Discussion Writing the answer as 109.0766 m would give the false impression that we know the answer to a precision of about 0.0001 m, whereas we actually have a precision of only about 1 m.

Note that although both factors were known to two decimal places, our solution is properly given with no decimal places. It is the number of significant figures that matters in multiplication or division. In scientific notation, we write 1.09×10^2 m.

Practice Problem 1.5 Significant Figures in Division

Write the solution to 28.84 m divided by 6.2 s with the correct number of significant figures.

When an integer, or a fraction of integers, is used in an equation, the precision of the result is not affected by the integer or the fraction; the number of significant figures is limited only by the measured values in the problem. The fraction $\frac{1}{2}$ in an equation is *exact;* it does not reduce the number of significant figures to one. In an equation such as $C = 2\pi r$ for the circumference of a circle of radius r, the factors 2 and π are exact. We use as many digits for π as we need to maintain the precision of the other quantities.

Order-of-Magnitude Estimates Sometimes a problem may be too complicated to solve precisely, or information may be missing that would be necessary for a precise calculation. In such a case, an **order-of-magnitude** solution is the best we can do. By *order of magnitude,* we mean "roughly what power of ten?" An order of magnitude calculation is done to at most one significant figure. Even when a more precise solution is feasible, it is often a good idea to start with a quick, "**back-of-the-envelope estimate**" (a calculation so short that it could easily fit on the back of an envelope). Why? Because we can often make a good guess about the correct order of magnitude of the answer to a problem, even before we start solving the problem. If the answer comes out with a different order of magnitude, we go back and search for an error. Suppose a problem concerns a vase that is knocked off a fourth-story window ledge. We can guess by experience the order of magnitude of the time it takes the vase to hit the ground. It might be 1 s, or 2 s, but we are certain that it is *not* 1000 s or 0.000 01 s.

CHECKPOINT 1.4

What are some of the reasons for making order-of-magnitude estimates?

1.5 UNITS

A **metric system** of units has been used for many years in scientific work and in European countries. The metric system is based on powers of ten (Fig. 1.1). In 1960, the General Conference of Weights and Measures, an international authority on units, proposed a revised metric system called the *Système International d'Unités* in French (abbreviated **SI**), which uses the meter (m) for length, the kilogram (kg) for mass, the second (s) for time, and four more base units (Table 1.1). **Derived units** are constructed from combinations of the base units. For example, the SI unit of force is $kg{\cdot}m/s^2$ (which can also be written $kg{\cdot}m{\cdot}s^{-2}$); this combination of units is given a special name, the newton (N), in honor of Isaac Newton. The newton is a derived unit because it is composed of a combination of base units. When units are named after famous scientists, the name of the unit is written with a lowercase letter, even though it is based on a proper name; the *symbol* for the unit is written with an uppercase letter. The inside front cover of the book has a complete listing of the derived SI units used in this book.

As an alternative to explicitly writing powers of ten, SI uses prefixes for units to indicate power of ten factors. Table 1.2 shows some of the powers of ten and the SI prefixes used for them. These are also listed on the inside front cover of the book. Note that when an SI unit with a prefix is raised to a power, the prefix is *also* raised to that power. For example, $8\ cm^3 = 2\ cm \times 2\ cm \times 2\ cm$.

SI units are preferred in physics and are emphasized in this book. Since other units are sometimes used, we must know how to convert units. Various scientific fields, even in physics, sometimes use units other than SI units, whether for historical or practical

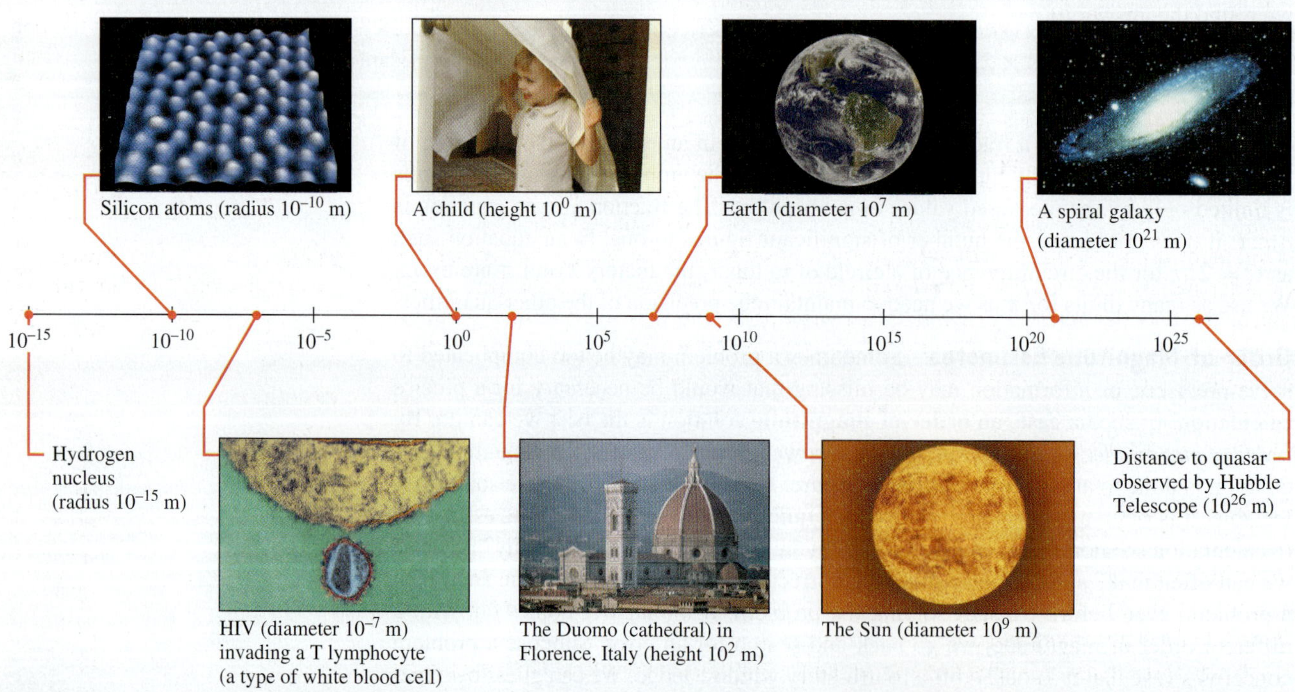

Figure 1.1 Scientific notation uses powers of ten to express quantities that have a wide range of values.

Table 1.1 SI Base Units

Quantity	Unit Name	Symbol	Definition
Length	meter	m	The distance traveled by light in vacuum during a time interval of 1/299 792 458 s.
Mass	kilogram	kg	The mass of the international prototype of the kilogram.
Time	second	s	The duration of 9 192 631 770 periods of the radiation corresponding to the transition between the two hyperfine levels of the ground state of the cesium-133 atom.
Electric current	ampere	A	The constant current in two long, thin, straight, parallel conductors placed 1 m apart in vacuum that would produce a force on the conductors of 2×10^{-7} newtons per meter of length.
Temperature	kelvin	K	The fraction 1/273.16 of the thermodynamic temperature of the triple point of water.
Amount of substance	mole	mol	The amount of substance that contains as many elementary entities as there are atoms in 0.012 kg of carbon-12.
Luminous intensity	candela*	cd	The luminous intensity, in a given direction, of a source that emits radiation of frequency 540×10^{12} Hz and that has a radiant intensity in that direction of 1/683 watts per steradian.

*Not used in this book

reasons. For example, in atomic and nuclear physics, the SI unit of energy (the joule, J) is rarely used; instead the energy unit used is usually the electron-volt (eV). Biologists and chemists use units that are not ordinarily used by physicists. One reason that SI is preferred is that it provides a common denominator—all scientists are familiar with the SI units.

In most of the world, SI units are used in everyday life and in industry. In the United States, however, the U.S. customary units—sometimes called English units—are still used. The base units for this system are the foot, the second, and the pound. The pound is legally defined in the United States as a unit of mass, but it is also commonly used as a unit of force (in which case it is sometimes called *pound-force*). Since mass and force are entirely different concepts in physics, this inconsistency is one good reason to use SI units.

What happened to the *Mars Climate Orbiter?*

In the autumn of 1999, to the chagrin of NASA, a \$125 million spacecraft was destroyed as it was being maneuvered into orbit around Mars. The company building the booster rocket provided information about the rocket's thrust in U.S. customary units, but the NASA scientists who were controlling the rocket thought the figures provided were in metric units. Arthur Stephenson, chairman of the *Mars Climate Orbiter* Mission Failure Investigation Board, stated that, "The 'root cause' of the loss of the spacecraft was the failed translation of English units into metric units in a segment of ground-based, navigation-related mission software." After a journey of 122 million miles, the *Climate Orbiter* dipped about 15 miles too deep into the Martian atmosphere, causing the propulsion system to overheat. The discrepancy in units unfortunately caused a dramatic failure of the mission.

Table 1.2 SI Prefixes

Prefix (abbreviation)	Power of Ten
peta- (P)	10^{15}
tera- (T)	10^{12}
giga- (G)	10^{9}
mega- (M)	10^{6}
kilo- (k)	10^{3}
deci- (d)	10^{-1}
centi- (c)	10^{-2}
milli- (m)	10^{-3}
micro- (μ)	10^{-6}
nano- (n)	10^{-9}
pico- (p)	10^{-12}
femto- (f)	10^{-15}

Converting Units If the statement of a problem includes a mixture of different units, the units must be converted to a single, consistent set before the problem is solved. Quantities to be added or subtracted *must be expressed in the same units*. Usually the best way is to convert everything to SI units. Common conversion factors are listed on the inside front cover of this book.

Examples 1.6 and 1.7 illustrate the technique for converting units. The quantity to be converted is multiplied by one or more conversion factors written as a fraction equal to 1. The units are multiplied or divided as algebraic quantities.

Some conversions are exact by definition. One meter is defined to be *exactly* equal to 100 cm; all SI prefixes are exactly a power of ten. The use of an exact conversion factor such as 1 m = 100 cm, or 1 foot = 12 inches, does not affect the precision of the result; the number of significant figures is limited only by the other quantities in the problem.

Example 1.6

Buying Clothes in a Foreign Country

Michel, an exchange student from France, is studying in the United States. He wishes to buy a new pair of jeans, but the sizes are all in *inches*. He does remember that 1 m = 3.28 ft and that 1 ft = 12 in. If his waist size is 82 cm, what is his waist size in inches?

Strategy Each conversion factor can be written as a fraction. If 1 m = 3.28 ft, then

$$\frac{3.28\ \text{ft}}{1\ \text{m}} = 1$$

We can multiply any quantity by 1 without changing its value. We arrange each conversion factor in a fraction and multiply one at a time to get from centimeters to inches.

Solution We first convert cm to meters.

$$82\ \cancel{\text{cm}} \times \frac{1\ \text{m}}{100\ \cancel{\text{cm}}}$$

Now, we convert meters to feet.

$$82\ \cancel{\text{cm}} \times \frac{1\ \cancel{\text{m}}}{100\ \cancel{\text{cm}}} \times \frac{3.28\ \text{ft}}{1\ \cancel{\text{m}}}$$

Finally, we convert feet to inches.

$$82\ \cancel{\text{cm}} \times \frac{1\ \cancel{\text{m}}}{100\ \cancel{\text{cm}}} \times \frac{3.28\ \cancel{\text{ft}}}{1\ \cancel{\text{m}}} \times \frac{12\ \text{in}}{1\ \cancel{\text{ft}}} = 32\ \text{in.}$$

In each case, the fraction is written so that the unit we are converting *from* cancels out.

As a check:

$$\cancel{\text{cm}} \times \frac{\cancel{\text{m}}}{\cancel{\text{cm}}} \times \frac{\cancel{\text{ft}}}{\cancel{\text{m}}} \times \frac{\text{in}}{\cancel{\text{ft}}} = \text{in.}$$

Discussion This problem could have been done in one step using a direct conversion factor from inches to centimeters (1 in. = 2.54 cm). One of the great advantages of SI units is that all the conversion factors are powers of ten (see Table 1.2); there is no need to remember that there are 12 inches in a foot, 4 quarts in a gallon, 16 ounces in a pound, 5280 feet in a mile, and so on.

Practice Problem 1.6 Driving on the Autobahn

A BMW convertible travels on the German Autobahn at a speed of 128 km/h. What is the speed of the car (a) in meters per second? (b) in miles per hour?

Example 1.7

Area of the Alveoli

The total area of the alveoli in the human lung is about 70 m^2. What is the area in (a) square centimeters and (b) square inches?

Strategy Look up the conversion factors between meters, centimeters, and inches. Since there are *two* powers of meters to convert, we need to square the conversion factors.

Solution (a) 1 m = 100 cm, so

$$70\ \text{m}^2 \times \left(100\ \tfrac{\text{cm}}{\text{m}}\right)^2 = 7.0 \times 10^5\ \text{cm}^2$$

(b) Using 1 in. = 2.54 cm,

$$7.0 \times 10^5\ \text{cm}^2 \times \left(\frac{1\ \text{in.}}{2.54\ \text{cm}}\right)^2 = 1.1 \times 10^5\ \text{in.}^2$$

Discussion Be careful when a unit is raised to a power other than 1; the conversion factor must be raised to the same power. Writing out the units to make sure they cancel prevents mistakes. When a quantity is raised to a power, both the number and the unit must be raised to the same power. The quantity $(100\ \text{cm})^3$ is equal to $100^3\ \text{cm}^3 = 10^6\ \text{cm}^3$; it is *not* equal to $100\ \text{cm}^3$, nor is it equal to 10^6 cm.

Practice Problem 1.7 Surface Area of Earth

The radius of Earth is 6.4×10^3 km. Find the surface area of Earth in square meters and in square miles. (Surface area of a sphere = $4\pi r^2$.)

Whenever a calculation is performed, always write out the units with each quantity. Combine the units algebraically to find the units of the result. This small effort has three important benefits:

1. It shows what the units of the result are. A common mistake is to get the correct numerical result of a calculation but to write it with the wrong units, making the answer wrong.
2. It shows where unit conversions must be done. If units that should have canceled do not, we go back and perform the necessary conversion. When a distance is calculated and the result comes out with units of meter-seconds per hour (m·s/h), we should convert hours to seconds.
3. It helps locate mistakes. If a distance is calculated and the units come out as meters per second (m/s), we know to look for an error.

CHECKPOINT 1.5

If 1 fluid ounce (fl oz) is approximately 30 mL, how many liters are in a half gallon (64 fl oz) of milk?

1.6 DIMENSIONAL ANALYSIS

Dimensions are basic *types* of units, such as time, length, and mass. (Warning: The word *dimension* has several other meanings, such as in "three-dimensional space" or "the dimensions of a soccer field.") Many different units of length exist: meters, inches, miles, nautical miles, fathoms, leagues, astronomical units, angstroms, and cubits, just to name a few. All have dimensions of length; each can be converted into any other.

We can add, subtract, or equate quantities only if they have the same dimensions (although they may not necessarily be given in the same units). It is possible to add 3 meters to 2 inches (after converting units), but it is not possible to add 3 meters to 2 kilograms. To analyze dimensions, treat them as algebraic quantities, just as we did with units in Section 1.5. Usually [M], [L], and [T] are used to stand for mass, length, and time dimensions, respectively. Equivalently, we can use the SI base units: kg for mass, m for length, and s for time.

Example 1.8

Dimensional Analysis for a Distance Equation

Analyze the dimensions of the equation $d = vt$, where d is distance traveled, v is speed, and t is elapsed time.

Strategy Replace each quantity with its dimensions. Distance has dimensions [L]. Speed has dimensions of length per unit time [L/T]. The equation is dimensionally consistent if the dimensions are the same on both sides.

Solution The right side has dimensions

$$\frac{[\text{L}]}{[\text{T}]} \times [\text{T}] = [\text{L}]$$

Since both sides of the equation have dimensions of length, the equation is dimensionally consistent.

Discussion If, by mistake, we wrote $d = v/t$ for the relation between distance traveled and elapsed time, we could quickly catch the mistake by looking at the dimensions. On the right side, v/t would have dimensions [L/T^2], which is not the same as the dimensions of d on the left side.

A quick dimensional analysis of this sort is a good way to catch algebraic errors. Whenever we are unsure whether an equation is correct, we can check the dimensions.

Practice Problem 1.8 Testing Dimensions of Another Equation

Test the dimensions of the following equation:

$$d = \frac{1}{2}at$$

where d is distance traveled, a is acceleration (which has SI units m/s^2), and t is the elapsed time. If incorrect, can you suggest what might have been omitted?

Applying Dimensional Analysis Dimensional analysis is good for more than just checking equations. In some cases, we can completely solve a problem—up to a dimensionless factor like $1/(2\pi)$ or $\sqrt{3}$—using dimensional analysis. To do this, first list all the relevant quantities on which the answer might depend. Then determine what combinations of them have the same dimensions as the answer for which we are looking. If only one such combination exists, then we have the answer, except for a possible dimensionless multiplicative constant.

Example 1.9

Violin String Frequency

While it is being played, a violin string produces a tone with frequency f in s^{-1}; the frequency is the number of vibrations *per second* of the string. The frequency depends only on the string's mass m, length L, and tension T. If the tension is increased 5.0%, how does the frequency change? Tension has SI unit kg·m/s^2.

Strategy We could make a study of violin strings, but let us see what we can find out by dimensional analysis. We want to find out how the frequency f can depend on m, L, and T. We won't know if there is a dimensionless constant involved, but we can work by proportions so any such constant will divide out.

Solution The unit of tension T is kg·m/s^2. The units of f do not contain kg or m; we can get rid of them from T by dividing the tension by the length and the mass:

$$\frac{T}{mL} \text{ has SI units } s^{-2}$$

That is almost what we want; all we have to do is take the square root:

$$\sqrt{\frac{T}{mL}} \text{ has SI units } s^{-1}$$

Therefore,

$$f = C\sqrt{\frac{T}{mL}}$$

where C is some dimensionless constant. To answer the question, let the original frequency and tension be f and T and the new frequency and tension be f' and T', where $T' = 1.050T$. Frequency is proportional to the square root of tension, so

$$\frac{f'}{f} = \sqrt{\frac{T'}{T}} = \sqrt{1.050} = 1.025$$

The frequency increases 2.5%.

Discussion We'll learn in Chapter 11 how to calculate the value of C, which is 1/2. That is the *only* thing we cannot get by dimensional analysis. There is *no* other way to combine T, m, and L to come up with a quantity that has the units of frequency.

Practice Problem 1.9 Increase in Kinetic Energy

When an object of mass m is moving with a speed v, it has kinetic energy associated with its motion. Energy is measured in kg·m^2·s^{-2}. If the speed of a moving object is increased by 25% while its mass remains constant, by what percentage does the kinetic energy increase?

CHECKPOINT 1.6

If two quantities have different dimensions, is it possible to (a) multiply; (b) divide; (c) add; (d) subtract them?

1.7 PROBLEM-SOLVING TECHNIQUES

No single method can be used to solve every physics problem. We demonstrate useful problem-solving techniques in the examples in every chapter of this text. Even for a particular problem, there may be more than one correct way to approach the solution. Problem-solving techniques are *skills* that must be *practiced* to be learned.

Think of the problem as a puzzle to be solved. Only in the easiest problems is the solution method immediately apparent. When you do not know the entire path to a solution, see where you can get by using the given information—find whatever you can. Exploration of this sort may lead to a solution by suggesting a path that had not been considered. Be willing to take chances. You may even find the challenge enjoyable!

When having some difficulty, it helps to work with a classmate or two. One way to clarify your thoughts is to put them into words. After you have solved a problem, try to explain it to a friend. If you can explain the problem's solution, you really do understand it. Both of you will benefit. But do not rely too much on help from others; the goal is for each of you to develop your own problem-solving skills.

General Guidelines for Problem Solving

1. Read the problem *carefully* and *all the way through*. Identify the goal of the problem: What are you trying to find?
2. Reread the problem and draw a sketch or diagram to help you visualize what is happening. If the problem involves motion or change, sketch it at different times (especially the initial and final situations).
3. Write down and organize the given information. Some of the information can be written in labels on the diagram. Be sure that the labels are unambiguous. Identify in the diagram the object, the position, the instant of time, or the time interval to which the quantity applies. Sometimes information might be usefully written in a table beside the diagram. Look at the wording of the problem again for information that is implied or stated indirectly. Decide on algebraic symbols to stand for each quantity and make sure your notation is clear and unambiguous.
4. Identify the units appropriate for the answer. If possible, make an estimate to determine the order of magnitude of the answer. This estimate is useful as a check on the final result to see if it is reasonable.
5. Think about how to get from the given information to the final desired information. Do not rush this step. Which principles of physics can be applied to the problem? Which will help get to the solution? How are the known and unknown quantities related? Are all of the known quantities relevant, or might some of them not affect the answer? Which equations are *relevant* and may lead to the solution to the problem?
6. Frequently, the solution involves more than one step. Intermediate quantities might have to be found first and then used to find the final answer. Try to map out a path from the given information to the solution. Whenever possible, a good strategy is to divide a complex problem into several simpler subproblems.
7. Perform algebraic manipulations with algebraic symbols (letters) as far as possible. Substituting the numbers in too early has a way of hiding mistakes.
8. Finally, if the problem requires a numerical answer, substitute the known numerical quantities, *with their units,* into the appropriate equation. Leaving out the units is a common source of error. Writing the units shows when a unit conversion needs to be done—and also may help identify an algebra mistake.
9. Once the solution is found, don't be in a hurry to move on. Check the answer—is it reasonable? Test your solution in special cases or with limiting values of quantities to see if the solution makes sense. (For example, what happens if the mass is very large? What happens as it approaches zero?) Try to think of other ways to solve the same problem. Many problems can be solved in several different ways. Besides providing a check on the answer, finding more than one method of solution deepens our understanding of the principles of physics and develops problem-solving skills that will help solve other problems.

1.8 APPROXIMATION

Physics is about building conceptual and mathematical models and comparing observations of the real world with the model. Simplified models help us to analyze complex situations. In various contexts we assume there is no friction, or no air resistance, no heat loss, or no wind blowing, and so forth. If we tried to take all these things into consideration with every problem, the problems would become vastly more complicated to solve. We never can take account of *every* possible influence. We freely make approximations whenever possible to turn a complex problem into an easier one, as long as the answer will be accurate enough for our purposes.

A valuable skill to develop is the ability to know when an assumption or approximation is reasonable. It might be permissible to ignore air resistance when dropping a stone, but not when dropping a beach ball. Why? We must always be prepared to justify any approximation we make by showing the answer is not changed very much by its use.

As well as making simplifying approximations in models, we also recognize that measurements are approximate. Every measured quantity has some uncertainty; it is impossible for a measurement to be exact to an arbitrarily large number of significant figures. Every measuring device has limits on the precision and accuracy of its measurements.

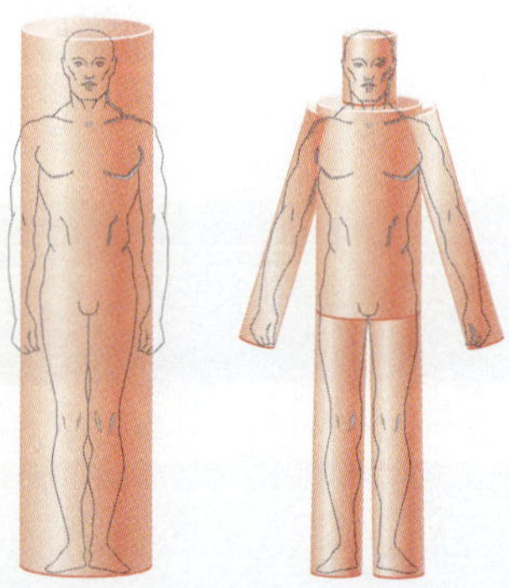

Figure 1.2 Approximation of human body by one or more cylinders to estimate the body's surface area.

Approximating the Surface Area of the Human Body Sometimes it is difficult or impossible to measure precisely a quantity that is needed for a problem. Then we have to make a reasonable estimate. Suppose we need to know the surface area of a human being to determine the heat loss by radiation in a cold room. We can estimate the height of an average person. We can also estimate the average distance around the waist or hips. Approximating the shape of a human body as a cylinder, we can estimate the surface area by calculating the surface area of a cylinder with the same height and circumference (Fig. 1.2a).

If we need a better estimate, we use a slightly more refined model. For instance, we might approximate the arms, legs, trunk, and head and neck as cylinders of various sizes (Fig. 1.2b). How different is the sum of these areas from the original estimate? The difference gives an idea of how close the first estimate is.

Example 1.10

Number of Cells in the Human Body

Average-sized cells in the human body are about 10 μm in diameter (Fig. 1.3). How many cells are in the human body? Make an order-of-magnitude estimate.

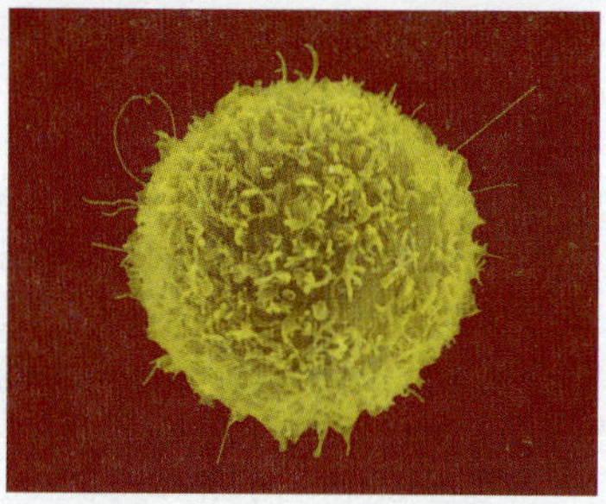

Figure 1.3 Scanning electron micrograph of a precursor T lymphocyte (a type of white blood cell in the human body). The cell is approximately 10 μm in diameter.

Strategy We divide this problem into three subproblems: estimating the volume of a human, estimating the volume of the average cell, and finally estimating the number of cells.

To find the volume of a human body, we approximate the body as a cylinder, as previously discussed. Next we assume the cells are spherical to find the volume of a cell. Third, the ratio of the two volumes (volume of the body to volume of the cell) shows how many cells are in the body.

Solution Model the body as a cylinder. A typical height is about 2 m. A typical *maximum* circumference (think hip size) is about 1 m. The corresponding radius is $1/(2\pi)$ m, or about 1/6 m. The *average* radius is somewhat smaller; say 0.1 m. The volume of a cylinder is the height times the cross-sectional area:

$$V = Ah = \pi r^2 h \approx 3 \times (0.1\text{ m})^2 \times (2\text{ m}) = 0.06\text{ m}^3$$

The volume of a sphere is $V = \frac{4}{3}\pi r^3$. Then the volume of an average cell is about

$$V_{\text{cell}} \approx \frac{4}{3} \times 3 \times (5 \times 10^{-6}\text{ m})^3 \approx 5 \times 10^{-16}\text{ m}^3$$

continued on next page

Example 1.10 continued

The number of cells is the ratio of the two volumes:

$$N = \frac{\text{volume of body}}{\text{average volume of cell}} \approx \frac{6 \times 10^{-2}\ \text{m}^3}{5 \times 10^{-16}\ \text{m}^3} \approx 1 \times 10^{14}$$

Discussion Based on this rough estimate, we cannot rule out the possibility that a better estimate might be 3×10^{13}. On the other hand, we *can* rule out the possibility that the number of cells is, say, 100 million ($= 10^8$).

Practice Problem 1.10 Drinking Water Consumed in the United States

How many liters of water are swallowed by the people living in the United States in one year? This is a type of problem made famous by the physicist Enrico Fermi (1901–1954), who was a master at this sort of back-of-the-envelope calculation. Such problems are often called *Fermi problems* in his honor. (Note: 1 liter = 10^{-3} m^3 $\approx$ 1 quart.)

1.9 GRAPHS

Graphs are used to help us see a pattern in the relationship between two quantities. It is much easier to see a pattern on a graph than to see it in a table of numerical values. When we do experiments in physics, we change one quantity (the **independent variable**) and see what happens to another (the **dependent variable**). We want to see how one variable *depends on* another. The value of the independent variable is usually plotted along the horizontal axis of the graph. In a plot of p versus q, which means p is plotted on the vertical axis and q on the horizontal axis, normally p is the dependent variable and q is the independent variable.

Some general guidelines for recording data and making graphs are given next.

Recording Data and Making Data Tables

1. Label columns with the names of the data being measured and be sure to include the units for the measurements. Do not erase any data, but just draw a line through data that you think are erroneous. Sometimes you may decide later that the data were correct after all.
2. Try to make a realistic estimate of the precision of the data being taken when recording numbers. For example, if the timer says 2.3673 s, but you know your reaction time can vary by as much as 0.1 s, the time should be recorded as 2.4 s. When doing calculations using measured values, remember to round the final answer to the correct number of significant figures.
3. Do not wait until you have collected all of your data to start a graph. It is much better to graph each data point as it is measured. By doing so, you can often identify equipment malfunction or measurement mistakes. You can also spot where something interesting happens and take data points closer together there. Graphing as you go means that you need to find out the range of values for both the independent and dependent variables.

Graphing Data

1. Make *large, neat* graphs. A tiny graph is not very illuminating. Use at least half a page. A graph made carelessly obscures the pattern between the two variables.
2. Label axes with the name of the quantities graphed and their units. Write a meaningful title.
3. When a linear relation is expected, use a ruler or straightedge to draw the best-fit straight line. Do not *assume* that the line must go through the origin—make a measurement to find out, if possible. Some of the data points will probably fall above the line and some will fall below the line.

4. Determine the slope of a best-fit line by measuring the ratio $\Delta y/\Delta x$ using as large a range of the graph as possible. (The notation Δy is read aloud as "delta y" and represents a change in the value of y.) Do not choose two data points to calculate the slope; instead, read values from two points on the best-fit line. Show the calculations. Do not forget to write the units; slopes of graphs in physics have units, since the quantities graphed have units.
5. When a nonlinear relationship is expected between the two variables, the best way to test that relationship is to manipulate the data algebraically so that a linear graph is expected. The human eye is a good judge of whether a straight line fits a set of data points. It is not so good at deciding whether a curve is parabolic, cubic, or exponential. To test the relationship $x = \frac{1}{2}at^2$, where x and t are the quantities measured and a is a constant, graph x versus t^2 instead of x versus t.
6. If one data point does not lie near the line or smooth curve connecting the other data points, that data point should be investigated to see whether an error was made in the measurement or whether some interesting event is occurring at that point. If something unusual is happening there, obtain additional data points in the vicinity.
7. When the slope of a graph is used to calculate some quantity, pay attention to the equation of the line and the units along the axes. The quantity to be found may be the inverse of the slope or twice the slope or one half the slope. The equation of the line will tell you how to interpret the slope and intercept of the line. For example, if the expected relationship is $v^2 = v_0^2 + 2ax$ and you plot v^2 versus x, rewrite the equation as $(v^2) = (2a)x + (v_0^2)$. This shows that the slope of the line is $2a$ and the vertical intercept is v_0^2.

Example 1.11

Length of a Spring

In an introductory physics laboratory experiment, students are investigating how the length of a spring varies with the weight hanging from it. Various weights (accurately calibrated to 0.01 N) ranging up to 6.00 N can be hung from the spring; then the length of the spring is measured with a meterstick (Fig. 1.4). The goal is to see if the weight F and length L are related by

$$F = kx$$

where L_0 is the length of the spring when no weight is hanging from it, $x = (L - L_0)$, and k is called the *spring constant* of the spring. Graph the data in the table and calculate k for this spring.

F (N):	0	0.50	1.00	2.50	3.00	3.50	4.00	5.00	6.00
L (cm):	9.4	10.2	12.5	17.9	19.7	22.5	23.0	28.8	29.5

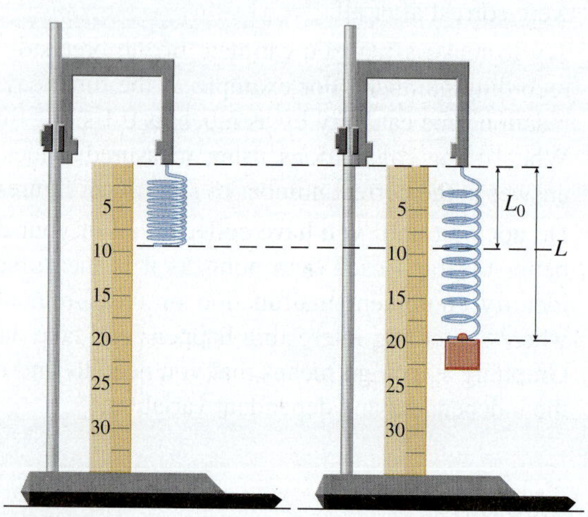

Figure 1.4
A hanging weight makes a spring stretch. In this experiment, students measure the length L of the spring when different weights are hung from it.

Strategy Weight is the independent variable, so it is plotted on the horizontal axis. After plotting the data points, we draw the best-fit straight line. Then we calculate the slope of the line, using two points on the line that are widely separated and that cross gridlines of the graph (so the values are easy to read). The slope of the graph is not k; we must solve the equation for L, since length is plotted on the vertical axis.

Solution Figure 1.5 shows a graph with data points and a best-fit straight line. There is some scatter in the data, but a linear relationship is plausible.

continued on next page

Example 1.11 continued

Two points where the line crosses gridlines of the graph are (0.80 N, 12.0 cm) and (4.40 N, 25.0 cm). From these, we calculate the slope:

$$\text{slope} = \frac{\Delta L}{\Delta F} = \frac{25.0 \text{ cm} - 12.0 \text{ cm}}{4.40 \text{ N} - 0.80 \text{ N}} = 3.61 \frac{\text{cm}}{\text{N}}$$

By analyzing the units of the equation $F = k(L - L_0)$, it is clear that the slope cannot be the spring constant; k has the same units as weight divided by length (N/cm). Is the slope equal to $1/k$? The units would be correct for that case. To be sure, we solve the equation of the line for L:

$$L = \left(\frac{1}{k}\right) F + L_0$$

We recognize the equation of a line in the familiar form $y = mx + b$, where the dependent variable L replaces y and the independent variable F replaces x. The intercept is $b = L_0$ and the slope is $m = 1/k$. Therefore,

$$k = \frac{1}{3.61 \text{ cm/N}} = 0.277 \text{ N/cm}$$

Discussion As discussed in the graphing guidelines, the slope of the straight-line graph is calculated from two widely spaced values *along the best-fit line*. We do not subtract values of actual data points. We are looking for an average value from the data; using two data points to find the slope would defeat the purpose of plotting a graph or of taking more than two data measurements. The values read from the graph, including the units, are indicated in Fig. 1.5. The units for the slope are cm/N, since we plotted centimeters versus newtons. For this particular problem the *inverse* of the slope is the quantity we seek, the spring constant in N/cm.

Practice Problem 1.11 Another Weight on Spring

What is the length of the spring of Example 1.11 when a weight of 8.00 N is suspended? Assume that the relationship found in Example 1.11 still holds for this weight.

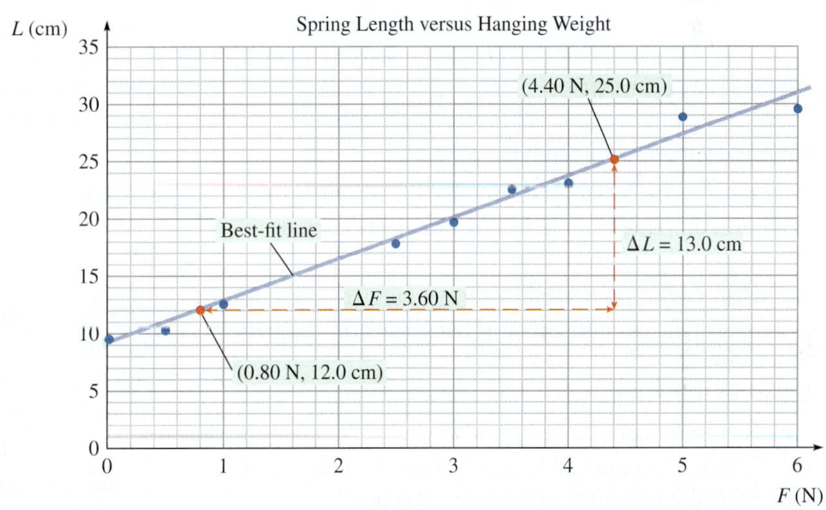

Figure 1.5
The students' graph of spring length L versus hanging weight F. After drawing a best-fit line, they calculate the slope using two points on the line.

CHECKPOINT 1.9

What value of k would you calculate by using only the first and last data points? Why is it better to use the value obtained from the best-fit line?

Master the Concepts

- Terms used in physics must be precisely defined. A term may have a different meaning in physics from the meaning of the same word in other contexts.
- A working knowledge of algebra, geometry, and trigonometry is essential in the study of physics.
- The *factor* by which a quantity is increased or decreased is the ratio of the new value to the original value.
- When we say that A is *proportional* to B (written $A \propto B$), we mean that if B increases by some factor, then A must increase by the same factor.
- In *scientific notation*, a number is written as the product of a number between 1 and 10 and a whole-number power of ten.
- *Significant figures* are the basic *grammar* of precision. They enable us to communicate quantitative information

continued on next page

Master the Concepts continued

and indicate the precision to which that information is known.

- When two or more quantities are added or subtracted, the result is as precise as the *least precise* of the quantities. When quantities are multiplied or divided, the result has the same number of significant figures as the quantity with the *smallest number of significant figures*.
- Order-of-magnitude estimates and calculations are made to be sure that the more precise calculations are realistic.
- The units used for scientific work are those from the *Système International* (*SI*). SI uses seven *base units*, which include the meter (m), the kilogram (kg), and the second (s) for length, mass, and time, respectively. Using combinations of the base units, we can construct other *derived units*.
- If the statement of a problem includes a mixture of different units, the units should be converted to a single, consistent set before the problem is solved. Usually the best way is to convert everything to SI units.
- Dimensional analysis is used as a quick check on the validity of equations. Whenever quantities are added, subtracted, or equated, they must have the same dimensions (although they may not necessarily be given in the same units).
- Mathematical approximations aid in simplifying complicated problems.
- Problem-solving techniques are *skills* that must be *practiced* to be learned.
- A graph is plotted to give a picture of the data and to show how one variable changes with respect to another. Graphs are used to help us see a pattern in the relationship between two variables.
- Whenever possible, make a careful choice of the variables plotted so that the graph displays a linear relationship.

Conceptual Questions

1. Give a few reasons for studying physics.
2. Why must words be carefully defined for scientific use?
3. Why are simplified models used in scientific study if they do not exactly match real conditions?
4. By what factor does tripling the radius of a circle increase (a) the circumference of the circle? (b) the area of the circle?
5. What are some of the advantages of scientific notation?
6. After which numeral is the decimal point usually placed in scientific notation? What determines the number of numerical digits written in scientific notation?
7. Are all the digits listed as "significant figures" precisely known? Might any of the significant digits be less precisely known than others? Explain.
8. Why is it important to write quantities with the correct number of significant figures?
9. List three of the base units used in SI.
10. What are some of the differences between the SI and the customary U.S. system of units? Why is SI preferred for scientific work?
11. Sort the following units into three groups of dimensions and identify the dimensions: fathoms, grams, years, kilometers, miles, months, kilograms, inches, seconds.
12. What are the first two steps to be followed in solving almost any physics problem?
13. Why do scientists plot graphs of their data instead of just listing values?
14. A student's lab report concludes, "The speed of sound in air is 327." What is wrong with that statement?
15. Once the solution of a problem has been found, what should be done before moving on to solve another problem?

Multiple-Choice Questions

Student Response System Questions

1. One kilometer is approximately
(a) 2 miles (b) 1/2 mile (c) 1/10 mile (d) 1/4 mile
2. By what factor does the volume of a cube increase if the length of the edges are doubled?
(a) 16 (b) 8 (c) 4 (d) 2 (e) $\sqrt{2}$
3. 55 mi/h is approximately
(a) 90 km/h (b) 30 km/h (c) 10 km/h (d) 2 km/h
4. If the length of a box is reduced to one third of its original value and the width and height are doubled, by what factor has the volume changed?
(a) 2/3 (b) 1 (c) 4/3 (d) 3/2
(e) depends on relative proportion of length to height and width
5. If the area of a circle is found to be half of its original value after the radius is multiplied by a certain factor, what was the factor used?
(a) $1/(2\pi)$ (b) 1/2 (c) $\sqrt{2}$ (d) $1/\sqrt{2}$ (e) 1/4
6. An equation for potential energy states $U = mgh$. If U is in $\text{kg}\cdot\text{m}^2\cdot\text{s}^{-2}$, m is in kg, and g is in $\text{m}\cdot\text{s}^{-2}$, what are the units of h?
(a) s (b) s^2 (c) m^{-1} (d) m (e) g^{-1}
7. In terms of the original diameter d, what new diameter will result in a new spherical volume that is a factor of eight times the original volume?
(a) $8d$ (b) $2d$ (c) $d/2$ (d) $d \times \sqrt[3]{2}$ (e) $d/8$

8. How many significant figures should be written in the sum 4.56 g + 9.032 g + 580.0078 g + 540.439 g?
(a) 3 (b) 4 (c) 5 (d) 6 (e) 7

9. The equation for the speed of sound in a gas states that $v = \sqrt{\gamma k_B T/m}$. Speed v is measured in m/s, γ is a dimensionless constant, T is temperature in kelvins (K), and m is mass in kg. What are the units for the Boltzmann constant, k_B?
(a) $kg \cdot m^2 \cdot s^2 \cdot K$ (b) $kg \cdot m^2 \cdot s^{-2} \cdot K^{-1}$ (c) $kg^{-1} \cdot m^{-2} \cdot s^2 \cdot K$
(d) $kg \cdot m/s$ (e) $kg \cdot m^2 \cdot s^{-2}$

10. How many significant figures should be written in the product 0.0078406 m × 9.45020 m?
(a) 3 (b) 4 (c) 5 (d) 6 (e) 7

Problems

- Ⓒ Combination conceptual/quantitative problem
- Biomedical application
- ✦ Challenging
- Blue # Detailed solution in the Student Solutions Manual
- [1, 2] Problems paired by concept
- connect Text website interactive or tutorial

1.3 The Use of Mathematics

1. The gardener is told that he must increase the height of his fences 37% if he wants to keep the deer from jumping in to eat the foliage and blossoms. If the current fence is 1.8 m high, how high must the new fence be?

2. What is the ratio of the number of seconds in a day to the number of hours in a day?

3. A spherical balloon expands when it is taken from the cold outdoors to the inside of a warm house. If its surface area increases 16.0%, by what percentage does the radius of the balloon change?

4. A spherical balloon is partially blown up and its surface area is measured. More air is then added, increasing the volume of the balloon. If the surface area of the balloon expands by a factor of 2.0 during this procedure, by what factor does the radius of the balloon change? (connect tutorial: car on curve)

5. A study finds that the metabolic rate of mammals is proportional to $m^{3/4}$, where m is total body mass. By what factor does the metabolic rate of a 70 kg human exceed that of a 5.0 kg cat?

6. Samantha is 1.50 m tall on her eleventh birthday and 1.65 m tall on her twelfth birthday. By what factor has her height increased? By what percentage?

7. The "scale" of a certain map is 1/10000. This means the length of, say, a road as represented on the map is 1/10000 the actual length of the road. What is the ratio of the *area* of a park as represented on the map to the actual area of the park? (connect tutorial: scaling)

8. Ⓒ On Monday, a stock market index goes up 5.00%. On Tuesday, the index goes down 5.00%. What is the net percentage change in the index for the two days? Explain why it is not zero.

9. According to Kepler's third law, the orbital period T of a planet is related to the radius R of its orbit by $T^2 \propto R^3$. Jupiter's orbit is larger than Earth's by a factor of 5.19. What is Jupiter's orbital period? (Earth's orbital period is 1 yr.)

10. If the radius of a circular garden plot is increased by 25%, by what percentage does the area of the garden increase?

11. A poster advertising a student election candidate is too large according to the election rules. The candidate is told she must reduce the length and width of the poster by 20.0%. By what percentage must the area of the poster be reduced?

12. An architect is redesigning a rectangular room on the blueprints of the house. He decides to double the width of the room, increase the length by 50%, and increase the height by 20%. By what factor has the volume of the room increased?

13. In cleaning out the artery of a patient, a doctor increases the radius of the opening by a factor of 2.0. By what factor does the cross-sectional area of the artery change?

14. ✦ A scanning electron micrograph of xylem vessels in a corn root shows the vessels magnified by a factor of 600. In the micrograph the xylem vessel is 3.0 cm in diameter. (a) What is the diameter of the vessel itself? (b) By what factor has the cross-sectional area of the vessel been increased in the micrograph?

15. ✦ The electric power P drawn from a generator by a lightbulb of resistance R is $P = V^2/R$, where V is the line voltage. The resistance of bulb B is 42% greater than the resistance of bulb A. What is the ratio P_B/P_A of the power drawn by bulb B to the power drawn by bulb A if the line voltages are the same?

1.4 Scientific Notation and Significant Figures

16. Rank these measurements of surface area in order of the number of significant figures, from fewest to greatest:
(a) 20 145 m^2; (b) 1.750×10^3 cm^2; (c) 0.00036 mm^2; (d) 8.0×10^{-2} mm^2; (e) 0.200 cm^2.

17. Perform these operations with the appropriate number of significant figures.
(a) 3.783×10^6 kg + 1.25×10^8 kg
(b) $(3.783 \times 10^6 \text{ m})/(3.0 \times 10^{-2} \text{ s})$

18. Write these numbers in scientific notation: (a) the U.S. population, 310000000; (b) the diameter of a helium nucleus, 0.0000000000000038 m.

19. In the following calculations, be sure to use an appropriate number of significant figures.
(a) $3.68 \times 10^7 \text{ g} - 4.759 \times 10^5 \text{ g}$
(b) $\dfrac{6.497 \times 10^4 \text{ m}^2}{5.1037 \times 10^2 \text{ m}}$

20. Rank the results of the following calculations in order of the number of significant figures, from least to greatest.
(a) 6.85×10^{-5} m + 2.7×10^{-7} m
(b) 702.35 km + 1897.648 km
(c) 5.0 m × 4.3 m
(d) $(0.04/\pi)$ cm
(e) $(0.040/\pi)$ m

21. Solve the following problem and express the answer in scientific notation with the appropriate number of significant figures:

$$(3.2 \text{ m}) \times (4.0 \times 10^{-3} \text{ m}) \times (1.3 \times 10^{-8} \text{ m})$$

22. Rank these measurements in order of the number of significant figures, from least to greatest.
(a) 7.68 g (b) 0.420 kg
(c) 0.073 m (d) 7.68×10^5 g
(e) 4.20×10^3 kg (f) 7.3×10^{-2} m
(g) 2.300×10^4 s

23. Solve the following problem and express the answer in meters per second (m/s) with the appropriate number of significant figures: (3.21 m)/(7.00 ms) = ? [*Hint*: Note that ms stands for milliseconds.]

24. Solve the following problem and express the answer in meters with the appropriate number of significant figures and in scientific notation:

$$3.08 \times 10^{-1} \text{ km} + 2.00 \times 10^3 \text{ cm}$$

25. Given these measurements, identify the number of significant figures and rewrite in standard scientific notation.
(a) 0.005 74 kg (b) 2 m (c) 0.450×10^{-2} m
(d) 45.0 kg (e) 10.09×10^4 s (f) $0.095 00 \times 10^5$ mL

1.5 Units

26. Rank the following lengths from smallest to greatest:
(a) 1 μm; (b) 1000 nm; (c) 100 000 pm;
(d) 0.01 cm; (e) 0.000 000 000 1 km.

27. Rank these speed measurements from smallest to greatest:
(a) 55 mi/h; (b) 82 km/h; (c) 33 m/s;
(d) 3.0 cm/ms; (e) 1.0 mi/min.

28. The density of body fat is 0.9 g/cm^3. Find the density in kg/m^3.

29. A cell membrane is 7.0 nm thick. How thick is it in inches?

30. The label on a small soda bottle lists the volume of the drink as 355 mL. Use the conversion factor 1 gal = 128 fl oz. (a) How many fluid ounces are in the bottle? (b) A competitor's drink is labeled 16.0 fl oz. How many milliliters are in that drink?

31. The length of the river span of the Brooklyn Bridge is 1595.5 ft. The total length of the bridge is 6016 ft. Find the length in meters of (a) the river span and (b) the total bridge length.

32. A beaker contains 255 mL of water. (1 mL = 1 milliliter; 1 L = 1000 cm^3.) What is the volume of the water in (a) cubic centimeters? (b) cubic meters?

33. A nerve impulse travels along a myelinated neuron at 80 m/s. What is this speed in (a) mi/h and (b) cm/ms?

34. The first modern Olympics in 1896 had a marathon distance of 40 km. In 1908, for the Olympic marathon in London, the length was changed to 42.195 km to provide the British royal family with a better view of the race. This distance was adopted as the official marathon length in 1921 by the International Amateur Athletic Federation. What is the official length of the marathon in miles?

35. At the end of 2006 an expert economist from the Global Economic Institute in Kiel, Germany, predicted a drop in the value of the U.S. dollar against the euro of 10% over the next five years. If the exchange rate was $1.27 to 1 euro on November 5, 2006, and was $1.45 to 1 euro on November 5, 2007, what was the actual drop in the value of the dollar over the first year?

36. The intensity of the Sun's radiation that reaches Earth's atmosphere is 1.4 kW/m^2 (kW = kilowatt; W = watt). Convert this to W/cm^2.

37. Blood flows through the aorta at an average speed of v = 18 cm/s. The aorta is roughly cylindrical with a radius r = 12 mm. The volume rate of blood flow through the aorta is $\pi r^2 v$. Calculate the volume rate of blood flow through the aorta in L/min.

38. A molecule in air is moving at a speed of 459 m/s. How many meters would the molecule move during 7.00 ms (milliseconds) if it didn't collide with any other molecules?

39. Express this product in units of km^3 with the appropriate number of significant figures: (3.2 km) × (4.0 m) × (13×10^{-3} mm).

40. (a) How many square centimeters are in 1 square foot? (1 in. = 2.54 cm.) (b) How many square centimeters are in 1 square meter? (c) Using your answers to parts (a) and (b), but without using your calculator, determine roughly how many square feet are in one square meter.

41. A snail crawls at a pace of 5.0 cm/min. Express the snail's speed in (a) ft/s and (b) mi/h.

42. An average-sized capillary in the human body has a cross-sectional area of about 150 μm^2. What is this area in square millimeters (mm^2)?

1.6 Dimensional Analysis

43. An equation for potential energy states $U = mgh$. If U is in joules (J), with m in kg, h in m, and g in m/s^2, find the combination of SI base units that are equivalent to joules.

44. One equation involving force states that $F_{net} = ma$, where F_{net} is in newtons (N), m is in kg, and a is in m·s^{-2}. Another equation states that $F = -kx$, where F is in newtons, k is in kg·s^{-2}, and x is in m. (a) Analyze the dimensions of ma and kx to show they are equivalent. (b) What are the dimensions of the force unit newton?

45. An equation for the period T of a planet (the time to make one orbit about the Sun) is $T^2 = 4\pi^2 r^3/(GM)$, where T is in s, r is in m, G is in $m^3/(kg \cdot s^2)$, and M is in kg. Show that the equation is dimensionally correct.

46. The relationship between kinetic energy K (SI unit $kg \cdot m^2 \cdot s^{-2}$) and momentum p is $K = p^2/(2m)$, where m stands for mass. What is the SI unit of momentum?

47. An expression for buoyant force is $F_B = \rho g V$, where F_B has dimensions $[MLT^{-2}]$, ρ (density) has dimensions $[ML^{-3}]$, and g (gravitational field strength) has dimensions $[LT^{-2}]$. (a) What must be the dimensions of V? (b) Which could be the correct interpretation of V: velocity or volume?

48. ✦ An object moving at constant speed v around a circle of radius r has an acceleration a directed toward the center of the circle. The SI unit of acceleration is m/s^2. (a) Use dimensional analysis to find a as a function of v and r. (b) If the speed is increased 10.0%, by what percentage does the radial acceleration increase?

1.8 Approximation

49. What is the approximate distance from your eyes to a book you are reading?

50. What is the approximate volume of your physics textbook in cubic centimeters (cm^3)?

51. Estimate the average mass of a person's leg.

52. Estimate the number of times a human heart beats during its lifetime.

53. What is the order of magnitude of the height (in meters) of a 40 story building?

54. Estimate the surface area of skin covering a human body by assuming that the body can be approximated as a 2.0 m tall cylinder with a radius of 15 cm with two cylindrical arms of length 1.0 m and radius of 5.0 cm.

1.9 Graphs

55. A nurse recorded the values shown in the following chart for a patient's temperature. Plot a graph of temperature versus elapsed time. From the graph, find (a) an estimate of the temperature at noon and (b) the slope of the graph. (c) Would you expect the graph to follow the same trend over the next 12 hours? Explain.

Time	Temp (°F)
10:00 A.M.	100.00
10:30 A.M.	100.45
11:00 A.M.	100.90
11:30 A.M.	101.35
12:45 P.M.	102.48

56. A patient's temperature was 97.0°F at 8:05 A.M. and 101.0°F at 12:05 P.M. If the temperature change with respect to elapsed time was linear throughout the day, what would the patient's temperature be at 3:35 P.M.?

57. A physics student plots results of an experiment as v versus t. The equation that describes the line is given by $at = v - v_0$. (a) What is the slope of this line? (b) What is the vertical axis intercept of this line?

58. A linear plot of speed versus elapsed time has a slope of 6.0 m/s^2 and a vertical intercept of 3.0 m/s. (a) What is the change in speed in the time interval between 4.0 s and 6.0 s? (b) What is the speed when the elapsed time is equal to 5.0 s?

59. An object is moving in the x-direction. A graph of the distance it has moved as a function of time is shown. (a) What are the slope and vertical axis intercept? (Be sure to include units.) (b) What physical significance do the slope and intercept on the vertical axis have for this graph?

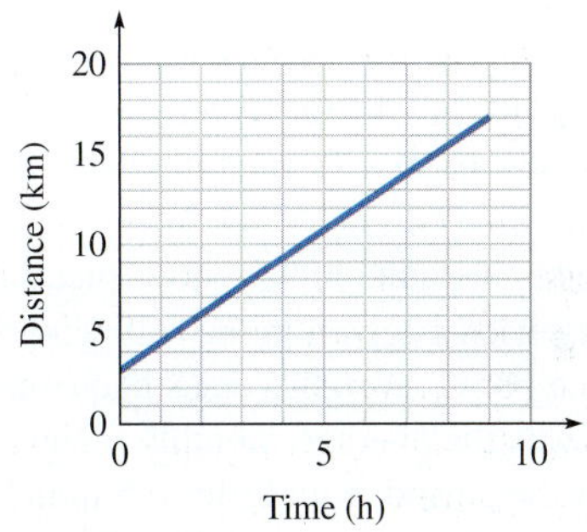

60. You have just performed an experiment in which you measured many values of two quantities, A and B. According to theory, $A = cB^3 + A_0$. You want to verify that the values of c and A_0 are correct by making a graph of your data that enables you to determine their values from a slope and a vertical axis intercept. What quantities do you put on the vertical and horizontal axes of the plot?

61. A graph of x versus t^4, with x on the vertical axis and t^4 on the horizontal axis, is linear. Its slope is 25 m/s^4 and its vertical axis intercept is 3 m. Write an equation for x as a function of t.

62. In a physics lab, students measure the sedimentation velocity v of spheres with radius r falling through a fluid. The expected relationship is $v = 2r^2 g(\rho - \rho_f)/(9\eta)$. (a) How should the students plot the data to test this relationship? (b) How could they determine the value of η from their plot, assuming values of the other constants are known?

63. In a laboratory you measure the decay rate of a sample of radioactive carbon. You write down the following measurements:

Time (min)	0	15	30	45	60	75	90
Decay rate (decays/s)	405	237	140	90	55	32	19

(a) Plot the decay rate versus time. (b) Plot the natural logarithm of the decay rate versus time. Explain why the presentation of the data in this form might be useful.

Collaborative Problems

64. (a) Estimate the number of breaths you take in one year. (b) Estimate the volume of air you breathe in during one year.

65. Use dimensional analysis to determine how the linear speed (v in m/s) of a particle traveling in a circle depends on some, or all, of the following properties: r is the radius of the circle; ω is an angular frequency in s^{-1} with which the particle orbits about the circle, and m is the mass of the particle. There is no dimensionless constant involved in the relation.

66. Estimate the number of automobile repair shops in your city by considering its population, how often an automobile needs repairs, and how many cars each shop can service per day. Then look in the yellow pages of your phone directory to see how accurate your estimate is. By what percentage was your estimate off?

67. The weight of a baby measured over an 11 month period is given in the following weight chart. (a) Plot the baby's weight versus age over the 11 mon. (b) What was the average monthly weight gain for this baby over the period from birth to 5 mon? How do you find this value from the graph? (c) What was the average monthly weight gain for the baby over the period from 5 mon to 10 mon? (d) If a baby continued to grow at the same rate as in the first 5 mon of life, what would the child weigh at age 12 yr?

Weight (lb)	Age (mon)
6.6	0 (birth)
7.4	1.0
9.6	2.0
11.2	3.0
12.0	4.0
13.6	5.0
13.8	6.0
14.8	7.0
15.0	8.0
16.6	9.0
17.5	10.0
18.4	11.0

68. It is useful to know when a small number is negligible. Perform the following computations: (a) 186.300 + 0.0030, (b) 186.300 − 0.0030, (c) 186.300 × 0.0030, (d) 186.300/0.0030. (e) For cases (a) and (b), what percent error will result if you ignore the 0.0030? Explain why you can never ignore the smaller number, 0.0030, for case (c) and case (d). (f) What rule can you make about ignoring small values?

69. ✦ Estimate the number of hairs on the average human head. [*Hint*: Consider the number of hairs in an area of 1 in.^{2} or 1 cm^2 and then consider the area covered by hair on the head.]

Comprehensive Problems

70. You are given these approximate measurements: (a) the radius of Earth is 6×10^6 m, (b) the length of a human body is 6 ft, (c) a cell's diameter is 10^{-6} m, (d) the width of the hemoglobin molecule is 3×10^{-9} m, and (e) the distance between two atoms (carbon and nitrogen) is 3×10^{-10} m. Write these measurements in the simplest possible metric prefix forms (in either nm, Mm, μm, or whatever works best).

71. A typical virus is a packet of protein and DNA (or RNA) and can be spherical in shape. The influenza A virus is a spherical virus that has a diameter of 85 nm. If the volume of saliva coughed onto you by your friend with the flu is 0.010 cm^3 and 10^{-9} is the fraction of that volume that consists of viral particles, how many influenza viruses have just landed on you?

72. The smallest "living" thing is probably a type of infectious agent known as a viroid. Viroids are plant pathogens that consist of a circular loop of single-stranded RNA, containing about 300 bases. (Think of the bases as beads strung on a circular RNA string.) The distance from one base to the next (measured along the circumference of the circular loop) is about 0.35 nm. What is the diameter of a viroid in (a) m, (b) μm, and (c) in.?

73. The largest living creature on Earth is the blue whale, which has an average length of 70 ft. The largest blue whale on record was 1.10×10^2 ft long. (a) Convert this length to meters. (b) If a double-decker London bus is 8.0 m long, how many double-decker-bus lengths is the record whale?

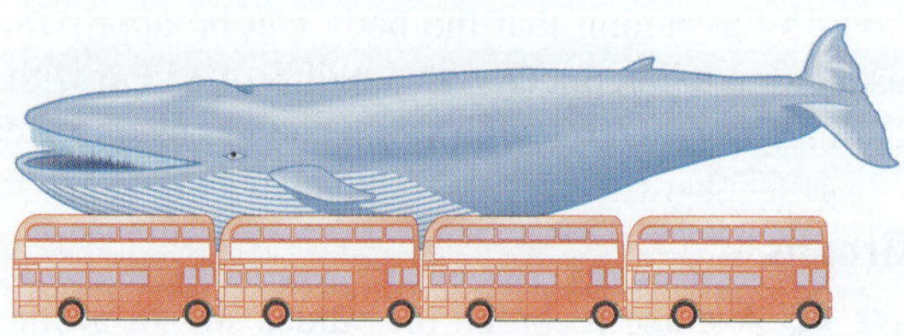

74. The record blue whale in Problem 73 had a mass of 1.9×10^5 kg. Assuming that its average density was 0.85 g/cm^3, as has been measured for other blue whales, what was the volume of the whale in cubic meters (m^3)? (Average density is the ratio of mass to volume.)

75. The total length of the blood vessels in the body is roughly 100 000 km. Most of this length is due to the capillaries, which have an average diameter of 8 μm. Estimate the total volume of blood in the human body by assuming that all the blood is found in the capillaries and that they are always full of blood.

76. A sheet of paper has length 27.95 cm, width 8.5 in., and thickness 0.10 mm. What is the volume of a sheet of paper in m^3? (Volume = length × width × thickness.)

77. ✦ The average speed of a nitrogen molecule in air is proportional to the square root of the temperature in

kelvins (K). If the average speed is 475 m/s on a warm summer day (temperature = 300.0 K), what is the average speed on a frigid winter day (250.0 K)?

78. A furlong is 220 yd; a fortnight is 14 d. How fast is 1 furlong per fortnight (a) in μm/s? (b) in km/d?

79. In the United States, we often use miles per hour (mi/h) when discussing speed, but the SI unit of speed is m/s. What is the conversion factor for changing m/s to mi/h? If you want to make a quick approximation of the speed in mi/h given the speed in m/s, what might be the easiest conversion factor to use?

80. Two thieves, escaping after a bank robbery, drop a sack of money on the sidewalk. (a) Estimate the mass of the sack if it contains \$5000 in quarter-dollar coins. (b) Estimate the mass if the sack contains \$1 000 000 in \$20 bills.

81. The weight W of an object is given by $W = mg$, where m is the object's mass and g is the gravitational field strength. The SI unit of field strength g, expressed in SI base units, is m/s^2. What is the SI unit for weight, expressed in base units?

82. Kepler's third law of planetary motion says that the square of the period of a planet (T^2) is proportional to the cube of the distance of the planet from the Sun (r^3). Mars is about twice as far from the Sun as Venus. How does the period of Mars compare with the period of Venus?

83. Ⓒ One morning you read in the *New York Times* that the net worth of the richest man in the world, Carlos Slim Helú of Mexico, is \$59 000 000 000. Later that day you see him on the street, and he gives you a \$100 bill. What is his net worth now? (Think of significant figures.)

84. The average depth of the oceans is about 4 km, and oceans cover about 70% of Earth's surface. Make an order-of-magnitude estimate of the volume of water in the oceans. Do not look up any data. (Use your ingenuity to estimate the radius or circumference of Earth. One method is to estimate the distance between two cities and then estimate what fraction of Earth's circumference that distance represents by visualizing the two cities on a globe.)

85. Suppose you have a pair of Seven League Boots. These are magic boots that enable you to stride along a distance of 7.0 leagues with each step. (a) If you march along at a military march pace of 120 paces/min, what will be your speed in km/h? (b) Assuming you could march on top of the oceans when you step off the continents, how long (in minutes) will it take you to march around the Earth at the equator? (1 league = 3 mi = 4.8 km.)

86. A car has a gas tank that holds 12.5 U.S. gal. Using the conversion factors from the inside front cover, (a) determine the size of the gas tank in cubic inches. (b) A cubit is an ancient measurement of length that was defined as the distance from the elbow to the tip of the finger, about 18 in. long. What is the size of the gas tank in cubic cubits?

87. ✦ The weight of an object at the surface of a planet is proportional to the planet's mass and inversely proportional to the square of the radius of the planet. Jupiter's radius is 11 times Earth's, and its mass is 320 times Earth's. An apple weighs 1.0 N on Earth. How much would it weigh on Jupiter?

88. ✦ The speed of ocean waves depends on their wavelength l (measured in meters) and the gravitational field strength l (measured in m/s^2) in this way:

$$v = K\lambda^p g^q$$

where K is a dimensionless constant. Find the values of the exponents p and q.

89. ✦ Without looking up any data, make an order-of-magnitude estimate of the annual consumption of gasoline (in gallons) by passenger cars in the United States. Make reasonable estimates for any quantities you need. Think in terms of average quantities. (1 gal ≈ 4 L.)

90. ✦ Find the order of magnitude of the number of cups of water required to fill a bathtub.

91. ✦ Three of the fundamental constants of physics are the speed of light $c = 3.0 \times 10^8$ m/s, the universal gravitational constant $G = 6.7 \times 10^{-11}\ m^3{\cdot}kg^{-1}{\cdot}s^{-2}$, and Planck's constant $h = 6.6 \times 10^{-34}\ kg{\cdot}m^2{\cdot}s^{-1}$.

(a) Find a combination of these three constants that has the dimensions of time. This time is called the *Planck time* and represents the age of the universe before which the laws of physics as presently understood cannot be applied. (b) Using the formula for the Planck time derived in part (a), what is the time in seconds?

92. ✦ Use dimensional analysis to determine how the period T of a swinging pendulum (the elapsed time for a complete cycle of motion) depends on some, or all, of these properties: the length L of the pendulum, the mass m of the pendulum bob, and the gravitational field strength g (in m/s^2). Assume that the amplitude of the swing (the maximum angle that the string makes with the vertical) has no effect on the period.

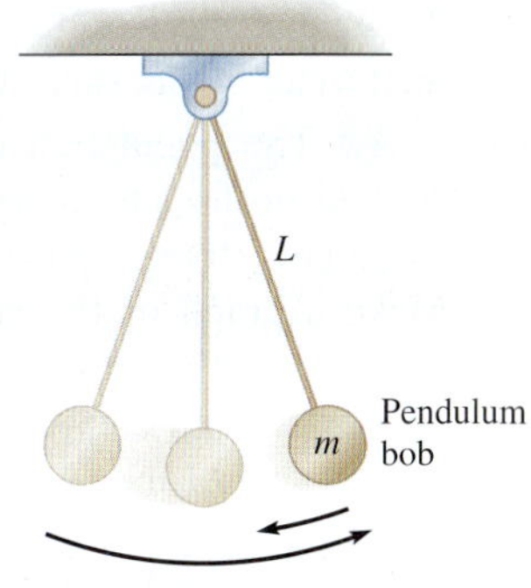

93. ✦ Astronauts aboard the International Space Station use a *massing chair* to measure their mass. The chair is attached to a spring and is free to oscillate back and forth. The frequency of the oscillation is measured and is used to calculate the total mass m attached to the spring. If the spring constant of the spring k is measured in kg/s^2 and the chair's frequency f is 0.50 s^{-1} for a 62 kg astronaut, what is the chair's frequency for a

75 kg astronaut? The chair itself has a mass of 10.0 kg. [*Hint*: Use dimensional analysis to find out how *f* depends on *m* and *k*.]

94. (a) How many center-stripe road reflectors, separated by 17.6 yd, are required along a 2.20 mile section of curving mountain roadway? (b) Solve the same problem for a road length of 3.54 km with the markers placed every 16.0 m. Would you prefer to be the highway engineer in a country with a metric system or U.S. customary units?

95. A baby was persistently spitting up after nursing and the pediatrician prescribed Zantac syrup to reduce the baby's stomach acid. The prescription called for 0.75 mL to be taken twice a day for a month. The pharmacist printed a label for the bottle of syrup that said "3/4 tsp. twice a day." By what factor was the baby overmedicated before the error was discovered at the baby's next office visit 2 weeks later? [*Hint*: 1 tsp = 4.9 mL.]

96. On April 15, 1999, a South Korean cargo plane crashed due to a confusion over units. The plane was to fly from Shanghai, China, to Seoul, Korea. After takeoff the plane climbed to 900 m. Then the first officer was instructed by the Shanghai tower to climb to 1500 m and maintain that altitude. The captain, after reaching 1450 m, twice asked the first officer at what altitude they should fly. He was twice told incorrectly they were to be at 1500 ft. The captain pushed the control column quickly forward and started a steep descent. The plane could not recover from the dive and crashed. How much above the correct altitude did the captain think they were when he started his rapid descent and lost control? (It turns out that aircraft altitudes are given in feet throughout the world except in China, Mongolia, and the former Soviet states where meters are used.)

97. ✦ The population of a culture of yeast cells is studied in the laboratory to see the effects of limited resources (food, space) on population growth. (a) Make a graph of the yeast population (measured as total mass of yeast cells) versus elapsed time. Draw a best-fit smooth curve. (b) After a long time, the population approaches a maximum known as the *carrying capacity*. Estimate the carrying capacity for this population. (c) When the population is much smaller than the carrying capacity, the growth is expected to be exponential: $m(t) = m_0 e^{rt}$, where m is the population at any time t, m_0 is the initial population, r is the *intrinsic growth rate* (i.e., the growth rate in the absence of limits), and e is the base of natural logarithms (see Appendix A.3). To obtain a straight-line graph from this exponential relationship, we can plot the natural logarithm of m/m_0:

$$\ln \frac{m}{m_0} = \ln e^{rt} = rt$$

Make a graph of $\ln (m/m_0)$ versus t from $t = 0$ to $t = 6.0$ h, and use it to estimate the intrinsic growth rate r for the yeast population.

Time (h)	Mass (g)
0.0	3.2
2.0	5.9
4.0	10.8
6.0	19.1
8.0	31.2
10.0	46.5
12.0	62.0
14.0	74.9
16.0	83.7
18.0	89.3
20.0	92.5
22.0	94.0
24.0	95.1

Answers to Practice Problems

1.1 a 4% decrease

1.2 48.6 W

1.3 (a) five; 1.0544×10^{-4} kg; (b) four; 5.800×10^{-3} cm; (c) ambiguous, three to six; if three, 6.02×10^5 s

1.4 The least precise value is to the nearest hundredth of a meter, so we round the result to the nearest hundredth of a meter: 564.50 m or, in scientific notation, 5.6450×10^2 m; five significant figures.

1.5 4.7 m/s

1.6 (a) 35.6 m/s; (b) 79.5 mi/h

1.7 5.1×10^{14} m^2; 2.0×10^8 mi^2

1.8 The equation is dimensionally inconsistent; the right side has dimensions [L/T]. To have matching dimensions we must multiply the right side by [T]; the equation must involve time squared, as in: $d = \frac{1}{2}at^2$.

1.9 kinetic energy = (constant) × mv^2; kinetic energy increases by 56%.

1.10 10^{11} L (Make a rough estimate of the population to be about 3×10^8 people, each drinking about 1.5 L/day.)

1.11 38.0 cm

Answers to Checkpoints

1.3 The volume increases by a factor of 27.

1.4 Order-of-magnitude estimates provide a quick method for obtaining limited precision solutions to problems. Even if greater accuracy is required, order-of-magnitude calculations are still useful as they provide a check as to the accuracy of the higher precision calculation.

1.5 1.9 L

1.6 (a) and (b) It is possible to multiply or divide quantities with different dimensions. (c) and (d) To be added or subtracted, quantities *must* have the same dimensions.

1.9 0.299 N/cm. The value from the best-fit line takes *all* the data into account. Using just two data points would ignore all the rest of the data and would magnify the effect of measurement errors in those two data points.

CHAPTER

2

Motion Along a Line

Despite its enormous mass (425–900 kg), the Cape buffalo is capable of running at a top speed of about 55 km/h (34 mi/h). Since the top speed of the African lion is about the same, how is it ever possible for a lion to catch the buffalo, especially since the lion typically makes its move from a distance of 20 to 30 m from the buffalo? (See p. 35 for the answer.)

BIOMEDICAL APPLICATIONS

- Doppler echocardiography (Ex. 2.6)
- Speed and acceleration of animals (Sec. 2.3; Probs. 7, 27, 75, 76)
- Action potentials in neurons (Prob. 87)

Concepts & Skills to Review

- scientific notation and significant figures (Section 1.4)
- converting units (Section 1.5)
- problem-solving techniques (Section 1.7)
- meaning of *velocity* in physics (Section 1.2)

2.1 POSITION AND DISPLACEMENT

CONNECTION:

The topic of Chapters 2 and 3 is **kinematics**: the mathematical description of motion. Beginning in Chapter 4, we will learn the principles of physics that predict and explain *why* objects move the way they do.

Position

To describe motion unambiguously, we need a way to say *where* an object is located. Suppose that at 3:00 P.M. a train stops on an east-west track as a result of an engine problem. The engineer wants to call the railroad office to report the problem. How can he tell them where to find the train? He might say something like "three kilometers east of the old trestle bridge." Notice that he uses a point of reference: the old trestle bridge. Then he states how far the train is from that point and in what direction. If he omits any of the three pieces (the reference point, the distance, or the direction), then his description of the train's whereabouts is ambiguous.

The same thing is done in physics. First, we choose a reference point, called the **origin**. Then, to describe the location of something, we give its distance from the origin and the direction. For motion along a line, we can choose the line of motion to be the x-axis of a coordinate system. The origin is the point $x = 0$. The position of an object can be described by its x-coordinate, which tells us both how far the object is from the origin and on which side. For the train in Fig. 2.1, we choose the origin at the center of the bridge and the $+x$-direction to the east. Then $x = +3$ km means the train is 3 km east of the bridge and $x = -26$ km means the train is 26 km west of the bridge.

Displacement

Once the train is underway, we might want to describe its motion. At 3:14 P.M., it leaves its initial position, 3 km east of the origin (see Fig. 2.1). At 3:56 P.M., the train is 26 km west of the origin, which is 29 km to the west of its initial position. **Displacement** is defined as the change of the position—the final position minus the initial position. The displacement is written Δx, where the symbol Δ (the uppercase Greek letter delta) means *the change in* the quantity that follows.

Displacement

$$\Delta x = x_f - x_i \quad (2\text{-}1)$$

Figure 2.1 Initial (x_i) and final (x_f) positions of a train. (Train not to scale.)

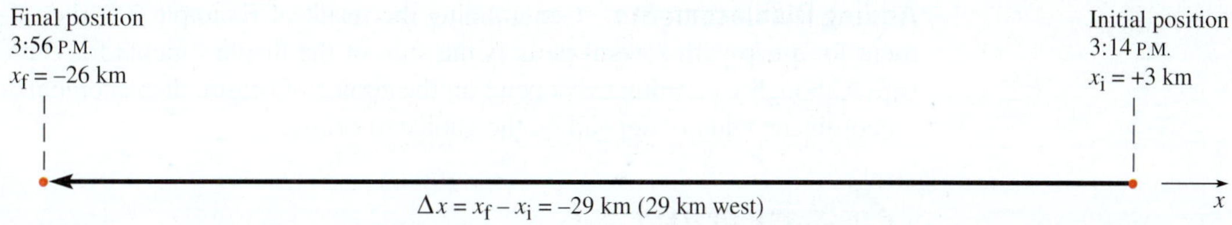

Figure 2.2 With the x-axis pointing east, $\Delta x = x_f - x_i = -26\text{ km} - (+3\text{ km}) = -29\text{ km}$. The train's displacement is 29 km west.

We can subtract x-coordinates to find the displacement of the train. If we choose the x-axis to the east, then $x_i = +3$ km and $x_f = -26$ km. The displacement is

$$\Delta x = x_f - x_i = (-26\text{ km}) - (+3\text{ km}) = -29\text{ km}$$

The displacement is 29 km in the $-x$-direction (west) (Fig. 2.2).

Displacement Versus Distance Notice that the magnitude of the displacement is not necessarily equal to the *distance traveled*. Suppose the train first travels 7 km to the east, putting it 10 km east of the origin, and then reverses direction and travels 36 km to the west. The total distance traveled in that case is (7 km + 36 km) = 43 km, but the magnitude of the displacement—which is the distance between the initial and final positions—is 29 km. The displacement depends only on the starting and ending positions, not on the path taken.

Example 2.1

A Mule Hauling Corn to Market

A mule hauls the farmer's wagon along a straight road for 4.3 km directly east to the neighboring farm where a few bushels of corn are loaded onto the wagon. Then the farmer drives the mule back along the same straight road, heading west for 7.2 km to the market. Find the displacement of the mule from the starting point to the market.

Strategy The problem gives us two successive displacements along a straight line. Let's choose the $+x$-axis to point east and an arbitrary point along the road to be the origin. Suppose the mule starts at position x_1 (Fig. 2.3). It goes east until it reaches the neighbor's farm at position x_2. The displacement to the neighbor's farm is $x_2 - x_1 = 4.3$ km east.

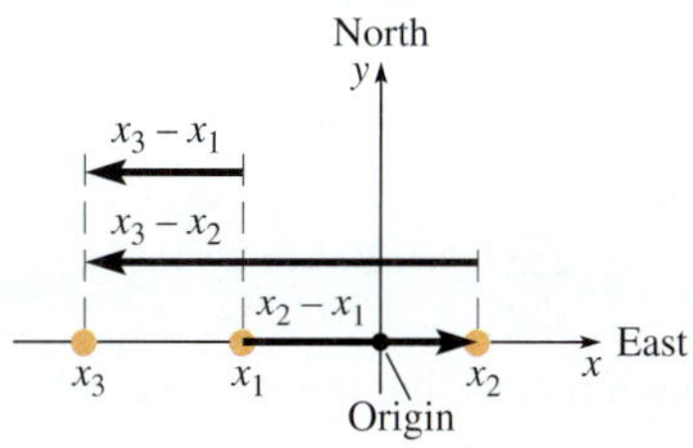

Figure 2.3
The total displacement is the sum of two successive displacements.

Then the mule goes 7.2 km west to reach the market at position x_3. The displacement from the neighbor's farm to the market is $x_3 - x_2 = -7.2$ km (negative because the displacement is in the $-x$-direction). The problem asks for the displacement of the mule from x_1 to x_3.

Solution We can eliminate x_2, the intermediate position, by adding the two displacements:

$$(x_3 - x_2) + (x_2 - x_1) = -7.2\text{ km} + 4.3\text{ km}$$
$$x_3 - x_1 = -2.9\text{ km}$$

The displacement is 2.9 km west.

Discussion When we added the two displacements, the intermediate position x_2 dropped out, as it must since the displacement is independent of the path taken from the initial position to the final position. The result does not depend on the choice of origin.

Practice Problem 2.1 A Nervous Squirrel

A nervous squirrel, trying to cross a road, first moves 3.0 m east, then 4.0 m west, then 1.2 m west, then 6.0 m east. What is the squirrel's total displacement?

Adding Displacements Generalizing the result of Example 2.1, the total displacement for a trip with several parts is the sum of the displacements for each part of the trip. Although x-coordinates depend on the choice of origin, displacements (*changes* in x-coordinates) do *not* depend on the choice of origin.

CHECKPOINT 2.1

In Example 2.1, is the magnitude of the displacement equal to the distance traveled? Explain.

2.2 VELOCITY: RATE OF CHANGE OF POSITION

We introduced *velocity* as a quantity with magnitude and direction in Section 1.2. The magnitude is the speed with which the object moves and the direction is the direction of motion. Now we develop a mathematical definition of velocity that fits that description. Note that displacement indicates by how much and in what direction the position has changed, but implies nothing about *how long* it took to move from one point to the other. Velocity depends on both the displacement and the time interval.

Average Velocity

When a displacement Δx occurs during a time interval Δt, the **average velocity** during that time interval is

Average velocity

$$v_{av,x} = \frac{\Delta x}{\Delta t} \tag{2-2}$$

Since Δt is always positive, the direction of the average velocity is the same as the direction of the displacement. The symbol Δ does not stand alone and cannot be canceled in equations because it *modifies* the quantity that follows it; $\frac{\Delta x}{\Delta t}$ means $\frac{x_f - x_i}{t_f - t_i}$, which is *not* the same as x/t.

Example 2.2

Average Velocity of a Train

Find the average velocity in kilometers per hour of the train shown in Fig. 2.1 during the time interval between 3:14 P.M., when the train is 3 km east of the origin, and 3:56 P.M., when it is 26 km west of the origin.

Strategy We choose the $+x$-axis to the east, as before. Then the displacement is $\Delta x = -29$ km, which means 29 km to the west. The average velocity is also to the west, so $v_{av,x}$ is negative. We convert Δt to hours to find the average velocity in kilometers per hour.

Solution The time interval is $\Delta t = 56 \text{ min} - 14 \text{ min} = 42$ min. Converting to hours,

$$\Delta t = 42 \text{ min} \times \frac{1 \text{ h}}{60 \text{ min}} = 0.70 \text{ h}$$

The average velocity is

$$v_{av,x} = \frac{\Delta x}{\Delta t} = \frac{-29 \text{ km}}{0.70 \text{ h}} = -41 \text{ km/h}$$

The negative sign means that the average velocity is directed along the negative x-axis, or to the west.

continued on next page

Example 2.2 continued

Discussion If the train had started at the same instant of time, 3:14 P.M., and had traveled directly west at a constant 41 km/h, it would have ended up in exactly the same place—26 km west of the trestle bridge—at 3:56 P.M.

Had we started measuring time from when we first spotted the motionless train at 3:00 P.M., instead of 3:14 P.M., we would have found the average velocity over a different time interval, changing the average velocity. The average velocity depends on the time interval considered.

The magnitude of the train's average velocity is *not* equal to the total distance traveled divided by the time interval for the complete trip. The latter quantity is called the average *speed*. The total distance is 7 km + 36 km = 43 km, so we have

$$\text{average speed} = \frac{\text{distance traveled}}{\text{total time}} = \frac{43\text{ km}}{0.70\text{ h}} = 61\text{ km/h}$$

The distinction arises because the average velocity is the constant velocity that would result in the same *displacement* (during the given time interval), while the average speed is the constant speed that would result in the same *distance traveled* (during the same time interval).

Practice Problem 2.2 Average Velocity for a Different Time Interval

What is the average velocity of the same train during the time interval from 3:28 P.M., when it is at $x = 10$ km, to 3:56 P.M., when it is at $x = -26$ km?

Average Speed Versus Average Velocity The *average* velocity does not convey detailed information about the motion during the corresponding time interval Δt. The average velocity would be the same for any other motion that takes the object through the same displacement in the same amount of time. However, the average *speed*, defined as the total *distance* traveled divided by the time interval, depends on the path traveled.

CHECKPOINT 2.2A

Can average speed ever be greater than the magnitude of the average velocity? Explain.

Instantaneous Velocity

The speedometer of a car does not indicate the average speed for an entire trip. When a speedometer reads 55 mi/h, it does *not* necessarily mean that the car travels 55 miles in the next hour; the car could change its speed or direction or stop during that hour. The speedometer reading can be used to calculate how far the car travels during a *very short time interval*—short enough that the speed does not change appreciably. For instance, at 55 mi/h (= 25 m/s), we can calculate that in 0.010 s the car moves 25 m/s × 0.010 s = 0.25 m, as long as the speed does not change significantly during that 0.010 s interval.

Similarly, the **instantaneous velocity** is a quantity whose magnitude is the speed and whose direction is the direction of motion. When we refer simply to *the velocity*, we always mean the *instantaneous* velocity. The velocity can be used to calculate the *displacement* of the object *during a very short time interval*, as long as *neither the speed nor the direction of motion* change significantly during that time interval. The sign of the velocity v_x indicates the direction of motion (positive for the $+x$-direction or negative for the $-x$-direction).

Thus, the velocity at some instant of time t is the average velocity during a *very short* time interval:

CONNECTION:

Couldn't we omit "x" subscripts in average ($v_{av,x}$) and instantaneous (v_x) velocity? If we wanted to understand only motion along a line, then we certainly would. However, in Chapter 3 we generalize the definitions of position, displacement, velocity, and acceleration as vector quantities in three dimensions. Using the "x" subscripts now lets us carry forward everything in Chapter 2 *without requiring a change in notation*. Then, when you look back to review Chapter 2, you won't have to remember different definitions for the same symbol. For example, in Chapter 3 we'll learn that v (without the subscript) stands for the *magnitude* of the velocity (the speed), which can never be negative.

Instantaneous velocity

$$v_x = \lim_{\Delta t \to 0} \frac{\Delta x}{\Delta t} \tag{2-3}$$

(Δx is the displacement during a *very short* time interval Δt)

The notation $\lim_{\Delta t \to 0}$ is read "the limit, as Δt approaches zero, of . . . ." In other words, let the time interval get smaller and smaller, *approaching*—but never reaching—zero. This notation in Eq. (2-3) reminds you that Δt must be a *very short* time interval. How short a time interval is short enough? If you use a shorter time interval and the calculation of v_x always gives the same value (to within the precision of your measurements), then Δt is short enough. In other words, Δt must be short enough that we can treat the velocity as constant during that time interval. When v_x is constant, cutting Δt in half also cuts the displacement in half, giving the same value for $\Delta x/\Delta t$.

Motion Diagrams

Suppose a cart moves to the right with increasing speed. Imagine shooting a video of the cart. Each frame in the video shows the positions of the cart at equally spaced times. If the images are compiled from successive frames into a single image, the resulting **motion diagram** (Fig. 2.4a) shows the position of the cart at equally spaced times. It's easier and just as useful to draw motion diagrams as a series of dots (Fig. 2.4b). This motion diagram shows that the speed is increasing because the displacements are getting larger. In Fig. 2.4c, black and red arrows represent the cart's displacement and velocity, respectively.

A motion diagram is closely related to a graph of x versus t. Let's choose the x-direction to the right in Fig. 2.4. The notation $x(t)$ represents position x as a function of time t. On a *graph* of $x(t)$, the x-axis is vertical, so let's rotate the motion diagram so the x-axis points up the page (Fig. 2.5). Then each dot on the motion diagram shows the vertical position of a point on the graph; these points are equally spaced along the time axis because Δt is constant.

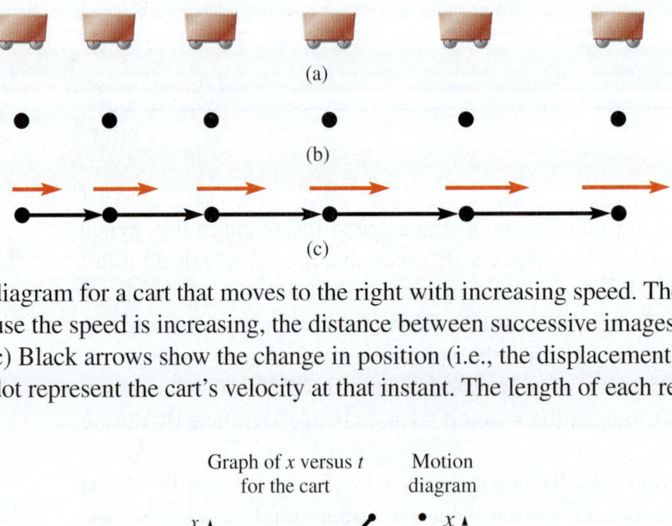

Figure 2.4 (a) The motion diagram for a cart that moves to the right with increasing speed. The cart's position is shown at equally spaced times. Because the speed is increasing, the distance between successive images increases. (b) A simplified motion diagram for the cart. (c) Black arrows show the change in position (i.e., the displacement) from one "frame" to the next. Red arrows above each dot represent the cart's velocity at that instant. The length of each red arrow represents the speed.

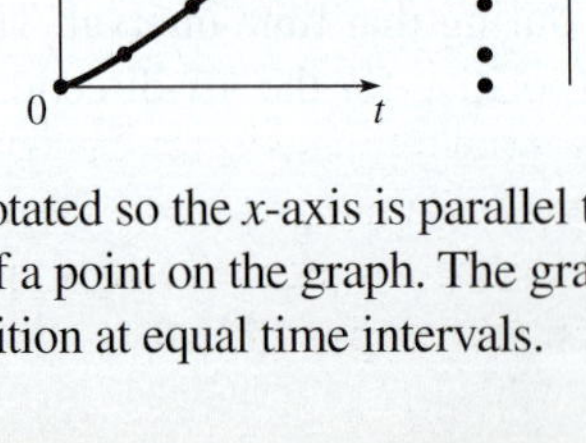

Figure 2.5 The motion diagram for the cart, rotated so the x-axis is parallel to the x-axis on the graph of $x(t)$. Each dot on the motion diagram shows the vertical position of a point on the graph. The graph points are equally spaced along the time axis because the motion diagram shows the cart's position at equal time intervals.

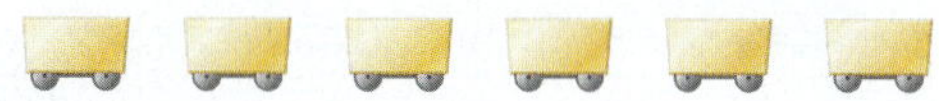

Figure 2.6 Motion diagram for a cart moving to the right.

CHECKPOINT 2.2B

The motion diagram (Fig. 2.6) shows a cart moving to the right. Describe the motion in words, and sketch a graph of $x(t)$.

Graphical Relationships Between Position and Velocity

For motion along the x-axis, the displacement is Δx. The average velocity can be represented on the graph of $x(t)$ as the slope of a line connecting two points (called a *chord*). In Fig. 2.7a, the displacement $\Delta x = x_3 - x_1$ is the *rise* of the graph (the change along the vertical axis) and the time interval $\Delta t = t_3 - t_1$ is the *run* of the graph (the change along the horizontal axis). The slope of the chord is the rise over the run:

$$\text{slope of chord} = \frac{\text{rise}}{\text{run}} = \frac{\Delta x}{\Delta t} = v_{\text{av},x} \qquad (2\text{-}4)$$

The slope of the chord is the average velocity for that time interval.

Finding v_x on a Graph of $x(t)$ To find the *instantaneous* velocity at some time $t = t_2$, we draw lines showing the average velocity for shorter and shorter time intervals. As the time interval is reduced (Fig. 2.7b), the average velocity changes. As Δt gets shorter and shorter, the chord approaches a tangent line to the graph at t_2. Thus,

> v_x is the *slope of the line tangent to the graph of* $x(t)$ at the chosen time.

In Fig. 2.8, the position of the train considered in Example 2.2 is graphed as a function of time, where 3:00 P.M. is chosen as $t = 0$. The graph of position versus time shows a curving line, but that does not mean the train travels along a curved path. The motion of the train is along a straight line since the track runs in an east-west direction.

A horizontal portion of the graph indicates that the position is not changing during that time interval and, therefore, it is at rest (its velocity is zero). Sloping portions of the graph indicate that the train is moving. The steeper the graph, the larger the speed of the train. The sign of the slope indicates the direction of motion. A positive slope indicates

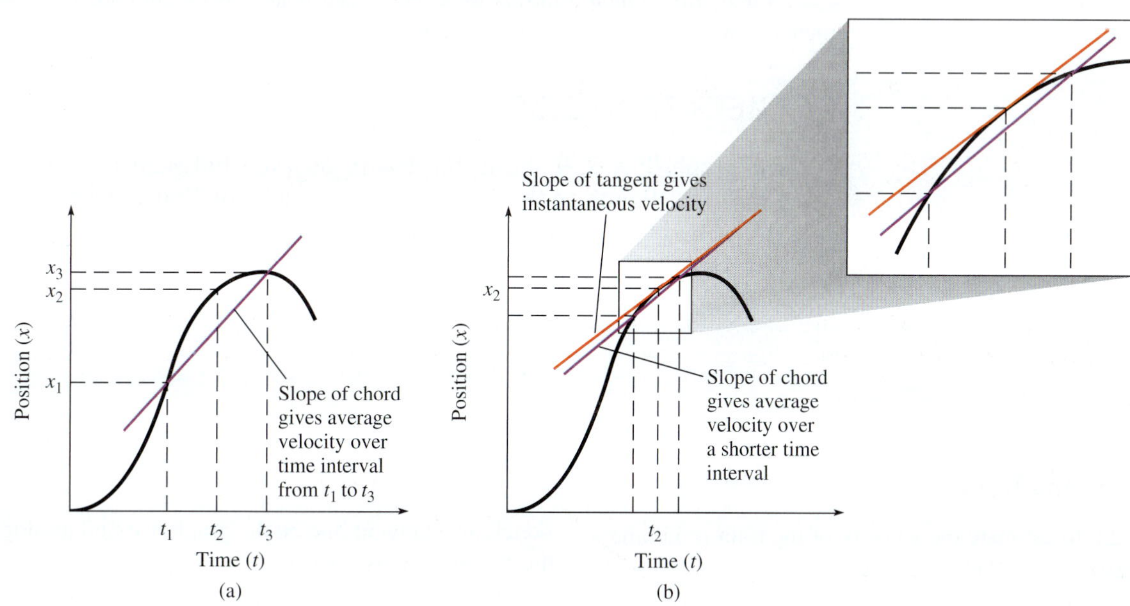

Figure 2.7 A graph of $x(t)$ for an object moving along the x-axis. (a) The average velocity $v_{x,\text{av}}$ for the time interval t_1 to t_3 is the slope of the chord connecting those two points on the graph. (b) The average velocity measured over a shorter time interval. As the time interval gets shorter and shorter, the average velocity approaches the *instantaneous* velocity v_x at the instant t_2. The slope of the *tangent line* to the graph is v_x at that instant.

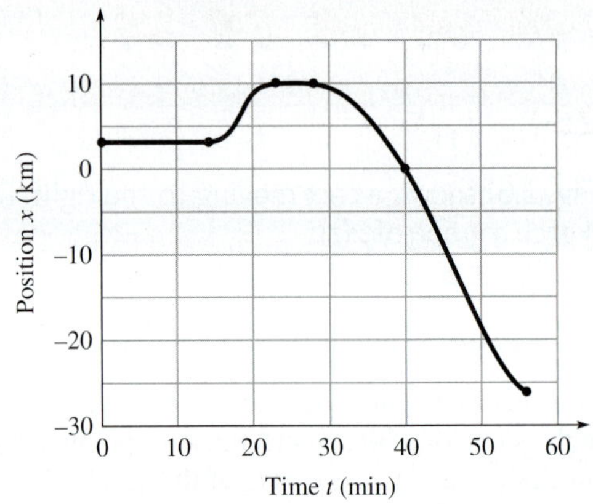

x (km)	t (min)
+3	0
+3	14
+10	23
+10	28
0	40
−26	56

Figure 2.8 Graph of position x versus time t for the train. The positions of the train at various times are marked with a dot. The position of the train would have to be measured at more frequent time intervals to accurately trace out the shape of the graph.

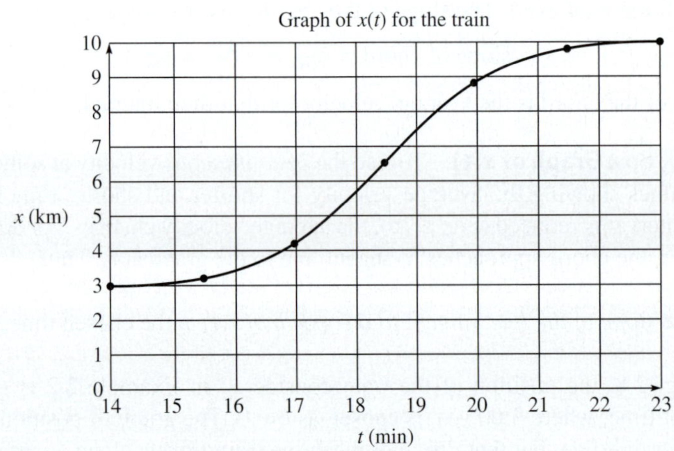

Figure 2.9 A more detailed graph of the train's position from $t = 14$ min to $t = 23$ min. Points on the graph show the train's position at equal time intervals of 1.5 min.

motion in the $+x$-direction, and a negative slope indicates motion in the $-x$-direction. In Fig. 2.8, the train is at rest from $t = 0$ to $t = 14$ min. Then it moves east, speeding up at first and then slowing down until it comes to rest at $t = 23$ min. It remains at rest until $t = 28$ min, after which it moves west. The speed is increasing until about $t = 45$ min; then it slows slightly while still moving west.

CHECKPOINT 2.2C

The graph (Fig. 2.9) shows the train's position between $t = 14$ min and $t = 23$ min. Points on the graph show the train's position at equal time intervals of 1.5 min. Draw a motion diagram and sketch a qualitative graph of $v_x(t)$. (Don't worry about numerical values—just sketch the shape of the graph.)

Example 2.3

Velocity of the Train

Use Fig. 2.8 to estimate the velocity of the train in kilometers per hour at $t = 40$ min.

Strategy Figure 2.8 is a graph of $x(t)$. The slope of a line tangent to the graph at $t = 40$ min is v_x at that instant. After sketching a tangent line on the graph, we find its slope from the rise divided by the run.

Solution Figure 2.10 shows a tangent line drawn on the graph. Using the endpoints of the tangent line, the rise is

continued on next page

Example 2.3 continued

(−25 km) − (15 km) = −40 km. The run is approximately (57 min) − (30 min) = 27 min = 0.45 h. Then

$$v_x \approx -40\ \text{km}/(0.45\ \text{h}) \approx -89\ \text{km/h}$$

The velocity is approximately 89 km/h in the $-x$-direction (west).

Discussion Since the slope of a line is constant, any two points *on the tangent line* would give the same value for the slope. Using widely spaced points gives a more precise value for the slope.

Practice Problem 2.3 Maximum Eastward Velocity

Use Fig. 2.9 to estimate the maximum velocity of the train in kilometers per hour during the time it moves east (t = 14 min to t = 23 min).

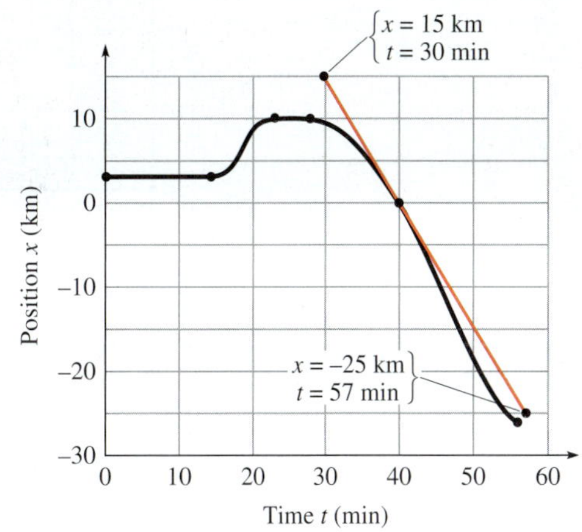

Figure 2.10
On the graph of $x(t)$, the slope of a line tangent to the graph at t = 40 min is v_x at t = 40 min.

Finding Δx with Constant Velocity What about the other way around? Given a graph of $v_x(t)$, how can we determine the displacement (change in position)? If v_x is constant during a time interval, then the average velocity is equal to the instantaneous velocity:

$$v_x = v_{\text{av},x} = \frac{\Delta x}{\Delta t} \quad \text{(for constant } v_x\text{)} \tag{2-5}$$

and therefore

$$\Delta x = v_x\,\Delta t \quad \text{(for constant } v_x\text{)} \tag{2-6}$$

The graph of Fig. 2.11 shows v_x versus t for an object moving along the x-axis with constant velocity v_1 from time t_1 to t_2. The displacement Δx during the time interval $\Delta t = t_2 - t_1$ is $v_1\,\Delta t$. The shaded rectangle has "height" v_1 and "width" Δt. Since the area of a rectangle is the product of the height and width, the displacement Δx is represented by the area of the rectangle between the graph of $v_x(t)$ and the time axis for the time interval considered.

When we speak of the area under a graph, we are not talking about the literal number of square centimeters of paper or computer screen. The figurative area under a graph usually does not have dimensions of an ordinary area [L^2]. In a graph of $v_x(t)$, v_x has dimensions [L/T] and time has dimensions [T]; areas on such a graph have dimensions [L/T] × [T] = [L], which is correct for a displacement. The *units* of Δx are determined by the units used on the axes of the graph. If v_x is in meters per second and t is in seconds, then the displacement is in meters.

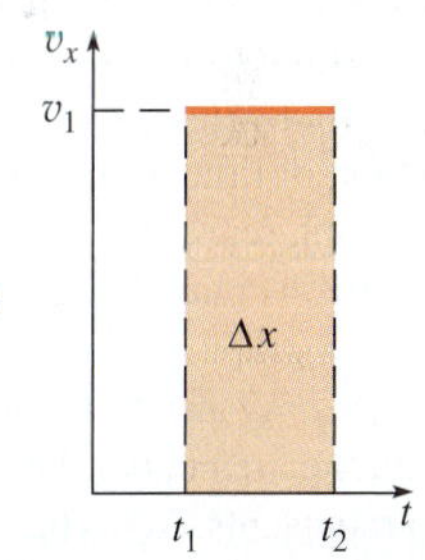

Figure 2.11 Displacement Δx between t_1 and t_2 is represented by the shaded area under the red $v_x(t)$ graph because $\Delta x = v_x\,\Delta t$.

Finding Δx with Changing Velocity What if the velocity is not constant? The displacement Δx during a *very small* time interval Δt can be found in the same way as for constant velocity since, during a short enough time interval, the velocity does not change appreciably. Then v_x and Δt are the height and width of a narrow rectangle (Fig. 2.12a)

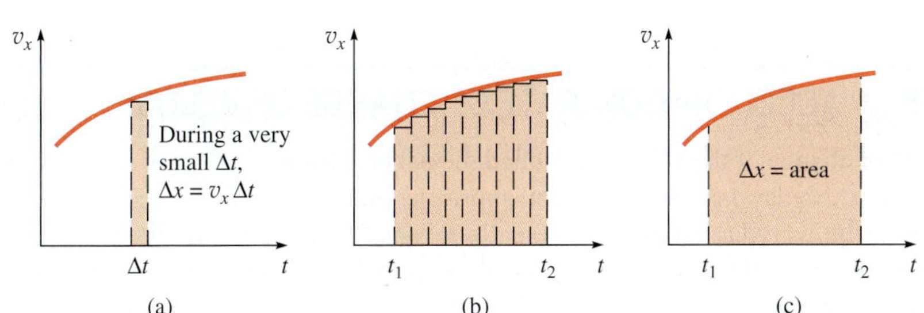

Figure 2.12 (a) Displacement Δx during a short time interval is approximately the area of a rectangle of height v_x and width Δt. (b) During a longer time interval, the displacement is approximately the sum of the areas of the rectangles. (c) The area under the v_x versus t graph for any time interval represents the displacement during that interval.

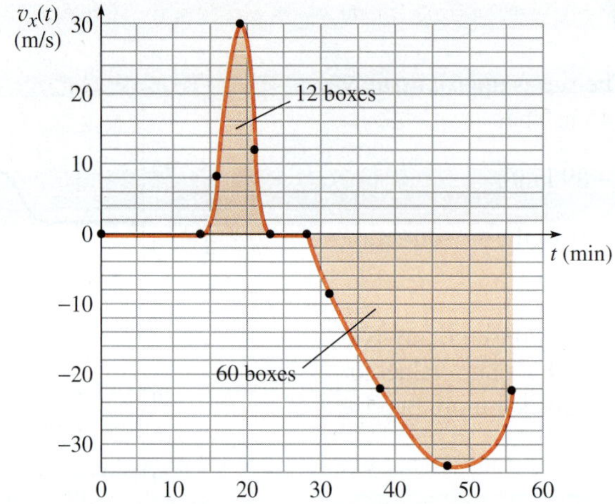

Figure 2.13 A graph of train velocity versus time. The train's displacement from $t = 14$ min to $t = 23$ min is the shaded area under the graph during that time interval. To estimate the area, count the number of grid boxes under the curve, estimating the fraction of the boxes that are only partly below the curve. Each box is 2 m/s in height and 5 min (= 300 s) in width, so each box represents an "area" (displacement) of 2 m/s × 300 s = 600 m = 0.60 km. The total number of shaded boxes for this time interval is about 12, so the displacement is about $\Delta x \approx 12 \times 0.60$ km = +7.2 km, which is close to the actual value of 7 km (during this time interval the train went from +3 km to +10 km). The shaded area for the time interval $t = 28$ min to $t = 56$ min is below the time axis; this negative area represents displacement in the $-x$-direction (west). The number of shaded grid boxes in this interval is about 60, so the displacement during this time interval is $\Delta x \approx -(60) \times 0.60$ km = −36 km.

CONNECTION:

The slope of a graph and the area under a graph have a consistent interpretation: On a graph of *any* quantity Q as a function of time, the slope of the graph represents the instantaneous rate of change of Q. On a graph of the *rate of change of Q* as a function of time, the area under the graph represents ΔQ. The slope of a graph of $x(t)$ is v_x, the rate of change of x; the area under a graph of $v_x(t)$ is Δx. In Section 2.3, we'll learn a similar graphical relationship between velocity and acceleration. The slope of a graph of $v_x(t)$ is a_x, the rate of change of v_x; the area under a graph of $a_x(t)$ is Δv_x.

and the displacement during that short time interval is the area of the rectangle. To find the total displacement during any time interval, the areas of all the narrow rectangles are added together (Fig. 2.12b). To improve the approximation, we let the time interval Δt approach zero and find that the displacement Δx during any time interval equals the area under the graph of $v_x(t)$ (Fig. 2.12c). When v_x is negative, x is decreasing and the displacement is in the $-x$-direction, so we must count the area as negative when it is below the time axis.

> Δx is the area under the graph of $v_x(t)$. The area is negative when the graph is beneath the time axis ($v_x < 0$).

The magnitude of the train's displacement is represented as the shaded areas in Fig. 2.13. The train's displacement from $t = 14$ min to $t = 23$ min is +7 km (area *above* the t-axis means displacement in the $+x$-direction) and from $t = 28$ min to $t = 56$ min it is −36 km (area *below* the t-axis means displacement in the $-x$-direction). The total displacement from $t = 0$ to $t = 56$ min is Δx = (+7 km) + (−36 km) = −29 km.

2.3 ACCELERATION: RATE OF CHANGE OF VELOCITY

The rate of change of the velocity is called the **acceleration**. The use of the word *acceleration* in everyday language is often imprecise and not in accord with its scientific definition. In everyday language, it usually means "an increase in speed". In physics, acceleration can indicate any kind of change in velocity, including decreasing speed and change in direction.

The concept of acceleration is much less intuitive for most people than the concept of velocity. Keep reminding yourself that the acceleration tells you how the velocity *is changing*. The direction of the *change* in velocity is not necessarily the same as the direction of either the initial or final velocities.

Average Acceleration

The **average acceleration** during a time interval Δt is:

$$a_{av,x} = \frac{\Delta v_x}{\Delta t} \qquad (2\text{-}7)$$

Since average acceleration is the change in velocity divided by the corresponding time interval, the SI units of acceleration are (m/s)/s = m/s^2, read as "meters per second squared." Thinking of m/s^2 as (m/s)/s can help you develop an understanding of what acceleration is. Suppose an object has a constant acceleration $a_x = +3.0$ m/s^2. Then v_x increases 3.0 m/s during every second of elapsed time (the change in v_x is +3.0 m/s per second). If $a_x = -2.0$ m/s^2, then v_x would decrease 2.0 m/s during every second (the change in v_x is −2.0 m/s per second).

For example, suppose it takes 30 s for a truck to slow down from 25 m/s to 10 m/s while traveling east. With the x-axis pointing east, the truck's average acceleration during that time interval is

$$a_{av,x} = \frac{\Delta v_x}{\Delta t} = \frac{-15\text{ m/s}}{30\text{ s}} = -0.50\text{ m/s}^2$$

or 0.50 m/s^2 to the west.

CONNECTION:

Compare average acceleration [Eq. (2-7)] and average velocity [Eq. (2-2)]. Each is the change in a quantity divided by the time interval during which the change occurs. Each can have different values for different time intervals.

Instantaneous Acceleration

To find the **instantaneous acceleration**, we calculate the average acceleration during a *very short time interval*:

Definition of instantaneous acceleration

$$a_x = \lim_{\Delta t \to 0} \frac{\Delta v_x}{\Delta t} \qquad (2\text{-}8)$$

(Δv_x is the change in velocity during a *very short* time interval Δt)

CONNECTION:

The rate of change of any quantity Q is

$$\lim_{\Delta t \to 0} \frac{\Delta Q}{\Delta t}$$

Velocity is the rate of change of position and acceleration is the rate of change of velocity.

The time interval Δt must be short enough that we can treat the acceleration as constant during that time interval. Just as with instantaneous velocity, the word *instantaneous* is not always repeated. *Acceleration* without the adjective means *instantaneous* acceleration.

The chapter opener asked how an African lion can ever catch a Cape buffalo. Although Cape buffaloes and African lions have about the same top *speed*, lions are capable of much larger *accelerations* than are buffaloes. Starting from rest, it takes a buffalo much longer to get to its top speed. On the other hand, lions have much less stamina. Once the buffalo reaches its top speed, it can maintain that speed much longer than the lion can. Thus, a Cape buffalo is capable of outrunning a lion unless the stalking lion can get fairly close before charging.

Can the lion catch the buffalo?

Conceptual Example 2.4

Direction of Acceleration While Slowing Down

Damon moves in the $-x$-direction on his motor scooter. He slows down as he approaches a stop sign. While slowing down, is the scooter's acceleration component a_x positive or negative? Sketch a graph of $v_x(t)$ to illustrate how v_x changes.

Strategy The acceleration has the same direction as the *change* in the velocity.

Solution and Discussion Damon is moving in the $-x$-direction, so v_x is negative. He is slowing down, so v_x is

continued on next page

Example 2.4 continued

getting smaller in magnitude (i.e., closer to zero). Therefore, the change in v_x is positive ($\Delta v_x > 0$). Since Δv_x is positive, a_x is positive. The acceleration is in the $+x$-direction.

We know that v_x starts out negative (because Damon moves in the $-x$-direction), decreases *in magnitude* (because he slows down), and reaches zero when he comes to rest. A *plausible* graph is shown in Fig. 2.14. (We don't have enough information to determine the precise shape.)

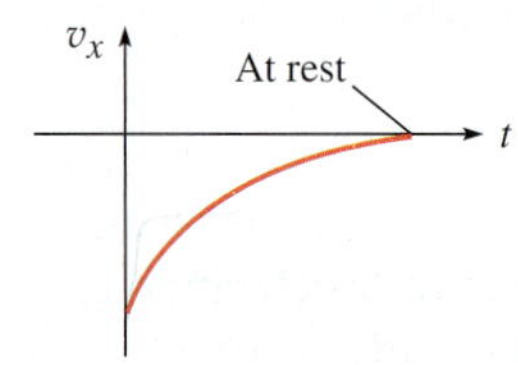

Figure 2.14
A graph of $v_x(t)$, showing that v_x starts out negative and increases until it is zero.

Conceptual Practice Problem 2.4 Continuing on His Way

As Damon pulls away from the stop sign, continuing in the $-x$-direction, his speed gradually increases. What is the sign of a_x? What is the direction of the acceleration? Sketch a motion diagram.

The Direction of the Acceleration

Generalizing Example 2.4, suppose an object moves along the x-axis. When the acceleration is in the same direction as the velocity, the object is speeding up. If v_x and a_x are both positive, the object is moving in the $+x$-direction and is speeding up. If they are both negative, the object is moving in the $-x$-direction and is speeding up.

When the acceleration and velocity are in opposite directions, the object is slowing down. When v_x is positive and a_x is negative, the object is moving in the positive x-direction and is slowing down. When v_x is negative and a_x is positive, the object is moving in the negative x-direction and is slowing down.

In straight-line motion, the acceleration is always in the same direction as the velocity, in the direction opposite to the velocity, or zero.

Graphical Relationships Between Velocity and Acceleration

Both velocity and acceleration measure rates of change: velocity is the rate of change of position and acceleration is the rate of change of velocity. Therefore, the graphical relationship of acceleration to velocity is the same as the graphical relationship of velocity to position:

a_x is the slope on a graph of $v_x(t)$ and Δv_x is the area under a graph of $a_x(t)$.

Figure 2.15 shows a graph of v_x versus t for Damon slowing down on his scooter. He is moving in the $-x$-direction, so $v_x < 0$, and his speed is decreasing, so $|v_x|$ is decreasing. The slope of a tangent line to the graph is a_x at that instant. Three tangent lines are drawn, showing that a_x is positive (the slopes are positive) and is not constant (the slopes are not all the same).

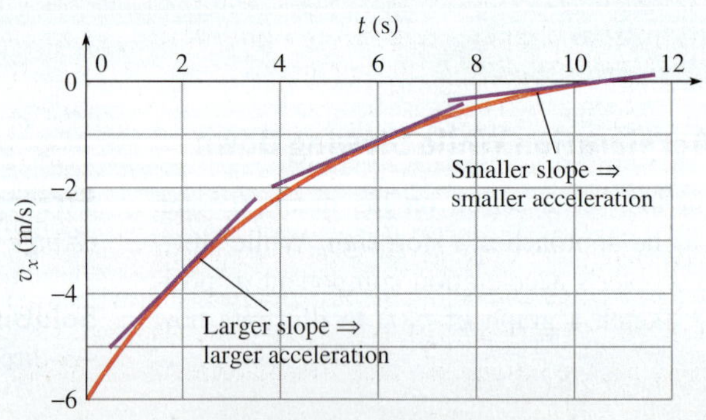

Figure 2.15 In this graph of v_x versus t, as Damon is stopping, v_x is negative, but a_x (the slope) is positive. The value of v_x is increasing, but—since it is less than zero to begin with and is getting closer to zero as time goes on—the speed is *decreasing*. The slopes of the three tangent lines shown represent the instantaneous accelerations (a_x) at three different times.

Example 2.5

Acceleration of a Sports Car

Figure 2.16 shows data for v_x as a function of time as a sports car starts from rest and travels in a straight line in the $+x$-direction, with the driver speeding up as quickly as possible. (a) What is the average acceleration of the sports car from 0 to 30.0 m/s? (b) What is the maximum acceleration of the car? (c) What is the car's displacement from $t = 0$ to $t = 19.1$ s (when it reaches 60.0 m/s)? (d) What is the car's average velocity during the entire 19.1 s interval?

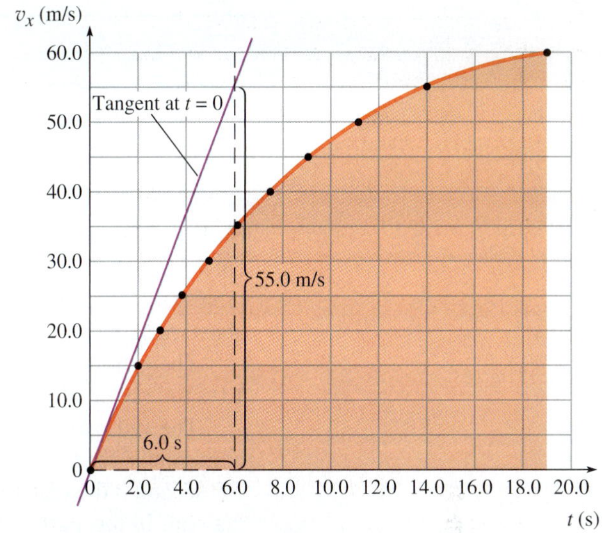

v_x (m/s)	0	15.0	20.0	25.0	30.0	35.0	40.0	45.0	50.0	55.0	60.0
t (s)	0	2.0	2.9	3.8	4.9	6.2	7.6	9.1	11.2	14.0	19.1

Figure 2.16

Data table and graph of $v_x(t)$ for a sports car.

Strategy (a) To find the average acceleration, the change in velocity for the time interval is divided by the time interval. (b) The instantaneous acceleration is the slope of the velocity graph, so it is maximum where the graph is steepest. At that point, the velocity is changing at a high rate. We expect the maximum acceleration to take place early on; the magnitude of acceleration must decrease as the velocity gets higher and higher—there is a maximum velocity for the car, after all. (c) The displacement Δx is the area under the $v_x(t)$ graph. The graph is not a simple shape such as a triangle or rectangle, so an estimate of the area is made. (d) Once we have a value for the displacement, we can apply the definition of average velocity.

Given: Graph of $v_x(t)$ in Fig. 2.16.

To find: (a) $a_{av,x}$ for $v_x = 0$ to 30.0 m/s; (b) maximum value of a_x; (c) Δx from $v_x = 0$ to 60.0 m/s; (d) $v_{av,x}$ from $t = 0$ to 19.1 s

Solution (a) The car starts from rest, so $v_{xi} = 0$. It reaches $v_x = 30.0$ m/s at $t = 4.9$ s, according to the data table. Then for this time interval,

$$a_{av,x} = \frac{\Delta v_x}{\Delta t} = \frac{30.0\text{ m/s} - 0\text{ m/s}}{4.9\text{ s} - 0\text{ s}} = 6.1\text{ m/s}^2$$

The average acceleration for this time interval is 6.1 m/s^2 in the $+x$-direction.

(b) The acceleration a_x, at any instant of time, is the slope of the tangent line to the $v_x(t)$ graph at that time. To find the maximum acceleration, we look for the steepest part of the graph. In this case, the largest slope occurs near $t = 0$, just as the car is starting out. In Fig. 2.16 a tangent line to the $v_x(t)$ graph at $t = 0$ is drawn and labeled. Values for the rise and run to calculate the slope of the tangent line are read from the graph. The tangent line passes through the two points ($t = 0$, $v_x = 0$) and ($t = 6.0$ s, $v_x = 55.0$ m/s) on the graph, so the rise is 55.0 m/s for a run of 6.0 s. The slope of this line is

$$a_x = \frac{\text{rise}}{\text{run}} = \frac{55.0\text{ m/s} - 0\text{ m/s}}{6.0\text{ s} - 0\text{ s}} = +9.2\text{ m/s}^2$$

The maximum acceleration is 9.2 m/s^2 in the $+x$-direction.

(c) Δx is the area under the $v_x(t)$ graph shown shaded in Fig. 2.16. The area can be estimated by counting the number of grid boxes under the curve. Each box is 5.0 m/s in height and 2.0 s in width, so each represents an "area" (displacement) of 10 m. When counting the number of boxes under the curve, a best estimate is made for the fraction of the boxes that are only partly below the curve. Approximately 75 boxes lie below the curve, so the displacement is $\Delta x = 75 \times 10\text{ m} = 750$ m. Since the car travels along a straight line and does not change direction, 750 m is also the distance traveled. (d) The average velocity during the 19.1 s interval is

$$v_{av,x} = \frac{\Delta x}{\Delta t} = \frac{750\text{ m}}{19.1\text{ s}} = 39\text{ m/s}$$

Discussion The graph of velocity as a function of time is often the most helpful graph to have when solving a problem. If that graph is not given in the problem, it is useful to sketch one. The $v_x(t)$ graph shows displacement, velocity, and acceleration at once: the velocity v_x is given by the points or the curve graphed, the displacement Δx is the area under the curve, and the acceleration a_x is the slope of the curve.

Why is the average velocity 39 m/s? Why is it not halfway between the initial velocity (0 m/s) and the final velocity (60 m/s)? If the acceleration were constant, the average velocity would indeed be $\frac{1}{2}(0 + 60\text{ m/s}) = 30$ m/s. The actual average velocity is somewhat higher than that—the acceleration is greater at the start, so less of the time interval is spent going (relatively) slow and more is spent going fast. The speed is less than 30 m/s for only 4.9 s, but is greater than 30 m/s for 14.2 s.

Practice Problem 2.5 Acceleration at a Later Time

What is the instantaneous acceleration of the car at $t = 14.0$ s?

CHECKPOINT 2.3

What physical quantity does the slope of the tangent to a graph of v_x versus time represent?

2.4 VISUALIZING MOTION ALONG A LINE WITH CONSTANT ACCELERATION

Motion Diagrams In Fig. 2.17, three carts move in the same direction with three different values of constant acceleration. The position of each cart is depicted in a motion diagram with a time interval of 1.0 s. Red arrows representing the cart's velocity are shown above each position.

The yellow cart has zero acceleration and, therefore, constant velocity. During each 1.0 s time interval, its displacement is the same: 1.0 m/s × 1.0 s = 1.0 m to the right.

The red cart has a constant acceleration of 0.2 m/s^2 to the right. Although m/s^2 is normally read "meters per second squared," it can be useful to think of it as "m/s per second": the cart's velocity component v_x increases 0.2 m/s during each 1.0 s time interval. In this case, acceleration is in the same direction as the velocity, so the speed increases. The motion diagram shows that displacement of the cart during successive 1.0 s time intervals gets larger and larger.

The blue cart has a constant acceleration of 0.2 m/s^2 in the $-x$-direction, which is the direction *opposite* to the velocity. The velocity component v_x decreases by 0.2 m/s during each 1.0 s interval. The velocity and acceleration are in opposite directions, so the speed is decreasing. Now the motion diagram shows that displacements during 1.0 s intervals get smaller and smaller.

Graphs Figure 2.18 shows graphs of $x(t)$, $v_x(t)$, and $a_x(t)$ for each of the carts. The acceleration graphs are horizontal since each of the carts has a constant acceleration. All three v_x graphs are straight lines. Since a_x is the rate of change of v_x, the slope of the $v_x(t)$ graph is a_x at that time t. With constant acceleration, the slope is the same everywhere and the graph is linear. Remember that a positive a_x does mean that v_x is increasing, but not necessarily that the *speed* is increasing. If v_x is negative, then a positive a_x indicates a *decreasing* speed. (See Conceptual Example 2.4.) Speed is increasing when the acceleration and velocity are in the same direction (a_x and v_x both positive *or* both negative). Speed is decreasing when acceleration and velocity are in opposite directions—when a_x and v_x have opposite signs.

The position graph is linear for the yellow cart because it has constant velocity. For the red cart, the $x(t)$ graph curves with increasing slope, showing that v_x is increasing. For the blue cart, the $x(t)$ graph curves with decreasing slope, showing that v_x is decreasing.

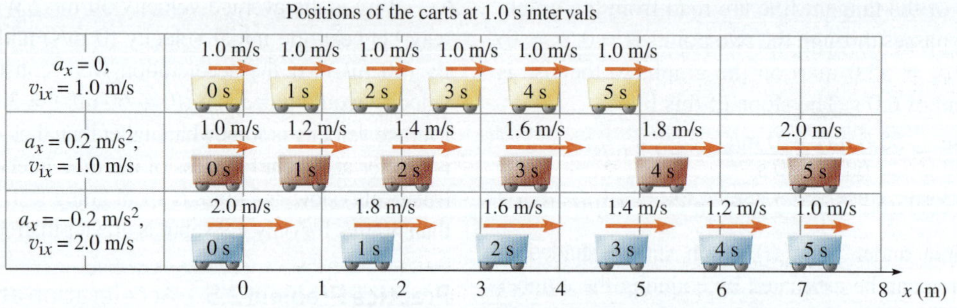

Figure 2.17 Each cart is shown in a motion diagram at 1.0 s time intervals. The arrows above each cart indicate the instantaneous velocities. All three carts move with constant acceleration.

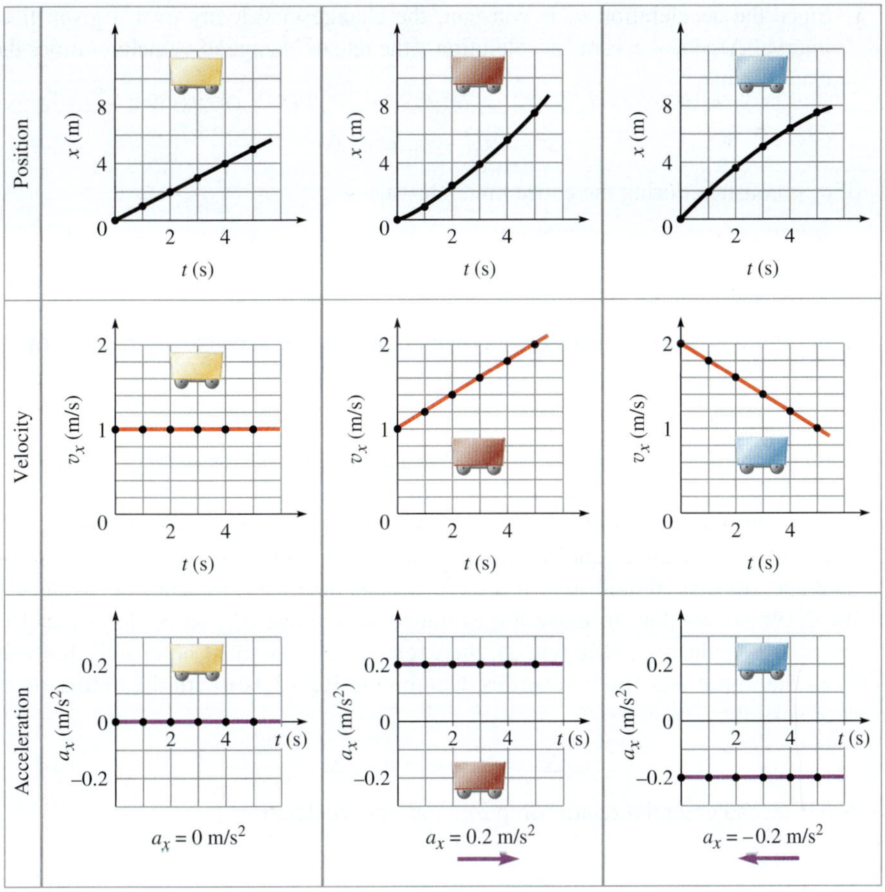

Figure 2.18 Graphs of position, velocity, and acceleration as functions of time for the carts of Fig. 2.17. The slope of the $x(t)$ graph at any time is the velocity v_x at that time. The slope of the $v_x(t)$ graph at any time is the acceleration a_x at that time.

CHECKPOINT 2.4

Do the red and blue carts in Fig. 2.17 ever have the same velocity? If so, when? If the motion diagram didn't use red arrows to indicate the velocity of each cart, how could you still answer the question?

2.5 KINEMATIC EQUATIONS FOR MOTION ALONG A LINE WITH CONSTANT ACCELERATION

The *kinematic equations* we introduce next are relationships between position, velocity, acceleration, and time that apply to the important special case of an object whose acceleration is *constant* (both in magnitude and direction). Although a graph of $v_x(t)$ can be used to find the position of an object as a function of time, having an algebraic method at our disposal will be very convenient. First, let us agree on a consistent notation:

- Choose an origin and a direction for the positive axis. (For vertical motion, it is conventional to use the y-axis instead of the x-axis, where the $+y$-direction is up.)
- At an initial time t_i, the initial position and velocity are x_i and v_{ix}.
- At a later time $t_f = t_i + \Delta t$, the final position and velocity are x_f and v_{fx}.

Two essential relationships between position, velocity, and acceleration enable us to find the position of an object moving along a line with constant acceleration:

1. Since the acceleration a_x is constant, the change in velocity over a given time interval $\Delta t = t_f - t_i$ is the acceleration—the rate of change of velocity—times the elapsed time:

$$\Delta v_x = v_{fx} - v_{ix} = a_x \, \Delta t \qquad \text{(2-9)}$$

(if a_x is constant during the entire time interval)

Equation (2-9) is the definition of a_x [Eq. (2-8)] with the assumption that a_x is constant.

2. Since the velocity changes linearly with time, the average velocity is given by:

$$v_{av,x} = \tfrac{1}{2}(v_{fx} + v_{ix}) \text{ (constant } a_x) \qquad \text{(2-10)}$$

Equation (2-10) is *not* true *in general*, but it is true for constant acceleration. To see why, refer to the $v_x(t)$ graph in Fig. 2.19a. The graph is linear because the acceleration—the slope of the graph—is constant. The displacement during any time interval is represented by the area under the graph. The average velocity is found by forming a rectangle with an area equal to the area under the curve in Fig. 2.19a, because the average velocity should give the same displacement in the same time interval. Figure 2.19b shows that, to make the excluded area above $v_{av,x}$ (triangle 1) equal to the extra area under $v_{av,x}$ (triangle 2), the average velocity must be exactly halfway between the initial and final velocities. Combining Eq. (2-10) with the definition of average velocity,

$$\Delta x = x_f - x_i = v_{av,x} \, \Delta t \qquad \text{(2-2)}$$

gives our second essential relationship for constant acceleration:

$$\Delta x = \tfrac{1}{2}(v_{fx} + v_{ix}) \, \Delta t \qquad \text{(2-11)}$$

(if a_x is constant during the entire time interval)

If the acceleration is *not* constant, there is no reason why the average velocity has to be exactly halfway between the initial and the final velocity. As an illustration, imagine a trip where you drive along a straight highway at 80 km/h for 50 min and then at 60 km/h for 30 min. Your acceleration is zero for the entire trip *except* during the few seconds while you slowed from 80 km/h to 60 km/h. The magnitude of your average velocity is *not* 70 km/h. You spent more time going 80 km/h than you did going 60 km/h, so the magnitude of your average velocity would be greater than 70 km/h.

Other Useful Relationships for Constant Acceleration Two more useful relationships can be formed between the various quantities (displacement, initial and final velocities, acceleration, and time interval) by eliminating some quantity from Eqs. (2-9)

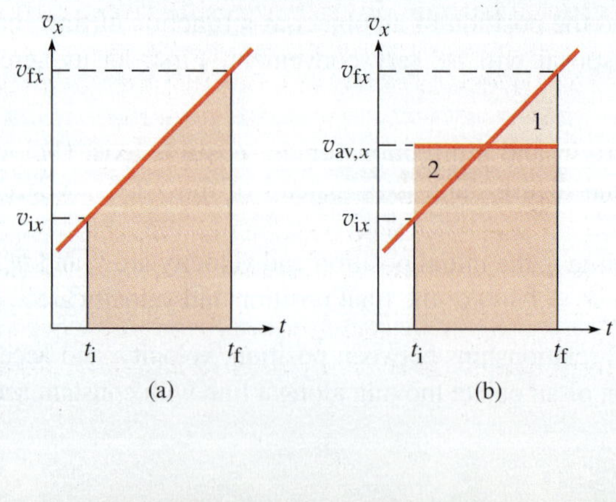

Figure 2.19 Finding the average velocity when the acceleration is constant. (a) On a graph of $v_x(t)$, the area under the graph is the displacement during that time interval. (b) The average velocity is the value of v_x that would produce the same displacement during the same time interval, so the areas under the two graphs are equal.

and (2-11). For example, suppose we don't know the final velocity v_{fx}. Then we can solve Eq. (2-9) for v_{fx}, substitute into Eq. (2-11), and simplify:

$$\Delta x = \tfrac{1}{2}(v_{fx} + v_{ix})\,\Delta t = \tfrac{1}{2}[(v_{ix} + a_x\,\Delta t) + v_{ix}]\,\Delta t$$

$$\Delta x = v_{ix}\,\Delta t + \tfrac{1}{2}a_x(\Delta t)^2 \quad \text{(constant } a_x\text{)} \qquad \text{(2-12)}$$

We can interpret Eq. (2-12) graphically. Figure 2.20 shows a $v_x(t)$ graph for motion with constant acceleration. The displacement that occurs between t_i and a later time t_f is the area under the graph for that time interval. Partition this area into a rectangle plus a triangle. The area of the rectangle is

$$\text{base} \times \text{height} = v_{ix}\,\Delta t$$

The height of the triangle is the change in velocity, which is equal to $a_x\,\Delta t$. The area of the triangle is

$$\tfrac{1}{2}\text{base} \times \text{height} = \tfrac{1}{2}\Delta t \times a_x\,\Delta t = \tfrac{1}{2}a_x(\Delta t)^2$$

Adding these areas gives Eq. (2-12).

Another useful relationship comes from eliminating the time interval Δt:

$$\Delta x = \tfrac{1}{2}(v_{fx} + v_{ix})\,\Delta t = \tfrac{1}{2}(v_{fx} + v_{ix})\left(\frac{v_{fx} - v_{ix}}{a_x}\right) = \frac{v_{fx}^2 - v_{ix}^2}{2a_x}$$

Rearranging terms,

$$v_{fx}^2 - v_{ix}^2 = 2a_x\,\Delta x \quad \text{(constant } a_x\text{)} \qquad \text{(2-13)}$$

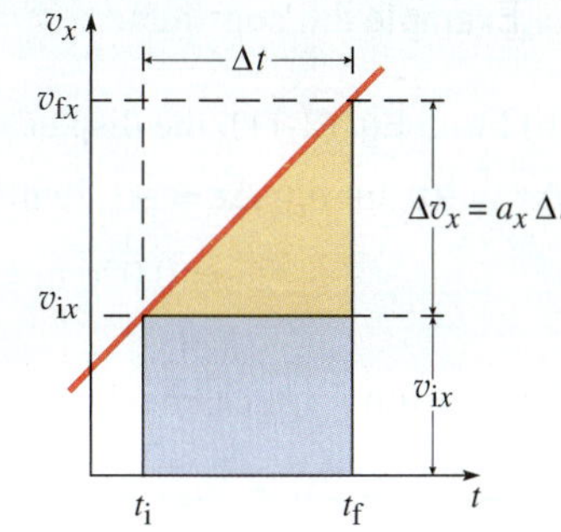

Figure 2.20 Graphical interpretation of Eq. (2-12). The area of the blue rectangle is $v_{ix}\,\Delta t$. The area of the yellow triangle is $\frac{1}{2}$base × height = $\frac{1}{2}(\Delta t)(a_x\,\Delta t)$.

CHECKPOINT 2.5

At 3:00 P.M., an airplane is moving due west at 460 km/h. At 3:05 P.M., it is moving due west at 480 km/h. Is its average velocity during the time interval necessarily 470 km/h west? Explain.

Example 2.6

Doppler Echocardiography

The maximum acceleration of blood in the aorta can be used to test ventricular function. The period during which maximum acceleration of the blood in the aorta occurs is during the first portion of the left ventricle's pumping action. During this period, the acceleration is essentially constant. Doppler echocardiography uses ultrasound to measure blood speeds in the aorta. The results for one patient show that the blood in the aorta begins at a speed of 0.10 m/s and undergoes constant acceleration for 38 ms, reaching a peak speed of 1.29 m/s. (a) What is the acceleration reflected in these data? (b) How far does the blood travel during this period?

Strategy Choose the x-axis in the direction of blood flow. Then the given information is: $\Delta t = 38\text{ ms} = 38 \times 10^{-3}\text{ s}$; $v_{ix} = 0.10\text{ m/s}$; $v_{fx} = 1.29\text{ m/s}$. The goal of the problem is to find a_x and Δx. Equation (2-9) can be solved for a_x in terms of the three given quantities. Equation (2-11) can be solved for Δx in terms of the three given quantities. Equations (2-12) and (2-13) contain both of the unknowns, so using them is correct but would lead to more complicated algebra.

Solution (a) The velocity change is $\Delta v_x = v_{fx} - v_{ix} = 1.29\text{ m/s} - 0.10\text{ m/s} = 1.19\text{ m/s}$. The time interval is $\Delta t = 38 \times 10^{-3}$ s. The acceleration is then

$$a_x = \frac{\Delta v_x}{\Delta t} = \frac{1.19\text{ m/s}}{38 \times 10^{-3}\text{ s}} = 31.32\text{ m/s}^2$$

Rounded to two significant figures, the acceleration is 31 m/s^2.

continued on next page

Example 2.6 continued

(b) From Eq. (2-11), the displacement is

$$\Delta x = \tfrac{1}{2}(v_{fx} + v_{ix})\,\Delta t = \tfrac{1}{2}(1.29\ \text{m/s} + 0.10\ \text{m/s})(38 \times 10^{-3}\ \text{s})$$
$$= 0.026\ \text{m} = 2.6\ \text{cm}$$

Discussion Quick check using Eq. (2-12):

$$\Delta x = v_{ix}\,\Delta t + \tfrac{1}{2}a_x(\Delta t)^2 = 0.10\ \text{m/s} \times 38 \times 10^{-3}\ \text{s} + \tfrac{1}{2} \times 31.32\ \text{m/s}^2 \times (38 \times 10^{-3}\ \text{s})^2 = 2.6\ \text{cm}$$

This isn't an *independent* check because Eq. (2-12) is derived from Eqs. (2-9) and (2-11); it's just a quick check to see if we made any algebra mistakes.

The value of a_x calculated from the velocity data could be used by a cardiologist to draw conclusions about the forces acting on the blood and thus to evaluate cardiac function.

Practice Problem 2.6 Ejection of Moss Spores

Measurements of sphagnum moss spores indicate that they undergo accelerations up to 360 000 m/s^2 as they are ejected from the parent moss plant. (a) Assuming a constant acceleration of this magnitude, how far will a sphagnum moss spore travel in 0.40 ms, starting from rest? (b) How fast will it be moving at that time?

Example 2.7

A Sliding Brick

Starting from rest, a brick slides along a straight line down an icy roof with a constant acceleration of magnitude 4.9 m/s^2 (Fig. 2.21). How fast is the brick moving when it reaches the edge of the roof 0.90 s later?

Strategy What is the direction of the acceleration? It has to be downward along the roof, in the same direction as the brick's velocity. An acceleration opposite the velocity would make the brick slow down, but since it starts from rest, a constant acceleration can only make it speed up. We choose the $+x$-axis in the direction of the acceleration. Then we use the acceleration to find how the velocity changes during the time interval.

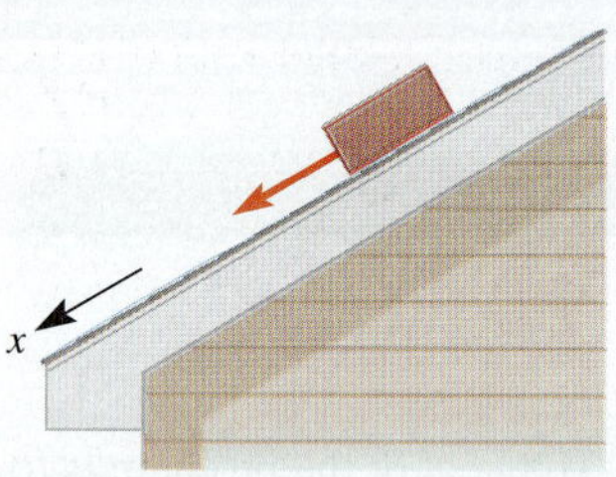

Figure 2.21 A brick sliding down an icy roof.

Solution With the x-axis in the direction of the acceleration, $a_x = +4.9$ m/s^2. The brick is initially at rest so $v_{ix} = 0$. We want to know v_{fx} at the end of the time interval $\Delta t = 0.90$ s. Since a_x is constant, v_x changes at a constant rate:

$$\Delta v_x = v_{fx} - v_{ix} = a_x\,\Delta t = (+4.9\ \text{m/s}^2) \times (0.90\ \text{s}) = 4.4\ \text{m/s}$$

At the edge of the roof, the brick is moving at 4.4 m/s parallel to the roof.

Discussion Conceptual check: $a_x = +4.9$ m/s^2 means that v_x increases 4.9 m/s every second. The brick slides for a bit less than 1 s, so the increase in v_x is a bit less than 4.9 m/s.

Practice Problem 2.7 Displacement of the Brick

How far from the edge of the roof was the brick when it started sliding?

Example 2.8

Two Spaceships

Two spaceships are moving from the same starting point in the $+x$-direction with constant accelerations. The silver spaceship has an initial velocity of +2.00 km/s and an acceleration of +0.400 km/s^2. The black spaceship has an initial velocity of +6.00 km/s and an acceleration of −0.400 km/s^2. (a) Find the time at which the silver spaceship just overtakes the black spaceship. (b) Sketch graphs of $v_x(t)$ for the two spaceships. (c) Sketch a motion diagram showing the positions of the two spaceships at 1.0 s intervals.

Strategy We can find the positions of the spaceships at later times from the initial velocities and the accelerations.

continued on next page

Example 2.8 continued

At first, the black spaceship is moving faster, so it pulls out ahead. Later, the silver ship overtakes the black ship at the instant their *positions are equal.*

Solution (a) The position of either spaceship at a later time is given by Eq. (2-12):

$$x_f = x_i + \Delta x = x_i + v_{ix}\,\Delta t + \tfrac{1}{2}a_x\,(\Delta t)^2$$

We set the final position of the silver spaceship equal to that of the black spaceship ($x_{fs} = x_{fb}$):

$$x_{is} + v_{isx}\,\Delta t + \tfrac{1}{2}a_{sx}\,(\Delta t)^2 = x_{ib} + v_{ibx}\,\Delta t + \tfrac{1}{2}a_{bx}\,(\Delta t)^2$$

Subscripts are useful for preventing you from mixing up similar quantities. The subscripts s and b stand for silver and black, respectively. The subscripts i and f stand for initial and final, respectively. A skilled problem-solver must be able to come up with algebraic symbols that are explicit and unambiguous.

The initial positions are the same: $x_{is} = x_{ib}$. Subtracting the initial positions from each side, moving all terms to one side, and factoring out one power of Δt yields

$$(\Delta t)(v_{isx} + \tfrac{1}{2}a_{sx}\,\Delta t - v_{ibx} - \tfrac{1}{2}a_{bx}\,\Delta t) = 0$$

This equation has two solutions—there are two times at which the spaceships are at the same position. One solution is $\Delta t = 0$. We already knew that the two spaceships started at the same *initial* position. The other solution, which gives the time at which one spaceship overtakes the other, is found by setting the expression in parentheses equal to zero. Solving for Δt,

$$\Delta t = \frac{2(v_{isx} - v_{ibx})}{a_{bx} - a_{sx}} = \frac{2 \times (2.00\text{ km/s} - 6.00\text{ km/s})}{-0.400\text{ km/s}^2 - 0.400\text{ km/s}^2} = 10.0\text{ s}$$

The silver spaceship overtakes the black spaceship 10.0 s after they leave the starting point.

(b) Figure 2.22 shows the $v_x(t)$ graphs with $t_i = 0$. Note that the area under the graphs from t_i to t_f is the same in the two graphs: the spaceships have the same displacement during that interval.

(c) Equation (2-12) can be used to find the position of each spaceship as a function of time. Choosing $x_i = 0$, $t_i = 0$, and $t = t_f$, the position at time t is

$$x(t) = 0 + v_{ix}t + \tfrac{1}{2}a_xt^2$$

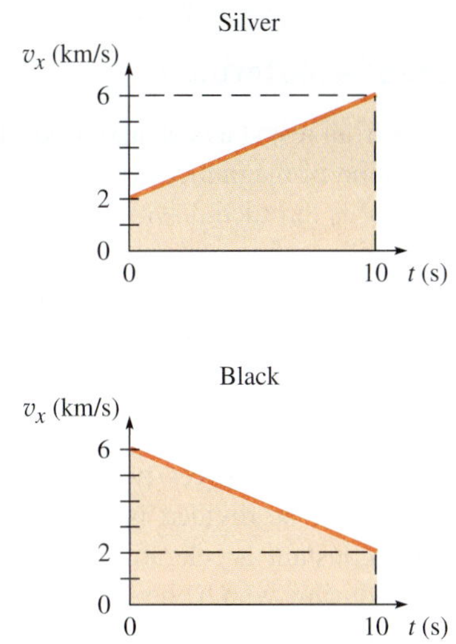

Figure 2.22

Graphs of v_x versus t for the silver and black spaceships. The shaded area under each graph represents the displacement Δx during the time interval.

Figure 2.23 shows the data table calculated this way and the corresponding motion diagram.

Discussion Quick check: the two ships must have the same displacement at $\Delta t = 10.0$ s.

$$\begin{aligned}\Delta x_s &= v_{isx}\,\Delta t + \tfrac{1}{2}a_{sx}(\Delta t)^2\\ &= 2.00\text{ km/s} \times 10.0\text{ s} + \tfrac{1}{2} \times 0.400\text{ km/s}^2 \times (10.0\text{ s})^2\\ &= 40.0\text{ km}\end{aligned}$$

$$\begin{aligned}\Delta x_b &= v_{ibx}\,\Delta t + \tfrac{1}{2}a_{bx}(\Delta t)^2\\ &= 6.00\text{ km/s} \times 10.0\text{ s} + \tfrac{1}{2} \times (-0.400\text{ km/s}^2) \times (10.0\text{ s})^2\\ &= 40.0\text{ km}\end{aligned}$$

Practice Problem 2.8 Time to Reach the Same Velocity

When do the two spaceships have the same *velocity*? What is the value of the velocity then?

t (s)	0	1.0	2.0	3.0	4.0	5.0	6.0	7.0	8.0	9.0	10.0
x_s (km)	0	2.2	4.8	7.8	11.2	15.0	19.2	23.8	28.8	34.2	40.0
x_b (km)	0	5.8	11.2	16.2	20.8	25.0	28.8	32.2	35.2	37.8	40.0

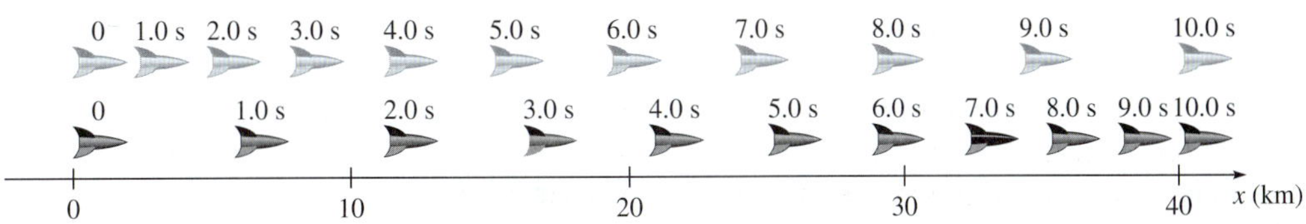

Figure 2.23

Calculated positions of the spaceships at 1.0 s time intervals and a motion diagram.

Example 2.9

Displacement of a Motorboat

A motorboat starts from rest at a dock and heads due east with a constant acceleration of magnitude 2.8 m/s^2. After traveling for 140 m, the motor is throttled down to slow down the boat at 1.2 m/s^2 (while still moving east) until its speed is 16 m/s. Just as the boat attains the speed of 16 m/s, it passes a buoy due east of the dock. (a) Sketch a qualitative graph of $v_x(t)$ for the motorboat from the dock to the buoy. Let the $+x$-axis point east. (b) What is the distance between the dock and the buoy?

Strategy This problem involves two different values of acceleration, so it must be divided into two subproblems. The equations for constant acceleration cannot be applied to a time interval during which the acceleration changes. But for each of two time intervals, the acceleration of the boat is constant: from t_1 to t_2, $a_{1x} = +2.8$ m/s^2; from t_2 to t_3, $a_{2x} = -1.2$ m/s^2. The two subproblems are connected by the position and velocity of the boat at the instant the acceleration changes. This is reflected in the graph of $v_x(t)$: It consists of two different straightline segments with different slopes that connect with the same value of v_x at time t_2.

For subproblem 1, the boat speeds up with a constant acceleration of 2.8 m/s^2 to the east. We know the acceleration, the displacement (140 m east), and the initial velocity: the boat starts from rest, so the initial velocity v_{1x} is zero. We need to calculate the final velocity v_{2x}, which then becomes the initial velocity for the second subproblem. The boat is always headed to the east, so we choose east as the positive x-direction.

Subproblem 1

Known: $v_{1x} = 0$; $a_{1x} = +2.8$ m/s^2;
$\Delta x_{21} = x_2 - x_1 = 140$ m.

To find: v_{2x}.

For subproblem 2, we know the acceleration and final velocity v_{3x}, and we have just found the initial velocity v_{2x} from subproblem 1. Because the boat is slowing down, its acceleration is in the direction opposite its velocity; therefore, $a_{2x} < 0$. From these three quantities we can find the displacement of the boat during the second time interval.

Subproblem 2

Known: v_{2x} from subproblem 1;
$a_{2x} = -1.2$ m/s^2; $v_{3x} = +16$ m/s.

To find: $\Delta x_{32} = x_3 - x_2$.

Adding the displacements for the two time intervals gives the total displacement. The magnitude of the total displacement is the distance between the dock and the buoy.

Solution (a) The graph starts with $v_x = 0$ at $t = t_1$. We choose $t_1 = 0$ for simplicity. The graph is a straight line with slope $+2.8$ m/s^2 until $t = t_2$. Then, starting from where the graph left off, the graph continues as a straight line with slope -1.2 m/s^2 until the graph reaches $v_x = 16$ m/s at $t = t_3$. Figure 2.24 shows the $v_x(t)$ graph. It is not quantitatively accurate because we have not calculated the values of t_2 and t_3.

(b1) To find v_{2x} without knowing the time interval, we eliminate Δt from Eqs. (2-9) and (2-11) for constant acceleration:

$$\Delta x_{21} = \frac{1}{2}(v_{2x} + v_{1x})\,\Delta t = \frac{1}{2}(v_{2x} + v_{1x})\left(\frac{v_{2x} - v_{1x}}{a_{1x}}\right) = \frac{v_{2x}^2 - v_{1x}^2}{2a_{1x}}$$

Solving for v_{2x},

$$v_{2x} = \pm\sqrt{v_{1x}^2 + 2a_{1x}\,\Delta x} = \pm\sqrt{0 + 2 \times 2.8\text{ m/s}^2 \times 140\text{ m}}$$
$$= \pm 28\text{ m/s}$$

The boat is moving east, in the $+x$-direction, so the correct sign here is positive: $v_{2x} = +28$ m/s.

(b2) The final velocity for the first interval (v_{2x}) is the *initial* velocity for the second interval. The final velocity is v_{3x}. Using the same equation just derived for this time interval,

$$\Delta x_{32} = \frac{v_{3x}^2 - v_{2x}^2}{2a_{2x}} = \frac{(16\text{ m/s})^2 - (28\text{ m/s})^2}{2 \times (-1.2\text{ m/s}^2)} = +220\text{ m}$$

The total displacement is

$$x_3 - x_1 = (x_3 - x_2) + (x_2 - x_1) = 220\text{ m} + 140\text{ m} = +360\text{ m}$$

The buoy is 360 m from the dock.

Discussion The natural division of the problem into two parts occurs because the boat has two different constant accelerations during two different time periods. In problems that can be subdivided in this way, the final velocity and position found in the first part becomes the initial velocity and position for the second part.

Practice Problem 2.9 Time to Reach the Buoy

What is the time required by the boat in Example 2.9 to reach the buoy?

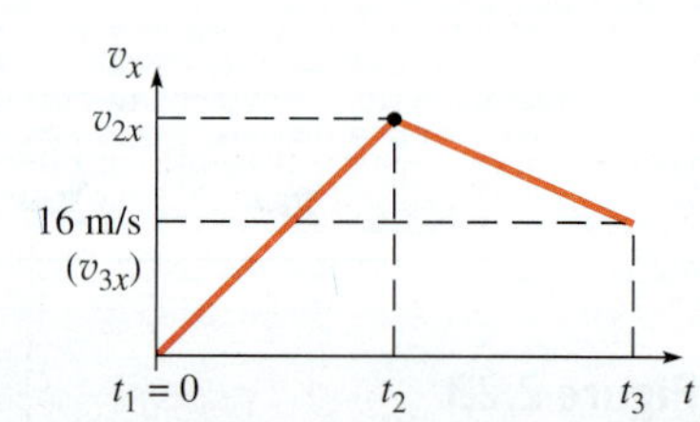

Figure 2.24
Graph of v_x versus t for the motorboat.

2.6 FREE FALL

Suppose you are standing on a bridge over a deep gorge. If you drop a stone into the gorge, how fast does it fall? You know from experience that it does not fall at a constant velocity; the longer it falls, the faster it goes. A better question is: What is the stone's acceleration?

First, let us simplify the problem. If the stone were moving very fast, air resistance would oppose its motion. When it is not falling so fast, the effect of air resistance is negligibly small. In **free fall,** no forces act on an object other than the gravitational force that makes the object fall. On Earth, free fall is an idealization since there is always *some* air resistance. We also assume that the stone's change in altitude is small enough that Earth's gravitational pull on it is constant.

CONNECTION:

Free fall is an example of motion with constant acceleration.

Free-fall Acceleration An object in free fall has a constant downward acceleration, called the *free-fall acceleration.* The magnitude of this acceleration varies a little from one place to another near Earth's surface, but at any given place, it has the same value for every object, regardless of the mass of the object. The symbol g represents the magnitude of the free-fall acceleration. Unless another value is given in a particular problem, please assume that the magnitude of the free-fall acceleration near Earth's surface is

$$g = 9.80 \text{ m/s}^2 \qquad (2\text{-}14)$$

When dealing with vertical motion, the y-axis is usually chosen to be positive pointing upward. The direction of the free-fall acceleration is down, so $a_y = -g$. The same techniques and equations used for other constant acceleration situations are used with free fall.

Earth's gravity always pulls downward, so the acceleration of an object in free fall is always downward and constant in magnitude, *regardless of whether the object is moving up, down, or is at rest, and independent of its speed.* If the object is moving downward, the downward acceleration makes it speed up; if it is moving upward, the downward acceleration makes it slow down.

Acceleration at Highest Point If an object is thrown straight up, its velocity is zero at the highest point of its flight. Why? On the way up, its velocity v_y is positive (if the positive y-axis is pointing up). On the way down, v_y is negative. Since v_y changes continuously, it must pass through zero to change sign (Fig. 2.25). At the highest point, the velocity is zero but the *acceleration* is *not* zero. If the acceleration were to suddenly become zero at the top of flight, the velocity would no longer change; the object would get *stuck at the top* rather than fall back down. The velocity is zero at the top but it does not *stay* zero: it keeps changing at the same rate.

CHECKPOINT 2.6

Is it possible for an object in free fall to be moving upward? Explain.

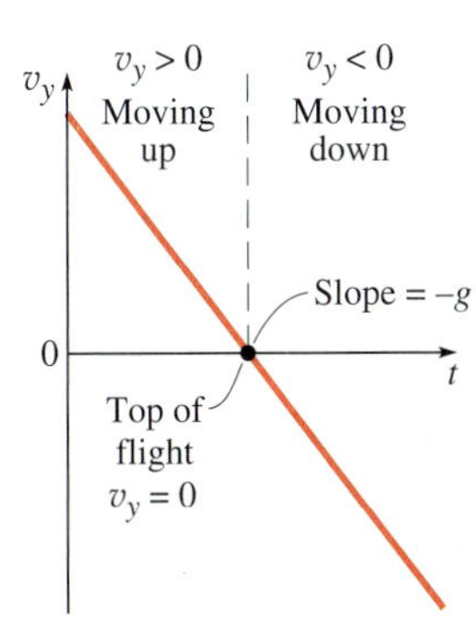

Figure 2.25 Graph of v_y versus t for an object thrown upward. The slope of the graph is the same at every point, whether the object moves up, down, or is at the top of flight: it is the acceleration $a_y = -g$.

Example 2.10

Throwing Stones

Standing on a bridge, you throw a stone straight upward. The stone hits a stream, 44.1 m below the point at which you release it, 4.00 s later. (a) Sketch graphs of $y(t)$ and $v_y(t)$. The positive y-axis points up. (b) What is the velocity of the stone just after it leaves your hand? (c) What is the velocity of the stone just before it hits the water? (d) Draw a motion diagram for the stone, showing its position at 0.1 s intervals during the first 0.9 s of its motion.

Strategy Ignoring air resistance, the stone is in free fall once your hand releases it and until it hits the water. For the time interval during which the stone is in free fall, the initial velocity is the velocity of the stone *just after* it leaves your hand and the final velocity is the velocity *just before* it hits the water. During free fall, the stone's acceleration is constant and equal to 9.80 m/s^2 downward. Known: $a_y = -9.80$ m/s^2; $\Delta y = -44.1$ m at $\Delta t = 4.00$ s. To find: v_{iy} and v_{fy}.

Solution (a) Let's choose the origin at the release point so the stone starts at $y = 0$. As the stone moves up, y increases until it reaches the maximum height. Then it moves downward until it hits the stream at a point below $y = 0$. The graph of $v_y(t)$ is a straight line because the acceleration is constant. The stone initially moves upward ($v_y > 0$). At the top of flight, $v_y = 0$. Then $v_y < 0$ as the stone moves downward. The value of v_y is the slope of the $y(t)$ graph: initially positive, steadily decreasing until it is zero at the top of flight; then the slope continues to decrease, becoming negative. From these observations, we can sketch the graphs of $y(t)$ and $v_y(t)$ (Fig. 2.26).

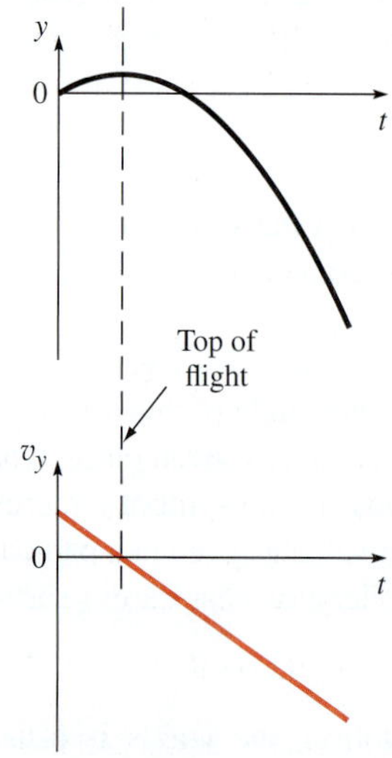

Figure 2.26
Graphs of $y(t)$ and $v_y(t)$ for the stone.

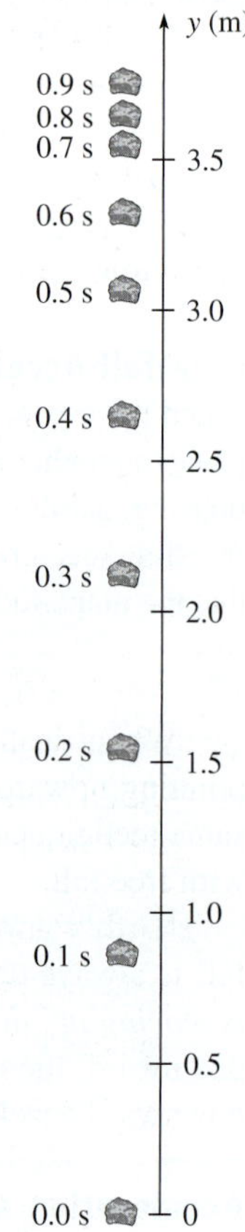

Figure 2.27
Motion diagram for a stone moving straight up.

(b) Equation (2-12) can be used to solve for v_{iy} since all the other quantities in it (Δy, Δt, and a_y) are known.

$$\Delta y = v_{iy}\,\Delta t + \frac{1}{2}a_y\,(\Delta t)^2$$

Solving for v_{iy},

$$v_{iy} = \frac{\Delta y}{\Delta t} - \frac{1}{2}a_y\,\Delta t = \frac{-44.1\text{ m}}{4.00\text{ s}} - \frac{1}{2}(-9.80\text{ m/s}^2 \times 4.00\text{ s}) \qquad (1)$$
$$= -11.0\text{ m/s} + 19.6\text{ m/s} = 8.6\text{ m/s}$$

The initial velocity is 8.6 m/s upward.

(c) The change in v_y is $a_y\,\Delta t$ from Eq. (2-9):

$$v_{fy} = v_{iy} + a_y\,\Delta t$$

Substituting the expression for v_{iy} found previously,

$$v_{fy} = \left(\frac{\Delta y}{\Delta t} - \frac{1}{2}a_y\,\Delta t\right) + a_y\,\Delta t = \frac{\Delta y}{\Delta t} + \frac{1}{2}a_y\,\Delta t \qquad (2)$$
$$= \frac{-44.1\text{ m}}{4.00\text{ s}} + \frac{1}{2}(-9.80\text{ m/s}^2 \times 4.00\text{ s})$$
$$= -11.0\text{ m/s} - 19.6\text{ m/s} = -30.6\text{ m/s}$$

The final velocity is 30.6 m/s downward.

(d) Choosing $y_i = 0$ and $t_i = 0$, the position of the stone as a function of time is

$$y(t) = v_{iy}t + \frac{1}{2}a_y t^2$$

The motion diagram is shown in Fig. 2.27.

Discussion The final speed is greater than the initial speed, as expected. Equations (1) and (2) have a direct interpretation, which is a good check on their validity. The first term, $\Delta y/\Delta t$, is the average velocity of the stone during the 4.00 s of free fall. The second term, $-\frac{1}{2}a_y\Delta t$, is *half* the change in v_y since $\Delta v_y = a_y\,\Delta t$. Because the acceleration is constant, the average velocity is halfway between the initial and final velocities. Therefore, the initial velocity is the average velocity minus half of the change, while the final velocity is the average velocity plus half of the change.

Practice Problem 2.10 Height Attained by Stone

(a) How high above the bridge does the stone go? [*Hint*: What is v_y at the highest point?] (b) If you dropped the stone instead of throwing it, how long would it take to hit the water?

Master the Concepts

- Displacement is the change in position: $\Delta x = x_f - x_i$. The displacement depends only on the starting and ending positions, not on details of the motion. The magnitude of the displacement is not necessarily equal to the total distance traveled; it is the straight-line distance from the initial position to the final position.
- Average velocity is the constant velocity that would cause the same displacement in the same amount of time.

$$v_{av,x} = \frac{\Delta x}{\Delta t} \text{ (for any time interval } \Delta t) \qquad (2\text{-}2)$$

- Velocity is a measure of how fast and in what direction something moves. Its direction is the direction of the object's motion and its magnitude is the instantaneous speed. It is the instantaneous rate of change of the position.

$$v_x = \lim_{\Delta t \to 0} \frac{\Delta x}{\Delta t} \text{ (for a \textit{very short} time interval } \Delta t) \qquad (2\text{-}3)$$

- Average acceleration is the constant acceleration that would cause the same velocity change in the same amount of time.

$$a_{av,x} = \frac{\Delta v_x}{\Delta t} \text{ (for any time interval } \Delta t) \qquad (2\text{-}7)$$

- Acceleration is the instantaneous rate of change of the velocity.

$$a_x = \lim_{\Delta t \to 0} \frac{\Delta v_x}{\Delta t} \text{ (for a \textit{very short} time interval } \Delta t) \qquad (2\text{-}8)$$

  Acceleration does not necessarily mean speeding up. A velocity can change by decreasing speed or by changing direction.
- Interpreting graphs: On a graph of $x(t)$, the slope at any point is v_x. On a graph of $v_x(t)$, the slope at any point is a_x, and the area under the graph during any time interval is the displacement Δx during that time interval. If v_x is negative, the displacement is also negative, so we must count the area as negative when it is below the time axis.

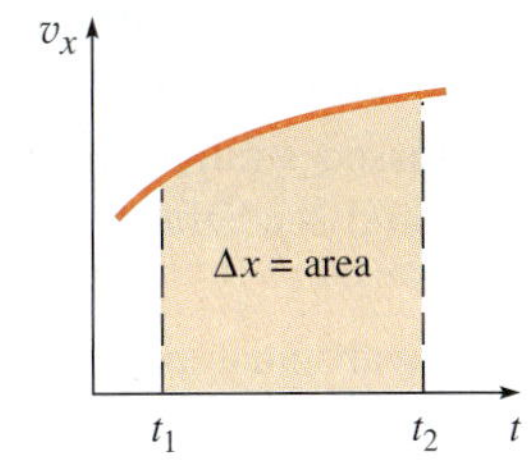

On a graph of $a_x(t)$, the area under the curve is Δv_x, the change in v_x during that time interval.

- Essential relationships for constant acceleration problems: if a_x is constant during the entire time interval Δt from t_i until a later time $t_f = t_i + \Delta t$,

$$\Delta v_x = v_{fx} - v_{ix} = a_x \Delta t \qquad (2\text{-}9)$$

$$\Delta x = \tfrac{1}{2}(v_{fx} + v_{ix})\Delta t \qquad (2\text{-}11)$$

$$\Delta x = v_{ix}\Delta t + \tfrac{1}{2}a_x(\Delta t)^2 \qquad (2\text{-}12)$$

$$v_{fx}^2 - v_{ix}^2 = 2a_x \Delta x \qquad (2\text{-}13)$$

These same relationships hold for position, velocity, and acceleration along the y-axis if a_y is constant.

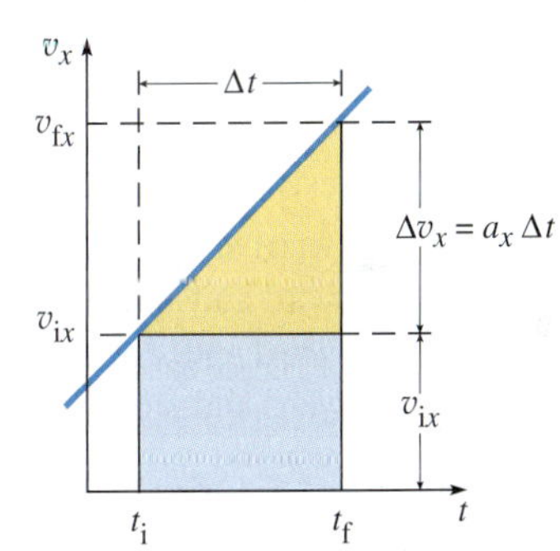

- An object in free fall has a constant downward acceleration. The magnitude of the acceleration g varies a little from place to place near Earth's surface. A typical value is $g = 9.80\ \text{m/s}^2$.

Conceptual Questions

1. Explain how these quantities differ: distance traveled, displacement, and displacement magnitude.
2. Explain the difference between speed and velocity.
3. On a graph of v_x versus time, what quantity does the area under the graph represent?
4. On a graph of v_x versus time, what quantity does the slope of the graph represent?
5. On a graph of a_x versus time, what quantity does the area under the graph represent?
6. On a graph of x versus time, what quantity does the slope of the graph represent?
7. What is the relationship between average velocity and instantaneous velocity? An object can have different instantaneous velocities at different times. Can the same object have different average velocities? Explain.
8. Can the velocity of an object be zero and the acceleration be nonzero at the same time? Explain.
9. You are bicycling along a straight north-south road. Let the x-axis point north. Describe your motion in each of the following cases. Example: $a_x > 0$ and $v_x > 0$ means

you are moving north and speeding up. (a) $a_x > 0$ and $v_x < 0$. (b) $a_x = 0$ and $v_x < 0$. (c) $a_x < 0$ and $v_x = 0$. (d) $a_x < 0$ and $v_x < 0$. (e) Based on your answers, explain why it is not a good idea to use the expression "negative acceleration" to mean slowing down.

10. When a coin is tossed straight up, what can you say about its velocity and acceleration at the highest point of its motion?
11. You throw a ball up with initial speed v_i, and when it reaches its high point at height h, you throw another ball into the air with the same initial speed v_i. Will the two balls cross at half the height h, more than half, or less than half? Explain.
12. When a coin is tossed directly upward, what can you say about its velocity and acceleration at the high point of the toss?

Multiple-Choice Questions

1. A ball is thrown straight up into the air. Ignore air resistance. While the ball is in the air its acceleration
   (a) increases.
   (b) is zero.
   (c) remains constant.
   (d) decreases on the way up and increases on the way down.
   (e) changes direction.
2. Which car has a westward acceleration?
   (a) a car traveling westward at constant speed
   (b) a car traveling eastward and speeding up
   (c) a car traveling westward and slowing down
   (d) a car traveling eastward and slowing down
   (e) a car starting from rest and moving toward the east

Questions 3 and 4. A toy rocket is propelled straight upward from the ground and reaches a height H. After an elapsed time Δt, measured from the time the rocket was first fired off, the rocket has fallen back down to the ground, landing at the same spot from which it was launched.

Answer choices:

(a) zero (b) $2\frac{H}{\Delta t}$

(c) $\frac{H}{\Delta t}$ (d) $\frac{1}{2}\frac{H}{\Delta t}$

3. What is the magnitude of the average velocity of the rocket during this time?
4. What is the average speed of the rocket during this time?
5. A leopard starts from rest at $t = 0$ and runs in a straight line with a constant acceleration until $t = 3.0$ s. The distance covered by the leopard between $t = 1.0$ s and $t = 2.0$ s is
   (a) the same as the distance covered during the first second.
   (b) twice the distance covered during the first second.
   (c) three times the distance covered during the first second.
   (d) four times the distance covered during the first second.

Multiple-Choice Questions 6–15. A jogger is exercising along a long, straight road that runs north-south. She starts out heading north. Her motion is described by the following graph of $v_x(t)$.

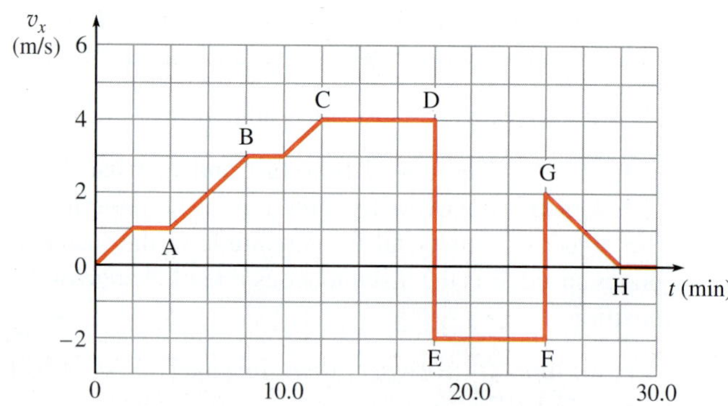

Multiple-Choice Questions 6–15

6. What is the displacement of the jogger from $t = 18.0$ min to $t = 24.0$ min?
   (a) 720 m, south (b) 720 m, north
   (c) 2160 m, south (d) 3600 m, north
7. What is the displacement of the jogger for the entire 30.0 min?
   (a) 3120 m, south (b) 2400 m, north
   (c) 2400 m, south (d) 3840 m, north
8. What is the total distance traveled by the jogger in 30.0 min?
   (a) 3840 m (b) 2340 m (c) 2400 m (d) 3600 m
9. What is the average velocity of the jogger during the 30.0 min?
   (a) 1.3 m/s, north (b) 1.7 m/s, north
   (c) 2.1 m/s, north (d) 2.9 m/s, north
10. What is the average speed of the jogger for the 30 min?
    (a) 1.4 m/s (b) 1.7 m/s (c) 2.1 m/s (d) 2.9 m/s
11. In what direction is she running at time $t = 20$ min?
    (a) south (b) north (c) not enough information
12. In which region of the graph is a_x positive?
    (a) A to B (b) C to D (c) E to F (d) G to H
13. In which region is a_x negative?
    (a) A to B (b) C to D (c) E to F (d) G to H
14. In which region is the velocity directed to the south?
    (a) A to B (b) C to D (c) E to F (d) G to H
15. What distance does the jogger travel during the first 10.0 min ($t = 0$ to 10.0 min)?
    (a) 8.5 m (b) 510 m (c) 900 m (d) 1020 m

16. The figure shown here has four graphs of x versus time. Which graph shows a constant, positive, nonzero velocity?

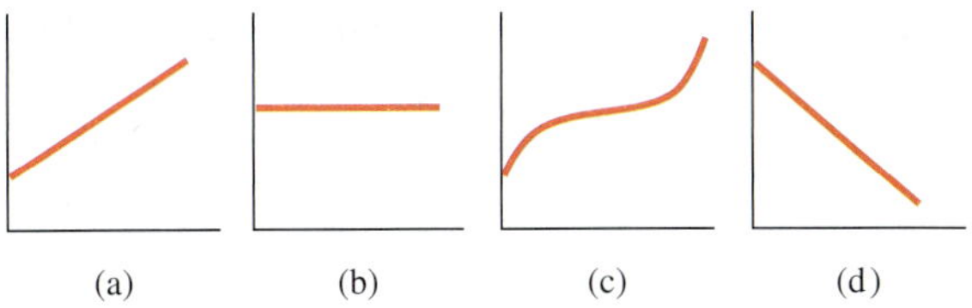

Multiple-Choice Questions 16–20

Questions 17–20. The four graphs show v_x versus time.

17. Which graph shows a constant velocity?

18. Which graph shows a_x constant and positive?

19. Which graph shows a_x constant and negative?

20. Which graph shows a changing a_x that is always positive?

Multiple-Choice Questions 21–30. Each row of the table describes an object moving along the x-axis. Based on the information given in two of the columns, choose the correct entry for the other columns. The question number is in parentheses.

Sign of v_x	Sign of a_x	Moving in what direction?	Change in *speed*
? (21)	+	? (22)	increasing
–	0	? (23)	? (24)
+	? (25)	? (26)	decreasing
? (27)	? (28)	$-x$	not changing
–	+	? (29)	? (30)

Problems

- Combination conceptual/quantitative problem
- Biological or medical application
- ✦ Challenging problem
- Blue # Detailed solution in the Student Solutions Manual
- [1, 2] Problems paired by concept
- connect Text website interactive or tutorial

2.1 Position and Displacement

1. A displacement of 32 cm east is followed by displacements of 48 cm east and then 64 cm west. What is the total displacement?

2. A squirrel is trying to locate some nuts he buried for the winter. He moves 4.0 m to the right of a stone and digs unsuccessfully. Then he moves 1.0 m to the left of his hole, changes his mind, and moves 6.5 m to the right of that position and digs a second hole. No luck. Then he moves 8.3 m to the left and digs again. He finds a nut at last. What is the squirrel's total displacement from its starting point?

3. A runner, jogging along a straight-line path, starts at a position 60 m east of a milestone marker and heads west. After a short time interval he is 20 m west of the mile marker. Choose east to be the positive x-direction. (a) What is the runner's displacement from his starting point? (b) What is his displacement from the milestone? (c) The runner then turns around and heads east. If at a later time the runner is 140 m east of the milestone, what is his displacement from the starting point at this time? (d) What is the total distance traveled from the starting point if the runner stops at the final position listed in part (c)?

4. At 3:00 P.M. a car is located 20 km south of its starting point. One hour later it is 96 km farther south. After two more hours, it is 12 km south of the original starting point. (a) What is the displacement of the car between 3:00 P.M. and 6:00 P.M.? (b) What is the displacement of the car from the starting point to the location at 4:00 P.M.? (c) What is the displacement of the car between 4:00 P.M. and 6:00 P.M.?

2.2 Velocity: Rate of Change of Position

5. For the train of Example 2.2, find the average velocity between 3:14 P.M. when the train is at 3 km east of the origin and 3:28 P.M. when it is 10 km east of the origin.

6. A cyclist travels 10.0 km east in a time of 11 min 40 s. What is his average velocity in meters per second?

7. One of the fastest known animals is the Indian spine-tailed swift. If a swift flies 3.2 km due north in a time of 32.8 s, what is its average velocity? Express your answer in both m/s and mi/h.

8. Jason drives due west with a speed of 35.0 mi/h for 30.0 min, then continues in the same direction with a speed of 60.0 mi/h for 2.00 h, and then drives still farther west at 25.0 mi/h for 10.0 min. What is Jason's average velocity for the entire trip?

9. Two cars, a Porsche Boxster convertible and a Toyota Scion xB, are traveling at constant speeds in the same direction, although the Boxster is 186 m behind the Scion. The speed of the Boxster is 24.4 m/s and the speed of the Scion is 18.6 m/s. Sketch graphs of $x(t)$ for the two cars on the same axes. How much time does it take for the Boxster to catch the Scion? [*Hint*: What must be true about the displacement of the two cars when they meet?] (connect tutorial: catchup)

10. Speedometer readings are obtained and graphed as a car skids to a stop along a straight-line path. How far does the car move between $t = 0$ and $t = 16$ s? Sketch a motion diagram showing the position of the car at 2 s intervals and sketch a graph of $x(t)$. (connect tutorial: start/stop traffic)

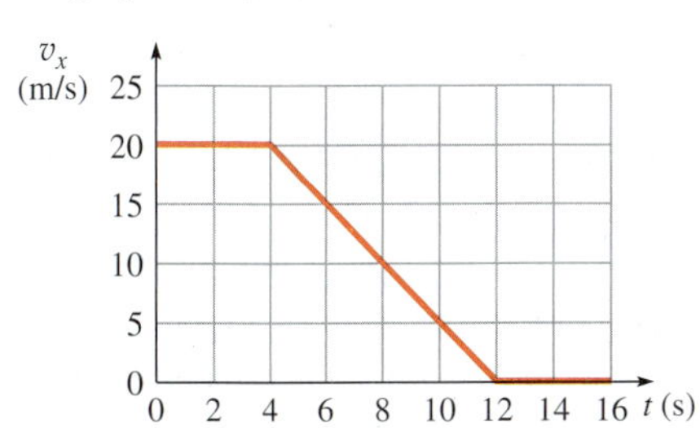

Problems 10 and 36

11. A graph is plotted of the vertical velocity v_y of an elevator versus time. The y-axis points up. (a) How high is the elevator above the starting point ($t = 0$) after 20 s has elapsed? (b) When is the elevator at its highest location above the starting point? (c) Describe the motion in words. (d) Sketch a graph of $y(t)$.

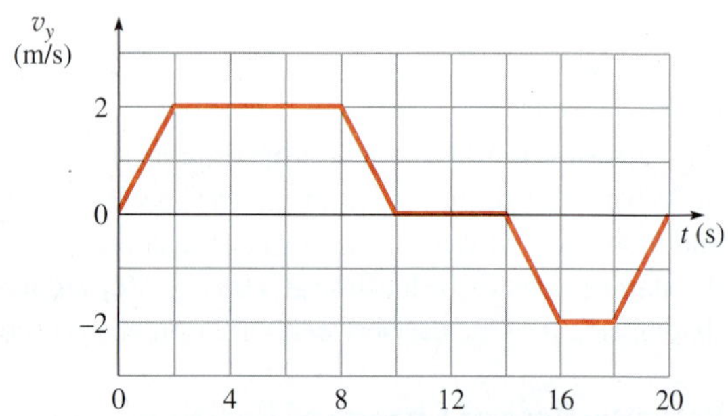

Problems 11 and 30

Problems 12–15. A bicycle is moving along a straight line. The graph in the figure shows its position from the starting point as a function of time. Consider the 1 s time intervals 0–1 s, 1–2 s, and so on.

12. Rank the time intervals in order of increasing displacement, from largest negative to largest positive.
13. Rank the time intervals in order of increasing speed.
14. Rank the time intervals in order of decreasing velocity v_x, from largest positive to largest negative.
15. How far does the object move from $t = 0$ to $t = 3$ s?

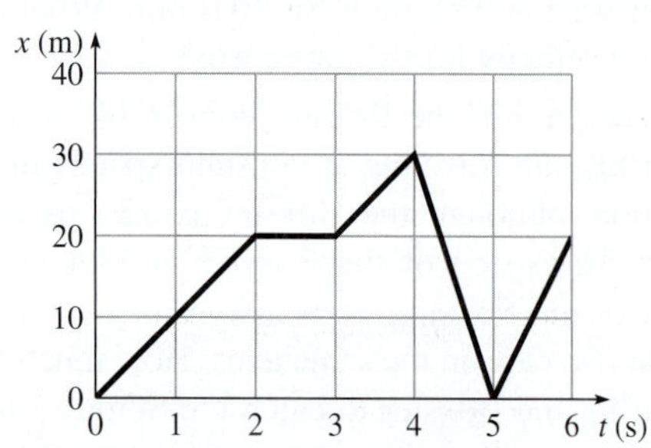

Problems 12–15

16. A ball thrown by a pitcher on a women's softball team is timed at 65.0 mi/h. The distance from the pitching rubber to home plate is 43.0 ft. In major league baseball the corresponding distance is 60.5 ft. If the batter in the softball game and the batter in the baseball game are to have equal times to react to the pitch, with what speed must the baseball be thrown? Assume the ball travels with a constant velocity. [*Hint*: There is no need to convert units; set up a ratio.]
17. A motor scooter travels east at a speed of 12 m/s. The driver then reverses direction and heads west at 15 m/s. What is the change in velocity of the scooter? Give magnitude and direction.
18. Ⓒ To pass a physical fitness test, Massimo must run 1000 m at an average rate of 4.0 m/s. He runs the first 900 m in 250 s. Is it possible for Massimo to pass the test? If so, how fast must he run the last 100 m to pass the test? Explain.
19. The graph shows v_x in meters per second versus t in seconds for a skateboard moving along the x-axis. How far does the board move between $t = 3.00$ s and $t = 8.00$ s? Sketch a motion diagram and a graph of $x(t)$ for the same time interval.
20. The graph shows $x(t)$ for a skater traveling along the x-axis. (a) What is $v_{av,x}$ for the interval from $t = 0$ to $t = 4.0$ s? (b) from $t = 0$ to $t = 5.0$ s?
21. The graph shows $x(t)$ for a skater traveling along the x-axis. What is v_x at $t = 2.0$ s?
22. The graph shows $x(t)$ for an object traveling along the x-axis. Plot v_x as a function of time for this object from $t = 0$ to $t = 8$ s.

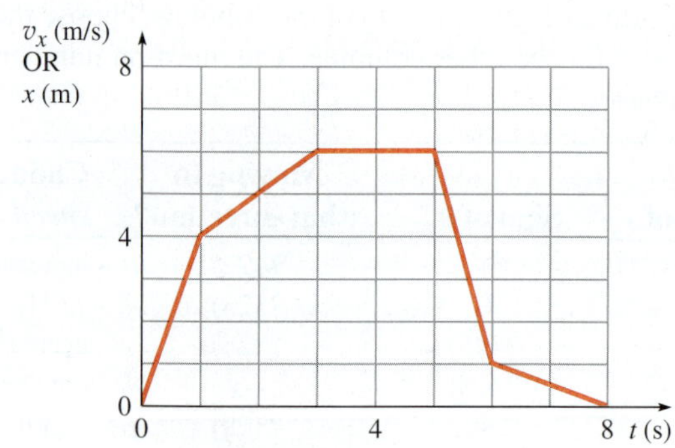

Problems 19–22 and 29

23. A chipmunk, trying to cross a road, first moves 80 cm to the right, then 30 cm to the left, then 90 cm to the right, and finally 310 cm to the left. (a) What is the chipmunk's total displacement? (b) If the elapsed time was 18 s, what was the chipmunk's average speed? (c) What was its average velocity?
24. Rita Jeptoo of Kenya was the first female finisher in the 110th Boston Marathon. She ran the first 10.0 km in a time of 0.5689 h. Assume the race course to be along a straight line. (a) What was her average speed during the first 10.0 km segment of the race? (b) She completed the entire race, a distance of 42.195 km, in a time of 2.3939 h. What was her average speed for the race?
25. A relay race is run along a straight-line track of length 300.0 m running south to north. The first runner starts at the south end of the track and passes the baton to a teammate at the north end of the track. The second runner races back to the start line and passes the baton to a third runner who races 100.0 m northward to the finish line. The magnitudes of the average velocities of the first, second, and third runners during their

parts of the race are 7.30 m/s, 7.20 m/s, and 7.80 m/s, respectively. What is the average velocity of the baton for the entire race? [*Hint*: You will need to find the time spent by each runner in completing her portion of the race.]

26. Using Fig. 2.8, estimate the train's maximum speed.

2.3 Acceleration: Rate of Change of Velocity

27. One of the fastest land animals of North America is the pronghorn antelope. If a pronghorn antelope accelerates from rest in a straight line with a constant acceleration of 1.7 m/s^2, how much time does it take for the antelope to reach a speed of 22 m/s?

28. If a car traveling at 28 m/s is brought to a full stop in 4.0 s after the brakes are applied, find the average acceleration during braking.

29. The graph with Problem 19 shows $v_x(t)$ for a skateboard moving along the x-axis. Rank the times $t = 0.5$ s, 1.5 s, 2.5 s, 3.5 s, 4.5 s, and 5.5 s, in order of the magnitude of the acceleration, from largest to smallest.

30. Sketch the acceleration of the elevator in Problem 11 as a function of time.

31. An 1100 kg airplane starts from rest; 8.0 s later it reaches its takeoff speed of 35 m/s. What is the average acceleration of the airplane during this time?

32. (a) In Fig. 2.16, what is the instantaneous acceleration of the sports car of Example 2.5 at the time of 14 s from the start? (b) What is the displacement of the car from $t = 12.0$ s to $t = 16.0$ s? (c) What is the average velocity of the car in the 4.0 s time interval from 12.0 s to 16.0 s?

33. An automobile is traveling along a straight road heading to the southeast at 24 m/s when the driver sees a deer begin to cross the road ahead of her. She steps on the brake and brings the car to a complete stop in an elapsed time of 8.0 s. A data recording device, triggered by the sudden braking action, records the following velocities and times as the car slows. Let the positive x-axis be directed to the southeast. Plot a graph of v_x versus t and find (a) the average acceleration as the car comes to a stop and (b) the instantaneous acceleration at $t = 2.0$ s.

v_x (m/s)	24	17.3	12.0	8.7	6.0	3.5	2.0	0.75	0
t (s)	0	1.0	2.0	3.0	4.0	5.0	6.0	7.0	8.0

34. Draw a motion diagram and sketch graphs of $x(t)$, $v_x(t)$, and $a_x(t)$ for a sprinter running a short race on a straight track, from just before the start of the race until the sprinter has stopped after finishing the race.

35. In each motion diagram, the dots are labeled with the "frame" number (frame 1 is the first position of the object). Choose the x-axis pointing to the right. For each diagram, sketch graphs of $x(t)$, $v_x(t)$, and $a_x(t)$ and describe the motion in words.

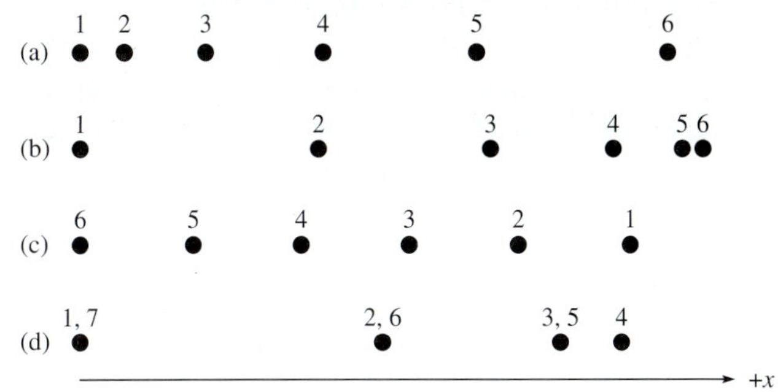

36. The graph with Problem 10 shows speedometer readings as a car skids to a stop. What is the magnitude of the acceleration at $t = 7.0$ s? Sketch a graph of $a_x(t)$.

37. The figure shows a plot of $v_x(t)$ for a car traveling in a straight line. (a) What is $a_{av,x}$ between $t = 6$ s and $t = 11$ s? (b) What is $v_{av,x}$ for the same time interval? (c) What is $v_{av,x}$ for the interval $t = 0$ to $t = 20$ s? (d) What is the increase in the car's speed between 10 s and 15 s? (e) How far does the car travel from time $t = 10$ s to time $t = 15$ s?

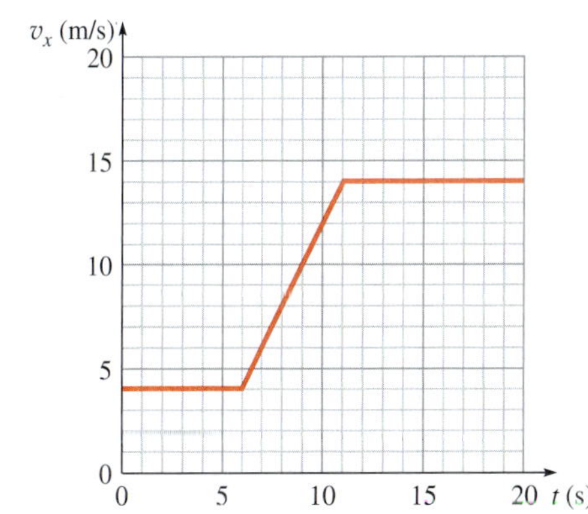

Problems 37 and 38

38. Sketch a graph of $a_x(t)$ for the car in Problem 37.

39. The graph shows v_x versus t for an object moving along a straight line. (a) What is a_x at $t = 11$ s? (b) What is a_x at $t = 3$ s? (c) Sketch a graph of $a_x(t)$. (d) How far does the object travel from $t = 12$ s to $t = 14$ s? (connect tutorial: x, v, a)

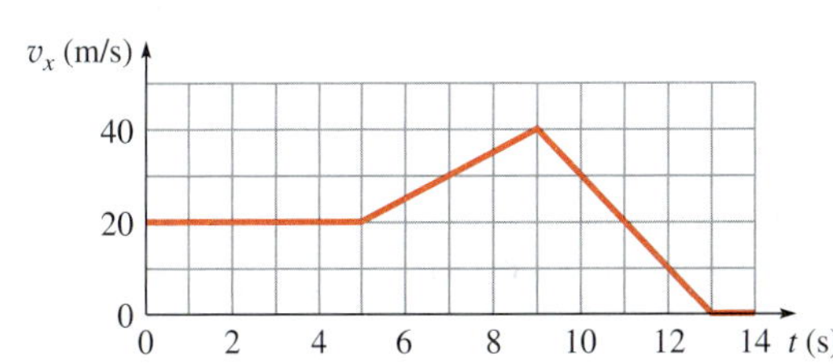

40. (a) Using Fig. 2.15, estimate Damon's acceleration at $t = 2.0$ s. (b) Estimate his average velocity between $t = 0$ and $t = 10.0$ s.

2.4 Visualizing Motion Along a Line with Constant Acceleration; 2.5 Kinematic Equations for Motion Along a Line with Constant Acceleration

41. Four objects move to the right with constant acceleration. Rank the motion diagrams in order of the magnitude of the acceleration, from greatest to least. The time interval between dots is the same in each diagram.

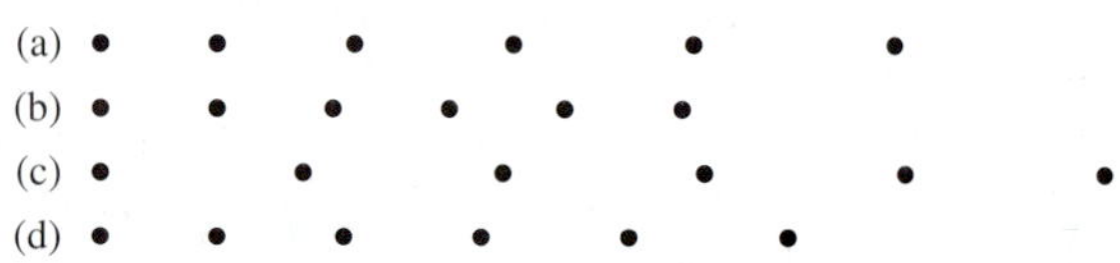

42. A toboggan is sliding in a straight line down a snowy slope. The table shows the speed of the toboggan at various times during its trip. (a) Make a graph of the speed as a function of time. (b) Judging by the graph, is it plausible that the toboggan's acceleration is constant? If so, what is the acceleration?

Time Elapsed, t (s)	Speed of Toboggan, v (m/s)
0	0
1.14	2.8
1.62	3.9
2.29	5.6
2.80	6.8

43. A pilot without special training or equipment can tolerate a horizontal acceleration of up to about $9g$ for a short period of time (about a minute) without losing consciousness. (a) How long would it take a supersonic jet in horizontal flight to accelerate from 200 m/s to 700 m/s at an acceleration of $9.0g$? (b) How far would the jet travel during this time?

44. The St. Charles streetcar in New Orleans starts from rest and has a constant acceleration of 1.20 m/s^2 for 12.0 s. (a) Draw a graph of v_x versus t. (b) How far has the train traveled at the end of the 12.0 s? (c) What is the speed of the train at the end of the 12.0 s? (d) Draw a motion diagram, showing the streetcar's position at 2.0 s intervals.

45. An airplane lands and starts down the runway with a southwest velocity of 55 m/s. What constant acceleration allows it to come to a stop in 1.0 km? Sketch a graph of $v_x(t)$.

46. A train is traveling south at 24.0 m/s when the brakes are applied. It slows down with constant acceleration to a speed of 6.00 m/s in a time of 9.00 s. (a) Draw a graph of v_x versus t for a 12 s interval (starting 2 s before the brakes are applied and ending 1 s after the brakes are released). Let the x-axis point to the north. (b) What is the acceleration of the train during the 9.00 s interval? (c) How far does the train travel during the 9.00 s?

47. ✦ A 1200 kg airplane starts from rest and moves forward with a constant acceleration of magnitude 5.00 m/s^2 along a runway that is 250 m long. (a) How long does it take the plane to reach a speed of 46.0 m/s? (b) How far along the runway has the plane moved when it reaches 46.0 m/s?

48. A car is speeding up and has an instantaneous velocity of 1.0 m/s in the $+x$-direction when a stopwatch reads 10.0 s. It has a constant acceleration of 2.0 m/s^2 in the $+x$-direction. (a) What is the speed when the stopwatch reads 12.0 s? (b) How far does the car move between t = 10.0 s and t = 12.0 s?

49. You are driving your car along a country road at a speed of 27.0 m/s. As you come over the crest of a hill, you notice a farm tractor 25.0 m ahead of you on the road, moving in the same direction as you at a speed of 10.0 m/s. You immediately slam on your brakes and slow down with a constant acceleration of magnitude 7.00 m/s^2. Will you hit the tractor before you stop? How far will you travel before you stop or collide with the tractor? If you stop, how far is the tractor in front of you when you finally stop?

50. A typical sneeze has a maximum speed of 44 m/s. Suppose the material emitted in the sneeze begins inside the nose at rest, 2.0 cm from the nostrils. It has a constant acceleration for the first 0.25 cm and then moves at constant velocity for the remainder of the distance. (a) What is the acceleration as it moves the first 0.25 cm? (b) How long does it take to move the 2.0 cm distance in the nose? (c) Sketch a graph of $v_x(t)$.

51. A train is traveling along a straight, level track at 26.8 m/s. Suddenly the engineer sees a truck stalled on the tracks 184 m ahead. If the maximum possible braking acceleration has magnitude 1.52 m/s^2, can the train be stopped in time?

52. In a cathode ray tube in an old TV, electrons are accelerated from rest with a constant acceleration of magnitude 7.03×10^{13} m/s^2 during the first 2.0 cm of the tube's length; then they move at essentially constant velocity another 45 cm before hitting the screen. (a) Find the speed of the electrons when they hit the screen. (b) How long does it take them to travel the length of the tube?

53. The graph is of v_x versus t for an object moving along the x-axis. Sketch a motion diagram between t = 9.0 s and t = 13.0 s and describe the motion in words. How far does the object move between t = 9.0 s and t = 13.0 s? Solve using two methods: a graphical analysis and an algebraic solution.

54. The graph is of v_x versus t for an object moving along the x-axis. Sketch a motion diagram between t = 5.0 s and t = 9.0 s and describe the motion in words. What is the acceleration between t = 5.0 s and t = 9.0 s?

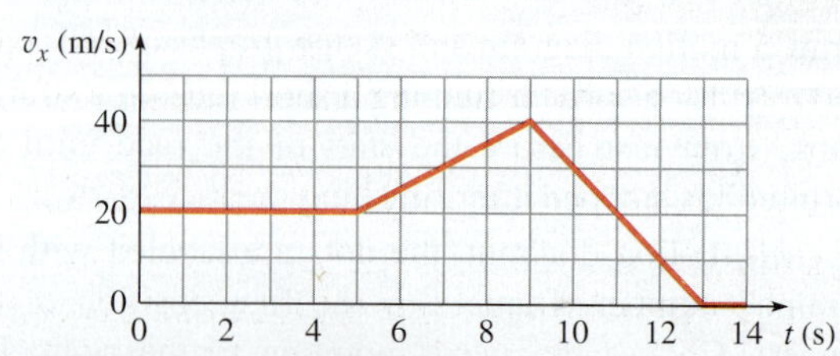

Problems 53–54

55. A train, traveling at a constant speed of 22 m/s, comes to an incline with a constant slope. While going up the incline, the train slows down with a constant acceleration of magnitude 1.4 m/s^2. (a) Draw a graph of v_x versus t where the x-axis points up the incline. (b) What is the speed of the train after 8.0 s on the incline? (c) How far has the train traveled up the incline after 8.0 s? (d) Draw a motion diagram, showing the train's position at 2.0 s intervals.

2.6 Free Fall

In the problems, please assume the free-fall acceleration $g = 9.80$ m/s^2 unless a different value is given in the problem statement. Ignore air resistance.

56. A brick is thrown vertically upward with an initial speed of 3.00 m/s from the roof of a building. If the building is 78.4 m tall, how much time passes before the brick lands on the ground?

57. A penny is dropped from the observation deck of the Empire State building (369 m above ground). With what velocity would it strike the ground if air resistance were negligible?

58. (a) How long does it take for a golf ball to fall from rest for a distance of 12.0 m? (b) How far would the ball fall in twice that time?

59. Grant jumps 1.3 m straight up into the air to slam-dunk a basketball into the net. With what speed did he leave the floor?

60. During a walk on the Moon, an astronaut accidentally drops his camera over a 20.0 m cliff. It leaves his hands with zero speed, and after 2.0 s it has attained a velocity of 3.3 m/s downward. How far has the camera fallen after 4.0 s?

61. Glenda drops a coin from ear level down a wishing well. The coin falls a distance of 7.00 m before it strikes the water. If the speed of sound is 343 m/s, how long after Glenda releases the coin will she hear a splash?

62. A stone is launched straight up by a slingshot. Its initial speed is 19.6 m/s and the stone is 1.50 m above the ground when launched. (a) How high above the ground does the stone rise? (b) How much time elapses before the stone hits the ground?

63. A 55 kg lead ball is dropped from the leaning tower of Pisa. The tower is 55 m high. (a) How far does the ball fall in the first 3.0 s of flight? (b) What is the speed of the ball after it has traveled 2.5 m downward? (c) What is the speed of the ball 3.0 s after it is released? (d) If the ball is thrown vertically upward from the top of the tower with an initial speed of 4.80 m/s, where will it be after 2.42 s?

64. ✦ A balloonist, riding in the basket of a hot air balloon that is rising vertically with a constant velocity of 10.0 m/s, releases a sandbag when the balloon is 40.8 m above the ground. What is the bag's speed when it hits the ground?

65. ✦ Superman is standing 120 m horizontally away from Lois Lane. A villain throws a rock vertically downward with a speed of 2.8 m/s from 14.0 m directly above Lois. (a) If Superman is to intervene and catch the rock just before it hits Lois, what should be his minimum constant acceleration? (b) How fast will Superman be traveling when he reaches Lois?

66. ✦ A student, looking toward his fourth-floor dormitory window, sees a flowerpot with nasturtiums (originally on a window sill above) pass his 1.0 m high window in 0.051 s. The distance between floors in the dormitory is 4.0 m. From a window on which floor did the flowerpot fall?

67. ✦ You drop a stone into a deep well and hear it hit the bottom 3.20 s later. This is the time it takes for the stone to fall to the bottom of the well, plus the time it takes for the sound of the stone hitting the bottom to reach you. Sound travels about 343 m/s in air. How deep is the well?

Collaborative Problems

68. ✦ A rocket engine can accelerate a rocket launched from rest vertically up with an acceleration of 20.0 m/s^2. However, after 50.0 s of flight the engine fails. Ignore air resistance. (a) What is the rocket's altitude when the engine fails? (b) When does it reach its maximum height? (c) What is the maximum height reached? [*Hint*: A graphical solution may be easiest.] (d) What is the velocity of the rocket just before it hits the ground?

69. An unmarked police car starts from rest just as a speeding car passes at a speed of v. If the police car speeds up with a constant acceleration of magnitude a, what is the speed of the police car when it catches up to the speeder, who does not realize she is being pursued and does not vary her speed?

70. ✦ Find the point of no return for an airport runway 1.50 mi in length if a jet plane can speed up at 10.0 ft/s^2 and slow down at 7.00 ft/s^2. The point of no return is the point where the pilot can no longer abort the takeoff without running out of runway. How much time is available from the start of the motion to decide on a course of action?

71. Your current case as an FBI investigator involves a possible assassination attempt. The crime scene is a tall building, about 150 m high. As a foreign official was walking into the building, a flowerpot fell from above him and nearly landed on his head. One witness says that someone accidentally knocked the flowerpot from a twenty-fourth-story window, 94 m above the ground. Another witness was inside the building looking out of an eighteenth-story window when she saw the flowerpot fall. She claims that the flowerpot took exactly 0.044 s to fall from the top of the window to the bottom of the window, a distance of 1.5 m. (She knows this because she was videotaping her pet dachshund doing tricks and the flowerpot falling past the window was filmed in the background.) The top of

the eighteenth-story window is 75 m above the ground. Could the flowerpot have fallen with zero initial velocity from the twenty-fourth-story window?

72. ✦ Ⓒ An elevator starts at rest on the ninth floor. At $t = 0$, a passenger pushes a button to go to another floor. The graph for this problem shows the acceleration a_y of the elevator as a function of time. Let the y-axis point upward. (a) Has the passenger gone to a higher or lower floor? (b) Sketch a graph of the velocity v_y of the elevator versus time. (c) Sketch a graph of the position y of the elevator versus time.

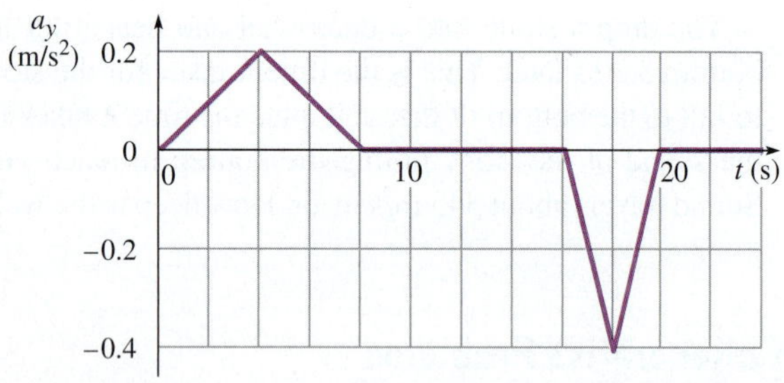

Comprehensive Problems

In the problems, please assume the free-fall acceleration $g = 9.80\ m/s^2$ unless a more precise value is given in the problem statement. Ignore air resistance.

73. (a) If a freestyle swimmer traveled 1500 m in a time of 14 min 53 s, how fast was his average speed? (b) If the pool was rectangular and 50 m in length, how does the speed you found compare with his sustained swimming speed of 1.54 m/s during one length of the pool after he had been swimming for 10 min? What might account for the difference?

74. While passing a slower car on the highway, you accelerate uniformly from 17.4 m/s to 27.3 m/s in a time of 10.0 s. (a) How far do you travel during this time? (b) What is your acceleration magnitude?

75. Ⓒ Fish don't move as fast as you might think. A small trout has a top swimming speed of only about 2 m/s, which is about the speed of a brisk walk (for a human, not a fish!). It may seem to move faster because it is capable of large *accelerations*—it can dart about, changing its speed or direction very quickly. If a trout starts from rest and accelerates to 2.0 m/s in 0.050 s, what is the trout's average acceleration?

76. A cheetah can accelerate from rest to 24 m/s in 2.0 s. Assuming the acceleration is constant over the time interval, (a) what is the magnitude of the acceleration of the cheetah? (b) What is the distance traveled by the cheetah in these 2.0 s? (c) A runner can accelerate from rest to 6.0 m/s in the same time, 2.0 s. What is the magnitude of the acceleration of the runner? By what factor is the cheetah's average acceleration magnitude greater than that of the runner?

77. A rocket is launched from rest. After 8.0 min, it is 160 km above the Earth's surface and is moving at a speed of 7.6 km/s. Assuming the rocket moves up in a straight line, what are its (a) average velocity and (b) average acceleration?

78. A streetcar named Desire travels between two stations 0.60 km apart. Leaving the first station, it accelerates for 10.0 s at 1.0 m/s^2 and then travels at a constant speed until it is near the second station, when it brakes at 2.0 m/s^2 in order to stop at the station. Sketch a graph of $v_x(t)$. How long did this trip take? [*Hint*: What's the average velocity?]

79. A stone is thrown vertically downward from the roof of a building. It passes a window 16.0 m below the roof with a speed of 25.0 m/s. It lands on the ground 3.00 s after it was thrown. What was (a) the initial velocity of the stone and (b) how tall is the building?

80. A car traveling at 29 m/s (65 mi/h) runs into a bridge abutment after the driver falls asleep at the wheel. (a) If the driver is wearing a seat belt and comes to rest within a 1.0 m distance, what is his acceleration (assumed constant)? (b) A passenger who isn't wearing a seat belt is thrown into the windshield and comes to a stop in a distance of 10.0 cm. What is the acceleration of the passenger?

81. To pass a physical fitness test, Marcella must run 1.00 km at an average speed of 3.33 m/s. She runs the first 0.500 km at an average of 4.20 m/s. (a) How much time does she have to run the remaining 0.500 km? (b) What should be her average speed over the last 500 m in order to finish with an overall average speed of 3.33 m/s?

82. At 3:00 P.M., a bank robber is spotted driving north on I-15 at milepost 126. His speed is 112.0 mi/h. At 3:37 P.M., he is spotted at milepost 185 doing 105.0 mi/h. During this time interval, what are the bank robber's displacement, average velocity, and average acceleration? (Assume a straight highway.)

83. ✦ Ⓒ The graph for this problem shows the vertical velocity v_y of a bouncing ball as a function of time. The y-axis points up. Answer these questions based on the data in the graph. (a) At what time does the ball reach its maximum height? (b) For how long is the ball in contact with the floor? (c) What is the maximum height of the ball? (d) What is the acceleration of the ball while in the air? (e) What is the average acceleration of the ball while in contact with the floor?

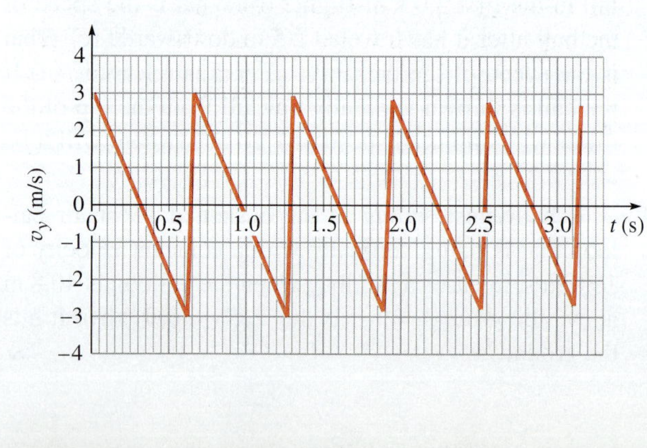

84. ✦ A motorcycle is speeding on a straight, level highway at constant speed. At $t = 0$, the motorcycle passes a police car that is initially at rest. The officer gives chase, but the motorcyclist doesn't notice and keeps moving at constant speed. The graph below shows $v_x(t)$ for both. (a) When are the motorcycle and police car moving at the same speed? (b) At $t = 16$ s, has the police car caught up with the speeder? Explain.

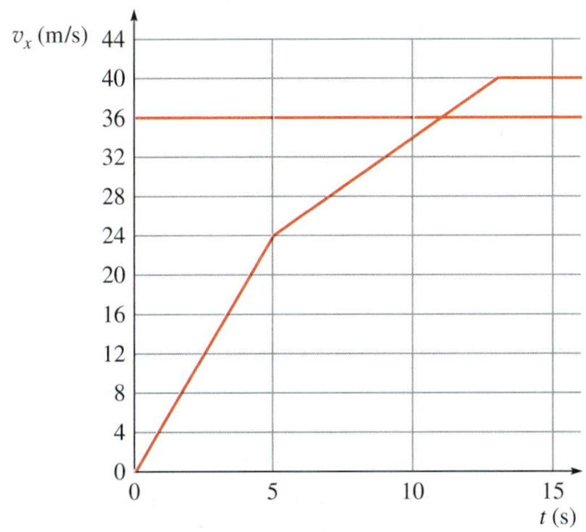

85. ✦ The graph shows the position x of a switch engine in a rail yard as a function of time t. At which of the labeled times t_0 to t_7 is (a) $a_x < 0$, (b) $a_x = 0$, (c) $a_x > 0$, (d) $v_x = 0$, (e) the speed decreasing?

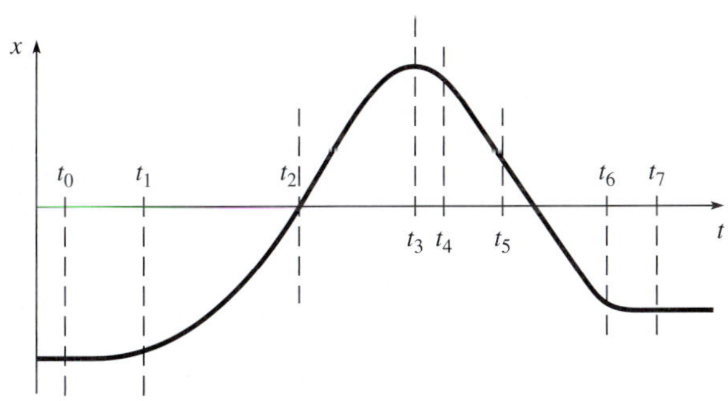

86. ✦ An airtrack glider, 8.0 cm long, blocks light as it goes through a photocell gate. The glider is released from rest on a frictionless inclined track and the gate is positioned so that the glider has traveled 96 cm when it is in the middle of the gate. The timer gives a reading of 333 ms for the glider to pass through this gate. Friction is negligible. What is the acceleration (assumed constant) of the glider along the track?

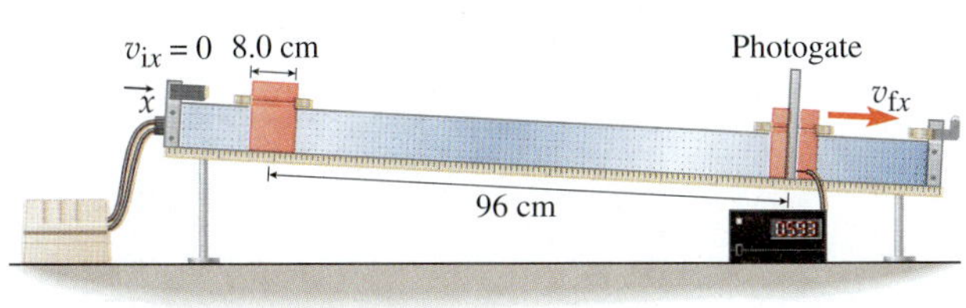

87. ✦ In the human nervous system, signals are transmitted along neurons as *action potentials* that travel at speeds of up to 100 m/s. (An action potential is a traveling influx of sodium ions through the membrane of a neuron.) The signal is passed from one neuron to another by the release of neurotransmitters in the synapse. Suppose someone steps on your toe. The pain signal travels along a 1.00 m long sensory neuron to the spinal column, across a synapse to a second 1.00 m long neuron, and across a second synapse to the brain. Suppose that the synapses are each 100 nm wide, that it takes 0.10 ms for the signal to cross each synapse, and that the action potentials travel at 100.0 m/s. (a) At what average speed does the signal cross a synapse? (b) How long does it take the signal to reach the brain? (c) What is the average speed of propagation of the signal?

Answers to Practice Problems

2.1 3.8 m east

2.2 $v_{\text{av},x} = \Delta x/\Delta t = (-36\text{ km})/(28\text{ min}) = -1.29\text{ km/min} = -77\text{ km/h}$. The average velocity is 77 km/h in the $-x$-direction (west).

2.3 About 100 to 110 km/h in the $+x$-direction (east)

2.4 The velocity is increasing in magnitude, so the acceleration is in the same direction as the velocity (the $-x$-direction). Thus, a_x is negative; the acceleration is in the $-x$-direction.

2.5 A tangent line at $t = 14.0$ s intersects the axes at approximately $(t, v_x) = (0, 35\text{ m/s})$ and $(17\text{ s}, 60\text{ m/s})$. The acceleration is the slope of this tangent line:
$a_x = (60\text{ m/s} - 35\text{ m/s})/(17\text{ s} - 0) = 1.5\text{ m/s}^2$.

2.6 (a) $\Delta x = v_{ix}\,\Delta t + \frac{1}{2}a_x\,(\Delta t)^2$
$= 0 + \frac{1}{2} \times 360\,000\text{ m/s}^2 \times (0.40 \times 10^{-3}\text{ s})^2 = 2.9\text{ cm}$
(b) $\Delta v_x = a_x \Delta t = 360\,000\text{ m/s}^2 \times 0.40 \times 10^{-3}\text{ s} = 140\text{ m/s}$; with $v_{ix} = 0$, $v_{fx} = 140\text{ m/s}$.

2.7 $\Delta x = \frac{1}{2}a_x(\Delta t)^2 = 2.0\text{ m}$

2.8 Set the velocities equal: $v_{isx} + a_{sx}t = v_{ibx} + a_{bx}t$. Solving, $t = \dfrac{v_{ibx} - v_{isx}}{a_{sx} - a_{bx}} = \dfrac{4.00\text{ km/s}}{0.800\text{ km/s}^2} = 5.00\text{ s}$. At that time, $v_{sx} = v_{bx} = 4.00\text{ km/s}$ (in the $+x$-direction).

2.9 From subproblem 1, solving $\Delta x_{21} = v_{1x}t_2 + \frac{1}{2}a_{1x}t_2^{\,2}$ yields $t_2 = 10.0$ s. From subproblem 2, solving $v_{3x} = v_{2x} + a_{2x}(t_3 - t_2)$ yields $t_3 - t_2 = 10.0$ s. Then $t_3 = 20$ s; it takes 20 s to reach the buoy.

2.10 (a) From release to the top of flight,
$\Delta y = (v_y^{\,2} - v_{iy}^{\,2})/(2a_y) = -(8.6\text{ m/s})^2/(-19.6\text{ m/s}^2) = 3.8\text{ m}$;
(b) With an initial velocity of zero, $-44.1\text{ m} = \Delta y = -\frac{1}{2}gt^2$ and $t = \sqrt{-2\Delta y/g} = 3.00\text{ s}$.

Answers to Checkpoints

2.1 No. The magnitude of the displacement is the shortest distance between two points. The distance traveled can be greater than or equal to the displacement, depending on the path taken. In Example 2.1 the displacement is 2.9 km to the west, and the distance traveled is 11.5 km.

2.2A Yes. Average speed is the distance traveled divided by the time interval in moving from point A to point B. Average velocity is the displacement from point A to point B divided by the same time interval. The magnitude of the displacement is the shortest possible distance from A to B. Thus the average velocity magnitude is less than or equal to the average speed.

2.2B The cart is moving to the right at constant speed because the distance from one "frame" to the next stays the same.

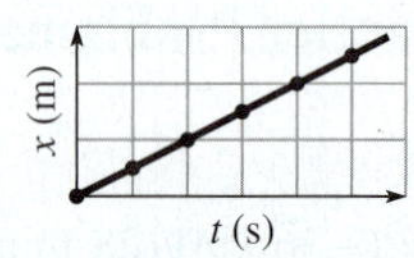

2.2C The slope of the $x(t)$ graph represents the value of v_x. The slope begins at zero, increases to a maximum around $t = 18.5$ min, and then decreases back to zero. The motion diagram shows dots closely spaced when the speed is low and widely spaced where it is high.

Motion diagram for the train from $t = 14$ min to $t = 23$ min.

Graph of $v_x(t)$

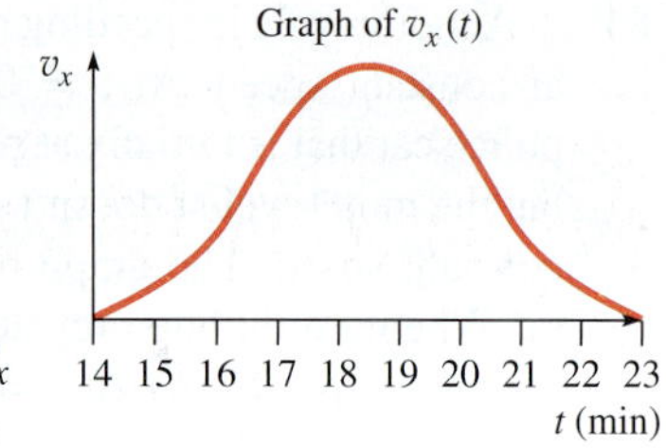

2.3 The slope of the tangent to a graph of v_x versus time is the instantaneous acceleration a_x at the time.

2.4 Yes; between 2 s and 3 s, the red cart's velocity component v_x increases from 1.4 m/s to 1.6 m/s while the blue cart's decreases from 1.6 m/s to 1.4 m/s, so they must be equal sometime during that interval.

The successive displacements Δx on the motion diagram tell us the average velocity during the time interval ($v_{av,x} = \Delta x/\Delta t$), so we can tell that the red cart is slower from $t = 1$ s to $t = 2$ s, faster from $t = 3$ s to $t = 4$ s, and the average velocities are equal between $t = 2$ s and $t = 3$ s. The velocities change continuously, so the carts must have the same instantaneous velocity at some time between $t = 2$ s and $t = 3$ s.

2.5 Only if the plane's acceleration is constant must its average velocity be 470 km/h west. If its acceleration is not constant, the average velocity is not necessarily 470 km/h west. To find the average velocity, we would divide the plane's displacement by the time interval.

2.6 Yes. If you throw a ball upward, it is in free fall as soon as it loses contact with your hand.

CHAPTER

3

Motion in a Plane

A gull scoops up a clam and takes it high above the ground. While flying parallel to the ground, the gull lets go of the clam. The clam lands on a rock below and cracks open. Then the gull alights and enjoys lunch. A beachcomber on the beach sees the clam fall along a parabolic path, just as a projectile would. Why does the clam not drop straight down? What does the path of the falling clam look like to the gull? (See pp. 71–72 and 79 for the answers.)

BIOMEDICAL APPLICATIONS

- Jumping locusts and snow leopards (Probs. 59, 103)
- Spitting archer fish (Prob. 62)
- Seed dispersal (Prob. 64)
- Acceleration in a centrifuge (Prob. 107)
- Fish ladders (Prob. 113)

Concepts & Skills to Review

- trigonometric functions: sine, cosine, and tangent (Appendix A.7)
- Pythagorean theorem (Appendix A.6)
- position, displacement, velocity, and acceleration (Sections 2.1–2.3)
- average and instantaneous quantities (Sections 2.2–2.3)
- motion along a line with constant acceleration (Sections 2.4–2.6)

3.1 GRAPHICAL ADDITION AND SUBTRACTION OF VECTORS

CONNECTION:

Vector quantities must be added and subtracted according to special rules that take their directions into account. All vector quantities follow the *same* rules of addition and subtraction.

Chapter 2 introduced the quantities position, displacement, velocity, and acceleration to describe motion along a line—that is, motion in one dimension of space. To describe motion in more than one dimension, we need a full treatment of vector addition and subtraction because position, displacement, velocity, and acceleration are vectors. (Other vectors you will study in this book include force, momentum, angular momentum, torque, and the electric and magnetic fields.)

Vectors and Scalars All **vectors** have a direction in space as well as a magnitude. The direction of any vector is always a *physical direction in space* such as up, down, north, or 35° south of west.

Vector quantities are usually drawn as arrows pointing in the direction of the vector; the length of the arrow is proportional to the magnitude of the vector. By contrast, a **scalar** quantity can have magnitude, algebraic sign (positive or negative), and units, but not a direction in space. It wouldn't make sense to draw an arrow to represent a scalar such as mass!

In this book, an arrow over a boldface symbol indicates a vector quantity ($\vec{\mathbf{r}}$). (Some books use boldface without the arrow or the arrow without boldface.) When writing by hand, always draw an arrow over a vector symbol to distinguish it from a scalar. When the symbol for a vector is written without the arrow and in italics rather than boldface (r), it stands for the *magnitude* of the vector (which is a scalar). Absolute value bars are also used to stand for the magnitude of a vector, so $r = |\vec{\mathbf{r}}|$. The magnitude of a vector may have units and is never negative; it can be positive or zero.

Conceptual Example 3.1

Body Temperature

Normal body temperature is 37°C. An adult with the flu might have a body temperature of around 39°C. Is temperature a vector quantity or a scalar quantity?

Strategy If a quantity is a vector, it must have both a magnitude and a physical direction in space.

Solution and Discussion Does temperature have a direction? A temperature in Fahrenheit (°F) or Celsius (°C) can be above or below zero—is that a direction? No. A vector must have a *physical direction in space*. It does not make sense to say that the body temperature of a patient is "38.4°C in the southwest direction." "The patient's temperature is up 1.4°C today," means that it has increased, not that it is pointing vertically upward. Temperature is a scalar. If we need to subtract temperatures to find the change in temperature, we subtract them like ordinary numbers. If the patient's temperature changes from 38.4°C to 37.7°C, the temperature change is

$$\Delta T = T_{\text{final}} - T_{\text{initial}} = 37.7°\text{C} - 38.4°\text{C} = -0.7°\text{C}$$

We define the change in a quantity as the final value minus the original value—not as the larger value minus the smaller—so a decrease is a negative change.

Conceptual Practice Problem 3.1 Bank Balance

When you deposit a paycheck, the balance of your checking account "goes up." When you pay a bill, it "goes down." Is the balance of your account a vector quantity?

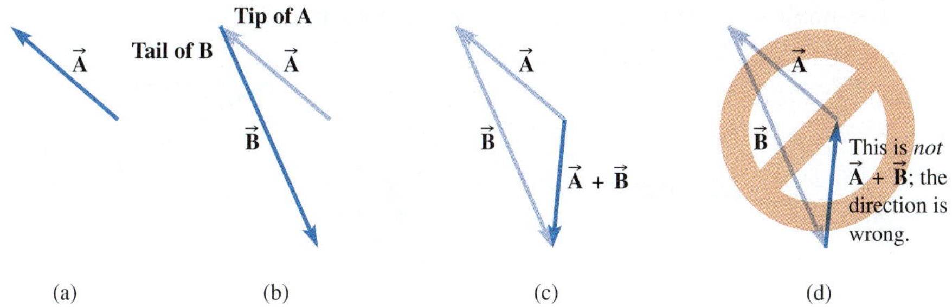

Figure 3.1 Adding two vectors graphically. (a) Draw one vector arrow. (b) Draw the second, starting where the first arrow ended. (c) The sum of the two is represented by an arrow drawn from the start of the first to the end of the second. (d) Be careful to avoid this common mistake.

When scalars are added or subtracted, they do so in the usual way: 3 kg of water plus 2 kg of water is equal to 5 kg of water. Adding or subtracting vectors is different. Vectors follow rules of addition and subtraction that take into account the *directions* of the vectors as well as their magnitudes. Whenever you need to add or subtract quantities, check whether they are vectors. If so, be sure to add or subtract them correctly *as vectors. Do not just add or subtract their magnitudes.* A plus sign (+) between vector quantities indicates *vector addition*, not ordinary addition. An equals sign (=) between vector quantities means that the vectors are identical in magnitude *and* direction, *not* simply that their magnitudes are equal.

Graphical Vector Addition We start with a graphical method to help develop your intuition. To add two vectors graphically, first draw an arrow to represent one of them (Fig. 3.1a). (It does not matter in what order vectors are added; $\vec{\mathbf{A}} + \vec{\mathbf{B}} = \vec{\mathbf{B}} + \vec{\mathbf{A}}$.) The arrow points in the direction of the vector and its length is proportional to the magnitude of the vector. It doesn't matter where you start drawing the arrow. The value of a vector is not changed by moving it as long as its direction and magnitude are not changed.

Now draw the second vector arrow starting where the first ends. In other words, place the "tail" of the second arrow at the "tip" of the first (Fig 3.1b). Finally, draw an arrow starting from the *tail* of the first and ending at the *tip* of the second. This arrow represents the sum of the two vectors (Fig. 3.1c). A common error is to draw the sum from the tip of the second to the tail of the first (Fig. 3.1d). If the lengths and directions of the vectors are drawn accurately to scale, using a ruler and a protractor, then the length and direction of the sum can be determined with the ruler and protractor. To add more than two vectors, continue drawing them tip to tail. The sum of two or more vectors is called the **resultant**.

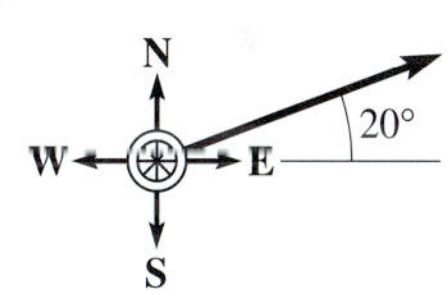

Figure 3.2 The direction of this vector is 20° north of east (20° N of E).

Vector Subtraction *To subtract a vector is to add its opposite* (a vector with the same magnitude but opposite direction): $\vec{\mathbf{r}}_f - \vec{\mathbf{r}}_i = \vec{\mathbf{r}}_f + (-\vec{\mathbf{r}}_i)$. Multiplying a vector by the scalar −1 reverses the vector's direction while leaving its magnitude unchanged, so $-\vec{\mathbf{r}}_i = -1 \times \vec{\mathbf{r}}_i$ is a vector equal in magnitude and opposite in direction to $\vec{\mathbf{r}}_i$.

Using the Cardinal Directions of the Compass Any direction in the horizontal plane can be specified by giving an angle with respect to north, south, east, or west. For example, the direction of the vector in Fig. 3.2 is "20° north of east," which means that the vector makes a 20° angle with the east direction and is on the north (rather than the south) side of east. The same direction could be described as "70° east of north," although it is customary to use the smaller angle. Northeast means "45° north of east" or, equivalently, "45° east of north."

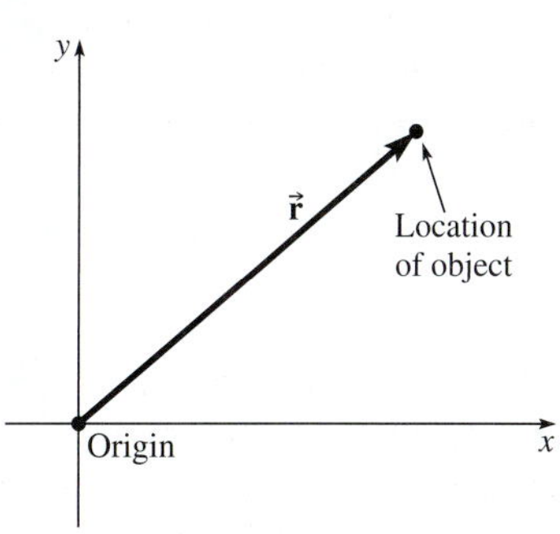

Figure 3.3 The position vector $\vec{\mathbf{r}}$ is drawn starting from the origin of the coordinate system and ending at the object's location.

Position and Displacement

The position $\vec{\mathbf{r}}$ of an object can be represented as a vector arrow drawn from the origin to the location of the object (Fig. 3.3). Its magnitude is the distance from the origin.

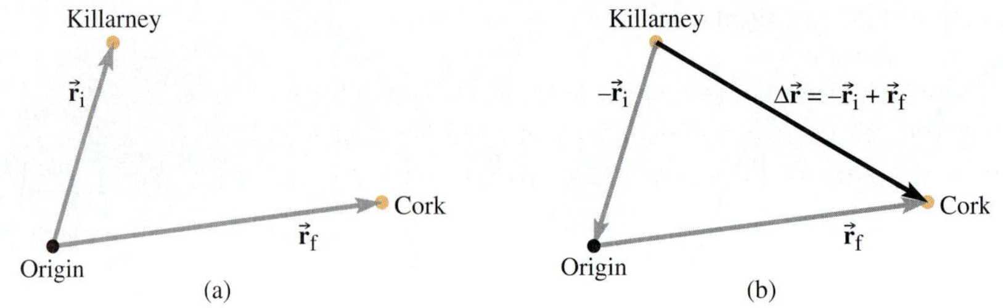

Figure 3.4 (a) Two position vectors, $\vec{\mathbf{r}}_i$ and $\vec{\mathbf{r}}_f$ drawn from an arbitrary origin to the starting point (Killarney) and to the ending point (Cork) of a trip. (b) The final position vector minus the initial position vector is the displacement $\Delta\vec{\mathbf{r}}$, found by adding $-\vec{\mathbf{r}}_i + \vec{\mathbf{r}}_f$.

The displacement is literally the *change in position* (the final position vector minus the initial position vector):

Displacement

$$\Delta\vec{\mathbf{r}} = \vec{\mathbf{r}}_f - \vec{\mathbf{r}}_i \qquad (3\text{-}1)$$

When the symbol Δ is used with a vector quantity, it always represents *vector* subtraction. Figure 3.4 shows the graphical subtraction of two position vectors to illustrate the displacement for a trip from Killarney to Cork. This same procedure is used to subtract any kind of vector quantity (velocity, acceleration, etc.). Note that the order matters: $\vec{\mathbf{B}} - \vec{\mathbf{A}} = -(\vec{\mathbf{A}} - \vec{\mathbf{B}})$.

Successive Displacements Can Be Added as Vectors As in Example 2.1, the total displacement for a trip with several parts is the vector sum of the displacements for each part of the trip because

$$\vec{\mathbf{r}}_3 - \vec{\mathbf{r}}_1 = (\vec{\mathbf{r}}_3 - \cancel{\vec{\mathbf{r}}_2}) + (\cancel{\vec{\mathbf{r}}_2} - \vec{\mathbf{r}}_1) \qquad (3\text{-}2)$$

Note that position vectors are *subtracted* to find the displacement, whereas successive displacements are *added* to find the total displacement. Example 3.2 explores this idea further.

Example 3.2

An Irish Adventure (1)

On a trip from Killarney to Cork, Charlotte and Shona drive 27° west of south for 18 km to Kenmare, then directly south for 17 km to Glengariff, and then finally 13° north of east for 48 km to Cork. Find the displacement vector for the entire trip by adding the three displacements graphically.

Blarney castle.

Strategy To add the displacement vectors, place the tail of each successive vector at the tip of the preceding vector. The value of a vector is not changed by moving it as long as its direction and magnitude are not changed, so a vector can be drawn starting at any point. The sum of the three displacements is then drawn from the tail of the first vector to the tip of the last vector. To add vectors graphically and get an accurate result, we use a ruler and a protractor. The protractor is used to draw the vector arrows in the correct directions and the ruler is used to draw them with the correct lengths. Then the length and direction of the sum can be determined with the ruler and protractor.

Solution Let's call the four positions $\vec{\mathbf{r}}_1$ (Killarney), $\vec{\mathbf{r}}_2$ (Kenmare), $\vec{\mathbf{r}}_3$ (Glengariff), and $\vec{\mathbf{r}}_4$ (Cork). The displacement for the whole trip is $\vec{\mathbf{r}}_4 - \vec{\mathbf{r}}_1$. The problem gives the displacements for the three parts of the trip; let's call them $\vec{\mathbf{A}} = \vec{\mathbf{r}}_2 - \vec{\mathbf{r}}_1 = 18$ km, 27° west of south; $\vec{\mathbf{B}} = \vec{\mathbf{r}}_3 - \vec{\mathbf{r}}_2 = 17$ km, south; and $\vec{\mathbf{C}} = \mathbf{r}_4 - \mathbf{r}_3 = 48$ km, 13° north of east. The sum of these three displacements is the total displacement because

$$\vec{\mathbf{A}} + \vec{\mathbf{B}} + \vec{\mathbf{C}} = (\cancel{\vec{\mathbf{r}}_2} - \vec{\mathbf{r}}_1) + (\cancel{\vec{\mathbf{r}}_3} - \cancel{\vec{\mathbf{r}}_2}) + (\vec{\mathbf{r}}_4 - \cancel{\vec{\mathbf{r}}_3}) = \vec{\mathbf{r}}_4 - \vec{\mathbf{r}}_1$$

continued on next page

Example 3.2 continued

Next we choose a convenient scale for the lengths of the vector arrows. Here we choose to represent 1 km as an arrow length of 0.2 cm, so the length of the vector arrow for $\vec{\mathbf{A}}$ should be

$$18 \text{ km} \times \frac{0.2 \text{ cm}}{1 \text{ km}} = 3.6 \text{ cm}$$

Similarly, the arrows for $\vec{\mathbf{B}}$ and $\vec{\mathbf{C}}$ should be 3.4 cm and 9.6 cm long, respectively.

After drawing the three vector arrows tip to tail, the arrow from the tail of the first vector to the tip of the last vector represents the sum (Fig. 3.5). This arrow is measured to have length 8.9 cm and its direction is 30° south of east. The total displacement has magnitude

$$8.9 \text{ cm} \times \frac{1 \text{ km}}{0.2 \text{ cm}} = 44.5 \text{ km}$$

Rounding to two significant figures, the total displacement $\vec{\mathbf{A}} + \vec{\mathbf{B}} + \vec{\mathbf{C}}$ has magnitude 45 km and is directed 30° south of east.

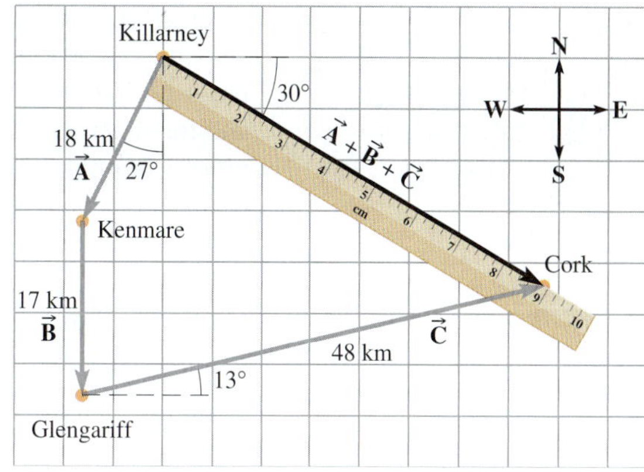

Figure 3.5

Graphical addition of the displacement vectors for the trip from Killarney to Cork via Kenmare and Glengariff. The gridlines on the graph paper are 1 cm apart.

Discussion Note that the answer includes both the magnitude and direction of the displacement. If a homework or exam question has you calculate a vector quantity such as position or velocity, don't forget to specify the direction as well as the magnitude in your answer. One without the other is incomplete.

Although the magnitude and direction of a position vector depends on the choice of origin, the magnitude and direction of a displacement (*change* of position) does *not* depend on the choice of origin.

The total *distance* traveled by Charlotte and Shona is 18 km + 17 km + 48 km = 83 km, which is *not* equal to the magnitude of the total displacement. Finding the total distance involves adding three *scalars*, while finding the total displacement involves adding three *vectors*. The magnitude of the total displacement is the *straight-line* distance from Killarney to Cork.

Practice Problem 3.2 A Traveling Executive

An executive flies from Kansas City to Chicago (displacement = 400 mi in the direction 30° north of east) and then from Chicago to Tulsa (600 mi, 45° south of west). Add the two displacements graphically to find the total displacement from Kansas City to Tulsa.

3.2 VECTOR ADDITION AND SUBTRACTION USING COMPONENTS

Components of a Vector

Any vector can be expressed as the sum of vectors parallel to the x-, y-, and (if needed) z-axes. The x-, y-, and z-components of a vector indicate the magnitude and direction of the three vectors along the three perpendicular axes. The sign of a component indicates the direction along that axis. The x-, y-, and z-components of vector $\vec{\mathbf{A}}$ are written with subscripts as follows: A_x, A_y, and A_z. One exception to this otherwise consistent notation is that the x-, y-, and z-components of a position vector $\vec{\mathbf{r}}$ are usually written x, y, and z (instead of r_x, r_y, and r_z). For now we will deal only with vectors in the xy-plane.

The x-component of a position vector $\vec{\mathbf{r}}$ is x, the x-coordinate. For all other vectors, the x-component is designated by a subscript x. For example, the x-component of a velocity vector $\vec{\mathbf{v}}$ is written v_x. Components of vectors have magnitude, units, and an algebraic sign. The sign indicates the direction: a positive x-component indicates the direction of the positive x-axis, while a negative x-component indicates the opposite direction (the negative x-axis).

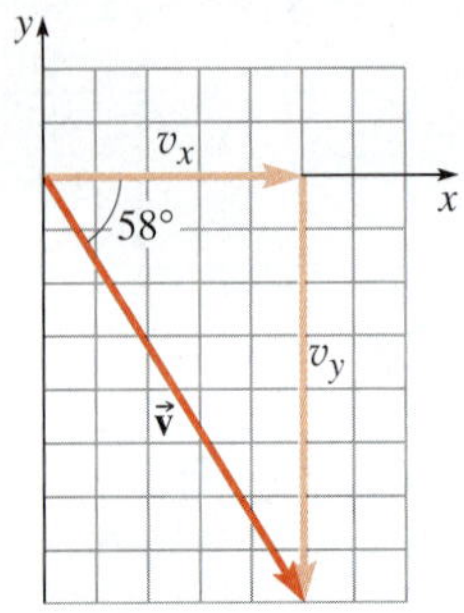

Figure 3.6 Resolving a velocity vector $\vec{v}$ into x- and y-components by drawing a right triangle with the vector arrow as the hypotenuse and the sides parallel to the x- and y-axes.

Finding Components The process of finding the components of a vector is called **resolving** the vector into its components. Before resolving a vector into components, we must choose a coordinate system (the directions of the x- and y-axes). Consider the velocity vector $\vec{v}$ in Fig. 3.6, which has magnitude 9.4 m/s and is directed 58° below the $+x$-axis. We can think of $\vec{v}$ as the sum of two vectors, one parallel to the x-axis and the other parallel to the y-axis. The magnitudes of these two vectors are the *magnitudes* (absolute values) of the x- and y-components of $\vec{v}$. We can find the magnitudes of the components using the right triangle in Fig. 3.6 and the trigonometric functions in Fig. 3.7. The length of the arrow represents the magnitude of the vector ($v = 9.4$ m/s), so

$$\cos 58° = \frac{\text{adjacent}}{\text{hypotenuse}} = \frac{|v_x|}{v} \quad \text{and} \quad \sin 58° = \frac{\text{opposite}}{\text{hypotenuse}} = \frac{|v_y|}{v} \qquad (3\text{-}3)$$

Now we must determine the correct algebraic sign for each of the components. From Fig. 3.6, the vector along the x-axis points in the *positive* x-direction and the vector along the y-axis points in the *negative* y-direction, so in this case,

$$v_x = +v \cos 58° = 5.0 \text{ m/s} \quad \text{and} \quad v_y = -v \sin 58° = -8.0 \text{ m/s} \qquad (3\text{-}4)$$

Using the right triangle in Fig. 3.8 gives the same values for the x- and y-components of $\vec{v}$ since $\cos 32° = \sin 58°$ and $\sin 32° = \cos 58°$.

Problem-Solving Strategy: Finding the x- and y-Components of a Vector from its Magnitude and Direction

1. Draw a right triangle with the vector as the hypotenuse and the other two sides parallel to the x- and y-axes.
2. Determine one of the unknown angles in the triangle.
3. Use trigonometric functions to find the magnitudes of the components. Make sure your calculator is in "degree mode" to evaluate trigonometric functions of angles in degrees and "radian mode" for angles in radians.
4. Determine the correct algebraic sign for each component.

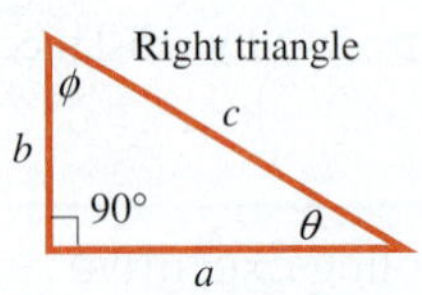

$$\phi = 90° - \theta$$

$$\sin\theta = \frac{\text{side opposite } \angle\theta}{\text{hypotenuse}} = \frac{b}{c}$$

$$\cos\theta = \frac{\text{side adjacent } \angle\theta}{\text{hypotenuse}} = \frac{a}{c}$$

$$\tan\theta = \frac{\text{side opposite } \angle\theta}{\text{side adjacent } \angle\theta} = \frac{b}{a}$$

Figure 3.7 Review of the trigonometric functions (see Appendix A.7 for more information).

Sometimes a vector is written as a list of its components in order, separated by a comma, inside parentheses. The velocity vector of Fig. 3.6 can be written: $\vec{v} = (5.0 \text{ N}, -8.0 \text{ N})$.

Finding Magnitude and Direction We must also know how to reverse the process to find a vector's magnitude and direction from its component.

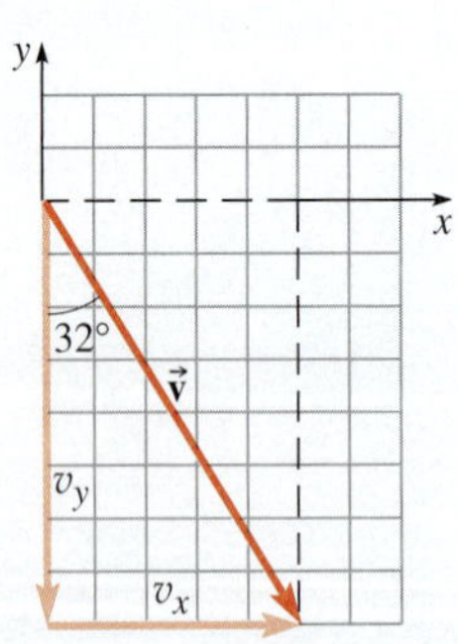

Figure 3.8 Resolving the velocity vector into components using a different right triangle. Note that the vector arrow is still the hypotenuse, and the sides are still parallel to the x- and y-axes.

Problem-Solving Strategy: Finding the Magnitude and Direction of a Vector $\vec{A}$ from its x- and y-Components

1. Sketch the vector on a set of x- and y-axes in the correct quadrant, according to the signs of the components.
2. Draw a right triangle with the vector as the hypotenuse and the other two sides parallel to the x- and y-axes.
3. In the right triangle, choose which of the unknown angles you want to determine.
4. Use the inverse tangent function to find the angle. The lengths of the sides of the triangle represent $|A_x|$ and $|A_y|$. If θ is opposite the side parallel to the x-axis, then $\tan\theta$ = opposite/adjacent = $|A_x/A_y|$. If θ is opposite the side parallel to the

continued on next page

y-axis, then $\tan \theta =$ opposite/adjacent $= |A_y/A_x|$. If your calculator is in "degree mode," then the result of the inverse tangent operation will be in degrees. [In general, the inverse tangent has two possible values between 0 and 360° because $\tan \alpha = \tan (\alpha + 180°)$. However, when the inverse tangent is used to find one of the angles in a right triangle, the result can never be greater than 90°, so the value the calculator returns is the one you want.]

5. Interpret the angle: specify whether it is the angle below the horizontal, or the angle west of south, or the angle clockwise from the negative y-axis, etc.
6. Use the Pythagorean theorem to find the magnitude of the vector.

$$A = \sqrt{A_x^2 + A_y^2} \tag{3-5}$$

Suppose we knew the components of the velocity vector in Fig. 3.6, but not the magnitude and direction. Let us find the angle θ between $\vec{\mathbf{v}}$ and the $+x$-axis:

$$\theta = \tan^{-1} \frac{\text{opposite}}{\text{adjacent}} = \tan^{-1} \frac{|v_y|}{|v_x|} = \tan^{-1} \frac{8.0 \text{ m/s}}{5.0 \text{ m/s}} = 58° \tag{3-6}$$

From the Pythagorean theorem, the magnitude of $\vec{\mathbf{v}}$ is

$$v = \sqrt{v_x^2 + v_y^2} = \sqrt{(+5.0 \text{ m/s})^2 + (-8.0 \text{ m/s})^2} = 9.4 \text{ m/s}$$

Adding Vectors Using Components

It is generally easier and more accurate to add vectors algebraically rather than graphically. The algebraic method relies on adding the components of the vectors. Remember that each vector is thought of as the sum of vectors parallel to the axes (Fig. 3.9a). When adding vectors, we can add them in any order and group them as we please. So we can sum the x-components to find the x-component of the sum (Fig. 3.9b) and then do the same with the y-components (Fig. 3.9c):

$$\vec{\mathbf{C}} = \vec{\mathbf{A}} + \vec{\mathbf{B}} \quad \text{if and only if} \quad C_x = A_x + B_x \quad \text{and} \quad C_y = A_y + B_y \tag{3-7}$$

In Eq. (3-7), $A_x + B_x$ represents ordinary addition since the signs of the components carry the direction information.

Problem-Solving Strategy: Adding Vectors Using Components

1. Find the x- and y-components of each vector to be added.
2. Add the x-components (*with their algebraic signs*) of the vectors to find the x-component of the sum. (If the signs are not correct, the sum will not be correct.)
3. Add the y-components (with their algebraic signs) of the vectors to find the y-component of the sum.
4. If necessary, use the x- and y-components of the sum to find the magnitude and direction of the sum.

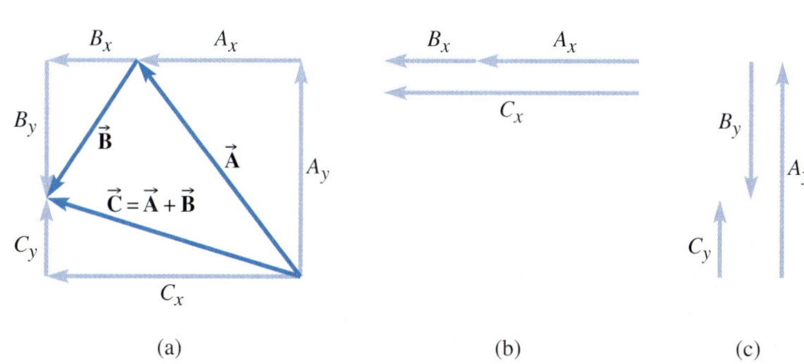

Figure 3.9 (a) $\vec{\mathbf{C}} = \vec{\mathbf{A}} + \vec{\mathbf{B}}$, shown graphically with the x- and y-components of each vector illustrated. (b) $C_x = A_x + B_x$; (c) $C_y = A_y + B_y$. This illustrates the fact that vector addition can be done by components.

Estimation Using Graphical Addition Even when using the component method to add vectors, the graphical method is an important first step. A rough sketch of vector addition, even one made without carefully measuring the lengths or the angles, has important benefits. Sketching the vectors makes it much easier to get the signs of the components correct. The graphical addition also serves as a check on the answer—it provides an estimate of the magnitude and direction of the sum, which can be used to check the algebraic answer. Graphical addition gives you a mental picture of what is going on and an intuitive feel for the algebraic calculations.

CHECKPOINT 3.2A

Two displacements $\vec{\mathbf{A}}$ and $\vec{\mathbf{B}}$ have x- and y-components as follows: $A_x = +3.0$ km, $A_y = -6.0$ km, $B_x = -8.5$ km, $B_y = -1.2$ km. The total displacement is $\vec{\mathbf{C}} = \vec{\mathbf{A}} + \vec{\mathbf{B}}$. What are the x- and y-components of $\vec{\mathbf{C}}$?

Choosing x- and y-Axes

A problem can be made easier to solve with a good choice of axes. We can choose any direction we want for the *x*- and *y*-axes, as long as they are perpendicular to one another. Three common choices are

- *x*-axis horizontal and *y*-axis vertical, when the vectors all lie in a vertical plane;
- *x*-axis east and *y*-axis north, when the vectors all lie in a horizontal plane; and
- *x*-axis parallel to an inclined surface and *y*-axis perpendicular to it.

Example 3.3

An Irish Adventure (2)

In the trip of Example 3.2, Charlotte and Shona drive 27° west of south for 18 km to Kenmare, then directly south for 17 km to Glengariff, and then finally 13° north of east for 48 km to Cork. Use the component method to find the magnitude and direction of the displacement vector for the entire trip.

Strategy As before, let's call the three successive displacements $\vec{\mathbf{A}}$, $\vec{\mathbf{B}}$, and $\vec{\mathbf{C}}$ respectively. To add the vectors using components, we first choose directions for the *x*- and *y*-axes. Then we find the *x*- and *y*-components of the three displacements. Adding the *x*- or *y*-components of the three displacements gives the *x*- or *y*-component of the total displacement. Finally, from the components we find the magnitude and direction of the total displacement.

Solution A good choice is the conventional one: *x*-axis to the east and the *y*-axis to the north. The first displacement ($\vec{\mathbf{A}}$) is directed 27° west of south. Both of its components are negative since west is the $-x$-direction and south is the $-y$-direction. To find the components, we draw a right triangle with the vector as the hypotenuse and the sides parallel to the *x*- and *y*-axes (Fig. 3.10). Using the right triangle in Fig. 3.10, the side of the triangle opposite the 27° angle is parallel to the *x*-axis. The sine function relates the opposite side to the hypotenuse:

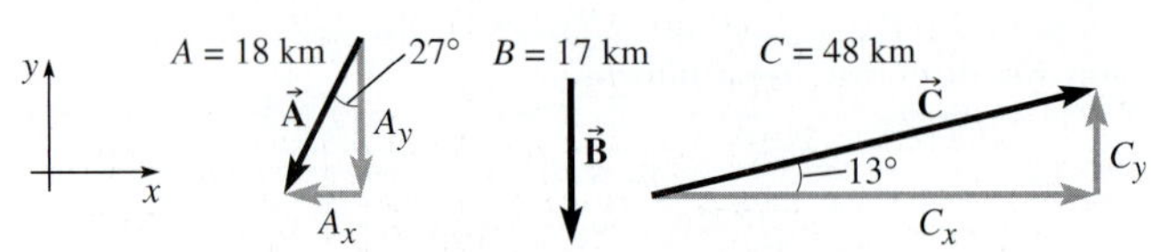

Figure 3.10
Resolving $\vec{\mathbf{A}}$, $\vec{\mathbf{B}}$, and $\vec{\mathbf{C}}$ into *x*- and *y*-components.

$$A_x = -A \sin 27° = -18 \text{ km} \times 0.454 = -8.17 \text{ km}$$

where *A* is the magnitude of $\vec{\mathbf{A}}$. The cosine relates the adjacent side to the hypotenuse:

$$A_y = -A \cos 27° = -18 \text{ km} \times 0.891 = -16.0 \text{ km}$$

Displacement $\vec{\mathbf{B}}$ has no *x*-component since its direction is south. Therefore,

$$B_x = 0 \quad \text{and} \quad B_y = -17 \text{ km}$$

The direction of $\vec{\mathbf{C}}$ is 13° north of east. Both its components are positive. From Fig. 3.10, the side of the right triangle opposite the 13° angle is parallel to the *y*-axis, so

$$C_x = +C \cos 13° = +48 \text{ km} \times 0.974 = +46.8 \text{ km}$$
$$C_y = +C \sin 13° = +48 \text{ km} \times 0.225 = +10.8 \text{ km}$$

continued on next page

Example 3.3 continued

Now we sum the x- and y-components separately to find the x- and y-components of the total displacement:

$$\Delta x = A_x + B_x + C_x$$
$$= (-8.17\text{ km}) + 0 + 46.8\text{ km} = +38.63\text{ km}$$
$$\Delta y = A_y + B_y + C_y$$
$$= (-16.0\text{ km}) + (-17\text{ km}) + 10.8\text{ km} = -22.2\text{ km}$$

The magnitude and direction of $\Delta\vec{\mathbf{r}}$ can be found from the right triangle in Fig. 3.11. The magnitude is represented by the hypotenuse. Using the Pythagorean theorem:

$$\Delta r = \sqrt{(\Delta x)^2 + (\Delta y)^2} = \sqrt{(38.63\text{ km})^2 + (-22.2\text{ km})^2}$$
$$= 45\text{ km}$$

The angle θ is found from the inverse tangent of opposite over adjacent.

$$\theta = \tan^{-1}\frac{\text{opposite}}{\text{adjacent}} = \tan^{-1}\frac{22.2\text{ km}}{38.63\text{ km}} = 30°$$

Since $+x$ is east and $-y$ is south, the direction of the displacement is 30° south of east. The magnitude and direction of the displacement found using components agree with the displacement found graphically in Fig. 3.5.

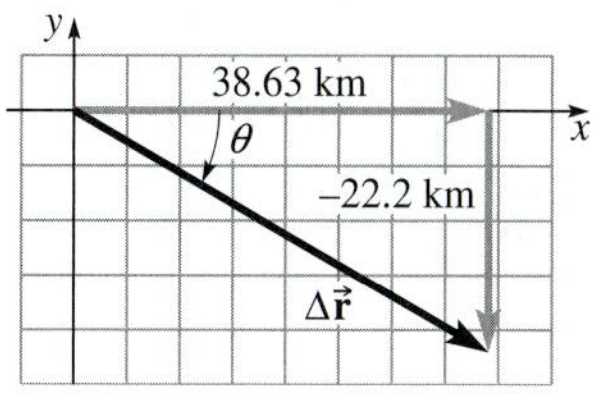

Figure 3.11
Using a right triangle to find the magnitude and direction of $\Delta\vec{\mathbf{r}}$. The magnitude is calculated using the Pythagorean theorem. The angle θ is calculated from the inverse tangent function.

Discussion Note that the x-component of one displacement was found using the sine function while another was found using the cosine. The x-component (or the y-component) of the vector can be related to *either* the sine or the cosine, depending on which angle in the right triangle is used.

Practice Problem 3.3 Changing the Coordinate Axes

Find the x- and y-components of the displacements for the three legs of the trip if the x-axis points south and the y-axis points east.

CHECKPOINT 3.2B

Sketch a vector arrow representing a displacement with x-component −6.0 m and y-component +2.0 m.

Unit Vector Notation

The same concept of vector components may be used to write vectors in a compact way. The **unit vectors** $\hat{\mathbf{x}}$ (read aloud as "x hat"), $\hat{\mathbf{y}}$, and $\hat{\mathbf{z}}$ are defined as vectors of magnitude 1 that point in the $+x$-, $+y$-, and $+z$-directions, respectively. (In some books, you may see them written as $\hat{\mathbf{i}}$, $\hat{\mathbf{j}}$, and $\hat{\mathbf{k}}$, respectively.) They are called *unit* vectors because the magnitude of each is the pure number 1—they do *not* have physical units such as kilograms or meters. Any vector $\vec{\mathbf{A}}$ can be written as the sum of three vectors along the coordinate axes:

$$\vec{\mathbf{A}} = A_x\hat{\mathbf{x}} + A_y\hat{\mathbf{y}} + A_z\hat{\mathbf{z}} \quad (3\text{-}8)$$

Here A_x is the x-component of $\vec{\mathbf{A}}$ which has physical units and can be positive or negative. $A_x\hat{\mathbf{x}}$ is a vector of magnitude $|A_x|$ directed in the $+x$-direction if $A_x > 0$ and in the $-x$-direction if $A_x < 0$. For example, consider the velocity vector $\vec{\mathbf{v}}$ of Fig. 3.8. $\vec{\mathbf{v}}$ has x-component $v_x = +5.0$ m/s and y-component $v_y = -8.0$ m/s, so $\vec{\mathbf{v}} = (+5.0\text{ m/s})\,\hat{\mathbf{x}} + (-8.0\text{ m/s})\hat{\mathbf{y}}$.

Using unit vector notation is one way to keep track of vector components in vector addition and subtraction without writing separate equations for each component. Adding two vectors in the xy-plane looks like this:

$$\vec{\mathbf{A}}_1 + \vec{\mathbf{A}}_2 = (A_{1x}\hat{\mathbf{x}} + A_{1y}\hat{\mathbf{y}}) + (A_{2x}\hat{\mathbf{x}} + A_{2y}\hat{\mathbf{y}}) \quad (3\text{-}9)$$

Regrouping the terms shows that the x-component of the sum is the sum of the x-components and likewise for the y-components:

$$\vec{\mathbf{A}}_1 + \vec{\mathbf{A}}_2 = (A_{1x} + A_{2x})\,\hat{\mathbf{x}} + (A_{1y} + A_{2y})\,\hat{\mathbf{y}} \quad (3\text{-}10)$$

3.3 VELOCITY

The definitions of average velocity, instantaneous velocity, average acceleration, and instantaneous acceleration from Chapter 2 still apply when the motion is not in a straight line as long as we add and subtract them as vectors. Suppose we want to know the instantaneous velocity of a race car at point P as it goes around a curved section of a racetrack (Fig. 3.12a). At a slightly later time the race car is at point Q. Let $\vec{\mathbf{r}}_i$ be the position of the car at P and $\vec{\mathbf{r}}_f$ be the position at point Q.

Average Velocity The displacement $\Delta\vec{\mathbf{r}} = \vec{\mathbf{r}}_f - \vec{\mathbf{r}}_i$ is represented as an arrow from P to Q. Alternatively, to subtract $\vec{\mathbf{r}}_i$ from $\vec{\mathbf{r}}_f$ the two vectors can be drawn with their tails at the same point. After reversing the direction of $\vec{\mathbf{r}}_i$ to represent $-\vec{\mathbf{r}}_i$ (Fig. 3.12b), the arrows are tip to tail and ready to add: $\Delta\vec{\mathbf{r}} = \vec{\mathbf{r}}_f + (-\vec{\mathbf{r}}_i)$. The average velocity during this time interval is the displacement $\Delta\vec{\mathbf{r}}$ divided by the time interval:

$$\vec{\mathbf{v}}_{av} = \frac{\vec{\mathbf{r}}_f - \vec{\mathbf{r}}_i}{t_f - t_i} = \frac{\Delta\vec{\mathbf{r}}}{\Delta t} \tag{3-11}$$

The direction of the average velocity is the direction of the displacement $\Delta\vec{\mathbf{r}}$.

Instantaneous Velocity The instantaneous velocity at P is the limit of the average velocity as Δt approaches zero. As we shorten the time interval between the initial and final positions by moving point Q closer and closer to P, the direction of the displacement vector $\Delta\vec{\mathbf{r}}$ gradually changes, approaching the tangent to the curved path at P (Fig. 3.12c). Expressed in mathematical terminology, the instantaneous velocity is the limit of $\Delta\vec{\mathbf{r}}/\Delta t$ as the time interval approaches zero:

CONNECTION

The rate of change of any vector quantity $\vec{\mathbf{Q}}$ is

$$\lim_{\Delta t \to 0} \frac{\Delta\vec{\mathbf{Q}}}{\Delta t}$$

Velocity is the rate of change of the position vector, and acceleration is the rate of change of the velocity vector.

Velocity

$$\vec{\mathbf{v}} = \lim_{\Delta t \to 0} \frac{\Delta\vec{\mathbf{r}}}{\Delta t} \tag{3-12}$$

($\Delta\vec{\mathbf{r}}$ is the change in velocity during a *very short* time interval Δt)

With this definition, the instantaneous velocity at P becomes tangent to the curve at P (Fig. 3.12d). Here we are talking about a tangent to the actual path through space, *not* a tangent line on a graph of position versus time. The magnitude of the velocity vector is the speed at which the object moves and the direction of the velocity vector is the direction of motion.

Component Equations A vector equation is always equivalent to a set of equations, one for each component. The x- and y-components of the average velocity are

$$v_{av,x} = \frac{\Delta x}{\Delta t} \quad \text{and} \quad v_{av,y} = \frac{\Delta y}{\Delta t} \tag{3-13}$$

The x- and y-components of the instantaneous velocity are

$$v_x = \lim_{\Delta t \to 0} \frac{\Delta x}{\Delta t} \quad \text{and} \quad v_y = \lim_{\Delta t \to 0} \frac{\Delta y}{\Delta t} \tag{3-14}$$

To put Eq. (3-14) into words, the x-component of an object's velocity is the rate of change of its x-coordinate and the y-component of its velocity is the rate of change of its y-coordinate.

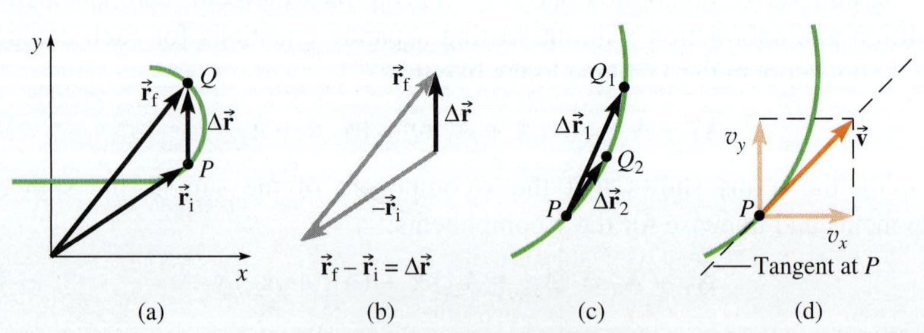

Figure 3.12 (a) Position vectors for two points on the curve. (b) The displacement $\Delta\vec{\mathbf{r}}$ from point P to point Q. (c) As the time interval is decreased, the final point moves closer and closer to P; the direction of the displacement $\Delta\vec{\mathbf{r}}$ approaches the tangent to the curve at P. (d) Instantaneous velocity can be resolved into components along perpendicular axes.

Example 3.4

An Irish Adventure (3)

In their trip from Kenmare to Cork via Glengariff, Charlotte and Shona travel a total distance of 83 km in 1.4 h. The total displacement for the trip is 45 km, 30° south of east. What is their average velocity? Contrast it with their average speed, defined as the total distance divided by the time interval.

Strategy The average velocity is calculated from the displacement—not from the distance traveled.

Solution The magnitude of the average velocity is

$$|\vec{\mathbf{v}}_{av}| = \frac{|\Delta\vec{\mathbf{r}}|}{\Delta t} = \frac{45\text{ km}}{1.4\text{ h}} = 32\text{ km/h}$$

The average velocity has the same direction as the displacement, so $\vec{\mathbf{v}}_{av} = 32$ km/h, 30° south of east. The average speed is

$$\text{average speed} = \frac{83\text{ km}}{1.4\text{ h}} = 59\text{ km/h}$$

Therefore, $|\vec{\mathbf{v}}_{av}|$ is not equal to the average speed. Furthermore, average velocity is a vector quantity with a direction in space, and average speed is a scalar.

Practice Problem 3.4 Average Velocity Versus Average Speed

In Example 3.4, $|\vec{\mathbf{v}}_{av}|$ was less than the average speed. Can $|\vec{\mathbf{v}}_{av}|$ ever be greater than the average speed? Can $|\vec{\mathbf{v}}_{av}|$ ever be equal to the average speed? Explain.

3.4 ACCELERATION

The average acceleration $\vec{\mathbf{a}}_{av}$ is the change in velocity divided by the elapsed time:

$$\vec{\mathbf{a}}_{av} = \frac{\vec{\mathbf{v}}_f - \vec{\mathbf{v}}_i}{t_f - t_i} = \frac{\Delta\vec{\mathbf{v}}}{\Delta t} \qquad (3\text{-}15)$$

For motion in a plane, this vector equation is equivalent to two component equations:

$$a_{av,x} = \frac{\Delta v_x}{\Delta t} \quad \text{and} \quad a_{av,y} = \frac{\Delta v_y}{\Delta t} \qquad (3\text{-}16)$$

The direction of $\vec{\mathbf{a}}_{av}$ is the same as the direction of $\Delta\vec{\mathbf{v}}$ (Fig. 3.13).

CHECKPOINT 3.4A

Four airplanes are initially all moving south at 200 m/s. Ten minutes later, plane A is moving south at 200 m/s, plane B is moving east at 200 m/s, plane C is moving south at 300 m/s, and plane D is moving north at 200 m/s. Rank the four planes in decreasing order of $|\vec{\mathbf{a}}_{av}|$, the magnitude of the average acceleration.

Instantaneous acceleration is the limit of the average acceleration as the time interval approaches zero:

Acceleration

$$\vec{\mathbf{a}} = \lim_{\Delta t \to 0} \frac{\Delta\vec{\mathbf{v}}}{\Delta t} \qquad (3\text{-}17)$$

($\Delta\vec{\mathbf{v}}$ is the change in velocity during a *very short* time interval Δt)

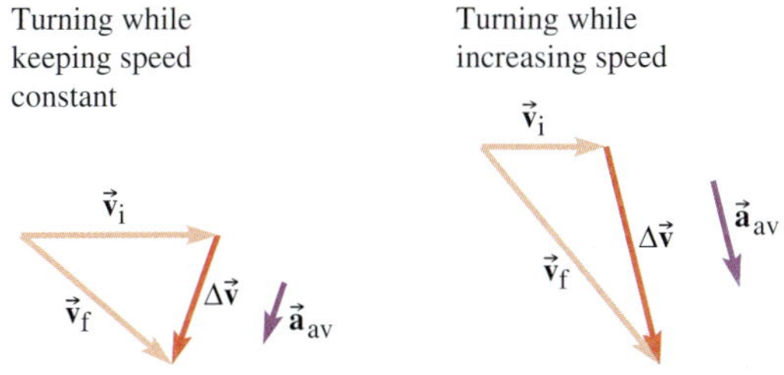

Figure 3.13 Two examples to illustrate that the average acceleration is always in the same direction as the change in velocity $\Delta\vec{\mathbf{v}}$ during the same time interval.

In component form,

$$a_x = \lim_{\Delta t \to 0} \frac{\Delta v_x}{\Delta t} \quad \text{and} \quad a_y = \lim_{\Delta t \to 0} \frac{\Delta v_y}{\Delta t} \tag{3-18}$$

In straight-line motion the acceleration is always along the same line as the velocity. For motion in two dimensions, the acceleration vector can make any angle with the velocity vector because the velocity vector can change in magnitude, in direction, or both. The direction of the acceleration is the direction of the *change* in velocity $\Delta\vec{\mathbf{v}}$ during a *very short* time interval.

The concept of the acceleration vector is much less intuitive for most people than the concept of the velocity vector. Always stop to think: the acceleration vector tells you how the velocity vector *is changing*.

CHECKPOINT 3.4B

An airplane is initially moving due north at 400 km/h. After making a slight course correction, it is moving at the same speed but in a direction 2.0° east of north. Is the plane's average acceleration during this time interval zero? Explain.

Example 3.5

Skating Uphill

An inline skater is traveling on a level road with a speed of 8.94 m/s; 120.0 s later she is climbing a hill with a 15.0° angle of incline at a speed of 7.15 m/s. (a) What is the change in her velocity? (b) What is her average acceleration during the 120.0 s time interval?

Strategy The change in velocity is *not* 8.94 m/s − 7.15 m/s = 1.79 m/s. That is the change in *speed*. The change in velocity is found by subtracting the initial velocity *vector* from the final velocity *vector*. After first making a graphical sketch, we use the component method. The average acceleration is the change in velocity divided by the elapsed time.

Solution (a) Figure 3.14a shows the initial and final velocity vectors and the slope of the hill. The initial velocity is

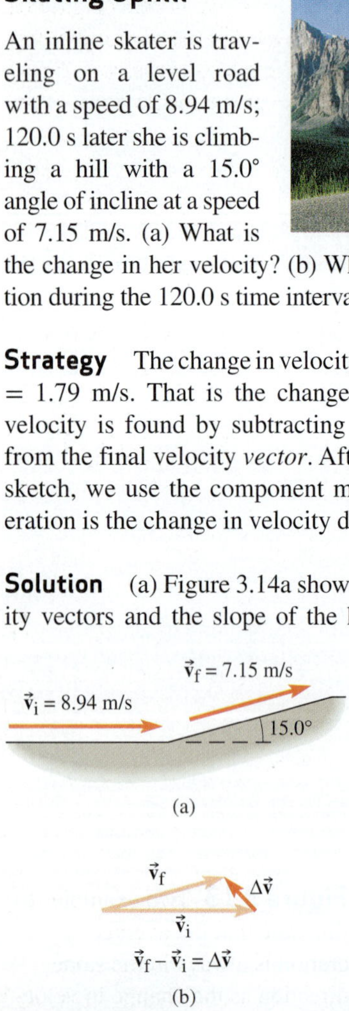

Figure 3.14
(a) Change in velocity as the skater slows going uphill and (b) graphical subtraction of velocity vectors.

horizontal as the skater skates on level ground. The final velocity is 15.0° above the horizontal. To subtract the two velocity vectors graphically, we place the tails of the vectors together. The change in velocity $\Delta\vec{\mathbf{v}}$ is found by drawing a vector arrow from the tip of $\vec{\mathbf{v}}_i$ to the tip of $\vec{\mathbf{v}}_f$. Judging by the graphical subtraction in Fig. 3.14b, the change in velocity is roughly at a 45° angle above the $-x$-axis. Its magnitude is smaller than the magnitudes of the initial and final velocity vectors—something like 2 to 3 m/s.

The components v_{fx} and v_{fy} can be found from a right triangle (Fig. 3.15):

$$v_{fx} = v_f \cos\theta = 7.15 \text{ m/s} \times 0.9659 = 6.91 \text{ m/s}$$
$$v_{fy} = v_f \sin\theta = 7.15 \text{ m/s} \times 0.2588 = 1.85 \text{ m/s}$$

Since v_i has only an x-component,

$$v_{iy} = 0 \quad \text{and} \quad v_{ix} = v_i = 8.94 \text{ m/s}$$

Now we subtract the components to find the components of $\Delta\vec{\mathbf{v}}$:

$$\Delta v_x = v_{fx} - v_{ix} = (6.91 - 8.94) \text{ m/s} = -2.03 \text{ m/s}$$

and

$$\Delta v_y = v_{fy} - v_{iy} = (1.85 - 0) \text{ m/s} = +1.85 \text{ m/s}$$

To find the magnitude of $\Delta\vec{\mathbf{v}}$, we apply the Pythagorean theorem (Fig. 3.16):

$$|\Delta\vec{\mathbf{v}}|^2 = (\Delta v_x)^2 + (\Delta v_y)^2 = (-2.03 \text{ m/s})^2 + (1.85 \text{ m/s})^2 = 7.54 \text{ (m/s)}^2$$
$$|\Delta\vec{\mathbf{v}}| = 2.75 \text{ m/s}$$

continued on next page

Example 3.5 continued

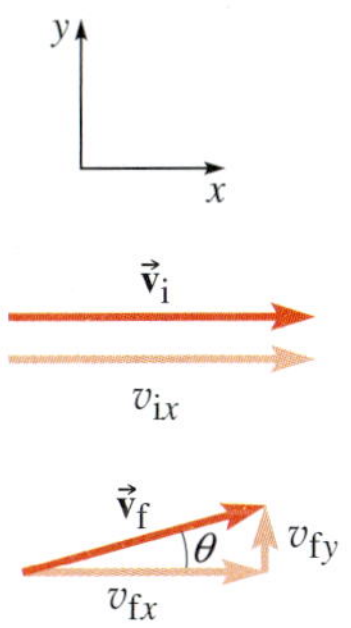

Figure 3.15
Initial and final velocity vectors resolved into components.

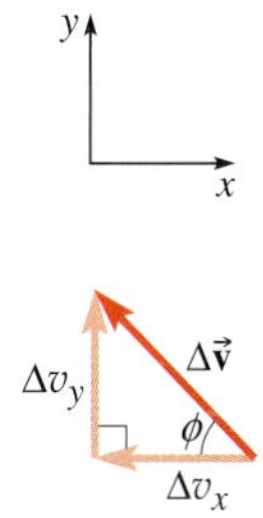

Figure 3.16
Reconstruction of $\Delta\vec{\mathbf{v}}$ from its components (not to scale).

The angle is found from

$$\tan\phi = \frac{\text{opposite}}{\text{adjacent}} = \left|\frac{\Delta v_y}{\Delta v_x}\right| = \frac{1.85\text{ m/s}}{2.03\text{ m/s}} = 0.9113$$

$$\phi = \tan^{-1} 0.9113 = 42.3°$$

The direction of the change in velocity $\Delta\vec{\mathbf{v}}$ is 42.3° above the negative x-axis.

(b) The magnitude of the average acceleration is

$$|\vec{\mathbf{a}}_{av}| = \frac{|\Delta\vec{\mathbf{v}}|}{\Delta t} = \frac{2.75\text{ m/s}}{120.0\text{ s}} = 0.0229\text{ m/s}^2$$

The direction of the average acceleration is the same as the direction of $\Delta\vec{\mathbf{v}}$: 42.3° above the negative x-axis.

Discussion Checking back with the graphical subtraction in Fig. 3.14b, the magnitude of $\Delta\vec{\mathbf{v}}$ appears to be roughly $\frac{1}{4}$ to $\frac{1}{3}$ the magnitude of $\Delta\vec{\mathbf{v}}$. Since $\frac{1}{4} \times 8.94\text{ m/s} = 2.24\text{ m/s}$ and $\frac{1}{3} \times 8.94\text{ m/s} = 2.98\text{ m/s}$, the answer of 2.75 m/s is reasonable.

Figure 3.14b also shows the direction of $\Delta\vec{\mathbf{v}}$ to be roughly midway between the $+y$- and $-x$-axes. We found the direction of $\Delta\vec{\mathbf{v}}$ to be 42.3° above the $-x$-axis and, therefore, 47.7° from the $+y$-axis. So the direction we calculated is also reasonable based on the graphical subtraction.

Practice Problem 3.5 Change in Sailboat Velocity

A C&C 30 sailboat is sailing at 12.0 knots (6.17 m/s) directly east across the harbor. When a gust of wind comes up, the boat changes its direction to 11.0° north of east and its speed increases to 14.0 knots (7.20 m/s). [A boat's speed is customarily expressed in knots, which means nautical miles per hour. A nautical mile (6076 ft) is a little longer than a statute mile (5280 ft).] (a) What is the magnitude and direction of the change in velocity of the sailboat in meters per second? (b) If this velocity change occurs during a 2.0 s time interval, what is the average acceleration of the sailboat during that interval?

CHECKPOINT 3.4C

For an object moving in a straight line, how does the direction of the acceleration vector compare with that of the velocity vector?

3.5 MOTION IN A PLANE WITH CONSTANT ACCELERATION

If an object moves in the xy-plane with constant acceleration, then both a_x and a_y are constant. By looking separately at the motion along two perpendicular axes, the y-direction and the x-direction, each component becomes a one-dimensional problem, which we studied in Chapter 2. We can apply any of the constant acceleration relationships from Section 2.5 separately to the x-components and to the y-components.

It is generally easiest to choose the axes so that the acceleration has only one nonzero component. Suppose we choose the axes so that the acceleration is in the positive or negative y-direction. Then $a_x = 0$ and v_x is constant. With this choice, the constant acceleration relationships [Eqs. (2-9) through (2-13)] become

***x*-axis: $a_x = 0$**	***y*-axis: constant a_y**	
$\Delta v_x = 0$ (v_x is constant)	$\Delta v_y = a_y\,\Delta t$	(3-19)
$\Delta x = v_x\,\Delta t$	$\Delta y = \frac{1}{2}(v_{fy} + v_{iy})\,\Delta t$	(3-20)
	$\Delta y = v_{iy}\,\Delta t + \frac{1}{2}a_y(\Delta t)^2$	(3-21)
	$v_{fy}^2 - v_{iy}^2 = 2a_y\,\Delta y$	(3-22)

Why are only two equations shown in the column for the x-axis? The other two are redundant when $a_x = 0$.

Note that there is no mixing of components in Eqs. (3-19) through (3-22). Each equation pertains either to the x-components or to the y-components; none contains the x-component of one vector quantity and the y-component of another. The only quantity that appears in both x- and y-component equations is the time interval—a scalar.

CONNECTION:

Projectile motion is free fall for objects with a nonzero horizontal velocity component.

Motion of Projectiles

An object in free fall near the Earth's surface has a constant acceleration. As long as air resistance is negligible, the constant downward pull of gravity gives the object a constant downward acceleration with magnitude g. In Section 2.6 we considered objects in free fall, but only when they had no horizontal velocity component, so they moved straight up or straight down. Now we consider objects (called **projectiles**) in free fall that have a *nonzero* horizontal velocity component. The motion of a projectile takes place in a vertical plane.

Suppose some medieval marauders are attacking a castle. They have a catapult that propels large stones into the air to bombard the walls of the castle (Fig. 3.17). Picture a stone leaving the catapult with initial velocity $\vec{\mathbf{v}}_i$. ($\vec{\mathbf{v}}_i$ is the *initial* velocity for the interval during which the stone moves *as a projectile*. In other words, it is the velocity of the stone just as it loses contact with the catapult.) The **angle of elevation** is the angle of the initial velocity above the horizontal. Once the stone is in the air, the only force acting on it is the downward gravitational force, provided that the air resistance has a negligible effect on the motion. The **trajectory** (path) of the stone is shown in Fig. 3.18. The positive x-axis is chosen in the horizontal direction (to the right) and the positive y-axis is upward.

If the initial velocity $\vec{\mathbf{v}}_i$ is at an angle θ above the horizontal, then resolving it into components gives

$$v_{ix} = v_i \cos\theta \quad \text{and} \quad v_{iy} = v_i \sin\theta \tag{3-23}$$

($+y$-axis up, θ measured from the horizontal x-axis)

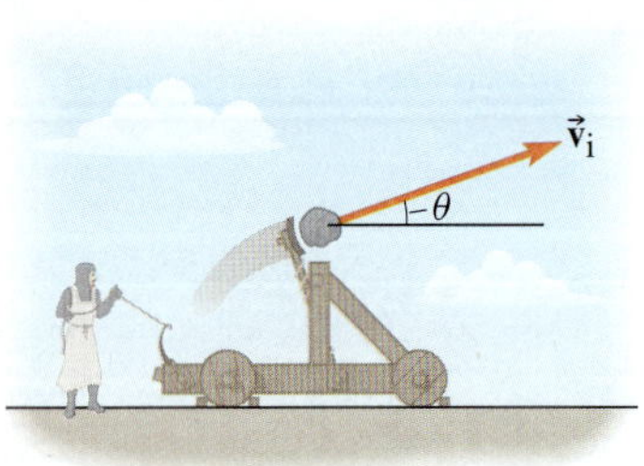

Figure 3.17 A medieval catapult projects a stone into the air. The velocity of the stone when it loses contact with the catapult is $\vec{\mathbf{v}}_i$.

With the y-axis pointing up, $a_y = -g$ because the acceleration is downward (in the $-y$-direction). The acceleration has no x-component ($a_x = 0$), so the stone's horizontal velocity component v_x is *constant*. The vertical velocity component v_y changes at a constant rate, exactly as if the stone were propelled straight up with an initial speed of v_{iy}. The initially positive v_y decreases until, at the top of flight, $v_y = 0$. Then the pull of gravity makes the projectile fall back downward. During the downward trip, v_y is still changing at the same constant rate with which it changed on the way up and at the top of the path. The acceleration has the same constant value—magnitude and direction—for the entire path.

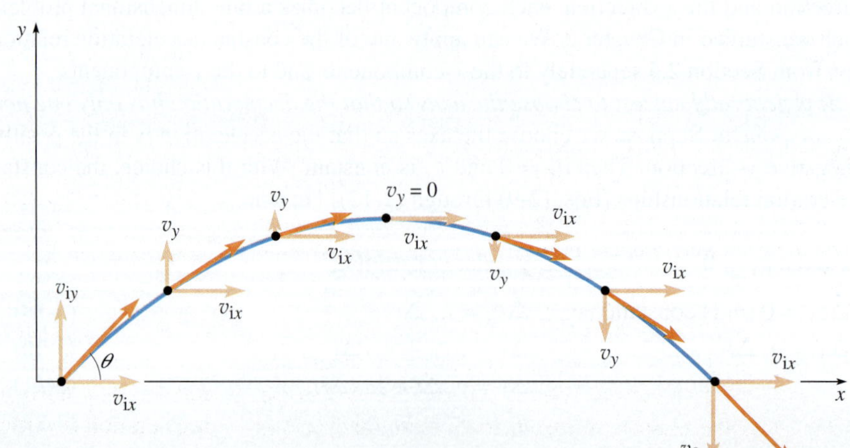

Figure 3.18 Motion diagram showing the trajectory of a projectile. The position is drawn at equal time intervals. Superimposed are the velocity vectors along with their x- and y-components. The horizontal velocity component is constant because no horizontal force acts. The vertical component changes due to the downward gravitational force.

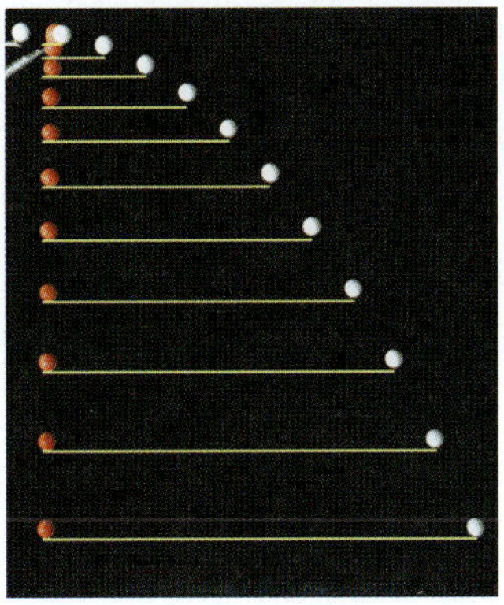

Figure 3.19 Independence of horizontal and vertical motion of a projectile in the absence of air resistance. The vertical motion of the projectile (white) is the same as that of an object (red) that falls straight down.

The motion of a projectile when air resistance is negligible is the superposition of horizontal motion with constant velocity and vertical motion with constant acceleration. The vertical and horizontal motions each proceed independently, as if the other motion were not present. In the experiment of Fig. 3.19, one ball was dropped and, at the same instant, another was projected horizontally. The photo shows the two balls at equally spaced times, just as a motion diagram does. The *vertical* motion of the two is identical; at every instant, the two are at the same height. The fact that they have different horizontal motion does not affect their vertical motion. (This statement would *not* be true if air resistance were significant.)

EVERYDAY PHYSICS DEMO

Take two coins with different masses to a room with a high table or countertop. Place the lighter one at the edge of the table and then slide the more massive one so they collide. Listen for the sound of the two coins hitting the floor. The coins will slide off the table with different horizontal velocities but will land at the same time.

Example 3.6

Lunchtime for a Gull

A gull scoops up a clam and takes it high above the ground (Fig. 3.20). While flying parallel to the ground, the gull lets go of the clam. The clam lands on a rock below and cracks open. Then the gull alights and enjoys lunch. (Ignore air resistance.) (a) Show that the path of the clam as viewed by a beachcomber on the beach is a parabola (i.e., it can be described by an equation of the form $y = Ax^2 + Bx + C$). (b) Explain why the clam does not drop straight down.

Strategy The clam is a projectile with a constant downward acceleration. Choosing the y-axis up, we can apply Eq. (3-21) with $a_y = -g$. We need to find y as a function of x. The quantity that appears in both the x- and y-component equations is the time, so we plan to start with equations for $y(t)$ and $x(t)$ and then eliminate t.

Solution (a) Choose the origin at the point where the clam has just been released. Then $x_i = 0$ and $y_i = 0$. Choose $t = 0$ just as the clam is released. Its velocity at $t = 0$ is the same as that of the gull, which is horizontal. Therefore, $v_{iy} = 0$. Now we find its position at time t:

$$\Delta x = x - x_i = x - 0 = v_{ix}t \qquad (3\text{-}20)$$

$$\Delta y = y - y_i = y - 0 = v_{iy}t + \tfrac{1}{2}a_y t^2 = 0 + \tfrac{1}{2}a_y t^2 \qquad (3\text{-}21)$$

Now we solve the first equation for t ($t = x/v_{ix}$) and substitute into the second.

$$y = \tfrac{1}{2}a_y\left(\frac{x}{v_{ix}}\right)^2 = \left(\frac{a_y}{2v_{ix}^2}\right)x^2$$

This is the equation of a parabola with $A = \dfrac{a_y}{2v_{ix}^2}$, $B = 0$, $C = 0$. The path of the clam is parabolic as viewed by the beachcomber.

(b) While it is carried by the gull, the clam has the same horizontal velocity as the gull. Once it is released, the clam retains that horizontal velocity component because the acceleration is vertical.

continued on next page

Example 3.6 continued

Figure 3.20
After digging up a clam, a gull takes it high above the ground and drops it to try to crack open the shell.

Figure 3.21
The path of the clam after being released is parabolic.

Discussion If we made a different choice for the origin in (a), the path of the clam would still have the same shape (parabolic). The equation for y as a function of x would still come out in the form of a parabola, $y = Ax^2 + Bx + C$. The values of B and C would be different, but this only shifts the location of the parabola; it doesn't change the shape.

Conceptual Practice Problem 3.6 Throwing Stones

You stand at the edge of a cliff and throw stones horizontally into the river below. To double the horizontal displacement of a stone from the cliff to where it lands, by what factor must you increase the stone's initial speed? Ignore air resistance.

Graphing Projectile Motion Figure 3.22 shows graphs of the x- and y-components of the position and velocity of a projectile as functions of time. In this case, the projectile is launched above flat ground at $t = 0$ and returns to the same elevation at a later time t_f. The y-component of velocity decreases linearly from its initial value; the slope of the line is $a_y = -g$. When $v_y = 0$, the projectile is at the apex of its trajectory. Then v_y continues to decrease at the same rate and is now negative with its magnitude getting larger and larger. At t_f, when the projectile has returned to its original altitude, the y-component of the velocity has the same magnitude as at $t = 0$ but with the opposite sign ($v_y = -v_{iy}$).

The graph of $y(t)$ indicates that the projectile moves upward, quickly at first and then gradually slowing, until it reaches the maximum height. The slope of the tangent to

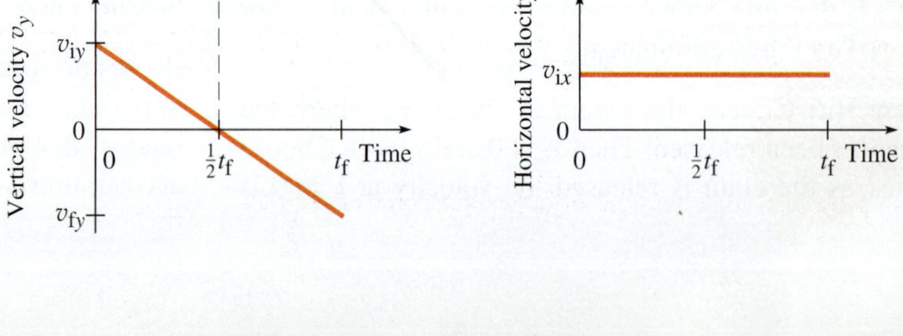

Figure 3.22 Projectile motion: graphs of y, v_y, x, and v_x as functions of time.

the $y(t)$ graph at any particular moment of time is v_y at that instant. At the highest point of the $y(t)$ graph, the tangent is horizontal and $v_y = 0$. After that, gravity makes the projectile start to fall downward. The shape of the graph of $y(t)$ is parabolic, but remember that this is y as a function of time, not the trajectory (y as a function of x).

The horizontal velocity is constant, so the graph of $v_x(t)$ is a horizontal line. The horizontal position x increases uniformly in time because the object is moving with a constant v_x.

CHECKPOINT 3.5A

When a basketball is thrown in an arc toward the net, what can you say about its velocity and acceleration at the highest point of the arc?

Example 3.7

Attacking the Castle Walls

The catapult used by the marauders hurls a stone with a velocity of 50.0 m/s at a 30.0° angle of elevation (Fig. 3.23). (a) What is the maximum height reached by the stone? (b) What is its *range* (defined as the horizontal distance traveled when the stone returns to its original height)? (c) How long has the stone been in the air when it returns to its original height?

Strategy The problem gives both the magnitude and direction of the initial velocity of the stone. Ignoring air resistance, the stone has a constant downward acceleration once it has been launched—until it hits the ground or some obstacle. We choose the positive y-axis upward and the positive x-axis in the direction of horizontal motion of the stone (toward the castle). When the stone reaches its maximum height, the velocity component in the y-direction is zero since the stone goes no higher. When the stone returns to its original height, $\Delta y = 0$ and $v_y = -v_{iy}$. The range can be found once the time of flight t_f is known—time is the quantity that connects the x-component equations to the y-component equations. Therefore, we solve (c) before (b). One way to find t_f is to find the time to reach maximum height and then double it (see Fig. 3.22). (Other methods include setting $\Delta y = 0$ or setting $v_y = -v_{iy}$.)

Solution (a) First we find the x- and y-components of the initial velocity for an angle of elevation $\theta = 30.0°$.

$$v_{iy} = v_i \sin\theta \quad \text{and} \quad v_{ix} = v_i \cos\theta$$

The maximum height is the vertical displacement Δy when $v_{fy} = 0$.

$$\Delta y = \frac{1}{2}(v_{fy} + v_{iy})\,\Delta t = \frac{1}{2}(0 + v_i \sin\theta)\,\Delta t$$

Eliminating the time interval using $v_{fy} - v_{iy} = a_y \Delta t$ yields

$$\Delta y = \frac{1}{2}(v_i \sin\theta)\left(\frac{0 - v_i \sin\theta}{a_y}\right) = -\frac{(v_i \sin\theta)^2}{2a_y}$$

$$= \frac{-(50.0\text{ m/s} \times \sin 30.0°)^2}{2 \times (-9.80\text{ m/s}^2)} = 31.9\text{ m}$$

The maximum height of the projectile is 31.9 m above its launch height.

(c) The initial and final heights are the same. Due to this symmetry, the time of flight (t_f) is *twice* the time it takes the

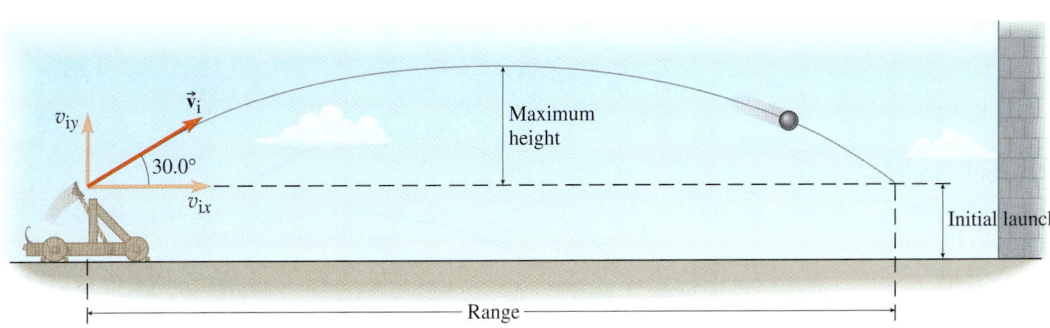

Figure 3.23
A catapult projects a stone into the air in an attack on a castle wall.

continued on next page

Example 3.7 continued

projectile to reach its maximum height. The time to reach the maximum height can be found from

$$v_{fy} = 0 = v_{iy} + a_y \Delta t$$

Solving for Δt,

$$\Delta t = \frac{-v_{iy}}{a_y}$$

The time of flight is

$$t_f = 2\,\Delta t = 2 \times \frac{-50.0\text{ m/s} \times \sin 30.0°}{-9.80\text{m/s}^2} = 5.10\text{ s}$$

(b) The range is

$$\Delta x = v_{ix}\, t_f = (50.0\text{ m/s} \times \cos 30.0°) \times 5.10\text{ s} = 221\text{ m}$$

Discussion Quick check: using

$$y_f - y_i = v_{iy}\,\Delta t + \tfrac{1}{2}a_y\,(\Delta t)^2$$

we can check that $\Delta y = 31.9$ m when $\Delta t = \frac{1}{2} \times 5.10$ s and that $\Delta y = 0$ when $\Delta t = 5.10$ s. Here we check the first of these:

$$\Delta y = (50.0\text{ m/s} \times \sin 30.0°) \times 2.55\text{ s} + \tfrac{1}{2} \times (-9.80\text{ m/s}^2) \times (2.55\text{ s})^2$$
$$= 63.8\text{ m} + (-31.9\text{ m}) = 31.9\text{ m}$$

which is correct. This is not an *independent* check, since this equation can be derived from the others, but it can reveal algebra or calculation errors.

Since we analyze the horizontal motion independently from the vertical motion, we start by resolving the given initial velocity into *x*- and *y*-components. Time is what connects the horizontal and vertical motions.

Practice Problem 3.7 Maximum Height for Arrows

Archers have joined in the attack on the castle and are shooting arrows over the walls. If the angle of elevation for an arrow is 45°, find an expression for the maximum height of the arrow in terms of v_i and g. [*Hint*: Simplify the expression using $\sin 45° = \cos 45° = 1/\sqrt{2}$.]

EVERYDAY PHYSICS DEMO

On a warm day, take a garden hose (or squirt gun) and aim the nozzle so that the water streams upward at an angle above the horizontal. Set the nozzle for a fast, narrow stream for best effect. Once the water leaves the nozzle, it becomes a projectile with a constant downward acceleration (ignoring the small effect of air resistance). The continuous stream of water lets us see the parabolic path easily. Stand in one place and try aiming the nozzle at different angles of elevation to find an angle that gives the maximum range. Aim for a particular spot on the ground (at a distance less than the maximum range) and see if you can find two different angles of elevated nozzle position that allow the stream to hit the target spot (Fig. 3.24). If you don't have a hose or squirt gun handy, try tossing a ball at different angles, with the same initial speed each time.

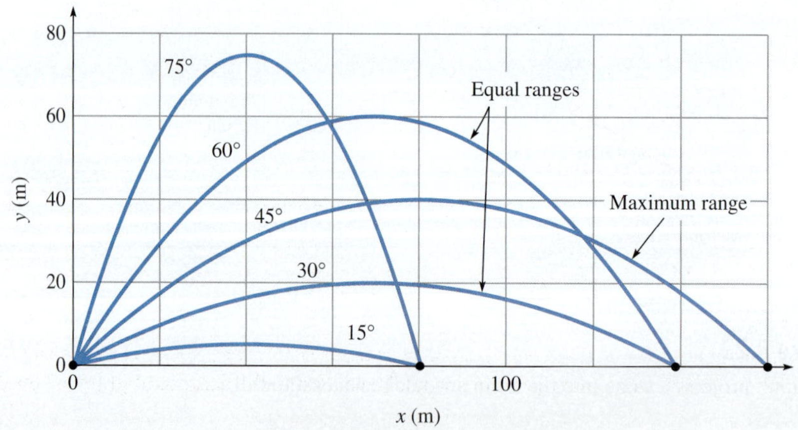

Figure 3.24 Parabolic trajectories of projectiles launched over level ground with the same initial speed ($v_i = 34.3$ m/s) at five different angles. The ranges of projectiles launched at angles θ and $90° - \theta$ are the same. The maximum range occurs for $\theta = 45°$.

CHECKPOINT 3.5B

Two projectiles launched at 30° and at 60° with the same initial speed will land at the same point (see Fig. 3.24). If they are launched simultaneously, do they land at the same time? If not, which lands first?

Conceptual Example 3.8

Monkey and Hunter

An inexperienced hunter aims and shoots an arrow straight at a coconut that is being held by a monkey in a tree (Fig. 3.25). At the same instant that the arrow leaves the bow, the monkey drops the coconut. Ignoring air resistance, does the arrow hit the coconut, the monkey, or neither?

Strategy and Solution If there were no gravity, the arrow would fly straight to the coconut (along the dashed blue line in Fig. 3.25). Since gravity gives the dropped coconut and the released arrow the same constant acceleration downward, they each fall the same vertical distance below the positions they would have had with no gravity. The coconut falls along the dashed red line; the distance fallen at 0.25 s intervals is marked. The arrow falls below the blue dashed line by the same distances, marked along its trajectory at 0.25 s intervals.

The arrow ends up hitting the coconut no matter what the initial speed of the arrow (as long as the arrow's range is at least as large as the horizontal distance to the coconut). The higher the speed of the arrow, the sooner they meet and the shorter the vertical distance that the coconut falls before being hit.

Discussion An experienced hunter would have aimed *above* the initial position of the coconut to compensate for gravity; he would have missed the coconut but might have hit the monkey unless the monkey jumped down to retrieve the coconut.

Conceptual Practice Problem 3.8 Changes in Position and Velocity for Consecutive Arrows

An arrow is shot into the air. One second later, a second arrow is shot with the same initial velocity. While the two are both in the air, does the difference in their positions ($\vec{\mathbf{r}}_2 - \vec{\mathbf{r}}_1$) stay constant or does it change with time? Does the difference in their velocities ($\vec{\mathbf{v}}_2 - \vec{\mathbf{v}}_1$) stay constant or does it change with time?

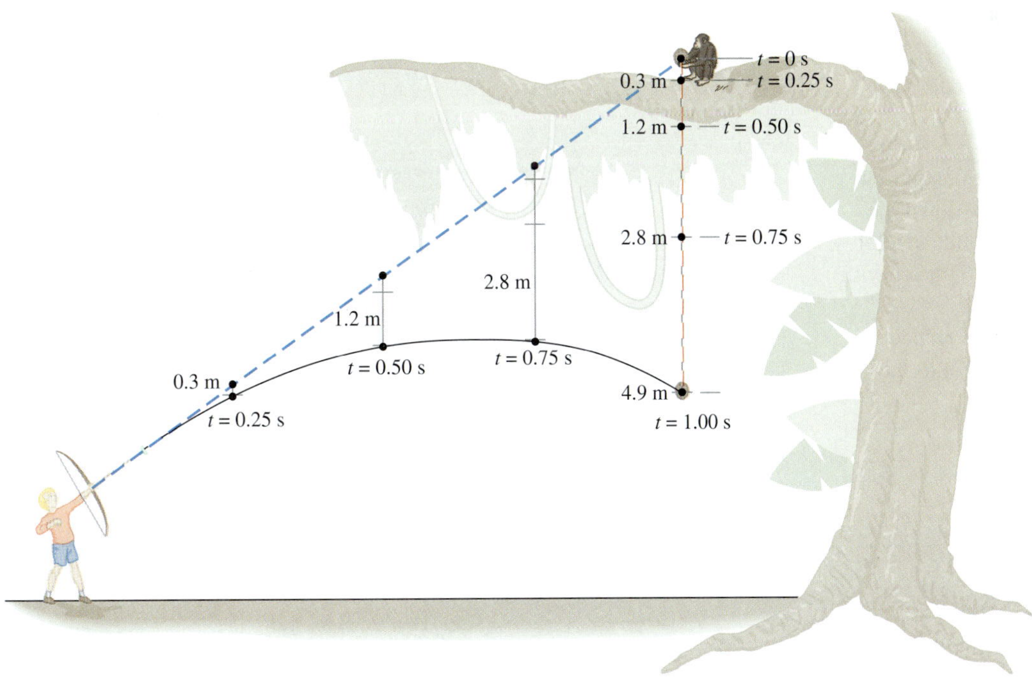

Figure 3.25

A monkey drops a coconut at the very instant an arrow is shot toward the coconut. At any instant, the coconut and arrow have fallen the same distance below where their positions would have been in the absence of gravity. This distance is $\frac{1}{2}gt^2$.

3.6 VELOCITY IS RELATIVE; REFERENCE FRAMES

CONNECTION:

In Chapter 4, we will learn the central idea of relativity: that the same principles of physics valid in one reference frame are also valid in any other reference frame that moves at constant velocity with respect to the first.

Until now, we have tacitly assumed in most situations that displacements, velocities, and accelerations should be measured in a **reference frame** attached to Earth's surface—that is, using a coordinate system in which the origin is a fixed point relative to Earth's surface and in which the coordinate axes have fixed directions relative to Earth's surface. This choice of reference frame is one of convenience, not necessity. The principles of physics presented starting in Chapter 4 are not restricted to one particular choice of reference frame; they have exactly the same form in any two reference frames whose relative velocity is constant.

Some of the ideas about relativity arose centuries before Einstein's theory. In the fourteenth century, the prevailing view was that the Sun and Moon revolve around a fixed Earth. The scholastic philosopher Nicole Oresme (1323–1382) argued instead that the Earth rotates around its axis. He wrote that the motion of one object can only be perceived relative to some other object. Therefore, we don't notice Earth's rotation because everything around us is moving with us.

Relative Velocity

Suppose Wanda is walking down the aisle of a train moving along the track at a constant velocity (Fig. 3.26). Imagine asking, "How fast is Wanda moving?" This question is not well defined. Do we mean her speed as measured by Tim, a passenger on the train, or her speed as measured by Greg, who is standing on the ground and looking into the train as it passes by? The answer to the question "How fast?" depends on the observer.

Figure 3.27 shows Wanda walking from one end of the car to the other during a time interval Δt. The displacement of Wanda as measured by Tim—her displacement *relative to the train*—is $\Delta\vec{\mathbf{r}}_{WT} = \vec{\mathbf{v}}_{WT}\Delta t$. During the same time interval, the *train's* displacement *relative to Greg* is $\Delta\vec{\mathbf{r}}_{TG} = \vec{\mathbf{v}}_{TG}\Delta t$. As measured by Greg, Wanda's displacement is partly due to her motion relative to the train and partly due to the motion of the train relative to the ground. Figure 3.27 shows that $\Delta\vec{\mathbf{r}}_{WT} + \Delta\vec{\mathbf{r}}_{TG} = \Delta\vec{\mathbf{r}}_{WG}$. Dividing by the time interval Δt gives the relationship between the three velocities:

$$\vec{\mathbf{v}}_{WT} + \vec{\mathbf{v}}_{TG} = \vec{\mathbf{v}}_{WG} \tag{3-24a}$$

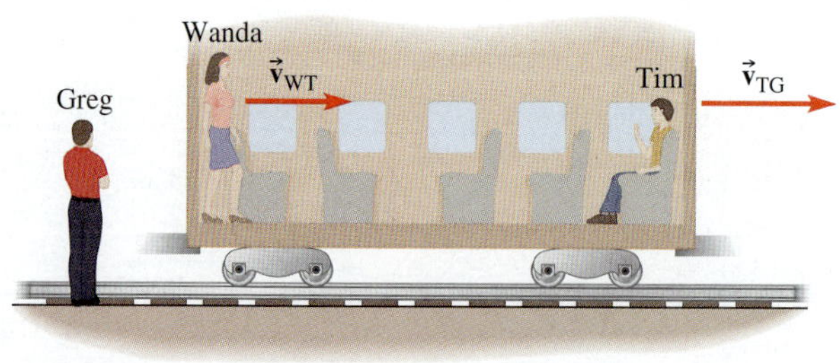

Figure 3.26 Tim and Greg watch Wanda walk down the aisle of a train. Wanda's velocity with respect to Tim (or with respect to the train) is $\vec{\mathbf{v}}_{WT}$; Tim's velocity with respect to Greg (or with respect to the ground) is $\vec{\mathbf{v}}_{TG}$.

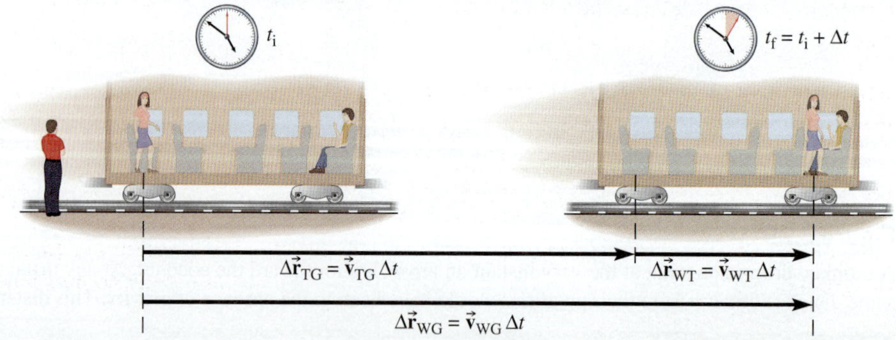

Figure 3.27 Wanda's displacement relative to the ground is the sum of her displacement relative to the train and the displacement of the train relative to the ground.

Solving for $\vec{\mathbf{v}}_{WT}$, we find that the velocity of Wanda relative to the train is her velocity minus the train's velocity, both measured in the *same reference frame*.

Relative Velocity

$$\vec{\mathbf{v}}_{WT} = \vec{\mathbf{v}}_{WG} - \vec{\mathbf{v}}_{TG} \quad (3\text{-}24b)$$

Generalizing Eq. (3-24b), the velocity of A relative to B is the vector difference of the two velocities as measured in a common reference frame. When working with relative velocities, we sometimes need to change the order of the subscripts, which *reverses the direction*. For example, Tim moves to the right relative to Greg, while Greg moves to the left relative to Tim: $\vec{\mathbf{v}}_{GT} = -\vec{\mathbf{v}}_{TG}$.

Sailing was one of Einstein's favorite pastimes throughout his life. Perhaps he started thinking about relativity while out on his sailboat?

Applications: Relative Velocities for Pilots and Sailors Relative velocities are of enormous practical interest to pilots of aircraft, sailors, and captains of ocean freighters. The pilot of an airplane is ultimately concerned with the motion of the plane with respect to the ground—the takeoff and landing points are fixed points on the ground. However, the controls of the plane (engines, rudder, ailerons, and spoilers) affect the motion of the plane *with respect to the air*. Pilots refer to *airspeed*, the speed of the plane with respect to the air, and *groundspeed*, the speed of the plane with respect to the ground. The *course* of a plane is the intended direction of its motion with respect to the ground, while the *heading* is the direction of its motion with respect to the air. The direction that the nose of the plane is pointing is its heading, not its course.

A sailor has to consider three different velocities of the boat: with respect to shore (for launching and landing), with respect to the air (for the behavior of the sails), and with respect to the water (for the behavior of the rudder). The heading of a boat is the direction of its motion with respect to the water, which is not the same as its course if there is a current. As for an airplane, the direction that a boat is pointing is its heading, not its course.

CHECKPOINT 3.6

In Fig. 3.26, if the train is moving at 18.0 m/s with respect to the ground and Wanda walks at 1.5 m/s with respect to the train, how fast is Wanda moving (a) with respect to Greg and (b) with respect to Tim?

Example 3.9

Flight from Denver to Chicago

An airplane flies from Denver to Chicago (1770 km) in 4.4 h when no wind blows. On a day with a tailwind, the plane makes the trip in 4.0 h. (a) What is the wind speed? (b) If a headwind blows with the same speed, how long does the trip take?

Strategy We assume the plane has the same *airspeed*—the same speed relative to the air—in both cases. Once the plane is up in the air, the behavior of the wings, control surfaces, and so on, depends on how fast the air is rushing by; the ground speed is irrelevant. But it is not irrelevant for the passengers, who are interested in a displacement relative to the ground.

Solution Let $\vec{\mathbf{v}}_{PG}$ and $\vec{\mathbf{v}}_{PA}$ represent the velocity of the plane relative to the ground and the velocity of the plane relative to the air, respectively. The wind velocity—the velocity of the air relative to the ground—can be written $\vec{\mathbf{v}}_{AG}$. Then $\vec{\mathbf{v}}_{PA} = \vec{\mathbf{v}}_{PG} - \vec{\mathbf{v}}_{AG}$. With no wind,

$$v_{PA} = v_{PG} = \frac{\Delta x}{\Delta t} = \frac{1770 \text{ km}}{4.4 \text{ h}} = 400 \text{ km/h}$$

(a) On the day with the tailwind,

$$v_{PG} = \frac{\Delta x}{\Delta t} = \frac{1770 \text{ km}}{4.0 \text{ h}} = 440 \text{ km/h}$$

We expect v_{PA} to be the same regardless of whether there is a wind or not. Since we are dealing with a tailwind, $\vec{\mathbf{v}}_{PA}$

continued on next page

Example 3.9 continued

and $\vec{\mathbf{v}}_{AG}$ are in the same direction, which we label as the $+x$-direction in Fig. 3.28. Then,

$$v_{PAx} = v_{PGx} - v_{AGx}$$
$$v_{AGx} = v_{PGx} - v_{PAx}$$
$$= 440\text{ km/h} - 400\text{ km/h} = 40\text{ km/h}$$

$v_{AGy} = 0$, so the wind speed is $v_{AG} = 40$ km/h.

(b) With a 40 km/h headwind, $\vec{\mathbf{v}}_{PA}$ and $\vec{\mathbf{v}}_{AG}$ are in opposite directions (Fig. 3.29). The velocity of the plane with respect to the ground is

$$v_{PGx} = v_{PAx} + v_{AGx} = 400\text{ km/h} + (-40\text{ km/h}) = 360\text{ km/h}$$

The ground speed of the plane is 360 km/h and the trip takes

$$\Delta t = \frac{\Delta x}{v_{PG}} = \frac{1770\text{ km}}{360\text{ km/h}} = 4.9\text{ h}$$

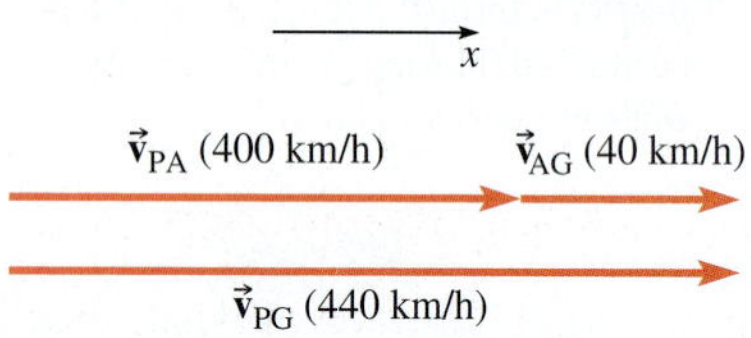

Figure 3.28 Addition of velocity vectors in the case of a tailwind. Lengths of vectors are not to scale.

Discussion Quick check: the trip takes longer with a headwind (4.9 h) than with no wind (4.4 h), as we expect.

Practice Problem 3.9 Rowing Across the Bay

Jamil, practicing to get on the crew team at school, rows a one-person racing shell to the north shore of the bay for a distance of 3.6 km to his friend's dock. On a day when the water is still (no current flowing), it takes him 20 min (1200 s) to reach his friend. On another day when a current flows southward, it takes him 30 min (1800 s) to row the same course. Ignore air resistance. (a) What is the speed of the current in meters per second? (b) How long does it take Jamil to return home with that same current flowing?

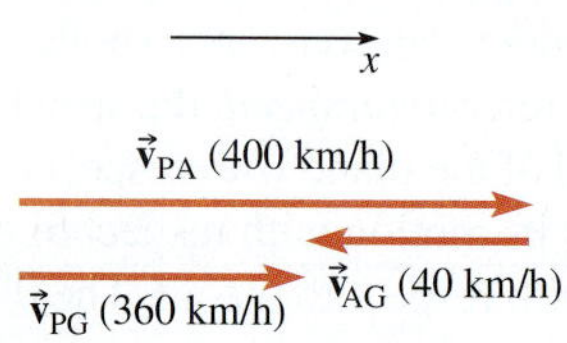

Figure 3.29 Addition of velocity vectors in the case of a headwind. Lengths of vectors are not to scale.

Relative Velocities in Two Dimensions The vector equations (3-24) apply to situations where the velocities are not all along the same line, as illustrated in Example 3.10.

Example 3.10

Rowing Across a River

Jack wants to row directly across a river from the east shore to a point on the west shore. The width of the river is 250 m and the current flows from north to south at 0.61 m/s. The trip takes Jack 4.2 min. In what direction did he head his rowboat to follow a course due west across the river? At what speed with respect to still water is Jack able to row?

Strategy We start with a sketch of the situation (Fig. 3.30). To keep the various velocities straight, we choose subscripts as follows: R = rowboat; W = water; S = shore. The velocity of the current given is the velocity of the water relative to the shore: $\vec{\mathbf{v}}_{WS} = 0.61$ m/s, south. The velocity of the rowboat relative to shore ($\vec{\mathbf{v}}_{RS}$) is due west. The magnitude of $\vec{\mathbf{v}}_{RS}$ can be found from the displacement relative to shore and the time interval, both of which are given. The question asks for the magnitude and direction of the velocity of the rowboat relative to the water ($\vec{\mathbf{v}}_{RW}$). The three velocities are related by

$$\vec{\mathbf{v}}_{RW} = \vec{\mathbf{v}}_{RS} - \vec{\mathbf{v}}_{WS}$$

or

$$\mathbf{v}_{RW} + \mathbf{v}_{WS} = \mathbf{v}_{RS}$$

To compensate for the current carrying the rowboat south with respect to shore, Jack heads (points) the rowboat upstream (against the current) at some angle to the north of west.

Solution In a sketch of the vector addition (Fig. 3.31), the velocity of the rowboat with respect to the water is at an angle θ north of west. With respect to shore, Jack travels 250 m in 4.2 min, so his speed with respect to shore is

$$v_{RS} = \frac{250\text{ m}}{4.2\text{ min} \times 60\text{ s/min}} = 0.992\text{ m/s}$$

We can find the angle at which the rowboat should be headed by finding the tangent of the angle between $\vec{\mathbf{v}}_{RW}$ and $\vec{\mathbf{v}}_{RS}$:

$$\tan\theta = \frac{v_{WS}}{v_{RS}} = \frac{0.61\text{ m/s}}{0.992\text{ m/s}}$$
$$\theta = 32°\text{N of W}$$

The speed at which Jack is able to row with respect to still water is the magnitude of $\vec{\mathbf{v}}_{RW}$. Since $\vec{\mathbf{v}}_{RS}$ and $\vec{\mathbf{v}}_{WS}$ are *perpendicular*, the Pythagorean theorem yields

$$|\vec{\mathbf{v}}_{RW}| = \sqrt{v_{WS}^2 + v_{RS}^2} = \sqrt{(0.61\text{ m/s})^2 + (0.992\text{ m/s})^2}$$
$$= 1.16\text{ m/s}$$

Jack rows at a speed of 1.16 m/s with respect to the water. (This result has three significant figures due to the addition inside the square root.)

Discussion If $\vec{\mathbf{v}}_{RS}$ and $\vec{\mathbf{v}}_{WS}$ had not been perpendicular, we could not have used the Pythagorean

continued on next page

Example 3.10 continued

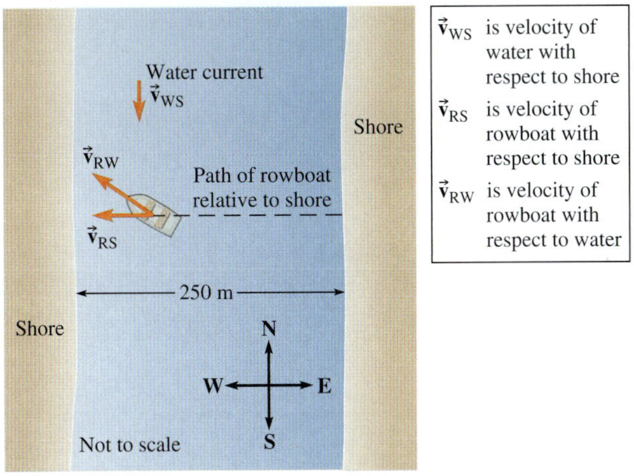

Figure 3.30

Rowing across a river. To go due west across the river, Jack must head his boat at some angle upstream (north of west); otherwise the current would carry him downstream.

theorem in this way. Rather, we would use the component method to add the two vectors.

If Jack had headed the rowboat directly west, the current would have carried him south, so he would have traveled in a direction south of west relative to shore. He has to compensate by heading upstream at just such an angle that his velocity relative to shore is directed west.

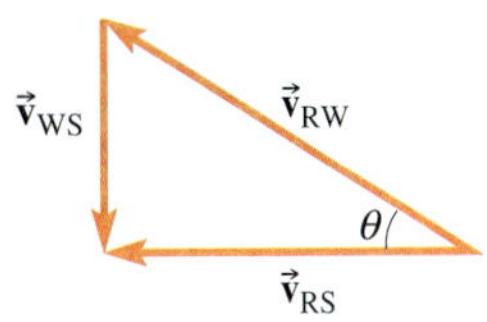

Figure 3.31

Graphical addition of the velocity vectors: $\vec{v}_{RW} + \vec{v}_{WS} = \vec{v}_{RS}$

Practice Problem 3.10 Heading Straight Across

If Jack were to head straight across the river, in what direction with respect to shore would he travel? How long would it take him to cross? How far downstream would he be carried? Assume that he rows at the same speed with respect to the water as in Example 3.10.

EVERYDAY PHYSICS DEMO

The next time you're on the escalator in a department store, watch someone on the neighboring escalator and visualize his position *relative to you*. From the change in his position relative to you, what is the direction of his velocity *relative to you*? Think about his and your velocities relative to the building and see if Eq. (3-24) supports your conclusion.

What does the path of the falling clam look like to the gull?

At the beginning of this chapter, we asked what the path followed by the falling clam looks like as seen by the gull flying through the air. If the gull continues to fly at the same horizontal velocity after dropping the clam, it is directly overhead when the clam hits the rock because they both have the same constant horizontal component of velocity with respect to the ground (Fig. 3.32a). The clam drop illustrates that the vertical motion of a projectile (in the absence of air resistance) is independent of the horizontal motion. A motion diagram for the clam in the reference frame of the ground would look like the white object in Fig. 3.19; in the reference frame of the gull it would look like the red object.

In the gull's reference frame—that is, using its own position as the origin of the coordinate axes—the velocity of the clam just after being released is zero. Therefore, the gull sees the clam fall straight down; it sees the rocks and other objects on the beach moving horizontally (Fig. 3.32b). Both observers agree that when the clam hits the rocks, the gull is directly overhead. At any instant, if the velocity of the clam with respect to the gull is $\vec{v}_{CG}$, the velocity of the gull with respect to the rocks is $\vec{v}_{GR}$, and the velocity of the clam with respect to the rocks is $\vec{v}_{CR}$, then $\vec{v}_{CG} = \vec{v}_{CR} - \vec{v}_{GR}$.

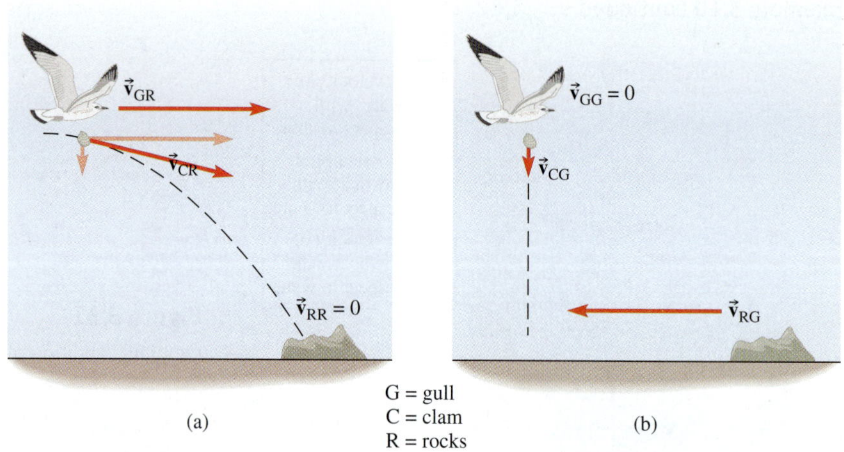

Figure 3.32 Trajectory of the clam in two different reference frames. (a) Beachcomber view: The gull flies along a horizontal line while the clam follows a parabolic path. (b) Bird's eye view: The gull sees the rocks moving while the clam drops straight down, landing on the rocks just as the rocks move under the clam.

Master the Concepts

- Vectors are added graphically by drawing each vector so that its tail is placed at the tip of the previous vector. The sum is drawn as a vector arrow from the tail of the first vector to the tip of the last. Addition of vectors is commutative: $\vec{\mathbf{A}} + \vec{\mathbf{B}} = \vec{\mathbf{B}} + \vec{\mathbf{A}}$.

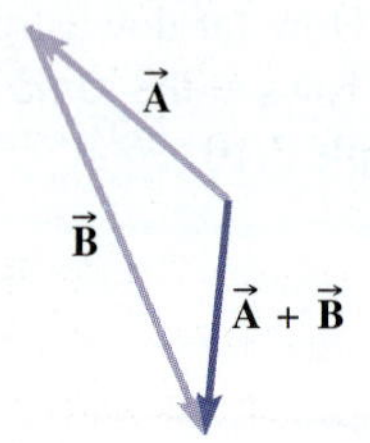

- Vectors are subtracted by adding the opposite of the second vector: $\vec{\mathbf{A}} - \vec{\mathbf{B}} = \vec{\mathbf{A}} + (-\vec{\mathbf{B}})$.
- Addition and subtraction of vectors algebraically using components is generally easier and more precise than the graphical method. The graphical method is still a useful first step to get an approximate answer.
- To find the components of a vector, first draw a right triangle with the vector as the hypotenuse and the other two sides parallel to the x- and y-axes. Then use the trigonometric functions to find the magnitudes of the components. The correct algebraic sign must be determined for each component. The same triangle can be used to find the magnitude and direction of a vector if its components are known.

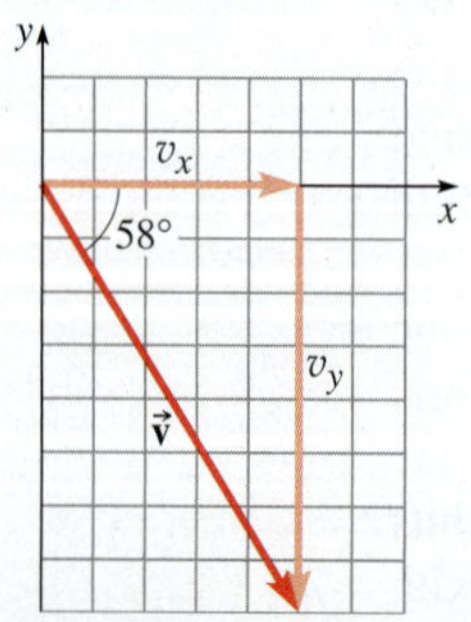

- To add vectors algebraically, add their components to find the components of the sum:

$$\vec{\mathbf{A}} + \vec{\mathbf{B}} = \vec{\mathbf{C}} \text{ if and only if}$$
$$A_x + B_x = C_x \text{ and } A_y + B_y = C_y$$

- The x- and y-axes are chosen to make the problem easiest to solve. Any choice is valid as long as the two are perpendicular. If the direction of the acceleration is known, choose x- and y-axes so that the acceleration vector is parallel to one of the axes.
- Position, displacement, velocity, and acceleration are vector quantities with both magnitude and direction. They must be added and subtracted as vectors.
- The equations for position, displacement, average velocity, instantaneous velocity, average acceleration, and instantaneous acceleration in Chapter 2 apply to *each perpendicular component* of the corresponding vector quantities for motion in two or three dimensions.
- The instantaneous velocity vector is tangent to the path of motion.

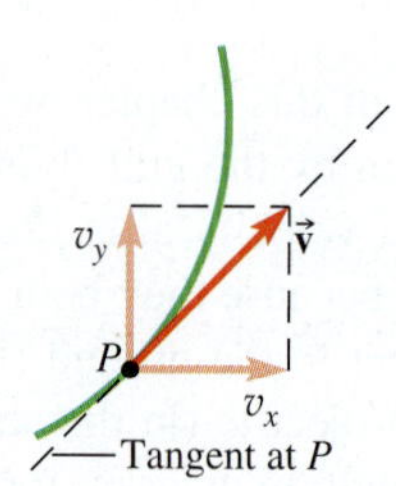

- The instantaneous acceleration vector does *not* have to be tangent to the path of motion, since velocities can change both in direction and in magnitude.
- For a projectile or any object moving with constant acceleration in the $\pm y$-direction, the motion in the x- and y-directions can be treated separately. Since $a_x = 0$, v_x is constant. Thus, the motion is a superposition of constant

continued on next page

Master the Concepts continued

velocity motion in the x-direction and constant acceleration motion in the y-direction.

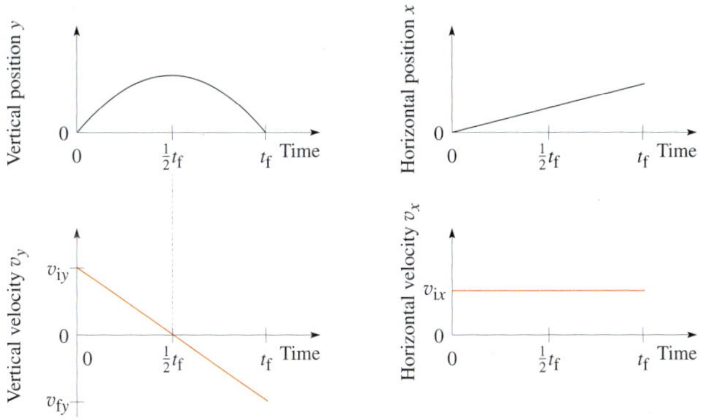

- The kinematic equations for an object moving in two dimensions with constant acceleration along the y-axis are

x-axis: $a_x = 0$	y-axis: constant a_y	
$\Delta v_x = 0$	$\Delta v_y = a_y \Delta t$	(3-19)
$\Delta x = v_x \Delta t$	$\Delta y = \frac{1}{2}(v_{fy} + v_{iy}) \Delta t$	(3-20)
	$\Delta y = v_{iy} \Delta t + \frac{1}{2} a_y (\Delta t)^2$	(3-21)
	$v_{fy}^2 - v_{iy}^2 = 2a_y \Delta y$	(3-22)

- To relate the velocities of objects measured in different reference frames, use the vector equation

$$\vec{\mathbf{v}}_{AB} = \vec{\mathbf{v}}_{AC} - \vec{\mathbf{v}}_{BC} \quad \text{(3-24b)}$$

where $\vec{\mathbf{v}}_{AB}$ represents the velocity of A relative to B, and so forth.

Conceptual Questions

1. If two vectors have the same magnitude, are they necessarily equal? If not, why not? Can two vectors with different magnitudes ever be equal?
2. (a) Is it possible for the sum of two vectors to be smaller in magnitude than the magnitude of either vector? (b) Is it possible for the magnitude of the sum of two vectors to be larger than the sum of the magnitudes of the two vectors?
3. What is the distinction between a vector and a scalar quantity? Give two examples of each.
4. Is it possible for two identical projectiles with identical initial speeds, but with two different angles of elevation, to land in the same spot? Explain. Ignore air resistance and sketch the trajectories.
5. If the trajectory is parabolic in one reference frame, is it always, never, or sometimes parabolic in another reference frame that moves at constant velocity with respect to the first reference frame? If the trajectory can be other than parabolic, what else can it be?
6. You are standing on a balcony overlooking the beach. You throw a ball straight up into the air with speed v_i and throw an identical ball straight down with speed v_i. Ignoring air resistance, how do the speeds of the balls compare just before they hit the ground?
7. You throw a ball up with initial speed v_i and when it reaches its high point at height h, you throw another ball into the air with the same initial speed v_i. Will the two balls cross at half the height h, or more than half, or less than half? Explain.
8. If an object is traveling at a constant velocity, is it necessarily traveling in a straight line? Explain.
9. Can the average speed and the magnitude of the average velocity ever be equal? If so, under what circumstances?
10. Give an example of an object whose acceleration is (1) in the same direction as its velocity, (2) opposite its velocity, and (3) perpendicular to its velocity.
11. Name a situation where the speed of an object is constant while the velocity is not.
12. Tell whether or not each of the following objects has a constant velocity and explain your reasoning. (a) A car driving around a curve at constant speed on a flat road. (b) A car driving straight up a 6° incline at constant speed. (c) The Moon.
13. Explain how to add two displacement vectors of magnitudes $3L$ and $4L$ so that the vector sum has magnitude (a) L; (b) $7L$; (c) $5L$.
14. Compare the advantages and disadvantages of the two methods of vector addition (graphical and algebraic).
15. Can the x-component of a vector ever be greater than the magnitude of the vector? Explain.
16. Why is the muzzle of a rifle not aimed directly at the center of the target? Why is this more important at longer ranges?
17. Does the monkey, coconut, and hunter demonstration still work if the hunter is in a higher tree and the arrow is pointed *downward* at the monkey and coconut?

Multiple-Choice Questions

1. Vector $\vec{\mathbf{A}}$ in the drawing is equal to
(a) $\vec{\mathbf{C}} + \vec{\mathbf{D}}$ (b) $\vec{\mathbf{C}} + \vec{\mathbf{D}} + \vec{\mathbf{E}}$ (c) $\vec{\mathbf{C}} + \vec{\mathbf{F}}$
(d) $\vec{\mathbf{B}} + \vec{\mathbf{C}}$ (e) $\vec{\mathbf{B}} + \vec{\mathbf{F}}$

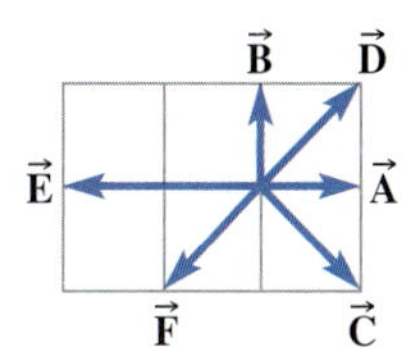

2. Which sum is *not* equal to zero?
(a) $\vec{C} + \vec{D} + \vec{E}$ (b) $\vec{B} + \vec{C} + \vec{F}$
(c) $\vec{D} + \vec{F}$ (d) $\vec{A} + \vec{B} + \vec{F}$

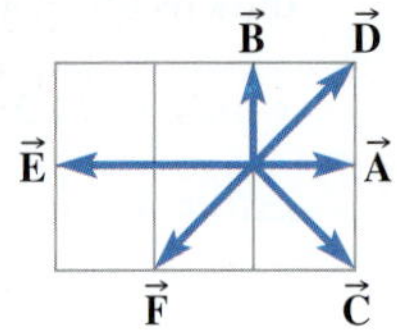

3. A hunter spots a pheasant flying along horizontally. If he shoots the pheasant, the time interval between the bird being shot and the dead bird hitting the ground depends on
(a) the speed with which the bird was flying.
(b) the height of the bird above the ground.
(c) the speed of the bird and its height above the ground.

4. A runner moves along a circular track at a constant speed.
(a) Her acceleration is zero.
(b) Her velocity is constant.
(c) Both (a) and (b) are true.
(d) Both her acceleration and her velocity are changing.

5. A boy plans to cross a river in a rubber raft. The current flows from north to south at 1 m/s. In what direction should he head to get across the river to the east bank in the least amount of time if he is able to paddle the raft at 1.5 m/s in still water?
(a) directly to the east
(b) south of east
(c) north of east
(d) The three directions require the same time to cross the river.

6. A boy plans to paddle a rubber raft across a river to the east bank while the current flows downriver from north to south at 1 m/s. He is able to paddle the raft at 1.5 m/s in still water. In what direction should he head the raft to go straight east across the river to the opposite bank?
(a) directly to the east (b) south of east
(c) north of east (d) north (e) south

7. A kicker kicks a football from the 5 yard line to the 45 yard line (both on the same half of the field). Ignoring air resistance, where along the trajectory is the speed of the football a minimum?
(a) at the 5 yard line, just after the football leaves the kicker's foot
(b) at the 45 yard line, just before the football hits the ground
(c) at the 15 yard line, while the ball is still going higher
(d) at the 35 yard line, while the ball is coming down
(e) at the 25 yard line, when the ball is at the top of its trajectory

8. Two balls, identical except for color, are projected horizontally from the roof of a tall building at the same instant. The initial speed of the red ball is twice the initial speed of the blue ball. Ignoring air resistance,
(a) the red ball reaches the ground first.
(b) the blue ball reaches the ground first.
(c) both balls land at the same instant with different speeds.
(d) both balls land at the same instant with the same speed.

9. A person stands on the roof garden of a tall building with one ball in each hand. If the red ball is thrown horizontally off the roof and the blue ball is simultaneously dropped over the edge, which statement is true?
(a) Both balls hit the ground at the same time, but the red ball has a higher speed just before it strikes the ground.
(b) The blue ball strikes the ground first, but with a lower speed than the red ball.
(c) The red ball strikes the ground first with a higher speed than the blue ball.
(d) Both balls hit the ground at the same time with the same speed.

10. A ball is thrown into the air and follows a parabolic trajectory. At the highest point in the trajectory,
(a) the velocity is zero, but the acceleration is not zero.
(b) both the velocity and the acceleration are zero.
(c) the acceleration is zero, but the velocity is not zero.
(d) neither the acceleration nor the velocity are zero.

11. A ball is thrown into the air and follows a parabolic trajectory. Point *A* is the highest point in the trajectory and point *B* is a point as the ball is falling back to the ground. Choose the correct relationship between the speeds and the magnitudes of the acceleration at the two points.
(a) $v_A > v_B$ and $a_A = a_B$ (b) $v_A < v_B$ and $a_A > a_B$
(c) $v_A = v_B$ and $a_A \neq a_B$ (d) $v_A < v_B$ and $a_A = a_B$

Questions 12–14. Two projectiles launched with the same initial speed but at different launch angles 30° and 60° land at the same spot (see Fig. 3.24). Ignore air resistance. Answer choices:
(a) projectile launched at 30°
(b) projectile launched at 60°
(c) They are equal.

12. Which has the larger horizontal velocity component v_x?

13. Which has a longer time of flight Δt (time interval between launch and hitting the ground)?

14. For which is the product $v_x \Delta t$ larger?

15. A sailor climbs the mast in a bosun's chair to make an emergency repair while the boat is moving forward at a steady 3.0 m/s (5.8 knots). At the top of the mast, he drops his wrench. If air resistance is negligible, the wrench lands on the deck
(a) significantly in front of the mast.
(b) significantly behind the mast.
(c) at the base of the mast.

16. A stone is thrown at an angle of 20° below the horizontal from the top of a cliff. Assume no air resistance. One second after being thrown, the stone's acceleration makes an angle θ below the horizontal. Which is true?
(a) $\theta = 0$ (b) $\theta = 20°$ (c) $0 < \theta < 20°$
(d) $20° < \theta < 90°$ (e) $\theta = 90°$

17. A stone is thrown at an angle of 20° below the horizontal from the top of a cliff. Assume no air resistance. One second after being thrown, the stone's velocity is at angle θ below the horizontal. Which is true?
(a) $\theta = 0$ (b) $\theta = 20°$ (c) $0 < \theta < 20°$
(d) $20° < \theta < 90°$ (e) $\theta = 90°$

Problems

- Combination conceptual/quantitative problem
- Biological or medical application
- ✦ Challenging problem
- Blue # Detailed solution in the Student Solutions Manual
- [1, 2] Problems paired by concept
- connect Text website interactive or tutorial

3.1 Graphical Addition and Subtraction of Vectors

1. Displacement vector $\vec{A}$ is directed to the west and has magnitude 2.56 km. A second displacement vector is also directed to the west and has magnitude 7.44 km. (a) What are the magnitude and direction of $\vec{A} + \vec{B}$? (b) What are the magnitude and direction of $\vec{A} - \vec{B}$ (c) What are the magnitude and direction of $\vec{B} - \vec{A}$?
2. Vector $\vec{A}$ is directed along the positive x-axis and has magnitude 1.73 units. Vector $\vec{B}$ is directed along the negative x-axis and has magnitude 1.00 unit. (a) What are the magnitude and direction of $\vec{A} + \vec{B}$? (b) What are the magnitude and direction of $\vec{A} - \vec{B}$? (c) What are the magnitude and direction of $\vec{B} - \vec{A}$?
3. Two vectors have magnitudes 3.0 and 4.0. How are the directions of the two vectors related if (a) the sum has magnitude 7.0, or (b) if the sum has magnitude 5.0? (c) What relationship between the directions gives the smallest magnitude sum and what is this magnitude?
4. A runner is practicing on a circular track that is 300 m in circumference. From the point farthest to the west on the track, he starts off running due north and follows the track as it curves around toward the east. (a) If he runs halfway around the track and stops at the farthest eastern point of the track, what is the distance he traveled? (b) What is his displacement?
5. Two displacement vectors each have magnitude 20 km. One is directed 60° above the $+x$-axis; the other is directed 60° below the $+x$-axis. What is the vector sum of these two displacements?
6. Orville walks 320 m due east. He then continues walking along a straight line, but in a different direction, and stops 200 m northeast of his starting point. How far did he walk during the second portion of the trip and in what direction?
7. Rank the vectors $\vec{A}$, $\vec{B}$, and $\vec{C}$ in order of increasing magnitude. Explain your reasoning.
8. Vectors $\vec{A}$, $\vec{B}$, and $\vec{C}$ are shown in the figure. Draw vectors $\vec{D}$ and $\vec{E}$, where $\vec{D} = \vec{A} + \vec{B}$ and $\vec{E} = \vec{A} + \vec{C}$. (b) Show that $\vec{A} + \vec{B} = \vec{B} + \vec{A}$ by graphical means.

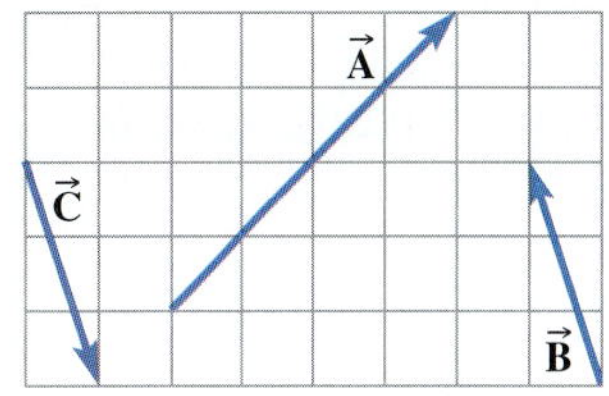

Problems 7, 8, 15, and 17

9. In the drawing, what is the vector sum of forces $\vec{D} + \vec{E} + \vec{F}$ if each grid square is 2 N on a side?

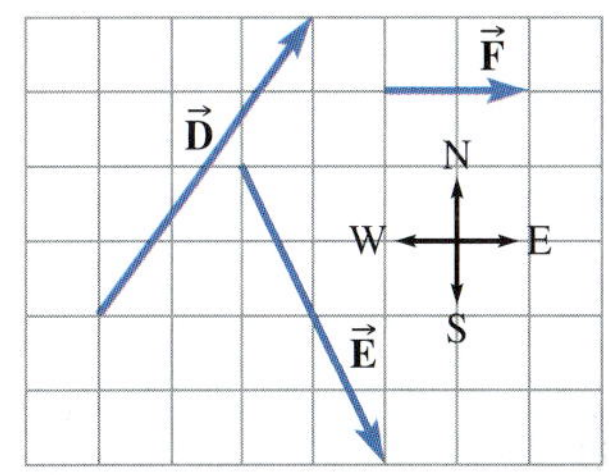

Problems 9, 10, and 16

10. Rank the vectors $\vec{D}$, $\vec{E}$, and $\vec{F}$ in order of increasing magnitude. Explain your reasoning.
11. Two vectors, each of magnitude 4.0 cm, are directed at a small angle α below the horizontal as shown. (The grid is 1 cm on a side.) (a) Let $\vec{C} = \vec{A} + \vec{B}$. Sketch the direction of $\vec{C}$ and estimate its magnitude. (b) Let $\vec{D} = \vec{A} - \vec{B}$. Sketch the direction of $\vec{D}$ and estimate its magnitude. (connect tutorial: vectors)

Problems 11, 22, and 23

12. Michaela is planning a trip in Ireland from Killarney to Cork to visit Blarney Castle. (See Example 3.2.) She also wants to visit Mallow, which is located 39 km due east of Killarney and 22 km due north of Cork. Draw the displacement vectors for the trip when she travels from Killarney to Mallow to Cork. (a) What is the magnitude of her displacement once she reaches Cork? Use graph paper, ruler, and protractor to find a graphical solution. (b) How much additional distance does Michaela travel in going to Cork by way of Mallow instead of going directly from Killarney to Cork?
13. A scout troop is practicing its orienteering skills with map and compass. First they walk due east for 1.2 km. Next, they walk 45° west of north for 2.7 km. In what direction must they walk to go directly back to their starting point? How far will they have to walk? Use graph paper, ruler, and protractor to find a graphical solution.
14. A sailboat sails from Marblehead Harbor directly east for 45 nautical miles, then 60° south of east for 20.0 nautical miles, east for 30.0 nautical miles, 30° east of north

for 10.0 nautical miles, and finally west for 62 nautical miles. At that time the wind dies, and the auxiliary engine fails to start. The crew decides to notify the Coast Guard of their position. Using graph paper, ruler, and protractor, sketch a graphical addition of the displacement vectors to find the sailboat's displacement from the harbor.

3.2 Vector Addition and Subtraction Using Components

15. Rank vectors $\vec{\mathbf{A}}$, $\vec{\mathbf{B}}$, and $\vec{\mathbf{C}}$ in Problem 7 in order of increasing x-component if the x-axis points to the right. Explain your reasoning.
16. With the y-axis pointing north, rank vectors $\vec{\mathbf{D}}$, $\vec{\mathbf{E}}$, and $\vec{\mathbf{F}}$ in Problem 9 in order of increasing y-component. Explain your reasoning.
17. Rank, in order of increasing x-component, $\vec{\mathbf{A}}+\vec{\mathbf{B}}$, $\vec{\mathbf{B}}+\vec{\mathbf{C}}$, and $\vec{\mathbf{A}}+\vec{\mathbf{C}}$ in Problem 7. The x-axis points to the right. Explain your reasoning.
18. A vector is 20.0 m long and makes an angle of 60.0° counterclockwise from the y-axis (on the side of the $-x$-axis). What are the x- and y-components of this vector?
19. Vector $\vec{\mathbf{A}}$ has magnitude 4.0 units; vector $\vec{\mathbf{B}}$ has magnitude 6.0 units. The angle between $\vec{\mathbf{A}}$ and $\vec{\mathbf{B}}$ is 60.0°. What is the magnitude of $\vec{\mathbf{A}}+\vec{\mathbf{B}}$?
20. Vector $\vec{\mathbf{A}}$ is directed along the positive y-axis and has magnitude $\sqrt{3.0}$ units. Vector $\vec{\mathbf{B}}$ is directed along the negative x-axis and has magnitude 1.0 unit. (a) What are the magnitude and direction of $\vec{\mathbf{A}}+\vec{\mathbf{B}}$? (b) What are the magnitude and direction of $\vec{\mathbf{A}}-\vec{\mathbf{B}}$? (c) What are the x- and y-components of $\vec{\mathbf{B}}-\vec{\mathbf{A}}$?
21. Vector $\vec{\mathbf{a}}$ has components $a_x = -3.0\ \text{m/s}^2$ and $a_y = +4.0\ \text{m/s}^2$. (a) What is the magnitude of $\vec{\mathbf{a}}$? (b) What is the direction of $\vec{\mathbf{a}}$? Give an angle with respect to one of the coordinate axes.
22. In Problem 11, let $\alpha = 10°$ and find the magnitude of vector $\vec{\mathbf{C}}$ using the component method.
23. In Problem 11, let $\alpha = 10°$ and find the magnitude of vector $\vec{\mathbf{D}}$ using the component method.
24. Find the x- and y-components of the four vectors shown in the drawing.

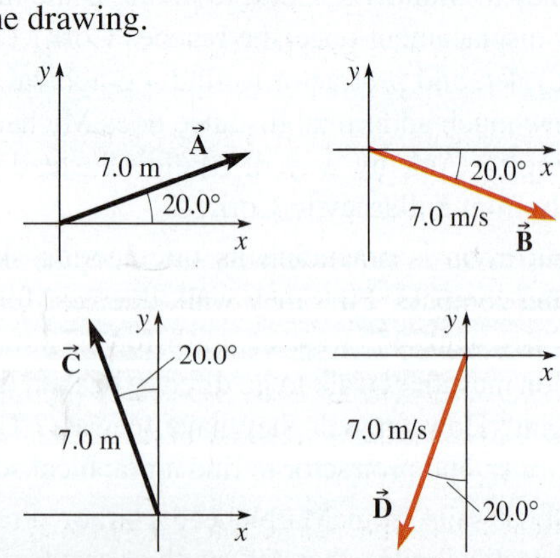

Problems 24–26

25. Suppose the vector $\vec{\mathbf{B}}$ is doubled in magnitude without changing its direction. What happens to its x- and y-components? Explain your reasoning.
26. Sketch a vector that has the same y-component as $\vec{\mathbf{B}}$ but with an x-component that is reversed in sign.
27. The velocity vector of a sprinting cheetah has x- and y-components $v_x = +16.4$ m/s and $v_y = -26.3$ m/s. (a) What is the magnitude of the velocity vector? (b) What angle does the velocity vector make with the $+x$- and $-y$-axes?
28. In each case, the x- and y-components of a vector are given. Find the magnitude and direction of the vector. (a) $A_x = -5.0$ m/s, $A_y = +8.0$ m/s; (b) $B_x = +120$ m, $B_y = -60.0$ m; (c) $C_x = -13.7$ m/s, $C_y = -8.8$ m/s; (d) $D_x = 2.3\ \text{m/s}^2$, $D_y = 6.5\ \text{cm/s}^2$.
29. A vector $\vec{\mathbf{A}}$ has a magnitude of 22.2 cm and makes an angle of 130.0° with the positive x-axis. What are the x- and y-components of this vector?
30. Vector $\vec{\mathbf{B}}$ has magnitude 7.1 and direction 14° below the $+x$-axis. Vector $\vec{\mathbf{C}}$ has x-component $C_x = -1.8$ and y-component $C_y = -6.7$. Compute (a) the x- and y-components of $\vec{\mathbf{B}}$; (b) the magnitude and direction of $\vec{\mathbf{C}}$; (c) the magnitude and direction of $\vec{\mathbf{C}}+\vec{\mathbf{B}}$; (d) the magnitude and direction of $\vec{\mathbf{C}}-\vec{\mathbf{B}}$; (e) the x- and y-components of $\vec{\mathbf{C}}-\vec{\mathbf{B}}$.
31. Margaret walks to the store using the following path: 0.500 mi west, 0.200 mi north, 0.300 mi east. What is her total displacement? That is, what is the length and direction of the vector that points from her house directly to the store? Use vector components to find the answer.
32. Jerry bicycles from his dorm to the local fitness center: 3.00 mi east and 2.00 mi north. Cindy's apartment is located 1.50 mi west of Jerry's dorm. If Cindy is able to meet Jerry at the fitness center by bicycling in a straight line, what is the length and direction she must travel?
33. Repeat Problem 13 using the component (algebraic) method.
34. Use the component method to obtain a more precise description of the sailboat's location in Problem 14.
35. You will be hiking to a lake with some of your friends by following the trails indicated on a map at the trailhead. The map says that you will travel 1.6 mi directly north, then 2.2 mi in a direction 35° east of north, then finally 1.1 mi in a direction 15° north of east. At the end of this hike, how far will you be from where you started, and what direction will you be from your starting point?

3.3 Velocity

36. A runner times his speed around a circular track with a circumference of 0.478 mi. At the start he is running toward the east and the track starts bending toward the north. If he goes halfway around, he will be running

toward the west. He finds that he has run a distance of 0.750 mi in 4.00 min. What is his (a) average speed and (b) average velocity in m/s?

37. A runner times his speed around a track with a circumference of 0.50 mi. He finds that he has run a distance of 1.00 mi in 4.0 min. What is his (a) average speed and (b) average velocity magnitude in m/s?

38. Peggy drives from Cornwall to Atkins Glen in 45 min. Cornwall is 73.6 km from Illium in a direction 25° west of south. Atkins Glen is 27.2 km from Illium in a direction 15° south of west. Using Illium as your origin, (a) draw the initial and final position vectors, (b) find the displacement during the trip, and (c) find Peggy's average velocity for the trip.

39. To get to a concert in time, a harpsichordist has to drive 122 mi in 2.00 h. (a) If he drove at an average speed of 55.0 mi/h in a due west direction for the first 1.20 h, what must be his average speed if he drives 30.0° south of west for the remaining 48.0 min? (b) What is his average velocity for the entire trip?

40. A bicycle travels 3.2 km due east in 0.10 h, then 4.8 km at 15.0° east of north in 0.15 h, and finally another 3.2 km due east in 0.10 h to reach its destination. The time lost in turning is negligible. What is the average velocity for the entire trip?

41. A car travels east at 96 km/h for 1.0 h. It then travels 30.0° east of north at 128 km/h for 1.0 h. (a) What is the average speed for the trip? (b) What is the average velocity for the trip?

42. A speedboat moves west at 108 km/h for 20.0 min. It then moves at 60.0° south of west at 90.0 km/h for 10.0 min. (a) What is the average speed for the trip? (b) What is the average velocity for the trip?

43. See Problem 12. During Michaela's travel from Killarney to Cork via Mallow, her actual travel time in the car is 48 min. (a) What is her average speed in m/s? (b) What is the magnitude of her average velocity in m/s?

44. ✦ Geoffrey drives from his home town due east at 90.0 km/h for 80.0 min. After visiting a friend for 15.0 min, he drives in a direction 30.0° south of west at 76.0 km/h for 45.0 min to visit another friend. (a) How far is it to his home from the second town? (b) If it takes him 45.0 min to drive directly home, what is his average velocity on the third leg of the trip? (c) What is his average velocity during the first two legs of his trip? (d) What is his average velocity over the entire trip? (e) What is his average speed during the entire trip if he spent 55.0 min visiting the second friend?

3.4 Acceleration

45. A hawk is flying north at 2.0 m/s with respect to the ground; 10.0 s later, it is flying south at 5.0 m/s. What is its average acceleration during this time interval?

46. A skydiver is falling straight down at 55 m/s when he opens his parachute and slows to 8.3 m/s in 3.5 s. What is the average acceleration of the skydiver during those 3.5 s?

47. A car travels three quarters of the way around a circle of radius 20.0 m in a time of 8.50 s at a constant speed. The initial velocity is west and the final velocity is south. (a) Find its average velocity for this trip. (b) What is the car's average acceleration during these 8.50 s? (c) Explain how a car moving at constant speed has a nonzero average acceleration.

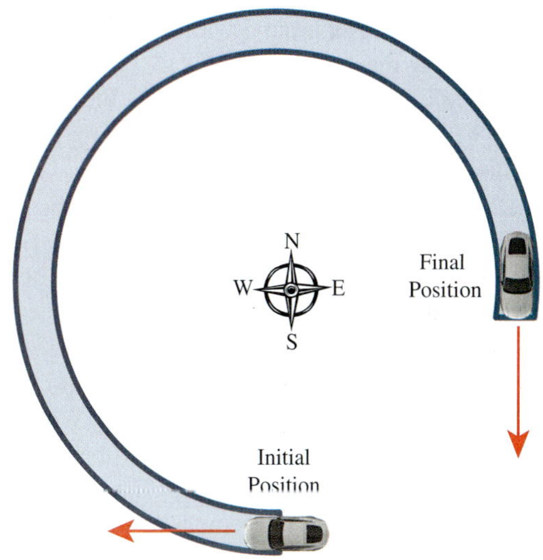

48. At $t = 0$, an automobile traveling north begins to make a turn. It follows one-quarter of the arc of a circle with a radius of 10.0 m until, at $t = 1.60$ s, it is traveling east. The car does not alter its speed during the turn. Find (a) the car's speed, (b) the change in its velocity during the turn, and (c) its average acceleration during the turn.

49. At the beginning of a 3.0 h plane trip, you are traveling due north at 192 km/h. At the end, you are traveling 240 km/h in the northwest direction (45° west of north). (a) Draw your initial and final velocity vectors. (b) Find the change in your velocity. (c) What is your average acceleration during the trip?

50. John drives 16 km directly west from Orion to Chester at a speed of 90 km/h, amd then directly south for 8.0 km to Seiling at a speed of 80 km/h, and then finally 34 km southeast to Oakwood at a speed of 100 km/h. Assume he travels at constant velocity during each of the three segments. (a) What was the change in velocity during this trip? [*Hint*: Do not assume he starts from rest and stops at the end.] (b) What was the average acceleration during this trip?

51. A particle's constant acceleration is south at 2.50 m/s^2. At $t = 0$, its velocity is 40.0 m/s east. What is its velocity at $t = 8.00$ s?

52. A particle's constant acceleration is north at 100 m/s^2. At $t = 0$, its velocity vector is 60 m/s east. At what time will the magnitude of the velocity be 100 m/s?

3.5 Motion in a Plane with Constant Acceleration

53. Rank the projectiles in Fig. 3.24 in order of increasing time of flight.

54. A baseball is thrown horizontally from a height of 9.60 m above the ground with a speed of 30.0 m/s. Where is the ball after 1.40 s has elapsed?

55. A clump of soft clay is thrown horizontally from 8.50 m above the ground with a speed of 20.0 m/s. Where is the clay after 1.50 s? Assume it sticks in place when it hits the ground.

56. A tennis ball is thrown horizontally from an elevation of 14.0 m above the ground with a speed of 20.0 m/s. (a) Where is the ball after 1.60 s? (b) If the ball is still in the air, how long before it hits the ground and where will it be with respect to the starting point once it lands?

57. A ball is thrown from a point 1.0 m above the ground. The initial velocity is 19.6 m/s at an angle of 30.0° above the horizontal. (a) Find the maximum height of the ball above the ground. (b) Calculate the speed of the ball at the highest point in the trajectory.

58. An arrow is shot into the air at an angle of 60.0° above the horizontal with a speed of 20.0 m/s. (a) What are the x- and y-components of the velocity of the arrow 3.0 s after it leaves the bowstring? (b) What are the x- and y-components of the displacement of the arrow during the 3.0 s interval?

59. The snow leopard (*Uncia uncia*) is an endangered species that lives in the mountains of central Asia. It is thought to be the longest jumper in the animal kingdom. If a snow leopard jumps at 35° above the horizontal and lands 15.0 m away on flat ground, what was its initial speed?

60. You have been employed by the local circus to plan their human cannonball performance. For this act, a spring-loaded cannon will shoot a human projectile, the Great Flyinski, across the big top to a net below. The net is located 5.0 m lower than the muzzle of the cannon from which the Great Flyinski is launched. The cannon will shoot the Great Flyinski at an angle of 35.0° above the horizontal and at a speed of 18.0 m/s. The ringmaster has asked that you decide how far from the cannon to place the net so that the Great Flyinski will land in the net and not be splattered on the floor, which would greatly disturb the audience. What do you tell the ringmaster? (connect interactive: projectile motion)

61. ✦ A cannonball is catapulted toward a castle. The cannonball's velocity when it leaves the catapult is 40 m/s at an angle of 37° with respect to the horizontal and the cannonball is 7.0 m above the ground at this time. (a) What is the maximum height above the ground reached by the cannonball? (b) Assuming the cannonball makes it over the castle walls and lands back down on the ground, at what horizontal distance from its release point will it land? (c) What are the x- and y-components of the cannonball's velocity just before it lands? The y-axis points up.

62. An archer fish spies a meal of a grasshopper sitting on a long stalk of grass at the edge of the pond in which he is swimming. If the fish is to successfully spit at and strike the grasshopper, which is 0.200 m away horizontally and 0.525 m above his mouth, what is the minimum speed at which the archer fish must spit? What angle above the horizontal must he spit?

63. After being assaulted by flying cannonballs, the knights on the castle walls (12 m above the ground) respond by propelling flaming pitch balls at their assailants. One ball lands on the ground at a distance of 50 m from the castle walls. If it was launched at an angle of 53° above the horizontal, what was its initial speed?

64. The orange jewelweed (*Impatiens capensis*) has seed pods that explode when lightly touched, launching the seeds as projectiles to disperse them. Suppose a seed is launched at 1.2 m/s from a height of 1.1 m. Assume air resistance is negligible and that the seed follows a clear path to the ground. (a) If the seed is launched horizontally, at what horizontal distance from the seed pod does the seed hit the ground? (b) If the seed is launched at 17° above the horizontal, at what horizontal distance does the seed hit the ground? (c) In the second case, the horizontal distance measured is 0.44 m. Was air resistance negligible?

65. The range R of a projectile is defined as the magnitude of the horizontal displacement of the projectile *when it returns to its original altitude*. (In other words, the range is the distance between the launch point and the impact point on flat ground.) A projectile is launched at $t = 0$ with initial speed v_i at an angle θ above the horizontal. (a) Find the time t at which the projectile returns to its original altitude. (b) Show that the range is

$$R = \frac{v_i^2 \sin 2\theta}{g}$$

[*Hint*: Use the trigonometric identity $\sin 2\theta = 2 \sin\theta \cos\theta$.]

66. Use the expression in Problem 65 to find (a) the maximum range of a projectile with launch speed v_i and (b) the launch angle θ at which the maximum range occurs.

67. A projectile is launched at $t = 0$ with initial speed v_i at an angle θ above the horizontal. (a) What are v_x and v_y at the projectile's highest point? (b) Find

the time t at which the projectile reaches its maximum height. (c) Show that the maximum height H of the projectile is

$$H = \frac{(v_i \sin\theta)^2}{2g}$$

68. ✦ A ballplayer standing at home plate hits a baseball that is caught by another player at the same height above the ground from which it was hit. The ball is hit with an initial velocity of 22.0 m/s at an angle of 60.0° above the horizontal. (a) How high will the ball rise? (b) How much time will elapse from the time the ball leaves the bat until it reaches the fielder? (c) At what distance from home plate will the fielder be when he catches the ball? (connect tutorial: projectile)

69. A suspension bridge is 60.0 m above the level base of a gorge. A stone is thrown or dropped from the bridge. Ignore air resistance. At the location of the bridge g has been measured to be 9.83 m/s^2. (a) If you drop the stone, how long does it take for it to fall to the base of the gorge? (b) If you *throw* the stone straight down with a speed of 20.0 m/s, how long before it hits the ground? (c) If you throw the stone with a velocity of 20.0 m/s at 30.0° above the horizontal, how far from the point directly below the bridge will it hit the level ground?

70. ✦ A circus performer is shot out of a cannon and flies over a net that is placed horizontally 6.0 m from the cannon. When the cannon is aimed at an angle of 40° above the horizontal, the performer is moving in the horizontal direction and just barely clears the net as he passes over it. What is the muzzle speed of the cannon and how high is the net?

71. ✦ Show that for a projectile launched at an angle of 45° the maximum height of the projectile is one quarter of the range (the distance traveled on flat ground).

3.6 Velocity Is Relative; Reference Frames

72. Two cars are driving toward each other on a straight, flat Kansas road. The Jeep Wrangler is traveling at 82 km/h north and the Ford Taurus is traveling at 48 km/h south, both measured relative to the road. What is the velocity of the Jeep relative to an observer in the Ford?

73. Two cars are driving toward each other on a straight and level road in Alaska. The BMW is traveling at 100.0 km/h north and the VW is traveling at 42 km/h south, both velocities constant and measured relative to the road. At a certain instant, the distance between the cars is 10.0 km. After what time interval, starting from that moment, will the two cars meet? [*Hint*: Consider a reference frame in which one of the cars is at rest.]

74. A car is driving directly north on the freeway at a speed of 110 km/h and a truck is leaving the freeway driving 85 km/h in a direction that is 35° west of north. What is the velocity of the truck relative to the car?

75. A Nile cruise ship takes 20.8 h to go upstream from Luxor to Aswan, a distance of 208 km, and 19.2 h to make the return trip downstream. Assuming the ship's speed relative to the water is the same in both cases, calculate the speed of the current in the Nile.

76. An airplane has a velocity relative to the ground of 210 m/s toward the east. The pilot measures his airspeed (the speed of the plane relative to the air) to be 160 m/s. What is the minimum wind velocity possible?

77. A small plane is flying directly west with an airspeed of 30.0 m/s. The plane flies into a region where the wind is blowing at 10.0 m/s at an angle of 30° to the south of west. (a) If the pilot does not change the heading of the plane, what will be the ground speed of the airplane? (b) What will be the new course, relative to the ground, of the airplane? (connect tutorial: flight of crow)

78. A small plane is flying directly west with an airspeed of 30.0 m/s. The plane flies into a region where the wind is blowing at 10.0 m/s at an angle of 30° to the south of west. In that region, the pilot changes the heading to maintain her due west course. (a) What is the change she makes in the heading to compensate for the wind? (b) After the heading change, what is the ground speed of the airplane?

79. A boat that can travel at 4.0 km/h in still water crosses a river with a current of 1.8 km/h. At what angle must the boat be pointed upstream to travel straight across the river? In other words, in what direction is the velocity of the boat relative to the water?

80. At an antique car rally, a Stanley Steamer automobile travels north at 40 km/h and a Pierce Arrow automobile travels east at 50 km/h. Relative to an observer riding in the Stanley Steamer, what are the x- and y-components of the velocity of the Pierce Arrow car? The x-axis is to the east and the y-axis is to the north.

81. Sheena can row a boat at 3.00 mi/h in still water. She needs to cross a river that is 1.20 mi wide with a current flowing at 1.60 mi/h. Not having her calculator ready, she guesses that to go straight across, she should head 60.0° upstream. (a) What is her speed with respect to the starting point on the bank? (b) How long does it take her to cross the river? (c) How far upstream or downstream from her starting point will she reach the

opposite bank? (d) In order to go straight across, what angle upstream should she have headed?

82. A dolphin wants to swim directly back to its home bay, which is 0.80 km due west. It can swim at a speed of 4.00 m/s relative to the water, but a uniform water current flows with speed 2.83 m/s in the southeast direction. (a) What direction should the dolphin head? (b) How long does it take the dolphin to swim the 0.80 km distance home?

83. ✦ A boy swims across a river in the shortest time he can, by always heading straight for the opposite bank. He can swim at a speed of 0.500 m/s relative to the water. The river is 25.0 m wide and the boy ends up at 50.0 m downstream from his starting point. (a) How fast is the current flowing in the river? (b) What is the speed of the boy relative to a friend standing on the riverbank?

84. ✦ An aircraft has to fly between two cities, one of which is 600.0 km north of the other. The pilot starts from the southern city and encounters a steady 100.0 km/h wind that blows from the northeast. The plane has a cruising speed of 300.0 km/h in still air. (a) In what direction (relative to east) must the pilot head her plane? (b) How long does the flight take?

Collaborative Problems

85. From the edge of the rooftop of a building, a boy throws a stone at an angle 25.0° above the horizontal. The stone hits the ground 4.20 s later, 105 m away from the base of the building. (Ignore air resistance.) (a) For the stone's path through the air, sketch graphs of x, y, v_x, and v_y as functions of time. These need to be only *qualitatively* correct—you need not put numbers on the axes. (b) Find the initial velocity of the stone. (c) Find the initial height h from which the stone was thrown. (d) Find the maximum height H reached by the stone.

86. You are serving as a consultant for the newest James Bond film. In one scene, Bond must fire a projectile from a cannon and hit the enemy headquarters located on the top of a cliff 75.0 m above and 350 m from the cannon. The cannon will shoot the projectile at an angle of 40.0° above the horizontal. The director wants to know what the speed of the projectile must be when it is fired from the cannon so that it will hit the enemy headquarters. What do you tell him? [*Hint*: Don't assume the projectile will hit the headquarters at the highest point of its flight.]

87. A helicopter is flying horizontally at 8.0 m/s and an altitude of 18 m when a package of emergency medical supplies is ejected horizontally backward with a speed of 12 m/s *relative to the helicopter*. Ignoring air resistance, what is the horizontal distance between the package and the helicopter when the package hits the ground?

88. ✦ A spotter plane sees a school of tuna swimming at a steady 5.00 km/h northwest. The pilot informs a fishing trawler, which is just then 100.0 km due south of the fish. The trawler sails along a straight-line course and intercepts the tuna after 4.0 h. How fast did the trawler move? [*Hint*: First find the velocity of the trawler relative to the tuna.]

89. ✦ Ⓒ One of the tricky things about learning to sail is distinguishing the "true wind" from the "apparent wind." The true wind is the velocity of the air relative to the water, whereas the apparent wind is the velocity of the air relative to the *sailboat*. The figure shows three different directions for the true wind. (a) In each case, draw a vector diagram to establish the magnitude and direction of the apparent wind. (b) In which of the three cases is the apparent wind speed greater than the true wind speed? (Assume that the speed of the boat relative to the water is less than the true wind speed.) (c) In which of the three cases is the direction of the apparent wind direction forward of the true wind? ["Forward" means coming from a direction more nearly straight ahead. For example, (1) is forward of (2), which is forward of (3).]

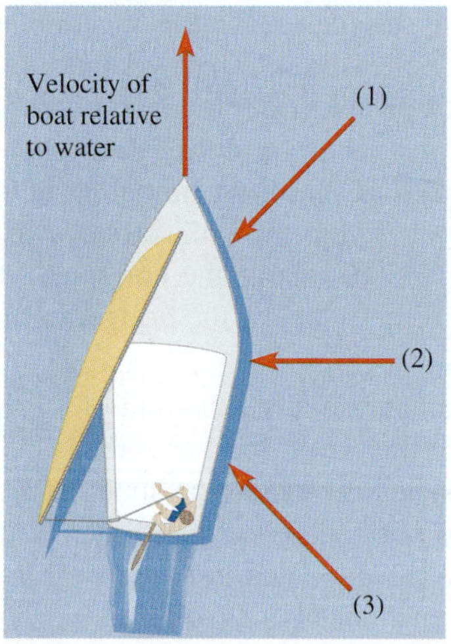

90. (*Note*: *In this problem you practice thinking about how the same events look in different reference frames. You won't need to use the "relative velocity formula."*) Samantha goes kayaking on a straight river. After she has paddled upstream for a while, she realizes she dropped a lifejacket overboard when she launched so she turns around and paddles downstream to retrieve it. The lifejacket has been drifting along with the current the whole time, but eventually Samantha catches up to it. Assume the river current flows at a constant speed and that Samantha uses the same paddling effort upstream and downstream, so her speeds *relative to the water* are the same. (a) Draw vectors to represent Samantha's upstream displacement (from launch to turnaround) and her downstream displacement (from turnaround to lifejacket retrieval) in the reference frame of the riverbank. If the magnitudes are unequal, be sure to show which is larger. (b) Now draw vectors representing Samantha's total (upstream + downstream) displacement and the total displacement of the lifejacket, both in the reference frame of the riverbank. (c) In the reference frame of the *water*, the lifejacket is at rest the whole time. What does

that tell you about Samantha's total displacement relative to the water? Sketch vector arrows showing her upstream and downstream displacements in the reference frame of the water. (d) Does Samantha spend more time paddling upstream, more time paddling downstream, or the same time each way? Explain your reasoning.

Comprehensive Problems

91. Harrison traveled 2.00 km west, then 5.00 km in a direction 53.0° south of west, then 1.00 km in a direction 60.0° north of west. (a) In what direction, and for how far, should Harrison travel to return to his starting point? (b) If Harrison returns directly to his starting point with a speed of 5.00 m/s, how long will the return trip take?
92. Paula swims across a river that is 10.2 m wide. She can swim at 0.833 m/s in still water, but the river flows with a speed of 1.43 m/s. If Paula swims in such a way that she crosses the river in as short a time as possible, how far downstream is she when she gets to the opposite shore?
93. You are watching the Super Bowl where your favorite team is leading by a score of 21 to 20. The other team is lining up to try to kick the winning field goal. You watched their kicker warm up and you saw that he could kick the football with a velocity of 21 m/s. He lines up for a 45 yd kick. You watch as he kicks the ball at an angle of 35° above the horizontal. Assuming he kicks the ball straight and with the same speed as during the warmup, will the ball clear the 10 ft high goal post, or will your favorite team win the Super Bowl?
94. Jason is practicing his tennis stroke by hitting balls against a wall. The ball leaves his racquet at a height of 60 cm above the ground at an angle of 80° with respect to the *vertical*. (a) The speed of the ball as it leaves the racquet is 20 m/s and it must travel a distance of 10 m before it reaches the wall. How far above the ground does the ball strike the wall? (b) Is the ball on its way up or down when it hits the wall?
95. Imagine a trip where you drive along an east-west highway at 80.0 km/h for 45.0 min and then you turn onto a highway that runs 38.0° north of east and travel at 60.0 km/h for 30.0 min. (a) What is your average velocity for the trip? (b) What is your average velocity on the return trip when you drive 38.0° south of west at 60.0 km/h for the first 30.0 min and then west at 80.0 km/h for the last 45.0 min?
96. A jetliner flies east for 600.0 km, then turns 30.0° toward the south and flies another 300.0 km. (a) How far is the plane from its starting point? (b) In what direction could the jetliner have flown directly to the same destination (in a straight-line path)? (c) If the jetliner flew at a constant speed of 400.0 km/h, how long did the trip take? (d) Moving at the same speed, how long would the direct flight have taken?
97. An African swallow carrying a very small coconut is flying horizontally with a speed of 18 m/s. (a) If it drops the coconut from a height of 100 m above the Earth, how long will it take before the coconut strikes the ground? (b) At what horizontal distance from the release point will the coconut strike the ground?
98. A pilot starting from Athens, New York, wishes to fly to Sparta, New York, which is 320 km from Athens in the direction 20.0° N of E. The pilot heads directly for Sparta and flies at an airspeed of 160 km/h. After flying for 2.0 h, the pilot expects to be at Sparta, but instead he finds himself 20 km due west of Sparta. He has forgotten to correct for the wind. (a) What is the velocity of the plane relative to the air? (b) Find the velocity (magnitude and direction) of the plane relative to the ground. (c) Find the wind speed and direction.
99. The citizens of Paris were terrified during World War I when they were suddenly bombarded with shells fired from a long-range gun known as Big Bertha. The barrel of the gun was 36.6 m long and it had a muzzle speed of 1.46 km/s. When the gun's angle of elevation was set to 55°, what would be the range? For the purposes of solving this problem, neglect air resistance. (The actual range at this elevation was 121 km; air resistance cannot be ignored for the high muzzle speed of the shells.)
100. The pilot of a small plane finds that the airport where he intended to land is fogged in. He flies 55 mi west to another airport to find that conditions there are too icy for him to land. He flies 25 mi at 15° east of south and is finally able to land at the third airport. (a) How far and in what direction must he fly the next day to go directly to his original destination? (b) How many extra miles beyond his original flight plan has he flown?
101. A particle has a constant acceleration of 5.0 m/s^2 to the east. At time $t = 0$, it is 2.0 m east of the origin and its velocity is 20 m/s north. What are the components of its position vector at $t = 2.0$ s?
102. A baseball batter hits a long fly ball that rises to a height of 44 m. An outfielder on the opposing team can run at 7.6 m/s. What is the farthest the fielder can be from where the ball will land so that it is possible for him to catch the ball?
103. A locust jumps at an angle of 55.0° and lands 0.800 m from where it jumped. (a) What is the maximum height of the locust during its jump? Ignore air resistance. (b) If it jumps with the same initial speed at an angle of 45.0°, would the maximum height be larger or smaller? (c) What about the range? (d) Calculate the maximum height and range for this angle.
104. An airplane is traveling from New York to Paris, a distance of 5.80×10^3 km. Ignore the curvature of the Earth. (a) If the cruising speed of the airplane is 350.0 km/h, how much time will it take for the airplane to make the round-trip on a calm day? (b) If a steady wind blows from New York to Paris at 60.0 km/h, how much time will the round-trip take? (c) How much time will it take if there is a crosswind of 60.0 km/h?

105. A gull is flying horizontally 8.00 m above the ground at 6.00 m/s. The bird is carrying a clam in its beak and plans to crack the clamshell by dropping it on some rocks below. Ignoring air resistance, (a) what is the horizontal distance to the rocks at the moment that the gull should let go of the clam? (b) With what speed relative to the rocks does the clam smash into the rocks? (c) With what speed relative to the gull does the clam smash into the rocks?

106. A beanbag is thrown horizontally from a dorm room window a height h above the ground. It hits the ground a horizontal distance h (the *same* distance h) from the dorm directly below the window from which it was thrown. Ignoring air resistance, find the direction of the beanbag's velocity just before impact.

107. ✦ A sample in a centrifuge moves in a circle of radius 8.0 cm at a constant speed of 500 m/s. (a) How much time does it take for the velocity's direction to change by 45° (1/8 of a revolution)? (b) What is the magnitude of the average acceleration during that time?

108. ✦ The invention of the cannon in the fourteenth century made the catapult unnecessary and ended the safety of castle walls. Stone walls were no match for balls shot from cannons. Suppose a cannonball of mass 5.00 kg is launched from a height of 1.10 m, at an angle of elevation of 30.0° with an initial velocity of 50.0 m/s, toward a castle wall of height 30 m and located 215 m away from the cannon. (a) The range of a projectile is defined as the horizontal distance traveled when the projectile returns to its original height. What will be the range reached by the projectile if it is not intercepted by the wall? (b) If the cannonball travels far enough to hit the wall, find the height at which it strikes.

109. ✦ In a plate glass factory, sheets of glass move along a conveyor belt at a speed of 15.0 cm/s. An automatic cutting tool descends at preset intervals to cut the glass to size. Since the assembly belt must keep moving at constant speed, the cutter is set to cut at an angle to compensate for the motion of the glass. If the glass is 72.0 cm wide and the cutter moves across the width at a speed of 24.0 cm/s, at what angle should the cutter be set?

110. ✦ A pilot wants to fly from Dallas to Oklahoma City, a distance of 330 km at an angle of 10.0° west of north. The pilot heads directly toward Oklahoma City with an airspeed of 200 km/h. After flying for 1.0 h, the pilot finds that he is 15 km off course to the west of where he expected to be after 1.0 h, assuming there was no wind. (a) What is the velocity and direction of the wind? (b) In what direction should the pilot have headed his plane to fly directly to Oklahoma City without being blown off course?

111. A ball is thrown horizontally off the edge of a cliff with an initial speed of 20.0 m/s. (a) How long does it take for the ball to fall to the ground 20.0 m below? (b) How long would it take for the ball to reach the ground if it were dropped from rest off the cliff edge? (c) How long would it take the ball to fall to the ground if it were thrown at an initial velocity of 20.0 m/s but 18° below the horizontal?

112. ✦ A marble is rolled so that it is projected horizontally off the top landing of a staircase. The initial speed of the marble is 3.0 m/s. Each step is 0.18 m high and 0.30 m wide. Which step does the marble strike first?

113. ✦ When fish head upstream to spawn, they often must make their way up a waterfall. If the water is not moving too fast, the fish can swim right up through the falling water. If the water is falling with too great a speed, the fish jump out of the water to get to a place in the waterfall where the water is not falling so fast. When humans build dams that interrupt the usual route followed by the fish, artificial fish ladders must also be built to allow the fish to get back uphill to the spawning area. These fish ladders consist of a series of small waterfalls with still pools of water in between them (see the figure). Assume that the water is at rest in the pools at the top and bottom of one "rung" of the fish ladder, that water falls straight down from one pool to the next, and that fish can swim at 5.0 m/s with respect to the water. (a) What is the maximum height of a waterfall up which the fish can swim without having to jump? (b) If a waterfall is 1.5 m high, how high must the fish jump to get to water through which it can swim? Assume that they jump straight up. (c) What initial speed must a fish have to jump the height found in part (b)? (d) For a 1.0 m high waterfall, how fast will the fish be swimming with respect to the ground when it starts swimming up the waterfall?

114. ✦ A motor scooter rounds a curve on the highway at a constant speed of 20.0 m/s. The original direction of the scooter was due east; after rounding the curve the

scooter is going 36° north of east. The radius of curvature of the road at the location of the curve is 150 m. What is the average acceleration of the scooter as it rounds the curve?

115. ✦ You want to make a plot of the trajectory of a projectile. That is, you want to make a plot of the height y of the projectile as a function of horizontal distance x. The projectile is launched from the origin with an initial speed v_i at an angle θ above the horizontal. Show that the equation of the trajectory followed by the projectile is

$$y = \left(\frac{v_{iy}}{v_{ix}}\right)x + \left(\frac{-g}{2v_{ix}^2}\right)x^2$$

116. ✦ A person climbs from a Paris metro station to the street level by walking up a stalled escalator in 94 s. It takes 66 s to ride the same distance when standing on the escalator when it is operating normally. How long would it take for him to climb from the station to the street by walking up the moving escalator?

Answers to Practice Problems

3.1 No; the checkbook balance may increase or decrease, but there is no spatial direction associated with it. When we say it "goes down," we do not mean that it moves in a direction toward the center of Earth! Rather, we really mean that it decreases. The balance is a scalar.

3.2 240 mi 20° W of S

3.3 $A_x = +16$ km; $A_y = -8.2$ km; $B_x = +17$ km; $B_y = 0$ km; $C_x = -11$ km; $C_y = +47$ km

3.4 $|\vec{\mathbf{v}}_{av}|$ can never be greater than the average speed because the magnitude of the displacement cannot be greater than the distance traveled. $|\vec{\mathbf{v}}_{av}|$ can be equal to the average speed if the magnitude of the displacement is equal to the distance traveled, which is true when the motion is along a straight line with no change in direction.

3.5 (a) 1.64 m/s directed 33° east of north; (b) 0.82 m/s^2 directed 33° east of north

3.6 2

3.7 $v_i^2/(4g)$

3.8 Ignoring air resistance, the two arrows have the same constant horizontal velocity component: $v_{2x} - v_{1x} = 0$ (choosing the x-axis horizontal and the y-axis up). Their vertical velocity components are different, but they *change at the same rate*, so $v_{2y} - v_{1y}$ stays constant. The difference in their velocities $(\vec{\mathbf{v}}_2 - \vec{\mathbf{v}}_1)$ stays constant. This constant difference in their velocities makes the difference in their positions $(\vec{\mathbf{r}}_2 - \vec{\mathbf{r}}_1)$ change with time

3.9 (a) 1.0 m/s; (b) 15 min

3.10 28° south of west; 3.6 min; 130 m

Answers to Checkpoints

3.2A $C_x = -5.5$ km and $C_y = -7.2$ km

3.2B

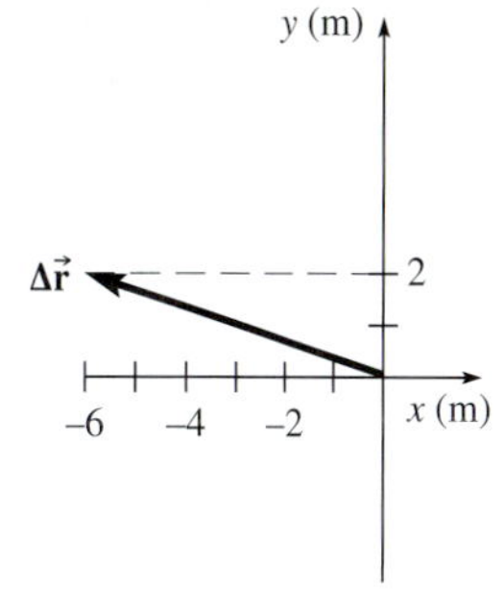

3.4A The average acceleration is the change in velocity divided by the time interval. Δt is the same for all four, so the largest change in velocity produces the largest average acceleration. The change in velocity of plane A is zero. The vector diagrams illustrate the change in velocity for the other planes. In decreasing order: D, B, C, A.

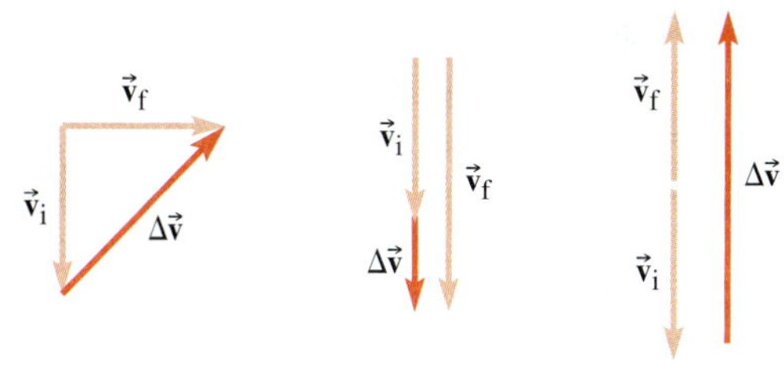

3.4B Velocity is a vector quantity. The plane's speed does not change, but its velocity does. Therefore, $\Delta\vec{\mathbf{v}} \neq 0$ and $\vec{\mathbf{a}}_{av} = \Delta\vec{\mathbf{v}}/\Delta t \neq 0$.

3.4C For straight-line motion, the acceleration vector is either in the same or in the opposite direction to the velocity vector. If the speed is increasing, $\vec{\mathbf{a}}$ is in the same direction as $\vec{\mathbf{v}}$; if the speed is decreasing, $\vec{\mathbf{a}}$ is in the direction opposite to $\vec{\mathbf{v}}$. If the velocity is constant, then $\vec{\mathbf{a}} = 0$.

3.5A The horizontal velocity component does not change. The vertical component is zero at the highest point, so the velocity vector is directed horizontally. The acceleration is constant and directed vertically downward throughout the flight, including at the highest point.

3.5B No, the projectile launched at 30° lands first because its initial v_y is smaller, so it doesn't go as high. (Remember that we can treat the horizontal and vertical motions independently.) It has a larger initial v_x, which allows it to go the same horizontal distance as the 60° projectile but in a shorter time.

3.6 (a) 19.5 m/s (b) 1.5 m/s

CHAPTER

4 Force and Newton's Laws of Motion

A sailplane (or "glider") is a small, unpowered, high-performance aircraft. A sailplane must be initially towed a few thousand feet into the air by a small airplane, after which it relies on regions of upward-moving air, such as thermals and ridge currents, to ascend further. Suppose a small plane requires about 120 m of runway to take off by itself. When it is towing a sailplane, how much runway does it need? (See p. 126 for the solution.)

BIOMEDICAL APPLICATIONS

- Tensile and contact forces in the body (Sec. 4.7; Probs. 6, 21, 121, 138, 160, 164)
- Traction apparatus (Ex. 4.1; Prob. 132)
- Newton's third law: swimming, walking, skiing (Sec. 4.4)
- Peak force on a runner's foot (Prob. 42)
- Effects of acceleration on the human body (Sec. 4.10)

Concepts & Skills to Review

- addition of vectors (Sections 3.1 and 3.2)
- vector components (Section 3.2)
- acceleration (Sections 2.3 and 3.4)
- motion with constant acceleration (Sections 2.4, 2.5, and 3.5)
- motion diagrams (Section 2.4)
- problem-solving techniques (Section 1.7)
- meanings of *velocity* and *mass* in physics (Section 1.2)

4.1 INTERACTIONS AND FORCES

This chapter begins our study of **mechanics**, the branch of physics that considers how interactions between objects affect the motion of those objects. Just as human life would be dull without social interactions, the physical universe would be dull without physical interactions. Social interactions with friends and family change our behavior; physical interactions change the "behavior" (motion, temperature, etc.) of matter.

An interaction between two objects can be described and measured in terms of two *forces*, one exerted on each of the two interacting objects. A **force** is a push or a pull. When you play soccer, your foot exerts a force on the ball while the two are in contact, thereby changing the speed and direction of the ball's motion. At the same time, the ball exerts a force on your foot, the effect of which you can feel. To understand the motion of an object, whether it be a soccer ball or the International Space Station, we need to analyze the forces acting on the object.

CONNECTION:

Newton's third law (Sec. 4.4) tells us not only that forces always come in *interaction pairs* but also how the magnitudes and directions of the forces are related.

To correctly identify forces, you should be able to describe them as (*type of force*) exerted on (*object*) by (*object*). For example: contact force exerted on the ball by the foot; gravitational force exerted on the Space Station by Earth.

Long-Range Forces Forces exerted on macroscopic objects—objects that are large enough for us to observe without instrumentation—can be either long-range forces or contact forces. **Long-range forces** do not require the two objects to be touching. These forces can exist even if the two objects are far apart and even if there are other objects between the two. For example, gravity is a long range force. The gravitational force exerted on the Earth by the Sun keeps the Earth in orbit around the Sun, despite the great distance between them and despite other planets that occasionally come between them. The Earth also exerts a long-range gravitational force on objects on or near its surface. We call the size of the gravitational force (also called the strength, or **magnitude**, of the force) that a planet or moon exerts on a nearby object the object's **weight**.

Part 3 of this book treats electromagnetic forces in detail. Until then, you can safely assume that gravity is the only significant long-range interaction unless the statement of a problem indicates otherwise.

EVERYDAY PHYSICS DEMO

Besides gravity, other long-range forces are electric or magnetic in nature. On a dry day, run a plastic comb vigorously through your hair or rub it on a wool sweater until you hear some crackling. Now hold the comb close to small pieces of a torn paper napkin. Observe the long-range electrical interaction between the paper and the comb.

Now take a refrigerator magnet. Hold it near but not touching the refrigerator door or another magnet. You can feel the effect of a long-range magnetic interaction.

Contact Forces All forces exerted on macroscopic objects, other than long-range gravitational and electromagnetic forces, involve contact. **Contact forces** exist only as long as the objects are touching one another. Your foot has no noticeable effect on a

Figure 4.1 A soccer player's foot exerts a force on the ball only when they are touching. The ball also exerts a force on the foot, but only while they touch. Once the ball loses contact with the foot, the only forces acting on it are a long-range gravitational force due to the Earth and a contact force due to the air.

soccer ball's motion until the two come into contact, and the force lasts only as long as they are in contact (Fig. 4.1). Once the ball moves away from your foot, your foot has no further influence over the ball's motion.

The idea of contact is a useful simplification for macroscopic objects. What we call a single contact force is really the net effect of enormous numbers of electromagnetic forces between atoms on the surfaces of the two objects. On an atomic scale, the idea of "contact" breaks down. There is no way to define "contact" between two atoms—in other words, there is no unique distance between the atoms at which the forces they exert on one another suddenly become zero.

CHECKPOINT 4.1A

Identify the forces acting on the soccer player in Fig. 4.1. Describe each as (*type of force*) exerted on the player by (*object*).

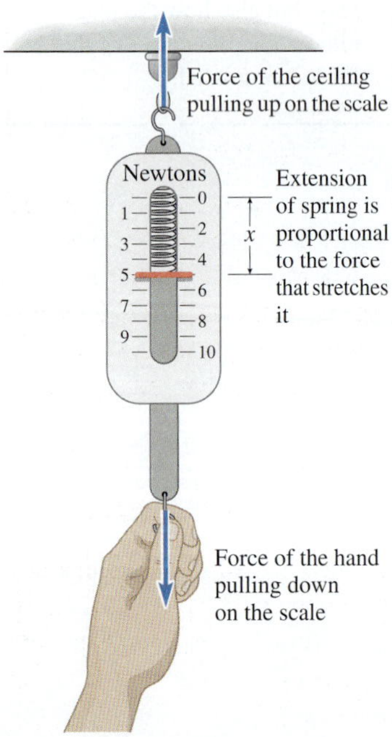

Figure 4.2 As the bottom of a spring scale is pulled downward, the spring stretches. We can measure the force by measuring the extension of the spring. For many springs, the extension is approximately proportional to the force, which makes calibration easy. Note that there is a pull on *both* ends of the scale. The ceiling pulls up on the scale and supports the scale from above. (Bathroom scales are similar, but they measure the *compression* of a spring.)

Measuring Forces

If the concept of force is to be useful in physics, there must be a way to measure forces. Consider a simple spring scale (Fig. 4.2). As the scale's pan is pulled down, a spring is stretched. The harder you pull, the more the spring stretches. As the spring stretches, an attached pointer moves. Then all we have to do to measure the applied force is to calibrate the scale so the amount of stretch measures the magnitude of the force. For many springs, the extension is approximately proportional to the force, which makes calibration easy.

In the United States, supermarket scales are generally calibrated to measure forces in pounds (lb). In the SI system, the unit of force is the **newton** (N). To convert pounds to newtons, use the approximate conversion factors

$$1 \text{ lb} = 4.448 \text{ N} \quad \text{or} \quad 1 \text{ N} = 0.2248 \text{ lb} \tag{4-1}$$

There are more sophisticated means for measuring forces than a supermarket scale. Even so, many operate on the same principle as the supermarket scale: a force is measured by the deformation—change of size or shape—it produces in some object.

Force Is a Vector Quantity

The magnitude of a force is *not* a complete description of the force. The *direction* of the force is equally important. The direction of the brief contact force exerted by a soccer player's foot on the ball can make the difference between scoring a goal or not. Force is a vector quantity that must be added (or subtracted) using the same methods used for other vector quantities such as position, velocity, and acceleration.

Example 4.1

Traction on a Foot

In a traction apparatus, three segments of a cord pull on the central pulley, each with magnitude 22.0 N, in the directions shown in Fig. 4.3. What is the sum of the forces exerted on the central pulley by the three cord segments? Give the magnitude and direction of the sum.

Strategy First, we sketch the graphical addition of the three forces to get an estimate of the magnitude and direction of the sum. Then, to get a precise answer, we resolve the three forces into their x- and y-components, sum the components, and then calculate the magnitude and direction of the sum.

Solution Figure 4.4 shows the graphical addition of the three forces exerted on the central pulley by the cord segments. From this sketch, we can tell that the sum of the three forces is at a relatively small angle above the horizontal (roughly half of 45°) and has a magnitude a bit larger than 44 N.

To find an algebraic solution, we find the components along the x- and y-axes and add them (Fig. 4.5). The x-components of the forces are

$$F_{1x} = F_{2x} = (22.0\text{ N}) \cos 45.0°$$
$$F_{3x} = (22.0\text{ N}) \cos 30.0°$$

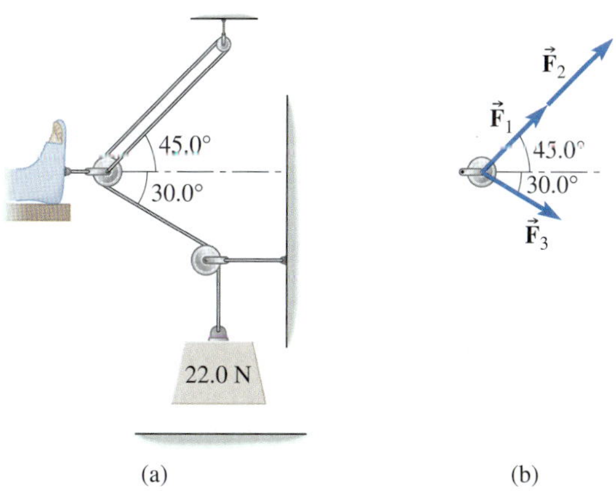

Figure 4.3
(a) A foot in traction; (b) the three forces exerted on the central pulley by the cord segments.

The y-components of the forces are

$$F_{1y} = F_{2y} = (22.0\text{ N}) \sin 45.0°$$
$$F_{3y} = (-22.0\text{ N}) \sin 30.0°$$

The sum of the x-components is

$$\begin{aligned} F_x &= F_{1x} + F_{2x} + F_{3x} \\ &= 2 \times (22.0\text{ N}) \cos 45.0° + (22.0\text{ N}) \cos 30.0° \\ &= 31.11\text{ N} + 19.05\text{ N} = 50.16\text{ N} \end{aligned}$$

We keep an extra decimal place for now to minimize round-off error. The sum of the y-components is

$$\begin{aligned} F_y &= F_{1y} + F_{2y} + F_{3y} \\ &= 2 \times (22.0\text{ N}) \sin 45.0° + (-22.0\text{ N}) \sin 30.0° \\ &= 31.11\text{ N} - 11.00\text{ N} = 20.11\text{ N} \end{aligned}$$

The magnitude of the sum is (Fig. 4.6):

$$F = \sqrt{F_x^2 + F_y^2} = \sqrt{(50.16\text{ N})^2 + (20.11\text{ N})^2} = 54.0\text{ N}$$

and the direction of the sum is

$$\theta = \tan^{-1}\frac{\text{opposite}}{\text{adjacent}} = \tan^{-1}\frac{20.11\text{ N}}{50.16\text{ N}} = 21.8°$$

The sum of the forces exerted on the pulley by the three cord segments is 54.0 N at an angle 21.8° above the $+x$-axis.

Discussion To check the answer, look back at the graphical estimate. The magnitude of the sum (54 N) is somewhat larger than 44 N and the direction is at an angle very nearly half of 45° above the horizontal.

Practice Problem 4.1 Changing the Pulley Angles

The pulleys are moved, after which $\vec{\mathbf{F}}_1$ and $\vec{\mathbf{F}}_2$ are at an angle of 30.0° above the x-axis and $\vec{\mathbf{F}}_3$ is 60.0° below the x-axis. (a) What is the sum of these three forces in component form? (b) What is the magnitude of the sum? (c) At what angle with the horizontal is the sum?

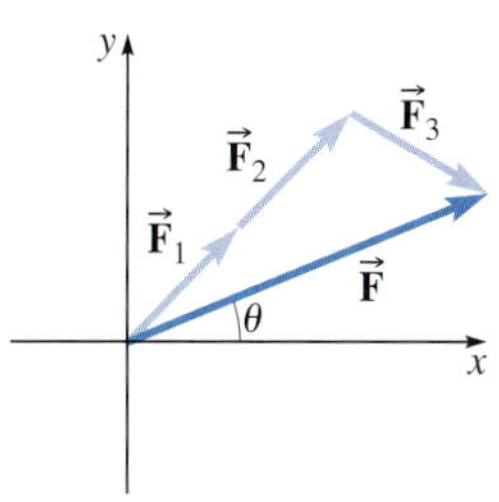

Figure 4.4
Graphical sum of the forces on the pulley due to the cord segments: $\vec{\mathbf{F}} = \vec{\mathbf{F}}_1 + \vec{\mathbf{F}}_2 + \vec{\mathbf{F}}_3$.

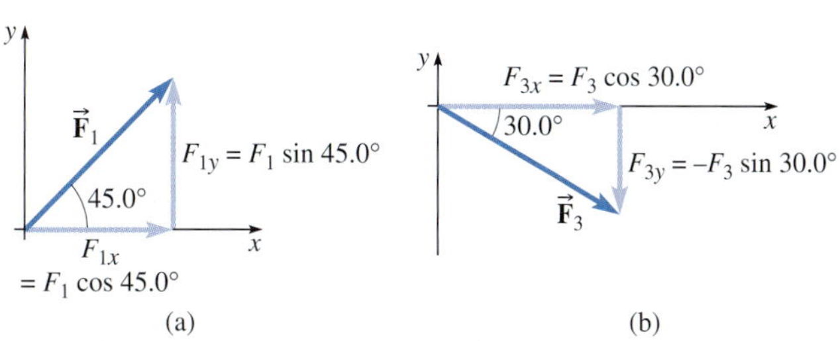

Figure 4.5
Using right triangles to find the components of (a) $\vec{\mathbf{F}}_1$ and (b) $\vec{\mathbf{F}}_3$. For clarity, the vector arrows are drawn twice as long as they were in Fig. 4.4.

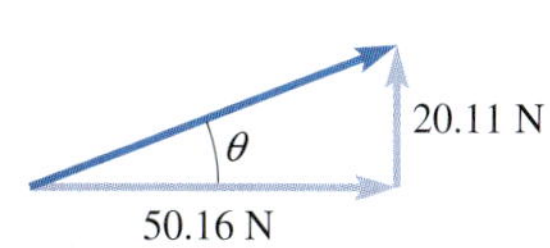

Figure 4.6
Finding the sum from its components. The magnitude is found using the Pythagorean theorem; the angle θ is found from the inverse tangent of opposite over adjacent.

CONNECTION:

In this chapter, we learn about a few kinds of forces. Later, when we learn about other forces, we always treat them the same: we add up *all* the forces acting on an object to find the net force.

Net Force

When more than one force acts on an object, the subsequent motion of the object is determined by the *net force* acting on the object. The **net force** is the vector sum of all the forces acting on an object.

Definition of net force

If $\vec{\mathbf{F}}_1, \vec{\mathbf{F}}_2, \ldots, \vec{\mathbf{F}}_n$ are *all* the forces acting on an object, then the net force $\vec{\mathbf{F}}_{\text{net}}$ acting on that object is the vector sum of those forces:

$$\vec{\mathbf{F}}_{\text{net}} = \sum \vec{\mathbf{F}} = \vec{\mathbf{F}}_1 + \vec{\mathbf{F}}_2 + \cdots + \vec{\mathbf{F}}_n \quad (4\text{-}2)$$

The symbol Σ is a capital Greek letter sigma that stands for "sum."

CHECKPOINT 4.1B

In Example 4.1, is the sum of the forces due to the three cord segments the *net* force on the central pulley?

Free-Body Diagrams

An essential tool used to find the net force acting on an object is a **free-body diagram** (FBD): a simplified sketch of a single object with force vectors drawn to represent *every* force *acting on that object*. (For example, the sum of three forces calculated in Example 4.1 is *not* the net force on the central pulley because the forces on the pulley due to the patient's leg and due to gravity are not included.) The net force must *not* include any forces that act on other objects. To draw an FBD:

- Draw the object in a simplified way—you don't have to be Michelangelo to solve physics problems! Almost any object can be represented as a box or a circle, or even a dot.
- Identify all the forces that are exerted on the object. Take care not to omit any forces that are exerted on the object. Consider that everything touching the object may exert one or more contact forces. Then identify long-range forces (for now, just gravity unless electric or magnetic forces are specified in the problem).
- Check your list of forces to make sure that each force is exerted *on* the object of interest *by* some other object. Make sure you have not included any forces that are exerted *on other objects*.
- Draw vector arrows representing all the forces acting on the object. We usually draw the vectors as arrows that start on the object and point away from it. Draw the arrows so they correctly illustrate the directions of the forces. If you have enough information to do so, draw the lengths of the arrows so they are proportional to the magnitudes of the forces.

Example 4.2

Net Force on an Airplane

The forces on an airplane in flight heading eastward are as follows: gravity = 16.0 kN (kilonewtons), downward; lift = 16.0 kN, upward; thrust = 1.8 kN, east; and drag = 0.8 kN, west. (Lift, thrust, and drag are three forces that the air exerts on the plane.) What is the net force on the plane?

continued on next page

Example 4.2 continued

Strategy All the forces acting on the plane are given in the statement of the problem. After drawing these forces in the FBD for the plane, we add the forces to find the net force. To resolve the force vectors into components, we choose x- and y-axes pointing east and up respectively. All four forces are then lined up with the axes, so each will have only one nonzero component, with a sign that indicates the direction along that axis. For example, the drag force points in the $-x$-direction, so its x-component is negative and its y-component is zero.

Solution Figure 4.7a is the FBD for the plane, using $\vec{\mathbf{L}}$, $\vec{\mathbf{T}}$, and $\vec{\mathbf{D}}$ for the lift, thrust, and drag, respectively. $\vec{\mathbf{W}}$ stands for the gravitational force on the plane; its magnitude is the plane's weight W. The sum of the x-components of the forces is

$$\begin{aligned}\sum F_x &= L_x + T_x + W_x + D_x \\ &= 0 + (1.8\text{ kN}) + 0 + (-0.8\text{ kN}) = 1.0\text{ kN}\end{aligned}$$

The sum of the y-components of the forces is

$$\begin{aligned}\sum F_y &= L_y + T_y + W_y + D_y \\ &= (16\text{ kN}) + 0 + (-16\text{ kN}) + 0 = 0\end{aligned}$$

The net force is 1.0 kN east.

Discussion A graphical check of the vector addition is a good idea. Figure 4.7b shows that the sum of the four forces is indeed in the $+x$-direction (east).

Practice Problem 4.2 New Forces on the Airplane

Find the net force on the airplane if the forces are gravity = 16.0 kN, downward; lift = 15.5 kN, upward; thrust = 1.2 kN, north; drag = 1.2 kN, south.

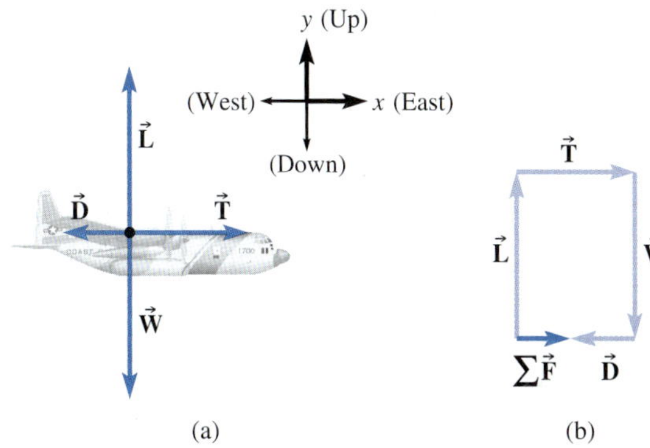

Figure 4.7
(a) Free-body diagram for the airplane. (b) Graphical addition of the four force vectors yields the net force, $\sum\vec{\mathbf{F}}$.

4.2 INERTIA AND EQUILIBRIUM: NEWTON'S FIRST LAW OF MOTION

In 1687, Isaac Newton (1643–1727) published one of the greatest scientific works of all time, his *Philosophiae Naturalis Principia Mathematica* (or *Principia* for short). The Latin title translates as *The Mathematical Principles of Natural Philosophy*. In the *Principia*, Newton stated three laws of motion that form the basis of classical mechanics. These laws describe how one or more forces acting on an object affect its motion and how the forces that interacting objects exert on one another are related.

Together with his law of universal gravitation, Newton's laws of motion showed for the first time that the motion of the heavenly bodies (the Sun, the planets, and their satellites) and the motion of earthly bodies can be understood using the same physical principles. To pre-Newtonian thinkers, it seemed that there must be two different sets of physical laws: one set to describe the motion of the heavenly bodies, thought to be perfect and enduring, and another to describe the motion of earthly bodies that always come to rest.

Newton's First Law of Motion

Newton's first law says that an object acted on by zero net force moves in a straight line with constant speed, or, if it is at rest, remains at rest. Using the concept of the velocity vector, which is a measure of both the speed *and the direction of motion* of an object, we can state the first law:

Newton's First Law of Motion

An object's velocity vector $\vec{\mathbf{v}}$ remains constant if and only if the net force acting on the object is zero.

This concise statement of Newton's first law includes both the case of an object at rest (zero velocity) and a moving object (nonzero velocity). Certainly it makes sense that an object at rest remains at rest unless some force acts on it to make it start to move. On the other hand, it may not be obvious that an object can continue to move without forces acting to keep it moving. In our experience, most moving objects come to rest because of forces that oppose motion, such as friction and air resistance. A hockey puck can slide the entire length of a rink with very little change in speed or direction because the ice is slippery (frictional forces are small). If we could remove *all* the resistive forces, including friction and air resistance, the puck would slide without changing its speed or direction at all.

No force is required to keep an object in motion if there are no forces opposing its motion. When a hockey player strikes the puck with his stick, the brief contact force exerted on the puck by the stick changes the puck's velocity, but once the puck loses contact with the stick, it continues to slide along the ice even though the stick no longer exerts a force on it.

Inertia Newton's first law is also called the **law of inertia.** In physics, **inertia** means resistance to *changes* in velocity. It does *not* mean resistance to the continuation of motion (or the tendency to come to rest). Newton based the law of inertia on the ideas of some of his predecessors, including Galileo Galilei (1564–1642) and René Descartes (1596–1650). In a series of clever experiments in which he rolled a ball up inclines of different angles, Galileo postulated that, if he could eliminate all resistive forces, a ball rolling on a horizontal surface would never stop (Fig. 4.8). Galileo made a brilliant conceptual leap from the real world with friction to an imagined, ideal world, free of friction. The law of inertia contradicted the view of the Greek philosopher Aristotle (384–322 B.C.E.). Almost 2000 years before Galileo, Aristotle had formulated his view that the natural state of an object is to be at rest; and, for an object to remain in motion, a force would have to act on it continuously. Galileo conjectured that, in the absence of friction and other resistive forces, an object in motion will continue to move even though no force is pushing or pulling it.

However, Galileo thought that the sustained motion of an object would be in a great circle around the Earth. Shortly after Galileo's death, Descartes argued that the motion of an object free of any forces should be along a straight line rather than a circle. Newton acknowledged his debt to Galileo, Descartes, and others when he wrote: "If I have seen farther, it is because I was standing on the shoulders of giants."

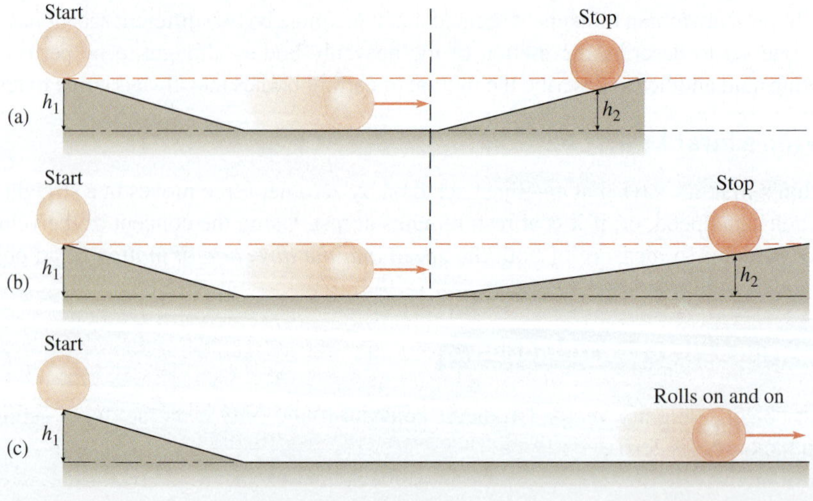

Figure 4.8 (a) Galileo found that a ball rolled down an incline stops when it reaches *almost* the same height on the second incline. He decided that it would reach the *same* height if resistive forces could be eliminated. (b) As the second incline is made less and less steep, the ball rolls farther and farther before stopping. (c) If the second incline is horizontal and there are no resistive forces, the ball would never stop.

Conceptual Example 4.3

Snow Shoveling

The task of shoveling newly fallen snow from the driveway can be thought of as a struggle against the inertia of the snow. Without the application of a net force, the snow remains at rest on the ground. However, there is an important way that the inertia of the snow makes it *easier* to shovel. Explain.

Strategy Think about the physical motions used when shoveling snow. (If you live where there is no snow, think about shoveling gravel from a wheelbarrow to line a garden path.) In order for the shoveling to be facilitated by the snow's inertia, there must be a time when the snow is moving on its own, without the shovel pushing it.

Solution and Discussion Imagine scooping up a shovelful of snow and swinging the shovel forward toward the side of the driveway. The snow and the shovel are both in motion. Then suddenly the forward motion of the shovel stops, but the snow continues to move forward because of its inertia; it slides forward off the shovel, to be pulled down to the ground by gravity. The snow does not stop moving forward when the forward force due to the shovel is removed.

This procedure works best with fairly dry snow. Wet sticky snow tends to cling to the shovel. The frictional force on the snow due to the shovel keeps it from moving forward and makes the job far more difficult. In this case, it might help to give the shovel a thin coating of cooking oil to reduce the frictional force the shovel exerts on the snow.

Conceptual Practice Problem 4.3 Inertia on the Subway

Emma, a college student, stands on a subway car, holding on to an overhead strap. As the train starts to pull out of the station, she feels thrust toward the rear of the car; as the train comes to a stop at the next station, she feels thrust forward. Explain the role played by inertia in this situation.

CHECKPOINT 4.2

The *Voyager* 1 and *Voyager* 2 space probes were launched in 1977. Between them, they explored all the large planets of the outer solar system (Jupiter, Saturn, Uranus, and Neptune) and 48 of their moons (Fig. 4.9).

The *Voyager* probes are now exploring the outer reaches of the solar system more than 14 billion kilometers from the Sun. They are heading out of the solar system at speeds of about 16 km/s, without being propelled by rockets or any other kind of engine. How can they continue to move at such high speeds for many years without an engine to drive them?

Figure 4.9 Io, one of the moons of Jupiter, as photographed by *Voyager* 1 from a distance of 862 000 km.

EVERYDAY PHYSICS DEMO

For an easy demonstration of inertia, place a quarter on top of an index card, or a credit card, balanced on top of a drinking glass (Fig. 4.10a). With your thumb and forefinger, flick the card so it flies out horizontally from under the quarter. What happens to the quarter? The horizontal force on the coin due to friction is small. With a negligibly small horizontal force, the coin tends to remain motionless while the card slides out from under it (Fig. 4.10b). Once the card is gone, gravity pulls the coin down into the glass (Fig. 4.10c).

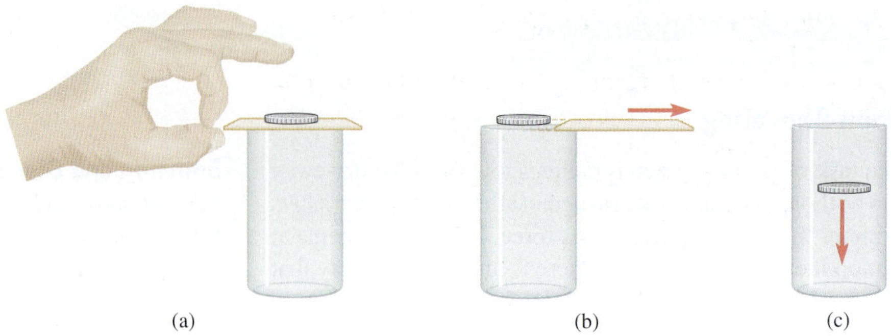

Figure 4.10 A demonstration of inertia. A similar demonstration that you may have seen is pulling a tablecloth out from under a table setting, leaving all the dishes and glasses in place. (If you want to try this, please practice first with plastic dishes, not your grandmother's china.)

An Object in Equilibrium Moves with Constant Velocity

When the net force acting on an object is zero, the object is said to be in **translational equilibrium**:

For an object in equilibrium,

$$\sum \vec{\mathbf{F}} = 0 \qquad \text{(4-3a)}$$

Equilibrium conveys the idea that the forces are in balance; there is as much force upward as there is downward, as much to the right as to the left, and so forth. *Translation* refers to motion without rotation. Any object moving with a constant velocity, whether at rest or moving in a straight line at constant speed, is in translational equilibrium. A vector can only have zero magnitude if all of its components are zero, so

For an object in equilibrium,

$$\sum F_x = 0 \quad \text{and} \quad \sum F_y = 0 \quad \left(\text{and} \quad \sum F_z = 0\right) \qquad \text{(4-3b)}$$

In an equilibrium problem, choose x- and y-axes so the fewest number of force vectors have both x- and y-components. It is always good practice to make a conscious *choice* of axes and then to draw them in the FBDs and any other sketches that you make in solving the problem.

Example 4.4

Sliding a Chest

In order to slide a chest that weighs 750 N across the floor at constant velocity, you must push it horizontally with a force of 450 N (Fig. 4.11). Find the contact force that the floor exerts on the chest.

Figure 4.11 Sliding a chest across the floor.

Strategy The chest moves with constant velocity, so it is in equilibrium. The net force acting on it is zero. We will identify all the forces acting on the chest, draw an FBD, do a graphical addition of the forces, choose x- and y-axes, resolve the forces into their x- and y-components, and then set $\Sigma F_x = 0$ and $\Sigma F_y = 0$.

Solution There are three forces acting on the chest. The gravitational force $\vec{\mathbf{W}}$ has magnitude 750 N and is directed downward. Your push $\vec{\mathbf{F}}$ has magnitude 450 N and its direction is horizontal. The contact force due to the floor $\vec{\mathbf{C}}$ has unknown magnitude and direction. However, remembering that the chest is in equilibrium, upward and downward force components must balance, as must the horizontal force

continued on next page

Example 4.4 continued

components. Therefore, $\vec{\mathbf{C}}$ must be roughly in the direction shown in the FBD (Fig. 4.12a), as is confirmed by adding the three forces graphically (Fig. 4.12b). The sum is zero because the tip of the last vector ends up at the tail of the first one.

Choosing the x-axis to the right and the y-axis up means that two of the three force vectors, $\vec{\mathbf{W}}$ and $\vec{\mathbf{F}}$, have one component that is zero:

$$W_x = 0 \quad \text{and} \quad W_y = -750\text{ N}$$
$$F_x = 450\text{ N} \quad \text{and} \quad F_y = 0$$

Now we set the x- and y-components of the net force each equal to zero because the chest is in equilibrium.

$$\sum F_x = W_x + F_x + C_x = 0 + 450\text{ N} + C_x = 0$$
$$\sum F_y = W_y + F_y + C_y = -750\text{ N} + 0 + C_y = 0$$

These equations tell us the components of $\vec{\mathbf{C}}$: $C_x = -450$ N and $C_y = +750$ N. Then the magnitude of the contact force is (Fig. 4.12c)

$$C = \sqrt{C_x^2 + C_y^2} = \sqrt{(-450\text{ N})^2 + (750\text{ N})^2} = 870\text{ N}$$

$$\theta = \tan^{-1}\frac{\text{opposite}}{\text{adjacent}} = \tan^{-1}\frac{750\text{ N}}{450\text{ N}} = 59°$$

The contact force due to the floor is 870 N, directed 59° above the leftward horizontal ($-x$-axis).

Note that we didn't need to know any details about contact forces to solve this problem. We explore contact forces in more detail in Section 4.6.

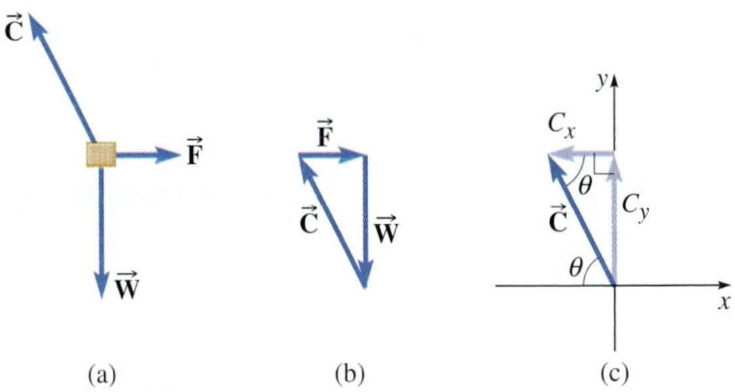

Figure 4.12

(a) An FBD for the chest; (b) graphical addition of the three forces showing that the sum is zero. (c) Finding the magnitude and direction of the contact force.

Discussion The x- and y-components of the contact force and its magnitude and direction are all reasonable based on the graphical addition, so we can be confident that we did not make an error such as a sign error with one of the components.

Practice Problem 4.4 The Chest at Rest

Suppose the same chest is at rest. You push it horizontally with a force of 110 N but it does not budge. What is the contact force on the chest due to the floor during the time you are pushing?

Application: Spring Scale

Using Newton's first law, we can understand how a spring scale can be used to measure weight (the magnitude of the gravitational force exerted on an object). If a melon remains at rest in the pan of the scale, the net force on the melon must be zero. There are only two forces acting on the melon: gravity pulls down and the scale pulls up. Then these two forces must be equal in magnitude and opposite in direction. The scale measures the magnitude of the force it exerts on the melon, which is equal to the weight of the melon.

4.3 NET FORCE, MASS, AND ACCELERATION: NEWTON'S SECOND LAW OF MOTION

When a *nonzero* net force acts on an object, the object's velocity changes. Newton's second law says that the *rate of change of the object's velocity*—that is, the object's acceleration—is proportional to the net force acting on it and inversely proportional to its mass:

Newton's Second Law of Motion

$$\vec{\mathbf{a}} = \frac{1}{m}\sum\vec{\mathbf{F}} \quad \text{or} \quad \sum\vec{\mathbf{F}} = m\vec{\mathbf{a}} \tag{4-4}$$

If the net force is zero, then the acceleration is zero, in accordance with Newton's first law. If the net force is not zero, then the acceleration has the same direction as the net force. When the net force is constant, the acceleration is also constant. In component form, Newton's second law is

$$\sum F_x = ma_x \quad \text{and} \quad \sum F_y = ma_y \tag{4-5}$$

If all the forces acting on an object are known, then Eq. (4-4) can be used to calculate its acceleration. Alternatively, sometimes we know the object's acceleration but we have incomplete information about the forces acting on it; then Eq. (4-4) provides information about the unknown forces.

SI Unit of Force

The SI unit of force, the newton, is *defined* so that a net force of 1 N gives a 1 kg mass an acceleration of 1 m/s^2:

$$1 \text{ N} = 1 \text{ kg·m/s}^2 \tag{4-6}$$

Defining the unit of force in this way makes it possible to write Eqs. (4-4) and (4-5) without needing a constant of proportionality to convert between the force unit and kg·m/s^2.

What Is Mass?

The acceleration of an object is proportional to the net force on it and is in the same direction (Fig. 4.13). A larger net force causes a more rapid change in the velocity vector. Newton's second law also says that the acceleration is inversely proportional to the object's mass. The same net force acting on two different objects causes a smaller acceleration on the object with greater mass (Fig. 4.14). Mass is a measure of an object's inertia—the amount of resistance to *changes in velocity*. Newton's second law serves as our *definition* of mass.

In everyday language mass and weight are sometimes used as synonyms, but in physics, mass and weight are different physical properties. The *mass* of an object is a measure of its inertia, but its *weight* is the magnitude of the gravitational force acting on it. Imagine taking a shuffleboard puck to the Moon. Since the Moon's surface gravity is weaker than the Earth's, the puck's weight would be smaller on the Moon, but the puck's *mass* would be the same as on Earth. Ignoring the effects of friction, an astronaut playing shuffleboard on the Moon would have to exert the same horizontal force on the puck as on Earth to give it the same acceleration (Fig. 4.15).

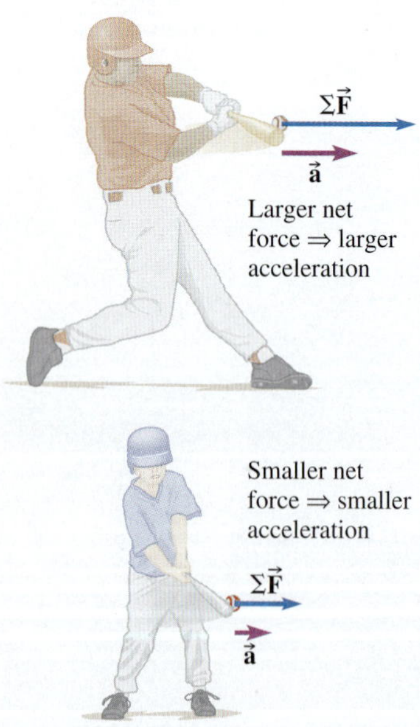

Figure 4.13 The acceleration of a baseball is proportional to the net force acting on it.

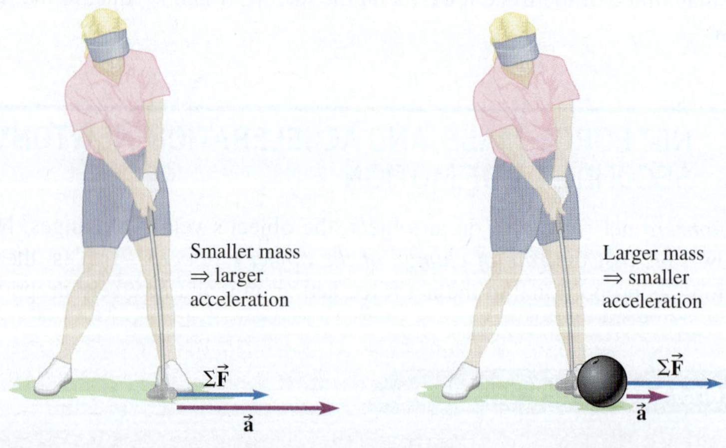

Figure 4.14 The same net force acting on two different objects produces accelerations in inverse proportion to the masses.

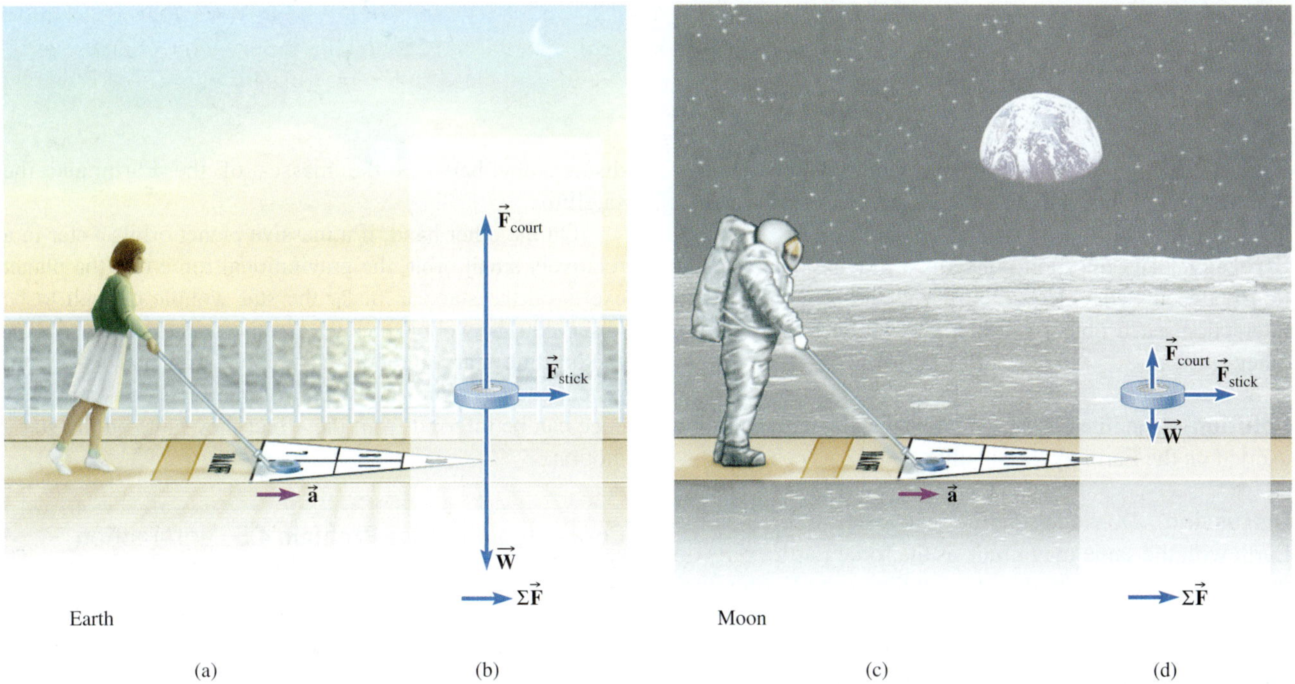

Figure 4.15 An astronaut playing shuffleboard (a) on Earth and (c) on the Moon. FBDs for a puck of mass m being given the same push ($\vec{F}_{stick}$) on a *frictionless* court on (b) Earth and (d) on the Moon. The acceleration ($\vec{a}$) of the puck must be the same since the mass of the puck is the same: $\Sigma\vec{F} = \vec{F}_{stick} = m\vec{a}$.

4.4 INTERACTION PAIRS: NEWTON'S THIRD LAW OF MOTION

In Section 4.1, we learned that forces always exist in pairs. Every force is part of an interaction between two objects and each of the interacting objects exerts a force on the other. We call the two forces an **interaction pair**; each force is the **interaction partner** of the other. When you push open a door, the door pushes you. When two cars collide, each exerts a force on the other. Note that interaction partners *always act on different objects*—the two objects that are interacting.

Newton's third law of motion says that interaction partners always have the *same magnitude* and are in *opposite directions*.

Newton's Third Law of Motion

In an interaction between two objects, each object exerts a force on the other. These two forces are equal in magnitude and opposite in direction.

Equivalently, we can write

$$\vec{F}\text{ (on } B \text{ by } A) = -\vec{F}\text{ (on } A \text{ by } B) \tag{4-7}$$

Do not assume that Newton's third law is involved *every* time two forces *happen* to be equal and opposite—*it ain't necessarily so!* You will encounter many situations in which two equal and opposite forces act *on a single object*. These forces cannot be *interaction partners* because they act on the same object. Interaction partners act on *different objects*, one on each of the two objects that are interacting. Note also that interaction partners are always of the same type (both gravitational, or both magnetic, or both frictional, etc.).

We will use Newton's third law frequently when analyzing forces. For instance, in Conceptual Example 4.9, Newton's third law is used to analyze forces that act when a horse pulls a sleigh.

Conceptual Example 4.5

An Orbiting Satellite

Earth exerts a gravitational force on an orbiting communications satellite. What is the interaction partner of this force?

Strategy The question concerns a gravitational interaction between two objects: Earth and the satellite. In this interaction, each object exerts a gravitational force on the other.

Solution The interaction partner is the gravitational force exerted on the Earth by the satellite.

Discussion Does the satellite really exert a force on the Earth with the same magnitude as the force Earth exerts on the satellite? If so, why does the satellite orbit Earth rather than Earth orbiting the satellite? Newton's third law says that the interaction partners are equal in magnitude, but does not say that these two forces have equal *effects*. The effect of a net force on an object's motion depends on the object's mass. These two forces of equal magnitude have vastly different effects due to the great discrepancy between the masses of the Earth and the satellite.

On the other hand, if a massive planet orbits a star in a relatively small orbit, the gravitational force that the planet exerts on the star can make the star wobble enough to be observed. The wobble enables astronomers to discover planets orbiting stars other than the Sun. The planets do not reflect enough light toward Earth to be seen, but their presence can be inferred from the effect they have on the star's motion.

Conceptual Practice Problem 4.5 Interaction Partner of a Surface Contact Force

In Example 4.4, the contact force exerted on the chest by the floor was 870 N, directed 59° above the leftward horizontal ($-x$-axis). Describe the interaction partner of this force—in other words, what object exerts it on what other object? What are the magnitude and direction of the interaction partner?

CHECKPOINT 4.4

In the photo, two children are pulling on a toy. If they are exerting equal and opposite forces on the toy, are these two forces interaction partners? Why or why not?

EVERYDAY PHYSICS DEMO

The next time you go swimming, notice that you use Newton's third law to get the water to push you forward. When you push down and backward on the water with your arms and legs, the water pushes up and forward on you. The various swimming strokes are devised so that you exert as large a force as possible backward on the water during the power part of the stroke, and then as small a force as possible forward on the water during the return part of the stroke. Notice a similar effect when you are walking, skating, or skiing. To get the ground to push you forward, your feet push backward on the ground. Conceptual Example 4.9 explores these forces in more detail.

The forces exerted by these two children on a toy cannot be interaction partners because they act on the same object (the toy). The interaction of the force exerted by a child on the toy is the force that the *toy* exerts *on that child.*

Internal and External Forces

When we say that a soccer ball interacts with the Earth (gravity), with a player's foot, and with the air, we are treating the ball as a single entity. But the ball really consists of an enormous number of protons, neutrons, and electrons, all interacting with each other. The protons and neutrons interact with each other to form atomic nuclei; the nuclei interact with electrons to form atoms; interactions between atoms form molecules; and the molecules interact to form the structure of the thing we call a soccer ball. It would be difficult to have to deal with all of these interactions to predict the motion of a soccer ball.

Defining a System Let us call the set of particles that constitute the soccer ball a **system**. Once we have defined a system, we can classify all the interactions that affect the system as either **internal** or **external** to the system. For an internal interaction, *both* interacting objects are part of the system. When we add up all the forces acting on the system to find the net force, every internal interaction contributes two forces—an interaction pair—that always add to zero. For an external interaction, *only one of the two interaction partners is exerted on the system*. The other partner is exerted on an object outside the system and does not contribute to the net force on the system. Therefore, to find the net force on the system, we can ignore all the internal forces and just add the external forces.

The insight that internal forces always add to zero is particularly powerful because the choice of what constitutes a system is completely arbitrary. We can choose *any* set of objects and define it to be a system. In one problem, it may be convenient to think of the soccer ball as a system; in another, we may choose a system consisting of both the soccer ball and the player's foot. The second choice might be useful if we do not have detailed information about the interaction between the foot and the ball.

4.5 GRAVITATIONAL FORCES

Newton's Law of Universal Gravitation

Now we turn our attention to learning about a few forces in more detail, beginning with gravity. According to **Newton's law of universal gravitation**, any two objects exert gravitational forces on each other that are proportional to the masses (m_1 and m_2) of the two objects and inversely proportional to the square of the distance (r) between their centers. Strictly speaking, the law of gravitation as presented here applies only to point particles and symmetrical spheres. (The *point particle* is a common model in physics used when the size of an object is negligibly small and the internal structure is irrelevant.) Nevertheless, the law of gravitation is *approximately* true for any two objects if the distance between their centers is large compared with their sizes.

In mathematical language, the magnitude of the gravitational force is written:

Magnitude of the Gravitational Force

$$F = \frac{Gm_1m_2}{r^2} \qquad \text{(4-8)}$$

The constant of proportionality ($G = 6.674 \times 10^{-11}\ \text{N}\cdot\text{m}^2/\text{kg}^2$) is called the **universal gravitational constant**. Equation (4-8) is only part of the law of universal gravitation because it gives only the magnitudes of the gravitational forces that each object exerts on the other. The directions are equally important: each object is pulled toward the other's center (Fig. 4.16). In other words, gravity is an attractive force. The forces on the two objects are equal in magnitude and the directions are opposite, as they must be since they form an interaction pair.

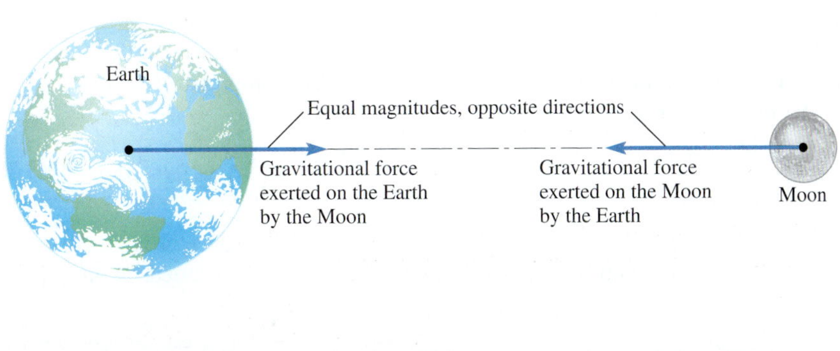

Figure 4.16 Gravity is always an attractive force. The force that each body exerts on the other is equal in magnitude, even though the masses may be very different. The force exerted *on the Moon by the Earth* is of the same magnitude as the force exerted on the Earth by the Moon. The directions are opposite.

Gravitational forces exerted *by* ordinary objects on each other are so small as to be negligible in most cases (see Practice Problem 4.6). Gravitational forces exerted by Earth, on the other hand, are much larger due to Earth's large mass.

Example 4.6

Weight at High Altitude

When you are in a commercial airliner cruising at an altitude of 6.4 km, by what percentage has your weight (as well as the weight of the airplane) changed compared with your weight on the ground?

Strategy Your weight is the magnitude of Earth's gravitational force exerted on you. Newton's law of universal gravitation gives the magnitude of the gravitational force at a distance r from the center of the Earth. For your weight on the ground W_1, we can use the mean radius of the Earth R_E (listed in the inside back cover of the book) as the distance between the Earth's center and you: $r_1 = R_E = 6.37 \times 10^6$ m (Fig. 4.17). At an altitude of $h = 6.4 \times 10^3$ m above the surface, your weight is W_2 and your distance from Earth's center is $r_2 = R_E + h$. Your mass m, the mass of the Earth M_E ($= 5.97 \times 10^{24}$ kg), and G are the same in the two cases, so it is efficient to write a ratio of the weights and let those factors cancel out.

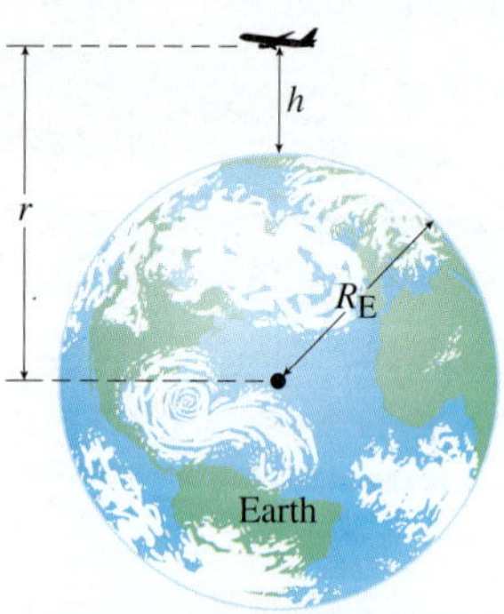

Figure 4.17 The gravitational force depends on the distance r to the *center* of the Earth. At an altitude h, $r = R_E + h$.

Solution The ratio of your weight in the airplane to your weight on the ground is

$$\frac{W_2}{W_1} = \frac{\dfrac{GM_E m}{r_2^2}}{\dfrac{GM_E m}{r_1^2}} = \frac{r_1^2}{r_2^2} = \frac{R_E^{\ 2}}{(R_E + h)^2} = \frac{1}{(1 + h/R_E)^2}$$

$$= \frac{1}{(1 + 0.001\,004\,7)^2} = 0.9980$$

Since $0.9980 = 1 - 0.0020$ and $0.0020 = 0.20/100$, your weight decreases by 0.20%.

Discussion Although 6400 m may seem like a significant altitude to us, it's a small fraction of the Earth's radius (0.10%), so the weight change is a small percentage. When judging whether a quantity is small or large, always ask: "Small (or large) compared to what?"

Practice Problem 4.6 A Creative Defense

After an automobile collision, one driver claims that the gravitational force between the two cars caused the collision. Estimate the magnitude of the gravitational force exerted by one car on another when they are driving side-by-side in parallel lanes and comment on the driver's claim.

Gravitational Field Strength

For an object near Earth's surface, the distance between the object and the Earth's center is very nearly equal to the Earth's mean radius, $R_E = 6.37 \times 10^6$ m. The mass of the Earth is $M_E = 5.97 \times 10^{24}$ kg, so the weight of an object of mass m near Earth's surface is

$$W = \frac{GM_E m}{R_E^2} = m\left(\frac{GM_E}{R_E^2}\right) \qquad (4\text{-}9)$$

Notice that for objects near Earth's surface, the constants in the parentheses are always the same and the weight of the object is proportional to its mass. Rather than recalculate that combination of constants over and over, we call the combination the **gravitational field strength** g near Earth's surface:

$$g = \frac{GM_E}{R_E^2} = \frac{6.674 \times 10^{-11}\ \text{N}\cdot\text{m}^2\cdot\text{kg}^{-2} \times (5.97 \times 10^{24}\ \text{kg})}{(6.37 \times 10^6\ \text{m})^2} \approx 9.8\ \text{N/kg} \qquad (4\text{-}10)$$

The units *newtons per kilogram* reinforce the conclusion that weight is proportional to mass: g tells us how many newtons of gravitational force are exerted on an object for every kilogram of the object's mass. The weight of a 1.0 kg object near Earth's surface is 9.8 N (2.2 lb). Using g, the weight of an object of mass m near Earth's surface is usually written

Relationship between mass and weight

$$W = mg \tag{4-11a}$$

In vector form:

$$\vec{\mathbf{W}} = m\vec{\mathbf{g}} \tag{4-11b}$$

In Eq. (4-11b), $\vec{\mathbf{W}}$ stands for the gravitational force and $\vec{\mathbf{g}}$ is called the **gravitational field**; the direction of both is downward. The italic (scalar) symbol g is the *magnitude* of a vector, so its value is *never negative*.

Variations in Earth's Gravitational Field The Earth is not a perfect sphere; it is slightly flattened at the poles. Since the distance from the surface to the center of the Earth is smaller there, the field strength at sea level is greatest at the poles (9.832 N/kg) and smallest at the equator (9.814 N/kg). Altitude also matters; as you climb above sea level, your distance from Earth's center increases and the field strength decreases. Tiny local variations in the field strength are also caused by geologic formations. On top of dense bedrock, g is a little greater than above less dense rock. Geologists and geophysicists measure these variations to study Earth's structure and also to locate deposits of various minerals, water, and oil. The device they use, a *gravimeter*, is essentially a mass hanging on a spring. As the gravimeter is carried from place to place, the extension of the spring increases where g is larger and decreases where g is smaller. The mass hanging from the spring does not change, but its weight does ($W = mg$).

Furthermore, due to Earth's rotation, the *effective* value of g that we measure in a coordinate system attached to Earth's surface is slightly less than the true value of the field strength. This effect is greatest at the equator, where the effective value of g is 9.784 N/kg, about 0.3% smaller than the true value of g. The effect gradually decreases with latitude to zero at the poles. We learn more about this effect in Chapter 5.

The most important thing to remember from this discussion is that, unlike G, g is *not* a universal constant. The value of g is a function of position. Near Earth's surface, the variations are small, so we can adopt an average value $g = 9.80$ N/kg as a default.

Gravitational Field and Free-Fall Acceleration

An object in free fall is assumed to have only one force acting on it: gravity. Other forces, such as air resistance, must be negligibly small for this approximation to be valid. We can write the gravitational force on the object as $\vec{\mathbf{W}} = m\vec{\mathbf{g}}$ where the gravitational field vector $\vec{\mathbf{g}}$ has magnitude g and is directed downward (in the direction of the gravitational force). From Newton's second law,

$$\vec{\mathbf{F}}_{\text{net}} = m\vec{\mathbf{g}} = m\vec{\mathbf{a}}$$

Dividing by the mass yields

$$\vec{\mathbf{a}} = \vec{\mathbf{g}} \tag{4-12}$$

Therefore, the acceleration of an object in free fall is $\vec{\mathbf{g}}$ regardless of the object's mass. Since 1 N = 1 kg·m/s^2, 9.80 N/kg = 9.80 m/s^2—the magnitude of the free-fall acceleration near Earth's surface has average value 9.80 m/s^2.

More massive objects have the same free-fall acceleration as less massive objects. True, a more massive object is harder to accelerate: the acceleration of an object subjected to a given force is inversely proportional to its mass. However, the stronger gravitational force on a more massive object compensates for its greater inertia, giving it the same free-fall acceleration as a less massive object.

Gravitational Field Strength on Other Planets

Equation (4-11) can be used to find the weight of an object at or above the surface of *any* planet or moon, but the value of g will be different due to the different mass M of the planet or moon and the different distance r from the planet's center:

$$g = \frac{GM}{r^2} \tag{4-13}$$

For instance, by substituting the mass and radius of Mars into Eq. (4-13), we find that $g = 3.7$ N/kg on the surface of Mars.

CHECKPOINT 4.5

If you climb Mt. McKinley, what happens to the weight of your gear? What happens to its mass?

Example 4.7

"Weighing" Figs in Kilograms

In most countries other than the United States, produce is sold in mass units (grams or kilograms) rather than in force units (pounds or newtons). The scale still measures a force, but the scale is calibrated to show the mass of the produce instead of its weight. What is the weight of 350 g of fresh figs, in newtons and in pounds?

Strategy Weight is mass times the gravitational field strength. We will assume $g = 9.80$ N/kg. The weight in newtons can be converted to pounds using the conversion factor 1 N = 0.2248 lb.

Solution The weight of the figs in newtons is

$$W = mg = 0.35 \text{ kg} \times 9.80 \text{ N/kg} = 3.43 \text{ N}$$

Converting to pounds,

$$W = 3.43 \text{ N} \times 0.2248 \text{ lb/N} = 0.771 \text{ lb}$$

The figs weigh 3.4 N or 0.77 lb.

Discussion This is the weight of the figs at a location where g has its average value of 9.80 N/kg. The figs would weigh a little more in the northern city of St. Petersburg, Russia, where g is larger, and a little less in Quito, Ecuador, where g is smaller.

Practice Problem 4.7 Figs on the Moon

What would those figs weigh on the surface of the Moon, where $g = 1.62$ N/kg?

CONNECTION:

In Example 4.4, we resolved the contact force on a sliding chest into components perpendicular to and parallel to the contact surface. It is often convenient to think of these components as two separate but related contact forces: the *normal force* and the *frictional force*.

4.6 CONTACT FORCES

We have already solved some problems involving forces exerted between two solid objects in contact. Now we look at contact forces in more detail.

Normal Force

A contact force perpendicular to the contact surface that prevents two solid objects from passing through one another is called the **normal force**. (In geometry, the word *normal* means *perpendicular*.) Consider a book resting on a horizontal table surface. The normal force due to the table must have just the right magnitude to keep the book from falling through the table. If no other vertical forces act, the normal force on the book is equal in magnitude to the book's weight because the book is in equilibrium (Fig. 4.18a).

According to Newton's third law, two objects in contact exert equal and opposite normal forces on one another; each pushes the other away. In our example, a downward normal force is exerted on the table by the book. In *everyday* language, we might say that the table "feels the book's weight." That is not an accurate statement in the

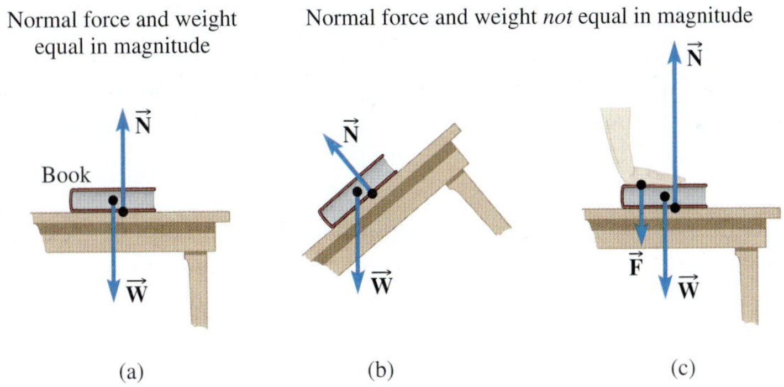

Figure 4.18 (a) The normal force is equal in magnitude to the weight of the book; the two forces sum to zero. (b) On an incline, the normal force is smaller than the weight of the book and is *not vertical*. (c) If you push down on the book ($\vec{F}$) the normal force on the book due to the table is larger than the book's weight.

language of physics. The table cannot "feel" the gravitational force on the book; the table can only feel forces exerted *on the table*. What the table does "feel" is the normal force—a *contact* force—exerted on the table by the book.

If the table's surface is horizontal, the normal force on the book will be vertical and equal in magnitude to the book's weight. If the surface of the table is *not* horizontal, the normal force is not vertical and is not equal in magnitude to the weight of the book. Remember that the normal force is *perpendicular to the contact surface* (Fig. 4.18b). Even on a horizontal surface, if there are other vertical forces acting on the book, then the normal force is *not* equal in magnitude to the book's weight (Fig. 4.18c). Never *assume* anything about the magnitude of the normal force. In general, we can figure out what the magnitude of the normal force must be in various situations if we have enough information about other forces.

What Causes Normal Forces How does the table "know" how hard to push on the book? First imagine putting the book on a bathroom scale instead of the table. A spring inside the scale provides the upward force. The spring "knows" how hard to push because, as it is compressed, the force it exerts increases. When the book reaches equilibrium, the spring is exerting just the right amount of force, so there is no tendency to compress it further. The spring is compressed until it pushes up with a force equal to the book's weight. If the spring were stiffer, it would exert the same upward force but with less compression.

The forces that bind atoms together in a rigid solid, like the table, act like extremely stiff springs that can provide large forces with little compression—so little that it's usually not noticed. The book makes a tiny indentation in the surface of the table (Fig. 4.19); a heavier book would make a slightly larger indentation. If the book were to be placed on a soft foam surface, the indentation would be much more noticeable.

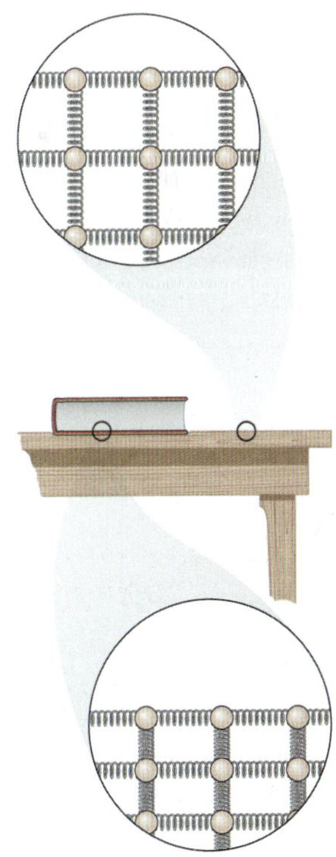

Figure 4.19 The book compresses the "atomic springs" in the table until they push up on the book to hold it up. The slight decrease in the distance between atoms is greatly exaggerated here.

CHECKPOINT 4.6

Your laptop is resting on the surface of your desk, which stands on four legs on the floor. Identify the normal forces acting on the desk and give their directions.

Friction

A contact force *parallel* to the contact surface is called **friction**. We distinguish two types: **static friction** and **kinetic** (or **sliding**) **friction**. When the two objects are slipping or sliding across one another, as when a loose shingle slides down a roof, the friction is kinetic. When no slipping or sliding occurs, such as between the tires of a car parked on a hill and the road surface, the friction is called static. Static friction acts to prevent objects from *starting* to slide; kinetic friction acts to try to make sliding objects stop sliding. Note that two objects in contact with one another that move with the same velocity exert *static* frictional forces on one another, because there is no *relative* motion between the two. For example, if a conveyor belt carries an air freight package up an incline and the package is not sliding, the two move with the same velocity and the friction is *static*.

Static Friction Frictional forces are complicated on the microscopic level and are an active field of current research. Despite the complexities, we can make some approximate statements about the frictional forces between dry, solid surfaces. In a simplified model, the maximum magnitude of the force of static friction $f_{s,max}$ that can occur in a particular situation is proportional to the magnitude of the normal force N acting between the two surfaces.

$$f_{s,max} \propto N$$

If you want better traction between the tires of a rear-wheel-drive car and the road, it helps to put something heavy in the trunk to increase the normal force between the tires and the road.

The constant of proportionality is called the **coefficient of static friction** (symbol μ_s):

Maximum force of static friction

$$f_{s,max} = \mu_s N \tag{4-14}$$

Since $f_{s,max}$ and N are both magnitudes of forces, μ_s is a dimensionless number. Its value depends on the condition and nature of the surfaces. Equation (4-14) provides only an *upper limit* on the force of static friction in a particular situation. The actual force of friction in a given situation is not necessarily the maximum possible. It tells us only that, if sliding does not occur, the magnitude of the static frictional force is less than or equal to this upper limit:

$$f_s \le \mu_s N \tag{4-15}$$

Kinetic (Sliding) Friction For sliding or kinetic friction, the force of friction is only weakly dependent on the speed and is roughly proportional to the normal force. In the simplified model we will use, the force of kinetic friction is assumed to be proportional to the normal force and independent of speed:

Force of kinetic (sliding) friction

$$f_k = \mu_k N \tag{4-16}$$

where f_k is the magnitude of the force of kinetic friction and μ_k is called the **coefficient of kinetic friction**. The coefficient of static friction is always larger than the coefficient of kinetic friction for an object on a given surface. On a horizontal surface, a larger force is required to start the object moving than is required to keep it moving at a constant velocity.

Direction of Frictional Forces Equations (4-14) through (4-16) relate only the *magnitudes* of the frictional and normal forces on an object. Remember that the frictional force is perpendicular to the normal force between the same two surfaces. Friction is always parallel to the contact surface, but there are many directions parallel to a given contact surface. Here are some rules of thumb for determining the direction of a frictional force.

- The static frictional force acts in whatever direction necessary to prevent the objects from beginning to slide or slip relative to each other.

- Kinetic friction acts in a direction that tends to make the sliding or slipping stop. If a book slides to the left along a table, the table exerts a kinetic frictional force on the book to the right, in the direction opposite to the motion of the book.
- From Newton's third law, frictional forces come in interaction pairs. If the table exerts a frictional force on the sliding book to the right, the book exerts a frictional force on the table to the *left* with the same magnitude.

Example 4.8

Coefficient of Kinetic Friction for the Sliding Chest

Example 4.4 involved sliding a 750 N chest to the right at constant velocity by pushing it with a horizontal force of 450 N. We found that the contact force on the chest due to the floor had components $C_x = -450$ N and $C_y = +750$ N, where the x-axis points to the right and the y-axis points up (Fig. 4.20). What is the coefficient of kinetic friction for the chest-floor surface?

Strategy To find the coefficient of friction, we need to know what the normal and frictional forces are. They are the components of the contact force that are perpendicular and parallel to the contact surface. Since the surface is horizontal (in the x-direction), the x-component of the contact force is friction and the y-component is the normal force.

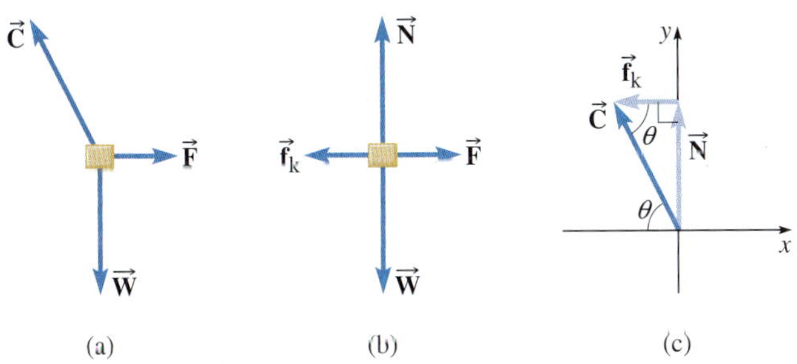

Figure 4.20

(a) FBD for the chest. $\vec{C}$ is the contact force due to the floor.
(b) FBD in which the contact force is replaced by two perpendicular forces, the normal force $\vec{N}$ and the kinetic frictional force $\vec{f}_k$.
(c) Resolving $\vec{C}$ into normal and frictional components.

Solution The magnitude of the force due to sliding friction is $f_k = |C_x| = 450$ N. The magnitude of the normal force is $N = |C_y| = 750$ N. Now we can calculate the coefficient of kinetic friction from $f_k = \mu_k N$:

$$\mu_k = \frac{f_k}{N} = \frac{450\text{ N}}{750\text{ N}} = 0.60$$

Discussion If we had written $f_k = C_x = -450$ N, we would have ended up with a negative coefficient of friction. The coefficient of friction is a relationship between the *magnitudes* of two forces, so it cannot be negative.

Practice Problem 4.8 Chest at Rest

Suppose the same chest is at rest. You push to the right with a force of 110 N but the chest does not budge. What are the normal and frictional forces on the chest due to the floor while you are pushing? Explain why you do not need to know the coefficient of static friction to answer this question.

Conceptual Example 4.9

Horse, Sleigh, and Newton's Third Law

A horse pulls a sleigh to the right at constant velocity on level ground. The horse exerts a horizontal force $\vec{F}_{sh}$ on the sleigh. (The subscripts indicate the force on the sleigh due to the horse.) (a) Draw three FBDs, one for the horse, one for the sleigh, and one for the system horse + sleigh. (b) To make the sleigh increase its velocity, there must be a nonzero net force to the right acting on the sleigh. Suppose the horse pulls harder (F_{sh} increases in magnitude). According to Newton's third law, the sleigh always pulls back on the horse with a force of *the same* magnitude as the force with which the horse pulls the sleigh. Does this mean that no matter how hard it pulls, the horse can never make the net force on the sleigh nonzero? Explain. (c) Identify the interaction partner of each force acting on the sleigh.

Strategy (a) In each FBD, we include only the *external* forces acting on that system. All three systems move with constant velocity, so the net force on each is zero. (b) Looking at the FBD for the sleigh, we can determine the conditions under which the net force on the sleigh can be nonzero. (c) For a force exerted on the sleigh by X, its interaction partner must be the force exerted on X by the sleigh.

continued on next page

Conceptual Example 4.9 continued

Solution and Discussion (a) If we think of the normal and frictional forces as separate forces, then there are four forces acting on the sleigh: the force exerted by the horse $\vec{\mathbf{F}}_{sh}$, the gravitational force due to Earth $\vec{\mathbf{F}}_{sE}$, the normal force on the sleigh due to the ground $\vec{\mathbf{N}}_{sg}$, and kinetic (sliding) friction due to the ground $\vec{\mathbf{f}}_{sg}$. Figure 4.21 shows the FBD for the sleigh. The net force is zero, so its horizontal and vertical components must each be zero: $\vec{\mathbf{F}}_{sh} + \vec{\mathbf{f}}_{sg} = 0$ and $\vec{\mathbf{N}}_{sg} + \vec{\mathbf{F}}_{sE} = 0$.

Similarly, four forces are acting on the horse: the force exerted by the sleigh $\vec{\mathbf{F}}_{hs}$, the gravitational force $\vec{\mathbf{F}}_{hE}$, the normal force due to the ground $\vec{\mathbf{N}}_{hg}$, and friction due to the ground $\vec{\mathbf{f}}_{hg}$. Newton's third law says that $\vec{\mathbf{F}}_{hs} = -\vec{\mathbf{F}}_{sh}$; the sleigh pulls back on the horse with a force equal in magnitude to the forward pull of the horse on the sleigh. Therefore, $\vec{\mathbf{F}}_{hs}$ is to the left and has the same magnitude as $\vec{\mathbf{F}}_{sh}$. The horse is in equilibrium, so $\vec{\mathbf{F}}_{hs} + \vec{\mathbf{f}}_{hg} = 0$ and $\vec{\mathbf{N}}_{hg} + \vec{\mathbf{F}}_{hE} = 0$. The first of these equations means that the frictional force has to be to the *right*. How does the horse get friction to push it *forward*? By pushing *backward* on the ground with its feet. We all do the same thing when taking a step; by pushing backward on the ground, we get the ground to push forward on us. This is *static* friction because the horse's hoof is not sliding along the ground. If there were no friction (imagine the ground to be icy), the hoof might slide backward. Static friction acts to prevent sliding, so the frictional force on the hoof is forward. Figure 4.22 shows the FBD for the horse.

Of the eight forces acting either on the horse or on the sleigh, two are internal forces for the horse + sleigh system: $\vec{\mathbf{F}}_{sh}$ and $\vec{\mathbf{F}}_{hs}$. They add to zero since they are interaction partners, so we can omit them from the FBD for the system (Fig. 4.23). The two frictional forces on the system horse + sleigh are *not* interaction partners, but they are equal in magnitude and opposite in direction. From the FBDs, $\vec{\mathbf{f}}_{hg} = -\vec{\mathbf{F}}_{hs}$ and $\vec{\mathbf{f}}_{sg} = -\vec{\mathbf{F}}_{sh}$. Because $\vec{\mathbf{F}}_{hs}$ and $\vec{\mathbf{F}}_{sh}$ are interaction partners, they are equal and opposite. Therefore, $\vec{\mathbf{f}}_{hg}$ and $\vec{\mathbf{f}}_{sg}$ are equal and opposite. The system is in equilibrium.

(b) The FBD for the sleigh (see Fig. 4.21) shows that if the horse pulls the sleigh with a force greater in magnitude than the force of friction on the sleigh ($F_{sh} > f_{sg}$), then the net force on the sleigh is nonzero and to the right. From Fig. 4.22, we need $f_{hg} > F_{hs}$ to have a nonzero net force to the right on the horse. So the frictional force on the horse would have to increase to enable it to pull the sleigh with a greater force. Then in Fig. 4.23, the two frictional forces are no longer equal in magnitude. The forward frictional force on the horse is greater than the backward frictional force on the sleigh, so the net force on the system horse + sleigh is to the right.

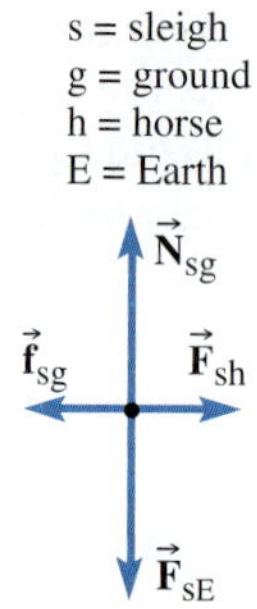

Figure 4.21
Free-body diagram for the sleigh. The subscripts identify the objects involved in the interaction. For example, $\vec{\mathbf{F}}_{sh}$ stands for the force on the sleigh due to the horse. Since the FBD is for the sleigh, we include only forces exerted *on* the sleigh, so the first subscript is always "s."

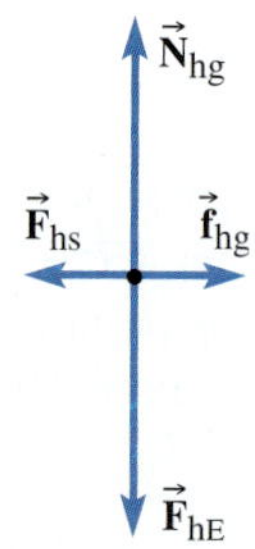

Figure 4.22
Free-body diagram for the horse. $\vec{\mathbf{F}}_{hs}$, the force exerted on the horse by the sleigh, is the interaction partner of $\vec{\mathbf{F}}_{sh}$ in Fig. 4.21, the force exerted on the sleigh by the horse. Therefore, $\vec{\mathbf{F}}_{hs} = -\vec{\mathbf{F}}_{sh}$.

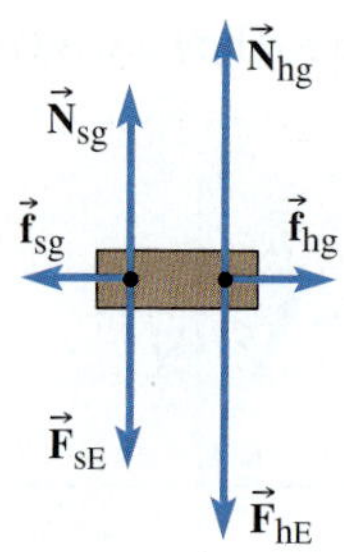

Figure 4.23
FBD for the system horse + sleigh. The internal forces $\vec{\mathbf{F}}_{sh}$ and $\vec{\mathbf{F}}_{hs}$ are omitted—they form an interaction pair, so they add to zero.

(c)

Force Exerted on Sleigh	Interaction Partner
Force on the sleigh due to the horse $\vec{\mathbf{F}}_{sh}$	Force on the horse due to the sleigh $\vec{\mathbf{F}}_{hs}$
Gravitational force on the sleigh due to Earth $\vec{\mathbf{F}}_{sE}$	Gravitational force on Earth due to the sleigh $\vec{\mathbf{F}}_{Es}$
Normal force on the sleigh due to the ground $\vec{\mathbf{N}}_{sg}$	Normal force on the ground due to the sleigh $\vec{\mathbf{N}}_{gs}$
Friction on the sleigh due to the ground $\vec{\mathbf{f}}_{sg}$	Friction on the ground due to the sleigh $\vec{\mathbf{f}}_{gs}$

Practice Problem 4.9 Passing a Truck

A car is moving north and speeding up to pass a truck on a level road. The combined contact force exerted *on the road by all four tires* has vertical component 11.0 kN downward and horizontal component 3.3 kN southward. The drag force exerted on the car by the air is 1.2 kN southward. (a) Draw the FBD for the car. (b) What is the weight of the car? (c) What is the net force acting on the car?

Microscopic Origin of Friction What looks like the smooth surface of a solid to the unaided eye is generally quite rough on a microscopic scale (Fig. 4.24). Friction is caused by atomic or molecular bonds between the "high points" on the surfaces of the two objects. These bonds are formed by microscopic electromagnetic forces that hold the atoms or molecules together. If the two objects are pushed together harder, the surfaces deform a little more, enabling more "high points" to bond. That is why the force of kinetic friction and the maximum force of static friction are proportional to the normal force. A bit of lubricant drastically decreases the frictional forces, because the two surfaces can float past one another without many of the "high points" coming into contact.

In static friction, when these molecular bonds are stretched, they pull back harder. The bonds have to be broken before sliding can begin. Once sliding begins, molecular bonds are continually made and broken as "high points" come together in a hit-or-miss fashion. These bonds are generally not as strong as those formed in the absence of sliding, which is why $\mu_s > \mu_k$.

For dry, solid surfaces, the amount of friction depends on how smooth the surfaces are and how many contaminants are present on the surface. Does polishing two steel surfaces decrease the frictional forces when they slide across each other? Not necessarily. In an extreme case, if the surfaces are extremely smooth and all surface contaminants are removed, the steel surfaces form a "cold weld"—essentially, they become one piece of steel. The atoms bond as strongly with their new neighbors as they do with the old.

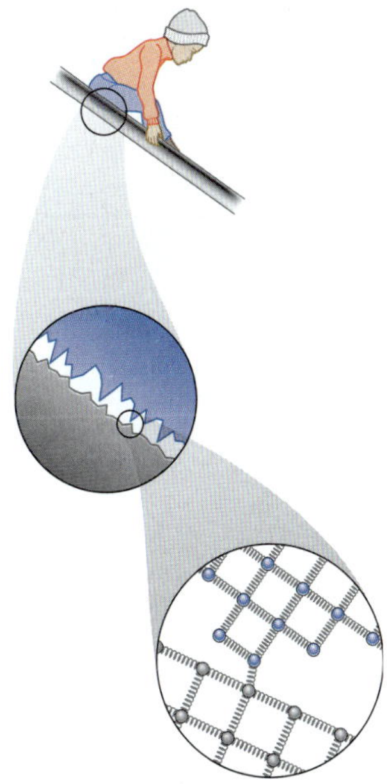

Figure 4.24 Friction is caused by bonds between atoms that form between the "high points" of the two surfaces that come into contact.

Application: Equilibrium on an Inclined Plane

Suppose we wish to pull a large box up a *frictionless* incline to a loading dock platform. Figure 4.25 shows the three forces acting on the box. $\vec{\mathbf{F}}_a$ represents the applied force with which we pull. The force is parallel to the incline. If we choose the x- and y-axes to be horizontal and vertical, respectively, then two of the three forces have both x- and y-components. On the other hand, if we choose the x-axis parallel to the incline and the y-axis perpendicular to it, then only one of the three forces has both x- and y-components (the gravitational force).

With axes chosen, the weight of the box is then resolved into two perpendicular components (Fig. 4.26a). To find the x- and y-components of the gravitational force $\vec{\mathbf{W}}$, we must determine the angle that $\vec{\mathbf{W}}$ makes with one of the axes. The right triangle of Fig. 4.26b shows that $\alpha + \phi = 90°$, since the interior angles of a triangle add up to 180°. The x- and y-axes are perpendicular, so $\alpha + \beta = 90°$. Therefore, $\beta = \phi$.

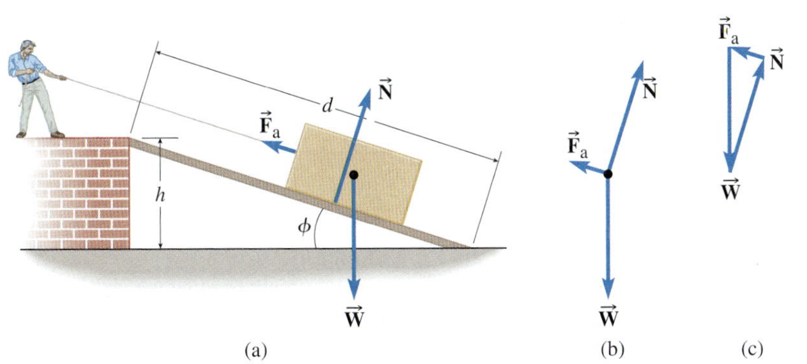

Figure 4.25 (a) Forces acting on a box of mass m as it is pulled up an incline. (b) Free-body diagram for the box. (c) Graphical addition showing that, *if the box moves with constant velocity*, the net force is zero.

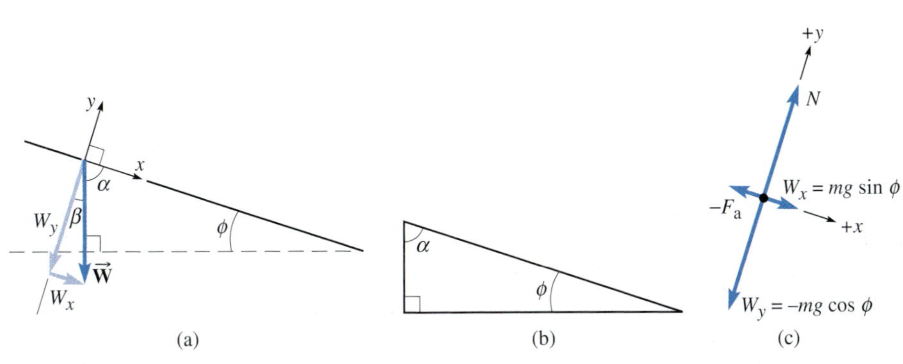

Figure 4.26 (a) Resolving the weight into components parallel to and perpendicular to the incline. (b) A right triangle shows that $\alpha + \phi = 90°$. (c) Free-body diagram for the box on the incline with the gravitational force separated into x- and y-components.

The y-component of $\vec{\mathbf{W}}$ is perpendicular to the surface of the incline. From Fig. 4.26a, the side parallel to the y-axis is adjacent to angle β, so

$$\cos \beta = \frac{\text{adjacent}}{\text{hypotenuse}} = \frac{|W_y|}{|W|}$$

Since W_y is in the $-y$-direction and $W = mg$,

$$W_y = -mg \cos \beta = -mg \cos \phi$$

The x-component of the weight tends to make the box slide down the incline (in the positive x-direction). Using the same triangle,

$$W_x = +mg \sin \phi$$

When the box is pulled with a force equal in magnitude to W_x up the incline (in the negative x-direction), it will slide up with constant velocity. The component of the box's weight perpendicular to the incline is supported by the normal force $\vec{\mathbf{N}}$ that pushes the box away from the incline. Figure 4.26c is an FBD in which the gravitational force is separated into its x- and y-components.

If the box is in equilibrium, whether at rest or moving along the incline at constant velocity, the force components along each axis sum to zero:

$$\sum F_x = (-F_a) + mg \sin \phi = 0$$

and

$$\sum F_y = N + (-mg \cos \phi) = 0$$

On an incline, the normal force is *not* equal in magnitude to the weight and it does not point straight up. If the applied force has magnitude $mg \sin \phi$, we can pull the box up the incline at constant velocity. If friction acts on the box, we must pull with a force greater than $mg \sin \phi$ to slide the box up the incline at constant velocity.

Example 4.10

Pushing a Safe up an Incline

A new safe is being delivered to the First National Bank. It is to be placed in the wall at a height of 1.5 m above the floor. The delivery people have a portable ramp, which they plan to use to help them push the safe up and into position. The mass of the safe is 510 kg, the coefficient of static friction along the incline is $\mu_s = 0.42$, and the coefficient of kinetic friction along the incline is $\mu_k = 0.33$. The ramp forms an angle $\theta = 15°$ above the horizontal. (a) How hard do the movers have to push to start the safe moving up the incline? Assume that they push in a direction parallel to the incline. (b) To slide the safe up at a constant speed, with what magnitude force must the movers push?

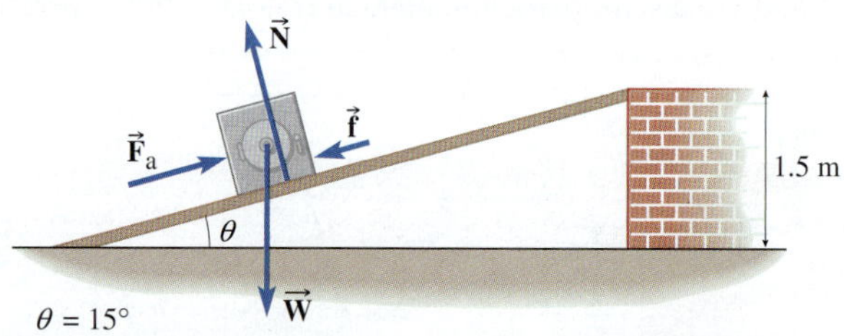

Figure 4.27

Forces acting on the safe as it is moved up the incline.

Strategy (a) When the safe *starts* to move, its velocity is changing, so the safe is *not* in equilibrium. Nevertheless, to find the minimum applied force to start the safe moving, we can find the *maximum* applied force for which the safe *remains at rest*—an equilibrium situation. (b) The safe is in equilibrium as it slides with a constant velocity. Both parts of the problem can be solved by drawing the FBD, choosing axes, and setting the x- and y-components of the net force equal to zero.

Solution First we draw a diagram to show forces acting (Fig. 4.27). When the crate is in equilibrium, these forces must add to zero. Figure 4.28a is a free-body diagram for the crate. Figure 4.28b shows the graphical addition of the four forces giving a net force of zero.

Before resolving the forces into components, we must choose x- and y-axes. To use the coefficient of friction, we have to resolve the contact force on the safe due to the incline into components *parallel and perpendicular to the incline*—friction and the normal force, respectively—rather

continued on next page

Example 4.10 continued

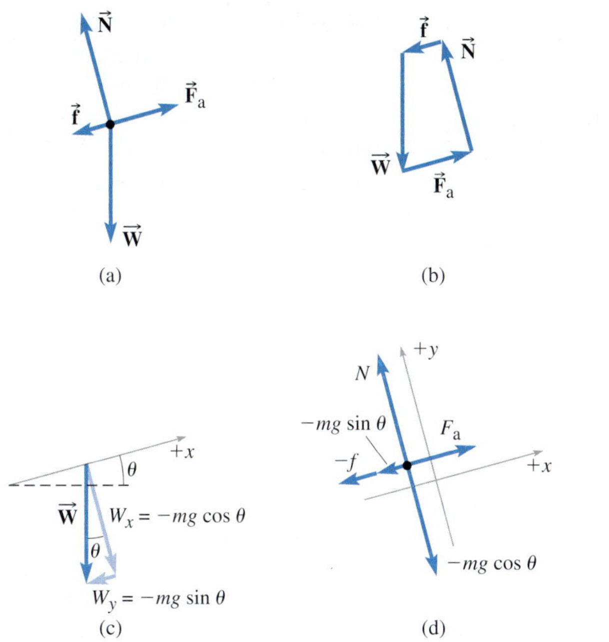

Figure 4.28

(a) Free-body diagram for the safe. (b) If the safe is in equilibrium, the forces must add to give a zero net force. (c) Resolving the weight into x- and y-components. (d) An FBD in which the forces are replaced with their x- and y components.

than into horizontal and vertical components. Therefore, we choose x- and y-axes parallel and perpendicular to the incline so friction is along the x-axis and the normal force is along the y-axis.

The gravitational force $\vec{\mathbf{W}}$ can be resolved into its components: $W_x = -mg \sin\theta$ and $W_y = -mg\cos\theta$ (Fig. 4.28c). Now we draw the FBD with $\vec{\mathbf{W}}$ replaced by its components (Fig. 4.28d).

(a) Suppose that the safe is initially at rest. As the movers start to push, F_a gets larger and the force of static friction gets larger to "try" to keep the safe from sliding. Eventually, at some value of F_a, static friction reaches its maximum possible value $\mu_s N$. If the movers continue to push harder, increasing F_a further, the force of static friction cannot increase past its maximum value $\mu_s N$, so the safe starts to slide. The direction of the frictional force is along the incline and downward since friction is "trying" to keep the safe from sliding *up* the incline.

The normal force is *not* equal in magnitude to the weight of the safe. To find the normal force, sum the y-components of the forces:

$$\sum F_y = N + (-mg\cos\theta) = 0$$

Then $N = mg\cos\theta$. The normal force is *less than the weight* since $\cos\theta < 1$.

When the movers push with the largest force for which the safe does *not* slide,

$$\sum F_x = F_{ax} + f_x + W_x = 0$$

The applied force is in the $+x$-direction, so $F_{ax} = +F_a$. The frictional force has its maximum magnitude and is in the $-x$-direction, so $f_x = -f_{s,max} = -\mu_s N = -\mu_s mg\cos\theta$. From the FBD, $W_x = -mg\sin\theta$. Then,

$$\sum F_x = F_a - \mu_s mg\cos\theta - mg\sin\theta = 0$$

Solving for F_a gives

$$\begin{aligned} F_a &= mg\,(\mu_s\cos\theta + \sin\theta) \\ &= 510\text{ kg} \times 9.80\text{ m/s}^2 \times (0.42 \times \cos 15° + \sin 15°) \\ &= 3300\text{ N} \end{aligned}$$

An applied force that *exceeds* 3300 N starts the box moving up the incline.

(b) Once the safe is sliding, the movers need only push hard enough to make the net force on the safe equal to zero if they want the safe to slide at constant velocity. We are now dealing with sliding friction, so the frictional force is now $f_x = -\mu_k N = -\mu_k mg\cos\theta$.

$$\begin{aligned} \sum F_x &= F_{ax} + f_x + W_x \\ &= F_a - \mu_k mg\cos\theta - mg\sin\theta \\ &= 0 \end{aligned}$$

$$\begin{aligned} F_a &= mg\,(\mu_k\cos\theta + \sin\theta) \\ &= 510\text{ kg} \times 9.80\text{ m/s}^2 \times (0.33 \times \cos 15° + \sin 15°) \\ &= 2900\text{ N} \end{aligned}$$

The movers push with a force $\vec{\mathbf{F}}_a$ of magnitude 2900 N directed up the incline. Despite the friction that opposes the safe's motion, the force exerted by the movers is still less than what they would need to exert to lift the safe straight up (5000 N).

Discussion In (b), the expression $F_a = mg\,(\mu_k\cos\theta + \sin\theta)$ shows that the applied force up the incline has to balance the sum of two forces down the incline: the frictional force ($\mu_k mg\cos\theta$) and the component of the gravitational force down the incline ($mg\sin\theta$). This balance of forces is shown graphically in the FBD (Fig. 4.28d).

Practice Problem 4.10 Smoothing the Infield Dirt

During the seventh-inning stretch of a baseball game, groundskeepers drag mats across the infield dirt to smooth it. A groundskeeper is pulling a mat at a constant velocity by applying a force of 120 N at an angle of 22° above the horizontal. The coefficient of kinetic friction between the mat and the ground is 0.60. Find (a) the magnitude of the frictional force between the dirt and the mat and (b) the weight of the mat.

EVERYDAY PHYSICS DEMO

To estimate the coefficient of static friction between a coin and the cover of your physics book, place the coin on the book and slowly lift the cover. Note the angle of the cover when the coin starts to slide. Explain how you can use this angle to find the coefficient of static friction. Can you devise an experiment to find the coefficient of kinetic friction?

Now try two different coins with different masses. Do they start to slide at about the same angle? If not, which one starts to slide first—the more massive coin or the less massive one?

4.7 TENSION

Consider a heavy chandelier hanging by a chain from the ceiling (Fig. 4.29a). The chandelier is in equilibrium, so the upward force on it due to the chain is equal in magnitude to the chandelier's weight. With what force does the chain pull downward on the ceiling? The ceiling has to pull up with a force equal to the total weight of the chain and the chandelier. The interaction partner of this force—the force the chain exerts on the ceiling—is equal in magnitude and opposite in direction. Therefore, if the weight of the chain is negligibly small compared with the weight of the chandelier, then the chain exerts forces of equal magnitude at its two ends. The forces at the ends would *not* be equal, however, if you grabbed the chain in the middle and pulled it up or down or if we could not neglect the weight of the chain. We can generalize this observation:

An *ideal* cord (or rope, string, tendon, cable, or chain) pulls in the direction of the cord with forces of equal magnitude on the objects attached to its ends as long as no external force is exerted on it anywhere between the ends. An ideal cord has zero mass and zero weight.

A single link of the chain (Fig. 4.29b) is pulled at both ends by the neighboring links. The magnitude of these forces is called the **tension** in the chain. Similarly, a little segment of a cord is pulled at both its ends by the tension in the neighboring pieces of the cord. If the segment is in equilibrium, then the net force acting on it is zero. As long as there are no other forces exerted on the segment, the forces exerted by its neighbors must be equal in magnitude and opposite in direction. Therefore, the tension has the same value everywhere and is equal to the force that the cord exerts on the objects attached to its ends.

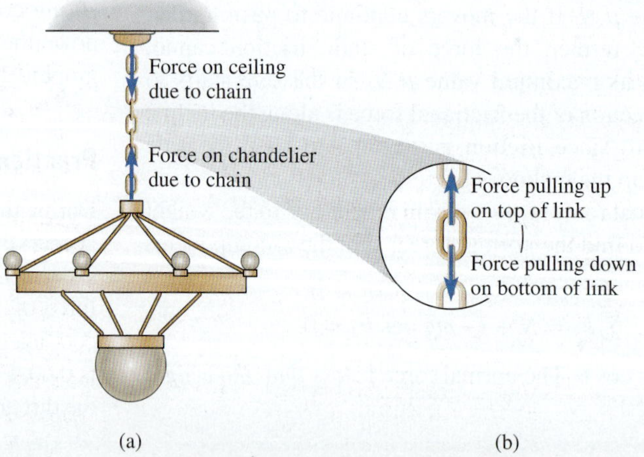

Figure 4.29 (a) The chain pulls up on the chandelier at one end and pulls down on the ceiling at the other. If the weight of the chain itself is negligibly small, these forces are equal in magnitude, because the net force *on* the chain is zero. The magnitude of these forces is the tension in the chain. (b) The chain is under tension. Each link is pulled in opposite directions by its neighbors.

Example 4.11

Archery Practice

Figure 4.30 shows the bowstring of a bow and arrow just before it is released. The archer is pulling back on the midpoint of the bowstring with a horizontal force of 162 N. What is the tension in the bowstring?

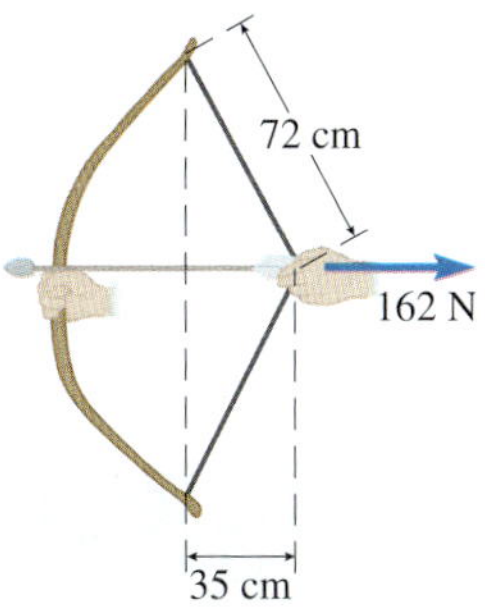

Figure 4.30

The force applied to the bowstring by an archer.

Strategy Consider a small segment of the bowstring that touches the archer's finger. That piece of the string is in equilibrium, so the net force acting on it is zero. We draw the FBD, choose coordinate axes, and apply the equilibrium condition: $\Sigma F_x = 0$ and $\Sigma F_y = 0$. We know the force exerted on the segment of string by the archer's fingers. That segment is also pulled on each end by the tension in the string. Can we assume the tension in the string is the same everywhere? The weight of the string is small compared with the other forces acting on it. The archer pulls sideways on the bowstring, exerting little or no *tangential* force, so we can assume the tension is the same everywhere.

Solution Figure 4.31a is an FBD for the segment of bowstring being considered. The two tension forces have the same magnitude T. Figure 4.31b shows these three forces adding to zero. From this sketch, we expect the tension T to be roughly the same as F_a. We choose the x-axis to the right and the y-axis upward. To find the components of the forces due to tension in the string, we draw a triangle (Fig. 4.31c). From the measurements given, we can find the angle θ.

$$\sin\theta = \frac{\text{opposite}}{\text{hypotenuse}} = \frac{35\text{ cm}}{72\text{ cm}} = 0.486$$

$$\theta = \sin^{-1} 0.486 = 29.1°$$

The x-component of the tension force exerted on the upper end of the segment is

$$T_{1x} = -T\sin\theta$$

The x-component of the force exerted on the lower end of the string is the same. Therefore,

$$\sum F_x = -2T\sin\theta + F_a = 0$$

Solving for T gives

$$T = \frac{F_a}{2\sin\theta} = \frac{162\text{ N}}{2\times 0.486} = 170\text{ N}$$

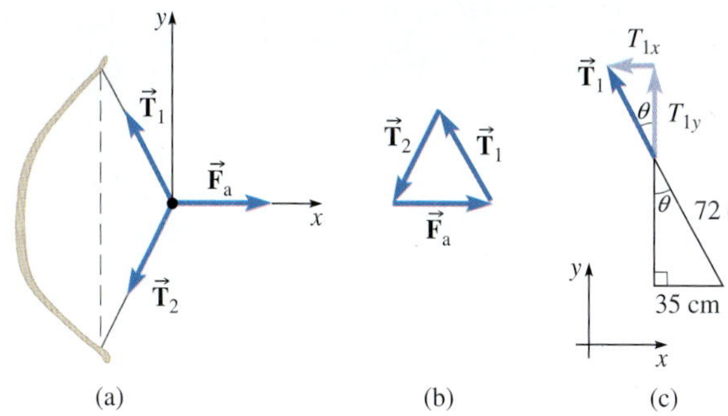

Figure 4.31

(a) FBD for a point on the bowstring. (b) Graphical addition of the three forces showing that the sum is zero. (c) The angle θ is used to find the x- and y-components of the forces exerted at each end of the bowstring.

Discussion The tension is only slightly larger than F_a, a reasonable result given the picture of graphical vector addition in Fig. 4.31b.

In this problem, only the x-components of the forces had to be used. The y-components must also add to zero. At the upper end of the string, the y-component of the force exerted by the bow is $+T\cos\theta$, while at the lower end it is $-T\cos\theta$. Therefore, $\Sigma F_y = 0$.

The expression $T = F_a/(2\sin\theta)$ can be evaluated for limiting values of θ to make sure that the expression is correct. As θ approaches 90°, the tension approaches

$$\frac{F_a}{2\sin 90°} = \frac{1}{2}F_a$$

That is correct because the archer would be pulling to the right with a force F_a, while each side of the bowstring would pull to the left with a force of magnitude T. For equilibrium, $F_a = 2T$ or $T = \frac{1}{2}F_a$.

As θ gets smaller, $\sin\theta$ decreases and the tension increases (for a fixed value of F_a). That agrees with our intuition. The larger the tension, the smaller the angle the string needs to make in order to supply the necessary horizontal force.

Practice Problem 4.11 Tightrope Practice

Jorge decides to rig up a tightrope in the backyard so his children can develop a good sense of balance (Fig. 4.32). For safety reasons, he positions a horizontal cable only 0.60 m above the ground. If the 6.00 m long cable sags by 0.12 m from its taut horizontal position when Denisha (weight 250 N) is standing on the middle of it, what is the tension in the cable? Ignore the weight of the cable.

continued on next page

Example 4.11 continued

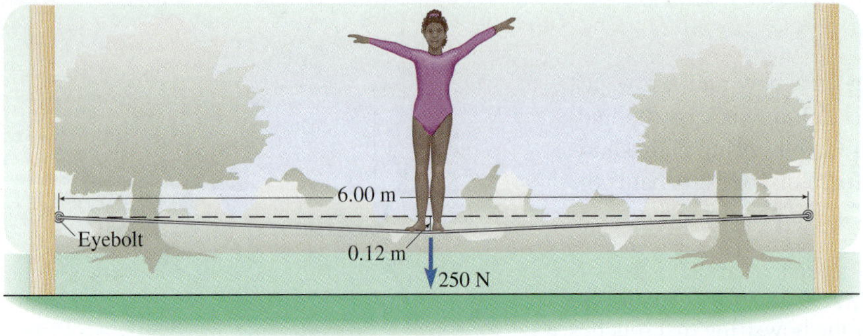

Figure 4.32 Tightrope for balancing practice. What is the tension in the cable?

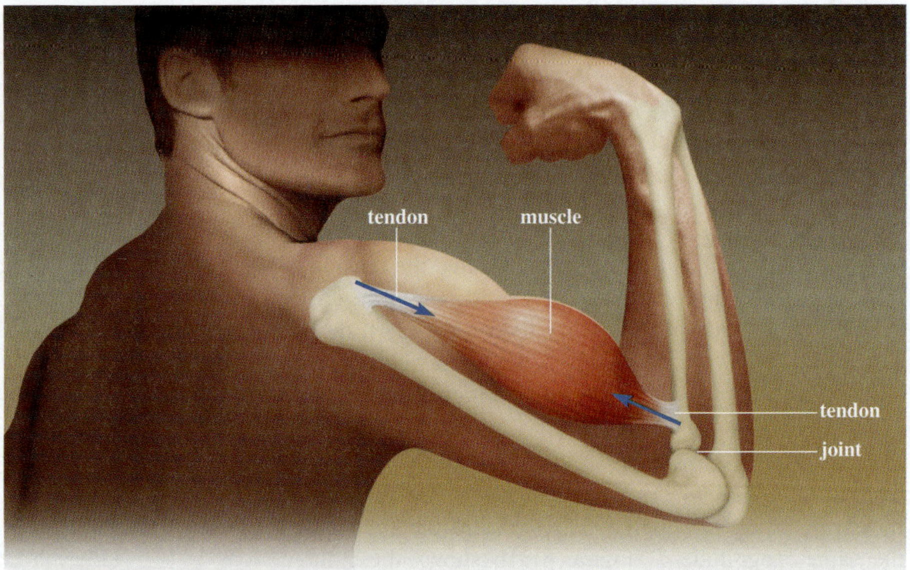

Figure 4.33 A muscle contracts, increasing the tension in the attached tendons. The tendons exert forces on two different bones.

Application: Tensile Forces in the Body Tensile forces are central in the study of animal motion, or biomechanics. Muscles are usually connected by tendons, one at each end of the muscle, to two different bones, which in turn are linked at a joint (Fig. 4.33). When the muscle contracts, the tension in the tendons increases, pulling on both of the bones.

EVERYDAY PHYSICS DEMO

Sit with your arm bent at the elbow with a heavy object on the palm of your hand. You can feel the contraction of the biceps muscle. With your other hand, feel the tendon that connects the biceps muscle to your forearm.

Now place your hand palm down on the desktop and push down. Now it is the triceps muscle that contracts, pulling up on the bone on the other side of the elbow joint. Muscles and tendons cannot push; they can only pull. The biceps muscle cannot push the forearm downward, but the triceps muscle can pull on the other side of the joint. In both cases, the arm acts as a lever.

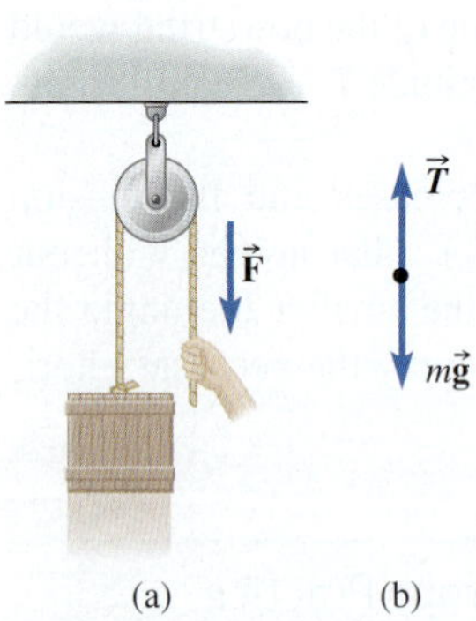

Figure 4.34 (a) Using a pulley to lift a crate by pulling *downward* on a rope with force $\vec{F}$. (b) A free-body diagram for the crate. If the crate is in equilibrium, then the tension T must be equal to the weight of the crate mg.

Application: Ideal Pulleys A pulley can change the direction of the force exerted by a cord under tension. To lift something heavy, it is easier to stand on the ground and pull *down* on the rope than to get above the weight on a platform and pull up on the rope (Fig. 4.34).

An *ideal* pulley has no mass and turns with no friction. An ideal pulley exerts no forces on the cord that are *tangent* to the cord—it is not pulling in either direction along the cord. As a result, the tension of an ideal cord that runs through an ideal pulley is the same on both sides of the pulley. An ideal pulley changes the direction of the force exerted by a cord without changing its magnitude. As long as a real pulley has a small mass and negligible amount of friction, we can approximate it as an ideal pulley.

Look back at Example 4.1. The three cord segments are sections of a single cord wrapped around some pulleys. If these pulleys are ideal, the tension in the cord must be the same everywhere. At its lowest end, the cord holds up a 22.0 N weight. Since the weight is in equlibrium, we know the tension must be 22.0 N.

Example 4.12

A Two-Pulley System

A 1804 N engine is hauled upward at constant speed (Fig. 4.35). What are the tensions in the three ropes labeled A, B, and C? Assume the ropes and the pulleys labeled L and R are ideal.

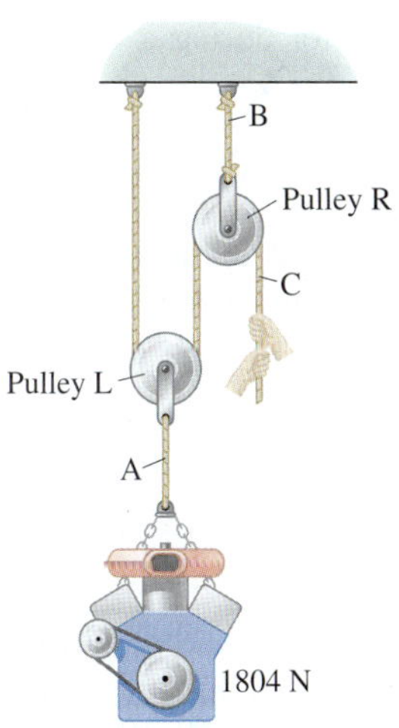

Figure 4.35
A system of pulleys used to raise a heavy engine.

Strategy The engine and pulley L move up at constant speed, so the net force on each of them is zero. Pulley R remains at rest, so the net force on it is also zero. We can draw the FBD for any or all of these objects and then apply the equilibrium condition. If the pulleys are ideal, the tension in the rope is the same on both sides of the pulley. Therefore, rope C—which is attached to the ceiling, passes around both pulleys, and is pulled downward at the other end—has the same tension throughout. Call the tensions in the three ropes T_A, T_B, and T_C. To analyze the forces exerted on a pulley, we define our system so the part of the rope wrapped around the pulley is considered part of the pulley. Then there are two cord segments pulling on the pulley, each with the same tension.

Solution There are two forces acting on the engine: the gravitational force (1804 N, downward) and the upward pull of rope A. These must be equal and opposite (Fig. 4.36a), since the net force is zero. Therefore $T_A = 1804$ N.

The FBD for pulley L (Fig. 4.36b) shows rope A pulling down with a force of magnitude T_A and rope C pulling upward on *each side*. The rope has the same tension throughout, so all forces labeled T_C in Fig. 4.36b,c have the same magnitude. For the net force to equal zero,

$$2T_C = T_A$$

$$T_C = \tfrac{1}{2}T_A = 902.0 \text{ N}$$

Figure 4.36c is the FBD for pulley R. Rope B pulls upward on it with a force of magnitude T_B. On *each side* of the pulley, rope C pulls downward. For the net force to equal zero,

$$T_B = 2T_C = 1804 \text{ N}$$

Discussion The engine is raised by pulling *down* on a rope—the pulleys change the direction of the applied force needed to lift the engine. In this case they also change the *magnitude* of the required force. They do that by making the rope pull up on the engine twice, so the person pulling the rope only needs to exert a force equal to half the engine's weight.

Practice Problem 4.12 System of Ropes, Pulleys, and Engine

Consider the entire collection of ropes, pulleys, and the engine to be a single system. Draw the FBD for this system and show that the net force on the system is zero. [*Hint*: Remember that only forces exerted by objects *external* to the system are included in the FBD.]

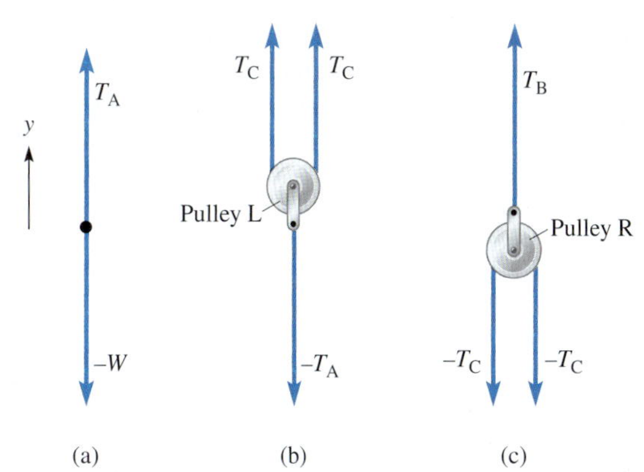

Figure 4.36
FBDs for the (a) engine, (b) pulley L, and (c) pulley R. The force arrows are labeled with their *y*-components, where the *y*-axis is up.

4.8 APPLYING NEWTON'S LAWS

We can now apply Newton's second law to a great variety of situations involving the forces we have encountered so far—gravity, contact forces, and tension. The following steps are helpful in most problems that involve Newton's second law.

Problem-Solving Strategy for Newton's Second Law

- Decide what objects (or systems of objects) will have Newton's second law applied to them.
- Identify all the *external* forces acting on the object(s).
- Use Newton's third law to relate the magnitudes and directions of interaction partners.
- Draw an FBD to show all the forces acting on the object(s).
- Choose a coordinate system. If the direction of the acceleration is known, choose axes so that the net force and the acceleration are along one of the axes.
- Find the net force by adding the forces as vectors.
- Use Newton's second law to relate the net force to the acceleration.
- Relate the acceleration to the change in the velocity vector during a time interval of interest.

CONNECTION:

If the net force acting on an object is *constant*, then the object moves with *constant acceleration*. Then all of the techniques from Chapters 2 and 3 for analyzing motion with constant acceleration can be applied.

Example 4.13 illustrates how to use Newton's second law to find unknown forces and the acceleration; it then uses the acceleration to find the change in velocity.

Example 4.13

The Broken Suitcase

The wheels fall off Beatrice's suitcase, so she ties a rope to it and drags it along the floor of the airport terminal (Fig. 4.37). The rope makes a 40.0° angle with the horizontal. The suitcase has a mass of 36.0 kg and Beatrice pulls on the rope with a force of 65.0 N. (a) What is the magnitude of the normal force acting on the suitcase due to the floor? (b) If the coefficient of kinetic friction between the suitcase and the marble floor is $\mu_k = 0.13$, find the frictional force acting on the suitcase. (c) What is the acceleration of the suitcase while Beatrice pulls with a 65.0 N force at 40.0°? (d) Starting from rest, for how long a time must she pull with this force until the suitcase reaches a comfortable walking speed of 0.5 m/s?

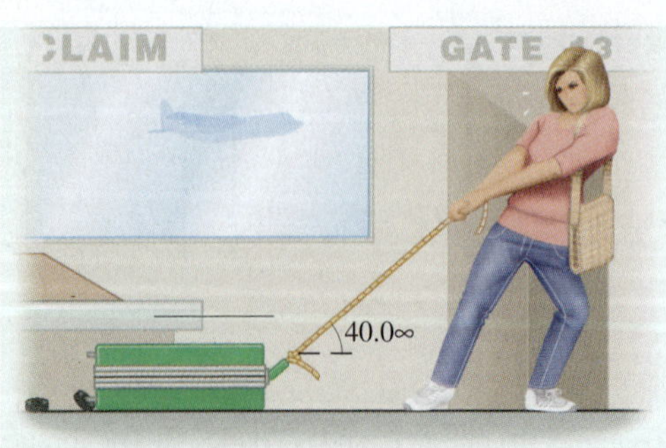

Figure 4.37
Beatrice dragging her suitcase.

Strategy Since the suitcase is dragged horizontally along the floor, the vertical component of its velocity is always zero. The vertical acceleration component of the suitcase is zero because the vertical velocity component does not change. (If it did have a vertical acceleration component, the suitcase would begin to move either down through the floor or up into the air.) If we choose the $+y$-axis up and the $+x$-axis to be horizontal, then $a_y = 0$. We resolve the forces acting on the suitcase into their components, draw a free-body diagram for the suitcase, and apply Newton's second law.

Solution (a) Figure 4.38 shows the forces acting on the suitcase, where $\vec{\mathbf{F}}$ is the force exerted by Beatrice. All the other forces are either parallel or perpendicular to the floor, so only $\vec{\mathbf{F}}$ needs to be resolved into x- and y-components.

$$F_x = F \cos 40.0° = 65.0 \text{ N} \times 0.766 = 49.8 \text{ N}$$
$$F_y = F \sin 40.0° = 65.0 \text{ N} \times 0.643 = 41.8 \text{ N}$$

Figure 4.39 is an FBD in which $\vec{\mathbf{F}}$ is replaced by its components. The vertical force components add to zero since $a_y = 0$.

continued on next page

Example 4.13 continued

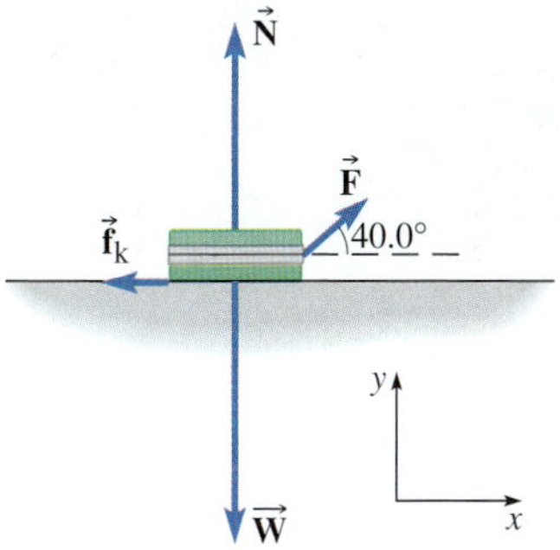

Figure 4.38
Forces acting on a suitcase dragged along the floor. The lengths of the vector arrows are not to scale.

$$\sum F_y = ma_y = 0$$
$$N + F\sin 40.0° - W = 0$$

We can solve this equation for the magnitude of the normal force. The magnitude of the gravitational force is $W = mg$, so

$$N = mg - F\sin 40.0°$$
$$= (36.0\text{ kg} \times 9.80\text{ N/kg}) - (65.0\text{ N} \times \sin 40.0°)$$
$$= 352.8\text{ N} - 41.8\text{ N} = 311\text{ N}$$

(b) The magnitude of the kinetic frictional force is

$$f_k = \mu_k N = 0.13 \times 311\text{ N} = 40.43\text{ N}$$

Rounded to two significant figures, the frictional force is 40 N in the $-x$-direction (opposite the motion of the suitcase).

(c) The y-component of the acceleration is zero. To find the x-component, we apply Newton's second law to the x-components of the forces acting on the suitcase:

$$\sum F_x = +F\cos 40.0° + (-f_k)$$
$$= 49.79\text{ N} - 40.43\text{ N} = 9.36\text{ N}$$
$$a_x = \frac{\sum F_x}{m} = \frac{9.36\text{ N}}{36.0\text{ kg}} = 0.260\text{ m/s}^2$$

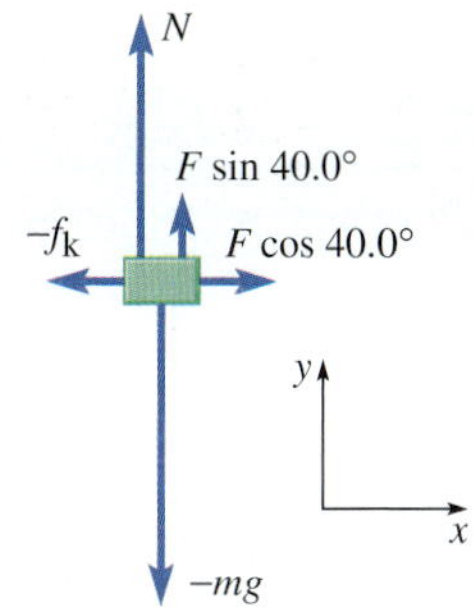

Figure 4.39
FBD for the suitcase, with the forces represented by their x- and y-components.

Here we have replaced newtons per kilogram with the equivalent meters per second squared, the usual way to write the SI units of acceleration. The acceleration is 0.3 m/s² in the $+x$-direction.

(d) With constant a_x,

$$\Delta v_x = a_x \Delta t$$

The suitcase starts from rest so $v_{ix} = 0$ and $\Delta v_x = v_{fx} - v_{ix} = v_{fx}$. Then,

$$\Delta t = \frac{v_{fx}}{a_x} = \frac{0.5\text{ m/s}}{0.260\text{ m/s}^2} = 2\text{ s}$$

Discussion What Beatrice probably wants to do is to drag the suitcase along at constant velocity. To do that, she must first accelerate the suitcase from rest. Once the suitcase is moving at the desired velocity, she pulls a little less hard, so the net force is zero and the suitcase slides at constant speed. She would do so without thinking much about it, of course!

Practice Problem 4.13 The Continuing Story . . .

How hard does Beatrice pull at a 40.0° angle while the suitcase slides along the floor at constant velocity? [*Hint*: Do *not* assume that the normal force is the same as in the previous discussion.]

Connected Objects Sometimes two or more objects are constrained to have the same acceleration by the way they are connected. In Example 4.14, we look at a train engine pulling five freight cars. The couplings maintain a fixed distance between the cars, so at any instant the cars move with the same velocity; if they didn't, the distance between them would change. The velocities don't have to be constant; they just have to change in exactly the same way, which implies that the accelerations must also be the same at any instant.

Example 4.14

Coupling Force on First and Last Freight Cars

A train engine pulls out of a station along a straight horizontal track with five identical freight cars behind it, each of which weighs 90.0 kN. The train reaches a speed of 15.0 m/s within 5.00 min of starting out. If the engine pulls with a constant force during this interval, with what magnitude of force does the coupling between cars pull forward on the first and last of the freight cars? Ignore air resistance and friction on the freight cars.

Strategy A sketch of the situation is shown in Fig. 4.40. To find the force exerted by the first coupling, we consider all five cars to be one system so we do not have to worry

continued on next page

Example 4.14 continued

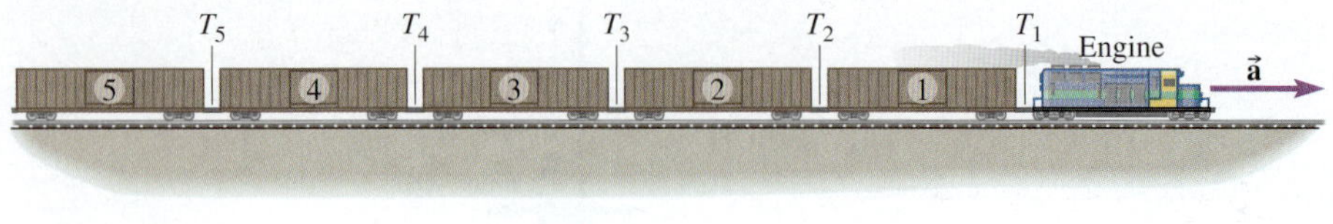

Figure 4.40
An engine pulling five identical freight cars. The entire train has a constant acceleration $\vec{\mathbf{a}}$ to the right.

about the forces exerted on one car by another: these *internal* forces add to zero by Newton's third law. For example, car 1 pulls on car 2 and car 2 pulls on car 1 with an equal but opposite force, so the two add to zero. The only *external* forces on the group of five cars are the normal force, gravity, and the pull of the first coupling. To find the force exerted by the fifth coupling, we consider car five by itself to be a system. In each case, once we identify a system, we draw a free-body diagram, choose a coordinate system, and then apply Newton's second law.

As discussed previously, the engine and the cars must all have the same acceleration at any instant. We expect the acceleration to be *constant* because the engine pulls with a constant force. We can calculate the acceleration of the train from the initial and final velocities and the elapsed time.

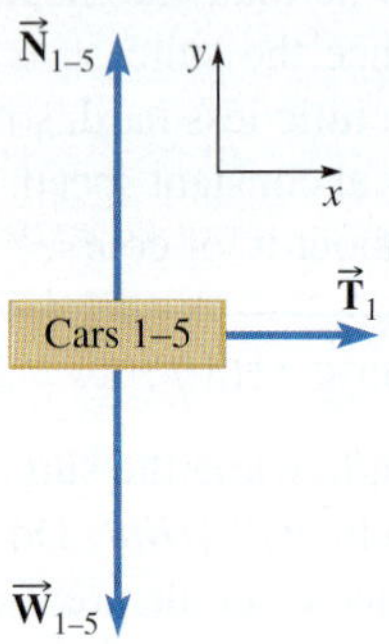

Figure 4.41
FBD for the system consisting of cars 1–5 (but not the engine). Only *external* forces are shown.

Solution For the tension T_1 in the first coupling, we consider the five cars as *one system* of mass M. Figure 4.41 shows the FBD in which cars 1 to 5 are treated as a single object. We choose the x-axis in the direction of motion of the train and the y-axis up. Since the train moves along the x-axis, the acceleration vector is along the x-axis. Therefore, $a_y = 0$. Using the y-component of Newton's second law, the vertical forces add to zero:

$$\sum F_y = Ma_y = N_{1\text{-}5} - W_{1\text{-}5} = 0$$

The only external horizontal force is the force $\vec{\mathbf{T}}_1$ due to the tension in the first coupling. This force is constant according to the problem statement, so we know that the acceleration a_x is constant:

$$\sum F_x = T_1 = Ma_x$$

The mass of the system M is five times the mass of one car m. We are given the *weight* of one car ($W = 90.0$ kN $= 9.00 \times 10^4$ N). From the relation between mass and weight, $W = mg$, the mass of one car is $m = W/g$ and the mass of five cars is $M = 5W/g$.

The constant acceleration of the train is

$$a_x = \frac{\Delta v_x}{\Delta t} = \frac{v_{\text{fx}} - v_{\text{ix}}}{t_\text{f} - t_\text{i}} = \frac{15.0\ \text{m/s} - 0}{300\ \text{s} - 0} = 0.0500\ \text{m/s}^2$$

Therefore,

$$T_1 = Ma_x = \frac{5W}{g} \times \frac{\Delta v_x}{\Delta t} = \frac{5 \times 9.00 \times 10^4\ \text{N}}{9.80\ \text{m/s}^2} \times \frac{15.0\ \text{m/s}}{300\ \text{s}}$$
$$= 2.30\ \text{kN}$$

Now consider the last freight car (car 5). If we ignore friction and air resistance, the only external forces acting are the force $\vec{\mathbf{T}}_5$ due to the tension in the fifth coupling, the normal force $\vec{\mathbf{N}}_5$, and the gravitational force $\vec{\mathbf{W}}_5$; the FBD is shown in Fig. 4.42. Since $\vec{\mathbf{N}}_5 + \vec{\mathbf{W}}_5 = 0$, the net force is equal to $\vec{\mathbf{T}}_5$. From Newton's second law,

$$\sum F_x = T_5 = ma_x = \frac{W}{g}a_x$$

$$T_5 = \frac{W}{g} \times \frac{\Delta v_x}{\Delta t} = \frac{9.00 \times 10^4\ \text{N}}{9.80\ \text{m/s}^2} \times \frac{15.0\ \text{m/s}}{300\ \text{s}} = 459\ \text{N}$$

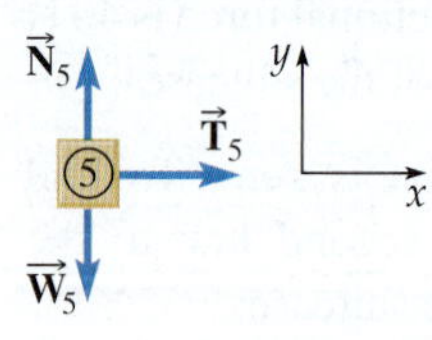

Figure 4.42
FBD for car 5. (Vector lengths are not to the same scale as those in Fig. 4.41.)

Discussion We considered two systems (cars 1 to 5 and car 5) that have the same acceleration and different masses. As expected, the net force is proportional to the mass: the net force on five cars is five times the net force on one car.

The solution to this problem is much simpler when Newton's second law is applied to a system comprising all five cars, rather than to each car individually. Although the problem can be solved by looking at individual cars, to find the tension in the first coupling you would have to draw five FBDs (one for each car) and apply Newton's second law five times. That's because each car, except the fifth, is acted on by the unequal tensions in the couplings on either side. You'd have to first find the tension in the fifth coupling, then the fourth, then the third, and so on.

Practice Problem 4.14 Coupling Force Between First and Second Freight Cars

With what force does the coupling between the first and second cars pull forward on the second car? [*Hint*: Try two methods. One of them is to draw the FBD for the first car and apply Newton's *third* law as well as the second.]

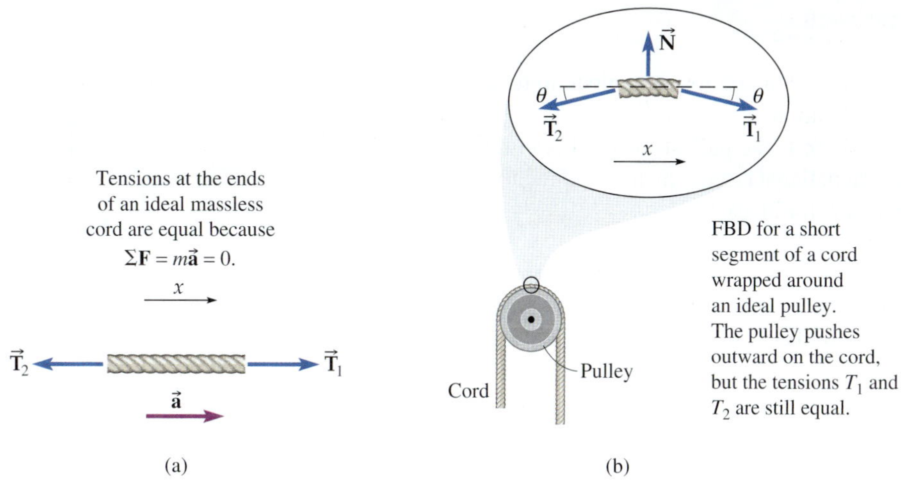

Figure 4.43 (a) FBD for an ideal cord with acceleration $\vec{\mathbf{a}}$. Applying Newton's second law along the x-axis: $\Sigma F_x = T_1 - T_2 = ma_x$. The ideal cord has mass $m = 0$, so $T_1 = T_2$: the tensions at the ends are equal. (b) An ideal cord passing around an ideal pulley and the FBD for a short segment of the cord at the top of the pulley. Choosing the x-axis to be horizontal, the normal force has no x-component. Applying Newton's second law along the x-axis: $\Sigma F_x = T_1 \cos\theta - T_2 \cos\theta = ma_x$. With $m = 0$, $T_1 = T_2$. The same reasoning can be applied to any segment of cord in contact with the pulley to show that the tensions are the same on either side of an ideal pulley.

Example 4.15 deals with two objects connected by an ideal cord. Although it may have a nonzero acceleration, the net force on an *ideal* cord is still zero because it has *zero mass*: if $m = 0$, then $\Sigma\vec{\mathbf{F}} = m\vec{\mathbf{a}} = 0$. As a result, the tension is the same at the two ends as long as no external force acts on the cord between the ends (Fig. 4.43a). An ideal cord that passes over an ideal pulley (having negligible mass and turning with negligible friction) has the same tension at its ends. The pulley exerts an external force on part of the cord, but this force is everywhere *perpendicular to the cord*. As Fig. 4.43b shows, an external force that has no component tangent to the cord does not affect the tension in the cord.

Example 4.15

Two Blocks Hanging on a Pulley

In Fig. 4.44, two blocks are connected by an ideal cord that does not stretch; the cord passes over an ideal pulley. If the masses are $m_1 = 26.0$ kg and $m_2 = 42.0$ kg, what are the accelerations of each block and the tension in the cord?

Strategy Since m_2 is greater than m_1, the downward force of gravity is stronger on the right side than on the left. We expect block 2's acceleration to be downward and block 1's to be upward.

The cord does not stretch, so blocks 1 and 2 move at the same speed at any instant (in opposite directions). Therefore, the accelerations of the two blocks are equal in magnitude and opposite in direction. If the accelerations had different magnitudes, then soon the two blocks would be moving with different speeds. That could only happen if the cord either stretches or contracts.

The tension in the cord must be the same everywhere along the cord since the masses of the cord and pulley are negligible and the pulley turns without friction.

We treat each block as a separate system, draw FBDs for each, and then apply Newton's second law to each. It is convenient to choose the positive y-direction differently for the two blocks since we know their accelerations are in opposite directions. For each block, we choose the $+y$-axis in the direction of the acceleration of that block: upward for block m_1 and downward for m_2. Doing so means that a_y has the same magnitude *and sign* (both positive) for the two blocks. (As an alternative, we could choose the $+y$-axis upward for both. Then we would write $a_{2y} = -a_{1y}$. Either choice is valid.)

Solution Figure 4.45 shows FBDs for the two blocks. Two forces act on each: gravity and the pull of the cord. The acceleration vectors are drawn *next to* the FBDs. Thus, we know the direction of the net force: it is always the same as the direction of the acceleration. Then we know that the tension must be greater than m_1g to give block 1 an upward acceleration and less than m_2g to give block 2 a downward

continued on next page

Example 4.15 continued

acceleration. The $+y$-axes are drawn for each block to be in the direction of the acceleration.

Figure 4.44
Two hanging blocks connected on either side of a frictionless pulley by a massless, flexible cord that does not stretch.

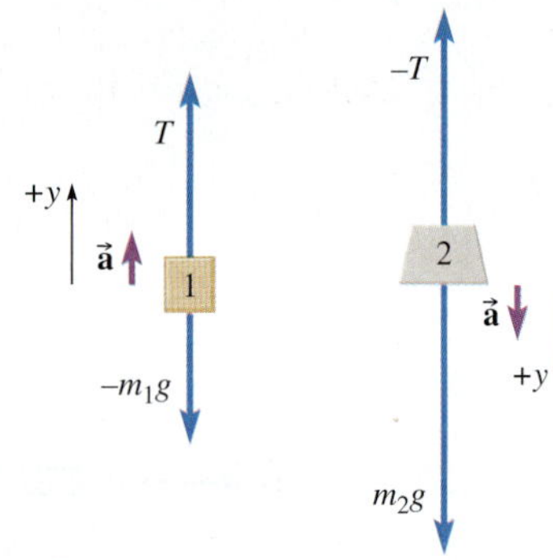

Figure 4.45
FBDs for the hanging blocks labeled with the y-components of the forces. We draw the acceleration vector *next to* each FBD as a guide—the net force has to be in the direction of the acceleration. However, the acceleration vector is not *part of* the FBD (it is not a force to be added to the others).

From the FBD of block 1, the pull of the cord is in the $+y$-direction and the gravitational force is in the $-y$-direction. Then Newton's second law for block 1 is

$$\sum F_{1y} = T - m_1 g = m_1 a_{1y}$$

For block 2, the pull of the cord is in the $-y$-direction and the gravitational force is in the $+y$-direction. Newton's second law for block 2 is

$$\sum F_{2y} = m_2 g - T = m_2 a_{2y}$$

The tension T in the cord is the same in the two equations. Also a_{1y} and a_{2y} are identical, so we write them simply as a_y. We then have a system of two equations with two unknowns. We can add the equations to obtain

$$m_2 g - m_1 g = m_2 a_y + m_1 a_y$$

Solving for a_y, we find

$$a_y = \frac{(m_2 - m_1)g}{m_2 + m_1}$$

Substituting numerical values,

$$a_y = \frac{(42.0\text{ kg} - 26.0\text{ kg}) \times 9.80\text{ N/kg}}{42.0\text{ kg} + 26.0\text{ kg}}$$

$$= 2.31\,\frac{\text{N}}{\text{kg}} \times \frac{1\text{ kg}\cdot\text{m/s}^2}{1\text{ N}} = 2.31\text{ m/s}^2$$

The blocks have the same magnitude acceleration. For block 1 the acceleration points upward and for block 2 it points downward.

To find T we can substitute the expression for a_y into either of the two original equations. Using the first equation,

$$T - m_1 g = m_1 \frac{(m_2 - m_1)g}{m_2 + m_1}$$

Solving for T yields

$$T = \frac{2m_1 m_2}{m_1 + m_2} g$$

Substituting,

$$T = \frac{2 \times 26.0\text{ kg} \times 42.0\text{ kg}}{68.0\text{ kg}} \times 9.80\text{ N/kg} = 315\text{ N}$$

Discussion A few quick checks:

- a_y is positive, which means that the accelerations are in the directions we expect.
- The tension (315 N) is between $m_1 g$ (255 N) and $m_2 g$ (412 N), as it must be for the accelerations to be in opposite directions.
- The units and dimensions are correct for all equations.
- We can check algebraic expressions in special cases for which we have some intuition. For example, if the masses had been *equal*, we expect the blocks to hang in equilibrium (either at rest or moving at constant velocity) due to the equal pull of gravity on the two blocks. Substituting $m_1 = m_2$ into the expressions for a_y and T gives $a_y = 0$ and $T = m_1 g = m_2 g$, which is just what we expect.

Note that we did *not* find out which way the blocks move. We found the directions of their *accelerations*. If the blocks start out at rest, then the block of mass m_2 moves downward and the block of mass m_1 moves upward. However, if initially m_2 is moving up and m_1 down, they continue to move in those directions, slowing down since their accelerations are opposite to their velocities. Eventually they come to rest and then reverse directions.

Practice Problem 4.15 Another Check

Using the numerical values of the tension and the acceleration calculated in Example 4.15, verify Newton's second law directly for each of the two blocks.

Examples 4.16, 4.17, and 4.18 illustrate how different concepts and problem-solving techniques from Chapters 2–4 can be brought together to find the solution to a physics problem.

Example 4.16

Hauling a Crate up to a Third-Floor Window

A student is moving into a dorm room on the third floor and he decides to use a block and tackle arrangement (Fig. 4.46) to move a crate of mass 91 kg from the ground up to his window. If the breaking strength of the available rope is 550 N, what is the minimum time required to haul the crate to the level of the window, 30.0 m above the ground, without breaking the rope?

Strategy The tension in the rope is T and is the same at both ends or anywhere along the rope, assuming the rope and pulleys are ideal. Two pieces of rope support the lower pulley, each pulling upward with a force of magnitude T. The gravitational force acts downward. We draw an FBD for the system consisting of the crate and the lower pulley and set the tension equal to the breaking force of the rope to find the maximum possible acceleration of the crate. Then we use the maximum acceleration to find the minimum time to move the required distance to the third-floor window. We choose the y-axis to be upward. Known: $m = 91$ kg; $\Delta y = 30.0$ m; $T_{max} = 550$ N; $v_{iy} = 0$. To find: Δt, the time to raise the crate 30.0 m with the maximum tension in the cable.

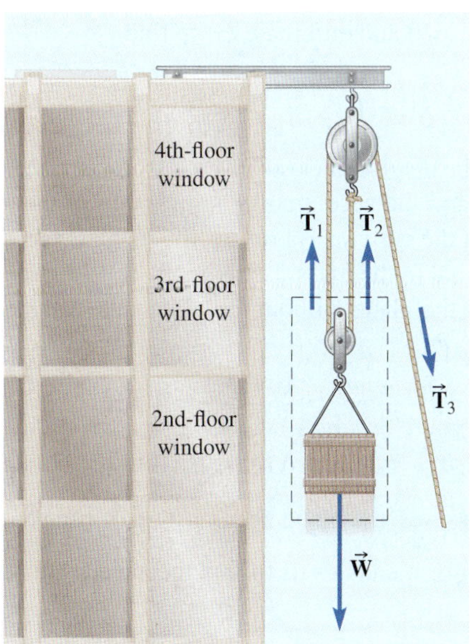

Figure 4.46 Block and tackle setup.

Solution From the FBD (Fig. 4.47), if the forces acting up are greater than the force acting down, the net force is upward and the crate's acceleration is upward. In terms of components, with the $+y$-direction chosen to be upward,

$$\sum F_y = T + T - mg = ma_y$$

Solving for the acceleration,

$$a_y = \frac{T + T - mg}{m}$$

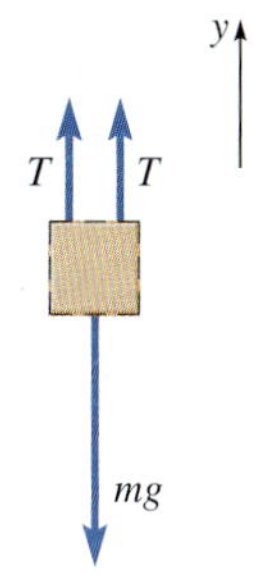

Figure 4.47 FBD for the crate and lower pulley. (This system is outlined by dashed lines in Fig. 4.46.)

Setting $T = 550$ N, the maximum possible value before the cable breaks, and substituting the other known values:

$$a_y = \frac{550\text{ N} + 550\text{ N} - 91\text{ kg} \times 9.80\text{ m/s}^2}{91\text{ kg}} = 2.288\text{ m/s}^2$$

The time to move the crate up a distance Δy starting from rest can be found from

$$\Delta y = v_{iy}\,\Delta t + \tfrac{1}{2}a_y(\Delta t)^2 \qquad (3\text{-}21)$$

Setting $v_{iy} = 0$ and solving for Δt, we find

$$\Delta t = \pm\sqrt{\frac{2\,\Delta y}{a_y}}$$

Our equation applies only for $\Delta t \geq 0$ (the crate reaches the window *after* it leaves the ground). Taking the positive root and substituting numerical values,

$$\Delta t = \sqrt{\frac{2 \times 30.0\text{ m}}{2.288\text{ m/s}^2}} = 5.1\text{ s}$$

This is the minimum possible to haul the crate up without breaking the rope.

Discussion In reality, the student is not likely to achieve this *minimum possible* time. To do so would mean pulling the rope at an unrealistic speed. At the end of the 5.1 s interval, $v_{fy} = 2.288\text{ m/s}^2 \times 5.1\text{ s} = 12$ m/s! More likely, the student would hoist the crate at a roughly constant velocity (except at the beginning, to get it moving, and at the end, to let it come to rest). For motion with a constant velocity, the tension in the rope would be equal to half the weight of the crate (450 N).

Practice Problem 4.16 Hauling the Crate with a Single Pulley

If only a single pulley, attached to the beam above the fourth floor, were available and if the student had a few friends to help him pull on the cable, could they haul the crate up to the third-floor window using the same rope? If so, what is the minimum time required to do so?

Example 4.17

Towing a Glider

What length runway does the plane need?

A small plane of mass 760 kg requires 120 m of runway to take off by itself. (120 m is the horizontal displacement of the plane just before it lifts off the runway, not the entire length of the runway.) As a simplified model, ignore friction and drag forces and assume the plane's engine makes the air exert a constant forward force on the plane. (a) When the plane is towing a 330 kg glider, how much runway does it need? (b) If the final speed of the plane just before it lifts off the runway is 28 m/s, what is the tension in the tow cable while the plane and glider are moving along the runway?

Strategy We draw FBDs for the two cases: plane alone, then plane + glider. The motion in both cases is horizontal (along the runway), because we are told the displacement *before it lifts off the runway*. Until the plane begins to lift off the runway, its vertical acceleration component is zero. We need not be concerned with the vertical forces (gravity, the normal force, and lift—the upward force on the plane's wings due to the air) since they cancel one another to produce zero vertical acceleration. We use Newton's second law to compare the accelerations in the two cases and then use the accelerations to compare the displacements.

Solution (a) When the plane takes off by itself, four forces act on it (Fig. 4.48). Three are vertical and the third—the thrust due to the engine—is horizontal. Choosing the x-axis to be horizontal, Newton's second law says

$$\sum F_{1x} = F = m_1 a_{1x}$$

where F is the thrust, m_1 is the plane's mass, and a_{1x} is its horizontal acceleration component.

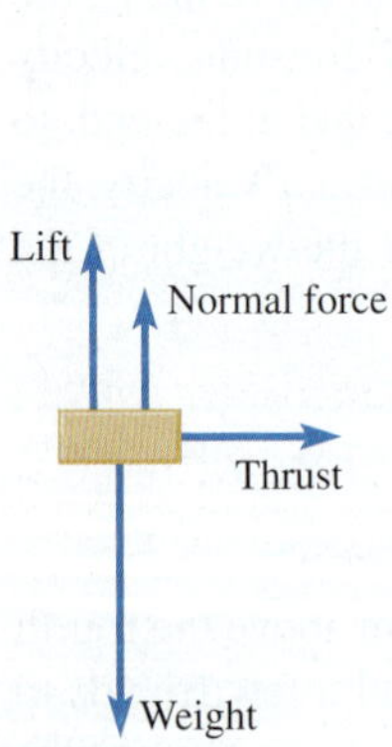

Figure 4.48
FBD for the plane.

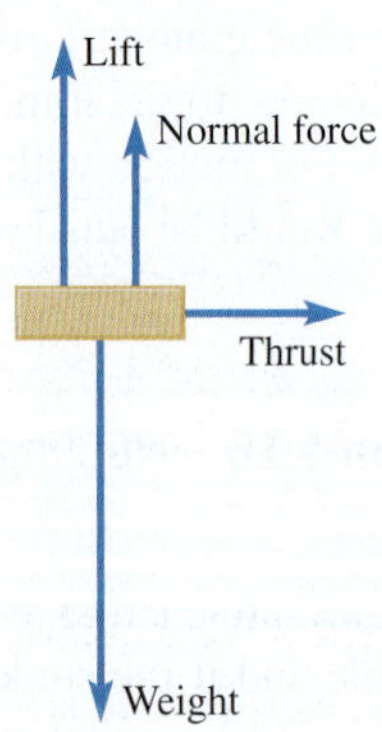

Figure 4.49
FBD for the system plane + glider.

When the glider is towed, we can consider the plane, glider, and cable to be a single system (Fig. 4.49). There is still only one horizontal external force and it is the same thrust as before. The tension in the cable is an *internal* force. Therefore,

$$\sum F_{2x} = F = (m_1 + m_2)a_x$$

where $m_1 + m_2$ is the total mass of the system (plane mass m_1 plus glider mass m_2) and a_x is the horizontal acceleration component of plane and glider. We ignore the mass of the cable.

The problem statement gives neither the thrust nor either of the accelerations. We can continue by setting the thrusts equal and finding the ratio of the accelerations:

$$m_1 a_{1x} = (m_1 + m_2)a_x \qquad \Rightarrow \qquad \frac{a_x}{a_{1x}} = \frac{m_1}{m_1 + m_2}$$

The magnitude of the acceleration is inversely proportional to the mass of the system for the same net force.

How is the acceleration related to the runway distance? The plane must get to the same final speed in order to lift off the runway. From our two basic constant acceleration equations

$$\Delta v_x = v_{fx} - v_{ix} = a_x\,\Delta t \qquad (2\text{-}9)$$

$$\Delta x = \tfrac{1}{2}(v_{fx} + v_{ix})\,\Delta t \qquad (2\text{-}11)$$

we can substitute $v_{ix} = 0$ and eliminate Δt to find

$$\Delta x = \frac{1}{2}(v_{fx} + 0)\left(\frac{v_{fx}}{a_x}\right) = \frac{v_{fx}^2}{2a_x}$$

In both cases, the displacement is inversely proportional to the acceleration and the acceleration is inversely proportional to the mass of the system. Therefore, the displacement is *directly* proportional to the mass. Letting $\Delta x_1 = 120$ m be the displacement of the plane without the glider, we can set up a proportion:

$$\frac{\Delta x}{\Delta x_1} = \frac{a_{1x}}{a_x} = \frac{m_1 + m_2}{m_1} = \frac{1090\text{ kg}}{760\text{ kg}} = 1.434$$

$$\Delta x = 1.434 \times 120\text{ m} = 172.08\text{ m}$$

To two significant figures, the plane needs 170 m of runway.

(b) The final speed given enables us to find the acceleration:

$$\Delta x = \frac{v_{fx}^2}{2a_x} \quad \text{or} \quad a_x = \frac{v_{fx}^2}{2\,\Delta x}$$

With $v_{fx} = 28$ m/s, $v_{ix} = 0$, and $\Delta x = 172.08$ m,

$$a_x = \frac{(28\text{ m/s})^2}{2 \times 172.08\text{ m}} = 2.278\text{ m/s}^2$$

The tension in the cable is the only horizontal force acting on the glider. Therefore,

$$\sum F_x = T = m_2 a_x = 330\text{ kg} \times 2.278\text{ m/s}^2 = 750\text{ N}$$

The tension is 750 N.

continued on next page

Example 4.17 continued

Discussion This solution is based on a simplified model, so we can only regard the answers as approximate. Nevertheless, it illustrates Newton's second law. The same net force produces an acceleration inversely proportional to the mass of the object upon which it acts. Here we have the same net force acting on two different objects: first the plane alone, then the plane and glider together.

Alternatively, we can look at forces acting only on the plane. When towing the glider, the cable pulls backward on the plane. The net force *on the plane* is smaller, so its acceleration is smaller. The smaller acceleration means that it takes more time to reach takeoff speed and travels a longer distance before lifting off the runway.

Practice Problem 4.17 Engine Thrust

What is the thrust provided by the airplane's engines in Example 4.17?

Example 4.18

A Pulley, an Incline, and Two Blocks

A block of mass $m_1 = 2.60$ kg rests on an incline that is angled at 30.0° above the horizontal (Fig. 4.50). An ideal cord is connected from block 1 over an ideal, frictionless pulley to another block of mass $m_2 = 2.20$ kg that is hanging 2.00 m above the ground. The coefficient of kinetic friction between the incline and block 1 is 0.180. The blocks are initially at rest. (a) How long does it take for block 2 to reach the ground? (b) Sketch a motion diagram for block 2 with a time interval of 0.5 s.

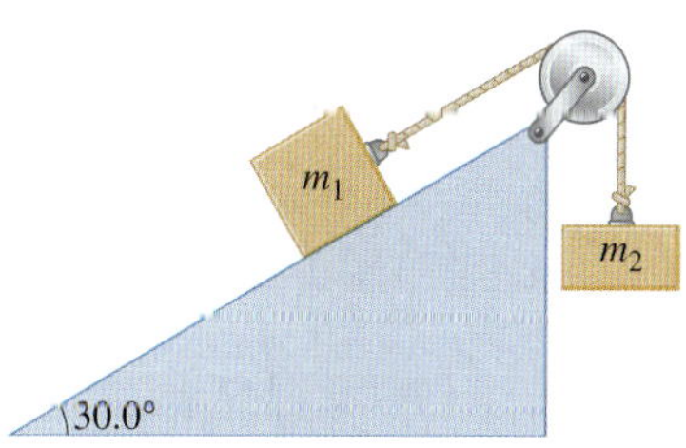

Figure 4.50 Block on an incline connected to a hanging block by a cord passing over a pulley.

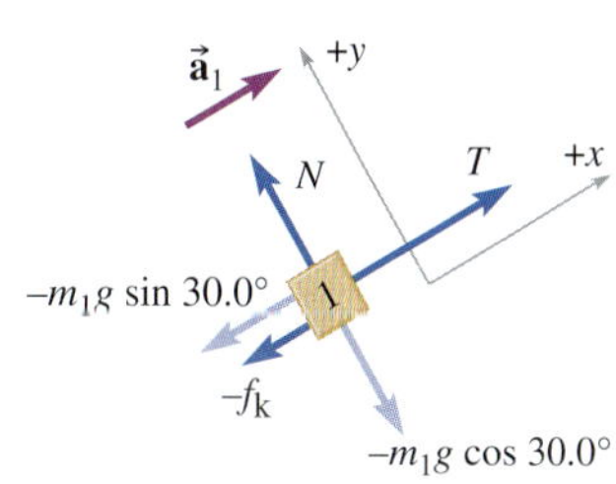

Figure 4.51 FBD for block 1. The forces are labeled with their x- and y-components.

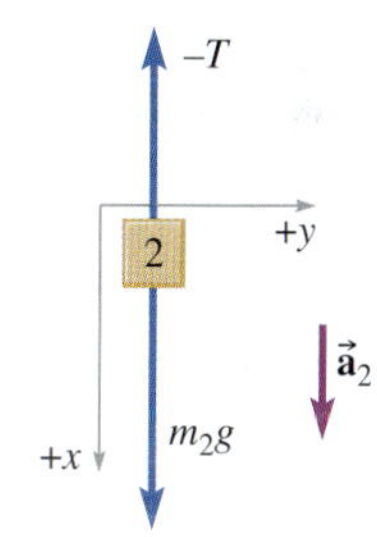

Figure 4.52 FBD for block 2 with the downward direction chosen as $+x$. The forces are labeled with their x-components.

Strategy The problem says that the blocks start from rest and that block 2 hits the floor, so block 2's acceleration is downward and block 1's is up the incline. For block 1, we choose axes parallel and perpendicular to the incline so that its acceleration has only one nonzero component. The magnitudes of the accelerations of the two blocks are equal since they are connected by an ideal cord that does not stretch. Since the cord and pulley are ideal, the tension is the same at the two ends.

Solution (a) We start by drawing separate FBDs for each block (Figs. 4.51 and 4.52). Since block 1 slides up the incline, the frictional force $\vec{f}_k$ acts down the incline to oppose the sliding. The gravitational force on block 1 is resolved into two components, one along the incline and one perpendicular to the incline.

Using the FBDs, we write Newton's second law in component form for each block. Block 1 has no acceleration component perpendicular to the incline. It does not sink into the incline or rise above it; it can only slide along the incline. Thus, the net force on block 1 in the direction perpendicular to the incline—the direction we have chosen as the y-axis for block 1—is zero.

$$\sum F_y = N - m_1 g \cos\theta = 0$$

or

$$N = m_1 g \cos\theta$$

Here $\theta = 30.0°$. Along the incline, in the x-direction for block 1, the acceleration is nonzero:

$$\sum F_x = T - m_1 g \sin\theta - f_k = m_1 a_x$$

The kinetic frictional force is related to the normal force:

$$f_k = \mu_k N = \mu_k m_1 g \cos\theta$$

By substitution,

$$T - m_1 g \sin\theta - \mu_k m_1 g \cos\theta = m_1 a_x \qquad (1)$$

continued on next page

Example 4.18 continued

For block 2, we choose an x-axis pointing downward. Doing so simplifies the solution, since then the two blocks have the same a_x. Applying Newton's second law,

$$\sum F_x = m_2 g - T = m_2 a_x \qquad (2)$$

The tension in the cord T and the x-component of acceleration a_x are both unknown in Eqs. (1) and (2). We solve for T in Eq. (2) and substitute into Eq. (1):

$$T = m_2 g - m_2 a_x = m_2(g - a_x)$$

$$m_2(g - a_x) - m_1 g \sin\theta - \mu_k m_1 g \cos\theta = m_1 a_x$$

Rearranging and solving for a_x yields

$$a_x = \frac{m_2 - m_1(\sin\theta + \mu_k \cos\theta)}{m_1 + m_2} g \qquad (3)$$

Substituting the known and given values,

$$a_x = \frac{2.20\text{ kg} - 2.60\text{ kg} \times (0.50 + 0.180 \times 0.866)}{2.60\text{ kg} + 2.20\text{ kg}} \times 9.80\text{ m/s}^2$$

$$= 1.01\text{ m/s}^2$$

Block 2 has a distance of 2.00 m to travel starting from rest with a constant downward acceleration of 1.01 m/s^2. From Eq. (2-12) with $v_{ix} = 0$,

$$\Delta x = \tfrac{1}{2} a_x (\Delta t)^2$$

The time to travel that distance is

$$\Delta t = \sqrt{\frac{2\Delta x}{a_x}} = \sqrt{\frac{2 \times 2.00\text{ m}}{1.01\text{ m/s}^2}} = 2.0\text{ s}$$

(b) Figure 4.53 shows the motion diagram for block 2. Choosing $x_i = 0$ and $t_i = 0$, the position as a function of time is $x = \frac{1}{2} a_x t^2$.

Discussion One advantage to solving for a_x algebraically in Eq. (3) before substituting numerical values is that dimensional analysis can easily be used to check for errors. In Eq. (3), the quantity in parentheses is dimensionless—the values of trigonometric functions are pure numbers as are coefficients of friction. Therefore, the numerator is the sum of two quantities with dimensions of force, the denominator is the sum of two masses, and force divided by mass gives an acceleration.

What if the problem did not tell us the directions of the blocks' accelerations? We could figure it out by comparing the force with which gravity pulls down on block 2 ($m_2 g$) with the component of the gravitational force pulling block 1 down the incline ($m_1 g \sin\theta$). Whichever is greater "wins the tug-of-war," assuming that static friction doesn't prevent the blocks from starting to slide. Once we know the direction of block 1's acceleration, we can determine the direction of the kinetic frictional force. If block 1 is not initially at rest, the kinetic frictional force opposes the direction of sliding, even though that may be opposite to the direction of the acceleration.

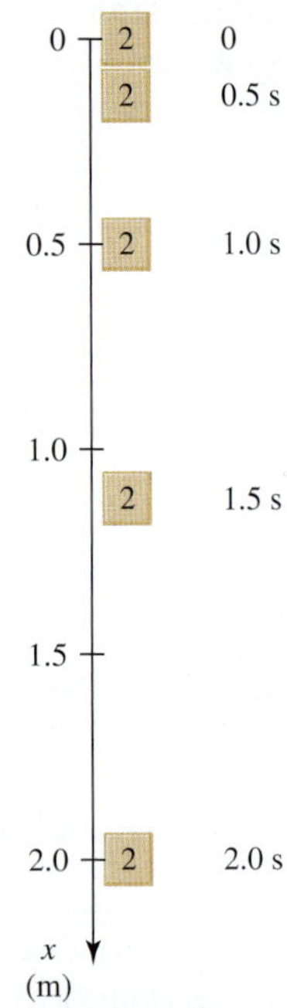

Figure 4.53 Motion diagram for block 2.

t (s)	x (m)
0	0
0.5	0.125
1.0	0.50
1.5	1.125
2.0	2.0

Practice Problem 4.18 More Fun with a Pulley and an Incline

Suppose that $m_1 = 3.8$ kg and $m_2 = 1.2$ kg and the coefficient of kinetic friction is 0.18. The blocks are released from rest and block 1 starts to slide. (a) Does block 1 slide up or down the incline? (b) In which direction does the kinetic frictional force act? (c) Find the acceleration of block 1.

CHECKPOINT 4.8

Is it ever useful to choose the x- and y-axes so the x-axis is not horizontal? If yes, give an example.

4.9 REFERENCE FRAMES

Imagine a train moving at constant velocity with respect to the ground (Fig. 4.54). Suppose Tim does some experiments using the train's reference frame for his measurements. Greg does similar experiments using the reference frame of the ground. Tim and Greg disagree about the numerical value of an object's velocity, but since their velocity

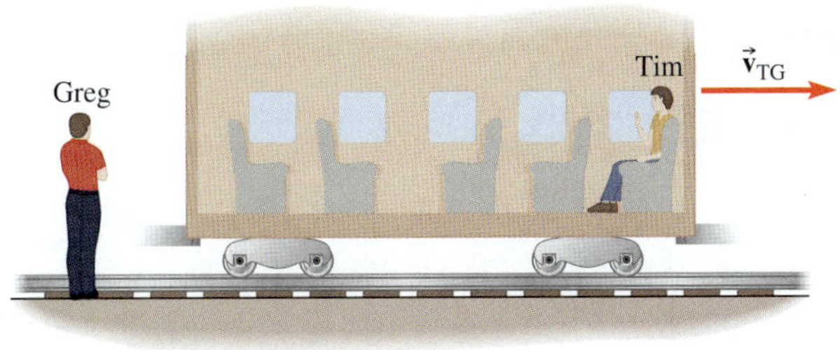

Figure 4.54 Greg's frame of reference is that of the ground; Tim's is that of the train, which moves at constant velocity $\vec{\mathbf{v}}_{TG}$ with respect to the ground.

measurements *differ by a constant*, they will always agree about *changes* in velocity and about accelerations. Both observers can use Newton's second law to relate the net force to the acceleration. *The basic laws of physics, such as Newton's laws of motion, work equally well in any two reference frames if they move with a constant relative velocity.*

CONNECTION:

The principle that the *laws* of physics are the same in different inertial reference frames is important in Newtonian mechanics, but is more general than that; it is one of the two postulates of Einstein's theory of relativity (see Chapter 26).

Newton's First Law Defines an Inertial Reference Frame You might wonder why we need Newton's first law—isn't it just a special case of the second law when $\Sigma\vec{\mathbf{F}} = 0$? No, the first law *defines* what kind of reference frame we can use when applying the second law. For the second law to be valid, we must use an *inertial reference frame*—a reference frame in which the law of inertia holds—to observe the motion of objects.

Is a reference frame attached to Earth's surface truly inertial? No, but it is close enough in many circumstances. When analyzing the motion of a soccer ball, the fact that Earth rotates about its axis does not have much effect. But if we want to analyze the motion of a meteor falling from a great distance toward Earth, Earth's rotation must be considered. We will take a closer look at the effect of Earth's rotation in Chapter 5.

4.10 APPARENT WEIGHT

Imagine being in an elevator when the cable snaps. Assume that some safety mechanism brings you to rest after you have been in free fall for a while. While you are in free fall, you *seem* to be "weightless," but your weight has not changed; the Earth still pulls downward with the same gravitational force. In free fall, gravity gives the elevator and everything in it a downward acceleration equal to $\vec{\mathbf{g}}$. If you jump up from the elevator floor, you seem to "float" up to the ceiling of the elevator. Your *weight* hasn't changed, but your *apparent* weight is zero while you are in free fall.

Similarly, astronauts in a space station in orbit around the Earth are in free fall (their acceleration is equal to the local value of $\vec{\mathbf{g}}$). Earth exerts a gravitational force on them so they are not weightless; their *apparent* weight is zero.

Imagine an object that appears to be resting on a bathroom scale. The scale measures the object's *apparent* weight W', which is equal to the true weight only if the object and the scale have zero acceleration. Newton's second law requires that

$$\sum\vec{\mathbf{F}} = \vec{\mathbf{N}} + m\vec{\mathbf{g}} = m\vec{\mathbf{a}}$$

where $\vec{\mathbf{N}}$ is the normal force of the scale pushing up. The apparent weight W' is the reading of the scale—that is, the magnitude of $\vec{\mathbf{N}}$:

$$W' = |\vec{\mathbf{N}}| = N$$

In Fig. 4.55a, the acceleration of the elevator is upward. The normal force must be larger than the weight for the net force to be upward (Fig. 4.55b). Writing the forces in component form where the $+y$-direction is upward

$$\sum F_y = N - mg = ma_y$$

or

$$N = mg + ma_y$$

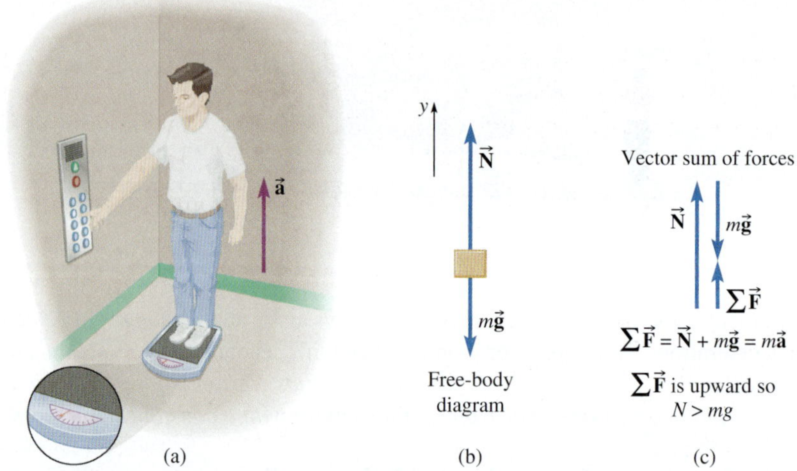

Figure 4.55 (a) Apparent weight in an elevator with acceleration upward. (b) FBD for the passenger. (c) The normal force must be greater than the weight to have an upward net force.

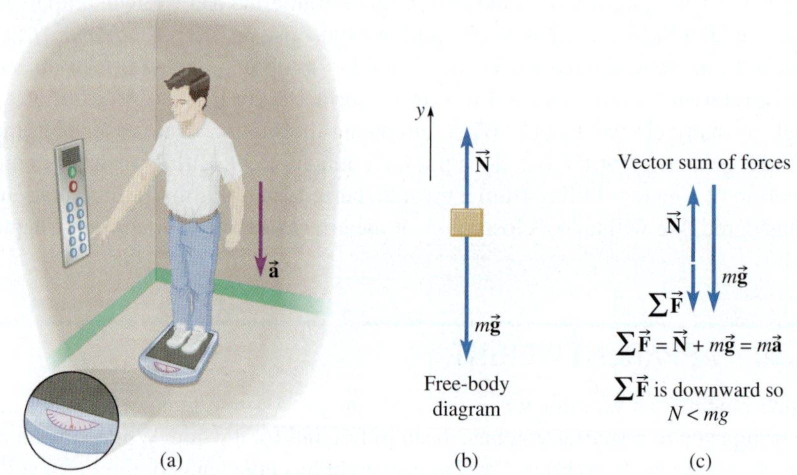

Figure 4.56 (a) Apparent weight in an elevator with acceleration downward. (b) FBD for the passenger. (c) The normal force must be less than the weight to have a downward net force.

Therefore,

$$W' = N = m(g + a_y) \tag{4-17}$$

Since the elevator's acceleration is upward, $a_y > 0$; the apparent weight is greater than the true weight (Fig. 4.55c).

In Fig. 4.56a, the acceleration is downward. Then the net force must also point downward. The normal force is still upward, but it must be smaller than the weight in order to produce a downward net force (Fig. 4.56b). It is still true that $W' = m(g + a_y)$, but now the acceleration is downward ($a_y < 0$). The apparent weight is less than the true weight (Fig. 4.56c). If the elevator is in free fall, then $a_y = -g$ and the apparent weight of the unfortunate passenger is zero.

Example 4.19

Apparent Weight in an Elevator

A passenger weighing 598 N rides in an elevator. What is the apparent weight of the passenger in each of the following situations? In each case, the magnitude of the elevator's acceleration is 0.500 m/s^2. (a) The passenger is on the first floor and has pushed the button for the fifteenth floor; the elevator is beginning to move upward. (b) The elevator is slowing down as it nears the fifteenth floor.

continued on next page

Example 4.19 continued

Strategy In each case, we sketch the FBD for the passenger. The apparent weight is equal to the magnitude of the normal force exerted by the floor on the passenger. The only other force acting is gravity. Newton's second law lets us find the normal force from the weight and the acceleration. Known: $W = 598$ N; magnitude of the acceleration is $a = 0.500\ \text{m/s}^2$. To find: W'.

Solution (a) Let the $+y$-axis be upward. When the elevator starts up from the first floor it has acceleration in the upward direction as its speed increases. Since the elevator's acceleration is upward, $a_y > 0$ (as in Fig. 4.55). We expect the apparent weight $W' = N$ to be greater than the true weight—the floor must push up with a force greater than W to cause an upward acceleration. Figure 4.57 is the FBD. Newton's second law says

$$\sum F_y = N - W = ma_y$$

Since $W = mg$, we can substitute $m = W/g$.

$$W' = N = W + ma_y = W + \frac{W}{g}a_y = W\left(1 + \frac{a_y}{g}\right)$$

$$= 598\ \text{N} \times \left(1 + \frac{0.500\ \text{m/s}^2}{9.80\ \text{m/s}^2}\right) = 629\ \text{N}$$

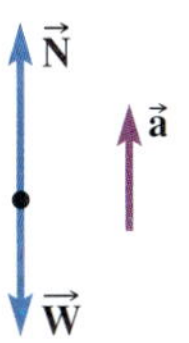

Figure 4.57
FBD for the passenger in an elevator with upward acceleration.

(b) When the elevator approaches the fifteenth floor, it is slowing down while still moving upward; its acceleration is downward ($a_y < 0$) as in Fig. 4.56. The apparent weight is less than the true weight. Figure 4.58 is the FBD. Again, $\Sigma F_y = N - W = ma_y$, but this time $a_y = -0.500\ \text{m/s}^2$.

$$N = W\left(1 + \frac{a_y}{g}\right)$$

$$= 598\ \text{N} \times \left(1 + \frac{-0.500\ \text{m/s}^2}{9.80\ \text{m/s}^2}\right) = 567\ \text{N}$$

Discussion The apparent weight is greater when the direction of the elevator's acceleration is upward. That can happen in two cases: either the elevator is moving up with increasing speed, or it is moving down with decreasing speed.

Practice Problem 4.19 Elevator Descending

What is the apparent weight of a passenger of mass 42.0 kg traveling in an elevator in each of the following situations? In each case, the magnitude of the elevator's acceleration is $0.460\ \text{m/s}^2$. (a) The passenger is on the fifteenth floor and has pushed the button for the first floor; the elevator is beginning to move downward. (b) The elevator is slowing down as it nears the first floor.

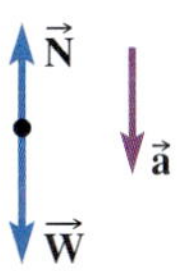

Figure 4.58
FBD for the passenger in an elevator with downward acceleration.

EVERYDAY PHYSICS DEMO

Take a bathroom scale to an elevator. Stand on the scale inside the elevator and push a button for a higher floor. When the elevator's acceleration is upward, you can feel the increase in your apparent weight and can see the increase by the reading on the scale. When the elevator slows down to stop, the elevator's acceleration is downward and your apparent weight is less than your true weight.

What is happening in your body while the elevator accelerates? The inertia principle means that your blood and internal organs cannot have the same acceleration as the elevator until the correct net force acts on them. Blood tends to collect in the lower extremities during acceleration upward and in the upper body during acceleration downward until the forces exerted on the blood by the body readjust to give the blood the same acceleration as the elevator. Likewise, the internal organs shift position within the body cavity, resulting in a funny feeling in the gut as the elevator starts and stops. To avoid this problem, high-speed express elevators in skyscrapers keep the acceleration relatively small, but maintain that acceleration long enough to reach high speeds. That way, the elevator can travel quickly to the upper floors without making the passengers feel too uncomfortable.

CHECKPOINT 4.10

You are standing on a bathroom scale in an elevator that is moving downward. Nearing your stop, the elevator's speed is decreasing. Is the scale reading greater or less than your weight?

4.11 AIR RESISTANCE

So far we have ignored the effect of air resistance on falling objects and projectiles. A skydiver relies on a parachute to provide a large force of air resistance (also called **drag**). Even with the parachute closed, drag is not negligible when the skydiver is falling rapidly. The drag force is similar to friction between two solid surfaces in that the direction of the force *opposes the motion* of the object through the air. However, in contrast to the force of friction, the magnitude of the drag force is strongly dependent on the speed of the object. In many cases, air drag is proportional to the square of the speed. Drag also depends on the size and shape of the object.

Since the drag force increases as the speed increases, a falling object approaches an equilibrium situation in which the drag force is equal in magnitude to the weight but opposite in direction. The velocity at which this equilibrium occurs is called the object's *terminal velocity*. (See the text website for a more detailed treatment of drag.)

connect

EVERYDAY PHYSICS DEMO

Drop a basket-style paper coffee filter (or a cupcake paper) and a coin simultaneously from as high above the floor as you can safely do so. Air resistance on the coin is negligible unless it is dropped from a great height. At the other extreme, the effect of air resistance on the coffee filter is very noticeable; it reaches its terminal speed almost immediately. Stack several (two to four) coffee filters together and drop them simultaneously with a single coffee filter. Why is the terminal speed higher for the stack? Crumple a coffee filter into a ball and drop it simultaneously with the coin. Air resistance on the coffee filter is now reduced, but still noticeable.

4.12 FUNDAMENTAL FORCES

One of the main goals of physics has been to understand the immense variety of forces in the universe in terms of the fewest number of fundamental laws. Physics has made great progress in this quest for *unification*; today all forces are understood in terms of just four fundamental interactions (Fig. 4.59). At the high temperatures present in the early universe, two of these interactions—the electromagnetic and weak forces—are now understood as the effects of a single electroweak interaction. The ultimate goal is to describe all forces in terms of a single interaction.

Gravity You may be surprised to learn that gravity is by far the *weakest* of the fundamental forces. Any two objects exert gravitational forces on one another, but the force is tiny unless at least one of the masses is large. We tend to notice the relatively

Figure 4.59 All forces result from just four fundamental forces: gravity, electromagnetism, and the weak and strong forces.

large gravitational forces exerted by planets and stars, but not the feeble gravitational forces exerted by smaller objects, such as the gravitational force this book exerts on your body.

Gravity has an unlimited range. The force gets weaker as the distance between two objects increases, but it never drops to zero, no matter how far apart the objects get.

Newton's law of gravity is an early example of unification. Before Newton, people did not understand that the same kind of force that makes an apple fall from a tree also keeps the planets in their orbits around the Sun. A single law—Newton's law of universal gravitation—describes both.

Electromagnetism The electromagnetic force is unlimited in range, like gravity. It acts on particles with electric charge. The electric and magnetic forces were unified into a single theoretical framework in the nineteenth century. We study electromagnetic forces in detail in Part 3 of this book.

Electromagnetism is the fundamental interaction that binds electrons to nuclei to form atoms and binds atoms together in molecules and solids. It is responsible for the properties of solids, liquids, and gases and forms the basis of the sciences of chemistry and biology. It is the fundamental interaction behind all macroscopic contact forces such as the frictional and normal forces between surfaces and forces exerted by cords, springs, muscles, and the wind.

The electromagnetic force is *much* stronger than gravity. For example, the electrical repulsion of two electrons at rest is about 10^{43} times as strong as the gravitational attraction between them. Macroscopic objects have a nearly perfect balance of positive and negative electric charge, resulting in a nearly perfect balance of attractive and repulsive electromagnetic forces between the objects. Therefore, despite the fundamental strength of the electromagnetic forces, the sum of the electromagnetic forces between two macroscopic objects is often negligibly small except when atoms on the two surfaces come very close to each other—what we think of as *in contact*. On a microscopic level, there is no fundamental difference between contact forces and other electromagnetic forces.

The Strong Force The strong force holds protons and neutrons together in the atomic nucleus. The same force binds quarks (a family of elementary particles) in combinations so they can form protons and neutrons and many more exotic subatomic particles. The strong force is the strongest of the four fundamental forces—hence its name—but its range is short: its effect is negligible at distances much larger than the size of an atomic nucleus (about 10^{-15} m).

The Weak Force The range of the weak force is even shorter than that of the strong force (about 10^{-17} m). It is manifest in many radioactive decay processes.

Master the Concepts

- An interaction between two objects consists of two forces, one on each of the objects. Loosely speaking, a *force* is a push or a pull. Gravity and electromagnetic forces have unlimited range. All other forces exerted on macroscopic objects involve contact. Force is a vector quantity.
- The SI unit of force is the newton: $1\ \text{N} = 1\ \text{kg·m/s}^2$.
- The *net force* on a system is the vector sum of all the forces acting on it:

$$\vec{\mathbf{F}}_{net} = \sum\vec{\mathbf{F}} = \vec{\mathbf{F}}_1 + \vec{\mathbf{F}}_2 + \cdots + \vec{\mathbf{F}}_n \quad (4\text{-}2)$$

Since all the internal forces form interaction pairs, we need only sum the external forces.

- *Newton's first law of motion*: If zero net force acts on an object, then the object's velocity does not change. Velocity is a vector whose magnitude is the speed at which the object moves and whose direction is the direction of motion.
- *Newton's second law of motion* relates the net force acting on an object to the object's acceleration and its mass:

$$\vec{\mathbf{a}} = \frac{\sum\vec{\mathbf{F}}}{m} \quad \text{or} \quad \sum\vec{\mathbf{F}} = m\vec{\mathbf{a}} \quad (4\text{-}4)$$

The acceleration is always in the same direction as the net force. Many problems involving Newton's second law—whether equilibrium or nonequilibrium—can be solved by treating the x- and y-components of the forces and the acceleration separately:

$$\sum F_x = ma_x \quad \text{and} \quad \sum F_y = ma_y \quad (4\text{-}5)$$

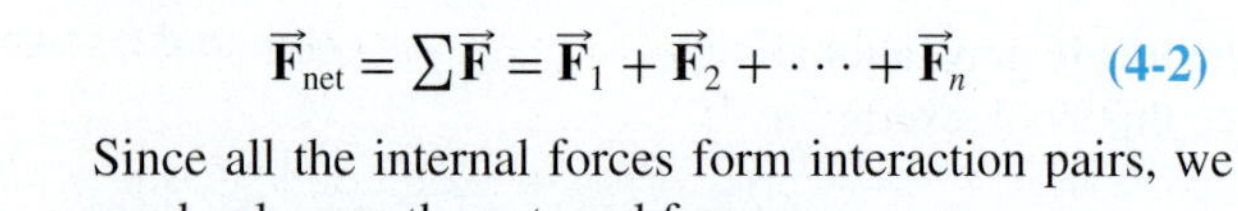

- *Newton's third law of motion*: In an interaction between two objects, each object exerts a force on the other. These two forces are equal in magnitude and opposite in direction:

$$\vec{\mathbf{F}}\,(\text{on } B \text{ by } A) = -\vec{\mathbf{F}}\,(\text{on } A \text{ by } B) \quad (4\text{-}7)$$

- A *free-body diagram* (FBD) includes vector arrows representing every force acting on the chosen object due to some other object, but no forces acting on other objects.

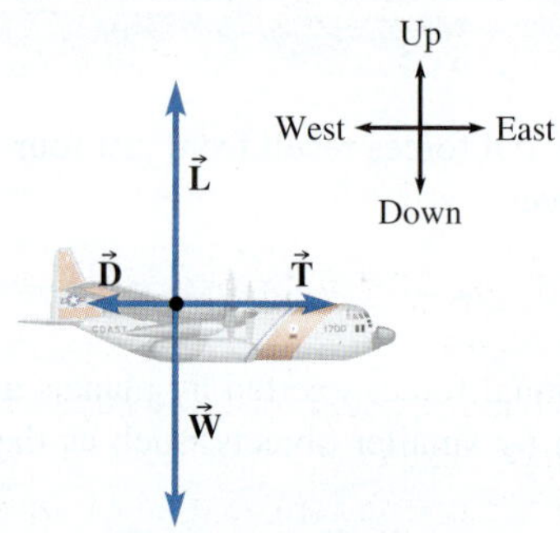

- The magnitude of the *gravitational force* between two objects is

$$F = \frac{Gm_1m_2}{r^2} \quad (4\text{-}8)$$

where r is the distance between their centers. Each object is pulled toward the other's center.

- The *weight* of an object is the magnitude of the gravitational force acting on it. An object's weight is proportional to its mass: $W = mg$ [Eq. (4-11a)], where g is the gravitational field strength. Near Earth's surface, $g \approx 9.80$ N/kg.
- The *normal force* is a contact force perpendicular to the contact surfaces that pushes each object away from the other.

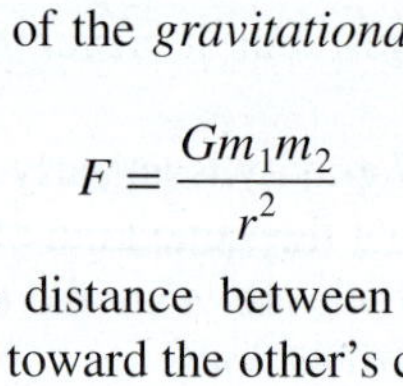

- *Friction* is a contact force parallel to the contact surfaces. In a simplified model, the kinetic frictional force and the maximum static frictional force are proportional to the normal force acting between the same contact surfaces.

$$f_s \le \mu_s N \quad (4\text{-}15)$$

$$f_k = \mu_k N \quad (4\text{-}16)$$

continued on next page

Master the Concepts continued

The static frictional force acts in the direction that tends to keep the surfaces from beginning to slide. The direction of the kinetic frictional force is in the direction that would tend to make the sliding stop.

- An ideal cord pulls in the direction of the cord with forces of equal magnitude on the objects attached to its ends as long as no external force tangent to the cord is exerted on it anywhere between the ends. The tension of an ideal cord that runs through an ideal pulley is the same on both sides of the pulley.
- An object that is accelerating has an apparent weight that differs from its true weight. The apparent weight is equal to the normal force exerted by a supporting surface with the same acceleration. A helpful trick is to think of the apparent weight as the reading of a bathroom scale that supports the object.
- The drag force exerted on an object moving through air opposes the motion of the object but, unlike kinetic friction, is strongly dependent on the object's speed. When an object falls at its terminal velocity, the drag force is equal and opposite to the gravitational force, so the acceleration is zero.
- At the fundamental level, there are four interactions: gravity, the strong and weak interactions, and the electromagnetic interaction. Contact forces are large-scale manifestations of many microscopic electromagnetic interactions.

Conceptual Questions

1. Explain the need for automobile seat belts in terms of Newton's first law.
2. An American visitor to Finland is surprised to see heavy metal frames outside of all the apartment buildings. On Saturday morning the purpose of the frames becomes evident when several apartment dwellers appear, carrying rugs and carpet beaters to each frame. What role does the principle of inertia play in the rug beating process? Do you see a similarity to the role the principle of inertia plays when you throw a baseball?
3. You are lying on the beach after a dip in the ocean where the waves were buffeting you around. Is it true that there are now no forces acting on you? Explain.
4. A dog goes swimming at the beach and then shakes himself all over to get dry. What principle of physics aids in the drying process? Explain.
5. In an attempt to tighten the loosened steel head of a hammer, a carpenter holds the hammer vertically, raises it up, and then brings it down rapidly, hitting the bottom end of the wood handle on a two-by-four board. Explain how this tightens the head back onto the handle.

6. When a car begins to move forward, what force makes it do so? Remember that it has to be an *external* force; the internal forces all add to zero. How does the engine, which is part of the car, cause an external force to act on the car?
7. Two cars are headed toward each other in opposite directions along a narrow country road. The cars collide head-on, crumpling up the hoods of both. Describe what happens to the car bodies in terms of the principle of inertia. Does the rear end of the car stop at the same time as the front end?
8. Can an object in free fall be in equilibrium? Explain.
9. (a) What assumptions do you make when you call the reading of a bathroom scale your "weight"? What does the scale really tell you? (b) Under what circumstances might the reading of the scale *not* be equal to your weight?
10. A freight train consists of an engine and several identical cars on level ground. Determine whether each of these statements is correct or incorrect and explain why. (a) If the train is moving at constant speed, the engine must be pulling with a force greater than the train's weight. (b) If the train is moving at constant speed, the engine's pull on the first car must exceed that car's backward pull on the engine. (c) If the train is coasting, its inertia makes it slow down and eventually stop.
11. (a) Does a man weigh more at the North Pole or at the equator? (b) Does he weigh more at the top of Mt. Everest or at the base of the mountain?
12. What is the acceleration of an object thrown straight up into the air at the highest point of its motion? Does the answer depend on whether air resistance is negligible or not? Explain.
13. If a wagon starts at rest and pulls back on you with a force equal to the force you pull on it, as required by Newton's third law, how is it possible for you to make the wagon start to move? Explain.
14. You are standing on a bathroom scale in an elevator. In which of these situations must the scale read the same as when the elevator is at rest? Explain. (a) Moving up at constant speed. (b) Moving up with increasing speed.

(c) In free fall (after the elevator cable has snapped and before the safety brakes have engaged).

15. A heavy ball hangs from a string attached to a sturdy wooden frame. A second string is attached to a hook on the bottom of the lead ball. You pull slowly and steadily on the lower string. Which string do you think will break first? Explain.

16. An SUV collides with a Mini Cooper convertible. Is the force exerted on the Mini by the SUV greater than, equal to, or less than the force exerted on the SUV by the Mini? Explain.

17. You are standing on one end of a light wooden raft that has floated 3 m away from the pier. If the raft is 6 m long by 2.5 m wide and you are standing on the raft end nearest to the pier, can you propel the raft back toward the pier where a friend is standing with a pole and hook trying to reach you? You have no oars. Make suggestions of what to do without getting yourself wet.

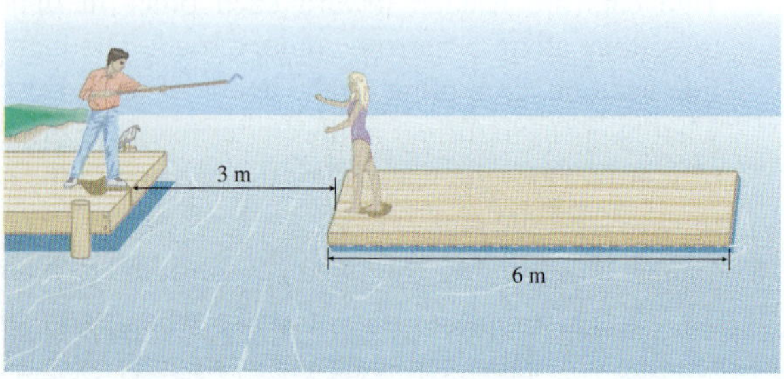

18. What does it mean when we refer to a cord as an "ideal cord" and a pulley as an "ideal pulley"?

19. If a feather and a lead brick are dropped simultaneously from the top of a ladder, the lead brick hits the ground first. What would happen if the experiment is repeated on the surface of the Moon?

20. Two boys are trying to break a cord. Gerardo says they should each pull in opposite directions on the two ends; Stefan says they should tie the cord to a pole and both pull together on the opposite end. Which plan is more likely to work?

21. Why might an elevator cable break during acceleration when lifting a lighter load than it normally supports at rest or at constant velocity?

22. If air resistance is ignored, what force(s) act on an object in free fall?

23. The net force acting on an object is constant. Under what circumstances does the object move along a straight line? Under what circumstances does the object move along a curved path?

24. Pulleys and inclined planes are examples of *simple machines*. Explain what these machines do in Examples 4.10, 4.12, and 4.16 to make a task easier to perform.

25. For a problem about a crate sliding along an inclined plane, is it possible to choose the x-axis so that it is parallel to the incline?

26. A bird sits on a stretched clothesline, causing it to sag slightly. Is the tension in the line greatest where the bird sits, greater at either end of the line where it is attached to poles, or the same everywhere along the line? Treat the line as an ideal cord with negligible weight.

27. You decide to test your physics knowledge while going over a waterfall in a barrel. You take a baseball into the barrel with you and as you are falling vertically downward, you let go of the ball. What do you expect to see for the motion of the ball relative to the barrel? Will the ball fall faster than you and move toward the bottom of the barrel? Will it move slower than you and approach the top of the barrel, or will it hover apparently motionless within the falling barrel? Explain. [*Warning*: Do not try this.]

28. A person is standing on a bathroom scale. Which of the following is *not* a force exerted *on the scale*: a contact force due to the floor, a contact force due to the person's feet, the weight of the person, the weight of the scale?

29. Does the concept of a contact force apply to both a macroscopic scale and an atomic scale? Explain.

30. If an object is acted on by a single constant force, is it possible for the object to remain at rest? Is it possible for the object to move with constant velocity? Is it possible for the object's speed to be decreasing? Is it possible for it to change direction? In each case that is possible, describe the direction of the force.

31. If an object is acted on by two constant forces is it possible for the object to move at constant velocity? If so, what must be true about the two forces? Give an example.

32. A parked automobile slips out of gear, rolls unattended down a slight incline, and then along a level road until it hits a stone wall. Draw an FBD to show the forces acting on the car while it is in contact with the wall.

Multiple-Choice Questions

1. Interaction partners
   (a) are equal in magnitude and opposite in direction and act on the same object.
   (b) are equal in magnitude and opposite in direction and act on different objects.
   (c) appear in an FBD for a given object.
   (d) always involve gravitational force as one partner.
   (e) act in the same direction on the same object.

2. Within a given system, the internal forces
   (a) are always balanced by the external forces.
   (b) all add to zero.
   (c) are determined only by subtracting the external forces from the net force on the system.
   (d) determine the motion of the system.
   (e) can never add to zero.

3. A friction force is
(a) a contact force that acts parallel to the contact surfaces.
(b) a contact force that acts perpendicular to the contact surfaces.
(c) a scalar quantity since it can act in any direction along a surface.
(d) always proportional to the weight of an object.
(e) always equal to the normal force between the objects.

4. When a force is called a "normal" force, it is
(a) the usual force expected given the arrangement of a system.
(b) a force that is perpendicular to the surface of the Earth at any given location.
(c) a force that is always vertical.
(d) a contact force perpendicular to the contact surfaces between two solid objects.
(e) the net force acting on a system.

5. Your car won't start, so you are pushing it. You apply a horizontal force of 300 N to the car, but it doesn't budge. What force is the interaction partner of the 300 N force you exert?
(a) the frictional force exerted on the car by the road
(b) the force exerted on you by the car
(c) the frictional force exerted on you by the road
(d) the normal force on you by the road
(e) the normal force on the car by the road

6. Which of these is *not* a long-range force?
(a) the force that makes raindrops fall to the ground
(b) the force that makes a compass point north
(c) the force that a person exerts on a chair while sitting
(d) the force that keeps the Moon in its orbital path around the Earth

7. When an object is in translational equilibrium, which of these statements is *not* true?
(a) The vector sum of the forces acting on the object is zero.
(b) The object must be stationary.
(c) The object has a constant velocity.
(d) The speed of the object is constant.

8. To make an object start moving on a surface with friction requires
(a) less force than to keep it moving on the surface.
(b) the same force as to keep it moving on the surface.
(c) more force than to keep it moving on the surface.
(d) a force equal to the weight of the object.

9. A thin string that can support a weight of 35.0 N, but breaks under any larger weight, is attached to the ceiling of an elevator. How large a mass can be attached to the string if the initial acceleration as the elevator starts to ascend is 3.20 m/s^2?
(a) 3.57 kg (b) 2.69 kg (c) 4.26 kg
(d) 2.96 kg (e) 5.30 kg

10. A woman stands on a bathroom scale in an elevator that is not moving. The scale reads 500 N. The elevator then moves downward at a constant velocity of 4.5 m/s. What does the scale read while the elevator descends with constant velocity?
(a) 100 N (b) 250 N (c) 450 N
(d) 500 N (e) 750 N

11. A 70.0 kg man stands on a bathroom scale in an elevator. What does the scale read if the elevator is slowing down at a rate of 3.00 m/s^2 while descending?
(a) 70 kg (b) 476 N (c) 686 N
(d) 700 N (e) 896 N

12. A space probe leaves the solar system to explore interstellar space. Once it is far from any stars, when must it fire its rocket engines?
(a) All the time, in order to keep moving.
(b) Only when it wants to speed up.
(c) When it wants to speed up or slow down.
(d) Only when it wants to turn.
(e) When it wants to speed up, slow down, or turn.

13. A small plane climbs with a constant velocity of 250 m/s at an angle of 28° with respect to the horizontal. Which statement is true concerning the magnitude of the net force on the plane?
(a) It is equal to zero.
(b) It is equal to the weight of the plane.
(c) It is equal to the magnitude of the force of air resistance.
(d) It is less than the weight of the plane but greater than zero.
(e) It is equal to the component of the weight of the plane in the direction of motion.

14. Two blocks are connected by a light string passing over a pulley (see the figure and connect tutorial: pulley). The block with mass m_1 slides on a frictionless horizontal surface, and the block with mass m_2 hangs vertically. If $m_1 > m_2$ and the pulley and string are assumed to be ideal, the tension in the string is
(a) zero.
(b) less than m_2g.
(c) equal to m_2g.
(d) greater than m_2g, but less than m_1g.
(e) equal to m_1g.
(f) greater than m_1g.

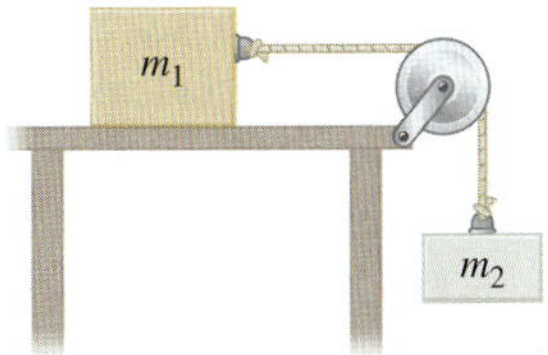

15. You place two different coins on the cover of your physics book and then slowly lift the cover. Assuming the coefficients of static friction are the same, which is true?
(a) The more massive coin starts to slide first.
(b) The less massive coin starts to slide first.
(c) The two coins start to slide at the same time.

16. A crate containing a new water heater weighs 800 N. The crate rests on the basement floor. Tim pushes horizontally on it with a force of 400 N, but it doesn't budge. What can you conclude about the coefficient of static friction between the crate and the floor?
   (a) $\mu_s = 0.5$ (b) $\mu_s \geq 0.5$ (c) $\mu_s \leq 0.5$
   (d) Not enough information is given to draw any of these conclusions.
17. A crate containing a new water heater weighs 800 N. Tim and a friend push horizontally on the water heater with a force of 600 N as it slides across the floor with constant velocity. What can you conclude about the coefficient of kinetic friction between the crate and the floor?
   (a) $\mu_k = 0.75$ (b) $\mu_k \geq 0.75$ (c) $\mu_k \leq 0.75$
   (d) Not enough information is given to draw any of these conclusions.
18. A woman stands on an airport's moving sidewalk and moves due west at constant velocity. The frictional force on the woman is _____.
   (a) zero
   (b) kinetic and to the west
   (c) kinetic and to the east
   (d) static and to the west
   (e) static and to the east

Questions 19–22. For each situation, how does the magnitude of the normal force *N* compare with the object's weight *W*? Answer choices:
   (a) equal to *W*
   (b) greater than *W*
   (c) less than *W*
   (d) The given information is insufficient to determine the relative magnitude of the normal force.
19. A child (weight *W*) sits on a level floor. The normal force on the child is _____.
20. A car (weight *W*) is parked on an incline. The magnitude of the normal force on the car is _____.
21. A weightlifter (weight *W*) holds a 400 N barbell above his head. The magnitude of the normal force on the weightlifter due to the floor is _____.
22. A passenger (weight *W*) rides in an elevator. The magnitude of the normal force on the passenger due to the floor is _____.
23. Two blocks of unequal masses are connected by an ideal cord that passes over an ideal pulley. Which is true concerning the accelerations of the blocks and the net forces acting on the blocks?
   (a) accelerations equal in magnitude, net forces unequal in magnitude
   (b) accelerations unequal in magnitude, net forces equal in magnitude
   (c) accelerations equal in magnitude, net forces equal in magnitude
   (d) accelerations unequal in magnitude, net forces unequal in magnitude

Questions 24–26. A ball is tossed straight up. Air resistance is *not* negligible; the force of air resistance is opposite in direction to the ball's velocity. Assume its magnitude is less than the ball's weight. Answer choices:
   (a) less than *g*
   (b) equal to *g*
   (c) greater than *g*
24. On the way up, the magnitude of the ball's acceleration is _____.
25. At the top the magnitude of the ball's acceleration is _____.
26. On the way down, the magnitude of the ball's acceleration is _____.

Problems

C Combination conceptual/quantitative problem
Biological or medical application
✦ Challenging problem
Blue # Detailed solution in the Student Solutions Manual
[1, 2] Problems paired by concept
connect Text website interactive or tutorial

4.1 Force

1. A sack of flour has a weight of 19.8 N. What is its weight in pounds?
2. An astronaut weighs 175 lb. What is his weight in newtons?
3. A force of 20 N is directed at an angle of 60° above the *x*-axis. A second force of 20 N is directed at an angle of 60° below the *x*-axis. What is the vector sum of these two forces?
4. Two draft horses, Sam and Bob, are dragging a sled loaded with jugs of maple syrup. They pull with horizontal forces of equal magnitude 1.50 kN on the front of the sled. The force due to Sam is in the direction 15° north of east, and the force due to Bob is 15° south of east. Use the graphical method of vector addition to find the magnitude and direction of the sum of the forces exerted on the sled by the two horses.

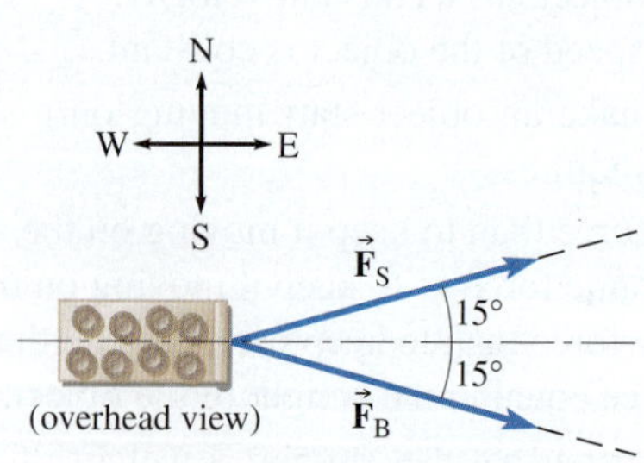

Problems 4 and 5

5. In Problem 4, if Sam pulls at 10° north of east while Bob pulls at 15° south of east, is it still possible for the sum of the two forces to be due east if their magnitudes are not the same? Which force must have the larger magnitude? Illustrate with a sketch.

6. Suppose you are standing on the floor doing your daily exercises. For one exercise, you lift your arms up and out until they are horizontal. In this position, assume that the deltoid muscle exerts a force of 270 N at an angle of 15° above the horizontal on the humerus, as shown in the figure. What are the *x*- and *y*-components of this force?

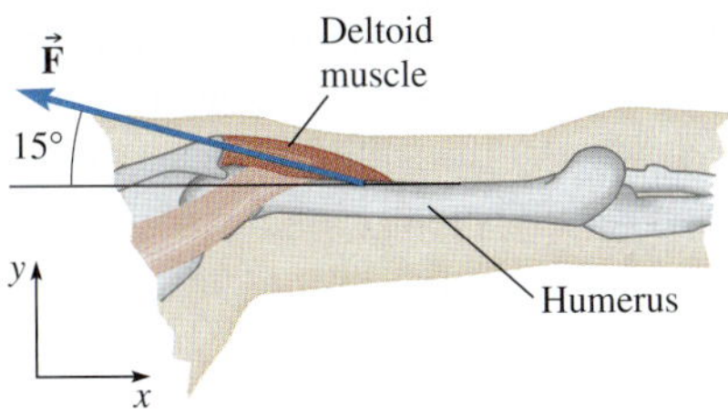

7. While tilling your garden, you exert a force on the handles of the tiller that has components $F_x = +85$ N and $F_y = -132$ N. The *x*-axis is horizontal and the *y*-axis points up. What are the magnitude and direction of this force?

8. Juan is helping his mother rearrange the living room furniture. Juan pushes on the armchair with a force of 30 N directed at an angle of 15° above a horizontal line while his mother pushes with a force of 40 N directed at an angle of 20° below the same horizontal. What is the vector sum of these two forces?

9. In the drawing, what is the vector sum of forces $\vec{A} + \vec{B} + \vec{C}$ if each grid square is 2 N on a side?

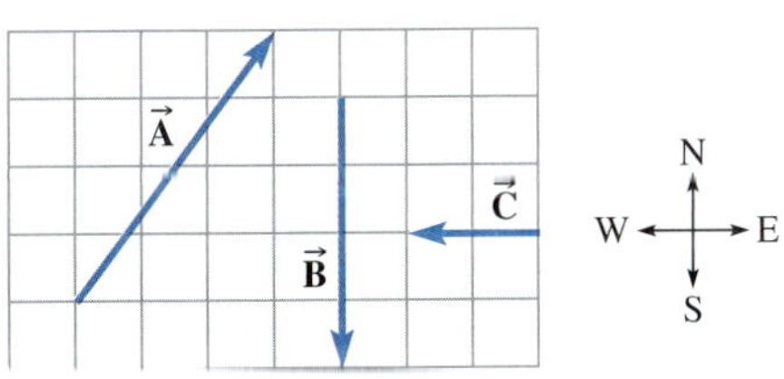

10. In the drawing, what is the vector sum of forces $\vec{D} + \vec{E} + \vec{F}$ if each grid square is 2 N on a side?

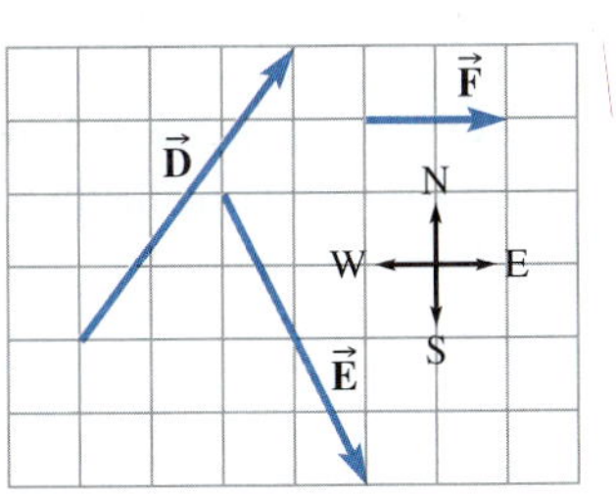

11. Two of Robin Hood's men are pulling a sledge loaded with some gold along a path that runs due north to their hideout. One man pulls his rope with a force of 62 N at an angle of 12° east of north and the other pulls with the same force at an angle of 12° west of north. Assume the ropes are parallel to the ground. What is the sum of these two forces on the sledge?

12. A barge is hauled along a straight-line section of canal by two horses harnessed to tow ropes and walking along the tow paths on either side of the canal. Each horse pulls with a force of 560 N at an angle of 15° with the centerline of the canal. Find the sum of the two forces exerted by the horses on the barge.

13. On her way to visit Grandmother, Red Riding Hood sat down to rest and placed her basket of goodies beside her. A wolf came along, spotted the basket, and began to pull on the handle with a force of 6.4 N at an angle of 25° with respect to vertical. Red was not going to let go easily, so she pulled on the handle with a force of 12 N. If the sum of these two forces on the basket is straight up, at what angle was Red Riding Hood pulling?

14. Two objects, *A* and *B*, are acted on by the forces shown in the FBDs. Is the magnitude of the net force acting on object *B* greater than, less than, or equal to the magnitude of the net force acting on object *A*? Explain.

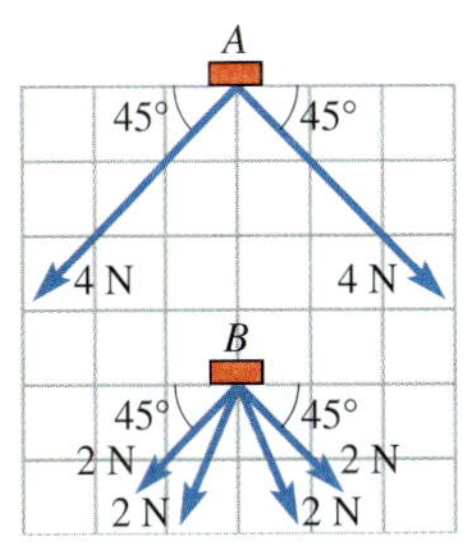

15. Find the magnitude and direction of the net force on the object in each of the FBDs for this problem.

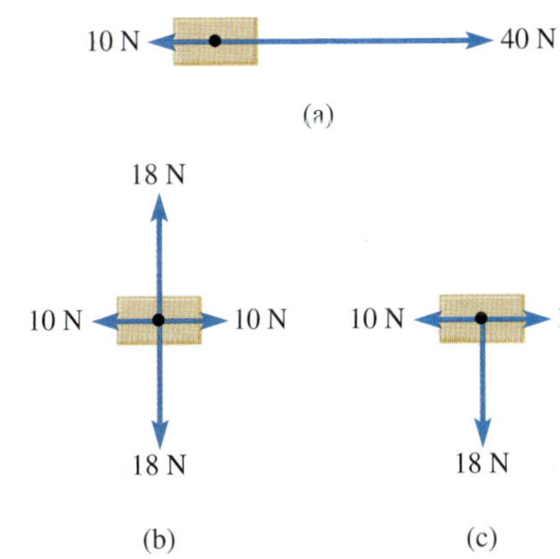

16. A truck driving on a level highway is acted on by the following forces: a downward gravitational force of 52 kN (kilonewtons); an upward contact force due to the road of 52 kN; another contact force due to the road of 7 kN, directed east; and a drag force due to air resistance of 5 kN, directed west. What is the net force acting on the truck?

4.2 Inertia and Equilibrium: Newton's First Law of Motion; 4.3 Net Force, Mass, and Acceleration: Newton's Second Law of Motion

17. A tennis ball (mass 57.0 g) moves toward the player's racquet at 47.5 m/s. It is in contact with the racquet for 3.60 ms, after which it moves in the opposite direction

at 50.2 m/s. What is the average force on the ball during this time interval?

18. A red-tailed hawk that weighs 8 N is gliding due north at constant speed. What is the total force acting on the hawk due to the air? Draw a free-body diagram for the hawk.

19. An 80 N crate of apples sits at rest on the horizontal bed of a parked pickup truck. What is the total contact force exerted on the crate by the bed of the pickup? Draw a free-body diagram for the crate.

20. Forces of magnitudes 2000 N and 3000 N act on five objects. The directions of the forces are shown in the sketches. Rank the objects according to the magnitude of the net force, from smallest to largest. Explain your reasoning.

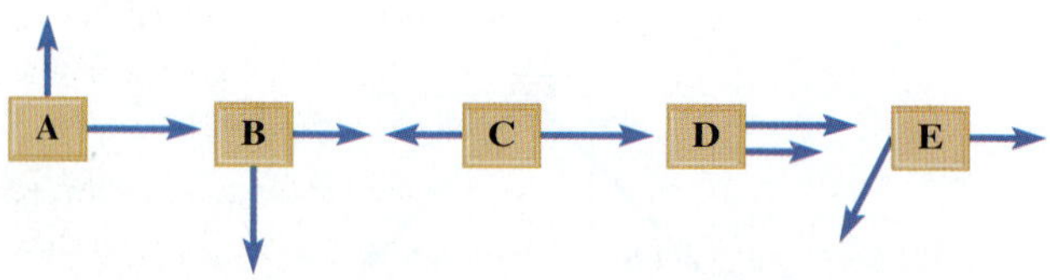

21. A person stands on the ball of one foot. The force due to the ground pushing up on the ball of the foot has magnitude 750 N. Ignore the weight of the foot itself. The other significant forces acting on the foot are the Achilles tendon pulling up and the tibia pushing down on the ankle joint. If the force due to the Achilles tendon is 2230 N, what is the force exerted on the foot by the tibia?

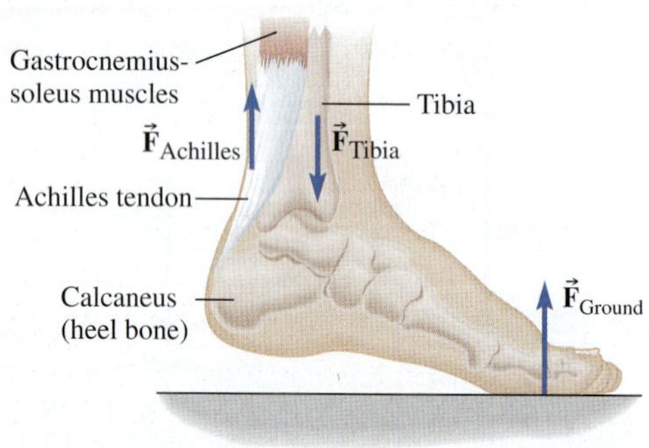

22. A sailboat, tied to a mooring with a line, weighs 820 N. The mooring line pulls horizontally toward the west on the sailboat with a force of 110 N. The sails are stowed away and the wind blows from the west. The boat is moored on a still lake—no water currents push on it. Draw an FBD for the sailboat and indicate the magnitude of each force.

23. A hummingbird is hovering motionless beside a flower. The blur of its wings shows that they are rapidly beating up and down. If the air pushes upward on the bird with a force of 0.30 N, what is the weight of the hummingbird?

24. You are pulling a suitcase through the airport at a constant speed. The handle of the suitcase makes an angle of 60° with respect to the horizontal direction. If you pull with a force of 5.0 N parallel to the handle, what is the horizontal component of the contact force due to the floor acting on the suitcase?

25. ✦ A model sailboat is slowly sailing west across a pond at 0.33 m/s. A gust of wind gives the sailboat a constant acceleration of 0.30 m/s^2 directed 28° south of west during a time interval of 2.0 s. (a) If the net force on the sailboat during the 2.0 s interval has magnitude 0.375 N, what is the sailboat's mass? (b) What is the new velocity of the boat after the 2.0 s gust of wind?

26. A man is lazily floating on an air mattress in a swimming pool. If the weight of the man and air mattress together is 806 N, what is the upward force of the water acting on the mattress?

27. What is the acceleration of an automobile of mass 1.40×10^3 kg when it is subjected to a net forward force of 3.36×10^3 N?

28. A bag of potatoes with weight 39.2 N is suspended from a string that exerts a force of 46.8 N. If the bag's acceleration is upward at 1.90 m/s^2, what is the mass of the potatoes?

29. A large wooden crate is pushed along a horizontal, frictionless surface by a force of 100 N. The acceleration of the crate is measured to be 2.5 m/s^2. What is the mass of the crate?

4.4 Interaction Pairs: Newton's Third Law of Motion

30. A hanging plant is suspended by a cord from a hook in the ceiling. Draw an FBD for each of these: (a) the system consisting of plant, soil, and pot; (b) the cord; (c) the hook; (d) the system consisting of plant, soil, pot, cord, and hook. Label each force arrow using subscripts (for example, $\vec{\mathbf{F}}_{ch}$ would represent the force exerted on the cord by the hook).

31. A bike is hanging from a hook in a garage. Consider the following forces: (1) the force of the Earth pulling down on the bike, (2) the force of the bike pulling up on the Earth, and (3) the force of the hook pulling up on the bike. (a) Which two forces are equal and opposite because of Newton's third law? (b) Which two forces are equal and opposite because of Newton's *first* law? Explain.

32. A woman who weighs 600 N sits on a chair with her feet on the floor and her arms resting on the chair's armrests. The chair weighs 100 N. Each armrest exerts an upward force of 25 N on her arms. The seat of the chair exerts an upward force of 500 N. (a) What force does the floor exert on her feet? (b) What force does the floor exert on the chair? (c) Consider the woman and the chair to be a single system. Draw an FBD for this system that includes only the *external* forces acting on it.

33. A fisherman is holding a fishing rod with a large fish hanging from the line. Identify the forces acting on the rod and describe the interaction partner of each.

34. In Problem 33, identify the forces acting on the fish and describe the interaction partner of each.

Problems 33 and 34

Problems 35–37. A skydiver, who weighs 650 N, is falling at a constant speed with his parachute open. Consider the apparatus that connects the parachute to the skydiver to be part of the parachute. The parachute pulls upward on the skydiver with a force of 620 N.

35. (a) Identify the forces acting on the skydiver. Describe each force as: (*type of force*) exerted on (*object 1*) by (*object 2*). (b) Draw an FBD for the skydiver. (c) Find the magnitude of the force on the skydiver due to the air. (d) Identify the interaction partner for each force acting on the skydiver. For each interaction partner, describe it as (*type of force*) exerted on (*object 1*) by (*object 2*) and determine its magnitude and direction.

36. (a) Identify the forces acting on the parachute. Describe each force as: (*type of force*) exerted on (*object 1*) by (*object 2*). (b) Draw an FBD for the parachute. (c) What are the magnitude and direction of the force on the parachute due to the skydiver? (d) Identify the interaction partner for each force acting on the parachute. For each interaction partner, describe it as (*type of force*) exerted on (*object 1*) by (*object 2*).

37. Consider the skydiver and parachute to be a single system. Identify the external forces acting on this system and draw an FBD.

38. Margie, who weighs 543 N, is standing on a bathroom scale that weighs 45 N. (a) With what magnitude force does the scale push up on Margie? (b) What is the interaction partner of that force? (c) With what magnitude force does the floor push up on the scale? (d) Identify the interaction partner of that force.

4.5 Gravitational Forces

39. (a) Calculate your weight in newtons. (b) What is the weight in newtons of 250 g of cheese? (c) Name a common object whose weight is about 1 N.

40. A young South African girl has a mass of 40.0 kg. (a) What is her weight in newtons? (b) If she came to the United States, what would her weight be in pounds as measured on an American scale? Assume $g = 9.80$ N/kg in both locations.

41. A man weighs 0.80 kN on Earth. What is his mass in kilograms?

42. The peak force on a runner's foot during a race is found to be vertical and three times his weight. What is the peak force on the foot of a runner whose mass is 85 kg?

43. In a binary star system, two stars orbit their common center of mass. In one such system, star A has 4.0 times the mass of star B. (a) Draw and label vector arrows for the gravitational forces that each star exerts on the other, showing how their directions and magnitudes are related. (b) Draw and label vector arrows to illustrate the accelerations of the stars, showing how their directions and magnitudes are related.

44. An astronaut stands at a position on the Moon such that Earth is directly overhead and releases a Moon rock that was in her hand. (a) Which way will it fall? (b) What is the gravitational force exerted by the Moon on a 1.0 kg rock resting on the Moon's surface? (c) What is the gravitational force exerted by the Earth on the same 1.0 kg rock resting on the surface of the Moon?

45. Find the ratio of the Earth's gravitational force on a satellite when it is on the ground to the gravitational force exerted when the satellite is orbiting at an altitude of 320 km.

46. How far above the surface of the Earth does an object have to be in order for it to have the same weight as it would have on the surface of the Moon? (Ignore any effects from the Earth's gravity for the object on the Moon's surface or from the Moon's gravity for the object above the Earth.)

47. Find and compare the weight of a 65 kg man on Earth with the weight of the same man on (a) Mars, where $g = 3.7$ N/kg; (b) Venus, where $g = 8.9$ N/kg; and (c) Earth's Moon, where $g = 1.6$ N/kg.

48. Find the altitudes above the Earth's surface where Earth's gravitational field strength would be (a) two thirds and (b) one third of its value at the surface. [*Hint*: First find the radius for each situation; then recall that the altitude is the distance from the *surface* to a point above the surface. Use proportional reasoning.]

49. During a balloon ascension, wearing an oxygen mask, you measure the weight of a calibrated 5.00 kg mass and find that the value of the gravitational field strength at your location is 9.792 N/kg. How high above sea level, where the gravitational field strength was measured to be 9.803 N/kg, are you located?

50. At what altitude above the Earth's surface would your weight be half of what it is at the Earth's surface?

51. (a) What is the magnitude of the gravitational force that the Earth exerts on the Moon? (b) What is the magnitude of the gravitational force that the Moon exerts on the Earth? See the inside back cover for necessary information.

52. What is the approximate magnitude of the gravitational force exerted by the Sun on the Voyager spacecraft when they are separated by 15 billion km? Each spacecraft has a mass of approximately 722 kg during the mission.

53. In free fall, we assume the acceleration to be constant. Not only is air resistance ignored, but the gravitational field strength is assumed to be constant. From what height can an object fall to the Earth's surface such that the gravitational field strength changes less than 1.000% during the fall?

54. A solar sailplane is going from Earth to Mars. Its sail is oriented to give a solar radiation force of 8.00×10^2 N. The gravitational force due to the Sun is 173 N and the gravitational force due to the Earth is 1.00×10^2 N. All forces are in the plane formed by Earth, Sun, and sailplane. The mass of the sailplane is 14500 kg. (a) What is the net force (magnitude and direction) acting on the sailplane? (b) What is the acceleration of the sailplane?

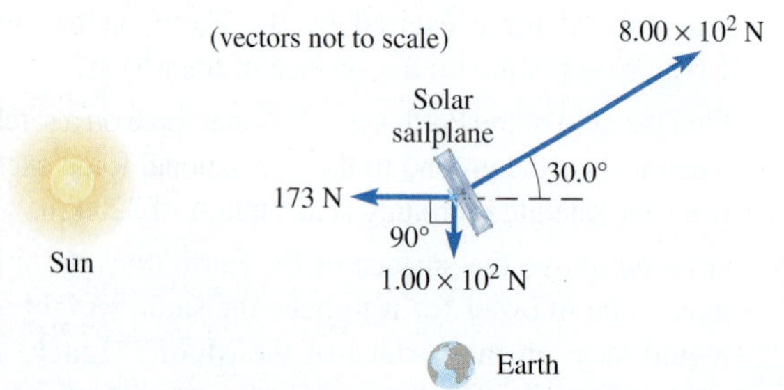

Problems 55–57. Assume the elevator is supported by a single cable. Forces exerted by the guide rails and air resistance are negligible.

55. A 2010 kg elevator moves with an upward acceleration of 1.50 m/s^2. What is the force exerted by the cable on the elevator?

56. While an elevator of mass 2530 kg moves upward, the force exerted by the cable is 27.6 kN. (a) What is the acceleration of the elevator? (b) If at some point in the motion the velocity of the elevator is 1.20 m/s upward, what is the elevator's velocity 4.00 s later?

57. While an elevator of mass 832 kg moves downward, the upward force due to the supporting cable is a constant 7730 N. Between $t = 0$ and $t = 4.00$ s, the elevator's displacement is 5.00 m downward. What is the elevator's speed at $t = 4.00$ s?

58. The vertical component of the acceleration of a sailplane is zero when the air pushes up against its wings with a force of 3.0 kN. (a) Assuming that the only vertical forces on the sailplane are that due to gravity and that due to the air pushing up against its wings, find the gravitational force on the Earth due to the sailplane. (b) If the wing stalls and the upward force decreases to 2.0 kN, what is the vertical acceleration of the sailplane?

59. A man lifts a 2.0 kg stone vertically with his hand at a constant upward velocity of 1.5 m/s. What is the magnitude of the force of the man's hand on the stone?

60. A man lifts a 2.0 kg stone vertically with his hand at a constant upward *acceleration* of 1.5 m/s^2. What is the magnitude of the force of the man's hand on the stone?

61. Using the masses and mean distances found on the inside back cover, calculate the net gravitational force on the Moon (a) during a lunar eclipse (Earth between Moon and Sun) and (b) during a solar eclipse (Moon between Earth and Sun).

62. ✦ A binary star consists of two stars of masses M_1 and $4.0M_1$ a distance d apart. Is there any point where the gravitational field due to the two stars is zero? If so, where is that point?

4.6 Contact Forces

63. Mechanical advantage is the ratio of the force required without the use of a simple machine to that needed when using the simple machine. Compare the force to lift an object with that needed to slide the same object up a frictionless incline and show that the mechanical advantage of the inclined plane is the length of the incline divided by the height of the incline (d/h in Fig. 4.25).

64. A book rests on the surface of the table. Consider the following four forces that arise in this situation: (a) the force of the Earth pulling on the book, (b) the force of the table pushing on the book, (c) the force of the book pushing on the table, and (d) the force of the book pulling on the Earth. The book is not moving. Which pair of forces must be equal in magnitude and opposite in direction even though they are *not* an interaction pair? Explain.

65. A crate of artichokes is on a ramp that is inclined 10.0° above the horizontal. Give the direction of the normal force and the friction force acting on the crate in each of these situations. (a) The crate is at rest. (b) The crate is sliding up the ramp. (c) The crate is sliding down the ramp.

66. An 80.0 N crate of apples sits at rest on a ramp that runs from the ground to the bed of a truck. The ramp is inclined at 20.0° to the ground. (a) What is the normal force exerted on the crate by the ramp? (b) The interaction partner of this normal force has what magnitude and direction? It is exerted *by* what object *on* what object? Is it a contact or a long-range force? (c) What is the static frictional force exerted on the crate by the ramp? (d) What is the minimum possible value of the coefficient of static friction? (e) The normal and frictional forces are perpendicular components of the contact force exerted on the crate by the ramp. Find the magnitude and direction of the contact force.

67. An 85 kg skier is sliding down a ski slope at a constant velocity. The slope makes an angle of 11° above the horizontal direction. Ignore air resistance. (a) What is

the force of kinetic friction acting on the skier? (b) What is the coefficient of kinetic friction between the skis and the snow?

68. A book that weighs 10 N is at rest in six different situations. Blue arrows indicate forces exerted on the book by an object that is not shown. Rank the situations according to the magnitude of the normal force on the 10 N book *due to the table*, from smallest to greatest. Explain your reasoning.

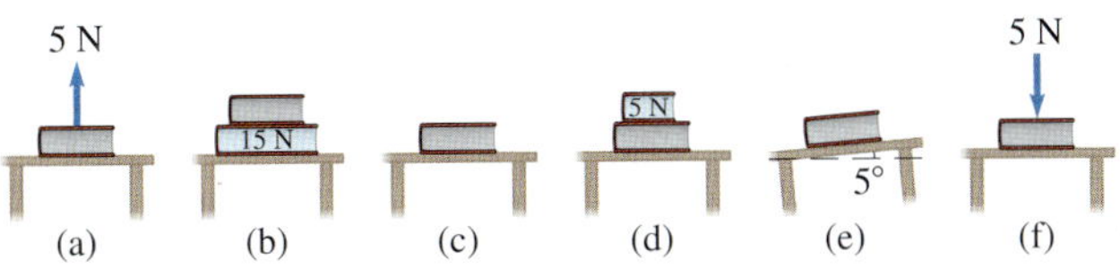

Problems 69–72. A crate of potatoes of mass 18.0 kg is on a ramp with angle of incline 30° to the horizontal. The coefficients of friction are $\mu_s = 0.75$ and $\mu_k = 0.40$. Find the frictional force (magnitude and direction) on the crate if

69. the crate is at rest.
70. the crate is sliding down the ramp.
71. the crate is sliding *up* the ramp.
72. the crate is being carried up the ramp at constant velocity by a conveyor belt (without sliding).
73. (a) In Example 4.10, if the movers stop pushing on the safe, can static friction hold the safe in place without having it slide back down? (b) If not, what minimum force needs to be applied to hold the safe in place?
74. ✦ A 3.0 kg block is at rest on a horizontal floor. If you push horizontally on the 3.0 kg block with a force of 12.0 N, it just starts to move. (a) What is the coefficient of static friction? (b) A 7.0 kg block is stacked on top of the 3.0 kg block. What is the magnitude F of the force, acting horizontally on the 3.0 kg block as before, that is required to make the two blocks start to move?
75. A horse is trotting along pulling a sleigh through the snow. To move the sleigh, of mass m, straight ahead at a constant speed, the horse must pull with a force of magnitude T. (a) What is the net force acting on the sleigh? (b) What is the coefficient of kinetic friction between the sleigh and the snow?
76. ✦ Before hanging new William Morris wallpaper in her bedroom, Brenda sanded the walls lightly to smooth out some irregularities on the surface. The sanding block weighs 2.0 N and Brenda pushes on it with a force of 3.0 N at an angle of 30.0° with respect to the vertical, and angled toward the wall. Draw an FBD for the sanding block as it moves straight up the wall at a constant speed. What is the coefficient of kinetic friction between the wall and the block?
77. A conveyor belt carries apples up an incline to a cider press. The apples neither slide nor roll as they move up the incline. (a) Does the belt exert a *static* or *kinetic* frictional force on an apple? Explain. (b) If the normal and frictional forces exerted on an apple have magnitudes 1.0 N and 0.40 N, respectively, what (if anything) can you conclude about the static and kinetic coefficients of friction?
78. A box sits on a horizontal wooden ramp. The coefficient of static friction between the box and the ramp is 0.30. You grab one end of the ramp and lift it up, keeping the other end of the ramp on the ground. What is the angle between the ramp and the horizontal direction when the box begins to slide down the ramp? (connect tutorial: crate on ramp)
79. ✦ In a playground, two slides have different angles of incline θ_1 and θ_2 ($\theta_2 > \theta_1$). A child slides down the first at constant speed; on the second, his acceleration down the slide is a. Assume the coefficient of kinetic friction is the same for both slides. (a) Find a in terms of θ_1, θ_2, and g. (b) Find the numerical value of a for $\theta_1 = 45°$ and $\theta_2 = 61°$.

4.7 Tension

80. A sailboat is tied to a mooring with a horizontal line. The wind is from the southwest. Draw an FBD and identify all the forces acting on the sailboat.
81. A towline is attached between a car and a glider. As the car speeds due east along the runway, the towline exerts a horizontal force of 850 N on the glider. What is the magnitude and direction of the force exerted by the glider on the towline?
82. In Example 4.14, find the tension in the coupling between cars 2 and 3. (connect tutorial: towing a train)
83. A 200.0 N sign is suspended from a horizontal strut of negligible weight. The force exerted on the strut by the wall is horizontal. Draw an FBD to show the forces acting on the strut. Find the tension T in the diagonal cable supporting the strut.

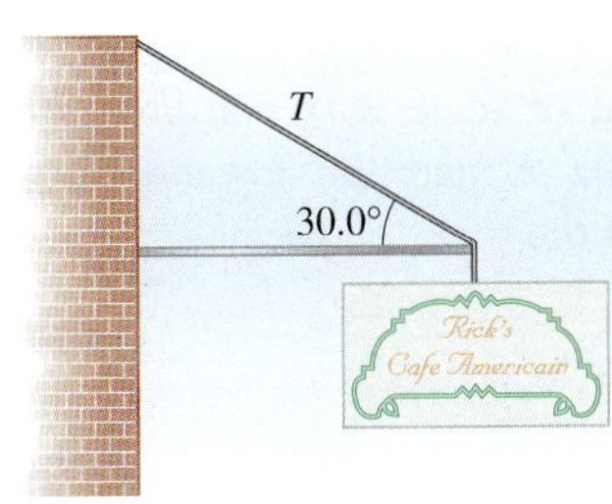

84. Two boxes with different masses are tied together on a frictionless ramp surface. What is the tension in each of the cords?

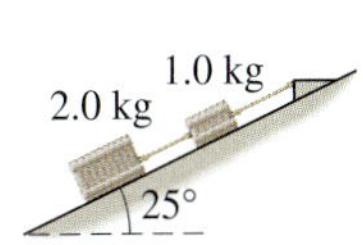

85. A pulley is attached to the ceiling. Spring scale A is attached to the wall and a rope runs horizontally from it and over the pulley. The same rope is then attached to spring scale B. On the other side of scale B hangs a 120 N

weight. What are the readings of the two scales A and B? Neglect the weights of the ropes, pulley, and scales.

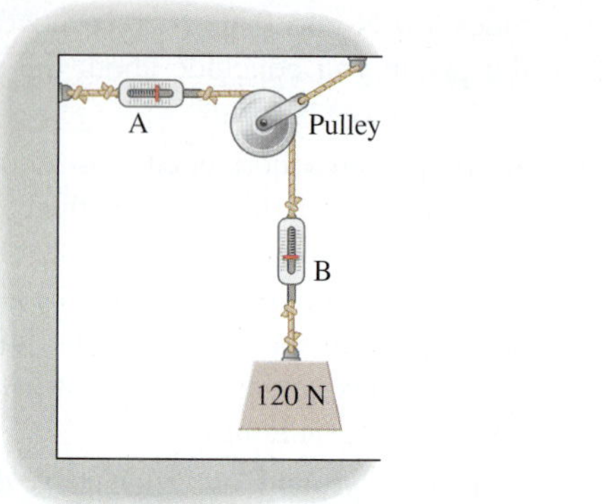

86. Spring scale A is attached to the floor and a rope runs vertically upward, loops over the pulley, and runs down on the other side to a 120 N weight. Scale B is attached to the ceiling and the pulley is hung below it. What are the readings of the two spring scales, A and B? Neglect the weights of the rope, pulley, and scales.

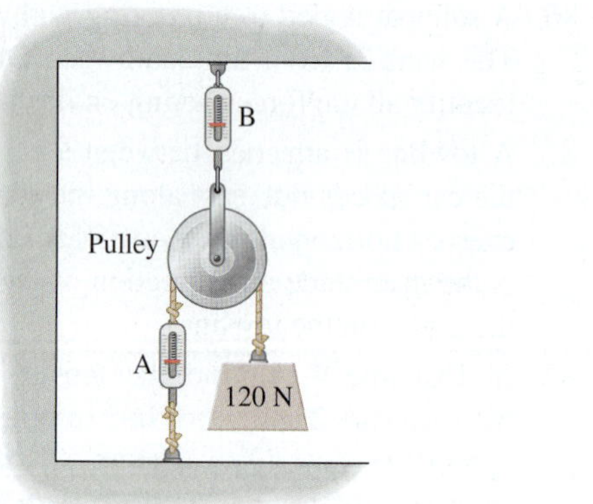

87. Two springs are connected in series so that spring scale A hangs from a hook on the ceiling and a second spring scale, B, hangs from the hook at the bottom of scale A. Apples weighing 120 N hang from the hook at the bottom of scale B. What are the readings on the upper scale A and the lower scale B? Ignore the weights of the ropes and scales.

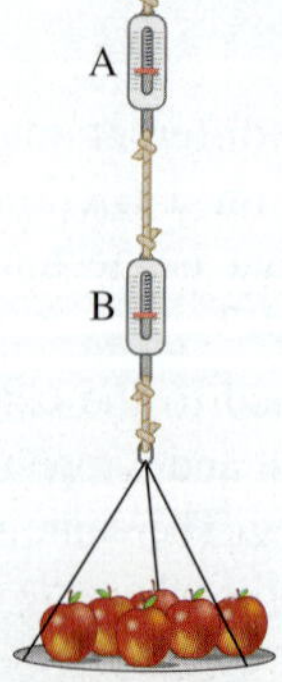

88. A pulley is hung from the ceiling by a rope. A block of mass M is suspended by another rope that passes over the pulley and is attached to the wall. The rope fastened to the wall makes a right angle with the wall. Ignore the masses of the rope and the pulley. Find (a) the tension in the rope from which the pulley hangs and (b) the angle θ that the rope makes with the ceiling.

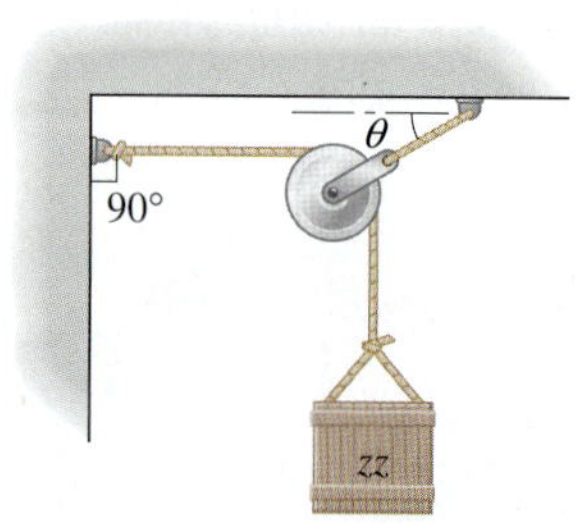

89. A 2.0 kg ball tied to a string fixed to the ceiling is pulled to one side by a force $\vec{\mathbf{F}}$. Just before the ball is released and allowed to swing back and forth, (a) how large is the force $\vec{\mathbf{F}}$ that is holding the ball in position and (b) what is the tension in the string?

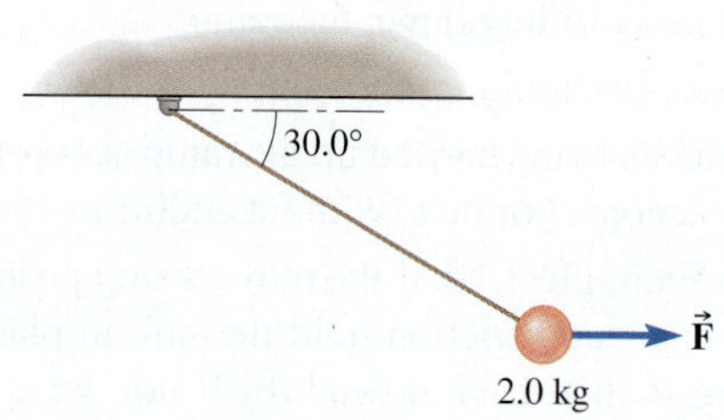

90. Two wooden crates with masses as shown are tied together by a horizontal cord. Another cord is tied to the first crate, and it is pulled with a force of 195 N at an angle of 20.0°, as shown. Each crate has a coefficient of kinetic friction of 0.550. Find the tension in the rope between the crates and the magnitude of the acceleration of the system.

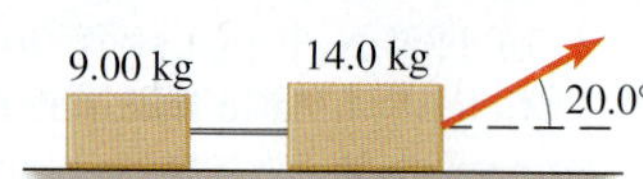

91. A 45 N lithograph is supported by two wires. One wire makes a 25° angle with the vertical and the other makes a 15° angle with the vertical. Find the tension in each wire. (connect tutorial: hanging picture)

92. ✦ A crow perches on a clothesline midway between two poles. Each end of the rope makes an angle of θ below the horizontal where it connects to the pole. If the weight of the crow is W, what is the tension in the rope? Ignore the weight of the rope.

93. The drawing shows a wire attached to two back teeth and stretched across a front tooth. The purpose of this arrangement is to apply a force $\vec{\mathbf{F}}$ to the front tooth. (The figure has been simplified by running the wire straight from the front tooth to the back teeth.) If the tension in the wire is 12 N, what are the magnitude and direction of the force $\vec{\mathbf{F}}$ applied to the front tooth?

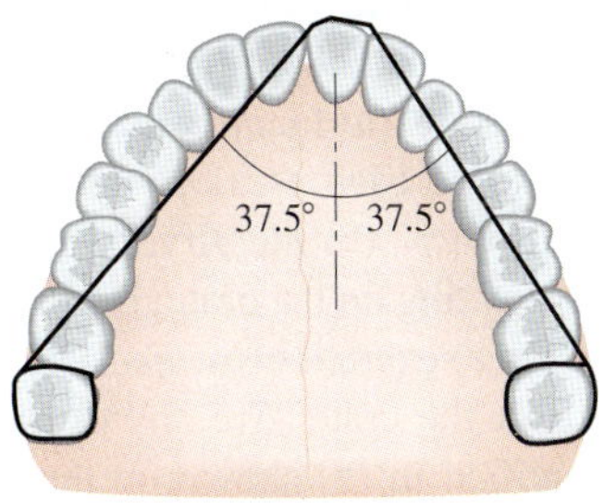

94. A cord cut into two equal sections, with a spring balance to measure forces inserted between them, is hanging from a hook attached to the ceiling. A mass of 10 kg is hanging from the lower end of the cord. The spring balance indicates a reading of 98 N for the force. Then two people hold the opposite ends of the same cord and pull against each other horizontally until the balance in the middle again reads 98 N. With what force must each person pull to attain this result?

95. Two blocks, masses m_1 and m_2, are connected by a massless cord. If the two blocks are pulled with a constant tension on a frictionless surface by applying a force of magnitude T_2 to a second cord connected to m_2, what is the ratio of the tensions in the two cords T_1/T_2 in terms of the masses?

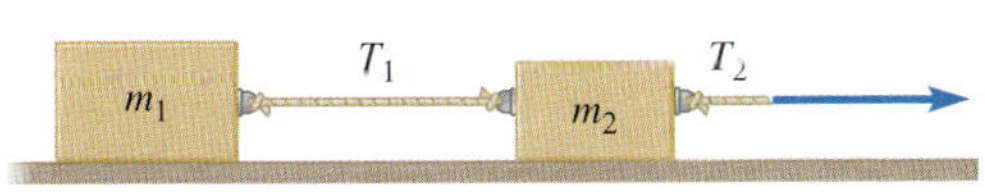

4.8 Applying Newton's Second Law

96. The coefficient of static friction between a block and a horizontal floor is 0.40, while the coefficient of kinetic friction is 0.15. The mass of the block is 5.0 kg. A horizontal force is applied to the block and slowly increased. (a) What is the value of the applied horizontal force at the instant that the block starts to slide? (b) What is the net force on the block after it starts to slide?

97. A 2.0 kg toy locomotive is pulling a 1.0 kg caboose. The frictional force of the track on the caboose is 0.50 N backward along the track. If the train's acceleration forward is 3.0 m/s^2, what is the magnitude of the force exerted by the locomotive on the caboose?

98. An engine pulls a train of 20 freight cars, each having a mass of 5.0×10^4 kg with a constant force. The cars move from rest to a speed of 4.0 m/s in 20.0 s on a straight track. Ignoring friction, find the force with which the 10th car pulls the 11th one (at the middle of the train). (connect tutorial: school bus)

99. In Fig. 4.44, two blocks are connected by a lightweight, flexible cord that passes over a frictionless pulley. (a) If $m_1 = 3.0$ kg and $m_2 = 5.0$ kg, what are the accelerations of each block? (b) What is the tension in the cord?

100. A rope is attached from a truck to a 1400 kg car. The rope will break if the tension is greater than 2500 N. Ignoring friction, find the maximum possible acceleration of the truck if the rope does not break. Should the driver of the truck be concerned that the rope might break?

101. An accelerometer—a device to measure acceleration—can be as simple as a small pendulum hanging in the cockpit. A similar accelerometer is found in the inner ear of vertebrates. Suppose you are flying a small plane in a straight, horizontal line and your accelerometer hangs at a constant angle of 12° behind the vertical shown in the figure. What is your acceleration?

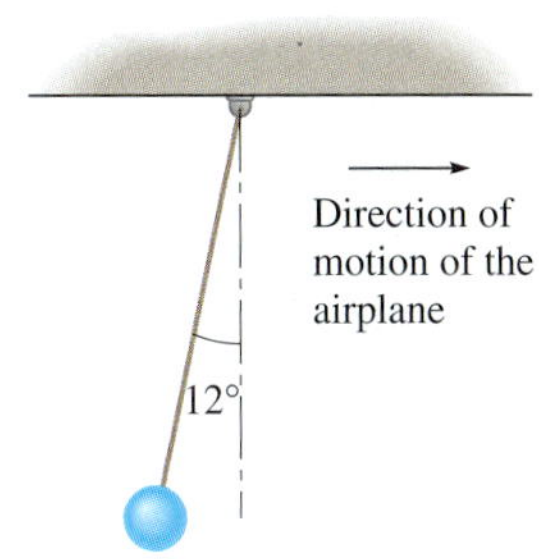

102. A box full of books rests on a wooden floor. The normal force the floor exerts on the box is 250 N. (a) You push horizontally on the box with a force of 120 N, but it refuses to budge. What can you say about the coefficient of static friction between the box and the floor? (b) If you must push horizontally on the box with a force of at least 150 N to start it sliding, what is the coefficient of static friction? (c) Once the box is sliding, you only have to push with a force of 120 N to keep it sliding at constant speed. What is the coefficient of kinetic friction?

103. A helicopter is lifting two crates simultaneously. One crate with a mass of 200 kg is attached to the helicopter by a cable. The second crate with a mass of 100 kg is hanging below the first crate and attached to the first crate by a cable. As the helicopter accelerates upward at a rate of 1.0 m/s^2, what is the tension in each of the two cables?

4.10 Apparent Weight

104. A person stands on a bathroom scale in an elevator. Rank the scale readings from highest to lowest based on the given information about the speed v or the magnitude of the acceleration a: (a) ascending with increasing speed (a = 1.0 m/s^2); (b) descending at constant speed (v = 2.0 m/s); (c) descending at constant speed (v = 4.0 m/s) (d) descending with increasing speed (a = 2.0 m/s^2); (e) ascending with decreasing speed (a = 2.0 m/s^2).

105. Oliver has a mass of 76.2 kg. He is riding in an elevator that has a downward acceleration of 1.37 m/s^2. With what magnitude force does the elevator floor push upward on Oliver?

106. While on an elevator, Jaden's apparent weight is 550 N. When he is on the ground, the scale reading is 600 N. What is Jaden's acceleration?

107. When on the ground, Ian's weight is measured to be 640 N. When Ian is on an elevator, his apparent weight is 700 N. What is the net force on the system (Ian and the elevator) if their combined mass is 1050 kg?

108. Refer to Example 4.19. What is the apparent weight of the same passenger (weighing 598 N) in the following situations? In each case, the magnitude of the elevator's acceleration is 0.50 m/s^2. (a) After having stopped at the 15th floor, the passenger pushes the 8th floor button; the elevator is beginning to move downward. (b) The elevator is moving downward and is slowing down as it nears the 8th floor.

109. Ⓒ You are standing on a bathroom scale inside an elevator. Your weight is 140 lb, but the reading of the scale is 120 lb. (a) What is the magnitude and direction of the acceleration of the elevator? (b) Can you tell whether the elevator is speeding up or slowing down?

110. Yolanda, whose mass is 64.2 kg, is riding in an elevator that has an upward acceleration of 2.13 m/s^2. What force does she exert on the floor of the elevator?

111. Felipe is going for a physical before joining the swim team. He is concerned about his weight, so he carries his scale into the elevator to check his weight while heading to the doctor's office on the 21st floor of the building. If his scale reads 750 N while the elevator has an upward acceleration of 2.0 m/s^2, what does the nurse measure his weight to be?

112. ✦ Luke stands on a scale in an elevator that has a constant acceleration upward. The scale reads 0.960 kN. When Luke picks up a box of mass 20.0 kg, the scale reads 1.200 kN. (The acceleration remains the same.) (a) Find the acceleration of the elevator. (b) Find Luke's weight.

4.12 Fundamental Forces

113. Which of the fundamental forces has the shortest range?

114. Which of the fundamental forces governs the motion of planets in the solar system? Is this the strongest or the weakest of the fundamental forces? Explain.

115. Which of the following forces have an unlimited range: strong force, contact force, electromagnetic force, gravitational force?

116. Which of the following forces bind electrons to nuclei to form atoms: strong force, contact force, electromagnetic force, gravitational force?

117. Which of the fundamental forces binds quarks together to form protons, neutrons, and many exotic subatomic particles?

Collaborative Problems

118. Ⓒ A box containing a new TV weighs 350 N. Phineas is pushing horizontally on it with a force of 150 N, but it doesn't budge. (a) Identify all the forces acting on the crate. Describe each as: (*type of force*) exerted on the crate by (*object*). (b) Identify the interaction partner of each force acting on the crate. Describe each partner as: (*type of force*) exerted on (*object*) by (*object*). (c) Draw an FBD for the crate. Are any of the interaction partners identified in (b) shown on the FBD? (d) What is the *net* force acting on the crate? Use your answer to determine the magnitude of all the forces acting on the crate. (e) If there are pairs of forces on the FBD that are equal in magnitude and opposite in direction, are these *interaction pairs*? Explain.

119. Ⓒ A truck is towing a 1250 kg car at a constant speed up a hill that makes an angle of $\alpha = 10.0°$ with respect to the horizontal. A rope is attached from the truck to the car at an angle of $\beta = 15.0°$ with respect to horizontal. Ignore friction on the car in this problem. (a) Draw an FBD showing all the forces on the car. (b) What choice of x- and y-axes will make it easiest to find the tension in the rope? Explain. (c) Find the tension.

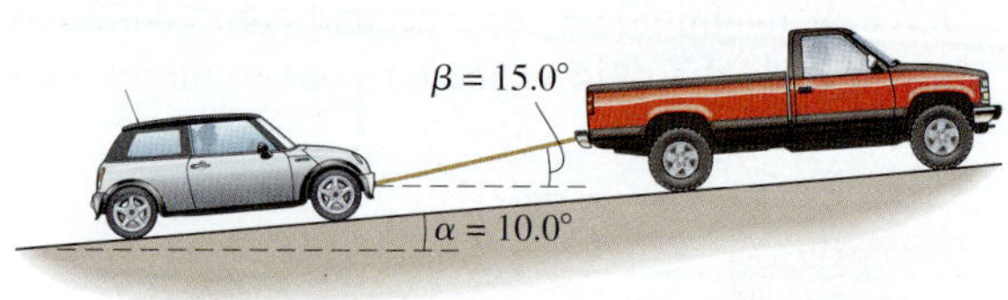

120. Ⓒ You grab a book and give it a quick push across the top of a horizontal table. After a short push, the book slides across the table, and because of friction, comes to a stop. (a) Draw an FBD of the book while you are pushing it. (b) Draw an FBD of the book after you have stopped pushing it, while it is sliding across the table. (c) Draw an FBD of the book after it has stopped sliding. (d) In which of the preceding cases is the net force on the book not equal to zero? (e) If the book has a mass of 0.50 kg and the coefficient of friction between the book and the table is 0.40, what is the net force acting on the book in part (b)? (f) If there were no friction between the table and the book, what would the free-body diagram for part (b) look like? Would the book slow down in this case? Why or why not?

121. Ⓒ When you hold up a 50 N weight in your hand, with your forearm horizontal and your palm up, the upward force exerted by your biceps on your

forearm is much larger than 50 N—perhaps as much as 5000 N. How can that be? What other forces are acting on your forearm? Draw an FBD for the forearm, showing all of the forces. Assume that all the forces exerted on the forearm are purely vertical—either up or down.

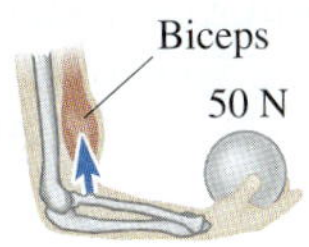

122. Ⓒ The readings of the two spring scales shown in the drawing are the same. (a) Explain why they are the same. [*Hint*: Draw free-body diagrams.] (b) What is the reading?

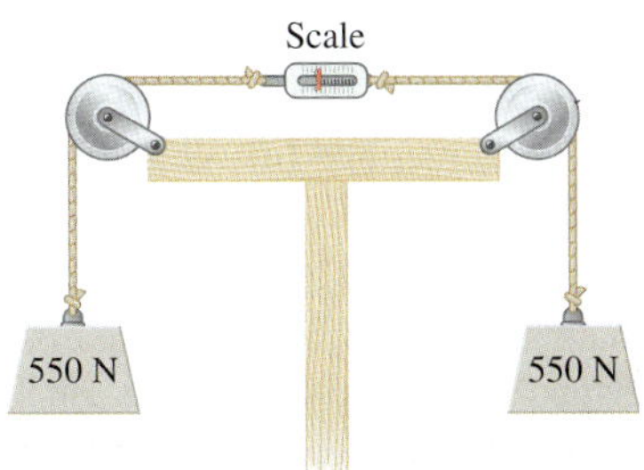

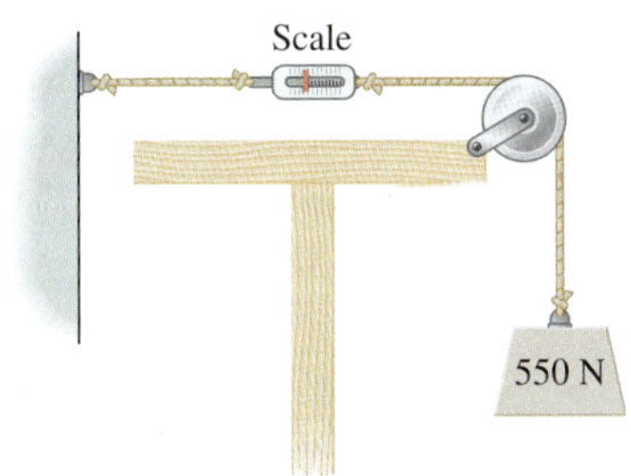

123. The coefficient of static friction between a block and a horizontal floor is 0.35, while the coefficient of kinetic friction is 0.22. The mass of the block is 4.6 kg and it is initially at rest. (a) What is the minimum horizontal applied force required to make the block start to slide? (b) Once the block is sliding, if you keep pushing on it with the same minimum starting force as in part (a), does the block move with constant velocity or does it accelerate? (c) If it moves with constant velocity, what is its velocity? If its velocity is changing, what is its acceleration?

124. ✦ You want to hang a 15 N picture as in part (a) using some very fine twine that will break with more than 12 N of tension. Can you do this? What if you have it as illustrated in part (b) of the figure?

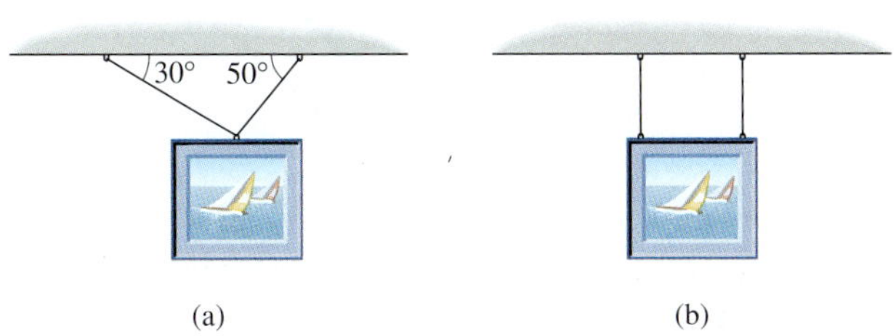

125. ✦ The coefficient of static friction between block A and a horizontal floor is 0.45, and the coefficient of static friction between block B and the floor is 0.30. The mass of each block is 2.0 kg and they are connected together by a horizontal cord. (a) A horizontal force $\vec{\mathbf{F}}$ pulling on block B is slowly increased until the blocks start to slide. What is the magnitude of $\vec{\mathbf{F}}$ just before they start to slide? (b) What is the tension in the cord connecting blocks A and B just before they start to slide?

126. ✦ Ⓒ While trying to decide where to hang a framed picture, you press it against the wall to keep it from falling. The picture weighs 5.0 N, and you press against the frame with a force of 6.0 N at an angle of 40° from the vertical. (a) What is the normal force exerted on the picture by the wall? (b) What is the minimum coefficient of static friction between the wall and the picture? (c) Depending on the force you exert, the frictional force exerted on the picture by the wall could have either of two possible directions. Explain why.

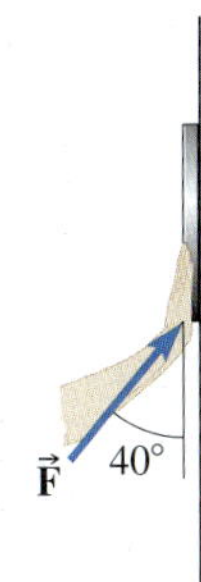

127. ✦ A block of mass $m_1 = 3.0$ kg rests on a frictionless horizontal surface. A second block of mass $m_2 = 2.0$ kg hangs from an ideal cord of negligible mass that runs over an ideal pulley and then is connected to the first block. The blocks are released from rest. (a) Find the acceleration of the two blocks after they are released. (b) What is the velocity of the first block 1.2 s after the release of the blocks, assuming the first block does not run out of room on the table and the second block does not land on the floor? (c) How far has block 1 moved during the 1.2 s interval? (d) What is the displacement of the blocks from their initial positions 0.40 s after they are released?

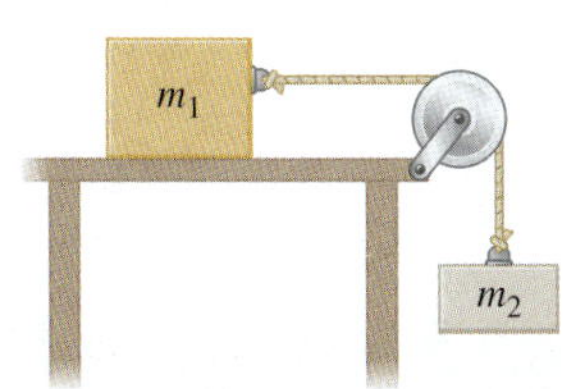

Problems 127 and 128

128. ✦ A block of mass m_1 slides to the right with coefficient of kinetic friction μ_k on a horizontal surface. The block is connected to a hanging block of mass m_2 by a light cord that passes over a light, frictionless pulley. (a) Find the acceleration of each of the blocks and the tension in the cord. (b) Check your answers in the special cases $m_1 \ll m_2$, $m_1 \gg m_2$, and $m_1 = m_2$. (c) For what value of m_2 (if any) do the two blocks slide at constant velocity? What is the tension in the cord in that case?

Comprehensive Problems

129. A car is driving on a straight, level road at constant speed. Draw an FBD for the car, showing the significant forces that act upon it.

130. You want to push a 65 kg box up a 25° ramp. The coefficient of kinetic friction between the ramp and the box is 0.30. With what magnitude force parallel to the ramp should you push on the box so that it moves up the ramp at a constant speed?

131. An airplane is cruising along in a horizontal level flight at a constant velocity, heading due west. (a) If the weight of the plane is 2.6×10^4 N, what is the net force on the plane? (b) With what force does the air push upward on the plane?

132. A young boy with a broken leg is undergoing traction. (a) Find the magnitude of the total force of the traction apparatus applied to the leg, assuming the weight of the leg is 22 N and the weight hanging from the traction apparatus is also 22 N. (b) What is the horizontal component of the traction force acting on the leg? (c) What is the magnitude of the force exerted on the femur by the lower leg?

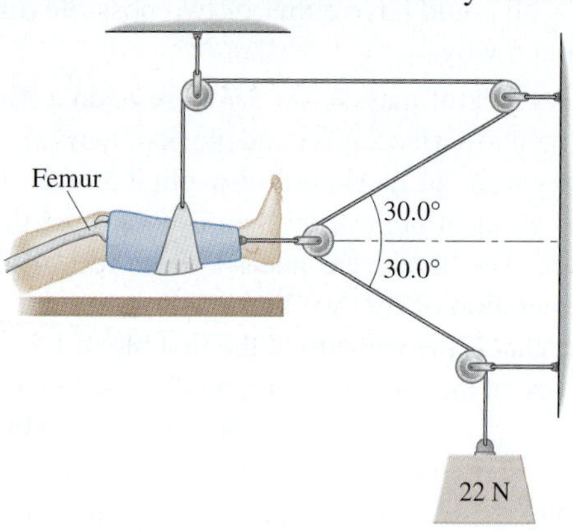

133. In the sport of curling, a player slides a 20.0 kg granite stone down a 38 m long ice rink. Draw FBDs for the stone (a) while it sits at rest on the ice; (b) while it slides down the rink; (c) during a head-on collision with an opponent's stone that was at rest on the ice.

134. The tallest spot on Earth is Mt. Everest, which is 8850 m above sea level. If the radius of the Earth to sea level is 6370 km, how much does the gravitational field strength change between the sea level value at that location (9.826 N/kg) and the top of Mt. Everest?

135. By what percentage does the weight of an object change when it is moved from the equator at sea level, where the effective value of g is 9.784 N/kg, to the North Pole where $g = 9.832$ N/kg?

136. Two canal workers pull a barge along the narrow waterway at a constant speed. One worker pulls with a force of 105 N at an angle of 28° with respect to the forward motion of the barge and the other worker, on the opposite tow path, pulls at an angle of 38° relative to the barge motion. Both ropes are parallel to the ground. (a) With what magnitude force should the second worker pull to make the sum of the two forces be in the forward direction? (b) What is the magnitude of the force on the barge from the two tow ropes?

137. A large wrecking ball of mass m is resting against a wall. It hangs from the end of a cable that is attached at its upper end to a crane that is just touching the wall. The cable makes an angle of θ with the wall. Ignoring friction between the ball and the wall, find the tension in the cable.

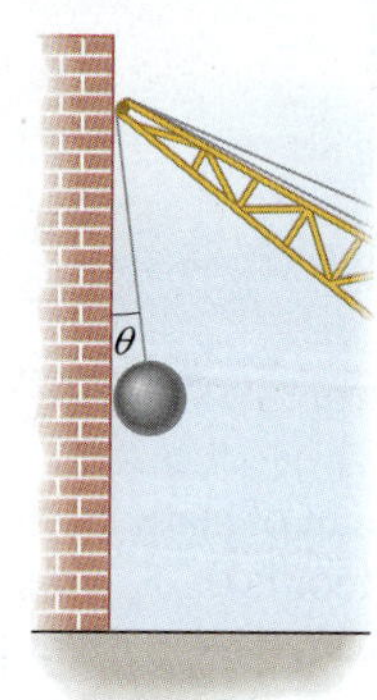

138. The figure shows the quadriceps and the patellar tendons attached to the patella (the kneecap). If the tension T in each tendon is 1.30 kN, what are the magnitude and direction of the contact force $\vec{\mathbf{F}}$ exerted on the patella by the femur?

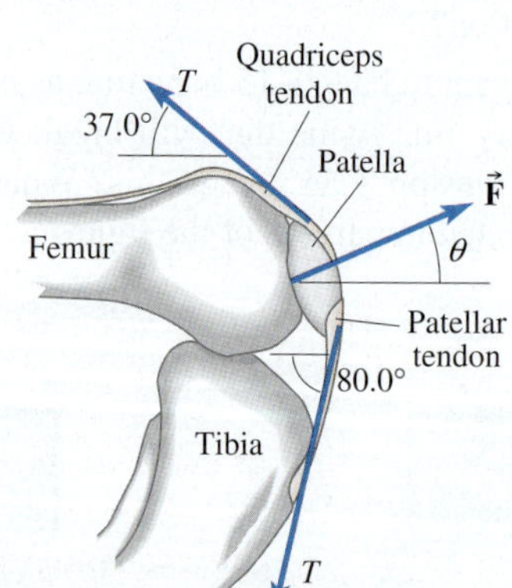

139. Two blocks lie side by side on a frictionless table. The block on the left is of mass m; the one on the right is of mass $2m$. The block on the right is pushed to the left with a force of magnitude F, pushing the other block in turn. What force does the block on the left exert on the block to its right?

140. A locomotive pulls a train of 10 identical cars, on a track that runs east-west, with a force of 2.0×10^6 N directed east. What is the force with which the *last* car to the west pulls on the rest of the train?

141. The coefficient of static friction between a brick and a wooden board is 0.40, and the coefficient of kinetic friction between the brick and board is 0.30. You place the brick on the board and slowly lift one end of the board off the ground until the brick starts to slide down the board. (a) What angle does the board make with the ground when the brick starts to slide? (b) What is the acceleration of the brick as it slides down the board?

142. In Fig. 4.15 an astronaut is playing shuffleboard on Earth. The puck has a mass of 2.0 kg. Between the board and puck the coefficient of static friction is 0.35 and that of kinetic friction is 0.25. (a) If she pushes the puck with a force of 5.0 N in the forward direction, does the puck move? (b) As she is pushing, she trips and the force in the forward direction suddenly becomes 7.5 N. Does the puck move? (c) If so, what is the acceleration of the puck along the board if she maintains contact between puck and stick as she regains her footing while pushing steadily with a force of 6.0 N on the puck? (d) She carries her game to the Moon and again pushes a moving puck with a force of 6.0 N forward. Is the acceleration of the puck during contact more, the same, or less than on Earth? Explain. (connect tutorial: rough table)

143. ✦ A roller coaster is towed up an incline at a steady speed of 0.50 m/s by a chain parallel to the surface of the incline. The slope is 3.0%, which means that the elevation increases by 3.0 m for every 100.0 m of horizontal distance. The mass of the roller coaster is 400.0 kg. Ignoring friction, find the magnitude of the force exerted on the roller coaster by the chain.

144. A 320 kg satellite is in orbit around the Earth 16000 km above the Earth's surface. (a) What is the weight of the satellite when in orbit? (b) What was its weight when it was on the Earth's surface, before being launched? (c) While it orbits the Earth, what force does the satellite exert on the Earth?

145. ✦ The mass of the Moon is 0.0123 times that of the Earth. A spaceship is traveling along a line connecting the centers of the Earth and the Moon. At what distance from the Earth does the spaceship find the gravitational pull of the Earth equal in magnitude to that of the Moon? Express your answer as a percentage of the distance between the centers of the two bodies.

146. ✦ Ⓒ A toy freight train consists of an engine and three identical cars. The train is moving to the right at constant speed along a straight, level track. Three spring scales are used to connect the cars as follows: spring scale A is located between the engine and the first car; scale B is between the first and second cars; scale C is between the second and third cars. (a) If air resistance and friction are negligible, what are the relative readings on the three spring scales A, B, and C? (b) Repeat part (a), taking air resistance and friction into consideration this time. [*Hint*: Draw an FBD for the car in the middle.] (c) If air resistance and friction together cause a force of magnitude 5.5 N on each car, directed toward the left, find the readings of scales A, B, and C.

147. ✦ Four *identical* spring scales, A, B, C, and D are used to hang a 220.0 N sack of potatoes. (a) Assume the scales have negligible weights and all four scales show the same reading. What is the reading of each scale? (b) Suppose that each scale has a weight of 5.0 N. If scales B and D show the same reading, what is the reading of each scale?

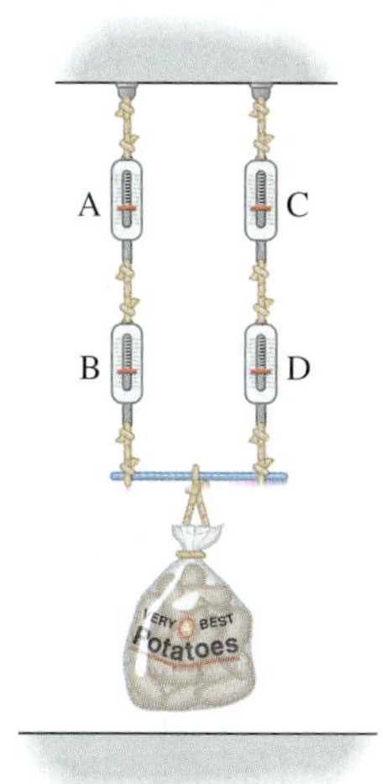

148. A computer weighing 87 N rests on the horizontal surface of your desk. The coefficient of static friction between the computer and the desk is 0.60. (a) Draw an FBD for the computer. (b) What is the magnitude of the frictional force acting on the computer? (c) How hard would you have to push on it to get it to start to slide across the desk?

149. ✦ A refrigerator magnet weighing 0.14 N is used to hold up a photograph weighing 0.030 N. The magnet attracts the refrigerator door with a magnetic force of 2.10 N. (a) Identify the interactions between the magnet and other objects. (b) Draw an FBD for the magnet, showing all the forces that act on it. (c) Which of these forces are long-range and which are contact forces? (d) Find the magnitudes of all the forces acting on the magnet.

150. ✦ A 50.0 kg crate is suspended between the floor and the ceiling using two spring scales, one attached to the ceiling and one to the floor. If the lower scale reads

120 N, what is the reading of the upper scale? Ignore the weight of the scales.

151. ✦ Spring scale A is attached to the ceiling. A 10.0 kg mass is suspended from the scale. A second spring scale, B, is hanging from a hook at the bottom of the 10.0 kg mass and a 4.0 kg mass hangs from the second spring scale. (a) What are the readings of the two scales if the masses of the scales are negligible? (b) What are the readings if each scale has a mass of 1.0 kg?

152. A tire swing hangs at a 12° angle to the vertical when a stiff breeze is blowing. In terms of the tire's weight W, (a) what is the magnitude of the horizontal force exerted on the tire by the wind? (b) What is the tension in the rope supporting the tire? Ignore the weight of the rope.

153. ✦ A boy has stacked two blocks on the floor so that a 5.00 kg block is on top of a 2.00 kg block. (a) If the coefficient of static friction between the two blocks is 0.400 and the coefficient of static friction between the bottom block and the floor is 0.220, with what minimum force should the boy push horizontally on the upper block to make both blocks start to slide together along the floor? (b) If he pushes too hard, the top block starts to slide off the lower block. What is the maximum force with which he can push without that happening if the coefficient of kinetic friction between the bottom block and the floor is 0.200?

154. Anthony is going to drive a flat-bed truck up a hill that makes an angle of 10° with respect to the horizontal direction. A 36.0 kg package sits in the back of the truck. The coefficient of static friction between the package and the truck bed is 0.380. What is the maximum acceleration the truck can have without the package falling off the back?

155. You want to lift a heavy box with a mass of 98.0 kg using the two-pulley system as shown. With what minimum force do you have to pull down on the rope in order to lift the box at a constant velocity? One pulley is attached to the ceiling and one to the box.

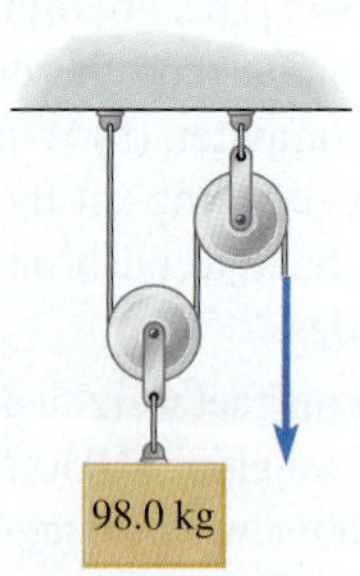

156. ✦ A crate of oranges weighing 180 N rests on a flatbed truck 2.0 m from the back of the truck. The coefficients of friction between the crate and the bed are $\mu_s = 0.30$ and $\mu_k = 0.20$. The truck drives on a straight, level highway at a constant 8.0 m/s. (a) What is the force of friction acting on the crate? (b) If the truck speeds up with an acceleration of 1.0 m/s^2, what is the force of the friction on the crate? (c) What is the maximum acceleration the truck can have without the crate starting to slide?

157. A crate of books is to be put on a truck by rolling it up an incline of angle θ using a dolly. The total mass of the crate and the dolly is m. Assume that rolling the dolly up the incline is the same as sliding it up a frictionless surface. (a) What is the magnitude of the *horizontal* force that must be applied just to hold the crate in place on the incline? (b) What horizontal force must be applied to roll the crate up at constant speed? (c) In order to start the dolly moving, it must be accelerated from rest. What horizontal force must be applied to give the crate an acceleration up the incline of magnitude a? (connect tutorial: cart on ramp)

158. ✦ A toy cart of mass m_1 moves on frictionless wheels as it is pulled by a string under tension T. A block of mass m_2 rests on top of the cart. The coefficient of static friction between the cart and the block is μ_s. Find the maximum tension T that will not cause the block to slide on the cart if the cart rolls on (a) a horizontal surface or (b) up a ramp of angle θ above the horizontal. In both cases, the string is parallel to the surface on which the cart rolls.

159. ✦ A helicopter of mass M is lowering a truck of mass m onto the deck of a ship. (a) At first, the helicopter and the truck move downward together (the length of the cable doesn't change). If their downward speed is decreasing at a rate of $0.10g$, what is the tension in the cable? (b) As the truck gets close to the deck, the helicopter stops moving downward. While it hovers, it lets out the cable so that the truck is still moving downward. If the truck's downward speed is decreasing at a rate of $0.10g$, while the helicopter is at rest, what is the tension in the cable?

160. ✦ A student's head is bent over her physics book. The head weighs 50.0 N and is supported by the muscle force $\vec{\mathbf{F}}_m$ exerted by the neck extensor muscles and by the contact force $\vec{\mathbf{F}}_c$ exerted at the atlantooccipital joint. Given that the magnitude of $\vec{\mathbf{F}}_m$ is 60.0 N and is directed 35° below the horizontal, find (a) the magnitude and (b) the direction of $\vec{\mathbf{F}}_c$.

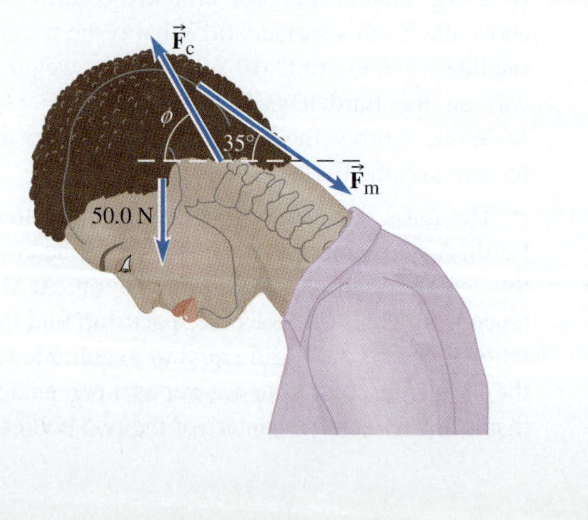

161. ✦ (a) If a spacecraft moves in a straight line between the Earth and the Sun, at what point would the force of gravity on the spacecraft due to the Sun be as large as that due to the Earth? (b) If the spacecraft is close to, but not at, this equilibrium point, does the net force on the spacecraft tend to push it toward or away from the equilibrium point? [*Hint*: Imagine the spacecraft a small distance d closer to the Earth and find out which gravitational force is stronger.]

162. ✦ In a movie, a stuntman places himself on the vertical front of a truck as the truck accelerates. The coefficient of friction between the stuntman and the truck is 0.65. The stuntman is not standing on anything but can "stick" to the front of the truck as long as the truck continues to accelerate. What minimum forward acceleration will keep the stuntman on the front of the truck?

163. ✦ An airplane of mass 2800 kg has just lifted off the runway. It is gaining altitude at a constant 2.3 m/s while the horizontal component of its velocity is increasing at a rate of 0.86 m/s^2. Assume $g = 9.81$ m/s^2. (a) Find the direction of the force exerted on the airplane by the air. (b) Find the horizontal and vertical components of the plane's acceleration if the force due to the air has the same magnitude but has a direction 2.0° closer to the vertical than its direction in part (a).

164. A person is doing leg lifts with 3.00 kg ankle weights. The lower leg itself has a mass of 5.00 kg. When the leg is held still at an angle of 30.0° with respect to the horizontal, the patellar tendon pulls on the tibia with a force of 337 N at an angle of 20.0° *with respect to the lower leg*. Find the magnitude and direction of the force exerted on the tibia by the femur, assuming it is the only other significant force acting on the lower leg.

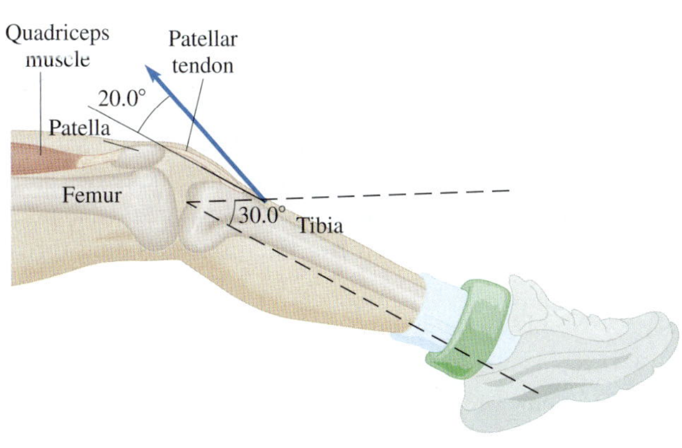

Review and Synthesis

165. The forces on a small airplane (mass 1160 kg) in horizontal flight heading eastward are as follows: weight = 11.37 kN downward, lift = 11.37 kN upward, thrust = 1.800 kN eastward, and drag = 1.400 kN westward. At $t = 0$, the plane's speed is 60.0 m/s. If the forces remain constant, how far does the plane travel in the next 60.0 s?

166. Two blocks are connected by a lightweight, flexible cord that passes over a frictionless pulley. If $m_1 = 3.6$ kg and $m_2 = 9.2$ kg, and block 2 is initially at rest 140 cm above the floor, how long does it take block 2 to reach the floor?

167. ✦ A 10.0 kg watermelon and a 7.00 kg pumpkin are attached to each other via a cord that wraps over a pulley, as shown. Friction is negligible everywhere in this system. (connect tutorial: pulley) (a) Find the accelerations of the pumpkin and the watermelon. Specify magnitude and direction. (b) If the system is released from rest, how far along the incline will the pumpkin travel in 0.30 s? (c) What is the speed of the watermelon after 0.20 s?

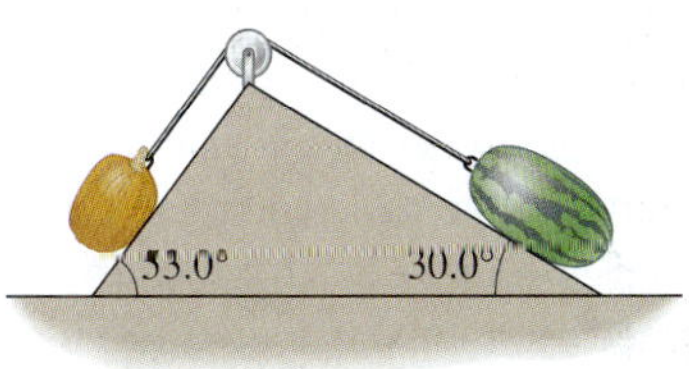

168. In the physics laboratory, a glider is released from rest on a frictionless air track inclined at an angle. If the glider has gained a speed of 25.0 cm/s in traveling 50.0 cm from the starting point, what was the angle of inclination of the track? Draw a graph of $v_x(t)$ when the positive x-axis points down the track.

169. A 10.0 kg block is released from rest on a frictionless track inclined at an angle of 55°. (a) What is the net force on the block after it is released? (b) What is the acceleration of the block? (c) If the block is released from rest, how long will it take for the block to attain a speed of 10.0 m/s? (d) Draw a motion diagram for the block. (e) Draw a graph of $v_x(t)$ for values of velocity between 0 and 10 m/s. Let the positive x-axis point down the track.

170. A skier with a mass of 63 kg starts from rest and skis down an icy (frictionless) slope that has a length of 50 m at an angle of 32° with respect to the horizontal. At the bottom of the slope, the path levels out and becomes horizontal, the snow becomes less icy, and the skier begins to slow down, coming to rest in a distance of 140 m along the horizontal path. (a) What is the

speed of the skier at the bottom of the slope? (b) What is the coefficient of kinetic friction between the skier and the horizontal surface?

171. A woman of mass 51 kg is standing in an elevator. (a) If the elevator floor pushes up on her feet with a force of 408 N, what is the acceleration of the elevator? (b) If the elevator is moving at 1.5 m/s as it passes the fourth floor on its way down, what is its speed 4.0 s later?

172. A model rocket is fired vertically from rest. It has a constant acceleration of 17.5 m/s^2 for the first 1.5 s. Then its fuel is exhausted, and it is in free fall with negligible air resistance. The rocket has a mass of 87 g; the mass of the fuel is much less than 87 g. (a) What was the net force on the rocket during the first 1.5 s after liftoff? (b) What force was exerted on the rocket by the burning fuel? (c) How high does the rocket travel? (d) How long after liftoff does the rocket return to the ground? (e) Sketch a graph of the rocket's vertical velocity vs. time from launch until it returns to the ground. (f) What was the net force on the rocket after its fuel was spent?

173. ✦ An astronaut of mass 60.0 kg and a small asteroid of mass 40.0 kg are initially at rest with respect to the space station. The astronaut pushes the asteroid with a constant force of magnitude 250 N for 0.35 s. Gravitational forces are negligible. (a) How far apart are the astronaut and the asteroid 5.00 s after the astronaut stops pushing? (b) What is their relative speed at this time?

174. Marie slides a paper plate with a slice of pizza across a horizontal table to her friend Jaden. The coefficient of friction between the table and plate is 0.32. If the pizza must travel 44 cm to get from Marie to Jaden, what initial speed should Marie give the plate of pizza so that it just stops when it gets to Jaden?

175. Julia is delivering newspapers. Suppose she is driving at 15 m/s along a straight road and wants to drop a paper out the window from a height of 1.00 m so it slides along the shoulder and comes to rest in the customer's driveway. At what horizontal distance before the driveway should she drop the paper? The coefficient of kinetic friction between the newspaper and the ground is 0.40. Ignore air resistance and assume no bouncing or rolling.

176. ✦ Carlos and Shannon are sledding down a snow-covered slope that is angled at 12° below the horizontal. When sliding on snow, Carlos's sled has a coefficient of friction $\mu_k = 0.10$; Shannon has a "supersled" with $\mu_k = 0.010$. Carlos takes off down the slope starting from rest. When Carlos is 5.0 m from the starting point, Shannon starts down the slope from rest. (a) How far have they traveled when Shannon catches up to Carlos? (b) How fast is Shannon moving with respect to Carlos as she passes by?

177. ✦ At time $t = 0$, block A of mass 0.225 kg and block B of mass 0.600 kg rest on a horizontal frictionless surface a distance 3.40 m apart, with block A located to the left of block B. A horizontal force of 2.00 N directed to the right is applied to block A for a time interval $\Delta t = 0.100$ s. During the same time interval, a 5.00 N horizontal force directed to the left is applied to block B. How far from B's initial position do the two blocks meet? How much time has elapsed from $t = 0$ until the blocks meet?

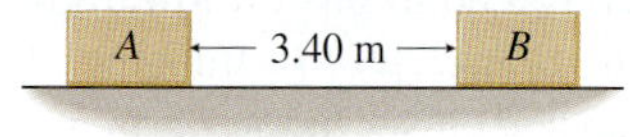

178. A crate is sliding down a frictionless ramp that is inclined at 35.0°. (a) If the crate is released from rest, how far does it travel down the incline in 2.50 s if it does not get to the bottom of the ramp before the time has elapsed? (b) How fast is the crate moving after 2.50 s of travel?

179. You are watching a television show about Navy pilots. The narrator says that when a Navy jet takes off, it accelerates because the engines are at full throttle and because there is a catapult that propels the jet forward. You begin to wonder how much force is supplied by the catapult. You look on the Web and find that the flight deck of an aircraft carrier is about 90 m long, that an F-14 has a mass of 33 000 kg, that each of the two engines supplies 27 000 lb of thrust and that the takeoff speed of such a plane is about 160 mi/h. Estimate the average force on the jet due to the catapult.

Answers to Practice Problems

4.1 (a) $F_x = 49.1$ N, $F_y = 2.9$ N; (b) $F = 49.2$ N; (c) 3.4° above the horizontal

4.2 0.5 kN downward

4.3 In the first case, the principle of inertia says that Emma tends to stay at rest with respect to the ground as the subway car begins to move forward, until forces acting on her (exerted by the strap and the floor) make her move forward. In the second case, Emma keeps moving forward with respect to the ground with constant speed as the subway car slows down, until forces acting on her make her slow down as well. The strap pulls backward on her and she pulls forward on the strap, which she interprets as being thrust forward.

4.4 760 N, 8.3° to the left of the $+y$-axis or 81.7° above the $-x$-axis

4.5 The contact force exerted on the floor by the chest; 870 N, 59° below the rightward horizontal ($+x$-axis)

4.6 For $m_1 = m_2 = 1000$ kg and $r = 4$ m, $F \approx 4$ μN, which is about the same magnitude as the weight of a mosquito. The claim that this tiny force caused the collision is ridiculous.

4.7 0.57 N or 0.13 lb

4.8 The chest is in equilibrium, so the net force on it is zero. Setting the net force equal to zero separately for the horizontal and vertical components gives the answer: the normal force is 750 N, up, and the frictional force is 110 N, to the left. The quantity $\mu_s N$ is the *maximum* possible magnitude of the force of static friction for a surface. In this problem, the frictional force does not necessarily have the maximum possible magnitude.

4.9 (a)

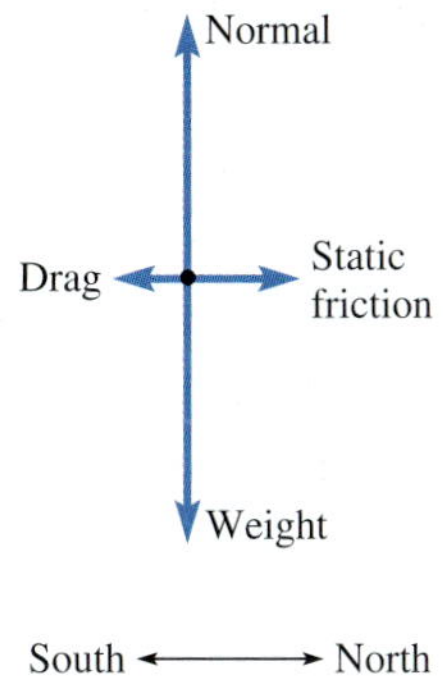

(b) Weight of the car = 11.0 kN; (c) 2.1 kN northward

4.10 (a) 110 N; (b) 230 N

4.11 3100 N

4.12

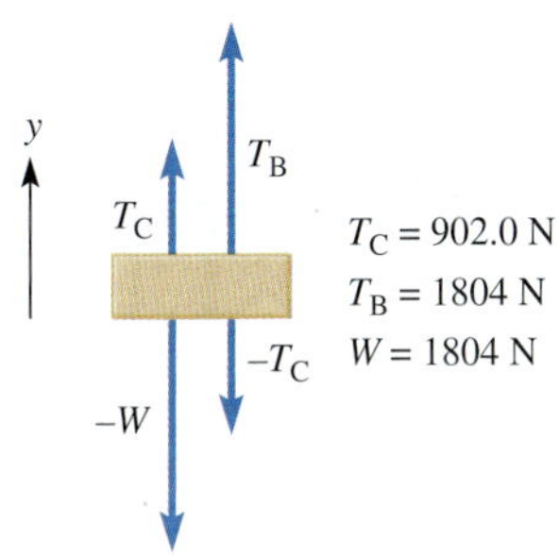

$$\Sigma F_y = 902\text{ N} + 1804\text{ N} - 1804\text{ N} - 902\text{ N} = 0$$

4.13 54 N

4.14 1.84 kN

4.15 Block 1: $\Sigma F_{1y} = T - m_1 g = 315\text{ N} - 255\text{ N} = 60\text{ N}$; $m_1 a_{1y} = 60\text{ N}$
Block 2: $\Sigma F_{2y} = m_2 g - T = 412\text{ N} - 315\text{ N} = 97\text{ N}$; $m_2 a_{2y} = 97\text{ N}$

4.16 Impossible to pull the crate up with a single pulley. The entire weight of the crate would be supported by a single strand of cable and that weight exceeds the breaking strength of the cable.

4.17 2500 N

4.18 (a) down the incline; (b) up the incline; (c) 0.2 m/s^2 down the incline

4.19 (a) 392 N; (b) 431 N

Answers to Checkpoints

4.1A Contact force exerted on the player by the ball; contact force exerted on the player by the ground; contact force exerted on the player by the air; gravitational force exerted on the player by Earth.

4.1B No, the net force is the sum of *all* the forces acting on the pulley. The patient's foot also exerts a force on the pulley.

4.2 The *Voyager* space probes are so far from the Sun that the gravitational forces exerted on them due to the Sun are negligibly small. To a very good approximation, we can say that the net force acting on them is zero. Therefore, the probes continue moving at constant speed along a straight line. No applied force has to be maintained by an engine to keep them moving because there are no forces that oppose their motion.

4.4 The two forces exerted by the two children on a toy cannot be interaction partners because they act on the *same* object (the toy), not on two different objects. Interaction partners act on different objects, one on each of the two objects that are interacting. The interaction partner of the force exerted by one child on the toy is the force that the toy exerts on that child.

4.5 The weight of the gear decreases as the value of g decreases. The mass of the gear does not change.

4.6 One upward normal force on each leg due to the floor and one downward normal force on the desktop due to the laptop.

4.8 Yes. For motion along an incline, it simplifies the problem to choose one axis parallel to the incline and the other perpendicular to the incline.

4.10 Your velocity is downward and decreasing in magnitude, so your acceleration is upward. Then the upward normal force exerted on you by the scale must be greater than your weight.

CHAPTER

5 Circular Motion

German athlete Susanne Keil throws the hammer during the German Athletics championships. Keil qualified for the 2004 Olympics in Athens with a 67.77 m throw.

In the track and field event called the *hammer throw*, the "hammer" is actually a metal ball (mass 4.00 kg for women or 7.26 kg for men) attached by a cable to a grip. The athlete whirls the hammer several times around while not leaving a circle of radius 2.1 m and then releases it. The winner is the athlete whose hammer lands the farthest distance away. How large a force does an athlete have to exert on the grip to whirl the massive hammer around in a circle? What kind of path does the hammer follow once it is released? (See pp. 163–164 for the answer.)

BIOMEDICAL APPLICATIONS

- Centrifuges (Exs. 5.2, 5.4; Probs. 14, 53, 54, 84)
- Effects of acceleration on organisms (Sec. 5.7; Ex. 5.4; Ex. 5.14; Probs. 14, 15, 59, 60)
- Dung beetles (Prob. 8)
- Flagellum of a bacterium (Prob. 79)

Concepts & Skills to Review

- gravitational forces (Section 4.5)
- Newton's second law: force and acceleration (Section 4.3)
- velocity and acceleration (Sections 3.3 and 3.4)
- apparent weight (Section 4.10)
- normal and frictional forces (Section 4.6)

5.1 DESCRIPTION OF UNIFORM CIRCULAR MOTION

Ask someone to name the most important machine ever invented by humans and you are likely to get the *wheel* as a response. Rotating objects are so important to modern—and even not-so-modern—technology that we barely notice them. Examples include wheels on cars, bicycles, trains, and lawnmowers; propellers on airplanes and helicopters; DVDs and Blu-ray discs; computer hard drives; the gears and hands of an analog clock; amusement park rides and centrifuges—the list is endless.

Rotation of a Rigid Object To describe circular motion, we could use the familiar definitions of displacement, velocity, and acceleration. But much of the circular motion around us occurs in the rotation of a rigid object. A **rigid object** is one for which the distance between any two points of the object remains the same when the object is translated or rotated. When such an object rotates, every point on the object moves in a circular path. The radius of the path for any point is the distance between that point and the axis of rotation. When a DVD spins inside a DVD player, different points on the DVD have different velocities and accelerations. The velocity and acceleration of a given point keep changing direction as the DVD spins. It would be clumsy to describe the rotation of the DVD by talking about the motion of arbitrary points on it. However, some quantities are the *same* for every point on the DVD. It is much simpler, for instance, to say "the DVD spins at 210 rev/min" instead of saying "a point 6.0 cm from the rotation axis of the DVD is moving at 1.3 m/s."

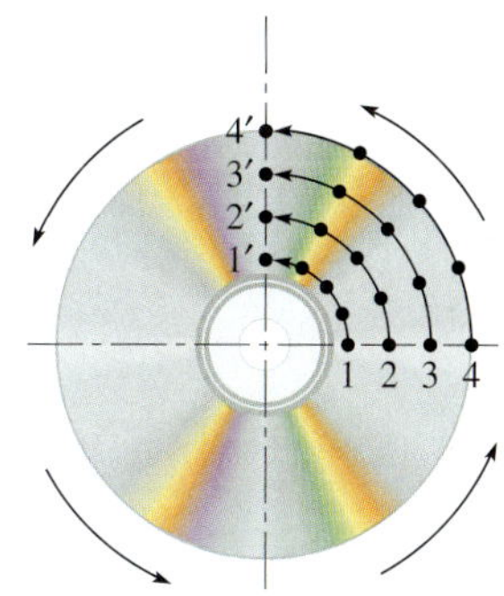

Figure 5.1 Motion diagrams for four points on a DVD as it rotates through $\frac{1}{4}$ turn. Points 1, 2, 3, and 4 travel through the same angle but different distances to reach their new positions, marked 1′, 2′, 3′, and 4′, respectively.

Angular Displacement and Angular Velocity To simplify the description of circular motion, we concentrate on *angles* instead of distances. If a DVD spins through $\frac{1}{4}$ of a turn, every point moves through the same angle (90°), but points at different radii move different linear distances. On the DVD shown in Fig. 5.1, point 1 near the axis of rotation moves through a smaller distance than point 4 on the circumference. For this reason we define a set of variables that are analogous to displacement, velocity, and acceleration, but use angular measure instead of linear distance. Instead of displacement, we speak of **angular displacement** $\Delta\theta$, the angle through which the DVD turns. A point on the DVD moves along the circumference of a circle. As the point moves from the angular position θ_i to the angular position θ_f, a radial line drawn between the center of the circle and that point sweeps out an angle $\Delta\theta = \theta_f - \theta_i$, which is the angular displacement of the DVD during that time interval (Fig. 5.2).

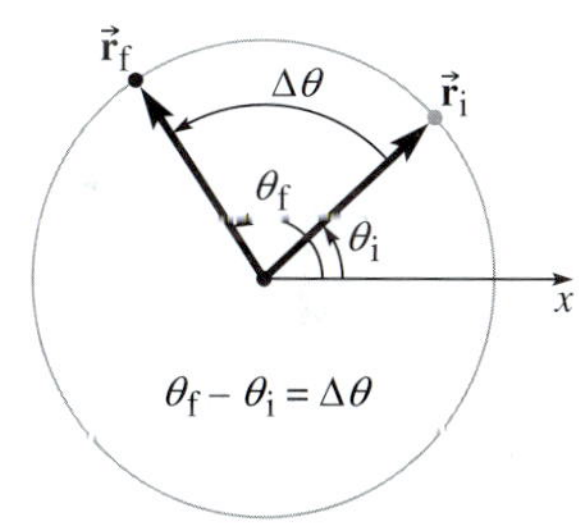

Figure 5.2 Angular positions such as θ_i and θ_f are measured counterclockwise from a reference axis (usually the x-axis).

Definition of angular displacement

$$\Delta\theta = \theta_f - \theta_i \qquad (5\text{-}1)$$

CONNECTION:

Equations (5-1) through (5-3) have a familiar form because ω is the *rate of change* of θ, just as velocity is the rate of change of position.

The sign of the angular displacement indicates the sense of the rotation. The usual convention is that a positive angular displacement represents counterclockwise rotation and a negative angular displacement represents clockwise rotation.

+ means Counterclockwise
− means Clockwise

Counterclockwise and clockwise are only well defined for a particular viewing direction; counterclockwise rotation viewed from above is clockwise when viewed from below.

The average angular velocity ω_{av} is the average rate of change of the angular displacement.

Definition of average angular velocity

$$\omega_{av} = \frac{\Delta\theta}{\Delta t} \quad (5\text{-}2)$$

If we let the time interval Δt become shorter and shorter, we are averaging over smaller and smaller time intervals. In the limit $\Delta t \to 0$, ω_{av} becomes the instantaneous angular velocity ω.

Definition of instantaneous angular velocity

$$\omega = \lim_{\Delta t \to 0} \frac{\Delta\theta}{\Delta t} \quad (5\text{-}3)$$

Remember that the notation $\lim_{\Delta t \to 0}$ indicates that $\Delta\theta$ is the angular displacement during a *very short* time interval Δt (short enough that the ratio $\Delta\theta/\Delta t$ doesn't change significantly if we make the time interval even shorter).

The angular velocity indicates—through its algebraic sign—in what direction the DVD is spinning. Since angular displacements can be measured in degrees or radians, angular velocities have units such as degrees/second, radians/second, degrees/day, and the like.

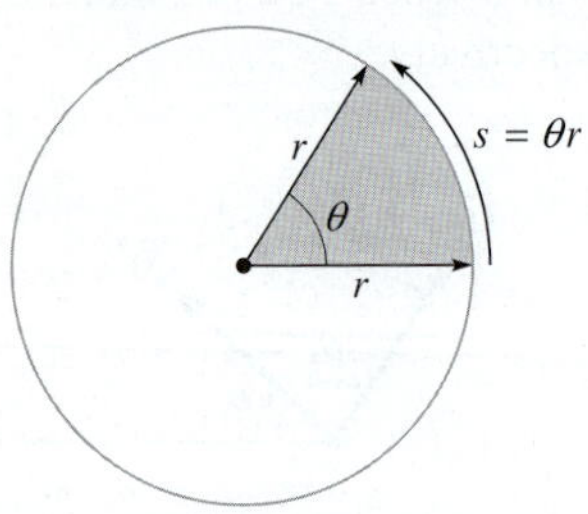

Figure 5.3 Definition of the radian: angle θ in radians is the arc length s divided by the radius r. The angle shown is 1 rad ≈ 57.3°.

Radian Measure You may be most familiar with measuring angles in degrees, but in many situations the most convenient measure is the **radian**. One such situation is when we relate the angular displacement or angular velocity of a rotating object with the distance traveled by, or the speed of, some point on the object.

In Fig. 5.3, an angle θ between two radii of a circle define an arc of length s. We say that θ is the angle *subtended* by the arc. The arc length is proportional to both the radius of the circle and to the angle subtended. The angle θ in radians is *defined* as

$$\theta \text{ (in radians)} = \frac{s}{r} \quad (5\text{-}4)$$

where r is the radius of the circle. Since an angle in radians is defined by the ratio of two lengths, it is dimensionless (a pure number). We use the term radians, abbreviated "rad," to keep track of the angular measure used. Since "rad" is not a physical unit like meters or kilograms, it does not have to balance in Eq. (5-4). For the same reason, we can drop "rad" whenever there is no chance of being misunderstood. We can write $\omega = 23\ \text{s}^{-1}$ as long as context makes it clear that we mean 23 radians per second.

In equations that relate linear variables to angular variables [e.g., Eq. (5-4)], think of r as the number of meters of arc length per radian of angle subtended. In other words, think of r as having units of meters per radian. Doing so, the radians cancel out in these equations. For example, if $\theta = 2.0$ rad and $r = 1.2$ m, then the arc length is

$$s = \theta r = 2.0\ \cancel{\text{rad}} \times 1.2\ \frac{\text{m}}{\cancel{\text{rad}}} = 2.4\ \text{m}$$

Since the arc length for an angle of 360° is the circumference of the circle, the radian measure of an angle of 360° is

$$\theta = \frac{s}{r} = \frac{2\pi r}{r} = 2\pi\ \text{rad}$$

Therefore, the conversion between degrees and radians is

$$360° = 2\pi\ \text{rad} \quad (5\text{-}5)$$

Example 5.1

Angular Speed of Earth

Earth is rotating about its axis. What is its angular speed in rad/s? (The question asks for angular *speed*, so we do not have to worry about the direction of rotation.)

Strategy The Earth's angular velocity is constant, or nearly so. Therefore, we can calculate the average angular velocity for any convenient time interval and, in turn, the Earth's instantaneous angular speed $|\omega|$.

Solution It takes the Earth 1 day to complete one rotation, during which the angular displacement is 2π rad. More formally, during a time interval $\Delta t = 1$ day, the angular displacement of the Earth is $\Delta\theta = 2\pi$ rad. So the angular speed of the Earth is 2π rad/day, and then convert days to seconds.

$$1 \text{ day} = 24 \text{ h} = 24 \text{ h} \times 3600 \text{ s/h} = 86\,400 \text{ s}$$

$$|\omega| = \frac{2\pi \text{ rad}}{86\,400 \text{ s}} = 7.3 \times 10^{-5} \text{ rad/s}$$

Discussion Notice that this problem is analogous to a problem in linear motion such as: "A car travels in a straight line at constant speed. In 3 h, it has traveled 192 mi. What is its velocity in m/s?" Just about everything in circular motion and rotation has this kind of analog—which means we can draw heavily on what we have already learned.

Earth actually completes one rotation in 23.9345 h (see inside back cover) rather than in 24 h due to Earth's motion around the Sun. This distinction would be important only if we needed a more precise value of $|\omega|$ (more than two significant figures).

Practice Problem 5.1 Angular Speed of Venus

Venus completes one rotation about its axis every 5816 h. What is the angular speed of the rotation of Venus in rad/s?

Relation Between Linear and Angular Speed

For a point moving in a circular path of radius r, the linear distance traveled along the circular path during an angular displacement of $\Delta\theta$ (in radians) is the arc length s where

$$s = r|\Delta\theta| = r|\theta_f - \theta_i| \quad \text{(angles in radians)} \qquad (5\text{-}6)$$

The point in question could be a point particle moving in a circular path, or it could be any point on a rotating rigid object. Since Eq. (5-6) comes directly from the definition of the radian, any equation derived from Eq. (5-6) is valid only when the angles are measured in radians.

What is the linear speed at which the point moves? The average linear speed is the distance traveled divided by the time interval,

$$v_{av} = \frac{s}{\Delta t} = \frac{r|\Delta\theta|}{\Delta t} \quad (\Delta\theta \text{ in radians})$$

We recognize $\Delta\theta/\Delta t$ as the average angular velocity ω_{av}. If we take the limit as Δt approaches zero, both average quantities (v_{av} and ω_{av}) become instantaneous quantities (v and ω). Therefore, the relationship between v and ω is

Relationship between linear speed and angular speed

$$v = r|\omega| \quad (\omega \text{ in radians per unit time}) \qquad (5\text{-}7)$$

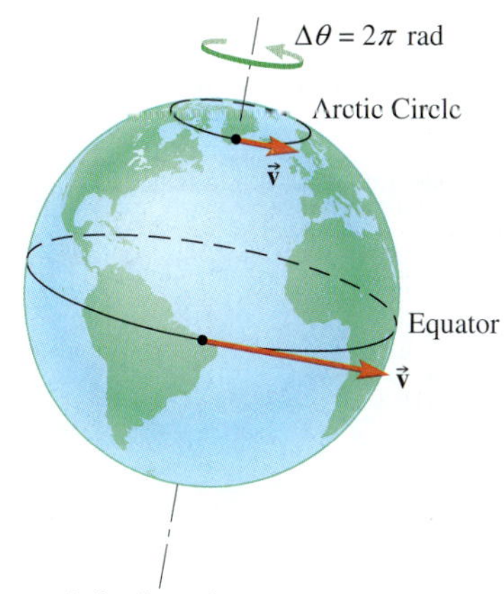

Figure 5.4 A person standing at the equator is moving much faster than another person standing at the Arctic Circle, but their *angular* speeds are the same.

Equation (5-7) relates only the *magnitudes* of the linear and angular speeds. The direction of the velocity vector $\vec{v}$ is tangent to the circular path. For a rotating object, points farther from the axis move at higher linear speeds; they have a circle of bigger radius to travel and, therefore, cover more distance in the same time interval. For example, a person standing at the equator has a much higher linear speed due to Earth's rotation than does a person standing at the Arctic Circle (Fig. 5.4).

Period and Frequency

When the speed of a point moving in a circle is constant, its motion is called **uniform circular motion**. Even though the speed of the point is constant, the velocity is not: the direction of the velocity is changing. This distinction is important when we find the acceleration of an object in uniform circular motion (see Section 5.2). The time for the point to travel completely around the circle is called the **period** of the motion, T. The **frequency** of the motion, which is the number of revolutions per unit time, is defined as

$$f = \frac{1}{T} \tag{5-8}$$

For example, suppose that a wind turbine turns steadily and completes 24 revolutions in 120 seconds. Its period we compute as T = (120 s)/(24 rev) = 5.0 s (meaning 5.0 seconds per revolution). Its frequency, defined as the number of revolutions per unit time, we can compute from the same data as (24 rev)/(120 s) = 0.20 rev/s.

CHECKPOINT 5.1

If a computer hard drive spins at 7200 rev/min, what is its period?

The speed is the total distance traveled divided by the time taken,

$$v = \frac{2\pi r}{T} = 2\pi r f \tag{5-9a}$$

Then, for uniform circular motion

$$|\omega| = \frac{v}{r} = 2\pi f \quad (\omega \text{ in radians per unit time}) \tag{5-9b}$$

where, in SI units, angular velocity ω is measured in rad/s and frequency f is measured in hertz (Hz). The hertz is a derived unit equal to 1 rev/s. The dimensions of Eq. (5-9b) are correct since both revolutions and radians are pure numbers. The physical dimensions on both sides are a number per second (s^{-1}).

Example 5.2

Speed in a Centrifuge

A centrifuge is spinning at 5400 rev/min. (a) Find the period (in s) and frequency (in Hz) of the motion. (b) If the radius of the centrifuge is 14 cm, how fast (in m/s) is an object at the outer edge moving?

Strategy 5400 rev/min *is* the frequency, but in a unit other than Hz. After a unit conversion, the other quantities can be found using the relations already discussed.

Solution (a) First convert rev/min to Hz:

$$f = 5400 \frac{\text{rev}}{\text{min}} \times \frac{1 \text{ min}}{60 \text{ s}} = 90 \text{ rev/s}$$

so the frequency is $f = 90$ Hz $= 90\ s^{-1}$. The period is

$$T = 1/f = 0.011 \text{ s}$$

(b) To find the linear speed, we first find the angular speed in rad/s:

$$|\omega| = 90 \frac{\text{rev}}{\text{s}} \times 2\pi \frac{\text{rad}}{\text{rev}} = 180\pi \text{ rad/s}$$

So $|\omega| = 2\pi f = 180\ \pi$ rad/s. The linear speed is

$$v = |\omega| r = 180\pi\ \text{s}^{-1} \times 0.14 \text{ m} = 79 \text{ m/s}$$

Discussion Notice that much of this problem was done with unit conversions. Instead of memorizing a formula such as $|\omega| = 2\pi f$, an understanding of where the formula came from (in this case, that 2π radians correspond to one revolution) is more useful and less prone to error.

Practice Problem 5.2 Clothing in the Drier

An automatic clothing drier spins at 51.6 rev/min. If the radius of the drier drum is 30.5 cm, how fast is the outer edge of the drum moving?

Rolling Without Slipping: Rotation and Translation Combined

When an object is rolling, it is both rotating and translating. The wheel rotates about an axle, but the axle is not at rest; it moves forward or backward. What is the relationship between the angular speed of the wheel and the linear speed of the axle? You might guess that $v = |\omega|r$ is the answer. You would be right, as long as the object rolls without slipping or skidding.

There is no fixed relationship between the linear and angular speeds of a wheel if it is allowed to skid or slip. When an impatient driver guns the engine the instant a traffic light turns green, the automobile wheels are likely to slip. The rubber sliding against the road surface makes the squealing sound and leaves tracks on the road. The driver could actually make the acceleration of the car greater by giving the engine *less* gas. When the wheels are skidding or slipping, *kinetic* friction propels the car forward instead of the potentially larger force of *static* friction.

For a wheel that rolls *without* slipping, as the wheel turns through one complete rotation, the axle moves a distance equal to the circumference of the wheel (Fig. 5.5). Think of a paint roller leaving a line of paint as it rolls along a wall. After one complete rotation, the same point on the roller wheel is touching the wall as was initially touching it. The length of the line of paint is $2\pi r$. The elapsed time is T, so the axle's speed is

$$v_{\text{axle}} = \frac{2\pi r}{T}$$

while the angular speed of the roller is

$$|\omega| = \frac{2\pi}{T}$$

Thus,

$$v_{\text{axle}} = |\omega|r \quad (\omega \text{ in radians per unit time}) \tag{5-10}$$

Figure 5.5 (a) As a wheel of radius r that rolls without slipping turns through one complete revolution, the distance its axle moves is equal to the circumference of the wheel ($2\pi r$). (b) As a wheel rolls without slipping through an angle $\Delta\theta$, the distance the axle moves is equal to the arc length s.

Example 5.3

Angular Speed of a Rolling Wheel

Kevin is riding his motorcycle at a speed of 13.0 m/s. If the diameter of the rear tire is 65.0 cm, what is the angular speed of the rear wheel? Assume that it rolls without slipping.

Strategy The given diameter of the tire enables us to find the circumference and, thus, the distance traveled in one revolution of the wheel. From the speed of the motorcycle we can find how many revolutions the tire must make per second.

Solution During one revolution of the wheel, the motorcycle travels a distance equal to the tire's circumference $2\pi r$ (see Fig. 5.5). Then the time to make one revolution is T, and the speed v is

$$v = \frac{\text{distance}}{\text{time}} = \frac{2\pi r}{T}$$

Therefore, $T = 2\pi r/v$. For each revolution there is an angular displacement of $\Delta\theta = 2\pi$ radians, so

$$|\omega| = \frac{|\Delta\theta|}{\Delta t} = \frac{2\pi}{T}$$

Substituting $T = 2\pi r/v$ and remembering that the radius is half the diameter,

$$|\omega| = \frac{2\pi}{2\pi r/v} = \frac{v}{r} = \frac{13.0 \text{ m/s}}{(0.650 \text{ m})/2} = 40.0 \frac{\text{rad}}{\text{s}}$$

Discussion Check: Time for one revolution is

$$\frac{2\pi \text{ rad}}{40.0 \text{ rad/s}} = 0.157 \text{ s.}$$

Time to travel a distance $2\pi r = 2.04$ m is

$$\frac{2.04 \text{ m}}{13.0 \text{ m/s}} = 0.157 \text{ s.}$$

Looks good.

You could have obtained this answer immediately by looking back through the text for the equation $|\omega| = v/r$ and plugging in numbers, but the solution here shows that you can re-create that equation. Here, and in many cases, there is no need to memorize a formula if you understand the concepts behind the formula. You are then less apt to make a mistake by forgetting a factor or constant in the equation, or by using an inappropriate formula. For another example, if an object moves along a straight line at a constant velocity, you know instantly that the displacement is the velocity times the time interval—not because you have memorized an equation ($\Delta\vec{\mathbf{r}} = \vec{\mathbf{v}}\Delta t$), but because you understand the concepts of displacement and velocity. This is the sort of internalization of scientific thinking that you will develop with more and more practice in problem solving.

Practice Problem 5.3 Rolling Drum

A cylindrical steel drum is tipped over and rolled along the floor of a warehouse. If the drum has a radius of 0.40 m and makes one complete turn every 8.0 s, how long does it take to roll the drum 36 m?

5.2 RADIAL ACCELERATION

For a particle undergoing uniform circular motion, the *magnitude* of the velocity vector is constant, but its direction is continuously changing. At any instant of time, the direction of the instantaneous velocity is tangent to the path, as discussed in Section 3.3. Since the *direction* of the velocity continually changes, the particle has a nonzero acceleration.

In Fig. 5.6a, two velocity vectors of equal magnitude are drawn tangent to a circular path of radius r, representing the velocity at two different times of an object moving around a circular path with constant speed. At any instant, the velocity vector is perpendicular to a radius drawn from the center of the circle to the position of the object. As the time between velocity measurements approaches zero, the radii become closer together (Fig. 5.6b). For circular motion, just as for any other kind of motion, the acceleration is defined as

$$\vec{\mathbf{a}} = \lim_{\Delta t \to 0} \frac{\Delta\vec{\mathbf{v}}}{\Delta t} \tag{3-17}$$

To find the acceleration, we must first find the change in velocity $\Delta\vec{\mathbf{v}}$ for a very short time interval. Figure 5.6c shows that as the time interval Δt approaches zero, the angle between the two velocities also approaches zero and $\Delta\vec{\mathbf{v}}$ becomes perpendicular to the velocity.

Since $\Delta\vec{\mathbf{v}}$ is perpendicular to the velocity, it is directed along a radius of the circle. Inspection of Figs. 5.6b and 5.6c shows that $\Delta\vec{\mathbf{v}}$ is radially *inward* (toward the center of

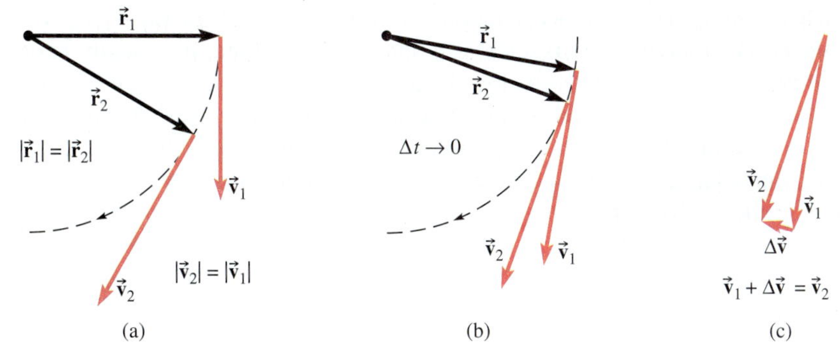

Figure 5.6 Uniform circular motion at constant speed. (a) The velocity vector is always tangent to the circular path and perpendicular to the radius at that point. (b) As the time interval between two velocity measurements decreases, the angle between the velocity vectors decreases. (c) The change in velocity ($\Delta\vec{v}$) is found by placing the tails of the two velocity vectors together. Then $\Delta\vec{v}$ is drawn from the tip of the initial velocity ($\vec{v}_1$) to the tip of the final velocity ($\vec{v}_2$) so that $\vec{v}_1 + \Delta\vec{v} = \vec{v}_2$.

CONNECTION:

Radial acceleration is not a new kind of acceleration. The acceleration vector for an object moving in uniform circular motion is directed radially inward toward the center of the circle.

the circle). Since the acceleration $\vec{a}$ has the same direction as $\Delta\vec{v}$ (in the limit $\Delta t \to 0$), the acceleration is also directed radially inward (Fig. 5.7)—that is, along a radius of the circular path toward the center of the circle. The acceleration of an object undergoing *uniform* circular motion is often called the **radial acceleration $\vec{a}_r$**. The word *radial* here just reminds us of the direction of the acceleration. (A synonym for radial acceleration is *centripetal acceleration*. *Centripetal* means "toward the center.")

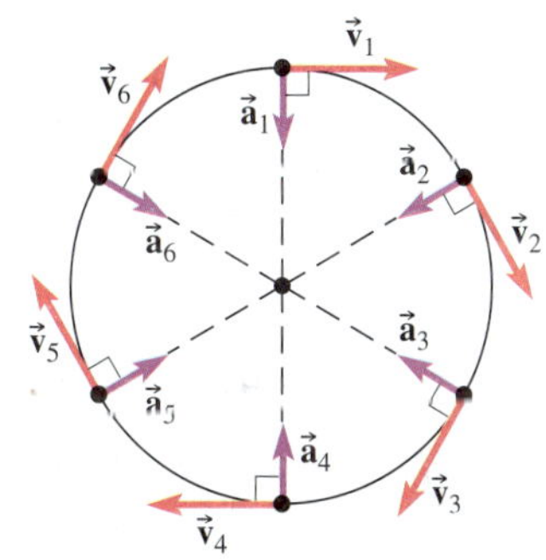

Figure 5.7 Motion diagram for an object in uniform circular motion, with acceleration and velocity vectors drawn at each of the six points. The acceleration is always directed toward the center of the circle, perpendicular to the velocity (see connect interactive: circular motion).

CHECKPOINT 5.2

Does a radial acceleration mean that the speed of the object is changing?

Magnitude of the Radial Acceleration

To find the magnitude of the radial acceleration for uniform circular motion, we must find the change in velocity $\Delta\vec{v}$ for a time interval Δt in the limit $\Delta t \to 0$. The velocity keeps the same magnitude but changes direction at a steady rate, equal to the angular velocity ω. In a time interval Δt, the velocity $\vec{v}$ rotates through an angle equal to the angular displacement $\Delta\theta = \omega\,\Delta t$. During this time interval, the velocity vector sweeps out an arc of a circle of "radius" v (Fig. 5.8). In the limit $\Delta t \to 0$, the magnitude of $\Delta\vec{v}$ becomes equal to the arc length, since a very short arc approaches a straight line. Then

$$|\Delta\vec{v}| = \text{arc length} = \text{radius of circle} \times \text{angle subtended}$$
$$= v\,|\Delta\theta| = v\,|\omega|\,\Delta t$$

Acceleration is the rate of change of velocity, so the magnitude of the radial acceleration is

$$a_r = |\vec{a}| = \frac{|\Delta\vec{v}|}{\Delta t} = v|\omega| \quad (\omega \text{ in radians per unit time}) \qquad (5\text{-}11)$$

where absolute value symbols are used with the vector quantities to indicate their magnitudes. Velocity and angular velocity are not independent; $v = |\omega|\,r$. It is usually most convenient to write the magnitude of the radial acceleration in terms of one or the other of these two quantities. So we write the radial acceleration in two other equivalent ways using $v = |\omega|\,r$:

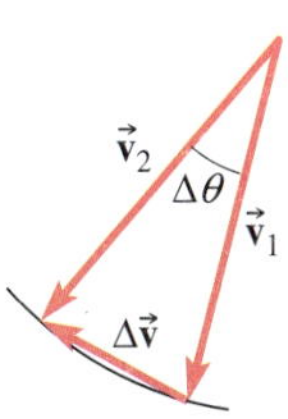

Figure 5.8 The velocity vector sweeps out an arc of a circle whose "length" is nearly equal to that of the chord $\Delta\vec{v}$.

Radial acceleration

$$a_r = \frac{v^2}{r} \quad \text{or} \quad a_r = \omega^2 r \quad (\omega \text{ in radians per unit time}) \qquad (5\text{-}12)$$

When finding the radial acceleration, use whichever form of Eq. (5-12) is more convenient. For rotating objects such as a spinning centrifuge, it's usually easiest to think in terms of the angular velocity. For an object moving around a circle, such as a satellite in orbit whose speed is known, it might be easier to use v^2/r. Since the two equations are equivalent, either can be used in any situation. Note that Eqs. (5-11) and (5-12) assume that ω is expressed in *radians* per unit time (normally rad/s, but rad/min or rad/h would also be correct).

Example 5.4

Testing the Acceleration That a Pilot Can Withstand

Centrifuges are used to establish the maximum acceleration a pilot can withstand without "blacking out" (Fig. 5.9). If the pilot undergoes a radial acceleration of 4.00*g* (as measured at her head) and the radial distance from her head to the axis of rotation is 12.5 m, what is the period of rotation of the centrifuge?

Figure 5.9
The 20 G Centrifuge at NASA's Ames Research Center, Moffett Field, California, can spin a pilot with an acceleration of up to 20*g*.

Strategy The radial acceleration can be found from the radius of the circular path and either the linear or the angular speed. The period is the time for one complete revolution; in one revolution the distance traveled is the circumference of the circle.

Solution The radial acceleration is

$$a_r = \frac{v^2}{r} \quad (5\text{-}12)$$

Therefore, the linear speed is $v = \sqrt{a_r r}$. The linear speed is the distance traveled in one revolution $(2\pi r)$ divided by the period T:

$$v = \frac{2\pi r}{T} \quad (5\text{-}9a)$$

Solving for the period, we obtain

$$T = \frac{2\pi r}{v} = \frac{2\pi r}{\sqrt{a_r r}} = 2\pi\sqrt{\frac{r}{a_r}} = 2\pi\sqrt{\frac{12.5\text{ m}}{4.00 \times 9.80\text{ m/s}^2}} = 3.55\text{ s}$$

Discussion For a quick check, let's calculate how fast the pilot is moving.

$$v = \sqrt{a_r r} = \sqrt{4.00 \times 9.80\text{ m/s}^2 \times 12.5\text{ m}} = 22.1\text{ m/s } (\approx 50\text{ mi/h})$$

This seems like a reasonable order of magnitude for the conditions; 22 m/s would be too fast to take a curve of radius 12.5 m on a motorcycle or in a car, but you wouldn't want an acceleration as large as 4*g* then! Now let's verify the period using the speed: $T = 2\pi r/v = (78.5\text{ m}) / (22.1\text{ m/s}) = 3.55\text{ s}$.

The problem can be solved using angular speed ω instead of linear speed v. The radial acceleration is $a_r = \omega^2 r$ and the period is

$$T = \frac{2\pi}{\omega} = \frac{2\pi}{\sqrt{a_r/r}} = 2\pi\sqrt{\frac{r}{a_r}}$$

the same result as before.

Practice Problem 5.4 A Spinning Blu-ray Disc

If a Blu-ray disc spins at 7200 rev/min, what is the radial acceleration of a point on the outer rim of the disc? The disc is 12 cm in diameter.

Applying Newton's Second Law to Uniform Circular Motion

Now that we know the magnitude and direction of the acceleration of any object in uniform circular motion, we can use Newton's second law to relate the net force acting on the object to the speed and radius of its motion. The net force is found in the usual way: each of the individual forces acting on the object is identified and then the forces are added as vectors. Every force acting must be exerted *by some other object*. Resist the temptation to add in a new, separate force just because something moves in a circle. For an object to move in a circle at constant speed, real, physical forces such as gravity, tension, normal forces, and friction must act on it; these forces combine to produce a net force that has the correct magnitude and is always perpendicular to the velocity of the object.

Problem-Solving Strategy for an Object in Uniform Circular Motion

1. Begin as for any Newton's second law problem: identify all the forces acting on the object and draw an FBD.
2. Choose perpendicular axes at the point of interest so that one is radial and the other is tangent to the circular path.
3. Find the radial component of each force.
4. Apply Newton's second law in the radial direction:

$$\sum F_r = ma_r$$

Here $\sum F_r$ is the radial component of the net force and the radial component of the acceleration is

$$a_r = \frac{v^2}{r} = \omega^2 r \tag{5-12}$$

(For uniform circular motion, neither the net force nor the acceleration has a tangential component because the speed is constant.)

Example 5.5

The Hammer Throw

What force does the athlete exert on the grip? What path does the hammer follow after release?

An athlete whirls a 4.00 kg hammer six or seven times around and then releases it. Although the purpose of whirling it around several times is to increase the hammer's speed, assume that *just before* the hammer is released, it moves at constant speed along a circular arc of radius 1.7 m. At the instant she releases the hammer, it is 1.0 m above the ground and its velocity is directed 40° above the horizontal. The hammer lands a horizontal distance of 74.0 m away. What force does the athlete apply to the grip just before she releases it? Ignore air resistance.

Strategy After release, the only force acting on the hammer is gravity. The hammer moves in a parabolic trajectory like any other projectile. By analyzing the projectile motion of the hammer, we can find the speed of the hammer just after its release. Just *before* release, the forces acting on the hammer are the tension in the cable and gravity. We can relate the net force on the hammer to its radial acceleration, calculated from the speed and radius of its path. The problem becomes two subproblems, one dealing with circular motion and the other with projectile motion. The final velocity for the circular motion is the initial velocity for the projectile motion.

Solution During its projectile motion, the initial velocity has magnitude v_i (to be determined) and direction $\theta = 40°$ above the horizontal. Choosing the $+y$-axis pointing up, the displacement of the hammer (in component form) is $\Delta x = 74.0$ m and $\Delta y = -1.0$ m (Fig. 5.10), the acceleration of the hammer is $a_x = 0$ and $a_y = -g$, and the initial velocity is $v_{ix} = v_i \cos\theta$ and $v_{iy} = v_i \sin\theta$. Then, from Eqs. (3-20) and (3-21),

$$\Delta x = (v_i \cos\theta)\,\Delta t \quad \text{and} \quad \Delta y = (v_i \sin\theta)\,\Delta t - \tfrac{1}{2}g(\Delta t)^2$$

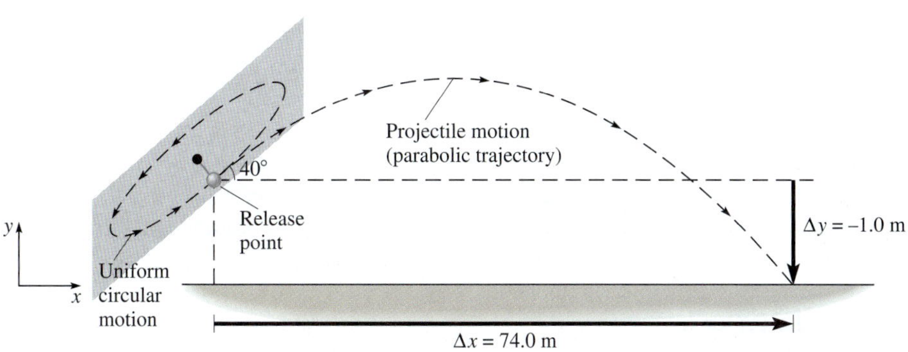

Figure 5.10 Path of the hammer from just before its release until it hits the ground. (Distances are *not* to scale.)

continued on next page

Example 5.5 continued

Solving the left equation for Δt and substituting into the right equation gives

$$\Delta y = v_i \sin\theta \frac{\Delta x}{\cancel{v_i} \cos\theta} - \frac{1}{2} g\left(\frac{\Delta x}{v_i \cos\theta}\right)^2$$

After a bit of algebra, we can solve for v_i. First we multiply through by $2v_i^2 \cos^2\theta$:

$$2v_i^2 \cos^2\theta\, \Delta y = 2v_i^2 \cos^2\theta \frac{\Delta x \sin\theta}{\cos\theta} - \frac{\cancel{2} \cancel{v_i^2 \cos^2\theta}}{\cancel{2}} g\left(\frac{\Delta x}{\cancel{v_i \cos\theta}}\right)^2$$

Subtracting the first term on the right side from both sides and factoring out v_i^2,

$$v_i^2(2\,\Delta y \cos^2\theta - 2\,\Delta x \cos\theta \sin\theta) = -g(\Delta x)^2$$

Now we solve for v_i:

$$v_i = \sqrt{\frac{g(\Delta x)^2}{2\,\Delta x \cos\theta \sin\theta - 2\,\Delta y \cos^2\theta}}$$

$$= \sqrt{\frac{9.80\text{ m/s}^2 \times (74.0\text{ m})^2}{2(74.0\text{ m}) \cos 40° \sin 40° - 2(-1.0\text{ m}) \cos^2 40°}}$$

$$= 26.9\text{ m/s}$$

The net force on the hammer can be found from Newton's second law. The two forces acting on the hammer are due to the tension in the cable and to gravity (Fig. 5.11). We ignore the gravitational force, assuming that the hammer's weight is small compared with the tension in the cable. Then the tension in the cable is the only significant force acting on the hammer. Assuming uniform circular motion, the cable pulls radially inward and causes a radial acceleration of magnitude v^2/r. Newton's second law in the radial direction is

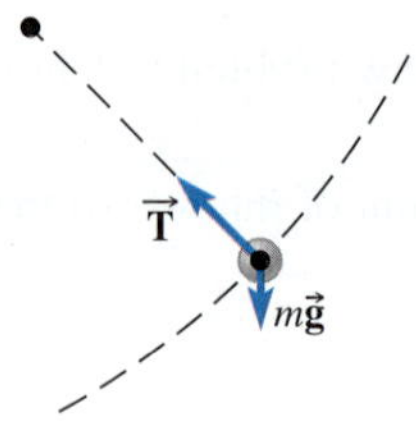

Figure 5.11 FBD for the hammer just before its release. (Not to scale.)

$$\sum F_r = T = ma_r = \frac{mv^2}{r}$$

Substituting numerical values,

$$T = \frac{4.00\text{ kg} \times (26.9\text{ m/s})^2}{1.7\text{ m}} = 1700\text{ N}$$

The tension is much larger than the weight of the hammer ($\approx$ 40 N), so the assumption that we could ignore the weight is justified. The athlete must apply a force of magnitude 1700 N—almost 400 lb—to the grip.

Discussion This example demonstrates the cumulative nature of physics concepts. The basic concepts keep reappearing, to be used over and over and to be extended for use in new contexts. Part of the problem involves new concepts (radial acceleration); the rest of the problem involves old material (Newton's second law, projectile motion, and tension in a cord).

Practice Problem 5.5 Rotating Carousel

A horse located 8.0 m from the central axis of a rotating carousel moves at a speed of 6.0 m/s. The horse is at a fixed height (it does not move up and down). What is the net force acting on a child seated on this horse? The child's weight is 130 N.

Example 5.6

Conical Pendulum

Suppose you whirl a stone in a horizontal circle at a slow speed so that the weight of the stone is *not* negligible compared with the tension in the cord. Then the cord cannot be horizontal—the tension must have a vertical component to cancel the weight and leave a horizontal net force (Fig. 5.12). If the cord has length L, the stone has mass m, and the cord makes an angle ϕ with the vertical direction, what is the constant angular speed of the stone?

Strategy The net force must point toward the center of the circle, since the stone is in uniform circular motion. With the stone in the position depicted in Fig. 5.12a, the direction of the net force is along the $+x$-axis. This time the tension in the cord does not pull toward the center, but the *net* force does.

Solution Start by drawing an FBD (Fig. 5.12b). Now apply Newton's second law in component form. The acceleration has components $a_x = \omega^2 r$ and $a_y = 0$. For the x-components,

$$\sum F_x = T \sin\phi = ma_x = m\omega^2 r$$

continued on next page

Example 5.6 continued

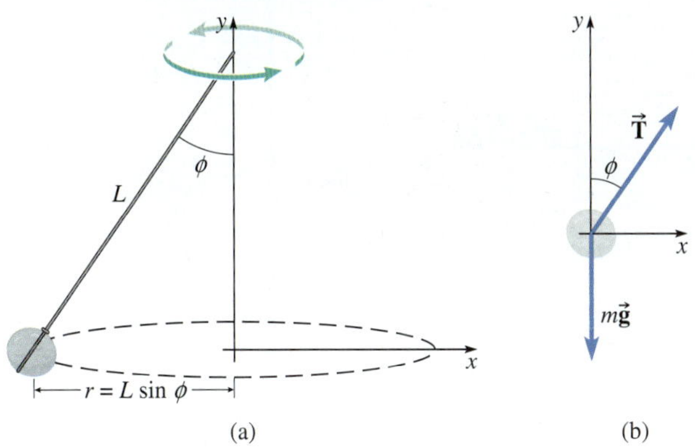

Figure 5.12
(a) A stone is whirled in a horizontal circle of radius $r = L \sin \phi$.
(b) An FBD for the stone.

Since the problem does not specify r, we must express r in terms of L and ϕ. In Fig. 5.12a, the radius forms a right triangle with the cord and the y-axis. Then

$$r = L \sin \phi$$

and

$$\sum F_x = T \sin \phi = m\omega^2 L \sin \phi$$

Therefore, $T = m\omega^2 L$. For the y-components,

$$\sum F_y = T \cos \phi - mg = ma_y = 0 \quad \Rightarrow \quad T \cos \phi = mg$$

Now we eliminate the tension:

$$(m\omega^2 L) \cos \phi = mg$$

Solving for $|\omega|$,

$$|\omega| = \sqrt{\frac{g}{L \cos \phi}}$$

Discussion We should check the dimensions of the final expression. Since $\cos \phi$ is dimensionless,

$$\sqrt{\frac{[\text{L/T}^2]}{[\text{L}]}} = \frac{1}{[\text{T}]}$$

which is correct for ω (SI unit rad/s).

Another check is to ask how ω and ϕ are related for a given length cord. As ϕ increases toward 90°, the cord gets closer to horizontal and the radius increases. In our expression, as ϕ increases, $\cos \phi$ decreases and, therefore, ω increases, in accordance with experience: the stone would have to be whirled faster and faster to make the cord more nearly horizontal.

Conceptual Practice Problem 5.6 Conical Pendulum on the Moon

Examine the result of Example 5.6 to see how ω depends on g, all other things being equal. Where the gravitational field is weaker, do you have to whirl the stone faster or more slowly to keep the cord at the same angle ϕ? Is that in accord with your intuition?

5.3 UNBANKED AND BANKED CURVES

Unbanked Curves When you drive an automobile in a circular path along an unbanked roadway, friction acting on the tires due to the pavement acts to keep the automobile moving in a curved path. This frictional force acts *sideways*, toward the center of the car's circular path (Fig. 5.13). The frictional force might also have a tangential component; for example, if the car is braking, a component of the frictional force makes the car slow down by acting backward (opposite to the car's velocity). For now we assume that the car's speed is constant and that the forward or backward component of the frictional force is negligibly small.

As long as the tires roll without slipping, there is no relative motion between the bottom of the tires and the road, so it is the force of *static* friction that acts (see Section 4.6). If the car is in a skid, then it is the smaller force of kinetic friction that acts as the bottom portion of the tire slides along the pavement. As the speed of the car increases; or for slippery surfaces with low coefficients of friction, the static frictional force may not be enough to hold the car in its curved path.

Application of Radial Acceleration and Contact Forces: Banked Curves To help prevent cars from going into a skid or losing control, the roadway is often banked (tilted at a slight angle) around curves so that the outer portion of the road—the part farthest from the center of curvature—is higher than the inner portion. Banking changes the angle and magnitude of the normal force, $\vec{\mathbf{N}}$, so that it has a horizontal component N_x

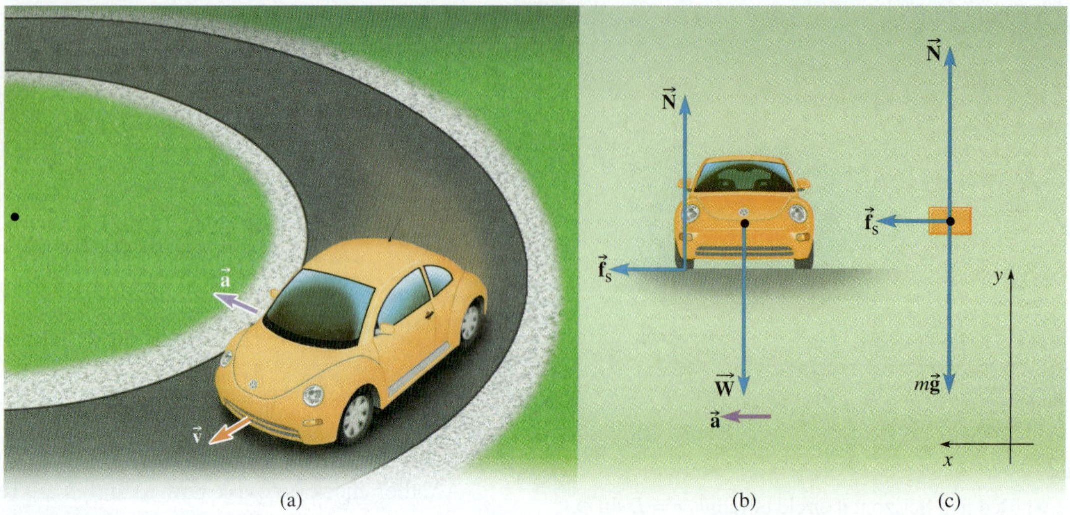

Figure 5.13 (a) A car negotiating a curve at constant speed on an *unbanked* roadway. The car's acceleration is toward the center of the circular path. (b) A head-on view of the same car. The center of the circular path is to the left as viewed here. The force vectors $\vec{\mathbf{N}}$ and $\vec{\mathbf{f}}_s$ are shown acting on one tire, but they represent the *total* normal and frictional forces acting on all four tires. (c) FBD for the car.

directed toward the center of curvature (in the radial direction—see Fig. 5.14). Then we need no longer rely solely on friction to keep the car moving in a circular path as it negotiates the curve; this component of the normal force acts to help the car remain on the curved path. Figure 5.14 shows a banked road with the normal force, the gravitational force, and, in parts (b) and (c), the radial component of the normal force N_x. We choose the axes so that the x-axis is in the direction of the acceleration, which is to the left; the axes are *not* parallel and perpendicular to the incline.

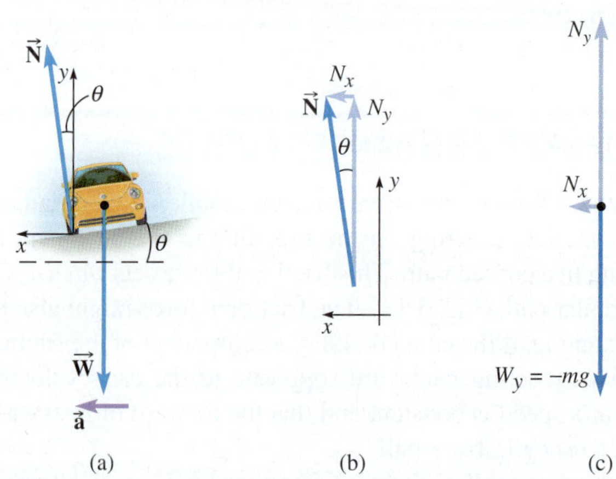

Figure 5.14 (a) Head-on view of a car negotiating a curve at constant speed on a *banked* roadway. The car's acceleration is toward the center of the circular path (to the left as viewed here). $\vec{\mathbf{N}}$, represents the *total* normal force acting on all four tires. The car moves at just the right speed so that the frictional force is zero. (b) Resolving the normal force into x- and y-components. (c) FBD for the car with the normal force represented by its components.

Example 5.7

Safe Speeds on Unbanked and Banked Curves

A car is going around an unbanked curve at the recommended speed of 11 m/s (see Fig. 5.13). (a) If the radius of curvature of the path is 25 m and the coefficient of static friction between the rubber and the road is $\mu_s = 0.70$, does the car skid as it goes around the curve? (b) What happens if the driver ignores the highway speed limit sign and travels at 18 m/s? (c) What maximum speed is safe for traveling around the curve if the road surface is wet from a recent rainstorm

continued on next page

Example 5.7 continued

and the coefficient of static friction between the wet road and the rubber tires is $\mu_s = 0.50$? (d) For a car to safely negotiate the curve in icy conditions at a speed of 13 m/s, what banking angle would be required (see Fig. 5.14)?

Strategy The force of static friction is the only horizontal force acting on the car when the curve is not banked. The maximum force of static friction, which depends on road conditions, determines the maximum possible radial acceleration of the car. Therefore, we can compare the radial acceleration necessary to go around the curve at the specified speeds with the maximum possible radial acceleration determined by the coefficient of static friction. For part (d), in icy conditions we cannot rely much on friction, but the normal force has a horizontal component when the road is banked.

Solution (a) We find the radial acceleration required for a speed of 11 m/s:

$$a_r = \frac{v^2}{r} = \frac{(11\ \text{m/s})^2}{25\ \text{m}} = 4.8\ \text{m/s}^2$$

In order to have that acceleration, the component of the net force acting toward the center of curvature must be

$$\sum F_r = ma_r = m\frac{v^2}{r}$$

The only force with a horizontal component is the static frictional force acting on the tires due to the road (see the FBD in Fig. 5.13c). Therefore,

$$\sum F_r = f_s = m\frac{v^2}{r}$$

We must check to make sure that the maximum frictional force is not exceeded:

$$f_s \le \mu_s N$$

Since $N = mg$, the car can go around the curve without skidding as long as

$$\cancel{m}\frac{v^2}{r} \le \mu_s \cancel{m}g$$

Thus, the radial acceleration cannot exceed $\mu_s g$. That limits the car to speeds satisfying

$$v \le \sqrt{\mu_s g r}$$

Substituting numerical values,

$$v \le \sqrt{0.70 \times 9.80\ \text{m/s}^2 \times 25\ \text{m}} = 13\ \text{m/s}$$

Since 11 m/s is less than the maximum safe speed of 13 m/s, the car safely negotiates the curve without skidding.

(b) At 18 m/s, the car moves at a speed higher than the maximum safe speed of 13 m/s. The frictional force cannot supply the radial acceleration needed for the car to go around the curve—the car goes into a skid.

(c) In part (a), we found that the car is limited to speeds satisfying

$$v \le \sqrt{\mu_s g r}$$

With $\mu_s = 0.50$, the maximum safe speed is

$$v_{max} = \sqrt{\mu_s g r} = \sqrt{0.50 \times 9.80\ \text{m/s}^2 \times 25\ \text{m}} = 11\ \text{m/s}$$

which is the same maximum speed recommended by the road sign. The highway engineer knew what she was doing when she had the sign placed along the road.

(d) Finally, we find the banking angle that would enable cars to travel around the curve at 13 m/s in icy conditions. Assuming that friction is negligible, the horizontal component of the normal force is the only horizontal force. With the x-axis pointing toward the center of curvature and the y-axis vertical (Fig. 5.14),

$$\sum F_x = N \sin\theta = mv^2/r \qquad (1)$$

and

$$\sum F_y = N\cos\theta - mg = 0 \qquad (2)$$

Dividing Eq. (1) by Eq. (2) gives

$$\frac{N\sin\theta}{N\cos\theta} = \tan\theta = \frac{mv^2/r}{mg} = \frac{v^2}{rg}$$

$$\theta = \tan^{-1}\frac{v^2}{rg} = \tan^{-1}\frac{(13\ \text{m/s})^2}{25\ \text{m} \times 9.80\ \text{m/s}^2} = 35° \qquad (3)$$

Discussion Notice that the mass of the car does not appear in Eq. (3); the same banking angle holds for a scooter, motorcycle, car, or tractor-trailer. Notice also that the banking angle depends on the square of the speed. Automobile racetracks and bicycle racetracks have highly banked road surfaces at hairpin curves to minimize skidding of the high-speed vehicles. However, a banking angle of 35° is far greater than those used in practice along public roadways.

Practice Problem 5.7 Actual Banking Angle

The curve in Example 5.7 is actually banked at 4.0°. What is the safest speed to go around the curve in icy conditions?

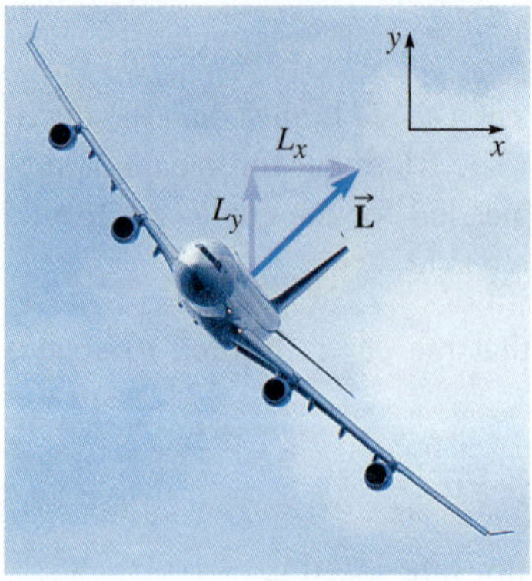

Figure 5.15 The lift force $\vec{L}$ is perpendicular to the wings of the plane. To turn, the pilot tilts the wings so a component of the lift force is directed toward the center of the circular path of the plane.

If there is *no* friction between the road and the tires, then there is only one speed at which it is safe to drive around a given curve. *With* friction, there is a *range* of safe speeds. The static frictional force can have any magnitude from 0 to $\mu_s N$, and it can be directed either up or down the bank of the road.

Application: Banking Angle of an Airplane When an airplane pilot makes a turn in the air, the pilot makes use of a banking angle. The airplane itself is tilted as if it were traveling over an inclined surface. Because of the shape of the wings, an aerodynamic force called *lift* acts upward when the plane is in level flight. To go around a turn, the wings are tilted; the lift force stays perpendicular to the wings and, therefore, now has a horizontal component (Fig. 5.15), just as the normal force has a horizontal component for a car on a banked curve. This component supplies the necessary radial acceleration, while the vertical component of the lift holds the plane up. Therefore,

$$L_x = ma_r = \frac{mv^2}{r} \text{ and } L_y = mg$$

where the x-axis is horizontal and the y-axis is vertical. The lift force is different in its physical origin from the normal force, but its components split up the same way, so a plane in a turn banks its wings at the same angle that a road would be banked for the same speed and radius of curvature. Of course, planes usually move much faster than cars and use large radii of curvature when they turn.

CHECKPOINT 5.3

A plane can't make a turn without tilting its wings. Why can a car turn on a flat road?

5.4 CIRCULAR ORBITS OF SATELLITES AND PLANETS

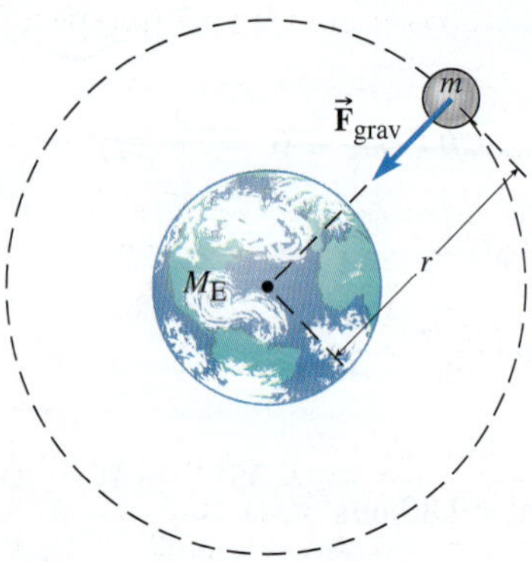

Figure 5.16 Satellite in orbit around Earth.

Application of Radial Acceleration: Circular Orbits A satellite can orbit Earth in a circular path because of the long-range gravitational force on the satellite due to the Earth. The magnitude of the gravitational force on the satellite is

$$F = \frac{Gm_1m_2}{r^2} \quad \text{(4-8)}$$

where the universal gravitational constant is $G = 6.67 \times 10^{-11}$ N·m²/kg². We can use Newton's second law to find the speed of a satellite in circular orbit at constant speed. Let m be the mass of the satellite, and M be the mass of the Earth. The direction of the gravitational force on the satellite is always toward the center of the Earth, which is the center of the orbit (Fig. 5.16). Since gravity is the only force acting on the satellite,

$$\sum F_r = G\frac{mM}{r^2}$$

where r is the distance from the *center* of the Earth to the satellite. Then, from Newton's second law,

$$\sum F_r = ma_r = \frac{mv^2}{r}$$

Setting these equal,

$$G\frac{\cancel{m}M}{r^{\cancel{2}}} = \frac{\cancel{m}v^2}{\cancel{r}}$$

Solving for the speed yields

$$v = \sqrt{\frac{GM}{r}} \quad \text{(5-13)}$$

Notice that the mass of the satellite does not appear in the equation for speed; it has been algebraically canceled. The greater inertia of a more massive satellite is overcome by a proportionally greater gravitational force acting on it. Thus, the speed of a satellite in a circular orbit does not depend on the mass of the satellite. Equation (5-13) also shows that satellites in lower orbits (smaller radii) have greater speeds.

We have been discussing satellites orbiting Earth, but the same principles apply to the circular orbits of satellites around other planets and to the orbits of the planets around the Sun. For planetary orbits, M in Eq. (5-13) would be the Sun's mass instead of the Earth's mass, because the *Sun's* gravitational pull keeps the planets in their orbits. The planetary orbits are actually ellipses (Fig. 5.17) instead of circles, although for most of the planets in the solar system the ellipses are nearly circular. Mercury is the exception; its orbit is markedly different from a circle.

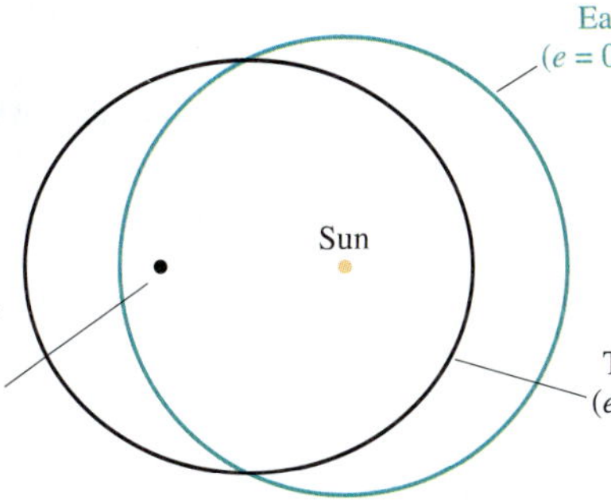

Figure 5.17 The shapes of two elliptical orbits around the Sun. (The ***sizes*** of the orbits are not to scale.) An ellipse looks like an elongated circle. The degree of elongation is measured by a quantity called the eccentricity e. A circle is a special case of an ellipse with $e = 0$. Most of the planetary orbits are nearly circular, with the exception of Mercury. The sum of the distances from any point on an ellipse to each of two fixed points (called the *foci*) is constant. The Sun is at one focus of each orbit. Since Earth's orbit is nearly circular, the second focus is very near the Sun.

Example 5.8

Speed of a Satellite

The Hubble Space Telescope is in a circular orbit 613 km above Earth's surface. The average radius of the Earth is 6.37×10^3 km and the mass of Earth is 5.97×10^{24} kg. What is the speed of the telescope in its orbit?

Strategy We first need to find the orbital radius of the telescope. It is not 613 km; that is the distance from the *surface* of Earth to the telescope. We must add the radius of the Earth to 613 km to find the orbital radius, which is measured from the center of the Earth to the telescope. Then we use Newton's second law, along with what we know about radial acceleration.

Solution The radius of the telescope's orbit is

$$r = 6.13 \times 10^2 \text{ km} + 6.37 \times 10^3 \text{ km} = (0.613 + 6.37) \times 10^3 \text{ km}$$
$$= 6.98 \times 10^3 \text{ km}$$

The net force on the telescope is equal to the gravitational force, given by Newton's law of gravity. Newton's second law relates the net force to the acceleration. Both are directed radially inward.

$$\sum F_r = \frac{GmM}{r^2} = \frac{mv^2}{r}$$

where m is the mass of the telescope. Solving for the speed, we find

$$v = \sqrt{\frac{GM}{r}}$$

$$v = \sqrt{\frac{6.67 \times 10^{-11} \text{ N·m}^2/\text{kg}^2 \times 5.97 \times 10^{24} \text{ kg}}{6.98 \times 10^6 \text{ m}}}$$

$$v = 7550 \text{ m/s} = 27\,200 \text{ km/h}$$

Discussion *Any* satellite orbiting Earth at an altitude of 613 km has this same speed, regardless of its mass.

Practice Problem 5.8 Speed of Earth in Its Orbit

What is the speed of Earth in its approximately circular orbit about the Sun? The average Earth-Sun distance is 1.50×10^{11} m and the mass of the Sun is 1.987×10^{30} kg. Once you find the speed, use it along with the distance traveled by the Earth during one revolution about the Sun to calculate the time in seconds for one orbit.

Kepler's Laws of Planetary Motion

At the beginning of the seventeenth century, Johannes Kepler (1571–1630) proposed three laws to describe the motion of the planets. These laws predated Newton's laws of motion and his law of gravity. They offered a far simpler description of planetary motion than anything that had been proposed previously. We turn history on its head and look at one of Kepler's laws as a consequence of Newton's laws. The fact that Newton could derive Kepler's laws from his own work on gravity was seen as a confirmation of Newtonian mechanics.

Kepler's laws of planetary motion are

- The planets travel in elliptical orbits (see Fig. 5.17) with the Sun at one focus of the ellipse.
- A line drawn from a planet to the Sun sweeps out equal areas in equal time intervals.
- The square of the orbital period is proportional to the cube of the average distance from the planet to the Sun.

Kepler's first law can be derived from the inverse square law of gravitational attraction. The derivation is a bit complicated, but for any two objects that have such an attraction, the orbit of one about the other is an ellipse, with the stationary object located at one focus. (Planetary orbits are also affected by gravitational interactions with other planets; Kepler's laws ignore these small effects.) The circle is a special case of an ellipse where the two foci coincide. We discuss Kepler's second law in Chapter 8.

Application of Radial Acceleration: Kepler's Third Law for a Circular Orbit We can derive Kepler's third law from Newton's law of universal gravitation for the special case of a circular orbit. The gravitational force gives rise to the radial acceleration:

$$\sum F_r = \frac{GmM}{r^2} = \frac{mv^2}{r}$$

Here, M is the mass of the Sun, m is the mass of the planet, r is the orbital radius, and v is the orbital speed. Solving for v yields

$$v = \sqrt{\frac{GM}{r}}$$

The distance traveled during one revolution is the circumference of the circle, which is equal to $2\pi r$. The speed is the distance traveled during one orbit divided by the period:

$$v = \sqrt{\frac{GM}{r}} = \frac{2\pi r}{T}$$

Now we solve for T:

$$T = 2\pi\sqrt{\frac{r^3}{GM}}$$

Squaring both sides yields

$$T^2 = \frac{4\pi^2}{GM}r^3 = \text{constant} \times r^3 \qquad (5\text{-}14)$$

Equation (5-14) is Kepler's third law: the square of the period of a planet is directly proportional to the cube of the average orbital radius.

Application of Radial Acceleration: Geostationary Orbits Although Kepler's laws were derived for the motion of planets around the Sun, they apply to satellites orbiting the Earth as well. In Eq. (5-14), M would then stand for the mass of Earth. Many satellites, such as those used for communications, are placed in a *geostationary orbit* (or *geosynchronous equatorial orbit*)—a circular orbit in Earth's equatorial plane whose period is equal to Earth's rotational period (Fig. 5.18). A satellite in geostationary orbit remains directly above a particular point on the equator; to observers on the ground, it seems to hover above that point without moving. Due to their fixed positions with respect to Earth's surface, geostationary satellites are used as relay stations for communication signals. In Example 5.9, we find the speed of a geostationary satellite.

CHECKPOINT 5.4

Do all geostationary satellites, no matter their masses, have to be at the same height above Earth? Explain.

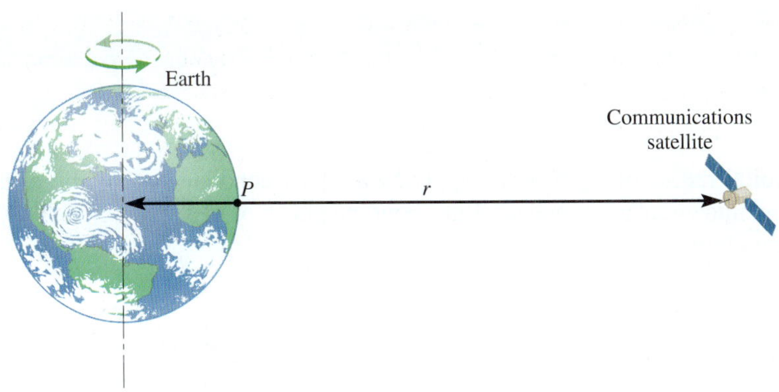

Figure 5.18 Geostationary satellite orbiting the Earth. The satellite has the same angular velocity as Earth, so it is always directly above point P.

Example 5.9

Geostationary Satellite

A 300.0 kg communications satellite is placed in a geostationary orbit 35 800 km above a relay station located in Kenya. What is the speed of the satellite in orbit?

Strategy The period of the satellite is 1 d or approximately 24 h. To find the speed of the satellite in orbit we use Newton's law of gravity and his second law of motion along with what we know about radial acceleration.

Solution Let m be the mass of the satellite and let M be the mass of the Earth. Gravity is the only force acting on the satellite in its orbit. From Newton's law of universal gravitation, Newton's second law, and the expression for radial acceleration,

$$\sum F_r = \frac{GmM}{r^2} = \frac{mv^2}{r}$$

Solving for the speed yields

$$v = \sqrt{\frac{GM}{r}}$$

We must add the mean radius of the Earth, $R_E = 6.37 \times 10^6$ m, to the height of the satellite above the Earth's surface to find the orbital radius.

$$r = h + R_E = 3.58 \times 10^7 \text{ m} + 0.637 \times 10^7 \text{ m}$$
$$= 4.217 \times 10^7 \text{ m}$$

Substituting numerical values into the speed equation,

$$v = \sqrt{\frac{6.67 \times 10^{-11} \text{ N·m}^2/\text{kg}^2 \times 5.97 \times 10^{24} \text{ kg}}{4.217 \times 10^7 \text{ m}}}$$

$$= \sqrt{9.443 \times 10^6 \text{ m}^2/\text{s}^2}$$

$$v = 3.07 \times 10^3 \text{ m/s}$$

Discussion This result, an orbital speed of 3.07 km/s and a distance above Earth's surface of 35 800 km, applies to *all* geostationary satellites. The mass of the satellite does not matter; it cancels out of the equations for orbital radius and for speed.

If we were actually putting a satellite into orbit, we would use a more accurate value for the period. We should use a time of 23 h and 56 min, which is the length of a *sidereal day*—the time for Earth to complete one rotation about its axis relative to the fixed stars. The solar day, 24 h, is the period of time between the daily appearances of the Sun at its highest point in the sky. The fact that Earth moves around the Sun is what causes the difference between these two ways of measuring the length of a day. The error introduced by using the longer time is negligible in this problem.

We can use Kepler's third law to check the result. Examples 5.8 and 5.9 both concern circular orbits around the Earth. Is the square of the period proportional to the cube of the orbital radius? From Example 5.8, $r_1 = 6.98 \times 10^3$ km and

$$T_1 = \frac{2\pi r_1}{v} = \frac{2\pi \times 6.98 \times 10^3 \text{ km}}{7.55 \text{ km/s}} = 5810 \text{ s}$$

From the present example, $r_2 = 4.22 \times 10^7$ m and

$$T_2 = 24 \text{ h} \times \frac{3600 \text{ s}}{1 \text{ h}} = 86\,400 \text{ s}$$

The ratio of the squares of the periods is

$$\left(\frac{T_2}{T_1}\right)^2 = \left(\frac{86\,400 \text{ s}}{5810 \text{ s}}\right)^2 = 221$$

The ratio of the cubes of the radii is

$$\left(\frac{r_2}{r_1}\right)^3 = \left(\frac{4.22 \times 10^7 \text{ m}}{6.98 \times 10^6 \text{ m}}\right)^3 = 221$$

Practice Problem 5.9 Orbital Radius of Venus

The period of the orbit of Venus around the Sun is 0.615 Earth years. Using this information, find the radius of its orbit in terms of R, the radius of Earth's orbit around the Sun.

Example 5.10

Orbiting Satellites

A satellite revolves about Earth with an orbital radius of r_1 and speed v_1. If an identical satellite were set into circular orbit with the same speed about a planet of mass three times that of Earth, what would its orbital radius be?

Strategy We can apply Newton's law of universal gravitation and set up a ratio to solve for the new orbital radius.

Solution From Newton's second law, the magnitude of the gravitational force on the satellite is equal to the satellite's mass times the magnitude of its radial acceleration:

$$\frac{GmM}{r_1^2} = \frac{mv_1^2}{r_1}$$

where M and m are the masses of Earth and of the satellite, respectively. Solving for r_1 yields

$$r_1 = \frac{GM}{v_1^2}$$

Now we apply Newton's second law to the orbit of the second satellite about the planet of mass $3M_E$:

$$\frac{Gm \times 3M}{r_2^2} = \frac{mv_1^2}{r_2}$$

$$r_2 = \frac{G \times 3M}{v_1^2}$$

The ratio of r_2 to r_1 is

$$\frac{r_2}{r_1} = \frac{G \times 3M/v_1^2}{GM_E/v_1^2} = 3$$

Thus, $r_2 = 3r_1$.

Discussion Notice that we did not rush to substitute numerical values for the constants G and M into the equations. We took the ratio r_2/r_1 so that these constants cancel.

Practice Problem 5.10 Period of Lunar Lander

A lunar lander is orbiting about the Moon. If the radius of its orbit is $\frac{1}{3}$ the radius of Earth, what is the period of its orbit?

5.5 NONUNIFORM CIRCULAR MOTION

So far we have focused on *uniform* circular motion. Now we can extend the discussion to nonuniform circular motion, where the angular velocity changes with time.

Figure 5.19a shows the velocity vectors $\vec{v}_1$ and $\vec{v}_2$ at two different times for an object moving in a circle with changing speed. In this case, the speed is increasing ($v_2 > v_1$). In Fig. 5.19b, we subtract $\vec{v}_1$ from $\vec{v}_2$ to find the change in velocity. In the limit $\Delta t \to 0$, $\Delta\vec{v}$ does *not* become perpendicular to the velocity, as it did for uniform circular motion. Thus, the direction of the acceleration is *not* radial if the speed is changing. However, we can resolve the acceleration into tangential and radial components (Fig. 5.19c). The radial component a_r changes the *direction* of the velocity,

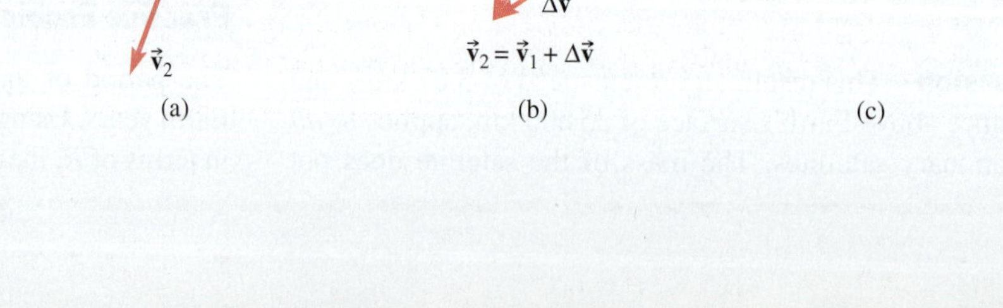

Figure 5.19 Motion along a circular path with a changing speed: (a) the magnitude of velocity $\vec{v}_2$ is greater than the magnitude of velocity $\vec{v}_1$, (b) the direction of $\Delta\vec{v}$ is not radial when the speed is changing, and (c) components of $\vec{a}$ can be taken along a tangent to the curved path (a_t) and along a radius (a_r).

and the tangential component a_t changes the *magnitude* of the velocity (the speed). Since these are perpendicular components of the acceleration, the magnitude of the acceleration is

$$a = \sqrt{a_r^2 + a_t^2}$$

Using the same method as in Section 5.2 to find the radial acceleration, but working here with only the radial *component* of the acceleration, we find that

$$a_r = \frac{v^2}{r} = \omega^2 r \quad (\omega \text{ in radians per unit time}) \qquad (5\text{-}12)$$

For circular motion, *whether uniform or nonuniform*, the radial component of the acceleration is given by Eq. (5-12). In *uniform* circular motion the radial component of the acceleration a_r is constant in magnitude, but for nonuniform circular motion a_r changes as the speed changes.

Also still true for nonuniform circular motion is the relationship between speed and angular speed:

$$v = r|\omega| \qquad (5\text{-}7)$$

CONNECTION:

Resolving a vector into perpendicular components is nothing new. Until now we've always found components along fixed x- and y-axes. Here we resolve the acceleration into radial and tangential components, which is useful because:

- the radial acceleration is always given by Eq. (5-12); and
- the tangential acceleration is zero if the speed is constant.

Many problems involving nonuniform circular motion are solved in the same way as for uniform circular motion. We find the *radial component of the net force* and then apply Newton's second law along the radial direction:

$$\sum F_r = ma_r$$

Problem-Solving Strategy for an Object in Nonuniform Circular Motion

1. Begin as for any Newton's second law problem: identify all the forces acting on the object and draw an FBD.
2. Choose perpendicular axes at the point of interest so that one axis is radial and the other is tangent to the circular path.
3. Find the radial component of each force.
4. Apply Newton's second law along the radial direction:

$$\sum F_r = ma_r$$

where

$$a_r = \frac{v^2}{r} = \omega^2 r$$

5. If necessary, apply Newton's second law to the tangential force components:

$$\sum F_t = ma_t$$

The tangential acceleration component a_t determines how the speed of the object changes.

CHECKPOINT 5.5

For an object in circular motion, what is it about the radial acceleration that distinguishes between uniform and nonuniform circular motion?

Example 5.11

Vertical Loop-the-Loop

A roller coaster includes a vertical circular loop of radius 20.0 m (Fig. 5.20a). What is the minimum speed at which the car must move at the top of the loop so that it doesn't lose contact with the track?

Strategy A roller coaster car moving around a vertical loop is in nonuniform circular motion; its speed decreases on the way up and increases on the way back down. Nevertheless, it is moving in a circle and has a radial acceleration component as given in Eq. (5-12) as long as it moves in a circle. The only forces acting on the car are gravity and the normal force of the track pushing the car. Even if frictional or drag forces are present, at the top of the loop they act in the tangential direction and, thus, do not contribute to the radial component of the net force. At the top of the loop, the track exerts a normal force on the car as long as the car moves with a speed great enough to stay on the track. If the car moves too slowly, it loses contact with the track and the normal force is then zero.

Solution The normal force exerted by the track on the car at the top pushes the car *away* from the track (downward); the normal force cannot pull up on the car. Then, at the top of the loop, the gravitational force and the normal force both point straight down toward the center of the loop. Figure 5.20b is an FBD for the car. From Newton's second law,

$$\sum F_r = N + mg = ma_r = \frac{mv_{top}^2}{r}$$

or

$$N = \frac{mv_{top}^2}{r} - mg$$

where v_{top} stands for the speed at the top. In this expression, N stands for the magnitude of the normal force. Since $N \geq 0$,

$$m\left(\frac{v_{top}^2}{r} - g\right) \geq 0$$

which simplifies to

$$v_{top} \geq \sqrt{gr}$$

continued on next page

Figure 5.20

(a) A roller coaster car on a vertical circular loop. At the bottom of the loop, the car's acceleration $\vec{a}_{bottom}$ points upward toward the center of the circle. At the top of the loop, the car's acceleration $\vec{a}_{top}$ points downward. The magnitude of $\vec{a}_{top}$ is smaller than that of $\vec{a}_{bottom}$ because the speed is smaller at the top than at the bottom. (b) FBD for the car at the top of the loop. The track is above the car, so the normal force on the car due to the track is *downward*. (c) FBD for the car at the bottom of the loop.

Example 5.11 continued

Imagine sending a roller coaster car around the loop many times with a slightly smaller speed at the top each time. As v_{top} approaches $\sqrt{gr}$, the normal force at the top gets smaller and smaller. When $v_{top} = \sqrt{gr}$, the normal force just becomes zero at the top of the loop. Any slower and the car loses contact with the track *before* getting to the highest point and would fall off the track unless prevented from falling by a backup safety mechanism. Therefore, the minimum speed at the top is

$$v_{top} = \sqrt{gr} = \sqrt{9.80 \text{ m/s}^2 \times 20.0 \text{ m}} = 14.0 \text{ m/s}$$

Discussion If the car is going faster than 14 m/s at the top, its radial acceleration is larger. The track pushing on the car provides the additional net force component that results in a larger radial acceleration. The minimum speed occurs when gravity alone provides the radial acceleration at the top of the loop. In other words, $a_r = g$ at the top of the loop for minimum speed.

Practice Problem 5.11 Normal Force at the Bottom of the Track

If the speed of the roller coaster at the *bottom* of the loop is 25 m/s, what is the normal force exerted on the car by the track in terms of the car's weight *mg*? (See Fig. 5.20c.)

EVERYDAY PHYSICS DEMO

Go outside on a warm day and fill a bucket with water. Swing the bucket around in a vertical circle over your head. What, if anything, keeps the water in the bucket when the bucket is upside down over your head? Why doesn't the water spill out? Do any upward forces act on the water at that point? [*Hint*: The FBD for the water when it is directly overhead is similar to the FBD for a roller coaster car at the top of a loop.]

Conceptual Example 5.12

Acceleration of a Pendulum Bob

A pendulum is released from rest at point *A* (Fig. 5.21). (a) Sketch a qualitative motion diagram from *B* to *D*. (b) Sketch an FBD and the acceleration vector for the pendulum bob at points *B* and *C*.

Strategy (a) The pendulum bob moves along the arc of a circle, but not at constant speed. The spacing between points on the motion diagram is larger where the bob is moving faster.

(b) Two forces appear on each FBD: gravity and the force due to the cord. The gravitational force is the same at both points (magnitude *mg*, direction down), but the force due to the cord varies in magnitude and in direction. Its direction is always along the cord. The net force on the bob is the sum of these two forces, and its direction is the same as the direction of the acceleration. We can use what we know about the acceleration to guide us in drawing the forces. At any point, the radial component of the acceleration is related to the speed at that point by $a_r = v^2/r$. The tangential acceleration is in the same direction as the velocity if the speed is increasing and in the opposite direction if the speed is decreasing.

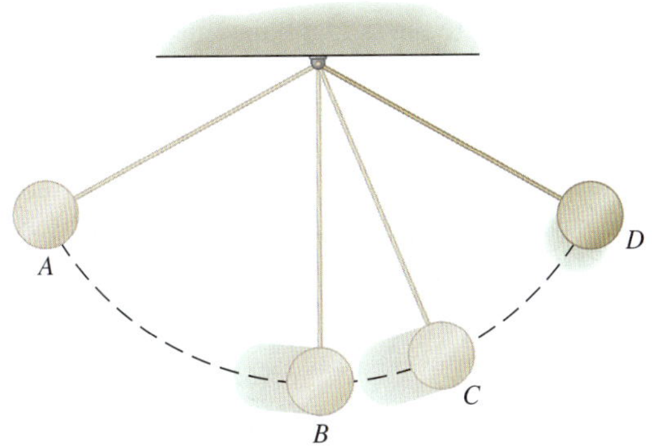

Figure 5.21
A pendulum swings to the right, starting from rest at point *A*.

Solution and Discussion (a) As the pendulum bob swings toward the bottom (from *A* to *B*), its speed is increasing; as it rises on the other side (from *B* to *D*), its speed is decreasing. A motion diagram from *B* to *D* is shown in Fig. 5.22.

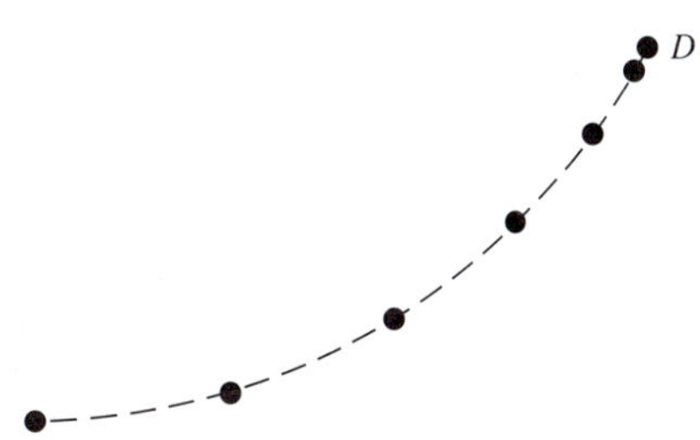

Figure 5.22
Motion diagram for the pendulum bob from point *B* to point *D*. The bob follows the arc of a circle. Its speed decreases as it rises, so the spacing between points decreases.

continued on next page

Conceptual Example 5.12 continued

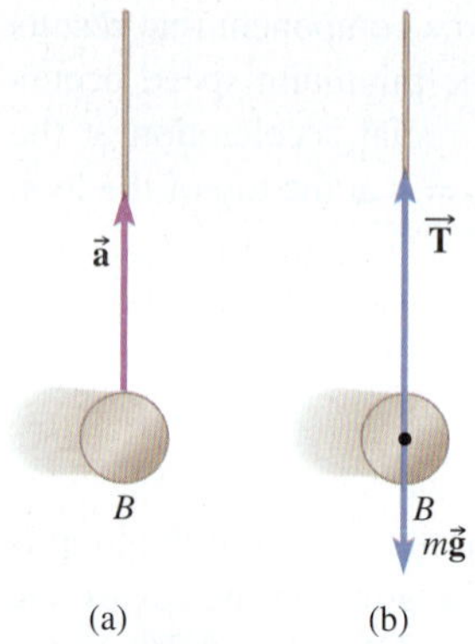

Figure 5.23
(a) Acceleration of the bob at point B. (b) FBD for the bob at B.

The spacing between points decreases because the speed is decreasing.

(b) At point B, the tension in the cord pulls straight up and gravity pulls down, so the tangential component of the net force is zero and the tangential acceleration is zero. Therefore, the acceleration points in the radial direction: straight up. The tension must be larger than the weight of the bob to give an upward net force. Figure 5.23 shows the acceleration and the FBD.

The acceleration at point C has both tangential and radial components. The tangential acceleration is opposite to the velocity because the bob is slowing down. Figure 5.24 shows the tangential and radial acceleration components

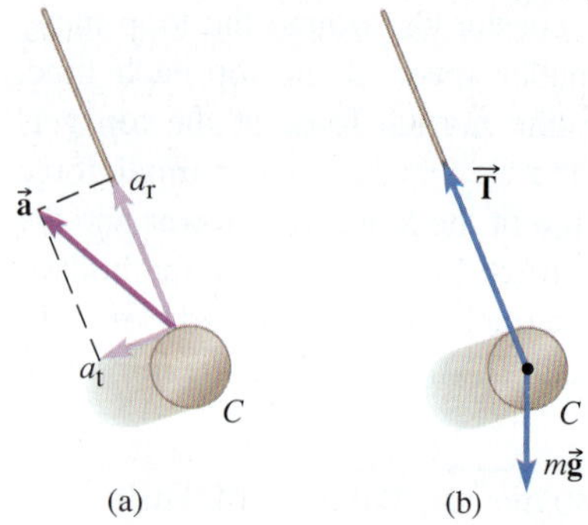

Figure 5.24
(a) At point C, the bob has both tangential and radial acceleration components. (b) FBD for the bob at C.

added to form the acceleration vector $\vec{a}$ and the FBD for the bob. When the two forces are added, they give a net force in the same direction as the acceleration vector.

Conceptual Practice Problem 5.12 Analysis of the Bob at Point D

Sketch the FBD and the acceleration vector for the pendulum bob at point D, the highest point in its swing to the right.

5.6 ANGULAR ACCELERATION

An object in nonuniform circular motion has a changing speed and a changing angular velocity. To describe how the angular velocity changes, we define an angular acceleration. If the angular velocity is ω_1 at time t_1 and is ω_2 at time t_2, the change in angular velocity is

$$\Delta\omega = \omega_2 - \omega_1$$

The time interval during which the angular velocity changes is $\Delta t = t_2 - t_1$. The average rate at which the angular velocity changes is called the **average angular acceleration,** α_{av}.

$$\alpha_{av} = \frac{\omega_2 - \omega_1}{t_2 - t_1} = \frac{\Delta\omega}{\Delta t} \quad (5\text{-}15)$$

As we let the time interval become shorter and shorter, α_{av} approaches the **instantaneous angular acceleration,** α.

$$\alpha = \lim_{\Delta t \to 0} \frac{\Delta\omega}{\Delta t} \quad (5\text{-}16)$$

If ω is in units of rad/s, α is in units of rad/s^2.

The angular acceleration is closely related to the tangential component of the acceleration. The tangential component of velocity is

$$v_t = r\,|\omega| \quad (5\text{-}7)$$

Equation (5-7) gives us a way to relate tangential acceleration to the angular acceleration. The tangential acceleration is the rate of change of the tangential velocity, so

$$a_t = \frac{\Delta v_t}{\Delta t} = r\left|\frac{\Delta\omega}{\Delta t}\right| \quad \text{(in the limit } \Delta t \to 0\text{)}$$

Therefore,

Relationship between tangential acceleration and angular acceleration

$$a_t = r\,|\alpha| \quad (5\text{-}17)$$

Table 5.1 Relationships Between θ, ω, and α for Constant Angular Acceleration

Constant Acceleration Along x-Axis		Constant Angular Acceleration	
$\Delta v_x = v_{fx} - v_{ix} = a_x \Delta t$	(3-19)	$\Delta\omega = \omega_f - \omega_i = \alpha \Delta t$	(5-18)
$\Delta x = \frac{1}{2}(v_{fx} + v_{ix}) \Delta t$	(3-20)	$\Delta\theta = \frac{1}{2}(\omega_f + \omega_i) \Delta t$	(5-19)
$\Delta x = v_{ix} \Delta t + \frac{1}{2}a_x(\Delta t)^2$	(3-21)	$\Delta\theta = \omega_i \Delta t + \frac{1}{2}\alpha (\Delta t)^2$	(5-20)
$v_{fx}^2 - v_{ix}^2 = 2a_x \Delta x$	(3-22)	$\omega_f^2 - \omega_i^2 = 2\alpha \Delta\theta$	(5-21)

CONNECTION:

Because α is the rate of change of ω, and ω is the rate of change of θ, the equations for constant α have the same form as those for constant α_x. Just take the Eqs. (3-19) through (3-22) and replace x with θ, v_x with ω, and a_x with α to get Eqs. (5-18) through (5-21).

Constant Angular Acceleration

The mathematical relationships between θ, ω, and α are the same as the mathematical relationships between x, v_x, and a_x that we developed in Chapters 2 and 3. Each quantity is the instantaneous rate of change of the preceding quantity. For example, a_x is the rate of change of v_x and α is the rate of change of ω. Because the mathematical relationships are the same, we can draw upon the skills and equations we developed to solve problems with constant acceleration a_x. All we have to do is take the equations for constant acceleration and replace x with θ, v_x with ω, and a_x with α (Table 5.1).

Equation (5-18) is the definition of average angular acceleration, with α_{av} replaced by α since the angular acceleration is constant. Constant α means that ω changes linearly with time; therefore, the average angular velocity is halfway between the initial and final angular velocities for any time interval $\omega_{av} = \frac{1}{2}(\omega_i + \omega_f)$. Using this form for ω_{av} along with the definition of ω_{av} ($\omega_{av} = \Delta\theta/\Delta t$) yields Eq. (5-19). Equations (5-20) and (5-21) can be derived from the preceding two relations in a manner analogous to the derivations of Eqs. (3-21) and (3-22) in Section 3.5.

CHECKPOINT 5.6

A centrifuge is "spinning up" with a constant angular acceleration. Can the radial acceleration of a sample in the centrifuge be constant? Explain.

Example 5.13

A Rotating Potter's Wheel

A potter's wheel rotates from rest to 210 rev/min in a time of 0.75 s. (a) What is the angular acceleration of the wheel during this time, assuming constant angular acceleration? (b) How many revolutions does the wheel make during this time interval? (c) Find the tangential and radial components of the acceleration of a point 12 cm from the rotation axis when the wheel is spinning at 180 rev/min.

Strategy We know the initial and final frequencies, so we can find the initial and final angular velocities. We also know the time it takes for the wheel to get to the final angular velocity. That is all we need to find the average angular acceleration that, for constant angular acceleration, is equal to the instantaneous angular acceleration. To find the number of revolutions, we can find the angular displacement $\Delta\theta$ in radians and then divide by 2π rad/rev. We can find the angular velocity at $t = 0.75$ s and use it to find the radial acceleration component. The tangential acceleration is calculated from α.

continued on next page

Example 5.13 continued

Solution (a) Initially the wheel is at rest, so the initial angular velocity is zero.

$$\omega_i = 0 \text{ rad/s}$$

Converting 210 rev/min to rad/s gives the final angular velocity:

$$\omega_f = 210 \frac{\text{rev}}{\text{min}} \times \frac{1}{60} \frac{\text{min}}{\text{s}} \times 2\pi \frac{\text{rad}}{\text{rev}} = 7.0\pi \text{ rad/s}$$

The angular acceleration is the rate of change of the angular velocity. Since α is constant, we can calculate it by finding the *average* angular acceleration for the time interval:

$$\alpha = \frac{\omega_f - \omega_i}{t_f - t_i} = \frac{7.0\pi \text{ rad/s} - 0}{0.75 \text{ s} - 0} = \frac{7.0\pi \text{ rad/s}}{0.75 \text{ s}} = 29 \text{ rad/s}^2$$

(b) The angular displacement is

$$\Delta\theta = \tfrac{1}{2}(\omega_f + \omega_i)\,\Delta t = \tfrac{1}{2}(7.0\pi \text{ rad/s} + 0)(0.75 \text{ s}) = 8.25 \text{ rad}$$

Since 2π rad = one revolution, the number of revolutions is

$$\frac{8.25 \text{ rad}}{2\pi \text{ rad/rev}} = 1.3 \text{ rev}$$

(c) At 180 rev/min, the angular velocity is

$$\omega = 180 \frac{\text{rev}}{\text{min}} \times \frac{1}{60} \frac{\text{min}}{\text{s}} \times 2\pi \frac{\text{rad}}{\text{rev}} = 6.0\pi \text{ rad/s}$$

The radial acceleration component is

$$a_r = \omega^2 r = (6.0\pi \text{ rad/s})^2 \times 0.12 \text{ m} = 43 \text{ m/s}^2$$

and the tangential acceleration component is

$$a_t = \alpha r = 29 \text{ rad/s}^2 \times 0.12 \text{ m} = 3.5 \text{ m/s}^2$$

Discussion A quick check involves another of the equations for constant acceleration:

$$\omega_f^2 - \omega_i^2 = 2\alpha\,\Delta\theta$$

Since $\omega_i = 0$, we can check

$$\omega_f = \sqrt{2\alpha\,\Delta\theta}$$

From the answers to (a) and (b),

$$\sqrt{2\alpha\,\Delta\theta} = \sqrt{2 \times 29 \text{ rad/s}^2 \times 8.25 \text{ rad}} = 22 \text{ rad/s}$$

The original value for ω_f in rad/s was 7.0π rad/s. Since $\pi \approx 22/7$, the check is successful.

Practice Problem 5.13 The London Eye

The London Eye, a Ferris wheel on the banks of the Thames, has radius 67.5 m. At its cruising angular speed, it takes 30.0 min to make one complete revolution. Suppose that it takes 20.0 s to bring the wheel from rest to its cruising speed and that the angular acceleration is constant during startup. (a) What is the angular acceleration during startup? (b) What is the angular displacement of the wheel during startup?

The London Eye

5.7 APPARENT WEIGHT AND ARTIFICIAL GRAVITY

Application: Apparent Weightlessness of Orbiting Astronauts You are no doubt familiar with pictures of astronauts "floating" while in orbit around the Earth. It seems as if the astronauts are weightless. To be truly weightless, the force of gravity acting on the astronauts due to Earth would have to be zero, or at least close to zero. Is it? We can calculate the weight of an astronaut in orbit. The orbital altitude for the International Space Station is about 420 km above the Earth. Then the orbital radius is 420 km + 6370 km = 6790 km. Comparing the astronaut's weight in orbit with his or her weight on Earth's surface,

$$\frac{W_{\text{orbit}}}{W_{\text{surface}}} = \frac{\dfrac{GMm}{(R_E + h)^2}}{\dfrac{GMm}{R_E^2}} = \frac{R_E^2}{(R_E + h)^2} = \frac{(6370 \text{ km})^2}{(6790 \text{ km})^2} = 0.88$$

The weight in orbit is 0.88 times the weight on the surface. The astronaut weighs less but certainly isn't *weightless*! Then why does the astronaut *seem* to be weightless?

Recall Section 4.10 on the apparent weightlessness of someone unfortunate enough to be in an elevator when the cable snaps. In that situation, the elevator and the passenger both have the same acceleration ($\vec{\mathbf{a}} = \vec{\mathbf{g}}$). Similarly, the astronaut has the same acceleration as the space station, which is equal to the *local* gravitational field $\vec{\mathbf{g}}$. Apparent weightlessness occurs when $\vec{\mathbf{a}} = \vec{\mathbf{g}}$, where $\vec{\mathbf{g}}$ is the *local* gravitational field.

CONNECTION:

In Section 4.10 we discussed apparent weight for motion along a line. The principle here is the same. Imagine the astronaut is standing on a scale; the apparent weight is the scale reading.

Application: Artificial Gravity In order for astronauts to spend long periods of time living in a space station without the deleterious effects of apparent weightlessness, *artificial gravity* would have to be created on the station. Many science fiction novels and movies feature ring-shaped space stations that rotate in order to create artificial gravity for the occupants. In a rotating space station, the acceleration of an astronaut is inward (toward the rotation axis), but the apparent gravitational field is outward. Therefore, the ceiling of rooms on the station are closest to the rotation axis and the floor is farthest away (Fig. 5.25).

The centrifuge is a device that creates artificial gravity on a smaller scale. Centrifuges are common not only in scientific and medical laboratories but also in everyday life. The first successful centrifuge was used to separate cream from milk in the 1880s. Water drips out of sopping wet clothes due to the pull of gravity when the clothes are hung on a clothesline, but the water is removed much faster by the artificial gravity created in the spin cycle of a washing machine.

The human body can be adversely affected not only by too little artificial gravity, but also by too much. Stunt pilots have to be careful about the accelerations to which they subject their bodies. An acceleration of about $3g$ can cause temporary blindness due to an inadequate supply of oxygen to the retina; the heart has difficulty pumping blood up to the head due to the blood's increased apparent weight. Larger accelerations can cause unconsciousness. Pressurized flight suits enable pilots to sustain accelerations up to about $5g$.

Figure 5.25 A rotating space station from the movie *2001: A Space Odyssey*. The apparent gravitational field is outward (away from the axis of rotation of the space station).

Example 5.14

Stunt Pilot

Dave wants to practice vertical circles for a flying show exhibition. (a) What must the minimum radius of the circle be to ensure that his acceleration at the bottom does not exceed $3.0g$? The speed of the plane is 78 m/s at the bottom of the circle. (b) What is Dave's apparent weight at the bottom of the circular path? Express your answer in terms of his true weight.

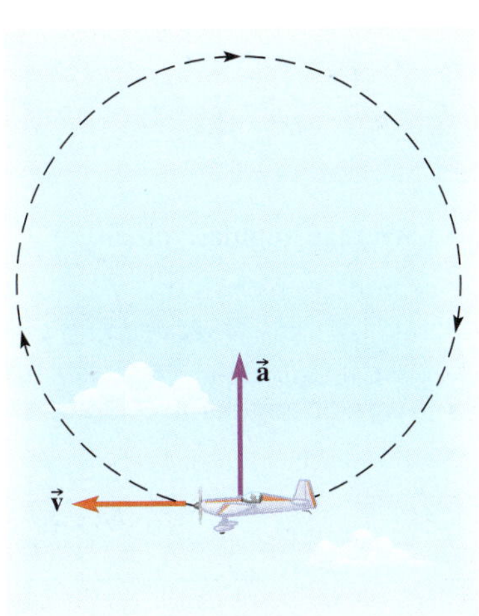

Figure 5.26

Velocity and acceleration vectors for the plane at the bottom of the circle.

Strategy For the *minimum* radius, we use the maximum possible radial acceleration since $a_r = v^2/r$. For the maximum radial acceleration, the *tangential* acceleration must be zero (Fig. 5.26)—the magnitude of the acceleration is $a = \sqrt{a_r^2 + a_t^2}$. Therefore, the radial acceleration component has magnitude $3.0g$ at the bottom. To find Dave's apparent weight, we do not need to use the numerical value of the radius found in part (a); we already know that his acceleration is upward and has magnitude $3.0g$.

Solution (a) The magnitude of the radial acceleration is

$$a_r = v^2/r$$

Solving for the radius,

$$r = \frac{v^2}{a_r} = \frac{v^2}{3.0g} = \frac{(78\ \text{m/s})^2}{3.0 \times 9.80\ \text{m/s}^2} = 210\ \text{m}$$

continued on next page

Example 5.14 continued

(b) Dave's apparent weight is the magnitude of the normal force of the plane pushing up on him. Let the y-axis point upward. The normal force is up, and the gravitational force is down (Fig. 5.27). Then

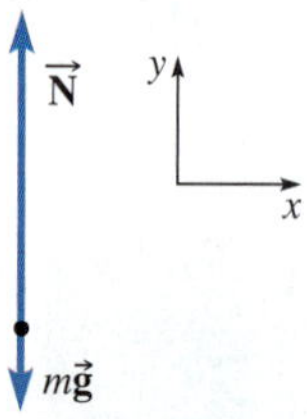

$$\sum F_y = N - mg = ma_y$$

where $a_y = +3.0g$. Therefore,

$$W' = N = m(g + a_y) = 4.0mg$$

His apparent weight is 4.0 times his true weight.

Figure 5.27 FBD for Dave.

Discussion It might have been tempting to jump to the conclusion that an acceleration of $3.0g$ means that his apparent weight is $3.0mg$. But is his apparent weight zero when his acceleration is zero? No.

Practice Problem 5.14 Astronaut's Apparent Weight

What is the apparent weight of a 730 N astronaut when her spaceship has an acceleration of magnitude $2.0g$ in the following two situations: (a) just above the surface of Earth, acceleration straight up; (b) far from any stars or planets?

Application of Apparent Weight to Objects at Rest with Respect to Earth's Surface Due to Earth's rotation, the *effective* value of g measured in a coordinate system attached to Earth's surface is slightly less than the true value of the gravitational field strength (see Section 4.5). The net force of an object placed on a scale is *not* zero because the object has a radial acceleration $a_r = \omega^2 r$ directed toward Earth's axis of rotation (Fig. 5.28). This relatively small effect is greatest where r is greatest—at the equator, where the effective value of g is about 0.3% smaller than the true value of g.

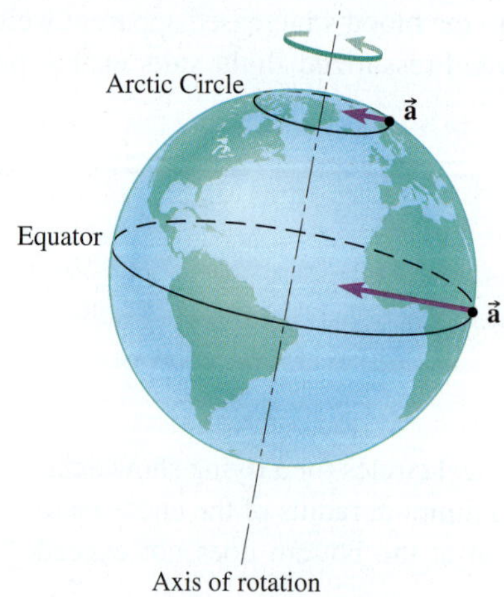

Figure 5.28 An object at rest with respect to Earth's surface has a radial acceleration due to Earth's rotation. The angular frequency ω is the same everywhere, so the radial acceleration $a_r = \omega^2 r$ is proportional to the distance from the axis of rotation.

Master the Concepts

- The angular displacement $\Delta\theta$ is the angle through which an object has turned. Positive and negative angular displacements indicate rotation in different directions. Conventionally, positive represents counterclockwise motion.

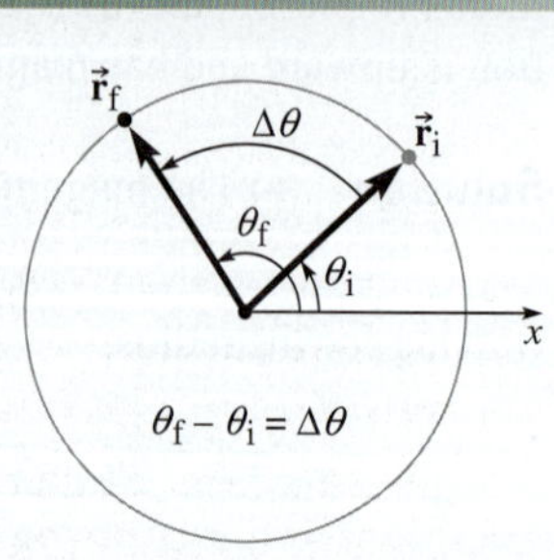

- Average angular velocity:

$$\omega_{av} = \frac{\theta_2 - \theta_1}{t_2 - t_1} = \frac{\Delta\theta}{\Delta t} \tag{5-2}$$

- Average angular acceleration:

$$\alpha_{av} = \frac{\omega_2 - \omega_1}{t_2 - t_1} = \frac{\Delta\omega}{\Delta t} \tag{5-15}$$

continued on next page

Master the Concepts continued

- The instantaneous angular velocity and acceleration are the limits of the average quantities as $\Delta t \to 0$.
- A useful measure of angle is the radian:

$$2\pi \text{ radians} = 360°$$

The arc length s of a circle of radius r subtended by an angle θ is

$$s = \theta r \quad (\theta \text{ in radian measure}) \qquad (5\text{-}4)$$

- The speed of an object in circular motion (including a point on a rotating object) is

$$v = r\,|\omega| \quad (\omega \text{ in radians per unit time}) \qquad (5\text{-}7)$$

- The tangential acceleration component is related to the angular acceleration by

$$a_t = r\,|\alpha| \quad (\alpha \text{ in radians per time}^2) \qquad (5\text{-}17)$$

- An object moving in a circle has a radial acceleration component given by

$$a_r = \frac{v^2}{r} = \omega^2 r \quad (\omega \text{ in radians per unit time}) \qquad (5\text{-}12)$$

- The tangential and radial acceleration components are two perpendicular components of the acceleration vector. The radial acceleration component changes the direction of the velocity, and the tangential acceleration component changes the speed.

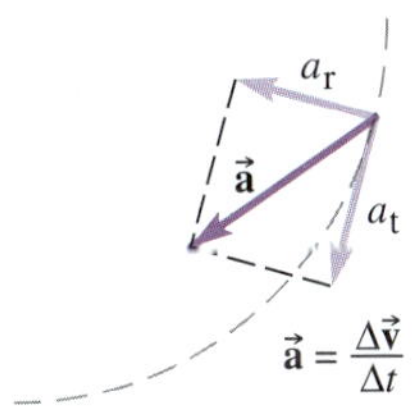

- Uniform circular motion means that v and ω are constant. In uniform circular motion, the time to complete one revolution is constant and is called the period T. The frequency f is the number of revolutions completed per second.

$$f = 1/T \qquad (5\text{-}8)$$

$$|\omega| = v/r = 2\pi f \qquad (5\text{-}9)$$

where the SI unit of angular velocity is rad/s and that of frequency is rev/s = Hz.

- A rolling object is both rotating and translating. If the object rolls without skidding or slipping, then

$$v_{\text{axle}} = r\,|\omega| \qquad (5\text{-}10)$$

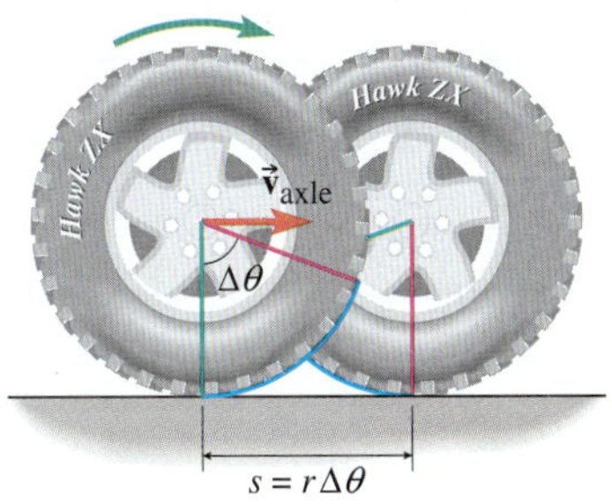

- Kepler's third law says that the square of the period of a planetary orbit is proportional to the cube of the orbital radius:

$$T^2 = \text{constant} \times r^3 \qquad (5\text{-}14)$$

- For constant angular acceleration, we can use equations analogous to those we developed for constant acceleration a_x:

$$\Delta\omega = \omega_f - \omega_i = \alpha\,\Delta t \qquad (5\text{-}18)$$

$$\Delta\theta = \tfrac{1}{2}(\omega_f + \omega_i)\,\Delta t \qquad (5\text{-}19)$$

$$\Delta\theta = \omega_i\,\Delta t + \tfrac{1}{2}\alpha\,(\Delta t)^2 \qquad (5\text{-}20)$$

$$\omega_f^2 - \omega_i^2 = 2\alpha\,\Delta\theta \qquad (5\text{-}21)$$

Conceptual Questions

1. Is depressing the "accelerator" (gas pedal) of a car the only way that the driver can make the car accelerate (in the physics sense of the word)? If not, what else can the driver do to give the car an acceleration?
2. Two children ride on a merry-go-round. One is 2 m from the axis of rotation and the other is 4 m from it. Which child has the larger (a) linear speed, (b) acceleration, (c) angular speed, and (d) angular displacement?
3. Explain why the orbital radius and the speed of a satellite in circular orbit are not independent.
4. In uniform circular motion, is the velocity constant? Is the acceleration constant? Explain.
5. In uniform circular motion, the net force is perpendicular to the velocity and changes the direction of the velocity but not the speed. If a projectile is launched horizontally, the net force (ignoring air resistance) is perpendicular to the initial velocity, and yet the projectile gains speed as it falls. What is the difference between the two situations?
6. The speed of a satellite in circular orbit around a planet does not depend on the mass of the satellite. Does it depend on the mass of the planet? Explain.
7. A flywheel (a massive disk) rotates with constant angular acceleration. For a point on the rim of the flywheel, is the tangential acceleration component constant? Is the radial acceleration component constant?
8. Explain why the force of gravity due to the Earth does not pull the Moon in closer and closer on an inward spiral until it hits Earth's surface.
9. When a roller coaster takes a sharp turn to the right, it feels as if you are pushed toward the left. Does a force push you to the left? If so, what is it? If not, why does there *seem* to be such a force?

10. Is there anywhere on Earth where a bathroom scale reads your true weight? If so, where? Where does your apparent weight due to Earth's rotation differ most from your true weight?

11. A physics teacher draws a cutaway view of a car rounding a banked curve as a rectangle atop a right triangle. A student draws a coordinate system based on the drawing. Is there another choice of axes that would make the problem easier to solve?

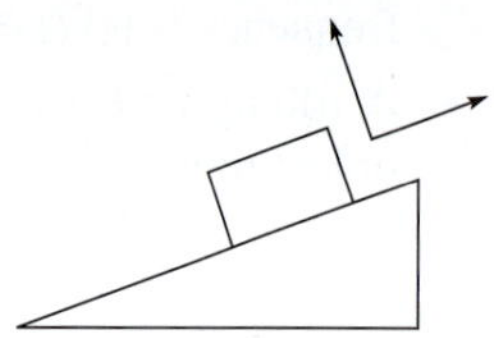

12. A bridal party is at a rehearsal dinner. The best man challenges the bridegroom to pick up an olive using only a brandy snifter. How does the groom accomplish this task?

Multiple-Choice Questions

Student Response System Questions

Questions 1–4: A satellite in orbit travels around the Earth in uniform circular motion. In the figure, the satellite moves counterclockwise (*ABCDA*). Answer choices:

(a) $+x$ (b) $+y$ (c) $-x$ (d) $-y$
(e) 45° above $+x$ (toward $+y$)
(f) 45° below $+x$ (toward $-y$)
(g) 45° above $-x$ (toward $+y$)
(h) 45° below $-x$ (toward $-y$)

1. What is the direction of the satellite's average velocity for one quarter of an orbit, starting at *C* and ending at *D*?
2. What is the direction of the satellite's instantaneous velocity at point *D*?
3. What is the direction of the satellite's average acceleration for one half of an orbit, starting at *C* and ending at *A*?
4. What is the direction of the satellite's instantaneous acceleration at point *C*?

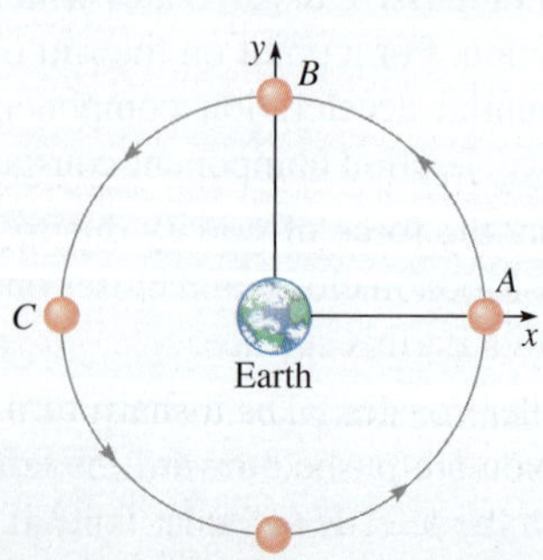

Multiple-Choice Questions 1–4 and Problem 90

5. An object moving in a circle at a constant speed has an acceleration that is
(a) in the direction of motion
(b) toward the center of the circle
(c) away from the center of the circle
(d) zero

6. A spider sits on a DVD that is rotating at a constant angular speed. The acceleration $\vec{\mathbf{a}}$ of the spider is
(a) greater the closer the spider is to the central axis.
(b) greater the farther the spider is from the central axis.
(c) nonzero and independent of the location of the spider on the DVD.
(d) zero.

7. Two satellites are in orbit around Mars with the same orbital radius. Satellite 2 has twice the mass of satellite 1. The radial acceleration of satellite 2 has
(a) twice the magnitude of the radial acceleration of satellite 1.
(b) the same magnitude as the radial acceleration of satellite 1.
(c) half the magnitude of the radial acceleration of satellite 1.
(d) four times the magnitude of the radial acceleration of satellite 1.

Questions 8–9: A boy swings in a tire swing. Answer choices:
(a) At the highest point of the motion
(b) At the lowest point of the motion
(c) At a point neither highest nor lowest
(d) It is constant.

8. When is the tangential acceleration the greatest?
9. When is the tension in the rope the greatest?

Questions 10–11 concern these three statements:
(1) Its acceleration is constant.
(2) Its radial acceleration component is constant in magnitude.
(3) Its tangential acceleration component is constant in magnitude.

10. An object is in uniform circular motion. Identify the correct statement(s).
(a) 1 only (b) 2 only (c) 3 only
(d) 1, 2, and 3 (e) 2 and 3 (f) 1 and 2
(g) 1 and 3 (h) None of them

11. An object is in nonuniform circular motion with constant angular acceleration. Identify the correct statement(s). (Use the same answer choices as in Question 10.)

12. An astronaut is out in space far from any large bodies. He uses his jets to start spinning, then releases a baseball he has been holding in his hand. Ignoring the gravitational force between the astronaut and the baseball, how would you describe the path of the baseball after it leaves the astronaut's hand?
(a) It continues to circle the astronaut in a circle with the same radius it had before leaving the astronaut's hand.
(b) It moves off in a straight line.
(c) It moves off in an ever-widening arc.

Problems

Combination conceptual/quantitative problem
Biomedical application
Challenging
Blue # Detailed solution in the Student Solutions Manual
[1, 2] Problems paired by concept
connect Text website interactive or tutorial

5.1 Description of Uniform Circular Motion

1. A carnival swing is fixed on the end of an 8.0 m long beam. If the swing and beam sweep through an angle of 120°, what is the distance through which the riders move?
2. Convert these to radian measure: (a) 30.0°, (b) 33.3 revolutions.
3. Find the average angular speed of the second hand of an analog clock. What is its angular displacement during 5.0 s?
4. An elevator cable winds on a drum of radius 90.0 cm that is connected to a motor. (a) If the elevator moves down at 0.50 m/s, what is the angular speed of the drum? (b) If the elevator moves down 6.0 m, how many revolutions has the drum made? (c) What is the drum's frequency of rotation?
5. A wheel of radius 30 cm is rotating at a rate of 2.0 revolutions every 0.080 s. (a) Through what angle, in radians, does the wheel rotate in 1.0 s? (b) What is the linear speed of a point on the wheel's rim? (c) What is the wheel's frequency of rotation?
6. A soccer ball of diameter 31 cm rolls without slipping at a linear speed of 2.8 m/s. (a) Through how many revolutions has the soccer ball turned as it moves a linear distance of 18 m? (b) What is the ball's angular speed?
7. A bicycle is moving at 9.0 m/s. What is the angular speed of its tires if their radius is 35 cm? (connect tutorial: car tire)
8. Dung beetles are renowned for building large (relative to their body size) balls of dung and rolling them on the ground. (a) If a dung beetle can roll (without slipping) a ball of dung whose radius is 2.5 cm at a linear speed of 3.5 cm/s, through what angle does the ball roll as the ball moves a distance of 15 cm? (b) What is the angular speed (assumed constant) of the ball's rotation?
9. ✦ In the construction of railroads, it is important that curves be gentle, so as not to damage passengers or freight. Curvature is not measured by the radius of curvature, but in the following way. First a 100.0 ft long chord is measured. Then the curvature is reported as the angle subtended by two radii at the endpoints of the chord. (The angle is measured by determining the angle between two tangents 100 ft apart; since each tangent is perpendicular to a radius, the angles are the same.) In modern railroad construction, track curvature is kept below 1.5°. What is the radius of curvature of a "1.5° curve"? [*Hint*: Since the angle is small, the length of the chord is approximately equal to the arc length along the curve.]

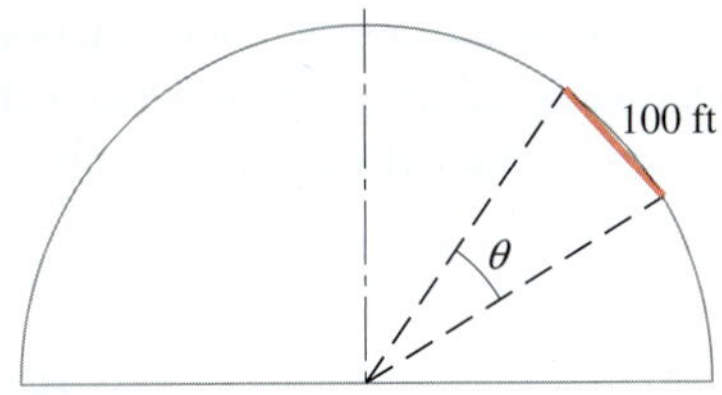

Problems 10–12. Five flywheels are spinning as follows: (a) radius 8.0 cm, period 4.0 ms; (b) radius 2.0 cm, period 4.0 ms; (c) radius 8.0 cm, period 1.0 ms; (d) radius 2.0 cm, period 1.0 ms; (e) radius 1.0 cm, period 4.0 ms.

10. Rank the flywheels in order of angular speed, largest to smallest. Explain.
11. Rank the flywheels in order of the linear speed at the rim, largest to smallest. Explain.

5.2 Radial Acceleration

12. Rank the flywheels of Problems 10 and 11 in order of the radial acceleration of a point on the rim, largest to smallest. Explain.
13. Objects that are at rest relative to Earth's surface are in circular motion due to Earth's rotation. What is the radial acceleration of an African baobab tree located at the equator?
14. An apparatus is designed to study insects at an acceleration of magnitude 980 m/s^2 (= 100g). The apparatus consists of a 2.0 m rod with insect containers at either end. The rod rotates about an axis perpendicular to the rod and at its center. (a) How fast does an insect move when it experiences a radial acceleration of 980 m/s^2? (b) What is the angular speed of the insect? (connect tutorial: centrifuge)

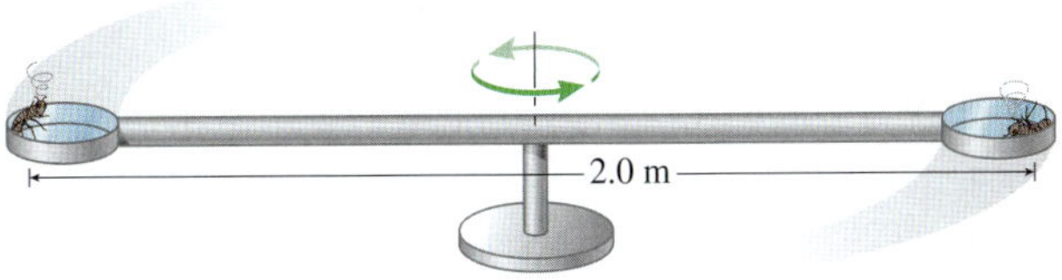

15. Medical testing has established that the maximum acceleration a pilot can be subjected to without losing consciousness is approximately 5.0g if the axis of acceleration is aligned with the spine. (See Ex. 5.4.) A pilot can avoid "blackout" at accelerations up to approximately 9.0g by wearing special "g-suits" that help keep blood pressure in the brain at a sufficient level. (a) Assuming this to be the case, what is the minimum safe radius of curvature for an unprotected pilot flying an F-15 in a horizontal circular loop at 750 km/h? (b) What does this radius become if the pilot is wearing a g-suit?

16. The rotor is an amusement park ride where people stand against the inside of a cylinder. Once the cylinder is spinning fast enough, the floor drops out. (a) What force keeps the people from falling out the bottom of the cylinder? (b) If the coefficient of friction is 0.40 and the cylinder has a radius of 2.5 m, what is the minimum angular speed of the cylinder so that the people don't fall out? (Normally the operator runs it considerably faster as a safety measure.)

17. Verify that all three expressions for radial acceleration ($v\omega$, v^2/r, and $\omega^2 r$) have the correct dimensions for an acceleration.

18. A 0.700 kg ball is on the end of a rope that is 1.30 m in length. The ball and rope are attached to a pole and the entire apparatus, including the pole, rotates about the pole's symmetry axis. The rope makes an angle of 70.0° with respect to the vertical. What is the tangential speed of the ball?

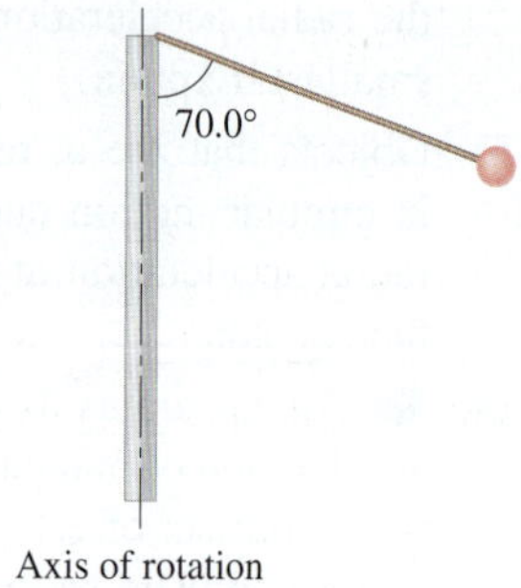

19. ✦ A child's toy has a 0.100 kg ball attached to two strings, A and B. The strings are also attached to a stick and the ball swings around the stick along a circular path in a horizontal plane. Both strings are 15.0 cm long and make an angle of 30.0° with respect to the horizontal. (a) Draw an FBD for the ball showing the tension forces and the gravitational force. (b) Find the magnitude of the tension in each string when the ball's angular speed is 6.00π rad/s.

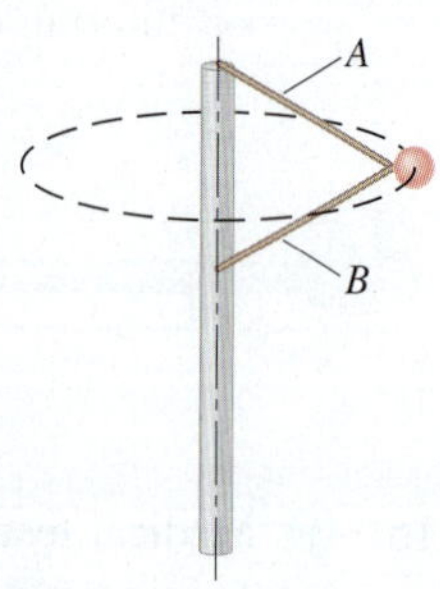

20. ✦ A child swings a rock of mass m in a horizontal circle using a rope of length L. The rock moves at constant speed v. (a) Ignoring gravity, find the tension in the rope. (b) Now include gravity (the weight of the rock is no longer negligible, although the weight of the rope still is negligible). What is the tension in the rope? Express the tension in terms of m, g, v, L, and the angle θ that the rope makes with the horizontal. (connect tutorial: skip rope)

21. ✦ A *conical pendulum* consists of a bob (mass m) attached to a string (length L) swinging in a horizontal circle (Fig. 5.12). As the string moves, it sweeps out the area of a cone. The angle that the string makes with the vertical is ϕ. (a) What is the tension in the string? (b) What is the period of the pendulum?

5.3 Unbanked and Banked Curves

22. Ⓒ A curve in a stretch of highway has radius R. The road is unbanked. The coefficient of static friction between the tires and road is μ_s. (a) What is the fastest speed that a car can safely travel around the curve? (b) Explain what happens when a car enters the curve at a speed greater than the maximum safe speed. Illustrate with an FBD. (connect interactive: banked curve)

23. A highway curve has a radius of 825 m. At what angle should the road be banked so that a car traveling at 26.8 m/s (60 mi/h) has no tendency to skid sideways on the road? [*Hint*: No tendency to skid means the frictional force is zero.]

24. A curve in a highway has radius of curvature 120 m and is banked at 3.0°. On a day when the road is icy, what is the safest speed to go around the curve?

25. A roller coaster car of mass 320 kg (including passengers) travels around a horizontal curve of radius 35 m. Its speed is 16 m/s. What is the magnitude and direction of the total force exerted on the car by the track?

26. A velodrome is built for use in the Olympics. The radius of curvature of the surface is 20.0 m. At what angle should the surface be banked for cyclists moving at 18 m/s? (Choose an angle so that no frictional force is needed to keep the cyclists in their circular path. Large banking angles *are* used in velodromes.)

27. Ⓒ A car drives around a curve with radius 410 m at a speed of 32 m/s. The road is not banked. The mass of the car is 1400 kg. (a) What is the frictional force on the car? (b) Does the frictional force necessarily have magnitude $\mu_s N$? Explain.

28. ✦ A car drives around a curve with radius 410 m at a speed of 32 m/s. The road is banked at 5.0°. The mass

of the car is 1400 kg. (a) What is the frictional force on the car? (b) At what speed could you drive around this curve so that the force of friction is zero?

29. A road with a radius of 75.0 m is banked so that a car can navigate the curve at a speed of 15.0 m/s without any friction. On a cold day when the street is icy, the coefficient of static friction between the tires and the road is 0.120. What is the *slowest* speed the car can go around this curve without sliding *down* the bank?

30. ✦ A curve in a stretch of highway has radius R. The road is banked at angle θ to the horizontal. The coefficient of static friction between the tires and road is μ_s. What is the fastest speed that a car can travel through the curve?

31. An airplane is flying at constant speed v in a horizontal circle of radius r. The lift force on the wings due to the air is perpendicular to the wings. At what angle to the vertical must the wings be banked to fly in this circle? (connect tutorial: plane in turn)

32. ✦ A road with a radius of 75.0 m is banked so that a car can navigate the curve at a speed of 15.0 m/s without any friction. When a car is going 20.0 m/s on this curve, what minimum coefficient of static friction is needed if the car is to navigate the curve without slipping?

5.4 Circular Orbits of Satellites and Planets

33. What is the average linear speed of the Earth about the Sun?

34. The orbital speed of Earth about the Sun is 3.0×10^4 m/s and its distance from the Sun is 1.5×10^{11} m. The mass of Earth is approximately 6.0×10^{24} kg and that of the Sun is 2.0×10^{30} kg. What is the magnitude of the force exerted by the Sun on Earth? [*Hint*: Two different methods are possible. Try both.]

35. Two satellites are in circular orbits around Jupiter. One, with orbital radius r, makes one revolution every 16 h. The other satellite has orbital radius $4.0r$. How long does the second satellite take to make one revolution around Jupiter?

36. The Hubble Space Telescope orbits Earth 613 km above Earth's surface. What is the period of the telescope's orbit?

37. Io, one of Jupiter's satellites, has an orbital period of 1.77 d. Europa, another of Jupiter's satellites, has an orbital period of about 3.54 d. Both moons have nearly circular orbits. Use Kepler's third law to find the distance of each satellite from Jupiter's center. Jupiter's mass is 1.9×10^{27} kg.

38. A spy satellite is in circular orbit around Earth. It makes one revolution in 6.00 h. (a) How high above Earth's surface is the satellite? (b) What is the satellite's acceleration?

5.5 Nonuniform Circular Motion

39. A roller coaster has a vertical loop with radius 29.5 m. With what minimum speed should the roller coaster car be moving at the top of the loop so that the passengers do not lose contact with the seats?

40. Ⓒ A pendulum is 0.80 m long, and the bob has a mass of 1.0 kg. At the bottom of its swing, the bob's speed is 1.6 m/s. (a) What is the tension in the string at the bottom of the swing? (b) Explain why the tension is greater than the weight of the bob.

41. A 35.0 kg child swings on a rope with a length of 6.50 m that is hanging from a tree. At the bottom of the swing, the child is moving at a speed of 4.20 m/s. What is the tension in the rope?

42. A car approaches the top of a hill that is shaped like a vertical circle with a radius of 55.0 m. What is the fastest speed that the car can go over the hill without losing contact with the ground?

5.6 Angular Acceleration

43. A child pushes a merry-go-round from rest to a final angular speed of 0.50 rev/s with constant angular acceleration. In doing so, the child pushes the merry-go-round 2.0 revolutions. What is the angular acceleration of the merry-go-round?

44. A cyclist starts from rest and pedals so that the wheels make 8.0 revolutions in the first 5.0 s. What is the angular acceleration of the wheels (assumed constant)?

45. During normal operation, a computer's hard disk spins at 7200 rev/min. If it takes the hard disk 4.0 s to reach this angular velocity starting from rest, what is the average angular acceleration of the hard disk in rad/s^2?

46. A hamster of mass 0.100 kg gets onto his 20.0 cm diameter exercise wheel and runs along inside the wheel for 0.800 s until its frequency of rotation is 1.00 Hz. (a) What is the tangential acceleration of the wheel, assuming it is constant? (b) What is the normal force on the hamster just before he stops? The hamster is at the bottom of the wheel during the entire 0.800 s.

47. A clothes washer reaches an angular speed of 1400 rev/min in 2.0 s, starting from rest, during the spin cycle. (a) Assuming the angular acceleration is constant, what is its magnitude? (b) How many revolutions does the washer make during this time interval?

48. A wheel's angular acceleration is constant. Initially its angular velocity is zero. During the first 1.0 s time interval, it rotates through an angle of 90.0°. (a) Through what angle does it rotate during the next 1.0 s time interval? (b) Through what angle during the third 1.0 s time interval?

49. A car that is initially at rest moves along a circular path with a constant tangential acceleration component of

2.00 m/s^2. The circular path has a radius of 50.0 m. The initial position of the car is at the far west location on the circle and the initial velocity is to the north. (a) After the car has traveled $\frac{1}{4}$ of the circumference, what is the speed of the car? (b) At this point, what is the radial acceleration component of the car? (c) At this same point, what is the total acceleration of the car?

50. A disk rotates with constant angular acceleration. The initial angular speed of the disk is 2.0π rad/s. After the disk rotates through 10.0π radians, the angular speed is 7.0π rad/s. (a) What is the magnitude of the angular acceleration? (b) How much time did it take for the disk to rotate through 10.0π radians? (c) What is the tangential acceleration of a point located at a distance of 5.0 cm from the center of the disk?

51. A "blink of an eye" is a time interval of about 150 ms for an average adult. The "closure" portion of the blink takes only about 55 ms. Let us model the closure of the upper eyelid as uniform angular acceleration through an angular displacement of 15°. (a) What is the value of the angular acceleration the eyelid undergoes while closing? (b) What is the tangential acceleration of the edge of the eyelid while closing if the radius of the eyeball is 1.25 cm?

52. A study was done observing the ability of the eye to rapidly rotate in order to follow a moving object by placing contact lenses that contain accelerometers on a subject's eye. The eyeball has radius 1.25 cm. Suppose that, while the subject watches a moving object, the eyeball rotates through 20.0° in a time interval of 75 ms. (a) What is the magnitude of the average angular velocity of the eye? (b) Assume that the eye starts at rest, rotates with a constant angular acceleration during the first half of the interval, then the rotation slows with a constant angular acceleration during the second half until it comes to rest. What is the magnitude of the angular acceleration of the eye? (c) What tangential acceleration would the contact-lens accelerometers record in this case?

53. In a Beams ultracentrifuge, the rotor is suspended magnetically in a vacuum. Since there is no mechanical connection to the rotor, the only friction is the air resistance due to the few air molecules in the vacuum. If the rotor is spinning with an angular speed of 5.0×10^5 rad/s and the driving force is turned off, its spinning slows down at an angular rate of 0.40 rad/s^2. (a) How long does the rotor spin before coming to rest? (b) During this time, through how many revolutions does the rotor spin?

54. The rotor of the Beams ultracentrifuge (see Problem 53) is 20.0 cm long. For a point at the end of the rotor, find the (a) initial speed, (b) tangential acceleration component, and (c) maximum radial acceleration component.

55. ✦ A pendulum is 0.800 m long, and the bob has a mass of 1.00 kg. When the string makes an angle of $\theta = 15.0°$ with the vertical, the bob is moving at 1.40 m/s. Find the tangential and radial acceleration components and the tension in the string. [*Hint*: Draw an FBD for the bob. Choose the x-axis to be tangential to the motion of the bob and the y-axis to be radial. Apply Newton's second law.]

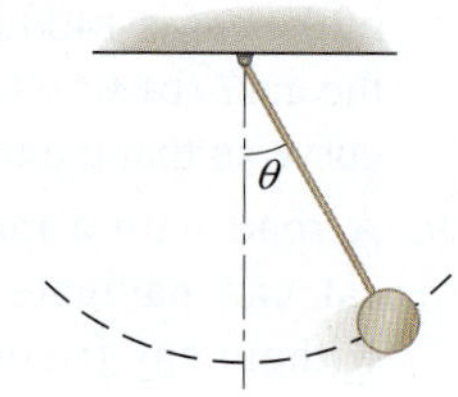

Problems 55 and 56

56. ✦ Find the tangential acceleration of a freely swinging pendulum when it makes an angle θ with the vertical.

5.7 Apparent Weight and Artificial Gravity

57. If a clothes washer's drum has a radius of 25 cm and spins at 4.0 rev/s, what is the strength of the artificial gravity to which the clothes are subjected? Express your answer as a multiple of g.

58. A space station is shaped like a ring and rotates to simulate gravity. If the radius of the space station is 120 m, at what frequency must it rotate so that it simulates Earth's gravity? [*Hint*: The apparent weight of the astronauts must be the same as their weight on Earth.] (connect tutorial: space station)

59. A biologist is studying growth in space. He wants to simulate Earth's gravitational field, so he positions the plants on a rotating platform in the spaceship. The distance of each plant from the central axis of rotation is $r = 0.20$ m. What angular speed is required?

60. ✦ A biologist is studying plant growth and wants to simulate a gravitational field twice as strong as Earth's. She places the plants on a horizontal rotating table in her laboratory on Earth at a distance of 12.5 cm from the axis of rotation. What angular speed will give the plants an effective gravitational field $\vec{\mathbf{g}}_{eff}$, whose magnitude is 2.0g? [*Hint*: Remember to account for Earth's gravitational field $\vec{\mathbf{g}}$ as well as the artificial gravity $\vec{\mathbf{g}}_{art}$.]

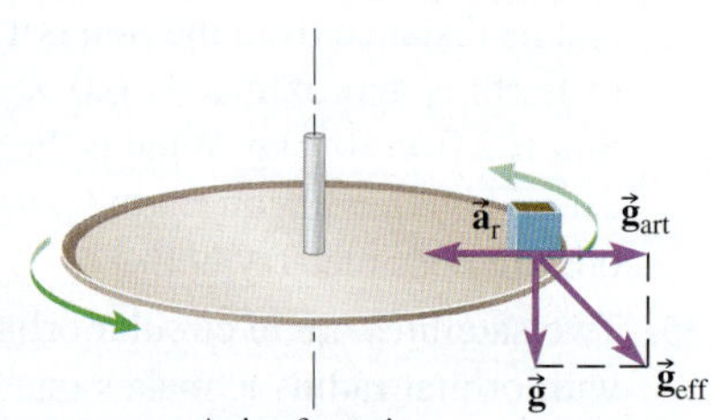

61. A person of mass M stands on a bathroom scale inside a Ferris wheel compartment. The Ferris wheel has radius R and angular velocity ω. What is the apparent weight of the person (a) at the top and (b) at the bottom?

62. A person rides a Ferris wheel that turns with constant angular velocity. Her weight is 520.0 N. At the top of the ride her apparent weight is 1.5 N different from her true weight. (a) Is her apparent weight at the top

521.5 N or 518.5 N? Why? (b) What is her apparent weight at the bottom of the ride? (c) If the angular speed of the Ferris wheel is 0.025 rad/s, what is its radius?

Collaborative Problems

63. Mars has a mass of about 6.42×10^{23} kg. The length of a day on Mars is 24 h and 37 min, a little longer than the length of a day on Earth. Your task is to put a satellite into a circular orbit around Mars so that it stays above one spot on the surface, orbiting Mars once each Mars day. At what distance from the center of the planet should you place the satellite?

64. ✦ A spacecraft is in orbit around Jupiter. The radius of the orbit is 3.0 times the radius of Jupiter (which is $R_J = 71\ 500$ km). The gravitational field at the surface of Jupiter is 23 N/kg. What is the period of the spacecraft's orbit? [*Hint*: You don't need to look up any more data about Jupiter to solve the problem.]

65. ✦ The time to sunset can be estimated by holding out your arm, holding your fingers horizontally in front of your eyes, and counting the number of fingers that fit between the horizon and the setting Sun. (a) What is the angular speed, in radians per second, of the Sun's apparent circular motion around the Earth? (b) Estimate the angle subtended by one finger held at arm's length. (c) How long in minutes does it take the Sun to "move" through this same angle?

66. ✦ What's the quickest way to make a U-turn at constant speed? Suppose that you need to make a 180° turn on a circular path. The minimum radius (due to the car's steering system) is 5.0 m, while the maximum (due to the width of the road) is 20.0 m. Your acceleration must never exceed 3.0 m/s^2 or else you will skid. Should you use the smallest possible radius, so the distance is small, or the largest, so you can go faster without skidding, or something in between? What is the minimum possible time for this U-turn?

67. ✦ You take a homemade "accelerometer" to an amusement park. This accelerometer consists of a metal nut attached to a string and connected to a protractor, as shown in the figure. While riding a roller coaster that is moving at uniform speed around a horizontal circular path, you hold up the accelerometer and notice that the string is making a constant angle of 55° with respect to the vertical with the nut pointing away from the center of the circle, as shown. (a) What is the radial acceleration of the roller coaster? (b) What is your radial acceleration expressed as a multiple of g? (c) If the roller coaster track is turning in a radius of 80.0 m, how fast are you moving?

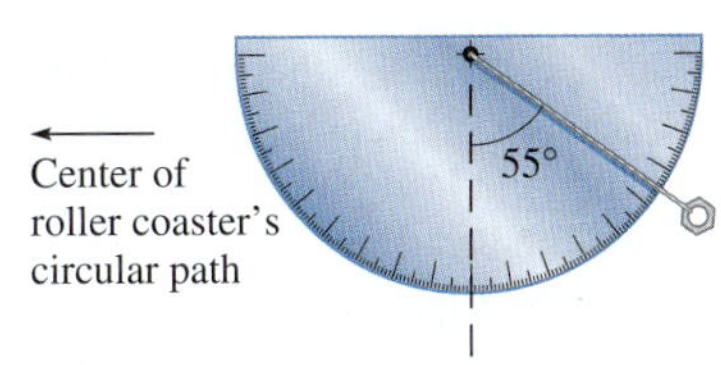

68. Your car's wheels are 65 cm in diameter, and the wheels are spinning at an angular velocity of 101 rad/s. How fast is your car moving in kilometers per hour (assume no slippage)?

Comprehensive Problems

69. The Earth rotates on its own axis once per day (24.0 h). What is the tangential speed of the summit of Mt. Kilimanjaro (elevation 5895 m above sea level), which is located approximately on the equator, due to the rotation of the Earth? The equatorial radius of Earth is 6378 km.

70. A trimmer for cutting weeds and grass near trees and borders has a nylon cord of 0.23 m length that whirls about an axle at 660 rad/s. What is the linear speed of the tip of the nylon cord?

71. A high-speed dental drill is rotating at 3.14×10^4 rad/s. Through how many degrees does the drill rotate in 1.00 s?

72. A jogger runs counterclockwise around a path of radius 90.0 m at constant speed. He makes 1.00 revolution in 188.4 s. At $t = 0$, he is heading due east. (a) What is the jogger's instantaneous velocity at $t = 376.8$ s? (b) What is his instantaneous velocity at $t = 94.2$ s?

73. Two gears A and B are turning in mesh. Gear A's radius to the point of contact between the gears is 8.0 cm and that of gear B is 4.0 cm. (a) What is the linear speed of the contact point when gear A's angular velocity is 6.0 rad/s counterclockwise? (b) What is B's angular velocity?

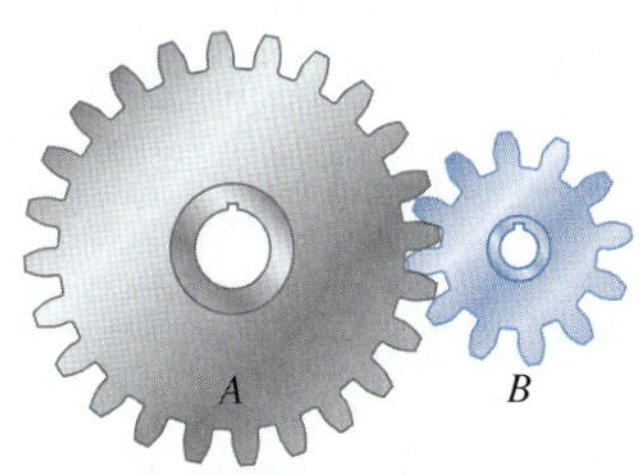

Problems 73 and 74

74. If gear A in Problem 73 has an initial frequency of 0.955 Hz and an angular acceleration of 3.0 rad/s^2, how many rotations does each gear go through in 2.0 s?

75. The Milky Way galaxy rotates about its center with a period of about 200 million yr. The Sun is 2×10^{20} m from the center of the galaxy. How fast is the Sun moving with respect to the center of the galaxy?

76. A small object of mass 0.50 kg is attached by a 0.50 m long cord to a pin set into the surface of a frictionless table top. The object moves in a circle on the horizontal

surface with a speed of 2.0π m/s. (a) What is the magnitude of the radial acceleration of the object? (b) What is the tension in the cord?

77. Two blocks, one with mass $m_1 = 0.050$ kg and one with mass $m_2 = 0.030$ kg, are connected to each other by a string. The inner block is connected to a central pole by another string as shown in the figure with $r_1 = 0.40$ m and $r_2 = 0.75$ m. When the blocks are spun around on a horizontal frictionless surface at an angular speed of 1.5 rev/s, what is the tension in each of the two strings?

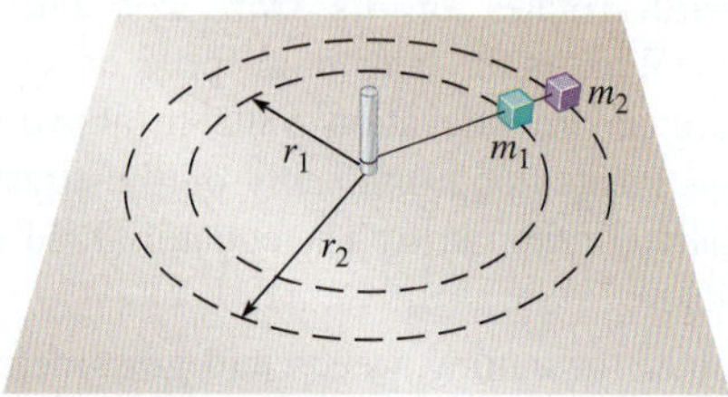

78. The Milky Way galaxy rotates about its center with a period of about 200 million yr. The Sun is 2×10^{20} m from the center of the galaxy. (a) What is the Sun's radial acceleration? (b) What is the net gravitational force on the Sun due to the other stars in the Milky Way?

79. Bacteria swim using a corkscrew-like helical flagellum that rotates. For a bacterium with a flagellum that has a pitch of 1.0 μm that rotates at 110 rev/s, how fast could it swim if there were no "slippage" in the medium in which it is swimming? The pitch of a helix is the distance between "threads."

80. You place a penny on an old turntable at a distance of 10.0 cm from the center. The coefficient of static friction between the penny and the turntable is 0.350. The turntable's angular acceleration is 2.00 rad/s^2. How long after you turn on the turntable will the penny begin to slide?

81. A coin is placed on an old turntable. If the coefficient of static friction between the coin and the turntable is 0.10, how far from the center of the turntable can the coin be placed without having it slip off when the turntable rotates at 33.3 rev/min?

82. ✦ Objects that are at rest relative to Earth's surface are in circular motion due to Earth's rotation. What is the radial acceleration of a painting hanging in the Prado Museum in Madrid, Spain, at a latitude of 40.2° North? (Note that the object's radial acceleration is not directed toward the center of the Earth.)

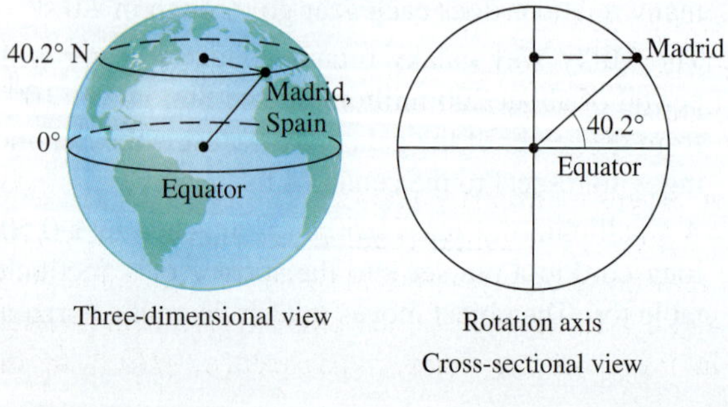

Three-dimensional view

Cross-sectional view

83. In an amusement park rocket ride, cars are suspended from 4.25 m cables attached to rotating arms at a distance of 6.00 m from the axis of rotation. The cables swing out at a constant angle of 45.0° when the ride is operating. What is the angular speed of rotation?

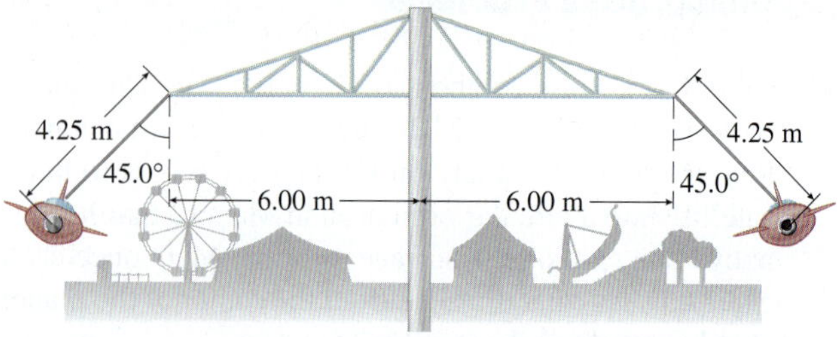

84. Centrifuges are commonly used in biological laboratories for the isolation and maintenance of cell preparations. For cell separation, the centrifugation conditions are typically 1.0×10^3 rev/min using an 8.0 cm radius rotor. (a) What is the radial acceleration of material in the centrifuge under these conditions? Express your answer as a multiple of g. (b) At 1.0×10^3 rev/min (and with an 8.0 cm rotor), what is the net force on a red blood cell whose mass is 9.0×10^{-14} kg? (c) What is the net force on a virus particle of mass 5.0×10^{-21} kg under the same conditions? (d) To pellet out virus particles and even to separate large molecules such as proteins, super-high-speed centrifuges called ultracentrifuges are used in which the rotor spins in a vacuum to reduce heating due to friction. What is the radial acceleration inside an ultracentrifuge at 75 000 rev/min with an 8.0 cm rotor? Express your answer as a multiple of g.

85. A proposed "space elevator" consists of a cable going all the way from the ground to a space station in geostationary orbit (always above the same point on Earth's surface). Elevator "cars" would climb the cable to transport cargo to outer space. Consider a cable connected between the equator and a space station at height H above the surface. Ignore the mass of the cable*. (a) Find the height H. (b) Suppose there is an elevator car of mass 100 kg sitting halfway up at height $H/2$. What tension T would be required in the cable to hold the car in place? Which part of the cable would be under tension (above the car or below it)?

86. A star near the visible edge of a galaxy travels in a uniform circular orbit. It is 40 000 ly (light-years) from the galactic center and has a speed of 275 km/s.

*More realistically, the mass of the cable is one of the primary engineering challenges of a space elevator. The cable is so long that it would have a very large mass and would have to withstand an enormous tension to support its own weight. The cable would need to be supported by a counterweight positioned beyond the geostationary orbit. Some believe *carbon nanotubes* hold the key to producing a cable with the required properties.

(a) Estimate the total mass of the galaxy based on the motion of the star. [*Hint*: For this estimate, assume the total mass to be concentrated at the galactic center and relate it to the gravitational force on the star.] (b) The total *visible* mass (i.e., matter we can detect via electromagnetic radiation) of the galaxy is 10^{11} solar masses. What fraction of the total mass of the galaxy is visible*, according to this estimate?

87. Massimo, a machinist, is cutting threads for a bolt on a lathe. He wants the bolt to have 18 threads per inch. If the cutting tool moves parallel to the axis of the would-be bolt at a linear velocity of 0.080 in./s, what must the rotational speed of the lathe chuck be to ensure the correct thread density? [*Hint*: One thread is formed for each complete revolution of the chuck.]

88. In Chapter 19 we will see that a charged particle can undergo uniform circular motion when acted on by a magnetic force and no other forces. (a) For that to be true, what must be the angle between the magnetic force and the particle's velocity? (b) The magnitude of the magnetic force on a charged particle is proportional to the particle's speed, $F = kv$. Show that two identical charged particles moving in circles at different speeds in the same magnetic field must have the same period. (c) Show that the radius of the particle's circular path is proportional to the speed.

89. A rotating flywheel slows down at a constant rate due to friction in its bearings. At $t = 0$, its angular velocity is ω_0. At $t = 60$ s, its angular velocity is $0.80\omega_0$. What is the angular velocity at $t = 180$ s?

Review and Synthesis

90. ✦ A satellite travels around Earth in uniform circular motion at an altitude of 35 800 km above Earth's surface. The satellite is in geosynchronous orbit. In the figure with Multiple-Choice Questions 1–4, the satellite moves counterclockwise (*ABCDA*). State directions in terms of the *x*- and *y*-axes. (a) What is the satellite's instantaneous velocity at point *C*? (b) What is the satellite's average velocity for one quarter of an orbit, starting at *A* and ending at *B*? (c) What is the satellite's average acceleration for one quarter of an orbit, starting at *A* and ending at *B*? (d) What is the satellite's instantaneous acceleration at point *D*?

*In many galaxies the stars appear to have roughly the *same orbital speed* over a large range of distances from the center. A popular hypothesis to explain such galaxy rotation velocities is the existence of *dark matter*—matter that we cannot detect via electromagnetic radiation. Dark matter is thought to account for the majority of the mass of some galaxies and nearly a fourth of the total mass of the universe.

91. ✦ Ⓒ Objects that are at rest relative to the Earth's surface are in circular motion due to Earth's rotation. (a) What is the radial acceleration of an object at the equator? (b) Is the object's apparent weight greater or less than its weight? Explain. (c) By what percentage does the apparent weight differ from the weight at the equator? (d) Is there any place on Earth where a bathroom scale reading is equal to your true weight? Explain.

92. ✦ Ⓒ Earth's orbit around the Sun is nearly circular. The period is 1 yr = 365.25 d. (a) In an elapsed time of 1 d, what is Earth's angular displacement? (b) What is the change in Earth's velocity, $\Delta\vec{\mathbf{v}}$? (c) What is Earth's average acceleration during 1 d? (d) Compare your answer for (c) to the magnitude of Earth's instantaneous radial acceleration. Explain.

93. ✦ Find the orbital radius of a geostationary satellite without using the speed found in Example 5.9. Start by writing an equation that relates the period, radius, and speed of the orbiting satellite. Then apply Newton's second law to the satellite. You will have two equations with two unknowns (the speed and radius). Eliminate the speed algebraically and solve for the radius.

94. Two blocks are connected by a light string passing over an ideal pulley. The block with mass $m_1 = 20.0$ kg slides on a frictionless horizontal surface, while the block with mass $m_2 = 1.0$ kg hangs vertically. The radius of the pulley is 6.0 cm. (a) Assuming that the pulley rotates such that the string doesn't slip, find the angular acceleration of the pulley. (b) If the block on the table is released from rest, calculate how many revolutions the pulley has made 2.0 s later, assuming the other block hasn't reached the floor.

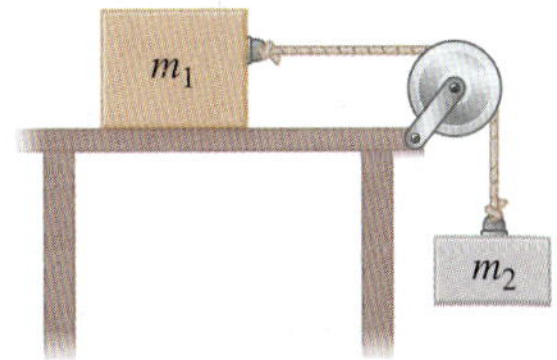

95. A ball weighing 20.0 N is tied to a string fixed to the ceiling. The string makes a 30.0° angle with the ceiling. Initially, the ball is held in place by a force $\vec{\mathbf{F}}$ that is perpendicular to the string. (a) What is the magnitude of the force $\vec{\mathbf{F}}$? (b) What is the tension in the string? (c) Just after the ball is released and allowed to swing back and forth, what are the tension in the string, the radial acceleration of the ball, and the tangential acceleration of the ball?

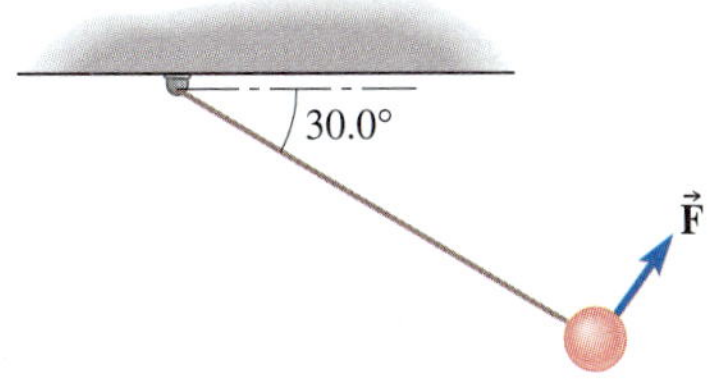

96. A wheel of radius r rolls to the right without slipping on a horizontal road. Its axle moves at a constant speed v_{axle}. (a) Find the velocities of points *A*, *B*, and *C* with respect to the axle. Express your answers in terms of v_{axle} and r, as needed. [*Hint*: In the

reference frame of the axle, the wheel is rotating in place at a constant angular speed ω.] (b) Find the velocities of points A, B, and C with respect to the road. (c) Comment on the velocity of point C with respect to the road.

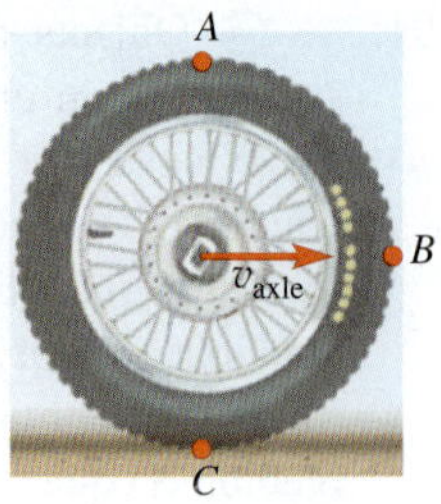

Answers to Practice Problems

5.1 3.001×10^{-7} rad/s

5.2 1.65 m/s

5.3 1.9 min

5.4 7200 rev/min $\times$ 2π rad/rev $\times$ (1/60) min/s = 240π rad/s; $a_r = \omega^2 r = (240\pi \text{ rad/s})^2 \times 0.060 \text{ m} = 34\ 000 \text{ m/s}^2$.

5.5 60 N toward the center of the circular path

5.6 More slowly

5.7 4.1 m/s

5.8 29.7 km/s; 3.17×10^7 s

5.9 $0.723R$

5.10 2.44 h

5.11 $4.2mg$

5.12 Acceleration is purely tangential:

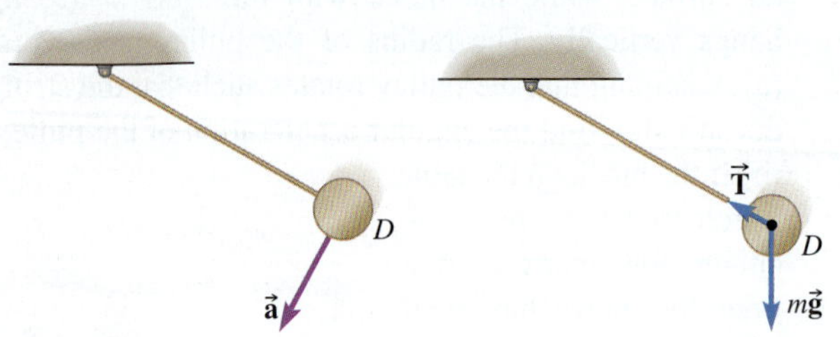

5.13 (a) 1.75×10^{-4} rad/s^2; (b) 0.0349 rad (2.00°)

5.14 (a) 2200 N; (b) 1500 N

Answers to Checkpoints

5.1 8.3 ms

5.2 No, for uniform circular motion the *direction* of the velocity vector is continuously changing but the magnitude of the velocity (the speed) is unchanged.

5.3 The car has friction between the road and the tires to exert a horizontal force that causes the radial acceleration.

5.4 To be geostationary the satellites must have an orbital period of 1 d. The only quantities that affect the period are the mass of Earth and the radial distance from Earth's center. These quantities are the same for all satellites no matter the mass.

5.5 For nonuniform circular motion, the direction and the magnitude of the velocity are both changing. There are tangential and radial components to the acceleration. The magnitude of the radial component changes as the speed changes. For uniform circular motion, the magnitude of the velocity is constant but the direction changes. The radial acceleration is constant in magnitude (and the tangential acceleration is zero).

5.6 The radial acceleration cannot be constant because the radius r is constant, but the angular velocity ω is changing $a_r = \omega^2 r$.

Conservation of Energy

CHAPTER

6

As a kangaroo hops along, the maximum height of each hop might be around 2.8 m. This height is only slightly higher than that achieved by an Olympic high jumper, but the kangaroo is able to achieve this height hop after hop as it travels with a horizontal velocity of 15 m/s or more. What features of kangaroo anatomy make this feat possible? It cannot simply be a matter of having more powerful leg muscles. If it were, the kangaroo would have to consume large amounts of energy-rich food to supply the muscles with enough chemical energy for each jump, but in reality a kangaroo's diet consists largely of grasses that are poor in energy content. (See pp. 215–216 for the answer.)

BIOMEDICAL APPLICATIONS

- Stored elastic energy in jumping, walking, and running animals (Sec. 6.7; Ex. 6.12; PP 6.12; Probs. 62, 69, 117)
- Power supplied by molecular motors in bacteria and in muscles (Ex. 6.13; PP 6.13)
- Metabolism (Probs. 8, 33, 86, 112–114)
- Elastic properties of virus capsid (Prob. 70)

Concepts & Skills to Review

- gravitational forces (Section 4.5)
- Newton's second law: force and acceleration (Section 4.3)
- components of vectors (Section 3.2)
- circular orbits (Section 5.4)
- area under a graph (Sections 2.2 and 2.3)

6.1 PREVIEW OF THE LAW OF CONSERVATION OF ENERGY

Until now, we have relied on Newton's laws of motion to be the fundamental physical laws used to analyze the forces that act on objects and to predict the motion of objects. Now we introduce another physical principle: the conservation of energy. A **conservation law** is a physical principle that identifies some quantity that does not change with time. Conservation of energy means that every physical process leaves the total energy in the universe unchanged. Energy can be converted from one form to another, or transferred from one place to another. If we are careful to account for all the energy transformations, we find that the total energy remains the same.

The Law of Conservation of Energy

The total energy in the universe is unchanged by any physical process:

total energy before = total energy after.

"Turn down the thermostat—we're trying to conserve energy!" In ordinary language, *conserving energy* means trying not to waste useful energy resources. In the scientific meaning of *conservation*, energy is *always* conserved no matter what happens. When we "produce" or "generate" electric energy, for instance, we aren't creating any new energy; we're just converting energy from one form into another that's more useful to us.

Conservation of energy is one of the few universal principles of physics. No exceptions to the law of conservation of energy have been found. Conservation of energy is a powerful tool in the search to understand nature. It applies equally well to radioactive decay, the gravitational collapse of a star, a chemical reaction, a biological process such as respiration, and to the generation of electricity by a wind turbine (Fig. 6.1). Think about the energy conversions that make life possible. Green plants use photosynthesis to convert the energy they receive from the Sun into stored chemical energy. When animals eat the plants, that stored energy enables motion, growth, and maintenance of body temperature. Energy conservation governs every one of these processes.

Figure 6.1 At a California wind farm, these wind turbines convert the energy of motion of the air into electric energy.

Problem-Solving Strategy: Choosing Between Alternative Solution Methods

Some problems can be solved using either energy conservation *or* Newton's second law, so it always pays to consider both methods. If both methods can be used to answer the question, think about which is easier to apply. Sometimes that won't be clear until you've gotten started—if the solution starts to get complicated, consider trying the other method. When time permits, solve the problem both ways. Doing so is a way to check your answer and can lead to insights you might not gain by using only one method.

Historical Development of the Principle of Energy Conservation Although many scientists contributed to the development of the law of conservation of energy, the law's first clear statement was made in 1842 by the German surgeon Julius Robert von Mayer (1814–1878). As a ship's physician on a voyage to what is now Indonesia, Mayer had noticed that the sailors' venous blood was a much deeper red in the tropics

Table 6.1 Some Common Forms of Energy

Form of Energy	Brief Description
Translational kinetic	Energy of translational motion (Chapter 6)
Elastic	Energy stored in a "springy" object or material when it is deformed (Chapter 6)
Gravitational	Energy of gravitational interactions (Chapter 6)
Rotational kinetic	Energy of rotational motion (Chapter 8)
Vibrational, acoustic, seismic	Energy of the oscillatory motions of atoms and molecules in a substance caused by a mechanical wave passing through it (Chapters 11 and 12)
Internal	Energies of motion and interaction of atoms and molecules in solids, liquids, and gases, related to our sensation of temperature (Chapters 13–15)
Electromagnetic	Energy of interaction of electric charges and currents; energy of electromagnetic fields, including electromagnetic waves such as light (Chapters 14, 17–22)
Rest	The total energy of a particle of mass m when it is at rest, given by Einstein's famous equation $E = mc^2$ (Chapters 26, 29, and 30)
Chemical	Energies of motion and interaction of electrons in atoms and molecules (Chapter 28)
Nuclear	Energies of motion and interaction of protons and neutrons in atomic nuclei (Chapters 29 and 30)

Figure 6.2 The stored chemical energy in food enables a weightlifter to lift the barbell over her head.

than it was in Europe. He concluded that less oxygen was being used because they didn't need to "burn" as much fuel to keep the body warm in the warmer climate.

In 1843, the English physicist James Prescott Joule (1818–1889), whose "day job" was running the family brewery, performed precise experiments to show that gravitational potential energy could be converted into a previously unrecognized form of energy (internal energy). It had previously been thought that forces such as friction "use up" energy. Thanks to Mayer, Joule, and others, we now know that friction converts mechanical forms of energy into internal energy and that total energy is always conserved.

Forms of Energy

Energy comes in many different forms (Fig. 6.2). Table 6.1 summarizes the main forms of energy discussed in this text and indicates the principal chapters that discuss each one. At the most *fundamental* level, there are only three types of energy: energy due to motion (**kinetic energy**), stored energy due to interaction (**potential energy**), and rest energy. Every form of energy listed in Table 6.1 can be understood as one or more of these three types.

To apply the energy conservation principle, we need to learn how to calculate the amount of each form of energy. There isn't one formula that applies to all. Fortunately, we don't have to learn about all of them at once. This chapter focuses on three forms of macroscopic mechanical energy (kinetic energy, gravitational potential energy, and elastic potential energy). For now, we use energy conservation as a tool to understand the **translational** motion of objects, but we do not consider rotational motion or changes in the *internal* energy of an object. We assume that these moving objects are perfectly rigid, so every point on the object moves through the same displacement.

6.2 WORK DONE BY A CONSTANT FORCE

To apply the principle of energy conservation, we need to learn how energy can be converted from one form to another. We begin with an example. Suppose the trunk in Fig. 6.3a weighs 220 N and must be lifted a height $h = 4.0$ m. To lift it at constant speed, Rosie must exert a force of 220 N on the rope, assuming an ideal pulley and

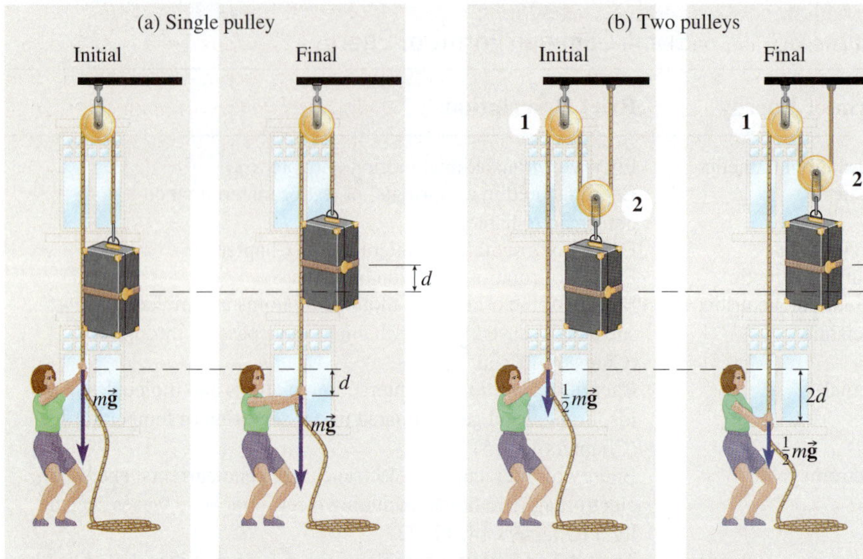

Figure 6.3 (a) Rosie moves a trunk into her dorm room through the window. (b) The two-pulley system makes it easier for Rosie to lift the trunk: the *force* she must exert is halved. Is she getting something for nothing, or does she still have to do the same amount of *work* to lift the trunk?

rope. (We ignore for now the brief initial time when she pulls with more than 220 N to accelerate the trunk from rest to its constant speed and the brief time she pulls with less than 220 N to let it come to rest.)

Rosie would only have to exert half the force (110 N) if she were to use the two-pulley system of Fig. 6.3b (see Example 4.12). She doesn't get something for nothing, though. To lift the trunk 4.0 m, the sections of rope on *both* sides of pulley 2 must be shortened by 4.0 m, so Rosie must pull an 8.0 m length of rope. The two-pulley system enables her to pull with half the force, but now she must pull the rope through *twice the distance*.

Notice that the *product* of the magnitude of the force and the distance is the same in both cases:

$$220 \text{ N} \times 4.0 \text{ m} = 110 \text{ N} \times 8.0 \text{ m} = 880 \text{ N·m} = W$$

This product is called the **work** (W) done by Rosie on the rope. Work is a scalar quantity; it does not have a direction. The same symbol W is often used for the weight of an object. To avoid confusion, we write mg for weight and let W stand for work.

Don't be misled by the many different meanings the word *work* has in ordinary conversation. We talk about doing homework, or going to work, or having too much work to do. Not everything we call "work" in conversation is *work* as defined in physics.

The SI unit of work and energy is the newton-meter (N·m), which is given the name joule (symbol: J) in honor of James Prescott Joule.

$$1 \text{ J} = 1 \text{ N·m}$$

Using either method, Rosie must do 880 J of work on the rope to lift the trunk. When we say that Rosie does 880 J of work, we mean that Rosie supplies 880 J of energy—the amount of energy required to lift the trunk 4.0 m. **Work** *is an energy transfer that occurs when a force acts on an object that moves.*

Rosie does no work on the rope while she holds it in one place because the displacement is zero. She can just as well fasten it and walk away (Fig. 6.4). *If there is no displacement, no work is done and no energy is transferred.* Why then does she get tired if she holds the rope in place for a long time? Although Rosie does no work *on the rope* when holding it in place, work *is* done inside her body by muscle fibers, which have to do work internally to maintain tension in the muscle. This internal work converts chemical energy into internal energy—the muscle warms up—but no energy is transferred *to the trunk.*

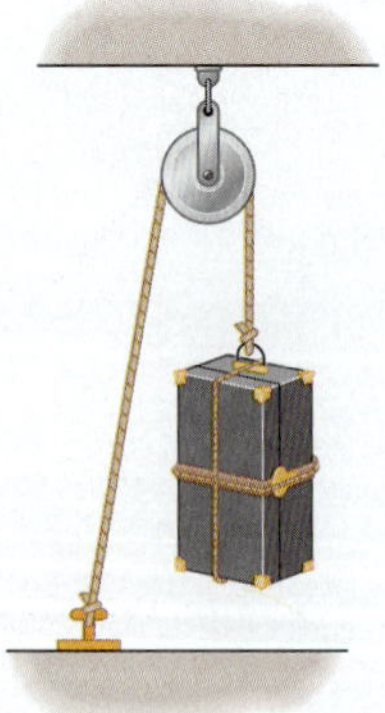

Figure 6.4 While the trunk is held in place by tying the rope, no work is done and no energy transfers occur.

Work Done by a Force not Parallel to the Displacement The force that Rosie exerts on the rope is in the same direction as the displacement of that end of the rope. More generally, how much work is done by a constant force that is at some angle to the displacement? It turns out that only the *component* of the force *in the direction of the displacement* does work. So, in general, the work done by a constant force is defined as the product of the magnitude of the displacement and the *component* of the force *in the direction of the displacement*. If θ represents the angle between the force and displacement vectors when they are drawn starting at the same point, then the force component in the direction of the displacement is $F \cos \theta$ (Fig. 6.5). Therefore, work done by a constant force on an object can be written $W = F\,\Delta r \cos \theta$, where F is the magnitude of the force and Δr is the magnitude of the displacement of the object.

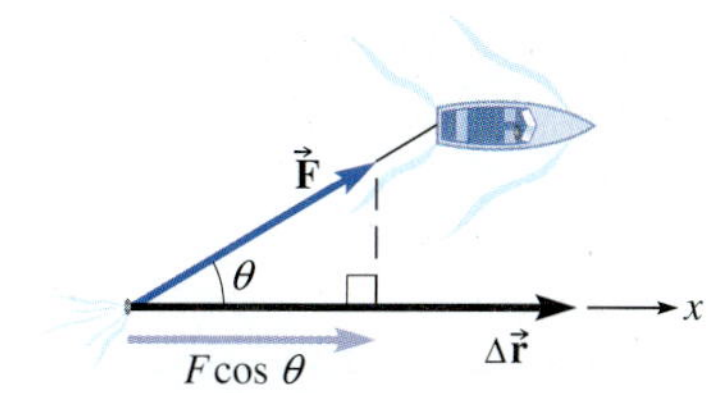

Figure 6.5 The work done by the force of the towrope on the water-skier during a displacement $\Delta\vec{\mathbf{r}}$ is $(F \cos \theta)\,\Delta r$, where $(F \cos \theta)$ is the component of $\vec{\mathbf{F}}$ in the direction of $\Delta\vec{\mathbf{r}}$.

Work done by a constant force $\vec{\mathbf{F}}$ acting on an object whose displacement is $\Delta\vec{\mathbf{r}}$

$$W = F\,\Delta r \cos \theta \qquad (6\text{-}1)$$

(θ is the angle between $\vec{\mathbf{F}}$ and $\Delta\vec{\mathbf{r}}$)

Work can also be expressed as the scalar product of the force and the displacement: $W = \vec{\mathbf{F}} \cdot \Delta\vec{\mathbf{r}}$. The **scalar product** (or **dot product**) of two vectors is defined by the equation $\vec{\mathbf{A}} \cdot \vec{\mathbf{B}} = AB \cos \theta$, where θ is the angle between $\vec{\mathbf{A}}$ and $\vec{\mathbf{B}}$ when they are drawn starting at the same point. The special name and notation are used because this pattern occurs often in physics and mathematics. See Appendix A.8 for more information on the scalar product.

If we choose the x-axis parallel to the displacement, then the component of the force in the direction of the displacement is $F_x = F \cos \theta$, so $W = F_x\,\Delta x$. Alternatively, we can identify $\Delta r \cos \theta$ in Eq. (6-1) as the component of the *displacement* in the direction of the *force* (Fig. 6.6). Therefore, if we choose the x-axis parallel to the *force*, then the component of the displacement in the direction of the force is Δx and $W = F_x\,\Delta x$, as before:

Work done by a constant force $\vec{\mathbf{F}}$ acting on an object whose displacement is $\Delta\vec{\mathbf{r}}$

$$W = F_x\,\Delta x \qquad (6\text{-}2)$$

($\vec{\mathbf{F}}$ and/or $\Delta\vec{\mathbf{r}}$ parallel to the x-axis)

Work Can Be Positive, Negative, or Zero When the angle between $\vec{\mathbf{F}}$ and $\Delta\vec{\mathbf{r}}$ is less than 90°, $\cos \theta$ in Eq. (6-1) is positive, so the work done by the force is positive ($W > 0$). If the angle between $\vec{\mathbf{F}}$ and $\Delta\vec{\mathbf{r}}$ is greater than 90°, $\cos \theta$ is negative and the work done by the force is negative ($W < 0$). Pay careful attention to the algebraic sign when calculating work. For example, the rope pulls Rosie's trunk in the direction of its displacement, so $\theta = 0$ and $\cos \theta = 1$; the rope does positive work on the trunk. At the same time, gravity pulls downward in the direction opposite to the displacement, so $\theta = 180°$ and $\cos \theta = -1$; gravity does *negative* work on the trunk.

If the force is perpendicular to the displacement, $\theta = 90°$ and $\cos 90° = 0$, so the work done is zero. For example, the normal force exerted by a stationary surface on a sliding object does no work because it is perpendicular to the displacement of the object (Fig. 6.7a). Even if the surface is curved, at any instant the normal force is perpendicular to the velocity of the object. During a short time interval, then, the normal force is perpendicular to the displacement $\Delta\vec{\mathbf{r}} = \vec{\mathbf{v}}\,\Delta t$ (Fig. 6.7b), so the normal force still does zero work.

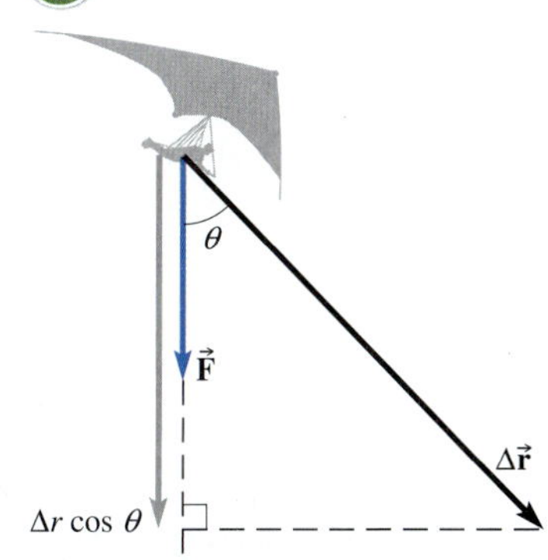

Figure 6.6 The work done by the force of gravity on the hang glider during a displacement $\Delta\vec{\mathbf{r}}$ is $F(\Delta r \cos \theta)$. F is the magnitude of the force and $\Delta r \cos \theta$ is the component of $\Delta\vec{\mathbf{r}}$ in the direction of $\vec{\mathbf{F}}$.

On the other hand, if the surface exerting the normal force is moving, then the normal force can do work. In Fig. 6.7c, the normal force exerted by the forklift on the pallet does positive work as it lifts the pallet.

No work is done by the tension in the string on a swinging pendulum bob because the tension is always perpendicular to the velocity of the bob (Fig. 6.8a). Similarly, no work is done by the Earth's gravitational force on a satellite in circular orbit (Fig. 6.8b). In a circular orbit, the gravitational force is always directed along a radius from the satellite to the center of the Earth. At every point in the orbit, the gravitational force is perpendicular to the velocity of the satellite (which is tangent to the circular orbit).

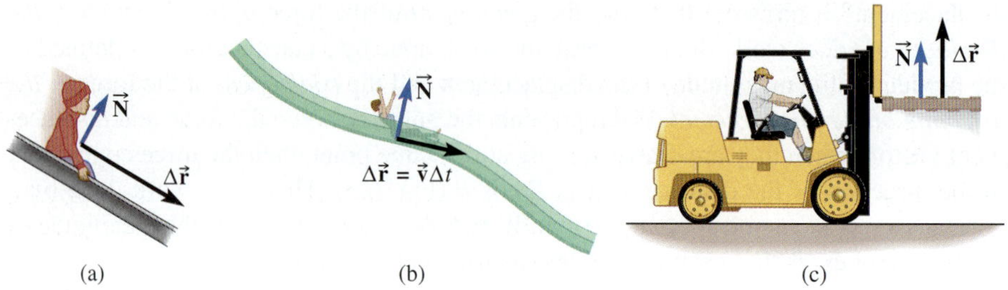

Figure 6.7 (a) The normal force does no work because it is perpendicular to the displacement. (b) Even while sliding on a curved surface, the direction of the normal force is always perpendicular to the displacement during a short Δt, so it does no work. (c) The normal force that the forklift exerts on the pallet does work; it is not perpendicular to the displacement.

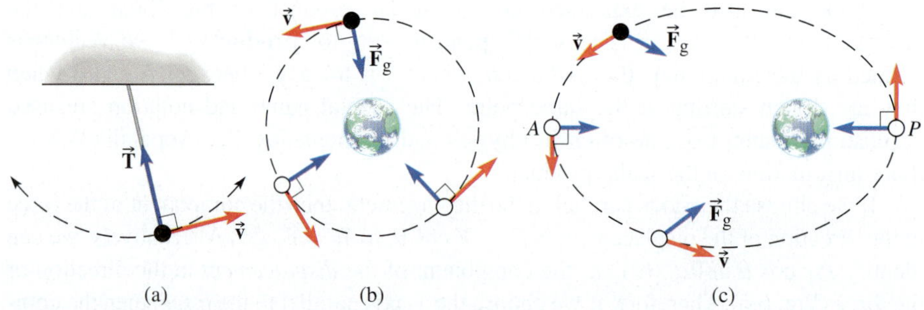

Figure 6.8 (a) The tension in the string of a pendulum is always perpendicular to the velocity of the pendulum bob, so the string does no work on the bob. (b) A satellite in a circular orbit around Earth. No matter where the satellite is in its circular orbit, it experiences a gravitational force directed toward the center of the Earth. This force is always perpendicular to the satellite's velocity; thus, gravity does no work on the satellite. (c) A satellite in an elliptical orbit around Earth. In an elliptical orbit, the gravitational force is *not* always perpendicular to the velocity. As the satellite moves counterclockwise in its orbit from point P to point A, gravity does negative work; from A to P, gravity does positive work.

Application of Work: Elliptical Orbits By contrast, gravity does work on a satellite in a noncircular orbit (Fig. 6.8c). Only at points A and P are the gravitational force and the satellite's velocity perpendicular. Wherever the angle between the gravitational force and the velocity is less than 90°, gravity is doing positive work, increasing the satellite's kinetic energy by making it move faster. Wherever the angle between the gravitational force and the velocity is greater than 90°, gravity is doing negative work, decreasing the satellite's kinetic energy by slowing it down.

CHECKPOINT 6.2

A force is applied to a moving object, but no work is done. How is that possible?

Example 6.1

Antique Chest Delivery

A valuable antique chest is to be moved into a truck. The weight of the chest is 1400 N. To get the chest from the ground onto the truck bed, which is 1.0 m higher, the movers must decide what to do. Should they lift it straight up, or should they push it up their 4.0 m long ramp? Assume they push the chest on a wheeled dolly, which in a simplified model is equivalent to sliding it up a *frictionless* ramp.

continued on next page

Example 6.1 continued

(a) Find the work done by the movers on the chest if they lift it straight up 1.0 m at constant speed.
(b) Find the work done by the movers on the chest if they slide the chest up the 4.0 m long *frictionless* ramp at constant speed by pushing parallel to the ramp.
(c) Find the work done by gravity on the chest in each case.
(d) Find the work done by the normal force of the ramp on the chest. Assume that all the forces are constant.

Strategy To calculate work, we use either Eq. (6-1) or Eq. (6-2), whichever is easier. For (a) and (b), we must calculate the force exerted by the movers. Drawing the FBD helps us calculate the forces. The ramp is a simple machine—just as for Rosie's pulleys, the ramp cannot reduce the amount of work that must be done, so we expect the work done by the movers to be the same in both cases (ignoring friction). We expect the work done by gravity to be negative in both cases, since the chest is moving up while gravity pulls down. The normal force due to the ramp is perpendicular to the displacement, so it does zero work on the chest. Since more than one force does work on the chest, we use subscripts to clarify which work is being calculated.

Given: Weight of chest $mg = 1400$ N; length of ramp $d = 4.0$ m; height of ramp $h = 1.0$ m

To find: Work done by movers on the chest W_m and work done by gravity on the chest W_g in the two cases; work done by the normal force on the chest W_N.

Solution (a) The displacement is 1.0 m straight up. The movers must exert an upward force $\vec{\mathbf{F}}_m$ equal in magnitude to the weight of the chest to move it at constant speed (Fig. 6.9). The work done to lift it 1.0 m is

$$W_m = F_m \Delta r \cos\theta = 1400 \text{ N} \times 1.0 \text{ m} \times \cos 0 = +1400 \text{ J}$$

where $\theta = 0$ because $\vec{\mathbf{F}}_m$ and $\Delta\vec{\mathbf{r}}$ are in the same direction (upward).

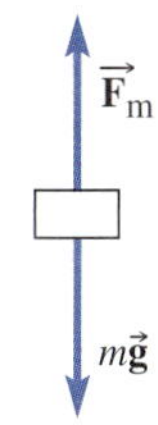

Figure 6.9
FBD for the chest as the movers lift it straight up at constant speed.

(b) Figure 6.10 shows a sketch of the situation. We take the x-axis along the inclined ramp and the y-axis perpendicular to the ramp and resolve the gravitational force into its x- and y-components (Fig. 6.11a). Figure 6.11b is the FBD for the chest. Sliding along at constant speed, the chest's acceleration is zero, so the x-components of the forces add to zero.

Figure 6.10
An antique chest is pushed up a ramp into a truck.

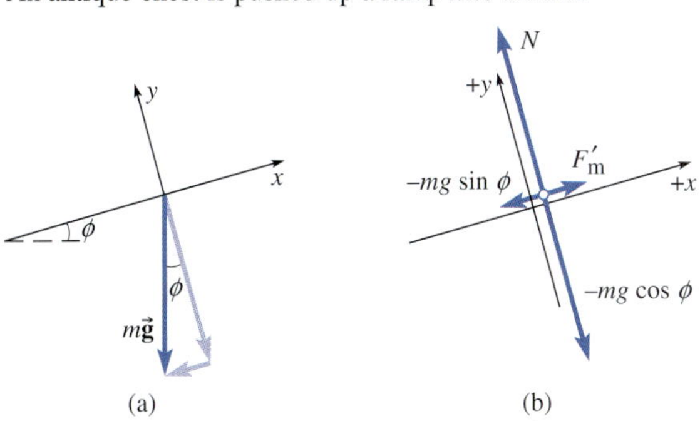

Figure 6.11
(a) Resolving $m\vec{\mathbf{g}}$ into x- and y-components; (b) FBD for the chest.

The x-component of the gravitational force acts in the $-x$-direction and the force exerted by the movers $\vec{\mathbf{F}}'_m$ acts in the $+x$-direction. [The *prime* symbol indicates that the force exerted by the movers is different from what it was in part (a).]

$$\sum F_x = F'_m - mg \sin\phi = 0$$

From the right triangle formed by the ramp, the ground, and the truck bed in Fig. 6.12:

$$\sin\phi = \frac{\text{height of truck bed}}{\text{distance along ramp}} = \frac{h}{d}$$

continued on next page

Example 6.1 continued

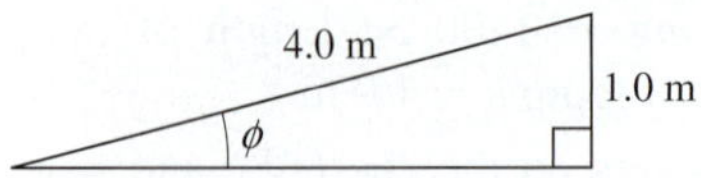

Figure 6.12 Finding the angle of the incline.

We can now solve for F'_m

$$F'_m = mg \sin \phi = \frac{mgh}{d}$$

The force and displacement are in the same direction, so $\theta = 0$:

$$W_m = F'_m d \cos 0 = \frac{mgh}{d} \times d \times 1 = mgh = +1400 \text{ J}$$

The work done by the movers is the same as in (a).

(c) In both cases, the force of gravity has magnitude mg and acts downward. Choosing the y-axis so it now points upward, $F_{gy} = -mg$. In both cases, the component of the displacement along the y-axis is $\Delta y = h = 1.0$ m. The work done by gravity is the same for the two cases. Using Eq. (6-2),

$$W_g = F_{gy} \Delta y = -mg \Delta y = -1400 \text{ N} \times 1.0 \text{ m} = -1400 \text{ J}$$

(d) The normal force of the ramp on the chest does no work because it acts in a direction perpendicular to the displacement of the chest.

$$W_N = N \Delta r \cos 90° = 0$$

Discussion Since d, the length of the ramp, cancels when multiplying the force times the distance, the work done by the movers is the same for *any* length ramp (as long as the height is the same). Using the ramp, the movers apply one quarter the force over a displacement that is four times larger. With a *real* ramp, friction acts to oppose the motion of the chest, so the movers would have to do *more* than 1400 J of work to slide the chest up the ramp. There's no getting around it; if the movers want to get that chest into the truck, they're going to have to do *at least* 1400 J of work.

Practice Problem 6.1 Bicycling Uphill

A bicyclist climbs a 2.0 km long hill that makes an angle of 2.6° with the horizontal. The total weight of the bike and the rider is 750 N. How much work is done on the bike and rider by gravity?

Total Work

When several forces act on an object, the total work is the sum of the work done by each force individually:

$$W_{total} = W_1 + W_2 + \cdots + W_N \quad (6\text{-}3)$$

Work is a scalar, not a vector. It can be positive, negative, or zero, but does not have a direction. Because we assume a rigid object with no rotational or internal motion, another way to calculate the total work is to find the work done by the *net* force as if there were a single force acting:

$$W_{total} = F_{net} \Delta r \cos \theta \quad (6\text{-}4)$$

To show that these two methods give the same result, let's choose the x-axis in the direction of the displacement. Then the work done by each individual force is the x-component of the force times Δx. From Eq. (6-3),

$$W_{total} = F_{1x} \Delta x + F_{2x} \Delta x + \cdots + F_{Nx} \Delta x$$

Factoring out the Δx from each term,

$$W_{total} = (F_{1x} + F_{2x} + \cdots + F_{Nx}) \Delta x = (\sum F_x) \Delta x$$

ΣF_x is the x-component of the net force. In Eq. (6-4), $F_{net} \cos \theta$ is the component of the net force in the direction of the displacement, which is the x-component of the net force. The two methods give the same total work.

Example 6.2

Fun on a Sled

Diane pulls a sled along a snowy path on level ground with her little brother Jasper riding on the sled (Fig. 6.13). The total mass of Jasper and the sled is 26 kg. The cord makes a 20.0° angle with the ground. As a simplified model, assume that the force of friction on the sled is determined by $\mu_k = 0.16$, even though the surfaces are not dry (some

continued on next page

Example 6.2 continued

snow melts as the runners slide along it). Find (a) the work done by Diane and (b) the work done by the ground on the sled while the sled moves 120 m along the path at a constant 3 km/h. (c) What is the total work done on the sled?

Strategy (a,b) To find the work done by a force on an object, we need to know the magnitudes and directions of the force and of the displacement of the object. The sled's acceleration is zero, so the vector sum of all the external forces (gravity, friction, rope tension, and the normal force) is zero. We draw the FBD and use Newton's second law to find the tension in the rope and the force of kinetic friction on the sled. Then we apply Eq. (6-1) or Eq. (6-2) to find the work done by each. (c) We have two methods to find the total work. We'll use Eq. (6-3) to calculate the total work and Eq. (6-4) as a check.

Solution (a) The FBD is shown in Fig. 6.14. The x- and y-axes are parallel and perpendicular to the ground, respectively. After resolving the tension into its components (Fig. 6.15), Newton's second law with zero acceleration yields

$$\sum F_x = +T\cos\theta - f_k = 0 \quad (1)$$

$$\sum F_y = +T\sin\theta - mg + N = 0 \quad (2)$$

where T is the tension and $\theta = 20.0°$. The force of kinetic friction is

$$f_k = \mu_k N$$

Substituting this into Eq. (1)

$$T\cos\theta - \mu_k N = 0 \quad (3)$$

To find the tension, we need to eliminate the unknown normal force N. Equation (2) also involves the normal force N. We multiply Eq. (2) by μ_k,

$$\mu_k T\sin\theta - \mu_k mg + \mu_k N = 0 \quad (4)$$

Adding Eqs. (3) and (4) eliminates N. Then we solve for T.

$$T\cos\theta + \mu_k T\sin\theta - \mu_k mg = 0$$

$$T = \frac{\mu_k mg}{\mu_k\sin\theta + \cos\theta}$$

$$= \frac{0.16 \times 26\text{ kg} \times 9.80\text{ m/s}^2}{0.16 \times \sin 20.0° + \cos 20.0°} = 41\text{ N}$$

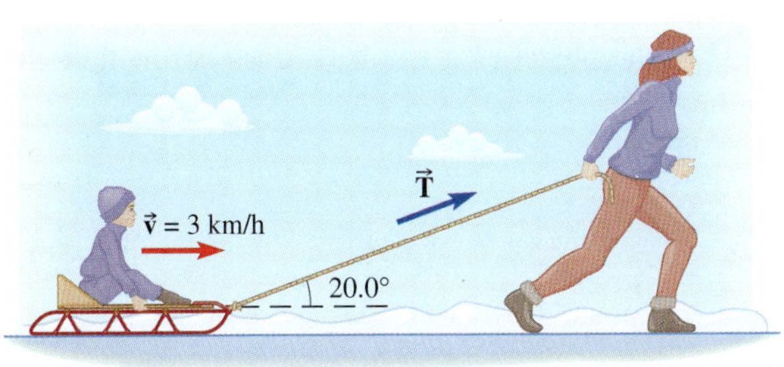

Figure 6.13
Jasper being pulled on a sled.

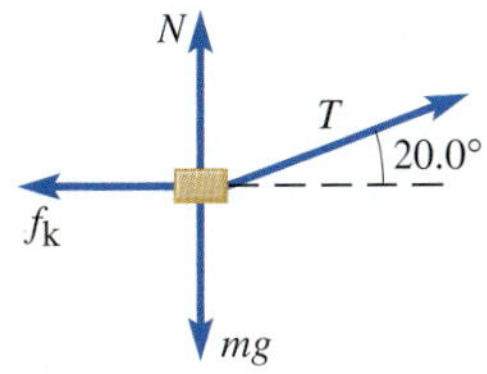

Figure 6.14
FBD for Jasper and the sled.

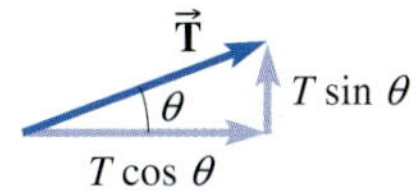

Figure 6.15
Resolving the tension into x- and y-components.

Now that we know the tension, we find the work done by Diane. The component of the tension T acting parallel to the displacement is $Tx = T\cos\theta$ and the displacement is $\Delta x = 120$ m. The work done by Diane is

$$W_T = (T\cos\theta)\Delta x$$

$$= 41\text{ N} \times \cos 20.0° \times 120\text{ m} = +4600\text{ J}$$

(b) The force on the sled due to the ground has two components: N and f_k. The normal force does no work since it is perpendicular to the displacement of the sled. Friction acts in a direction opposite to the displacement, so the angle between the force and displacement is 180°. The work done by friction is

$$W_f = f_k\,\Delta x\cos 180° = -f_k\,\Delta x$$

From Eq. (1),

$$f_k = T\cos\theta$$

Therefore, the work done by the ground—the work done by the frictional force—is

$$W_f = -f_k\,\Delta x = -(T\cos\theta)\,\Delta x$$

Except for the negative sign, W_f is the same as W_T: $W_f = -4600$ J.

(c) The tension and friction are the only forces that do work on the sled. The normal force and gravity are both perpendicular to the displacement, so they do zero work.

$$W_{total} = W_T + W_f = 4600\text{ J} + (-4600\text{ J}) = 0$$

Discussion To check (c), note that the sled travels with constant velocity, so the net force acting on it is zero. $W_{total} = F_{net}\,\Delta r\cos\theta = 0$.

The speed (3 km/h) was not used in the solution. Assuming that the frictional force on the sled is independent of speed, Diane exerts the same force to pull the sled at *any* constant speed. Then the work she does is the same for a 120 m displacement. At a higher speed, though, she has to do that amount of work in a shorter time interval.

Practice Problem 6.2 A Different Angle

Find the tension if Diane pulls at an angle $\theta = 30.0°$ instead of 20.0°, assuming the same coefficient of friction. What is the work done by Diane on the sled in this case for a 120 m displacement? Explain how the tension can be greater but the work done by Diane smaller.

Work Done by Dissipative Forces

The work done by kinetic friction was calculated in Example 6.2 according to a simplified model of friction. In this model, when friction truly does −4600 J of work on the sled, it transfers 4600 J of energy from the sled to the ground's internal energy—the ground warms up a bit. In reality, 4600 J of energy is converted into internal energy *shared* between the ground and the sled—both the ground and the sled warm up a little. So the 4600 J is not all transferred to the ground; some stays in the sled but is converted to a different form of energy.

Rather than saying friction does −4600 J of work, a more accurate statement is that friction *dissipates* 4600 J of energy. **Dissipation** is the conversion of energy from an organized form to a disorganized form such as the kinetic energy associated with the random motions of the atoms and molecules within an object, which is part of the object's internal energy. As a practical matter, we usually are not concerned with *where* the internal energy appears. When we can calculate the work done by friction using Eq. (6-1), we get the correct amount of energy dissipated; we just don't know how much of it is transferred to the stationary surface and how much remains in the sliding object. This is how we apply the term *work* to kinetic friction or to other dissipative forces such as air resistance. (In Chapters 13–15, we study internal energy in detail.)

6.3 KINETIC ENERGY

Suppose a constant net force $\vec{\mathbf{F}}_{\text{net}}$ acts on a rigid object of mass m during a displacement $\Delta\vec{\mathbf{r}}$. Choosing the x-axis in the direction of the net force, the total work done on the object is

$$W_{\text{total}} = F_{\text{net}}\,\Delta x$$

where Δx is the x-component of the displacement. Newton's second law tells us that $\vec{\mathbf{F}}_{\text{net}} = m\vec{\mathbf{a}}$, so

$$W_{\text{total}} = ma_x\,\Delta x \qquad (6\text{-}5)$$

Since the acceleration is constant, we can use any of the equations for constant acceleration from Chapter 2. From Eq. (2-13), $v_{\text{fx}}^2 - v_{\text{ix}}^2 = 2a_x\,\Delta x$ or

$$a_x\,\Delta x = \tfrac{1}{2}(v_{\text{fx}}^2 - v_{\text{ix}}^2)$$

Substituting into Eq. (6-5) yields

$$W_{\text{total}} = \tfrac{1}{2}m(v_{\text{fx}}^2 - v_{\text{ix}}^2)$$

Since the net force is in the x-direction, a_y and a_z are both zero. Only the x-component of the velocity changes; v_y and v_z are constant. As a result,

$$v_{\text{f}}^2 - v_{\text{i}}^2 = (v_{\text{fx}}^2 + \cancel{v_{\text{fy}}^2} + \cancel{v_{\text{fz}}^2}) - (v_{\text{ix}}^2 + \cancel{v_{\text{iy}}^2} + \cancel{v_{\text{iz}}^2}) = v_{\text{fx}}^2 - v_{\text{ix}}^2$$

Therefore, the total work done is

$$W_{\text{total}} = \tfrac{1}{2}m(v_{\text{f}}^2 - v_{\text{i}}^2) = \tfrac{1}{2}mv_{\text{f}}^2 - \tfrac{1}{2}mv_{\text{i}}^2$$

The total work done is equal to the change in the quantity $\frac{1}{2}mv^2$, which is called the object's **translational kinetic energy** (symbol K). (Often we just say *kinetic energy* if it is understood that we mean translational kinetic energy.) Translational kinetic energy is the energy associated with motion of the object as a whole; it does not include the energy of rotational or internal motion.

Translational kinetic energy

$$K = \tfrac{1}{2}mv^2 \qquad (6\text{-}6)$$

Work-kinetic energy theorem

$$W_{\text{total}} = \Delta K \qquad (6\text{-}7)$$

In the examples we've considered so far, the object moves at constant speed so $\Delta K = 0$; that's why the total work was zero.

Kinetic energy is a scalar quantity and is always positive if the object is moving or zero if it is at rest. Kinetic energy is never negative, although a *change* in kinetic energy can be negative. The kinetic energy of an object moving with speed v is equal to the work that must be done on the object to accelerate it to that speed starting from rest. When the total work done is positive, the object's speed increases, increasing the kinetic energy. When the total work done is negative, the object's speed decreases, decreasing the kinetic energy.

Conceptual Example 6.3

Collision Damage

Why is the damage caused by an automobile collision so much worse when the vehicles involved are moving at high speeds?

Strategy When a collision occurs, the kinetic energy of the automobiles gets converted into other forms of energy. We can use the kinetic energy as a rough measure of how much damage can be done in a collision.

Solution and Discussion Suppose we compare the kinetic energy of a car at two different speeds: 60.0 mi/h and 72.0 mi/h (which is 20.0% greater than 60.0 mi/h). If kinetic energy were proportional to speed, then a 20.0% increase in speed would mean a 20.0% increase in kinetic energy. However, since kinetic energy is proportional to the *square* of the speed, a 20.0% speed increase causes an increase in kinetic energy greater than 20.0%. Working by proportions, we can find the percent increase in kinetic energy:

$$\frac{K_2}{K_1} = \frac{\frac{1}{2}\cancel{m}v_2^2}{\frac{1}{2}\cancel{m}v_1^2} = \left(\frac{72.0 \text{ mi/h}}{60.0 \text{ mi/h}}\right)^2 = 1.44$$

Therefore, a 20.0% increase in speed causes a 44% increase in kinetic energy. What seems like a relatively modest difference in speed makes a lot of difference when a collision occurs.

Practice Problem 6.3 Two Different Cars Collide with a Stone Wall

Suppose a sports utility vehicle and a small electric car both collide with a stone wall and come to a dead stop. If the SUV mass is 2.5 times that of the small car and the speed of the SUV is 60.0 mi/h while that of the other car is 40.0 mi/h, what is the ratio of the kinetic energy changes for the two cars (SUV to small car)?

Example 6.4

Bungee Jumping

A bungee jumper makes a jump in the Gorge du Verdon in southern France. The jumping platform is 182 m above the bottom of the gorge. The jumper weighs 780 N. If the jumper falls to within 68 m of the bottom of the gorge, how much work is done by the bungee cord on the jumper during his descent? Ignore air resistance.

Strategy Ignoring air resistance, only two forces act on the jumper during the descent: gravity and the tension in the cord. Since the jumper has zero kinetic energy at both the highest and lowest points of the jump, the change in kinetic energy for the descent is zero. Therefore, the total work done by the two forces on the jumper must equal zero.

Solution Let W_g and W_c represent the work done on the jumper by gravity and by the cord. Then

$$W_{total} = W_g + W_c = \Delta K = 0$$

The work done by gravity is

$$W_g = F_y\,\Delta y = -mg\,\Delta y$$

where the weight of the jumper is $mg = 780$ N. With $y = 0$ at the bottom of the gorge, the vertical component of the displacement is

$$\Delta y = y_f - y_i = 68 \text{ m} - 182 \text{ m} = -114 \text{ m}$$

Then the work done by gravity is

$$W_g = -(780 \text{ N}) \times (-114 \text{ m}) = +89 \text{ kJ}$$

continued on next page

Example 6.4 continued

The work done by the cord is $W_c = W_{total} - W_g = -89$ kJ.

Discussion The work done by gravity is positive, since the force and the displacement are in the same direction (downward). If not for the negative work done by the cord, the jumper would have a kinetic energy of 89 kJ after falling 114 m.

The length of the bungee cord is not given, but it does not affect the answer. At first the jumper is in free fall as the cord plays out to its full length; only then does the cord begin to stretch and exert a force on the jumper, ultimately bringing him to rest again. Regardless of the length of the cord, the total work done by gravity and by the cord must be zero since the change in the jumper's kinetic energy is zero.

Practice Problem 6.4 The Bungee Jumper's Speed

Suppose that during the jumper's descent, at a height of 111 m above the bottom of the gorge, the cord has done −21.7 kJ of work on the jumper. What is the jumper's speed at that point?

CHECKPOINT 6.3A

Kinetic energy and work are related. Can kinetic energy ever be negative? Can work ever be negative?

CHECKPOINT 6.3B

Rank these objects in order of increasing kinetic energy: (a) a 5000 kg elephant walking at 3 m/s; (b) a 100 kg man skateboarding at 15 m/s; (c) a 100 000 kg whale drifting along at 0.5 m/s; (d) a 30 kg eagle diving at 50 m/s; and (e) a 50 kg cheetah running at 30 m/s.

6.4 GRAVITATIONAL POTENTIAL ENERGY (1)

Gravitational Potential Energy When Gravitational Force Is Constant

Toss a stone up with initial speed v_i. Ignoring air resistance, how high does the stone go? We can solve this problem with Newton's second law, but let's use work and energy instead. The stone's initial kinetic energy is $K_i = \frac{1}{2}mv_i^2$. For an upward displacement Δy, gravity does negative work $W_{grav} = -mg\,\Delta y$. No other forces act, so this is the total work done on the stone.

$$W_{grav} = -mg\,\Delta y = K_f - K_i$$

From the standpoint of energy conservation, where did the stone's initial kinetic energy go? If total energy cannot change, it must be "stored" somewhere. Furthermore, the stone gets its kinetic energy back as it falls from its highest point to its initial position, so the energy is stored in a way that is easily recovered as kinetic energy. Stored energy due to the interaction of an object with something else (here, Earth's gravitational field) that can easily be recovered as kinetic energy is called **potential energy** (symbol U).

The stone's loss of kinetic energy ($\Delta K = -mg\,\Delta y$) is accompanied by an increase in gravitational potential energy ($\Delta U = +mg\,\Delta y$). In general, the change in gravitational potential energy when an object moves up or down is the *negative* of the work done by gravity:

Change in gravitational potential energy

$$\Delta U_{grav} = -W_{grav} \quad (6\text{-}8)$$

If the gravitational field is uniform, the work done by gravity is

$$W_{grav} = F_y\,\Delta y = -mg\,\Delta y$$

where the y-axis points up. Therefore,

Change in gravitational potential energy

$$\Delta U_{grav} = mg\,\Delta y \tag{6-9}$$

(uniform $\vec{\mathbf{g}}$, y-axis up)

Equation (6-9) holds *even if the object does not move in a straight-line path.*

Significance of the Negative Sign in Eq. (6-8) The work done by gravity is the amount of stored (potential) energy gravity *gives to* the stone. On the way up (Fig. 6.16a), gravity takes energy away from the stone ($W_{grav} < 0$), so the amount of potential energy increases ($\Delta U > 0$). On the way down (Fig. 6.16b), gravity gives the stone energy ($W_{grav} > 0$) and the amount of stored energy decreases ($\Delta U < 0$). W_{grav} and ΔU have opposite signs because energy is being transformed from one form to another without changing the total amount.

CHECKPOINT 6.4A

A stone is tossed straight up in the air and is moving upward starting at $y = 0$. The y-axis is up. Ignore air resistance. (a) Does the gravitational potential energy increase, decrease, or stay the same? (b) What about the kinetic energy? (c) Sketch graphs of the kinetic and potential energies as functions of y, the height, on the same axes.

Other Forms of Potential Energy In addition to gravitational potential energy, other kinds of potential energy include elastic potential energy (Section 6.7) and electric potential energy (Chapter 17). Forces that have potential energies associated with them are called **conservative forces**, for reasons we explain shortly. Not every force has an associated potential energy. For instance, there is no such thing as "frictional potential energy." When kinetic friction does work, it converts energy into a disorganized form that is not easily recoverable as kinetic energy.

Mechanical Energy The total work done on an object can always be written as the sum of the work done by conservative forces (W_{cons}) plus the work done by nonconservative forces (W_{nc}). Since the total work is equal to the change in the object's kinetic energy [Eq. (6-7)],

$$W_{total} = W_{cons} + W_{nc} = \Delta K \quad \Rightarrow \quad W_{nc} = \Delta K - W_{cons} \tag{6-10}$$

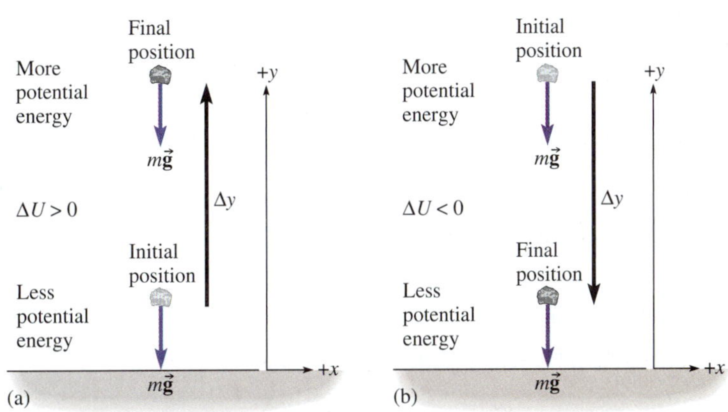

Figure 6.16 (a) As the stone moves up, the gravitational force and the stone's displacement are in opposite directions, so the work done by gravity is negative: $W_{grav} < 0$. Gravity takes kinetic energy away from the stone and stores it as gravitational potential energy, so the potential energy *increases*: $\Delta U = -W_{grav} > 0$. (b) As the stone moves down, the force and the displacement are in the same direction, so the work done by gravity is positive: $W_{grav} > 0$. Gravity gives kinetic energy to the stone, decreasing the stored potential energy, so the potential energy *decreases*: $\Delta U = -W_{grav} < 0$.

Following the same reasoning we used for gravity [see Eq. (6-8)], the change in the total potential energy is equal to the negative of the work done by the conservative forces:

$$\Delta U = -W_{\text{cons}} \tag{6-11}$$

Combining Eqs. (6-10) and (6-11) yields

$$W_{\text{nc}} = \Delta K + \Delta U = \Delta E_{\text{mech}} \tag{6-12}$$

The sum of the kinetic and potential energies ($K + U$) is called the **mechanical energy** E_{mech}. W_{nc} is equal to the change in mechanical energy. Conservative forces such as gravity do *not* change the mechanical energy; they just change one form of mechanical energy into another. Work done by conservative forces is already accounted for by the change in potential energy.

The term *conservative force* comes from a time before the general law of conservation of energy was understood and when no forms of energy other than mechanical energy were recognized. Back then, it was thought that certain forces conserved energy and others did not. Now we believe that *total* energy is *always* conserved. Nonconservative forces do not conserve *mechanical* energy, but they do conserve *total* energy.

Conservation of Mechanical Energy

When nonconservative forces do no work, mechanical energy is conserved:

$$K_{\text{i}} + U_{\text{i}} = K_{\text{f}} + U_{\text{f}} \tag{6-13}$$

CHECKPOINT 6.4B

You toss a basketball straight up and then catch it at the same height. Due to air resistance, its speed is a bit smaller just before you catch it than it was just after you tossed it. Compare the initial and final values of the ball's kinetic energy. What about the ball's gravitational potential energy? Its mechanical energy? If the mechanical energy has changed, what caused it to change?

Choosing Where the Potential Energy Is Zero

Notice that when we apply Eq. (6-12), only the *change* in potential energy enters the calculation. Therefore, we can always assign the value of the potential energy for any *one* position. Most often, we choose some convenient position and assign it to have zero potential energy. Once that choice is made, the potential energy of every other configuration is determined by Eq. (6-11).

For gravitational potential energy in a uniform gravitational field, we usually choose the potential energy to be zero at some convenient place: on the floor, on a table, or at the top of a ladder. After assigning $y = 0$ to that place, the potential energy at any other place is $U = mgy$.

Gravitational potential energy

$$U_{\text{grav}} = mgy \tag{6-14}$$

(uniform $\vec{\mathbf{g}}$, y-axis up, assign $U = 0$ to $y = 0$)

Potential energy is then positive above $y = 0$ and negative below it. There is no special significance to the sign of the potential energy. What matters is the sign of the potential energy *change*.

Example 6.5

Rock Climbing in Yosemite

A team of climbers is rappelling down steep terrain in the Yosemite valley (Fig. 6.17). Mei-Ling (mass 60.0 kg) slides down a line starting from rest 12.0 m above a horizontal shelf. If she lands on the shelf below with a speed of 2.0 m/s, calculate the energy dissipated by the kinetic frictional forces acting between her and the line. The local value of g is 9.78 N/kg. Ignore air resistance.

Strategy The forces acting on Mei-Ling are gravity and kinetic friction (Fig. 6.18). The only force whose work is not included in the change in potential energy is the work done by kinetic friction. Therefore, the change in the mechanical energy, $\Delta K + \Delta U$, is equal to the work done by friction. Since we know Mei-Ling's initial and final speeds as well as her mass, we can calculate the change in her kinetic energy. From the change in height, we can calculate the change in potential energy.

Figure 6.17

Mei-Ling rappelling downward from a position 12.0 m above a shelf.

Given: $m = 60.0$ kg;
$\Delta y = -12.0$ m;
$v_i = 0$;
$v_f = 2.0$ m/s;
$g = 9.78$ N/kg.

To find: change in mechanical energy $\Delta K + \Delta U$.

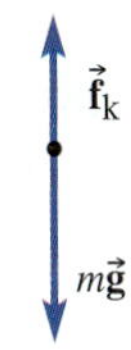

Figure 6.18

FBD for Mei-Ling.

Solution $W_{nc} = \Delta K + \Delta U$, so we need to calculate the changes in kinetic and potential energy. Mei-Ling's kinetic energy is initially zero since she starts at rest. The change in her kinetic energy is

$$\Delta K = \tfrac{1}{2}mv_f^2 - \tfrac{1}{2}mv_i^2 = \tfrac{1}{2}mv_f^2 - 0 = \tfrac{1}{2}(60.0 \text{ kg}) \times (2.0 \text{ m/s})^2$$
$$= +120 \text{ J}$$

The change in her potential energy is

$$\Delta U = mg\,\Delta y = 60.0 \text{ kg} \times 9.78 \text{ m/s}^2 \times (0 - 12.0 \text{ m}) = -7040 \text{ J}$$

The work done by friction is equal to the change in mechanical energy:

$$\Delta K + \Delta U = 120 \text{ J} + (-7040 \text{ J}) = -6920 \text{ J}$$

The amount of energy dissipated by friction (converted from mechanical energy into internal energy) is 6920 J. Fortunately, Mei-Ling is wearing gloves, so her hands don't get burned.

Discussion If the line had broken when Mei-Ling was at the top, her final kinetic energy would have been +7040 J—disastrously large since it corresponds to a final speed of

$$v = \sqrt{\frac{K}{\frac{1}{2}m}} = \sqrt{\frac{7040 \text{ J}}{30.0 \text{ kg}}} = 15.3 \text{ m/s}$$

Instead, kinetic friction reduces her final kinetic energy to a manageable +120 J (which corresponds to a final speed of 2.0 m/s). Mei-Ling can absorb this much kinetic energy safely by landing on the shelf while bending her knees.

Practice Problem 6.5 Energy Dissipated by Air Resistance

A ball thrown straight up at an initial speed of 14.0 m/s reaches a maximum height of 7.6 m. What fraction of the ball's initial kinetic energy is dissipated by air resistance as the ball moves upward?

Example 6.6

A Quick Descent

A ski trail makes a vertical descent of 78 m. A novice skier, unable to control his speed, skis down this trail and is lucky enough not to hit any trees. What is his speed at the bottom of the trail, ignoring friction and air resistance?

Strategy When nonconservative forces do no work, mechanical energy does not change. A skilled skier can control his speed by, in effect, controlling how much work the frictional force does on the skis. Here we assume *no* friction or air resistance. Then the only forces acting on the skier are the normal force and gravity (Fig. 6.19). The normal force does no work, since it is always perpendicular to the skier's velocity, so $W_{nc} = 0$.

Solution Because $W_{nc} = 0$, the mechanical energy does not change:

$$K_i + U_i = K_f + U_f$$

If we choose the y-axis up and $y = 0$ at the bottom of the hill, $y_i = 78$ m and $y_f = 0$. Then

$$U_i = mgy_i \quad \text{and} \quad U_f = 0$$

If the skier starts with zero kinetic energy, then $K_i = 0$ and $K_f = \frac{1}{2}mv_f^2$. Setting the mechanical energies equal,

$$0 + mgy_i = \tfrac{1}{2}mv_f^2 + 0$$

Solving for the final speed v_f,

$$v_f = \sqrt{2gy_i} = \sqrt{2 \times 9.80 \text{ m/s}^2 \times 78 \text{ m}} = 39 \text{ m/s}$$

Discussion Notice that the solution did not depend on the detailed shape of the path. If the slope were constant (Fig. 6.20), we could use Newton's second law to find the skier's acceleration and then the change in velocity:

$$\sum F_x = mg \sin\theta = ma_x \quad \Rightarrow \quad a_x = g \sin\theta$$

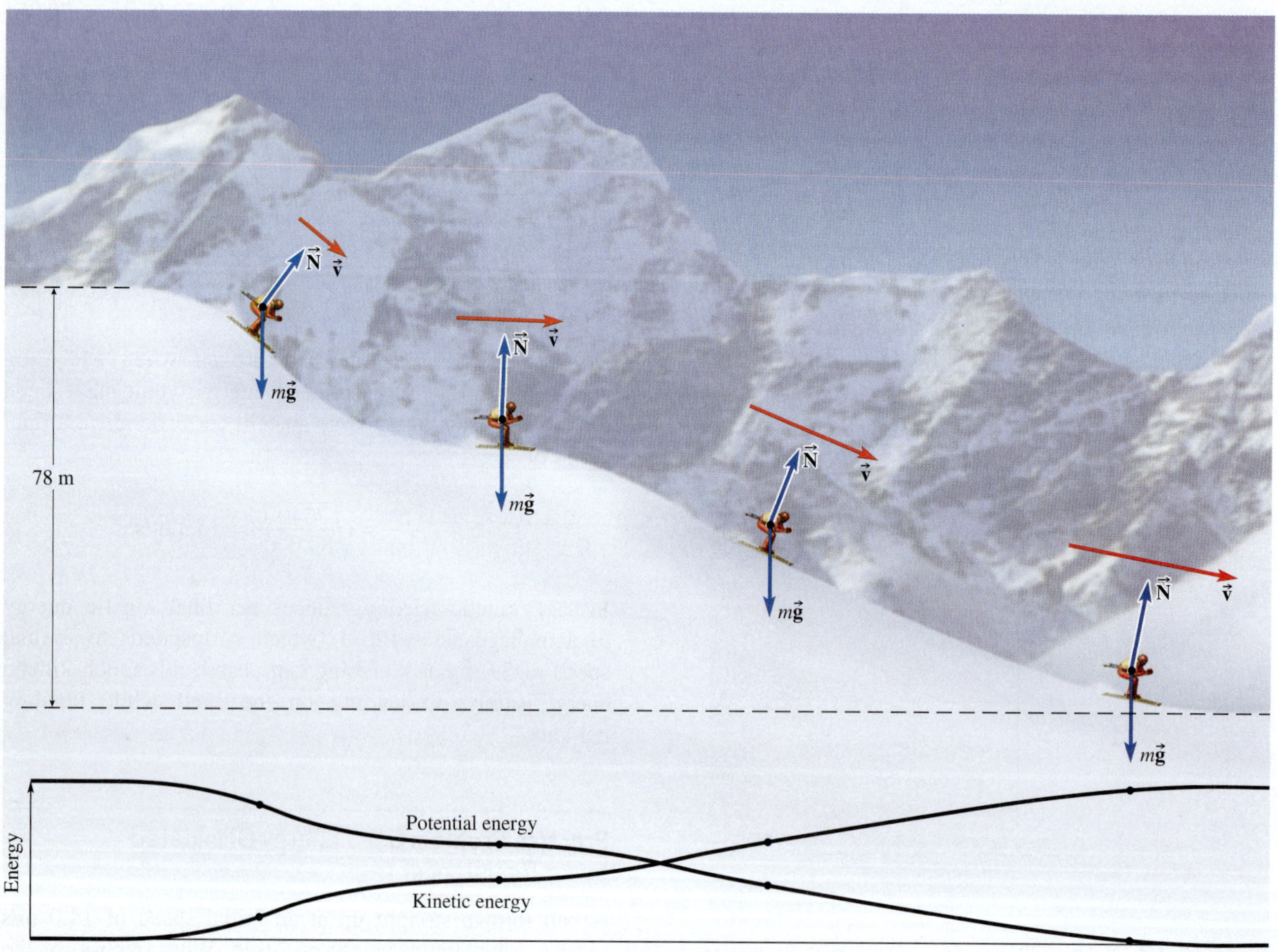

Figure 6.19
The final speed of the skier depends only on the initial and final altitudes if no friction acts.

continued on next page

Example 6.6 continued

From Eq. (2-13),

$$\Delta x = \frac{v_{fx}^2 - v_{ix}^2}{2a_x} = \frac{v_{fx}^2}{2g \sin \theta} = \frac{h}{\sin \theta} \quad \Rightarrow \quad v_{fx} = \sqrt{2gh}$$

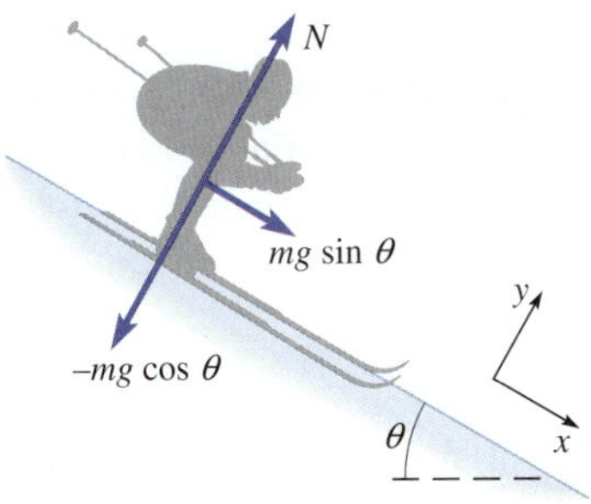

Figure 6.20
FBD for the skier on a constant slope.

where $h = 78$ m.

This method shows that the final speed does not depend on the angle of the slope, but the energy method shows that the final speed is the same for *any* shape path, not just for constant slopes. On the other hand, the *time* that it takes the skier to reach the bottom *does* depend on the length and contour of the trail.

A final speed of 39 m/s (87 mi/h) is dangerously fast. In reality, friction and air resistance do negative work on the skier, so the final speed would be smaller.

Practice Problem 6.6 Speeding Roller Coaster

A roller coaster is hauled to the top of the first hill of the ride by a motorized chain drive. After that, the train of cars is released and no more energy is supplied by an external motor. The cars are moving at 4.0 m/s at the top of the first hill, 35.0 m above the ground. How fast are they moving at the top of the second hill, 22.0 m above the ground? Ignore friction and air resistance.

Recognizing a Conservative Force

In Example 6.6, the final speed doesn't depend on the shape of the trail: it could have been a steep descent, or a long gradual one, or have a complicated profile with varying slope. It could even be a vertical descent—the final speed is the same for free fall off a 78 m high building. The change in gravitational potential energy depends on the initial and final positions but *not* on the path taken. That's why we can write $\Delta U = mg\,\Delta y$ [Eq. (6-9)].

Any time the work done by a force is *independent of path*—that is, the work depends only on the initial and final positions—the force is conservative. Energy stored as potential energy by a conservative force during a displacement from point A to point B can be recovered as kinetic energy. We can simply reverse displacement to get all of the energy back: $\Delta U_{B \to A} = -\Delta U_{A \to B}$.

The work done by friction, air resistance, and other contact forces *does* depend on path, so these forces cannot have potential energies associated with them. We cannot use friction to store energy in a form that is completely recoverable as kinetic energy.

6.5 GRAVITATIONAL POTENTIAL ENERGY (2)

The expressions for gravitational potential energy developed in Section 6.4 apply when the gravitational force is *constant* (or nearly constant). If the gravitational force is not constant, such as when a satellite is placed into orbit around the Earth, Eqs. (6-9) and (6-14) cannot be used. Instead, we need to use an expression for gravitational potential energy that corresponds to Newton's law of universal gravitation. Recall that the magnitude of the gravitational force that one object exerts on another is

$$F = \frac{Gm_1m_2}{r^2} \qquad (4\text{-}8)$$

where r is the distance between the centers of the objects. The corresponding expression for gravitational potential energy in terms of the distance between two objects is

Gravitational potential energy

$$U = -\frac{Gm_1m_2}{r} \qquad (6\text{-}15)$$

(assign $U = 0$ when $r = \infty$)

A graph showing the gravitational potential energy as a function of r is shown in Fig. 6.21. Note that we have assigned the potential energy to be zero at infinite

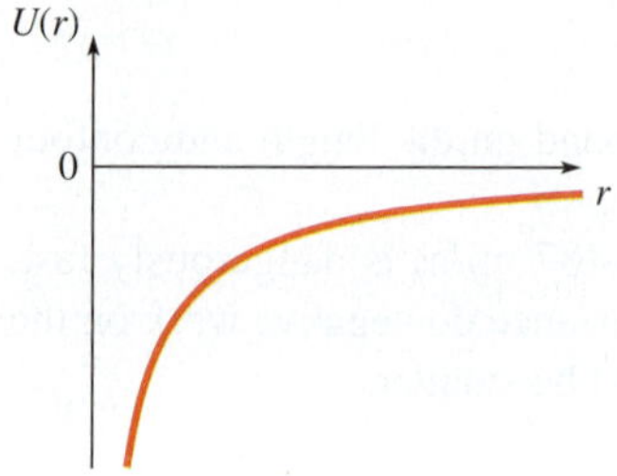

Figure 6.21 Gravitational potential energy as a function of r, the distance between the centers of two objects. The potential energy increases as the distance increases.

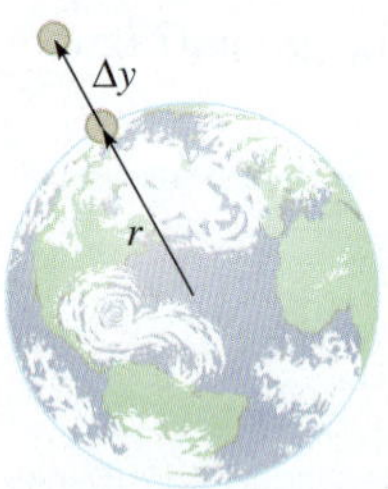

Figure 6.22 An object at a distance r from Earth's center moves up a small distance Δy (greatly exaggerated in the figure).

separation ($U = 0$ when $r = \infty$). Why this choice? Simply put, any other choice would mean adding a constant term to the expression for U. This constant term would *always subtract out* of our equations, which involve only *changes* in potential energy.

This choice ($U = 0$ when $r = \infty$) means that the gravitational potential energy is *negative* for any finite value of r, because potential energy decreases as the objects get closer together and increases as they get farther apart.

Does Eq. (6-15) Contradict Eq. (6-9)? Calculus is used to derive Eq. (6-15) but we can *verify* that it is consistent with Eq. (6-9) without using calculus. For a *very small* displacement from r_i to $r_f = r_i + \Delta y$ (Fig. 6.22), the potential energy change given by Eq. (6-15) must reduce to the constant-force case:

$$\Delta U = U_f - U_i = \left(-\frac{GM_E m}{r_i + \Delta y}\right) - \left(-\frac{GM_E m}{r_i}\right)$$

Rearranging and factoring out the common factors $GM_E m$ and then rewriting with a common denominator,

$$\Delta U = GM_E m\left(\frac{1}{r_i} - \frac{1}{r_i + \Delta y}\right) = GM_E m\frac{\cancel{r_i} + \Delta y - \cancel{r_i}}{r_i\,(r_i + \Delta y)}$$

For values of Δy that are small compared with r_i, $r_i + \Delta y \approx r_i$. Making that approximation in the denominator, we obtain

$$\Delta U = m\left(\frac{GM_E}{r_i^2}\right)\Delta y \quad (\Delta y \ll r_i) \qquad \text{(6-16)}$$

The quantity in the parentheses in Eq. (6-16) is the gravitational field strength g, the gravitational force on the object divided by its mass m. Then, $\Delta U = mg\,\Delta y$, in agreement with Eq. (6-9).

CHECKPOINT 6.5

As Mercury travels in its elliptical orbit about the Sun, how does its mechanical energy at its nearest point (*perihelion*) to the Sun compare with that at its farthest point (*aphelion*) from the Sun? How does its potential energy compare at the same two points?

Example 6.7

Orbital Speed of Mercury

The orbit of the planet Mercury around the Sun is an ellipse. At its perihelion ($r_p = 4.60 \times 10^7$ km), its orbital speed is 59 km/s. What is its orbital speed at aphelion ($r_a = 6.98 \times 10^7$ km)?

Strategy Ignoring the small gravitational forces exerted by other planets, the only force acting on Mercury is the gravitational force due to the Sun. Gravity is a conservative force, so the mechanical energy is constant. Figure 6.23 is a sketch of the orbit. At aphelion, Mercury is farther from the Sun than at perihelion, so the potential energy is greater. Then the kinetic energy must be smaller, so the answer must be less than 59 km/s.

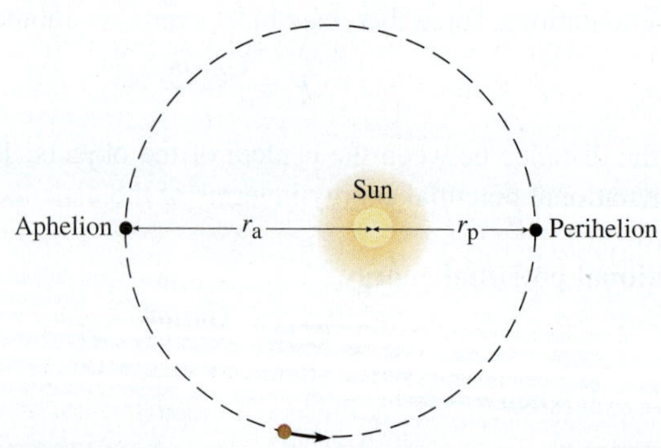

Figure 6.23
Sketch of Mercury's orbit.

continued on next page

Example 6.7 continued

Given: $v_p = 5.9 \times 10^4$ m/s, $r_p = 4.60 \times 10^{10}$ m, $r_a = 6.98 \times 10^{10}$ m.

To find: v_a.

Solution Mechanical energy is constant:

$$K_p + U_p = K_a + U_a$$

The kinetic energy of Mercury at perihelion is $K_p = \frac{1}{2}mv_p^2$, where m is the mass of Mercury; the kinetic energy at aphelion is $K_a = \frac{1}{2}mv_a^2$. The potential energies at perihelion and at aphelion are

$$U_p = -\frac{GM_S m}{r_p} \quad \text{and} \quad U_a = -\frac{GM_S m}{r_a}$$

respectively, where $M_S = 1.99 \times 10^{30}$ kg is the mass of the Sun. From conservation of energy:

$$\frac{1}{2}mv_p^2 + \left(-\frac{GM_S m}{r_p}\right) = \frac{1}{2}mv_a^2 + \left(-\frac{GM_S m}{r_a}\right)$$

The mass of Mercury cancels out. Solving for v_a,

$$\frac{1}{2}v_a^2 = \frac{1}{2}v_p^2 + \left(-\frac{GM_S}{r_p}\right) - \left(-\frac{GM_S}{r_a}\right)$$

$$v_a = \sqrt{v_p^2 + 2GM_S\left(\frac{1}{r_a} - \frac{1}{r_p}\right)}$$

Substituting numerical values yields $v_a = 39$ km/s.

Discussion The speed at aphelion is less than the speed at perihelion, as expected.

Practice Problem 6.7 Speed at a Different Distance

What is Mercury's orbital speed when its distance from the Sun is 5.80×10^7 km?

Example 6.8

Escape Speed

(a) Ignoring air resistance, find the minimum initial speed a projectile must have at Earth's surface if the projectile is to escape Earth's gravitational pull. (b) Sketch a graph of the kinetic and potential energies as functions of r, the distance from Earth's center.

Strategy What does "escape Earth's gravitational pull" mean? The gravitational force on the projectile due to Earth *approaches* zero at large distances, but never *reaches* zero. We are looking for the initial speed so that, even though Earth's gravity keeps pulling the projectile back, the projectile can keep moving away from Earth. The gravitational force is not constant, and the trajectory of the projectile may be complicated, so using $\Sigma\vec{\mathbf{F}} = m\vec{\mathbf{a}}$ is impractical. We try an energy approach.

The only force acting on the projectile is gravity, so the mechanical energy is constant. To escape, the projectile must have enough initial kinetic energy so that it can reach an unlimited distance from Earth.

Solution (a) The mechanical energy is constant:

$$K_i + U_i = K_f + U_f$$

Initially the projectile is at a distance R, Earth's radius, from Earth's center and is moving at initial speed v_i. At some later time, the projectile has speed v_f at distance r_f from Earth. Then

$$\frac{1}{2}mv_i^2 + \left(-\frac{GMm}{R}\right) = K_f + U_f$$

where m is the projectile's mass and M is Earth's mass. To escape, the projectile must be able to reach any value of r_f, no matter how large. As r_f gets larger and larger, the potential energy approaches its maximum value, which is zero. (Mathematically, as $r_f \to \infty$, $U_f \to 0$.) The *minimum* value of v_i gives the projectile *just enough* energy. So we assume that the projectile can reach its maximum potential energy without any kinetic energy left over ($K_f = 0$):

$$\frac{1}{2}mv_i^2 + \left(-\frac{GMm}{R}\right) = 0 + 0$$

Solving for v_i,

$$\frac{1}{2}mv_i^2 = \frac{GMm}{R} \quad \Rightarrow \quad v_i = \sqrt{\frac{2GM}{R}} = 11.2 \text{ km/s}$$

(b) As the projectile moves away from Earth, the potential energy increases and the kinetic energy decreases. Their sum $K + U$ (the mechanical energy) remains constant because no nonconservative forces are acting. The potential energy approaches zero as the distance increases, and because the projectile is launched at escape speed, so does the kinetic energy. The graph of $U(r)$ (Fig. 6.24) is the same as Fig. 6.21. The graph of $K(r)$ looks like a "mirror image" because $K + U = 0$.

Discussion The speed found in part (a) is called the **escape speed** of Earth. Note that the escape speed is independent of the mass of the projectile because both the kinetic energy and the potential energy are proportional to the projectile's mass.

continued on next page

Example 6.8 continued

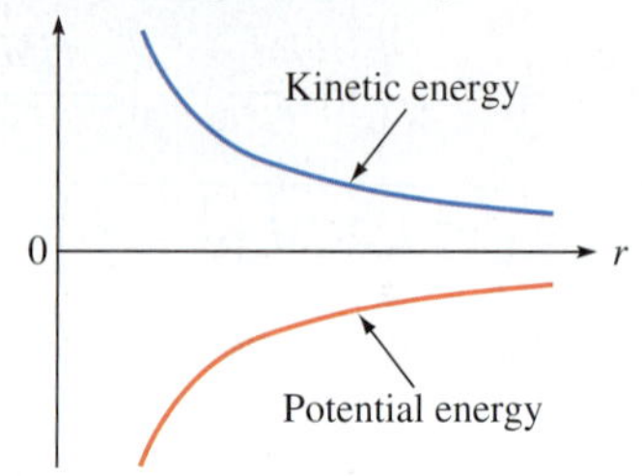

Figure 6.24

The kinetic and potential energies of a projectile launched from Earth's surface at escape speed, ignoring air resistance, as functions of r, the distance from Earth's center. The graphs start at $r = R$, where R is Earth's radius. As the projectile moves away from Earth, the potential energy *increases* and the kinetic energy *decreases* but their sum (the mechanical energy) remains constant.

The concept of escape speed helps explain why there is little hydrogen gas (H_2) or helium gas (He) in Earth's atmosphere. We will see in Chapter 13 that the molecules in a gas have an average kinetic energy determined by the temperature of the gas. In a mixture of gases, the molecules with the smallest mass have the highest average speeds. The average speeds of hydrogen and helium in our atmosphere are large enough that they can escape the atmosphere. A negligible fraction of the nitrogen, oxygen, or water molecules have speeds greater than the escape speed, so they persist in the atmosphere.

Practice Problem 6.8 Protons Streaming Away from the Sun

Particles such as protons and electrons are continually streaming away from the Sun in all directions. They carry off some of the energy released in the thermonuclear reactions occurring in the Sun. How fast must a proton be moving at a distance of 7.00×10^9 m from the center of the Sun for it to escape the Sun's gravitational pull and leave the solar system?

Figure 6.25 Application of work done by a variable force: drawing a compound bow.

CONNECTION:

See Sections 2.2 and 2.3 to review how we found that the area under a graph of $v_x(t)$ is Δx and that the area under a graph of $a_x(t)$ is Δv_x.

6.6 WORK DONE BY VARIABLE FORCES

So far we have considered only constant forces when calculating work. The advantage of using energy methods really shines in problems dealing with variable forces, where it's difficult to use Newton's second law. How can we calculate the work done by a variable force? Consider an archer drawing back a compound bow (Fig. 6.25). The compound bow is designed to make it easier to draw the string back and hold it back because, at a certain point, the force required to draw the string farther stops increasing. A convenient way to describe how the force varies with string position is to plot a graph. Figure 6.26 shows the force that must be applied to hold the string back as a function of distance. How can we calculate the work done by the archer as he draws the string back 40 cm?

We've asked analogous questions in previous chapters. Recall how we find the displacement Δx when the velocity v_x is not constant (Section 2.2). We divide the time interval into a series of *short* time intervals and sum up the displacements that occur during each one.

To approximate the work done by a variable force F_x, we divide the overall displacement into a series of small displacements Δx. During each small displacement, the work done is

$$\Delta W = F_x \, \Delta x \qquad (6\text{-}17)$$

On a graph of $F_x(x)$, each ΔW is the area of a rectangle of height F_x and width Δx (Fig. 6.27). The total work done is the sum of the areas of these rectangles. This approximation gets better as we make the rectangles thinner and thinner, so *the total work done is the area under the graph of* $F_x(x)$ from x_i to x_f.

In Fig. 6.26, the "area" of each rectangle represents (0.050 m × 20.0 N) = 1.0 J of work. There are approximately 36 rectangles under the graph between $x = 0$ and $x = 40$ cm, so the work done by the archer is +36 J.

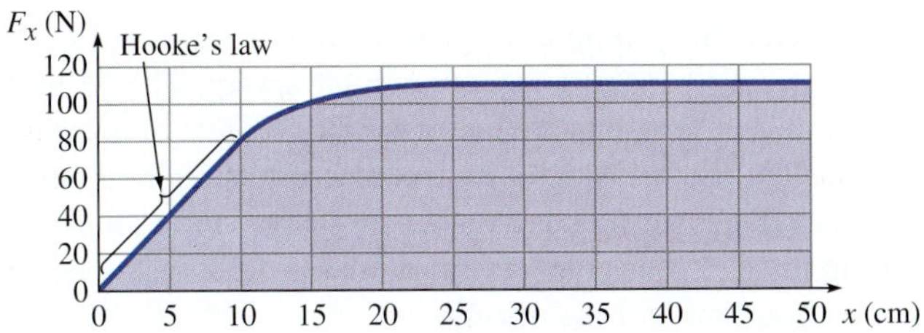

Figure 6.26 The force to draw back the compound bow depends on how far it is drawn. In this graph, the "area" represented by each rectangle is 0.050 m × 20.0 N = 1.0 J.

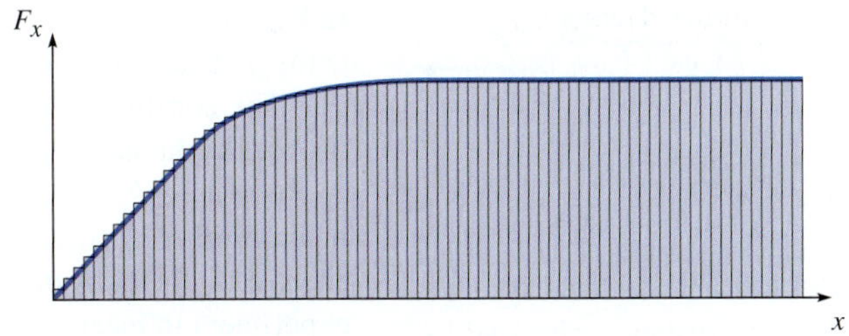

Figure 6.27 Each rectangle's area approximates the work done during a small displacement. The total area of the rectangles approximates the total work done.

Example 6.9

Archery Practice

To draw back a *simple* bow, the force the archer exerts on the string continues to increase as the displacement of the string increases and the bow bends slightly. The force-versus-position graph of Fig. 6.28 describes such a bow. Calculate the work done by the archer on the string as he draws the string back 40.0 cm.

Strategy The work done by the archer is the area under the force-versus-position graph. This time, instead of counting rectangles, we can calculate the triangular area formed by the force-versus-position graph.

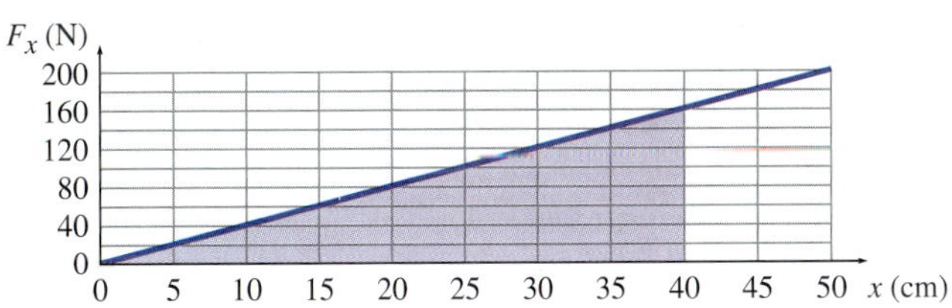

Figure 6.28
A simple bow requires a force proportional to the displacement of the string.

Solution We want to find the work done by the archer to draw the string back 40.0 cm, so the base of the triangle is 40.0 cm. The altitude of the triangle is the force at 40.0 cm: 160 N. The area of a triangle is $\frac{1}{2}$(base × altitude), so

$$W = \tfrac{1}{2}(0.400 \text{ m} \times 160 \text{ N}) = +32 \text{ J}$$

Discussion To check, we can count the number of rectangles (including the half rectangles) that lie under the graph. There are 32 rectangles and each represents 20 N × 0.05 m = 1 J of work, so the answer is correct.

By doing 32 J of work on the bowstring, the archer stores this much energy in the bow. When the arrow is released, the bowstring does 32 J of work on the arrow, giving the arrow a kinetic energy of 32 J.

Practice Problem 6.9 A Gentle Pull

How much work would you do to draw the string of the *compound* bow (see Fig. 6.26) back 10.0 cm?

Hooke's Law and Ideal Springs

In Example 6.9, the displacement of the bowstring is proportional to the force exerted by the archer. Robert Hooke (1635–1703) observed that, for many objects, the deformation—change in size or shape—of the object is proportional to the magnitude of the force that causes the deformation. This observation, called **Hooke's law**, is an approximation and is valid only within limits. For example, the compound bow of Fig. 6.26 is described by Hooke's law for an applied force less than 80 N.

Many springs are described by Hooke's law as long as they are not stretched or compressed too far. That is, the extension or compression—the increase or decrease in length from the relaxed length—is proportional to the force applied to the ends of the spring. When we refer to an **ideal spring**, we mean a spring that is described by Hooke's law and is also massless.

$$F = k\,\Delta L \qquad (6\text{-}18)$$

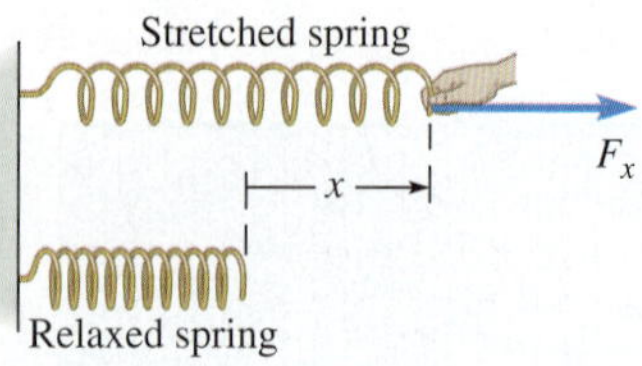

Figure 6.29 An ideal spring is stretched a distance x beyond its relaxed length.

In Eq. (6-18), F is the *magnitude* of the force exerted *on each end* of the spring and ΔL is the distance that the spring is stretched or compressed *from its relaxed length.*

The constant k is called the **spring constant** for a particular spring. The SI unit of force is the newton and the SI unit of length is the meter, so the SI units of a spring constant are N/m. The spring constant is a measure of how hard it is to stretch or compress a spring. A stiffer spring has a larger spring constant because larger forces must be exerted on the ends of the spring to stretch or compress it. Example 1.11 describes an experiment to measure the spring constant of a real spring.

In many situations, we are more interested in the forces exerted *by* the spring than in the forces exerted *on* it. From Newton's third law, the forces exerted *by* the spring on whatever is attached to its ends are equal in magnitude and opposite in direction to the forces exerted *by* those objects *on* the ends of the spring. Suppose that an ideal spring is aligned with the x-axis. One end is fixed in place and the other end can move along the x-axis (Fig. 6.29). For convenience, choose the origin so the moveable end is at $x = 0$ when the spring is relaxed. Then the force exerted by the moveable end of the spring on whatever is attached to it is

Force exerted *by* an ideal spring (Hooke's law)

$$F_x = -kx \qquad (6\text{-}19)$$

(F_x is the force exerted *by* the moveable end when its position is x; the spring is relaxed at $x = 0$.)

The negative sign in Eq. (6-19) indicates the direction of the force. The moveable end of the spring always pushes or pulls toward its relaxed position. If it is displaced in the $+x$-direction, the force it exerts is in the $-x$-direction (back toward $x = 0$). If it is displaced in the $-x$-direction, the force it exerts is in the $+x$-direction (again, back toward $x = 0$).

Example 6.10

Getting Down to Nuts and Bolts

In many hardware stores, bulk nuts and bolts are sold by weight. A spring scale in the store stretches 4.8 cm when 24.0 N of bolts are weighed. On the scale, what is the distance in centimeters between calibration marks that are marked in increments of 1 N? Assume an ideal spring.

Strategy The bolts are in equilibrium, so the spring scale is pulling upward on them with a force of 24.0 N (Fig. 6.30). Using Hooke's law and the data given, we can find the spring constant k. Then we can use Hooke's law again to find out how much the spring stretches when the applied force is increased by 1 N.

Solution Let the x-axis point up. When the pan of the scale is at $x = -4.8$ cm, it exerts a force $Fx = +24.0$ N on the bolts. From Hooke's law, $Fx = -kx$ and the spring constant is

$$k = -\frac{F_x}{x} = \frac{-24.0\text{ N}}{-4.8\text{ cm}} = 5.0\text{ N/cm}$$

Now let $F_x = 1.00$ N and solve for x:

$$x = -\frac{F_x}{k} = -\frac{1.00\text{ N}}{5.0\text{ N/cm}} = -0.20\text{ cm}$$

Since the relation between F and x is linear, the spring stretches an additional 0.20 cm for each additional newton of force. Therefore, the 1 N marks should be 0.20 cm apart.

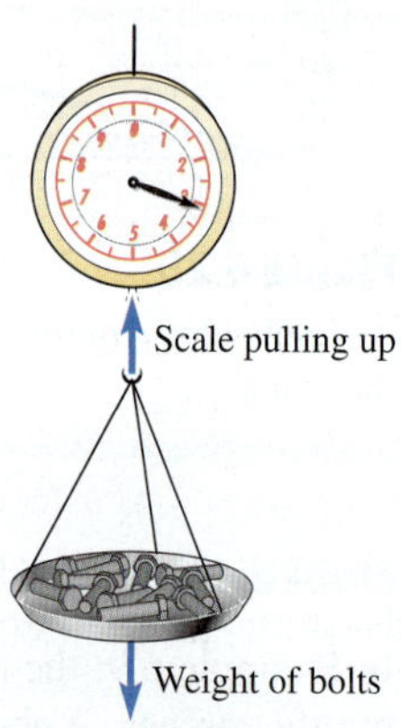

Figure 6.30 FBD for the pan of the scale.

Discussion A variation on the solution is to look back at the question and notice that we are asked how many centimeters the spring stretches for each newton of force, which is the *reciprocal* of the spring constant. The reciprocal of the spring constant is

$$\frac{1}{k} = -\frac{x}{F} = -\frac{-4.8\text{ cm}}{24.0\text{ N}} = 0.20\text{ cm/N}$$

The answer is reasonable: since it takes 5 N to make the spring stretch 1 cm, 1 N makes the spring stretch $\frac{1}{5}$ cm.

Practice Problem 6.10 Stretching a Spring

16.0 N of nuts are placed in the pan of the scale of Example 6.10. How far does the spring stretch?

Work Done by an Ideal Spring

To find the work done by an ideal spring, first we draw the $F_x(x)$ graph (Fig. 6.31). The unstretched position of the moveable end is $x = 0$. The work done by the spring as its moveable end moves from equilibrium ($x_i = 0$) to the final position x_f is the area of the shaded right triangle whose base is x and altitude is $-kx$:

$$W = \tfrac{1}{2}(\text{base} \times \text{altitude}) = -\tfrac{1}{2}kx^2 \qquad (6\text{-}20)$$

The area is negative because the graph is underneath the x-axis. Think of $-\frac{1}{2}kx^2$ as the average force $(-\frac{1}{2}kx)$ times the displacement (x).

More generally, if the moveable end starts at position x_i, not necessarily at the equilibrium point, the work done by the spring is

$$W_{\text{spring}} = (-\tfrac{1}{2}kx_f^2) - (-\tfrac{1}{2}kx_i^2) = -\tfrac{1}{2}kx_f^2 + \tfrac{1}{2}kx_i^2 \qquad (6\text{-}21)$$

Imagine the spring starting at equilibrium and ultimately ending up at a displacement x_f after passing through x_i. The total work done by the spring is $-\frac{1}{2}kx_f^2$; then we subtract the work that was done to get the spring to position x_i from equilibrium $(-\frac{1}{2}kx_i^2)$ to get the work done from x_i to x_f. Equation (6.21) is valid regardless of whether the spring is stretched $(x > 0)$ or compressed $(x < 0)$.

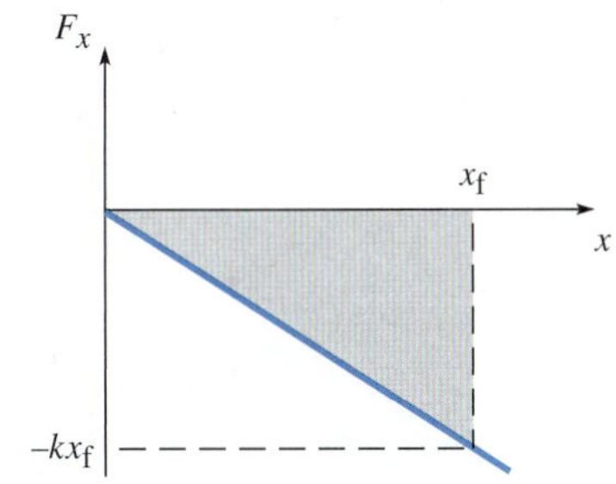

Figure 6.31 A spring is stretched to a final position x_f. The work done by the spring is the (negative) area between the $F_x(x)$ graph and the x-axis.

6.7 ELASTIC POTENTIAL ENERGY

The work done by an ideal spring [Eq. (6-21)] depends on the initial and final positions of the moveable end, but *not* on the path that was taken. Therefore, the force exerted by an ideal spring is *conservative*, and we can associate a potential energy with it. The kind of potential energy stored in a spring is called **elastic potential energy**.

Just as for gravity [see Eqs. (6-8) and (6-11)], the change in elastic potential energy is the *negative* of the work done by the spring:

$$\Delta U_{\text{elastic}} = -W_{\text{spring}} \qquad (6\text{-}22)$$

For example, if you increase the elastic energy stored in a spring by compressing it, the spring does *negative* work because the force its end exerts on your hand is in the direction opposite to its displacement. This stored elastic energy can be recovered as kinetic energy by, say, using the spring to shoot a stone. As the spring expands back to its original length, it does positive work on the stone to increase the stone's kinetic energy and the stored elastic energy decreases.

CONNECTION:

The change in potential energy is always equal to the negative of the work done by the associated force. See Eq. (6-11).

From Eqs. (6-21) and (6-22),

$$\Delta U_{\text{elastic}} = \tfrac{1}{2}kx_f^2 - \tfrac{1}{2}kx_i^2 \qquad (6\text{-}23)$$

Remember that only changes in potential energy enter our calculations, so we can assign $U = 0$ to any convenient position. The most convenient choice is to assign $U = 0$ when the spring is relaxed $(x = 0)$:

Elastic potential energy stored in an ideal spring

$$U_{\text{elastic}} = \tfrac{1}{2}kx^2 \qquad (6\text{-}24)$$

$U = 0$ when $x = 0$ (relaxed spring)

Conservation of Energy with More than One Form of Potential Energy When applying conservation of energy using $W_{nc} = \Delta K + \Delta U$ [Eq. (6-12)], ΔU must include the change in all forms of potential energy. For now, with two forms of potential energy,

$$\Delta U = \Delta U_{\text{grav}} + \Delta U_{\text{elastic}} \qquad (6\text{-}25)$$

W_{nc} is the work done by all forces *other than* those included in the potential energy. When $W_{nc} = 0$, the mechanical energy $K + U$ is constant.

CHECKPOINT 6.7

If a spring is compressed horizontally on a table and then released so it expands to its original relaxed position, where does the spring have the greatest elastic potential energy?

Example 6.11

The Dart Gun

In a dart gun (Fig. 6.32), a spring with $k = 400.0$ N/m is compressed 8.0 cm when the dart (mass $m = 20.0$ g) is loaded (Fig. 6.32a). What is the muzzle speed of the dart when the spring is released (Fig. 6.32b)? Ignore friction.

Strategy The elastic energy initially stored in the spring is converted into the kinetic energy of the dart as the spring expands. There is no change in gravitational potential energy since the motion of the dart is horizontal. The vertical normal forces do no work because they are perpendicular to the displacement of the dart. The spring pushes the dart to the right until it reaches its relaxed length. Assuming the spring can't pull the dart to the left (as it would if they stick together), the dart loses contact with the spring when the spring is at its relaxed length. We choose the origin at the relaxed position of the spring; therefore, $x_f = 0$. Using the x-axis in Fig. 6.32, $x_i = -8.0$ cm. The dart starts from rest, so $v_i = 0$. To find: v_f.

Solution Since we ignore friction, no work is done by nonconservative forces. Therefore, the mechanical energy is constant:

$$K_i + U_i = K_f + U_f$$

We can ignore the gravitational potential energy because it does not change. Using Eq. (6-24) for the elastic potential energy in the spring,

$$\tfrac{1}{2}mv_i^2 + \tfrac{1}{2}kx_i^2 = \tfrac{1}{2}mv_f^2 + \tfrac{1}{2}kx_f^2$$

After setting $x_f = 0$ and $v_i = 0$,

$$0 + \tfrac{1}{2}kx_i^2 = \tfrac{1}{2}mv_f^2 + 0$$

Solving for v_f,

$$v_f = \sqrt{\frac{k}{m}}\, x_i = \sqrt{\frac{400.0\ \text{N/m}}{0.0200\ \text{kg}}} \times 0.080\ \text{m} = 11\ \text{m/s}$$

Discussion Checking the units,

$$\sqrt{\frac{\text{N/m}}{\text{kg}}} \times \text{m} = \sqrt{\frac{(\cancel{\text{kg}}\cdot\cancel{\text{m}}/\text{s}^2)/\cancel{\text{m}}}{\cancel{\text{kg}}}} \times \text{m} = \frac{\text{m}}{\text{s}}$$

Notice that the muzzle speed is proportional to the distance the spring is compressed when the gun is cocked. If the spring is compressed halfway, it stores only *one quarter* as much elastic energy. The dart then acquires one quarter the kinetic energy, which means its speed is half as much. A more massive dart fired from the same gun would have a smaller muzzle speed, but the *same* kinetic energy.

Practice Problem 6.11 A Misfire

The same dart gun is cocked by compressing the spring the same distance (8.0 cm). This time the spring gets caught inside the gun, stopping at the point where it is still compressed by 4.0 cm. The dart is not caught inside the gun, but is released. Find the muzzle speed of the dart. [*Hint*: What is x_f in this case?]

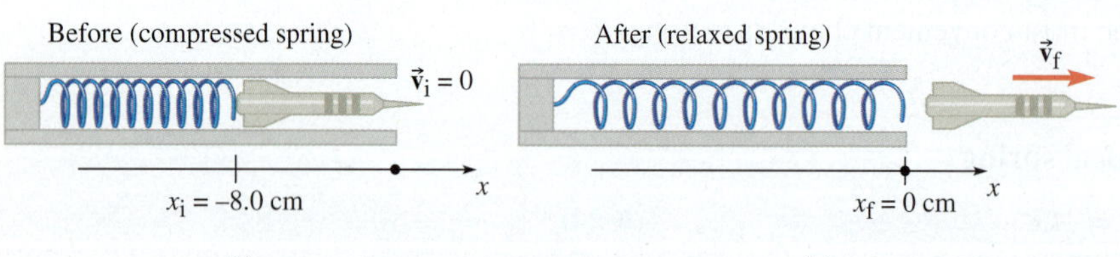

Figure 6.32
Dart gun (a) before and (b) after firing. The spring was compressed by 8.0 cm when the gun was cocked.

Application of Energy Conversion: Jumping When a human jumps, the muscles supply the energy to propel the body upward. Try jumping as high as you can from a standing start. You no doubt start by crouching down. Then you accelerate upward, straightening your legs and your body; your muscles convert chemical energy into the mechanical energy of your jump. If you are very athletic, you might be able to jump about 1 m above the floor.

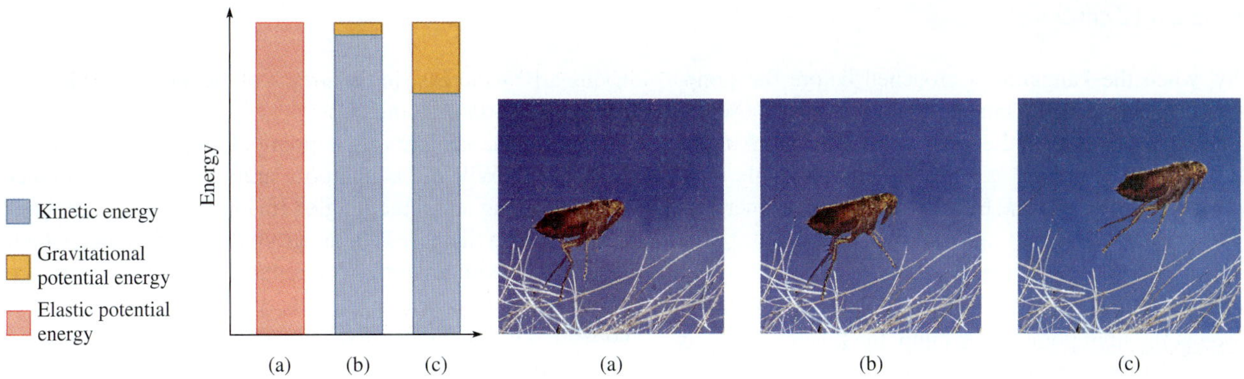

Figure 6.33 Energy transformations in the jump of a flea.

The kangaroo uses a different mechanism. It has long, elastic tendons and small muscles in its hind legs, in contrast to the relatively large muscles and short, stiffer tendons found in humans. The kangaroo folds its legs before a jump, using its muscles to stretch the tendons and converting chemical energy into elastic potential energy. The kangaroo then quickly extends its legs, relaxing the tendons like a released spring. The elastic energy stored in the tendons supplies much of the energy needed for the jump; the rest is supplied by the kangaroo's leg muscles, which convert some more chemical energy into mechanical energy.

How does the kangaroo keep jumping?

When the kangaroo lands on the ground, the tendons are stretched again as its legs bend. Thus, rather than dissipating all of the energy from the previous jump, a large fraction of it is recaptured as elastic energy in the tendons and then released to assist the next jump. This process reduces the amount of energy the muscles must supply for subsequent jumps and makes the kangaroo one of the most energy-efficient travelers among animals. The human body also stores some elastic energy in stretched tendons and in flexed foot bones when we run or jump, but not to the extent that its specialized anatomy enables the kangaroo to do.

Some insects jump using a catapult technique. The knee joint of a flea contains an elastic material called resilin (a rubber-like protein). The flea slowly bends its knee, stretching out the resilin and storing elastic energy, and then locks its knee in place (Fig. 6.33a). When the flea is ready to jump, the knee is unlocked and the resilin quickly contracts with a sudden conversion of the stored elastic energy into kinetic energy (Fig. 6.33b). Some of this kinetic energy is then converted into gravitational potential energy as the flea moves higher and higher (Fig. 6.33c). Ignoring air resistance and other dissipative forces, the total mechanical energy (kinetic energy + gravitational potential energy + elastic potential energy) does not change during the jump.

Example 6.12

The Hopping Kangaroo

Suppose the height h of a kangaroo's hop (Fig. 6.34) after it stretches its tendons a distance x_1 (beyond their unstretched length) is 2.0 m. How high would the hop be after it stretched the tendons 10% more than before (i.e., a distance $1.10x_1$ beyond their unstretched length)? In a simplified model, we assume that all the energy for a kangaroo's hop comes from the elastic energy stored in the tendons, which behave as ideal springs. Ignore air resistance and other energy dissipation.

Strategy Ignoring dissipation, the mechanical energy does not change. We have to include both gravitational and elastic potential energies in the mechanical energy. At first we consider a kangaroo jumping straight up. Then we try to generalize to more typical hopping with forward motion as well as upward motion.

Solution The mechanical energy does not change:

$$K_i + U_{i,grav} + U_{i,elastic} = K_f + U_{f,grav} + U_{f,elastic}$$

continued on next page

Example 6.12 continued

Initially, when the kangaroo is crouched before the jump, it has zero kinetic energy. For convenience, we choose the initial gravitational potential energy to be zero. Thinking of the elastic potential energy as being stored in a single ideal spring with spring constant k, the initial mechanical energy is

$$K_i + U_{i,grav} + U_{i,elastic} = 0 + 0 + \tfrac{1}{2}kx_i^2$$

where x_i represents the initial stretch of the tendons. With the kangaroo at the high point of the jump, the kinetic energy is again zero if it jumped straight up. The tendons are no longer stretched, so the elastic potential energy is zero. But now there is gravitational potential energy. At a height h above the initial point, the final mechanical energy is

$$K_f + U_{f,grav} + U_{f,elastic} = 0 + mgh + 0$$

where m is the kangaroo's mass. Setting the mechanical energies equal,

$$\frac{1}{2}kx_i^2 = mgh \quad \Rightarrow \quad h = \frac{kx_i^2}{2mg}$$

We don't know all of the constants (mass, spring constant, initial amount of stretch), so we set up a ratio:

$$\frac{h_2}{h_1} = \frac{\cancel{k}x_2^2/(\cancel{2mg})}{\cancel{k}x_1^2/(\cancel{2mg})} = \frac{x_2^2}{x_1^2}$$

For a 10% increase in stretch, $x_2 = 1.10x_1$ and

$$h_2 = \left(\frac{x_2}{x_1}\right)^2 h_1 = (1.10)^2 h_1 = 1.21 \times 2.0 \text{ m} = 2.4 \text{ m}$$

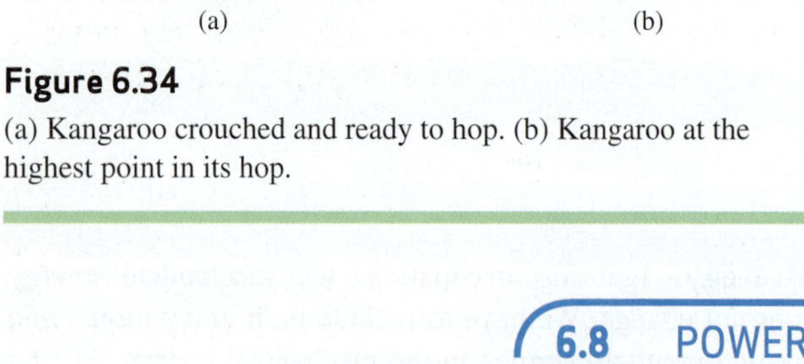

Figure 6.34

(a) Kangaroo crouched and ready to hop. (b) Kangaroo at the highest point in its hop.

Using a 10% increase in the stretch of the tendon, the kangaroo jumps about 21% higher.

When the kangaroo is hopping along, it does not jump straight up. Will the kangaroo's jump still be 21% higher when jumping at another angle? Imagine the kangaroo hopping along so that it leaves the ground at a 45° angle, which gives the maximum horizontal range per hop in the absence of air resistance. The elastic energy in the tendon is first converted to kinetic energy. This time, not all of the kinetic energy is converted to gravitational potential energy. The kinetic energy at the highest point of the jump is *not* zero because the kangaroo is still moving forward. The initial velocity can be resolved into components:

$$v^2 = v_x^2 + v_y^2 = 2v_x^2 \quad \text{(since } v_x = v_y \text{ for a 45° angle)}$$

At the highest point of the jump, the kinetic energy is $\frac{1}{2}mv_x^2$, which is half of the initial kinetic energy. Overall, *half* of the elastic energy of the tendon is converted to gravitational potential energy:

$$\tfrac{1}{2} \times \left(\tfrac{1}{2}kx_i^2\right) = mgh$$

Since h is still proportional to x_i^2, the height of the jump still increases by 21% if the stretch of the tendon is increased by 10%.

Discussion The storage of elastic energy in the tendon is a clever way for the kangaroo to get more "miles per gallon." Without such an energy storage system, most of the kangaroo's mechanical energy would be converted to an unrecoverable form of energy at the end of each hop. The tendons store some of the energy that would otherwise be lost and then release it to help the next jump. Since less mechanical energy is "lost" on each landing, the energy supplied by the kangaroo's muscles is less than it would otherwise be. Humans use a similar energy-saving mechanism when running (see Problem 117).

Practice Problem 6.12 Jumping with Joey

Suppose the kangaroo has a baby kangaroo (a *joey*) riding in her pouch. If the joey has grown to be one sixth the mass of its mother, how high can the kangaroo jump with the additional load? Assume that, without the joey, she can jump 2.8 m.

6.8 POWER

Sometimes the *rate* of energy conversion is important. When shopping for a sports car, you wouldn't ask the salesman how much work the engine can do. A tiny economy car like the Toyota Prius does more work than a Ferrari if the Prius is used for daily commuting while the Ferrari sits in the garage most of the time. But the Ferrari can do work *at a much faster rate* than the Prius can. In other words, it can change chemical energy in the gasoline into mechanical energy of the car at a much faster rate—it has a larger maximum power output. The higher power output enables the Ferrari to accelerate to high speeds much faster than the Prius. We give the name **power** (symbol P) to the rate of energy transfer or of energy conversion. The average power is the amount of energy transferred (ΔE) divided by the time the transfer takes (Δt):

Average power

$$P_{av} = \frac{\Delta E}{\Delta t} \tag{6-26}$$

The SI unit of power, the joule per second, is given the name watt (1 W = 1 J/s), after James Watt (1736–1819), a Scottish inventor who greatly improved the efficiency of steam engines. Remember that the unit symbol W stands for *watt*, not *work*.

In the United States, the maximum power output of an electric motor or automobile engine is often specified in horsepower, which is a non-SI unit of power (1 hp = 746 W).

The *kilowatt-hour* (kW·h) is a unit of energy, *not* a unit of power. One kilowatt-hour is the amount of energy transferred at a constant rate of 1 kW during a time interval of 1 h. The kilowatt-hour is commonly used by utility companies to measure the amount of electric energy delivered to consumers.

The work done by a force during a small time interval Δt is

$$W = F\,\Delta r \cos\theta \tag{6-1}$$

The magnitude of the displacement is

$$\Delta r = v\,\Delta t$$

Hence, the power—the rate at which the force does work—can be found from the force and the velocity.

$$P = \frac{W}{\Delta t} = \frac{F\,\Delta r \cos\theta}{\Delta t} = F\frac{\Delta r}{\Delta t}\cos\theta = Fv\cos\theta$$

Instantaneous power (rate at which work is done)

$$P = Fv\cos\theta \tag{6-27}$$

(θ is the angle between $\vec{\mathbf{F}}$ and $\vec{\mathbf{v}}$)

Equation (6.27) can be written using the scalar product: $P = \vec{\mathbf{F}}\cdot\vec{\mathbf{v}}$.

Example 6.13

Germ Power

A bacterium spins its helical flagellum like a rotary motor to overcome the drag force that opposes its motion in order to propel itself through water. If the bacterium is moving at a constant velocity of 80 μm/s and the drag force is 0.125 μN, what is this motor's power output?

Strategy Moving at constant velocity, the bacterium's kinetic energy is constant. Therefore the motor must do work at the same rate that the drag force dissipates energy.

Solution The drag force dissipates energy at a rate:

$$P_{drag} = F_{drag}v\cos\theta$$

The drag force is opposite to the velocity, so $\theta = 180°$ and $P_{drag} = -F_{drag}v$. The kinetic energy is not changing, so

$$P_{total} = P_{drag} + P_{motor} = 0$$

$$P_{motor} = +F_{drag}v = 0.125\ \mu\text{N} \times 80\ \mu\text{m/s} = 1.0 \times 10^{-11}\ \text{W}$$

Discussion Another strategy would be to find the force on the bacterium due to the motor. The net force is zero, so $\vec{\mathbf{F}}_{drag} + \vec{\mathbf{F}}_{motor} = 0$. Thus the force due to the motor is 0.125 μN in the direction of the velocity. The power is

$$P_{motor} = F_{motor}\,v\cos 0 = 1.0 \times 10^{-11}\ \text{W}$$

The result may seem like a tiny power output, but the energy released by the decomposition of one molecule of adenosine triphosphate (ATP) is approximately 5×10^{-20} J, so the motor would requires decomposition of more than 2×10^{8} ATP molecules per second.

Practice Problem 6.13 Muscle Power

To generate tension in a muscle, a myosin molecule pulls on an actin filament with a force of about 1 pN. If the actin filament moves at 2 μm/s, what is the power output of this molecular motor?

Example 6.14

Air Resistance on a Hill-Climbing Car

A 1000 kg car climbs a hill with a 4.0° incline at a constant 12.0 m/s (Fig. 6.35). (a) At what rate is the gravitational potential energy increasing? (b) If the mechanical power output of the engine is 20.0 kW, find the force of air resistance on the car. (Assume that air resistance is responsible for all of the energy dissipation.)

Strategy (a) We can find the rate of gravitational potential energy increase in two ways. One is to find the potential energy change during a time interval Δt and divide it by the time interval, which is equivalent to using the definition of average power [Eq. (6-26)]. The other possibility is to use Eq. (6-27) to find the rate at which the gravitational force does work.

(b) The car moves at constant speed, so its kinetic energy is not changing. Therefore, during any time interval, the work done by the engine (W_e) plus the (negative) work done by air resistance (W_a) is equal to the increase in the gravitational potential energy.

Given: car mass = 1000 kg; v = 12.0 m/s; 4.0° incline.
To find: (a) rate of potential energy change, $\Delta U/\Delta t$;
(b) force due to air resistance, $\vec{\mathbf{F}}_a$.

Solution (a) For a small change in elevation Δy, the change in potential energy is

$$\Delta U = mg\,\Delta y$$

The *rate* of potential energy change is

$$\frac{\Delta U}{\Delta t} = \frac{mg\,\Delta y}{\Delta t} = mg\frac{\Delta y}{\Delta t} = mgv_y$$

where $v_y = \Delta y/\Delta t$ is the y-component of the velocity. From Fig. 6.36, $v_y = v\sin\phi$, where $\phi = 4.0°$. Then,

$$\frac{\Delta U}{\Delta t} = mgv\sin\phi = 1000\text{ kg}\times 9.80\text{ m/s}^2\times 12.0\text{ m/s}\times\sin 4.0°$$
$$= 8200\text{ W}$$

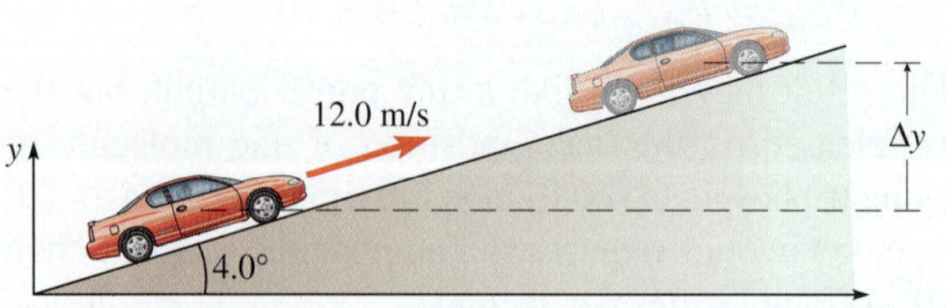

Figure 6.35
Car climbing a hill at constant speed.

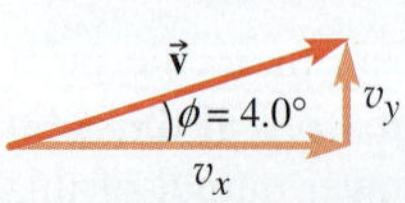

Figure 6.36
Resolving the velocity into x- and y-components.

(b) During any time interval Δt, the (positive) work done by the engine plus the (negative) work done by air resistance must equal the increase in the gravitational potential energy:

$$W_{total} = W_e + W_a = \Delta U$$

Dividing each term by Δt, we find

$$\frac{W_e}{\Delta t} + \frac{W_a}{\Delta t} = \frac{\Delta U}{\Delta t} \quad\Rightarrow\quad P_e + P_a = \frac{\Delta U}{\Delta t}$$

where P_e and P_a represent the power output of the engine and the rate at which air resistance does (negative) work on the car, respectively. Then,

$$P_a = \frac{\Delta U}{\Delta t} - P_e = 8.2\text{ kW} - 20.0\text{ kW} = -11.8\text{ kW}$$

So, of the 20.0 kJ of mechanical work that the engine does each second, 8.2 kJ goes into gravitational potential energy and 11.8 kJ goes into pushing air out of the way and stirring it up in the process.

The direction of the force of air resistance $\vec{\mathbf{F}}_a$ on the car is opposite to the car's velocity, so

$$P_a = F_a v\cos 180° = -F_a v$$

Solving for F_a,

$$F_a = -\frac{P_a}{v} = -\frac{-11\,800\text{ W}}{12.0\text{ m/s}} = 983\text{ N}$$

Discussion We can check (a) by using Eq. (6-27) to find the rate at which the gravitational force does work: $P = Fv\cos\theta$, where $F = mg$. The angle θ is *not* the same as ϕ. In Eq. (6-27), θ is the angle between the force and velocity vectors, which is 94.0° (Fig. 6.37). Then,

$$P = mgv\cos 94.0°$$
$$= 1000\text{ kg}\times 9.80\text{ m/s}^2\times 12.0\text{ m/s}\times\cos 94.0°$$
$$= -8200\text{ W}$$

Gravity does work on the car at a rate of −8200 W, which means the potential energy is *increasing* at a rate of +8200 W.

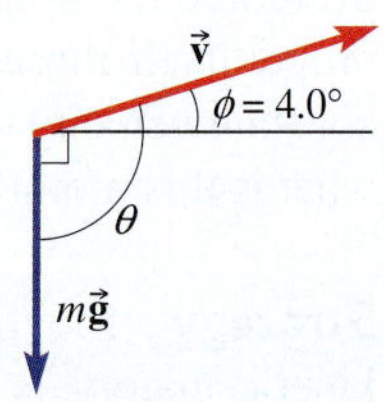

Figure 6.37
The angle between the force and the velocity is $\theta = 94.0°$. (The angle is exaggerated for clarity.)

We can also figure out what mechanical power the engine must supply to go 12.0 m/s on level ground. With no change in potential energy, all of the mechanical power output of the engine goes into stirring up the air, so $P_e + P_a = 0$. The magnitude of the force of air resistance is the same (983 N) since the speed is the same. Then air resistance dissipates energy at the same rate as before:

$$P_a = -F_a v = -983\text{ N}\times 12.0\text{ m/s} = -11.8\text{ kW}$$

continued on next page

Example 6.14 continued

Therefore, $P_e = 11.8$ kW. On level ground, the gravitational potential energy isn't increasing, so the engine only needs to do enough work to counteract the tendency of air resistance to slow down the car.

In this example, we have assumed that all of the mechanical power output of the engine is delivered to the wheels to propel the car forward. In reality, some of the engine's power output is used to run auxiliary devices such as headlights, radios, and windshield wipers. Friction (in the moving parts of the engine, transmission, and drivetrain) also reduces the amount of power that is actually delivered to the wheels.

Practice Problem 6.14 Mechanical Power Output on Flat Ground or Going Downhill

What mechanical power must the engine supply (a) to drive on level ground at 12.0 m/s and (b) to go down a 4.0° incline at 12.0 m/s? (Since this is the same speed as in Example 6.14, the force of air resistance is the same.)

Master the Concepts

- Conservation law: a physical law phrased in terms of a quantity that does not change with time.
- The law of conservation of energy: the total energy of the universe is unchanged by any physical process.
- Work is an energy transfer due to the application of a force. The work done by a constant force $\vec{\mathbf{F}}$ acting on an object during a displacement $\Delta\vec{\mathbf{r}}$ is

$$W = F\,\Delta r \cos\theta \quad (6\text{-}1)$$

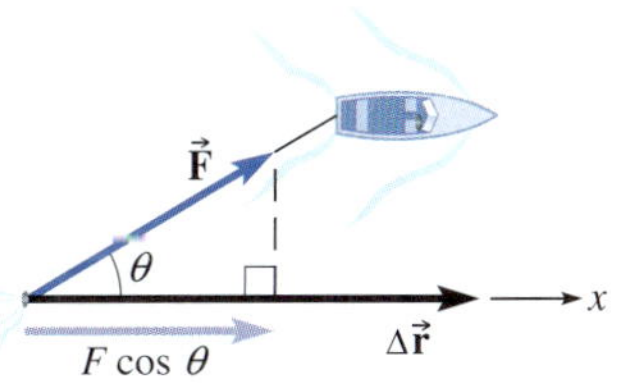

where θ is the angle between $\vec{\mathbf{F}}$ and $\Delta\vec{\mathbf{r}}$. If $\vec{\mathbf{F}}$ or $\Delta\vec{\mathbf{r}}$ is parallel to the x-axis,

$$W = F_x\,\Delta x \quad (6\text{-}2)$$

- When several forces act on an object, the total work is the sum of the work done by each force individually.
- Translational kinetic energy is the energy associated with motion of the object as a whole. The translational kinetic energy of an object of mass m moving with speed v is

$$K = \tfrac{1}{2}mv^2 \quad (6\text{-}6)$$

- The gravitational potential energy for an object of mass m in a *uniform* gravitational field is

$$U_{\text{grav}} = mgy \quad (6\text{-}14)$$

where the $+y$-axis points up and we assign $U = 0$ to $y = 0$.
- The gravitational potential energy for two objects of masses m_1 and m_2 whose centers are separated by a distance r is

$$U = -\frac{Gm_1m_2}{r} \quad (6\text{-}15)$$

where we assign $U = 0$ to infinite separation ($r = \infty$).

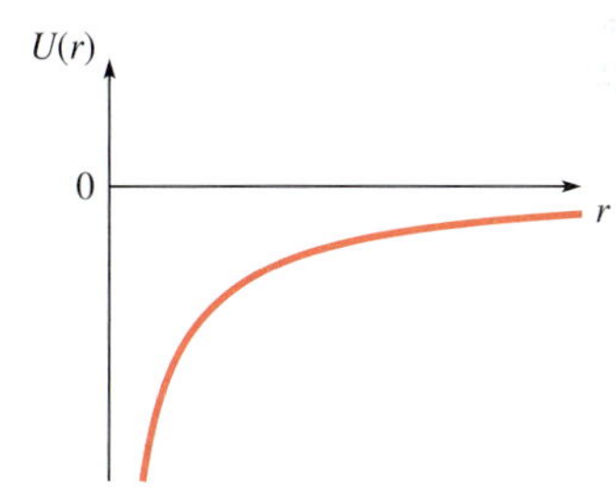

- There is no special significance to the sign of the potential energy. What matters is the sign of the potential energy *change*. Only *changes* in potential energy enter our calculations.
- The work done by a variable force directed along the x-axis during a displacement Δx is the area under the $F_x(x)$ graph from x_i to x_f.

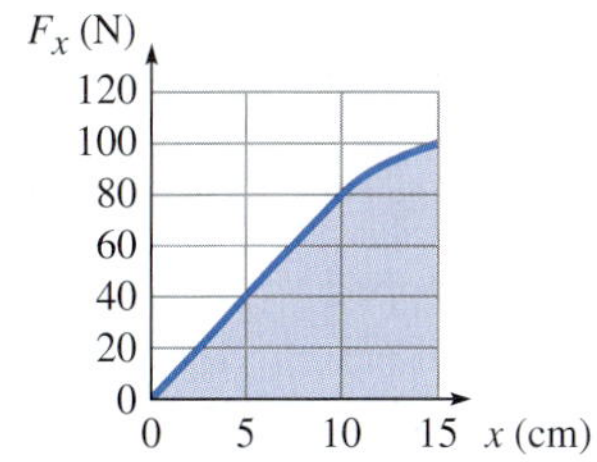

- Hooke's law: for many objects, the deformation is proportional to the magnitude of the force that causes the deformation. An ideal spring is massless and follows Hooke's law. The force exerted *by* the moveable end of an ideal spring when it is at position x is

$$F_x = -kx \quad (6\text{-}19)$$

where the origin is chosen so the spring is relaxed at $x = 0$ and k is called the spring constant.

continued on next page

Master the Concepts continued

- If we assign $U = 0$ to the relaxed spring ($x = 0$), the elastic potential energy stored in an ideal spring of spring constant k is

$$U_{\text{elastic}} = \tfrac{1}{2}kx^2 \quad \text{(6-24)}$$

- Mechanical energy is the sum of the kinetic and potential energies. The change in potential energy accounts for the work done by the forces associated with the potential energy. The work done by nonconservative forces is equal to the change in the mechanical energy:

$$W_{\text{nc}} = \Delta K + \Delta U \quad \text{(6-12)}$$

When the work done by nonconservative forces is zero, the mechanical energy does not change.

$$\text{If } W_{\text{nc}} = 0, \quad \Delta K + \Delta U = 0$$

- Average power is the average rate of energy conversion.

$$P_{\text{av}} = \frac{\Delta E}{\Delta t} \quad \text{(6-26)}$$

The instantaneous rate at which a force $\vec{\mathbf{F}}$ does work when the object it acts on moves with velocity $\vec{\mathbf{v}}$ is

$$P = Fv\cos\theta \quad \text{(6-27)}$$

where θ is the angle between $\vec{\mathbf{F}}$ and $\vec{\mathbf{v}}$

- The SI unit of work and energy is the joule. 1 J = 1 N·m. The SI unit of power is the watt. 1 W = 1 J/s.

Conceptual Questions

1. An object moves in a circle. Is the total work done on the object by external forces necessarily zero? Explain.
2. You are walking to class with a backpack full of books. As you walk at constant speed on flat ground, does the force exerted on the backpack by your back and shoulders do any work? If so, is it positive or negative? Answer the same questions in two other situations: (1) you are walking down some steps at constant speed; (2) you start to run faster and faster on a level sidewalk to catch a bus.
3. Why do roads leading to the top of a mountain wind back and forth? [*Hint*: Think of the road as an inclined plane.]
4. A mango falls to the ground. During the fall, does the Earth's gravitational field do positive or negative work W_{m} on the mango? Does the mango's gravitational field do positive or negative work W_{E} on the Earth? Compare the signs and the magnitudes of W_{m} and W_{E}.
5. Can static friction do work? If so, give an example. [*Hint*: Static friction acts to prevent *relative* motion along the contact surface.]
6. In the design of a roller coaster, is it possible for any hill of the ride to be higher than the first one? If so, how?
7. When a ball is dropped to the floor from a height h, it strikes the ground and briefly undergoes a change of shape before rebounding to a maximum height less than h. Explain why it does not return to the same height h.
8. A gymnast is swinging in a vertical circle about a crossbar. In terms of energy conservation, explain why the speed of the gymnast's body is slowest at the top of the circle and fastest at the bottom.
9. A bicycle rider notices that he is approaching a steep hill. Explain, in terms of energy, why the bicyclist pedals hard to gain as much speed as possible on level road before reaching the hill.
10. You need to move a heavy crate by sliding it across a smooth floor. The coefficient of sliding friction is 0.2. You can either push the crate horizontally or pull the crate using an attached rope. When you pull on the rope, it makes a 30° angle with the floor. Which way should you choose to move the crate so that you do the least amount of work? How can you answer this question without knowing the weight of the crate or the displacement of the crate?
11. The main effort of running is the work done by the muscles to accelerate the legs. When a foot strikes the ground, it is momentarily brought to rest while the remainder of the animal's body continues to move forward. When the foot is picked up, it is accelerated forward by one set of muscles in order to move ahead of the rest of the body. Then the foot is slowed down by a second set of muscles until it is brought to rest on the ground again. The muscles expend energy both when speeding up the leg and when slowing it down. How are thoroughbred horses, deer, and greyhounds adapted so that they can run at great speed?

12. Explain why an ideal spring *must* exert forces of equal magnitude on the objects attached to each end, even if the spring itself has a nonzero acceleration. [*Hint*: Use one of Newton's laws of motion and remember that an ideal spring has zero mass.] Is the amount of work done by the spring on the two objects necessarily the same? Explain. If the answer is no, give an example to illustrate.
13. Zorba and Boris are at a water park. There are two water slides with straight slopes that start at the same height and end at the same height. Slide A has a more gradual slope than slide B. Boris says he likes slide B better because you reach a faster speed, and he notes that he

got to the bottom level in less time on slide B as measured with his stop watch. His brother Zorba says you reach the same speed with either slide. Who is correct and why? Both slides have negligible friction.

Multiple-Choice Questions

Student Response System Questions

1. After getting on the Santa Monica Freeway, a sports car accelerates from 30 mi/h to 90 mi/h. Its kinetic energy
   (a) increases by a factor of $\sqrt{3}$.
   (b) increases by a factor of 3.
   (c) increases by a factor of 9.
   (d) increases by a factor that depends on the car's mass.
2. If a kangaroo on Earth can jump from a standing start so that its feet reach a height h above the surface, approximately how high can the same kangaroo jump from a standing start on the Moon's surface? $g_{Moon} \approx \frac{1}{6} g_{Earth}$. (Assume the kangaroo has an oxygen tank and pressure suit with negligible mass.)
   (a) h (b) $6h$ (c) $\frac{1}{6}h$
   (d) $36h$ (e) $\frac{1}{36}h$ (f) $\sqrt{6}h$

Questions 3–6. The orbit of Mercury is much more eccentric than the orbits of the other planets. That is, instead of being nearly circular, the orbit is noticeably elliptical.

Answer choices for Questions 3–5:
(a) its maximum value. (b) its minimum value.
(c) the same value as at every other point in the orbit.

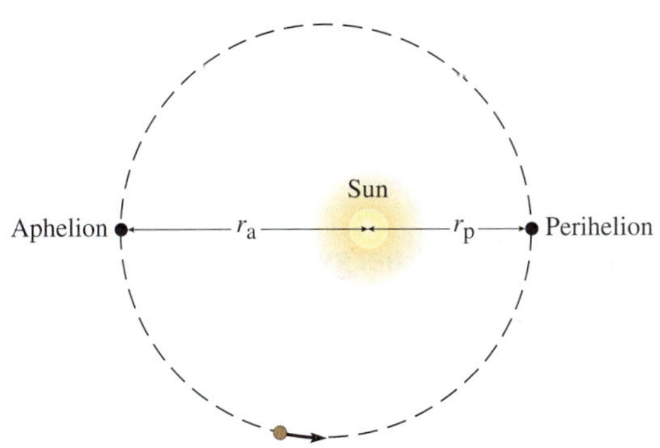

Multiple-Choice Questions 3–6

3. At perihelion, the gravitational potential energy of Mercury's orbit has
4. At perihelion, the kinetic energy of Mercury has
5. At perihelion, the mechanical energy of Mercury's orbit has
6. As Mercury moves from the perihelion to the aphelion, the work done by gravity on Mercury is
   (a) zero. (b) positive. (c) negative.
7. A hiker descends from the South Rim of the Grand Canyon to the Colorado River. During this hike, the work done by gravity on the hiker is
   (a) positive and depends on the path taken.
   (b) negative and depends on the path taken.
   (c) positive and independent of the path taken.
   (d) negative and independent of the path taken.
   (e) zero.
8. Two balls are thrown from the roof of a building with the same initial speed. One is thrown horizontally while the other is thrown at an angle of 20° above the horizontal. Which hits the ground with the greatest speed? Ignore air resistance.
   (a) The one thrown horizontally
   (b) The one thrown at 20°
   (c) They hit the ground with the same speed.
   (d) The answer cannot be determined with the given information.

Questions 9 and 10. A simple catapult, consisting of a leather pouch attached to rubber bands tied to two prongs of a wooden Y, has a spring constant k and is used to shoot a pebble horizontally. When the catapult is stretched by a distance d, it gives a pebble of mass m a launch speed v. Answer choices for Questions 9 and 10:
(a) $\sqrt{3}v$ (b) $3v$ (c) $3\sqrt{3}v$ (d) $9v$ (e) $27v$

9. What speed does the catapult give a pebble of mass m when stretched to a distance $3d$?
10. What speed does the catapult give a pebble of mass $m/3$ when stretched to a distance d?
11. A projectile is launched at an angle θ above the horizontal. Ignoring air resistance, what fraction of its initial kinetic energy does the projectile have at the top of its trajectory?
    (a) $\cos\theta$ (b) $\sin\theta$ (c) $\tan\theta$ (d) $\frac{1}{\tan\theta}$ (e) $\frac{1}{2}$
    (f) $\cos^2\theta$ (g) $\sin^2\theta$ (h) 0 (i) 1

Problems

Combination conceptual/quantitative problem
Biomedical application
Challenging
Blue # Detailed solution in the Student Solutions Manual
[1, 2] Problems paired by concept
connect Text website interactive or tutorial

Section 6.2 Work Done by a Constant Force

1. How much work must Denise do to drag her basket of laundry of mass 5.0 kg a distance of 5.0 m along a floor, if the force she exerts is a constant 30.0 N at an angle of 60.0° with the horizontal?
2. A sled is dragged along a horizontal path at a constant speed of 1.5 m/s by a rope that is inclined at an angle of 30.0° with respect to the horizontal. The total weight of the sled is 470 N. The tension in the rope is 240 N. How

much work is done by the rope on the sled in a time interval of 10.0 s?

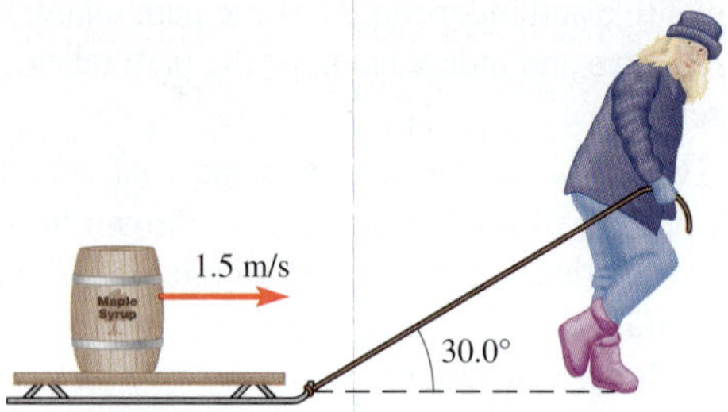

3. Hilda holds a gardening book of weight 10 N at a height of 1.0 m above her patio for 50 s. How much work does she do *on the book* during that 50 s?
4. A horizontal towrope exerts a force of 240 N due west on a water-skier while the skier moves due west a distance of 54 m. How much work does the towrope do on the water-skier?
5. A barge of mass 5.0×10^4 kg is pulled along the Erie Canal by two mules, walking along towpaths parallel to the canal on either side of it. The ropes harnessed to the mules make angles of 45° to the canal. Each mule is pulling on its rope with a force of 1.0 kN. How much work is done on the barge by both of these mules together as they pull the barge 150 m along the canal?
6. A 402 kg pile driver is raised 12 m above ground. (a) How much work must be done to raise the pile driver? (b) How much work does gravity do on the driver as it is raised? (c) The driver is now dropped. How much work does gravity do on the driver as it falls?

7. Jennifer lifts a 2.5 kg carton of cat litter from the floor to a height of 0.75 m. (a) How much *total* work is done on the carton during this operation? Jennifer then pours 1.2 kg of the litter into the cat's litter box on the floor. (b) How much work is done by gravity on the 1.2 kg of litter as it falls into the litter box?
8. Starting from rest, a horse pulls a 250 kg cart for a distance of 1.5 km. It reaches a speed of 0.38 m/s by the time it has walked 50.0 m and then walks at constant speed. The frictional force on the rolling cart is a constant 260 N. Each gram of oats the horse eats releases 9.0 kJ of energy; 10.0% of this energy can go into the work the horse must do to pull the cart. How many grams of oats must the horse eat to pull the cart?
9. Dirk pushes on a packing box with a horizontal force of 66.0 N as he slides it along the floor. The average friction force acting on the box is 4.80 N. How much *total* work is done on the box in moving it 2.50 m along the floor?
10. Juana slides a crate along the floor of the moving van. The coefficient of kinetic friction between the crate and the van floor is 0.120. The crate has a mass of 56.8 kg and Juana pushes with a horizontal force of 124 N. If 74.4 J of total work are done on the crate, how far along the van floor does it move?

Problems 11–14. A crate of mass $m_1 = 12.4$ kg is pulled by a massless rope up a 36.9° ramp. The rope passes over an ideal pulley and is attached to a hanging crate of mass $m_2 = 16.3$ kg. The crates move 1.40 m, starting from rest.

11. Find the work done by gravity on the hanging crate.
12. Find the work done by gravity on the sliding crate.

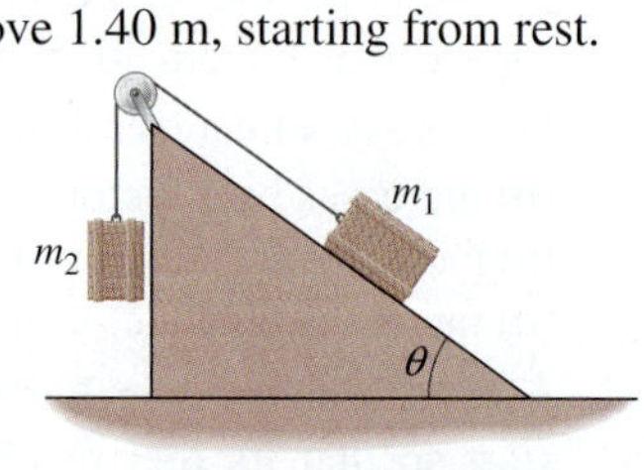

Problems 11-14, 46, and 47

13. If the incline is frictionless, find the total work done on the sliding crate. The tension in the rope is 110.5 N.
14. If the frictional force on the sliding crate has magnitude 19.4 N and the tension in the rope is 121.5 N, find the total work done on the sliding crate.
15. A 75.0 kg skier starts from rest and slides down a 32.0 m frictionless slope that is inclined at an angle of 15.0° with the horizontal. Ignore air resistance. Calculate the work done by gravity on the skier and the work done by the normal force on the skier.
16. In Problem 15, if the slope is not frictionless so that the skier has a final velocity of 10.0 m/s, calculate the work done by gravity, the work done by the normal force, the work done by friction, the force of friction (assuming it is constant), and the coefficient of kinetic friction. (connect tutorial: water slide)

Section 6.3 Kinetic Energy

17. An automobile with a mass of 1600 kg has a speed of 30.0 m/s. What is its kinetic energy?
18. A record company executive is on his way to a TV interview carrying his briefcase. The mass of the briefcase is 5.00 kg. The executive realizes that he is going to be late. Starting from rest, he starts to run, reaching a speed of 2.50 m/s. What is the work done by the executive on the briefcase during this time? Ignore air resistance.
19. In 1899, Charles M. "Mile a Minute" Murphy set a record for speed on a bicycle by pedaling for a mile at an average of 62.3 mi/h (27.8 m/s) on a 3 mi track of plywood planks set over railroad ties in the draft of a Long Island Railroad train. In 1985, a record was set for this type of "motor pacing" by Olympic cyclist John Howard who raced at 152.2 mi/h (68.04 m/s) in the wake of a race car at Bonneville Salt Flats. The race car had a modified tail assembly designed to

reduce the air drag on the cyclist. What was the kinetic energy of the bicycle plus rider in each of these feats? Assume that the mass of bicycle plus rider is 70.5 kg in each case.

20. Sam pushes a 10.0 kg sack of bread flour on a frictionless horizontal surface with a constant horizontal force of 2.0 N starting from rest. (a) What is the kinetic energy of the sack after Sam has pushed it a distance of 35 cm? (b) What is the speed of the sack after Sam has pushed it a distance of 35 cm?

21. Josie and Charlotte push a 12 kg bag of playground sand for a sandbox on a frictionless, horizontal, wet polyvinyl surface with a constant, horizontal force for a distance of 8.0 m, starting from rest. If the final speed of the sand bag is 0.40 m/s, what is the magnitude of the force with which they pushed?

22. A ball of mass 0.10 kg moving with speed of 2.0 m/s hits a wall and bounces back with the same speed in the opposite direction. What is the change in the ball's kinetic energy?

23. Jim rides his skateboard down a ramp that is in the shape of a quarter circle with a radius of 5.00 m. At the bottom of the ramp, Jim is moving at 9.00 m/s. Jim and his skateboard have a mass of 65.0 kg. How much work is done by friction as the skateboard goes down the ramp? (connect tutorial: energy, parts (a) and (b))

24. A 69.0 kg short-track ice skater is racing at a speed of 11.0 m/s when he falls down and slides across the ice into a padded wall that brings him to rest. Assuming that he doesn't lose any speed during the fall or while sliding across the ice, how much work is done by the wall while stopping the ice skater?

25. A plane weighing 220 kN (25 tons) lands on an aircraft carrier. The plane is moving horizontally at 67 m/s (150 mi/h) when its tailhook grabs hold of the arresting cables. The cables bring the plane to a stop in a distance of 84 m. (a) How much work is done on the plane by the arresting cables? (b) What is the force (assumed constant) exerted on the plane by the cables? (Both answers will be *underestimates*, since the plane lands with the engines full throttle forward; in case the tailhook fails to grab hold of the cables, the pilot must be ready for immediate takeoff.)

26. A shooting star is a meteoroid that burns up when it reaches Earth's atmosphere. Many of these meteoroids are quite small. Calculate the kinetic energy of a meteoroid of mass 5.0 g moving at a speed of 48 km/s and compare it to the kinetic energy of a 1100 kg car moving at 29 m/s (65 mi/h).

Section 6.4 Gravitational Potential Energy (1)

27. Sean climbs a tower that is 82.3 m high to make a jump with a parachute. The mass of Sean plus the parachute is 68.0 kg. If $U = 0$ at ground level, what is the potential energy of Sean and the parachute at the top of the tower? (connect tutorial: energy, parts (c) and (d))

28. Ⓒ Justin moves a desk 5.0 m across a level floor by pushing on it with a constant horizontal force of 340 N. (It slides for a negligibly small distance before coming to a stop when the force is removed.) Then, changing his mind, he moves it back to its starting point, again by pushing with a constant force of 340 N. (a) What is the change in the desk's gravitational potential energy during the round-trip? (b) How much work has Justin done on the desk? (c) If the work done by Justin is not equal to the change in gravitational potential energy of the desk, then where has the energy gone?

Problems 29–32. A skier passes through points *A–E* as shown. Points *B* and *D* are at the same height.

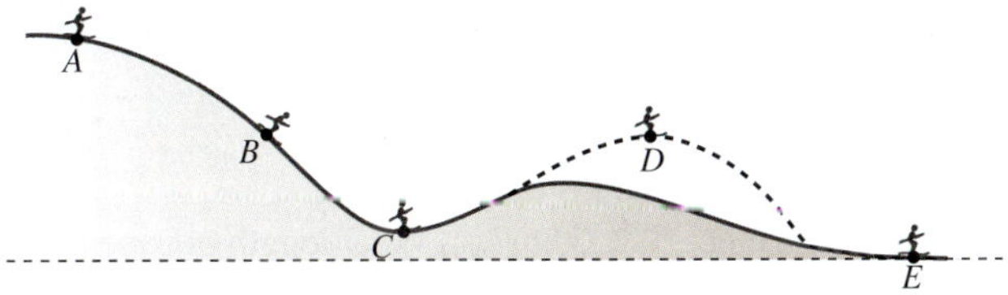

Problems 29–32

29. Rank the points in order of kinetic energy, from greatest to least, assuming no friction or air resistance.

30. Rank the points in order of gravitational potential energy, from greatest to least.

31. Rank the points in order of mechanical energy, from greatest to least, assuming no friction or air resistance.

32. Rank the points in order of mechanical energy, from greatest to least, taking friction and air resistance into consideration.

33. Brad tries out a weight-loss plan that involves repeatedly lifting a 50.0 kg barbell from the floor over his head to a height of 2.0 m. If he is able to complete three such lifts per minute, how long will it take for him to lose 0.50 kg of fat? "Burning" 1 g of fat supplies 39 kJ to the body; of this, 10% can be used by the muscles to lift the barbell. (Ignore the fat "burned" while he lowers the barbell to the floor.)

34. An airline executive decides to economize by reducing the amount of fuel required for long-distance flights. He orders the ground crew to remove the paint from the outer surface of each plane. The paint removed from a single plane has a mass of approximately 100 kg. (a) If the airplane cruises at an altitude of 12000 m, how much energy is saved in not having to lift the paint to that altitude? (b) How much energy is saved by not

having to move that amount of paint from rest to a cruising speed of 250 m/s?

35. Emil is tossing an orange of mass 0.30 kg into the air. (a) Emil throws the orange straight up and then catches it, throwing and catching it at the same point in space. What is the change in the potential energy of the orange during its trajectory? Ignore air resistance. (b) Emil throws the orange straight up, starting 1.0 m above the ground. He fails to catch it. What is the change in the potential energy of the orange during this flight?

36. A brick of mass 1.0 kg slides down an icy roof inclined at 30.0° with respect to the horizontal. (a) If the brick starts from rest, how fast is it moving when it reaches the edge of the roof 2.00 m away? Ignore friction. (b) Redo part (a) if the coefficient of kinetic friction is 0.10. (connect interactive: sliding brick)

37. An arrangement of two pulleys, as shown in the figure, is used to lift a 48.0 kg mass a distance of 4.00 m above the starting point. Assume the pulleys and rope are ideal and that all rope sections are essentially vertical. (a) What is the change in the potential energy of the weight when it is lifted a distance of 4.00 m? (b) How much work must be done to lift the 48.0 kg mass a distance of 4.00 m? (c) What length of rope must be pulled by the person lifting the weight 4.00 m higher in the air? (connect tutorial: block-and-tackle)

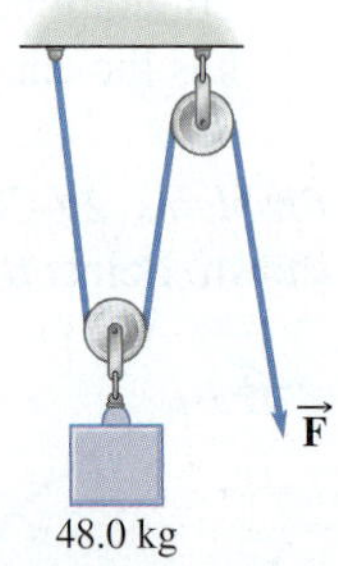

38. In Example 6.1, find the work done by the movers as they slide the chest up the ramp if the coefficient of friction between the chest and the ramp is 0.20. (connect tutorial: ramp)

39. A cart moving to the *right* passes point 1 at a speed of 20.0 m/s. Let $g = 9.81\ \text{m/s}^2$. (a) What is the speed of the cart as it passes point 3? (b) Will the cart reach position 4? Ignore friction.

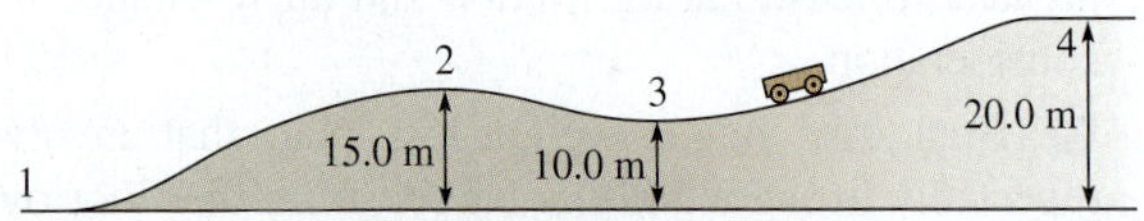

Problems 39 and 40

40. A cart starts from position 4 with a velocity of 15 m/s to the left. Find the speed with which the cart reaches positions 1, 2, and 3. Ignore friction.

41. Bruce stands on a bank beside a pond, grasps the end of a 20.0 m long rope attached to a nearby tree and swings out to drop into the water. If the rope starts at an angle of 35.0° with the vertical, what is Bruce's speed at the bottom of the swing?

42. The maximum speed of a child on a swing is 4.9 m/s. The child's height above the ground is 0.70 m at the lowest point in his motion. How high above the ground is he at his highest point?

43. If the skier of Example 6.6 is moving at 12 m/s at the bottom of the trail, calculate the total work done by friction and air resistance during the run. The skier's mass is 75 kg.

44. A 750 kg automobile is moving at 20.0 m/s at a height of 5.0 m above the bottom of a hill when it runs out of gasoline. The car coasts down the hill and then continues coasting up the other side until it comes to rest. Ignoring frictional forces and air resistance, what is the value of h, the highest position the car reaches above the bottom of the hill?

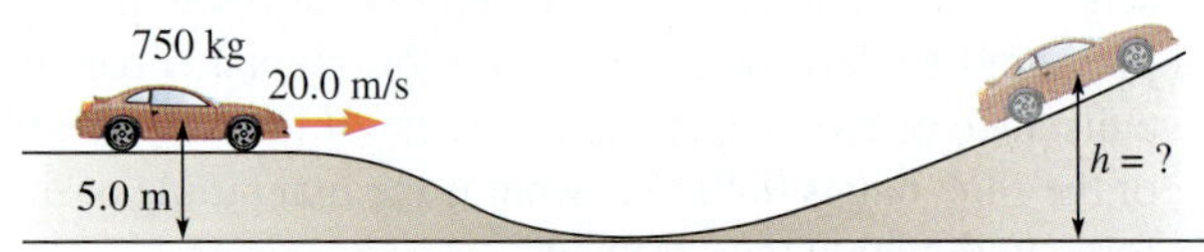

45. ✦ Ⓒ Rachel is on the roof of a building, h meters above ground. She throws a heavy ball into the air with a speed v, at an angle θ with respect to the horizontal. Ignore air resistance. (a) Find the speed of the ball when it hits the ground in terms of h, v, θ, and g. (b) For what value(s) of θ is the speed of the ball greatest when it hits the ground?

46. Refer to Problems 11–14. Use conservation of energy to find the final speed of the sliding crate if the incline is frictionless.

47. Refer to Problems 11–14. Find conservation of energy to find the final speed of the sliding crate if the frictional force on the sliding crate has magnitude 19.4 N.

Section 6.5 Gravitational Potential Energy (2)

48. You are on the Moon and would like to send a probe into space so that it does not fall back to the surface of the Moon. What launch speed do you need?

49. A planet with a radius of 6.00×10^7 m has a gravitational field of magnitude $30.0\ \text{m/s}^2$ at the surface. What is the escape speed from the planet?

50. The escape speed from the surface of Planet Zoroaster is 12.0 km/s. The planet has no atmosphere. A meteor far away from the planet moves at speed 5.0 km/s on a collision course with Zoroaster. How fast is the meteor going when it hits the surface of the planet?

51. The escape speed from the surface of the Earth is 11.2 km/s. What would be the escape speed from another planet of the same density (mass per unit volume) as Earth but with a radius twice that of Earth?

52. A satellite is placed in a noncircular orbit about the Earth. The farthest point of its orbit (*apogee*) is 4.0 Earth radii from the center of the Earth, while its nearest point (*perigee*) is 2.0 Earth radii from the Earth's center. If we define the gravitational potential energy U to be zero for an infinite separation of Earth and satellite, find the ratio $U_{\text{perigee}}/U_{\text{apogee}}$.

53. What is the minimum speed with which a meteor strikes the top of the Earth's stratosphere (about 40 km above Earth's surface), assuming that the meteor begins as a bit of interplanetary debris far from Earth? Assume the drag force is negligible until the meteor reaches the stratosphere.

54. A projectile with mass of 500 kg is launched straight up from the Earth's surface with an initial speed v_i. What magnitude of v_i enables the projectile to just reach a maximum height of $5R_E$, measured from the *center* of the Earth? Ignore air friction as the projectile goes through the Earth's atmosphere.

55. ✦ The orbit of Halley's comet around the Sun is a long thin ellipse. At its aphelion (point farthest from the Sun), the comet is 5.3×10^{12} m from the Sun and moves with a speed of 10.0 km/s. What is the comet's speed at its perihelion (closest approach to the Sun) where its distance from the Sun is 8.9×10^{10} m?

56. ✦ Suppose a satellite is in a circular orbit 3.0 Earth radii above the surface of the Earth (4.0 Earth radii from the center of the Earth). By how much does it have to increase its speed in order to be able to escape Earth? [*Hint*: You need to calculate the orbital speed and the escape speed.]

57. An asteroid hits the Moon and ejects a large rock from its surface. The rock has enough speed to travel to a point between the Earth and the Moon where the gravitational forces on it from the Earth and the Moon are equal in magnitude and opposite in direction. At that point the rock has a very small velocity toward Earth. What is the speed of the rock when it encounters Earth's atmosphere at an altitude of 700 km above the surface?

Section 6.6 Work Done by Variable Forces

58. How much work is done on the bowstring of Example 6.9 to draw it back by 20.0 cm? [*Hint*: Rather than recalculate from scratch, use proportional reasoning.]

59. An ideal spring has a spring constant $k = 20.0$ N/m. What is the amount of work that must be done to stretch the spring 0.40 m from its relaxed length?

60. The forces required to extend a spring to various lengths are measured. The results are shown in the following table. Using the data in the table, plot a graph that helps you to answer the following two questions: (a) What is the spring constant? (b) What is the relaxed length of the spring?

Force (N)	1.00	2.00	3.00	4.00	5.00
Spring length (cm)	14.5	18.0	21.5	25.0	28.5

61. The force that must be exerted to drive a nail into a wall is roughly as shown in the graph. The first 1.2 cm are through soft drywall; then the nail enters the solid wooden stud. How much work must be done to hammer the nail a horizontal distance of 5.0 cm into the wall?

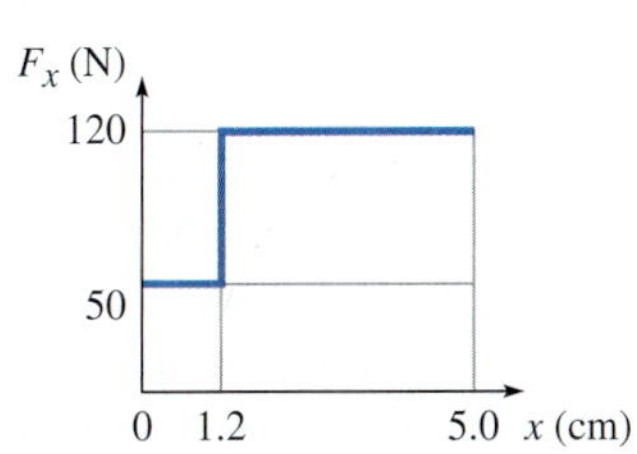

62. (a) If the length of the Achilles tendon increases 0.50 cm when the force exerted on it by the muscle increases from 3200 N to 4800 N, what is the "spring constant" of the tendon? (b) How much work is done by the muscle in stretching the tendon 0.50 cm as the force increases from 3200 N to 4800 N?

63. (a) If forces of magnitude 5.0 N applied to each end of a spring cause the spring to stretch 3.5 cm from its relaxed length, how far do forces of magnitude 7.0 N cause the same spring to stretch? (b) What is the spring constant of this spring? (c) How much work is done by the applied forces in stretching the spring 3.5 cm from its relaxed length? (connect tutorial: spring)

64. A block of wood is compressed 2.0 nm when inward forces of magnitude 120 N are applied to it on two opposite sides. (a) Assuming Hooke's law holds, what is the effective spring constant of the block? (b) Assuming Hooke's law still holds, how much is the same block compressed by inward forces of magnitude 480 N? (c) How much work is done by the applied forces during the compression of part (b)?

65. The length of a spring increases by 7.2 cm from its relaxed length when a mass of 1.4 kg is hanging in equilibrium from the spring. (a) What is the spring constant? (b) How much elastic potential energy is stored in the spring? (c) A different mass is suspended, and the spring length increases by 12.2 cm from its relaxed length to its new equilibrium position. What is the second mass?

66. A spring fixed at one end is compressed from its relaxed position by a distance of 0.20 m. See the graph of the applied external force, F_x, versus the position, x, of the spring. (a) Find the work done by the external force in compressing the spring 0.20 m starting from its relaxed position. (b) Find the work done by the external force to compress the spring from 0.10 m to 0.20 m.

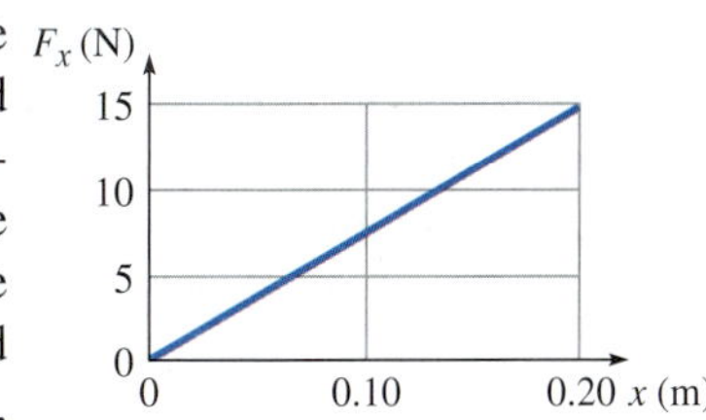

67. Rhonda keeps a 2.0 kg model airplane moving at constant speed in a horizontal circle at the end of a string of length 1.0 m. The tension in the string is 18 N. How much work does the string do on the plane during each revolution? (connect tutorial: circular motion)

68. The graph shows the force exerted on an object versus the position of that object along the x-axis. The force has no components other than along the x-axis. What is

the work done by the force on the object as the object is displaced from 0 to 3.0 m?

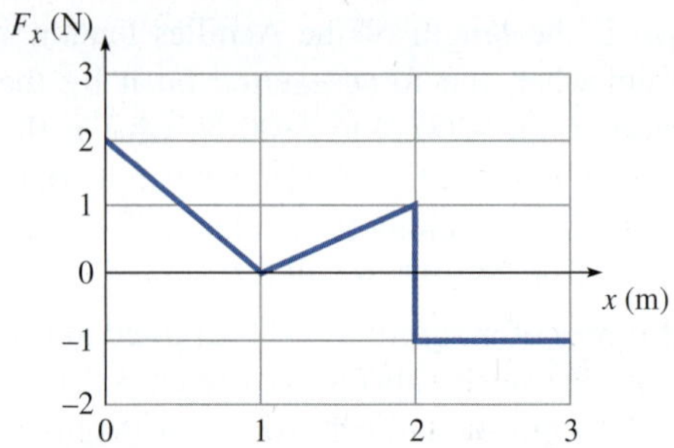

Section 6.7 Elastic Potential Energy

69. The tension in a ligament in the human knee is approximately proportional to the extension of the ligament, if the extension is not too large. If a particular ligament has an effective spring constant of 150 N/mm as it is stretched, (a) what is the tension in this ligament when it is stretched by 0.75 cm? (b) What is the elastic energy stored in the ligament when stretched by this amount?

70. An instrument known as an *atomic force microscope* (AFM) can be used to measure forces between atoms or molecules at the nanometer scale. Suppose you find the elasticity of biological membranes by measuring the indentation of the force probe into the membrane as a function of the applied force. You could then use an AFM to study the elastic properties of the capsid (outer shell) of a virus. The graph shows your data of the force applied to the capsid by the AFM (in nanonewtons) versus the indentation of the capsid (in nanometers). (a) What is the effective spring constant of the capsid for indentations of 0 to 14 nm? (b) How much elastic energy is stored in the membrane when it is indented 14 nm?

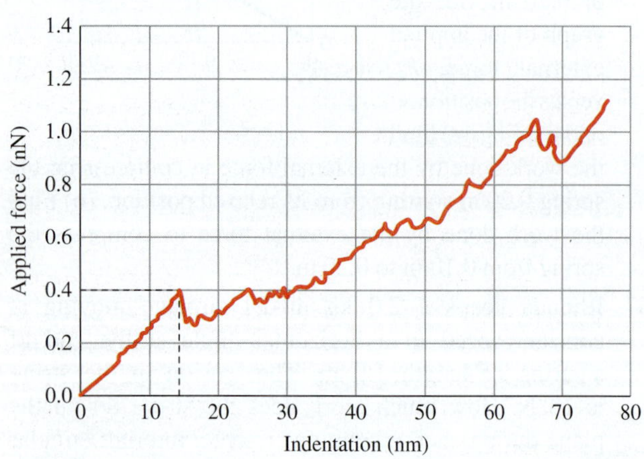

71. When the spring on a toy gun is compressed by a distance x, it will shoot a rubber ball straight up to a height of h. Ignoring air resistance, how high will the gun shoot the same rubber ball if the spring is compressed by an amount $2x$? Assume $x \ll h$.

72. You shoot a 51 g pebble straight up with a catapult whose spring constant is 320 N/m. The catapult is initially stretched by 0.20 m. How high above the starting point does the pebble fly? Ignore air resistance.

73. A gymnast of mass 52 kg is jumping on a trampoline. She jumps so that her feet reach a maximum height of 2.5 m above the trampoline and, when she lands, her feet stretch the trampoline down 75 cm. How far does the trampoline stretch when she stands on it at rest? [*Hint*: Assume the trampoline obeys Hooke's law when it is stretched.]

74. Jorge is going to bungee jump from a bridge that is 55.0 m over the river below. The bungee cord has an unstretched length of 27.0 m. To be safe, the bungee cord should stop Jorge's fall when he is at least 2.00 m above the river. If Jorge has a mass of 75.0 kg, what is the minimum spring constant of the bungee cord? (connect tutorial: spring scale)

75. A 2.0 kg block is released from rest and allowed to slide down a frictionless surface and into a spring. The far end of the spring is attached to a wall, as shown. The initial height of the block is 0.50 m above the lowest part of the slide and the spring constant is 450 N/m. (a) What is the block's speed when it is at a height of 0.25 m above the base of the slide? (b) How far is the spring compressed? (c) The spring sends the block back to the left. How high does the block rise?

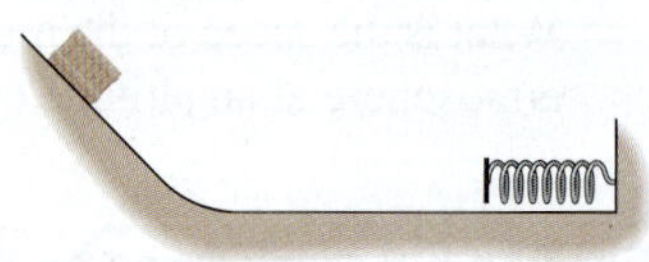

Problems 75 and 118

76. ✦ A block (mass m) hangs from a spring (spring constant k). The block is released from rest a distance d above its *equilibrium* position. (a) What is the speed of the block as it passes through the equilibrium point? (b) What is the maximum distance below the equilibrium point that the block will reach?

Section 6.8 Power

77. Lars, of mass 82.4 kg, can do work for about 2.0 min at the rate of 1.0 hp (746 W). How long will it take him to climb three flights of stairs, a vertical height of 12.0 m?

78. Show that 1 kilowatt-hour (kW·h) is equal to 3.6 MJ.

79. If a man has an average useful power output of 40.0 W, what minimum time would it take him to lift fifty 10.0 kg boxes to a height of 2.00 m?

80. In Section 6.2, Rosie lifts a trunk weighing 220 N up 4.0 m. If it takes her 40 s to lift the trunk, at what average rate does she do work?

81. A bicycle and its rider together have a mass of 75 kg. What power output of the rider is required to maintain a constant speed of 4.0 m/s (about 9 mi/h) up a 5.0% grade (a road that rises 5.0 m for every 100 m along the pavement)? Assume that frictional losses of energy are negligible.

82. The power output of a cyclist moving at a constant speed of 6.0 m/s on a level road is 120 W. (a) What is the force exerted on the cyclist and the bicycle by the air? (b) By bending low over the handlebars, the cyclist reduces the air resistance to 18 N. If she maintains a power output of 120 W, what will her speed be?

83. A patient's heart pumps 5.0 L of blood per minute into the aorta, which has a diameter of 1.8 cm. The average force exerted by the heart on the blood is 16 N. What is the average mechanical power output of the heart?

84. A motorist driving a 1200 kg car on level ground accelerates from 20.0 m/s to 30.0 m/s in a time of 5.0 s. Ignoring friction and air resistance, determine the *average* mechanical power in watts the engine must supply during this time interval.

85. A 62 kg woman takes 6.0 s to run up a flight of stairs. The landing at the top of the stairs is 5.0 m above her starting place. (a) What is the woman's average power output while she is running? (b) Would that be equal to her average power *input*—the rate at which chemical energy in food or stored fat is used? Why or why not?

86. How many grams of carbohydrate does a person of mass 74 kg need to metabolize to climb five flights of stairs (15 m height increase)? Each gram of carbohydrate provides 17.6 kJ of energy. Assume 10.0% efficiency—that is, 10.0% of the available chemical energy in the carbohydrate is converted to mechanical energy. What happens to the other 90% of the energy?

87. An object moves in the positive x-direction under the influence of a force F_x. A graph of F_x versus v_x is shown. (a) What is the instantaneous power on the object when its velocity is 11 m/s? (b) What is the instantaneous power on the object when its velocity is 16 m/s?

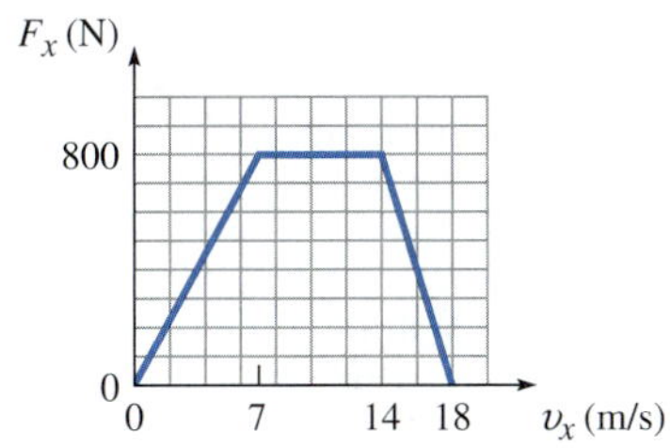

88. A top fuel drag racer with a mass of 500.0 kg completes a quarter-mile (402 m) drag race in a time of 4.2 s starting from rest. The car's final speed is 125 m/s. What is the engine's average power output? Ignore friction and air resistance.

89. (a) Calculate the change in potential energy of 1 kg of water as it passes over Niagara Falls (a vertical descent of 50 m). (b) At what rate is gravitational potential energy lost by the water of the Niagara River? The rate of flow is 5.5×10^6 kg/s. (c) If 10% of this energy can be converted into electric energy, how many households would the electricity supply? (An average household uses an average electrical power of about 1 kW.)

90. A car with mass of 1000.0 kg accelerates from 0 m/s to 40.0 m/s in 10.0 s. Ignore air resistance. The engine has a 22% efficiency, which means that 22% of the energy released by the burning gasoline is converted into mechanical energy. (a) What is the average mechanical power output of the engine? (b) What volume of gasoline is consumed? Assume that the burning of 1.0 L of gasoline releases 46 MJ of energy.

Collaborative Problems

91. You are driving a car through campus when a fellow student steps out in front of you. You slam on the brakes, creating a 9.0 m long skid mark as measured by the police officer standing on the corner. She also has a device that measures the coefficient of friction between rubber and asphalt as 0.60. Can she write you a ticket for speeding in this 25 mi/h zone?

92. A roller coaster car (mass = 988 kg including passengers) is about to roll down a track. The diameter of the circular loop is 20.0 m and the car starts out from rest 40.0 m above the lowest point of the track. Ignore friction and air resistance. (a) At what speed does the car reach the top of the loop? (b) What is the force exerted on the car by the track at the top of the loop? (c) From what minimum height above the bottom of the loop can the car be released so that it does not lose contact with the track at the top of the loop?

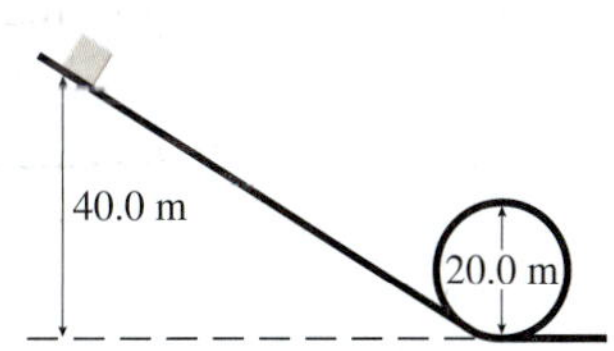

93. A 4.0 kg block is released from rest at the top of a frictionless plane of length 8.0 m that is inclined at an angle of 15° to the horizontal. A cord is attached to the block and trails along behind it. When the block reaches a point 5.0 m along the incline from the top, someone grasps the cord and pulls it parallel to the incline so that the tension is constant until the block comes to rest at the bottom of the incline. What is the tension? Solve the problem twice, once using work and energy and again using Newton's laws and

the equations for constant acceleration. Which method do you prefer?

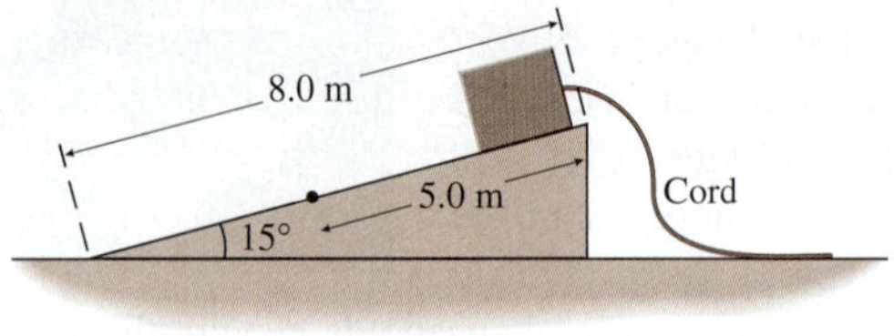

94. ✦ The bungee jumper of Example 6.4 made a jump into the Gorge du Verdon in southern France from a platform 182 m above the bottom of the gorge. The jumper weighed 780 N and came within 68 m of the bottom of the gorge. The cord's unstretched length is 30.0 m. (a) Assuming that the bungee cord follows Hooke's law when it stretches, find its spring constant. [*Hint*: The cord does not begin to stretch until the jumper has fallen 30.0 m.] (b) At what speed is the jumper falling when he reaches a height of 92 m above the bottom of the gorge?

95. ✦ A 1500 kg car coasts in neutral down a 2.0° hill. The car attains a terminal speed of 20.0 m/s. (a) How much power must the engine deliver to drive the car on a *level* road at 20.0 m/s? (b) If the maximum useful power that can be delivered by the engine is 40.0 kW, what is the steepest hill the car can climb at 20.0 m/s?

96. ✦ The escape speed from Earth is 11.2 km/s, but that is only the minimum speed needed to escape *Earth's* gravitational pull; it does not give the object enough energy to leave the solar system. What is the minimum speed for an object near the Earth's surface so that the object escapes both the Earth's and the Sun's gravitational pulls? Ignore drag due to the atmosphere and the gravitational forces due to the Moon and the other planets. Also ignore the rotation and the orbital motion of the Earth.

97. ✦ Ⓒ A wind turbine converts some of the kinetic energy of the wind into electric energy. Suppose that the blades of a small wind turbine have length $L = 4.0$ m. (a) When a 10 m/s (22 mi/h) wind blows head-on, what volume of air (in m^3) passes through the circular area swept out by the blades in 1.0 s? (b) What is the mass of this much air? Each cubic meter of air has a mass of 1.2 kg. (c) What is the translational kinetic energy of this mass of air? (d) If the turbine can convert 40% of this kinetic energy into electric energy, what is its electric power output? (e) What happens to the power output if the wind speed decreases to $\frac{1}{2}$ of its initial value? What can you conclude about electric power production by wind turbines?

Comprehensive Problems

98. A spring scale in a French market is calibrated to show the *mass* of vegetables in grams and kilograms. (a) If the marks on the scale are 1.0 mm apart for every 25 g, what maximum extension of the spring is required to measure up to 5.0 kg? (b) What is the spring constant of the spring? [*Hint*: Remember that the scale really measures *force*.]

99. Plot a graph of this data for a spring resting horizontally on a table. Use your graph to find (a) the spring constant and (b) the relaxed length of the spring.

Force (N)	0.200	0.450	0.800	1.500
Spring length (cm)	13.3	15.0	17.3	22.0

100. ✦ A particle, constrained to move along the x-axis, has a total mechanical energy of −100 J. The potential energy of the particle is shown in the graph. At time $t = 0$, the particle is located at $x = 5.5$ cm and is moving to the left. (a) What is the particle's potential energy at $t = 0$? What is its kinetic energy at this time? (b) What are the particle's total, potential, and kinetic energies when it is at $x = 1$ cm and moving to the right? (c) What is the particle's kinetic energy when it is at $x = 3$ cm and moving to the left? (d) Describe the motion of this particle starting at $t = 0$.

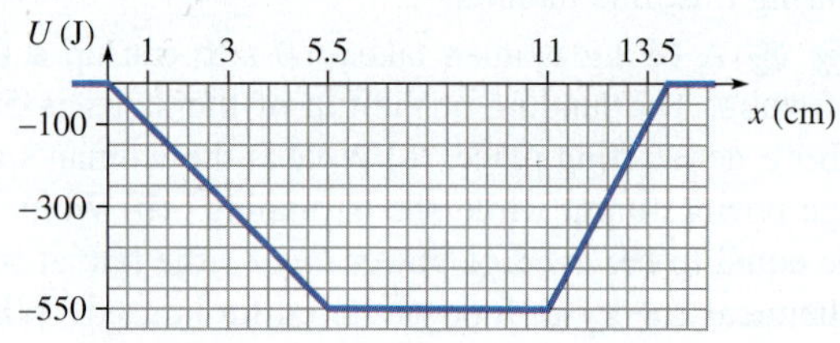

101. How much energy is expended by an 80.0 kg person in climbing a vertical distance of 15 m? Assume that muscles have an efficiency of 22%; that is, the work done by the muscles to climb is 22% of the energy expended.

102. Ugonna stands at the top of an incline and pushes a 100 kg crate to get it started sliding down the incline. The crate slows to a halt after traveling 1.50 m along the incline. (a) If the initial speed of the crate was 2.00 m/s and the angle of inclination is 30.0°, how much energy was dissipated by friction? (b) What is the coefficient of sliding friction?

103. A child's playground swing is supported by chains that are 4.0 m long. If the swing is 0.50 m above the ground and moving at 6.0 m/s when the chains are vertical, what is the maximum height of the swing?

104. A block slides down a plane that is inclined at an angle of 53° with respect to the horizontal. If the coefficient of kinetic friction is 0.70, what is the acceleration of the block?

105. If a high jumper needs to make his center of gravity rise 1.2 m, how fast must he be able to sprint? Assume all of his kinetic energy can be transformed into potential energy. For an extended object, the gravitational potential energy is $U = mgh$, where h is the height of the center of gravity.

106. A pole-vaulter converts the kinetic energy of running to elastic potential energy in the pole, which is then converted to gravitational potential energy. If a pole-vaulter's center of gravity is 1.0 m above the ground while he sprints at 10.0 m/s, what is the maximum height of his center of gravity during the vault? For an extended object, the gravitational potential energy is $U = mgh$, where h is the height of the center of gravity. (In 1988, Sergei Bubka was the first pole-vaulter ever to clear 6 m.)

107. A hang glider moving at speed 9.5 m/s dives to an altitude 8.2 m lower. Ignoring drag, how fast is it then moving?

108. A car moving at 30 mi/h is stopped by jamming on the brakes and locking the wheels. The car skids 50 ft before coming to rest. How far would the car skid if it were initially moving at 60 mi/h? [*Hint*: You will not have to do any unit conversions if you set up the problem as a proportion.]

109. A spring gun ($k = 28$ N/m) is used to shoot a 56 g ball horizontally. Initially the spring is compressed by 18 cm. The ball loses contact with the spring and leaves the gun when the spring is still compressed by 12 cm. What is the speed of the ball when it hits the ground, 1.4 m below the spring gun?

110. A spring with $k = 40.0$ N/m is at the base of a frictionless 30.0° inclined plane. A 0.50 kg object is pressed against the spring, compressing it 0.20 m from its equilibrium position. The object is then released. If the object is not attached to the spring, how far up the incline does it travel before coming to rest and then sliding back down?

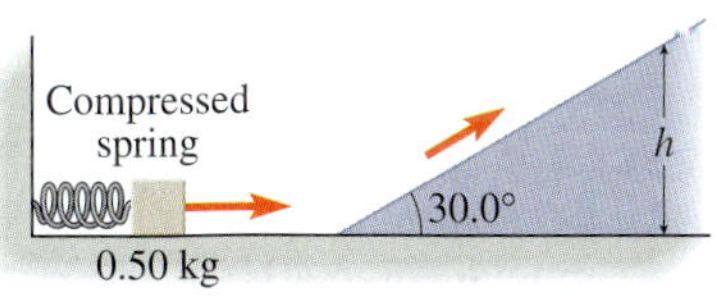

111. In an adventure movie, a 62.5 kg stunt woman falls 8.10 m and lands in a huge air bag. Her speed just before she hit the air bag was 10.5 m/s. (a) What is the total work done on the stunt woman during the fall? (b) How much work is done by gravity on the stunt woman? (c) How much work is done by air resistance on the stunt woman? (d) Estimate the magnitude of the average force of air resistance by assuming it is constant throughout the fall.

112. Yosemite Falls in California is about 740 m high. (a) What average power would it take for a 70 kg person to hike up to the top of Yosemite Falls in 1.5 h? (b) The human body is about 25% efficient at converting chemical energy to mechanical energy. How much chemical energy is used in this hike? (c) How many kilocalories of food energy would a person use in this hike?

113. (a) How much work does a major-league pitcher do on the baseball when he throws a 90.0 mi/h (40.2 m/s) fastball? The mass of a baseball is 153 g. (b) How many fastballs would a pitcher have to throw to "burn off" a 1520 kcal meal? Assume that 80.0% of the chemical energy in the food is converted to thermal energy and only 20.0% becomes the kinetic energy of the fastballs.

114. The amount of food energy per day required by a person resting under standard conditions is called the basal metabolic rate (BMR). (a) To generate 1 kcal, Jermaine's body needs approximately 0.010 mol of oxygen. If Jermaine's net intake of oxygen through breathing is 0.015 mol/min while he is resting, what is his BMR in kcal/day? (b) If Jermaine fasts for 24 h, how many pounds of fat does he lose? Assume that only fat is consumed. Each gram of fat consumed generates 9.3 kcal.

115. Tarzan is running along the ground and approaching a deep gully. A tree branch with a vine hangs over the gully. Tarzan must grab the vine and swing across the gully to the other side, where the ground surface is 1.7 m higher than the ground surface from which Tarzan starts. How fast does Tarzan have to be running to accomplish this feat?

116. Jane is running from the ivory hunters in the jungle. Cheetah throws a 7.0 m long vine toward her. Jane leaps onto the vine with a speed of 4.0 m/s. When she catches the vine, it makes an angle of 20° with respect to the vertical. (a) When Jane is at her lowest point, she has moved downward a distance h from the height where she originally caught the vine. Show that h is given by $h = L - L \cos 20°$, where L is the length of the vine. (b) How fast is Jane moving when she is at the lowest point in her swing? (c) How high can Jane swing above the lowest point in her swing?

117. Human feet and legs store elastic energy when walking or running. They are not nearly as efficient at doing so as kangaroo legs, but the effect is significant nonetheless. If not for the storage of elastic energy, a 70 kg man running at 4 m/s would lose about 100 J of mechanical energy each time he sets down a foot. Some of this energy is stored as elastic energy in the Achilles tendon and in the arch of the foot; the elastic energy is then converted back into the kinetic and gravitational potential energy of the leg, reducing the expenditure of metabolic energy. If the maximum tension in the Achilles tendon when the foot is set down is 4.7 kN and the tendon's spring constant is 350 kN/m, calculate how far the tendon stretches and how much elastic energy is stored in it.

118. A 0.50 kg block, starting at rest, slides down a 30.0° incline with a kinetic friction coefficient of 0.25 (see the figure with Problem 75). After sliding 85 cm down the incline, it slides across a frictionless horizontal surface and encounters a spring ($k = 35$ N/m). (a) What is the maximum compression of the spring? (b) After the

compression of part (a), the spring rebounds and shoots the block back up the incline. How far along the incline does the block travel before coming to rest?

119. The potential energy of a particle constrained to move along the x-axis is shown in the graph. At $x = 0$, the particle is moving in the $+x$-direction with a kinetic energy of 200 J. Can this particle get into the region 3 cm < x < 8 cm? Explain. If it can, what is its kinetic energy in that region? If it can't, what happens to it?

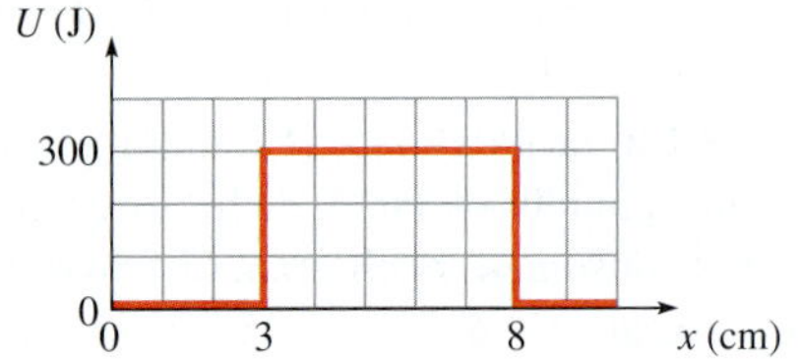

Problems 119 and 120

120. The potential energy of a particle constrained to move along the x-axis is shown in the graph. At $x = 0$, the particle is moving in the $+x$-direction with a kinetic energy of 400 J. Can this particle get into the region 3 cm < x < 8 cm? Explain. If it can, what is its kinetic energy in that region? If it can't, what happens to it?

121. ✦ Show that $U = -2K$ for any gravitational circular orbit. [*Hint*: Use Newton's second law to relate the gravitational force to the acceleration required to maintain uniform circular motion.]

122. ✦ Two springs with equal spring constants k are connected first in series (one after the other) and then in parallel (side by side) with a weight hanging from the bottom of the combination. What is the effective spring constant of the two different arrangements? In other words, what would be the spring constant of a single spring that would behave exactly as (a) the series combination and (b) the parallel combination? Ignore the weight of the springs. [*Hint* for (a): *each* spring stretches an amount $x = F/k$, but only one spring exerts a force on the hanging object. *Hint* for (b): *each* spring exerts a force $F = kx$.]

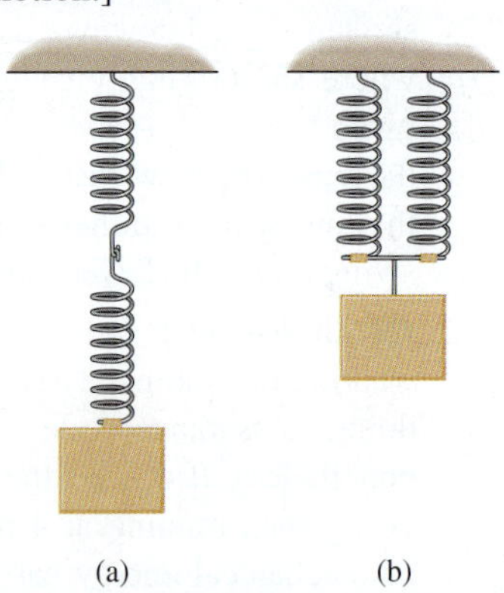

Problems 122, 126, and 127

123. ✦ When a 0.20 kg mass is suspended from a vertically hanging spring, it stretches the spring from its original length of 5.0 cm to a total length of 6.0 cm. The spring with the same mass attached is then placed on a horizontal frictionless surface. The mass is pulled so that the spring stretches to a *total* length of 10.0 cm; then the mass is released, and it oscillates back and forth. What is the maximum speed of the mass as it oscillates?

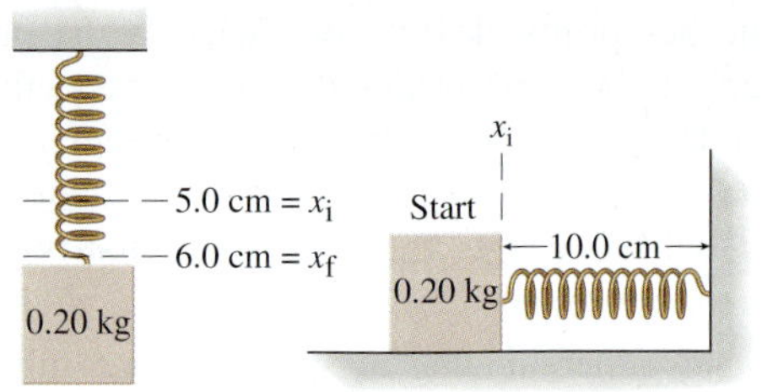

124. ✦ A spring used in an introductory physics laboratory stores 10.0 J of elastic potential energy when it is compressed 0.20 m. Suppose the spring is cut in half. When one of the halves is compressed by 0.20 m, how much potential energy is stored in it? [*Hint*: Does the half spring have the same k as the original uncut spring?]

125. ✦ Ⓒ An elevator can carry a maximum load of 1202 kg (including the mass of the elevator car). The elevator has an 801 kg counterweight that always moves with the same speed but in the *opposite direction* to the car. (a) What is the average power that must be delivered by the motor to carry the maximum load up 40.0 m in 60.0 s? (b) How would your answer be different if there were no counterweight?

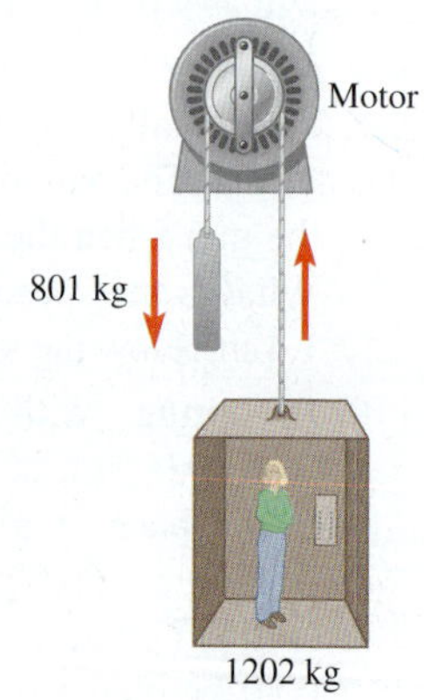

126. ✦ Two springs with spring constants k_1 and k_2 are connected in series. (a) What is the effective spring constant of the combination? (b) If a hanging object attached to the combination is displaced by 4.0 cm from the relaxed position, what is the potential energy stored in the spring for $k_1 = 5.0$ N/cm and $k_2 = 3.0$ N/cm? [See Problem 122(a).]

127. ✦ Two springs with spring constants k_1 and k_2 are connected in parallel. (a) What is the effective spring constant of the combination? (b) If a hanging object attached to the combination is displaced by 2.0 cm from the relaxed position, what is the potential energy stored in the spring for $k_1 = 5.0$ N/cm and $k_2 = 3.0$ N/cm? [See Problem 122(b).]

128. ✦ Ⓒ The graph shows the tension in a rubber band as it is first stretched and then allowed to contract. As you stretch a rubber band, the tension at a particular length (on the way to a maximum stretch) is larger than the tension at that same length as you let the rubber band contract. That is why the graph shows two separate lines, one for stretching and one for contracting; the

lines are not superimposed as you might have thought they would be. (a) Make a rough estimate of the total work done by the external force applied to the rubber band for the entire process. (b) If the rubber band obeyed Hooke's law, what would the answer to (a) have to be? (c) While the rubber band is stretched, is all of the work done on it accounted for by the increase in elastic potential energy? If not, what happens to the rest of it? [*Hint*: Take a rubber band and stretch it rapidly several times. Then hold it against your wrist or your lip.]

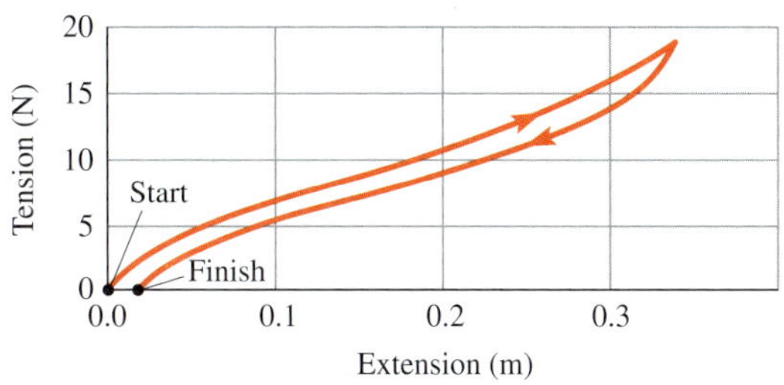

129. ✦ Use dimensional analysis to show that the electric power output of a wind turbine is proportional to the *cube* of the wind speed. The relevant quantities on which the power can depend are the length L of the rotor blades, the density ρ of air (SI units kg/m^3), and the wind speed v.

130. ✦ Use this method to find how the speed with which animals of similar shape can run up a hill depends on the size of the animal. Let L represent some characteristic length, such as the height or diameter of the animal. Assume that the maximum rate at which the animal can do work is proportional to the animal's surface area: $P_{max} \propto L^2$. Set the maximum power output equal to the rate of increase of gravitational potential energy and determine how the speed v depends on L.

Review and Synthesis

131. A block is released from rest and slides down an incline. The coefficient of sliding friction is 0.38, and the angle of inclination is 60.0°. (a) Use energy considerations to find how fast the block is sliding after it has traveled a distance of 30.0 cm along the incline. (b) Solve the same problem using Newton's second law instead of energy. Which method is easier?

132. A pendulum consists of a bob of mass m attached to the end of a cord of length L. The pendulum is released from a point at a height of $L/2$ above the lowest point of the swing. What is the tension in the cord as the bob passes the lowest point?

133. Two blocks of masses m_1 and m_2, resting on frictionless inclined planes, are connected by a massless rope passing over an ideal pulley. Angle $\phi = 45.0°$ and angle $\theta = 36.9°$; mass m_1 is 6.00 kg and mass m_2 is 4.00 kg. (a) Using energy conservation, find how fast the blocks are moving after they travel 2.00 m along the inclines. (b) Now solve the same problem using Newton's second law. [*Hint*: First find the acceleration of each of the blocks. Then find how fast either block is moving after it travels 2.00 m along the incline with constant acceleration.]

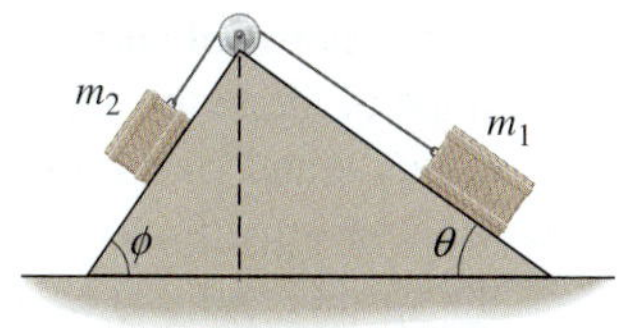

134. Tarzan wants to swing on a vine across a river. He is standing on a ledge above the water's edge, and the river is 5.00 m wide. The vine is attached to a tree branch that is 8.00 m directly above the opposite edge of the river. Initially the vine makes a 60.0° angle with the vertical as he is holding it. He swings across starting from rest, but unfortunately the vine breaks when the vine is 20.0° from the vertical. (a) Assuming Tarzan weighs 900.0 N, what was the tension in the vine just before it broke? (b) Does he land safely on the other side of the river?

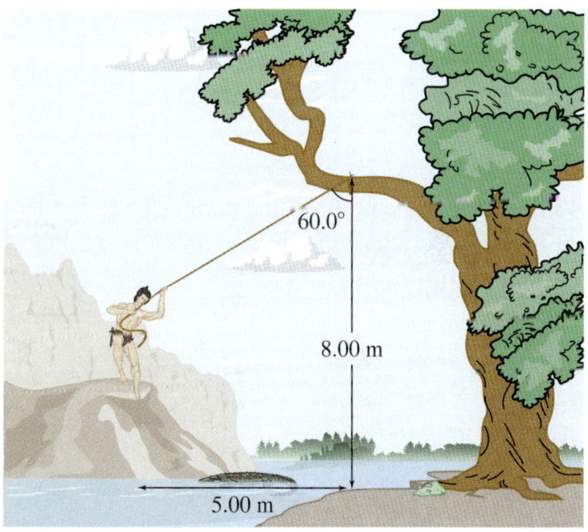

135. Three rocks are thrown from a cliff with the same initial speeds but in different directions: rock A is thrown straight down, rock B is thrown straight up, and rock C is thrown horizontally. Ignore air resistance. (a) Rank the speeds of the three rocks just before they hit the flat ground at the bottom of the cliff. (b) Compare two methods of answering this question: using constant-acceleration kinematics versus using energy. Which is easier? Why?

136. A packing carton slides down an inclined plane of angle 30.0° and of incline length 2.0 m. (a) If the initial speed of the carton is 4.0 m/s directed down the incline, what is the speed at the bottom? Ignore friction. (b) How long does it take the carton to slide down the incline?

137. ✦ A skier starts from rest at the top of a frictionless slope of ice in the shape of a hemispherical dome with radius R and slides down the slope. At a certain height h, the normal force becomes zero and the skier leaves the surface of the ice. What is h in terms of R?

138. A pendulum bob hung from a string of length L is released from height h above the bottom of its swing. The pendulum's swing is interrupted by a horizontal peg at height r above the bottom of the pendulum's swing (point B). After the string hits the peg, the bob swings around in a circular arc of radius r. Express your answers in terms of L, r, and g, as needed. (a) If the bob is to travel in a full circle of radius r around the peg without the string going slack, what is the minimum possible speed it can have at the top of that circle (point A)? (b) If the string breaks when the bob is at point A, moving at the minimum speed found in part (a), how far to the right of point B is the bob when it's at the same height as point B? (c) What is the minimum value of h for which the bob will make it to point A without going slack?

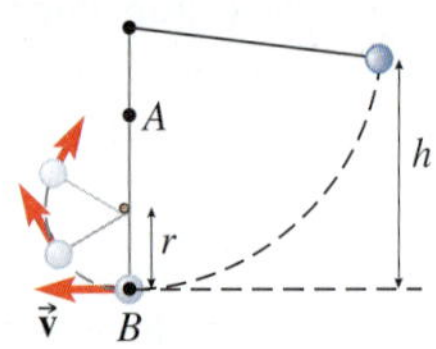

Answers to Practice Problems

6.1 −68 kJ

6.2 43 N; 4500 J; she pulls with a greater force but its component in the direction of the displacement is smaller.

6.3 $(2.5\,m)(1.50v)^2/(mv^2) = 5.6$

6.4 29 m/s

6.5 0.24

6.6 16.5 m/s

6.7 48 km/s

6.8 195 km/s

6.9 4.0 J

6.10 3.2 cm

6.11 9.8 m/s

6.12 2.4 m

6.13 2×10^{-18} W

6.14 (a) 11.8 kW (b) 3.6 kW

Answers to Checkpoints

6.2 The force is perpendicular to the displacement.

6.3A Kinetic energy is never negative. Work can be positive, negative, or zero, because kinetic energy can increase, decrease, or stay the same.

6.3B (b), (c), (a) = (e), (d)

6.4A (a) The gravitational potential energy increases until it reaches its maximum value when the stone reaches its highest point above the ground. (b) The kinetic energy decreases as the potential energy increases. It is zero at the highest point.

(c)

Kinetic and potential energies for a stone thrown upward as a function of height

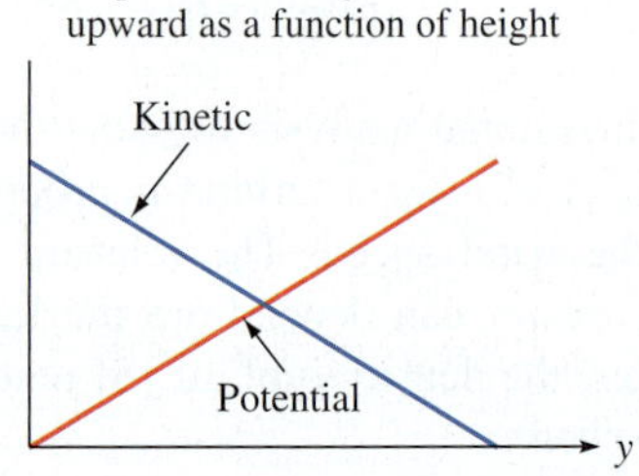

6.4B The final kinetic energy is smaller because the final speed is smaller. The initial and final potential energies are equal because the ball is at the same height. The final mechanical energy is less than the initial. The decrease in mechanical energy is caused by the nonconservative work done by the force of air resistance; some of the ball's mechanical energy has been converted into other forms as the ball stirs up the air.

6.5 The mechanical energy is the same throughout Mercury's orbit. The kinetic energy is greatest at the perihelion because the potential energy is smallest there.

6.7 The greatest elastic potential energy is at the maximum compression.

CHAPTER

7

Linear Momentum

After a collision, an accident investigator measures the lengths of skid marks on the road. How can the investigator use this information to figure out the velocities of the vehicles immediately before the collision? (See p. 253 for the answer.)

BIOMEDICAL APPLICATIONS

- Protecting the body from injury (Sec. 7.3; Ex. 7.2; PP 7.2)
- Jet propulsion in squid (Ex. 7.5)
- Ballistocardiography (Sec. 7.4)
- Center of mass of a pregnant woman (Prob. 33)
- Molecular motors (Prob. 97)

Concepts & Skills to Review

- conservation laws (Section 6.1)
- Newton's third law of motion (Section 4.4)
- Newton's second law of motion (Section 4.3)
- velocity (Section 3.3)
- components of vectors (Section 3.2)
- vector subtraction (Section 3.1)
- kinetic energy (Section 6.3)

7.1 A CONSERVATION LAW FOR A VECTOR QUANTITY

In Chapter 4 we learned how to determine the acceleration of an object by finding the net force acting on it and applying Newton's second law of motion. If the forces happen to be constant, then the resulting constant acceleration enables us to calculate changes in velocity and position. Calculating velocity and position changes when the forces are not constant is much more difficult. In many cases, the forces cannot even be easily determined. Conservation of energy is one tool that enables us to draw conclusions about motion without knowing all the details of the forces acting. Recall, for example, how easily we can calculate the escape speed of a projectile using conservation of energy, without even knowing the path the object takes. Now imagine how difficult the same calculation would be using Newton's second law, with a gravitational force that changes magnitude and direction depending on the path taken.

In this chapter we develop another conservation law. Conservation laws are powerful tools. If a quantity is conserved, then no matter how complicated the situation, we can set the value of the conserved quantity at one time equal to its value at a later time. The "before-and-after" aspect of a conservation law enables us to draw conclusions about the results of a complicated set of interactions without knowing all of the details.

CONNECTION:

Conservation laws can involve scalars, such as energy, or vectors, such as momentum.

The new conserved quantity, *momentum*, is a vector quantity, in contrast to energy, which is a scalar. When momentum is conserved, both the magnitude and the direction of the momentum must be constant. Equivalently, the x- and y-components of momentum are constant. When we find the total momentum of more than one object, we must add the momentum vectors according to the procedure by which vectors are always added.

7.2 MOMENTUM

The word *momentum* is often heard in broadcasts of sporting events. A sports broadcaster might say, "The home team has won five consecutive games; they have the momentum in their favor." The team with "momentum" is hard to stop; they are moving forward on a winning streak. A football player, running for the goal line with a football tucked under his arm, has momentum; he is hard to stop. This use of the word *momentum* is closer to the physics usage. In physics we would agree that the runner has momentum, but we have a precise definition in mind.

In everyday use, momentum has something to do with mass as well as with velocity. Would you rather have a running child bump into you, or a football player running with the same velocity? The child has much less momentum than the football player, even though their velocities are the same.

Could a quantity combining mass and velocity be useful in physics? Imagine a collision between two spaceships (Fig. 7.1). Let the spaceships be so far from planets and stars that we can ignore gravitational interactions with celestial bodies. The spaceships exert forces on one another while they are in contact. According to Newton's third law, these forces are equal and opposite. The force on ship 2 exerted by ship 1 is equal and opposite to the force exerted on ship 1 by ship 2:

$$\vec{\mathbf{F}}_{21} = -\vec{\mathbf{F}}_{12}$$

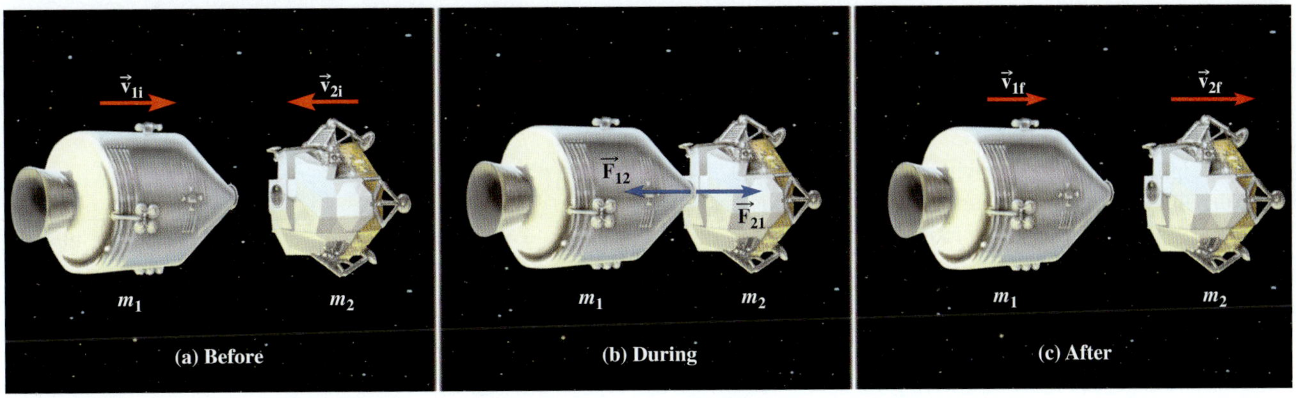

Figure 7.1 (a) Two spaceships about to collide. (b) During the collision, the spaceships exert forces on one another that are equal in magnitude and opposite in direction. (c) The velocities of the spaceships after the collision.

The changes in *velocities* of the two spaceships are *not* equal and opposite if the masses are different. Suppose a large spaceship (mass m_1) collides with a much smaller ship (mass $m_2 \ll m_1$). Assume for now that the forces are constant during the time interval Δt that the spaceships are in contact. Although the forces have the same magnitude, the magnitudes of the accelerations of the two ships are different because their masses are different. The ship with the larger mass has the smaller acceleration.

The acceleration of either spaceship causes its velocity to change by

$$\Delta\vec{\mathbf{v}} = \vec{\mathbf{a}}\,\Delta t = \frac{\vec{\mathbf{F}}}{m}\Delta t$$

The time interval Δt is the duration of the interaction between the two ships, so it must be the same for both ships.

Since the changes in velocity are inversely proportional to the masses, the changes in the *products* of mass and velocity are equal and opposite for the two objects involved in the interaction:

$$m_1\Delta\vec{\mathbf{v}}_1 = \vec{\mathbf{F}}_{12}\Delta t$$

$$m_2\Delta\vec{\mathbf{v}}_2 = \vec{\mathbf{F}}_{21}\Delta t = (-\vec{\mathbf{F}}_{12})\Delta t = -(m_1\Delta\vec{\mathbf{v}}_1)$$

This is a useful insight, so we give the product of mass and velocity a name and symbol: **linear momentum** (symbol $\vec{\mathbf{p}}$, SI unit kg·m/s). Linear momentum (or just *momentum*) is a vector quantity having the same direction as the velocity.

Definition of linear momentum

$$\vec{\mathbf{p}} = m\vec{\mathbf{v}} \qquad (7\text{-}1)$$

The collision of the two spaceships causes changes in their momenta that are equal in magnitude and opposite in direction:

$$\Delta\vec{\mathbf{p}}_2 = -\Delta\vec{\mathbf{p}}_1$$

In any interaction between two objects, momentum can be transferred from one object to the other. The momentum changes of the two objects are always equal and opposite, so the total momentum of the two objects is unchanged by the interaction. (By *total momentum* we mean the vector sum of the individual momenta of the objects.)

CONNECTION:

Newton's third law implies that during an interaction momentum is transferred from one object to another.

Example 7.1 gives some practice in finding the change in momentum of an object whose velocity changes. Remember that momentum is a vector quantity, so changes in momentum must be found by subtracting momentum vectors, not by subtracting the magnitudes of the momenta.

Example 7.1

Change of Momentum of a Moving Car

A car weighing 12 kN is driving due north at 30.0 m/s. After driving around a sharp curve, the car is moving east at 13.6 m/s. What is the change in momentum of the car?

Strategy The definition of momentum is $\vec{\mathbf{p}} = m\vec{\mathbf{v}}$. We can start by finding the car's mass. There are two potential pitfalls:

1. momentum depends not on weight but on mass, and
2. momentum is a vector, so we must take its direction into consideration as well as its magnitude. To find the change in momentum, we need to do a vector subtraction.

Solution The car's mass is

$$m = \frac{W}{g} = \frac{1.2 \times 10^4 \text{ N}}{9.8 \text{ m/s}^2} = 1220 \text{ kg}$$

The car's initial velocity is

$$\vec{\mathbf{v}}_i = 30.0 \text{ m/s, north}$$

The car's initial momentum is then

$$\vec{\mathbf{p}}_i = m\vec{\mathbf{v}}_i = 1220 \text{ kg} \times 30.0 \text{ m/s north}$$
$$= 3.66 \times 10^4 \text{ kg·m/s north}$$

After the curve, the final velocity is

$$\vec{\mathbf{v}}_f = 13.6 \text{ m/s, east}$$

The final momentum is

$$\vec{\mathbf{p}}_f = m\vec{\mathbf{v}}_f = 1220 \text{ kg} \times 13.6 \text{ m/s east}$$
$$= 1.66 \times 10^4 \text{ kg·m/s east}$$

Momentum vectors are added and subtracted according to the same methods used for other vectors. To find the change in the momentum, we draw vector arrows representing the addition of $\vec{\mathbf{p}}_f$ and $-\vec{\mathbf{p}}_i$ (Fig. 7.2). Since *in this case* the three vectors in Fig. 7.2 form a right triangle, the magnitude of $\Delta\vec{\mathbf{p}}$ can be found from the Pythagorean theorem

$$|\Delta\vec{\mathbf{p}}| = \sqrt{p_i^2 + p_f^2}$$
$$= \sqrt{(3.66 \times 10^4 \text{ kg·m/s})^2 + (1.66 \times 10^4 \text{ kg·m/s})^2}$$
$$= 4.02 \times 10^4 \text{ kg·m/s}$$

From the vector diagram, $\Delta\vec{\mathbf{p}}$ is directed at an angle θ east of south. Using trigonometry,

$$\tan\theta = \frac{\text{opposite}}{\text{adjacent}} = \frac{p_f}{p_i} = \frac{1.66 \times 10^4 \text{ kg·m/s}}{3.66 \times 10^4 \text{ kg·m/s}} = 0.454$$

$$\theta = \tan^{-1} 0.454 = 24.4°$$

Since the weight is given with two significant figures, we report the change in momentum of the car as 4.0×10^4 kg·m/s directed 24° east of south.

Discussion As with displacements, velocities, accelerations, and forces, it is crucial to remember that momentum is a vector. When finding changes in momentum, we must find the difference between final and initial momentum *vectors*. If the initial and final momenta had not been perpendicular, we would have had to resolve the vectors into x- and y-components in order to subtract them.

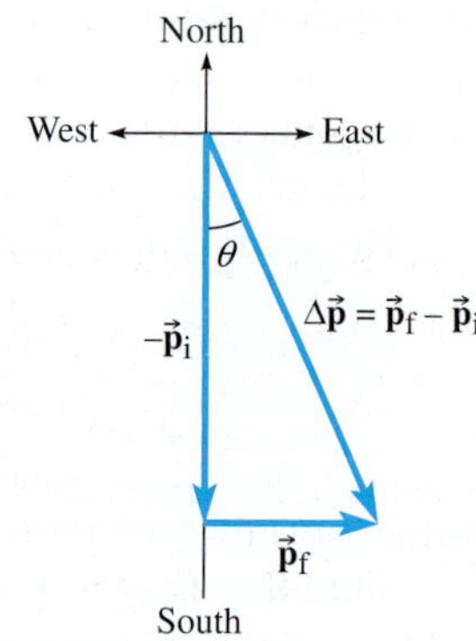

Figure 7.2 Vector subtraction to find the change in momentum.

Practice Problem 7.1 Falling Apple

(a) What is the momentum of an apple weighing 1.0 N just before it hits the ground if it falls out of a tree from a height of 3.0 m? (b) The apple falls because of the gravitational interaction between the apple and the Earth. How much does this interaction change *Earth's* momentum? How much does it change Earth's velocity?

CHECKPOINT 7.2

In Example 7.1, if the speed of the car had remained constant, would $\Delta\vec{\mathbf{p}}$ have been zero?

7.3 THE IMPULSE-MOMENTUM THEOREM

We found that the change in momentum of an object when a single force acts on it is equal to the product of the force acting on the object and the time interval during which the force acts:

$$\Delta\vec{\mathbf{p}} = \vec{\mathbf{F}}\,\Delta t$$

The product $\vec{\mathbf{F}}\,\Delta t$ is given the name **impulse** and the symbol $\vec{\mathbf{J}}$. Since the impulse is the product of a vector (the force) and a positive scalar (the time), impulse is a vector quantity having the same direction as that of the force. In words, $\Delta\vec{\mathbf{p}} = \vec{\mathbf{F}}\,\Delta t$ can be read as "*the change in momentum equals the impulse*." The SI units of impulse are newton-seconds (N·s) and those of momentum are kilogram-meters per second (kg·m/s). These are equivalent units, as can be demonstrated using the definition of the newton (Problem 3).

If an object is involved in more than one interaction, then its change in momentum during any time interval is equal to the *total* impulse during that time interval. The total impulse is the vector sum of the impulses due to each force. The total impulse is also equal to the net force times the time interval:

$$\vec{\mathbf{J}}_{\text{total}} = \vec{\mathbf{F}}_1\,\Delta t + \vec{\mathbf{F}}_2\,\Delta t + \cdots$$
$$= (\vec{\mathbf{F}}_1 + \vec{\mathbf{F}}_2 + \cdots)\,\Delta t = \sum\vec{\mathbf{F}}\,\Delta t$$

The total impulse on an object is equal to the change in the object's momentum during the same time interval. This relationship between total impulse and momentum change is called the impulse-momentum theorem and is especially useful in solving problems that involve collisions and impacts.

CONNECTION:

Impulse is a momentum transfer due to a force; work is an energy transfer due to a force.

	Impulse	Work
Definition	$\vec{\mathbf{F}}\,\Delta t$	$\vec{\mathbf{F}}\cdot\Delta\vec{\mathbf{r}}$
Vector or Scalar?	Vector	Scalar*
Physical meaning	Momentum transfer	Energy transfer

*The scalar or dot product of two vectors is introduced in Section 6.2.

Impulse-momentum theorem

$$\Delta\vec{\mathbf{p}} = \vec{\mathbf{J}}_{\text{total}} = \sum\vec{\mathbf{F}}\,\Delta t \tag{7-2}$$

Impulse When Forces Are Not Constant Our discussion so far has assumed that the forces acting are constant or that Δt is very small so the change in $\vec{\mathbf{F}}$ is negligible. That is a rather unusual situation; the concept of momentum would be of limited use if it were applicable only when forces are constant. However, everything we have said still applies to situations where the forces are not constant, as long as we use the *average* force to calculate the impulse.

$$\Delta\vec{\mathbf{p}} = \vec{\mathbf{J}}_{\text{total}} = \sum\vec{\mathbf{F}}_{\text{av}}\,\Delta t \tag{7-3}$$

Conceptual Example 7.2

Protecting the Body from Injury

Which causes the larger change in momentum of an object, an average force of 5 N acting for 4 s or an average force of 2 N acting for 10 s? How might this principle be used when designing products to protect the human body from injury? Give an example.

Solution and Discussion The change in momentum is equal to the impulse. The product of the force and the time interval gives the momentum change of the object. Over a period of 4 s, the 5 N force causes a momentum change of magnitude (5 N × 4 s) = 20 N·s, and the 2 N force acting for 10 s also causes a momentum change of magnitude (2 N × 10 s) = 20 N·s. The smaller force causes the same change in momentum because it acts for a longer time interval.

When designing products to protect the human body, one goal is to lengthen the time period during which a velocity change occurs. For example, when a movie stuntman falls from a great height, he lands on a large air bag (Fig. 7.3),

Figure 7.3

A stuntman lands safely on an air bag to break his fall. The air bag reduces the risk of injury in two ways. It changes the stuntman's momentum more gradually, so that forces of smaller magnitude act on his body. It also spreads these forces over a larger area so they are less likely to cause serious injury.

continued on next page

Conceptual Example 7.2 continued

which changes his momentum much more gradually than if he were to fall onto concrete. The average force exerted by the air bag on the stuntman is much smaller than the average force exerted by concrete would be. Nets used under circus acrobats serve the same purpose. The net gives and dips downward when the acrobat falls into it, gradually reducing the speed of the fall over a longer time interval than if she fell directly onto the ground.

Practice Problem 7.2 Pole-Vaulter Landing on a Padded Surface

A pole-vaulter vaults over the bar and falls onto thick padding. He lands with a speed of 9.8 m/s; the padding then brings him to a stop in a time of 0.40 s. What is the average force on his body *due to the padding* during that time interval? Express your answer as a fraction or multiple of his weight. [*Hint*: The force due to the padding is not the only force acting on the vaulter during the 0.40 s interval.]

Application: Designing a Safer Automobile One automotive design change implemented to minimize injury on collision is the foam padding built into automobile dashboards (Fig. 7.4). Automobile bumpers have shock absorbers built in to lessen damage to the car body in small collisions. The structure of the car itself is often a single piece of metal with reinforced supports (*unibody* construction) so that the entire body can crumple and absorb the change in momentum more slowly than it would if it were made of separate sections of metal that would slide into or over each other or fall off the car. The safety glass in a windshield has two advantages. One is that it does not shatter and send sharp shards of glass into human tissue, but the other is that it distorts when struck by solid objects such as human bones or a human head. The glass doesn't give much, but in a crash every little bit helps.

The use of seat belts plus the air bag is better than either alone. Without a seat belt, the body continues moving with the same speed the car had before the crash. The rapidly inflating air bag moves toward the body, and the effective velocity is then the sum of the two velocities (air bag velocity + body velocity) when the two collide. The body flying into the air bag can be injured more than a restrained body making more gradual contact with the air bag. An adult should sit at least 12 in. from the air bag container to avoid injury from the deploying air bag itself. The American Academy of Pediatrics recommends that children younger than age 13 should sit in the back seat, and they should sit in a car seat or booster seat appropriate for their size until they are at least 4 feet 9 inches (1.45 m) tall. Children should sit in a rear-facing car seat until age 2.

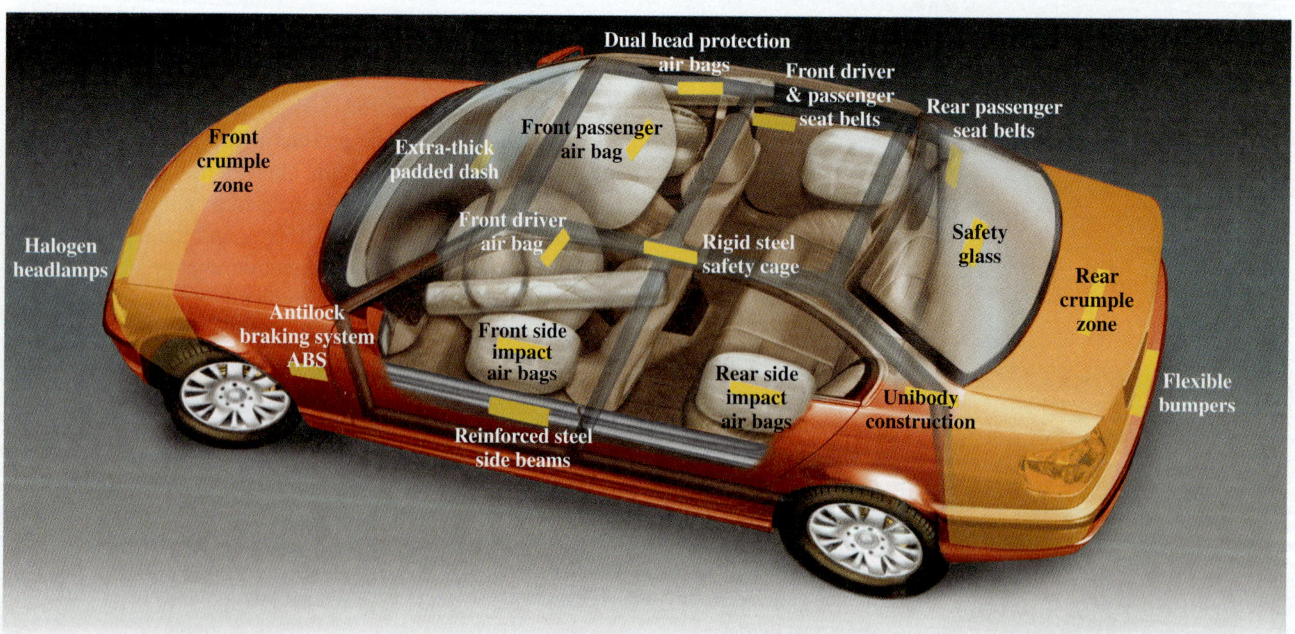

Figure 7.4 Some safety features of the modern automobile. Many of these features serve to lengthen the time interval during which a momentum change occurs in a crash, thereby lessening the forces acting on the passengers.

Example 7.3

Collision Between an Automobile and a Tree

A car moving at 20.0 m/s (44.7 mi/h) crashes into a tree. Find the magnitude of the average force acting on a passenger of mass 65 kg in each of the following cases. (a) The passenger is not wearing a seat belt. He is brought to rest by a collision with the windshield and dashboard that lasts 3.0 ms. (b) The car is equipped with a passenger-side air bag. The force due to the air bag acts for 30 ms, bringing the passenger to rest.

Strategy From the impulse-momentum theorem, $\Delta\vec{\mathbf{p}} = \vec{\mathbf{F}}_{av}\,\Delta t$, where $\vec{\mathbf{F}}_{av}$ is the average force acting on the passenger and Δt is the time interval during which the force acts. The change in the passenger's momentum is the same in the two cases. What differs is the time interval during which the change occurs. It takes a larger force to change the momentum in a shorter time interval.

Solution The magnitude of the passenger's initial momentum is

$$|\vec{\mathbf{p}}_i| = |m\vec{\mathbf{v}}_i| = 65\text{ kg} \times 20.0\text{ m/s} = 1300\text{ kg}\cdot\text{m/s}$$

His final momentum is zero, so the magnitude of the momentum change is

$$|\Delta\vec{\mathbf{p}}| = 1300\text{ kg}\cdot\text{m/s}$$

This momentum change divided by the time interval gives the magnitude of the average force in each case.

(a) No seat belt: $|\vec{\mathbf{F}}_{av}| = \dfrac{|\Delta\vec{\mathbf{p}}|}{\Delta t} = \dfrac{1300\text{ kg}\cdot\text{m/s}}{0.0030\text{ s}} = 4.3\times10^5\text{ N}$

(b) Air bag: $|\vec{\mathbf{F}}_{av}| = \dfrac{|\Delta\vec{\mathbf{p}}|}{\Delta t} = \dfrac{1300\text{ kg}\cdot\text{m/s}}{0.030\text{ s}} = 4.3\times10^4\text{ N}$

Discussion The average forces required to bring the passenger to rest are inversely proportional to the time interval over which those forces act. It is a far happier situation to have the momentum change over as long a period as possible to make the forces smaller. Automotive safety engineers design cars to minimize the average forces on the passengers during sudden stops and collisions.

The air bag also spreads the force over a much larger area than impact with a hard surface like the windshield, further reducing the risk of injury.

Practice Problem 7.3 Catching a Fastball

A baseball catcher is catching a fastball that is thrown at 43 m/s (96 mi/h) by the pitcher. If the mass of the ball is 0.15 kg and if the catcher moves his mitt backward toward his body by 8.0 cm as the ball lands in the glove, what is the magnitude of the average force acting on the catcher's mitt? Estimate the time interval required for the catcher to move his hands.

EVERYDAY PHYSICS DEMO

Try playing catch with a friend [outdoors] using a raw egg or a water balloon. How do you move your hands to minimize the chance of breaking the egg or balloon when you catch it? What is likely to happen if you forget and catch it as you would a ball?

Graphical Calculation of Impulse

When a force is changing, how can we find the impulse? We've asked similar questions in previous chapters. For simplicity we consider components along the x-axis. Recall:

- displacement $= \Delta x = v_{av,x}\,\Delta t =$ area under $v_x(t)$ graph
- change in velocity $= \Delta v_x = a_{av,x}\,\Delta t =$ area under $a_x(t)$ graph

In both cases, the mathematical relationship is that of a rate of change. Velocity is the rate of change of position with time, and acceleration is the rate of change of velocity with time. Now we have force as the rate of change of momentum with time. By analogy:

- impulse $= \vec{\mathbf{J}} = F_{av,x}\,\Delta t =$ area under $F_x(t)$ graph

So to find the impulse for a variable force, we find the area under the $F_x(t)$ graph. Then, if we wish to know the average force, we can divide the impulse by the time interval during which the force is applied.

CONNECTION:

See Sections 2.2, 2.3, and 6.6 to review how we used the area under a graph to find displacement, change in velocity, and work done by a force.

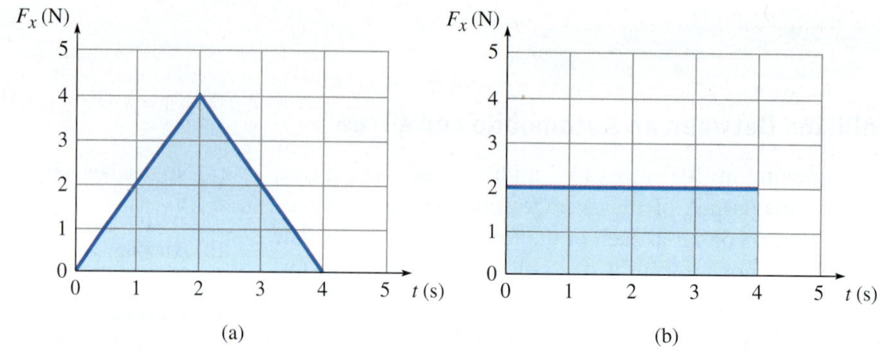

Figure 7.5 (a) The area under the $F_x(t)$ graph for a variable force is the impulse. (b) The average force for a given time interval is the constant force that would produce the same impulse.

The variable force of Fig. 7.5a increases linearly from 0 to 4 N in a time of 2 s; then it decreases from 4 N to 0 N in 2 s. The area under the $F_x(t)$ graph is found from the triangular area

$$J = \tfrac{1}{2}\text{ base} \times \text{height} = 2\text{ s} \times 4\text{ N} = 8\text{ N·s}$$

The average force during the 4 s time interval is

$$F_{\text{av}} = \frac{J}{\Delta t} = \frac{8\text{ N·s}}{4\text{ s}} = 2\text{ N}$$

Figure 7.5b shows the average force over the 4 s time interval; the area under the curve (the impulse) is the same as in Fig. 7.5a.

Example 7.4

Hitting the Wall

An experimental robotic car of mass 10.2 kg moving at 1.2 m/s in the $+x$-direction crashes into a brick wall and rebounds. A force sensor on the car's bumper records the force that the wall exerts on the car as a function of time. These data are shown in graphical form in Fig. 7.6. (a) What is the maximum magnitude of the force exerted on the car? (b) What is the average force on the car during the collision? (c) At what speed does the car rebound from the wall?

Strategy The maximum force can be read directly from the graph. To solve parts (b) and (c) of this problem, we must find the impulse exerted on the car. Since impulse is the area under the $F_x(t)$ curve, we'll make an estimate of the area. The impulse is then equal to the average force times the time interval and also to the car's change in momentum. Once we find the change in momentum, we use it to find the car's final speed.

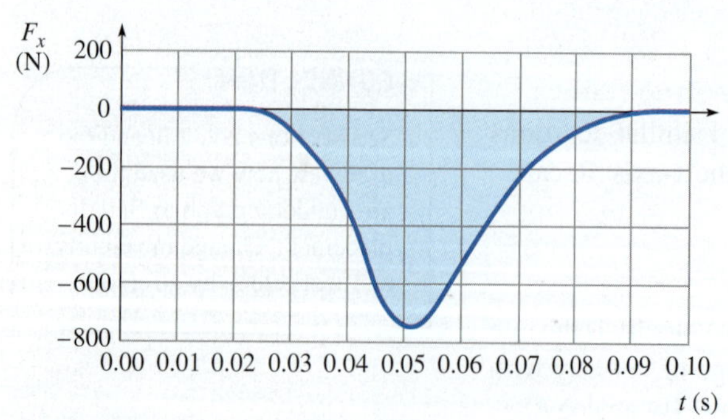

Figure 7.6
Force versus time for a car colliding with a wall.

Given: $m = 10.2$ kg; $v_{ix} = 1.2$ m/s; graph of $F_x(t)$
To find: (a) F_{max}; (b) F_{av}; (c) v_{fx}

Solution (a) From Fig. 7.6, the maximum force is approximately 750 N in magnitude.

(b) Each division on the horizontal axis represents 0.01 s, and each vertical division represents 200 N. Then the area of each grid box represents (200 N × 0.01 s) = 2 N·s. Counting the number of grid boxes between the $F_x(t)$ curve and the time axis, estimating fractions of boxes, yields about 10 boxes. Then the magnitude of the impulse is approximately

$$J = 10\text{ boxes} \times 2\text{ N·s/box} = 20\text{ N·s}$$

The collision is underway when the force is nonzero. So the collision begins at about $t = 0.025$ s and ends at about $t = 0.095$ s. The duration of the collision is

$$\Delta t = 0.07\text{ s}$$

The magnitude of the average force is approximately

$$F_{\text{av}} = \frac{J}{\Delta t} = \frac{20\text{ N·s}}{0.07\text{ s}} = 300\text{ N}$$

continued on next page

Example 7.4 continued

(c) The impulse gives us the momentum change. The force exerted by the wall is in the $-x$-direction. Thus, the x-component of the impulse is negative. In the graph of F_x versus t, the area lies under the time axis and so is counted as negative. So, working with x-components,

$$\Delta p_x = mv_{fx} - mv_{ix} = F_{av}\,\Delta t = -20\text{ N}\cdot\text{s}$$

Solving for v_{fx},

$$v_{fx} = \frac{\Delta p_x + mv_{ix}}{m} = \frac{\Delta p_x}{m} + v_{ix}$$

Substituting numerical values in this expression yields

$$v_{fx} = \frac{-20\text{ N}\cdot\text{s}}{10.2\text{ kg}} + 1.2\text{ m/s} = -0.8\text{ m/s}$$

The car rebounds at a speed of 0.8 m/s.

Discussion As a check, we compare the average force with the maximum force. The average force is a bit less than half of the maximum force. If the force were a linear function of time, the average would be exactly half the maximum. Here, the average force is less than that because more time is spent at smaller values of force than at the larger values.

Practice Problem 7.4 Car-Van Collision

A car weighing 13.6 kN is moving at 10.0 m/s in the $+x$-direction when it collides head-on with a van weighing 33.0 kN. The horizontal force exerted on the car before, during, and after the collision is shown in Fig. 7.7. What is the car's velocity just after the collision?

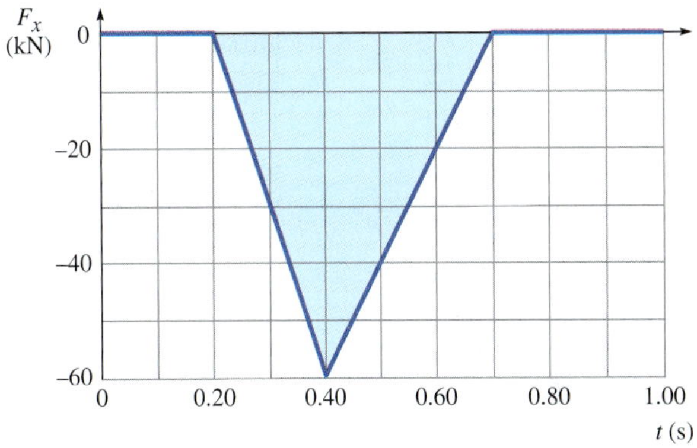

Figure 7.7
Varying force on a car during a car-van collision.

A Restatement of Newton's Second Law

We can use the relationship between impulse and momentum to find a new way to understand Newton's second law. Let's rewrite the impulse-momentum theorem this way:

$$\Sigma\vec{\mathbf{F}}_{av} = \frac{\Delta\vec{\mathbf{p}}}{\Delta t}$$

What happens if we let the time interval Δt get smaller and smaller, approaching zero? Then the average force is taken over a smaller and smaller time interval, approaching the instantaneous force:

Newton's Second Law

$$\Sigma\vec{\mathbf{F}} = \lim_{\Delta t\to 0}\frac{\Delta\vec{\mathbf{p}}}{\Delta t} \qquad (7\text{-}4)$$

CONNECTION:

Equation (7-4) is closer to Newton's original statement of his second law and is more general than $\Sigma\vec{\mathbf{F}} = m\vec{\mathbf{a}}$.

In words, *the net force is the rate of change of momentum.*

Equation (7-4) is *more general* than $\Sigma\vec{\mathbf{F}} = m\vec{\mathbf{a}}$, the form of Newton's second law used in Chapters 4 through 6, which holds only when mass is constant. One situation in which mass is not constant is the rocket engine. In a rocket engine, fuel combustion produces hot gases that are then expelled at high speeds (Fig. 7.8). The rocket's mass decreases as the exhaust gases are expelled.

When mass is constant, then it can be factored out:

$$\Sigma\vec{\mathbf{F}} = \lim_{\Delta t\to 0}\frac{\Delta\vec{\mathbf{p}}}{\Delta t} = \lim_{\Delta t\to 0}\frac{\Delta(m\vec{\mathbf{v}})}{\Delta t} = m\lim_{\Delta t\to 0}\frac{\Delta\vec{\mathbf{v}}}{\Delta t} = m\vec{\mathbf{a}}$$

Thus, Eq. (7-4) reduces to the familiar form of Newton's second law from Chapters 4–6 when mass is constant.

Figure 7.8 The Space Shuttle is propelled upward as hot gases are exhausted downward at high speeds from two solid-rocket boosters. The Shuttle program flew 135 missions during its 30 year existence (1981–2011). The five Shuttle orbiters were the first reusable spacecraft.

7.4 CONSERVATION OF MOMENTUM

Consider two pucks that bump into one another after sliding along a frictionless table. Figure 7.9 shows what happens to the two pucks before, during, and after their interaction. If we think of the two pucks as constituting a single system, then the gravitational interactions with the Earth and the contact interactions with the table are *external* interactions—interactions with objects external to the system. The force of gravity on each object is balanced by the normal force on the same object, and, thus, there is no net impulse up or down. Together, these forces produce a net external force of zero, so they leave the system's momentum unchanged. Since these two always cancel, we can ignore these external interactions and just focus on the interaction between the pucks. Therefore, we omit the normal and gravitational forces in Fig. 7.9.

Until contact is made, there is no interaction between the pucks (ignoring the small gravitational interaction between the two). During the collision, the pucks exert forces on each other. Force $\vec{\mathbf{F}}_{12}$ is the contact force acting on mass m_1, and force $\vec{\mathbf{F}}_{21}$ is the contact force acting on mass m_2. If we continue to regard the two pucks as parts of a single interacting system, then those forces are *internal* forces of this system. When they collide, some momentum is transferred from one puck to the other. The changes in momentum of the two are equal and opposite:

$$\Delta\vec{\mathbf{p}}_1 = -\Delta\vec{\mathbf{p}}_2$$

Since the change in momentum is the final momentum minus the initial momentum, we write:

$$\vec{\mathbf{p}}_{1\text{f}} - \vec{\mathbf{p}}_{1\text{i}} = -(\vec{\mathbf{p}}_{2\text{f}} - \vec{\mathbf{p}}_{2\text{i}})$$

Moving the initial momenta to the left side and the final momenta to the right:

$$\vec{\mathbf{p}}_{1\text{i}} + \vec{\mathbf{p}}_{2\text{i}} = \vec{\mathbf{p}}_{1\text{f}} + \vec{\mathbf{p}}_{2\text{f}} \quad (7\text{-}5)$$

Equation (7-5) says the sum of the momenta of the pucks before the interaction is equal to the sum of the momenta after the interaction; or, more simply, the total momentum of the objects is unchanged by the collision. This isn't surprising since, if some momentum is just transferred from one to the other, the total hasn't changed. We say that momentum is *conserved* for this collision. The interaction between the pucks changes the momentum of each puck, but the total momentum of the system is unchanged.

In a system composed of more than two objects, interactions between objects inside the system do not change the total momentum of the system—they just transfer some momentum from one part of the system to another. Only external interactions can change the total momentum of the system. To summarize:

- The total momentum of a system is the vector sum of the momenta of each object in the system.
- External interactions can change the total momentum of a system.
- Internal interactions do not change the total momentum of a system.

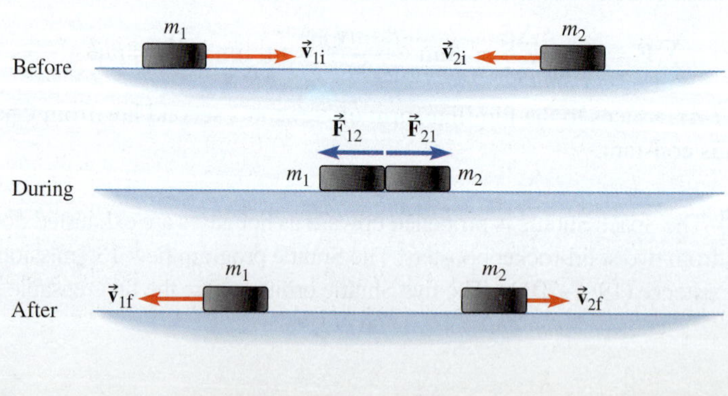

Figure 7.9 Two sliding pucks with different masses before, during, and after collision.

In the absence of external interactions, momentum is conserved:

Conservation of Linear Momentum

If the net external force acting on a system is zero, then the momentum of the system is conserved.

$$\text{If } \sum \vec{\mathbf{F}}_{\text{ext}} = 0, \quad \vec{\mathbf{p}}_i = \vec{\mathbf{p}}_f \qquad (7\text{-}6)$$

By definition, an *isolated*, or closed, system is subject to no external interactions; thus, *linear momentum is always conserved for an isolated system.* Remember that momentum is a vector quantity, so both the magnitude *and the direction* of the momentum at the beginning and end of the interaction must be the same. In component form, both p_x and p_y are unchanged by the interaction.

CHECKPOINT 7.4

When is the momentum of a system not conserved?

Application of Momentum Conservation: Recoil of a Rifle When a bullet is fired from a rifle, the system of rifle plus bullet must conserve momentum. Suppose the rifle is at rest before the bullet is fired. The momentum of the system is zero. When the bullet is fired, part of the system's mass breaks away and travels in one direction with a certain momentum. The rifle, which is the remaining mass of the system, moves in the exact opposite direction such that the total momentum of the system is still zero. The rifle has a much larger mass than the bullet, so it has a much smaller speed. The backward motion of the rifle is the *recoil* felt by anyone who has held a rifle against her shoulder and squeezed the trigger.

Application: Ballistocardiography Ballistocardiography is a diagnostic technique that measures the recoil of the human body due to the pumping of the heart. If the net external force is zero, a momentum change in one part of the body is accompanied by an equal and opposite momentum change in the rest of the body. A ballistocardiogram records the tiny recoil movements of the body that occur as the heart contracts, ejects blood into the aorta, and then is refilled with blood.

Application: Jets, Rockets, and Airplane Wings Jet engines and rockets operate by conservation of momentum. Hot combustion gases are forced out of nozzles at high speed by the engines. The increased backward momentum of the hot gases as they are expelled is accompanied by an increased forward momentum of the engines. Airplane wings generate lift by conservation of momentum. The main purpose of the wing is to deflect air downward, giving it a downward momentum component. (Exactly how the wing does this is the complicated part.) Since the wing pushes air downward, air pushes the wing upward.

Example 7.5

Jet Propulsion in Squid

Squid (Fig. 7.10) are the fastest swimmers among marine invertebrates. During a fast swim to evade a predator, some species can reach speeds of more than 10 m/s. A squid propels itself much as a jet or rocket does. It starts by filling an internal cavity with water. Then the *mantle*, a powerful muscle, squeezes the cavity and expels water through a narrow opening (the *siphon*) at high speed.

Suppose a squid of mass 182 g (including the water that will be expelled) is initially at rest. It then expels 54 g of water at an average speed of 62 cm/s (relative to the surrounding water). Ignoring drag forces, how fast is the squid moving immediately after expelling the water?

Strategy Consider the squid and the water inside its cavity to be a single system. Because we assume drag forces on the system are negligible, the net external force on the system is zero and the momentum of the system is conserved.

continued on next page

Example 7.5 continued

Figure 7.10 The Bigfin Reef Squid (*Sepioteuthis lessoniana*) is commonly found on coral reefs and in seagrass beds from Hawaii to the Red Sea. This photo, taken in Malaysia, clearly shows the siphon through which water is expelled to propel the squid forward.

Solution Initially the squid is at rest and the momentum of the system is zero. After expelling the water, the squid moves with velocity $\vec{\mathbf{v}}_s$ and the expelled water moves with average velocity $\vec{\mathbf{v}}_w$. The total momentum of the system is conserved:

$$\vec{\mathbf{p}}_i = \vec{\mathbf{p}}_f \quad \Rightarrow \quad 0 = m_s\vec{\mathbf{v}}_s + m_w\vec{\mathbf{v}}_w$$

Here $m_s = 182\text{ g} - 54\text{ g} = 128\text{ g}$ is the mass of the squid after expelling the water. Solving for $\vec{\mathbf{v}}_s$,

$$\vec{\mathbf{v}}_s = -\frac{m_w\vec{\mathbf{v}}_w}{m_s}$$

The minus sign means the squid moves in the opposite direction from the jet of water. The squid's speed is

$$v_s = \frac{m_w v_w}{m_s} = \frac{(54\text{ g}) \times (62\text{ cm/s})}{128\text{ g}} = 26\text{ cm/s}$$

Discussion Quick check: The mass of the squid is a bit more than twice the mass of the expelled water, so the speed of the water is a bit more than twice the speed of the squid.

A variation on the problem: Suppose the squid is not initially at rest. We can still apply conservation of momentum; the only difference is that the initial momentum is not zero. If the squid is initially moving at velocity $\vec{\mathbf{v}}_i$, then

$$(m_s + m_w)\vec{\mathbf{v}}_i = m_s\vec{\mathbf{v}}_s + m_w\vec{\mathbf{v}}_w$$

In this equation, all three velocities are measured with respect to the surrounding water. For the initial momentum, we write $(m_s + m_w)\vec{\mathbf{v}}_i$ because the water inside the squid is also moving at velocity $\vec{\mathbf{v}}_i$ before it is expelled.

Practice Problem 7.5 Skaters Pushing Apart

Two skaters on in-line skates, Lisa and Bart, are initially at rest. They push apart and start moving in opposite directions. If Lisa's speed just after they push apart is 2.0 m/s and her mass is 85% of Bart's mass, how fast is Bart moving at that time?

Conceptual Example 7.6

Escape on Slippery Ice

A pilot parachutes from his disabled aircraft and lands on the frozen surface of a lake. There is no breeze blowing and the lake surface is too slippery to walk on. What can the pilot do to reach the shore?

Strategy and Solution Since the person in jeopardy is a pilot, he begins to think about how hot gases forced backward from a jet engine cause the plane to move forward. That gives him an idea: he bundles the parachute into a package and pushes it as hard as possible in a direction away from the nearest point of the shore. If friction is negligible, the net external force on the system of pilot plus parachute is zero and the total momentum of the system cannot change. The momentum of the parachute plus the momentum of the pilot must still equal zero. By conservation of momentum, the pilot begins sliding in the opposite direction and glides toward the shore.

Discussion If friction brings the pilot to rest before he reaches the shore, he can search his pockets and belt loops for other items to throw away. Once he reaches shore, he can tie one end of a rope to a tree and, holding onto the other end, venture back out onto the ice to retrieve any essential items. The rope provides him with an external force so he can get back to shore.

Practice Problem 7.6 Recoil of a Rifle

During an afternoon of target practice, you fire a Winchester .308 rifle of mass 3.8 kg. The bullets have a mass of 9.72 g and leave the rifle at a muzzle velocity of 860 m/s. If you are sloppy and fire a round when the butt of the rifle is not firmly up against your shoulder, at what speed does the rifle butt smash into your shoulder? (Ouch!)

7.5 CENTER OF MASS

We have seen that the momentum of an isolated system is conserved even though parts of the system may interact with other parts; internal interactions transfer momentum between parts of the system but do not change the total momentum of the system. We can define a point called the **center of mass** (CM) that serves as an average location of the system. In Section 7.6, we prove that the center of mass of an isolated system must move with constant velocity, regardless of how complicated the motions of parts of the system may be. Then we can treat the mass of the system as if it were all concentrated at the CM, like a point particle. The CM of an object is not necessarily located within the object; for some objects, such as a boomerang, the center of mass is located outside of the object itself (Fig. 7.11a).

What if a system is not isolated, but has external interactions? Again imagine all of the mass of the system concentrated into a single point particle located at the CM. The motion of this fictitious point particle is determined by Newton's second law, where the net force is the sum of all of the external forces acting on *any part* of the system. In the case of a complex system composed of many parts interacting with one another, the motion of the CM is considerably simpler than the motion of an arbitrary particle of the system (Fig. 7.11b,c).

Location of Center of Mass For a system composed of two particles, the center of mass lies somewhere on a line between the two particles. In Fig. 7.12, particles of masses m_1 and m_2 are located at positions x_1 and x_2, respectively. We define the location of the CM for these two particles as

$$x_{\text{CM}} = \frac{m_1x_1 + m_2x_2}{m_1 + m_2} \tag{7-7}$$

The CM is a *weighted average* of the positions of the two particles. Here we use the word *weighted* in its statistical sense. The position of a particle with more mass counts

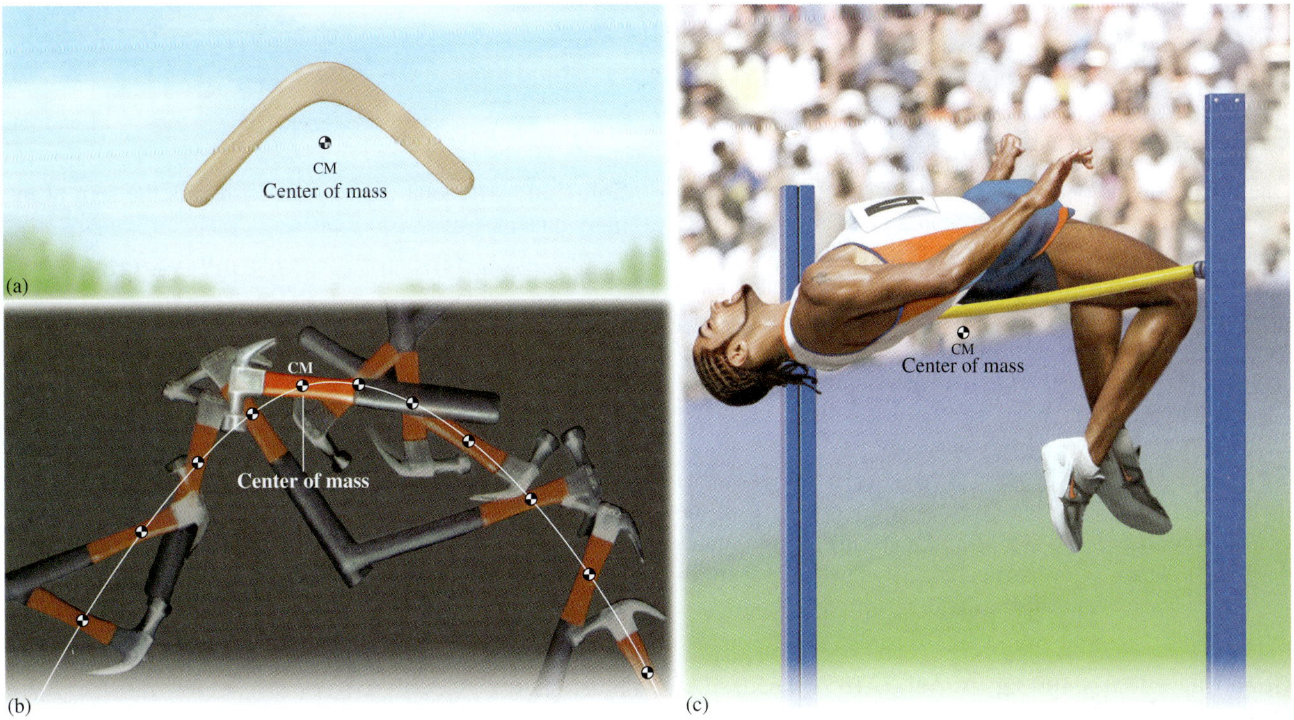

Figure 7.11 (a) The center of mass of a boomerang is a point outside of the boomerang. (b) The path followed by the center of mass when a hammer is tossed through the air. (c) British pole-vaulter Ben Challenger's center of mass actually passes *beneath* the bar as his body passes over the bar.

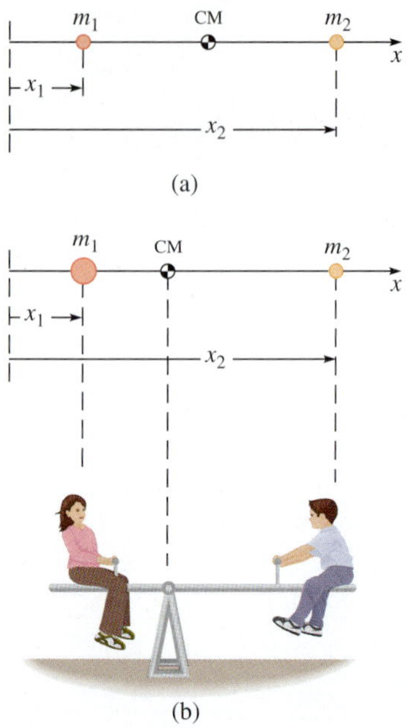

Figure 7.12 (a) Two particles of equal mass located at positions x_1 and x_2 from the origin. The CM is midway between the two. (b) Two particles of unequal mass. The CM is closer to the more massive particle. For two children balanced on a seesaw, the CM is at the fulcrum.

more—carries more *statistical* weight—than does the position of a particle with a smaller mass. We can rewrite Eq. (7-7) as a weighted average:

$$x_{CM} = \frac{m_1}{M}x_1 + \frac{m_2}{M}x_2 \tag{7-8}$$

Here $M = m_1 + m_2$ represents the total mass of the system. The statistical weight used for the location of each particle is the mass of that particle as a fraction of the total mass of the system.

Suppose masses m_1 and m_2 are equal. Then we expect the CM to be located midway between the two particles (Fig. 7.12a). If $m_1 = 2m_2$, as in Fig. 7.12b, then the CM is closer to the particle of mass m_1. Figure 7.12b shows that, in this case, the CM is twice as far from m_2 as from m_1.

For a system of N particles, at arbitrary locations in three-dimensional space, the definition of the CM is a generalization of Eq. (7-7).

Definition of center of mass

Vector form: $$\vec{\mathbf{r}}_{CM} = \frac{m_1\vec{\mathbf{r}}_1 + m_2\vec{\mathbf{r}}_2 + \cdots + m_N\vec{\mathbf{r}}_N}{M} = \frac{\sum m_n\vec{\mathbf{r}}_n}{M} \tag{7-9}$$

Component form: $$x_{CM} = \frac{m_1x_1 + m_2x_2 + \cdots + m_Nx_N}{M} = \frac{\sum m_nx_n}{M}$$

$$y_{CM} = \frac{\sum m_ny_n}{M} \qquad z_{CM} = \frac{\sum m_nz_n}{M}$$

where $M = m_1 + m_2 + \cdots + m_N = \sum m_n$

Remember that the symbol Σ stands for *sum.*

Example 7.7

Center of Mass of a Binary Star System

Due to the gravitational interaction between the two stars in a binary star system, each moves in a circular orbit around their CM. One star has a mass of 15.0×10^{30} kg; its center is located at $x = 1.0$ AU and $y = 5.0$ AU. The other has a mass of 3.0×10^{30} kg; its center is at $x = 4.0$ AU and $y = 2.0$ AU. Find the CM of the system composed of the two stars. (AU stands for *astronomical unit.* 1 AU = the average distance between the Earth and the Sun = 1.5×10^8 km.)

Strategy We treat the stars as point particles located at their centers. Since we are given x- and y-coordinates, the easiest way to proceed is to find the x- and y-coordinates of the CM. There is no particular advantage here in finding the position vector of the CM in terms of its length and direction.

Given: $m_1 = 15.0 \times 10^{30}$ kg $\quad x_1 = 1.0$ AU $\quad y_1 = 5.0$ AU
$m_2 = 3.0 \times 10^{30}$ kg $\quad x_2 = 4.0$ AU $\quad y_2 = 2.0$ AU

To find: x_{CM}; y_{CM}

Solution The total mass of the system is the sum of the individual masses:

$$M = m_1 + m_2 = 15.0 \times 10^{30}\text{ kg} + 3.0 \times 10^{30}\text{ kg} = 18.0 \times 10^{30}\text{ kg}$$

For the x-position, we find

$$x_{CM} = \frac{m_1}{M}x_1 + \frac{m_2}{M}x_2$$
$$= \frac{15.0 \times 10^{30}\text{ kg}}{18.0 \times 10^{30}\text{ kg}} \times 1.0\text{ AU} + \frac{3.0 \times 10^{30}\text{ kg}}{18.0 \times 10^{30}\text{ kg}} \times 4.0\text{ AU}$$
$$= 1.5\text{ AU}$$

and for the y-position, we find

$$y_{CM} = \frac{m_1}{M}y_1 + \frac{m_2}{M}y_2$$
$$= \frac{15.0}{18.0} \times 5.0\text{ AU} + \frac{3.0}{18.0} \times 2.0\text{ AU} = 4.5\text{ AU}$$

Discussion In Fig. 7.13, we mark the position of the CM. As we expect for the case of two particles, it is located closer to the

continued on next page

Example 7.7 continued

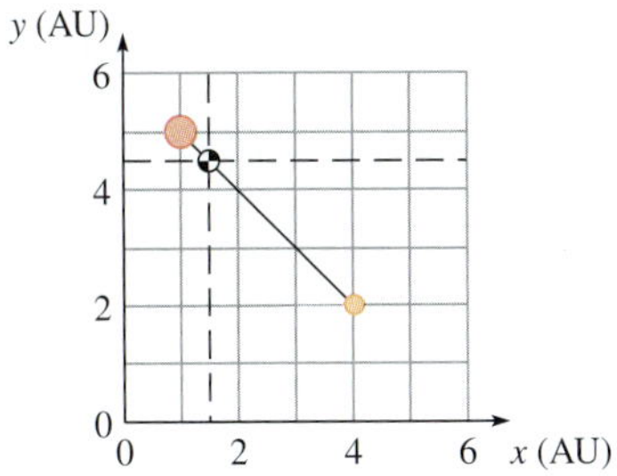

Figure 7.13

Finding the CM for the system of two stars.

larger mass and on a line connecting the two. Once the CM position is found in a problem, check to be sure its location is reasonable. Suppose we had made an error in this example and found the CM to be at $x = 1.5$ AU and $y = 1.7$ AU. This is not a reasonable location for the CM since it is not along the line connecting the two and is closer to the less massive star; we then would go back to look for the error.

Practice Problem 7.7 Three Balls with Unequal Masses

Three spherical objects are shown in Fig. 7.14. Their masses are $m_1 = m_3 = 1.0$ kg and $m_2 = 4.0$ kg. Find the location of the CM for the three objects.

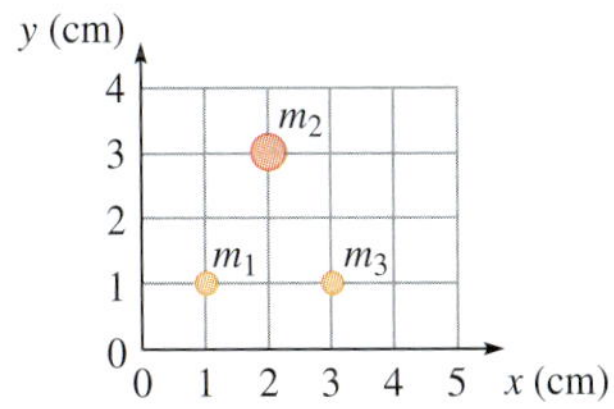

Figure 7.14

Three spheres located at x, y positions (1.0 cm, 1.0 cm), (2.0 cm, 3.0 cm), and (3.0 cm, 1.0 cm).

Using Symmetry to Locate the Center of Mass Most objects we deal with in real life are not composed of a small set of point particles or spherically symmetrical objects. In Example 7.7, we use the location of the center of each star to find the CM. Due to spherical symmetry, the CM of either star (by itself) is at its geometric center. The same technique can be applied to other shapes with symmetry. A standard 2 by 4, which is an 8 ft long uniform piece of lumber used in building 1.5 in. deep × 3.5 in. high, has its center of mass at its geometric center. By contrast, a "loaded" die does *not* have its CM at its geometric center, since a small metal plug has been inserted near one face to make the distribution of mass in the die asymmetrical. The definition of the CM [Eq. (7-9)] still holds as long as (x_n, y_n, z_n) are the coordinates of the CM of a part of the system with mass m_n.

7.6 MOTION OF THE CENTER OF MASS

Now that we know how to find the position of the CM of a system, we turn our attention to the motion of the CM. How is the velocity of the CM related to the velocities of the various parts of the system?

During a short time interval Δt, the displacement of the nth particle is $\Delta\vec{\mathbf{r}}_n = \vec{\mathbf{v}}_n \Delta t$ and the displacement of the center of mass is $\Delta\vec{\mathbf{r}}_{\text{CM}} = \vec{\mathbf{v}}_{\text{CM}} \Delta t$. From the definition of the CM [Eq. (7-9)], the displacements must be related as follows:

$$\Delta\vec{\mathbf{r}}_{\text{CM}} = \frac{\sum m_n \Delta\vec{\mathbf{r}}_n}{M} = \vec{\mathbf{v}}_{\text{CM}} \Delta t = \frac{\sum m_n \vec{\mathbf{v}}_n \Delta t}{M}$$

Dividing both sides by Δt and multiplying by M yields

$$M\vec{\mathbf{v}}_{\text{CM}} = \sum m_n \vec{\mathbf{v}}_n \tag{7-10}$$

The right side of Eq. (7-10) is the sum of the momenta of the particles that constitute the system—the total momentum of the system $\vec{\mathbf{p}}$. Therefore,

$$\vec{\mathbf{p}} = M\vec{\mathbf{v}}_{\text{CM}} \tag{7-11}$$

For two-dimensional motion, Eq. (7-11) is equivalent to two component equations

$$p_x = Mv_{\text{CM},x} \quad \text{and} \quad p_y = Mv_{\text{CM},y} \tag{7-12}$$

In Section 7.4, we showed that, for an isolated system, the total linear momentum is conserved. In such a system, Eq. (7-11) implies that the CM must move with constant velocity regardless of the motions of the individual particles. On the other hand, what if the system is not isolated? If a net external force acts on a system, the CM does not move with constant velocity. Instead, it moves as if all the mass were concentrated there into a fictitious point particle with all the external forces acting on that point. The motion of the CM obeys the following statement of Newton's second law:

$$\Sigma\vec{\mathbf{F}}_{ext} = M\vec{\mathbf{a}}_{CM} \tag{7-13}$$

where M is the total mass of the system, $\Sigma\vec{\mathbf{F}}_{ext}$ is the net external force, and $\vec{\mathbf{a}}_{CM}$ is the acceleration of the CM. [Eq. (7-13) is proved in Problem 43.]

CHECKPOINT 7.6

Turn back to Fig. 7.11b. Why does the CM of the hammer move along a parabolic path?

Example 7.8

An Exploding Rocket

A model rocket is fired from the ground in a parabolic trajectory. At the top of the trajectory, a horizontal distance of 260 m from the launch point, an explosion occurs within the rocket, breaking it into two fragments. One fragment, having one third of the mass of the rocket, falls straight down to Earth as if it had been dropped from rest at that point. At what horizontal distance from the launch point does the other fragment land? Ignore air resistance. [*Hint*: The two fragments land simultaneously.]

Strategy Two different strategies can be used to solve this problem.

Strategy 1: We apply conservation of momentum to the explosion. The momentum of the rocket *just before* the explosion is equal to the total momentum of the two fragments *just after* the explosion. Why can momentum conservation be assumed here? There is an external force—gravity—acting on the system. External forces change momentum. However, the explosion takes place in a *very short time interval*. From the impulse-momentum theorem [Eq. (7-2)], the momentum change of the system is the force of gravity multiplied by the time interval. As long as the time interval considered is sufficiently short, the momentum change of the system can be ignored.

Strategy 2: The explosion is caused by an *internal* interaction between two parts of the rocket. The motion of the CM of the system is unaffected by internal interactions, so it continues in the same parabolic path. Just before the explosion, the rocket is at the top of its trajectory, so it has $p_y = 0$ (with the y-axis pointing up). Just after the explosion, one fragment is at rest. Then the other fragment must have $p_y = 0$; otherwise, conservation of momentum would be violated. Then both fragments have $v_y = 0$ just after the explosion. Ignoring air resistance, they land simultaneously. At that same instant, the CM also reaches the ground.

Solution 1 First we make a sketch of the situation (Fig. 7.15). At the top of the trajectory, where the explosion occurs, $v_y = 0$; the rocket is moving in the x-direction. The initial momentum just before the explosion is entirely in the x-direction. If M is the mass of the rocket, then

$$p_{ix} = Mv_{ix}$$

Just after the explosion, one third of the mass of the rocket is at rest; it then drops straight down under the influence of the gravitational force. This piece has zero momentum just after the explosion. To conserve momentum, the other two thirds of the rocket must have a momentum equal to the momentum just before the explosion.

$$p_{ix} = p_{1x} + p_{2x}$$

$$Mv_{ix} = 0 + (\tfrac{2}{3}M)v_{2x}$$

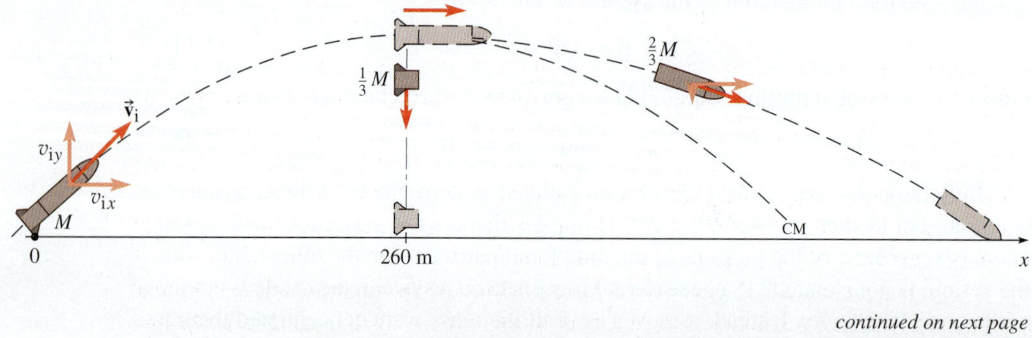

Figure 7.15
Rocket motion after explosion.

continued on next page

Example 7.8 continued

Solving for v_{2x}, we find

$$v_{2x} = \frac{3}{2}v_{ix}$$

The y-component of momentum must also be conserved:

$$p_{iy} = p_{1y} + p_{2y}$$

We know that both p_{iy} and p_{1y} are zero; therefore, p_{2y} is zero as well. Just after the explosion, both parts of the rocket have zero vertical components of velocity. Then both parts take the same time to fall to the ground as if the rocket had not exploded. With a horizontal velocity larger by a factor of $\frac{3}{2}$, the second piece of the rocket travels a horizontal distance from the explosion a factor of $\frac{3}{2}$ larger than 260 m (see Fig. 7.15). The distance from the launch point where this piece lands is

$$\Delta x = 260 \text{ m} + \tfrac{3}{2} \times 260 \text{ m} = 650 \text{ m}$$

Solution 2 The piece with mass $\frac{1}{3}M$ falls straight down and lands 260 m from the launch point. After the explosion, the CM continues to travel just as the rocket itself would have done if it had not broken apart. From the symmetry of the parabola, the CM touches the ground at a distance of $2 \times 260 \text{ m} = 520 \text{ m}$ from the launch point. Since we know the location of the CM and that of one of the pieces, we can find where the second piece lands:

$$Mx_{\text{CM}} = \tfrac{1}{3}Mx_1 + \tfrac{2}{3}Mx_2$$

After canceling the common factor of M,

$$x_{\text{CM}} = \tfrac{1}{3}x_1 + \tfrac{2}{3}x_2$$

Solving for x_2 yields

$$x_2 = \frac{3x_{\text{CM}} - x_1}{2} = \frac{3 \times 520 \text{ m} - 260 \text{ m}}{2} = 650 \text{ m}$$

which is the same answer that we found in Solution 1.

Discussion The insight that the motion of the CM is unaffected by internal interactions can be of enormous help. Note, however, that solution 2 would not be so simple if the two fragments did not land simultaneously. As soon as one fragment (fragment 1) hits the ground, the external force on the system is no longer due exclusively to gravity, so the CM doesn't continue to follow the same parabolic path. The normal and frictional forces acting on fragment 1 affect its subsequent motion and the subsequent motion of the CM even though the motion of fragment 2 is unaffected.

Practice Problem 7.8 Diana and the Raft

Diana (mass 55 kg) walks at 0.91 m/s (relative to the water) on a raft of mass 100.0 kg. The raft moves in the opposite direction at 0.50 m/s. Suppose it takes her 3.0 s to walk from one end of the raft to the other. (a) How far does Diana walk (relative to the water)? (b) How far does the raft move while Diana is walking? (c) How far does the CM of Diana and the raft move during the 3.0 s?

7.7 COLLISIONS IN ONE DIMENSION

What Is a Collision? In the macroscopic world, a moving object bumps into another object that may be at rest or in motion. The two objects exert forces on each other while they are in contact; as a result, their velocities change. In the microscopic and submicroscopic world, our picture of a collision is different. When atoms collide, they don't "touch" each other: the atom doesn't have a definite spatial boundary, so there are no surfaces to make "contact." However, the collision model is still useful for atoms and subatomic particles whenever there is an interaction in which the forces are strong over a short time interval, so that there is a clear "before collision" and a clear "after collision."

Analyzing Collisions Using Momentum Conservation We can often use conservation of momentum to analyze collisions even when external forces act on the colliding objects. If the net external force is small compared with the internal forces the colliding objects exert on each other during the collision, then the change in the total momentum of the two objects is small compared with the transfer of momentum from one object to the other. Then the total momentum after the collision is *approximately* the same as it was before the collision.

The same techniques that are used for collisions in the macroscopic world (car crashes, billiard ball collisions, baseball bats hitting balls) are also used in collisions in the microscopic world (gas molecules colliding with each other and with surfaces, radioactive decays of nuclei). First, we study collisions limited to motion along a line; later, we consider collisions limited to motion in a plane (in two dimensions).

Example 7.9

Collision in the Air

A krypton atom (mass 83.9 u) moving with a velocity of 0.80 km/s to the right and a water molecule (mass 18.0 u) moving with a velocity of 0.40 km/s to the left collide head-on. The water molecule has a velocity of 0.60 km/s to the right after the collision. What is the velocity of the krypton atom after the collision? (The symbol "u" stands for the atomic mass unit.)

Strategy Since we know both initial velocities and one of the final velocities, we can find the second final velocity by applying momentum conservation. Let the subscript "1" refer to the krypton atom and let the subscript "2" refer to the water molecule. Let the x-axis point to the right. Figure 7.16 shows before and after pictures of the collision.

Solution Momentum conservation requires that the final momentum be equal to the initial momentum:

$$\vec{\mathbf{p}}_{1i} + \vec{\mathbf{p}}_{2i} = \vec{\mathbf{p}}_{1f} + \vec{\mathbf{p}}_{2f}$$

Now we substitute $\vec{\mathbf{p}} = m\vec{\mathbf{v}}$ for each momentum. It is easiest to work in terms of components. For simplicity we drop the "x" subscripts, remembering that all quantities refer to x-components:

$$m_1 v_{1i} + m_2 v_{2i} = m_1 v_{1f} + m_2 v_{2f}$$

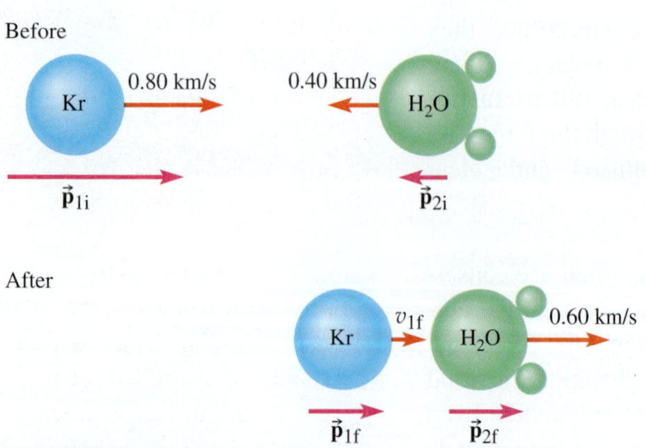

Figure 7.16

Before and after snapshots of a collision.

Since $m_1/m_2 = 83.9/18.0 = 4.661$, we can substitute $m_1 = 4.661 m_2$:

$$4.661 m_2 v_{1i} + m_2 v_{2i} = 4.661 m_2 v_{1f} + m_2 v_{2f}$$

The common factor m_2 divides out. Solving for v_{1f} gives

$$\begin{aligned} v_{1f} &= \frac{4.661 v_{1i} + v_{2i} - v_{2f}}{4.661} \\ &= \frac{4.661 \times 0.80\text{ km/s} + (-0.40\text{ km/s}) - 0.60\text{ km/s}}{4.661} \\ &= 0.59\text{ km/s} \end{aligned}$$

After the collision, the krypton atom moves to the right with a speed of 0.59 km/s.

Discussion To check this result, we calculate the total momentum (x-component) before and after the collision:

$$\begin{aligned} m_1 v_{1i} + m_2 v_{2i} &= (83.9\text{ u})(0.80\text{ km/s}) + (18.0\text{ u})(-0.40\text{ km/s}) \\ &= 60\text{ u·km/s} \end{aligned}$$

$$\begin{aligned} m_1 v_{1f} + m_2 v_{2f} &= (83.9\text{ u})(0.59\text{ km/s}) + (18.0\text{ u})(0.60\text{ km/s}) \\ &= 60\text{ u·km/s} \end{aligned}$$

Momentum is conserved. There is no need to convert u to kg since we only need to compare these two values.

If we made the mistake of thinking of momentum as a scalar, we would get the wrong answer. The sum of the *magnitudes* of the momenta before the collision is *not* equal to the sum of the *magnitudes* of the momenta after the collision. Conservation of energy is perhaps easier to understand intuitively since energy is a scalar quantity. Converting kinetic energy to potential energy is analogous to moving money from a checking account to a savings account; the total amount of money is the same before and after. This sort of analogy does *not* work with momentum!

Practice Problem 7.9 Head-On Collision

A 5.0 kg ball is at rest when it is struck head-on by a 2.0 kg ball moving along a track at 10.0 m/s. If the 2.0 kg ball is at rest after the collision, what is the speed of the 5.0 kg ball after the collision?

Elastic and Inelastic Collisions

Collisions are often classified based on what happens to the kinetic energy of the colliding objects. A ball dropped from a height h does not rebound to the same height. The kinetic energy of the ball just after the collision with the floor or ground is less than it was just before the collision; the amount of the kinetic energy decrease depends on the makeup of the ball and the ground. A racquetball dropped onto a hard wooden floor may rebound nearly to its original height, but a watermelon rebounds very little or not at all. Why do some objects rebound much better than others?

Imagine a racquetball colliding with the floor (Fig. 7.17). The bottom of the ball is flattened. What makes the ball rebound from the floor? The forces holding the ball together are like springs; the kinetic energy of the ball has been transformed largely into elastic potential energy stored in these springs. When the ball bounces back up, most of this energy is transformed back into kinetic energy. Then why does the watermelon not rebound? The watermelon, too, is deformed when it collides with the floor, but this deformation is not reversible. The kinetic energy of the watermelon is dissipated (changed into thermal energy).

A collision in which the *total* kinetic energy is the same before and after is called **elastic**. There is no *conservation law* for kinetic energy by itself. Total energy is always conserved, but that does not preclude some kinetic energy being transformed into another type of energy. The elastic collision is just a special kind of collision in which no kinetic energy is changed into other forms of energy.

It can be shown (see Problem 60) that for *any* elastic collision between two objects, the relative speed is the same before and after the collision. (This fact is most useful in one-dimensional collisions; in two-dimensional collisions the *direction* of the relative velocity changes due to the collision.) Since the relative velocity is in the opposite direction after a one-dimensional collision—first the objects move together, then they move apart—we can write:

$$v_{2ix} - v_{1ix} = -(v_{2fx} - v_{1fx}) \tag{7-14}$$

assuming the objects move along the x-axis. For a one-dimensional elastic collision, Eq. (7-14) is a useful alternative to setting the final kinetic energy equal to the initial kinetic energy.

When the final kinetic energy is less than the initial kinetic energy, the collision is said to be **inelastic**. Collisions between macroscopic objects are generally inelastic to some degree, but sometimes the change in kinetic energy is so small that we treat them as elastic. When a collision results in two objects sticking together, the collision is **perfectly inelastic**. The decrease of kinetic energy in a perfectly inelastic collision is as large as *possible* (consistent with the conservation of momentum).

Now that we have defined elastic and inelastic collisions, we can put together a problem-solving strategy for collision problems.

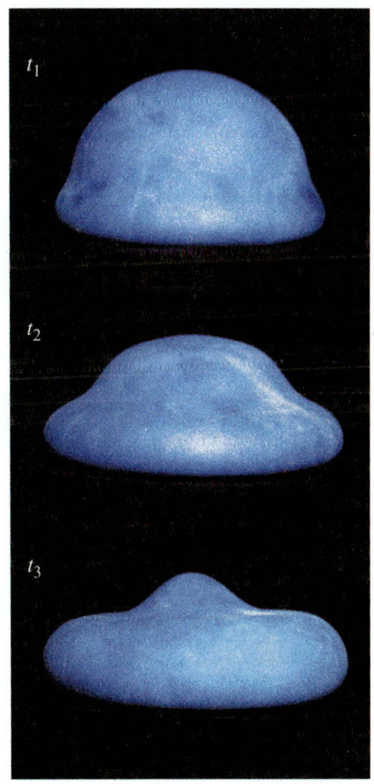

Figure 7.17 Three successive photos of a racquetball during its collision with the floor ($t_1 < t_2 < t_3$). As the ball is flattened, its kinetic energy is transformed into elastic potential energy. Then as the ball rebounds, the elastic potential energy is transformed back into kinetic energy.

Problem-Solving Strategy for Collisions Involving Two Objects

1. Draw before and after diagrams of the collision.
2. Collect and organize information on the masses and velocities of the two objects before and after the collision. Express the velocities in component form (with correct algebraic signs).
3. Set the sum of the momenta of the two before the collision equal to the sum of the momenta after the collision. Write one equation for each direction:

$$m_1 v_{1ix} + m_2 v_{2ix} = m_1 v_{1fx} + m_2 v_{2fx}$$

$$m_1 v_{1iy} + m_2 v_{2iy} = m_1 v_{1fy} + m_2 v_{2fy}$$

continued on next page

4. If the collision is known to be perfectly inelastic, set the final velocities equal:

$$v_{1fx} = v_{2fx} \quad \text{and} \quad v_{1fy} = v_{2fy}$$

5. If the collision is known to be perfectly elastic, then either set the final kinetic energy equal to the initial kinetic energy:

$$\tfrac{1}{2}m_1v_{1i}^2 + \tfrac{1}{2}m_2v_{2i}^2 = \tfrac{1}{2}m_1v_{1f}^2 + \tfrac{1}{2}m_2v_{2f}^2$$

or, for a one-dimensional collision, set the relative speeds equal:

$$v_{2ix} - v_{1ix} = -(v_{2fx} - v_{1fx})$$

6. Solve for the unknown quantities.

CHECKPOINT 7.7A

Is momentum conserved in a perfectly inelastic collision?

CHECKPOINT 7.7B

A bumper car traveling at speed v_i is moving toward a second car of equal mass that is at rest. Figure 7.18 shows two possible outcomes of the collision. (a, b) For each outcome, show that momentum is conserved and determine whether the collision is elastic, inelastic, or perfectly inelastic. (c) Think of another possible outcome that is consistent with momentum conservation.

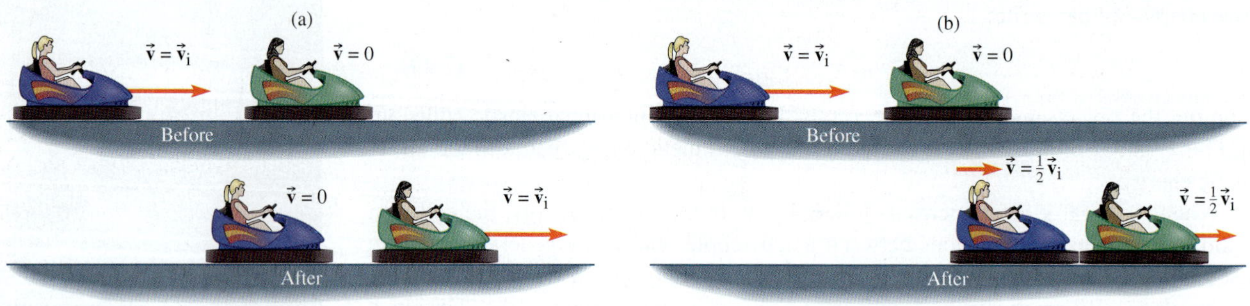

Figure 7.18 Two of the many possible outcomes of a collision between bumper cars of equal mass with one of them initially at rest.

Example 7.10

Collision at the Highway Entry Ramp

At a Route 3 highway on-ramp, a car of mass 1.50×10^3 kg is stopped at a stop sign, waiting for a break in traffic before merging with the cars on the highway. A pickup of mass 2.00×10^3 kg comes up from behind and hits the stopped car. Assuming the collision is elastic, how fast was the pickup going just before the collision if the car is pushed straight ahead onto the highway at 20.0 m/s just after the collision?

Strategy Conservation of momentum provides one equation relating the initial and final velocities. That the collision is elastic provides another equation. With two unknown velocities, these two equations enable us to solve for both. Let "1" refer to the car stopped at the stop sign and "2" refer to the pickup. All motions are in one direction, which we call the x-axis. To simplify the notation, we drop the x subscripts and let all p's and v's refer to x-components. Figure 7.19 shows a before and after diagram for the collision.

Given: $m_1 = 1.50 \times 10^3$ kg; $m_2 = 2.00 \times 10^3$ kg; before the collision, $v_{1i} = 0$; after the collision, $v_{1f} = 20.0$ m/s

To find: v_{2i} (speed of the pickup just before the collision)

continued on next page

Example 7.10 continued

Solution From conservation of momentum,

$$\cancel{m_1 v_{1i}} + m_2 v_{2i} = m_1 v_{1f} + m_2 v_{2f} \qquad (1)$$

where we cross out the first term because $v_{1i} = 0$. The collision is elastic, so the relative velocity after the collision is equal and opposite to the relative velocity before the collision [Eq. (7-14)]:

$$v_{2i} - \cancel{v_{1i}} = -(v_{2f} - v_{1f}) \qquad (2)$$

We want to solve these two equations for v_{2i}, so we can eliminate v_{2f}. Multiplying Eq. (2) through by m_2 and rearranging yields

$$m_2 v_{2i} = m_2 v_{1f} - m_2 v_{2f} \qquad (3)$$

Adding Eqs. (1) and (3) gives

$$2m_2 v_{2i} = (m_1 + m_2) v_{1f} \qquad (4)$$

Finally, we solve Eq. (4) for v_{2i}:

$$v_{2i} = \frac{m_1 + m_2}{2m_2} v_{1f} = \frac{1500 \text{ kg} + 2000 \text{ kg}}{4000 \text{ kg}} \times 20.0 \text{ m/s} = 17.5 \text{ m/s}$$

Discussion To check this answer, first solve for v_{2f}. Then you can verify that momentum is conserved [Eq. (1)] and that the relative velocity changes sign [Eq. (2)]. You can also calculate the total kinetic energy before and after the collision and show they are equal, as they must be for an elastic collision. We leave these checks to you for practice.

The road exerts frictional forces on the vehicles, so the net external force on the vehicles was not zero during the collision. We still use conservation of momentum because during the short time interval of the collision, friction doesn't have time to change the system's momentum significantly.

Practice Problem 7.10 Perfectly Inelastic Collision Between the Cars

Instead of colliding elastically, suppose the two vehicles lock bumpers when they collide. With the same initial conditions ($v_{1i} = 0$ and $v_{2i} = 17.5$ m/s), find the speed at which the car would be pushed out onto the highway.

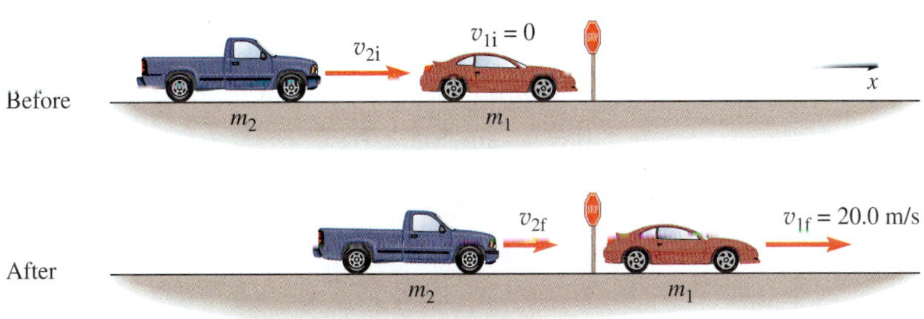

Figure 7.19 Before and after diagrams of the collision (side view).

Suppose in Example 7.10 that the entry ramp speed limit is 20 mi/h (8.94 m/s). By measuring the length of the skid marks from the stop sign and estimating the coefficient of friction, the accident investigator can determine that the car was pushed onto the highway at a speed of 20.0 m/s. Witnesses confirm that the car was stopped before the collision. Then the investigator calculates the speed of the pickup just before the collision using conservation of momentum. The duration Δt of the collision is so short that we can ignore momentum changes due to external forces and treat the two vehicles as an isolated system. Finding that the pickup exceeded the speed limit, the investigator adds speeding to the charges against the driver of the pickup.

How are skid marks used to find car velocities just before the collision?

7.8 COLLISIONS IN TWO DIMENSIONS

Most collisions are not limited to motion in one dimension in the absence of a track or other device to constrain motion to a single line. In a two-dimensional collision, we use the same techniques we used for one-dimensional collisions, as long as we remember that momentum is a vector. (connect interactive: the virtual pool table.) To apply conservation of momentum, it is usually easiest to work with x- and y-components.

CONNECTION:

See the Problem-Solving Strategy in Section 7.7. The same strategy applies to collisions in two or three dimensions.

Example 7.11

Colliding Pucks on an Air Table

A small puck (mass $m_1 = 0.10$ kg) is sliding to the right with an initial speed of 8.0 m/s on an air table (Fig. 7.20a). An air table has many tiny holes through which air is blown; the resulting air cushion allows objects to slide with very little friction. The puck collides with a larger puck (mass $m_2 = 0.40$ kg), which is initially at rest. Figure 7.20b shows the outcome of the collision: the pucks move off at angles $\phi_1 = 60.0°$ above and $\phi_2 = 30.0°$ below the initial direction of motion of the small puck. (a) What are the final speeds of the pucks? (b) Is this an elastic collision or an inelastic collision? (c) If inelastic, what fraction of the initial kinetic energy is converted to other forms of energy in the collision?

Strategy The system of two pucks is an isolated system because the net external force is zero. Therefore, we can apply conservation of momentum. Since motions in two dimensions are involved, we treat the horizontal and vertical components of momentum separately.

Figure 7.20 shows the pucks before and after the collision. Now we collect information on the known quantities, writing velocities in component form.

Masses:	$m_1 = 0.10$ kg; $m_2 = 0.40$ kg
Before collision:	$v_{1ix} = 8.0$ m/s; $v_{1iy} = v_{2ix} = v_{2iy} = 0$
After collision:	$v_{1fx} = v_{1f}\cos\phi_1$; $v_{1fy} = v_{1f}\sin\phi_1$; $v_{2fx} = v_{2f}\cos\phi_2$; $v_{2fy} = -v_{2f}\sin\phi_2$ ($\phi_1 = 60.0°$ and $\phi_2 = 30.0°$)
To find:	v_{1f} and v_{2f}; total kinetic energy before and after the collision

Solution (a) Working with components means that we set the total x-component of momentum before the collision equal to the total x-component of momentum after the collision. We treat the y-components in the same way. The initial momentum is in the x-direction only. Thus, the total momentum y-component after the collision must be zero.

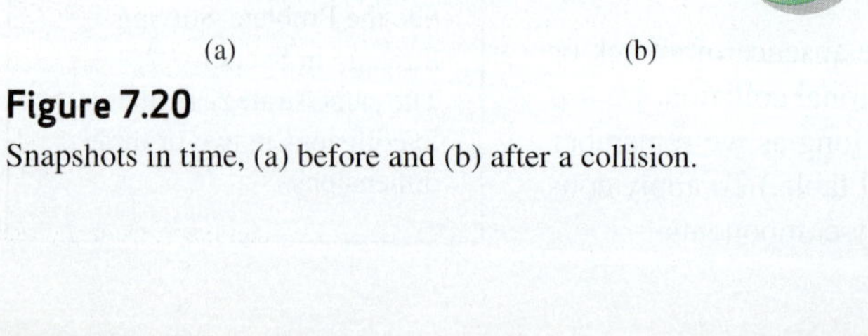

Figure 7.20
Snapshots in time, (a) before and (b) after a collision.

First we set the x-component of the total momentum before the collision equal to the x-component of the total momentum after the collision:

$$p_{1ix} + p_{2ix} = p_{1fx} + p_{2fx}$$

Each momentum component is now rewritten using $p_x = mv_x$:

$$m_1 v_{1ix} + 0 = m_1 v_{1f}\cos\phi_1 + m_2 v_{2f}\cos\phi_2$$

Since $m_2 = 4.0m_1$,

$$m_1 v_{1ix} = m_1 v_{1f}\cos 60.0° + 4.0m_1 v_{2f}\cos 30.0°$$

After canceling the common factor m_1 and substituting numerical values for cos 60.0° and cos 30.0°, this reduces to

$$v_{1ix} = 0.500v_{1f} + 3.46v_{2f} \qquad (1)$$

For conservation of the y-component of the momentum:

$$p_{1iy} + p_{2iy} = p_{1fy} + p_{2fy}$$

The y-component of $\vec{\mathbf{p}}_{2f}$ is negative because the y-component of $\vec{\mathbf{v}}_{2f}$ is negative.

$$0 = m_1 v_{1f}\sin\phi_1 + (-4.0m_1 v_{2f}\sin\phi_2)$$
$$0 = v_{1f}\sin 60.0° - 4.0v_{2f}\sin 30.0°$$

We solve for v_{2f} in terms of v_{1f}:

$$v_{2f} = \frac{\sin 60.0°}{4.0\sin 30.0°}v_{1f} = 0.433v_{1f} \qquad (2)$$

Equations (1) and (2) contain two unknowns. To eliminate one unknown, we substitute $0.433v_{1f}$ for v_{2f} in Eq. (1):

$$v_{1ix} = 0.500v_{1f} + 3.46(0.433v_{1f}) = 2.00v_{1f}$$

Solving this equation gives the value of v_{1f}:

$$v_{1f} = 4.0 \text{ m/s}$$

Then by substitution into Eq. (2), we find the value of v_{2f}:

$$v_{2f} = 0.433v_{1f} = 1.73 \text{ m/s, which rounds to } 1.7 \text{ m/s}$$

(b) Now that we have the final speeds, we can compare the initial and final kinetic energies.

$$K_i = \tfrac{1}{2}m_1 v_{1i}^2$$
$$K_i = \tfrac{1}{2}(0.10 \text{ kg}) \times (8.0 \text{ m/s})^2 = 3.2 \text{ J}$$

and

$$K_f = \tfrac{1}{2}m_1 v_{1f}^2 + \tfrac{1}{2}m_2 v_{2f}^2$$
$$= \tfrac{1}{2}(0.10 \text{ kg}) \times (4.0 \text{ m/s})^2 + \tfrac{1}{2}(0.40 \text{ kg}) \times (1.73 \text{ m/s})^2$$
$$= 0.80 \text{ J} + 0.60 \text{ J} = 1.40 \text{ J}$$

continued on next page

Example 7.11 continued

The final kinetic energy is less than the initial kinetic energy, so the collision is inelastic.

(c) The amount of kinetic energy converted to other forms of energy (primarily internal energy of the pucks) is

$$3.2\ \text{J} - 1.40\ \text{J} = 1.8\ \text{J}$$

We divide by the initial kinetic energy to find the fraction of the initial kinetic energy converted to other forms:

$$\frac{1.8\ \text{J}}{3.2\ \text{J}} = 0.56$$

Less than half of the kinetic energy of the incident puck therefore survives the collision as the kinetic energies of the two pucks.

Discussion Although a two-dimensional collision problem tends to require more complicated algebra than a one-dimensional problem, the physical principles are the same. As long as the net external force on the system is zero (or negligibly small), the total vector momentum must be conserved.

Practice Problem 7.11 Colliding Balls

A ball of mass m_1 moves at speed v_i along the $+x$-axis toward a second ball of mass m_2, which is initially at rest. The second ball has five times the mass of the first ball. After the collision between these two objects, m_1 moves along the $+y$-axis at a speed v_1, and m_2 moves at a speed $v_2 = \frac{1}{4}v_i$ at an angle of 36.9° below the $+x$-axis. Find v_1 in terms of v_i.

Conceptual Example 7.12

Eric at the Pool Table

Playing a game of pool, Eric is trying to decide whether to attempt a shot to sink the 4-ball in the pocket at corner B without *scratching* (sinking the cue ball "C" in corner A). He notices that the lines from the 4-ball to the two corner pockets happen to make a right angle (Fig. 7.21). The collision of the balls is nearly elastic. Assume Eric is a novice player and does not know how to put "English" (sidespin) on the cue ball. Should he attempt the shot?

Strategy We assume a perfectly elastic collision between the balls. They have the same mass. The cue ball moves with an initial velocity $\vec{\mathbf{v}}_i$ and strikes the 4-ball, which is initially at rest. The 4-ball falls in pocket B if its velocity after the collision, $\vec{\mathbf{v}}_4$, points toward the pocket. Assuming it does, we use conservation of momentum and kinetic energy to find the direction of the cue ball velocity, $\vec{\mathbf{v}}_c$, after the collision.

Solution Conservation of momentum requires that

$$m\vec{\mathbf{v}}_i = m\vec{\mathbf{v}}_c + m\vec{\mathbf{v}}_4$$

or

$$\vec{\mathbf{v}}_i = \vec{\mathbf{v}}_c + \vec{\mathbf{v}}_4$$

Figure 7.21
Should Eric try to sink the 4-ball?

continued on next page

Conceptual Example 7.12 continued

This vector addition is shown graphically in Fig. 7.22a. Since the collision is elastic, the total kinetic energy doesn't change:

$$\tfrac{1}{2}mv_i^2 = \tfrac{1}{2}mv_c^2 + \tfrac{1}{2}mv_4^2$$

or

$$v_i^2 = v_c^2 + v_4^2$$

Since v_i, v_c, and v_4 are the sides of a triangle, this is a statement of the Pythagorean theorem—the triangle must be a *right* triangle with v_i as the hypotenuse (Fig. 7.22b).

Figure 7.22

(a) Graphical addition of velocity vectors as required by the conservation of momentum. (b) Since $v_i^2 = v_c^2 + v_4^2$, the three velocities form a right triangle.

Therefore, the velocities of the 4-ball and the cue ball after the collision are perpendicular to each other.

If Eric sinks the 4-ball, the cue ball falls into pocket A. He shouldn't attempt this shot until he learns how to put some English on the ball.

Discussion Note that we did not resolve the velocities into *x*- and *y*-components. Doing so would have made the solution longer in this case.

We found that the two balls move at right angles after the collision. This result is true for any two-dimensional elastic collision between two objects of equal masses if one of them is initially at rest. In Example 7.11 the two pucks move at right angles after the collision, but the collision is inelastic—the masses are unequal.

Practice Problem 7.12 Finding the Speed Ratio

Suppose that the cue ball initially moves in the $-x$-direction. After the collision, the cue ball moves at 52.0° above the $-x$-axis and the 4-ball moves at 38.0° below the $-x$-axis. Find the ratio of the balls' speeds v_c/v_4 after the collision.

Master the Concepts

- Definition of linear momentum:

$$\vec{\mathbf{p}} = m\vec{\mathbf{v}} \quad (7\text{-}1)$$

- During an interaction, momentum is transferred from one object or system to another, but the total momentum of the two is unchanged.

$$\Delta\vec{\mathbf{p}}_2 = -\Delta\vec{\mathbf{p}}_1$$

- Impulse is the average force times the time interval.
- The total impulse equals the change in momentum:

$$\Delta\vec{\mathbf{p}} = \sum\vec{\mathbf{F}}\,\Delta t \quad (7\text{-}2)$$

- A conserved quantity is one that remains unchanged as time passes.
- Impulse is the area under a graph of force versus time.

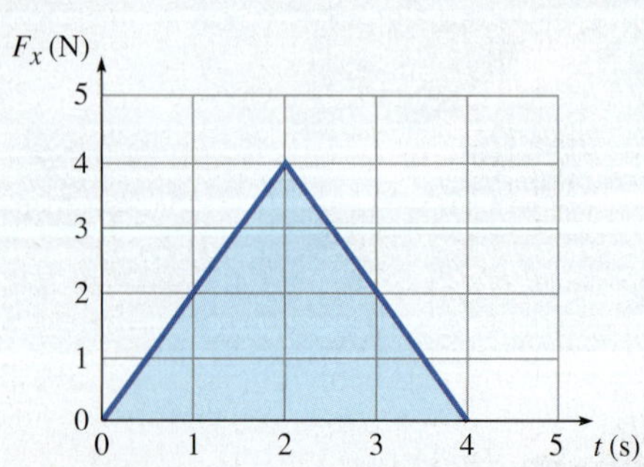

- The net force is the rate of change of momentum.

$$\sum\vec{\mathbf{F}} = \lim_{\Delta t \to 0} \frac{\Delta\vec{\mathbf{p}}}{\Delta t} \quad (7\text{-}4)$$

- External interactions may change the total momentum of a system.
- Internal interactions do not change the total momentum of a system.
- Conservation of linear momentum: if the net external force acting on a system is zero, then the momentum of the system is conserved.
- The position of the CM of a system of *N* particles is

$$x_{CM} = \frac{m_1x_1 + m_2x_2 + \cdots + m_Nx_N}{M}$$

and

$$y_{CM} = \frac{m_1y_1 + m_2y_2 + \cdots + m_Ny_N}{M} \quad (7\text{-}9)$$

where *M* is the total mass of the particles:

$$M = m_1 + m_2 + \cdots + m_N$$

- The total momentum of a system is equal to the total mass times the velocity of the center of mass:

$$\vec{\mathbf{p}} = \vec{\mathbf{p}}_1 + \vec{\mathbf{p}}_2 + \cdots + \vec{\mathbf{p}}_N = M\vec{\mathbf{v}}_{CM} \quad (7\text{-}11)$$

continued on next page

Master the Concepts continued

- No matter how complicated a system is, the CM moves as if all the mass of the system were concentrated to a point particle with all the external forces acting on it:

$$\sum \vec{\mathbf{F}}_{ext} = M\vec{\mathbf{a}}_{CM} \quad (7\text{-}13)$$

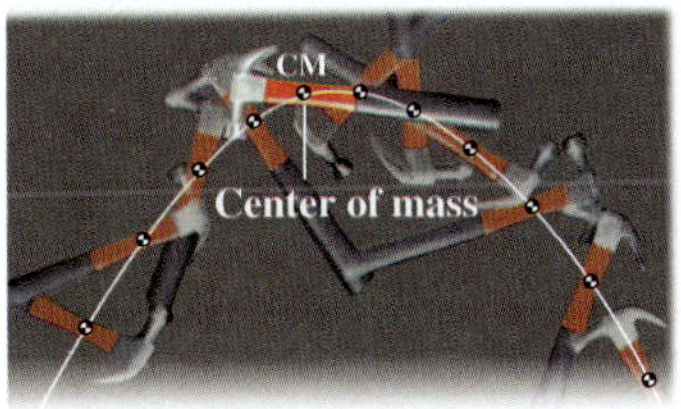

- The CM of an isolated system moves at constant velocity.
- Conservation of momentum is used to solve problems involving collisions, explosions, and the like. Even when external forces are acting, the momentum of the system just before a collision is nearly equal to the momentum just after if the collision interaction is brief. The impulse, and, therefore, the change in momentum of the system, is small since the time interval is small.

Conceptual Questions

1. You are trapped on the second floor of a burning building. The stairway is impassable, but there is a balcony outside your window. Describe what might happen in the following situations. (a) You jump from the second-story balcony to the pavement below, landing stiff-legged on your feet. (b) You jump into a privet hedge, landing on your back and rolling to your feet. (c) You jump into a firefighters' net, landing on your back. What happens to the net as you land in it? What do the firefighters do to cushion your fall even more?
2. A force of 30 N is applied for 5 s to each of two objects of different masses. (a) Which one has the greatest momentum change? (b) The greatest velocity change? (c) The greatest acceleration?
3. If you take a rifle and saw off part of the barrel, the muzzle speed (the speed at which bullets emerge from the barrel) will be smaller. Why?
4. A firecracker at rest explodes, sending fragments off in all directions. Initially the firecracker has zero momentum, but after the explosion the fragments flying off each have quite a lot of momentum. Hasn't momentum been created? If not, explain why not.
5. An astronaut in deep space is taking a space walk when the tether connecting him to his spaceship breaks. How can he get back to the ship? He doesn't have a rocket propulsion backpack, unfortunately, but he is carrying a big wrench.
6. An astronaut hits a golf ball on the surface of the Moon. Is the momentum of the ball conserved while it is in flight? Is there a *component* of its momentum that is conserved?
7. Which would be more effective: a hammer that collides *elastically* with a nail, or one that collides perfectly *inelastically*? Assume that the mass of the hammer is much larger than that of the nail.
8. Mary and Daryl are new to the sport of rock climbing. Mary says she wants a stiff rope because a stiff rope is a strong rope. Daryl insists that a good climbing rope must have some stretch. Who is correct, and why?
9. In your own words, phrase each of Newton's three laws of motion as a statement about momentum.
10. Two objects with different masses have the same kinetic energy. Which has the larger magnitude of momentum?
11. A woman is 1.60 m tall. When standing straight, is her CM necessarily 0.80 m above the floor? Explain.
12. The momentum of a system can only be changed by an external force. What is the external force that changes the momentum of a bicycle (with its rider) as it speeds up, slows down, or changes direction? Is it true that changes in the bicycle's kinetic energy must come from an external force? Explain.
13. In an egg toss, two people try to toss a raw egg back and forth without breaking it as they move farther and farther apart. Discuss a strategy in terms of impulse and momentum for catching the egg without breaking it.
14. In the "executive toy," two balls are pulled back and then released. After the collision, two balls move away on the opposite side. Why do we never see three balls move away following this action, although with a lower velocity so that linear momentum is still conserved?

15. A baseball batting coach emphasizes the importance of "follow-through" when a batter is trying for a home run. The coach explains that the follow-through keeps the bat in contact with the ball for a longer time so the ball will travel a greater distance. Explain the reasoning behind this statement in terms of the impulse-momentum theorem.

16. Micah is standing on his frictionless skateboard facing a concrete wall. He wants to project himself backward by throwing small balls at the wall. His friend Jeremy says that Micah need not throw the balls against the wall, he just needs to throw the balls away from himself, but Micah says the balls need something to push against if they are to propel him backward. Who is right and why?

Multiple-Choice Questions

Student Response System Questions

1. Two particles A and B of equal mass are located at some distance from each other. Particle A is at rest while B moves away from A at speed v. What happens to the center of mass of the system of two particles?
(a) It does not move.
(b) It moves with a speed v away from A.
(c) It moves with a speed v toward A.
(d) It moves with a speed $\frac{1}{2}v$ away from A.
(e) It moves with a speed $\frac{1}{2}v$ toward A.

2. A ball of mass m with initial speed v collides with another ball of mass M, initially at rest. After the collision the two balls stick together, moving with speed V. The ratio of the final speed V to the initial speed v is $V/v =$
(a) $\dfrac{M}{M+m}$ (b) $\dfrac{M+m}{M}$
(c) $\dfrac{m}{M+m}$ (d) $\dfrac{M+m}{m}$
(e) $\sqrt{\dfrac{M}{M+m}}$ (f) $\sqrt{\dfrac{m}{M+m}}$

3. Two uniform spheres with equal mass per unit volume are in contact with one another. The mass of sphere A is five times that of sphere B. The center of mass of the system is
(a) at the point where A and B touch.
(b) inside sphere B somewhere on the line joining the centers of A and B.
(c) inside sphere A somewhere on the line joining the centers.
(d) at the center of sphere A.
(e) outside of both spheres.

4. A 3.0 kg object is initially at rest. It then receives an impulse of magnitude 15 N·s. After the impulse, the object has
(a) a speed of 45 m/s.
(b) a momentum of magnitude 5.0 kg·m/s.
(c) a speed of 7.5 m/s.
(d) a momentum of magnitude 15 kg·m/s.

5. An object of mass m drops from rest a little above the Earth's surface for a time t. Ignore air resistance. After time t the magnitude of its momentum is
(a) mgt^2
(b) mgt
(c) $mg\sqrt{t}$
(d) $\sqrt{mgt}$
(e) $\dfrac{mgt^2}{2}$

6. An object at rest suddenly explodes into three parts of equal mass. Two of the parts move away at right angles to each other and with equal speeds v. What is the velocity of the third part just after the explosion?
(a) Direction of vector 1 and magnitude $2v$
(b) Direction of vector 2 and magnitude $\sqrt{2}v$
(c) Direction of vector 3 and magnitude $\dfrac{1}{\sqrt{2}}v$
(d) Direction of vector 2 and magnitude $\dfrac{1}{\sqrt{2}}v$
(e) Direction of vector 1 and magnitude $\dfrac{1}{\sqrt{2}}v$

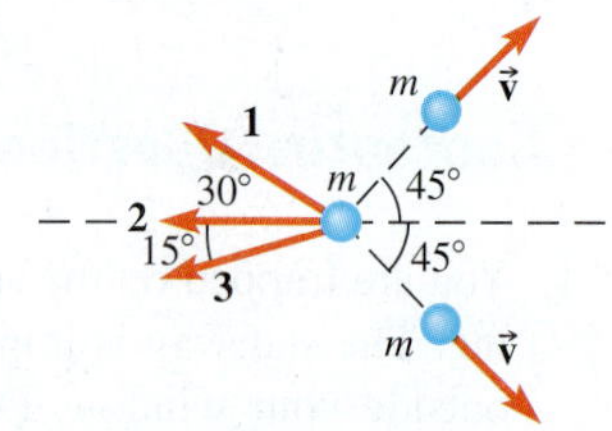

Multiple-Choice Questions 7–12 refer to a situation in which a golf ball is projected straight upward in the $+y$-direction. Ignore air resistance. The answer choices are found in the figures.

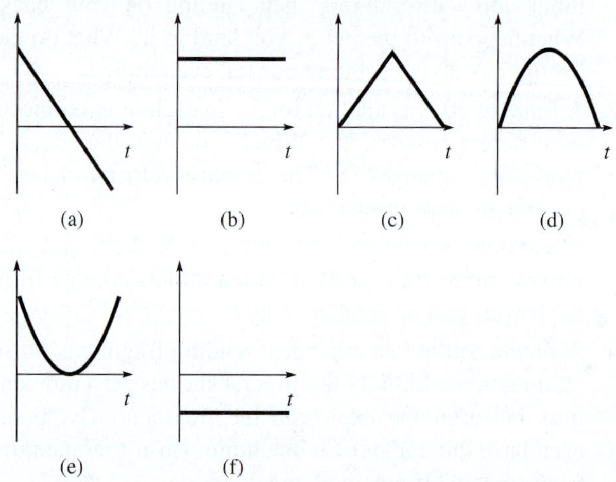

7. Which graph shows the acceleration a_y of the ball as a function of time?

8. Which graph shows the momentum p_y of the ball as a function of time?

9. Which graph shows the vertical position y of the ball as a function of time?

10. Which graph shows the total energy of the ball as a function of time?

11. Which graph shows the potential energy of the ball as a function of time?

12. Which graph shows the kinetic energy of the ball as a function of time?

Problems

- C Combination conceptual/quantitative problem
- Biomedical application
- ✦ Challenging
- Blue # Detailed solution in the Student Solutions Manual
- [1, 2] Problems paired by concept
- connect Text website interactive or tutorial

7.2 Momentum; 7.3 The Impulse-Momentum Theorem

1. Two cars, each of mass 1300 kg, are approaching each other on a head-on collision course. Each speedometer reads 19 m/s. What is the magnitude of the total momentum of the system?
2. What is the momentum of an automobile (weight = 9800 N) when it is moving at 35 m/s to the south?
3. Verify that the SI unit of impulse is the same as the SI unit of momentum.
4. A cue stick hits a cue ball with an average force of 24 N for a duration of 0.028 s. If the mass of the ball is 0.16 kg, how fast is it moving after being struck?
5. A system consists of three particles with these masses and velocities: mass 3.0 kg, moving north at 3.0 m/s; mass 4.0 kg, moving south at 5.0 m/s; and mass 7.0 kg, moving north at 2.0 m/s. What is the total momentum of the system?
6. A sports car traveling along a straight line increases its speed from 20.0 mi/h to 60.0 mi/h. (a) What is the ratio of the final to the initial magnitude of its momentum? (b) What is the ratio of the final to the initial kinetic energy?
7. At $t = 0$, six birds are flying south at 10 m/s. Their masses and their velocities at a later time are:
   (a) 200 g, 10 m/s north at $t = 30$ s
   (b) 200 g, 10 m/s east at $t = 30$ s
   (c) 200 g, 20 m/s north at $t = 60$ s
   (d) 400 g, 20 m/s north at $t = 60$ s
   (e) 400 g, 20 m/s south at $t = 10$ s
   (f) 400 g, 30 m/s west at $t = 90$ s
   Rank them in order of the magnitude of the momentum change, smallest to largest.
8. An object of mass 3.0 kg is projected into the air at a 55° angle. It hits the ground 3.4 s later. What is its change in momentum while it is in the air? Ignore air resistance.
9. A ball of mass 5.0 kg moving with a speed of 2.0 m/s in the $+x$-direction hits a wall and bounces back with the same speed in the $-x$-direction. What is the change of momentum of the ball?
10. Dynamite is being used to blast through rock to build a road. After one explosion, the masses and kinetic energies of several fragments of rock thrown up into the air are:
    (a) 8 kg, 400 J
    (b) 2 kg, 1600 J
    (c) 4 kg, 1600 J
    (d) 16 kg, 100 J
    (e) 1 kg, 1600 J
    Rank the fragments in order of the magnitude of their momentum, smallest to largest.
11. An object of mass 3.0 kg is allowed to fall from rest under the force of gravity for 3.4 s. What is the change in its momentum? Ignore air resistance.
12. What average force is necessary to bring a 50.0 kg sled from rest to a speed of 3.0 m/s in a period of 20.0 s? Assume frictionless ice.
13. Five cars are traveling on a highway. Their masses and initial speeds are:
    (a) 1500 kg, 30 m/s
    (b) 1500 kg, 20 m/s
    (c) 1000 kg, 30 m/s
    (d) 1000 kg, 20 m/s
    (e) 2000 kg, 40 m/s
    The cars use the same braking force to slow down and stop. Rank the cars in order of the time it takes them to stop, from smallest to greatest.
14. A bird (mass 31 g) is flying at 11.1 m/s when it flies into a glass window and bounces off at a speed of 4.1 m/s. The bird is in contact with the glass for 0.071 s. What is the average force on the bird during the collision?
15. For a safe reentry into the Earth's atmosphere, the pilots of a space capsule must reduce their speed from 2.6×10^4 m/s to 1.1×10^4 m/s. The rocket engine produces a backward force on the capsule of 1.8×10^5 N. The mass of the capsule is 3800 kg. For how long must they fire their engine? [*Hint*: Ignore the change in mass of the capsule due to the expulsion of exhaust gases.]
16. A 0.15 kg baseball traveling in a horizontal direction with a speed of 20 m/s hits a bat and is popped straight up with a speed of 15 m/s. (a) What is the change in momentum (magnitude and direction) of the baseball? (b) If the bat was in contact with the ball for 50 ms, what was the average force of the bat on the ball?
17. An automobile traveling at a speed of 30.0 m/s applies its brakes and comes to a stop in 5.0 s. If the automobile has a mass of 1.0×10^3 kg, what is the average horizontal force exerted on it during braking? Assume the road is level.
18. A 3.0 kg object is initially moving northward at 15 m/s. Then a force of 15 N, toward the east, acts on it for 4.0 s. (a) At the end of the 4.0 s, what is the object's final velocity? (b) What is the change in momentum during the 4.0 s?

19. ✦ A boy of mass 60.0 kg is rescued from a hotel fire by leaping into a firefighters' net. The window from which he leapt was 8.0 m above the net. The firefighters lower their arms as he lands in the net so that he is brought to a complete stop in a time of 0.40 s. (a) What is his change in momentum during the 0.40 s interval? (b) What is the impulse on the net due to the boy during the interval? [*Hint*: Do not ignore gravity.] (c) What is the average force on the net due to the boy during the interval?

20. ✦ A pole-vaulter of mass 60.0 kg vaults to a height of 6.0 m before dropping to thick padding placed below to cushion her fall. (a) Find the speed with which she lands. (b) If the padding brings her to a stop in a time of 0.50 s, what is the average force on her body due to the padding during that time interval?

7.4 Conservation of Momentum

21. A frog is sitting on a lily pad when it sees a delicious fly. He darts out his tongue at a speed of 3.7 m/s to catch the fly. The tongue has a mass of 0.41 g, and the rest of the frog plus the lily pad have a mass of 12.5 g. What is the recoil speed of the frog and lily pad? Ignore drag forces on the pad due to the water.

22. Diana is standing on a raft of mass 100.0 kg that is floating on a still lake. She decides to walk the length of the raft. If Diana's mass is 55 kg and she walks with a velocity of 0.91 m/s with respect to the shore, how fast and in what direction does the raft move while Diana is walking? Assume the raft is stationary with respect to the shore before Diana starts walking.

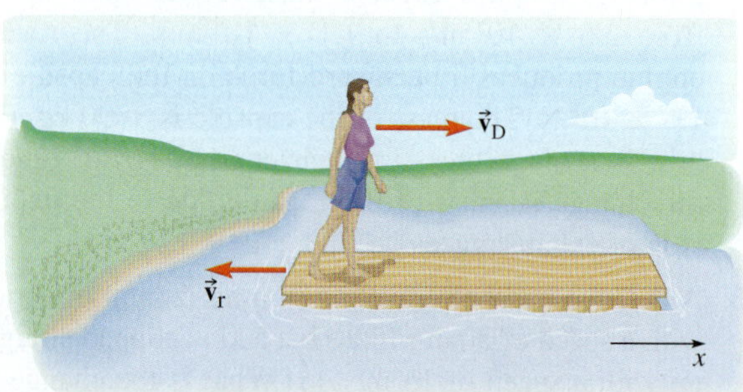

23. A rifle has a mass of 4.5 kg and it fires a bullet of mass 10.0 g at a muzzle speed of 820 m/s. What is the recoil speed of the rifle as the bullet leaves the gun barrel?

24. A 0.030 kg bullet is fired vertically at 200 m/s into a 0.15 kg baseball that is initially at rest. The bullet lodges in the baseball and, after the collision, the baseball/bullet rise to a height of 37 m. (a) What was the speed of the baseball/bullet right after the collision? (b) What was the average force of air resistance while the baseball/bullet was rising?

25. A submarine of mass 2.5×10^6 kg and initially at rest fires a torpedo of mass 250 kg. The torpedo has an initial speed of 100.0 m/s. What is the initial recoil speed of the submarine? Ignore the drag force of the water.

26. A uranium nucleus (mass 238 u), initially at rest, undergoes radioactive decay. After an alpha particle (mass 4.0 u) is emitted, the remaining nucleus is thorium (mass 234 u). If the alpha particle is moving at 0.050 times the speed of light, what is the recoil speed of the thorium nucleus? (Note: "u" is a unit of mass; it is *not* necessary to convert it to kg.)

27. Dash is standing on his frictionless skateboard with three balls, each with a mass of 100 g, in his hands. The combined mass of Dash and his skateboard is 60 kg. How fast should Dash throw the balls forward if he wants to move backward with a speed of 0.50 m/s? Do you think Dash can succeed? Explain.

28. A 58 kg astronaut is in space, far from any objects that would exert a significant gravitational force on him. He would like to move toward his spaceship, but his jet pack is not functioning. He throws a 720 g socket wrench with a velocity of 5.0 m/s in a direction away from the ship. After 0.50 s, he throws a 800 g spanner in the same direction with a speed of 8.0 m/s. After another 9.90 s, he throws a mallet with a speed of 6.0 m/s in the same direction. The mallet has a mass of 1200 g. How fast is the astronaut moving after he throws the mallet?

29. ✦ A cannon on a railroad car is facing in a direction parallel to the tracks. It fires a 98 kg shell at a speed of 105 m/s (relative to the ground) at an angle of 60.0° above the horizontal. If the cannon plus car have a mass of 5.0×10^4 kg, what is the recoil speed of the car if it was at rest before the cannon was fired? [*Hint*: A *component* of a system's momentum along an axis is conserved if the net external force acting on the system has no component along that axis.]

30. ✦ A marksman standing on a motionless railroad car fires a gun into the air at an angle of 30.0° from the horizontal. The bullet has a speed of 173 m/s (relative

to the ground) and a mass of 0.010 kg. The man and car move to the left at a speed of 1.0×10^{-3} m/s after he shoots. What is the mass of the man and car? (See the hint in Problem 29.)

7.5 Center of Mass; 7.6 Motion of the Center of Mass

31. Particle A is at the origin and has a mass of 30.0 g. Particle B has a mass of 10.0 g. Where must particle B be located if the coordinates of the CM are $(x, y) = (2.0\text{ cm}, 5.0\text{ cm})$?

32. Particle A has a mass of 5.0 g and particle B has a mass of 1.0 g. Particle A is located at the origin and particle B is at the point $(x, y) = (25\text{ cm}, 0)$. What is the location of the CM? (connect tutorial: center of mass)

33. Women often experience back pain when they are pregnant. Suppose a woman's mass before pregnancy is 68 kg and her center of mass when standing is located directly above the hips. By the thirty-fourth week of pregnancy, she has gained 8.0 kg (the combined mass of the fetus, the placenta, and the amniotic fluid). The center of mass of this 8.0 kg is 18 cm in front of the hips. If she hasn't changed her posture, how far in front of the hips is her center of mass? [Women have three wedge-shaped lumbar vertebrae, whereas men have only two. This evolutionary adaptation permits a greater curvature of the lumbar spine to keep a pregnant woman's CM above the hips. See *Nature* **450** (Dec 13, 2007) pp. 1075–1078.]

34. In an action movie, the hero dangles his archenemy over the edge of a cliff. The archenemy's mass is 68 kg and his center of mass is 44 cm horizontally past the edge of the cliff. The hero's center of mass is 15 cm horizontally from the edge of the cliff. What is the smallest value of the hero's mass so that the CM of the two is not out past the edge of the cliff (which would make them both fall into the ravine below)? Is this scenario reasonable?

35. The three objects in the figure each have the same mass. If one of the objects is moved 12 cm in the positive x-direction, by how much does the CM move?

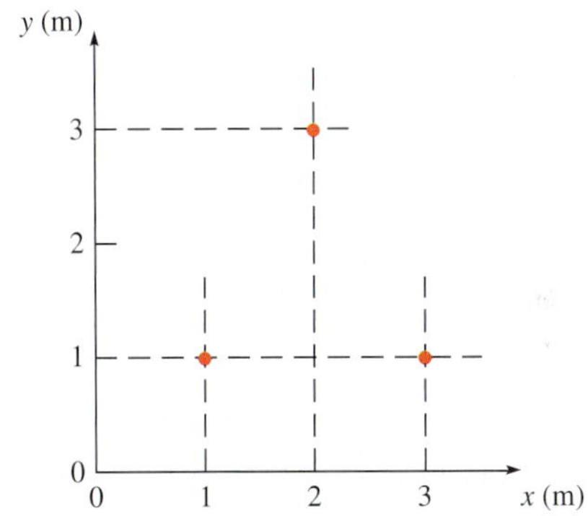

36. The positions of three particles, written as (x, y) coordinates, are: particle 1 (mass 4.0 kg) at (4.0 m, 0 m); particle 2 (mass 6.0 kg) at (2.0 m, 4.0 m); particle 3 (mass 3.0 kg) at (−1.0 m, −2.0 m). What is the location of the CM?

37. ✦ Belinda needs to find the CM of a sculpture she has made so that it will hang in a gallery correctly. The sculpture is all in one plane and consists of various shaped uniform objects with masses and sizes as shown. Where is the CM of this sculpture? Assume the thin rods connecting the larger pieces have no mass and place the reference frame origin at the top left corner of the sculpture.

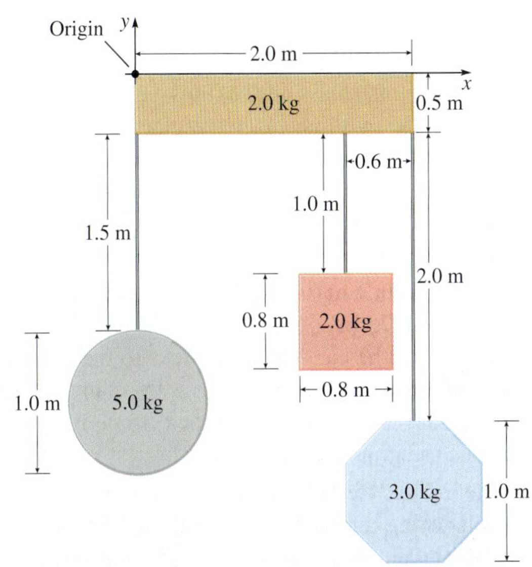

38. Find the x-coordinate of the CM of the composite object shown in the figure. The sphere, cylinder, and rectangular solid all have a uniform composition. Their masses and dimensions are: sphere: 200 g, diameter = 10 cm; cylinder: 450 g, length = 17 cm, radius = 5.0 cm; rectangular solid: 325 g, length in x-direction = 16 cm, height = 10 cm, depth = 12 cm.

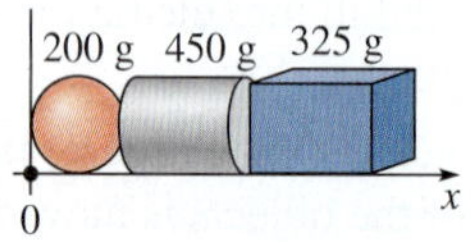

39. Consider two falling objects. Their masses are 3.0 kg and 4.0 kg. At time $t = 0$, the two are released from rest. What is the velocity of their CM at $t = 10.0$ s? Ignore air resistance.

40. Object A of mass 3 kg is moving in the $+x$-direction with a speed of 14 m/s. Object B of mass 4 kg is moving in the $-y$-direction with a speed of 7 m/s. What are the x- and y-components of the velocity of the CM of the two objects?

41. If a particle of mass 5.0 kg is moving east at 10 m/s and a particle of mass 15 kg is moving west at 10 m/s, what is the velocity of the CM of the pair?

42. An object located at the origin and having mass M explodes into three pieces having masses $M/4$, $M/3$, and $5M/12$. The pieces scatter on a horizontal frictionless xy-plane. The piece with mass $M/4$ flies away with velocity 5.0 m/s at 37° above the x-axis. The piece with mass $M/3$ has velocity 4.0 m/s directed at an angle of 45° above the $-x$-axis. (a) What are the velocity components of the third piece? (b) Describe the motion of the CM of the system after the explosion.

43. Prove Eq. (7-13) $\Sigma\vec{\mathbf{F}}_{ext} = M\vec{\mathbf{a}}_{CM}$. [*Hint*: Start with $\Sigma\vec{\mathbf{F}}_{ext} = \lim_{\Delta t \to 0} (\Delta\vec{\mathbf{p}}/\Delta t)$, where $\Sigma\vec{\mathbf{F}}_{ext}$ is the net external force acting on a system and $\vec{\mathbf{p}}$ is the total momentum of the system.]

7.7 Collisions in One Dimension

44. A helium atom (mass 4.00 u) moving at 618 m/s to the right collides with an oxygen molecule (mass 32.0 u) moving in the same direction at 412 m/s. After the collision, the oxygen molecule moves at 456 m/s to the right. What is the velocity of the helium atom after the collision?

45. A toy car with a mass of 120 g moves to the right with a speed of 0.75 m/s. A small child drops a 30.0 g piece of clay onto the car. The clay sticks to the car and the car continues to the right. What is the change in speed of the car? Consider the frictional force between the car and the ground to be negligible.

46. In the railroad freight yard, an empty freight car of mass m rolls along a straight level track at 1.0 m/s and collides with an initially stationary, fully loaded boxcar of mass $4.0m$. The two cars couple together on collision. (a) What is the speed of the two cars after the collision? (b) Suppose instead that the two cars are at rest after the collision. With what speed was the loaded boxcar moving before the collision if the empty one was moving at 1.0 m/s? (connect tutorial: sticking collision)

47. A 0.020 kg bullet traveling at 200.0 m/s east hits a motionless 2.0 kg block and bounces off it, retracing its original path with a velocity of 100.0 m/s west. What is the final velocity of the block? Assume the block rests on a perfectly frictionless horizontal surface.

48. A block of wood of mass 0.95 kg is initially at rest. A bullet of mass 0.050 kg traveling at 100.0 m/s strikes the block and becomes embedded in it. With what speed do the block of wood and the bullet move just after the collision?

49. A 0.020 kg bullet is shot horizontally and collides with a 2.00 kg block of wood. The bullet embeds in the block, and the block slides along a horizontal surface for 1.50 m. If the coefficient of kinetic friction between the block and surface is 0.400, what was the original speed of the bullet?

50. A 2.0 kg block is moving to the right at 1.0 m/s just before it strikes and sticks to a 1.0 kg block initially at rest. What is the total momentum of the two blocks after the collision?

51. A 75 kg man is at rest on ice skates. A 0.20 kg ball is thrown to him. The ball is moving horizontally at 25 m/s just before the man catches it. How fast is the man moving just after he catches the ball?

52. A BMW of mass 2.0×10^3 kg is traveling at 42 m/s. It approaches a 1.0×10^3 kg Volkswagen going 25 m/s in the same direction and strikes it in the rear. Neither driver applies the brakes. Ignore the relatively small frictional forces on the cars due to the road and due to air resistance. (a) If the collision slows the BMW down to 33 m/s, what is the speed of the VW after the collision? (b) During the collision, which car exerts a larger force on the other, or are the forces equal in magnitude? Explain.

53. A 100 g ball collides elastically with a 300 g ball that is at rest. If the 100 g ball was traveling in the positive x-direction at 5.00 m/s before the collision, what are the velocities of the two balls after the collision? (connect tutorial: elastic collision)

54. An object of 1.0 kg mass approaches a stationary object of 5.0 kg at 10.0 m/s and, after colliding, rebounds in the reverse direction along the same line with a speed of 5.0 m/s. What is the speed of the 5.0 kg object after the collision?

55. A 2.0 kg object is at rest on a perfectly frictionless surface when it is hit by a 3.0 kg object moving at 8.0 m/s. If the two objects are stuck together after the collision, what is the speed of the combination?

56. A spring of negligible mass is compressed between two blocks, A and B, which are at rest on a frictionless horizontal surface at a distance of 1.0 m from a wall on the

left and 3.0 m from a wall on the right. The sizes of the blocks and spring are small. When the spring is released, block A moves toward the left wall and strikes it at the same instant that block B strikes the right wall. The mass of A is 0.60 kg. What is the mass of B?

57. ✦ A 0.010 kg bullet traveling horizontally at 400.0 m/s strikes a 4.0 kg block of wood sitting at the edge of a table. The bullet is lodged into the wood. If the table height is 1.2 m, how far from the table does the block hit the floor?

58. ✦ Two objects with masses m_1 and m_2 approach each other with equal and opposite momenta so that the total momentum is zero. Show that, if the collision is elastic, the final *speed* of each object must be the same as its initial speed. (The final *velocity* of each object is *not* the same as its initial velocity, however.)

59. ✦ A 6.0 kg object is at rest on a perfectly frictionless surface when it is struck head-on by a 2.0 kg object moving at 10 m/s. If the collision is perfectly elastic, what is the speed of the 6.0 kg object after the collision? [*Hint*: You will need two equations.]

60. ✦ Use the result of Problem 58 to show that in *any* elastic collision between two objects, the relative speed of the two is the same before and after the collision. [*Hints*: Look at the collision in its CM *frame*—the reference frame in which the CM is at rest. The *relative* speed of two objects is the same in any inertial reference frame.]

7.8 Collisions in Two Dimensions

61. A firecracker is tossed straight up into the air. It explodes into three pieces of equal mass just as it reaches the highest point. Two pieces move off at 120 m/s at right angles to each other. How fast is the third piece moving?

62. Object A of mass M has an original velocity of 6.0 m/s in the $+x$-direction toward a stationary object (B) of the same mass. After the collision, A has velocity components of 1.0 m/s in the $+x$-direction and 2.0 m/s in the $+y$-direction. What is the magnitude of B's velocity after the collision?

63. ✦ (a) With reference to Practice Problem 7.11, find the momentum change of the ball of mass m_1 during the collision. Give your answer in x- and y-component form; express the components in terms of m_1 and v_i. (b) Repeat for the ball of mass m_2. How are the momentum changes related?

64. A hockey puck moving at 0.45 m/s collides with another puck that was at rest. The pucks have equal mass. The first puck is deflected 37° to the right and moves off at 0.36 m/s. Find the speed and direction of the second puck after the collision.

65. ✦ Puck 1 sliding along the x-axis strikes stationary puck 2 of the same mass. After the elastic collision, puck 1 moves off at speed v_{1f} in the direction 60.0° above the x-axis; puck 2 moves off at speed v_{2f} in the direction 30.0° below the x-axis. Find v_{2f} in terms of v_{1f}.

66. Block A, with a mass of 220 g, is traveling north on a frictionless surface with a speed of 5.0 m/s. Block B, with a mass of 300 g travels west on the same surface until it collides with A. After the collision, the blocks move off together with a velocity of 3.13 m/s at an angle of 42.5° to the north of west. What was B's speed just before the collision?

67. A 2.0 kg object (the "projectile") approaches a stationary object (the "target") at 5.0 m/s. The projectile is deflected through an angle of 60.0° and its speed after the collision is 3.0 m/s. What is the magnitude of the momentum of the target after the collision?

68. A 1500 kg car moving east at 17 m/s collides with a 1800 kg car moving south at 15 m/s, and the two cars stick together. (a) What is the velocity of the cars right after the collision? (b) How much kinetic energy was converted to another form during the collision?

69. A car with a mass of 1700 kg is traveling directly northeast (45° between north and east) at a speed of 14 m/s (31 mi/h), and collides with a smaller car with a mass of 1300 kg that is traveling directly south at a speed of 18 m/s (40 mi/h). The two cars stick together during the collision. With what speed and direction does the tangled mess of metal move right after the collision?

70. ✦ In a nuclear reactor, a neutron moving at speed v_i in the positive x-direction strikes a deuteron, which is at rest. The neutron is deflected by 90.0° and moves off with speed $0.577v_i$ in the positive y-direction. Find the x- and y-components of the deuteron's velocity after the collision. (The mass of the deuteron is twice the mass of the neutron.)

71. Two identical pucks are on an air table. Puck A has an initial velocity of 2.0 m/s in the $+x$-direction. Puck B is at rest. Puck A collides elastically with puck B, and A moves off at 1.0 m/s at an angle of 60° above the x-axis. What is the speed and direction of puck B after the collision?

72. A block of mass 2.00 kg slides eastward along a frictionless surface with a speed of 2.70 m/s. A chunk of clay with a mass of 1.50 kg slides southward on the same surface with a speed of 3.20 m/s. The two objects collide and move off together. What is their velocity after the collision?

73. In a circus trapeze act, two acrobats fly through the air and grab on to each other, then together grab a swinging bar. One acrobat, with a mass of 60 kg, is moving at 3.0 m/s at an angle of 10° above the horizontal, and the other, with a mass of 80 kg, is approaching her with a speed of 2.0 m/s at an angle of 20° above the horizontal. What is the direction and speed of the acrobats right after they grab on to each other?

74. ✦ Two African swallows fly toward each other, carrying coconuts. The first swallow is flying north horizontally with a speed of 20 m/s. The second swallow is flying at the same height as the first and in the opposite direction with a speed of 15 m/s. The mass of the first swallow is 0.270 kg and the mass of his coconut is

0.80 kg. The second swallow's mass is 0.220 kg and her coconut's mass is 0.70 kg. The swallows collide and lose their coconuts. Immediately after the collision, the 0.80 kg coconut travels 10° west of south with a speed of 13 m/s, and the 0.70 kg coconut moves 30° east of north with a speed of 14 m/s. The two birds are tangled up with each other and stop flapping their wings as they travel off together. What is the velocity of the birds immediately after the collision?

75. ✦ During a game of marbles, the "shooter," a marble with three times the mass of the other marbles, has a speed of 3.2 m/s just before it hits one of the marbles. The other marble bounces off the shooter in an elastic collision at an angle of 40°, as shown, and the shooter moves off at an angle θ. Determine (a) the speed of the shooter after the collision, (b) the speed of the marble after the collision, and (c) the angle θ.

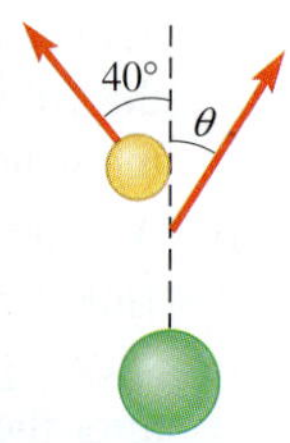

Collaborative Problems

76. Jane is sitting on a chair with her lower leg at a 30.0° angle with respect to the vertical, as shown. You need to develop a computer model of her leg to assist in some medical research. If you assume that her leg can be modeled as two uniform cylinders, one with mass $M = 20$ kg and length $L = 35$ cm and one with mass $m = 10$ kg and length $l = 40$ cm, where is the CM of her leg?

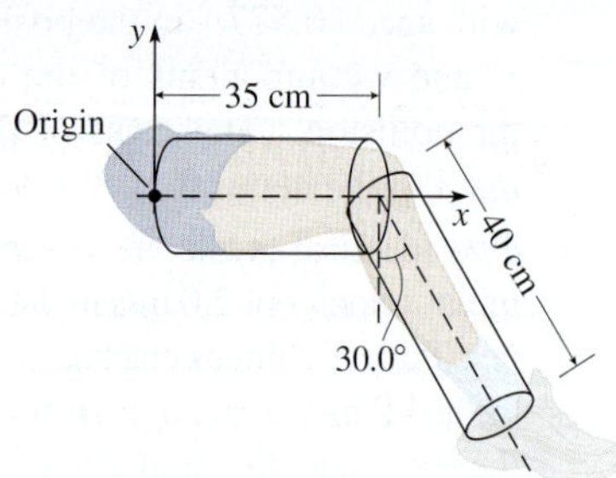

77. A 115 g ball is traveling to the left with a speed of 30 m/s when it is struck by a racket. The force on the ball, directed to the right and applied over 21 ms of contact time, is shown in the graph. What is the speed of the ball immediately after it leaves the racket? (connect tutorial: impulse)

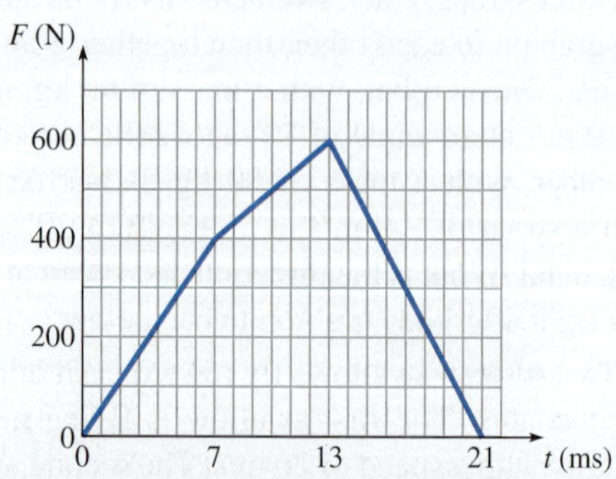

78. A man with a mass of 65 kg skis down a frictionless hill that is 5.0 m high. At the bottom of the hill the terrain levels out. As the man reaches the horizontal section, he grabs a 20 kg backpack and skis off a 2.0 m high ledge. At what horizontal distance from the edge of the ledge does the man land?

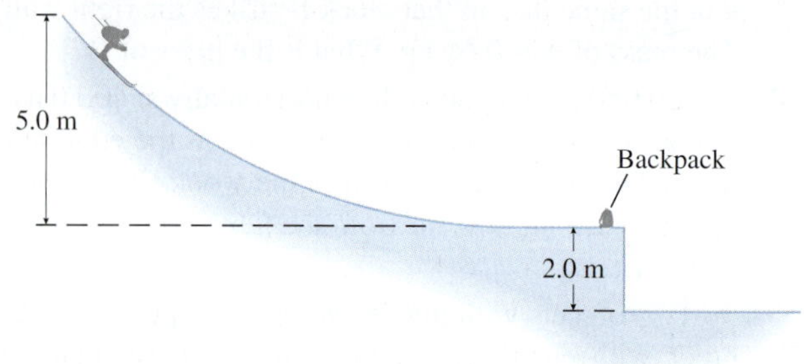

79. ✦ C In a lab experiment, two identical gliders on an air track are held together by a piece of string, compressing a spring between the gliders. While they are moving to the right at a common speed of 0.50 m/s, one student holds a match under the string and burns it, letting the spring force the gliders apart. One glider is then observed to be moving to the right at 1.30 m/s. (a) What velocity does the other glider have? (b) Is the total kinetic energy of the two gliders after the collision greater than, less than, or equal to the total kinetic energy before the collision? If greater, where did the extra energy come from? If less, where did the "lost" energy go?

80. A police officer is investigating the scene of an accident where two cars collided at an intersection. One car with a mass of 1100 kg moving west had collided with a 1300 kg car moving north. The two cars, stuck together, skid at an angle of 30° north of west for a distance of 17 m. The coefficient of kinetic friction between the tires and the road is 0.80. The speed limit for each car was 70 km/h. Was either car speeding?

Comprehensive Problems

81. A sled of mass 5.0 kg is coasting along on a frictionless ice-covered lake at a constant speed of 1.0 m/s. A 1.0 kg book is dropped vertically onto the sled. At what speed does the sled move once the book is on it?

82. An automobile weighing 13.6 kN is moving at 17.0 m/s when it collides with a stopped car weighing 9.0 kN. If they lock bumpers and move off together, what is their speed just after the collision?

83. For a system of three particles moving along a line, an observer in a laboratory measures the following masses and velocities. What is the velocity of the CM of the system?

Mass (kg)	v_x (m/s)
3.0	+290
5.0	−120
2.0	+52

84. An intergalactic spaceship is traveling through space far from any planets or stars, where no human has

gone before. The ship carries a crew of 30 people (of total mass 2.0×10^3 kg). If the speed of the spaceship is 1.0×10^5 m/s and its mass (excluding the crew) is 4.8×10^4 kg, what is the magnitude of the total momentum of the ship and the crew?

85. A baseball player pitches a fastball toward home plate at a speed of 41 m/s. The batter swings, connects with the ball of mass 145 g, and hits it so that the ball leaves the bat with a speed of 37 m/s. Assume that the ball is moving horizontally just before and just after the collision with the bat. (a) What is the magnitude of the change in momentum of the ball? (b) What is the impulse delivered to the ball by the bat? (c) If the bat and ball are in contact for 3.0 ms, what is the magnitude of the average force exerted on the ball by the bat?

86. ◆ A tennis ball of mass 0.060 kg is served. It strikes the ground with a velocity of 54 m/s (120 mi/h) at an angle of 22° below the horizontal. Just after the bounce it is moving at 53 m/s at an angle of 18° above the horizontal. If the interaction with the ground lasts 0.065 s, what average force did the ground exert on the ball?

87. A uniform rod of length 30.0 cm is bent into the shape of an inverted U. Each of the three sides is of length 10.0 cm. Find the location, in x- and y-coordinates, of the CM as measured from the origin.

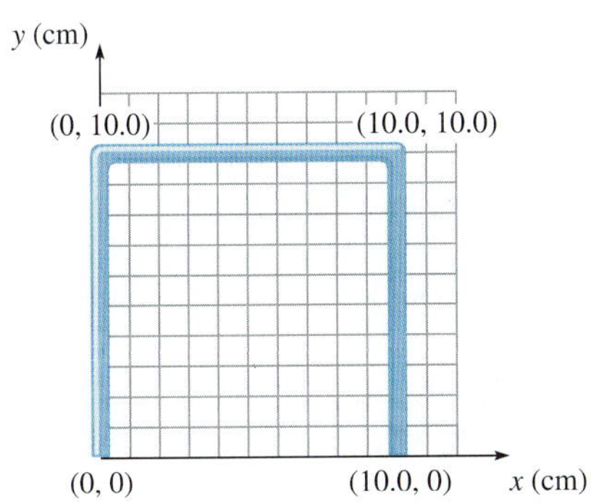

88. A child places 12 wooden blocks together, as shown in the figure. If each block has the same mass and density, where is the CM of these blocks? Each block is a cube with sides of 1.0 inch length. The origin of the coordinate system is at the center of the farthest block to the left.

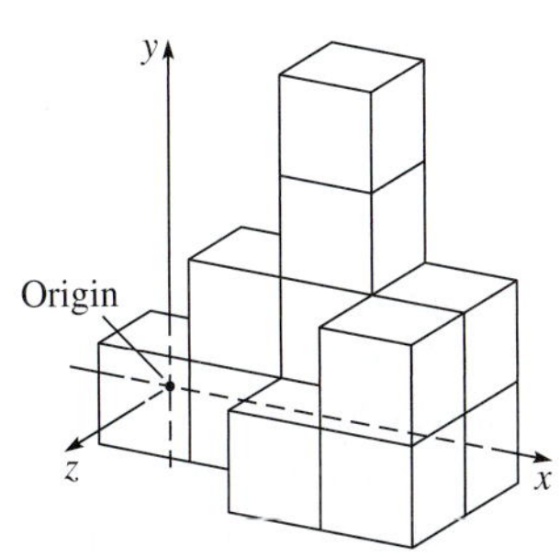

89. To contain some unruly demonstrators, the riot squad approaches with fire hoses. Suppose that the rate of flow of water through a fire hose is 24 kg/s, and the stream of water from the hose moves at 17 m/s. What force is exerted by such a stream on a person in the crowd? Assume that the water comes to a dead stop against the demonstrator's chest.

90. An inexperienced catcher catches a 130 km/h fastball of mass 140 g within 1 ms, whereas an experienced catcher slightly retracts his hand during the catch, extending the stopping time to 10 ms. What are the average forces imparted to the two gloved hands during the catches?

91. ◆ A stationary 0.1 g fly encounters the windshield of a 1000 kg automobile traveling at 100 km/h. (a) What is the change in momentum of the car due to the fly? (b) What is the change of momentum of the fly due to the car? (c) Approximately how many flies does it take to reduce the car's speed by 1 km/h?

92. A 0.15 kg baseball is pitched with a speed of 35 m/s (78 mi/h). When the ball hits the catcher's glove, the glove moves back by 5.0 cm (2 in.) as it stops the ball. (a) What was the change in momentum of the baseball? (b) What impulse was applied to the baseball? (c) Assuming a constant acceleration of the ball, what was the average force applied by the catcher's glove?

93. ◆ An object of mass 2.0 kg (the "projectile") approaches a stationary object (the "target") at 8.0 m/s. The projectile is deflected through an angle of 90.0° and its speed after the collision is 6.0 m/s. What is the speed of the target after the collision if the collision is perfectly elastic?

94. A radioactive nucleus is at rest when it spontaneously decays by emitting an electron and neutrino. The momentum of the electron is 8.20×10^{-19} kg·m/s, and it is directed at right angles to that of the neutrino. The neutrino's momentum has magnitude 5.00×10^{-19} kg·m/s. (a) In what direction does the newly formed ("daughter") nucleus recoil? (b) What is its momentum?

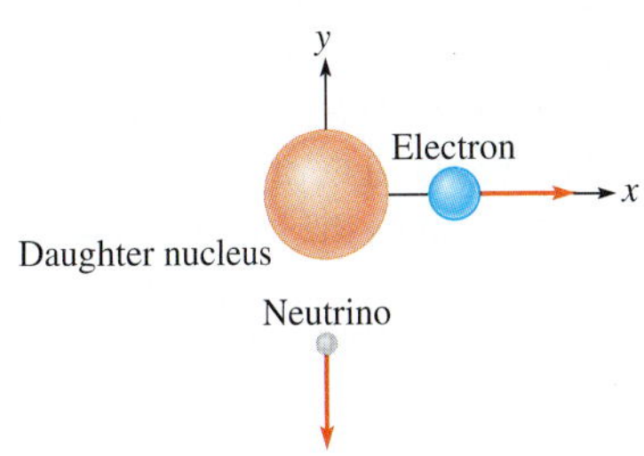

95. ◆ A 60.0 kg woman stands at one end of a 120 kg raft that is 6.0 m long. The other end of the raft is 0.50 m from a pier. (a) The woman walks toward the pier until she gets to the other end of the raft and stops there. Now what is the distance between the raft and the pier? (b) In (a), how far did the woman walk (relative to the pier)?

96. ◆ A jet plane is flying at 130 m/s relative to the ground. There is no wind. The engines take in 81 kg of air per second. Hot gas (burned fuel and air) is expelled from the engines at high speed. The engines provide a forward force on the plane of magnitude 6.0×10^4 N. At what speed relative to the ground is the gas being expelled?

[*Hint*: Look at the momentum change of the air taken in by the engines during a time interval Δt.] This calculation is approximate since we are ignoring the 3.0 kg of fuel consumed and expelled with the air each second.

97. ✦ Within cells, small organelles containing newly synthesized proteins are transported along microtubules by tiny molecular motors called kinesins. What force does a kinesin molecule need to deliver in order to accelerate an organelle with mass 0.01 pg (10^{-17} kg) from 0 to 1 μm/s within a time of 10 μs?

98. ✦ The pendulum bobs in the figure are made of soft clay so that they stick together after impact. The mass of bob A is half that of bob B. Bob B is initially at rest. What is the ratio of the kinetic energy of the combined bobs, just after impact, to the kinetic energy of bob A just before impact?

99. ✦ The pendulum bobs in the figure are made of soft clay so that they stick together after impact. The mass of bob A is half that of bob B. Bob B is initially at rest. If bob A is released from a height h above its lowest point, what is the maximum height attained by bobs A and B after the collision?

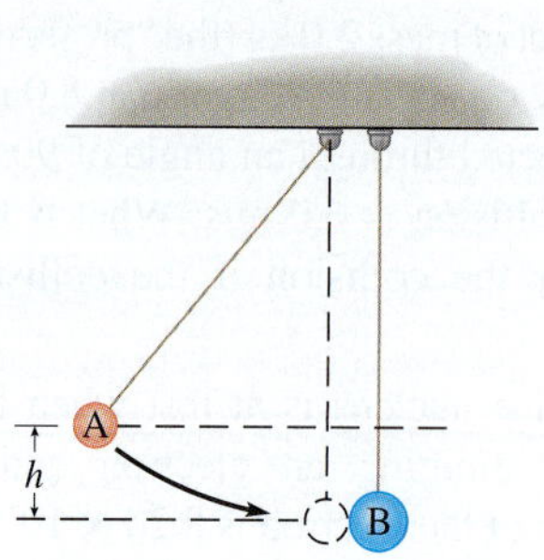

Problems 98 and 99

100. ✦ A flat, circular metal disk of uniform thickness has a radius of 3.0 cm. A hole is drilled in the disk that is 1.5 cm in radius. The hole is tangent to one side of the disk. Where is the CM of the disk now that the hole has been drilled? [*Hint*: The original disk (before the hole is drilled) can be thought of as having two pieces—the disk with the hole plus the smaller disk of metal drilled out. Write an equation that expresses x_{CM} of the original disk in terms of the centers of mass of the two pieces. Since the thickness is uniform, the mass of any piece is proportional to its area.]

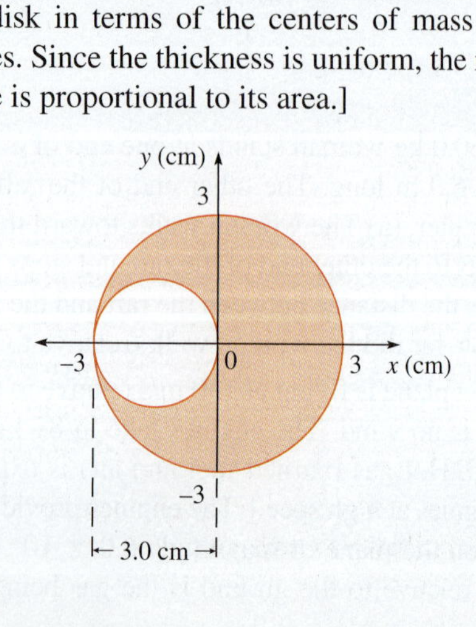

101. ✦ Two pendulum bobs have equal masses and lengths (5.1 m). Bob A is initially held horizontally while bob B hangs vertically at rest. Bob A is released and collides elastically with bob B. How fast is bob B moving immediately after the collision?

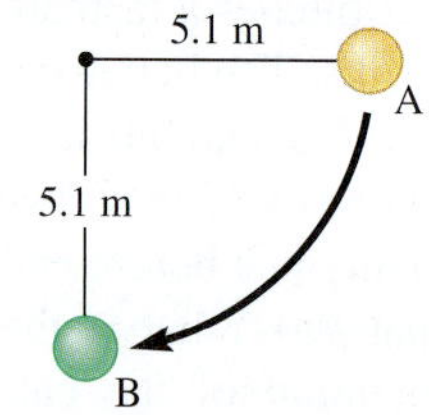

102. Two identical gliders, each with elastic bumpers and mass 0.10 kg, are on a horizontal air track. Friction is negligible. Glider 2 is stationary. Glider 1 moves toward glider 2 from the left with a speed of 0.20 m/s. They collide. After the collision, what are the velocities of glider 1 and glider 2?

103. ✦ A radium nucleus (mass 226 u) at rest decays into a radon nucleus (symbol Rn, mass 222 u) and an alpha particle (symbol α, mass 4 u). (a) Find the ratio of the speeds v_α/v_{Rn} after the decay. (b) Find the ratio of the magnitudes of the momenta p_α/p_{Rn}. (c) Find the ratio of the kinetic energies K_α/K_{Rn}. (Note: "u" is a unit of mass; it is *not* necessary to convert it to kg.)

Review and Synthesis

104. Gerald wants to know how fast he can throw a ball, so he hangs a 2.30 kg target on a rope from a tree. He picks up a 0.50 kg ball of putty and throws it horizontally against the target. The putty sticks to the target and the putty and target swing up a vertical distance of 1.50 m from its original position. How fast did Gerald throw the ball of putty?

105. It is the bottom of the ninth inning at a baseball game. The score is tied and there is a runner on second base when the batter gets a hit. The 85 kg base runner rounds third base and is heading for home with a speed of 8.0 m/s. Just before he reaches home plate, he crashes into the opposing team's catcher, and the two players slide together along the base path toward home plate. The catcher has a mass of 95 kg and the coefficient of friction between the players and the dirt on the base path is 0.70. How far do the catcher and base runner slide?

106. Pendulum bob A has half the mass of pendulum bob B. Each bob is tied to a string that is 5.1 m long. When bob A is held with its string horizontal and then released, it swings down and, once bob A's string is vertical, it collides elastically with bob B. How high do the bobs rise after the collision?

107. At the beginning of a scene in an action movie, the 78.0 kg star, Indianapolis Jones, stands on a ledge 3.70 m above the ground and the 55.0 kg heroine, Georgia

Smith, stands on the ground. Jones swings down on a rope, grabs Smith around the waist, and continues swinging until they come to rest on another ledge on the other side of the set. At what height above the ground should the second ledge be placed? Assume that Jones and Smith remain nearly upright during the swing so that their CMs are always the same distance above their feet.

108. A Vulcan spaceship has a mass of 65 000 kg and a Romulan spaceship is twice as massive. Both have engines that generate the same total force of 9.5×10^6 N. (a) If each spaceship fires its engine for the same amount of time, starting from rest, which will have the greater kinetic energy? Which will have the greater momentum? (b) If each spaceship fires its engine for the same *distance*, which will have the greater kinetic energy? Which will have the greater momentum? (c) Calculate the energy and momentum of each spaceship in parts (a) and (b), ignoring any change in mass due to whatever is expelled by the engines. In part (a), assume that the engines are fired for 100 s. In part (b), assume that the engines are fired for 100 m.

109. A boy of mass 60 kg is sledding down a 70 m slope starting from rest. The slope is angled at 15° below the horizontal. After going 20 m along the slope, he passes his friend, who jumps on the sled. The friend has a mass of 50 kg, and the coefficient of kinetic friction between the sled and the snow is 0.12. Ignoring the mass of the sled, find their speed at the bottom.

110. ✦ Ⓒ You want to throw a banana to a monkey hanging from a branch as shown in the figure. The banana has a mass of 200 g, and the monkey has a mass of 3.00 kg. The monkey is startled and drops from the branch the moment you throw the banana. Ignore air resistance. (a) In what direction should you aim the banana so the monkey catches it in the air? (b) Explain why your answer to part (a) is the same for different values of the banana's launch speed. (c) If the monkey catches the banana at the point indicated in the figure, what was the banana's initial speed? (d) What is the horizontal distance d to the spot where the monkey lands?

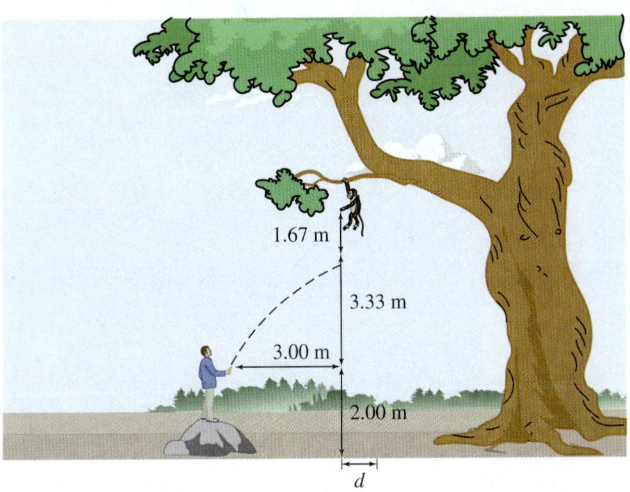

111. ✦ A 0.122 kg dart is fired from a gun with a speed of 132 m/s horizontally into a 5.00 kg wooden block. The block is attached to a spring with a spring constant of 8.56 N/m. The coefficient of kinetic friction between the block and the horizontal surface it is resting on is 0.630. After the dart embeds itself into the block, the block slides along the surface and compresses the spring. What is the maximum compression of the spring?

Answers to Practice Problems

7.1 (a) 0.78 kg·m/s downward; (b) 0.78 kg·m/s toward the apple; 1.3×10^{-25} m/s

7.2 3.5 times his weight

7.3 1700 N; 0.0037 s

7.4 0.8 m/s in the $-x$-direction

7.5 1.7 m/s

7.6 2.2 m/s

7.7 (2.0 cm, 2.3 cm)

7.8 (a) 2.7 m; (b) 1.5 m in the other direction; (c) the CM does not move

7.9 4.0 m/s

7.10 10.0 m/s

7.11 $0.751v_i$

7.12 0.781

Answers to Checkpoints

7.2 No, because the *direction* of the car's momentum would have changed.

7.4 When external forces act on a system, the momentum of the system is not conserved.

7.6 Despite the fact that the hammer is rotating, it is in free fall and its CM follows the same trajectory as a point particle in free fall.

7.7A Yes. Momentum is conserved in both elastic and inelastic collisions. In an inelastic collision, the initial and final *kinetic energies* are not equal.

7.7B (a) The total momentum is conserved: $mv_i + 0 = 0 + mv_i$. The total kinetic energies before and after are: $K_i = \frac{1}{2}mv_i^2 + 0$ and $K_f = 0 + \frac{1}{2}mv_i^2$. They are equal so the collision is elastic. (b) The total momentum is conserved: $mv_i + 0 = m\left(\frac{1}{2}v_i\right) + m\left(\frac{1}{2}v_i\right)$. The final velocities are the same so the collision is perfectly inelastic. (c) Suppose that the blue car's velocity x-component after the collision is $\frac{1}{4}v_i$ and the red car's is $\frac{3}{4}v_i$. This conserves momentum: $mv_i + 0 = m\left(\frac{1}{4}v_i\right) + m\left(\frac{3}{4}v_i\right)$ The total kinetic energies before and after are: $K_i = \frac{1}{2}mv_i^2 + 0$ and $K_f = \frac{1}{2}m\left(\frac{1}{4}v_i\right)^2 + \frac{1}{2}m\left(\frac{3}{4}v_i\right)^2 = \frac{5}{16}mv_i^2$. $K_f < K_i$ so the collision is inelastic.

CHAPTER

8 Torque and Angular Momentum

In gymnastics, the iron cross is a notoriously difficult feat, requiring incredible strength. Why does it require such great strength? (See pp. 290–291 for the answer.)

BIOMEDICAL APPLICATIONS

- Torque exerted by athletes, animals (PP 8.4; Probs. 42, 53, 94, and 113)
- Posture and center of gravity of animals, athletes (Sec. 8.4; PP 8.9, CQ 15 and 16, Probs. 90 and 91)
- Torque and equilibrium in the human body (Sec. 8.5; Ex. 8.10; PP 8.10; CQ 10–11, Probs. 18, 43–48, 118, and 124)
- Conservation of angular momentum in figure skaters, divers (Sec. 8.8; MCQ 10; Probs. 77–79, 82, and 83)

Concepts & Skills to Review

- translational equilibrium (Section 4.2)
- uniform circular motion and circular orbits (Sections 5.1 and 5.4)
- angular acceleration (Section 5.6)
- conservation of energy (Section 6.1)
- center of mass and its motion (Sections 7.5 and 7.6)
- rolling without slipping (Section 5.1)

8.1 ROTATIONAL KINETIC ENERGY AND ROTATIONAL INERTIA

When a rigid object is rotating about a fixed axis, it has kinetic energy because each particle other than those on the axis of rotation is moving in a circle around the axis. In principle, we can calculate the kinetic energy of rotation by summing the kinetic energy of each particle. To say the least, that sounds like a laborious task. We need a simpler way to express the rotational kinetic energy of such an object so that we don't have to calculate this sum over and over. Our simpler expression exploits the fact that the speed of each particle is proportional to the angular speed of rotation ω.

If a rigid object consists of N particles, the sum of the kinetic energies of the particles can be written mathematically using a subscript to label the mass and speed of each particle:

$$K_{\text{rot}} = \tfrac{1}{2}m_1v_1^2 + \tfrac{1}{2}m_2v_2^2 + \cdots + \tfrac{1}{2}m_Nv_N^2 = \sum_{n=1}^{N}\tfrac{1}{2}m_nv_n^2$$

where the notation $\sum_{n=1}^{N} Q_n$ stands for the sum $Q_1 + Q_2 + \ldots + Q_N$.

The speed of each particle is related to its distance from the axis of rotation. Particles that are farther from the axis move faster. In Section 5.1, we found that the speed of a particle moving in a circle is

$$v = r\omega \qquad (5\text{-}7)$$

where ω is the angular speed and r is the distance between the rotation axis and the particle (Fig. 8.1). By substitution, the rotational kinetic energy can be written

$$K_{\text{rot}} = \sum_{n=1}^{N} \tfrac{1}{2}m_nr_n^2\omega^2$$

The entire object rotates at the same angular velocity ω, so the constants $\frac{1}{2}$ and ω^2 can be factored out of each term of the sum:

$$K_{\text{rot}} = \tfrac{1}{2}\left(\sum_{n=1}^{N} m_nr_n^2\right)\omega^2$$

The quantity in the parentheses *cannot change* since the distance between each particle and the rotation axis stays the same if the object is rigid and doesn't change shape. However difficult it may be to compute the sum in the parentheses, we only need to do it *once* for any given mass distribution and axis of rotation.

Let's give the quantity in the parentheses the symbol I. In Chapter 5, we found it useful to draw analogies between translational variables and their rotational equivalents. By using the symbol I, we can see that translational and rotational kinetic energy have similar forms: translational kinetic energy is

$$K_{\text{tr}} = \tfrac{1}{2}mv^2$$

and rotational kinetic energy is

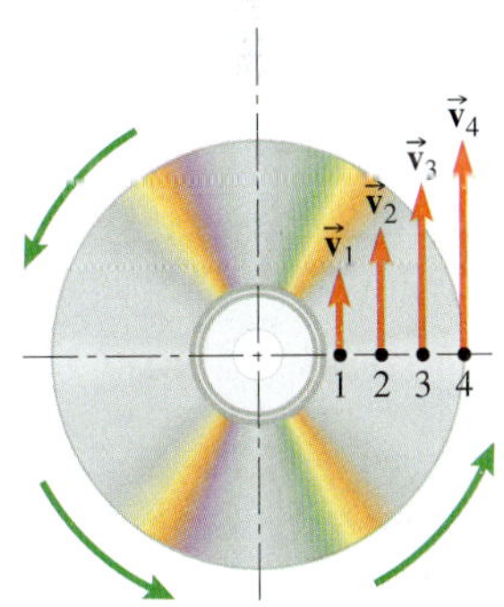

Figure 8.1 Four points on a spinning DVD. Points at greater distances from the center are moving faster than points closer to the center.

Rotational kinetic energy

$$K_{\text{rot}} = \tfrac{1}{2}I\omega^2 \qquad (8\text{-}1)$$

Since $v = r\omega$ was used to derive Eq. (8-1), ω must be expressed in radians per unit time (normally rad/s).

The quantity I is called the **rotational inertia**:

Rotational inertia

$$I = \sum_{n=1}^{N} m_n r_n^2 \qquad (8\text{-}2)$$

(SI unit: $kg \cdot m^2$)

CONNECTION:

Rotational and translational kinetic energies have the same form: $\frac{1}{2}$ inertia × speed^2.

Comparing the expressions for translational and rotational kinetic energies, we see that angular speed ω takes the place of speed v and rotational inertia I takes the place of mass m. Mass is a measure of the inertia of an object, or, in other words, how difficult it is to change the object's velocity. Similarly, for a rigid rotating object, I is a measure of its rotational inertia—how hard it is to change its angular velocity. That is why the quantity I is called the rotational inertia; it is also called the **moment of inertia**.

When a problem requires you to find a rotational inertia, there are three principles to follow.

Finding the Rotational Inertia

1. If the object consists of a *small* number of particles, calculate the sum $I = \sum_{n=1}^{N} m_n r_n^2$ directly.
2. For symmetrical objects with simple geometric shapes, calculus can be used to perform the sum in Eq. (8-2). Table 8.1 lists the results of these calculations for the shapes most commonly encountered.
3. Since the rotational inertia is a sum, you can always mentally deconstruct the object into several parts, find the rotational inertia of each part, and then add them. This is an example of the *divide-and-conquer* problem-solving technique.

connect

See the text website for more information on rotational inertia, including the parallel- and perpendicular-axis theorems.

Keep in mind that the rotational inertia of an object depends on the location of the rotation axis. For instance, imagine taking the hinges off the side of a door and putting them on the top so that the door swings about a horizontal axis like a cat flap door (Fig. 8.2b). The door now has a considerably larger rotational inertia than before the hinges were moved because the door's height is greater than its width. The door has the same mass as before, but its mass now lies on average much farther from the axis of rotation than that

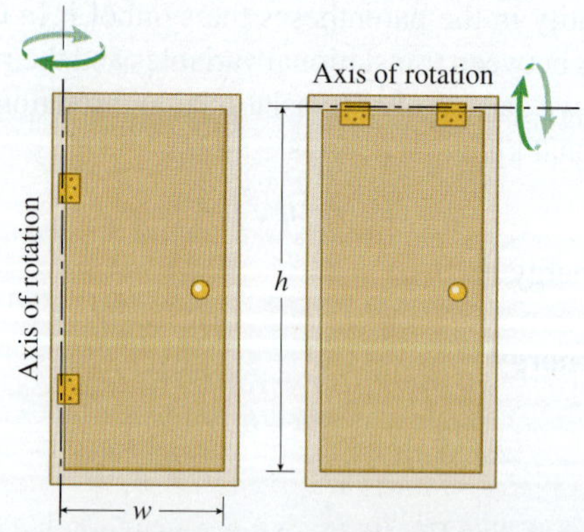

Figure 8.2 The rotational inertia of a door depends on the rotation axis. (a) The door with hinges at the side has a smaller rotational inertia, $I = \frac{1}{3}Mw^2$, than (b) the rotational inertia, $I = \frac{1}{3}Mh^2$, of the same door with hinges at the top, because the door is taller than it is wide.

Table 8.1 Rotational Inertia for Uniform Objects with Various Geometrical Shapes

Shape		Axis of Rotation	Rotational Inertia
Thin hollow cylindrical shell (or hoop)	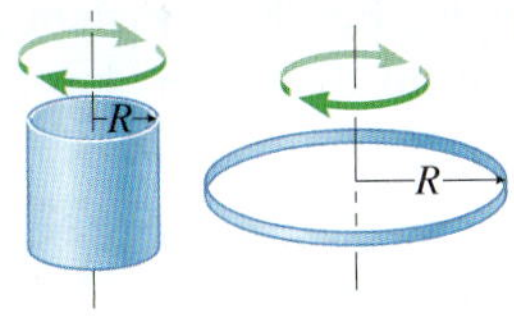	Central axis of cylinder	MR^2
Solid cylinder (or disk)	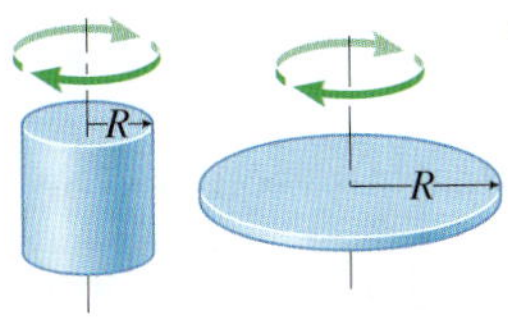	Central axis of cylinder	$\frac{1}{2}MR^2$
Hollow cylindrical shell or disk	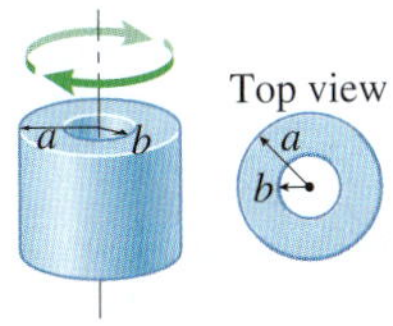	Central axis of cylinder	$\frac{1}{2}M(a^2 + b^2)$
Rectangular plate	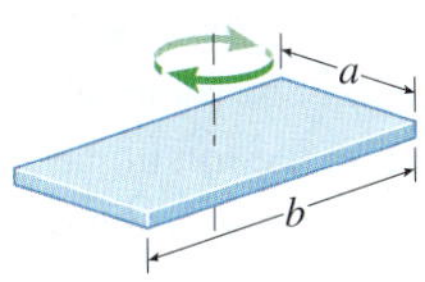	Perpendicular to plate through center	$\frac{1}{12}M(a^2 + b^2)$
Solid sphere	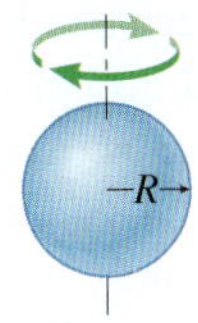	Through center	$\frac{2}{5}MR^2$
Thin hollow spherical shell	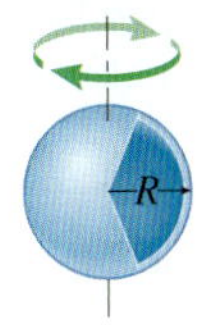	Through center	$\frac{2}{3}MR^2$
Thin rod (or rectangular plate)	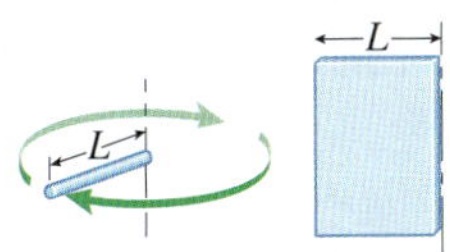	Perpendicular to rod through end (or along edge of plate)	$\frac{1}{3}ML^2$
Thin rod (or rectangular plate)	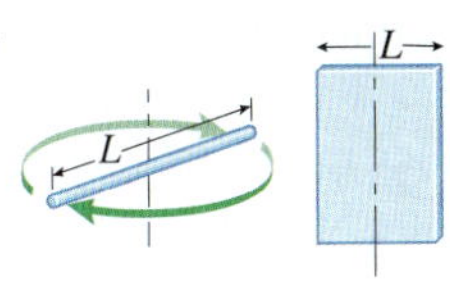	Perpendicular to rod through center (or parallel to edge of plate through center)	$\frac{1}{12}ML^2$

of the door in Fig. 8.2a. In applying Eq. (8-2) to find the rotational inertia of the door, the values of r_n range from 0 to the height of the door (h), whereas with the hinges in the normal position the values of r_n range from 0 only to the width of the door (w).

EVERYDAY PHYSICS DEMO

The change in rotational inertia of a rod as the rotation axis changes can be easily felt. Hold a baseball bat in the usual way, with your hands gripping the bottom of the bat. Swing the bat a few times. Now "choke up" on the bat—move your hands up the bat—and swing a few times. The bat is easier to swing because it now has a smaller rotational inertia. Children often choke up on a bat that is too massive for them. Even major league baseball players occasionally choke up on the bat when they want more control over their swing to place a hit in a certain spot (Fig. 8.3). On the other hand, choking up on the bat makes it impossible to hit a home run. To hit a long fly ball, you want the pitched baseball to encounter a bat that is swinging with a lot of rotational inertia.

Figure 8.3 Hank Aaron choking up on the bat.

CHECKPOINT 8.1

According to Table 8.1, the rotational inertia of a uniform cylinder or disk about its central axis depends only on the mass and radius. Why does it not depend on the height of the cylinder or thickness of the disk?

Example 8.1

Rotational Inertia of a Barbell

A barbell consists of two plates, each a uniform disk of mass 20 kg and radius 15 cm, attached 20 cm from each end of a uniform rod of mass 10 kg, radius 1.25 cm, and length 2.20 m (Fig. 8.4). Find the rotational inertia of the barbell about two different axes of rotation: (a) axis a, the central axis of the bar, and (b) axis b, perpendicular to the bar and through its midpoint. Ignore the thickness of the disks and the holes in the disks.

Strategy The rotational inertia of this composite object is the sum of the rotational inertias of the three parts (two disks and rod). Table 8.1 gives formulas for the rotational inertias of disks and rods, but *only for certain axes of rotation*. In particular, for axis b we have two disks rotating about an axis external to the disks, so none of the formulas in Table 8.1 apply; instead we'll return to the basic definition of rotational inertia [Eq. (8-2)] and make an approximation. Based on the distances between parts of the barbell and the two axes, we expect a smaller rotational inertia about axis a than about axis b. Let M and R be the mass and radius of each disk, and m, r, and L the mass, radius, and length of the rod, respectively.

Solution (a) Each of the three component parts, the two disks and the rod, are solid cylinders rotating about their central axes. (The two formulas in Table 8.1 for thin rods are for axes *perpendicular* to the rod, so they are not useful here.) From Table 8.1,

$$I = \tfrac{1}{2}MR^2 + \tfrac{1}{2}MR^2 + \tfrac{1}{2}mr^2$$

$$= 2 \times [\tfrac{1}{2} \times 20\text{ kg} \times (0.15\text{ m})^2] + \tfrac{1}{2} \times 10\text{ kg} \times (0.0125\text{ m})^2$$

$$= 2 \times 0.225\text{ kg}\cdot\text{m}^2 + 0.000\,78\text{ kg}\cdot\text{m}^2 = 0.45\text{ kg}\cdot\text{m}^2$$

(b) Table 8.1 gives the rotational inertia of the rod about axis b as $\frac{1}{12}mL^2$. The center of each disk (assumed to have negligible thickness) is a distance $d = \frac{1}{2}(2.20\text{ m} - 0.40\text{ m}) = 0.90\text{ m}$ from the midpoint of the rod. If we think of breaking a disk into tiny pieces and applying Eq. (8-2), each of the distances r_n is at least $d = 0.90$ m (to the center) but no more than $\sqrt{d^2 + R^2} \approx 0.91$ m (to the edge). Therefore, to a good approximation, we can assume each disk to be a point mass at a distance d from the axis. Then

$$I = Md^2 + Md^2 + \tfrac{1}{12}mL^2$$

$$= 2 \times [20\text{ kg} \times (0.90\text{ m})^2] + \tfrac{1}{12} \times 10\text{ kg} \times (2.20\text{ m})^2$$

$$= 2 \times 16.2\text{ kg}\cdot\text{m}^2 + 4.03\text{ kg}\cdot\text{m}^2 = 36\text{ kg}\cdot\text{m}^2$$

As expected, the rotational inertia is much smaller about axis a than about axis b.

Discussion The rod makes only a slight contribution to the rotational inertia about axis a because the radius of the rod is so much smaller than the radii of the disks, so its mass is on average much closer to the axis of rotation. The rod makes a more significant contribution to the rotational inertia about axis b because now the length, not radius, of the rod is relevant—its mass is distributed at distances from 0 to 1.10 m from the axis of rotation. Even if we account for the thickness of the disks, as long as their thicknesses are small relative to d, our estimate Md^2 of the contribution to I from each disk about axis b is still valid.

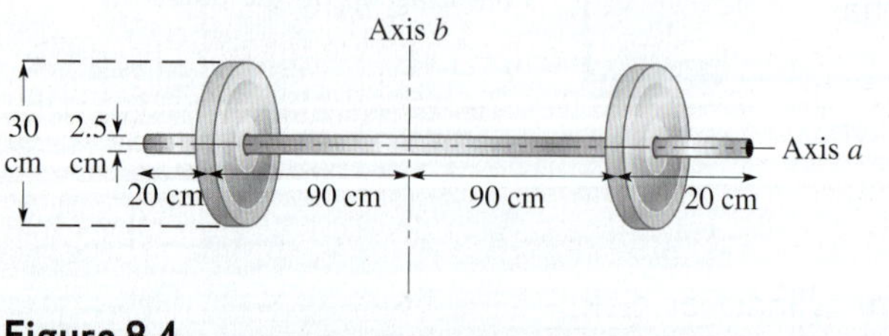

Figure 8.4
A barbell with two different rotation axes.

Practice Problem 8.1 Playground Merry-Go-Round

A playground merry-go-round is essentially a uniform disk that rotates about a vertical axis through its center (Fig. 8.5). Suppose the disk has a radius of 2.0 m and a mass of 160 kg; a child of mass 18.4 kg sits at the edge of the merry-go-round. What is the merry-go-round's rotational inertia, including the contribution due to the child? [*Hint*: Treat the child as a point mass at the edge of the disk.]

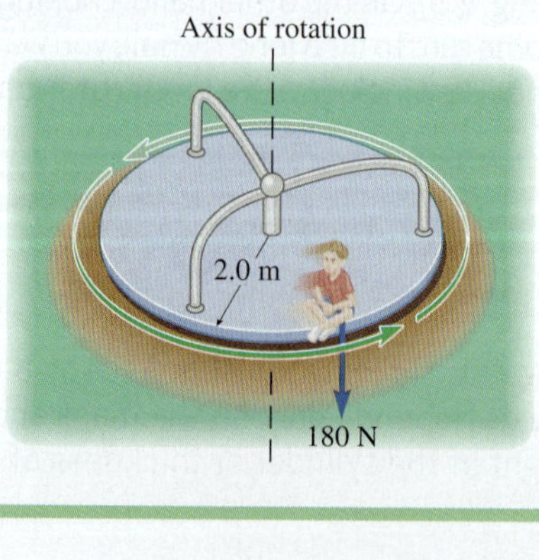

Figure 8.5
Child on a merry-go-round.

When applying conservation of energy to objects that rotate, the rotational kinetic energy is included in the mechanical energy. In Eq. (6-12),

$$W_{nc} = \Delta K + \Delta U \quad (6\text{-}12)$$

just as U stands for the sum of the elastic and gravitational potential energies, K stands for the sum of the translational and rotational kinetic energies:

$$K = K_{tr} + K_{rot}$$

CONNECTION:

We are applying the same principle of energy conservation to objects that can rotate.

Example 8.2

Atwood's Machine

Atwood's machine consists of a cord around a pulley of rotational inertia I, radius R, and mass M, with two blocks (masses m_1 and m_2) hanging from the ends of the cord as in Fig. 8.6. (Note that in Example 4.15 we analyzed Atwood's machine for the special case of a massless pulley; for a massless pulley $I = 0$.) Assume that the pulley is free to turn without friction and that the cord does not slip. Ignore air resistance. If the masses are released from rest, find how fast they are moving after they have moved a distance h (one up, the other down).

Strategy Ignoring both air resistance and friction means that no nonconservative forces act on the system; therefore, its mechanical energy is conserved:

$$\Delta U + \Delta K = 0$$

Gravitational potential energy is converted into the translational kinetic energies of the two blocks and the rotational kinetic energy of the pulley.

Solution For our convenience, we assume that $m_1 > m_2$. Mass m_1, therefore, moves down and m_2 moves up. After the masses have each moved a distance h, the changes in gravitational potential energy are

$$\Delta U_1 = -m_1 gh$$

$$\Delta U_2 = +m_2 gh$$

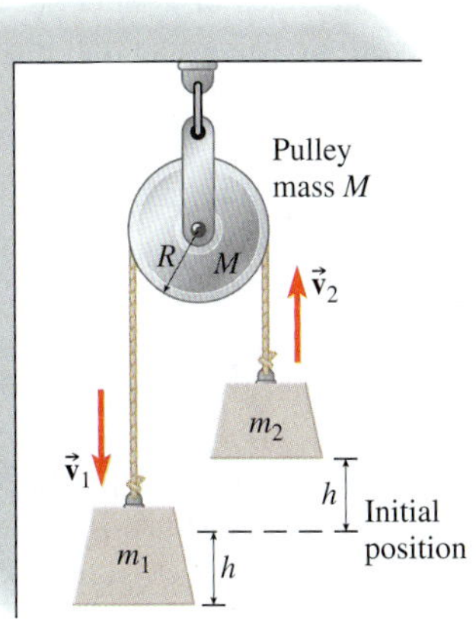

Figure 8.6
Atwood's machine.

The mechanical energy of the system includes the kinetic energies of three objects: the two masses and the pulley. All start with zero kinetic energy, so

$$\Delta K = \tfrac{1}{2}(m_1 + m_2)v^2 + \tfrac{1}{2}I\omega^2$$

The speed v of the masses is the same since the cord's length is fixed. The speed v and the angular speed of the pulley ω are related if the cord does not slip: the tangential speed of the pulley must equal the speed at which the cord moves. The tangential speed of the pulley is its angular speed times its radius:

$$v = \omega R$$

After substituting v/R for ω, the energy conservation equation becomes

$$\Delta U + \Delta K = [-m_1 gh + m_2 gh] + \left[\frac{1}{2}(m_1 + m_2)v^2 + \frac{1}{2}I\left(\frac{v}{R}\right)^2\right] = 0$$

or

$$\frac{1}{2}\left[(m_1 + m_2) + \frac{I}{R^2}\right]v^2 = (m_1 - m_2)gh$$

Solving this equation for v yields

$$v = \sqrt{\frac{2(m_1 - m_2)gh}{m_1 + m_2 + I/R^2}}$$

Discussion This answer is rich in information, in the sense that we can ask many "What if?" questions. Not only do these questions provide checks as to whether the answer is reasonable, they also enable us to perform thought experiments, which could then be checked by constructing an Atwood's machine and comparing the results.

For instance: What if m_1 is only slightly greater than m_2? Then the final speed v is small—as m_2 approaches m_1, v approaches 0. This makes intuitive sense: a small imbalance in weights produces a small acceleration. You should practice this kind of reasoning by making other such checks.

It is also enlightening to look at terms in an algebraic solution and connect them with physical interpretations.

continued on next page

Example 8.2 continued

The quantity $(m_1 - m_2)g$ is the imbalance in the gravitational forces pulling on the two sides. The denominator $(m_1 + m_2 + I/R^2)$ is a measure of the total inertia of the system—the sum of the two masses plus an inertial contribution due to the pulley. The pulley's contribution is *not* simply equal to its mass. If, for example, the pulley is a uniform disk with $I = \frac{1}{2}MR^2$, the term I/R^2 would be equal to *half* the mass of the pulley.

The same principles used to analyze Atwood's machine have many applications in the real world. One such application is in elevators, where one of the hanging masses is the elevator and the other is the counterweight. Of course, the elevator and counterweight are not allowed to hang freely from a pulley—we must also consider the energy supplied by the motor.

Practice Problem 8.2 Modified Atwood's Machine

Figure 8.7 shows a modified form of Atwood's machine where one of the blocks slides on a table instead of hanging from the pulley. The blocks are released from rest. Find the speed of the blocks after they have moved a distance h in terms of m_1, m_2, I, R, and h. Ignore friction.

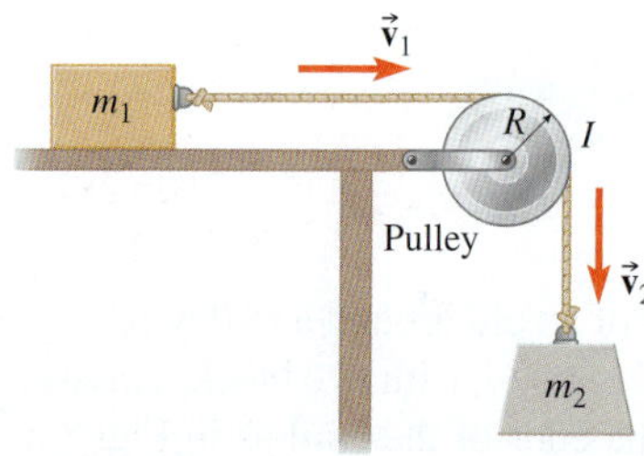

Figure 8.7 Modified Atwood's machine.

8.2 TORQUE

Suppose you place a bicycle upside down to repair it. First, you give one of the wheels a spin. If everything is working as it should, the wheel spins for quite a while; its angular acceleration is small. If the wheel doesn't spin for very long, then its angular velocity changes rapidly and the angular acceleration is large in magnitude; there must be excessive friction somewhere. Perhaps the brakes are rubbing on the rim or the bearings need to be repacked.

If we could eliminate *all* the frictional forces acting on the wheel, including air resistance, then we would expect the wheel to keep spinning without diminishing angular speed. In that case, its angular acceleration would be zero. The situation is reminiscent of Newton's first law: an object with no external interactions, or no net force acting on it, moves with constant velocity. We can state a "Newton's first law for rotation": a rotating object with no external interactions, and whose rotational inertia doesn't change, keeps rotating at constant angular velocity.

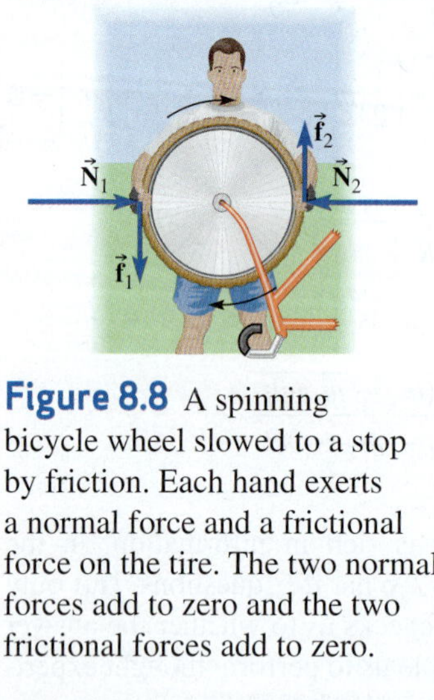

Figure 8.8 A spinning bicycle wheel slowed to a stop by friction. Each hand exerts a normal force and a frictional force on the tire. The two normal forces add to zero and the two frictional forces add to zero.

Of course, the hypothetical frictionless bicycle wheel does have external interactions. The Earth's gravitational field exerts a downward force and the axle exerts an upward force to keep the wheel from falling. Then is it true that, as long as there is no net external force, the angular acceleration is zero? No; it is easy to give the wheel an angular acceleration while keeping the net force zero. Imagine bringing the wheel to rest by pressing two hands against the tire on opposite sides. On one side, the motion of the rim of the tire is downward and the kinetic frictional force is upward (Fig. 8.8). On the other side, the tire moves upward and the frictional force is downward. In a similar way, we could apply equal and opposite forces to the opposite sides of a wheel at rest to make it start spinning. In either case, we exert equal magnitude forces, so that the net force is zero, and still give the wheel an angular acceleration.

Torque A quantity related to force, called **torque**, plays the role in rotation that force itself plays in translation. A torque is not separate from a force; it is impossible to exert a torque without exerting a force. Torque is a measure of how effective a given force is at twisting or turning something. For something rotating about a fixed axis such as the bicycle wheel, a torque can *change* the rotational motion either by making it rotate faster or by slowing it down.

When stopping the bicycle wheel with two equal and opposite forces, as in Fig. 8.8, the net applied force is zero and, thus, the wheel is in translational equilibrium; but the net torque is not zero, so it is not in rotational equilibrium. Both forces tend to give the wheel the same sign of angular acceleration; they are both making the wheel slow down. The two torques are in fact equal, with the same sign.

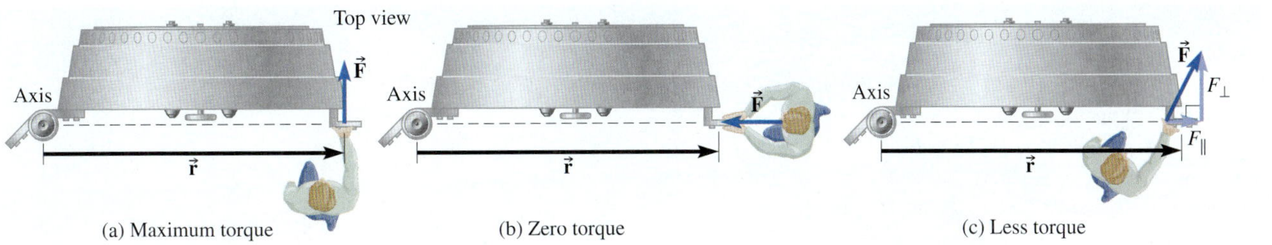

Figure 8.9 Torque on a bank vault door depends on the direction of the applied force. (a) Pushing tangentially gives the maximum torque. (b) Pushing radially with the same magnitude force gives zero torque. (c) The torque is proportional to the tangential component of the force ($F_\perp$).

Relationship Between Force and Torque What determines the torque produced by a particular force? Imagine trying to push open a massive bank vault door. Certainly you would push as hard as you can; the torque is proportional to the magnitude of the force. It also matters where and in what direction the force is applied. For maximum effectiveness, you push tangentially (Fig. 8.9a). If you pushed radially, straight in toward the axis of rotation that passes through the hinges, the door wouldn't rotate, no matter how hard you push (Fig. 8.9b). A force acting in any other direction could be decomposed into radial and tangential components, with the radial component contributing nothing to the torque (Fig. 8.9c). Only the tangential component of the force ($F_\perp$) produces a torque. Recall that the *radial* direction is directly toward or away from the axis of rotation. The *tangential* direction is perpendicular to both the radial direction and the axis of rotation; it is tangent to the circular path followed by a point on the object as the object rotates.

Furthermore, *where* you apply the force is critical (Fig. 8.10). Instinctively, you would push at the outer edge, as far from the rotation axis as possible. If you pushed close to the axis, it would be difficult to open the door. Torque is proportional to the distance between the rotation axis and the **point of application** of the force (the point at which the force is applied).

To satisfy the requirements of the previous paragraphs, we define the magnitude of the torque as the product of the distance between the rotation axis and the point of application of the force (r) with the perpendicular component of the force ($F_\perp$):

Definition of torque

$$\tau = \pm rF_\perp \tag{8-3}$$

where r is the shortest distance between the rotation axis and the point of application of the force and $F_\perp$ is the perpendicular component of the force.

The symbol for torque is τ, the Greek letter tau. The SI unit of torque is the N·m. The SI unit of *energy*, the joule, is equivalent to N·m, but we do not write torque in joules. Even though both energy and torque can be written using the same SI base units, the two quantities have different meanings; torque is not a form of energy. To help maintain the distinction, the joule is used for energy but *not* for torque.

CHECKPOINT 8.2

You are trying to loosen a nut, without success. Why might it help to switch to a wrench with a longer handle?

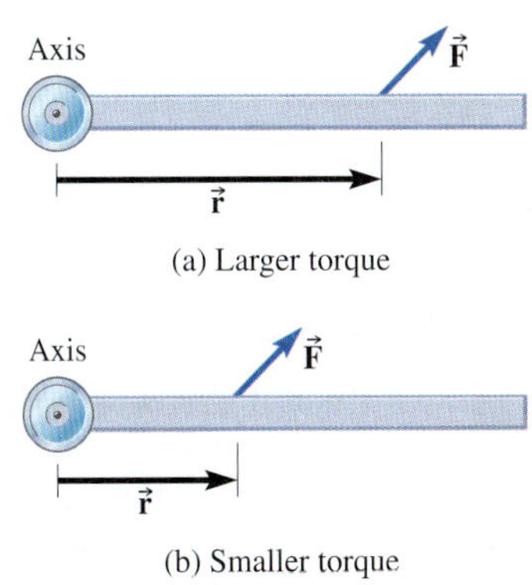

Figure 8.10 Torques; the same force at different distances from the axis.

Sign Convention for Torque The sign of the torque indicates the direction of the angular acceleration that torque would cause *by itself*. Recall from Section 5.1 that by convention a positive angular velocity ω means counterclockwise (CCW) rotation and a negative angular velocity ω means clockwise (CW) rotation. A positive angular acceleration α either increases the rate of CCW rotation (increases the magnitude of a positive ω) or decreases the rate of CW rotation (decreases the magnitude of a negative ω).

Figure 8.11 (a) When the cyclist climbs a hill, the top half of the chain exerts a large force $\vec{\mathbf{F}}$ on the sprocket attached to the rear wheel. As viewed here, the torque about the axis of rotation (the axle) due to this force is clockwise. By convention, we call this a negative torque. (b) When the brakes are applied, the brake pads are pressed onto the rim, giving rise to frictional forces on the rim. As viewed here, the frictional force $\vec{\mathbf{f}}$ causes a counterclockwise (positive) torque on the wheel about the axle.

We use the same sign convention for torque. A force whose perpendicular component tends to cause rotation in the CCW direction gives rise to a positive torque; if it is the only torque acting, it would cause a positive angular acceleration α (Fig. 8.11). A force whose perpendicular component tends to cause rotation in the CW direction gives rise to a negative torque. The symbol $\pm$ in Eq. (8-3) reminds us to assign the appropriate algebraic sign each time we calculate a torque.

The sign of the torque is *not* determined by the sign of the angular velocity (in other words, whether the wheel is spinning CCW or CW); rather, it is determined by the sign of the angular *acceleration* the torque would cause if acting alone. To determine the sign of a torque, imagine which way the torque would make the object begin to spin if it is initially not rotating.

In a more general treatment of torque, torque is a vector quantity defined as the cross product $\vec{\tau} = \vec{\mathbf{r}} \times \vec{\mathbf{F}}$. See Appendix A.8 for the definition of the cross product. For an object rotating about a fixed axis, Eq. (8-3) gives the component of $\vec{\tau}$ along the axis of rotation.

Example 8.3

A Spinning Bicycle Wheel

To stop a spinning bicycle wheel, suppose you push radially inward on opposite sides of the wheel, as shown in Fig. 8.8, with equal forces of magnitude 10.0 N. The radius of the wheel is 32 cm and the coefficient of kinetic friction between the tire and your hand is 0.75. The wheel is spinning in the CW sense. What is the net torque on the wheel?

Strategy The 10.0 N forces are directed radially toward the rotation axis, so they produce no torques themselves;

continued on next page

Example 8.3 continued

only perpendicular components of forces give rise to torques. The forces of kinetic friction between the hands and the tire are tangent to the tire, so they do produce torques. The normal force applied to the tire is 10.0 N on each side; using the coefficient of friction, we can find the frictional forces.

Solution The frictional force exerted by each hand on the tire has magnitude

$$f = \mu_k N = 0.75 \times 10.0 \text{ N} = 7.5 \text{ N}$$

The frictional force is tangent to the wheel, so $f_\perp = f$. Then the magnitude of each torque is

$$|\tau| = rf_\perp = 0.32 \text{ m} \times 7.5 \text{ N} = 2.4 \text{ N·m}$$

The two torques have the same sign, since they are both tending to slow down the rotation of the wheel. Is the torque positive or negative? The angular velocity of the wheel is negative since it rotates CW. The angular acceleration has the opposite sign because the angular speed is decreasing. Since $\alpha > 0$, the net torque is also positive. Therefore,

$$\sum \tau = +4.8 \text{ N·m}$$

Discussion The trickiest part of calculating torques is determining the sign. To check, look at the frictional forces in Fig. 8.8. Imagine which way the forces would make the wheel begin to rotate if the wheel were not originally rotating. The frictional forces point in a direction that would tend to cause a CCW rotation, so the torques are positive.

Practice Problem 8.3 Disc Brakes

In the disc brakes that slow down a car, a pair of brake pads squeeze a spinning rotor; friction between the pads and the rotor provides the torque that slows down the car. If the normal force that each pad exerts on a rotor is 85 N and the coefficient of friction is 0.62, what is the frictional force on the rotor due to each of the pads? If this force acts 8.0 cm from the rotation axis, what is the magnitude of the torque on the rotor due to the pair of brake pads?

Lever Arms

There is another, completely equivalent, way to calculate torques that is often more convenient than finding the perpendicular component of the force. Figure 8.12 shows a force $\vec{\mathbf{F}}$ acting at a distance r from an axis. The distance r is the length of a line perpendicular to the axis that runs from the axis to the force's point of application. The force makes an angle θ with that line. The torque is then

$$\tau = \pm rF_\perp = \pm r(F \sin \theta)$$

The factor $\sin \theta$ could be grouped with r instead of with F. Then $\tau = \pm(r \sin \theta)F$, or

Torque (using the lever arm)

$$\tau = \pm r_\perp F \quad (8\text{-}4)$$

The distance $r_\perp$ is called the **lever arm** (or **moment arm**). The magnitude of the torque is, therefore, the magnitude of the force times the lever arm.

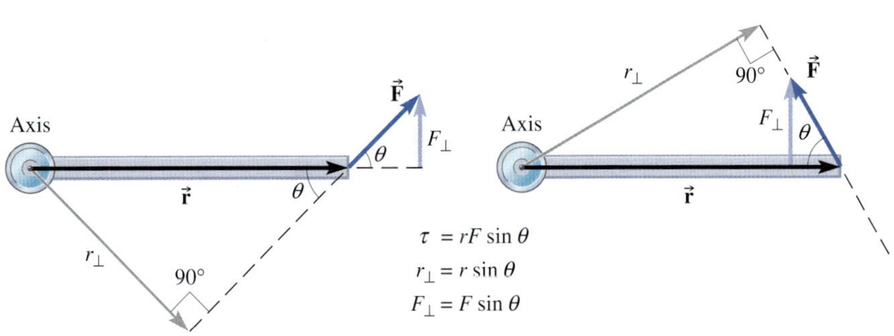

Figure 8.12 Finding the magnitude of a torque using the lever arm.

Finding Torques Using the Lever Arm

1. Draw a line parallel to the force through the force's point of application; this line (dashed in Fig. 8.12) is called the force's **line of action**.
2. Draw a line from the rotation axis to the line of action. This line must be perpendicular to both the axis and the line of action. The distance from the axis to the line of action along this perpendicular line is the lever arm ($r_\perp$). If the line of action of the force goes through the rotation axis, the lever arm and the torque are both zero (see Fig. 8.9b).
3. The magnitude of the torque is the magnitude of the force times the lever arm:

$$\tau = \pm r_\perp F \qquad \text{(8-4)}$$

4. Determine the algebraic sign of the torque as before.

Example 8.4

Screen Door Closer

An automatic screen door closer attaches to a door 47 cm away from the hinges and pulls on the door with a force of 25 N, making an angle of 15° with the door (Fig. 8.13). Find the magnitude of the torque exerted on the door due to this force about the rotation axis through the hinges using (a) the perpendicular component of the force and (b) the lever arm. (c) What is the sign of this torque as viewed from above?

Strategy For method (a), we must find the component of the 25 N force perpendicular to the radial direction. Then this component is multiplied by the length of the radial line. For method (b), we draw in the line of action of the force. Then the lever arm is the perpendicular distance from the line of action to the rotation axis. The torque is the magnitude of the force times the lever arm. We must be careful not to combine the two methods: the torque is *not* equal to the perpendicular force component times the lever arm. For (c), we determine whether this torque would tend to make the door rotate CCW or CW.

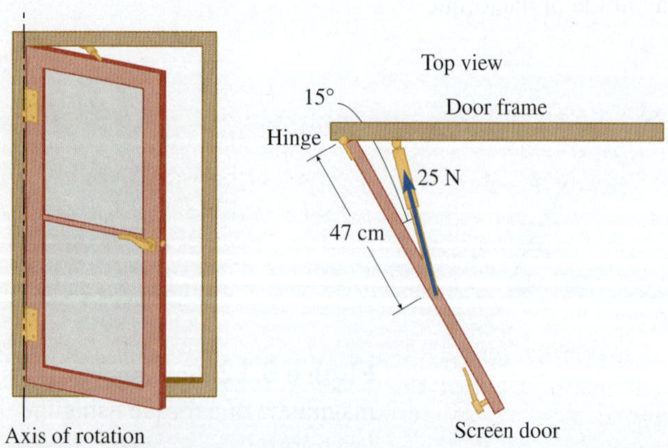

Figure 8.13

Screen door with automatic closing mechanism.

Solution (a) As shown in Fig. 8.14a, the radial component of the force ($F_\parallel$) passes through the rotation axis. The angle labeled 15° is actually a bit larger than 15°, but since the thickness of the door is much less than 47 cm, we approximate it as 15°. The perpendicular component is

$$F_\perp = F \sin 15°$$

The magnitude of the torque is

$$|\tau| = rF_\perp = 0.47\,\text{m} \times 25\,\text{N} \times \sin 15° = 3.0\,\text{N·m}$$

(b) Figure 8.14b shows the line of action of the force, drawn parallel to the force and passing through the point of application. The lever arm is the perpendicular distance between the rotation axis and the line of action. The distance r is approximately 47 cm (again ignoring the thickness of the door). Then the lever arm is

$$r_\perp = r \sin 15°$$

and the magnitude of the torque is

$$|\tau| = r_\perp F = 0.47\,\text{m} \times \sin 15° \times 25\,\text{N} = 3.0\,\text{N·m}$$

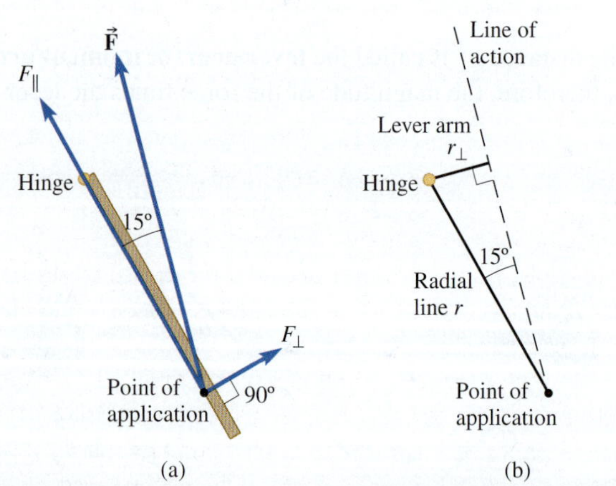

Figure 8.14

(a) Finding the perpendicular component of the force. (b) Finding the lever arm.

continued on next page

Example 8.4 continued

(c) Using the top view of Fig. 8.13, the torque tends to close the door by making it rotate counterclockwise (assuming the door is initially at rest and no other torques act). The torque is therefore positive as viewed from above.

Discussion The most common mistake to make in either solution method would be to use cosine instead of sine (or, equivalently, to use the complementary angle 75° instead of 15°). A check is a good idea. If the automatic closer were more nearly parallel to the door, the angle would be less than 15°. The torque would be smaller because the force is more nearly pulling straight in toward the axis. Since the sine function gets smaller for angles closer to zero, the expression checks out correctly.

It might seem silly for a door closer to pull at such an angle that the perpendicular component is relatively small. The reason it's done that way is so the door closer does not get in the way. A closer that pulled in a perpendicular direction would stick straight out from the door. As discussed in Section 8.5, the situation is much the same in our bodies. In order to not inhibit the motion of our limbs, our tendons and muscles are nearly parallel to the bones. As a result, the forces they exert must be much larger than we might expect.

Practice Problem 8.4 Exercise Is Good for You

A person is lying on an exercise mat and lifts one leg at an angle of 30.0° from the horizontal with an 89 N (20 lb) weight attached to the ankle (Fig. 8.15). The distance between the ankle weight and the hip joint (which is the rotation axis for the leg) is 84 cm. What is the torque due to the ankle weight on the leg?

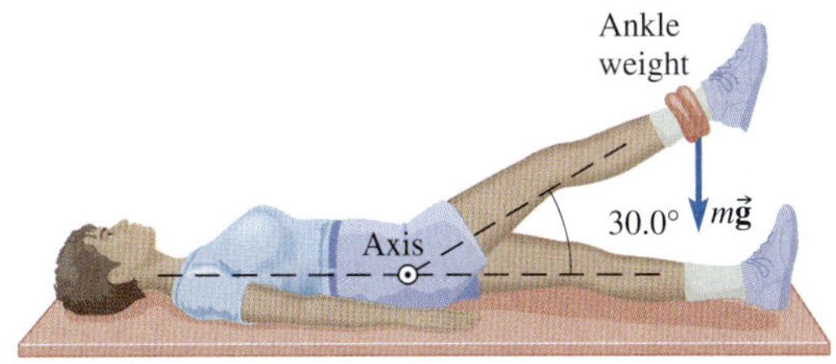

Figure 8.15
Exercise leg lifts.

Center of Gravity

We have seen that the torque produced by a force depends on the point of application of the force. What about gravity? The gravitational force on an object is not exerted at a single point, but is distributed throughout the volume of the object. When we talk of "the" force of gravity on something, we really mean the total force of gravity acting on each particle making up the system.

Fortunately, when we need to find the total torque due to the forces of gravity acting on an object, the total force of gravity can be considered to act at a single point. This point is called the **center of gravity**. The torque found this way is the same as finding all the torques due to the forces of gravity acting at every point in the object, and then adding them together. As you can verify in Problem 109, if the gravitational field is uniform in magnitude and direction, then the center of gravity of an object is located at the object's center of mass.

8.3 CALCULATING WORK DONE FROM THE TORQUE

Torques can do work, as anyone who has started a lawnmower with a pull cord can verify. Actually, it is the force that does the work, but in rotational problems it is often simpler to calculate the work done from the torque. Just as the work done by a constant force is the product of force and the parallel component of displacement, work done by a constant torque can also be calculated as the torque times the *angular* displacement.

CONNECTION:

We're not introducing a different kind of work, just a different way to calculate work.

Imagine a torque acting on a wheel that spins through an angular displacement $\Delta\theta$ while the torque is applied. The work done by the force that gives rise to the torque is the product of the perpendicular component of the force ($F_\perp$) with the arc length s through which the point of application of the force moves (Fig. 8.16). We use the perpendicular force component because that is the component parallel to the *displacement*, which is instantaneously tangent to the arc of the circle. Thus,

$$W = F_\perp s \quad (8\text{-}5)$$

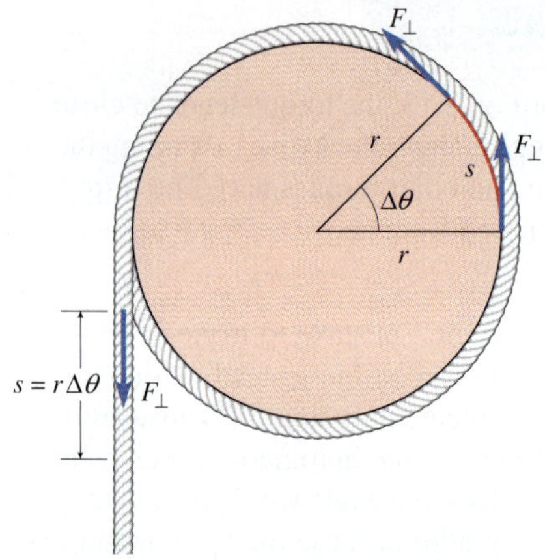

Figure 8.16 The work done by a torque is the product of the perpendicular force component $F_\perp$ and the arc length s.

To write the work in terms of torque, note that $\tau = rF_\perp$ and $s = r\Delta\theta$; then

$$W = F_\perp s = \frac{\tau}{r} \times r\Delta\theta = \tau\Delta\theta$$

$$W = \tau\Delta\theta \quad (\Delta\theta \text{ in radians}) \tag{8-6}$$

Work is indeed the product of torque and the angular displacement. If τ and $\Delta\theta$ have the same sign, the work done is positive; if they have opposite signs, the work done is negative. The *power* due to a constant torque—the rate at which work is done—is

$$P = \tau\omega \tag{8-7}$$

Example 8.5

Work Done on a Potter's Wheel

A potter's wheel is a heavy stone disk on which the pottery is shaped. Potter's wheels were once driven by the potter pushing on a foot treadle; today most potter's wheels are driven by electric motors. (a) If the potter's wheel is a uniform disk of mass 40.0 kg and diameter 0.50 m, how much work must be done by the motor to bring the wheel from rest to 80.0 rev/min? (b) If the motor delivers a constant torque of 8.2 N·m during this time, through how many revolutions does the wheel turn in coming up to speed?

Strategy Work is an energy transfer. In this case, the motor is increasing the rotational kinetic energy of the potter's wheel. Thus, the work done by the motor is equal to the change in rotational kinetic energy of the wheel, ignoring frictional losses. In the expression for rotational kinetic energy, we must express ω in rad/s; we cannot substitute 80.0 rev/min for ω. Once we know the work done, we use the torque to find the angular displacement.

Solution (a) The change in rotational kinetic energy of the wheel is

$$\Delta K = \tfrac{1}{2}I(\omega_f^2 - \omega_i^2) = \tfrac{1}{2}I\omega_f^2$$

Initially the wheel is at rest, so the initial angular velocity ω_i is zero. From Table 8.1, the rotational inertia of a uniform disk is

$$I = \tfrac{1}{2}MR^2$$

Substituting this for I,

$$\Delta K = \tfrac{1}{4}MR^2\omega_f^2$$

Before substituting numerical values, we need to convert 80.0 rev/min to rad/s:

$$\omega_f = 80.0\,\frac{\text{rev}}{\text{min}} \times 2\pi\,\frac{\text{rad}}{\text{rev}} \times \frac{1}{60}\,\frac{\text{min}}{\text{s}} = 8.38\text{ rad/s}$$

Substituting the known values for mass and radius,

$$\Delta K = \tfrac{1}{4} \times 40.0\text{ kg} \times (\tfrac{0.50}{2}\text{ m})^2 \times (8.38\text{ rad/s})^2 = 43.9\text{ J}$$

Therefore, the work done by the motor, rounded to two significant figures, is 44 J.

(b) The work done by a constant torque is

$$W = \tau\,\Delta\theta$$

Solving for the angular displacement $\Delta\theta$ gives

$$\Delta\theta = \frac{W}{\tau} = \frac{43.9\text{ J}}{8.2\text{ N·m}} = 5.35\text{ rad}$$

Since 2π rad = 1 revolution,

$$\Delta\theta = 5.35\text{ rad} \times \frac{1\text{ rev}}{2\pi\text{ rad}} = 0.85\text{ rev}$$

continued on next page

Example 8.5 continued

Discussion As always, work is an energy transfer. In this problem, the work done by the motor is the means by which the potter's wheel acquires its rotational kinetic energy. But work done by a torque does not *always* appear as a change in rotational kinetic energy. For instance, when you wind up a mechanical clock or a windup toy, the work done by the torque you apply is stored as elastic potential energy in some sort of spring.

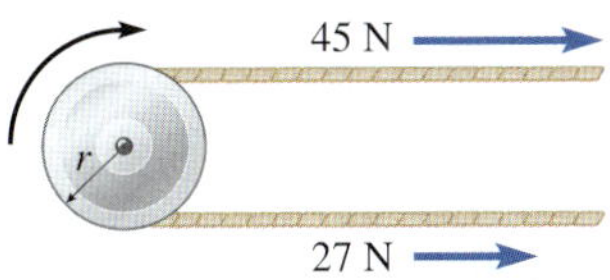

Figure 8.17
Air conditioner belt and pulley.

Practice Problem 8.5 Work Done on an Air Conditioner

A belt wraps around a pulley of radius 7.3 cm that drives the compressor of an automobile air conditioner. The tension in the belt on one side of the pulley is 45 N, and on the other side of the pulley it is 27 N (Fig. 8.17). How much work is done by the belt on the compressor during one revolution of the pulley?

8.4 ROTATIONAL EQUILIBRIUM

An object is in translational equilibrium when the net force acting on it is zero. It is quite possible for the net force acting to be zero, while the net torque is nonzero; the object would then have a nonzero angular acceleration. When designing a bridge or a new house, it would be unacceptable for any of the parts to have nonzero angular acceleration! Zero net force is sufficient to ensure *translational* equilibrium; if an object is also in *rotational* equilibrium, then the net torque acting on it must also be zero.

Conditions for equilibrium (both translational and rotational)

$$\sum \vec{\mathbf{F}} = 0 \quad \text{and} \quad \sum \tau = 0 \tag{8-8}$$

Choosing an Axis of Rotation in Equilibrium Problems Before tackling equilibrium problems, we must resolve a conundrum: if something is not rotating, then where is the axis of rotation? How can we calculate torques without knowing where the axis of rotation is? In some cases, perhaps involving axles or hinges, there may be a clear axis about which the object would rotate if the balance of forces and torques is disturbed. In many cases, though, it is not clear what the rotation axis would be, and in general it depends on exactly how the equilibrium is upset. Fortunately, the axis can be chosen *arbitrarily* when calculating torques *in equilibrium problems*.

In equilibrium, the net torque about *any* rotation axis must be zero. Does that mean that we have to write down an infinite number of torque equations, one for each possible axis of rotation? Fortunately, no. Although the proof is complicated, it can be shown that if the net force acting on an object is zero and the net torque about one rotation axis is zero, then the net torque about every other axis parallel to that axis must also be zero. Therefore, one torque equation is all we need.

Since the torque can be calculated about any desired axis, a judicious choice can greatly simplify the solution of the problem. The best place to choose the axis is usually at the point of application of an unknown force so that the unknown force does not appear in the torque equation.

CHECKPOINT 8.4

Is it possible for the net torque on an object to be zero and the net force nonzero? Is it possible for the net force to be zero and the net torque nonzero?

Problem-Solving Steps in Equilibrium Problems

- Identify an object or system in equilibrium. Draw a diagram showing all the forces acting on that object, each drawn at its point of application. Use the center of gravity as the point of application of any gravitational forces.
- To apply the force condition $\Sigma\vec{\mathbf{F}} = 0$, choose a convenient coordinate system and resolve each force into its x- and y-components.
- To apply the torque condition $\Sigma\tau = 0$, choose a convenient rotation axis—generally one that passes through the point of application of an unknown force. Then find the torque due to each force. Use whichever method is easier: either the lever arm times the magnitude of the force or the distance times the perpendicular component of the force. Determine the direction of each torque; then either set the sum of all the torques (with their algebraic signs) equal to zero or set the magnitude of the CW torques equal to the magnitude of the CCW torques.
- Not all problems require all three equations (two force component equations and one torque equation). Sometimes it is easier to use more than one torque equation, with a different axis. Before diving in and writing down all the equations, think about which approach is the easiest and most direct.

Example 8.6

Carrying a 6 × 6 Beam

Two carpenters are carrying a uniform 6 × 6 beam. The beam is 8.00 ft (2.44 m) long and weighs 425 N (95.5 lb). One of the carpenters, being a bit stronger than the other, agrees to carry the beam 1.00 m in from the end; the other carries the beam at its opposite end. What is the upward force exerted on the beam by each carpenter?

Strategy The conditions for equilibrium are that the net external force equal zero and the net external torque equal zero:

$$\sum\vec{\mathbf{F}} = 0 \quad \text{and} \quad \sum\tau = 0$$

Should we start with forces or with torques? In this problem, it is easiest to start with torques. If we choose the axis of rotation where one of the unknown forces acts, then that force has a lever arm of zero and its torque is zero. The torque equation can be solved for the other unknown force. Then with only one force still unknown, we set the sum of the y-components of the forces equal to zero.

Solution The first step is to draw a force diagram (Fig. 8.18). Each force is drawn at the point where it acts. Known distances are labeled.

We choose a rotation axis perpendicular to the xy-plane and passing through the point of application of $\vec{\mathbf{F}}_2$. The simplest way to find the torques for this example is to multiply each force by its lever arm. The lever arm for $\vec{\mathbf{F}}_1$ is

$$2.44 \text{ m} - 1.00 \text{ m} = 1.44 \text{ m}$$

and the magnitude of the torque due to this force is

$$|\tau| = Fr_\perp = F_1 \times 1.44 \text{ m}$$

Since the beam is uniform, its center of gravity is at its midpoint. We imagine the entire gravitational force to act at this point. Then the lever arm for the gravitational force is

$$\tfrac{1}{2} \times 2.44 \text{ m} = 1.22 \text{ m}$$

and the torque due to gravity has magnitude

$$|\tau| = Fr_\perp = 425 \text{ N} \times 1.22 \text{ m} = 518.5 \text{ N·m}$$

The torque due to $\vec{\mathbf{F}}_1$ is negative since, if it were the only torque, it would make the beam start to rotate clockwise about our chosen axis of rotation. The torque due to gravity is positive since, if it were the only torque, it would make the beam start to rotate counterclockwise. Therefore,

$$\sum\tau = -F_1 \times 1.44 \text{ m} + 518.5 \text{ N·m} = 0$$

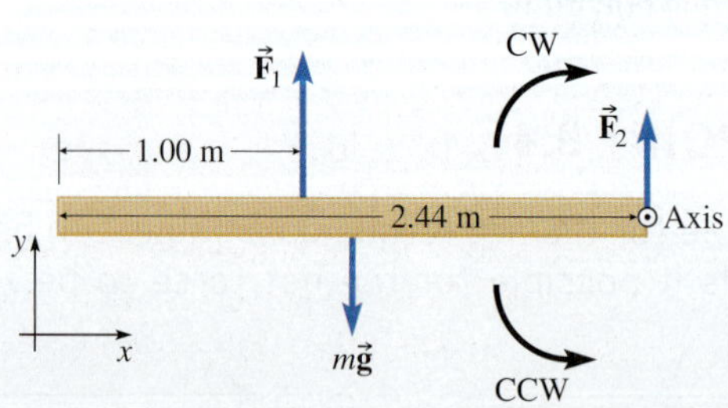

Figure 8.18 Diagram of the beam with rotation axis, forces, and distances shown.

continued on next page

Example 8.6 continued

Solving for F_1,

$$F_1 = \frac{518.5\ \text{N}\cdot\text{m}}{1.44\ \text{m}} = 360\ \text{N}$$

Since another condition for equilibrium is that the net force be zero,

$$\sum F_y = F_1 + F_2 - mg = 0$$

Solving for F_2,

$$F_2 = 425\ \text{N} - 360\ \text{N} = 65\ \text{N}$$

Discussion A good way to check this result is to make sure that the net torque about a *different axis* is zero—for an object in equilibrium, the net torque about any axis must be zero. Suppose we choose an axis through the point of application of $\vec{\mathbf{F}}_1$. Then the lever arm for $m\vec{\mathbf{g}}$ is 1.22 m − 1.00 m = 0.22 m and the lever arm for $\vec{\mathbf{F}}_2$ is 2.44 m − 1.00 m = 1.44 m. Setting the net torque equal to zero:

$$\sum \tau = -425\ \text{N} \times 0.22\ \text{m} + F_2 \times 1.44\ \text{m} = 0$$

Solving for F_2 gives

$$F_2 = \frac{425\ \text{N} \times 0.22\ \text{m}}{1.44\ \text{m}} = 65\ \text{N}$$

which agrees with the value calculated before. We could have used this second torque equation to find F_2 instead of setting ΣF_y equal to zero.

Practice Problem 8.6 A Diving Board

A uniform diving board of length 5.0 m is supported at two points; one support is located 3.4 m from the end of the board and the second is at 4.6 m from the end (Fig. 8.19). The supports exert vertical forces on the diving board. A diver stands at the end of the board over the water. Determine the directions of the support forces. (connect tutorial: plank) [*Hint*: In this problem, consider torques about different rotation axes.]

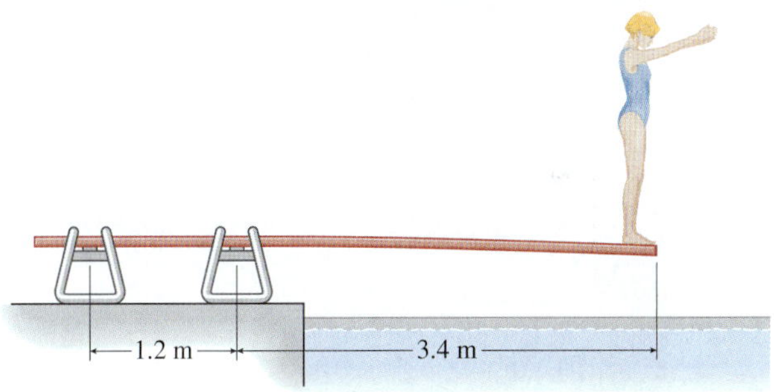

Figure 8.19
Diving board.

Application of Rotational Equilibrium: The Cantilever A diving board is an example of a cantilever—a beam or pole that extends beyond its support. The forces exerted by the supports on a diving board are considerably larger than if the same board were supported at both ends (see Problem 35). The advantage is that the far end of the board is left free to vibrate; as it does, the support forces adjust themselves to keep the board from tipping over. Dramatic effects can be achieved by architects using cantilevers (Fig. 8.20). Some practical reasons for cantilever construction include reaching out over a stream, rock outcropping, or steep hillside; creating usable outdoor space under the cantilever; and giving buildings a lighter and more spacious feel. Most airplane wings and some bridges are cantilevers.

Figure 8.20 Villa Méditerranée in Marseilles, France, designed by architect Stefano Boeri. A cantilevered exhibition space extends 40 m out over a 2000 m^2 pool and offers panoramic views of the sea.

Example 8.7

The Slipping Ladder

A 15.0 kg uniform ladder leans against a wall in the atrium of a large hotel (Fig. 8.21a). The ladder is 8.00 m long; it makes an angle $\theta = 60.0°$ with the floor. The coefficient of static friction between the floor and the ladder is $\mu_s = 0.45$. How far along the ladder can a 60.0 kg person climb before the ladder starts to slip? Assume that the wall is frictionless. (connect interactive: ladder and tutorial: ladder)

Strategy Consider the ladder and the climber as a single system. Until the ladder starts to slip, this system is in equilibrium. Therefore, the net external force and the net external torque acting on the system are both equal to zero:

$$\sum F_x = 0, \qquad \sum F_y = 0, \qquad \text{and} \qquad \sum \tau = 0$$

continued on next page

Example 8.7 continued

(a)

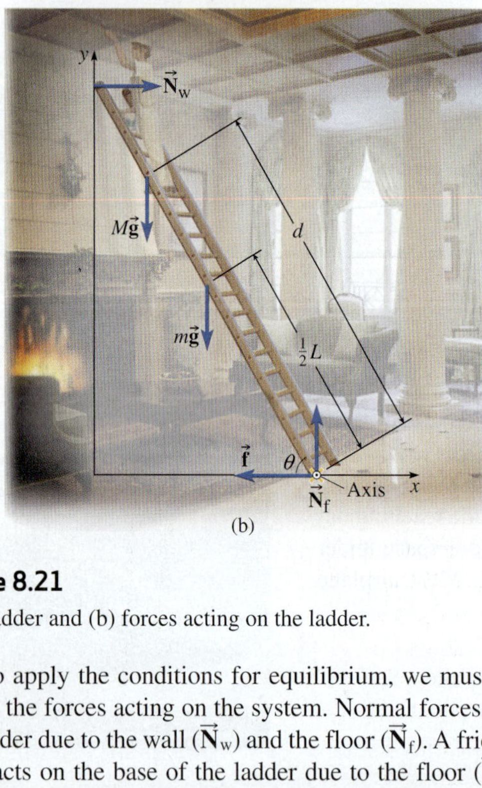

(b)

Figure 8.21

(a) A ladder and (b) forces acting on the ladder.

To apply the conditions for equilibrium, we must identify all the forces acting on the system. Normal forces act on the ladder due to the wall ($\vec{\mathbf{N}}_w$) and the floor ($\vec{\mathbf{N}}_f$). A frictional force acts on the base of the ladder due to the floor ($\vec{\mathbf{f}}$), but no frictional force acts on the top of the ladder since the wall is frictionless. Gravitational forces act on the ladder and on the person climbing it. As the person ascends the ladder, the frictional force $\vec{\mathbf{f}}$ has to increase to keep the ladder in equilibrium. The ladder begins to slip when the frictional force required to maintain equilibrium is larger than its maximum possible value $\mu_s N_f$. The ladder is about to slip when $f = \mu_s N_f$.

Solution The first step is to make a careful drawing of the ladder and label all distances and forces (Fig. 8.21b). Instead of cluttering the diagram with numerical values, we use L (= 8.00 m) for the length of the ladder, d for the unknown distance from the bottom of the ladder to the point where the person stands, and M (= 60.0 kg) and m (= 15.0 kg) for the masses of the person and ladder, respectively. The weight of the ladder acts at the ladder's center of gravity, which is the ladder's midpoint since it is uniform.

Now we apply the conditions for equilibrium. Starting with $\Sigma F_x = 0$, we find

$$N_w - f = 0$$

where, if the climber is at the highest point possible, the frictional force must have its maximum possible magnitude:

$$f = \mu_s N_f$$

Combining these two equations, we obtain a relationship between the magnitudes of the two normal forces:

$$N_w = \mu_s N_f$$

Next we use the condition $\Sigma F_y = 0$, which gives

$$N_f - Mg - mg = 0$$

The only unknown quantity in this equation is N_f, so we can solve for it:

$$N_f = Mg + mg = (M + m)g$$

Now we can find the other normal force, N_w:

$$N_w = \mu_s N_f = \mu_s (M + m)g$$

At this point, we know the magnitudes of all the forces:

$$\begin{aligned} Mg &= 588.0 \text{ N} \\ mg &= 147.0 \text{ N} \\ N_f &= Mg + mg = 735.0 \text{ N} \\ f = N_w &= \mu_s(Mg + mg) = 0.45 \times 735.0 \text{ N} = 330.75 \text{ N} \end{aligned}$$

We do not know the distance d, which is the goal of the problem. To find d we must set the net torque equal to zero.

First we choose a rotation axis. The most convenient choice is an axis perpendicular to the plane of Fig. 8.21 and passing through the bottom of the ladder. Since two of the five forces ($\vec{\mathbf{N}}_f$ and $\vec{\mathbf{f}}$) act at the bottom of the ladder, these two forces have zero lever arms and, thus, produce zero torque. Another reason why this is a convenient choice of axis is that the distance d is measured from the bottom of the ladder.

In this situation, with the forces either vertical or horizontal, it is probably easiest to use lever arms to find the torques. In three diagrams (Fig. 8.22), we first draw the line of action for each force; then the lever arm is the perpendicular distance between the axis and the line of action.

continued on next page

Example 8.7 continued

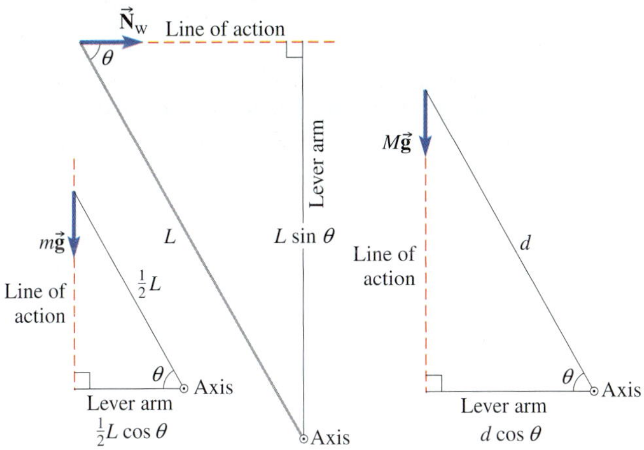

Figure 8.22
Finding the lever arm for each force.

The lever arms are:

For $m\vec{g}$, $r_\perp = \frac{1}{2} L \cos \theta = \frac{1}{2} (8.00 \text{ m}) \cos 60.0° = 2.00 \text{ m}$
For $\vec{N}_w$, $r_\perp = L \sin \theta = 8.00 \text{ m} \sin 60.0° = 6.928 \text{ m}$
For $M\vec{g}$, $r_\perp = d \cos \theta = 0.500d$

Using the usual convention that CCW torques are positive, the torque due to $\vec{N}_w$ is negative and the torques due to gravity are positive. The magnitude of each torque is the magnitude of the force times its lever arm:

$$\tau = Fr_\perp$$

Setting the net torque equal to zero yields

$$-N_w(L \sin \theta) + mg\,(\tfrac{1}{2}L \cos \theta) + Mg(d \cos \theta) = 0$$

Now we substitute values for the forces and lever arms and solve for d.

$$-(330.75 \text{ N})(6.928 \text{ m}) + (147.0 \text{ N})(2.000 \text{ m}) + (588.0 \text{ N})(0.5000d) = 0$$

$$-2291 \text{ N·m} + 294.0 \text{ N·m} + (294.0 \text{ N})d = 0$$

$$d = \frac{2291 \text{ N·m} - 294.0 \text{ N·m}}{294.0 \text{ N}} = 6.8 \text{ m}$$

The person can climb 6.8 m up the ladder without having it slip. (This is the distance *along the ladder*, not the height above the ground.)

Discussion If the person goes any higher, then his weight produces a larger CCW torque about our chosen rotation axis. To stay in equilibrium, the total CW torque would have to get larger. The only force providing a CW torque is the normal force due to the wall, which pushes to the right. However, if this force were to get larger, the frictional force would have to get larger to keep the net horizontal force equal to zero. Since friction already has its maximum magnitude, there is no way for the ladder to be in equilibrium if the person climbs any higher.

Practice Problem 8.7 Another Ladder Leaning on a Wall

A uniform ladder of mass 10.0 kg and length 3.2 m leans against a frictionless wall with its base located 1.5 m from the wall. If the ladder is not to slip, what must be the minimum coefficient of static friction between the bottom of the ladder and the ground? Assume the wall is frictionless.

EVERYDAY PHYSICS DEMO

Is it possible to *wind up* a spool by *pulling* on the thread? Take a dumbbell, spool of thread, or yo-yo and wrap some string around the center of its axle. Place the dumbbell on a table (or on the floor). Unwind a short length of string and try pulling perpendicularly to the axle at different angles to the horizontal (Fig. 8.23). Depending on the direction of your pull, the dumbbell can roll in either direction. Try to find the angle at which the rolling changes direction; at this angle the dumbbell does not roll at all. (Pulling at this angle, it is possible to make the spool *slide* along the table without rotating.)

What is special about this angle? Since the dumbbell is in equilibrium when pulling at this angle, we can analyze the torques using any rotation axis we choose. A convenient choice is the axis that passes through point *P*, the point of contact with the table. Then the contact force between the table and the dumbbell acts at the rotation axis, and its torque is zero. The torque due to gravity is also zero, since the line of action passes through point *P*. The dumbbell can only be in equilibrium if the torque due to the remaining force (the tension in the string) is zero. This torque is zero if the lever arm is zero, which means the line of action passes through point *P*.

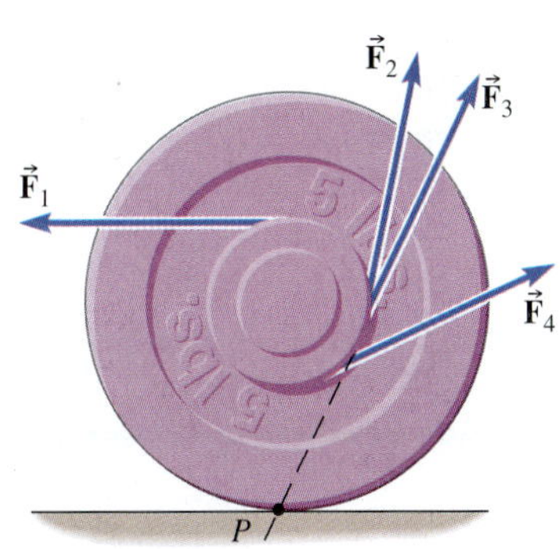

Figure 8.23 Forces $\vec{F}_1$ and $\vec{F}_2$ make the dumbbell roll to the left; $\vec{F}_4$ makes it roll to the right; $\vec{F}_3$ does not make it roll.

Example 8.8

The Sign and the Breaking Cord

A uniform beam of weight 196 N and of length 1.00 m is attached to a hinge on the outside wall of a restaurant. A cord is attached at the center of the beam and is attached to the wall, making an angle of 30.0° with the beam (Fig. 8.24a). The cord keeps the beam perpendicular to the wall. If the breaking tension of the cord is 620 N, how large can the mass of the sign be without breaking the cord?

Strategy The beam is in equilibrium; both the net force and the net torque acting on it must be zero. To find the maximum weight of the sign, we let the tension in the cord have its maximum value of 620 N. We do not know the force exerted by the hinge on the beam, so we choose an axis of rotation through the hinge. Then the force exerted by the hinge on the beam has a zero lever arm and does not enter the torque equation.

Before doing anything else, we draw a diagram showing each force acting on the beam and the chosen rotation axis. The FBD in previous chapters often placed all the force vectors starting from a single point. Now we draw each force vector starting at its point of application so that we can find the torque—either by finding the lever arm or by finding the perpendicular force component and the distance from the axis to the point of application.

Solution Figure 8.24b shows the forces acting on the beam; three of these contribute to the torque. The gravitational force on the beam can be taken to act at the midpoint of the beam since it is uniform. The force due to the cord has a perpendicular component (Fig. 8.24c) of

$$F_\perp = 620 \text{ N} \times \sin 30.0° = 310 \text{ N}$$

The two gravitational forces tend to rotate the beam CW, while the tension in the cord tends to rotate it CCW. The net torque must be equal to zero:

$$-0.50 \text{ m} \times 196 \text{ N} - 1.00 \text{ m} \times Mg + 0.50 \text{ m} \times 310 \text{ N} = 0$$

or

$$1.00 \text{ m} \times Mg = 0.50 \text{ m} \times (310 \text{ N} - 196 \text{ N})$$

Now we solve for the unknown mass M:

$$M = \frac{0.50 \text{ m} \times (310 \text{ N} - 196 \text{ N})}{1.00 \text{ m} \times 9.80 \text{ N/kg}} = 5.8 \text{ kg}$$

Discussion In this problem, we did not have to set the net force equal to zero. By placing the axis of rotation at the hinge, we eliminated two of the three unknowns from the torque equation: the horizontal and vertical components of the hinge force (or, equivalently, its magnitude and direction). If we wanted to find the hinge force as well, setting the net force equal to zero would be necessary.

Practice Problem 8.8 Hinge Forces

Find the vertical component of the force exerted by the hinge in two different ways: (a) setting the net force equal to zero and (b) using a torque equation about a different axis.

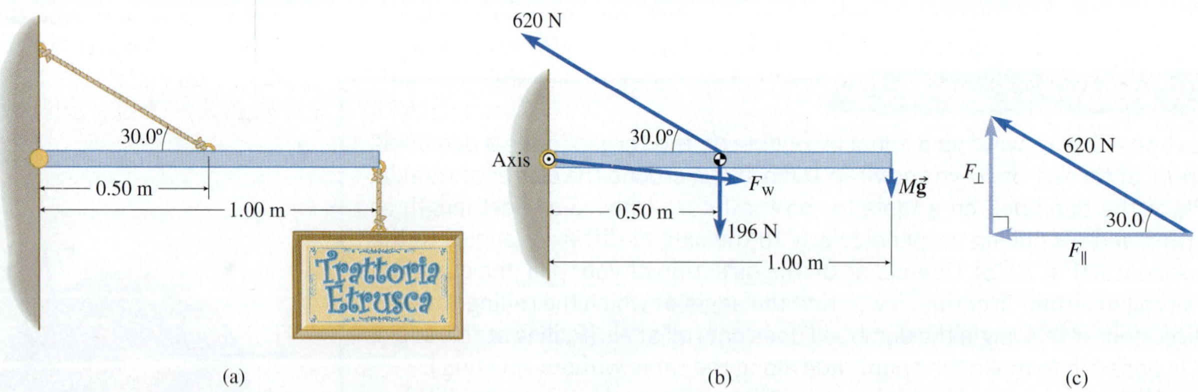

Figure 8.24
(a) A sign outside a restaurant. (b) Forces acting on the beam. (c) Finding the components of the tension in the cord.

Distributed Forces

Gravity is not the only force that is distributed rather than acting at a point. Contact forces, including both the normal component and friction, are spread over the contact surface. Just as for gravity, we can consider the contact force to act at a single point, but the location of that point is often not at all obvious. For a book sitting on a horizontal table, it seems reasonable that the normal force effectively acts at the geometric center of the book cover that touches the table. It is less clear where that effective point is if the book is on an incline or is sliding. As Example 8.9 shows, when something is about to topple over, contact is about to be lost everywhere except at the corner around which the toppling object is about to rotate. That corner then must be the location of the contact forces.

Example 8.9

The Toppling File Cabinet

A file cabinet of height a and width b is on a ramp at angle θ (Fig. 8.25a). The file cabinet is filled with papers in such a way that its center of gravity is at its geometric center. Find the largest θ for which the file cabinet does not tip over. Assume the coefficient of static friction is large enough to prevent sliding. (connect tutorial: file cabinet)

Strategy Until the file cabinet begins to tip over, it is in equilibrium; the net force acting on it must be zero and the total torque about any axis must also be zero. We first draw a force diagram showing the three forces (gravity, normal, friction) acting on the file cabinet. The point of application of the two contact forces (normal, friction) must be at the lower edge of the file cabinet if it is on the steepest possible incline, just about to tip over. In that case, contact has been lost over the rest of the bottom surface of the file cabinet so that only the lower edge makes good contact with the ramp.

As in all equilibrium problems, a good choice of rotation axis makes the problem easier to solve. We know that, at the maximum angle, the contact forces act at the bottom edge of the file cabinet. A good choice of rotation axis is along the bottom edge of the file cabinet, because then the normal and frictional forces have zero lever arm.

Solution Figure 8.25b shows the forces acting on the file cabinet at the maximum angle θ. The gravitational force is drawn at the center of gravity. Instead of drawing a single vector arrow for the gravitational force, we represent the gravitational force by its components parallel and perpendicular to the ramp. Then we find the lever arm for each of the components. The lever arm for the parallel component of the weight ($mg \sin \theta$) is $\frac{1}{2}a$ and the lever arm for the perpendicular component ($mg \cos \theta$) is $\frac{1}{2}b$. Setting the net torque equal to zero:

$$\sum \tau = -mg \cos \theta \times \tfrac{1}{2}b + mg \sin \theta \times \tfrac{1}{2}a = 0$$

After dividing out the common factors of $\frac{1}{2}mg$,

$$b \cos \theta = a \sin \theta$$

Solving for θ,

$$\theta = \tan^{-1} \frac{b}{a}$$

Discussion As a check, we can regard the normal and friction forces as two components of a single contact force. We can think of that contact force as acting at a single point—a "center of contact" analogous to the center of gravity. As the file cabinet is put on steeper and steeper surfaces, the effective point of application of the contact force moves toward the lower edge of the file cabinet (Fig. 8.26). If we take the rotation axis through the center of gravity so there is no gravitational torque, then the torque due to the contact force must be zero. The only way that can happen is

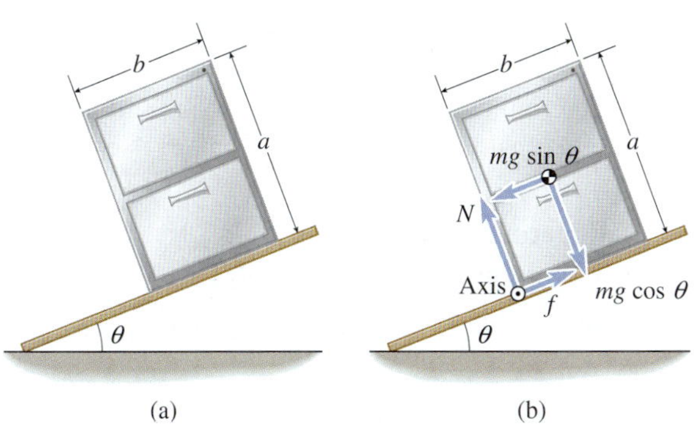

Figure 8.25
(a) File cabinet on an incline. (b) Forces acting on the file cabinet.

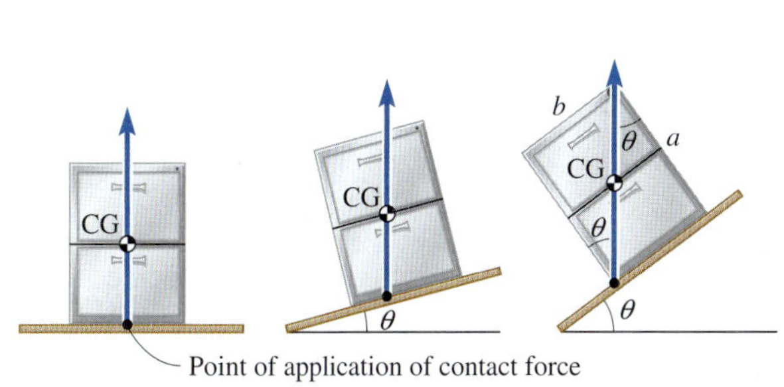

Figure 8.26
Contact force for various incline angles.

continued on next page

Example 8.9 continued

Figure 8.27
Yuri Chechi of Italy holds the pike position on the rings at the World Gymnastic Championships in Sabae, Japan.

if its lever arm is zero, which means that the contact force must point directly toward the center of gravity. If the angle θ has its maximum value, the contact force acts at the lower edge and $\tan \theta = b/a$. The file cabinet is about to tip when its center of gravity is directly above the lower edge. Any object supported only by contact forces can be in equilibrium only if the point of application of the total contact force is directly below the object's center of gravity.

Conceptual Practice Problem 8.9 Gymnast Holding a Pike Position

Figure 8.27 shows a gymnast holding a pike position. What can you say about the location of the gymnast's center of gravity?

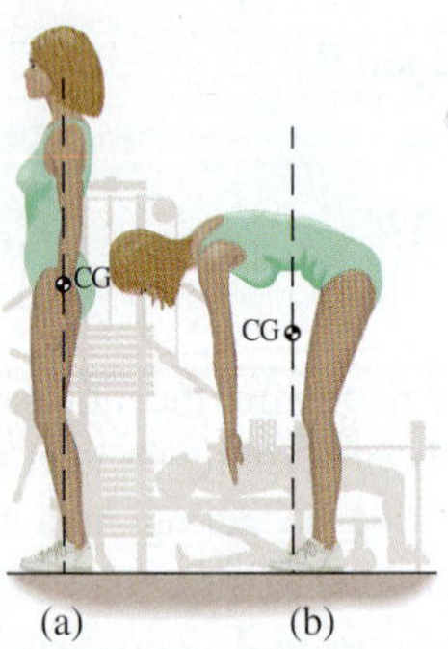

Figure 8.28 Location of the center of gravity when (a) standing and (b) reaching for the floor.

EVERYDAY PHYSICS DEMO

When a person stands up straight, the body's center of gravity lies directly above a point between the feet, about 3 cm in front of the ankle joint (Fig. 8.28a). When a person bends over to touch her toes, the center of gravity lies outside the body (Fig. 8.28b). Note that the lower half of the body must move backward to keep the center of gravity from moving out in front of the toes, which would cause the person to fall over.

An interesting experiment can be done that illustrates what happens to your balance when you shift your center of gravity. Stand against a wall with the heels of your feet touching the wall and your back pressed against the wall. Then carefully try to bend over as if to touch your toes, without bending your knees. Can you do this without falling over? Explain.

Mechanical Advantage

In its simplest form, a lever is a rigid bar that rotates about a fixed point (the fulcrum). The lever is an example of a simple machine that can be used to amplify a force (Fig. 8.29). The **mechanical advantage (MA)** is defined as the ratio of the output force (load) to the applied force:

$$\text{MA} = \frac{\text{load}}{\text{applied force}}$$

For an ideal lever, by using the fulcrum as the axis of rotation and setting the net torque on the lever equal to zero, we find that the mechanical advantage is equal to the ratio of the lever arms:

$$\text{MA} = \frac{\text{lever arm of applied force}}{\text{lever arm of load}}$$

The force amplification comes with a trade-off: the applied force must move through a larger distance than the load.

$$\text{MA} = \frac{\text{displacement of applied force}}{\text{displacement of load}}$$

connect See the text website for details about the mechanical advantage of other simple machines.

Figure 8.29 A wheelbarrow uses a lever to make it easier to lift heavy loads. The mechanical advantage is the ratio of the output force (load) to the applied force. For the wheelbarrow, the load is the weight of the wheelbarrow contents (W) and the applied force (F) is exerted by the gardener on the handles. If the rotational inertia of the lever itself is negligible, the net torque on the lever has to be zero. Using the fulcrum as the axis of rotation, we have $\Sigma\tau = Fd_2 - Wd_1 = 0$, where d_1 and d_2 are the lever arms of the applied force an load, respectively. The mechanical advantage is:

$$\text{MA} = (\text{load})/(\text{applied force}) = W/F = d_2/d_1.$$

8.5 APPLICATION: EQUILIBRIUM IN THE HUMAN BODY

We can use the concepts of torque and equilibrium to understand some of how the musculoskeletal system of the human body works. A muscle has tendons at each end that connect it to two different bones across a joint (the flexible connection between the bones). When the muscle contracts, it pulls the tendons, which in turn pull on the bones. Thus, the muscle produces a pair of forces of equal magnitude, one acting on each of the two bones. The biceps muscle (Fig. 8.30) in the upper arm attaches the scapula to the forearm (radius) across the inside of the elbow joint. When the biceps contracts, the forearm is pulled toward the upper arm. The biceps is a *flexor* muscle; it moves one bone closer to another.

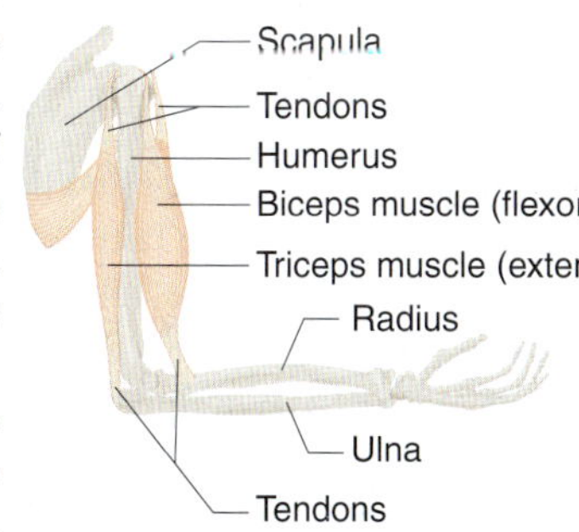

Figure 8.30 The biceps is a flexor muscle; the triceps is an extensor muscle.

A muscle can pull but not push, so a flexor muscle such as the biceps cannot reverse its action to push the forearm away from the upper arm. The *extensor* muscles make bones move apart from each other. In the upper arm (Fig. 8.30), an extensor muscle—the triceps—connects the scapula and humerus to the ulna (a bone in the forearm parallel to the radius) across the outside of the elbow. Since the biceps and triceps connect to the forearm on opposite sides of the elbow joint, they tend to cause rotation about the joint in opposite directions. When the triceps contracts it pulls the forearm away from the upper arm. Using flexor and extensor muscles on opposite sides of the joint, the body can produce both positive and negative torques, although both muscles pull in the same direction.

Suppose the arm is held in a horizontal position. The deltoid muscle (the muscle shown in Fig. 8.31) exerts a force $\vec{\mathbf{F}}_m$ on the humerus at an angle of about 15° above the horizontal. This force has to do two things. The vertical component (magnitude $F_m \sin 15° \approx 0.26F_m$) supports the weight of the arm, while the horizontal component (magnitude $F_m \cos 15° \approx 0.97F_m$) stabilizes the joint by pulling the humerus in against the shoulder (scapula). In Example 8.10, we estimate the magnitude of $\vec{\mathbf{F}}_m$.

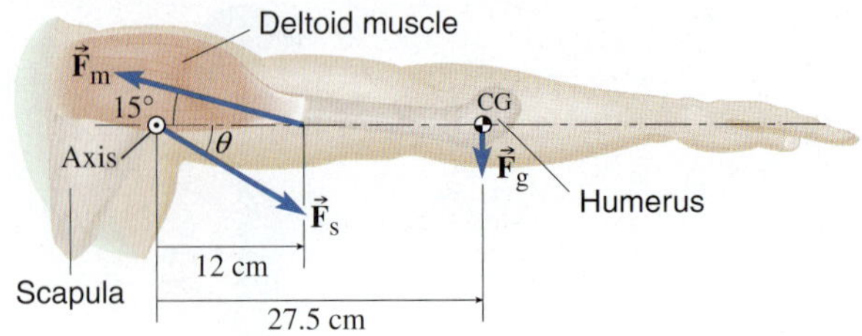

Figure 8.31 Forces exerted on an outstretched arm by the deltoid muscle ($\vec{\mathbf{F}}_m$), the scapula ($\vec{\mathbf{F}}_s$), and gravity ($\vec{\mathbf{F}}_g$).

Example 8.10

Force to Hold Arm Horizontal

A person is standing with his arm outstretched in a horizontal position. The weight of the arm is 30.0 N, and its center of gravity is at the elbow joint, 27.5 cm from the shoulder joint (see Fig. 8.31). The deltoid pulls on the upper arm at an angle of 15° above the horizontal and at a distance of 12 cm from the joint. What is the magnitude of the force exerted by the deltoid muscle on the arm?

Strategy The arm is in equilibrium, so we can apply the conditions for equilibrium: $\Sigma\vec{\mathbf{F}} = 0$ and $\Sigma\tau = 0$. When calculating torques, we choose the rotation axis at the shoulder joint because then the unknown force $\vec{\mathbf{F}}_s$, which acts on the arm at the joint, has a zero lever arm and produces zero torque. With only one unknown in the torque equation, we can solve immediately for F_m. We do not need to apply the condition $\Sigma\vec{\mathbf{F}} = 0$ unless we want to find $\vec{\mathbf{F}}_s$.

Solution The gravitational force is perpendicular to the line between its point of application and the rotation axis. Gravity produces a CW torque of magnitude

$$|\tau| = Fr = 30.0\text{ N} \times 0.275\text{ m} = 8.25\text{ N}\cdot\text{m}$$

For the torque due to $\vec{\mathbf{F}}_m$, we find the component of $\vec{\mathbf{F}}_m$ that is perpendicular to the line between its point of application and the rotation axis. Since this line is horizontal, we need the vertical component of $\vec{\mathbf{F}}_m$ which is $F_m \sin 15°$. Then the magnitude of the CCW torque due to $\vec{\mathbf{F}}_m$ is

$$|\tau| = F_\perp r = F_m \sin 15° \times 0.12\text{ m}$$

The sum of these torques is zero. With the usual sign convention (CCW is +),

$$F_m \sin 15° \times 0.12\text{ m} - 8.25\text{ N}\cdot\text{m} = 0$$

Solving for F_m,

$$F_m = \frac{8.25\text{ N}\cdot\text{m}}{\sin 15° \times 0.12\text{ m}} = 270\text{ N}$$

Discussion The force exerted by the muscle is much larger than the 30.0 N weight of the arm. The muscle must exert a larger force because the lever arm is small; the point of application is less than half as far from the joint as the center of gravity [0.12 m/(0.275 m) ≈ 4/9]. Also, the muscle cannot pull straight up on the arm; the vertical component of the muscle force is only about $\frac{1}{4}$ of the magnitude of the force. These two factors together make the weight supported (30.0 N) only $\frac{4}{9} \times \frac{1}{4} = \frac{1}{9}$ as large as the force exerted by the muscle.

Practice Problem 8.10 Holding a Juice Carton

Find the force exerted by the same person's deltoid muscle when holding a 1.0 L juice carton (weight 9.9 N) with the arm outstretched and parallel to the floor (as in Fig. 8.31). Assume that the juice carton is 60.0 cm from the shoulder.

Why does the iron cross require great strength?

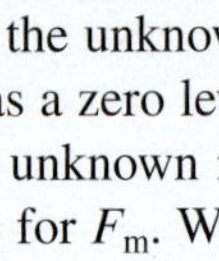

The Iron Cross When a gymnast does the iron cross (Fig. 8.32a), the primary muscles involved are the latissimus dorsi ("lats") and pectoralis major ("pecs"). Since the rings are supporting the gymnast's weight, they exert an upward force on the gymnast's arms. Thus, the task for the muscles is not to hold the arm up, but to pull it down. The lats pull on the humerus about 3.5 cm from the shoulder joint (Fig. 8.32b). The pecs pull on the humerus about 5.5 cm from the joint (Fig. 8.32c). The other ends of these two muscles connect to bone in many places, widely distributed over the back (lats) and chest (pecs). As a reasonable simplification, we can assume that these muscles pull at a 45° angle below the horizontal in the iron cross maneuver. We also assume that the two muscles exert equal forces, so we can replace the two with a single force acting at 4.5 cm from the joint.

To determine the force exerted, we look at the entire arm as a system in equilibrium. This time we can ignore the weight of the arm itself since the force exerted on the arm

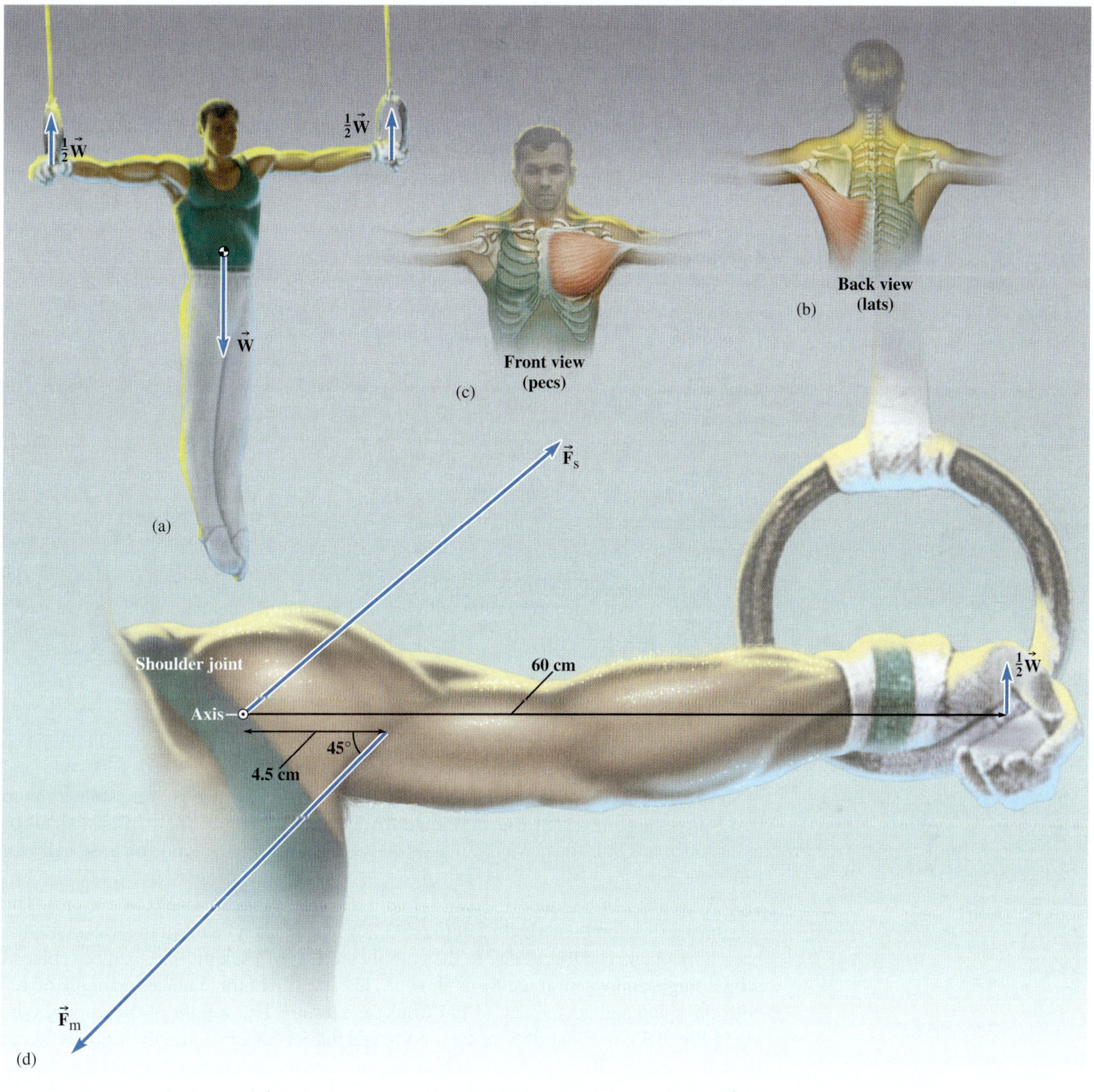

Figure 8.32 (a) Gymnast doing the iron cross. The principal muscles involved are (b) the "lats" and (c) the "pecs." (d) Simplified model of the forces acting on the arm of the gymnast.

by the ring is much larger—half the gymnast's weight is supported by each ring. The ring exerts an upward force that acts on the hand about 60 cm from the shoulder joint (see Fig. 8.32d). Taking torques about the shoulder, in equilibrium we have

$$|\text{CW torque}| = |\text{CCW torque}|$$

$$F_m \times 0.045 \text{ m} \times \sin 45° = \tfrac{1}{2}W \times 0.60 \text{ m}$$

$$F_m = \frac{\frac{1}{2}W \times 0.60 \text{ m}}{0.045 \text{ m} \times \sin 45°} = 9.4W$$

Thus, the force exerted by the lats and pecs *on one side* of the gymnast's body is more than nine times his weight.

Structure of Muscles and Bones in the Human Body The structure of the human body makes large muscular forces necessary. Are there advantages to the structure? Due to the small lever arms, the muscle forces are much larger than they would otherwise be, but the human body has traded this for a wide range of movement of the bones. The biceps and triceps muscles can move the lower arms through almost 180° while they change their lengths by only a few centimeters. The muscles also remain nearly parallel to the bones. If the biceps and triceps muscles were attached to the lower arm much farther from the elbow, there would have to be a large flap of skin to allow them to move so far away from the bones. The arrangement of our bones and muscles favors a wide range of movement.

Another advantage of the body structure is that it tends to minimize the rotational inertia of our limbs. For example, the muscles that control the motion of the lower arm are contained mostly within the *upper* arm. This keeps the rotational inertia of the lower arms about the elbow smaller. It also keeps the rotational inertia of the entire arm about the shoulder smaller. Smaller rotational inertia means that the energy we have to expend to move our limbs around is smaller.

The biceps muscle with its tendons is almost parallel to the humerus. One interesting observation is that the tendon connects to the radius at different points in different people. In one person this point may be 5.0 cm from the elbow joint, but in another person whose arm is the same length it may be 5.5 cm from the elbow. Thus, some people are naturally stronger than others because of their internal structure. Chimpanzees have an advantage over humans because their biceps muscle has a longer lever arm. Do not make the mistake of arm wrestling with an adult chimp; challenge the chimp to a game of chess instead.

Application of Equilibrium Conditions: Heavy Lifting

When lifting an object from the floor, our first instinct is to bend over and pick it up. This is not a good way to lift something heavy. The spine is an ineffective lever and is susceptible to damage when a heavy object is lifted with bent waist. It is much better to squat down and use the powerful leg muscles to do the lifting instead of using our back muscles. Analyzing torques in a simplified model of the back can illustrate why.

The spine can be modeled as a rod with an axis at the tailbone (the sacrum). The sacrum exerts a force, marked $\vec{\mathbf{F}}_s$ in Fig. 8.33, when a person bends at the waist with the back horizontal. The forces due to the complicated set of back muscles can be replaced with a single equivalent force $\vec{\mathbf{F}}_b$ as shown. This equivalent force makes an angle of 12° with the spine and acts about 44 cm from the sacrum. The weight of the upper body, $m\vec{\mathbf{g}}$ in Fig. 8.33, is about 65% of total body weight; its center of gravity is about 38 cm from the sacrum. By placing an axis at the sacrum, we can ignore the force $\vec{\mathbf{F}}_s$ in our torque equation. Since the vertical component of $\vec{\mathbf{F}}_b$ is $F_b \sin 12° \approx 0.21F_b$, only about $\frac{1}{5}$ the magnitude of the forces exerted by the back muscles is supporting the body weight. The much larger horizontal component is pressing the rod representing the spine into the sacrum.

If we put some numbers into this example, we can get an idea of the forces required for just supporting the upper body in this position. If the person's total weight is 710 N (160 lb), then the upper body weight is

$$mg = 0.65 \times 710 \text{ N}$$

Now we set the magnitude of the CCW torques about the axis equal to the magnitude of the CW torques:

$$F_b \times 0.44 \text{ m} \times \sin 12° = mg \times 0.38 \text{ m}$$

Substituting and solving,

$$F_b = \frac{0.65 \times 710 \text{ N} \times 0.38 \text{ m}}{0.44 \text{ m} \times \sin 12°} = 1920 \text{ N}$$

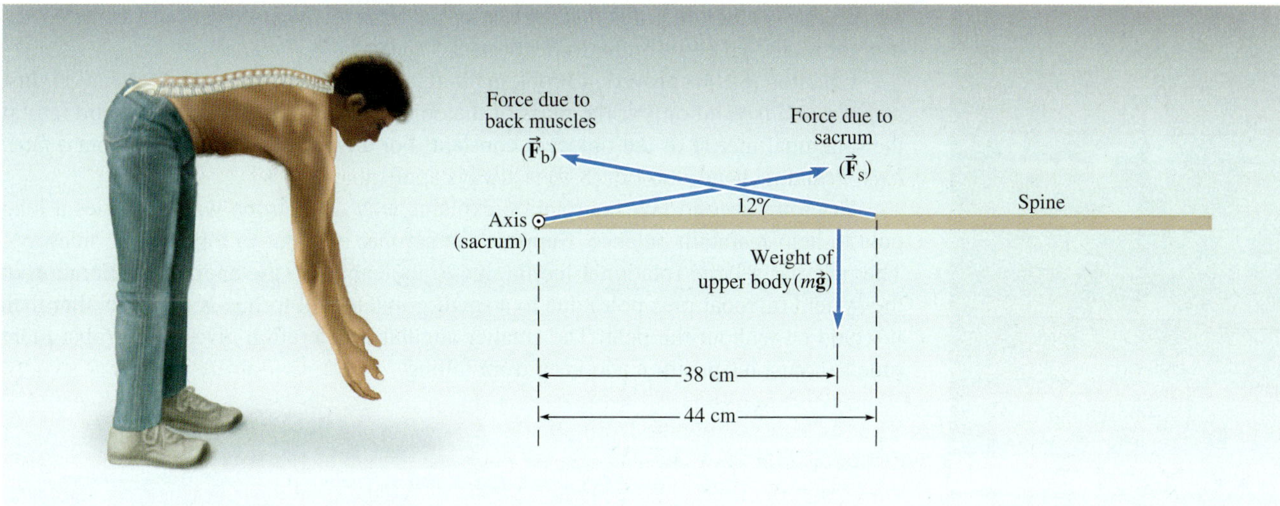

Figure 8.33 A simplified model of the human back when bent over.

The muscular force that compresses the spine is the horizontal component of $\vec{\mathbf{F}}_b$:

$$F_b \cos 12° = 1900 \text{ N}$$

or about 430 lb. This is over four times the weight of the upper body.

Now if the person tries to lift something with his arms in this position, the lever arm for the weight of the load is even longer than for the weight of the upper body. The back muscles must supply a much larger force. The spine is now compressed with a dangerously large force. A cushioning disk called the lumbosacral disk, at the bottom of the spine, separates the last vertebra from the sacrum. This disk can be ruptured or deformed, causing great pain when the back is misused in such a fashion.

If, instead of bending over, we bend our knees and lower our body, keeping it vertically aligned as much as possible while lifting a load, the centers of gravity of the body and load are positioned more closely in a line above the sacrum, as in Fig. 8.34. Then the lever arms of these forces with respect to an axis through the sacrum are relatively small, and the force on the lumbosacral disk is roughly equal to the upper body weight plus the weight being lifted.

Figure 8.34 A safer way to lift a heavy object.

8.6 ROTATIONAL FORM OF NEWTON'S SECOND LAW

The concepts of torque and rotational inertia can be used to formulate a "Newton's second law for rotation"—a law that fills the role of $\Sigma\vec{\mathbf{F}} = m\vec{\mathbf{a}}$ for rotation about a fixed axis:

Rotational form of Newton's second law

$$\sum \tau = I\alpha \qquad \text{(8-9)}$$

CONNECTION:

In Newton's second law for *rotation*, net torque takes the place of net force, rotational inertia takes the place of mass, and α takes the place of $\vec{\mathbf{a}}$.

When calculating the net torque $\Sigma\tau$, remember to assign the correct algebraic sign to each torque before adding them. The sum of the torques due to internal forces acting on a rigid object is always zero. Therefore, only *external* torques need be included in the net torque.

The angular acceleration of a rigid object is proportional to the net torque (more torque causes a larger α) and is inversely proportional to the rotational inertia (more inertia causes a smaller α). In rotational equilibrium, the angular acceleration must be

zero; Eq. (8-9) then requires that the net torque be zero. We used $\Sigma\tau = 0$ as the condition for rotational equilibrium in Sections 8.4 and 8.5.

Equation (8-9) is proved in Problem 60. It is subject to an important restriction. Just as $\Sigma\vec{\mathbf{F}} = m\vec{\mathbf{a}}$ is valid only if the mass of the object is constant, $\Sigma\tau = I\alpha$ is valid only if the rotational inertia of the object is constant. For a *rigid* object rotating about a *fixed axis*, I cannot change, so Eq. (8-9) is always applicable.

Newton's second law for rotation explains why a tightrope walker carries a long pole to help maintain balance. Suppose the acrobat is about to topple over sideways. The pole has a large rotational inertia due to its length, so the angular acceleration of the system (acrobat plus pole) due to a small gravitational torque is much smaller than it would be without the pole. The smaller angular acceleration gives the acrobat more time to adjust his position and keep from falling.

Example 8.11

The Grinding Wheel

A grinding wheel is a solid, uniform disk of mass 2.50 kg and radius 9.00 cm. Starting from rest, what constant torque must a motor supply so that the wheel attains a rotational speed of 126 rev/s in a time of 6.00 s?

Strategy Since the grinding wheel is a uniform disk, we can find its rotational inertia using Table 8.1. After converting the revolutions per second to radians per second, we can find the angular acceleration from the change in angular velocity over the given time interval. Once we have I and α, we can find the net torque from Newton's second law for rotation.

Solution The grinding wheel is a uniform disk, so its rotational inertia is

$$I = \tfrac{1}{2}mr^2$$

$$\tfrac{1}{2} \times 2.50\ \text{kg} \times (0.0900\ \text{m})^2 = 0.010125\ \text{kg}\cdot\text{m}^2$$

A single rotation of the wheel is equivalent to 2π radians, so

$$\omega = 126\,\frac{\text{rev}}{\text{s}} \times 2\pi\,\frac{\text{rad}}{\text{rev}}$$

The angular acceleration is

$$\alpha = \frac{\Delta\omega}{\Delta t}$$

Then the torque required is

$$\sum\tau = I\alpha = I\frac{\Delta\omega}{\Delta t}$$

$$= 0.010125\ \text{kg}\cdot\text{m}^2 \times \frac{126\ \text{rev/s} \times 2\pi\ \text{rad/rev}}{6.00\ \text{s}}$$

$$= 1.34\ \text{N}\cdot\text{m}$$

If there are no other torques on the wheel, the motor must supply a constant torque of 1.34 N·m.

Discussion We assumed that no other torques are exerted on the wheel. There is certain to be at least a small frictional torque on the wheel with a sign opposite to the sign of the motor's torque. Then the motor would have to supply a torque larger than 1.34 N·m. The *net* torque would still be 1.34 N·m.

Practice Problem 8.11 Another Approach

Verify the answer to Example 8.11 by: (a) finding the angular displacement of the wheel using equations for constant α; (b) finding the change in rotational kinetic energy of the wheel; and (c) finding the torque from $W = \tau\Delta\theta$.

8.7 THE MOTION OF ROLLING OBJECTS

A rolling object combines translational motion of the center of mass with rotation about an axis that passes through the center of mass (see Section 5.1). For an object that is rolling without slipping, $v_{\text{CM}} = \omega R$. As a result, there is a specific relationship between the rolling object's translational and rotational kinetic energies. The total kinetic energy of a rolling object is the sum of its translational and rotational kinetic energies.

A wheel with mass M and radius R has a rotational inertia that is some pure number times MR^2; it couldn't be anything else and still have the right units. We can write the rotational inertia about an axis through the CM as $I_{CM} = \beta MR^2$, where β is a pure number that measures how far from the axis of rotation the mass is distributed. Larger β means the mass is, on average, farther from the axis. From Table 8.1, a hoop has $\beta = 1$; a disk, $\beta = \frac{1}{2}$; and a solid sphere, $\beta = \frac{2}{5}$.

Using $I_{CM} = \beta MR^2$ and $v_{CM} = \omega R$, the rotational kinetic energy for a rolling object can be written

$$K_{rot} = \frac{1}{2}I_{CM}\omega^2 = \frac{1}{2}(\beta MR^2)\left(\frac{v_{CM}}{R}\right)^2 = \beta\left(\frac{1}{2}Mv_{CM}^2\right)$$

Since $\frac{1}{2}Mv_{CM}^2$ is the translational kinetic energy,

$$K_{rot} = \beta K_{tr} \tag{8-10}$$

This is convenient since β depends only on the shape, not on the mass or radius of the object. For a given shape rolling without slipping, the ratio of its rotational to translational kinetic energy is always the same (β).

The total kinetic energy can be written

$$K = K_{tr} + K_{rot} = \tfrac{1}{2}Mv_{CM}^2 + \tfrac{1}{2}I_{CM}\omega^2 \tag{8-11}$$

or in terms of β,

$$K = (1 + \beta)\,K_{tr} = (1 + \beta)\tfrac{1}{2}Mv_{CM}^2 \tag{8-12}$$

Thus, two objects of the same mass rolling at the same translational speed do *not* necessarily have the same kinetic energy. The object with the larger value of β has more rotational kinetic energy.

Conceptual Example 8.12

Hollow and Solid Rolling Balls

Starting from rest, two balls roll down a hill as in Fig. 8.35. One is solid, the other hollow. Which one is moving faster when it reaches the bottom of the hill? (connect tutorial: rolling)

Strategy and Solution Energy conservation is the best way to approach this problem. As a ball rolls down the hill, its gravitational potential energy decreases as its kinetic energy increases by the same amount. The total kinetic energy is the sum of the translational and rotational contributions.

We do not know the mass or the radius of either ball, and we cannot assume they are the same. Since both kinetic and potential energies are proportional to mass, mass does not affect the final speed. Also, the total kinetic energy does not depend on the radius of the ball [see Eq. (8-12)]. The final speeds of the two balls differ because different *fractions* of their total kinetic energies are translational.

One ball is a solid sphere and the other is approximately a spherical shell. The mass of a spherical shell is all concentrated on the surface of a sphere, while a solid sphere has its mass distributed throughout the sphere's volume. Therefore, the shell has a larger β than the solid sphere. When the shell rolls, it converts a bigger fraction of the lost potential energy into rotational kinetic energy; therefore, a smaller fraction becomes translational kinetic energy. The final speed of the solid sphere is larger since it puts a larger fraction of its kinetic energy into translational motion.

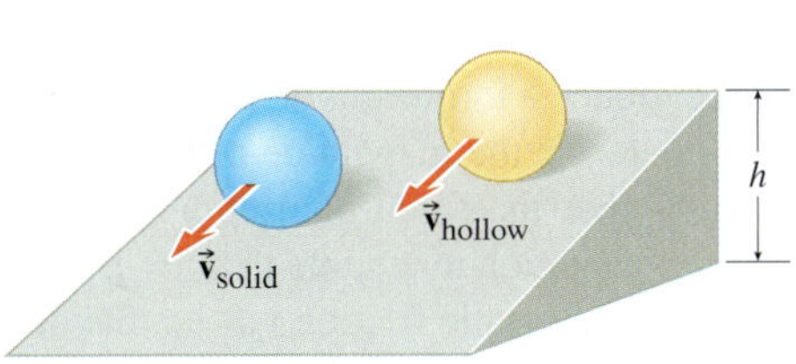

Figure 8.35
Rolling balls.

Discussion We can make this conceptual question into a quantitative one: what is the ratio of the speeds of the two balls at the bottom of the hill?

Let the height of the hill be h. Then for a ball of mass M, the loss of gravitational potential energy is Mgh. This

continued on next page

Conceptual Example 8.12 continued

amount of gravitational potential energy is converted into translational and rotational kinetic energy:

$$Mgh = K_{tr} + K_{rot} = (1 + \beta)K_{tr} = (1 + \beta)\frac{Mv_{CM}^2}{2}$$

Mass cancels out, as expected. We can solve for the final speed in terms of g, h, and β. The final speed is independent of the ball's mass and radius.

$$v_{CM} = \sqrt{\frac{2gh}{1 + \beta}}$$

The ratio of the final speeds for two balls rolling down the same hill is, therefore,

$$\frac{v_1}{v_2} = \sqrt{\frac{1 + \beta_2}{1 + \beta_1}}$$

To evaluate the ratio, we look up the rotational inertias in Table 8.1. The solid sphere has $\beta = \frac{2}{5}$ and the spherical shell has $\beta = \frac{2}{3}$. Then

$$\frac{v_{solid}}{v_{hollow}} = \sqrt{\frac{1 + \frac{2}{3}}{1 + \frac{2}{5}}} \approx 1.091$$

The solid ball's final speed is, therefore, 9.1% faster than that of the hollow ball. This ratio depends neither on the masses of the balls, the radii of the balls, the height of the hill, nor the slope of the hill.

Practice Problem 8.12 Fraction of Kinetic Energy That Is Rotational Energy

What fraction of a rolling ball's kinetic energy is rotational kinetic energy? Answer both for a solid ball and a hollow one.

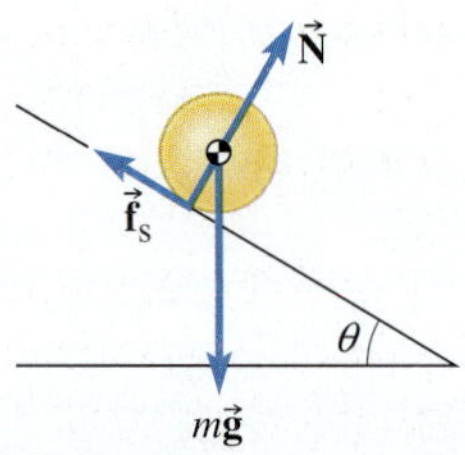

Figure 8.36 Forces acting on a ball rolling downhill.

CHECKPOINT 8.7

Give an example of how a marble can move so that (a) $K_{tr} > 0$ and $K_{rot} = 0$; (b) $K_{tr} = 0$ and $K_{rot} > 0$; (c) $K_{rot} = \frac{2}{5}K_{tr}$.

Acceleration of Rolling Objects What is the acceleration of a ball rolling down an incline? Figure 8.36 shows the forces acting on the ball. Static friction is the force that makes the ball rotate; if there were no friction, instead of rolling, the ball would just *slide* down the incline. This is true because friction is the only force acting that yields a nonzero torque about the rotation axis through the ball's center of mass. Gravity gives zero torque because it acts at the axis, so the lever arm is zero. The normal force points directly at the axis, so its lever arm is also zero.

The frictional force $\vec{\mathbf{f}}$ provides a torque

$$\tau = rf$$

where r is the ball's radius. An analysis of the forces and torques combined with Newton's second law in both forms enables us to calculate the acceleration of the ball in Example 8.13.

Example 8.13

Acceleration of a Rolling Ball

Calculate the acceleration of a solid ball rolling down a slope inclined at an angle θ to the horizontal (Fig. 8.37a).

Strategy The net torque is related to the angular acceleration by $\Sigma\tau = I\alpha$, Newton's second law for rotation. Similarly, the net force acting on the ball gives the acceleration of the center of mass: $\Sigma\vec{\mathbf{F}} = m\vec{\mathbf{a}}_{CM}$. The axis of rotation is through the ball's CM. As already discussed, neither gravity nor the normal force produce a torque about this axis; the net torque is $\Sigma\tau = rf$, where f is the magnitude of the frictional force. One problem is that the force of friction is unknown. We must resist the temptation to assume that $f = \mu_s N$; there is no reason to assume that static friction has its maximum possible magnitude. We do know that the two accelerations, translational and rotational, are related. We know that v_{CM} and ω are proportional since r is constant. To stay proportional,

continued on next page

Example 8.13 continued

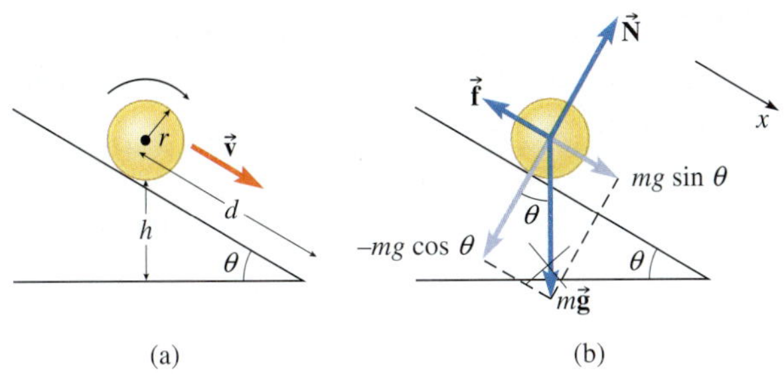

Figure 8.37
(a) A ball rolling downhill. (b) FBD for the ball, with the gravitational force resolved into components perpendicular and parallel to the incline.

they must change in lock step; their rates of change, a_{CM} and α, are proportional to each other by the same factor of r. Thus, $a_{CM} = \alpha r$. This connection should enable us to eliminate f and solve for the acceleration. Since the speed of a ball after rolling a certain distance was found to be independent of the mass and radius of the ball in Example 8.12, we expect the same to be true of the acceleration.

Solution Since the net torque is

$$\sum \tau = rf$$

the angular acceleration is

$$\alpha = \frac{\sum \tau}{I} = \frac{rf}{I} \qquad (1)$$

where I is the ball's rotational inertia about its CM.

Figure 8.37b shows the forces along the incline acting on the ball. The acceleration of the CM is found from Newton's second law. The component of the net force acting along the incline (in the direction of the acceleration) is

$$\sum F_x = mg \sin \theta - f = ma_{CM} \qquad (2)$$

Because the ball is rolling without slipping, the acceleration of the CM and the angular acceleration are related by

$$a_{CM} = \alpha r$$

Now we try to eliminate the unknown frictional force f from the previous equations. Solving Eq. (1) for f gives

$$f = \frac{I\alpha}{r}$$

Substituting this into Eq. (2), we get

$$mg \sin \theta - \frac{I\alpha}{r} = ma_{CM}$$

Now to eliminate α, we can substitute $\alpha = a_{CM}/r$:

$$mg \sin \theta - \frac{Ia_{CM}}{r^2} = ma_{CM}$$

Solving for a_{CM},

$$a_{CM} = \frac{g \sin \theta}{1 + I/(mr^2)}$$

For a solid sphere, $I = \frac{2}{5}mr^2$, so

$$a_{CM} = \frac{g \sin \theta}{1 + \frac{2}{5}} = \frac{5}{7} g \sin \theta$$

Discussion The acceleration of an object *sliding* down an incline without friction is $a = g \sin \theta$. The acceleration of the rolling ball is smaller than $g \sin \theta$ due to the frictional force directed up the incline.

We can check the answer using the result of Example 8.12. The ball's acceleration is constant. If the ball starts from rest as in Fig. 8.37a, after it has rolled a distance d, its speed v is

$$v = \sqrt{2ad} = \sqrt{2\left(\frac{g \sin \theta}{1 + \beta}\right)d}$$

where $\beta = \frac{2}{5}$. The vertical drop during this time is $h = d \sin \theta$, so

$$v = \sqrt{\frac{2gh}{1 + \beta}}$$

Practice Problem 8.13 Acceleration of a Hollow Cylinder

Calculate the acceleration of a thin hollow cylindrical shell rolling down a slope inclined at an angle θ to the horizontal.

8.8 ANGULAR MOMENTUM

Newton's second law for translational motion can be written in two ways:

$$\sum \vec{\mathbf{F}} = \lim_{\Delta t \to 0} \frac{\Delta \vec{\mathbf{p}}}{\Delta t} \text{ (general form)} \quad \text{or} \quad \sum \vec{\mathbf{F}} = m\vec{\mathbf{a}} \text{ (constant mass)}$$

In Eq. (8-9) we wrote Newton's second law for rotation as $\sum \tau = I\alpha$, which applies only when I is constant—that is, for a rigid object rotating about a fixed axis. A more general form of Newton's second law for rotation uses the concept of **angular momentum** (symbol L).

CONNECTION:

Note the analogy with

$$\sum \vec{\mathbf{F}} = \lim_{\Delta t \to 0} \frac{\Delta \vec{\mathbf{p}}}{\Delta t}$$

The net external torque acting on a system is equal to the rate of change of the angular momentum of the system.

$$\sum \tau = \lim_{\Delta t \to 0} \frac{\Delta L}{\Delta t} \quad \text{(8-13)}$$

The angular momentum of a rigid object rotating about a fixed axis is the rotational inertia times the angular velocity, which is analogous to the definition of linear momentum (mass times velocity):

CONNECTION:

Note the analogy with $\vec{\mathbf{p}} = m\vec{\mathbf{v}}$. See the Master the Concepts section at the end of the chapter for a complete table of these analogies.

Angular momentum

$$L = I\omega \quad \text{(8-14)}$$

(rigid object, fixed axis)

Either Eq. (8-13) or Eq. (8-14) can be used to show that the SI units of angular momentum are kg·m^2/s.

For a rigid object rotating around a fixed axis, angular momentum doesn't tell us anything new. The rotational inertia is constant for such an object since the distance r_n between every point on the object and the axis stays the same. Then any change in angular momentum must be due to a change in angular velocity ω:

$$\sum \tau = \lim_{\Delta t \to 0} \frac{\Delta L}{\Delta t} = \lim_{\Delta t \to 0} \frac{I \Delta \omega}{\Delta t} = I \lim_{\Delta t \to 0} \frac{\Delta \omega}{\Delta t} = I\alpha$$

Conservation of Angular Momentum However, Eq. (8-13) is *not* restricted to rigid objects or to fixed rotation axes. In particular, if the net external torque acting on a system is zero, then the angular momentum of the system cannot change. This is the **law of conservation of angular momentum**:

Conservation of angular momentum can be applied to any system if the net external torque on the system is zero (or negligibly small).

Conservation of angular momentum

$$\text{If} \quad \sum \tau = 0, \qquad L_i = L_f \quad \text{(8-15)}$$

Here L_i and L_f represent the angular momentum of the system at two different times. Conservation of angular momentum is one of the most basic and fundamental laws of physics, along with the two other conservation laws we have studied so far (energy and linear momentum). For an isolated system, the total energy, total linear momentum, and total angular momentum of the system are each conserved. None of these quantities can change unless some external agent causes the change.

CONNECTION:

Another conservation law

With conservation of energy, we add up the amounts of the different forms of energy (such as kinetic energy and gravitational potential energy) to find the *total* energy. The conservation law refers to the total energy. By contrast, linear momentum and angular momentum *cannot* be added to find the "total momentum." They are entirely different quantities, not two forms of the same quantity. They even have different dimensions, so it would be impossible to add them. Conservation of linear momentum and conservation of angular momentum are *separate* laws of physics.

Application of Angular Momentum: Figure Skaters In this section, we restrict our consideration to cases where the axis of rotation is fixed but where the rotational inertia is not necessarily constant. One familiar example of a changing rotational inertia occurs when a figure skater spins (Fig. 8.38). To start the spin, the skater glides along with her arms outstretched and then begins to rotate her body about a vertical axis by

Figure 8.38 Figure skater Lucinda Ruh at the (a) beginning and (b) end of a spin. Her angular velocity is much higher in (b) than in (a).

pushing against the ice with a skate. The push of the ice against the skate provides the external torque that gives the skater her initial angular momentum. Initially the skater's arms and the leg not in contact with the ice are extended away from her body. The mass of the arms and leg when extended contribute more to her rotational inertia than they do when held close to the body. As the skater spins, she pulls her arms and leg close and straightens her body to decrease her rotational inertia. As she does, her angular velocity increases dramatically in such a way that her angular momentum stays the same.

CHECKPOINT 8.8

If the skater then extends her arms and leg back to their initial configuration, does her angular velocity decrease back to its initial value, ignoring friction?

Applications of Angular Momentum: Hurricanes and Pulsars Many natural phenomena can be understood in terms of angular momentum. In a hurricane, circulating air is sucked inward by a low-pressure region at the center of the storm (the *eye*). As the air moves closer and closer to the axis of rotation, it circulates faster and faster. An even more dramatic example is the formation of a pulsar. Under certain conditions, a star can implode under its own gravity, forming a neutron star (a collection of tightly packed neutrons). If the Sun were to collapse into a neutron star, its radius would be only about 13 km. If a star is rotating before its collapse, then as its rotational inertia decreases dramatically, its angular velocity must increase to keep its angular momentum constant. Such rapidly rotating neutron stars are called pulsars because they emit regular pulses of x-rays, at the same frequency as their rotation, that can be detected when they reach Earth. Some pulsars rotate in only a few thousandths of a second per revolution.

Example 8.14

Mouse on a Wheel

A 0.10 kg mouse is perched at point B on the rim of a 2.00 kg wagon wheel that rotates freely in a horizontal plane at 1.00 rev/s (Fig. 8.39). The mouse crawls to point A at the center. Assume the mass of the wheel is concentrated at the rim. What is the frequency of rotation in rev/s when the mouse arrives at point A?

continued on next page

Example 8.14 continued

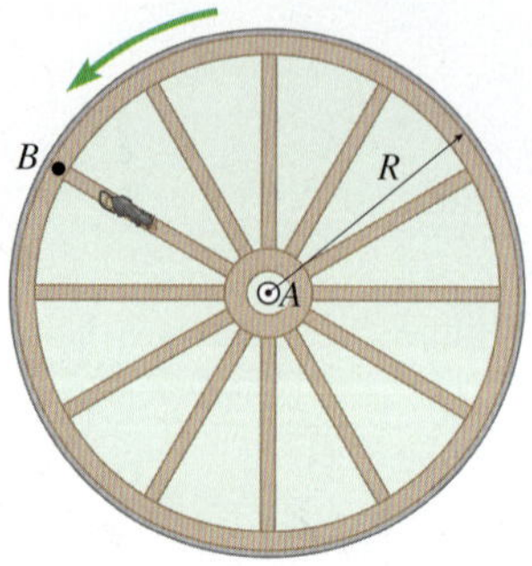

Figure 8.39
Mouse on a rotating wheel.

Strategy Assuming that frictional torques are negligibly small, there is no external torque acting on the mouse/wheel system. Then the angular momentum of the mouse/wheel system must be conserved; it takes an external torque to change angular momentum. The mouse and wheel exert torques on one another, but these *internal* torques only transfer some angular momentum between the wheel and the mouse without changing the total angular momentum. We can think of the system as initially being a rigid object with rotational inertia I_i. When the mouse reaches the center, we think of the system as a rigid object with a different rotational inertia I_f. The mouse changes the rotational inertia of the mouse/wheel system by moving from the outer rim, where its mass makes the maximum possible contribution to the rotational inertia, to the rotation axis, where its mass makes no contribution to the rotational inertia.

Solution Initially, all of the mass of the system is at a distance R from the rotation axis, where R is the radius of the wheel. Therefore,

$$I_i = (M + m)R^2$$

where M is the mass of the wheel and m is the mass of the mouse. After the mouse moves to the center of the wheel, its mass contributes nothing to the rotational inertia of the system:

$$I_f = MR^2$$

From conservation of angular momentum,

$$I_i\,\omega_i = I_f\,\omega_f$$

Substituting the rotational inertias and $\omega = 2\pi f$,

$$(M + m)R^2 \times 2\pi f_i = MR^2 \times 2\pi f_f$$

Factors of $2\pi R^2$ cancel from each side, leaving

$$(M + m)f_i = Mf_f$$

Solving for f_f,

$$f_f = \frac{M + m}{M} f_i = \frac{2.10 \text{ kg}}{2.00 \text{ kg}} (1.00 \text{ rev/s}) = 1.05 \text{ rev/s}$$

Discussion Conservation laws are powerful tools. We do not need to know the details of what happens as the mouse crawls along the spoke from the outer edge of the wheel; we need only look at the initial and final conditions.

A common mistake in this sort of problem is to assume that the initial rotational kinetic energy is equal to the final rotational kinetic energy. This is not true because the mouse crawling in toward the center expends energy to do so. In other words, the mouse converts some internal energy into rotational kinetic energy.

Practice Problem 8.14 Change in Rotational Kinetic Energy

What is the percentage change in the rotational kinetic energy of the mouse/wheel system?

Angular Momentum in Planetary Orbits

Conservation of angular momentum applies to planets orbiting the Sun in elliptical orbits. Kepler's second law says that the orbital speed varies in such a way that the line from the Sun to the planet sweeps out area at a constant rate (Fig. 8.40a). In Problem 117, you can show that Kepler's second law is a direct result of conservation

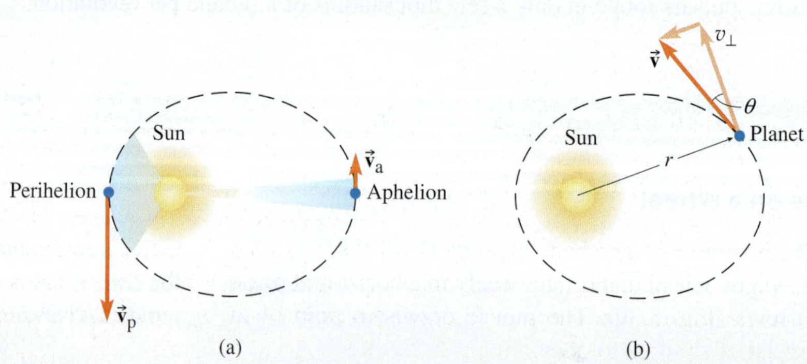

Figure 8.40 The planet's speed varies such that it sweeps out equal areas in equal time intervals. The eccentricity of the planetary orbit is exaggerated for clarity.

of angular momentum, where the angular momentum of the planet is calculated using an axis of rotation perpendicular to the plane of the orbit and passing through the Sun. When the planet is closer to the Sun, it moves faster; when it is farther away, it moves more slowly. Conservation of angular momentum can be used to relate the orbital speeds and radii at two different points in the orbit. The same applies to satellites and moons orbiting planets.

Example 8.15

Earth's Orbital Speed

At perihelion (closest approach to the Sun), Earth is 1.47×10^8 km from the Sun and its orbital speed is 30.3 km/s. What is Earth's orbital speed at aphelion (greatest distance from the Sun), when it is 1.52×10^8 km from the Sun? Note that at these two points Earth's velocity is perpendicular to a radial line from the Sun (see Fig. 8.40a).

Strategy We take the axis of rotation through the Sun. Then the gravitational force on Earth points directly toward the axis; with zero lever arm, the torque is zero. With no other external forces acting on the Earth, the net external torque is zero. Earth's angular momentum about the rotation axis through the Sun must therefore be conserved. To find Earth's rotational inertia, we treat it as a point particle since its radius is much less than its distance from the axis of rotation.

Solution The rotational inertia of the Earth is

$$I = mr^2$$

where m is Earth's mass and r is its distance from the Sun. The angular velocity is

$$\omega = \frac{v_\perp}{r}$$

where $v_\perp$ is the component of the velocity perpendicular to a radial line from the Sun. At the two points under consideration, $v_\perp = v$. As the distance from the Sun r varies, its speed v must vary to conserve angular momentum:

$$I_i\omega_i = I_f\omega_f$$

By substitution,

$$mr_i^2 \times \frac{v_i}{r_i} = mr_f^2 \times \frac{v_f}{r_f}$$

or

$$r_i v_i = r_f v_f \qquad (1)$$

Solving for v_f,

$$v_f = \frac{r_i}{r_f} v_i = \frac{1.47 \times 10^8 \text{ km}}{1.52 \times 10^8 \text{ km}} \times 30.3 \text{ km/s} = 29.3 \text{ km/s}$$

Discussion Earth moves slower at a point farther from the Sun. This is what we expect from energy conservation. The potential energy is greater at aphelion than at perihelion. Since the mechanical energy of the orbit is constant, the kinetic energy must be smaller at aphelion.

Equation (1) implies that the orbital speed and orbital radius are inversely proportional, but strictly speaking this equation only applies to the perihelion and aphelion. At a general point in the orbit, the *perpendicular component* $v_\perp$ is inversely proportional to r (see Fig. 8.40b). The orbits of Earth and most of the other planets are nearly circular so that $\theta \approx 0°$ and $v_\perp \approx v$.

Practice Problem 8.15 Puck on a String

A puck on a frictionless, horizontal air table is attached to a string that passes down through a hole in the table. Initially the puck moves at 12 cm/s in a circle of radius 24 cm. If the string is pulled through the hole, reducing the radius of the puck's circular motion to 18 cm, what is the new speed of the puck?

8.9 THE VECTOR NATURE OF ANGULAR MOMENTUM

Until now we have treated torque and angular momentum as scalar quantities. Such a treatment is adequate in the cases we have considered so far. However, the law of conservation of angular momentum applies to *all* systems, including rotating objects whose axis of rotation changes direction. Torque and angular momentum are actually vector quantities. Angular momentum is conserved in *both magnitude and direction* in the absence of external torques.

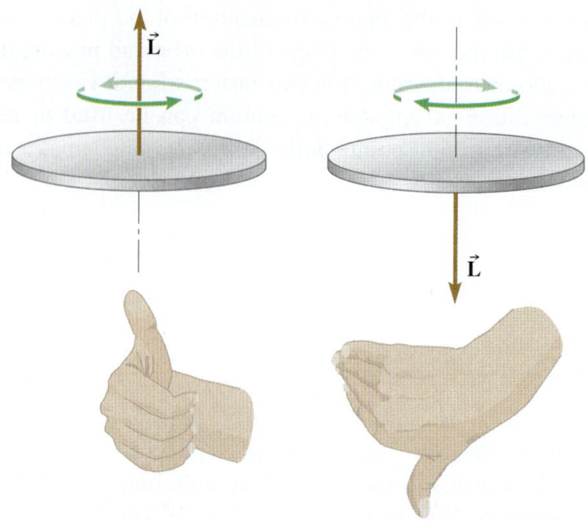

Figure 8.41 Right-hand rule for finding the direction of the angular momentum of a spinning disk.

An important special case is that of a symmetrical object rotating about an axis of symmetry, such as the spinning disk in Fig. 8.41. The magnitude of the angular momentum of such an object is $L = I\omega$. The direction of the angular momentum vector points along the axis of rotation. To choose between the two directions along the axis, a **right-hand rule** is used. Align your right hand so that, as you curl your fingers in toward your palm, your fingertips follow the object's rotation; then your thumb points in the direction of $\vec{\mathbf{L}}$.

For rotation about a *fixed* axis, the net torque is also along the axis of rotation, in the direction of the *change* in angular momentum it causes. The sign convention we have used up to now for angular momentum and torque gives the sign of the *z-component of the vector quantity*, where the z-axis points toward the viewer. For example, imagine viewing the disk of Figure 8.41 from above. If we choose the z-axis toward the viewer (up), then $L_z > 0$ for counterclockwise rotation and $L_z < 0$ for clockwise rotation.

Application of Angular Momentum: The Gyroscope A disk with a large rotational inertia can be used as a *gyroscope*. When the gyroscope spins at a large angular velocity, it has a large angular momentum. It is then difficult to change the orientation of the gyroscope's rotation axis because to do so requires changing its angular momentum. To change the direction of a large angular momentum requires a correspondingly large torque. Thus, a gyroscope can be used to maintain stability. Gyroscopes are used in guidance systems in airplanes, submarines, and space vehicles to maintain a constant direction in space.

Application of Angular Momentum: Rifle Bullets, Spinning Tops, and Earth's Rotation The same principle explains the great stability of rifle bullets and spinning tops. A rifle bullet is made to spin as it passes through the rifle's barrel. The spinning bullet then keeps its correct orientation—nose first—as it travels through the air. Otherwise, a small torque due to air resistance could make the bullet turn around randomly, greatly increasing air resistance and undermining accuracy. A properly thrown football is made to spin for the same reasons. A spinning top can stay balanced for a long time, while the same top soon falls over if it is not spinning.

The Earth's rotation gives it a large angular momentum. As the Earth orbits the Sun, the axis of rotation stays in a fixed direction in space. The axis points nearly at Polaris (the North Star), so even as the Earth rotates around its axis, Polaris maintains its position in the northern sky. The fixed direction of the rotation axis gives us the regular progression of the seasons (Fig. 8.42).

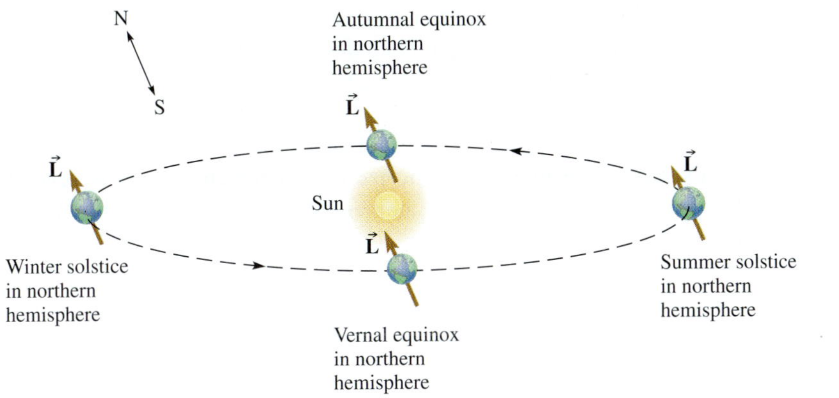

Figure 8.42 Spinning like a top, the Earth maintains the direction of its angular momentum due to rotation as it revolves around the Sun (not to scale).

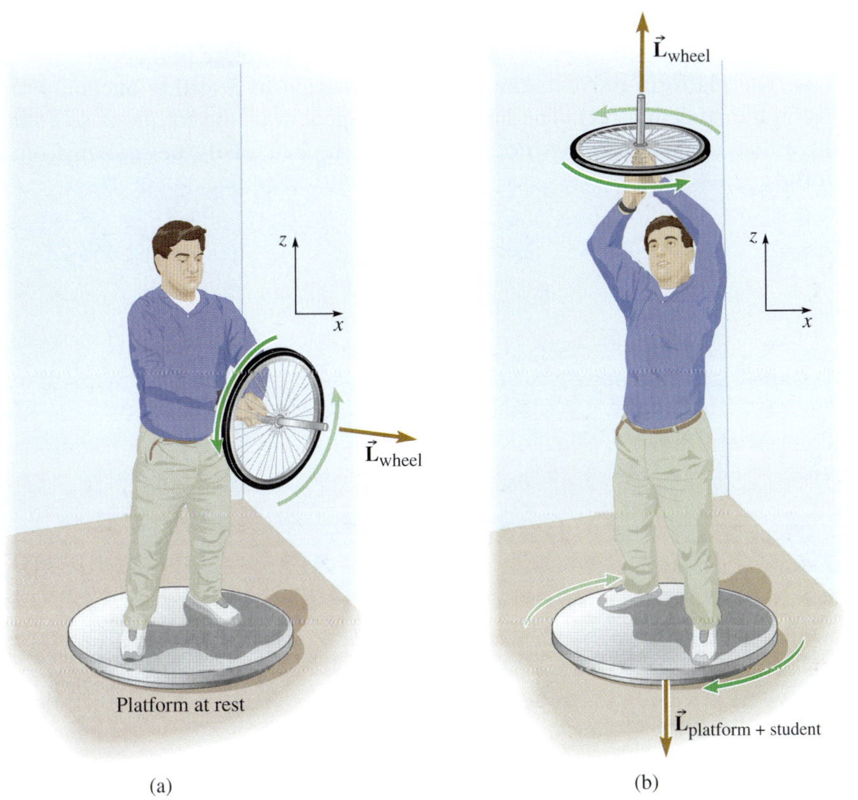

Figure 8.43 A demonstration of angular momentum conservation.

A Classic Demonstration of Angular Momentum

A demonstration often done in physics classes is for a student to hold a spinning bicycle wheel while standing on a platform that is free to rotate. The wheel's rotation axis is initially horizontal (Fig. 8.43a). Then the student repositions the wheel so that its axis of rotation is vertical (Fig. 8.43b). As he repositions the wheel, the platform begins to rotate opposite to the wheel's rotation. If we assume *no* friction acts to resist rotation of the platform, then the platform continues to rotate as long as the wheel is held with its axis vertical. If the student returns the wheel to its original orientation, the rotation of the platform stops.

The platform is free to rotate about a vertical axis. As a result, once the student steps onto the platform, *the vertical component* L_z of the angular momentum of the system (student + platform + wheel) is conserved. The horizontal components of $\vec{\mathbf{L}}$

are *not* conserved. The platform is not free to rotate about any horizontal axis since the floor can exert external torques to keep it from doing so. In vector language, we would say that only the vertical component of the external torque is zero, so only the vertical component of angular momentum is conserved.

Initially $L_z = 0$ since the student and the platform have zero angular momentum and the wheel's angular momentum is horizontal. When the wheel is repositioned so that it spins with an upward angular momentum ($L_z > 0$), the rest of the system (the student and the platform) must acquire an equal magnitude of downward angular momentum ($L_z < 0$) so that the vertical component of the total angular momentum is still zero. Thus, the platform and student rotate in the opposite sense from the rotation of the wheel. Since the platform and student have more rotational inertia than the wheel, they do not spin as fast as the wheel, but their vertical angular momentum is just as large.

The student and the wheel apply torques to each other to transfer angular momentum from one part of the system to the other. These torques are equal and opposite and they have both vertical and horizontal components. As the student lifts the wheel, he feels a strange twisting force that tends to rotate him about a horizontal axis. The platform prevents the horizontal rotation by exerting unequal normal forces on the student's feet. The horizontal component of the torque is so counterintuitive that, if the student is not expecting it, he can easily be thrown from the platform!

Master the Concepts

- The rotational kinetic energy of a rigid object with rotational inertia I and angular velocity ω is

$$K_{\text{rot}} = \tfrac{1}{2}I\omega^2 \quad (8\text{-}1)$$

In this expression, ω must be measured in *radians* per unit time.

- Rotational inertia is a measure of how difficult it is to change an object's angular velocity. It is defined as:

$$I = \sum_{n=1}^{N} m_n r_n^2 \quad (8\text{-}2)$$

where r_n is the perpendicular distance between a particle of mass m_n and the rotation axis. The rotational inertia depends on the location of the rotation axis.

- Torque measures the effectiveness of a force for twisting or turning an object. It can be calculated in two equivalent ways: either as the product of the perpendicular component of the force with the shortest distance between the rotation axis and the point of application of the force

$$\tau = \pm rF_\perp \quad (8\text{-}3)$$

or as the product of the magnitude of the force with its lever arm (the perpendicular distance between the line of action of the force and the axis of rotation)

$$\tau = \pm r_\perp F \quad (8\text{-}4)$$

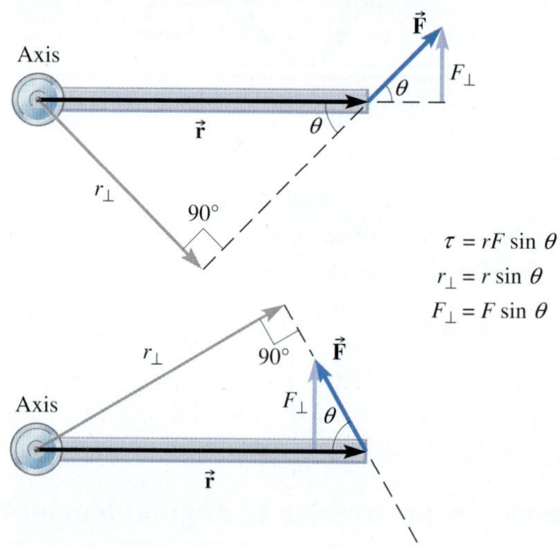

- A force whose perpendicular component tends to cause rotation in the CCW direction gives rise to a positive torque; a force whose perpendicular component tends to cause rotation in the CW direction gives rise to a negative torque.
- The work done by a constant torque is the product of the torque and the angular displacement:

$$W = \tau\,\Delta\theta \quad (\Delta\theta \text{ in radians}) \quad (8\text{-}6)$$

continued on next page

Master the Concepts continued

- The conditions for translational and rotational equilibrium are

$$\sum \vec{\mathbf{F}} = 0 \quad \text{and} \quad \sum \tau = 0 \qquad (8\text{-}8)$$

The rotation axis can be chosen *arbitrarily* when calculating torques in equilibrium problems. Generally, the best place to choose the axis is at the point of application of an unknown force so that the unknown force does not appear in the torque equation.

- Newton's second law for rotation is

$$\sum \tau = I\alpha \qquad (8\text{-}9)$$

where radian measure must be used for α. A more general form is

$$\sum \tau = \lim_{\Delta t \to 0} \frac{\Delta L}{\Delta t} \qquad (8\text{-}13)$$

where L is the angular momentum of the system.

- The total kinetic energy of an object that is rolling without slipping is the sum of the rotational kinetic energy about an axis through the CM and the translational kinetic energy:

$$K = \tfrac{1}{2}Mv_{\text{CM}}^2 + \tfrac{1}{2}I_{\text{CM}}\omega^2 \qquad (8\text{-}11)$$

- The angular momentum of a rigid object rotating about a fixed axis is the rotational inertia times the angular velocity:

$$L = I\omega \qquad (8\text{-}14)$$

- The law of conservation of angular momentum: if the net external torque acting on a system is zero, then the angular momentum of the system cannot change.

$$\text{If} \quad \sum \tau = 0, \qquad L_{\text{i}} = L_{\text{f}} \qquad (8\text{-}15)$$

- This table summarizes the analogous quantities and equations in translational and rotational motion.

Translation	**Rotation**
m	I
$\vec{\mathbf{F}}$	τ
$\vec{\mathbf{a}}$	α
$\sum \vec{\mathbf{F}} = m\vec{\mathbf{a}}$	$\sum \tau = I\alpha$
Δx	$\Delta\theta$
$W = F_x \Delta x$	$W = \tau\Delta\theta$
$\vec{\mathbf{v}}$	ω
$K = \frac{1}{2}mv^2$	$K = \frac{1}{2}I\omega^2$
$\vec{\mathbf{p}} = m\vec{\mathbf{v}}$	$L = I\omega$
$\sum \vec{\mathbf{F}} = \lim_{\Delta t \to 0} \frac{\Delta \vec{\mathbf{p}}}{\Delta t}$	$\sum \tau = \lim_{\Delta t \to 0} \frac{\Delta L}{\Delta t}$
If $\sum \vec{\mathbf{F}} = 0$, $\vec{\mathbf{p}}$ is conserved	If $\sum \tau = 0$, L is conserved

Conceptual Questions

1. In Fig. 8.2b, where should the doorknob be located to make the door easier to open?
2. Explain why it is easier to drive a wood screw using a screwdriver with a large-diameter handle rather than one with a thin handle.
3. Why is it easier to push open a swinging door from near the edge away from the hinges rather than in the middle of the door?
4. A book measures 3 cm by 16 cm by 24 cm. About which of the axes shown in the figure is its rotational inertia smallest?

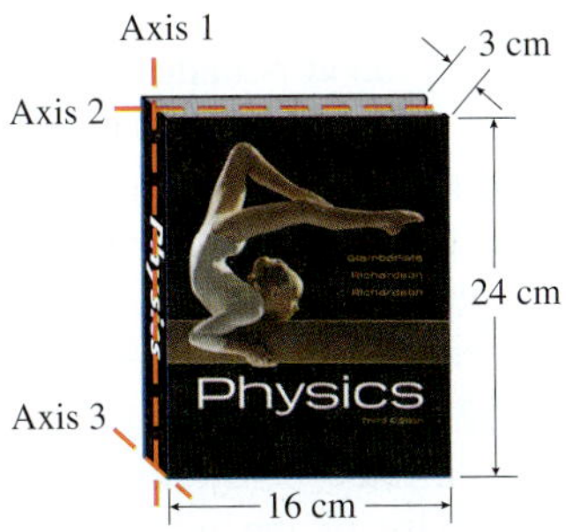

Conceptual Question 4

5. An object in equilibrium has only two forces acting on it. The forces must be equal in magnitude and opposite in direction in order to give a translational net force of zero. What else must be true of the two forces for the object to be in equilibrium? [*Hint*: Consider the lines of action of the forces.]
6. Why do many helicopters have a small propeller attached to the tail that rotates in a vertical plane? Why is this attached at the tail rather than somewhere else? [*Hint*: Most of the helicopter's mass is forward, in the cab.]
7. In the "Pinewood Derby," Cub Scouts construct cars and then race them down an incline. Some say that, everything else being equal (friction, drag coefficient, same wheels, etc.), a heavier car will win; others maintain that the weight of the car does not matter. Who is right? Explain. [*Hint*: Think about the fraction of the car's kinetic energy that is rotational.]
8. A large barrel lies on its side. In order to roll it across the floor, you apply a horizontal force, as shown in the figure. If the applied force points toward the axis of rotation,

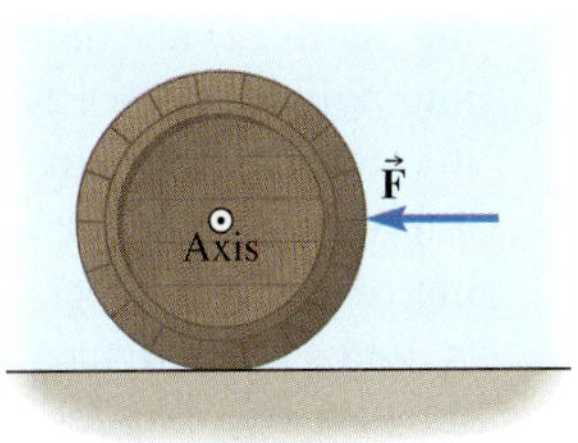

which runs down the center of the barrel through the center of mass, it produces zero torque about that axis. How then can this applied force make the barrel start to roll?

9. Animals that can run fast always have thin legs. Their leg muscles are concentrated close to the hip joint; only tendons extend into the lower leg. Using the concept of rotational inertia, explain how this helps them run fast.

10. Part (a) of the figure shows a simplified model of how the triceps muscle connects to the forearm. As the angle θ is changed, the tendon wraps around a nearly circular arc. Explain how this is much more effective than if the tendon is connected as in part (b) of the figure. [*Hint*: Look at the lever arm as θ changes.]

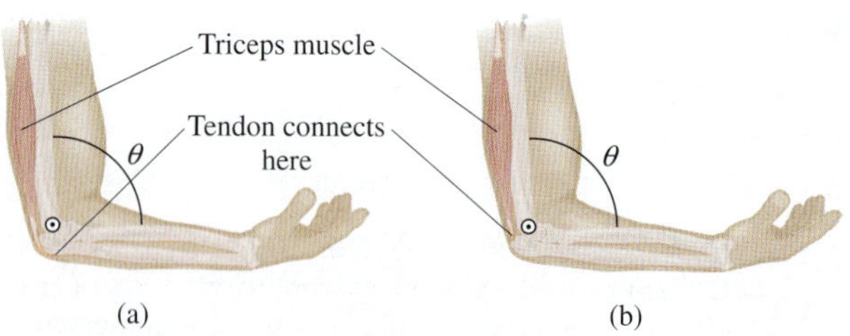

Question 10

11. Part (a) of the figure shows a simplified model of how the biceps muscle enables the forearm to support a load. What are the advantages of this arrangement as opposed to the alternative shown in part (b), where the flexor muscle is in the forearm instead of in the upper arm? Are the two equally effective when the forearm is horizontal? What about for other angles between the upper arm and the forearm? Consider also the rotational inertia of the forearm about the elbow and of the entire arm about the shoulder.

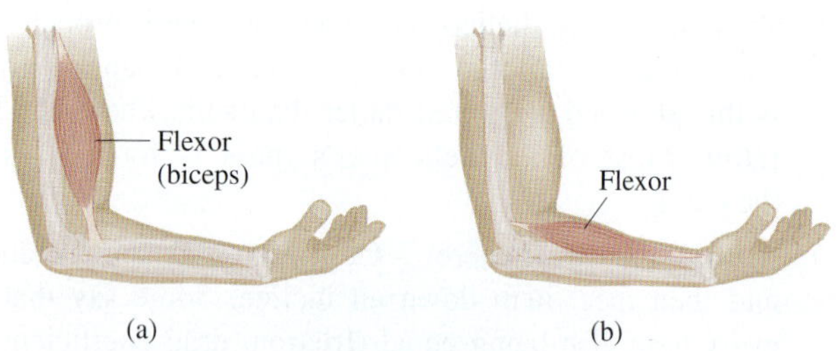

Question 11

12. In Section 8.6, it was asserted that the sum of all the internal torques (i.e., the torques due to internal forces) acting on a rigid object is zero. The figure shows two particles in a rigid object. The particles exert forces $\vec{\mathbf{F}}_{12}$ and $\vec{\mathbf{F}}_{21}$ on each other. These forces are directed along a line that joins the two particles. Explain why the torques due to these two forces must be equal and opposite even though the forces are applied at different points (and, therefore, possibly different distances from the axis).

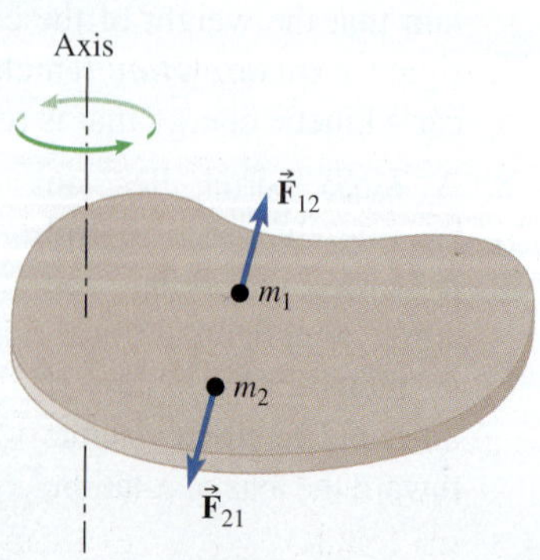

13. A playground merry-go-round (see Fig. 8.5) spins with negligible friction. A child moves from the center out to the rim of the merry-go-round platform. Let the system be the merry-go-round plus the child. Which of these quantities change: angular velocity of the system, rotational kinetic energy of the system, angular momentum of the system? Explain your answer.

14. The figure shows a balancing toy with weights extending on either side. The toy is extremely stable. It can be pushed quite far off center one way or the other but it does not fall over. Explain why it is so stable.

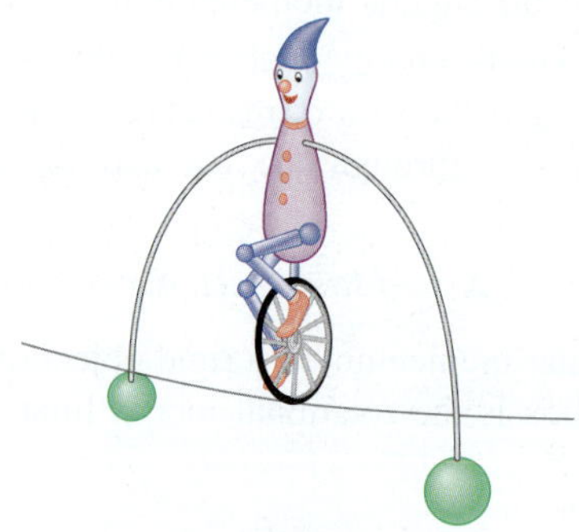

Question 14

15. Explain why the posture taken by defensive football linemen makes them more difficult to push out of the way. Consider both the height of the center of gravity and the size of the support base (the area on the ground bounded by the hands and feet touching the ground). In order to knock a person over, what has to happen to the center of gravity? Which do you think needs a more complex neurological system for maintaining balance: four legged animals or humans?

16. The center of gravity of the upper body of a bird is located below the hips; in a human, the center of gravity of the upper body is located well above the hips. Since the upper body is supported by the hips, are birds or humans more stable? Consider what happens if the upper body is displaced a little so that its center of gravity is not directly above or below the hips.

In what direction does the torque due to gravity tend to make the upper body rotate about an axis through the hips?

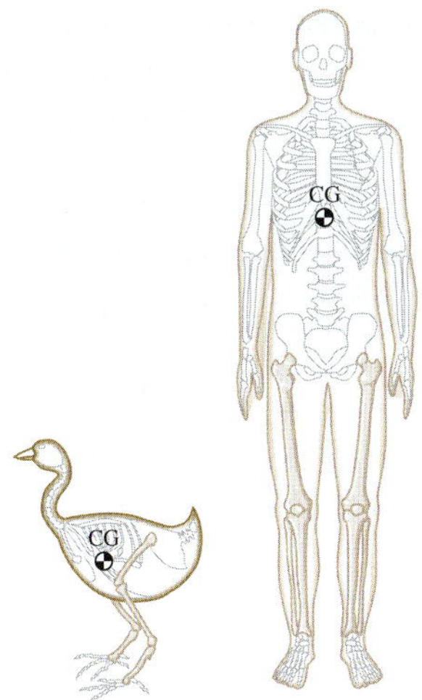

17. An astronaut wants to remove a bolt from a satellite in orbit. He positions himself so that he is at rest with respect to the satellite, then pulls out a wrench and attempts to remove the bolt. What is wrong with his method? How can he remove the bolt?

18. Your door is hinged to close automatically after being opened. Where is the best place to put a wedge-shaped door stopper on a slippery floor in order to hold the door open? Should it be placed close to the hinge or far from it?

19. You are riding your bicycle and approaching a rather steep hill. Which gear should you use to go uphill, a low gear or a high gear? With a low gear the wheel rotates less than with a high gear for one rotation of the pedals.

20. One way to find the center of gravity of an irregular flat object is to suspend it from various points so that it is free to rotate. When the object hangs in equilibrium, a vertical line is drawn downward from the support point. After drawing lines from several different support points, the center of gravity is the point where the lines all intersect. Explain how this works.

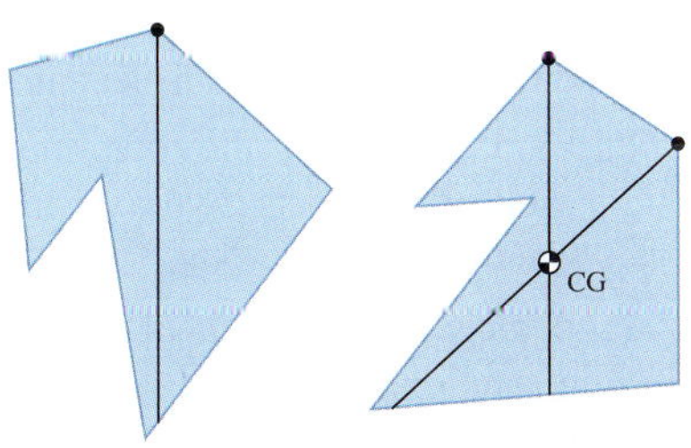

21. One of the effects of significant global warming would be the melting of part or all of the polar ice caps. This, in turn, would change the length of the day (the period of the Earth's rotation). Explain why. Would the day get longer or shorter?

Multiple-Choice Questions

Student Response System Questions

1. When both are expressed in terms of SI *base* units, torque has the same units as
(a) angular acceleration (b) angular momentum
(c) force (d) energy
(e) rotational inertia (f) angular velocity

2. A heavy box is resting on the floor. You would like to push the box to tip it over on its side, using the minimum force possible. Which of the arrows in the diagram shows the correct location and direction of the force? Assume enough friction so that the box does not slide; instead it rotates about point P.

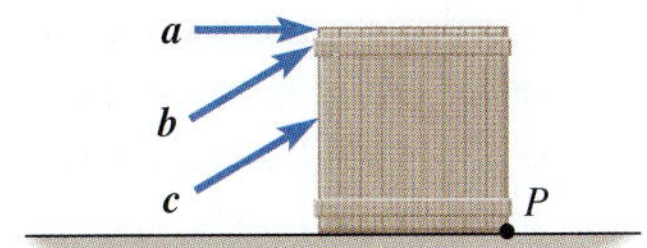

Questions 3–4: A uniform solid cylinder rolls without slipping down an incline. At the bottom of the incline, the speed, v, of the cylinder is measured and the translational and rotational kinetic energies (K_{tr}, K_{rot}) are calculated. A hole is drilled through the cylinder along its axis and the experiment is repeated; at the bottom of the incline the cylinder now has speed v' and translational and rotational kinetic energies K'_{tr} and K'_{rot}.

3. How does the speed of the cylinder compare with its original value?
(a) $v' < v$ (b) $v' = v$ (c) $v' > v$
(d) Answer depends on the radius of the hole drilled.

4. How does the ratio of rotational to translational kinetic energy of the cylinder compare with its original value?
(a) $\frac{K'_{rot}}{K'_{tr}} < \frac{K_{rot}}{K_{tr}}$ (b) $\frac{K'_{rot}}{K'_{tr}} = \frac{K_{rot}}{K_{tr}}$ (c) $\frac{K'_{rot}}{K'_{tr}} > \frac{K_{rot}}{K_{tr}}$
(d) Answer depends on the radius of the hole drilled.

5. The SI units of angular momentum are
(a) $\frac{\text{rad}}{\text{s}}$ (b) $\frac{\text{rad}}{\text{s}^2}$ (c) $\frac{\text{kg}\cdot\text{m}}{\text{s}^2}$
(d) $\frac{\text{kg}\cdot\text{m}^2}{\text{s}^2}$ (e) $\frac{\text{kg}\cdot\text{m}^2}{\text{s}}$ (f) $\frac{\text{kg}\cdot\text{m}}{\text{s}}$

6. Which of the forces in the figure produces the largest magnitude torque about the rotation axis indicated?
(a) 1 (b) 2 (c) 3 (d) 4

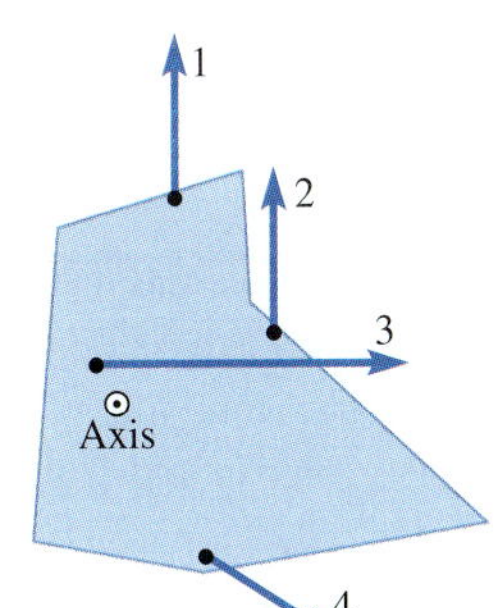

Multiple-Choice Questions 6–8

7. Which of the forces in the figure produces a CW torque about the rotation axis indicated?
(a) 3 only (b) 4 only (c) 1 and 2
(d) 1, 2, and 3 (e) 1, 2, and 4

8. Which pair of forces in the figure might produce equal magnitude torques with opposite signs?
(a) 2 and 3 (b) 2 and 4 (c) 1 and 2
(d) 1 and 3 (e) 1 and 4 (f) 3 and 4

9. A uniform bar of mass m is supported by a pivot at its top, about which the bar can swing like a pendulum. If a force F is applied perpendicularly to the lower end of the bar as in the diagram, how big must F be in order to hold the bar in equilibrium at an angle θ from the vertical?

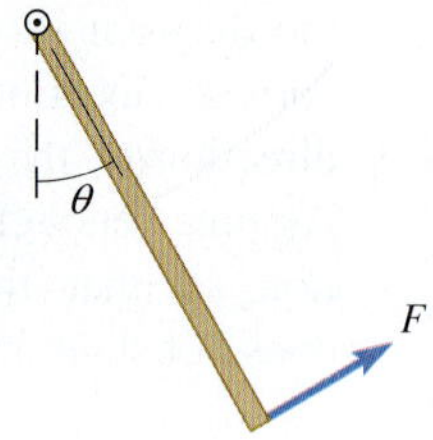

(a) $2mg$ (b) $2mg \sin \theta$
(c) $(mg/2) \sin \theta$ (d) $2mg \cos \theta$
(e) $(mg/2) \cos \theta$ (f) $mg \sin \theta$

10. A high diver in midair pulls her legs inward toward her chest in order to rotate faster. Doing so changes which of these quantities: her angular momentum L, her rotational inertia I, and her rotational kinetic energy K_{rot}?

(a) L only (b) I only
(c) K_{rot} only (d) L and I only
(e) I and K_{rot} only (f) all three

Problems

- C Combination conceptual/quantitative problem
- Biomedical application
- ✦ Challenging
- Blue # Detailed solution in the Student Solutions Manual
- [1, 2] Problems paired by concept
- connect Text website interactive or tutorial

8.1 Rotational Kinetic Energy and Rotational Inertia

1. Verify that $\frac{1}{2}I\omega^2$ has dimensions of energy.
2. What is the rotational inertia of a solid iron disk of mass 49 kg, with a thickness of 5.00 cm and radius of 20.0 cm, about an axis through its center and perpendicular to it?
3. A bowling ball made for a child has half the radius of an adult bowling ball. They are made of the same material (and therefore have the same mass *per unit volume*). By what factor is the (a) mass and (b) rotational inertia of the child's ball reduced compared with the adult ball?
4. Find the rotational inertia of the system of point particles shown in the figure assuming the system rotates about the (a) x-axis, (b) y-axis, (c) z-axis. The z-axis is perpendicular to the xy-plane and points out of the page. Point particle A has a mass of 200 g and is located at $(x, y, z) = (-3.0 \text{ cm}, 5.0 \text{ cm}, 0)$, point particle B has a mass of 300 g and is at (6.0 cm, 0, 0), and point particle C has a mass of 500 g and is at (−5.0 cm, −4.0 cm, 0). (d) What are the x- and y-coordinates of the center of mass of the system?

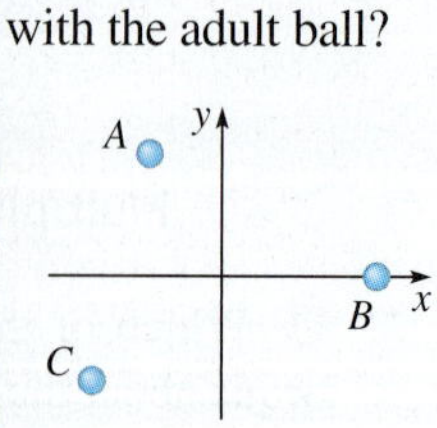

Problems 5–6. Four point masses of 3.0 kg each are arranged in a square on massless rods. The length of a side of the square is 0.50 m. Consider rotation of this square about three different axes: (a) an axis passing through masses B and C; (b) an axis passing through masses A and C; and (c) an axis passing through the center of the square and perpendicular to the plane of the square

5. Rank the three arrangements in increasing order of the rotational inertia.
6. Calculate the rotational inertia for each of the three arrangements.

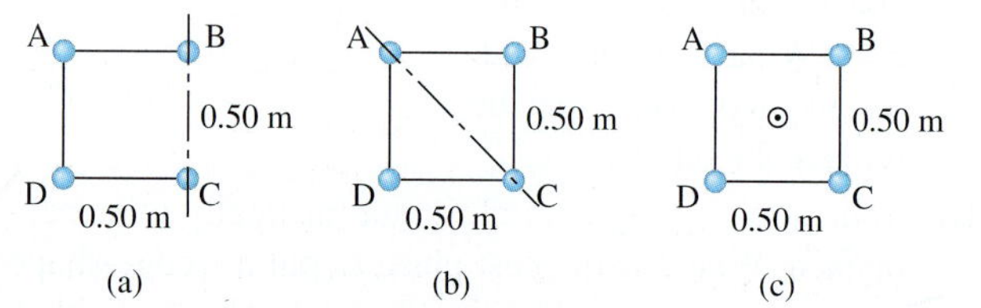

Problems 5 and 6

7. How much work is done by the motor in a CD player to make a CD spin, starting from rest? The CD has a diameter of 12.0 cm and a mass of 15.8 g. The laser scans at a constant tangential velocity of 1.20 m/s. Assume that the music is first detected at a radius of 20.0 mm from the center of the disc. Ignore the small circular hole at the CD's center.
8. Find the ratio of the rotational inertia of the Earth for rotation about its own axis to its rotational inertia for rotation about the Sun.
9. A bicycle has wheels of radius 0.32 m. Each wheel has a rotational inertia of 0.080 kg·m^2 about its axle. The total mass of the bicycle including the wheels and the rider is 79 kg. When coasting at constant speed, what fraction of the total kinetic energy of the bicycle (including rider) is the rotational kinetic energy of the wheels?
10. C In many problems in previous chapters, cars and other objects that roll on wheels were considered to act as if they were sliding without friction. (a) Can the same assumption be made for a wheel rolling *by itself*? Explain your answer. (b) If a moving car of total mass 1300 kg has four wheels, each with rotational inertia of 0.705 kg·m^2 and radius of 35 cm, what fraction of the total kinetic energy is rotational?
11. A centrifuge has a rotational inertia of 6.5×10^{-3} kg·m^2. How much energy must be supplied to bring it from rest to 420 rad/s (4000 rev/min)?

8.2 Torque

12. A mechanic turns a wrench using a force of 25 N at a distance of 16 cm from the rotation axis. The force is perpendicular to the wrench handle. What magnitude torque does she apply to the wrench?

13. The pull cord of a lawnmower engine is wound around a drum of radius 6.00 cm. While the cord is pulled with a force of 75 N to start the engine, what magnitude torque does the cord apply to the drum?

14. A child of mass 40.0 kg is sitting on a horizontal seesaw at a distance of 2.0 m from the supporting axis. What is the magnitude of the torque about the axis due to the weight of the child?

15. A 124 g mass is placed on one pan of a balance, at a point 25 cm from the support of the balance. What is the magnitude of the torque about the support exerted by the mass?

16. A uniform door weighs 50.0 N and is 1.0 m wide and 2.6 m high. What is the magnitude of the torque due to the door's own weight about a horizontal axis perpendicular to the door and passing through a corner?

17. In each of the five situations shown, a force is applied to a point on the handle of an old pump. Rank the situations in order of the magnitude of the torque applied to the handle, smallest to largest.

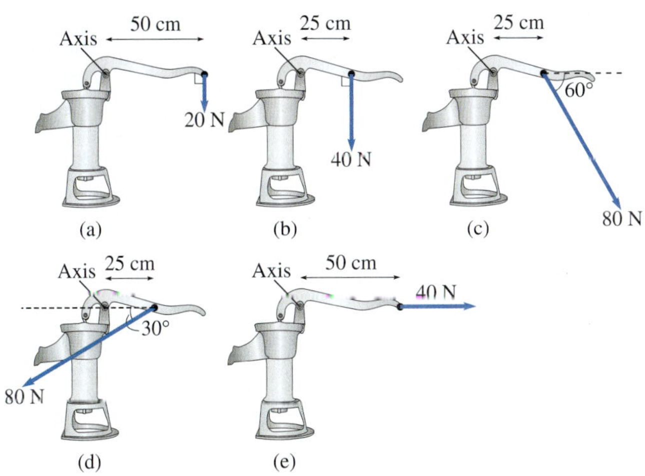

18. The human mandible (lower jaw) is attached to the temporomandibular joint (TMJ). The masseter muscle is largely responsible for pulling the mandible upward when you are talking or eating. It is attached at a horizontal distance of about 2.5 cm from the TMJ. If this muscle exerts a force of 220 N when you bite, what is the force your incisors exert? The horizontal distance from the TMJ to your incisors is 6.6 cm. Assume that both forces are vertical.

19. A tower outside the Houses of Parliament in London has a famous clock commonly referred to as Big Ben, the name of its 13 ton chiming bell. The hour hand of each clock face is 2.7 m long and has a mass of 60.0 kg. Assume the hour hand to be a uniform rod attached at one end. (a) What is the torque on the clock mechanism due to the weight of one of the four hour hands when the clock strikes noon? The axis of rotation is perpendicular to a clock face and through the center of the clock. (b) What is the torque due to the weight of one hour hand about the same axis when the clock tolls 9:00 A.M.?

20. ✦ Ⓒ Any pair of equal and opposite forces acting on the same object is called a *couple*. Consider the couple in part (a) of the figure. The rotation axis is perpendicular to the page through point P. (a) Show that the magnitude of the net torque is equal to Fd, where d is the distance between the lines of action. Because the distance d is independent of the location of the rotation axis, this shows that the torque is the same for any rotation axis. (b) Repeat for the couple in part (b) of the figure. Show that the torque magnitude is still Fd if d is the *perpendicular* distance between the lines of action of the forces.

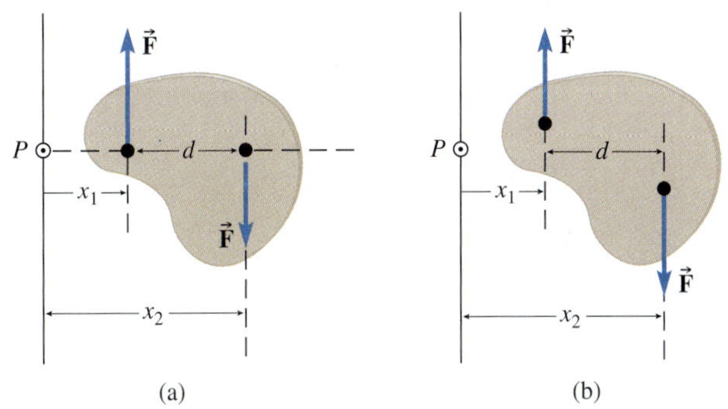

21. A 46.4 N force is applied to the outer edge of a door of width 1.26 m in such a way that it acts (a) perpendicular to the door, (b) at an angle of 43.0° with respect to the door surface, (c) so that the line of action of the force passes through the axis of the door hinges. Find the torque for these three cases.

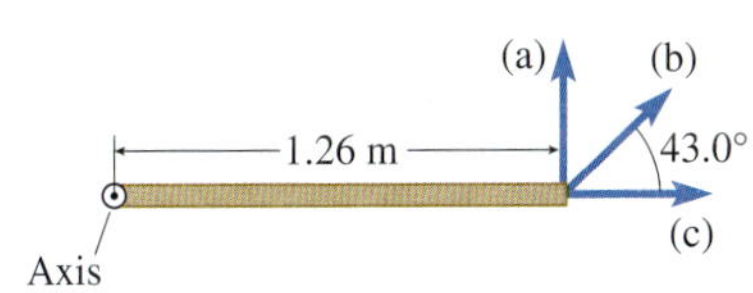

22. A trap door, of length and width 1.65 m, is held open at an angle of 65.0° with respect to the floor. A rope is attached to the raised edge of the door and fastened to the wall behind the door in such a position that the rope pulls perpendicularly to the trap door. If the mass of the trap door is 16.8 kg, what is the torque exerted on the trap door by the rope? (connect tutorial: deck hatch)

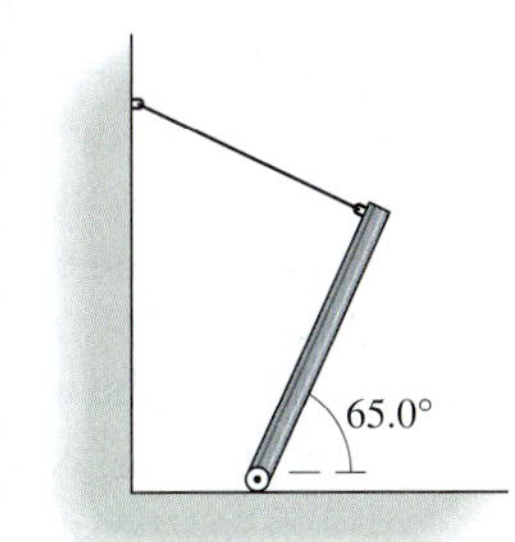

23. A weightless rod, 10.0 m long, supports three weights as shown. Where is its center of gravity?

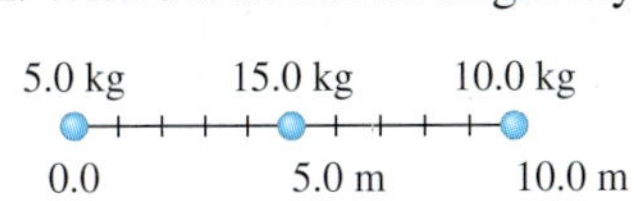

24. A door weighing 300.0 N measures 2.00 m × 3.00 m and is of uniform density; that is, the mass is uniformly distributed throughout the volume. A doorknob is

attached to the door as shown. Where is the center of gravity if the doorknob weighs 5.0 N and is located 0.25 m from the edge?

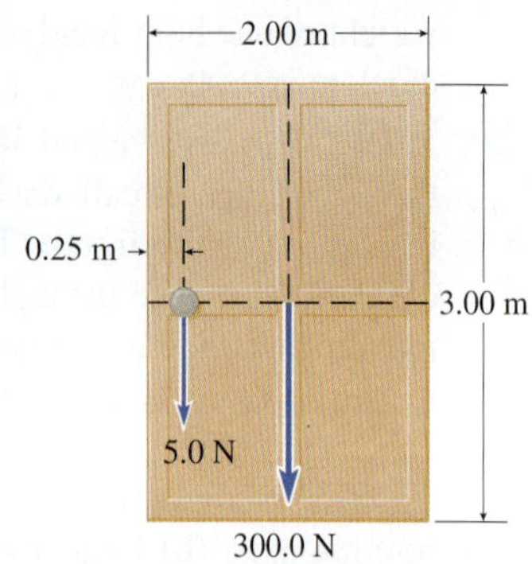

25. ✦ A plate of uniform thickness is shaped as shown. Where is the center of gravity? Assume the origin (0, 0) is located at the lower left corner of the plate; the upper left corner is at (0, s); and the upper right corner is at (s, s).

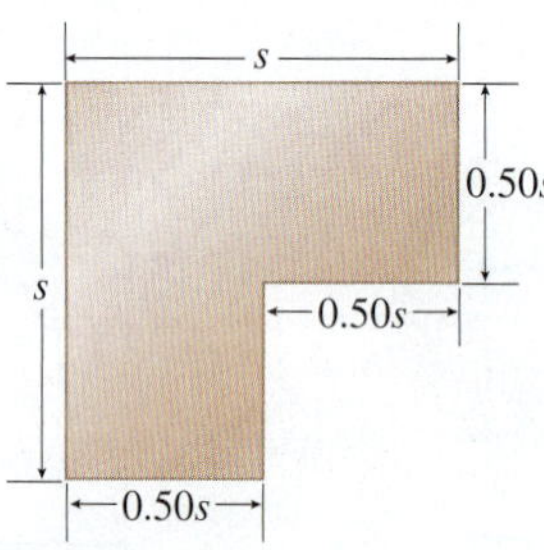

8.3 Calculating Work Done from the Torque

26. A stone used to grind wheat into flour is turned through 12 revolutions by a constant force of 20.0 N applied to the rim of a 10.0 cm radius shaft connected to the wheel. How much work is done on the stone during the 12 revolutions?

27. The radius of a wheel is 0.500 m. A rope is wound around the outer rim of the wheel. The rope is pulled with a force of magnitude 5.00 N, unwinding the rope and making the wheel spin CCW about its central axis. Ignore the mass of the rope. (a) How much rope unwinds while the wheel makes 1.00 revolution? (b) How much work is done by the rope on the wheel during this time? (c) What is the torque on the wheel due to the rope? (d) What is the angular displacement $\Delta\theta$, in radians, of the wheel during 1.00 revolution? (e) Show that the numerical value of the work done is equal to the product $\tau\Delta\theta$.

28. ✦ A flywheel of mass 182 kg has an effective radius of 0.62 m (assume the mass is concentrated along a circumference located at the effective radius of the flywheel). (a) How much work is done to bring this wheel from rest to a speed of 120 rev/min in a time interval of 30.0 s? (b) What is the applied torque on the flywheel (assumed constant)?

29. ✦ A Ferris wheel rotates because a motor exerts a torque on the wheel. The radius of the London Eye, a huge observation wheel on the banks of the Thames, is 67.5 m and its mass is 1.90×10^6 kg. The cruising angular speed of the wheel is 3.50×10^{-3} rad/s. (a) How much work does the motor need to do to bring the stationary wheel up to cruising speed? [*Hint*: Treat the wheel as a hoop.] (b) What is the torque (assumed constant) the motor needs to provide to the wheel if it takes 20.0 s to reach the cruising angular speed?

8.4 Rotational Equilibrium

30. A rod is being used as a lever as shown. The fulcrum is 1.2 m from the load and 2.4 m from the applied force. If the load has a mass of 20.0 kg, what force must be applied to lift the load?

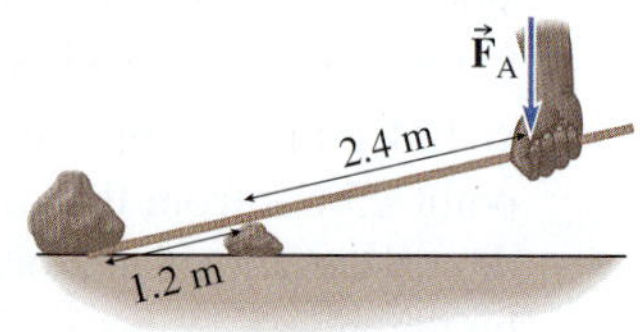

31. A weight of 1200 N rests on a lever at a point 0.50 m from a support. On the same side of the support, at a distance of 3.0 m from it, an upward force with magnitude F is applied. Ignore the weight of the board itself. If the system is in equilibrium, what is F?

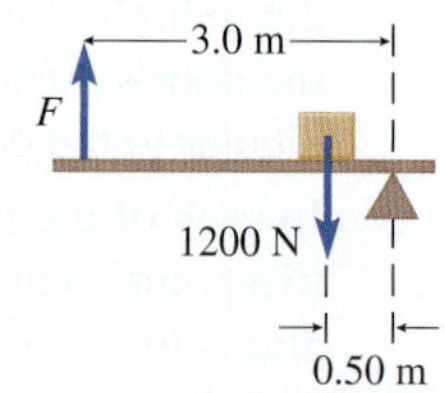

32. A sculpture is 4.00 m tall and has its center of gravity located 1.80 m above the center of its base. The base is a square with a side of 1.10 m. To what angle θ can the sculpture be tipped before it falls over? (connect tutorial: filing cabinet)

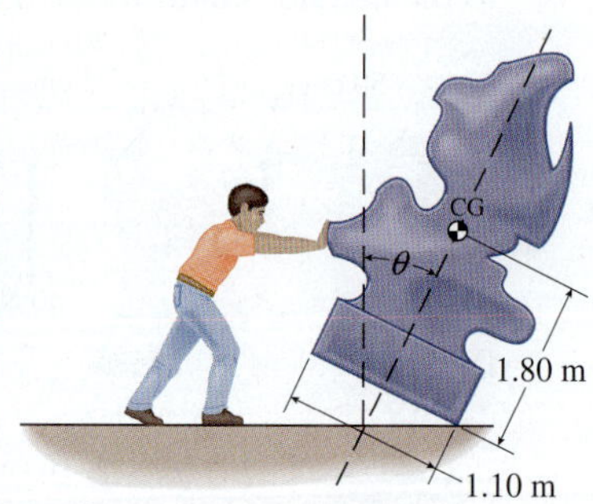

33. A house painter is standing on a uniform, horizontal platform that is held in equilibrium by two cables attached to supports on the roof. The painter has a mass of 75 kg, and the mass of the platform is 20.0 kg. The distance from the left end of the platform to where the painter is standing is d = 2.0 m, and the total length of the platform is 5.0 m. (a) How large is the force exerted by the left-hand cable on the platform? (b) How large is the force exerted by the right-hand cable?

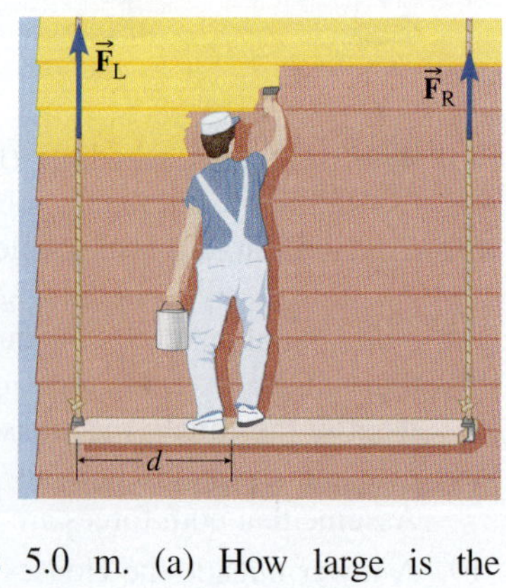

34. Ⓒ Four identical uniform metersticks are stacked on a table as shown. Where is the x-coordinate of the CM of the metersticks if the origin is

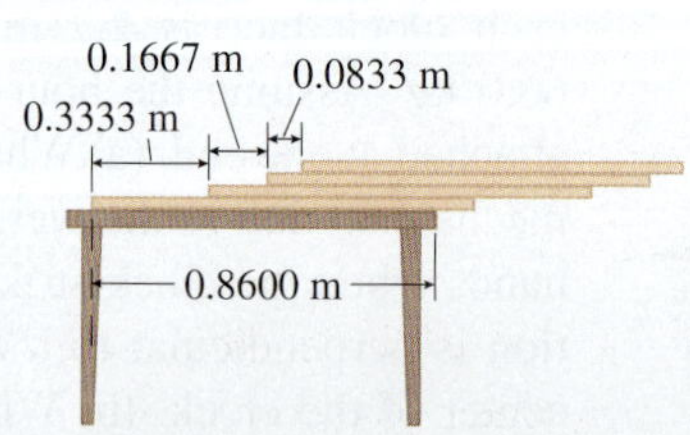

chosen at the left end of the lowest stick? Why does the system balance?

35. ✦ A uniform diving board, of length 5.0 m and mass 55 kg, is supported at two points; one support is located 3.4 m from the end of the board and the second is at 4.6 m from the end (see Fig. 8.19). What are the forces acting on the board due to the two supports when a diver of mass 65 kg stands at the end of the board over the water? Assume that these forces are vertical. (connect tutorial: plank) [*Hint*: In this problem, consider using two different torque equations about different rotation axes. This may help you determine the directions of the two forces.]

36. ✦ A house painter stands 3.0 m above the ground on a 5.0 m long ladder that leans against the wall at a point 4.7 m above the ground. The painter weighs 680 N and the ladder weighs 120 N. Assuming no friction between the house and the upper end of the ladder, find the force of friction that the driveway exerts on the bottom of the ladder. (connect interactive: ladder; tutorial: ladder)

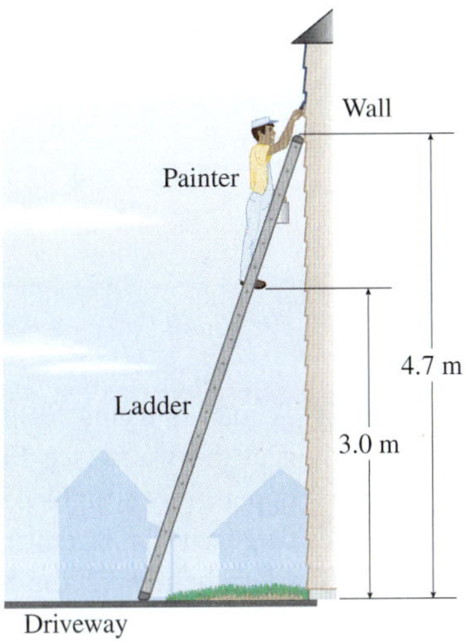

37. ✦ A mountain climber is rappelling down a vertical wall. The rope attaches to a buckle strapped to the climber's waist 15 cm to the right of his center of gravity. If the climber weighs 770 N, find (a) the tension in the rope and (b) the magnitude and direction of the contact force exerted by the wall on the climber's feet.

38. Ⓒ A sign is supported by a uniform horizontal boom of length 3.00 m and weight 80.0 N. A cable, inclined at an angle of 35° with the boom, is attached at a distance of 2.38 m from the hinge at the wall. The weight of the sign is 120.0 N. What is the tension in the cable, and what are the horizontal and vertical forces F_x and F_y exerted on the boom by the hinge? Comment on the magnitude of F_y.

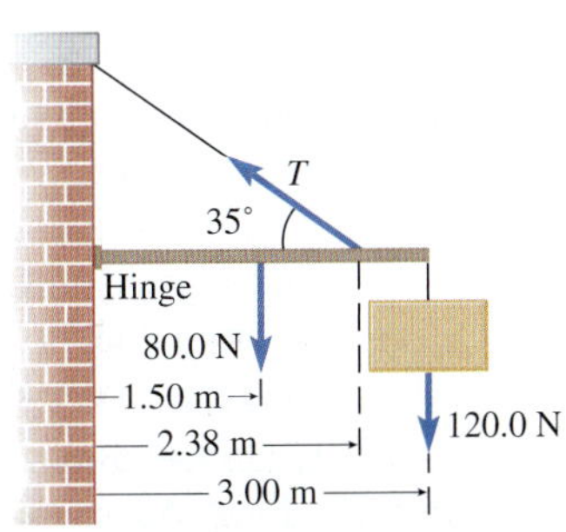

39. Ⓒ A boom of mass m supports a steel girder of weight W hanging from its end. One end of the boom is hinged at the floor; a cable attaches to the other end of the boom and pulls horizontally on it. The boom makes an angle θ with the horizontal. Find the tension in the cable as a function of m, W, θ, and g. Comment on the tension at $\theta = 0$ and $\theta = 90°$.

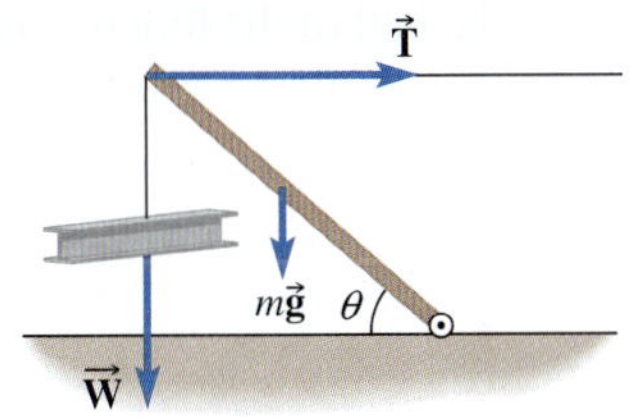

40. You are asked to hang a uniform beam and sign using a cable that has a breaking strength of 417 N. The store owner desires that it hang out over the sidewalk as shown. The sign has a weight of 200.0 N and the beam's weight is 50.0 N. The beam's length is 1.50 m and the sign's dimensions are 1.00 m horizontally × 0.80 m vertically. What is the minimum angle θ that you can have between the beam and cable?

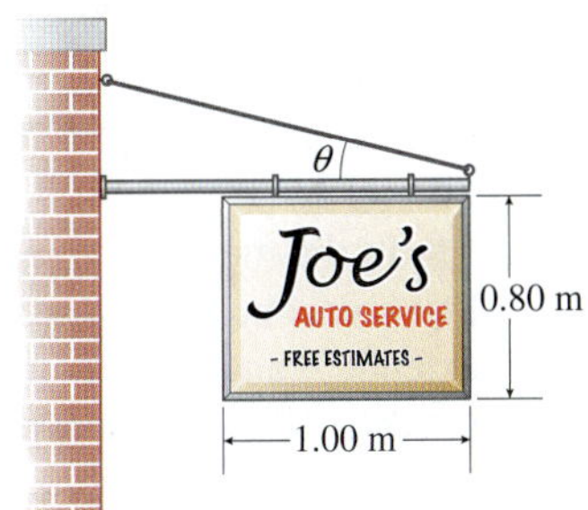

Problems 40 and 41

41. Refer to Problem 40. You chose an angle θ of 33.8°. An 8.7 kg raccoon has climbed onto the beam and is walking from the wall toward the point where the cable meets the beam. How far can the raccoon walk before the cable breaks?

42. A man is doing push-ups. He has a mass of 68 kg and his center of gravity is located at a horizontal distance of 0.70 m from his palms and 1.00 m from his feet. Find the forces exerted by the floor on his palms and feet.

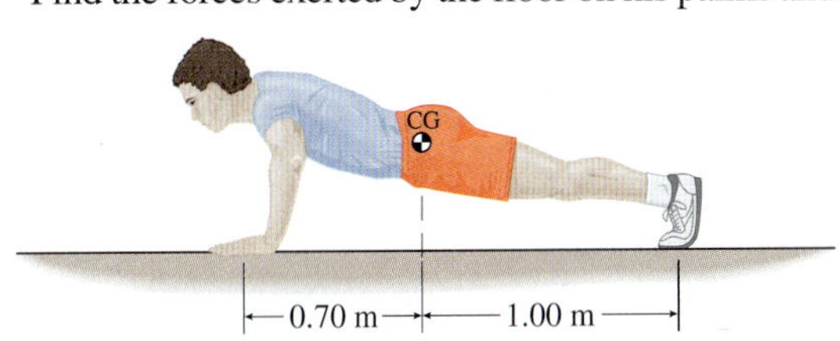

8.5 Application: Equilibrium in the Human Body

43. Your friend balances a package with mass $m = 10$ kg on top of his head while standing. The mass of his upper body is $M = 55$ kg (about 65% of his total mass). Because the spine is vertical rather than horizontal, the force exerted by the sacrum on the spine ($\vec{\mathbf{F}}_s$ in Fig 8.33) is directed approximately straight up and

the force exerted by the back muscles ($\vec{\mathbf{F}}_b$) is negligibly small. Find the magnitude of $\vec{\mathbf{F}}_s$.

44. Find the tension in the Achilles tendon and the force that the tibia exerts on the ankle joint when a person who weighs 750 N supports himself on the ball of one foot. The normal force $N = 750$ N pushes up on the ball of the foot on one side of the ankle joint, while the Achilles tendon pulls up on the foot on the other side of the joint.

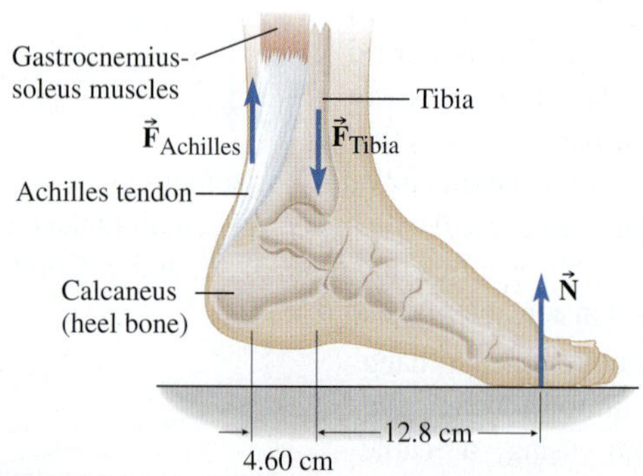

45. In the movie *Terminator*, Arnold Schwarzenegger lifts someone up by the neck and, with both arms fully extended and horizontal, holds the person off the ground. If the person being held weighs 700 N, is 60 cm from the shoulder joint, and Arnold has an anatomy analogous to that in Fig. 8.31, what force must *each* of the deltoid muscles exert to perform this task?

46. Find the force exerted by the biceps muscle in holding a 1.0 L milk carton (weight 9.9 N) with the forearm parallel to the floor. Assume that the hand is 35.0 cm from the elbow and that the upper arm is 30.0 cm long. The elbow is bent at a right angle, and one tendon of the biceps is attached to the forearm at a position 5.00 cm from the elbow, while the other tendon is attached at 30.0 cm from the elbow. The weight of the forearm and empty hand is 18.0 N, and the center of gravity of the forearm is at a distance of 16.5 cm from the elbow.

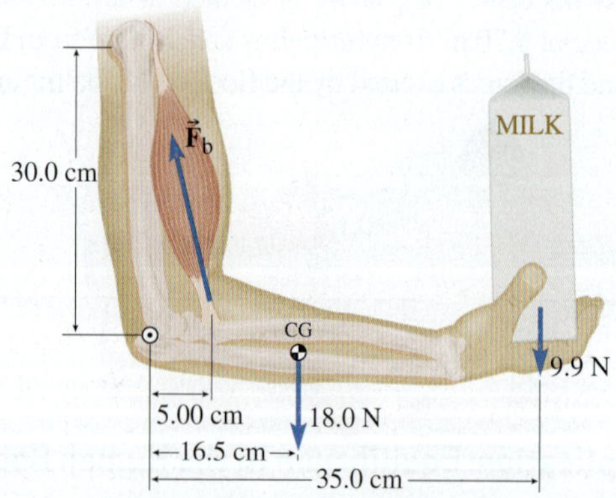

47. A person is doing leg lifts with 3.0 kg ankle weights. She is sitting in a chair with her legs bent at a right angle initially. The quadriceps muscles are attached to the patella via a tendon; the patella is connected to the tibia by the patellar tendon, which attaches to bone 10.0 cm below the knee joint. Assume that the tendon pulls at an angle of 20.0° *with respect to the lower leg*, regardless of the position of the lower leg. The lower leg has a mass of 5.0 kg, and its center of gravity is 22 cm below the knee. The ankle weight is 41 cm from the knee. If the person lifts one leg, find the force exerted by the patellar tendon to hold the leg at an angle of (a) 30.0° and (b) 90.0° with respect to the vertical.

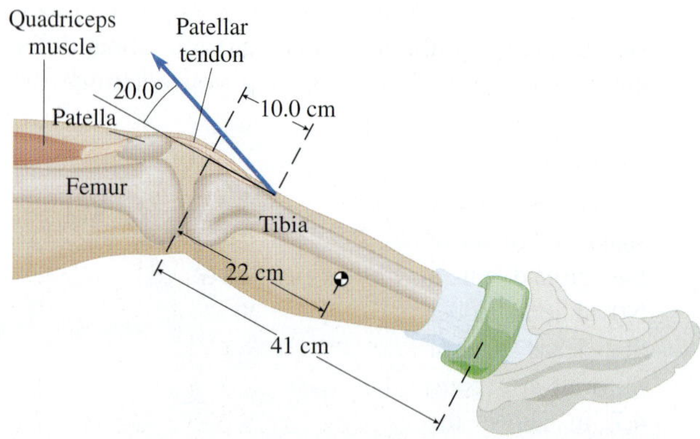

48. ✦ A man is trying to lift 60.0 kg off the floor by bending at the waist (see Fig. 8.33). Assume that the man's upper body weighs 455 N and the upper body's center of gravity is 38 cm from the sacrum (tailbone). (a) If, when bent over, the hands are a horizontal distance of 76 cm from the sacrum, what torque must be exerted by the erector spinae muscles to lift 60.0 kg off the floor? (The axis of rotation passes through the sacrum, as shown in Fig. 8.33.) (b) When bent over, the erector spinae muscles are a horizontal distance of 44 cm from the sacrum and act at a 12° angle above the horizontal. What force ($\vec{\mathbf{F}}_b$ in Fig. 8.33) do the erector spinae muscles need to exert to lift the weight? (c) What is the component of this force that compresses the spinal column?

8.6 Rotational Form of Newton's Second Law

49. Verify that the units of the rotational form of Newton's second law [Eq. (8-9)] are consistent. In other words, show that the product of a rotational inertia expressed in kg·m^2 and an angular acceleration expressed in rad/s^2 is a torque expressed in N·m.

50. A spinning flywheel has rotational inertia I = 400.0 kg·m^2. Its angular velocity decreases from 20.0 rad/s to zero in 300.0 s due to friction. What is the frictional torque acting?

51. A turntable must spin at 33.3 rev/min (3.49 rad/s) to play an old-fashioned vinyl record. How much torque must the motor deliver if the turntable is to reach its final angular speed in 2.0 revolutions, starting from rest? The turntable is a uniform disk of diameter 30.5 cm and mass 0.22 kg.

52. A lawn sprinkler has three spouts that spray water, each 15.0 cm long. As the water is sprayed, the sprinkler

turns around in a circle. The sprinkler has a total rotational inertia of 9.20×10^{-2} kg·m^2. If the sprinkler starts from rest and takes 3.20 s to reach its final speed of 2.2 rev/s, what force does each spout exert on the sprinkler?

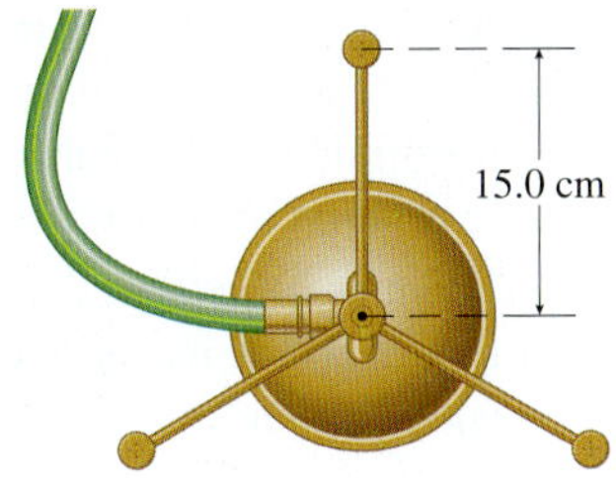

53. A discus thrower moves the discus in 1.5 complete revolutions in 1.4 s (starting from rest). The radius of the circular path of the discus is 0.90 m, and the mass of the discus is 2.0 kg. Assume a constant torque is applied to the discus by the athlete. (a) What is the angular speed of the discus just before release? (b) What torque does the athlete apply to the discus? (c) Approximately how far from the athlete does the discus land if it is released at a 45° angle to the horizontal?

54. A chain pulls tangentially on a 40.6 kg uniform cylindrical gear with a tension of 72.5 N. The chain is attached along the outside radius of the gear at 0.650 m from the axis of rotation. Starting from rest, the gear takes 1.70 s to reach its rotational speed of 1.35 rev/s. What is the total frictional torque opposing the rotation of the gear?

55. Four masses are arranged as shown. They are connected by rigid, massless rods of lengths 0.75 m and 0.50 m. What torque must be applied to cause an angular acceleration of 0.75 rad/s^2 about the axis shown?

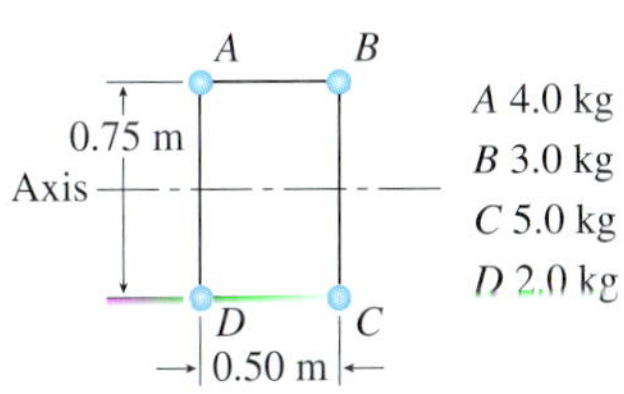

56. A bicycle wheel, of radius 0.30 m and mass 2 kg (concentrated on the rim), is rotating at 4.00 rev/s. After 50 s the wheel comes to a stop because of friction. What is the magnitude of the average torque due to frictional forces?

57. A playground merry-go-round (see Fig. 8.5), made in the shape of a solid disk, has a diameter of 2.50 m and a mass of 350.0 kg. Two children, each of mass 30.0 kg, sit on opposite sides at the edge of the platform. Approximate the children as point masses. (a) What torque is required to bring the merry-go-round from rest to 25 rev/min in 20.0 s? (b) If two other bigger children are going to push on the merry-go-round rim to produce this acceleration, with what force magnitude must each child push? (connect tutorial: roundabout)

58. Two children standing on opposite sides of a merry-go-round (see Fig. 8.5) are trying to rotate it. They each push in opposite directions with forces of magnitude 10.0 N. (a) If the merry-go-round has a mass of 180 kg and a radius of 2.0 m, what is the angular acceleration of the merry-go-round? (Assume the merry-go-round is a uniform disk.) (b) How fast is the merry-go-round rotating after 4.0 s?

59. ✦ Ⓒ Refer to Atwood's machine (Example 8.2). (a) Assuming that the cord does not slip as it passes around the pulley, what is the relationship between the angular acceleration of the pulley (α) and the magnitude of the linear acceleration of the blocks (a)? (b) What is the net torque on the pulley about its axis of rotation in terms of the tensions T_1 and T_2 in the left and right sides of the cord? (c) Explain why the tensions cannot be equal if $m_1 \neq m_2$. (d) Apply Newton's second law to each of the blocks and Newton's second law for rotation to the pulley. Use these three equations to solve for a, T_1, and T_2. (e) Since the blocks move with constant acceleration, use the result of Example 8.2 along with the constant acceleration equation $v_{fy}^2 - v_{iy}^2 = 2a_y\Delta y$ to check your answer for a.

60. ✦ Derive the rotational form of Newton's second law as follows. Consider a rigid object that consists of a large number N of particles. Let F_n, m_n, and r_n represent the tangential component of the net force acting on the nth particle, the mass of that particle, and the particle's distance from the axis of rotation, respectively. (a) Use Newton's second law to find a_n, the particle's tangential acceleration. (b) Find the torque acting on this particle. (c) Replace a_n with an equivalent expression in terms of the angular acceleration α. (d) Sum the torques due to all the particles and show that

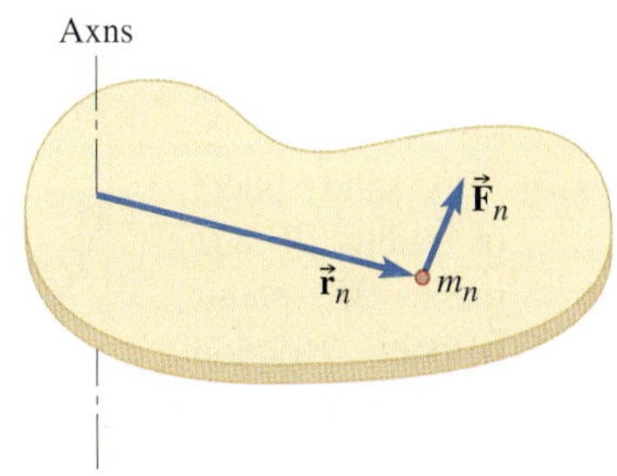

$$\sum_{n=1}^{N} \tau_n = I\alpha$$

8.7 The Motion of Rolling Objects

61. A solid sphere is rolling without slipping or sliding down a board that is tilted at an angle of 35° with respect to the horizontal. What is its acceleration?

62. A solid sphere is released from rest and allowed to roll down a board that has one end resting on the floor and is tilted at 30° with respect to the horizontal. If the sphere is released from a height of 60 cm above the floor, what is the sphere's speed when it reaches the lowest end of the board?

63. A hollow cylinder, a uniform solid sphere, and a uniform solid cylinder all have the same mass m. The three objects are rolling on a horizontal surface with identical translational speeds v. Find their total kinetic energies in terms of m and v and order them from smallest to largest.

64. A solid sphere of mass 0.600 kg rolls without slipping along a horizontal surface with a translational speed of 5.00 m/s. It comes to an incline that makes an angle of 30° with the horizontal surface. Ignoring energy losses due to friction, to what vertical height above the horizontal surface does the sphere rise on the incline?

65. A bucket of water with a mass of 2.0 kg is attached to a rope that is wound around a cylinder. The cylinder has a mass of 3.0 kg and is mounted horizontally on frictionless bearings. The bucket is released from rest. (a) Find its speed after it has fallen through a distance of 0.80 m. What are (b) the tension in the rope and (c) the acceleration of the bucket?

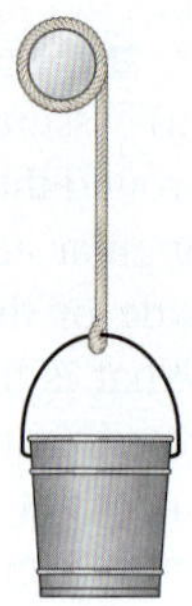

Problems 65 and 66

66. A 1.10 kg bucket is tied to a rope that is wrapped around a pole mounted horizontally on frictionless bearings. The cylindrical pole has a diameter of 0.340 m and a mass of 2.60 kg. When the bucket is released from rest, how long will it take to fall to the bottom of the well, a distance of 17.0 m?

67. ✦ A solid sphere of radius R and mass M slides without friction down a loop-the-loop track. The sphere starts from rest at a height of h above the horizontal. Assume that the radius of the sphere is small relative to the radius r of the loop. (a) Find the minimum value of h in terms of r so that the sphere remains on the track all the way around the loop. (b) Find the minimum value of h if, instead, the sphere rolls without slipping on the track.

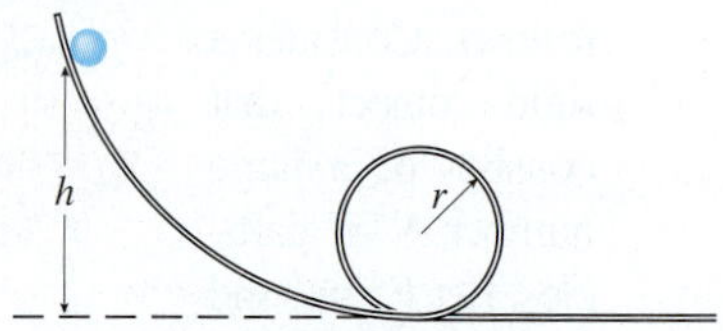

Problems 67–69

68. ✦ A hollow cylinder, of radius R and mass M, rolls without slipping down a loop-the-loop track of radius r. The cylinder starts from rest at a height h above the horizontal section of track. What is the minimum value of h so that the cylinder remains on the track all the way around the loop?

69. Ⓒ If the hollow cylinder of Problem 68 is replaced with a solid sphere, will the minimum value of h increase, decrease, or remain the same? Once you think you know the answer and can explain why, redo the calculation to find h.

70. ✦ The string in a yo-yo is wound around an axle of radius 0.500 cm. The yo-yo has both rotational and translational motion, like a rolling object, and has mass 0.200 kg and outer radius 2.00 cm. Starting from rest, it rotates and falls a distance of 1.00 m (the length of the string). Assume for simplicity that the yo-yo is a uniform circular disk and that the string is thin compared with the radius of the axle. (a) What is the speed of the yo-yo when it reaches the distance of 1.00 m? (b) How long does it take to fall? [*Hint*: The translational and rotational kinetic energies are related, but the yo-yo is *not* rolling on its outer radius.]

8.8 Angular Momentum

71. A uniform disk of mass 5.00 kg has a radius of 0.100 m and spins with a frequency of 0.550 rev/s. What is its angular momentum?

72. Assume the Earth is a uniform solid sphere with radius of 6.37×10^6 m and mass of 5.97×10^{24} kg. Find the magnitude of the angular momentum of the Earth due to rotation about its axis.

73. The mass of a flywheel is 5.6×10^4 kg. This particular flywheel has its mass concentrated at the rim of the wheel. If the radius of the wheel is 2.6 m and it is rotating at 350 rev/min, what is the magnitude of its angular momentum?

74. The angular momentum of a spinning wheel is 240 kg·m^2/s. After application of a constant braking torque for 2.5 s, it slows and has a new angular momentum of 115 kg·m^2/s. What is the torque applied?

75. How long would a braking torque of 4.00 N·m have to act to just stop a spinning wheel that has an initial angular momentum of 6.40 kg·m^2/s?

76. Six flywheels have masses, thicknesses, radii, and angular speeds as given in the table. Each flywheel is a solid disk. Rank the flywheels in order of their angular momentum, smallest to largest.

	Mass (kg)	Thickness (cm)	Radius (cm)	Angular Speed (rad/s)
A	10	1	20	30
B	20	2	20	30
C	20	2	40	15
D	20	2	40	30
E	20	8	10	60
F	5	0.5	20	60

77. A figure skater is spinning at a rate of 1.0 rev/s with her arms outstretched. She then draws her arms in to her chest, reducing her rotational inertia to 67% of its original value. What is her new rate of rotation?

78. A skater is initially spinning at a rate of 10.0 rad/s with a rotational inertia of 2.50 kg·m^2 when her arms are extended. What is her angular velocity after she pulls her arms in and reduces her rotational inertia to 1.60 kg·m^2?

79. A figure skater is spinning at 10.0 rad/s with her arms extended. Her rotational inertia is 2.50 kg·m^2. After pulling her arms in, her rotational inertia is 1.60 kg·m^2. How much work does she do to pull her arms in while spinning?

80. A uniform disk with a mass of 800 g and radius 17.0 cm is rotating on frictionless bearings with an angular speed of 18.0 Hz when Jill drops a 120 g clod of clay on a point 8.00 cm from the center of the disk, where it sticks. What is the new angular speed of the disk?

81. A spoked wheel with a radius of 40.0 cm and a mass of 2.00 kg is mounted horizontally on frictionless bearings. JiaJun puts his 0.500 kg guinea pig on the outer edge of the wheel. The guinea pig begins to run along the edge of the wheel with a speed of 20.0 cm/s with

respect to the ground. What is the angular velocity of the wheel? Assume the spokes of the wheel have negligible mass.

82. A diver can change his rotational inertia by drawing his arms and legs close to his body in the tuck position. After he leaves the diving board (with some unknown angular velocity), he pulls himself into a ball as closely as possible and makes 2.00 complete rotations in 1.33 s. If his rotational inertia decreases by a factor of 3.00 when he goes from the straight to the tuck position, what was his angular velocity when he left the diving board?

83. ✦ The rotational inertia for a diver in a pike position is about 15.5 kg·m^2; it is only 8.0 kg·m^2 in a tuck position. (a) If the diver gives himself an initial angular momentum of 106 kg·m^2/s as he jumps off the board, how many turns can he make when jumping off a 10.0 m platform in a tuck position? (b) How many in a pike position? [*Hint*: Gravity exerts no torque on the person as he falls; assume he is rotating throughout the 10.0 m dive.]

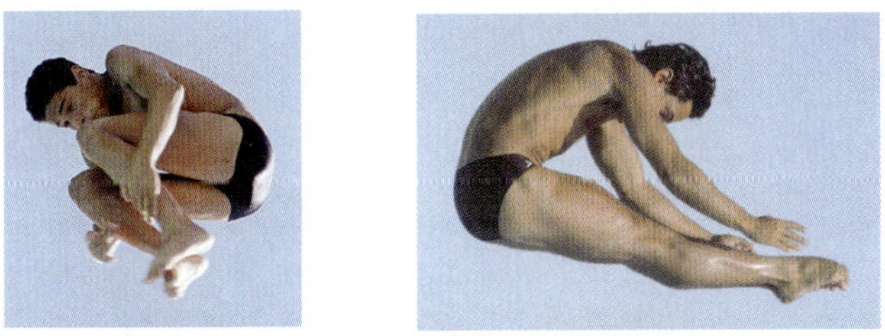

(a) (b)

Problem 83. (a) Mark Ruiz in the tuck position. (b) Gregory Louganis in the pike position.

84. Consider the merry-go-round of Practice Problem 8.1. The child is initially standing on the ground when the merry-go-round is rotating at 0.75 rev/s. The child then steps on the merry-go-round. How fast is the merry-go-round rotating now? By how much did the rotational kinetic energy of the merry-go-round and child change?

8.9 The Vector Nature of Angular Momentum

Problems 85 and 86. A solid cylindrical disk is to be used as a stabilizer in a ship. By using a massive disk rotating in the hold of the ship, the captain knows that a large torque is required to tilt its angular momentum vector. The mass of the disk to be used is 1.00×10^5 kg, and it has a radius of 2.00 m.

85. ✦ If the cylinder rotates at 300.0 rev/min, what is the magnitude of the average torque required to tilt its axis by 60.0° in a time of 3.00 s? [*Hint*: Draw a vector diagram of the initial and final angular momenta.]

86. Ⓒ How should the disk be oriented to prevent rocking from side to side and from bow to stern? Does this orientation make it difficult to steer the ship? Explain.

Collaborative Problems

87. ✦ One day when your friend from Problem 43 is picking up a package, you notice that he bends at the waist to pick it up rather than keeping his back straight and bending his knees. You suspect that the lower back pain he complains about is caused by the large force on his lower vertebrae ($\vec{\mathbf{F}}_s$ in Fig. 8.33) when he lifts objects in this way. Suppose that when the spine is horizontal, the back muscles exert a force $\vec{\mathbf{F}}_b$ as in Fig. 8.33 (44 cm from the sacrum and at an angle of 12° to the horizontal). Assume that the CM of his upper body (including the arms) is at its geometric center, 38 cm from the sacrum. Find the horizontal component of $\vec{\mathbf{F}}_s$ when your friend is holding a 10 kg package at a distance of 76 cm from his sacrum. Compare this with the magnitude of $\vec{\mathbf{F}}_s$ found in Problem 43.

88. ✦ Ⓒ A uniform solid cylinder rolls without slipping down an incline. A hole is drilled through the cylinder along its axis. The radius of the hole is 0.50 times the (outer) radius of the cylinder. (a) Does the cylinder take more or less time to roll down the incline now that the hole has been drilled? Explain. (b) By what percentage does drilling the hole change the time for the cylinder to roll down the incline? (connect tutorial: rolling)

89. ✦ (a) Assume the Earth is a uniform solid sphere. Find the kinetic energy of the Earth due to its rotation about its axis. (b) Suppose we could somehow extract 1.0% of the Earth's rotational kinetic energy to use for other purposes. By how much would that change the length of the day? (c) For how many years would 1.0% of the Earth's rotational kinetic energy supply the world's energy usage (assume a constant 1.0×10^{21} J per year)?

90. One way to determine the location of your center of gravity is shown in the diagram. A 2.2 m long uniform plank is supported by two bathroom scales, one at either end. Initially the scales each read 100.0 N. A 1.60 m tall student then lies on top of the plank, with the soles of his feet directly above scale B. Now scale A reads 394.0 N and scale B reads 541.0 N. (a) What is the student's weight? (b) How far is his center of gravity from the soles of his feet? (c) When standing, how far above the floor is his center of gravity, expressed as a fraction of his height?

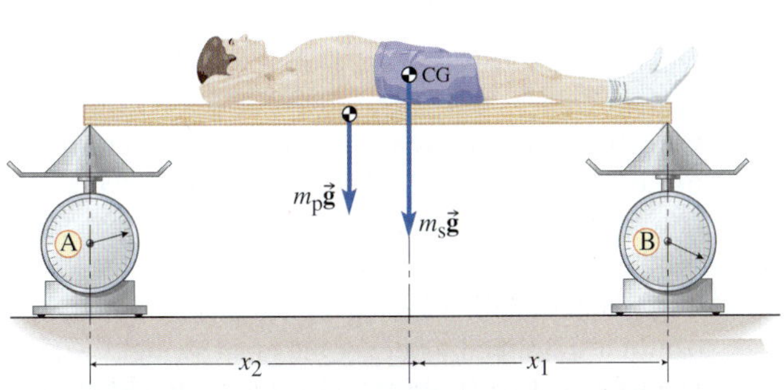

Problem 90

91. The posture of small animals may prevent them from being blown over by the wind. For example, with wind blowing from the side, a small insect stands with bent legs; the more bent the legs, the lower the body and the smaller the angle θ. The wind exerts a force on the insect, which causes a torque about the point where the downwind feet touch. The torque due to the weight of the insect must be equal and opposite to keep the insect from being blown over. For example, the drag force on a blowfly due to a sideways wind is $F_{\text{wind}} = cAv^2$, where v is the velocity of the wind, A is the cross-sectional area on which the wind is blowing, and $c \approx 1.3\ \text{N}\cdot\text{s}^2\cdot\text{m}^{-4}$. (a) If the blowfly has a cross-sectional side area of $0.10\ \text{cm}^2$, a mass of 0.070 g, and crouches such that $\theta = 30.0°$, what is the maximum wind speed in which the blowfly can stand? (Assume that the drag force acts at the center of gravity.) (b) How about if it stands such that $\theta = 80.0°$? (c) Compare with the maximum wind speed that a dog can withstand, if the dog stands such that $\theta = 80.0°$, has a cross-sectional area of $0.030\ \text{m}^2$, and weighs 10.0 kg. (Assume the same value of c.)

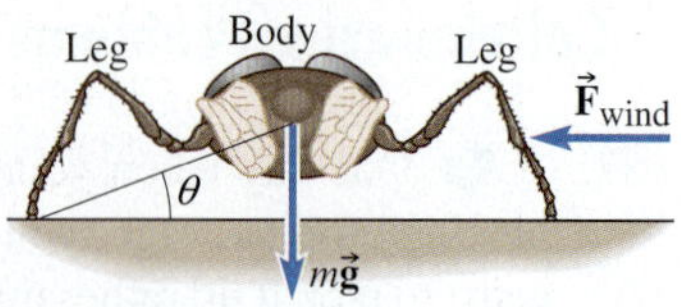

Comprehensive Problems

92. The Moon's distance from Earth varies between 3.56×10^5 km at perigee and 4.07×10^5 km at apogee. What is the ratio of its orbital speed around Earth at perigee to that at apogee?

93. A ceiling fan has four blades, each with a mass of 0.35 kg and a length of 60 cm. Model each blade as a rod connected to the fan axle at one end. When the fan is turned on, it takes 4.35 s for the fan to reach its final angular speed of 1.8 rev/s. What torque was applied to the fan by the motor? Ignore torque due to the air.

94. The distance from the center of the breastbone to a man's hand, with the arm outstretched and horizontal to the floor, is 1.0 m. The man is holding a 10.0 kg dumbbell, oriented vertically, in his hand, with the arm horizontal. What is the torque due to this weight about a horizontal axis through the breastbone perpendicular to his chest?

95. A uniform rod of length L is free to pivot around an axis through its upper end. If it is released from rest when horizontal, at what speed is the lower end moving at its lowest point? [*Hint*: The gravitational potential energy change is determined by the change in height of the center of gravity.]

96. ✦ A gymnast is performing a giant swing on the high bar. In a simplified model of the giant swing, treat the gymnast as a rigid object that swings around the bar without friction. With what angular speed should he be moving at the bottom of the giant swing in order to make it all the way around? The distance from the bar to his feet is 2.0 m and his center of gravity is 1.0 m from his feet. [*Note*: The bar can either push or pull on the gymnast, depending on the gymnast's speed and position.]

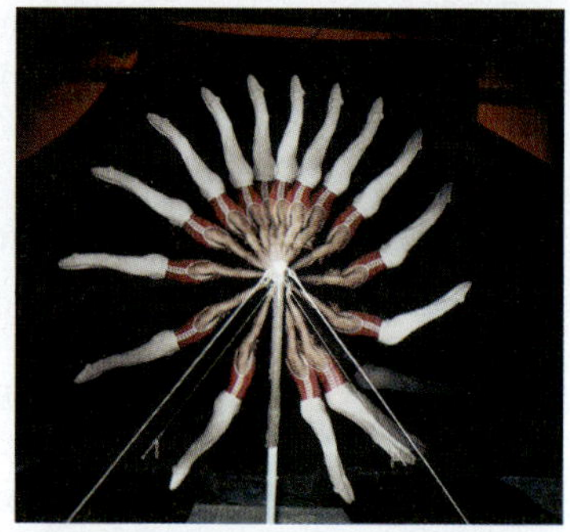

Problem 96. Notice that the angular speed is much greater at the bottom of the swing.

97. ✦ The 12.2 m crane weighs 18 kN and is lifting a 67 kN load. The hoisting cable (tension T_1) passes over a pulley at the top of the crane and attaches to an electric winch in the cab. The pendant cable (tension T_2), which supports the crane, is fixed to the top of the crane. Find the tensions in the two cables and the force $\vec{\mathbf{F}}_\text{p}$ at the pivot.

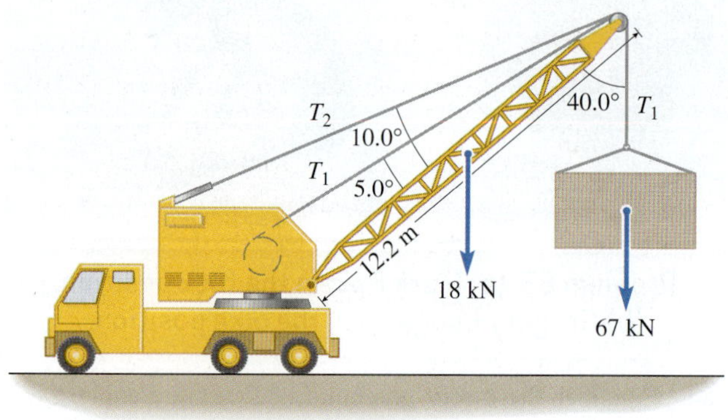

98. A collection of objects is set to rolling, without slipping, down a slope inclined at 30°. The objects are a solid sphere, a hollow sphere, a solid cylinder, and a hollow cylinder. A frictionless cube is also allowed to slide down the same incline. Rank the order in which they arrive at the finish line.

99. A uniform cylinder with a radius of 15 cm has been attached to two cords and the cords are wound around it and hung from the ceiling. The cylinder is released from rest, and the cords unwind as the cylinder descends. (a) What is the acceleration of the cylinder? (b) If the mass of the cylinder is 2.6 kg, what is the tension in each cord?

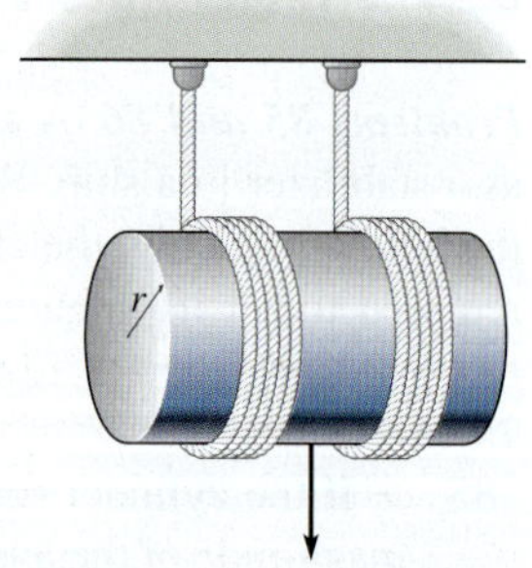

100. A grinding wheel, with a mass of 20.0 kg and a radius of 22.4 cm, is a uniform cylindrical disk. (a)

Find the rotational inertia of the wheel about its central axis. (b) When the grinding wheel's motor is turned off, friction causes the wheel to slow from 1200 rev/min to rest in 60.0 s. What torque must the motor provide to accelerate the wheel from rest to 1200 rev/min in 4.00 s? Assume that the frictional torque is the same regardless of whether the motor is on or off.

101. A 0.185 kg spherical steel ball is used in a pinball machine. The ramp is 1.35 m long and tilted at an angle of 5.00°. Just after a flipper hits the ball at the bottom of the ramp, the ball has an initial speed of 2.20 m/s. What is the speed of the ball when it reaches the top of the pinball machine?

102. A rotating star collapses under the influence of gravitational forces to form a pulsar. The radius of the pulsar is 1.0×10^{-4} times the radius of the star before collapse. There is no change in mass. In both cases, the mass of the star is uniformly distributed in a spherical shape. Find the ratios of the (a) angular momentum, (b) angular velocity, and (c) rotational kinetic energy of the star after collapse to the values before collapse. (d) If the period of the star's rotation before collapse is 1.0×10^7 s, what is its period after collapse?

103. A 5.60 kg uniform door is 0.760 m wide by 2.030 m high, and is hung by two hinges, one at 0.280 m from the top and one at 0.280 m from the bottom of the door. If the vertical components of the forces on each of the two hinges are identical, find the vertical and horizontal force components acting on each hinge due to the door.

104. A uniform solid cylinder rolls without slipping or sliding down an incline. The angle of inclination is 60.0°. Use energy considerations to find the cylinder's speed after it has traveled a distance of 30.0 cm along the incline.

105. Ⓒ In a motor, a flywheel (solid disk of radius R and mass M) is rotating with angular velocity ω_i. When the clutch is released, a second disk (radius r and mass m) initially not rotating is brought into frictional contact with the flywheel. The two disks spin around the same axle with frictionless bearings. After a short time, friction between the two disks brings them to a common angular velocity. (a) Ignoring external influences, what is the final angular velocity? (b) Does the total angular momentum of the two change? If so, account for the change. If not, explain why it does not. (c) Repeat (b) for the rotational kinetic energy.

106. Ⓒ A person on a bicycle (combined total mass 80.0 kg) starts from rest and coasts down a hill to the bottom 20.0 m below. Each wheel can be treated as a hoop with mass 1.5 kg and radius 40 cm. Ignore friction and air resistance. (a) Find the speed of the bike at the bottom. (b) Would the speed at the bottom be the same for a less massive rider? Explain.

107. ✦ A painter (mass 61 kg) is walking along a trestle, consisting of a uniform plank (mass 20.0 kg, length 6.00 m) balanced on two sawhorses. Each sawhorse is placed 1.40 m from an end of the plank. A paint bucket (mass 4.0 kg, diameter 0.28 m) is placed as close as possible to the right-hand edge of the plank while still having the whole bucket in contact with the plank. (a) How close to the right-hand edge of the plank can the painter walk before tipping the plank and spilling the paint? (b) How close to the left-hand edge can the same painter walk before causing the plank to tip? [*Hint*: As the painter walks toward the right-hand edge of the plank and the plank starts to tip clockwise, what is the force acting upward on the plank from the left-hand sawhorse support?]

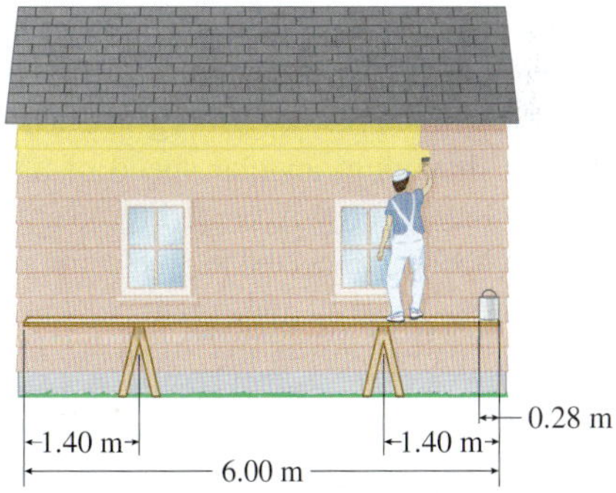

108. ✦ An experimental flywheel, used to store energy and replace an automobile engine, is a solid disk of mass 200.0 kg and radius 0.40 m. (a) What is its rotational inertia? (b) When driving at 22.4 m/s (50 mi/h), the fully energized flywheel is rotating at an angular speed of 3160 rad/s. What is the initial rotational kinetic energy of the flywheel? (c) If the total mass of the car is 1000.0 kg, find the ratio of the initial rotational kinetic energy of the flywheel to the translational kinetic energy of the car. (d) If the force of air resistance on the car is 670.0 N, how far can the car travel at a speed of 22.4 m/s (50 mi/h) with the initial stored energy? Ignore losses of mechanical energy due to means other than air resistance.

109. A flat object in the xy-plane is free to rotate about the z-axis. The gravitational field is uniform in the $-y$-direction. Think of the object as a large number of particles with masses m_n located at coordinates (x_n, y_n), as in the figure. (a) Show that the torques on the particles about the z-axis can be written $\tau_n = -x_n m_n g$. (b) Show that if the center of gravity is located at (x_{CG}, y_{CG}), the total torque due to gravity on the object must be $\Sigma \tau_n = -x_{CG} M g$, where M is the total mass of the object. (c) Show that $x_{CG} = x_{CM}$. (This same line of reasoning can be applied to objects that are not flat

and to other axes of rotation to show that $y_{CG} = y_{CM}$ and $z_{CG} = z_{CM}$.)

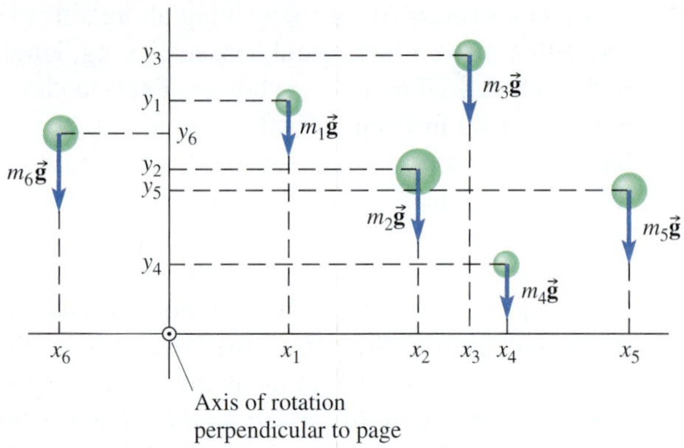

Problem 109

110. The operation of the Princeton Tokomak Fusion Test Reactor requires large bursts of energy. The power needed exceeds the amount that can be supplied by the utility company. Prior to pulsing the reactor, energy is stored in a giant flywheel of mass 7.27×10^5 kg and rotational inertia 4.55×10^6 kg·m^2. The flywheel rotates at a maximum angular speed of 386 rev/min. When the stored energy is needed to operate the reactor, the flywheel is connected to an electrical generator, which converts some of the rotational kinetic energy into electric energy. (a) If the flywheel is a uniform disk, what is its radius? (b) If the flywheel is a hollow cylinder with its mass concentrated at the rim, what is its radius? (c) If the flywheel slows to 252 rev/min in 5.00 s, what is the average power supplied by the flywheel during that time?

111. ✦ A box of mass 42 kg sits on top of a ladder. Ignoring the weight of the ladder, find the tension in the rope. Assume that the rope exerts horizontal forces on the ladder at each end. [*Hint*: Use a symmetry argument; then analyze the forces and torques on one side of the ladder.]

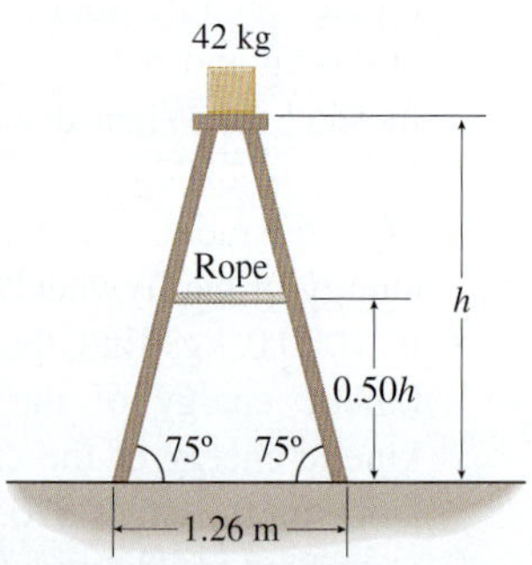

112. Ⓒ Nina wants to lean a ladder of mass 15 kg and length 8.0 m against a wall. She lifts one end over her head. Then she "walks" her hands from rung to rung toward the other end, which rests on the ground. (a) When she is holding the ladder 2.0 m from the end where she started, what vertical force does she exert on the ladder? (b) To "walk" more than 4.0 m along the ladder, she will need a helper. Explain why. What should the helper do?

113. A crustacean (*Hemisquilla ensigera*) rotates its anterior limb to strike a mollusk, intending to break it open. The limb reaches an angular velocity of 175 rad/s in 1.50 ms. We can approximate the limb as a thin rod rotating about an axis perpendicular to one end (the joint where the limb attaches to the crustacean). (a) If the mass of the limb is 28.0 g and the length is 3.80 cm, what is the rotational inertia of the limb about that axis? (b) If the extensor muscle is 3.00 mm from the joint and acts perpendicular to the limb, what is the muscular force required to achieve the blow?

114. A 2.0 kg uniform flat disk is thrown into the air with a linear speed of 10.0 m/s. As it travels, the disk spins at 3.0 rev/s. If the radius of the disk is 10.0 cm, what is the magnitude of its angular momentum?

115. ✦ Ⓒ A hoop of 2.00 m circumference is rolling down an inclined plane of length 10.0 m in a time of 10.0 s. It started out from rest. (a) What is its angular velocity when it arrives at the bottom? (b) If the mass of the hoop, concentrated at the rim, is 1.50 kg, what is the angular momentum of the hoop when it reaches the bottom of the incline? (c) What force(s) supplied the net torque to change the hoop's angular momentum? Explain. [*Hint*: Use a rotation axis through the hoop's center.] (d) What is the magnitude of this force?

116. A large clock has a second hand with a mass of 0.10 kg concentrated at the tip of the pointer. (a) If the length of the second hand is 30.0 cm, what is its angular momentum? (b) The same clock has an hour hand with a mass of 0.20 kg concentrated at the tip of the pointer. If the hour hand has a length of 20.0 cm, what is its angular momentum?

117. ✦ A planet moves around the Sun in an elliptical orbit (see Fig. 8.40). (a) Show that the external torque acting on the planet about an axis through the Sun is zero. (b) Since the torque is zero, the planet's angular momentum about this axis is constant. Write an expression for the planet's angular momentum in terms of its mass m, its distance r from the Sun, and its angular velocity ω. (c) Given r and ω, how much area is swept out during a short time Δt? [*Hint*: Think of the area as a fraction of the area of a *circle*, like a slice of pie; if Δt is short enough, the radius of the orbit during that time is nearly constant.] (d) Show that the area swept out per unit time is constant. You have just proved Kepler's second law!

118. ✦ A 68 kg woman stands straight with both feet flat on the floor. Her center of gravity is a horizontal distance of 3.0 cm in front of a line that connects her two ankle joints. The Achilles tendon attaches the calf muscle to the foot a distance of 4.4 cm behind the ankle joint. If the Achilles tendon is inclined at an angle of 81° with respect to the horizontal, find the force that each calf muscle needs to exert while she is standing. [*Hint*: Consider the equilibrium of the part of the body *above* the ankle joint.]

119. A merry-go-round (radius R, rotational inertia I_i) spins with negligible friction. Its initial angular velocity is ω_i. A child (mass m) on the merry-go-round moves from the center out to the rim. (a) Calculate the angular

velocity after the child moves out to the rim. (b) Calculate the rotational kinetic energy and angular momentum of the system (merry-go-round + child) before and after.

120. ✦ A spool of thread of mass m rests on a plane inclined at angle θ. The end of the thread is tied as shown. The outer radius of the spool is R, and the inner radius (where the thread is wound) is r. The rotational inertia of the spool is I. Give all answers in terms of m, θ, R, r, I, and g. (a) If there is no friction between the spool and the incline, describe the motion of the spool and calculate its acceleration. (b) If the coefficient of friction is large enough to keep the spool from slipping, calculate the magnitude and direction of the frictional force. (c) What is the minimum possible coefficient of friction to keep the spool from slipping in part (b)?

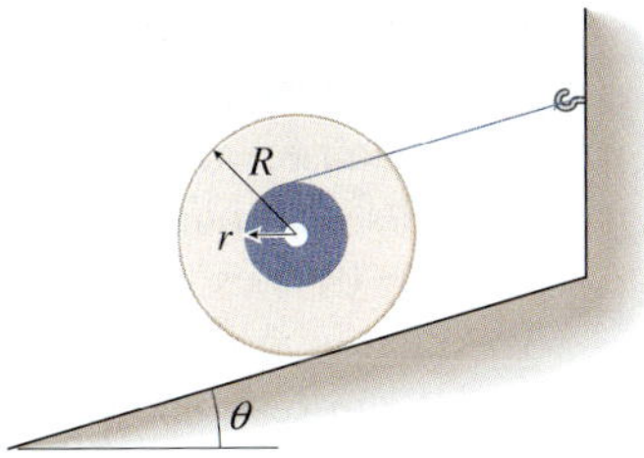

121. A bicycle travels up an incline at constant velocity. The magnitude of the frictional force due to the road on the rear wheel is $f = 3.8$ N. The upper section of chain pulls on the sprocket wheel, which is attached to the rear wheel, with a force $\vec{\mathbf{F}}_C$. The lower section of chain is slack. If the radius of the rear wheel is 6.0 times the radius of the sprocket wheel, what is the magnitude of the force $\vec{\mathbf{F}}_C$ with which the chain pulls?

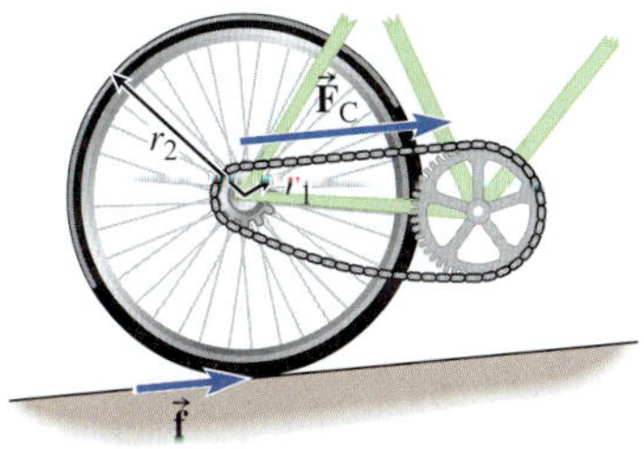

122. ✦ A circus roustabout is attaching the circus tent to the top of the main support post of length L when the post suddenly breaks at the base. The worker's weight is negligible relative to that of the uniform post. What is the speed with which the roustabout reaches the ground if (a) he jumps at the instant he hears the post crack or (b) if he clings to the post and rides to the ground with it? (c) Which is the safest course of action for the roustabout?

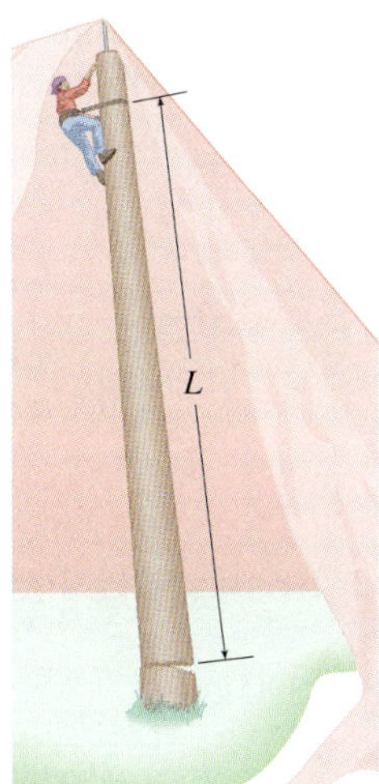

123. ✦ Ⓒ A student stands on a platform that is free to rotate and holds two dumbbells, each at a distance of 65 cm from his central axis. Another student gives him a push and starts the system of student, dumbbells, and platform rotating at 0.50 rev/s. The student on the platform then pulls the dumbbells in close to his chest so that they are each 22 cm from his central axis. Each dumbbell has a mass of 1.00 kg and the rotational inertia of the student, platform, and dumbbells is initially 2.40 kg·m^2. Model each arm as a uniform rod of mass 3.00 kg with one end at the central axis; the length of the arm is initially 65 cm and then is reduced to 22 cm. What is his new rate of rotation?

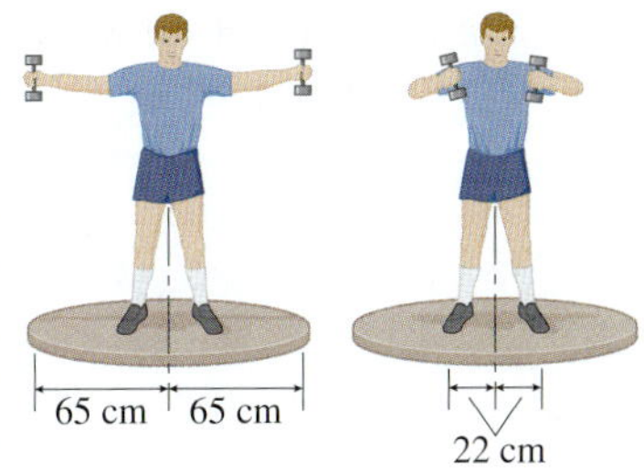

124. A person places his hand palm downward on a scale and pushes down on the scale until it reads 96 N. The triceps muscle is responsible for this arm extension force. Find the force exerted by the triceps muscle. The bottom of the triceps muscle is 2.5 cm to the left of the elbow joint, and the palm is pushing at approximately 38 cm to the right of the elbow joint.

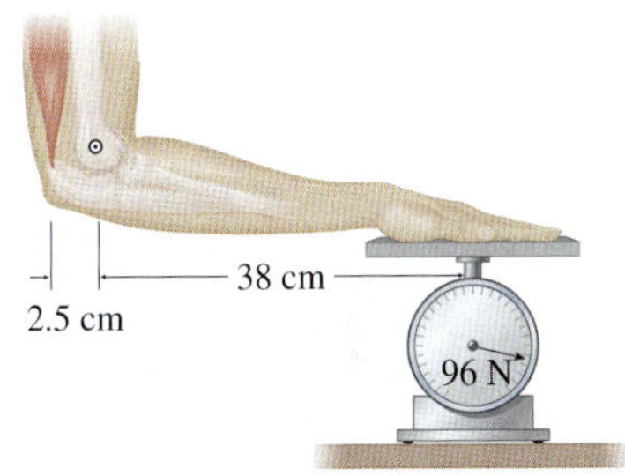

125. ✦ (a) Redo Example 8.7 to find an algebraic solution for d in terms of M, m, μ_s, L, and θ. (b) Use this expression to show that placing the ladder at a larger angle θ (that is, more nearly vertical) enables the person to climb farther up the ladder without having it slip, all other things being equal. (c) Using the numerical values from Example 8.7, find the minimum angle θ that enables the person to climb all the way to the top of the ladder.

Review and Synthesis

126. A block of mass m_2 hangs from a rope. The rope wraps around a pulley of rotational inertia I and then attaches to a second block of mass m_1, which sits on a frictionless table. What is the acceleration of the blocks when they are released?

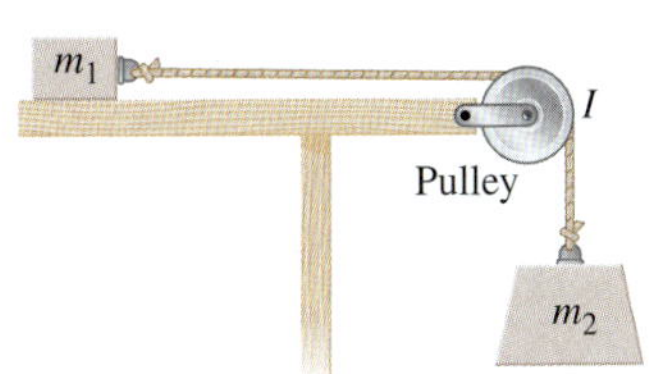

127. A modern sculpture has a large horizontal spring, with a spring constant of 275 N/m, that is attached to a 53.0 kg piece of uniform metal at its end and holds the metal at an angle of 50.0° above the horizontal direction.

The other end of the metal is wedged into a corner as shown. By how much has the spring stretched?

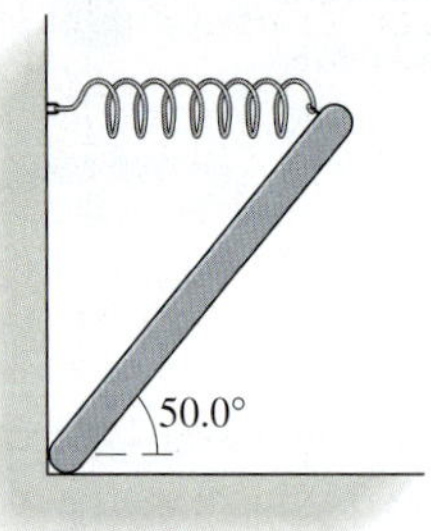

128. A hollow cylinder rolls without slipping or sliding along a horizontal surface toward an incline. (a) If the cylinder's speed is 3.00 m/s at the base of the incline and the angle of inclination is 37.0°, how far up the incline does the cylinder travel before coming to a stop? (b) What is the cylinder's acceleration?

129. ✦ A bicycle and its rider have a total mass of 74.0 kg. Each of its wheels can be modeled as a thin hoop with mass 1.30 kg and diameter of 70.0 cm. When the brakes are applied, two brake pads squeeze the rims of each wheel. Assume that the four brake pads exert normal forces on the wheels that are equal and constant in magnitude. The coefficient of kinetic friction between a brake pad and a wheel is 0.90. The bicycle is moving on level ground with a linear speed of 7.5 m/s. When the brakes are applied, the bicycle is stopped in 4.5 s. Find the magnitudes of (a) the static frictional force exerted by the road on each tire; (b) the angular acceleration of the wheels; (c) the net torque on each wheel; and (d) the normal force applied to a wheel by each of the brake pads.

130. Consider the apparatus shown in the figure (not to scale). The pulley, which can be treated as a uniform disk, has a mass of 60.0 g and a radius of 3.00 cm. The spool also has a radius of 3.00 cm. The rotational inertia of the spool, axle, and paddles about their axis of rotation is 0.00140 kg·m^2. The block has a mass of 0.870 kg and is released from rest. After the block has fallen a distance of 2.50 m, it has a speed of 3.00 m/s. How much energy has been delivered to the fluid in the beaker?

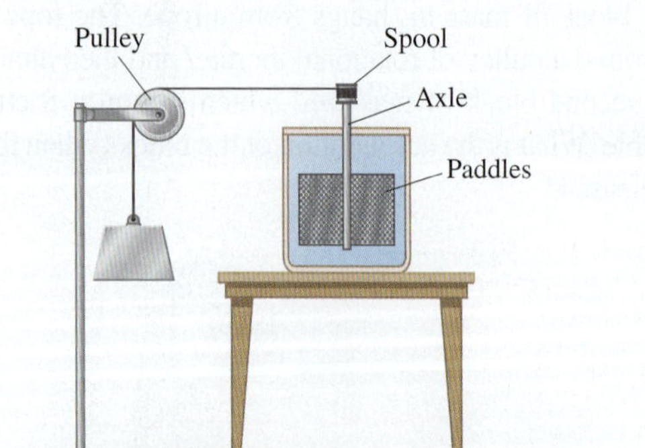

131. A uniform disk is rotated about its symmetry axis. The disk goes from rest to an angular speed of 11 rad/s in a time of 0.20 s with constant angular acceleration. The rotational inertia and radius of the disk are 1.5 kg·m^2 and 11.5 cm, respectively. (a) What is the angular acceleration during the 0.20 s interval? (b) What is the net torque on the disk during this time? (c) After the applied torque stops, a frictional torque remains. This torque causes an angular acceleration of magnitude 9.8 rad/s^2. Through what total angle θ (starting from time $t = 0$) does the disk rotate before coming to rest? (d) What is the speed of a point halfway between the rim of the disk and its rotation axis 0.20 s after the applied torque is removed?

132. A child's toy is made of a 12.0 cm radius rotating wheel that picks up 1.00 g pieces of candy in a pocket at its lowest point, brings the candy to the top, then releases it. The frequency of rotation is 1.60 Hz. (a) How far from its starting point does the candy land? (b) What is the radial acceleration of the candy when it is on the wheel?

133. ✦ You are mowing the lawn on a hill near your house when the lawnmower blade strikes a stone of mass 100 g and sends it flying horizontally toward a window. The lawnmower blade can be modeled as a thin rod with a mass of 2.0 kg and a length of 50 cm rotating about its center. The stone impacts the blade near one end and is ejected with a velocity perpendicular to the rotation axis and the blade at the moment of collision. As a result of the impact, the blade slows from 60 rev/s to 55 rev/s. The window is 1.00 m in height, and its center is located 10.0 m away and at the same height as the lawnmower. (a) With what speed is the stone shot out by the mower? [*Hint*: The external force due to the lawnmower's drive shaft on the system (blade + stone) cannot be ignored during the collision, but the external *torque* about the shaft *can* be ignored. The angular momentum of the stone just after impact can be calculated from its tangential velocity and its distance from the rotation axis.] (b) Ignoring air resistance, will the stone hit the window?

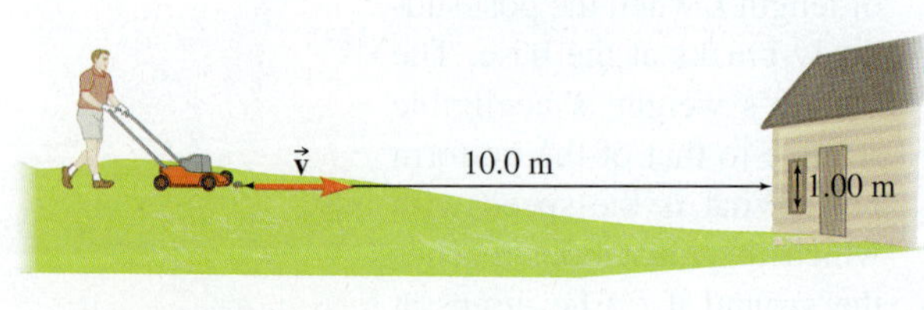

Answers to Practice Problems

8.1 390 kg·m^2

8.2 $v = \sqrt{\dfrac{2m_2gh}{m_1 + m_2 + I/R^2}}$

8.3 53 N; 8.4 N·m

8.4 −65 N·m

8.5 8.3 J

8.6 left support, downward; right support, upward

8.7 0.27

8.8 57 N, downward

8.9 It must lie in the same vertical plane as the two ropes holding up the rings. Otherwise, the gravitational force would have a nonzero lever arm with respect to a horizontal axis that passes through the contact points between his hands and the rings; thus, gravity would cause a net torque about that axis.

8.10 460 N

8.11 (a) 2380 rad; (b) 3.17 kJ; (c) 1.34 N·m

8.12 solid ball, $\frac{2}{7}$; hollow ball, $\frac{2}{5}$

8.13 $\frac{1}{28}g \sin \theta$

8.14 5% increase

8.15 16 cm/s

Answers to Checkpoints

8.1 Rotational inertia involves distances from masses to the rotation axis; distances *along* the rotation axis are irrelevant. Another way to see it: cut the cylinder or disk into a large number of thin disks with the same radius. Each thin disk has rotational inertia $I_i = \frac{1}{2}m_iR^2$. Now add up the rotational inertias of the thin disks: $I = \Sigma I_i = \Sigma \frac{1}{2}m_iR^2 = \frac{1}{2}R^2 \Sigma m_i = \frac{1}{2}MR^2$.

8.2 The longer handle lets you push at a greater distance from the rotation axis. Thus, you can exert a larger torque.

8.4 Yes in both cases. Torque depends not only on the magnitude and direction of the force but also on the point where the force is applied. Two forces that do not add to zero can produce torques that add to zero due to different lever arms. Then the net torque is zero and the net force nonzero; the object is in rotational equilibrium but not in translational equilibrium. Similarly, two forces that add to zero can have different lever arms and produce torques that do not add to zero. In this case the net force is zero and the net torque is nonzero; the object is in translational equilibrium but not in rotational equilibrium.

8.7 (a) falling without spinning; (b) spinning about a fixed axis; (c) rolling without slipping along a surface

8.8 Yes. If friction is negligible, the external torque is zero so her angular momentum does not change. Extending her arms and leg makes her rotational inertia increase back to its initial value, so her angular velocity decreases to its initial value.

CHAPTER

9 Fluids

A hippopotamus in Kruger National Park, South Africa, wants to feed on the vegetation growing on the bottom of a pond. When the hippo wades into the pond, it floats. How does a hippopotamus get its floating body to sink to the bottom of a pond? (See p. 336 for the answer.)

BIOMEDICAL APPLICATIONS

- Blood flow and blood pressure (Secs. 9.2, 9.5, 9.9; Exs. 9.9, 9.12; CQ 7; Probs. 24–26, 48, 66–68, 84–86, 97)
- Arterial flutter and aneurisms (Sec. 9.8; Prob. 98)
- IVs, syringes, and blood-sucking bugs (Probs. 61, 62, 68)
- Animals manipulating their densities to float or sink (Sec. 9.6; Ex. 9.8; PP 9.8; Probs. 40 and 41)
- Surface tension in the lungs (Sec. 9.11; Ex. 9.14; CQ 14)
- Specific-gravity measurements of blood and urine (Sec. 9.6)
- Pressure on divers and animals underwater (Ex. 9.3; Probs. 18, 19, and 99)

Concepts & Skills to Review

- conservation of energy (Chapter 6)
- force as rate of change of momentum (Section 7.3)
- conservation of momentum in collisions (Sections 7.7 and 7.8)
- equilibrium (Section 4.2)

9.1 STATES OF MATTER

Ordinary matter is usually classified into three familiar states or phases: solids, liquids, and gases. Solids tend to hold their shapes. Many solids are quite rigid; they are not easily deformed by external forces because each atom or molecule is held in a particular position by the forces exerted by its neighbors. Although the atoms or molecules vibrate around fixed equilibrium positions, they do not have enough energy to break the bonds with their neighbors. To bend an iron bar, for example, the arrangement of the atoms must be altered, which is not easy to do. A blacksmith heats iron in a forge to loosen the bonds between atoms so that he can bend the metal into the desired shape.

In contrast to solids, liquids and gases do not hold their shapes. A liquid flows and takes the shape of its container and a gas expands to fill its container. **Fluids**—both liquids and gases—are easily deformed by external forces. This chapter deals mainly with properties that are common to both liquids and gases.

The atoms or molecules in a fluid do not have fixed positions, so a fluid does not have a definite shape. An applied force can easily make a fluid flow; for instance, the squeezing of the heart muscle exerts a force that pumps blood through the blood vessels of the body. However, this squeezing does not change the *volume* of the blood by much. In many situations we can think of liquids as **incompressible**—that is, as having a fixed volume that is impossible to change. The shape of the liquid can be changed by pouring it from a container of one shape into a container of a different shape, but the volume of the liquid remains the same.

In most liquids, the atoms or molecules are almost as closely packed as those in the solid phase of the same material. The intermolecular forces in a liquid are almost as strong as those in solids, but the molecules are not locked in fixed positions as they are in solids. That is why the volume of the liquid can remain nearly constant while the shape is easily changed. Water is one of the exceptions: in cold water, the molecules in the liquid phase are actually *more* closely packed than those in the solid phase (ice).

Gases, on the other hand, cannot be characterized by a definite volume nor by a definite shape. A gas expands to fill its container and can easily be compressed. The molecules in a gas are very far apart compared to the molecules in liquids and solids. The molecules are almost free of interactions with each other except when they collide.

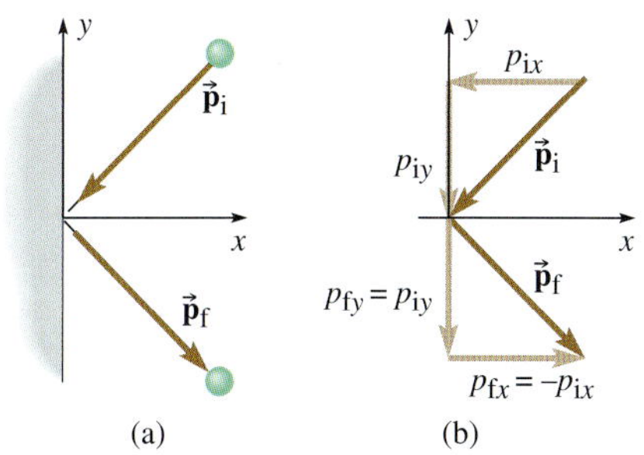

Figure 9.1 (a) A single fluid molecule bouncing off a container wall. (b) In this elastic collision, the *y*-component of the momentum is unchanged, while the *x*-component reverses direction.

9.2 PRESSURE

Microscopic Origin of Pressure A **static** fluid does not flow; it is everywhere at rest. In the study of fluid statics (*hydrostatics*), we also assume that any solid object in contact with the fluid—whether a vessel containing the fluid or an object submerged in the fluid—is at rest. The atoms or molecules in a static fluid are not themselves static; they are continually moving. The motion of people bouncing up and down and bumping into each other in a mosh pit gives you a rough idea of the motion of the closely packed atoms or molecules in a liquid; in gases, the atoms or molecules are much farther apart than in liquids, so they travel greater distances between collisions.

Fluid pressure is caused by collisions of the fast-moving atoms or molecules of a fluid. When a single molecule hits a container wall and rebounds, its momentum changes due to the force exerted on it by the wall. Figure 9.1a shows a molecule of

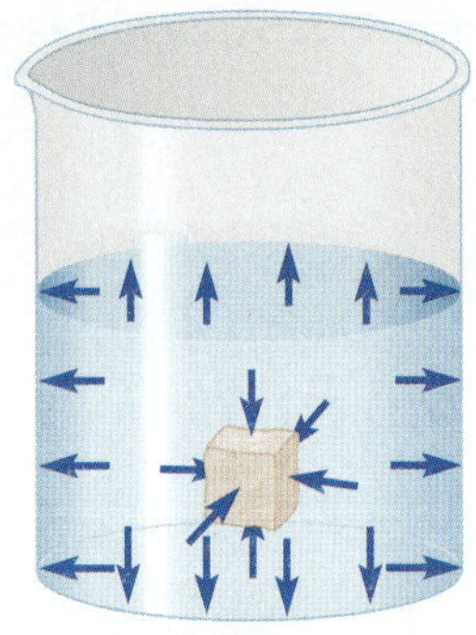

Figure 9.2 Forces due to a static fluid acting on the walls of the container and on a submerged object.

a fluid within a container making an elastic collision with one of the container walls. In this case, the y-component of momentum is unchanged, while the x-component reverses direction (Fig. 9.1b). The momentum change is in the $+x$-direction, which occurs because the wall exerts a force to the right on the molecule. By Newton's third law, the molecule exerts a force to the left on the wall during the collision. If we consider all the molecules colliding with this wall, *on average* they exert no force on the wall in the $\pm y$-direction, but all exert a force in the $-x$-direction. The frequent collisions of fluid molecules with the walls of the container cause a net force pushing outward on the walls.

Definition of Pressure A static fluid exerts a force on any surface with which it comes in contact; the direction of the force is perpendicular to the surface (Fig. 9.2). A static fluid *cannot* exert a force *parallel* to the surface. If it did, the surface would exert a force on the fluid parallel to the surface, by Newton's third law. This force would make the fluid flow along the surface, contradicting the premise that the fluid is static.

The *average pressure* of a fluid at points on a planar surface is

Average pressure

$$P_{av} = \frac{F}{A} \tag{9-1}$$

where F is the magnitude of the force acting perpendicularly to the surface, and A is the area of the surface. By imagining a tiny surface at various points within the fluid and measuring the force that acts on it, we can define the pressure at any point within the fluid. In the limit of a small area A, $P = F/A$ is the **pressure** of the fluid.

Pressure is a scalar quantity; it does not have a direction. The force acting on an object submerged in a fluid—or on some portion of the fluid itself—is a vector quantity; its direction is perpendicular to the contact surface. Pressure is defined as a scalar because, at a given location in the fluid, the magnitude of the force per unit area is the same for any orientation of the surface. The molecules in a static fluid are moving in random directions; there can be no preferred direction since that would constitute fluid flow. There is no reason that a surface would have a greater number of collisions, or collisions with more energetic molecules, for one particular surface orientation compared with any other orientation.

The SI unit for pressure is the newton per square meter (N/m^2), which is named the *pascal* (symbol Pa) after the French scientist Blaise Pascal (1623–1662). Another commonly used unit of pressure is the *atmosphere* (atm). One atmosphere is the *average* air pressure at sea level. The conversion factor between atmospheres and pascals is

$$1 \text{ atm} = 101.3 \text{ kPa}$$

Other units of pressure in common use are introduced in Section 9.5.

CHECKPOINT 9.2

A quarter (diameter 2.4 cm) and a dime (diameter 1.8 cm) rest on the bottom of a swimming pool. The water exerts a downward force on the upper surface of each coin. Assuming the water pressure is the same on both coins, by what factor is this downward force on the quarter larger than that on the dime?

Example 9.1

Pressure due to Stiletto-Heeled Shoes

A young woman weighing 534 N (120 lb) walks to her bedroom while wearing tennis shoes. She then gets dressed for her evening date, putting on her new stiletto-heeled dress shoes. The area of the heel section of her tennis shoe is 60.0 cm^2 and the area of the heel of her dress shoe is 1.00 cm^2. For each pair of shoes, find the average pressure caused by the heel making contact with the floor when her entire weight is supported by one heel.

Strategy The average pressure is the force applied to the floor divided by the contact area. The force that the heel exerts on the floor is 534 N. To keep the units straight, we convert the areas from square centimeters to square meters.

Solution To convert the area of the tennis shoe heel and the dress shoe heel from cm^2 to m^2, we use the conversion $(1\text{ m})^2 = (10^2\text{ cm})^2$. For the tennis shoe heel:

$$A = 60.0\text{ cm}^2 \times \left(\frac{1\text{ m}}{10^2\text{ cm}}\right)^2 = 6.00 \times 10^{-3}\text{ m}^2$$

For the dress shoe heel:

$$A = 1.00\text{ cm}^2 \times \left(\frac{1\text{ m}}{10^2\text{ cm}}\right)^2 = 1.00 \times 10^{-4}\text{ m}^2$$

The average pressure is the woman's weight divided by the area of the heel. For the tennis shoe:

$$P = \frac{F}{A} = \frac{534\text{ N}}{6.00 \times 10^{-3}\text{ m}^2} = 8.90 \times 10^4\text{ N/m}^2 = 89.0\text{ kPa}$$

For the stilettos:

$$P = \frac{534\text{ N}}{1.00 \times 10^{-4}\text{ m}^2} = 5.34 \times 10^6\text{ N/m}^2 = 5.34\text{ MPa}$$

Discussion In atmospheres, these pressures are 0.879 atm and 52.7 atm, respectively. The pressure due to the dress shoe is 60 times the pressure due to the tennis shoe since the same force is spread over $\frac{1}{60}$ the area.

Practice Problem 9.1 Pressure from an Ordinary Dress Shoe Heel

Fortunately for floor manufacturers, and for women's feet, stiletto heels are out of fashion more often than they are in fashion. Suppose that a woman's dress shoes have heels that are each 4.0 cm^2 in area. Find the pressure on the floor, when the entire weight is on a single heel, for such a shoe worn by the same woman as in Example 9.1. Find the factor by which this pressure exceeds the pressure from the tennis shoe heel.

Atmospheric Pressure

On the surface of the Earth, we live at the bottom of an ocean of fluid called air. The forces exerted by air on our bodies and on surfaces of other objects may be surprisingly large: 1 atm is approximately 10 N/cm^2 of surface area, or nearly 15 lb/in^2. We are not crushed by this pressure because most of the fluids in our bodies are at approximately the same pressure as the air around us. As an analogy, consider a sealed bag of potato chips. Why is the bag not crushed by the air pushing in on all sides? Because the air inside the bag is at the same pressure and pushes out on the sides of the bag. The pressure of the fluids inside our cells matches the pressure of the surrounding fluids pushing in on the cell membranes, so the cells do not rupture.

By contrast, the blood pressure in the arteries is as much as 20 kPa higher than atmospheric pressure. The strong, elastic arterial walls are stretched by the pressure of the blood inside; the walls squeeze the arterial blood to keep its higher pressure from being transmitted to other fluids in the body.

Changing weather conditions cause variations of approximately 5% in the actual value of air pressure at sea level; 101.3 kPa (1 atm) is only the *average* value. Air pressure also decreases with increasing elevation. (In Section 9.4, we study the effect of gravity on fluid pressure in detail.) The average air pressure in Leadville, Colorado, the highest incorporated city in the United States (elevation 3100 m), is 70 kPa. Some Tibetans live at altitudes of over 5000 m, where the average air pressure is only half its value at sea level. In problems, please assume that the atmospheric pressure is 1 atm unless the problem states otherwise.

9.3 PASCAL'S PRINCIPLE

If the weight of a static fluid is negligible (as, for example, in a hydraulic system under high pressure), then the pressure must be the same everywhere in the fluid. Why? In Fig. 9.3, imagine the submerged cube to be composed of the same fluid as its surroundings. Ignoring the fluid's weight, the only forces acting on the cubical piece of fluid are those due to the surrounding fluid pushing inward. The forces pushing on each pair of opposite sides of the cube must be equal in magnitude, since the fluid inside the cube is in equilibrium. Therefore, the pressure must be the same on both sides. Since we can extend this argument to any size and shape piece of fluid, *the fluid pressure must be the same everywhere in a weightless, static fluid.*

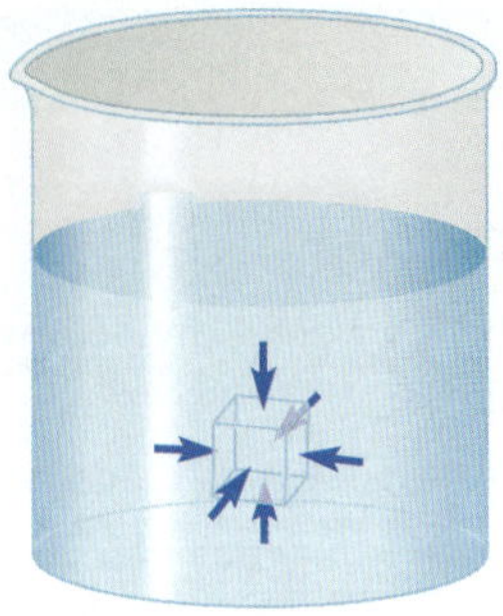

Figure 9.3 Forces acting on a cube of fluid.

More generally, when the weight of the fluid is *not* negligible, the pressure is not the same everywhere. In this case, analysis of the forces acting on a piece of fluid (see Conceptual Question 15) leads to a more general result called **Pascal's principle**.

Pascal's Principle

A *change in pressure* at any point in a confined fluid is transmitted everywhere throughout the fluid.

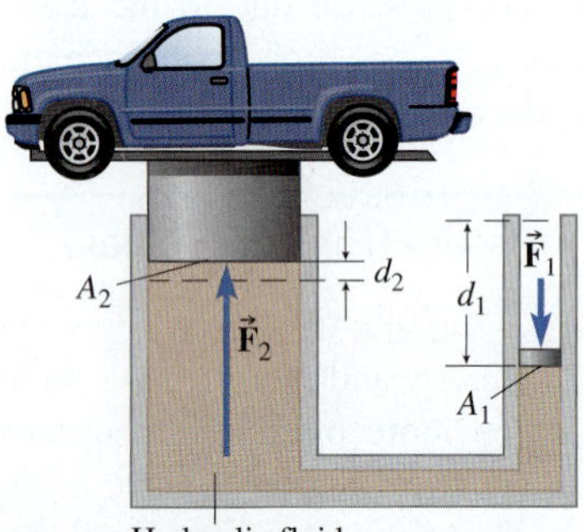

Figure 9.4 Simplified diagram of a hydraulic lift. Notice that piston 1 has to move a great distance (d_1) to lift the truck a much smaller distance (d_2). In a real hydraulic lift, piston 1 is usually replaced by a pump that draws fluid from a reservoir and pushes it into the hydraulic system.

Applications of Pascal's Principle: Hydraulic Lifts, Brakes, and Controls When a truck needs to have its muffler replaced, it is lifted into the air by a mechanism called a hydraulic lift (Fig. 9.4). A force is exerted on a liquid by a piston with a relatively small area; the resulting increase in pressure is transmitted everywhere throughout the liquid. Then the truck is lifted by the fluid pressure on a piston of much larger area. The upward force on the truck is much larger than the force applied to the small piston. Pascal's principle has many other applications, such as the hydraulic brakes in cars and trucks and the hydraulic controls in airplanes.

To analyze the forces in the hydraulic lift, let force F_1 be applied to the small piston of area A_1, causing a pressure increase:

$$\Delta P = \frac{F_1}{A_1}$$

A truck is supported by a piston of much larger area A_2 on the other side of the lift. The increase in pressure due to the small piston is transmitted everywhere in the liquid. Ignoring the weight of the fluid (or assuming the two pistons to be at the same height), the force F_2 exerted by the fluid on the large piston is related to F_1 by

$$\frac{F_1}{A_1} = \frac{F_2}{A_2}$$

CONNECTION:

As for levers, systems of pulleys, and other simple machines, the hydraulic lift can reduce the applied *force* needed to perform a task, but the *work* done is the same.

Since A_2 is larger than A_1, the force exerted on the large piston (F_2) is larger than the force applied to the small piston (F_1). We are not getting something for nothing; just as for the two-pulley systems discussed in Section 6.2, the advantage of the smaller force applied to the small piston is balanced by a greater distance it must be moved. The small piston has to move a long distance d_1 while the large piston moves a short distance d_2. Assuming the liquid to be incompressible, the volume of fluid displaced by each piston is the same, so

$$A_1 d_1 = A_2 d_2$$

The displacements of the pistons are inversely proportional to their areas, while the forces are directly proportional to the areas; then the product of force and displacement is the same:

$$\frac{F_1}{A_1} \times A_1 d_1 = \frac{F_2}{A_2} \times A_2 d_2 \quad \Rightarrow \quad F_1 d_1 = F_2 d_2$$

The work (force times displacement) done by moving the small piston equals the work done by the large piston in raising the truck.

Example 9.2

The Hydraulic Lift

In a hydraulic lift, if the radius of the smaller piston is 2.0 cm and the radius of the larger piston is 20.0 cm, what weight can the larger piston support when a force of 250 N is applied to the smaller piston?

Strategy According to Pascal's principle, the pressure increases the same amount at every point in the fluid. A natural way to work is in terms of proportions since the forces are proportional to the areas of the pistons.

Solution Since the pressure on the two pistons increases by the same amount,

$$\frac{F_1}{A_1} = \frac{F_2}{A_2}$$

Equivalently, the forces are proportional to the areas:

$$\frac{F_2}{F_1} = \frac{A_2}{A_1}$$

The ratio of the radii is $r_2/r_1 = 10$, so the ratio of the areas is $A_2/A_1 = (r_2/r_1)^2 = 100$. Then the weight that can be supported is

$$F_2 = 100F_1 = 25\,000\text{ N} = 25\text{ kN}$$

Discussion One common error in this sort of problem is to think of the area and the force as a *trade-off*—in other words, that the piston with the *large* area has the *small* force and vice versa. Since the pressures are the same, the force exerted by the fluid on either piston is proportional to the piston's area. We make the piston that lifts the truck large because we know the force on it will be large, *in direct proportion to* its area.

Practice Problem 9.2 Application of Pascal's Principle

Consider the hydraulic lift of Example 9.2. (a) What is the increase in pressure caused by the 250 N force on the small piston? (b) If the larger piston moves 5.0 cm, how far does the smaller piston move?

9.4 THE EFFECT OF GRAVITY ON FLUID PRESSURE

On a drive through the mountains or on a trip in a small plane, the feeling of our ears popping is evidence that pressure is not the same everywhere in a static fluid. Gravity makes fluid pressure increase as you move down and decrease as you move up. To understand more about this variation, we must first define the density of a fluid.

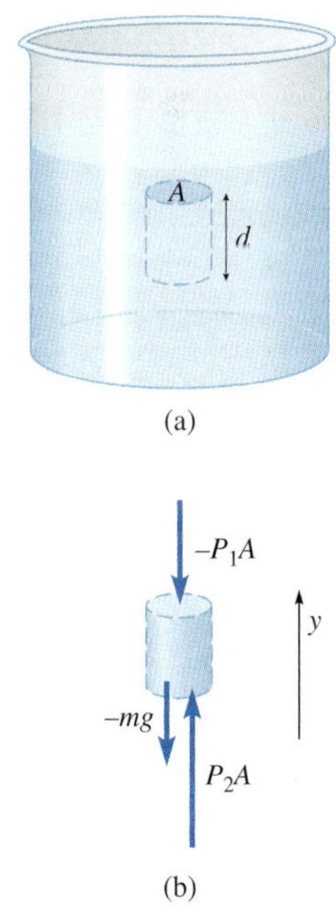

Figure 9.5 Applying Newton's second law to a cylinder of liquid tells us how pressure increases with increasing depth. (a) A cylinder of liquid of height d and area A. (b) Vertical forces on the cylinder of liquid.

Density The **density** of a substance is its mass per unit volume. The Greek letter ρ (rho) is used to represent density. The density of a uniform substance of mass m and volume V is

Density

$$\rho = \frac{m}{V} \quad (9\text{-}2)$$

The SI units of density are kilograms per cubic meter: kg/m^3. For a nonuniform substance, Eq. (9-2) defines the **average density**.

Table 9.1 lists the densities of some common substances. Note that temperatures and pressures are specified in the table. For solids and liquids, density is only weakly dependent on temperature and pressure. On the other hand, gases are highly compressible, so even a relatively small change in temperature or pressure can change the density of a gas significantly.

Pressure Variation with Depth due to Gravity Now, using the concept of density, we can find how pressure increases with depth due to gravity. Suppose we have a glass beaker containing a static liquid of uniform density ρ. Within this liquid, imagine a cylinder of liquid with cross-sectional area A and height d (Fig. 9.5a). The mass of the liquid in this cylinder is

$$m = \rho V$$

Table 9.1 Densities of Common Substances (at 0°C and 1 atm unless otherwise indicated)

Gases	Density (kg/m^3)	Liquids	Density (kg/m^3)	Solids	Density (kg/m^3)
Hydrogen	0.090	Gasoline	680	Polystyrene	100
Helium	0.18	Ethanol	790	Cork	240
Steam (100°C)	0.60	Oil	800–900	Wood (pine)	350–550
Nitrogen	1.25	Water (0°C)	999.84	Wood (oak)	600–900
Air (20°C)	1.20	Water (3.98°C)	999.98	Ice	917
Air (0°C)	1.29	Water (20°C)	998.21	Wood (ebony)	1000–1300
Oxygen	1.43	Seawater	1025	Bone	1500–2000
Carbon dioxide	1.98	Blood (37°C)	1060	Concrete	2000
		Mercury	13 600	Quartz, granite	2700
				Aluminum	2702
				Iron, steel	7860
				Copper	8920
				Lead	11 300
				Gold	19 300
				Platinum	21 500

where the volume of the cylinder is

$$V = Ad$$

The weight of the cylinder of liquid is therefore

$$mg = (\rho A d)g$$

The vertical forces acting on this column of liquid are shown in Fig. 9.5b. The pressure at the top of the cylinder is P_1 and the pressure at the bottom is P_2. Since the liquid in the column is in equilibrium, the net vertical force acting on it must be zero by Newton's second law:

$$\sum F_y = P_2 A - P_1 A - \rho A d g = 0$$

Dividing by the common factor A and rearranging yields:

Pressure variation with depth in a static fluid with uniform density

$$P_2 = P_1 + \rho g d \qquad (9\text{-}3)$$

where point 2 is a depth d below point 1

Since we can imagine a cylinder anywhere we choose within the liquid, Eq. (9-3) relates the pressure at any two points in a static liquid where point 2 is a depth d below point 1.

For gases, Eq. (9-3) can be applied as long as the depth d is small enough that changes in the density due to gravity are negligible. Since liquids are nearly incompressible, Eq. (9-3) holds to great depths in liquids.

For a liquid that is open to the atmosphere, suppose we take point 1 at the surface and point 2 a depth d below. Then $P_1 = P_{atm}$, so the pressure at a depth d below the surface is

Pressure at a depth *d* below the surface of a liquid open to the atmosphere

$$P = P_{atm} + \rho g d \qquad (9\text{-}4)$$

CHECKPOINT 9.4

Pressure in a static fluid depends on vertical position. Can it also depend on horizontal position? Explain.

Example 9.3

A Diver

A diver swims to a depth of 3.2 m in a freshwater lake. What is the increase in the force pushing in on her eardrum, compared to what it was at the lake surface? The area of the eardrum is 0.60 cm^2.

Strategy We can find the increase in pressure at a depth of 3.2 m and then find the corresponding increase in force on the eardrum. If the force on the eardrum at the surface is P_1A and the force at a depth of 3.2 m is P_2A, then the increase in the force is $(P_2 - P_1)A$.

Solution The increase in pressure depends on the depth d and the density of water. From Table 9.1, the density of water is 1000 kg/m^3 to two significant figures for any reasonable temperature.

$$P_2 - P_1 = \rho g d$$

$$\Delta P = 1000\ \text{kg/m}^3 \times 9.8\ \text{m/s}^2 \times 3.2\ \text{m}$$

$$= 31.4\ \text{kPa}$$

The increase in force on the eardrum is

$$\Delta F = \Delta P \times A$$

where $A = 0.60\ \text{cm}^2 = 6.0 \times 10^{-5}\ \text{m}^2$. Then

$$\Delta F = (3.14 \times 10^4\ \text{Pa}) \times (6.0 \times 10^{-5}\ \text{m}^2)$$

$$= 1.9\ \text{N}$$

Discussion A force also pushes *outward* on the eardrum due to the pressure inside the ear canal. If the diver descends rapidly so that the pressure inside the ear canal does not change, then a 1.9 N net force due to fluid pressure pushes inward on the eardrum. When the diver's ear "pops," the pressure inside the ear canal increases to equal the fluid pressure outside the eardrum, so that the net force due to fluid pressure on the eardrum is zero.

Practice Problem 9.3 Limits on Submarine Depth

A submarine is constructed so that it can safely withstand a pressure of 1.6×10^7 Pa. How deep may this submarine descend in the ocean if the average density of seawater is 1025 kg/m^3?

Conceptual Example 9.4

The Hydrostatic Paradox

Three vessels have different shapes, but the same base area and the same weight when empty (Fig. 9.6). The vessels are filled with water to the same level and then the air is pumped out. The top surface of the water is then at a low pressure that, for simplicity, we assume to be zero. (a) Are the water pressures at the bottom of each vessel the same? If not, which is largest and which is smallest? (b) If the three vessels containing water are weighed on a scale, do they give the same reading? If not, which weighs the most and which weighs the least? (c) If the water exerts the same downward force on the bottom of each vessel, is that force equal to the weight of water in the vessel? Is there a paradox here?

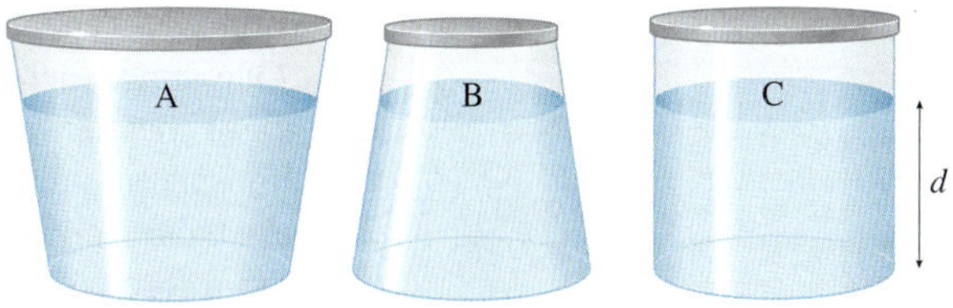

Figure 9.6
Three differently shaped vessels filled with water to same level.

[*Hint*: Think about the forces due to fluid pressure on the *sides* of the containers; do they have vertical components?]

continued on next page

Conceptual Example 9.4 continued

Solution and Discussion (a) The water at the bottom of each vessel is the same depth d below the surface. Water at the surface of each vessel is at a pressure $P_{\text{surface}} = 0$. Therefore, the pressures at the bottom must be equal:

$$P = P_{\text{surface}} + \rho g d = \rho g d$$

(b) The weight of each filled vessel is equal to the weight of the vessel itself plus the weight of the water inside. The vessels themselves have equal weights, but vessel A holds more water than C, whereas vessel B holds less water than C. Vessel A weighs the most and vessel B weighs the least.

(c) Each container supports the water inside by exerting an upward force equal in magnitude to the weight of the water. By Newton's third law, the water exerts a downward force on the container of the same magnitude. Figure 9.7 shows the forces acting on each container due to the water. In vessel C, the horizontal forces on any two diametrically opposite points on the walls of the container are equal and opposite; thus, the net force on the container walls is zero. The force on the bottom is

$$F = PA = (\rho g d)(\pi r^2)$$

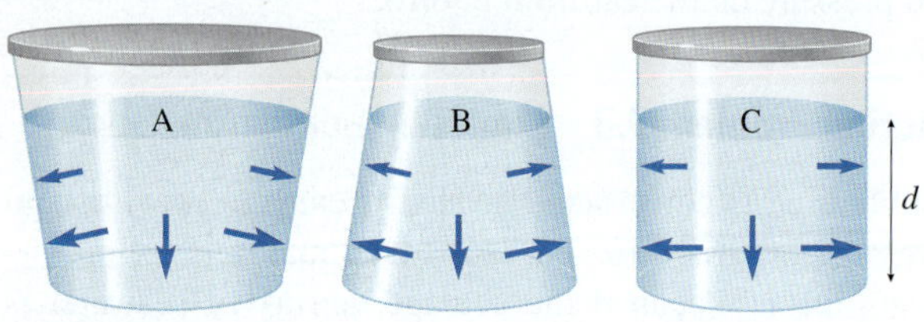

Figure 9.7
Forces exerted on the containers by the water.

The volume of water in the cylinder is $V = \pi r^2 d$, so

$$F = \rho g V = (\rho V)g = mg$$

The force on the bottom of vessel C is equal to the weight of the water, as expected. However, the force on the bottom of vessel A is less than the weight of the water in the container, while the force on the bottom of vessel B is greater than the weight of the water. Then how can the water be in equilibrium? In vessel A, the forces on the container walls have downward components as well as horizontal components. The horizontal components of the forces on any two diametrically opposite points are equal and opposite, so the horizontal components add to zero. The sum of the downward components of the forces on the walls and the downward force on the bottom of the container is equal to the weight of the water. Similarly, the forces on the walls of vessel B have upward components. In each case, the *total* force on the bottom *and sides* of the container due to the water is equal to the weight of the water.

Conceptual Practice Problem 9.4 Is Pressure Determined by Column Height?

Figure 9.8 shows a vessel with two points marked at the bottom of the water in the vessel. A narrow column of water is drawn above each point. (a) Is the pressure at point 2, P_2, the same as the pressure at point 1, P_1, even though the column of water above point 2 is not as tall? (b) Does $P = P_{\text{atm}} + \rho g d$ imply that $P_2 < P_1$? Explain.

Figure 9.8
Two different points on the bottom of an open vessel.

9.5 MEASURING PRESSURE

Many other units are used for pressure besides atmospheres and pascals. In the United States, the pressure in an automobile tire is measured in pounds per square inch (lb/in.^{2}). Weather bureaus record atmospheric pressure in bars or millibars. In the United States, television weather reports and home barometers measure pressure in inches of mercury. One atmosphere is equal to approximately 1.013 bar, 760.0 mm of mercury, or 29.9 in. of mercury. Blood pressure, the *difference* between the pressure in the blood and atmospheric pressure, is measured in millimeters of mercury (mm Hg), also called the torr. Inches or millimeters of mercury may seem like strange units for pressure: how can a force per unit area be equal to a *distance* (so many mm Hg)? There is an assumption inherent in using these pressure units that we can understand by studying the mercury manometer.

The Manometer

A mercury manometer consists of a vertical U-shaped tube, containing some mercury, with one side typically open to the atmosphere and the other connected to a vessel containing a gas whose pressure we want to measure. Figure 9.9 shows the manometer before it is connected to such a vessel. When both sides of the manometer are open to the atmosphere, the mercury levels are the same.

Now we connect an inflated balloon to the left side of the U-tube (Fig. 9.10). If the gas in the balloon is at a higher pressure than the atmosphere, the gas pushes the mercury down on the left side and forces it up on the right side. The density of a gas is small compared to the density of mercury, so every point within the gas is assumed to be at the same pressure no matter what the depth. At point B, the mercury pushes on the gas with the same magnitude force with which the gas pushes on the mercury, so point B is at the same pressure as the gas. Since point B' is at the same height within the mercury as point B, the pressure at B' is the same as at B. Point C is at atmospheric pressure.

The pressure at B is

$$P_B = P_C + \rho g d$$

where ρ is the density of mercury. The difference in the pressures on the two sides of the manometer is

$$\Delta P = P_B - P_C = \rho g d \tag{9-5}$$

Thus, the difference in mercury levels d is a measure of the pressure *difference*—commonly reported in millimeters of mercury (mm Hg).

The pressure measured when one side of the manometer is open is the *difference* between atmospheric pressure and the gas pressure rather than the absolute pressure of the gas. This difference is called the **gauge pressure**, since it is what most gauges (not just manometers) measure:

Gauge pressure

$$P_{\text{gauge}} = P_{\text{abs}} - P_{\text{atm}} \tag{9-6}$$

Since the density of mercury is 13 600 kg/m^3, 1.00 mm Hg can be converted to pascals by substituting $d = 1.00$ mm in Eq. (9-5):

$$1.00 \text{ mm Hg} = \rho g d = (13\,600 \text{ kg/m}^3)(9.80 \text{ m/s}^2)(0.001\,00 \text{ m}) = 133 \text{ Pa}$$

The liquid in a manometer may be something other than mercury, such as water or oil. Equation (9-5) still applies, as long as we use the correct density ρ of the liquid in the manometer.

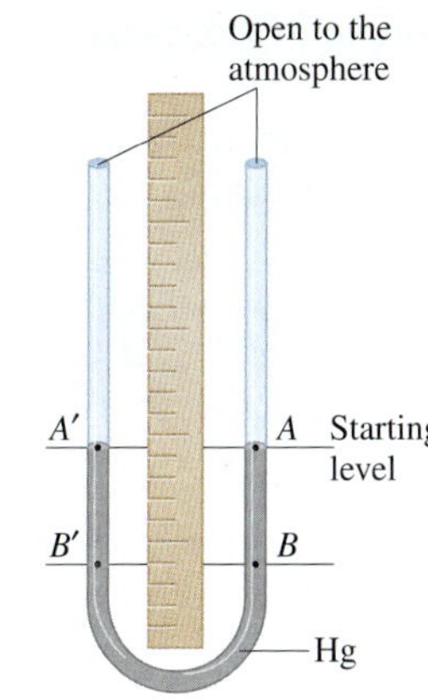

Figure 9.9 A mercury manometer open on both sides. Points A and A' are both at atmospheric pressure. Any two points (e.g., B and B') at the same height within the fluid are at the same pressure: $P_B = P_{B'}$.

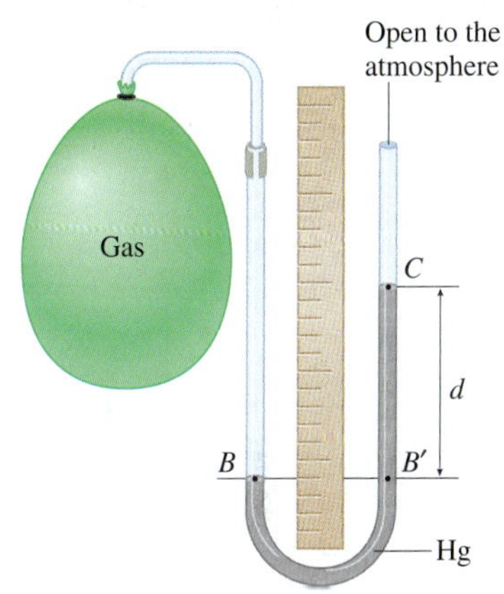

Figure 9.10 The manometer connected on one side to a container of gas at a pressure greater than atmospheric pressure.

Example 9.5

The Mercury Manometer

A manometer is attached to a container of gas to determine its pressure. Before the container is attached, both sides of the manometer are open to the atmosphere. After the container is attached, the mercury on the side attached to the gas container rises 12 cm above its previous level. (a) What is the gauge pressure of the gas in Pa? (b) What is the absolute pressure of the gas in Pa?

Strategy The mercury column is higher on the side connected to the container of gas, so we know that the pressure of the enclosed gas is lower than atmospheric pressure. We need to find the *difference* in levels of the mercury columns on the two sides. Careful: It is *not* 12 cm! If one side went up by 12 cm, then the other side has gone down by 12 cm, since the same volume of mercury is contained in the manometer.

Solution (a) The difference in the mercury levels is 24 cm (Fig. 9.11). Since the mercury on the gas side went *up*, the absolute pressure of the gas is *lower* than atmospheric

continued on next page

Example 9.5 continued

pressure. Therefore, the gauge pressure of the gas is *less than zero*. The gauge pressure in Pa is

$$P_{\text{gauge}} = \rho g d$$

where the "depth" is $d = -24$ cm (the mercury is 24 cm higher on the gas side). Then

$$P_{\text{gauge}} = 13\,600\ \text{kg/m}^3 \times 9.8\ \text{m/s}^2 \times (-0.24\ \text{m}) = -32\ \text{kPa}$$

(b) The absolute pressure of the gas is

$$P = P_{\text{gauge}} + P_{\text{atm}}$$
$$= -32\ \text{kPa} + 101\ \text{kPa} = 69\ \text{kPa}$$

Discussion As a check, the manometer tells us directly that the gauge pressure of the gas is −240 mm Hg. Converting to pascals gives

$$-240\ \text{mm Hg} \times 133\ \text{Pa/mm Hg} = -32\ \text{kPa}$$

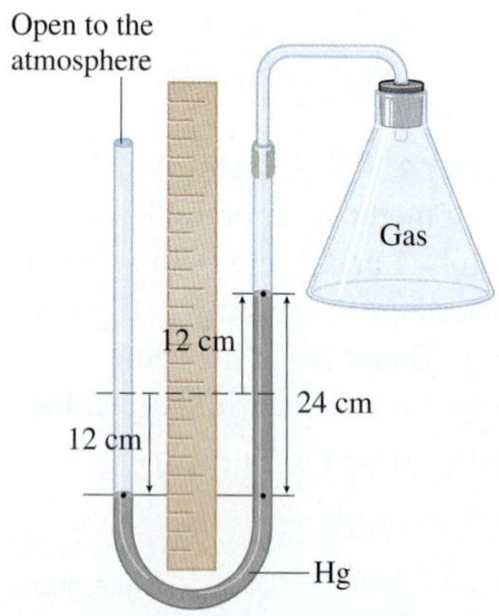

Figure 9.11 When a container of gas is attached to one side of the manometer, one side goes down 12 cm and the other side goes up 12 cm.

Practice Problem 9.5 Column Heights in Manometer

A mercury manometer is connected to a container of gas. (a) The height of the mercury column on the side connected to the gas is 22.0 cm (measured from the bottom of the manometer). What is the height of the mercury column on the open side if the gauge pressure is measured to be 13.3 kPa? (b) If the gauge pressure of the gas doubles, what are the new heights of the two columns?

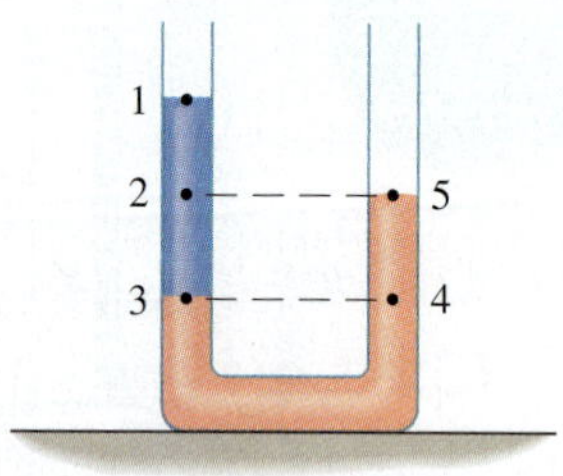

Figure 9.12 A manometer containing two different liquids. Both sides are open to the atmosphere.

CHECKPOINT 9.5

A manometer contains two different liquids of different densities (Fig. 9.12). Both sides are open to the atmosphere. Rank points 1-5 in order of the pressure, largest to smallest.

The Barometer

A manometer can act as a **barometer**—a device to measure atmospheric pressure. Instead of attaching a container with a gas to one end of the manometer, attach a container and a vacuum pump. Pump the air out of the container to get as close to a vacuum—zero pressure—as possible. Then the atmosphere pushes down on one side and pushes the fluid up on the other side toward the evacuated container.

Figure 9.13 shows a barometer in which the vacuum is not created by a vacuum pump. The barometer was invented by Evangelista Torricelli (1608–1647), an assistant to Galileo; in his honor, 1 mm Hg is called 1 torr.

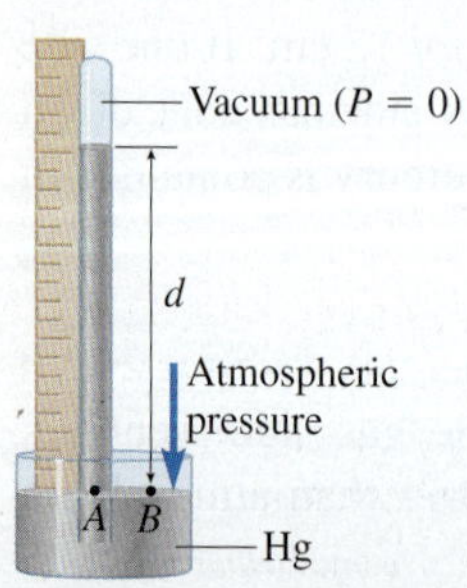

Figure 9.13 A simple barometer. A tube, of length greater than 76 cm and closed at one end, is filled with mercury. The tube is then inverted into an open container of mercury. Some mercury flows down from the tube into the bowl. The space left at the top of the tube is nearly a vacuum because nothing is left but a negligible amount of mercury vapor. Points *A* and *B* are at the same level in the mercury and, therefore, are both at atmospheric pressure since the bowl is open to the air. The distance *d* from *A* to the top of the mercury column in the closed tube is a measure of the atmospheric pressure (often called *barometric pressure* because it is measured with a barometer).

EVERYDAY PHYSICS DEMO

When you next have a drink with a straw, insert the straw into the drink and place your finger over the upper opening of the straw so that no more air can enter the straw. Raise the lower end of the straw up out of your drink. Does the liquid in the straw flow back down into your glass? What do you suppose is holding the liquid in place? Make an FBD on your paper napkin.

Some air is trapped between your finger and the top of the liquid in the straw; that air exerts a downward force on the liquid of magnitude P_1A (Fig. 9.14). A downward gravitational force mg also acts on the liquid. The air at the bottom of the straw exerts an upward force on the liquid of magnitude $P_{atm}A$; this upward force is what holds the liquid in place. Because the liquid does not pour out of the straw, but instead is in equilibrium,

$$P_{atm}A - P_1A - mg = 0$$

Thus, the pressure P_1 of the air trapped above the liquid must be less than atmospheric pressure.

How did P_1 become less than atmospheric pressure? As you pulled the straw up and out, the liquid in the straw falls a bit, expanding the volume available to the air trapped above the liquid. When a gas expands under conditions like this, its pressure decreases.

When you remove your finger from the top of the straw, air can get in at the top of the straw. Then the pressures above and below the liquid are equal, so the gravitational force pulls the liquid down and out of the straw.

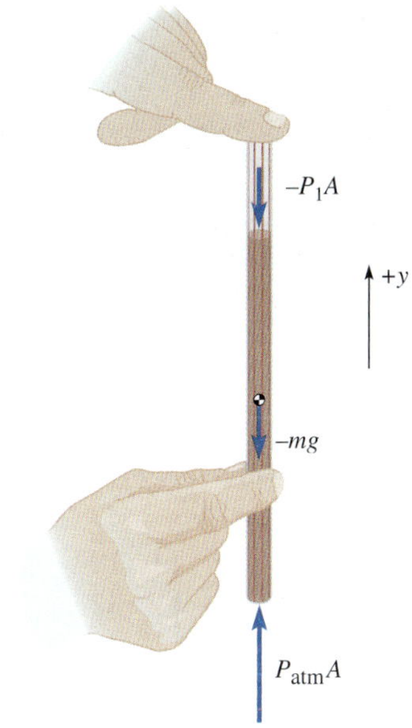

Figure 9.14 Force acting on the liquid inside a straw.

Application of the Manometer: Measuring Blood Pressure

Blood pressure is measured with a sphygmomanometer (Fig. 9.15). The oldest kind of sphygmomanometer consists of a mercury manometer on one side attached to a closed bag—the cuff. The cuff is wrapped around the upper arm at the level of the heart and is then pumped up with air. The manometer measures the gauge pressure of the air in the cuff.

At first, the pressure in the cuff is higher than the *systolic* pressure—the maximum pressure in the brachial artery that occurs when the heart contracts. The cuff pressure squeezes the artery closed, and no blood flows into the forearm. A valve on the cuff is then opened to allow air to escape slowly. When the cuff pressure decreases to just below the systolic pressure, a little squirt of blood flows past the constriction in the artery with each heartbeat. The sound of turbulent blood flow past the constriction can be heard through the stethoscope.

As air continues to escape from the cuff, the sound of blood flowing through the constriction in the artery continues to be heard. When the pressure in the cuff reaches the *diastolic* pressure in the artery—the minimum pressure that occurs when the heart muscle is relaxed—there is no longer a constriction in the artery, so the pulsing sounds cease. The *gauge* pressures for a healthy heart are nominally around 120 mm Hg (systolic) and 80 mm Hg (diastolic).

Figure 9.15 A sphygmomanometer being used to measure blood pressure.

9.6 THE BUOYANT FORCE

When an object is immersed in a fluid, the pressure on the lower surface of the object is higher than the pressure on the upper surface. The difference in pressures leads to an upward net force acting on the object due to the fluid pressure. If you try to push a beach ball underwater, you feel the effects of the buoyant force pushing the ball back up. It takes a rather large force to hold such an object completely underwater; the instant you let go, the object pops back up to the surface.

CONNECTION:

The buoyant force is not a new kind of force exerted by a fluid; it is the sum of forces due to fluid pressure.

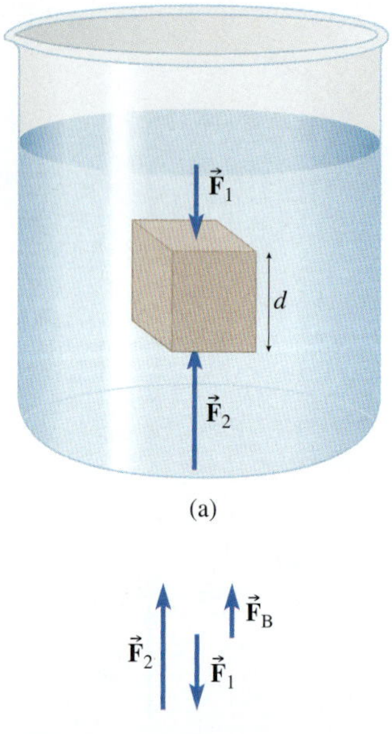

Figure 9.16 (a) Forces due to fluid pressure on the top and bottom of an immersed rectangular solid. (b) The buoyant force is the sum of $\vec{\mathbf{F}}_1$ and $\vec{\mathbf{F}}_2$. Since $|\vec{\mathbf{F}}_2| > |\vec{\mathbf{F}}_1|$, the net force due to fluid pressure is upward.

Consider a rectangular solid immersed in a fluid of uniform density ρ (Fig. 9.16a). For each vertical face (left, right, front, and back), there is a face of equal area opposite it. The forces on these two faces due to the fluid are equal in magnitude since the areas and the average pressures are the same. The directions are opposite, so the forces acting on the vertical faces cancel in pairs.

Let the top and bottom surfaces each have area A. The force on the lower face of the block is $F_2 = P_2A$; the force on the upper face is $F_1 = P_1A$. The total force on the block due to the fluid, called the **buoyant force** F_B, is upward since $F_2 > F_1$ (Fig. 9.16b).

$$\vec{\mathbf{F}}_B = \vec{\mathbf{F}}_1 + \vec{\mathbf{F}}_2$$

$$F_B = (P_2 - P_1)A$$

Since $P_2 - P_1 = \rho gd$, the magnitude of the buoyant force can be written as:

Buoyant force

$$F_B = \rho gdA = \rho gV \tag{9-7}$$

where $V = Ad$ is the volume of the block.

Note that ρV is the mass of the volume V of the fluid that the block displaces. Thus, the buoyant force on the submerged block is equal to the weight of an equal volume of fluid, a result called **Archimedes' principle**.

Archimedes' Principle

A fluid exerts an upward buoyant force on a submerged object equal in magnitude to the weight of the volume of fluid displaced by the object.

As expected from Newton's third law, the object exerts a force of equal magnitude downward on the fluid.

Archimedes' principle applies to a submerged object of *any shape* even though we derived it for a rectangular block. Why? Imagine replacing an irregular submerged object with enough fluid to fill the object's place. This "piece" of fluid is in equilibrium, so the buoyant force must be equal to its weight. The buoyant force is the net force exerted on the "piece" of fluid by the surrounding fluid, which is identical to the buoyant force on the irregular object since the two have the same shape and surface area.

The same argument can be used to show that if an object is only partly submerged, the buoyant force is still equal to the weight of fluid displaced. Equation (9-7) applies as long as V is the part of the object's volume below the fluid surface rather than the entire volume of the object.

Net Force due to Gravity and Buoyancy The net force due to gravity and buoyancy acting on an object totally or partially immersed in a fluid (Fig. 9.17) is

$$\vec{\mathbf{F}} = m\vec{\mathbf{g}} + \vec{\mathbf{F}}_B$$

The force of gravity on an object of volume V_o and average density ρ_o is

$$W = mg = \rho_o g V_o$$

and the buoyant force is

$$F_B = \rho_f g V_f$$

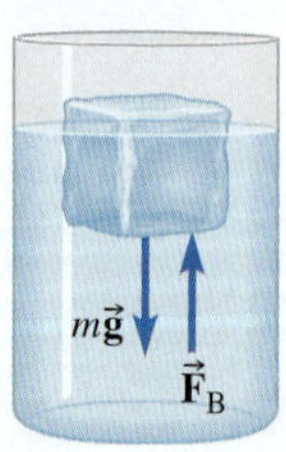

Figure 9.17 Forces acting on a floating ice cube. The ice cube is in equilibrium, so $\vec{\mathbf{F}}_B + m\vec{\mathbf{g}} = 0$.

where V_f and ρ_f are the volume of fluid displaced and the fluid density, respectively. Choosing up to be the $+y$-direction, the net force due to gravity and buoyancy is

$$F_y = \rho_f g V_f - \rho_o g V_o \tag{9-8}$$

Here F_y can be positive or negative, depending on which density is larger. Imagine releasing a pebble and an air bubble underwater. The pebble's average density is greater than the density of water, so the net force on it is downward; the pebble sinks. An air bubble's average density is less than the density of water, so the net force is upward, causing the bubble to rise toward the surface of the water.

If the object is completely submerged, the volumes of the object and the displaced fluid are the same and

$$F_y = (\rho_f - \rho_o)gV$$

If $\rho_o < \rho_f$, the object floats with only part of its volume submerged. In equilibrium, the object displaces a volume of fluid whose weight is equal to the object's weight. At that point the gravitational force equals the buoyant force and the object floats. Setting $F_y = 0$ in Eq. (9-8) yields

$$\rho_f g V_f = \rho_o g V_o$$

which can be rearranged as:

$$\frac{V_f}{V_o} = \frac{\rho_o}{\rho_f}$$

On the left side of this equation is the fraction of the object's volume that is submerged; it is equal to the ratio of the density of the object to the density of the fluid.

CHECKPOINT 9.6

Two identical pieces of wood are floating, one in a beaker of water and the other in a beaker of alcohol that has a density 0.8 times that of water. Does one piece of wood float higher above the liquid surface than the other? If so, which floats higher? Explain.

Specific Gravity The **specific gravity** of a substance is the ratio of its density to the density of water at 3.98°C. Water at 3.98°C is chosen as the reference material because at that temperature, the density of water is a maximum (at atmospheric pressure). To four significant figures, water at 3.98°C has a density of 1.000 g/cm^3 (1000 kg/m^3). If we say the specific gravity of seawater is 1.025, that means that seawater has a density of 1.025 g/cm^3 (1025 kg/m^3).

Specific gravity

$$\text{S.G.} = \frac{\rho}{\rho_{\text{water}}} = \frac{\rho}{1000\ \text{kg/m}^3} \qquad (9\text{-}9)$$

Applications of Specific-Gravity Measurements in Medicine Blood tests often include determination of the specific gravity of the blood—normally around 1.040 to 1.065. A reading that is too low may indicate anemia, since the presence of red blood cells increases the average density of the blood. Before taking blood from a donor, a drop of the blood is placed in a solution of known density. If the drop does not sink, it is not safe for the donor to give blood because the concentration of red blood cells is too low. Urinalysis also includes a specific-gravity measurement (normally 1.015 to 1.030); too high a value indicates an abnormally high concentration of dissolved salts, which can signal a medical problem.

Applications of Archimedes' Principle Freighters, aircraft carriers, and cruise ships float, although they are made from steel and other materials that are more dense than seawater. When a ship floats, the buoyant force acting on the ship is equal to the ship's weight. A ship is constructed so that it displaces a volume of seawater larger than

How can the hippo sink?

the volume of the steel and other construction materials. The *average* density of the ship is its weight divided by its total volume. A large part of a ship's interior is filled with air. All of the "empty" space contributes to the total volume; the resulting average density is less than that of seawater, allowing the ship to float.

Now we can understand how a hippopotamus can sink to the bottom of a pond: it can expel some of the air in its body by exhaling. Exhalation increases the average density of the hippopotamus so that it is just slightly above the density of the water; thus, it sinks. (An armadillo does just the opposite: it swallows air, inflating its stomach and intestines, to increase the buoyant force for a swim across a large lake. See Problem 41.) When the hippo needs to breathe, it swims back up to the surface.

Example 9.6

The Golden (?) Falcon

A small statue in the shape of a falcon has a weight of 24.1 N. The owner of the statue claims it is made of solid gold. When the statue is completely submerged in a container brimful of water, the weight of the water that spills over the top and into a bucket is 1.25 N. Find the density and specific gravity of the metal. Is the density consistent with the claim that the falcon is solid gold?

Strategy When the statue is completely submerged, it displaces a volume V of water equal to its own volume. The weight of the displaced water is equal to the buoyant force. Let $m_s g = 24.1$ N represent the weight of the statue (in terms of its mass m_s) and let $m_w g = 1.25$ N represent the weight of the water.

Solution The specific gravity (S.G.) of the statue is

$$\text{S.G.} = \frac{\rho_s}{\rho_w} = \frac{m_s/V}{m_w/V} = \frac{m_s}{m_w}$$

Rather than calculate the masses in kilograms, we recognize that a ratio of masses is equal to the ratio of the weights:

$$\text{S.G.} = \frac{m_s g}{m_w g} = \frac{24.1\text{ N}}{1.25\text{ N}} = 19.3$$

The density of the statue is

$$\rho_s = \text{S.G.} \times \rho_w = 19.3 \times 1000\text{ kg/m}^3 = 1.93 \times 10^4\text{ kg/m}^3$$

From Table 9.1, the statue has the correct density; it may *possibly* be gold.

Discussion According to legend, this method to determine the specific gravity of a solid was discovered by Archimedes in the third century B.C.E. King Hieron II asked Archimedes to find a way to check whether his crown was made of pure gold—without melting down the crown, of course! Archimedes came up with his method while he was taking a bath; he noticed the water level rising as he got in and connected the rising water level with the volume of water displaced by his body. In his excitement, he jumped out of the bath and ran naked through the streets of Siracusa (a city in Sicily) shouting "Eureka!"

Practice Problem 9.6 Identifying an Unknown Substance

An unknown solid substance has a weight of 142.0 N. The object is suspended from a scale and hung so that it is completely submerged in water (but not touching bottom). The scale reads 129.4 N. Find the specific gravity of the object and determine whether the substance could be anything listed in Table 9.1.

Example 9.7

Hidden Depths of an Iceberg

What percentage of a floating iceberg's volume is above water? The specific gravity of ice is 0.917 and the specific gravity of the surrounding seawater is 1.025.

Strategy The ratio of the density of ice to the density of seawater tells us the ratio of the volume of ice that is submerged in the seawater to the total volume of the iceberg. The rest of the ice is above the water.

Solution We could calculate the densities of seawater and of ice in SI units from their specific gravities, but that is

continued on next page

Example 9.7 continued

unnecessary; the ratio of the specific gravities is equal to the ratio of the densities:

$$\frac{\text{S.G.}_{\text{ice}}}{\text{S.G.}_{\text{seawater}}} = \frac{\rho_{\text{ice}}/\rho_{\text{water}}}{\rho_{\text{seawater}}/\rho_{\text{water}}} = \frac{\rho_{\text{ice}}}{\rho_{\text{seawater}}}$$

The fraction of the iceberg's volume that is submerged is equal to the ratio of the densities of ice and seawater. Thus, the ratio of the volume submerged to the total volume of ice is

$$\frac{V_{\text{submerged}}}{V_{\text{ice}}} = \frac{\rho_{\text{ice}}}{\rho_{\text{seawater}}} = \frac{\text{S.G.}_{\text{ice}}}{\text{S.G.}_{\text{seawater}}}$$

$$= \frac{0.917}{1.025} = 0.895$$

89.5% of the ice is below the surface of the water, leaving only 10.5% above the surface.

Discussion An alternative solution does not depend on remembering that the ratio of the volumes is equal to the ratio of the densities. The buoyant force is equal to the weight of a volume $V_{\text{submerged}}$ of water:

$$\text{buoyant force} = \rho_{\text{seawater}} V_{\text{submerged}} g$$

The weight of the iceberg is $mg = \rho_{\text{ice}} V_{\text{ice}} g$. From Newton's second law, the buoyant force must be equal in magnitude to the weight of the iceberg when it is floating in equilibrium:

$$\rho_{\text{seawater}} V_{\text{submerged}} g = \rho_{\text{ice}} V_{\text{ice}} g$$

or

$$\frac{V_{\text{submerged}}}{V_{\text{ice}}} = \frac{\rho_{\text{ice}}}{\rho_{\text{seawater}}}$$

The fact that ice floats is of great importance for the balance of nature. If ice were more dense than water, it would gradually fill up the ponds and lakes *from the bottom*. It would not form on top of lakes and remain there. The consequences for fish and other bottom dwellers of solidly frozen lakes would be catastrophic. The water below the surface layer of ice formed in winter remains just above freezing so that the fish are able to survive.

Practice Problem 9.7 Floating in Freshwater Versus Seawater

If the average density of a human body is 980 kg/m^3, what fraction of the body floats above water in a freshwater pond and what fraction floats above seawater in the ocean? The specific gravity of seawater is 1.025.

Conceptual Example 9.8

A Hovering Fish

How is it that a fish is able to hover almost motionless in one spot—until some attractive food is spotted and, with a flip of the tail, off it swims after the food? Fish have a thin-walled bladder, called a swim bladder, located under the spinal column. The swim bladder contains a mixture of oxygen and nitrogen obtained from the blood of the fish. How does the swim bladder help the fish keep the buoyant and gravitational forces balanced so that it can hover?

Solution and Discussion

If the fish's average density is greater than that of the surrounding water, it will sink; if its average density is smaller than that of the water, it will rise. By varying the volume of the swim bladder, the fish is able to vary its overall volume and, thus, its average density. By adjusting its average density to match the density of the surrounding water, the fish can remain suspended in position. The fish can also adjust the volume of the bladder when it wants to rise or sink. (See Problem 40.)

Conceptual Practice Problem 9.8 The Diving Beetle

A diving beetle traps a bubble of air under its wings. While under the water, the beetle uses the air in the bubble to breathe, gradually exchanging the oxygen for carbon dioxide. (a) What does the beetle do to the air bubble so that it can dive under the water? (b) Once under water, what does the beetle do so that it can rise to the surface? [*Hint*: Treat the beetle and the air bubble as a single system. How can the beetle change the buoyant force acting on the system?]

Buoyant Forces on Objects Immersed in a Gas Gases such as air are fluids and exert buoyant forces just as liquids do. The buoyant force due to air is often negligible if an object's average density is much larger than the density of air. To see a significant buoyant force in air, we must use an object with a small average density. A hot air

Figure 9.18 The buoyant force due to the outside air keeps these balloons aloft.

balloon has an opening at the bottom and a burner for heating the air within (Fig. 9.18). Many molecules of the heated air escape through the opening, decreasing the balloon's average density. When the balloon is less dense on average than the surrounding air, it rises because the buoyant force exceeds the weight of the balloon. At higher altitudes, the surrounding air becomes less and less dense, so at some particular altitude, the buoyant force is equal in magnitude to the weight of the balloon. Then, by Newton's second law, the net force on the balloon is zero. The balloon is in *stable* equilibrium at this altitude: if the balloon rises a bit, it experiences a net force downward, while if the balloon sinks down a bit, it is pushed back upward.

9.7 FLUID FLOW

Types of Fluid Flow The study of *moving* fluids is a wonderfully complex subject. To illustrate some important ideas in less complex situations, we limit our study at first to fluids flowing under special conditions.

One difference between moving fluids and static fluids is that a moving fluid can exert a force *parallel* to any surface over or past which it flows; a static fluid cannot. Since the moving fluid exerts a force against a surface, the surface must also exert a force on the fluid. This **viscous force** opposes the flow of the fluid; it is the counterpart to the kinetic frictional force between solids. An external force must act on a viscous fluid (and thereby do work) to keep it flowing. Viscosity is considered in Section 9.9. Until then, we consider only nonviscous fluids—fluid flow where the viscous forces are negligibly small. We also ignore surface tension, which is considered in Section 9.11.

Figure 9.19 A wind tunnel shows the streamlines in the laminar flow of air past a car.

Fluid flow can be characterized as steady or unsteady. When the flow is **steady**, the velocity of the fluid *at any point* is constant in time. The velocity is not necessarily the same everywhere, but at any particular point, the velocity of the fluid passing that point remains constant in time. The density and pressure at any point in a steadily flowing fluid are also constant in time.

Steady flow is **laminar**. The fluid flows in neat layers so that each small portion of fluid that passes a particular point follows the same path as every other portion of fluid that passes the same point. The path that the fluid follows, starting from any point, is called a **streamline** (Fig. 9.19). The streamlines may curve and bend, but they cannot cross each other; if they did, the fluid would have to "decide" which way to go when it gets to such a point. The direction of the fluid velocity at any point must be tangent to the streamline passing through that point. Streamlines are a convenient way to depict fluid flow in a sketch.

When the fluid velocity at a given point changes, the flow is *unsteady*. **Turbulence** is an extreme example of unsteady flow (Fig. 9.20). In turbulent flow, swirling vortices—whirlpools of fluid—appear. The vortices are not stationary; they move with the fluid. The flow velocity at any point changes erratically; prediction of the direction or speed of fluid flow under turbulent conditions is difficult. (See the text website for more information about turbulent flow.)

Figure 9.20 Turbulent flow of gas emerging from the nozzle of an aerosol can.

The Ideal Fluid The special case that we consider first is the flow of an **ideal fluid**. An ideal fluid is incompressible, undergoes laminar flow, and has no viscosity. Under some conditions, real fluids can be modeled as (nearly) ideal.

The flow of an ideal fluid is governed by two principles: the continuity equation and Bernoulli's equation. The continuity equation is an expression of conservation of mass for an incompressible fluid: since no fluid is created or destroyed, the total mass of the fluid must be constant. Bernoulli's equation, discussed in Section 9.8, is a form of the energy conservation law applied to fluid flow. Together, these two equations enable us to predict the flow of an ideal fluid.

The Continuity Equation

We start by deriving the continuity equation, which relates the speed of flow to the cross-sectional area of the fluid. Suppose an incompressible fluid flows into a pipe of nonuniform cross-sectional area under conditions of steady flow. In Fig. 9.21, the fluid on the left moves at speed v_1. During a time Δt, the fluid travels a distance

$$x_1 = v_1 \Delta t$$

If A_1 is the cross-sectional area of this section of pipe, then the mass of fluid moving past point 1 in time Δt is

$$\Delta m_1 = \rho V_1 = \rho A_1 x_1 = \rho A_1 v_1 \Delta t$$

During this same time interval, the mass of fluid moving past point 2 is

$$\Delta m_2 = \rho V_2 = \rho A_2 x_2 = \rho A_2 v_2 \Delta t$$

But, if the flow is steady, the mass passing through one section of pipe in time interval Δt must pass through any other section of the pipe in the same time interval. Therefore,

$$\Delta m_1 = \Delta m_2$$

or

$$\rho A_1 v_1 \Delta t = \rho A_2 v_2 \Delta t \tag{9-10}$$

The quantity $\rho A v$ is the *mass flow rate* of the fluid:

Mass flow rate

$$\frac{\Delta m}{\Delta t} = \rho A v \quad \text{(SI unit: kg/s)} \tag{9-11}$$

Since the time intervals Δt are the same, Eq. (9-11) says that the mass flow rate past any two points is the same. Since the density of an incompressible fluid is constant, the volume flow rate past any two points must also be the same:

Volume flow rate

$$\frac{\Delta V}{\Delta t} = A v \quad \text{(SI unit: m}^3\text{/s)} \tag{9-12}$$

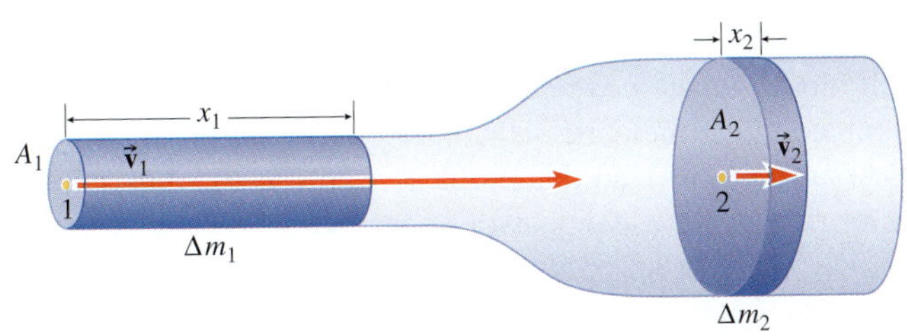

Figure 9.21 An incompressible fluid flowing horizontally through a nonuniform pipe.

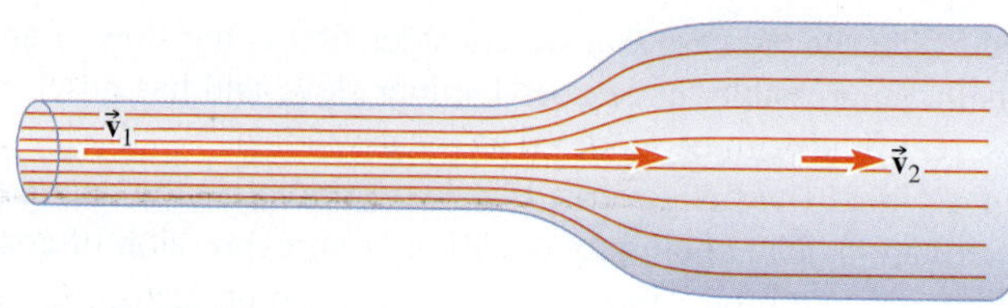

Figure 9.22 Streamlines in a pipe of varying cross-sectional area. Streamlines are closer together where the fluid velocity is larger and farther apart where the velocity is smaller.

The **continuity equation** for an incompressible fluid equates the volume flow rates past two different points:

Continuity equation for incompressible fluid

$$A_1v_1 = A_2v_2 \tag{9-13}$$

The same volume of fluid that enters the pipe in a given time interval exits the pipe in the same time interval. Where the radius of the tube is large, the speed of the fluid is small; where the radius is small, the fluid speed is large. A familiar example is what happens when you use your thumb to partially block the end of a garden hose to make a jet of water. The water moves past your thumb, where the cross-sectional area is smaller, at a greater speed than it moves in the hose. Similarly, water traveling along a river speeds up, forming rapids, when the riverbed narrows or is partially blocked by rocks and boulders.

Streamlines are closer together where the fluid flows faster and farther apart where it flows more slowly (Fig. 9.22). Thus, streamlines help us visualize fluid flow. The fluid velocity at any point is tangent to a streamline through that point.

Figure 9.23 Demonstrating the continuity equation at a bathroom sink. Notice that the stream of water is narrower where the flow speed is faster.

EVERYDAY PHYSICS DEMO

The continuity equation applies to an ideal fluid even if it is not flowing through a pipe. Turn on a faucet so that the water flows out in a moderate stream (Fig. 9.23). The falling water is in free fall, accelerated by gravity until it hits the sink below. As the water falls, its speed increases. The stream of water gradually narrows as it falls so that the product of speed and cross-sectional area is constant, as predicted by the continuity equation.

CHECKPOINT 9.7

An artery with an inner diameter of 1.20 cm narrows (due to plaque buildup) to an inner diameter of 1.00 cm. By what percentage does the speed of blood flow change when entering the narrower section?

Example 9.9

Speed of Blood Flow

The heart pumps blood into the aorta, which has an inner radius of 1.0 cm. The aorta feeds 32 major arteries. If blood in the aorta travels at a speed of 28 cm/s, at approximately what average speed does it travel in the arteries? Assume that blood can be treated as an ideal fluid and that the arteries each have an inner radius of 0.21 cm.

Strategy Since we have assumed blood to be an ideal fluid, we can apply the continuity equation. The main tube (the aorta) is connected to multiple tubes (the major arteries), so this problem seems to be more complicated than a single pipe with a constriction in it. What matters here is the total cross-sectional area into which the blood flows.

continued on next page

Example 9.9 continued

Solution We start by finding the cross-sectional area of the aorta

$$A_1 = \pi r_{\text{aorta}}^2$$

and then the total cross-sectional area of the arteries

$$A_2 = 32\pi r_{\text{artery}}^2$$

Now we apply the continuity equation and solve for the unknown speed.

$$A_1 v_1 = A_2 v_2$$

$$v_2 = v_1 \frac{A_1}{A_2} = 0.28 \text{ m/s} \times \frac{\pi \times (0.010 \text{ m})^2}{32\pi \times (0.0021 \text{ m})^2} = 0.20 \text{ m/s}$$

Discussion The blood flow slows in the arteries because the total cross-sectional area is greater than that of the aorta alone. From the arteries, the blood travels to the many capillaries of the body. Each capillary has a tiny cross-sectional area, but there are so many of them that the blood flow slows greatly once it reaches the capillaries (Problem 84). This allows time for the exchange of oxygen, carbon dioxide, and nutrients between the blood and the body tissues.

Practice Problem 9.9 Hosing Down a Wastebasket

A garden hose fills a 32 L wastebasket in 120 s. The opening at the end of the hose has a radius of 1.00 cm. (a) How fast is the water traveling as it leaves the hose? (b) How fast does the water travel if half the exit area is obstructed by placing a finger over the opening?

9.8 BERNOULLI'S EQUATION

The continuity equation relates the flow velocities of an ideal fluid at two different points based on the change in cross-sectional area of the pipe. According to the continuity equation, the fluid must speed up as it enters a constriction (Fig. 9.24) and then slow down to its original speed when it leaves the constriction. Using energy ideas, we will show that the pressure of the fluid in the constriction (P_2) cannot be the same as the pressure before or after the constriction (P_1). For horizontal flow *the speed is higher where the pressure is lower.* This principle is often called the **Bernoulli effect**.

The Bernoulli effect can seem counterintuitive at first; isn't rapidly moving fluid at *high* pressure? For instance, if you were hit with the fast-moving water out of a firehose, you would be knocked over easily. The force that knocks you over is indeed due to fluid pressure; you would justifiably conclude that the pressure was high. However, the pressure is not high *until you slow down the water* by getting in its way. The rapidly moving water in the jet is, in fact, approximately at atmospheric pressure (zero gauge pressure), but when you *stop* the water, its pressure increases dramatically.

Let's find the quantitative relationship between pressure changes and flow speed changes for an ideal fluid. In Fig. 9.25, the shaded volume of fluid flows to the right. If the left end moves a distance Δx_1, then the right end moves a distance Δx_2. Since the fluid is incompressible,

$$A_1 \, \Delta x_1 = A_2 \, \Delta x_2 = V$$

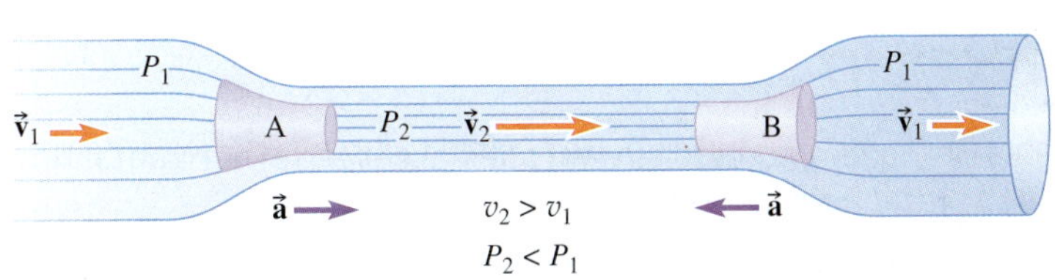

Figure 9.24 A small volume of fluid speeds up as it moves into a constriction (position A) and then slows down as it moves out of the constriction (position B).

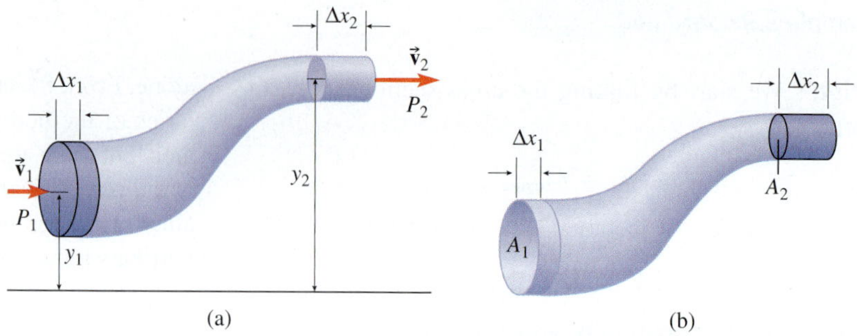

Figure 9.25 Applying conservation of energy to the flow of an ideal fluid. The shaded volume of fluid in (a) is flowing to the right; (b) shows the same volume of fluid a short time later.

Work is done by the neighboring fluid during this flow. Fluid behind (to the left) pushes forward, doing positive work, while fluid ahead pushes backward, doing negative work. The total work done on the shaded volume by neighboring fluid is

$$W = P_1 A_1 \,\Delta x_1 - P_2 A_2 \,\Delta x_2 = (P_1 - P_2)V$$

CONNECTION:

Bernoulli's equation is a restatement of the principle of energy conservation applied to the flow of an ideal fluid.

Since no dissipative forces act on an ideal fluid, the work done is equal to the total change in kinetic and gravitational potential energy. The net effect of the displacement is to move a volume V of fluid from height y_1 to height y_2 and to change its speed from v_1 to v_2. The energy change is therefore

$$\Delta E = \Delta K + \Delta U = \tfrac{1}{2}m(v_2^2 - v_1^2) + mg(y_2 - y_1)$$

where the $+y$-direction is up. Substituting $m = \rho V$ and equating the work done on the fluid to the change in its energy yields

$$(P_1 - P_2)\, V = \tfrac{1}{2}\rho V(v_2^2 - v_1^2) + \rho V g(y_2 - y_1)$$

Dividing both sides by V and rearranging yields Bernoulli's equation, named after Swiss mathematician Daniel Bernoulli (1700–1782), but first derived by fellow Swiss mathematician Leonhard Euler (pronounced like *oiler*, 1707–1783).

Bernoulli's equation (for ideal fluid flow)

$$P_1 + \rho g y_1 + \tfrac{1}{2}\rho v_1^2 = P_2 + \rho g y_2 + \tfrac{1}{2}\rho v_2^2$$

$$(\text{or } P + \rho g y + \tfrac{1}{2}\rho v^2 = \text{constant}) \qquad (9\text{-}14)$$

Bernoulli's equation relates the pressure, flow speed, and height at two points in an ideal fluid. Although we derived Bernoulli's equation in a relatively simple situation, it applies to the flow of any ideal fluid as long as points 1 and 2 are on the same streamline.

Each term in Bernoulli's equation has units of pressure, which in the SI system is Pa or N/m^2. Since a joule is a newton-meter, the pascal is also equal to a joule per cubic meter (J/m^3). Each term represents work or energy per unit volume. The pressure is the work done by the fluid on the fluid ahead of it per unit volume of flow. The kinetic energy per unit volume is $\frac{1}{2}\rho v^2$ and the gravitational potential energy per unit volume is $\rho g y$. (connect tutorial: energies)

CHECKPOINT 9.8

Discuss Bernoulli's equation in two special cases: (a) horizontal flow ($y_1 = y_2$) and (b) a static fluid ($v_1 = v_2 = 0$).

Example 9.10

Torricelli's Theorem

A barrel full of rainwater has a spigot near the bottom, at a depth of 0.80 m beneath the water surface. (a) When the spigot is directed horizontally (Fig. 9.26a) and is opened, how fast does the water come out? (b) If the opening points upward (Fig. 9.26b), how high does the resulting "fountain" go? (connect tutorial: waterfall)

Strategy The water at the surface is at atmospheric pressure. The water emerging from the spigot is *also* at atmospheric pressure since it is in contact with the air. If the pressure of the emerging water were different than that of the air, the stream would expand or contract until the pressures were equal. We apply Bernoulli's equation to two points: point 1 at the water surface and point 2 in the emerging stream of water.

Solution (a) Since $P_1 = P_2$, Bernoulli's equation is

$$\rho g y_1 + \tfrac{1}{2}\rho v_1^2 = \rho g y_2 + \tfrac{1}{2}\rho v_2^2$$

Point 1 is 0.80 m above point 2, so

$$y_1 - y_2 = 0.80 \text{ m}$$

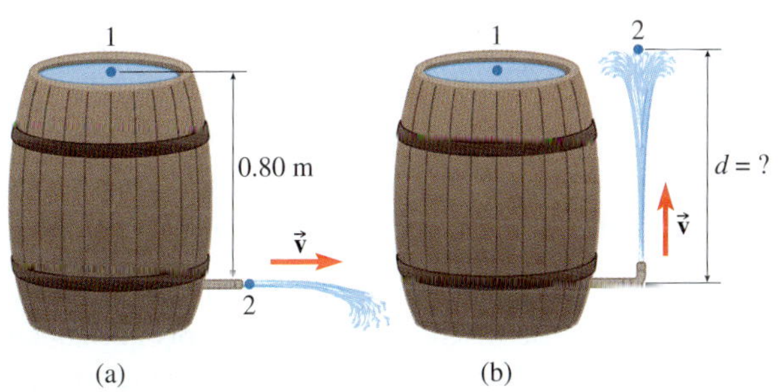

Figure 9.26

Full barrel of rainwater with open spigot (a) horizontal and (b) upward.

The speed of the emerging water is v_2. What is v_1, the speed of the water at the surface? The water at the surface is moving slowly, since the barrel is draining. The continuity equation requires that

$$v_1 A_1 = v_2 A_2$$

Since the cross-sectional area of the spigot A_2 is much smaller than the area of the top of the barrel A_1, the speed of the water at the surface v_1 is negligibly small compared with v_2. Setting $v_1 = 0$, Bernoulli's equation reduces to

$$\rho g y_1 = \rho g y_2 + \tfrac{1}{2}\rho v_2^2$$

After dividing through by ρ, we solve for v_2:

$$g(y_1 - y_2) = \tfrac{1}{2}v_2^2$$

$$v_2 = \sqrt{2g(y_1 - y_2)} = 4.0 \text{ m/s}$$

(b) Now take point 2 to be at the top of the fountain. Then $v_2 = 0$ and Bernoulli's equation reduces to

$$\rho g y_1 = \rho g y_2$$

The "fountain" goes right back up to the top of the water in the barrel!

Discussion The result of part (b) is called Torricelli's theorem. In reality, the fountain does not reach as high as the original water level; some energy is dissipated due to viscosity and air resistance.

Practice Problem 9.10 Fluid in Free Fall

Verify that the speed found in part (a) is the same as if the water just fell 0.80 m straight down. That shouldn't be too surprising since Bernoulli's equation is an expression of energy conservation.

Example 9.11

The Venturi Meter

A *Venturi meter* (Fig. 9.27) measures fluid speed in a pipe. A constriction (of cross-sectional area A_2) is put in a pipe of normal cross-sectional area A_1. Two vertical tubes, open to the atmosphere, rise from two points, one of which is in the constriction. The vertical tubes function like manometers, enabling the pressure to be determined. From this information the flow speed in the pipe can be determined.

Suppose that the pipe in question carries water, $A_1 = 2.0A_2$, and the fluid heights in the vertical tubes are

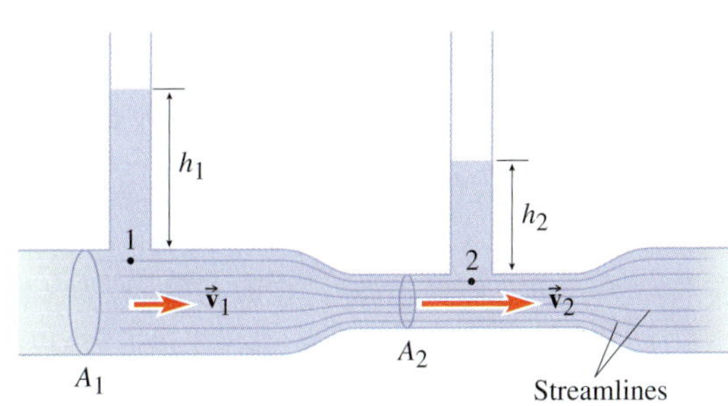

Figure 9.27

Venturi meter.

continued on next page

Example 9.11 continued

$h_1 = 1.20$ m and $h_2 = 0.80$ m. (a) Find the ratio of the flow speeds v_2/v_1. (b) Find the gauge pressures P_1 and P_2. (c) Find the flow speed v_1 in the pipe.

Strategy Neither of the two flow speeds is given. We need more than Bernoulli's equation to solve this problem. Since we know the ratio of the areas, the continuity equation gives us the ratio of the speeds. The height of the water in the vertical tubes enables us to find the pressures at points 1 and 2. The fluid pressure at the bottom of each vertical tube is the same as the pressure of the moving fluid just beneath each tube—otherwise, water would flow into or out of the vertical tubes until the pressure equalized. The water in the vertical tubes is static, so the gauge pressure at the bottom is $P = \rho g d$. Once we have the ratio of the speeds and the pressures, we apply Bernoulli's equation.

Solution (a) From the continuity equation, the product of flow speed and area must be the same at points 1 and 2. Therefore,

$$\frac{v_2}{v_1} = \frac{A_1}{A_2} = 2.0$$

The water flows twice as fast in the constriction as in the rest of the pipe.

(b) The gauge pressures are:

$$P_1 = \rho g h_1 = 1000 \text{ kg/m}^3 \times 9.80 \text{ N/kg} \times 1.20 \text{ m} = 11.8 \text{ kPa}$$

$$P_2 = \rho g h_2 = 1000 \text{ kg/m}^3 \times 9.80 \text{ N/kg} \times 0.80 \text{ m} = 7.8 \text{ kPa}$$

(c) Now we apply Bernoulli's equation. We can use gauge pressures as long as we do so on both sides—in effect we are just subtracting atmospheric pressure from both sides of the equation:

$$P_1 + \rho g y_1 + \tfrac{1}{2}\rho v_1^2 = P_2 + \rho g y_2 + \tfrac{1}{2}\rho v_2^2$$

Since the tube is horizontal, $y_1 \approx y_2$ and we can ignore the small change in gravitational potential energy density $\rho g y$. Then

$$P_1 + \tfrac{1}{2}\rho v_1^2 = P_2 + \tfrac{1}{2}\rho v_2^2$$

We are trying to find v_1, so we can eliminate v_2 by substituting $v_2 = 2.0v_1$:

$$P_1 + \tfrac{1}{2}\rho v_1^2 = P_2 + \tfrac{1}{2}\rho (2.0v_1)^2$$

Simplifying,

$$P_1 - P_2 = 1.5\rho v_1^2$$

$$v_1 = \sqrt{\frac{11\,800 \text{ Pa} - 7800 \text{ Pa}}{1.5 \times 1000 \text{ kg/m}^3}} = 1.6 \text{ m/s}$$

Discussion The assumption that $y_1 \approx y_2$ is fine as long as the pipe radius is small compared with the difference between the static water heights (40 cm). Otherwise, we would have to account for the different y values in Bernoulli's equation.

One subtle point: recall that we assumed that the fluid pressure at the bottom of the vertical tubes was the same as the pressure of the moving fluid just beneath. Does that contradict Bernoulli's equation? Since there is an abrupt change in fluid speed, shouldn't there be a significant difference in the pressures? No, because these points are *not on the same streamline*.

Practice Problem 9.11 Garden Hose

Water flows horizontally through a garden hose of radius 1.0 cm at a speed of 1.4 m/s. The water shoots horizontally out of a nozzle of radius 0.25 cm. What is the gauge pressure of the water inside the hose?

Application of Bernoulli's Principle: Arterial Flutter and Aneurisms Suppose an artery is narrowed due to buildup of plaque on its inner walls. The flow of blood through the constriction is similar to that shown in Fig. 9.24. Bernoulli's equation tells us that the pressure P_2 in the constriction is lower than the pressure elsewhere. The arterial walls are elastic rather than rigid, so the lower pressure allows the arterial walls to contract a bit in the constriction. Now the flow velocity is even higher and the pressure even lower. Eventually the artery wall collapses, shutting off the flow of blood. Then the pressure builds up, reopens the artery, and allows blood to flow. The cycle of arterial flutter then begins again.

The opposite may happen where the arterial wall is weak. Blood pressure pushes the artery walls outward, forming a bulge called an aneurism. The lower flow speed in the bulge is accompanied by a higher blood pressure, which enlarges the aneurism even more (see Problem 98). Ultimately the artery may burst from the increased pressure.

Application of Bernoulli's Principle: Airplane Wings How does an airplane wing generate lift? Figure 9.28 is a sketch of some streamlines for air flowing past an airplane wing in a wind tunnel. The streamlines bend, showing that the wing deflects air downward. By Newton's third law (or conservation of momentum), if the wing pushes downward on the air, the air also pushes upward on the wing. This upward force on the wing is lift. However, the situation is not as simple as air "bouncing" off the bottom of the wing—note that air passing above the wing is also deflected downward.

We can use Bernoulli's equation to get more insight into the generation of lift. (Bernoulli's equation applies in an approximate way to moving air. Even though air is not incompressible, for subsonic flight the density changes are small enough to be ignored.) If the air exerts a net upward force on the wing, the air pressure must be lower above the wing than beneath the wing. In Fig. 9.28, the streamlines above the wing are closer together than beneath the wing, showing that the flow speed above the wing is faster than it is beneath. This observation confirms that the pressure is lower above the wing, because where the pressure is lower, the flow speed is faster.

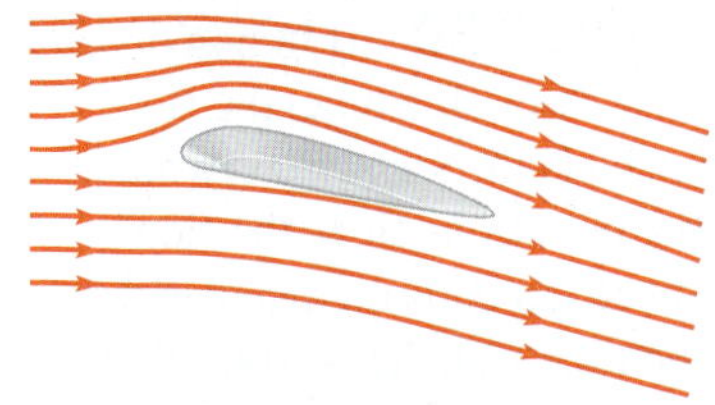

Figure 9.28 Streamlines showing the airflow past an airplane wing in a wind tunnel.

9.9 VISCOSITY

Bernoulli's equation ignores viscosity (fluid friction). According to Bernoulli's equation, an ideal fluid can continue to flow in a horizontal pipe at constant velocity on its own, just as a hockey puck would slide across frictionless ice at constant velocity without anything pushing it along. However, all real fluids have some viscosity. To keep a viscous fluid flowing, a net force due to fluid pressure must push the fluid forward to compensate for the viscous forces that oppose the flow (Fig. 9.29a). A pressure difference between the ends of the pipe must be maintained to keep the fluid moving. With the x-axis pointing in the direction of flow, the pressure is smaller at larger values of x (Fig. 9.29b). The pressure difference is important—in everything from blood flowing through arteries to oil pumped through a pipeline.

CONNECTION:

Kinetic friction makes a sliding object slow down unless an applied force balances the force of friction. Similarly, viscous forces oppose the flow of a fluid. Steady flow of a viscous fluid requires an applied force to balance the viscous forces. The applied force is due to the pressure difference.

To visualize viscous flow in a tube of circular cross section, imagine the fluid to flow in cylindrical layers, or shells. If there were no viscosity, all the layers would move at the same speed (Fig. 9.30a). In viscous flow, the fluid speed depends on the distance from the tube walls (Fig. 9.30b). The fastest flow is at the center of the tube. Layers closer to the wall of the tube move more slowly. The outermost layer of fluid, which is in contact with the tube, does not move. Each layer of fluid exerts viscous forces on the neighboring layers; these forces oppose the relative motion of the layers. The outermost layer exerts a viscous force on the tube.

A liquid is more viscous if the cohesive forces between molecules are stronger. The viscosity of a liquid decreases with increasing temperature because the molecules become less tightly bound. A decrease in the temperature of the human body is dangerous because the viscosity of the blood increases and the flow of blood through the body is hindered. Gases, on the other hand, have an increase in viscosity for an increase in

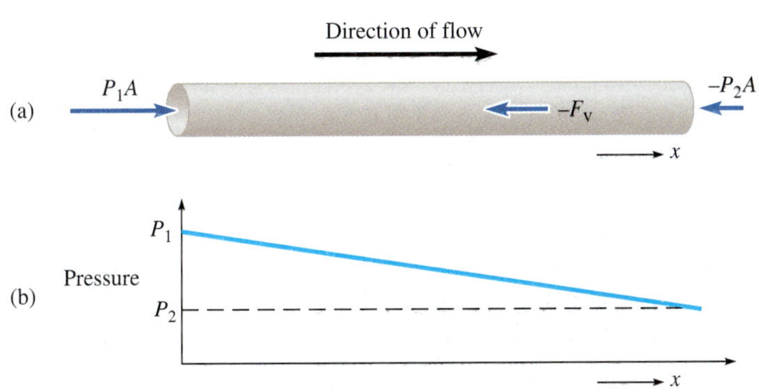

Figure 9.29 (a) To maintain viscous flow, a net force due to fluid pressure $(P_1 - P_2)A$ must be applied in the direction of flow to balance the viscous force F_v due to the pipe, which opposes flow. (b) The pressure in the fluid decreases from P_1 at the left end to P_2 at the right end.

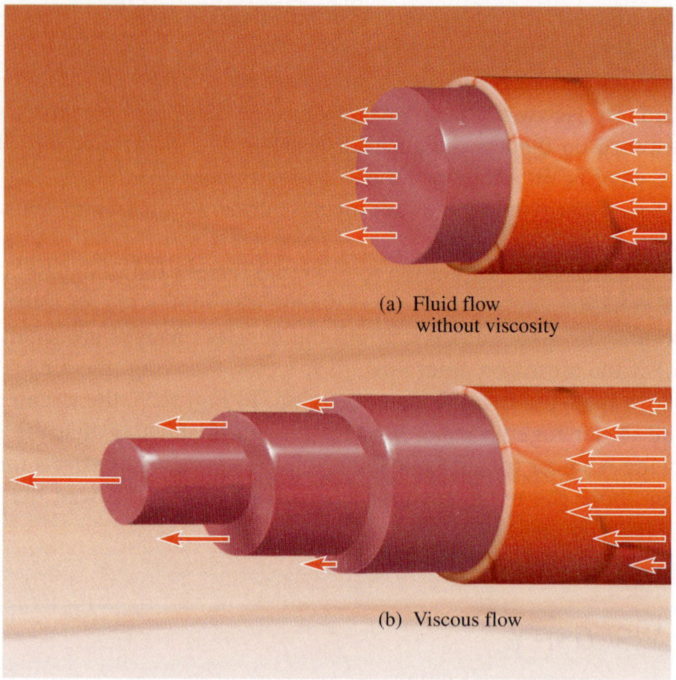

Figure 9.30 (a) In nonviscous flow through a tube, the flow speed is the same everywhere. (b) In viscous flow, the flow speed depends on distance from the tube wall. This simplified sketch shows layers of fluid each moving at a different speed, but in reality the flow speed increases continuously from zero for the outermost "layer" to a maximum speed at the center.

temperature. At higher temperatures the gas molecules move faster and collide more often with each other.

The coefficient of viscosity (or simply the *viscosity*) of a fluid is written as the Greek letter eta (η) and has units of pascal-seconds (Pa·s) in SI. Other viscosity units in common use are the poise (pronounced *pwäz*, symbol P; 1 P = 0.1 Pa·s) and the centipoise (1 cP = 0.01 P = 0.001 Pa·s). Table 9.2 lists the viscosities of some common fluids.

Poiseuille's Law

The volume flow rate $\Delta V/\Delta t$ for laminar flow of a viscous fluid through a horizontal, cylindrical pipe depends on several factors. First of all, the volume flow rate is proportional to the *pressure drop per unit length* ($\Delta P/L$)—also called the pressure gradient. If a pressure drop ΔP maintains a certain flow rate in a pipe of length L, then a similar pipe of length $2L$ needs twice the pressure drop to maintain the same flow rate (ΔP across the first half and another ΔP across the second half). Thus, the flow rate ($\Delta V/\Delta t$) must be proportional to the pressure drop per unit length ($\Delta P/L$).

Next, the flow rate is inversely proportional to the viscosity of the fluid. The more viscous the fluid, the smaller the flow rate, if all other factors are equal.

The only other consideration is the radius of the pipe. In the nineteenth century, during a study of flow in blood vessels, French physician Jean-Léonard Marie Poiseuille (1799–1869) discovered that the flow rate is proportional to the *fourth power* of the pipe radius:

Poiseuille's law (for viscous flow)

$$\frac{\Delta V}{\Delta t} = \frac{\pi}{8}\frac{\Delta P/L}{\eta} r^4 \qquad (9\text{-}15)$$

where $\Delta V/\Delta t$ is the volume flow rate, ΔP is the pressure difference between the ends of the pipe, r and L are the inner radius and length of the pipe, respectively, and η is

Table 9.2 Viscosities of Some Fluids

Substance	Temperature (°C)	Viscosity (Pa·s)
Gases		
Water vapor	100	1.3×10^{-5}
Air	0	1.7×10^{-5}
	20	1.8×10^{-5}
	30	1.9×10^{-5}
	100	2.2×10^{-5}
Liquids		
Acetone	30	0.30×10^{-3}
Methanol	30	0.51×10^{-3}
Ethanol	30	1.0×10^{-3}
Water	0	1.8×10^{-3}
	20	1.0×10^{-3}
	30	0.80×10^{-3}
	40	0.66×10^{-3}
	60	0.47×10^{-3}
	80	0.36×10^{-3}
	100	0.28×10^{-3}
Blood plasma	37	1.3×10^{-3}
Blood, whole	20	3.0×10^{-3}
	37	2.1×10^{-3}
Glycerin	20	0.83
	30	0.63
SAE 5W-30 motor oil	−30	≤ 6.6
	150	$\geq 2.9 \times 10^{-3}$

the viscosity of the fluid. Poiseuille's name is pronounced *pwahzoy*, in a rough English approximation.

It isn't often that we encounter a *fourth-power* dependence. Why such a strong dependence on radius? First of all, if fluids are flowing through two different pipes at the *same speed*, the volume flow rates are proportional to radius squared (flow rate = speed multiplied by cross-sectional area). But, in viscous flow, the average flow speed is larger for wider pipes; fluid farther away from the walls can flow faster. It turns out that the average flow speed for a given pressure gradient is also proportional to radius squared, giving the overall fourth power dependence on the pipe radius of Poiseuille's law.

Application of Viscous Flow: High Blood Pressure The strong dependence of flow rate on radius is important in blood flow. A person with cardiovascular disease has arteries narrowed by plaque deposits. To maintain the necessary blood flow to keep the body functioning, the blood pressure increases. If the diameter of an artery narrows to $\frac{1}{2}$ of its original value due to plaque deposits, the blood flow rate would decrease to $\frac{1}{16}$ of its original value if the pressure drop across it were to stay the same. To compensate for some of this decrease in blood flow, the heart pumps harder, increasing the blood pressure. (See Problem 85.) High blood pressure is not good either; it introduces its own set of health problems, not least of which is the increased demands placed on the heart muscle.

Example 9.12

Arterial Blockage

A cardiologist reports to her patient that the radius of the left anterior descending artery of the heart has narrowed by 10.0%. What percent increase in the blood pressure drop across the artery is required to maintain the normal blood flow through this artery?

Strategy We assume that the viscosity of the blood has not changed, nor has the length of the artery. To maintain normal blood flow, the volume flow rate must stay the same:

$$\frac{\Delta V_1}{\Delta t} = \frac{\Delta V_2}{\Delta t}$$

Solution If r_1 is the normal radius and r_2 is the actual radius, a 10.0% reduction in radius means $r_2 = 0.900r_1$. Then, from Poiseuille's law,

$$\frac{\pi(\Delta P_1/L)r_1^4}{8\eta} = \frac{\pi(\Delta P_2/L)r_2^4}{8\eta}$$

$$r_1^4\,\Delta P_1 = r_2^4\,\Delta P_2$$

We solve for the ratio of the pressure drops:

$$\frac{\Delta P_2}{\Delta P_1} = \frac{r_1^4}{r_2^4} = \frac{1}{(0.900)^4} = 1.52$$

Discussion A factor of 1.52 means there is a 52% increase in the blood pressure difference across that artery. The increased pressure must be provided by the heart. If the normal pressure drop across the artery is 10 mm Hg, then it is now 15.2 mm Hg. The person's blood pressure either must increase by 5.2 mm Hg, or blood flow will be reduced through this artery. The heart is under greater strain as it works harder, attempting to maintain an adequate flow of blood. (See Problem 85.)

Practice Problem 9.12 New Water Pipe

The town water supply is operating at nearly full capacity. The town board decides to replace the water main with a bigger one to increase capacity. If the maximum flow rate is to increase by a factor of 4.0, by what factor should they increase the radius of the water main?

9.10 VISCOUS DRAG

When an object moves through a fluid, the fluid exerts a drag force on it. When the relative velocity between the object and the fluid is low enough for the flow around the object to be laminar, the drag force derives from viscosity and is called **viscous drag**. The viscous drag force is proportional to the speed of the object ($F_D \propto v$). For larger relative speeds, the flow becomes turbulent and the drag force is proportional to the square of the object's speed ($F_D \propto v^2$). (For more information on turbulent drag, see the text website.)

connect

The viscous drag force depends also on the shape and size of the object. For a spherical object, the viscous drag force is given by Stokes's law:

Stokes's law (viscous drag on a sphere)

$$F_D = 6\pi\eta r v \quad \text{(9-16)}$$

where r is the radius of the sphere, η is the viscosity of the fluid, and v is the speed of the object with respect to the fluid.

CHECKPOINT 9.10

Compare and contrast the viscous drag force with the kinetic frictional force.

An object's **terminal velocity** is the velocity that produces just the right drag force so that the net force is zero. An object falling at its terminal velocity has zero acceleration, so it continues moving at that constant velocity. Using Stokes's law, we can find the terminal velocity of a spherical object falling through a viscous fluid. When the object moves at terminal velocity, the net force acting on it is zero. If $\rho_o > \rho_f$, the object sinks; the terminal velocity is downward, and the viscous drag force acts upward to oppose the motion. For an object, such as an air bubble in oil, that rises rather than sinks ($\rho_o < \rho_f$), the terminal velocity is *upward* and the drag force is *downward*. Example 9.13 shows how to calculate the terminal velocity by setting the net force on the falling object equal to zero.

Example 9.13

Falling Droplet

In an experiment to measure the electric charge of the electron, a fine mist of oil droplets is sprayed into the air and observed through a telescope as they fall. These droplets are so tiny that they soon reach their terminal velocity. If the radius of the droplets is 2.40 μm and the average density of the oil is 862 kg/m^3, find the terminal speed of the droplets. The density of air is 1.20 kg/m^3 and the viscosity of air is 1.8×10^{-5} Pa·s.

Strategy When the droplets fall at their terminal velocity, the net force on them is zero. We set the net force equal to zero and use Stokes's law for the drag force.

Solution We set the sum of the forces equal to zero when $v = v_t$.

$$\Sigma F_y = +F_D + F_B - W = 0$$

If m_{air} is the mass of displaced air, then

$$6\pi\eta r v_t + m_{air}g - m_{oil}g = 0$$

Solving for v_t,

$$v_t = \frac{g(m_{oil} - m_{air})}{6\pi\eta r} = \frac{g(\rho_{oil}\cdot\frac{4}{3}\pi r^3 - \rho_{air}\cdot\frac{4}{3}\pi r^3)}{6\pi\eta r}$$

$$= \frac{2(\rho_{oil} - \rho_{air})gr^2}{9\eta}$$

$$= \frac{2(862 \text{ kg/m}^3 - 1.20 \text{ kg/m}^3)(9.80 \text{ N/kg})(2.40 \times 10^{-6} \text{ m})^2}{9(1.8 \times 10^{-5} \text{ Pa·s})}$$

$$= 6.0 \times 10^{-4} \text{ m/s} = 0.60 \text{ mm/s}$$

Discussion We should check the units in the final expression:

$$\frac{(\text{kg/m}^3)\cdot(\text{N/kg})\cdot\text{m}^2}{\text{Pa·s}} = \frac{\text{N/m}}{(\text{N/m}^2)\cdot\text{s}} = \frac{\text{m}}{\text{s}}$$

Stokes's law was applied in this way by Robert Millikan (1868–1953) in his experiments in 1909–1913 to measure the charge of the electron. Using an atomizer, Millikan produced a fine spray of oil droplets. The droplets picked up electric charge as they were sprayed through the atomizer. Millikan kept a droplet suspended without falling by applying an upward electric force. After removing the electric force, he measured the terminal speed of the droplet as it fell through the air. He calculated the mass of the droplet from the terminal speed and the density of the oil using Stokes's law. By setting the magnitude of the electric force equal to the weight of a suspended droplet, Millikan calculated the electric charge of the droplet. He measured the charges of hundreds of different droplets and found that they were all multiples of the same quantity—the charge of an electron.

Practice Problem 9.13 Rising Bubble

Find the terminal velocity of an air bubble of 0.500 mm radius in a cup of vegetable oil. The specific gravity of the oil is 0.840, and the viscosity is 0.160 Pa·s. Assume the diameter of the bubble does not change as it rises.

EVERYDAY PHYSICS DEMO

A demonstration of terminal velocity can be done at home. Drop two objects at the same time: a coin and two or three nested cone-shaped paper coffee filters (or cupcake papers). You will see the effects of viscous drag on the coffee filters as they fall with a constant terminal velocity. Enlist the help of a friend so you can get a side view of the two objects falling. Why do the coffee filters work so well?

Application of Viscous Drag: Sedimentation Velocity and the Centrifuge For small particles falling in a liquid, the terminal velocity is also called the *sedimentation velocity*. The sedimentation velocity is often small for two reasons. First, if the particle isn't much more dense than the fluid, then the vector sum of the gravitational and buoyant forces is small. Second, the terminal velocity is proportional to r^2 (see Example 9.13); viscous drag is most important for small particles. Thus, it can take a long time for the particles to sediment out of solution. Because the sedimentation velocity is proportional to g, it can be increased by the use of a centrifuge, a rotating container that creates artificial gravity of magnitude $g_{eff} = \omega^2 r$ [see Eq. (5-12) and Section 5.7]. Ultracentrifuges are capable of rotating at 10^5 rev/min and produce artificial gravity approaching a million times g.

9.11 SURFACE TENSION

The surface of a liquid has special properties not associated with the interior of the liquid. The surface acts like a stretched membrane under tension. The **surface tension** (symbol γ, the Greek letter gamma) of a liquid is the force per unit *length* with which the surface pulls *on its edge*. The direction of the force is tangent to the surface at its edge. Surface tension is caused by the cohesive forces that pull the molecules toward each other. (See the text website for more details about surface tension.)

Application: How Insects Can Walk on the Surface of a Pond The high surface tension of water enables water striders and other small insects to walk on the surface of a pond. The foot of the insect makes a small indentation in the water surface (Fig. 9.31); the deformation of the surface enables the water to push upward on the foot as if the water surface were a thin sheet of rubber. Visually it looks similar to a person walking across the mat of a trampoline. Other small water creatures, such as mosquito larvae and planaria, hang from the surface of water, using surface tension to hold themselves up. In plants, surface tension aids in the transport of water from the roots to the leaves.

Figure 9.31 A water strider.

EVERYDAY PHYSICS DEMO

Place a needle (or a flat plastic-coated paper clip) gently on the surface of a glass of water. It may take some practice, but you should be able to get it to "float" on top of the water. Now add some detergent to the water and try again. The detergent reduces the surface tension of the water so it is unable to support the needle. Soaps and detergents are *surfactants*—substances that reduce the surface tension of a fluid. The reduced surface tension allows the water to spread out more, wetting more of a surface to be cleaned.

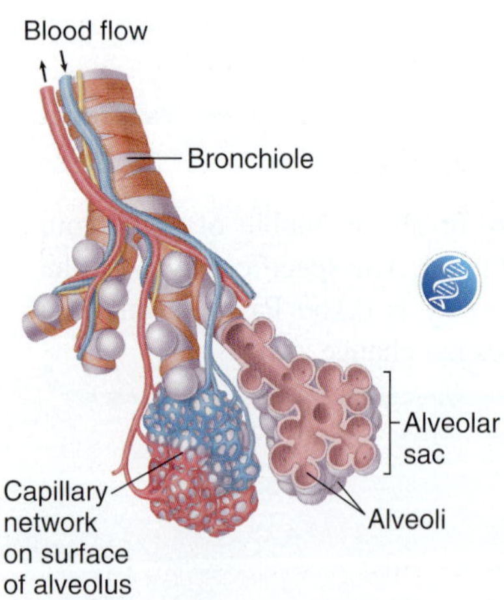

Figure 9.32 In the human lung, millions of tiny sacs called alveoli are inflated with each breath. Gas is exchanged between the air and the blood through the walls of the alveoli. The total surface area through which gas exchange takes place is about 80 m^2—about 40 times the surface area of the body.

Application: Surfactant in the Lungs The high surface tension of water is a hindrance in the lungs. The exchange of oxygen and carbon dioxide between inspired air and the blood takes place in tiny sacs called *alveoli*, 0.05 to 0.15 mm in radius, at the end of the bronchial tubes (Fig. 9.32). If the mucus coating the alveoli had the same surface tension as other body fluids, the pressure difference between the inside and outside of the alveoli would not be great enough for them to expand and fill with air. The alveoli secrete a surfactant that decreases the surface tension in their mucous coating so they can inflate during inhalation.

Bubbles

In an underwater air bubble, the surface tension of the water surface tries to contract the bubble while the pressure of the enclosed air pushes outward on the surface. In equilibrium, the air pressure inside the bubble must be larger than the water pressure outside so that the net outward force due to pressure balances the inward force due to surface tension. The excess pressure $\Delta P = P_{in} - P_{out}$ depends both on the surface tension and the size of the bubble. In Problem 79, you can show that the excess pressure is

$$\Delta P = \frac{2\gamma}{r} \qquad (9\text{-}17)$$

Look closely at a glass of champagne and you can see strings of bubbles rising, originating from the same points in the liquid. Why don't bubbles spring up from random locations? A very small bubble would require an insupportably large excess pressure. The bubbles need some sort of nucleus—a small dust particle, for instance—on which to form so they can start out larger, with excess pressures that aren't so large. The strings of bubbles in the glass of champagne are showing where suitable nuclei have been "found."

EVERYDAY PHYSICS DEMO

Blow up a balloon and notice that it's hard to get started and then gets easier as the balloon starts to inflate. As with bubbles [Eq. (9-17)], the excess pressure your lungs must supply is larger when the radius is smaller. Here the elastic forces of the balloon take the place of surface tension for a bubble. Eventually it gets harder again as these elastic forces increase (as if the surface tension were to increase).

Example 9.14

Lung Pressure

During inhalation the gauge pressure in the alveoli is about −400 Pa to allow air to flow in through the bronchial tubes. Suppose the mucous coating on an alveolus of initial radius 0.050 mm had the same surface tension as water (0.070 N/m). What lung pressure outside the alveoli would be required to begin to inflate the alveolus?

Strategy We model an alveolus as a sphere coated with mucus. Due to the surface tension of the mucus, the alveolus must have a lower pressure outside than inside, as for a bubble.

Solution The excess pressure is

$$\Delta P = \frac{2\gamma}{r} = \frac{2 \times 0.070\ \text{N/m}}{0.050 \times 10^{-3}\ \text{m}} = 2.8\ \text{kPa}$$

Thus, the pressure inside the alveolus would be 2.8 kPa higher than the pressure outside. The gauge pressure inside is −400 Pa, so the gauge pressure outside would be

$$P_{\text{out}} = -0.4\ \text{kPa} - 2.8\ \text{kPa} = -3.2\ \text{kPa}$$

Discussion The *actual* gauge pressure outside the alveoli is about −0.5 kPa rather than −3.2 kPa; then $\Delta P = P_{\text{in}} - P_{\text{out}} = -0.4\ \text{kPa} - (-0.5\ \text{kPa}) = 0.1\ \text{kPa}$ rather than 2.8 kPa. Here the surfactant comes to the rescue; by decreasing the surface tension in the mucus, it decreases ΔP to about 0.1 kPa and allows the expansion of the alveoli to take place. For a newborn baby, the alveoli are initially collapsed, making the required pressure difference about 4 kPa. That first breath is as difficult an event as it is significant.

Practice Problem 9.14 Champagne Bubbles

A bubble in a glass of champagne is filled with CO_2. When it is 2.0 cm below the surface of the champagne, its radius is 0.50 mm. What is the gauge pressure inside the bubble? Assume that champagne has the same average density as water and a surface tension of 0.070 N/m.

Master the Concepts

- Fluids are materials that flow and include both liquids and gases. A liquid is nearly incompressible, whereas a gas expands to fill its container.
- Pressure is the perpendicular force per unit area that a fluid exerts on any surface with which it comes in contact ($P = F/A$). The SI unit of pressure is the pascal ($1\ \text{Pa} = 1\ \text{N/m}^2$).
- The average air pressure at sea level is 1 atm = 101.3 kPa.
- Pascal's principle: A change in pressure at any point in a confined fluid is transmitted everywhere throughout the fluid.
- The average density of a substance is the ratio of its mass to its volume

$$\rho = \frac{m}{V} \quad (9\text{-}2)$$

- The specific gravity of a material is the ratio of its density to that of water at 3.98°C.
- Pressure variation with depth in a static fluid:

$$P_2 = P_1 + \rho g d \quad (9\text{-}3)$$

where point 2 is a depth d below point 1.

continued on next page

Master the Concepts continued

- Instruments to measure pressure include the manometer and the barometer. The barometer measures the pressure of the atmosphere. The manometer measures a pressure difference.
- Gauge pressure is the amount by which the absolute pressure exceeds atmospheric pressure:

$$P_{\text{gauge}} = P_{\text{abs}} - P_{\text{atm}} \qquad \text{(9-6)}$$

- Archimedes' principle: a fluid exerts an upward buoyant force on a completely or partially submerged object equal in magnitude to the weight of the volume of fluid displaced by the object:

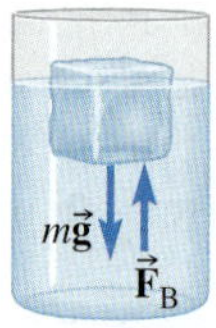

$$F_{\text{B}} = \rho g V \qquad \text{(9-7)}$$

where V is the volume of the part of the object that is submerged and ρ is the density of the fluid.

- In steady flow, the velocity of the fluid *at any point* is constant in time. In laminar flow, the fluid flows in neat layers so that each small portion of fluid that passes a particular point follows the same path as every other portion of fluid that passes the same point. The path that the fluid follows, starting from any point, is called a streamline. Laminar flow is steady. Turbulent flow is chaotic and unsteady. The viscous force opposes the flow of the fluid; it is the counterpart to the frictional force for solids.
- An ideal fluid exhibits laminar flow, has no viscosity, and is incompressible. The flow of an ideal fluid is governed by two principles: the continuity equation and Bernoulli's equation.
- The continuity equation states that the volume flow rate for an ideal fluid is constant:

$$\frac{\Delta V}{\Delta t} = A_1 v_1 = A_2 v_2 \qquad \text{(9-12, 9-13)}$$

- Bernoulli's equation relates pressure changes to changes in flow speed and height:

$$P_1 + \rho g y_1 + \tfrac{1}{2}\rho v_1^2 = P_2 + \rho g y_2 + \tfrac{1}{2}\rho v_2^2 \qquad \text{(9-14)}$$

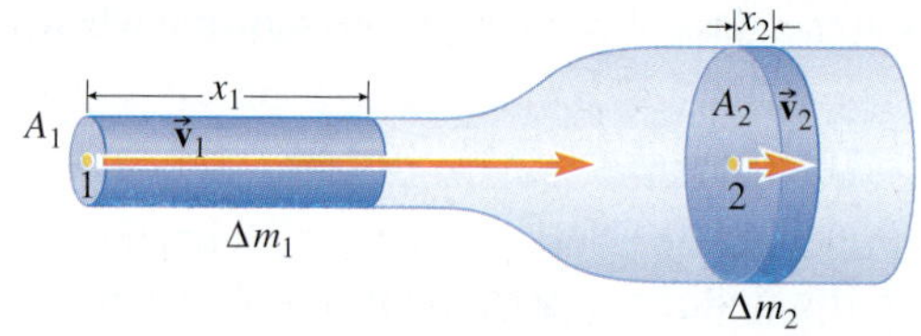

- Poiseuille's law gives the volume flow rate $\Delta V/\Delta t$ for viscous flow in a horizontal pipe:

$$\frac{\Delta V}{\Delta t} = \frac{\pi}{8}\frac{\Delta P/L}{\eta} r^4 \qquad \text{(9-15)}$$

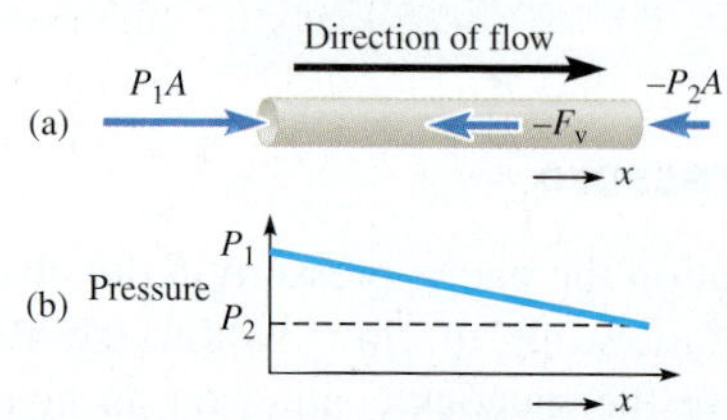

where ΔP is the pressure difference between the ends of the pipe, r and L are the inner radius and length of the pipe, respectively, and η is the viscosity of the fluid.

- Stokes's law gives the viscous drag force on a spherical object moving in a fluid:

$$F_{\text{D}} = 6\pi\eta r v \qquad \text{(9-16)}$$

- The surface tension γ (the Greek letter gamma) of a liquid is the force *per unit length* with which the surface pulls on its edge.

Conceptual Questions

1. Does a manometer (with one side open) measure absolute pressure or gauge pressure? How about a barometer? A tire pressure gauge? A sphygmomanometer?
2. A volunteer firefighter holds the end of a firehose as a strong jet of water emerges. (a) The hose exerts a large backward force on the firefighter. Explain the origin of this force. (b) If another firefighter steps on the hose, forming a constriction (a place where the area of the hose is smaller), the hose begins to pulsate wildly. Explain.
3. The weight of a boat is listed on specification sheets as its "displacement." Explain.
4. In tall buildings, the water supply system uses multiple pumping stations on different floors. At each station, water pumped up from below collects in a storage tank held at atmospheric pressure before it enters the pump. The storage tank supplies water to the floors below it. What are some of the reasons why these multiple pumping stations are used?
5. Can an astronaut on the Moon use a straw to drink from a normal drinking glass? How about if he pokes a straw through an otherwise sealed juice box? Explain.
6. It is commonly said that wood floats because it is "lighter than water" or that a stone sinks because it is "heavier than water." Are these accurate statements? If not, correct them.
7. Why must a blood pressure cuff be wrapped around the arm at the same vertical level as the heart?
8. A hot air balloon is floating in equilibrium with the surrounding air. (a) How does the pressure inside the balloon

compare with the pressure outside? (b) How does the density of the air inside compare to the density outside?

9. When helium weather balloons are released, they are purposely underinflated. Why? [*Hint*: The balloons go to very high altitudes.]
10. Bernoulli's equation applies only to *steady flow*. Yet Bernoulli's equation allows the fluid velocity at one point to be different than the velocity at another point. For the fluid velocity to change, the fluid must be accelerated as it moves from one point to another. In what way is the flow *steady*, then?
11. Before getting an oil change, it is a good idea to drive a few miles to warm up the engine. Why?
12. Your ears "pop" when you change altitude quickly—such as during takeoff or landing in an airplane, or during a drive in the mountains. Curiously, if you are a passenger in a high-speed train, your ears sometimes pop as the speed of the train increases rapidly—even though there is little or no change in altitude. Explain.
13. It is easier to get a good draft in a chimney on a windy day than when the outside air is still, all other things being equal. Why?
14. Two soap bubbles of *different radii* are formed at the ends of a tube with a closed valve in the middle. What happens to the bubbles when the valve is opened? (If the alveoli in the lung did not have a surfactant that reduces surface tension in the smaller alveoli, the same thing would happen in the lung, with disastrous results!)
15. *Pascal's principle: proof by contradiction.* Points A and B are near each other at the same height in a fluid. Suppose $P_A > P_B$. (a) Can both v_A and v_B be zero? Explain. (b) Point C is just above point D in a static fluid. Suppose the pressure at C increases by an amount ΔP. What would happen if the pressure at D did not increase by the same amount?
16. What are the advantages of using hydraulic systems rather than mechanical systems to operate automobile brakes or the control surfaces of an airplane?
17. In any hydraulic system, it is important to "bleed" air out of the line. Why?
18. Is it possible for a skin diver to dive to any depth as long as his snorkel tube is sufficiently long? (A snorkel is a face mask with a breathing tube that sticks above the surface of the water.)
19. Is the buoyant force on a soap bubble greater than the weight of the bubble? If not, why do soap bubbles sometimes appear to float in air?
20. A plastic water bottle open at the top is three-fourths full of water and is placed on a scale. The bottle has an indentation for a label midway up the side, and a strap has been placed around this indentation. If the strap is tightened, so the bottle is squeezed in at the middle and the water level is forced to rise, what happens to the reading on the scale? Is the water pressure at the bottom of the bottle the same?

Multiple-Choice Questions

Student Response System Questions

1. Bernoulli's equation applies to
   (a) any fluid.
   (b) an incompressible fluid, whether viscous or not.
   (c) an incompressible, nonviscous fluid, whether the flow is turbulent or not.
   (d) an incompressible, nonviscous, nonturbulent fluid.
   (e) a static fluid only.
2. A dam holding back the water in a reservoir exerts a horizontal force on the water. The magnitude of this force depends on
   (a) the maximum depth of the reservoir.
   (b) the depth of the water at the location of the dam.
   (c) the surface area of the reservoir.
   (d) both (a) and (b).
   (e) all three—(a), (b), and (c).
3. Bernoulli's equation is an expression of
   (a) conservation of mass.
   (b) conservation of energy.
   (c) conservation of momentum.
   (d) conservation of angular momentum.

Questions 4–5. Two spheres, A and B, fall through the same viscous fluid.

Answer choices for Questions 4 and 5:
(a) A has the larger terminal velocity.
(b) B has the larger terminal velocity.
(c) A and B have the same terminal velocity.
(d) Insufficient information is given to reach a conclusion.

4. A and B have the same radius; A has the larger mass. Which has the larger terminal velocity?
5. A and B have the same density; A has the larger radius. Which has the larger terminal velocity?
6. A glass of ice water is filled to the brim with water; the ice cubes stick up above the water surface. After the ice melts, which is true?
   (a) The water level is below the top of the glass.
   (b) The water level is at the top of the glass but no water has spilled.
   (c) Some water has spilled over the sides of the glass.
   (d) Impossible to say without knowing the initial densities of the water and the ice.
7. The continuity equation is an expression of
   (a) conservation of mass.
   (b) conservation of energy.
   (c) conservation of momentum.
   (d) conservation of angular momentum.

8. What is the gauge pressure of the gas in the closed tube in the figure? (Take the atmospheric pressure to be 76 cm Hg.)
(a) 20 cm Hg (b) −20 cm Hg (c) 96 cm Hg
(d) 56 cm Hg (e) −96 cm Hg (f) −56 cm Hg

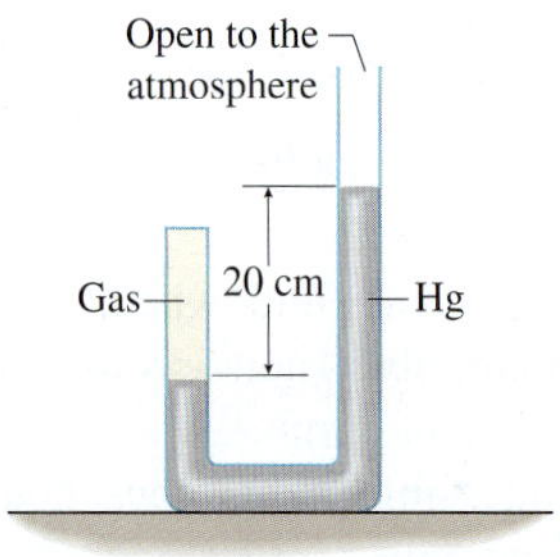

9. A manometer contains two different fluids of different densities. Both sides are open to the atmosphere. Which pair(s) of points in the figure have equal pressure?
(a) $P_1 = P_5$ (b) $P_2 = P_5$ (c) $P_3 = P_4$
(d) Both (a) and (c) (e) Both (b) and (c)

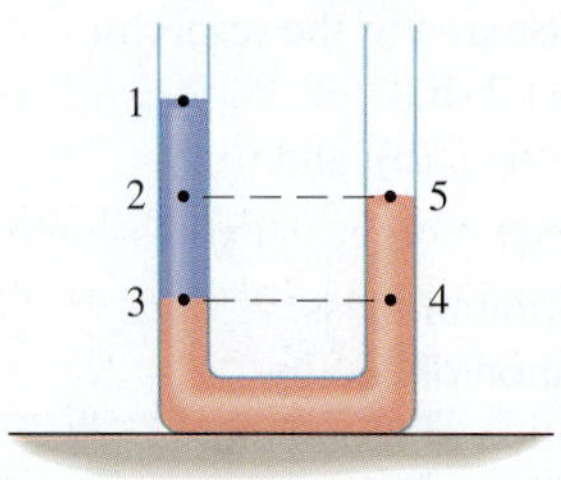

10. A Venturi meter is used to measure the flow speed of a *viscous* fluid. With reference to the figure, which is true?
(a) $h_3 = h_1$ (b) $h_3 > h_1$
(c) $h_3 < h_1$ (d) Insufficient information to determine

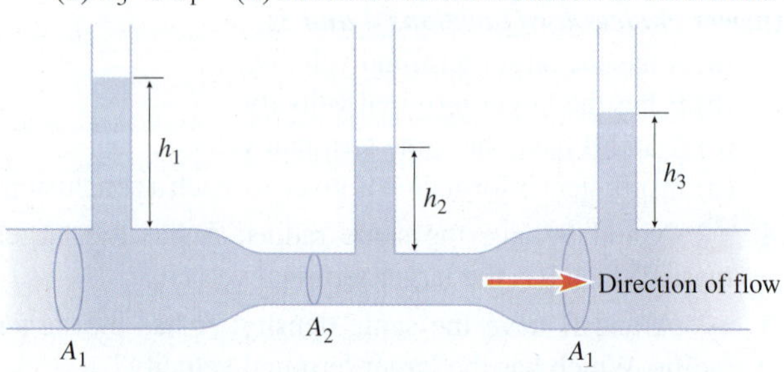

Problems

- C — Combination conceptual/quantitative problem
- Biomedical application
- ✦ Challenging
- Blue # — Detailed solution in the Student Solutions Manual
- [1, 2] — Problems paired by concept
- connect — Text website interactive or tutorial

9.2 Pressure

1. Someone steps on your toe, exerting a force of 500 N on an area of 1.0 cm^2. What is the average pressure on that area in atm?

2. What is the average pressure on the soles of the feet of a standing 90.0 kg person due to the contact force with the floor? Each foot has a surface area of 0.020 m^2.

3. C Atmospheric pressure is about 1.0×10^5 Pa on average. (a) What is the downward force of the air on a desktop with surface area 1.0 m^2? (b) Convert this force to pounds so you really understand how large it is. (c) Why does this huge force not cause the desk to collapse?

4. A 10 kg baby sits on a three-legged stool. The diameter of each of the stool's round feet is 2.0 cm. A 60 kg adult sits on a four-legged chair that has four circular feet, each with a diameter of 6.0 cm. Who applies the greater pressure to the floor and by how much?

5. A lid is put on a box that is 15 cm long, 13 cm wide, and 8.0 cm tall, and the box is then evacuated until its inner pressure is 0.80×10^5 Pa. How much force is required to lift the lid (a) at sea level; (b) in Denver, on a day when the atmospheric pressure is 67.5 kPa ($\frac{2}{3}$ the value at sea level)?

6. A container is filled with gas at a pressure of 4.0×10^5 Pa. The container is a cube, 0.10 m on a side, with one side facing south. What is the magnitude and direction of the force on the south side of the container due to the gas inside?

9.3 Pascal's Principle

7. A nurse applies a force of 4.40 N to the piston of a syringe. The piston has an area of 5.00×10^{-5} m^2. What is the pressure increase in the fluid within the syringe?

8. In a hydraulic lift, the radii of the pistons are 2.50 cm and 10.0 cm. A car weighing W = 10.0 kN is to be lifted by the force of the large piston. (a) What force F_a must be applied to the small piston? (b) When the small piston is pushed in by 10.0 cm, how far is the car lifted? (c) Find the mechanical advantage of the lift, which is the ratio W/F_a.

9.4 The Effect of Gravity on Fluid Pressure

9. At the surface of a freshwater lake the air pressure is 1.0 atm. At what depth under water in the lake is the water pressure 4.0 atm?

10. What is the pressure on a fish 10 m under the ocean surface?

11. The density of platinum is 21 500 kg/m^3. Find the ratio of the volume of 1.00 kg of platinum to the volume of 1.00 kg of aluminum.

12. In the Netherlands, a dike holds back the sea from a town below sea level. The dike springs a leak 3.0 m below the water surface. If the area of the hole in the dike is 1.0 cm^2, what force must the Dutch boy exert to save the town?

13. Each of five cylindrical drums with radius R is filled to a height h above the bottom with a liquid of density ρ. Rank the drums in order of the pressure at the bottom of the drum, largest to smallest.
(a) $R = 40$ cm, $h = 80$ cm, $\rho = 1000$ kg/m^3
(b) $R = 40$ cm, $h = 100$ cm, $\rho = 1000$ kg/m^3
(c) $R = 50$ cm, $h = 100$ cm, $\rho = 800$ kg/m^3
(d) $R = 50$ cm, $h = 80$ cm, $\rho = 800$ kg/m^3
(e) $R = 50$ cm, $h = 125$ cm, $\rho = 800$ kg/m^3

14. Each of six barrels is filled to a height h above the bottom with a liquid of density ρ. Each barrel has a hole of radius r in its side. The center of each hole is 20 cm above the barrel bottom. A plug in the hole keeps the liquid from escaping. Rank the barrels in order of the force on the plug due to the liquid in the barrel, from largest to smallest.
(a) $r = 1$ cm, $h = 100$ cm, $\rho = 1000$ kg/m^3
(b) $r = 1$ cm, $h = 120$ cm, $\rho = 1000$ kg/m^3
(c) $r = 1.25$ cm, $h = 120$ cm, $\rho = 800$ kg/m^3
(d) $r = 1.25$ cm, $h = 100$ cm, $\rho = 800$ kg/m^3
(e) $r = 1$ cm, $h = 145$ cm, $\rho = 1000$ kg/m^3

15. A giraffe's brain is approximately 3.4 m above its heart. Estimate the minimum gauge pressure that its heart must produce to move the blood to his brain. Ignore any effects from the blood flow through arteries of different area, and assume that giraffe blood is identical to human blood.

16. How high can you suck water up a straw? The pressure in the lungs can be reduced to about 10 kPa below atmospheric pressure.

17. Ⓒ A container has a large cylindrical lower part with a long thin cylindrical neck. The lower part of the container holds 12.5 m^3 of water and the surface area of the bottom of the container is 5.00 m^2. The height of the lower part of the container is 2.50 m, and the neck contains a column of water 8.50 m high. The total volume of the column of water in the neck is 0.200 m^3. (a) What is the magnitude of the force exerted by the water on the bottom of the container? (b) Explain why it is not equal to the weight of the water.

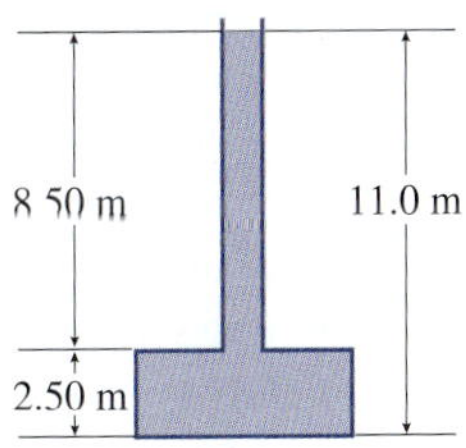

18. A sperm whale can reach depths of 2500 m below the surface of the ocean. What is the pressure on the whale's skin at that depth, assuming that the density of seawater is constant from the surface to that depth?

19. The maximum pressure most organisms can survive is about 1000 times atmospheric pressure. Only small, simple organisms such as tadpoles and bacteria can survive such high pressures. What then is the maximum depth at which these organisms can live under the sea (assuming that the density of seawater is 1025 kg/m^3)?

20. At the surface of a freshwater lake the pressure is 105 kPa. (a) What is the pressure increase in going 35.0 m below the surface? (b) What is the approximate pressure decrease in going 35 m above the surface? Air at 20°C has density of 1.20 kg/m^3.

9.5 Measuring Pressure

21. When a mercury manometer is connected to a gas main, the mercury stands 40.0 cm higher in the tube that is open to the air than in the tube connected to the gas main. A barometer at the same location reads 74.0 cm Hg. Determine the absolute pressure of the gas in cm Hg.

22. An experiment to determine the specific heat of a gas makes use of a water manometer attached to a flask. Initially the two columns of water are even. Atmospheric pressure is 1.0×10^5 Pa. After heating the gas, the water levels change to those shown. Find the change in pressure of the gas in Pa.

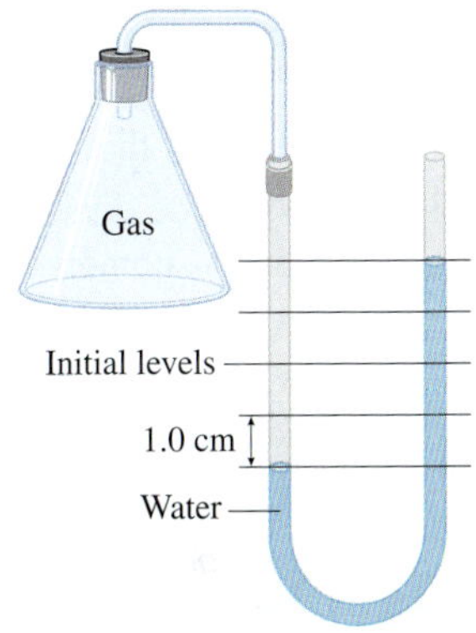

23. A manometer using oil (density 0.90 g/cm^3) as a fluid is connected to an air tank. Suddenly the pressure in the tank increases by 0.74 cm Hg. (a) By how much does the fluid level rise in the side of the manometer that is open to the atmosphere? (b) What would your answer be if the manometer used mercury instead?

24. An IV is connected to a patient's vein. The blood in the vein has a gauge pressure of 12 mm Hg. At least how far above the vein must the IV bag be hung in order for fluid to flow into the vein? Assume the fluid in the IV has the same density as blood.

25. Estimate the average blood pressure in a person's foot if the foot is 1.37 m below the aorta, where the average blood pressure is 104 mm Hg. For the purposes of this estimate, assume the blood isn't flowing.

26. A woman's systolic blood pressure when resting is 160 mm Hg. What is this pressure in (a) Pa, (b) lb/in.^{2}, (c) atm, (d) torr?

27. The gauge pressure of the air in an automobile tire is 32 lb/in.^{2}. Convert this to (a) Pa, (b) torr, (c) atm.

9.6 The Buoyant Force

28. Six wooden blocks (mass m) float in a barrel of water. The blocks are not all made from the same type of wood. The bottom of each block is submerged to a depth d below the water surface. Rank the blocks in order of the buoyant force on them, largest to smallest.

(a) $m = 20$ g, $d = 2.5$ cm (b) $m = 20$ g, $d = 2$ cm
(c) $m = 25$ g, $d = 2$ cm (d) $m = 25$ g, $d = 2.5$ cm
(e) $m = 10$ g, $d = 2$ cm (f) $m = 10$ g, $d = 1$ cm

29. (a) Which has the larger buoyant force acting on it when immersed in water, 1.0 kg of lead or 1.0 kg of aluminum? Explain. (b) Which has the larger buoyant force acting on it, 1.0 kg of steel that is sinking to the bottom of a lake or 1.0 kg of wood with a density of 500 kg/m^3 that is floating on the lake? Explain. (c) Once you have answered the qualitative questions, find the quantitative answers to parts (a) and (b).

30. A Canada goose floats with 25% of its volume below water. What is the average density of the goose?

31. A flat-bottomed barge, loaded with coal, has a mass of 3.0×10^5 kg. The barge is 20.0 m long and 10.0 m wide. It floats in freshwater. What is the depth of the barge below the waterline? (connect tutorial: boat)

32. (a) When ice floats in water at 0°C, what percent of its volume is submerged? (b) What is the specific gravity of ice?

33. (a) What is the density of an object that is 14% submerged when floating in water at 0°C? (b) What percentage of the object will be submerged if it is placed in ethanol at 0°C?

34. (a) What is the buoyant force on 0.90 kg of ice floating freely in liquid water? (b) What is the buoyant force on 0.90 kg of ice held completely submerged under water?

35. A block of birch wood floats in oil with 90.0% of its volume submerged. What is the density of the oil? The density of the birch is 0.67 g/cm^3.

36. When a block of ebony is placed in ethanol, what percentage of its volume is submerged?

37. A cylindrical disk has volume 8.97×10^{-3} m^3 and mass 8.16 kg. The disk is floating on the surface of some water with its flat surfaces horizontal. The area of each flat surface is 0.640 m^2. (a) What is the specific gravity of the disk? (b) How far below the water level is its bottom surface? (c) How far above the water level is its top surface?

38. An aluminum cylinder weighs 1.03 N. When this same cylinder is completely submerged in alcohol, the volume of the displaced alcohol is 3.90×10^{-5} m^3. If the cylinder is suspended from a scale while submerged in the alcohol, the scale reading is 0.730 N. What is the specific gravity of the alcohol? (connect tutorial: ball in beaker)

39. A solid piece of plastic, with a density of 890 kg/m^3, is placed in oil with a density of 830 kg/m^3 and the plastic sinks (1). Then the plastic is placed in water and it floats (2). (a) What percentage of the plastic is submerged in the water? (b) Finally, the same oil in which the plastic sinks is poured over the plastic and the water. Will less (3) or more (4) of the plastic be submerged in the water compared to 2? Explain. (c) Calculate the percentage of the plastic submerged in the water.

40. A fish uses a swim bladder to change its density so it is equal to that of water, enabling it to remain suspended under water. If a fish has an average density of 1080 kg/m^3 and mass 10.0 g with the bladder completely deflated, to what volume must the fish inflate the swim bladder in order to remain suspended in seawater of density 1060 kg/m^3?

41. Nine-banded armadillos (*Dasypus novemcinctus*) have a typical mass density of 1200 kg/m^3 and a mass of 7.0 kg (including their armor). When faced with a body of water to cross, the armadillo has two choices: to hold its breath and walk across the bottom or to swallow air into the stomach and intestine and float across. Approximately what volume of air does it need to swallow in order to float? Assume the swallowed air is at atmospheric pressure.

42. The average density of a fish can be found by first weighing it in air and then finding the scale reading for the fish completely immersed in water and suspended from a scale. If a fish has weight 200.0 N in air and scale reading 15.0 N in water, what is the average density of the fish?

Problems 43–44. While vacationing on the Outer Banks of North Carolina, you find an old coin that looks like it is made of gold. You know that there were many shipwrecks here, so you take the coin home to test it.

43. You suspend the coin from a spring scale and find that its mass is 49.7 g. You then let the coin hang completely submerged in a glass of water but not touching the bottom and find that the scale reads 47.1 g. Should you get excited about the possibility that this coin might really be gold? Explain.

44. On a kitchen scale you measure the coin's mass to be 49.7 g. You measure the mass of a small bowl of water as 314.8 g. You tie a slender thread around the coin and suspend it completely submerged in the water but not touching the bottom. The scale reading is now 317.4 g. Should you get excited about the possibility that this coin might really be gold? Explain.

45. ✦ (a) A piece of balsa wood with density 0.50 g/cm^3 is released under water. What is its initial acceleration? (b) Repeat for a piece of maple with density 0.750 g/cm^3. (c) Repeat for a ping-pong ball with an average density of 0.125 g/cm^3.

46. ✦ A piece of metal is released under water. The volume of the metal is 50.0 cm^3 and its specific gravity is 5.0. What is its initial acceleration?

9.7 Fluid Flow; 9.8 Bernoulli's Equation

47. A garden hose of inner radius 1.0 cm carries water at 2.0 m/s. The nozzle at the end has radius 0.20 cm. How fast does the water move through the nozzle?

48. If the average volume flow of blood through the aorta is 8.5×10^{-5} m^3/s and the cross-sectional area of the aorta is 3.0×10^{-4} m^2, what is the average speed of blood in the aorta?

49. A nozzle of inner radius 1.00 mm is connected to a hose of inner radius 8.00 mm. The nozzle shoots out water moving at 25.0 m/s. (a) At what speed is the water in the hose moving? (b) What is the volume flow rate? (c) What is the mass flow rate?

50. Water entering a house flows with a speed of 0.20 m/s through a pipe of 1.0 cm inside radius. What is the speed of the water at a point where the pipe tapers to a radius of 2.5 mm?

51. A horizontal segment of pipe tapers from a cross-sectional area of 50.0 cm^2 to 0.500 cm^2. The pressure at the larger end of the pipe is 1.20×10^5 Pa, and the speed is 0.040 m/s. What is the pressure at the narrow end of the segment?

52. In a tornado or hurricane, a roof may tear away from the house because of a difference in pressure between the air inside and the air outside. Suppose that air is blowing across the top of a 2000 ft^2 roof at 150 mi/h. What is the magnitude of the force on the roof?

53. Use Bernoulli's equation to estimate the upward force on an airplane's wing if the average flow speed of air is 190 m/s above the wing and 160 m/s below the wing. The density of the air is 1.3 kg/m^3, and the area of each wing surface is 28 m^2.

54. An airplane flies on a level path. There is a pressure difference of 500 Pa between the lower and upper surfaces of the wings. The area of each wing surface is about 100 m^2. The air moves below the wings at a speed of 80.5 m/s. Estimate (a) the weight of the plane and (b) the air speed above the wings.

55. A nozzle is connected to a horizontal hose. The nozzle shoots out water moving at 25 m/s. What is the gauge pressure of the water in the hose? Ignore viscosity and assume that the diameter of the nozzle is much smaller than the inner diameter of the hose.

56. Suppose air, with a density of 1.29 kg/m^3, is flowing into a Venturi meter. The narrow section of the pipe at point A has a diameter that is $\frac{1}{3}$ of the diameter of the larger section of the pipe at point B. The U-shaped tube is filled with *water* and the difference in height between the two sections of pipe is 1.75 cm. How fast is the air moving at point B?

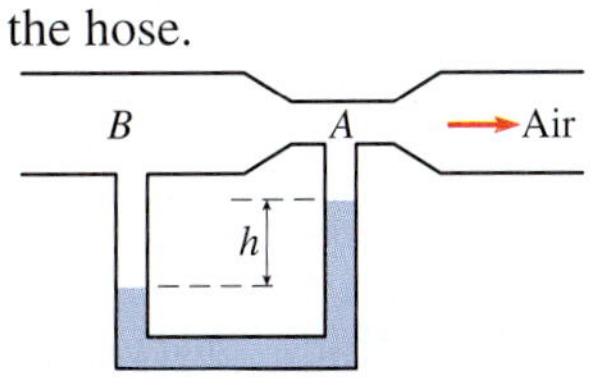

57. A water tower supplies water through the plumbing in a house. A 2.54 cm diameter faucet in the house can fill a cylindrical container with a diameter of 44 cm and a height of 52 cm in 12 s. How high above the faucet is the top of the water in the tower? (Assume that the diameter of the tower is so large compared to that of the faucet that the water at the top of the tower does not move.)

58. The volume flow rate of the water supplied by a well is 2.0×10^{-4} m^3/s. The well is 40.0 m deep. (a) What is the power output of the pump—in other words, at what rate does the well do work on the water? (b) Find the pressure difference the pump must maintain. (c) Can the pump be at the top of the well or must it be at the bottom? Explain.

9.9 Viscosity

59. Using Poiseuille's law [Eq. (9-15)], show that viscosity has SI units of pascal-seconds.

60. A viscous liquid is flowing steadily through a pipe of diameter D. Suppose you replace it by two parallel pipes, each of diameter $D/2$, but the same length as the original pipe. If the pressure difference between the ends of these two pipes is the same as for the original pipe, what is the total rate of flow in the two pipes compared to the original flow rate?

61. A hypodermic syringe is attached to a needle that has an internal radius of 0.300 mm and a length of 3.00 cm. The needle is filled with a solution of viscosity 2.00×10^{-3} Pa·s; it is injected into a vein at a gauge pressure of 16.0 mm Hg. Ignore the extra pressure required to accelerate the fluid from the syringe into the entrance of the needle. (a) What must the pressure of the fluid in the syringe be in order to inject the solution at a rate of 0.250 mL/s? (b) What force must be applied to the plunger, which has an area of 1.00 cm^2?

62. A bug from South America known as *Rhodnius prolixus* extracts the blood of animals. Suppose *Rhodnius prolixus* extracts 0.30 cm^3 of blood in 25 min from a human arm through its feeding tube of length 0.20 mm and radius 5.0 μm. What is the absolute pressure at the bug's end of the feeding tube if the absolute pressure at the other end (in the human arm) is 105 kPa? Assume the viscosity of blood is 0.0013 Pa·s. [*Note*: Negative absolute pressures are possible in liquids in very slender tubes.]

Problems 63–65. Four identical sections of pipe are connected in various ways to pumps that supply water at the pressures indicated in the figure. The water exits at the right at 1.0 atm. Assume viscous flow.

63. If the *total* volume flow rates in systems A and C are the same and the flow speed in each of the pipes in C is 3.0 m/s, what is the flow speed in system A?

64. If the total volume flow rate in system B is 0.020 m^3/s, what is the total volume flow rate in system C?

65. If the total volume flow rates in systems A and B are the same, at what pressure does the pump supply water in system A?

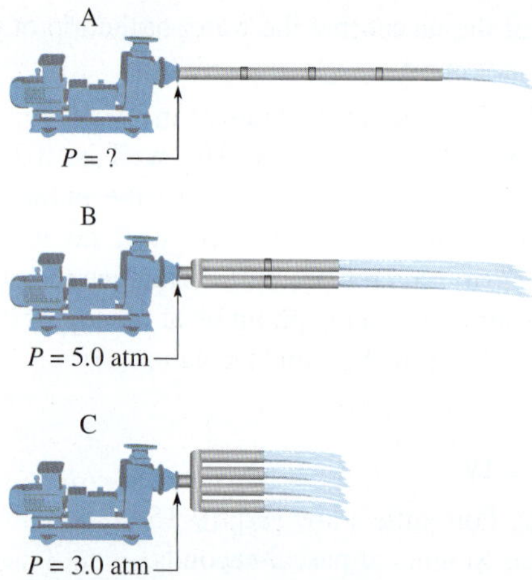

Problems 63–65

66. A capillary carries blood in the direction shown. Viscosity is *not* negligible. Points C and D are on the central axis of the capillary. Rank the points in order of decreasing fluid speed.

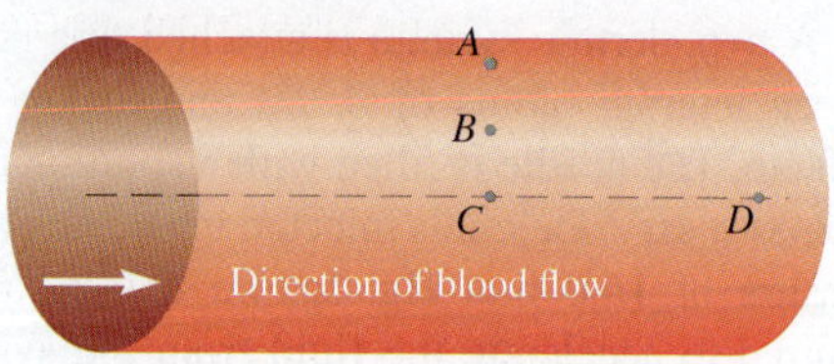

67. (a) What is the pressure difference required to make blood flow through an artery of inner radius 2.0 mm and length 0.20 m at a speed of 6.0 cm/s? (b) What is the pressure difference required to make blood flow at 0.60 mm/s through a capillary of radius 3.0 μm and length 1.0 mm? (c) Compare both answers to your average blood pressure, about 100 torr.

68. Blood plasma (at 37°C) is to be supplied to a patient at the rate of 2.8×10^{-6} m^3/s. If the tube connecting the plasma to the patient's vein has a radius of 2.0 mm and a length of 50 cm, what is the pressure difference between the plasma and the patient's vein?

69. (a) Since the flow rate is proportional to the pressure difference, show that Poiseuille's law can be written in the form $\Delta P = IR$, where I is the volume flow rate and R is a constant of proportionality called the fluid flow *resistance*. (Written this way, Poiseuille's law is analogous to *Ohm's law* for electric current to be studied in Chapter 18: $\Delta V = IR$, where ΔV is the potential drop across a conductor, I is the electric current flowing through the conductor, and R is the electrical resistance of the conductor.) (b) Find R in terms of the viscosity of the fluid and the length and radius of the pipe.

9.10 Viscous Drag

70. Five spheres are falling through the same viscous fluid, not necessarily at their terminal speeds. The radii r and speeds v of the spheres are given. Rank the spheres in order of decreasing viscous drag force on them.
(a) $r = 1.0$ mm, $v = 15$ mm/s
(b) $r = 1.0$ mm, $v = 30$ mm/s
(c) $r = 2.0$ mm, $v = 15$ mm/s
(d) $r = 2.0$ mm, $v = 30$ mm/s
(e) $r = 3.0$ mm, $v = 20$ mm/s

71. Two identical spheres are dropped into two different columns: one column contains a liquid of viscosity 0.5 Pa·s; the other contains a liquid of the same density but unknown viscosity. The sedimentation velocity in the second tube is 20% higher than the sedimentation velocity in the first tube. What is the viscosity of the second liquid?

72. A sphere of radius 1.0 cm is dropped into a glass cylinder filled with a viscous liquid. The mass of the sphere is 12.0 g, and the density of the liquid is 1200 kg/m^3. The sphere reaches a terminal speed of 0.15 m/s. What is the viscosity of the liquid?

73. An air bubble of 1.0 mm radius is rising in a container with vegetable oil of specific gravity 0.85 and viscosity 0.12 Pa·s. The container of oil and the air bubble are at 20°C. What is its terminal velocity?

74. *What keeps a cloud from falling?* A cumulus (fair-weather) cloud consists of tiny water droplets of average radius 5.0 μm. Find the terminal velocity for these droplets at 20°C, assuming viscous drag. (Besides the viscous drag force, there are also upward air currents called *thermals* that push the droplets upward. connect tutorial: rain drop)

75. C A flea is on the back of a squirrel climbing a tall tree. When the squirrel is near the top, the flea jumps off. (a) Assuming the drag force is viscous, estimate the terminal speed of the flea. Treat the flea as a drop of water of radius 1.0 mm falling through air at 20°C. (b) Does your result seem reasonable? If not, what do you think the problem is?

76. ✦ An aluminum sphere (specific gravity = 2.7) falling through water reaches a terminal speed of 5.0 cm/s. What is the terminal speed of an air bubble of the same radius rising through water? Assume viscous drag in both cases and ignore the possibility of changes in size or shape of the air bubble; the temperature is 20°C.

9.11 Surface Tension

77. An underwater air bubble has an excess inside pressure of 10 Pa. What is the excess pressure inside an air bubble with twice the radius?

78. Assume a water strider has a roughly circular foot of radius 0.02 mm. (a) What is the maximum possible upward force on the foot due to surface tension of the

water? (b) What is the maximum mass of this water strider so that it can keep from breaking through the water surface? The strider has six legs.

79. ✦ A hollow hemispherical object is filled with air as in part (a) of the figure. (a) Show that the magnitude of the force due to fluid pressure on the curved surface of the hemisphere has magnitude $F = \pi r^2 P$, where r is the radius of the hemisphere and P is the pressure of the air. Ignore the weight of the air. [*Hint*: First find the force on the *flat* surface. What is the net force on the hemisphere due to the air?] (b) Consider an underwater air bubble to be divided into two hemispheres along the circumference as in part (b) of the figure. The upper hemisphere of the water surface exerts a force of magnitude $2\pi r\gamma$ (circumference times force per unit length) on the lower hemisphere due to surface tension. Show that the air pressure inside the bubble must exceed the water pressure outside by $\Delta P = 2\gamma/r$.

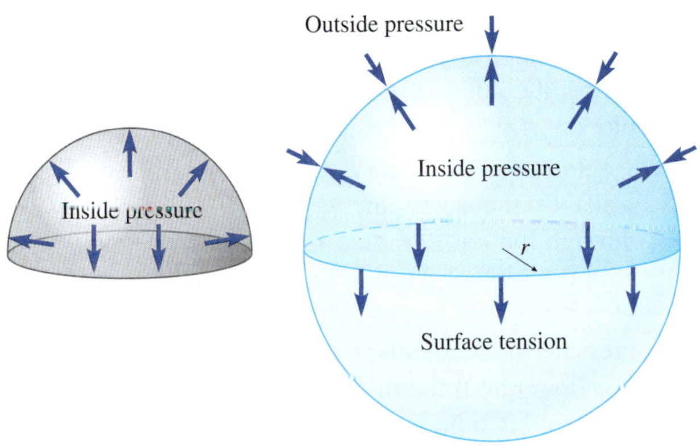

(a) (b)

Collaborative Problems

80. Ⓒ A wooden barrel full of water has a flat circular top of radius 25.0 cm with a small hole in it. A tube of height 8.00 m and inner radius 0.250 cm is suspended above the barrel with its lower end inserted snugly in the hole. Water is poured into the upper end of the tube until it is full. (a) What is the weight of the water in the tube? (b) What is the force with which the water in the barrel pushes up on the top of the barrel? (c) How can adding such a small weight of water lead to such a large force on the top of the barrel? (As a demonstration of the principle now named for him, Pascal astonished spectators by showing that the addition of a small amount of water to the tube could make the barrel burst.)

81. You are hiking through a lush forest with some of your friends when you come to a large river that seems impossible to cross. However, one of your friends notices an old metal barrel sitting on the shore. The barrel is shaped like a cylinder and is 1.20 m high and 0.76 m in diameter. One of the circular ends of the barrel is open and the barrel is empty. When you put the barrel in the water with the open end facing up, you find that the barrel floats with 33% of it under water. You decide that you can use the barrel as a boat to cross the river, as long as you leave about 30 cm sticking above the water. How much extra mass can you put in this barrel to use it as a boat?

82. ✦ Ⓒ On a nice day when the temperature outside is 20°C, you take the elevator to the top of the Willis Tower in Chicago, which is 440 m tall. (a) How much less is the air pressure at the top than the air pressure at the bottom? Express your answer both in Pa and atm. (connect tutorial: gauge) [*Hint*: The altitude change is small enough to treat the density of air as constant.] (b) How many pascals does the pressure decrease for every meter of altitude? (c) If the pressure gradient—the pressure decrease per meter of altitude—were uniform, at what altitude would the atmospheric pressure reach zero? (d) Atmospheric pressure does *not* decrease with a uniform gradient since the density of air decreases as you go up. Which is true: the pressure reaches zero at a *lower* altitude than your answer to (c), or the pressure is nonzero at that altitude and the atmosphere extends to a higher altitude? Explain.

83. ✦ A house with its own well has a pump in the basement with an output pipe of inner radius 6.3 mm. The pump can maintain a gauge pressure of 410 kPa in the output pipe. A showerhead on the second floor (6.7 m above the pump's output pipe) has 36 holes, each of radius 0.33 mm. The shower is on "full blast" and no other faucet in the house is open. (a) Ignoring viscosity, with what speed does water leave the showerhead? (b) With what speed does water move through the output pipe of the pump?

84. The average speed of blood in the aorta is 0.3 m/s, and the radius of the aorta is 1 cm. There are about 2×10^9 capillaries with an average radius of 6 μm. What is the approximate average speed of the blood flow in the capillaries?

85. ✦ The diameter of a certain artery has decreased by 25% due to arteriosclerosis. (a) If the same amount of blood flows through it per unit time as when it was unobstructed, by what percentage has the blood pressure difference between its ends increased? (b) If, instead, the pressure drop across the artery stays the same, by what factor does the blood flow rate through it decrease? (In reality we are likely to see a combination of some pressure increase with some reduction in flow.)

86. The average adult has about 5 L of blood, and a healthy adult heart pumps blood at a rate of about 80 cm^3/s. Estimate how long it takes for medicine delivered intravenously to travel throughout a person's body.

Comprehensive Problems

87. A block of aluminum that has dimensions 2.00 cm by 3.00 cm by 5.00 cm is suspended by a thread from a spring scale. (a) What is the scale reading in grams? (b) A beaker full of oil (density 850 kg/m^3) rests on a second scale, which reads 460.0 g. The block is then lowered so it is fully submerged in oil without touching the bottom of the beaker. What are the readings of the two scales now?

88. A 85.0 kg canoe made of thin aluminum has the shape of half of a hollowed-out log with a radius of 0.475 m and a length of 3.23 m. (a) When this is placed in the water, what percentage of the volume of the canoe is below the waterline? (b) How much additional mass can be placed in this canoe before it begins to sink? (connect interactive: buoyancy)

89. Two identical beakers are filled to the brim and placed on balance scales. The base area of the beakers is large enough that any water that spills out of the beakers will fall onto the table the scales are resting on. A block of pine (density = 420 kg/m^3) is placed in one of the beakers. The block has a volume of 8.00 cm^3. Another block of the same size, but made of steel, is placed in the other beaker. How does the scale reading change in each case?

90. A very large vat of water has a hole 1.00 cm in diameter located a distance 1.80 m below the water level. (a) How fast does water exit the hole? (b) How would your answer differ if the vat were filled with gasoline? (c) How would your answer differ if the vat contained water, but was on the Moon, where the gravitational field strength is 1.6 N/kg?

91. An atomizer is a device that delivers a fine mist of some liquid such as perfume by blowing air horizontally over the top of a tube immersed in the liquid. Suppose a perfume with density 800 kg/m^3 has a 3.0 cm tube extending vertically from the top of the liquid. What minimum speed does air flow over the top of the tube when the liquid just reaches the top of the tube?

92. The deepest place in the ocean is the Marianas Trench in the western Pacific Ocean, which is over 11.0 km deep. On January 23, 1960, the research sub *Trieste* went to a depth of 10.915 km, nearly to the bottom of the trench. This still is the deepest dive on record. The density of seawater is 1025 kg/m^3. (a) What is the water pressure at that depth? (b) What was the force due to water pressure on a flat section of area 1.0 m^2 on the top of the sub's hull?

93. The pressure in a water pipe in the basement of an apartment house is 4.10×10^5 Pa, but on the seventh floor it is only 1.85×10^5 Pa. What is the height between the basement and the seventh floor? Assume the water is not flowing; no faucets are opened.

94. The body of a 90.0 kg person contains 0.020 m^3 of body fat. If the density of fat is 890 kg/m^3, what percentage of the person's body weight is composed of fat?

95. Near sea level, how high a hill must you ascend for the reading of a barometer you are carrying to drop by 1.0 cm Hg? Assume the temperature remains at 20°C as you climb. The reading of a barometer on an average day at sea level is 76.0 cm Hg. (connect tutorial: gauge)

96. If you watch water falling from a faucet, you will notice that the flow decreases in radius as the water falls. This can be explained by the equation of continuity, since the cross-sectional area of the water decreases as the speed increases. If the water flows with an initial velocity of 0.62 m/s and a diameter of 2.2 cm at the faucet opening, what is the diameter of the water flow after the water has fallen 30 cm?

97. If the cardiac output of a small dog is 4.1×10^{-5} m^3/s, the radius of its aorta is 0.50 cm, and the aorta length is 40.0 cm, determine the pressure drop across the aorta of the dog. Assume the viscosity of blood is 4.0×10^{-3} Pa·s.

98. In an aortic aneurysm, a bulge forms where the walls of the aorta are weakened. If blood flowing through the aorta (radius 1.0 cm) enters an aneurysm with a radius of 3.0 cm, how much on average is the blood pressure higher inside the aneurysm than the pressure in the unenlarged part of the aorta? The average flow rate through the aorta is 120 cm^3/s. Assume the blood is nonviscous and the patient is lying down so there is no change in height.

99. Scuba divers are admonished not to rise faster than their air bubbles when rising to the surface. This rule helps them avoid the rapid pressure changes that cause the "bends." Air bubbles of 1.0 mm radius are rising from a scuba diver to the surface of the sea. Assume a water temperature of 20°C. (a) If the viscosity of the water is 1.0×10^{-3} Pa·s, what is the terminal velocity of the bubbles? (b) What is the largest rate of pressure change tolerable for the diver according to this rule?

100. A shallow well usually has the pump at the top of the well. (a) What is the deepest possible well for which a surface pump will work? [*Hint*: A pump maintains a pressure difference, keeping the outflow pressure higher than the intake pressure.] (b) Why is there not the same depth restriction on wells with the pump at the bottom?

101. ✦ A stone of weight W has specific gravity 2.50. (a) When the stone is suspended from a scale and submerged in water, what is the scale reading in terms of its weight in air? (b) What is the scale reading for the stone when it is submerged in oil (specific gravity = 0.90)?

102. ✦ A plastic beach ball has radius 20.0 cm and mass 0.10 kg, not including the air inside. (a) What is the weight of the beach ball including the air inside? Assume the air density is 1.3 kg/m^3 both inside and outside. (b) What is the buoyant force on the beach ball in air? The thickness of the plastic is about 2 mm—negligible compared with the radius of the ball. (c) The ball is thrown straight up in the air. At the top of its trajectory, what is its acceleration? [*Hint*: When $v = 0$, there is no drag force.]

103. ✦ A block of wood, with density 780 kg/m^3, has a cubic shape with sides 0.330 m long. A cord of negligible mass is used to tie a piece of lead to the bottom of the wood. The lead pulls the wood into the water until it is just completely covered with water. What is the mass of the lead? [*Hint*: Don't forget to consider the buoyant force on both the wood and the lead.]

104. ✦ Are evenly spaced specific-gravity markings on the cylinder of a hydrometer equal distances apart? In other words, is the depth d to which the cylinder is submerged linearly related to the density ρ of the fluid? To answer this question, assume that the cylinder has radius r and mass m. Find an expression for d in terms of ρ, r, and m, and see if d is a linear function of ρ.

105. A hydrometer is an instrument for measuring the specific gravity of a liquid. For example, vintners use a hydrometer to determine the density changes as wine is fermented, and producers of maple sugar and maple syrup use the hydrometer to find how much sugar is in the collected sap. Markings along a stem are calibrated to indicate the specific gravity for the level at which the hydrometer floats in a liquid. The weighted base ensures that the hydrometer floats vertically. Suppose the hydrometer has a cylindrical stem of cross-sectional area 0.400 cm^2. The total volume of the bulb and stem is 8.80 cm^3, and the mass of the hydrometer is 4.80 g. (a) How far from the top of the cylinder should a mark be placed to indicate a specific gravity of 1.00? (b) When the hydrometer is placed in alcohol, it floats with 7.25 cm of stem above the surface. What is the specific gravity of the alcohol? (c) What is the lowest specific gravity that can be measured with this hydrometer?

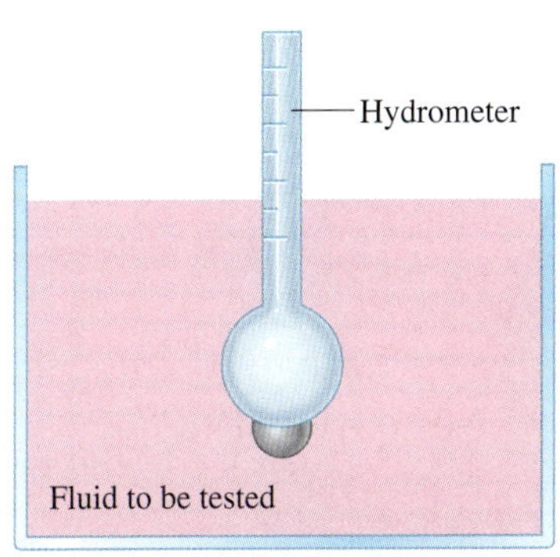

Problems 104 and 105

106. ✦ To measure the airspeed of a plane, a device called a Pitot tube is used. A simplified model of the Pitot tube is a manometer with one side connected to a tube facing directly into the "wind" (stopping the air that hits it head-on) and the other side connected to a tube so that the "wind" blows across its openings. If the manometer uses mercury and the levels differ by 25 cm, what is the plane's airspeed? The density of air at the plane's altitude is 0.90 kg/m^3.

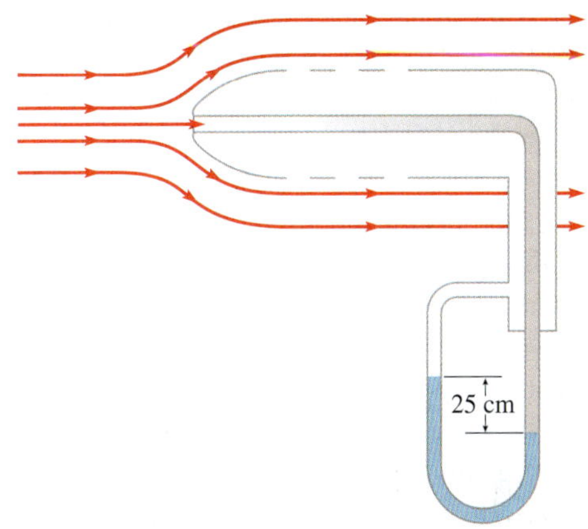

107. ✦ A U-shaped tube is partly filled with water and partly filled with a liquid that does not mix with water. Both sides of the tube are open to the atmosphere. What is the density of the liquid (in g/cm^3)?

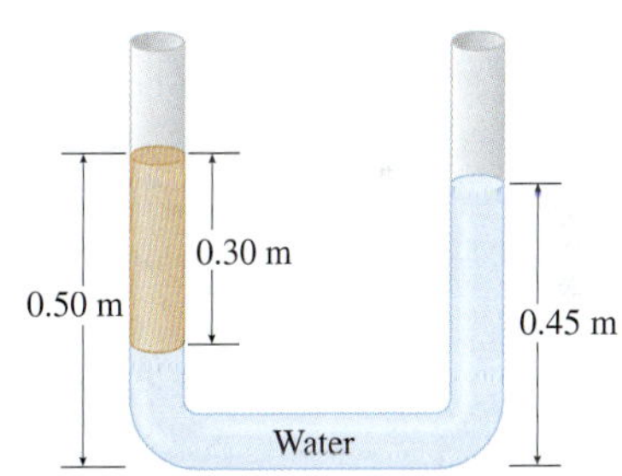

108. ✦ Water enters an apartment building 0.90 m below the street level with a gauge pressure of 52.0 kPa through the main pipe, which has a 5.00 cm radius. A second-story bathroom has an open faucet with a 1.20 cm radius that is located 4.20 m above the street. How fast is the water moving through the main pipe?

109. ✦ Atmospheric pressure is equal to the weight of a vertical column of air, extending all the way up through the atmosphere, divided by the cross-sectional area of the column. (a) Explain why that must be true. [*Hint*: Apply Newton's second law to the column of air.] (b) If the air all the way up had a uniform density of 1.29 kg/m^3 (the density at sea level at 0°C), how high would the column of air be? (c) In reality, the density of air decreases with increasing altitude. Does that mean that the height found in (b) is a lower limit or an upper limit on the height of the atmosphere?

Review and Synthesis

110. The pressure inside a bottle of champagne is 4.5 atm higher than the air pressure outside. The neck of the bottle has an inner radius of 1.0 cm. (a) What is the frictional force on the cork due to the neck of the bottle? (b) The cork (mass 7.5 g) is loosened until the friction force is zero and then released, allowing the cork to shoot out of the bottle. If the cork moves 1.0 cm along the neck of the bottle as the gauge pressure in the bottle decreases from 4.5 atm to zero, estimate the speed at which the cork shoots out of the bottle. [*Hint*: Estimate the average force on the cork to be half of the maximum.]

111. A hydraulic lift is lifting a car that weighs 12 kN. The area of the piston supporting the car is A, the area of the other piston is a, and the ratio A/a is 100.0. How far must the small piston be pushed down to raise the car a distance of 1.0 cm? [*Hint*: Consider the work to be done.]

112. ✦ Depressing the brake pedal in a car pushes on a piston with cross-sectional area 3.0 cm^2. The piston applies pressure to the brake fluid, which is connected to two pistons, each with area 12.0 cm^2. Each of these pistons presses a brake pad against one side of a rotor attached to one of the rotating wheels. See the figure for this problem. (a) When the force applied by the brake pedal to the small piston is 7.5 N, what is the normal force applied to each side of the rotor? (b) If the coefficient of kinetic friction between a brake pad and the rotor is 0.80 and each pad is (on average) 12 cm from the rotation axis of the rotor, what is the torque on the rotor due to the two pads?

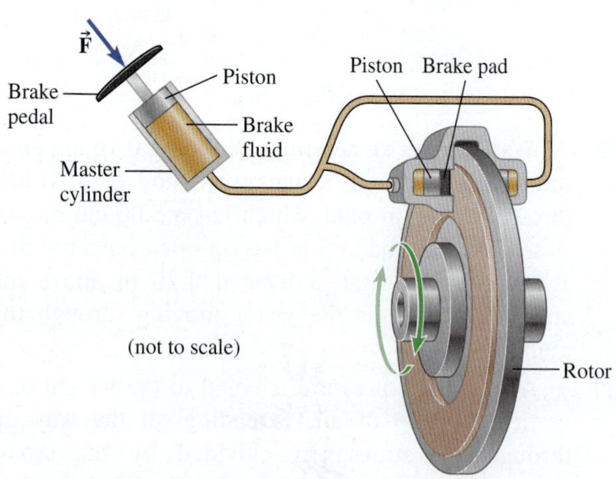

113. A dinoflagellate takes 5.0 s to travel 1.0 mm. Approximate a dinoflagellate as a sphere of radius 35.0 μm (ignoring the flagellum). (a) What is the drag force on the dinoflagellate in seawater of viscosity 0.0010 Pa·s? (b) What is the power output of the flagellate?

114. ✦ The potential energy associated with surface tension is much like the elastic potential energy of a stretched spring or a balloon. Suppose we do work on a puddle of liquid, spreading it out through a distance of Δs along a line L perpendicular to the force. (a) What is the work done on the fluid surface in terms of γ, L, and Δs? (b) The work done is equal to the increase in surface energy of the fluid. Show that the increase in energy is proportional to the increase in area. (c) Show that we can think of γ as the surface energy per unit area. (d) Show that the SI units of surface tension can be expressed either as N/m (force per unit length) or J/m^2 (energy per unit area).

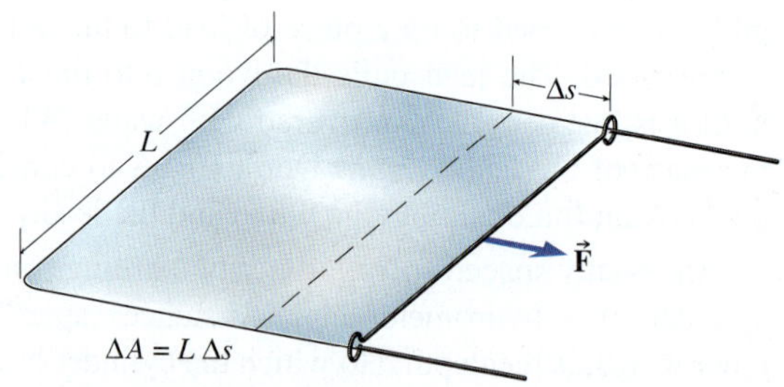

115. Ⓒ ✦ A beach ball is thrown straight up in the air. The graph shows its vertical velocity component as a function of time. The dashed line is tangent to the curve at the point where $v_y = 0$. (a) What shape would the $v_y(t)$ graph have if air drag on the beach ball were always negligibly small? (b) Describe feature(s) of the $v_y(t)$ graph that tell you that the force of air drag on the ball is *not* negligible. (c) Describe feature(s) of the graph that indicate that the buoyant force on the ball is significant. [*Hint*: Look at a *part* of the graph where air drag *is* negligible.] (d) Estimate the buoyant force on the ball as a fraction of its weight. (e) If the ball is thrown from a height of 1.0 m above the ground, approximately when does it hit the ground?

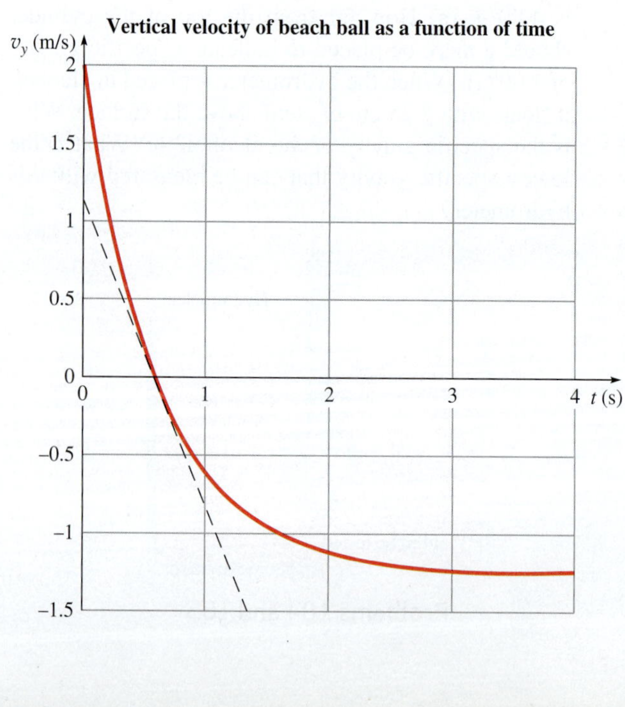

116. A cube that is 4.00 cm on a side and of density 8.00×10^2 kg/m^3 is attached to one end of a spring. The other end of the spring is attached to the base of a beaker. When the beaker is filled with water until the entire cube is submerged, the spring is stretched by 1.00 cm. What is the spring constant?

117. A section of pipe with an internal diameter of 10.0 cm tapers to an inner diameter of 6.00 cm as it rises through a height of 1.70 m at an angle of 60.0° with respect to the horizontal. The pipe carries water and its higher end is open to air. (a) If the speed of the water at the lower point is 15.0 cm/s, what are the pressure at the lower end and the speed of the water as it exits the pipe? (b) If the higher end of the pipe is 0.300 m above ground, at what horizontal distance from the pipe outlet does the water land?

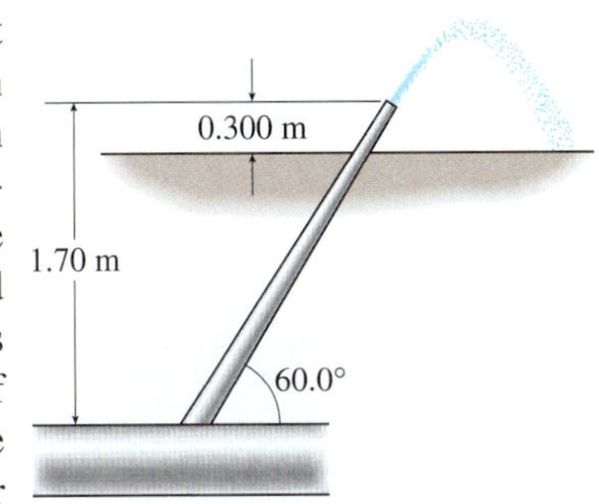

Answers to Practice Problems

9.1 1.3×10^6 N/m^2 = 1.3 MPa; the pressure is a factor of 15 greater than the pressure from the tennis shoe heel.

9.2 (a) 2.0×10^5 Pa; (b) 5.0 m

9.3 1.6 km

9.4 (a) Yes, $P_2 = P_1$. The column above point 2 is not as tall, but the pressure at the top of that column is *greater than* atmospheric pressure. (b) No, $P = P_{atm} + \rho g d$ gives the pressure at a depth d below a point where the pressure is P_{atm}.

9.5 (a) 32.0 cm; (b) 17.0 cm and 37.0 cm

9.6 S.G. = 11.3; could be lead

9.7 2% and 4%

9.8 (a) The beetle can squeeze the air bubble with its wings, compressing the air to reduce the bubble volume and decreasing the buoyant force. (b) When it is time to rise to the surface, the beetle relaxes the pressure on the bubble, allowing it to expand again.

9.9 (a) 0.85 m/s; (b) 1.7 m/s

9.10 $\sqrt{2gh} = 4.0$ m/ s

9.11 250 kPa

9.12 1.4

9.13 2.85 mm/s upward

9.14 480 Pa

Answers to Checkpoints

9.2 1.8

9.4 Pressure in a static fluid cannot depend on horizontal position. The net horizontal force on any part of the fluid must be zero—otherwise the horizontal acceleration would be nonzero and the fluid would begin to flow. The net vertical force *including the weight of the fluid* must also be zero, so pressure does depend on vertical position.

9.5 3 = 4, 2, 1 = 5. The pressures at 3 and 4 are equal because they are at the same depth in the red liquid. The pressures at 1 and 5 are equal to the air pressure in the room. Point 2 is intermediate in pressure; in the blue liquid, pressure increases with depth so $P_3 > P_2 > P_1$.

9.6 In both cases, the weight of the displaced liquid is equal to the weight of the wood. A smaller *volume* of water is displaced, due to its higher density, so the wood floats higher in water than in alcohol.

9.7 We expect the blood to flow faster in the narrower section because the volume flow rate must be the same. From the continuity equation, $v_2/v_1 = A_1/A_2 = (d_1/d_2)^2 = 1.20^2 = 1.44$. The speed increases 44%.

9.8 (a) For horizontal flow, Bernoulli's equation becomes $P_1 + \frac{1}{2}\rho v_1^2 = P_2 + \frac{1}{2}\rho v_2^2$; the pressure is lower where the flow speed is higher. (b) In a static fluid, Bernoulli's equation becomes $P_1 + \rho g y_1 = P_2 + \rho g y_2$. Letting $d = y_1 - y_2$, we have $P_2 - P_1 = \rho g y_1 - \rho g y_2 = \rho g d$, which is the pressure dependence with depth for a static fluid as discussed in Section 9.4.

9.10 Viscous drag and kinetic friction are both forces that oppose the motion of an object (relative to the surrounding fluid or relative to the surface on which the object slides, respectively). However, viscous drag depends strongly on the speed of the object ($F_D \propto v$), but kinetic friction does not.

CHAPTER

10 Elasticity and Oscillations

Near the top of the 241 m tall Hancock Tower in Boston, two steel boxes filled with lead are part of a system designed to reduce the swaying and twisting of the building caused by the wind. The mass of each box is nearly 300 000 kg (weight 300 tons). It might seem that adding a large mass to the top of the building would make it more "top heavy" and might increase the amount of swaying. Why is such a large mass used and how does it reduce the swaying of the building? (See p. 388 for the answer.)

BIOMEDICAL APPLICATIONS

- Structure of bone; elastic properties of bone, tendons, ligaments, and hair (Secs. 10.2, 10.3, 10.4; Ex. 10.2; CQ 10; Probs. 2, 3, 13–18, 110)
- Osteoporosis (Sec. 10.3)
- How walking speed depends on leg length (Ex. 10.10)
- Elastic energy storage in insects, scallops (Probs. 8–10)
- Vibration of eardrum (Probs. 39 and 40)
- Elastic properties of spider silk (Probs. 90 and 91)

Concepts & Skills to Review

- Hooke's law (Section 6.6)
- graphical relationship of position, velocity, and acceleration (Sections 2.2 and 2.3)
- elastic potential energy (Section 6.7)

10.1 ELASTIC DEFORMATIONS OF SOLIDS

If the net force and the net torque on an object are zero, the object is in equilibrium—but that does not mean that the forces and torques have no effect. An object is deformed when contact forces are applied to it (Fig. 10.1). A **deformation** is a change in the size or shape of the object. Many solids are stiff enough that the deformation cannot be seen with the human eye; a microscope or other sensitive device is required to detect the change in size or shape.

When the contact forces are removed, an **elastic** object returns to its original shape and size. Many objects are elastic as long as the deforming forces are not too large. On the other hand, any object may be permanently deformed or even broken if the forces acting are too large. An automobile that collides with a tree at a low speed may not be damaged; but at a higher speed the car suffers a permanent deformation, and the driver may suffer a broken bone.

CONNECTION:

The two topics of this chapter—elasticity and oscillations—may seem unrelated at first, but they are closely connected: many oscillations are caused by the kinds of elastic forces we study in Sections 10.1 through 10.4.

Figure 10.1 A tennis ball is flattened by the contact force exerted on it by the strings of the tennis racquet. Likewise, the strings of the racquet are deformed by the contact force exerted by the ball. The two forces are interaction partners.

10.2 HOOKE'S LAW FOR TENSILE AND COMPRESSIVE FORCES

Suppose we stretch a wire by applying tensile forces of magnitude F to each end. The length of the wire increases from L to $L + \Delta L$. How does the elongation ΔL depend on the original length L? Conceptual Example 10.1 helps answer this question.

Conceptual Example 10.1

Stretching Tendons

If a given tensile force stretches a tendon by an amount ΔL, how much would the same force stretch a tendon twice as long but identical in thickness and composition?

Strategy and Solution Think of the tendon of length $2L$ as two tendons of length L placed end-to-end (Fig. 10.2). Under the same tension, each of the two tendons stretches by an amount ΔL, so the total deformation of the long tendon is $2\Delta L$.

Practice Problem 10.1 Cutting a Spring in Half

If a spring (spring constant k) is cut in half, what is the spring constant of each of the two newly formed springs?

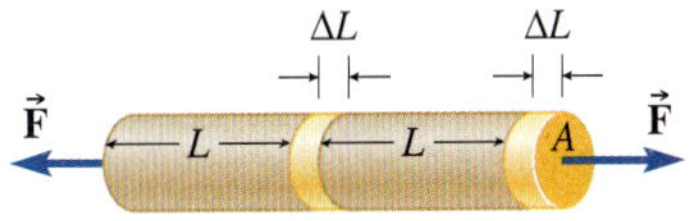

Figure 10.2 Two identical tendons are joined end-to-end and stretched by tensile forces. Each tendon stretches an amount ΔL.

Stress and Strain When stretched by the same tensile forces, the two tendons in Conceptual Example 10.1 get longer by an amount proportional to their original lengths: $\Delta L \propto L$. In other words, the two tendons have the same *fractional length change* $\Delta L/L$. The fractional length change is called the **strain**; it is a dimensionless measure of the degree of deformation.

$$\text{Strain} = \frac{\Delta L}{L} \text{ (fractional length change)} \qquad (10\text{-}1)$$

Suppose we had wires of the same composition and length but different thicknesses. It would require larger tensile forces to stretch the thicker wire the same amount as the thinner one; a thick steel cable is harder to stretch than the same length of a thin strand of steel. In Conceptual Question 13, we conclude that the tensile force required is proportional to the cross-sectional area of the wire ($F \propto A$). Thus, the same applied force *per unit area* produces the same deformation on wires of the same length and composition. The force per unit area is called the **stress**:

$$\text{Stress} = \frac{F}{A} \text{ (force per unit cross-sectional area)} \qquad (10\text{-}2)$$

The SI units of stress are the same as those of pressure: N/m^2 or Pa.

Hooke's Law Suppose that a solid object of initial length L is subjected to tensile or compressive forces of magnitude F. As a result of the forces, the length of the object is changed by magnitude ΔL. According to Hooke's law, the deformation is proportional to the deforming forces as long as they are not too large:

CONNECTION:

Hooke's law does not just apply to springs. The deformation of an object is often proportional to the forces applied to it.

$$F = k\Delta L \qquad (10\text{-}3)$$

In Eq. (10-3), k is a measure of the object's stiffness; it is analogous to the spring constant of a spring. This constant k depends on the length and cross-sectional area of the object. A larger cross-sectional area A makes k larger; a greater length L makes k smaller.

We can rewrite Hooke's law in terms of stress (F/A) and strain ($\Delta L/L$):

Hooke's law

$$\text{stress} \propto \text{strain}$$

$$\frac{F}{A} = Y\frac{\Delta L}{L} \qquad (10\text{-}4)$$

Equation (10-4) still says that the length change (ΔL) is proportional to the magnitude of the deforming forces (F). Stress and strain account for the effects of length and cross-sectional area; the proportionality constant Y depends only on the inherent stiffness of the material from which the object is composed; it is independent of the length and cross-sectional area. Comparing Eqs. (10-3) and (10-4), the "spring constant" k for the object is

$$k = \frac{YA}{L} \qquad (10\text{-}5)$$

The constant of proportionality Y in Eqs. (10-4) and (10-5) is called the **elastic modulus**, or **Young's modulus**; Y has the same units as those of stress (Pa), since strain is dimensionless. Young's modulus can be thought of as the inherent stiffness of a material; it measures the resistance of the material to elongation or compression. Material that is flexible and stretches easily (e.g., rubber) has a *low* Young's modulus. A stiff material (e.g., steel) has a high Young's modulus; it takes a larger stress to produce the same strain. Table 10.1 gives Young's modulus for a variety of common materials.

CHECKPOINT 10.2

Which stretches more when put under the same tension: a steel wire 2.0 m long or a copper wire 1.0 m long with the same diameter? (See Table 10.1.)

Table 10.1 Approximate Values of Young's Modulus for Various Substances

Substance	Young's Modulus (GPa)	Substance	Young's Modulus (GPa)
Rubber	0.002–0.008	Wood, along the grain	10–15
Human cartilage	0.024	Brick	14–20
Human vertebra	0.088 (compression); 0.17 (tension)	Concrete	20–30 (compression)
		Marble	50–60
Collagen, in bone	0.6	Aluminum	70
Human tendon	0.6	Cast iron	100–120
Wood, across the grain	1	Copper	120
Nylon	2–6	Wrought iron	190
Spider silk	4	Steel	200
Human femur	9.4 (compression); 16 (tension)	Diamond	1200

Application: Strength of Bone and of Concrete Hooke's law holds up to a maximum stress called the *proportional limit*. For many materials, Young's modulus has the same value for tension and compression. Some composite materials, such as bone and concrete, have significantly different Young's moduli for tension and compression. The components of bone include fibers of collagen (a protein found in all connective tissue) that give it strength under tension and hydroxyapatite crystals (composed of calcium and phosphate) that give it strength under compression. The different properties of these two substances lead to different values of Young's modulus for tension and compression.

Example 10.2

Compression of the Femur

A man whose weight is 0.80 kN is standing upright. By approximately how much is his femur (thighbone) shortened compared with when he is lying down? Assume that the compressive force on each femur is about half his weight (Fig. 10.3). The average cross-sectional area of the femur is 8.0 cm^2 and the length of the femur when lying down is 43.0 cm.

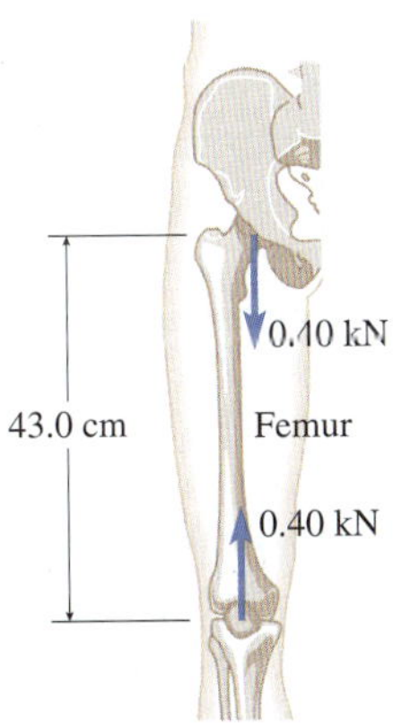

Figure 10.3 Compression of the femur.

Strategy A change in length of the femur involves a strain. After finding the stress and looking up the Young's modulus, we can find the strain using Hooke's law. We assume that each femur supports *half* the man's weight.

Solution The strain is proportional to the stress:

$$\frac{F}{A} = Y\frac{\Delta L}{L}$$

Solving this equation for ΔL gives

$$\Delta L = \frac{F/A}{Y} L$$

From Table 10.1, Young's modulus for a femur *in compression* is:

$$Y = 9.4 \text{ GPa}$$

We need to convert the cross-sectional area to m^2 since 1 Pa = 1 N/m^2:

$$A = 8.0 \text{ cm}^2 \times \left(\frac{1 \text{ m}}{100 \text{ cm}}\right)^2 = 0.000\,80 \text{ m}^2$$

The force on each leg is 0.40 kN, or 4.0×10^2 N. The length change is then

$$\Delta L = \frac{F/A}{Y} L = \frac{(4.0 \times 10^2 \text{ N})/(0.000\,80 \text{ m}^2)}{9.4 \times 10^9 \text{ Pa}} \times 43.0 \text{ cm}$$
$$= 5.3 \times 10^{-5} \times 43.0 \text{ cm} = 0.0023 \text{ cm}$$

Discussion The strain—or fractional length change—is 5.3×10^{-5}. Since the strain is much smaller than 1, we are justified in not worrying about whether the length is 43.0 cm with or without the compressive load; we would calculate the same value of ΔL (to two significant figures) either way.

Practice Problem 10.2 Fractional Length Change of a Cable

A steel cable of diameter 3.0 cm supports a load of 2.0 kN. What is the fractional length increase of the cable compared with the length when there is no load if $Y = 200$ GPa?

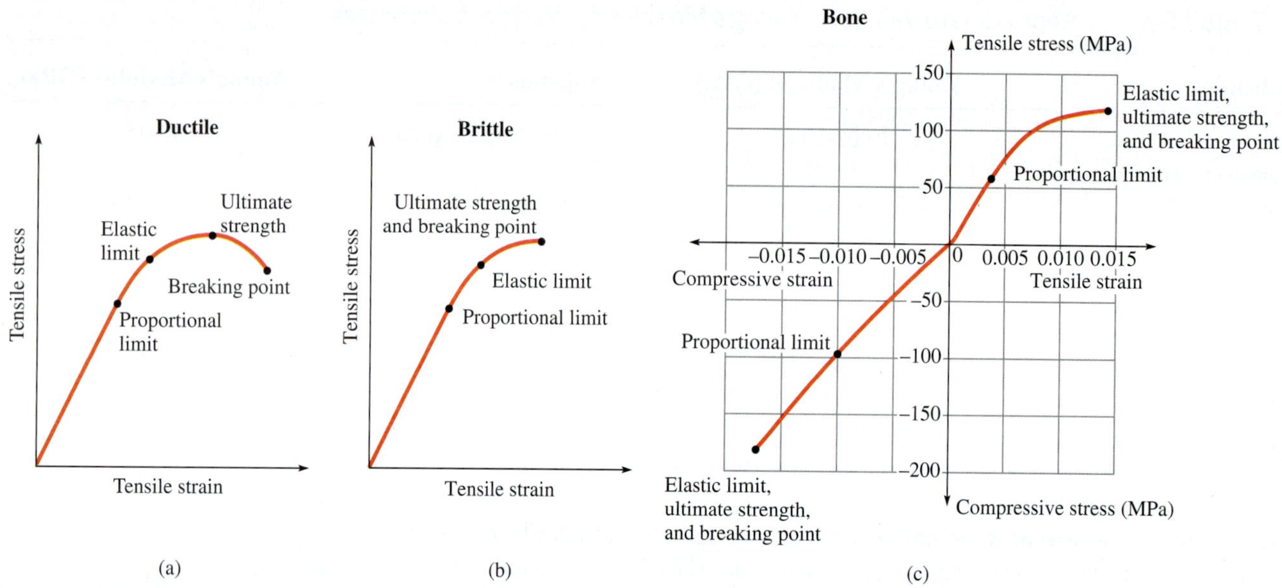

Figure 10.4 Stress-strain curves showing limits for (a) a ductile material, (b) a brittle material, and (c) compact bone. The elastic limit, ultimate strength, and breaking point are well separated for ductile materials, but close together for brittle materials.

10.3 BEYOND HOOKE'S LAW

If the tensile or compressive stress exceeds the proportional limit, the strain is no longer proportional to the stress (Fig. 10.4). The solid still returns to its original length when the stress is removed as long as the stress does not exceed the *elastic limit*. If the stress exceeds the elastic limit, the material is permanently deformed. For still larger stresses, the solid fractures when the stress reaches the *breaking point*. The maximum stress that can be withstood without breaking is called the *ultimate strength*. The ultimate strength can be different for compression and tension; then we refer to the compressive strength or the tensile strength of the material.

A *ductile* material continues to stretch beyond its ultimate tensile strength without breaking; the stress then *decreases* from the ultimate strength (Fig. 10.4a). Examples of ductile solids are the relatively soft metals, such as gold, silver, copper, and lead. These metals can be pulled like taffy, becoming thinner and thinner until finally reaching the breaking point. For a *brittle* substance, the ultimate strength and the breaking point are close together (Fig. 10.4b).

Figure 10.5 Load test of a floor joist made of prestressed concrete. Concrete itself does not have a high enough shear strength to withstand this much shear stress. In prestressed concrete, steel rods are stretched before the concrete is poured. After the concrete hardens, the frame holding the rods under tension is removed. The rods contract, compressing the concrete. The steel rods add the tensile and shear strength that concrete lacks.

Application: Elastic Properties of Bone; Osteoporosis Bone is an example of a brittle material; it fractures abruptly if the stress becomes too large (Fig. 10.4c). Under either tension or compression, its elastic limit, breaking point, and ultimate strength are approximately the same. Babies have more flexible bones than adults because they have built up less of the calcium compound hydroxyapatite. As people age, their bones become more brittle as the collagen fibers lose flexibility, and their bones also become weaker as calcium gets reabsorbed (a condition called osteoporosis).

Like bone, reinforced concrete has one component for tensile strength and another for compressive strength. Reinforced concrete contains steel rods that provide tensile and shear strength that concrete itself lacks (Fig. 10.5).

Application: The Human Vertebra Human anatomy has special features for adapting to the compressive stress associated with standing upright. For example, the vertebrae in the spinal column gradually increase in size from the neck to the tailbone. Such an arrangement places the stronger vertebrae in the lower positions, where they must support more weight. The vertebrae are separated by fluid-filled disks, which have a cushioning effect by spreading out the compressive forces.

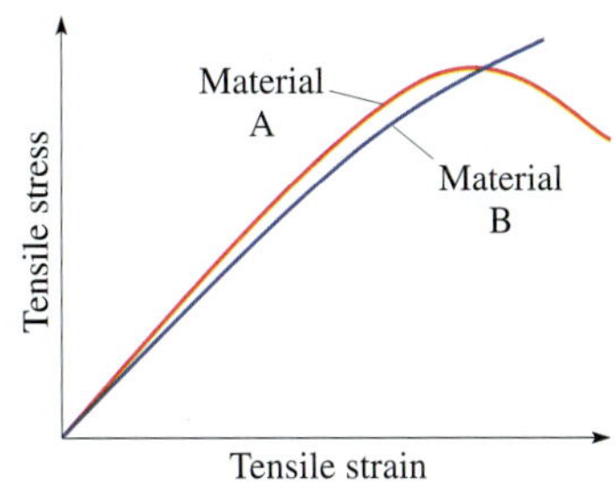

Figure 10.6 Stress-strain graphs for two materials.

CHECKPOINT 10.3

Stress-strain graphs for two different materials are shown in Fig. 10.6. Each graph ends at the breaking point for that material. (a) Which has the larger Young's modulus? Explain. (b) Which has the higher ultimate strength? Explain.

Example 10.3

Crane with Steel Cable

A crane is required to lift loads of up to 100 kN (11 tons). (a) What is the minimum diameter of the steel cable that must be used? (b) If a cable of twice the minimum diameter is used and it is 8.0 m long when no load is present, how much longer is it when supporting a load of 100 kN? (Data for steel: $Y = 200$ GPa; proportional limit = 0.20 GPa; elastic limit = 0.30 GPa; tensile strength = 0.50 GPa.)

Strategy The data given for steel consists of four quantities that all have the same units. It would be easy to mix them up if we didn't understand what each one means. Young's modulus is the proportionality constant between stress and strain. That will be useful in part (b) where we find the elongation of the cable; the elongation is the strain times the original length. However, we should first check that the stress is less than the proportional limit before using Young's modulus to find the strain.

The elastic limit is the maximum stress so that no permanent deformation occurs; the tensile strength is the maximum stress so that the cable does not break. We certainly don't want the cable to break, but it would be prudent to keep the stress under the elastic limit to give the cable a long useful life. Therefore, we choose a minimum diameter in (a) to keep the stress below the elastic limit.

Solution (a) We choose the minimum diameter to keep the stress less than the elastic limit:

$$\frac{F}{A} < \text{elastic limit} = 3.0 \times 10^8 \text{ Pa}$$

for $F = 1.0 \times 10^5$ N. Then

$$A > \frac{F}{\text{elastic limit}} = \frac{1.0 \times 10^5 \text{ N}}{3.0 \times 10^8 \text{ Pa}} = 3.33 \times 10^{-4} \text{ m}^2$$

The minimum cross-sectional area corresponds to the minimum diameter. The cross-sectional area of the cable is πr^2 or $\pi d^2/4$, so

$$d = \sqrt{\frac{4A}{\pi}} = \sqrt{\frac{4 \times 3.33 \times 10^{-4} \text{ m}^2}{\pi}} = 2.1 \text{ cm}$$

The minimum *diameter* is therefore 2.1 cm.

(b) If we double the diameter and keep the same load, the stress is reduced by a factor of 4 since the cross-sectional area is proportional to the square of the diameter. Therefore, the stress is

$$\frac{F}{A} = \frac{3.0 \times 10^8 \text{ Pa}}{4} = 7.5 \times 10^7 \text{ Pa}$$

The strain is then

$$\frac{\Delta L}{L} = \frac{F/A}{Y} = \frac{7.5 \times 10^7 \text{ Pa}}{2.0 \times 10^{11} \text{ Pa}} = 3.75 \times 10^{-4}$$

The strain is the fractional length change. Then the length change is

$$\Delta L = (3.75 \times 10^{-4})L = 3.75 \times 10^{-4} \times 8.0 \text{ m} = 3.0 \text{ mm}$$

Discussion By using a cable twice as thick as the minimum, we build in a safety factor. We don't want to be right at the edge of disaster! Since doubling the diameter of the cable increases the cross-sectional area of the cable by a factor of 4, the maximum stress on the cable is one fourth of the elastic limit.

continued on next page

Example 10.3 continued

Practice Problem 10.3 Tuning a Harpsichord String

A harpsichord string is made of yellow brass (Young's modulus 90 GPa, tensile strength 630 MPa). When tuned correctly, the tension in the string is 59.4 N, which is 93% of the maximum tension that the string can endure without breaking. What is the radius of the string?

Height Limits

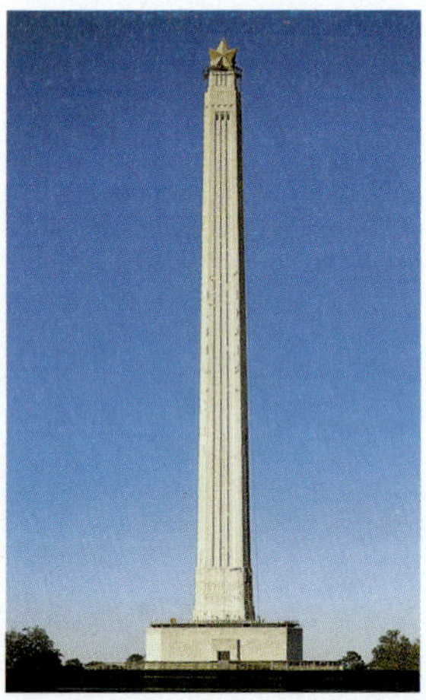

The San Jacinto monument in Texas is the tallest stone column in the world.

What limits the height of a stone column? If the column is too tall, it could be crushed under its own weight. The maximum height of a column is limited since the compressive stress at the bottom cannot exceed the compressive strength of the material (see Problem 82). However, the maximum height at which a vertical column buckles is generally less than the height at which it would be crushed.

Application: Bone Structure The bones of our limbs are hollow; the inside of the structural material is filled with marrow, which is structurally weak. A hollow bone is better able to resist fracture from bending and twisting forces than a solid bone with the same amount of structural material, although the hollow bone would buckle more easily under a compressive force along the central axis.

Application: Size Limitations on Organisms Why would the proportions of a giant's bones have to be different from a human's? If the giant's average density is the same as a human's, then his weight is larger by the same factor that his *volume* is larger. If the giant is five times as tall as a human, for instance, and has the same relative proportions, then his volume is $5^3 = 125$ times as large, since each of the three dimensions of any body part has increased by a factor of 5. On the other hand, the cross-sectional area of a bone is proportional to the *square* of its radius. So although the leg bones must support 125 times as much weight, the maximum compressive force they can withstand has only increased by a factor of 25. The giant would need much thicker legs (in relation to their length) to support his increased weight. Similar analysis can be applied to the twisting and bending forces that are more likely to break bones than are compressive forces. The result is the same: the bones of a giant could not have human proportions.

Some science fiction or horror movies portray giant insects as greatly magnified versions of a normal insect. Such a giant insect's legs would collapse under the weight of the insect.

Figure 10.7 A column made from a rolled sheet of paper can support a book.

EVERYDAY PHYSICS DEMO

Challenge a friend to use a single sheet of 8.5 in. × 11 in. paper and two paper clips to support a book at least 8 in. above a table. If your friend has no idea what to do, roll the sheet of paper into a narrow cylinder about 2 cm in diameter; then fasten the cylinder at the top and bottom with paper clips. Carefully place the book so that it is balanced on top of the cylinder (Fig. 10.7). If you have difficulty, try using thicker paper or a lighter book.

Use the same "apparatus" to get some insight into the buckling of columns. Try making the diameter of the paper cylinder twice as large. The walls of this column are thinner because there are fewer layers of the paper in the cylinder wall, although the same cross-sectional area of paper supports the book. If nothing happens, try a larger diameter. You will see the walls crumple in on themselves as the cylinder buckles and the book falls to the table.

10.4 SHEAR AND VOLUME DEFORMATIONS

In this section we consider two other kinds of deformation. In each case we define a stress (force per unit area), a strain (dimensionless), and a modulus (the constant of proportionality between stress and strain).

Shear Deformation

Unlike tensile and compressive forces, which are perpendicular to two opposite surfaces of an object, a **shear deformation** is the result of a pair of equal and opposite forces that act *parallel* to two opposite surfaces (Fig. 10.8). The **shear stress** is the magnitude of the shear force divided by the area of the surface on which the force acts:

$$\text{shear stress} = \frac{\text{shear force}}{\text{area of surface}} = \frac{F}{A} \quad (10\text{-}6)$$

Shear strain is the ratio of the relative displacement Δx to the separation L of the two surfaces:

$$\text{shear strain} = \frac{\text{displacement of surfaces}}{\text{separation of surfaces}} = \frac{\Delta x}{L} \quad (10\text{-}7)$$

The shear strain is proportional to the shear stress as long as the stress is not too large. The constant of proportionality is the **shear modulus** S.

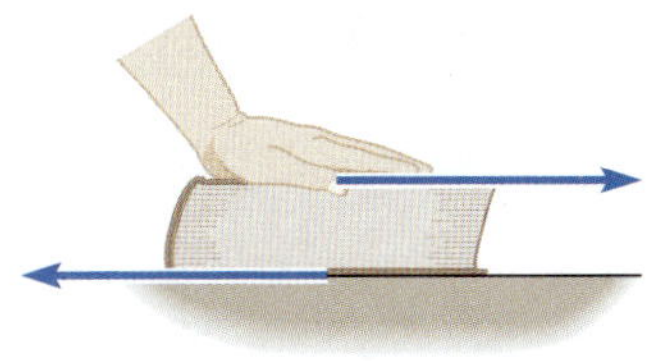

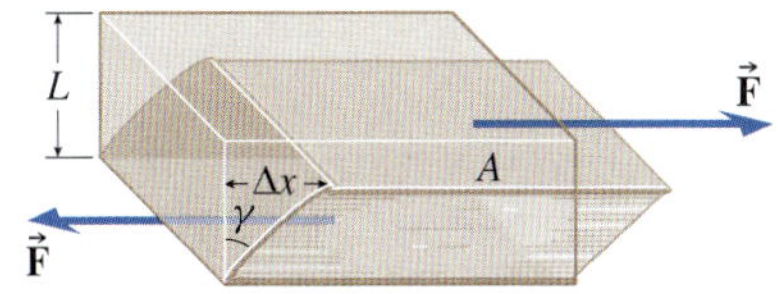

Figure 10.8 A book under shear stress. Shear forces produce the same kind of deformation in a solid block; the amount of the deformation is just smaller.

Hooke's law for shear deformation

$$\text{shear stress} \propto \text{shear strain}$$

$$\frac{F}{A} = S\frac{\Delta x}{L} \quad (10\text{-}8)$$

The units of shear stress and the shear modulus are the same as for tensile or compressive stress and Young's modulus, Pa. The strain is once again dimensionless. Table 10.2 lists shear moduli for various materials.

An example of shear stress is the cutting action of a pair of scissors (or "shears") on a piece of paper. The forces acting on the paper from above and below are offset from each other and act parallel to the cross-sectional surfaces of the paper (Fig. 10.9).

CONNECTION:

Hooke's law takes the same form for different kinds of stresses and strains. In each case, the strain is proportional to the stress.

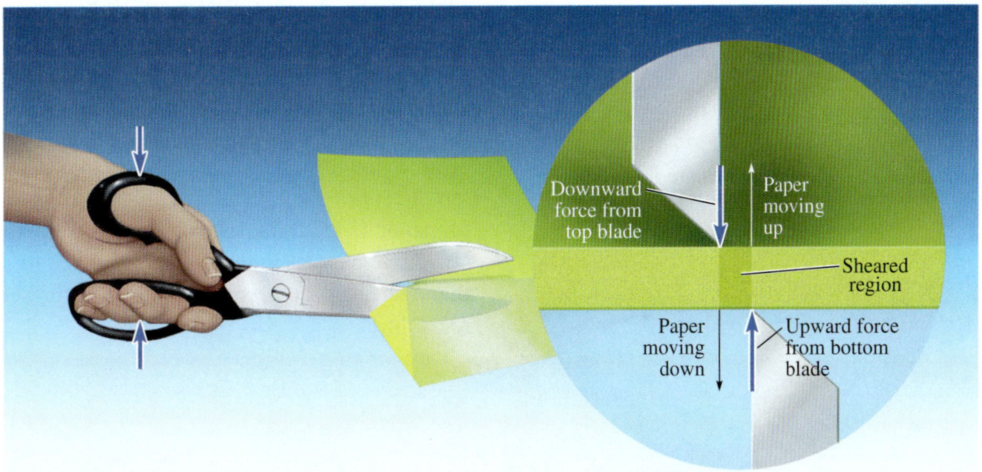

Figure 10.9 Scissors apply shear stress to a sheet of paper. The shear stress is the force exerted by a blade divided by the cross-sectional area of the paper—the thickness of the paper times the length of blade that is in contact with the paper.

Table 10.2 Shear and Bulk Moduli for Various Materials

Material	Shear Modulus *S* (GPa)	Bulk Modulus *B* (GPa)
Gases		
Air*		0.000 10
Air†		0.000 14
Liquids		
Ethanol		0.9
Water		2.2
Mercury		25
Solids		
Cast iron	40–50	60–90
Marble		70
Aluminum	25–30	70
Copper	40–50	120–140
Steel	80–90	140–160
Diamond		620

*At 0°C and 1 atm; constant temperature expansion or compression

†At 0°C and 1 atm; no heat flow during expansion or compression

Example 10.4

Cutting Paper

A sheet of paper of thickness 0.20 mm is cut with scissors that have blades of length 10.0 cm and width 0.20 cm. While cutting, the scissors blades each exert a force of 3.0 N on the paper; the length of each blade that makes contact with the paper is approximately 0.5 mm. What is the shear stress on the paper?

Strategy Shear stress is a force divided by an area. In this problem, identifying the correct area is tricky. The blades push two *cross-sectional* paper surfaces in opposite directions so they are displaced with respect to one another. The shear stress is the force exerted by each blade divided by this cross-sectional area—the thickness of the paper times the length of blade *in contact with the paper*. (Compare Figs. 10.8 and 10.9.) The total length and the width of the blades are irrelevant.

Solution The cross-sectional area is

$$A = \text{thickness} \times \text{contact length}$$
$$= 2.0 \times 10^{-4}\ \text{m} \times 5 \times 10^{-4}\ \text{m} = 1 \times 10^{-7}\ \text{m}^2$$

The shear stress is

$$\frac{F}{A} = \frac{3.0\ \text{N}}{1 \times 10^{-7}\ \text{m}^2} = 30\ \text{MPa}$$

Discussion To identify the correct area, remember that shear forces act *in the plane of* the surfaces that are displaced with respect to each other. By contrast, tensile and compressive forces are perpendicular to the area used to find tensile and compressive stresses.

Practice Problem 10.4 Shear Stress due to a Hole Punch

A hole punch has a diameter of 8.0 mm and presses onto ten sheets of paper with a force of 6.7 kN. If each sheet of paper is of thickness 0.20 mm, find the shear stress. [*Hint*: Be careful in deciding what area to use. Remember that a shear force acts *parallel* to the surface whose area is relevant.]

Application: Spiral Fractures When a bone is twisted, it is subjected to a shear stress. Shear stress is a more common cause of fracture than a compressive or tensile stress along the length of the bone. The twisting of a bone can result in a spiral fracture (Fig. 10.10).

(a) (b)

Figure 10.10 (a) A skier falls and his leg is subjected to a shear stress. (b) X-ray of a spiral fracture of the tibia.

Volume Deformation

As discussed in Chapter 9, a fluid exerts inward forces on an immersed solid object. These forces are perpendicular to the surfaces of the object. Since the fluid presses inward on all sides of the object (Fig. 10.11), the solid is compressed—its volume is reduced. The fluid pressure P is the force per unit surface area; it can be thought of as the **volume stress** on the solid object. Pressure has the same units as the other kinds of stress: Pa.

$$\text{volume stress} = \text{pressure} = \frac{F}{A} = P$$

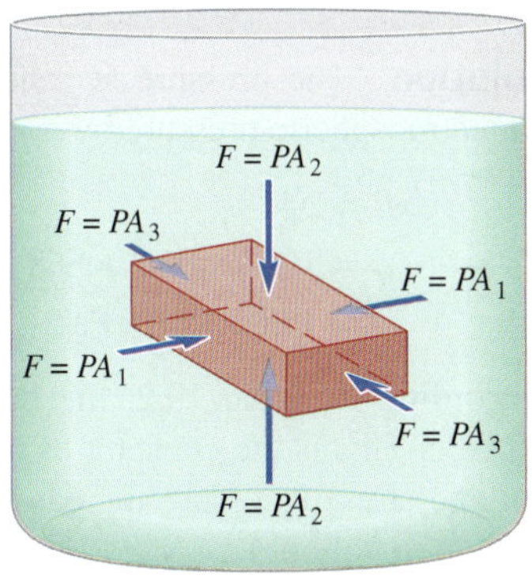

Figure 10.11 Forces on an object when submerged in a fluid.

The resulting deformation of the object is characterized by the **volume strain**, which is the fractional change in volume:

$$\text{volume strain} = \frac{\text{change in volume}}{\text{original volume}} = \frac{\Delta V}{V} \qquad (10\text{-}9)$$

Unless the stress is too large, the stress and strain are proportional within a constant of proportionality called the **bulk modulus** B. A substance with a large bulk modulus is more difficult to compress than a substance with a small bulk modulus.

An object at atmospheric pressure is already under volume stress: the air pressure already compresses the object slightly compared with what its volume would be in a vacuum. For solids and liquids, the volume strain due to atmospheric pressure is, for most purposes, negligibly small (5×10^{-5} for water). Since we are usually concerned with the deformation due to a *change* in pressure ΔP from atmospheric pressure, we can write Hooke's law as:

Hooke's law for volume deformation

$$\Delta P = -B\frac{\Delta V}{V} \qquad (10\text{-}10)$$

where V is the volume at atmospheric pressure. The negative sign in Eq. (10-10) allows the bulk modulus to be positive—an increase in the volume stress causes a *decrease* in volume, so ΔV is negative. Table 10.2 lists bulk moduli for various substances.

Unlike the stresses and strains discussed previously, volume stress can be applied to fluids (liquids and gases) as well as solids. The bulk moduli of liquids are generally not much less than those of solids, since the atoms in liquids are nearly as close together as those in solids. In Chapter 9 we assume that liquids are incompressible, which is often a good approximation since the bulk moduli of liquids are generally large. In gases, the atoms are much farther apart on average than in solids or liquids. Gases are much easier to compress than solids or liquids, so their bulk moduli are much smaller.

Example 10.5

Marble Statue Under Water

A marble statue of volume 1.5 m^3 is being transported by ship from Athens to Cyprus. The statue topples into the ocean when an earthquake-caused tidal wave sinks the ship; the statue ends up on the ocean floor, 1.0 km below the surface. Find the change in volume of the statue in cm^3 due to the pressure of the water. The density of seawater is 1025 kg/m^3.

continued on next page

Example 10.5 continued

Strategy The water pressure is the volume stress; it is the force per unit area pressing inward and perpendicular to all the surfaces of the statue. The water pressure at a depth d is greater than the pressure at the water surface; we can find the pressure using the given density of seawater. Then, using the bulk modulus of marble given in Table 10.2, we find the change in volume from Hooke's law.

Solution The pressure at a depth $d = 1.0$ km is larger than atmospheric pressure by

$$\Delta P = \rho g d$$
$$= 1025 \text{ kg/m}^3 \times 9.8 \text{ N/kg} \times 1000 \text{ m}$$
$$= 1.005 \times 10^7 \text{ Pa}$$

According to Table 10.2, the bulk modulus for marble is 70 GPa. This is the constant of proportionality between the volume stress (pressure increase) and the strain (fractional change in volume).

$$\Delta P = -B\frac{\Delta V}{V}$$

Solving for ΔV, we have

$$\Delta V = -\frac{\Delta P}{B}V = -\frac{1.005 \times 10^7 \text{ Pa}}{70 \times 10^9 \text{ Pa}} \times 1.5 \text{ m}^3$$
$$= -2.2 \times 10^{-4} \text{ m}^3 \times \left(\frac{100 \text{ cm}}{1 \text{ m}}\right)^3 = -220 \text{ cm}^3$$

The statue's volume decreases approximately 220 cm^3.

Discussion The fractional decrease in volume is

$$\frac{1.005 \times 10^7 \text{ Pa}}{70 \times 10^9 \text{ Pa}} \approx \frac{1}{7000}$$

or a reduction of 0.014%.

In calculating the pressure increase, we assumed that the density of seawater is constant—the equation $\Delta P = \rho g d$ is derived for a constant fluid density ρ. Should we worry that our calculation of ΔP is wrong? The result of Practice Problem 10.5 shows that the density of seawater at a depth of 1.0 km is only about 0.43% greater than its density at the surface. The calculation of ΔP is inaccurate by less than 0.5%—negligible here since we only know the depth to two significant figures.

Practice Problem 10.5 Compression of Water

Show that a pressure increase of 10 MPa (100 atm) on 1 m^3 of seawater causes a 0.43% decrease in volume. The bulk modulus of seawater is 2.3 GPa.

CONNECTION:

As shown in Sections 10.2–10.4, Hooke's law applies to small deformations of many kinds of objects, not just springs. Thus, simple harmonic motion occurs in many situations as long as the vibrations are not too large.

10.5 SIMPLE HARMONIC MOTION

Vibration, one of the most common kinds of motion, is repeated motion back and forth along the same path. Vibrations occur in the vicinity of a point of **stable equilibrium**. An equilibrium point is *stable* if the net force on an object when it is displaced a small distance from equilibrium points back toward the equilibrium point (Fig. 10.12). Such a force is called a **restoring force** since it tends to restore equilibrium. A special kind of vibratory motion—called **simple harmonic motion** (SHM)—occurs whenever the restoring force is proportional to the displacement from equilibrium.

Figure 10.13 shows a graph of F_x versus x for some restoring force. We choose $x = 0$ at the equilibrium position. Since the graph is not linear, the resulting oscillations are not SHM—unless the amplitude is small. For small amplitudes, we can approximate the graph near equilibrium by a straight line tangent to the curve at the equilibrium point. For small amplitude oscillations, the restoring force is approximately linear, so the resulting oscillations are (approximately) SHM. The ideal spring is a favorite model of physicists because the restoring force it provides is proportional to the displacement from equilibrium.

Consider a relaxed ideal spring with spring constant k and zero mass. The spring is fixed at one end and attached at the other to an object of mass m (Fig. 10.14) that slides without friction. Since the normal force is equal and opposite to the weight of the object, the net force on the object is that due to the spring. When the spring is relaxed, the net force is zero; the object is in equilibrium.

If the object is now pulled to the right to the position $x = A$ and then released, the net force on the object is

$$F_x = -kx \quad (10\text{-}11)$$

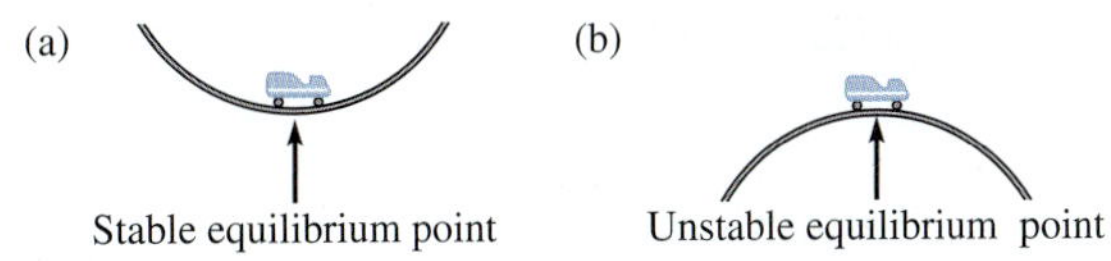

Figure 10.12 (a) A point of *stable* equilibrium for a roller-coaster car. If the car is displaced slightly from its position at the bottom of the track, the net force pulls the car back toward the equilibrium point. (b) A point of *unstable* equilibrium for a roller-coaster car. If the car is displaced slightly from the very top of the track, the net force pushes the car *away from* the equilibrium point.

where the negative sign tells us that the spring force is opposite in direction to the displacement from equilibrium. At first the object is to the right of the equilibrium position and the spring pulls to the left. Notice that the force exerted by the spring is in the correct direction to restore the object to the equilibrium position; it always pushes or pulls back toward the equilibrium point.

Imagine taking a series of photos at equal time intervals as the object oscillates back and forth. In Fig. 10.15 the blue dots are the positions of the object at equal time intervals over one-half of a full cycle, from one endpoint to the other. (A full cycle would include the return trip.)

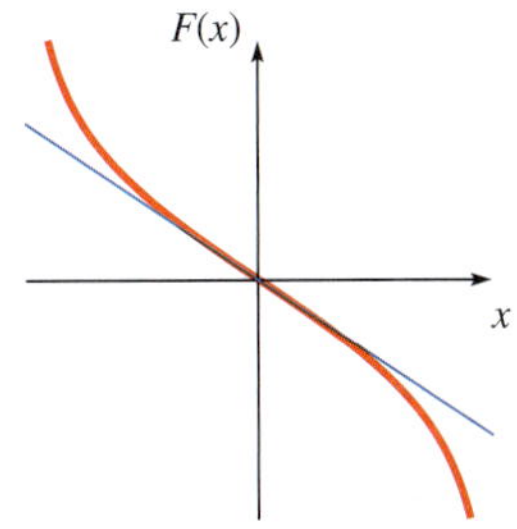

Figure 10.13 A nonlinear restoring force (red) can be approximated as a linear restoring force (blue) for small displacements.

Energy Analysis in SHM Figure 10.15 suggests that the speed is greatest as the object passes through the equilibrium position. The object slows as it approaches the endpoints and gains speed as it approaches the equilibrium point. At the endpoints ($x = \pm A$), the object is instantaneously at rest before heading back in the other direction. Conservation of energy supports these observations. The total mechanical energy of the mass and spring is constant.

$$E = K + U = \text{constant}$$

where K is the kinetic energy and U is the elastic potential energy stored in the spring. As the object oscillates back and forth, energy is converted from potential to kinetic and back to potential in the half-cycle shown in Fig. 10.15. From Section 6.7, the elastic potential energy of the spring is

$$U = \tfrac{1}{2}kx^2 \qquad (6\text{-}24)$$

The speed at any point x can be found from the energy equation

$$E = \tfrac{1}{2}mv_x^2 + \tfrac{1}{2}kx^2 \qquad (10\text{-}12)$$

The maximum displacement of the object is the **amplitude** A. At the maximum displacement, where the motion changes direction, the velocity is zero. Since the kinetic energy is zero at $x = \pm A$, all the energy is elastic potential energy at the endpoints. Therefore, the total energy E at the endpoints is

$$E_{\text{total}} = \tfrac{1}{2}kA^2 \qquad (10\text{-}13)$$

and, since energy is conserved, this must be the total energy at any point in the object's motion. The maximum speed v_m occurs at $x = 0$ where all the energy is kinetic. Thus, at $x = 0$, the total energy equals the kinetic energy

$$E_{\text{total}} = \tfrac{1}{2}mv_m^2$$

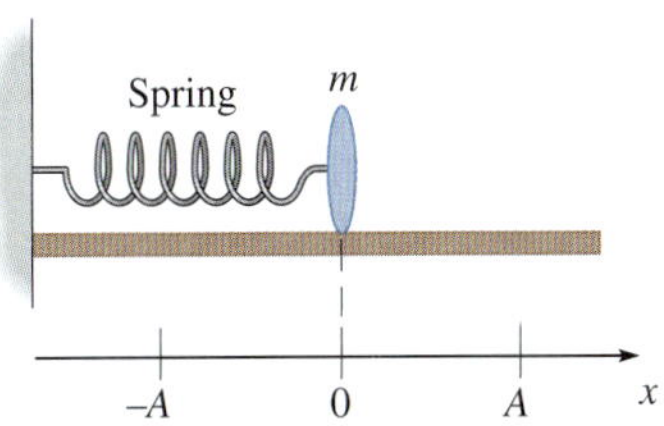

Figure 10.14 Spring in relaxed position. We choose the object's equilibrium position as the origin ($x = 0$).

CONNECTION:

Our study of SHM is based on familiar principles of energy conservation and Newton's second law, together with Hooke's law.

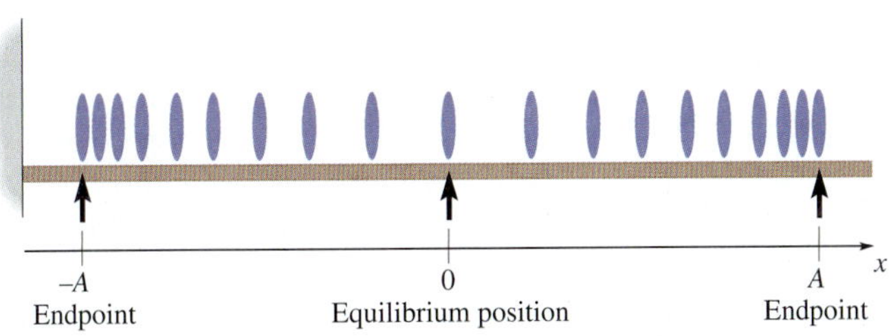

Figure 10.15 Positions of an oscillating object at equal time intervals over half a period. The spring is omitted for clarity.

and, from Eq. (10-13),

$$\tfrac{1}{2}mv_{\text{m}}^2 = \tfrac{1}{2}kA^2$$

Solving for v_{m} yields

$$v_{\text{m}} = \sqrt{\frac{k}{m}}A \qquad (10\text{-}14)$$

The maximum speed is proportional to the amplitude.

CHECKPOINT 10.5

What is the displacement of an object in SHM when the kinetic and potential energies are equal?

Acceleration in SHM The force on the object at any point x is given by Hooke's law; Newton's second law then gives the acceleration:

$$F_x = -kx = ma_x$$

Solving for the acceleration, we obtain:

Acceleration of an object in SHM

$$a_x(t) = -\frac{k}{m}x(t) \qquad (10\text{-}15)$$

Thus, the acceleration is a negative constant $(-k/m)$ times the displacement; the acceleration and displacement are always in opposite directions. Whenever the acceleration is a negative constant times the displacement, the motion is SHM.

The acceleration has its maximum magnitude a_{m}, where the force is largest, which is at the maximum displacement $x = \pm A$:

$$a_{\text{m}} = \frac{k}{m}A \qquad (10\text{-}16)$$

In SHM, the acceleration changes with time; Eq. (10-16) gives the *maximum* acceleration only. Equations derived for constant acceleration do not apply.

Example 10.6

Oscillating Model Rocket

A model rocket of 1.0 kg mass is attached to a horizontal spring with a spring constant of 6.0 N/cm. The spring is compressed by 18.0 cm and then released. The intent is to shoot the rocket horizontally, but the release mechanism fails to disengage, so the rocket starts to oscillate horizontally. Ignore friction and assume the spring to be ideal. (a) What is the amplitude of the oscillation? (b) What is the maximum speed? (c) What are the rocket's speed and acceleration when it is 12.0 cm from the equilibrium point?

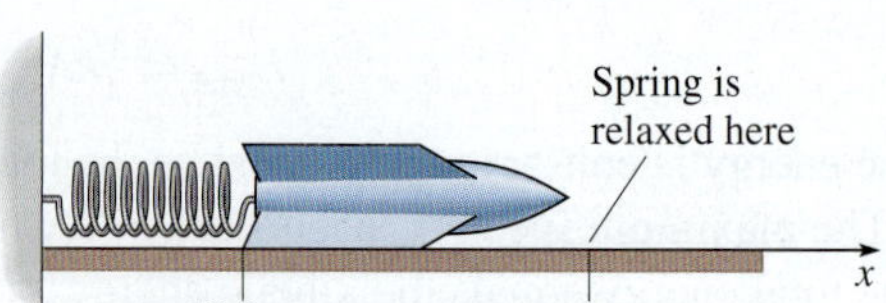

Figure 10.16

The model rocket before it is released.

Strategy First, we sketch the situation (Fig. 10.16). Initially all of the energy is elastic potential energy and the kinetic energy is zero. The initial displacement must be the maximum displacement—or amplitude—of the oscillations since to get farther from equilibrium would require more elastic energy than the total energy available. The speed at any position can be found using energy conservation ($\tfrac{1}{2}kx^2 + \tfrac{1}{2}mv_x^2 = \tfrac{1}{2}kA^2$). The maximum speed occurs when all of the energy is kinetic. The acceleration can be found from Newton's second law.

continued on next page

Example 10.6 continued

Solution (a) The amplitude of the oscillation is the maximum displacement, so $A = 18.0$ cm.

(b) From energy conservation, the maximum kinetic energy is equal to the maximum elastic potential energy:

$$K_m = \tfrac{1}{2}mv_m^2 = E = \tfrac{1}{2}kA^2$$

Solving for v_m,

$$v_m = \sqrt{\frac{k}{m}}\,A = \sqrt{\frac{6.0 \times 10^2 \text{ N/m}}{1.0 \text{ kg}}} \times 0.180 \text{ m} = 4.4 \text{ m/s}$$

(c) For the speed at a displacement of 0.120 m, we again use energy conservation.

$$\tfrac{1}{2}kx^2 + \tfrac{1}{2}mv^2 = \tfrac{1}{2}kA^2$$

Solving for v,

$$v = \sqrt{\frac{kA^2 - kx^2}{m}} = \sqrt{\frac{k}{m}(A^2 - x^2)}$$

$$= \sqrt{\frac{6.0 \times 10^2 \text{ N/m}}{1.0 \text{ kg}}[(0.180 \text{ m})^2 - (0.120 \text{ m})^2]} = 3.3 \text{ m/s}$$

From Newton's second law,

$$F_x = -kx = ma_x$$

At $x = \pm 0.120$ m,

$$a_x = -\frac{k}{m}x = \frac{6.0 \times 10^2 \text{ N/m}}{1.0 \text{ kg}} \times (\pm 0.120 \text{ m}) = \pm 72 \text{ m/s}^2$$

The magnitude of the acceleration is 72 m/s^2; the direction is toward the equilibrium point.

Discussion Note that at a given position (say $x = +0.120$ m), we can find the *speed* of the rocket, but the direction of the velocity can be either left or right; the rocket passes through each point (other than the endpoints) both on its way to the left and on its way to the right. By contrast, the *acceleration* at $x = +0.120$ m is always in the $-x$-direction, regardless of whether the rocket is moving to the left or to the right. If the rocket is moving to the right, then it is slowing down as it approaches $x = +A$; if it is moving to the left, then it is speeding up as it approaches $x = 0$.

Practice Problem 10.6 Maximum Acceleration of the Rocket

What is the maximum acceleration of the rocket in Example 10.6 and at what position(s) does it occur?

CONNECTION:

The period and frequency are defined exactly as for uniform circular motion, which is another kind of periodic motion.

10.6 THE PERIOD AND FREQUENCY FOR SHM

Definitions of Period and Frequency SHM is *periodic* motion because the same motion repeats over and over—a particle goes back and forth over the same path in exactly the same way. Each time the particle repeats its original motion, we say that it has completed another cycle. To complete one cycle of motion, the particle must be at the same point *and heading in the same direction* as it was at the start of the cycle. The **period** T is the time taken by one *complete* cycle. The **frequency** f is the number of cycles per unit time:

$$f = \frac{1}{T} \quad \text{(SI unit: Hz = cycles per second)} \qquad (5\text{-}8)$$

SHM is a special kind of periodic motion in which the restoring force is proportional to the displacement from equilibrium. Not all periodic vibrations are examples of simple harmonic motion since not all restoring forces are proportional to the displacement. Any restoring force can cause oscillatory motion. An electrocardiogram (Fig. 10.17) traces the periodic pattern of a beating heart, but the motion of the recorder needle is not simple harmonic motion. As we are about to show, in SHM the position is a *sinusoidal* function of time.

CHECKPOINT 10.6

The pendulum in a grandfather clock swings from its extreme leftmost position to its extreme rightmost position in 1.0 s. What is the frequency of its periodic motion?

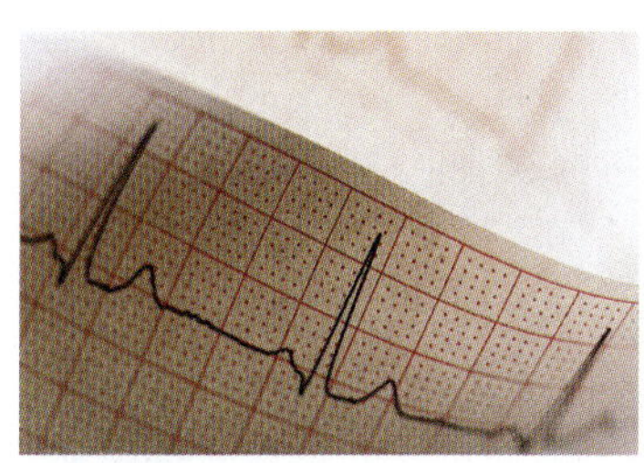

Figure 10.17 An electrocardiogram.

Circular Motion and SHM To learn more about SHM, imagine setting up an experiment (Fig. 10.18). We attach an object to an ideal spring, move the object away from the equilibrium position, and then release it. The object vibrates back and forth in simple

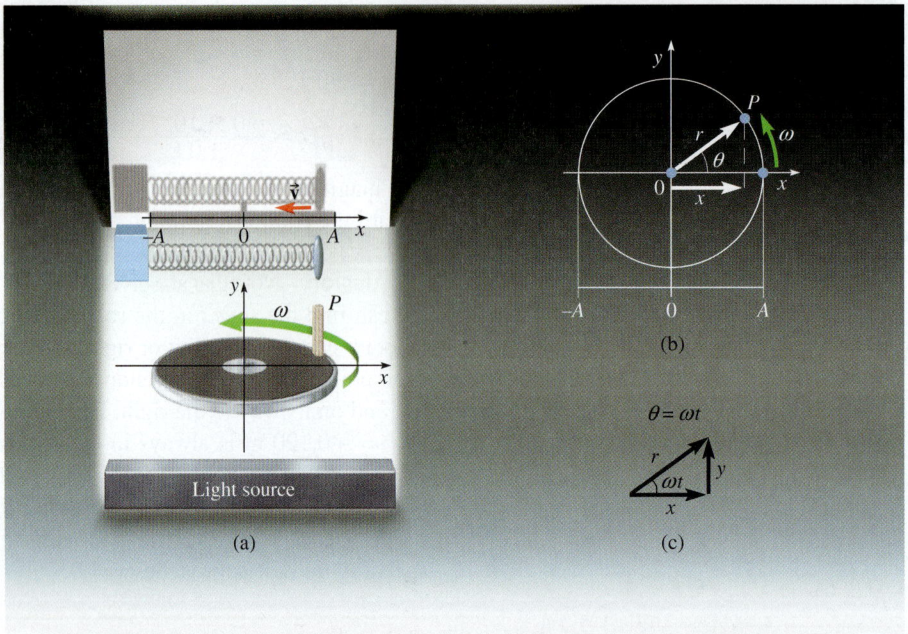

Figure 10.18 (a) An experiment to show the relation between uniform circular motion and SHM. (b) A pin P moving counterclockwise around a circle as a disk rotates with constant angular velocity ω. (c) Finding the x-component of the displacement.

harmonic motion with amplitude A. At the same time a horizontal circular disk, of radius $r = A$ and with a pin projecting vertically up from its outer edge, is set into rotation with uniform circular motion. Both the pin and the object attached to the spring are illuminated so that shadows of the vibrating object and of the pin on the rotating disk are seen on a screen. The speed of the disk is adjusted until the shadows oscillate with the same period. We will show that the motion of the two shadows is identical, so the mathematical description of one can be used for the other.

To find the mathematical description of SHM, we analyze the uniform circular motion of the pin. Figure 10.18b shows the pin P moving counterclockwise around a circle of radius A at a constant angular velocity ω in rad/s. For simplicity, let the pin start at $\theta = 0$ at time $t = 0$. The location of the pin at any time is then given by the angle θ:

$$\theta(t) = \omega t$$

The motion of the pin's shadow has the same x-component as the pin itself. Using a right triangle (Fig. 10.18c), we find that

$$x(t) = A \cos \theta = A \cos \omega t \quad \text{(10-17)}$$

Since the pin moves in uniform circular motion, its acceleration is constant in *magnitude* but not in direction; the acceleration is toward the center of the circle. In Section 5.2, the magnitude of the radial acceleration is shown to be

$$a_r = \omega^2 r = \omega^2 A \quad \text{(5-12)}$$

At any instant the direction of the acceleration vector is opposite to the direction of the displacement vector in Fig. 10.18b—that is, toward the center of the circle. Therefore,

$$a_x = -a_r \cos \theta = -\omega^2 A \cos \omega t \quad \text{(10-18)}$$

Comparing Eqs. (10-17) and (10-18), we see that, at any time t,

$$a_x(t) = -\omega^2 x(t) \quad \text{(10-19)}$$

In Eq. (10-15) we showed that in SHM the acceleration is proportional to the displacement:

$$a_x = -\frac{k}{m}x \tag{10-15}$$

Comparing the right-hand sides of Eqs. (10-15) and (10-19), the motions of the two shadows are identical as long as ω is given by

Angular frequency of a mass-spring system

$$\omega = \sqrt{\frac{k}{m}} \tag{10-20a}$$

In the context of SHM, the quantity ω is called the **angular frequency**. Note that the angular frequency is determined by the mass and the spring constant but is independent of the amplitude. Most of the equations involving ω are correct only if ω is measured in *radians* per unit time (e.g., rad/s). Don't forget to put your calculator into radian mode.

Equations (10-17) and (10-18) show that the position and acceleration of an object in SHM are sinusoidal functions of time (sine or cosine). In Problem 62, you can show that v_x is also a sinusoidal function of time. The term *harmonic* in *simple harmonic motion* refers to sinusoidal vibrations; this usage is related to similar usage in music and acoustics. In Chapter 12, we show that a complex vibration can be formed by combining harmonic (sinusoidal) vibrations at different frequencies, which is why the study of SHM is the basis for understanding more complex vibrations. The term *simple* in SHM means that no energy enters or leaves the system. In SHM, the amplitude of the vibration is constant.

Period and Frequency for an Ideal Mass-Spring System Since the object in SHM and the pin in circular motion have the same frequency and period, the relationships between ω, f, and T still apply. Therefore, the frequency and period of a mass-spring system are

$$f = \frac{\omega}{2\pi} = \frac{1}{2\pi}\sqrt{\frac{k}{m}} \tag{10-20b}$$

and

$$T = \frac{1}{f} = 2\pi\sqrt{\frac{m}{k}} \tag{10-20c}$$

With the identification of ω for a mass-spring system, we can write the maximum speed and acceleration from Eqs. (10-14) and (10-16):

$$v_m = \omega A \tag{10-21}$$

$$a_m = \omega^2 A \tag{10-22}$$

These expressions are more general than Eqs. (10-14) and (10-16)—they apply to any system in SHM, not just a mass-spring system.

To Find the Angular Frequency for Any Object in SHM

- Write down the restoring force as a function of the displacement from equilibrium. Since the restoring force is linear, it always takes the form $F = -kx$, where k is a constant.
- Use Newton's second law to relate the restoring force to the acceleration.
- Solve for ω using $a_x = -\omega^2 x$ [Eq. (10-19)].

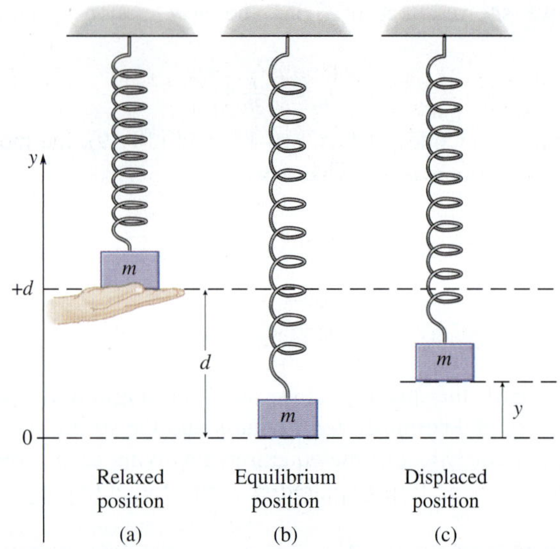

Figure 10.19 (a) A relaxed spring, of spring constant k, with mass m attached. (b) The same spring is extended to its equilibrium position, a distance d below the relaxed position, after mass m is allowed to hang freely. Note that we choose $y = 0$ at the equilibrium position, not at the relaxed position. (c) The spring is displaced from the equilibrium position.

A Vertical Mass and Spring

The mass and spring systems discussed so far oscillate horizontally. An oscillating mass on a vertical spring also exhibits SHM; the difference is that the equilibrium point is shifted downward by gravity. In our discussions, we assume ideal springs that obey Hooke's law and have a negligibly small mass of their own.

Suppose that an object of weight mg is hung from an ideal spring of spring constant k (Fig. 10.19). The object's equilibrium point is *not* the point at which the spring is relaxed. In equilibrium, the spring is stretched downward a distance d from its relaxed length so that the spring pulls up with a force equal to mg. Taking the $+y$-axis in the upward direction, the condition for equilibrium is

$$\sum F_y = +kd - mg = 0 \qquad \text{(at equilibrium)} \qquad (10\text{-}23)$$

Let us take the origin ($y = 0$) at the equilibrium point. If the object is displaced vertically from the equilibrium point to a position y, the spring force becomes

$$F_{\text{spring},y} = k(d - y)$$

If y is positive, the object is displaced upward and the spring force is less than kd. The y-component of the net force is then

$$\sum F_y = k(d - y) - mg = kd - ky - mg \qquad \text{(at displacement } y \text{ from equilibrium)}$$

From Eq. (10-23), we know that $kd = mg$; therefore,

$$\sum F_y = -ky$$

The restoring force provided by the spring and gravity together is $-k$ times the displacement from equilibrium. Therefore, the vertical mass-spring exhibits SHM with the same period and frequency as if it were horizontal.

Example 10.7

A Vertical Spring

A spring with spring constant k is suspended vertically. A model goose of mass m is attached to the unstretched spring and then released so that the bird oscillates up and down. (Ignore friction and air resistance; assume an ideal massless spring.) Calculate the kinetic energy, the elastic potential energy, the gravitational potential energy, and the total mechanical energy at (a) the point of release and (b) the equilibrium point. Take the gravitational potential energy to

continued on next page

Example 10.7 continued

be zero at the equilibrium point. (c) How long does it take the bird to move from its highest to its lowest position?

Strategy The bird oscillates in SHM about its equilibrium point $y = 0$ between two extreme positions $y = +A$ and $y = -A$ (Fig. 10.20). The amplitude A is equal to the distance the spring is stretched at the equilibrium point; it can be found by setting the net force on the bird equal to zero. The total mechanical energy is the sum of the kinetic energy, the elastic potential energy, and the gravitational potential energy. We expect the total energy to be the same at the two points; since no dissipative forces act, mechanical energy is conserved.

Solution The equilibrium point is where the net force on the bird is zero:

$$\sum F_y = +kd - mg = 0 \qquad (10\text{-}23)$$

In this equation, d is the extension of the spring at equilibrium. Since the bird is released where the spring is relaxed, d is also the amplitude of the oscillations:

$$A = d = \frac{mg}{k}$$

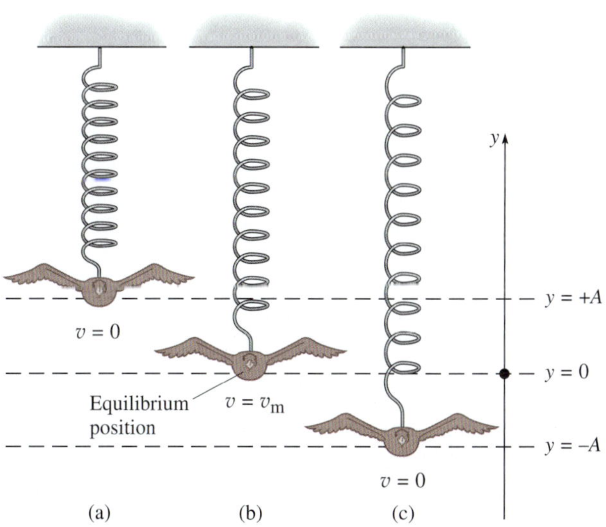

Figure 10.20

(a) The spring is unstretched before the model bird is released at position $y = +A$; (b) the model bird passes through the equilibrium position $y = 0$ with maximum speed; (c) the spring's maximum extension occurs when the bird is at $y = -A$.

(a) At the point of release, $v = 0$ and the kinetic energy is zero. The elastic energy is also zero—the spring is unstretched. The gravitational potential energy is

$$U_g = mgy = mgA = \frac{(mg)^2}{k}$$

The total mechanical energy is the sum of the kinetic and potential (elastic + gravitational) energies,

$$E = K + U_e + U_g = \frac{(mg)^2}{k}$$

(b) At the equilibrium point, the bird moves with its maximum speed $v_m = \omega A$. The angular frequency is the same as for a horizontal spring: $\omega = \sqrt{k/m}$ Then the kinetic energy is

$$K = \tfrac{1}{2}mv_m^2 = \tfrac{1}{2}m\omega^2A^2$$

Substituting $A = mg/k$ and $\omega^2 = k/m$,

$$K = \frac{1}{2}m\frac{k}{m}\frac{(mg)^2}{k^2} = \frac{1}{2}\frac{(mg)^2}{k}$$

The spring is stretched a distance A, so the elastic energy is

$$U_e = \frac{1}{2}kA^2 = \frac{1}{2}k\frac{(mg)^2}{k^2} = \frac{1}{2}\frac{(mg)^2}{k}$$

The gravitational potential energy is zero at $y = 0$. Therefore, the total mechanical energy is

$$E = K + U_e + U_g = \frac{1}{2}\frac{(mg)^2}{k} + \frac{1}{2}\frac{(mg)^2}{k} + 0 = \frac{(mg)^2}{k}$$

which is the same as at $y = +A$.

(c) The period is $2\pi\sqrt{m/k}$. Moving from $y = +A$ to $y = -A$ is *half* of a complete cycle, so the time it takes is $\pi\sqrt{m/k}$.

Discussion As the bird moves down from the release point toward the equilibrium point, gravitational potential energy is converted into elastic energy and kinetic energy. After the bird passes the equilibrium point, both kinetic and gravitational energy are converted into elastic energy. At the lowest point in the motion, the gravitational potential energy has its lowest value, while the elastic potential energy has its greatest value. The *total* potential energy (gravitational plus elastic) has its minimum value at the equilibrium point since the kinetic energy is maximum there.

Practice Problem 10.7 Energy at Maximum Extension

Calculate the energies at the lowest point in the oscillations in Example 10.7.

10.7 GRAPHICAL ANALYSIS OF SHM

We have shown that the position of a particle moving in SHM along the x-axis is

$$x(t) = A\cos\omega t \qquad (10\text{-}17)$$

Since the cosine function goes from -1 to $+1$, multiplying it by A gives us a displacement from $-A$ to $+A$. Figure 10.21a is a graph of the position as a function of time.

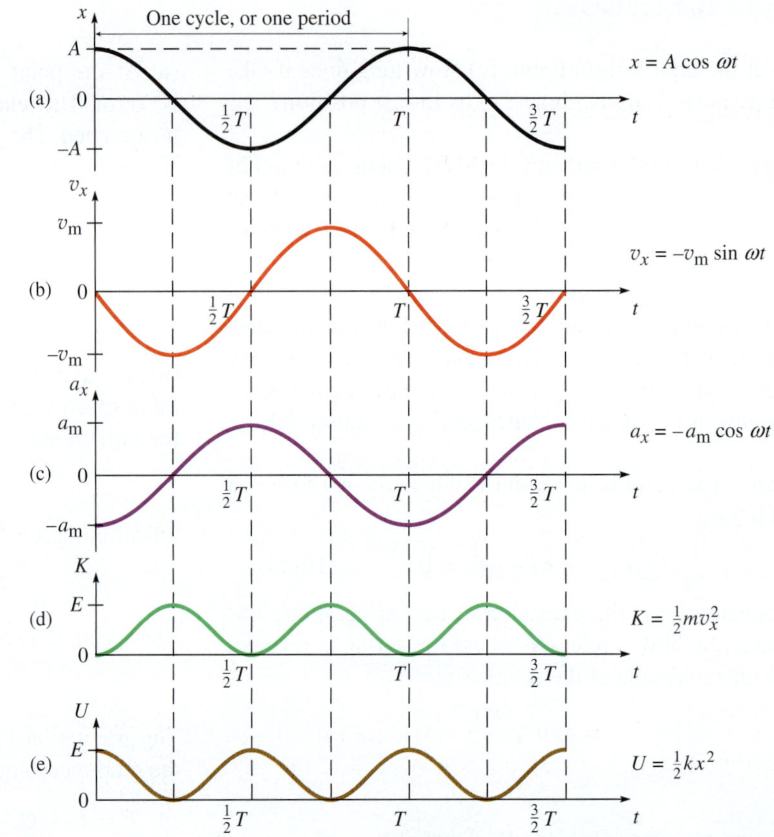

Figure 10.21 Graphs of (a) position, (b) velocity, and (c) acceleration as functions of time for a particle in simple harmonic motion. Observe the interrelationships between the three graphs. The velocity graph is one-quarter cycle ahead of the position graph; that is, $v_x(t)$ reaches its positive maximum one-quarter period before $x(t)$ reaches its positive maximum. Likewise, the acceleration is one-quarter cycle ahead of the velocity and one-half cycle ahead of the position. (d) Kinetic energy as a function of time. (e) Potential energy as a function of time.

The velocity at any time is the slope of the $x(t)$ graph. Note that the maximum slope in Fig. 10.21a occurs when $x = 0$, which confirms what we already know from energy conservation: the velocity is maximum at the equilibrium point. Note also that the velocity is zero when the displacement is a maximum ($+A$ or $-A$). Figure 10.21b shows a graph of $v_x(t)$. The equation describing this graph is (see Problem 62):

$$v_x(t) = -v_m \sin\omega t = -\omega A \sin\omega t \qquad \text{(10-24)}$$

The acceleration is the slope of the $v_x(t)$ graph. Figure 10.21c is a graph of $a_x(t)$, which is described by the equation

$$a_x(t) = -a_m \cos\omega t = -\omega^2 A \cos\omega t \qquad \text{(10-18)}$$

Figures 10.21d,e show the kinetic and potential energies as functions of time, respectively. The total energy $E = K + U = \frac{1}{2}kA^2$ is constant.

We have written the position as a function of time in terms of the cosine function, but we can just as correctly use the sine function. The difference between the two is the initial position at time $t = 0$. If the position is at a maximum ($x = A$) at $t = 0$, $x(t)$ is a cosine function. If the position is at the equilibrium point ($x = 0$) at $t = 0$, $x(t)$ is a sine function. By analyzing the slopes of the graphs, you can show (Problem 59) that if the position as a function of time is

$$x(t) = A\sin\omega t \qquad \text{(10-25a)}$$

then the velocity and acceleration are

$$v_x(t) = v_m \cos\omega t \qquad \text{(10-25b)}$$

$$a_x(t) = -a_m \sin\omega t \qquad \text{(10-25c)}$$

CHECKPOINT 10.7

(a) When the displacement of an object in SHM is zero, what is its speed? (b) When the speed is zero, what is the displacement?

Example 10.8

A Vibrating Loudspeaker Cone

A loudspeaker has a movable diaphragm (the *cone*) that vibrates back and forth to produce sound waves. The displacement of a loudspeaker cone playing a sinusoidal test tone is graphed in Fig. 10.22. Find (a) the amplitude of the motion, (b) the period of the motion, and (c) the frequency of the motion. (d) Write equations for $x(t)$ and $v_x(t)$.

Strategy The amplitude and period can be read directly from the graph. The frequency is the inverse of the period. Since $x(t)$ begins at the maximum displacement, it is described by a cosine function. By looking at the slope of $x(t)$, we can tell whether the velocity is a positive or negative sine function.

Solution (a) The amplitude is the maximum displacement shown on the graph: $A = 0.015$ m.

(b) The period is the time for one complete cycle. From the graph: $T = 0.040$ s.

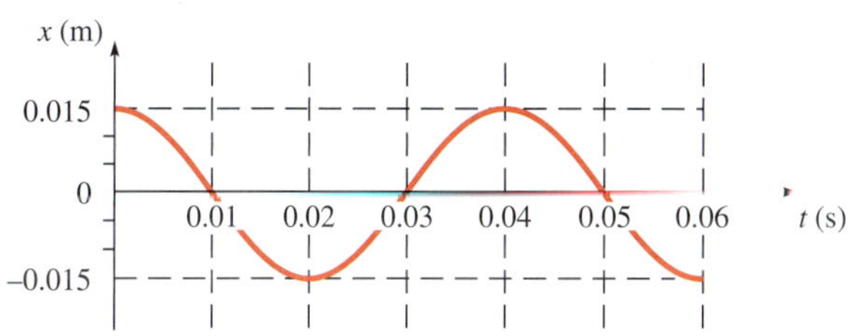

Figure 10.22
Horizontal displacement of a vibrating cone as a function of time.

(c) The frequency is the inverse of the period.

$$f = \frac{1}{T} = \frac{1}{0.040\ \text{s}} = 25\ \text{Hz}$$

(d) Since $x = +A$ at $t = 0$, we write $x(t)$ as a cosine function:

$$x(t) = A \cos \omega t$$

where $A = 0.015$ m and

$$\omega = 2\pi f = 160\ \text{rad/s}$$

The slope of $x(t)$ is initially zero and then goes negative. Therefore, $v_x(t)$ is a negative sine function:

$$v_x(t) = -v_\text{m} \sin \omega t$$

where $\omega = 2\pi f = 160$ rad/s and

$$v_\text{m} = \omega A = 160\ \text{rad/s} \times 0.015\ \text{m} = 2.4\ \text{m/s}$$

Discussion As a check, the velocity should be one-quarter cycle ahead of the position. If we imagine shifting the vertical axis to the right (ahead) by 0.01 s, the graph would have the shape of a negative sine function.

Practice Problem 10.8 Acceleration of the Speaker Cone

Sketch a graph and write an equation for $a_x(t)$.

10.8 THE PENDULUM

Simple Pendulum

When a pendulum swings back and forth, a string or thin rod constrains the bob to move along a circular arc. The motion is periodic for any amplitude. As we show here, when the amplitude is small, the motion is approximately SHM.

A *simple* (or *ideal*) pendulum consists of a bob (modeled as a point mass m) attached to a string or rod of length L and negligible mass. The restoring force is the tangential component of the weight, which has magnitude $mg \sin \theta$ when the tangential displacement along the circular arc is $s = L\theta$ (Fig. 10.23). The restoring force is *not* proportional to the displacement, so the motion is not SHM.

However, for small angles, $\sin \theta \approx \theta$, and then the restoring force is $F_\text{tan} \approx mg\theta$, which is (approximately) proportional to the displacement. Therefore, the motion of the pendulum is approximately SHM for small amplitudes. In this case, after substituting $\theta = s/L$, we have

$$F_\text{tan} \approx (mg/L)s$$

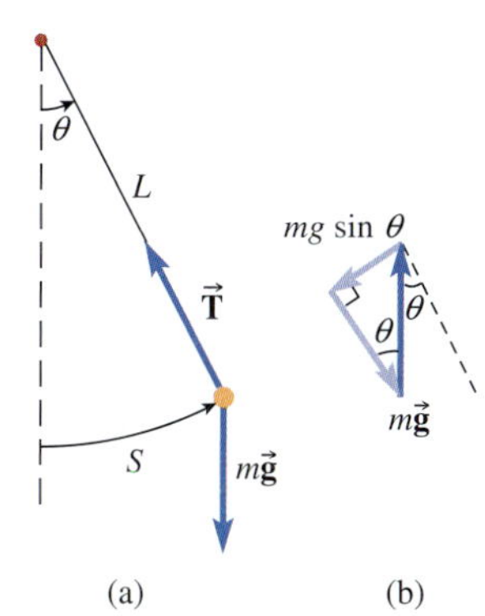

Figure 10.23 (a) A simple pendulum of length L is displaced so the string makes and angle θ with the vertical. The arc length s is equal to $L\theta$. The forces acting on the bob are the tension in the string and the weight. (b) The tangential component of the weight has magnitude $mg \sin \theta$. This is the restoring force; it always pulls the pendulum back toward the equilibrium position at $\theta = 0$.

The effective spring constant—that is, the constant of proportionality between the restoring force and the displacement—is

$$k_{eff} = mg/L$$

Then the angular frequency of the SHM is

$$\omega = \sqrt{\frac{k_{eff}}{m}} = \sqrt{\frac{g}{L}} \quad \text{(10-26a)}$$

and the period is

$$T = \frac{2\pi}{\omega} = 2\pi\sqrt{\frac{L}{g}} \quad \text{(10-26b)}$$

Note that the period depends on L and g but not on the mass of the pendulum.

Remember that Eq. (10-26b) is an approximation that assumes small amplitudes. The actual period of a pendulum is longer than the small-amplitude value of Eq. (10-26b). The discrepancy is less than 1% for amplitudes up to $\theta_{max} = 22°$ and rises to about 18% for an amplitude of $\theta_{max} = 90°$.

Be careful not to confuse the *angular frequency* of the pendulum with its *angular velocity*. The angular frequency of a given pendulum is constant, whereas the angular velocity (the rate of change of θ) changes with time between zero (at the extremes) and its maximum magnitude (at the equilibrium point).

EVERYDAY PHYSICS DEMO

The relation between the period and the length of the pendulum is easily tested. Make a simple pendulum by taping a thin string to a coin. Holding the end of the string, let the coin swing through a small arc and note the time for the coin to make ten complete oscillations, starting from one extreme position and returning to the same position ten times. Divide the time by ten to get the period. (This gives a more accurate value than timing a single period.) Measure the length of the pendulum and test Eq. (10-26b).

Repeat the experiment by holding the string at a position closer to the coin, shortening the length of the pendulum. What do you find? Is the period for the shorter pendulum longer, shorter, or the same as that measured for the longer pendulum?

The effect of a different mass on the period can also be tested by using two or three coins taped together, with the same length pendulum as used for the first measurement. Does a heavier coin oscillate with the same period as a lighter one (for the same length)?

Example 10.9

Grandfather Clock

A grandfather clock uses a pendulum with period 2.0 s to keep time. In one such clock, the pendulum bob has mass 150 g; the pendulum is set into oscillation by displacing it 33 mm to one side. (a) What is the length of the pendulum? (b) Does the initial displacement satisfy the small angle approximation?

Strategy The period depends on the length of the pendulum and on the gravitational field strength g. It does not depend on the mass of the bob. It also does not depend on the initial displacement, as long as it is small compared with the length.

Solution (a) Assuming small amplitudes, the period is

$$T = 2\pi\sqrt{\frac{L}{g}}$$

Solving for L,

$$L = \frac{T^2 g}{(2\pi)^2} = \frac{(2.0\ \text{s})^2 \times 9.80\ \text{m/s}^2}{(2\pi)^2} = 0.99\ \text{m}$$

continued on next page

Example 10.9 continued

(b) The small angle approximation is valid if the maximum displacement is small compared with the length of the pendulum.

$$\frac{x}{L} = \frac{33\text{ mm}}{990\text{ mm}} = 0.033$$

Is that small enough? If $\sin\theta = x/L = 0.033$, then

$$\theta = \sin^{-1} 0.033 = 0.033006$$

Sin θ and θ differ by less than 0.02%. Since we only know T to two significant figures, the approximation is good.

Discussion We should check that we didn't write the expression for the period "upside down," which is the most likely error we could make. Besides checking that the units work out, we know that a longer pendulum has a longer period, so L must go in the numerator. On the other hand, if g were larger, the restoring force would be larger and we would expect the period to shorten; thus, g belongs in the denominator.

Practice Problem 10.9 Pendulum on the Moon

A pendulum of length 0.99 m is taken to the Moon by an astronaut. The period of the pendulum is 4.9 s. What is the gravitational field strength on the surface of the Moon?

Physical Pendulum

Imagine that you have a simple pendulum of length L. Beside it you have a uniform metal bar of the same length, which is free to swing about an axis at one end. Would the two have the same period if they are set into oscillation?

For the simple pendulum, the bob is assumed to be a point mass; all the mass of the pendulum is at a distance L from the rotation axis. For the metal bar, however, the mass is uniformly distributed from the axis to a *maximum* distance L away from the axis. The center of mass is located at the midpoint, a distance $d = \frac{1}{2}L$ from the axis (Fig. 10.24). Since the mass is on average closer to the axis, the period is shorter than that of the simple pendulum.

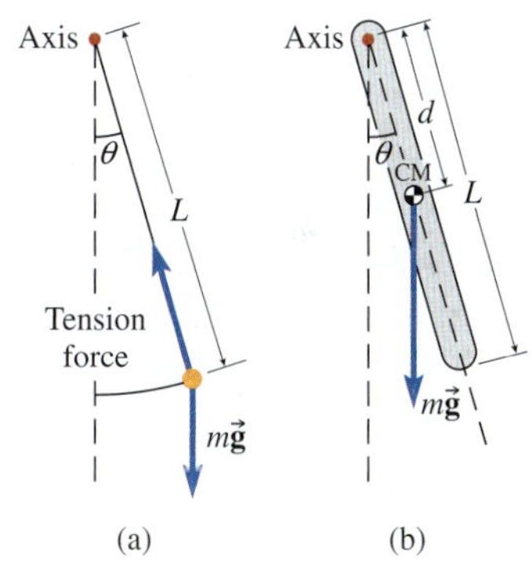

Figure 10.24 (a) A simple pendulum and (b) a physical pendulum.

Would this bar have a period equal to that of a simple pendulum of length $d = \frac{1}{2}L$? That is a good guess, since the center of mass of the bar is a distance $\frac{1}{2}L$ away from the rotation axis. Unfortunately, it isn't quite that easy. The gravitational force acts at the center of mass, but we *cannot* think of all the mass as being concentrated at that point—that would give the wrong rotational inertia. When set into oscillation, the bar, or any other rigid object free to rotate about a fixed axis, is called a **physical pendulum**. For small amplitudes, the period of a physical pendulum is

$$T = 2\pi\sqrt{\frac{I}{mgd}} \qquad (10\text{-}27)$$

where d is the distance from the rotation axis to the CM of the object and I is the rotational inertia about that axis. [See the text website for a derivation of Eq. (10-27).]

connect

For a uniform bar of length L, the CM is halfway down the bar:

$$d = \tfrac{1}{2}L$$

From Table 8.1, the rotational inertia of a uniform bar rotating about an axis through an endpoint is $I = \frac{1}{3}mL^2$. The period of oscillation is

$$T = 2\pi\sqrt{\frac{I}{mgd}} = 2\pi\sqrt{\frac{\frac{1}{3}mL^2}{(mg)\frac{1}{2}L}} = 2\pi\sqrt{\frac{2L}{3g}}$$

The bar has the same period as a simple pendulum of length $\frac{2}{3}L$.

CHECKPOINT 10.8

Is the expression for the period of a physical pendulum, Eq. (10-27), correct in the limiting case of a simple pendulum?

Example 10.10

Comparison of Walking Frequencies and Speeds for Various Creatures

During a relaxed walking pace, an animal's leg can be thought of as a physical pendulum of length L that pivots about the hip. (a) What is the relaxed walking frequency for a cat ($L = 30$ cm), dog (60 cm), human (1 m), giraffe (2 m), and a mythological titan (10 m)? (b) Derive an equation that gives the walking speed (amount of ground covered per unit time) for a given walking frequency f. [*Hint*: Start by drawing a picture of the leg position at the start of the swing (leg back) and the end of the swing (leg forward) and assume a comfortable angle of about 30° between these two positions. To how many steps does a complete period of the pendulum correspond?] (c) Find the walking speed for each of the animals listed in part (a).

Strategy We have to use an idealized model of the leg, since we don't know the exact location of the center of mass or the rotational inertia. The simple pendulum is not a good model, since it would assume all the mass of the leg at the foot! A much better model is to think of the leg as a uniform cylinder pivoting about one end.

Solution (a) For a uniform cylinder, the center of mass is a distance $d = \frac{1}{2}L$ from the pivot and the rotational inertia about an axis at one end is $I = \frac{1}{3}mL^2$. Then the period is

$$T = 2\pi\sqrt{\frac{I}{mgd}} = 2\pi\sqrt{\frac{\frac{1}{3}mL^2}{(mg)\frac{1}{2}L}} = 2\pi\sqrt{\frac{2L}{3g}}$$

and the frequency f is

$$f = \frac{1}{T} = \frac{1}{2\pi}\sqrt{\frac{3g}{2L}} \approx 0.2\sqrt{\frac{g}{L}}$$

Substituting the numerical values of L for each animal, we find the frequencies to be 1 Hz (cat), 0.8 Hz (dog), 0.6 Hz (human), 0.4 Hz (giraffe), and 0.2 Hz (titan).

(b) One period of the "pendulum" corresponds to two steps. In Fig. 10.25a, the right leg is about to step forward. The step occurs as the pendulum swings forward through half a cycle. In Fig. 10.25b, the right foot is about to touch the ground; in Fig. 10.25c, the right foot touches the ground and now the left leg is about to step forward. During this step, the right foot stays in place on the ground, but the right leg is swinging backward relative to the hip joint. During each step, the distance covered is approximately the length of a 30° arc of radius L, which is one twelfth the circumference of a circle of radius L. So during one period, the distance walked is

$$D = 2 \times \frac{1}{12} \times 2\pi L = \frac{\pi}{3}L \approx L$$

and the walking speed is

$$v = \frac{D}{T} = Lf = 0.2\sqrt{gL}$$

(c) The speeds are 0.3 m/s (cat), 0.5 m/s (dog), 0.6 m/s (human), 0.9 m/s (giraffe), and 2 m/s (titan).

Figure 10.25

The forward motion of a leg during walking is similar to the swing of a physical pendulum. From (a) to (b), the right leg swings forward like a pendulum. In (c), the right foot is on the ground and the left leg is about to swing forward.

continued on next page

Example 10.10 continued

Discussion You may be more familiar with walking speeds in mi/h. Converting the units, 0.6 m/s ≈ 1.3 mi/h, which is just about right for a leisurely walk. A brisk walk is about 3 mi/h for most people; to go much faster than that, you need to jog or run.

The solution says that longer legs walk faster, but the frequency of the steps is lower. You can verify that by walking beside a friend who is much taller or much shorter than you, or by taking your dog for a walk.

Practice Problem 10.10 Walking Speed for a Human

A more realistic model of a human leg of length 1.0 m has the center of mass 0.45 m from the hip and a rotational inertia of $\frac{1}{6}mL^2$. What is the walking speed predicted by this model?

EVERYDAY PHYSICS DEMO

Test the conclusion of Example 10.10 by walking beside a friend who is much taller or much shorter than you. Does the person with longer legs tend to walk faster but with a lower frequency of steps?

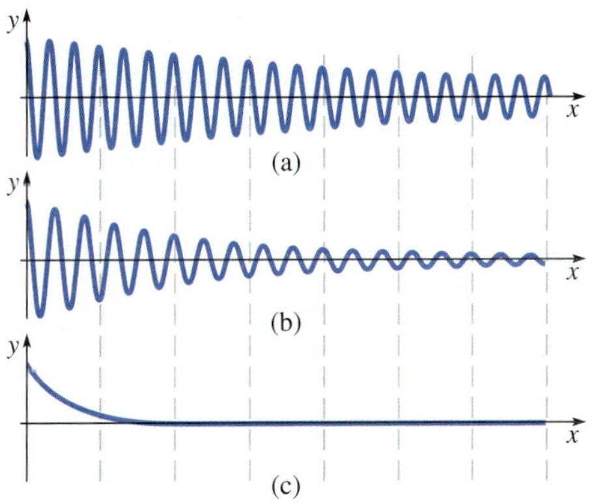

Figure 10.26 Graphs of $x(t)$ for a mass-spring system with increasing amounts of damping. In (c) the damping is sufficient to prevent oscillations from occurring.

10.9 DAMPED OSCILLATIONS

In SHM, we assume that no dissipative forces such as friction or viscous drag exist. Since the mechanical energy is constant, the oscillations continue forever with constant amplitude. SHM is a simplified model. The oscillations of a swinging pendulum or a vibrating tuning fork gradually die out as energy is dissipated. The amplitude of each cycle is a little smaller than that of the previous cycle (Fig. 10.26a). This kind of motion is called **damped oscillation**, where the word *damped* is used in the sense of *extinguished* or *restrained*. For a small amount of damping, oscillations occur at approximately the same frequency as if there were no damping. A greater degree of damping lowers the frequency slightly (Fig. 10.26b). Even more damping prevents oscillations from occurring at all (Fig. 10.26c).

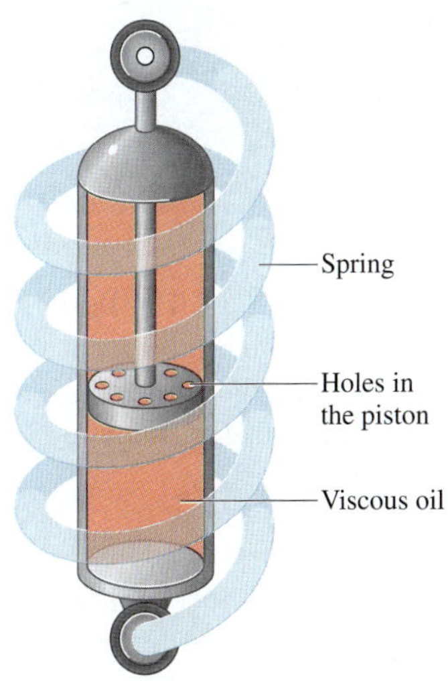

Figure 10.27 A shock absorber.

Application: Shock Absorbers Damping is not always a disadvantage. The suspension system of a car includes shock absorbers that cause the vibration of the body—a mass connected to the chassis by springs—to be quickly damped. The shock absorbers reduce the discomfort that passengers would otherwise experience due to the bouncing of an automobile as it travels along a bumpy road. Figure 10.27 shows how a shock absorber works. In order to compress or expand the shock absorber, a viscous oil must flow through the holes in the piston. The viscous force dissipates energy regardless of which direction the piston moves. The shock absorber enables the spring to smoothly

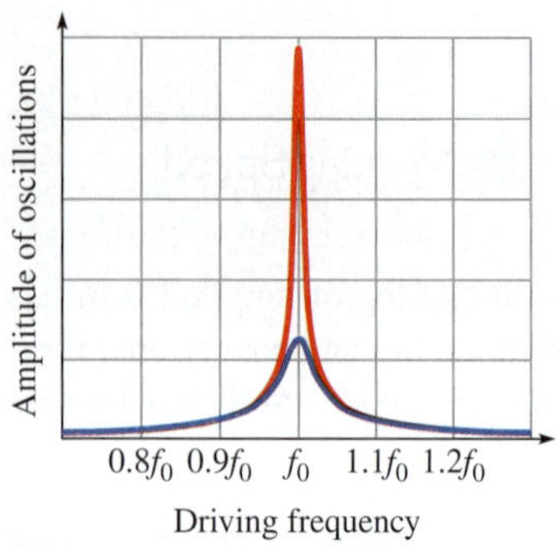

Figure 10.28 Two resonance curves for an oscillator with natural frequency f_0. The amplitude of the driving force is constant. In the red graph, the oscillator has one fourth as much damping as in the blue graph.

return to its equilibrium length without oscillating up and down (Fig. 10.26c). When the oil leaks out of the shock absorber, the damping is insufficient to prevent oscillations. After hitting a bump, the body of the car oscillates up and down (Fig. 10.26b).

10.10 FORCED OSCILLATIONS AND RESONANCE

When damping forces are present, the only way to keep the amplitude of oscillations from diminishing is to replace the dissipated energy from some other source. When a child is being pushed on a swing, the parent replaces the energy dissipated with a small push. In order to keep the amplitude of the motion constant, the parent gives a little push once per cycle, adding just enough energy each time to compensate for the energy dissipated in one cycle. The frequency of the *driving force* (the parent's push) matches the *natural frequency* of the system (the frequency at which it would oscillate on its own).

Forced oscillations (or driven oscillations) occur when a periodic external driving force acts on a system that can oscillate. The frequency of the driving force does not have to match the natural frequency of the system. Ultimately, the system oscillates at the driving frequency, even if it is far from the natural frequency. However, the amplitude of the oscillations is generally quite small unless the driving frequency f is close to the natural frequency f_0 (Fig. 10.28). When the driving frequency is equal to the natural frequency of the system, the amplitude of the motion is a maximum. This condition is called **resonance**.

At resonance, the driving force is always in the same direction as the object's velocity. Since the driving force is always doing positive work, the energy of the oscillator builds up until the energy dissipated balances the energy added by the driving force. For an oscillator with little damping, this requires a large amplitude. When the driving and natural frequencies differ, the driving force and velocity are no longer synchronized; sometimes they are in the same direction and sometimes in opposite directions. The driving force is not at resonance, so it sometimes does negative work. The net work done by the driving force decreases as the driving frequency moves away from resonance. Therefore, the oscillator's energy and amplitude are smaller than at resonance.

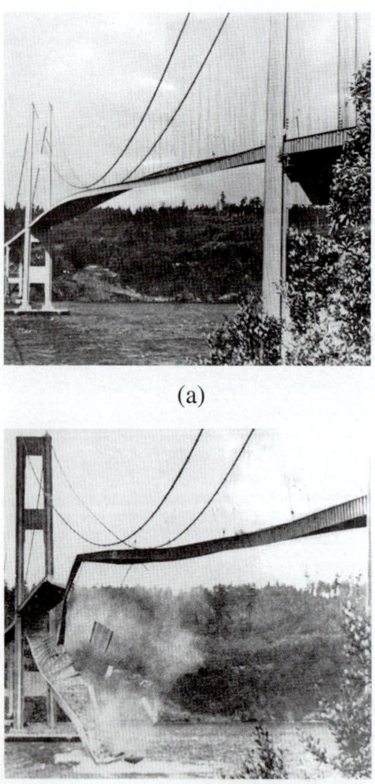

(a)

(b)

Figure 10.29 (a) The Tacoma Narrows Bridge begins to vibrate. (b) Ultimately the vibrations caused the bridge to collapse.

Applications of Resonance Large-amplitude vibrations due to resonance can be dangerous in some situations. Materials can be stressed past their elastic limits, causing permanent deformation or breaking. In 1940, the wind set the Tacoma Narrows Bridge in Washington state into vibration with increasing amplitude. Turbulence in the air as it flowed across the bridge caused the air pressure to fluctuate with a frequency matching one of the bridge's resonant frequencies. As the amplitude of the oscillations grew, the bridge was closed; soon after, the bridge collapsed (Fig. 10.29). Engineers now design bridges with much higher resonant frequencies so the wind cannot cause resonant vibrations.

In the nineteenth century, bridges were sometimes set into resonant vibration when the cadence of marching soldiers matched a resonant frequency of the bridge. After the collapse of several bridges due to resonance, soldiers were told to break step when crossing a bridge to eliminate the danger of their cadence setting the bridge into resonance.

How is the swaying of a tall building reduced?

Tall buildings sway back and forth at a particular resonant frequency determined by the structure. The vibration pattern is similar to what you see if you hold one end of a ruler to the edge of a desk and then pluck the other end. Engineers have many methods to reduce the amplitude of the swaying. One of the simplest and most widely used is the tuned mass damper (TMD). Building engineers attach a damped mass-spring system to the structure at a point where its vibration amplitude is largest—near the top. In the Hancock Tower, each of the 300 000 kg boxes is attached to the building frame with springs and shock absorbers and can slide back and forth, riding on a thin layer of oil that covers a 9 m long steel plate. The resonant frequency of the TMD is matched to the resonant frequency of the swaying building. When the swaying of the building drives the TMD into oscillation, energy is dissipated in the shock absorbers. The TMD in the Hancock Tower reduces the amplitude of its swaying by about 50%.

Master the Concepts

- A deformation is a change in the size or shape of an object.
- When deforming forces are removed, an *elastic* object returns to its original shape and size.
- Hooke's law, in a generalized form, says that the deformation of a material (measured by the strain) is proportional to the magnitude of the forces causing the deformation (measured by the stress). The definitions of stress and strain are as given in the following table.

Type of Deformation

	Tensile or Compressive	**Shear**	**Volume**
Stress	Force per unit cross-sectional area F/A	Shear force divided by the parallel area of the surface on which it acts F/A	Pressure P
Strain	Fractional length change $\Delta L/L$	Ratio of the relative displacement Δx to the separation L of the two parallel surfaces $\Delta x/L$	Fractional volume change $\Delta V/V$
Constant of proportionality	Young's modulus Y	Shear modulus S	Bulk modulus B

- If the tensile or compressive stress exceeds the *proportional limit*, the strain is no longer proportional to the stress. The solid still returns to its original length when the stress is removed as long as the stress does not exceed the *elastic limit*. If the stress exceeds the elastic limit, the material is permanently deformed. For larger stresses yet, the solid fractures when the stress reaches the *breaking point*. The maximum stress that can be withstood without breaking is called the *ultimate strength*.
- Vibrations occur in the vicinity of a point of stable equilibrium. An equilibrium point is *stable* if the net force on an object when it is displaced from equilibrium points back toward the equilibrium point. Such a force is called a restoring force since it tends to restore equilibrium.
- Simple harmonic motion is periodic motion that occurs whenever the restoring force is proportional to the displacement from equilibrium. In SHM, the position, velocity, and acceleration as functions of time are sinusoidal (i.e., sine or cosine functions). Any oscillatory motion is approximately SHM if the amplitude is small, because for small oscillations the restoring force is approximately linear.
- The period T is the time taken by one complete cycle of oscillation. The frequency f is the number of cycles per unit time:

$$f = \frac{1}{T} \qquad (5\text{-}8)$$

The angular frequency is measured in radians per unit time:

$$\omega = 2\pi f \qquad (5\text{-}9)$$

- The maximum velocity and acceleration in SHM are

$$v_m = \omega A \qquad \text{and} \qquad a_m = \omega^2 A \qquad (10\text{-}21, 10\text{-}22)$$

where ω is the angular frequency. The acceleration is proportional to and in the opposite direction from the displacement:

$$a_x(t) = -\omega^2 x(t) \qquad (10\text{-}19)$$

- The equations that describe SHM are

If $x = A$ at $t = 0$,	If $x = 0$ at $t = 0$,
$x = A\cos\omega t$	$x = A\sin\omega t$
$v_x = -v_m\sin\omega t$	$v_x = v_m\cos\omega t$
$a_x = -a_m\cos\omega t$	$a_x = -a_m\sin\omega t$

In either case, the velocity is one-quarter cycle ahead of the position, and the acceleration is one-quarter cycle ahead of the velocity.

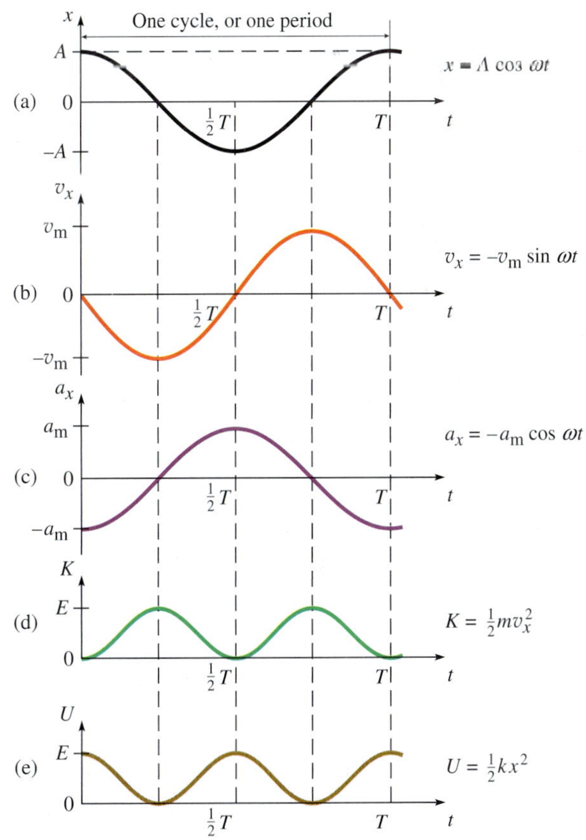

continued on next page

Master the Concepts continued

- The period of oscillation for a mass-spring system is

$$T = 2\pi\sqrt{\frac{m}{k}} \quad (10\text{-}20c)$$

For a simple pendulum it is

$$T = 2\pi\sqrt{\frac{L}{g}} \quad (10\text{-}26b)$$

and for a physical pendulum it is

$$T = 2\pi\sqrt{\frac{I}{mgd}} \quad (10\text{-}27)$$

- In the absence of dissipative forces, the total mechanical energy of a simple harmonic oscillator is constant and proportional to the square of the amplitude:

$$E = \tfrac{1}{2}kA^2 \quad (10\text{-}13)$$

where the potential energy has been chosen to be zero at the equilibrium point. At any point, the sum of the kinetic and potential energies is constant:

$$E = \tfrac{1}{2}mv_x^2 + \tfrac{1}{2}kx^2 = \tfrac{1}{2}kA^2 \quad (10\text{-}12)$$

Conceptual Questions

1. Young's modulus for diamond is about 20 times as large as that of glass. Does that tell you which is stronger? If not, what does it tell you?
2. A grandfather clock is running too fast. To fix it, should the pendulum be lengthened or shortened? Explain.
3. A karate student hits downward on a stack of concrete blocks supported at both ends. A block breaks. Explain where it starts to break first, at the bottom or at the top. (The block experiences shear, compressive, and tensile stresses. Recall that concrete has much less tensile strength than compressive strength. Which part of the block is stretched and which is compressed when the block bends in the middle?)

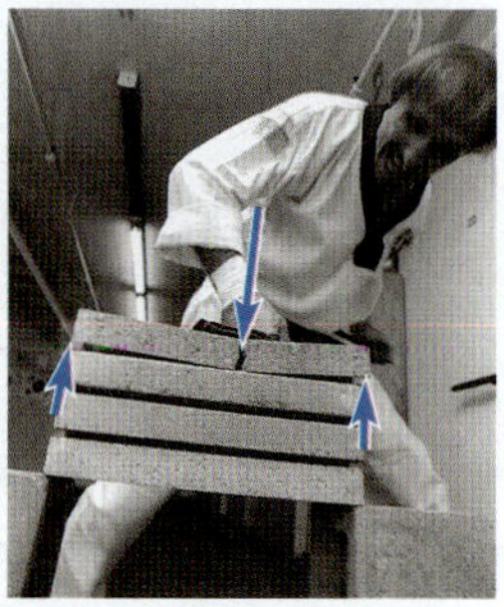

4. A cylindrical steel bar is compressed by the application of forces of magnitude F at each end. What magnitude forces would be required to compress by the same amount (a) a steel bar of the same cross-sectional area but one half the length? (b) a steel bar of the same length but one half the radius?
5. The columns built by the ancient Greeks and Romans to support temples and other structures are tapered; they are thicker at the bottom than at the top. This certainly has an aesthetic purpose, but is there an engineering purpose as well? What might it be?

6. Explain how the period of a mass-spring system can be independent of amplitude, even though the distance traveled during each cycle is proportional to the amplitude.
7. In a reciprocating saw, a *Scotch yoke* converts the rotation of the motor into the back-and-forth motion of the blade. The Scotch yoke is a mechanical device used to convert oscillatory motion to circular motion or vice versa. A wheel with a fixed knob rotates at constant angular velocity; the knob is constrained within a vertical slot causing the saw blade to move left and right without moving up and down. Is the motion of the saw blade SHM? Explain.

8. A mass hanging vertically from a spring and a simple pendulum both have a period of oscillation of 1 s on Earth. An astronaut takes the two devices to another planet where the gravitational field is stronger than that of Earth. For each of the two systems, state whether the period is now longer than 1 s, shorter than 1 s, or equal to 1 s. Explain your reasoning.
9. A bungee jumper leaps from a bridge and comes to a stop a few centimeters above the surface of the water below. At that lowest point, is the tension in the bungee cord equal to the jumper's weight? Explain why or why not.
10. Does it take more force to tear a longer tendon or a shorter tendon? Assume the tendons are identical except for their lengths and are ideal—there are no weak points. Does it take more *energy* to tear the long tendon or the short tendon? Explain.
11. A pilot is performing vertical loop-the-loops over the ocean at noon. The plane speeds up as it approaches the bottom of the circular loop and slows as it approaches the top of the loop. An observer in a helicopter is watching the shadow of the plane on the surface of the water. Does the shadow exhibit SHM? Explain.
12. Are you more likely to find steel rods in a horizontal concrete beam or in a vertical concrete column? Is concrete more in need of reinforcement under tensile or compressive stress?
13. Suppose that it takes tensile forces of magnitude F to produce a given strain $\Delta L/L$ in a steel wire of cross-sectional area A. If you had two such wires side by side and stretched them simultaneously, what magnitude tensile forces would be required to produce the same strain? By thinking of a thick wire as two (or more) thinner wires side

by side, explain why the force to produce a given strain must be proportional to the cross-sectional area. Thus, the strain depends on the stress—the force per unit area.

14. Think of a crystalline solid as a set of atoms connected by ideal springs. When a wire is stretched, how is the elongation of the wire related to the elongation of each of the interatomic springs? Use your answer to explain why a given tensile stress produces an elongation of the wire proportional to the wire's initial length—or, equivalently, that a given stress produces the same strain in wires of different lengths.

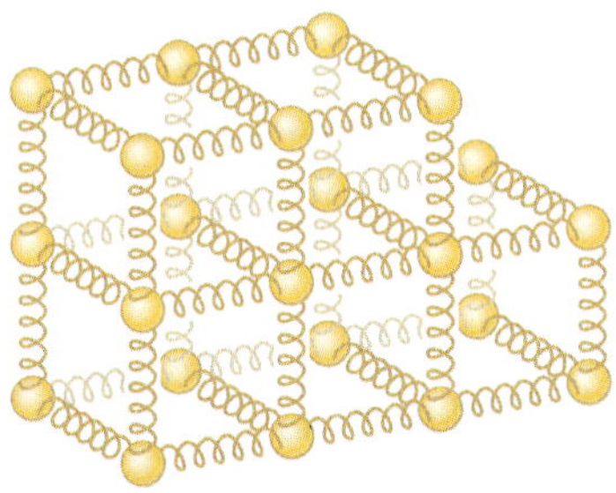

15. What are the advantages of using the concepts of stress and strain to describe deformations?
16. An old highway is built out of concrete blocks of equal length. A car traveling on this highway feels a little bump at the joint between blocks. The passengers in the car feel that the ride is uncomfortable at a speed of 45 mi/h, but much smoother at speeds either lower or higher than that. Explain.
17. The period of oscillation of a simple pendulum does not depend on the mass of the bob. By contrast, the period of a mass-spring system does depend on mass. Explain the apparent contradiction. [*Hint*: What provides the restoring force in each case? How does the restoring force depend on mass?]
18. A mass connected to an ideal spring is oscillating without friction on a horizontal surface. Sketch graphs of the kinetic energy, potential energy, and total energy as functions of time for one complete cycle.

Multiple-Choice Questions

Student Response System Questions

Questions 1–4. An object is suspended vertically from an ideal spring. The spring is initially in its relaxed position. The object is then released and oscillates about the equilibrium position. Answer choices for Questions 1–4:

(a) The spring is relaxed.
(b) The object is at the equilibrium point.
(c) The spring is at its maximum extension.
(d) The spring is somewhere between the equilibrium point and maximum extension.

1. The acceleration is greatest in magnitude and is directed upward when:
2. The speed is greatest when:
3. The acceleration is zero when:
4. The acceleration is greatest in magnitude and is directed downward when:
5. Two simple pendulums, A and B, have the same length, but the mass of A is twice the mass of B. Their vibrational amplitudes are equal. Their periods are T_A and T_B, respectively, and their energies are E_A and E_B. Choose the correct statement.
(a) $T_A = T_B$ and $E_A > E_B$ (b) $T_A > T_B$ and $E_A > E_B$
(c) $T_A > T_B$ and $E_A < E_B$ (d) $T_A = T_B$ and $E_A < E_B$
6. A force F applied to each end of a steel wire (length L, diameter d) stretches it by 1.0 mm. How much does F stretch another steel wire, of length $2L$ and diameter $2d$?
(a) 0.50 mm (b) 1.0 mm (c) 2.0 mm
(d) 4.0 mm (e) 0.25 mm
7. A stiff material is characterized by
(a) high ultimate strength.
(b) high breaking strength.
(c) high Young's modulus.
(d) high proportional limit.
8. A brittle material is characterized by
(a) high breaking strength and low Young's modulus.
(b) low breaking strength and high Young's modulus.
(c) high breaking strength and high Young's modulus.
(d) low breaking strength and low Young's modulus.
9. Which pair of quantities can be expressed in the same units?
(a) stress and strain
(b) Young's modulus and strain
(c) Young's modulus and stress
(d) ultimate strength and strain
10. Two wires have the same diameter and length. One is made of copper, the other brass. The wires are connected together end to end. When the free ends are pulled in opposite directions, the two wires *must* have the same
(a) stress. (b) strain. (c) ultimate strength.
(d) elongation. (e) Young's modulus.

Questions 11–20. See the graph of $v_x(t)$ for an object in SHM. Answer choices for each question:

(a) 1 s, 2 s, 3 s (b) 5 s, 6 s, 7 s (c) 0 s, 1 s, 7 s, 8 s
(d) 3 s, 4 s, 5 s (e) 0 s, 4 s, 8 s (f) 2 s, 6 s
(g) 3 s, 5 s (h) 1 s, 3 s (i) 5 s, 7 s
(j) 3 s, 7 s (k) 1 s, 5 s

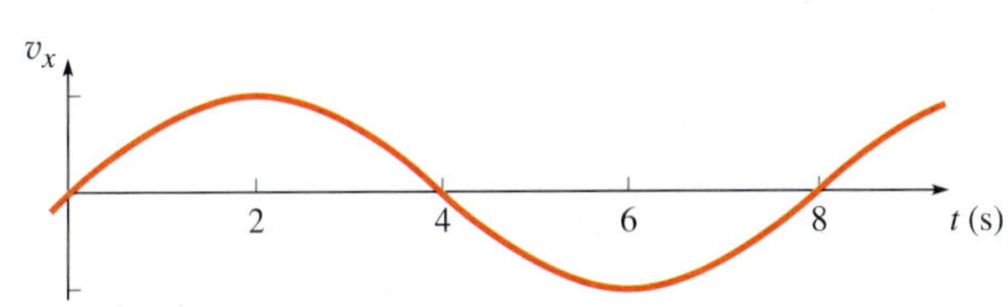

Multiple-Choice Questions 11–20

11. When is the kinetic energy maximum?
12. When is the kinetic energy zero?
13. When is the potential energy maximum?
14. When is the potential energy minimum?
15. When is the object at the equilibrium point?
16. When does the acceleration have its maximum magnitude?
17. Which answer specifies times when the net force is in the $+x$-direction?
18. Which answer specifies times when the object is on the $-x$-side of the equilibrium point ($x < 0$)?
19. Which answer specifies times when the object is moving away from the equilibrium point?
20. Which answer specifies times when the potential energy is decreasing?

Problems

C	Combination conceptual/quantitative problem
(icon)	Biomedical application
✦	Challenging
Blue #	Detailed solution in the Student Solutions Manual
[1, 2]	Problems paired by concept
connect	Text website interactive or tutorial

10.2 Hooke's Law for Tensile and Compressive Forces

1. A steel beam is placed vertically in the basement of a building to keep the floor above from sagging. The load on the beam is 5.8×10^4 N, the length of the beam is 2.5 m, and the cross-sectional area of the beam is 7.5×10^{-3} m^2. Find the vertical compression of the beam.
2. A 91 kg man's thighbone has a relaxed length of 0.50 m, a cross-sectional area of 7.0×10^{-4} m^2, and a Young's modulus of 11 GPa. By how much does the thighbone compress when the man is standing on both feet?
3. A man with a mass of 70 kg stands on one foot. His femur has cross-sectional area of 8.0 cm^2 and uncompressed length 50 cm. (a) How much shorter is the femur when he stands on one foot? (b) What is the fractional length change of the femur when the person moves from standing on two feet to standing on one foot?
4. A brass wire with Young's modulus of 92 GPa is 2.0 m long and has a cross-sectional area of 5.0 mm^2. If a weight of 5.0 kN is hung from the wire, by how much does it stretch?
5. A wire of length 5.00 m with a cross-sectional area of 0.100 cm^2 stretches by 6.50 mm when a load of 1.00 kN is hung from it. What is the Young's modulus for this wire?
6. Four brass wires are subjected to the same tensile stress. The wires have unstretched lengths and diameters as follows. Rank the four wires in decreasing order of the amount of stretch.
(a) length L, diameter d
(b) length $2L$, diameter d
(c) length $4L$, diameter $d/2$
(d) length $L/4$, diameter $d/2$
7. Two steel wires (of the same length and different radii) are connected together, end to end, and tied to a wall. An applied force stretches the combination by 1.0 mm. How far does the *midpoint* move?

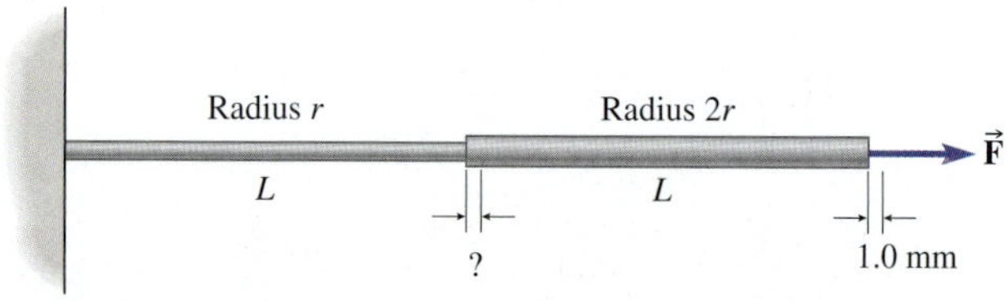

8. Abductin, an elastic protein found in the ligaments of scallops, has a Young's modulus of 4.0 MPa. The inner hinge ligament has a cross-sectional area of 0.78 mm^2 and a relaxed length of 1.0 mm. When the muscles in the shell relax, the shell opens. This increases efficiency as the muscles do not need to exert any force to open the shell, only to close it. If the muscles must exert a force of 1.5 N to keep the shell closed, by how much is the abductin ligament compressed?
9. ✦ Resilin is a rubber-like protein that helps insects to fly more efficiently. The resilin, attached from the wing to the body, is relaxed when the wing is down and is extended when the wing is up. As the wing is brought up, some elastic energy is stored in the resilin. The wing is then brought back down with little muscular energy, since the potential energy in the resilin is converted back into kinetic energy. Resilin has a Young's modulus of 1.7 MPa. (a) If an insect wing has resilin with a relaxed length of 1.0 cm and a cross-sectional area of 1.0 mm^2, how much force must the wings exert to extend the resilin to 4.0 cm? (b) How much energy is stored in the resilin?
10. ✦ It takes a flea 1.0×10^{-3} s to reach a peak speed of 0.74 m/s. (a) If the mass of the flea is 0.45×10^{-6} kg, what is the average power required? (b) Insect muscle has a maximum output of 60 W/kg. If 20% of the flea's weight is muscle, can the muscle provide the power needed? (c) The flea has a resilin pad at the base of the hind leg that compresses when the flea bends its leg to jump. If we assume the pad is a cube with a side of 6.0×10^{-5} m and the pad compresses fully, what is the energy stored in the compression of the pads of the two hind legs? The Young's modulus for resilin is 1.7 MPa. (d) Does this provide enough power for the jump?
11. A 0.50 m long guitar string, of cross-sectional area 1.0×10^{-6} m^2, has Young's modulus $Y = 2.0$ GPa. By how much must you stretch the string to obtain a tension of 20 N?

10.3 Beyond Hooke's Law

12. An acrobat of mass 55 kg is going to hang by her teeth from a steel wire and she does not want the wire to stretch beyond its elastic limit. The elastic limit for the wire is 250 MPa. What is the minimum diameter the wire should have to support her?

13. Using the stress-strain graph for bone (Fig. 10.4c), calculate Young's moduli for tension and for compression. Consider only small stresses.

14. A hair breaks under a tension of 1.2 N. What is the diameter of the hair? The tensile strength is 200 MPa.

15. The ratio of the tensile (or compressive) strength to the density of a material is a measure of how strong the material is "pound for pound." (a) Compare tendon (tensile strength 80.0 MPa, density 1100 kg/m^3) with steel (tensile strength 0.50 GPa, density 7700 kg/m^3): which is stronger "pound for pound" under tension? (b) Compare bone (compressive strength 160 MPa, density 1600 kg/m^3) with concrete (compressive strength 0.40 GPa, density 2700 kg/m^3): which is stronger "pound for pound" under compression?

16. Common sports injuries result in the tearing of tendons and ligaments due to overstretching. If the anterior cruciate ligament (ACL) in an athlete's knee has a length of 1.0 cm, a breaking point of 190 MPa, and a Young's modulus of 600 MPa, how far must it be stretched from its relaxed length to tear it?

17. The leg bone (femur) breaks under a compressive force of about 5×10^4 N for a human and 10×10^4 N for a horse. The human femur has a compressive strength of 160 MPa, whereas the horse femur has a compressive strength of 140 MPa. What is the effective cross-sectional area of the femur in a human and in a horse? (*Note*: Since the center of the femur contains bone marrow, which has essentially no compressive strength, the effective cross-sectional area is about 80% of the total cross-sectional area.)

18. The force of the tibia (shinbone) on the ankle joint for a person of weight 750 N standing on the ball of one foot is 2800 N. (See the following figure.) The ankle joint pushes upward on the bottom of the tibia with a force of 2800 N, while the top end of the tibia must feel a net downward force of approximately 2800 N (ignoring the weight of the tibia itself). The tibia has a length of 0.40 m, an average inner diameter of 1.3 cm, and an average outer diameter of 2.5 cm. (The central core of the bone contains marrow that has negligible compressive strength.) (a) Find the average cross-sectional area of the tibia. (b) Find the compressive stress in the tibia. (c) Find the change in length of the tibia due to the compressive forces.

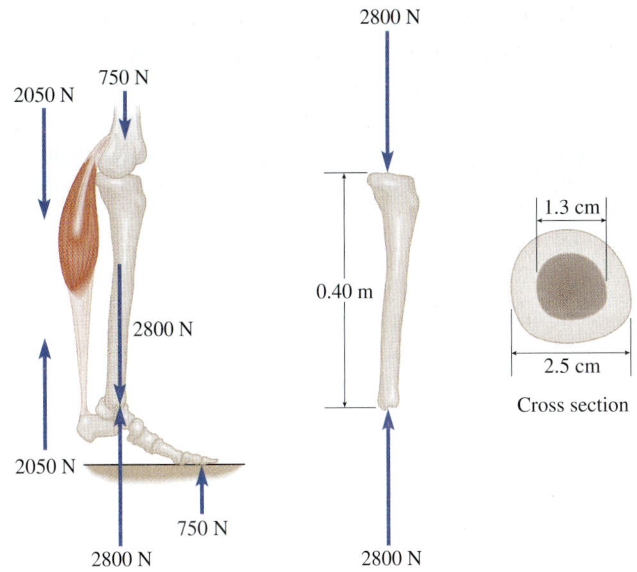

Problem 18

19. What is the maximum load that could be suspended from a copper wire of length 1.0 m and radius 1.0 mm without permanently deforming the wire? Copper has an elastic limit of 200 MPa and a tensile strength of 400 MPa.

20. What is the maximum load that could be suspended from a copper wire of length 1.0 m and radius 1.0 mm without breaking the wire? Copper has an elastic limit of 200 MPa and a tensile strength of 400 MPa.

21. The maximum strain of a steel wire with Young's modulus 200 GPa, just before breaking, is 0.20%. What is the stress at its breaking point, assuming that strain is proportional to stress up to the breaking point?

22. A marble column with a cross-sectional area of 25 cm^2 supports a load of 7.0×10^4 N. The marble has a Young's modulus of 60 GPa and a compressive strength of 200 MPa. (a) What is the stress in the column? (b) What is the strain in the column? (c) If the column is 2.0 m high, how much is its length changed by supporting the load? (d) What is the maximum weight the column can support?

23. A copper wire of length 3.0 m is observed to stretch by 2.1 mm when a weight of 120 N is hung from one end. (a) What is the diameter of the wire and what is the tensile stress in the wire? (b) If the tensile strength of copper is 400 MPa, what is the maximum weight that may be hung from this wire?

10.4 Shear and Volume Deformations

24. A sphere of copper is subjected to 100 MPa of pressure. The copper has a bulk modulus of 130 GPa. By what fraction does the volume of the sphere change? By what fraction does the radius of the sphere change?

25. Atmospheric pressure on Venus is about 90 times that on Earth. A steel sphere with a bulk modulus of 160 GPa has

a volume of 1.00 cm^3 on Earth. If it were put in a pressure chamber and the pressure were increased to that of Venus (9.12 MPa), how would its volume change?

26. How would the volume of 1.00 cm^3 of aluminum on Earth change if it were placed in a vacuum chamber and the pressure changed to that of the Moon (less than 10^{-9} Pa)?

27. Some claim that mountain climbers suffer from headaches due not only to a lack of oxygen in the brain, but also to the expansion of the brain in the cranium. Find the fractional change of the brain's volume due to a reduction in pressure from 101 kPa at sea level to 31 kPa high in the Himalayas. The bulk modulus is 2.1 GPa. (Another reason the brain expands is the dilation of the blood vessels in the brain in order to deliver more oxygen.)

28. Two steel plates are fastened together using four bolts. The bolts each have a shear modulus of 80 GPa and a shear strength of 600 MPa. The radius of each bolt is 1.0 cm. Normally, the bolts clamp the two plates together and the frictional forces between the plates keep them from sliding. If the bolts are loose, then the frictional forces are small and the bolts themselves would be subject to a large shear stress. What is the maximum shearing force F on the plates that the four bolts can withstand?

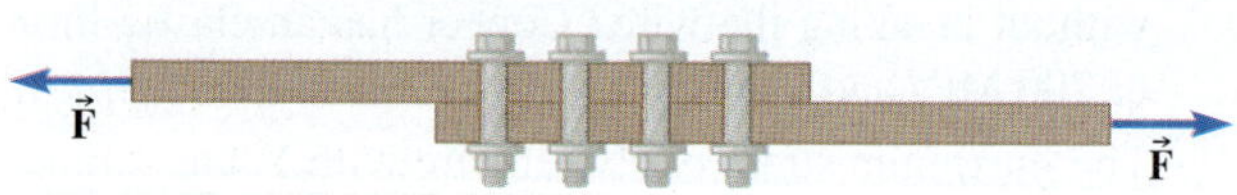

29. An anchor, made of cast iron of bulk modulus 60.0 GPa and of volume 0.230 m^3, is lowered over the side of the ship to the bottom of the harbor where the pressure is greater than sea level pressure by 1.75 MPa. Find the change in the volume of the anchor.

30. The upper surface of a cube of gelatin, 5.0 cm on a side, is displaced 0.64 cm by a tangential force. If the shear modulus of the gelatin is 940 Pa, what is the magnitude of the tangential force?

31. A large sponge has forces of magnitude 12 N applied in opposite directions to two opposite faces of area 42 cm^2 (see Fig. 10.8 for a similar situation). The thickness of the sponge (L) is 2.0 cm. The deformation angle (γ) is 8.0°. (a) What is Δx? (b) What is the shear modulus of the sponge?

10.5 Simple Harmonic Motion; 10.6 The Period and Frequency for SHM

32. The period of oscillation of a spring-and-mass system is 0.50 s and the amplitude is 5.0 cm. What is the magnitude of the acceleration at the point of maximum extension of the spring?

33. A sewing machine needle moves with a rapid vibratory motion, rather like SHM, as it sews a seam. Suppose the needle moves 8.4 mm from its highest to its lowest position and it makes 24 stitches in 9.0 s. What is the maximum needle speed?

34. The prong of a tuning fork moves back and forth when it is set into vibration. The distance the prong moves between its extreme positions is 2.24 mm. If the frequency of the tuning fork is 440.0 Hz, what are the maximum velocity and the maximum acceleration of the prong? Assume SHM.

35. The period of oscillation of an object in an ideal spring-and-mass system is 0.50 s and the amplitude is 5.0 cm. What is the speed at the equilibrium point?

36. Five ideal mass-spring systems are described by their masses, spring constants, and amplitudes of oscillation as follows. Rank them in decreasing order of the frequency of oscillations.
(a) mass m, spring constant k, amplitude A
(b) mass $2m$, spring constant k, amplitude A
(c) mass m, spring constant k, amplitude $2A$
(d) mass $2m$, spring constant $k/2$, amplitude A
(e) mass $2m$, spring constant $2k$, amplitude $A/2$

37. Five ideal mass-spring systems are described in Problem 36. Rank them in decreasing order of their total energy.

38. The frequency of vibration of a lithotriptor, an ultrasound generator used to destroy kidney stones, is 1.0 MHz. (a) What is the period of vibration? (b) What is the angular frequency?

39. The human eardrum responds to sound by vibrating. If the eardrum moves in simple harmonic motion at a frequency of 4.0 kHz and an amplitude of 0.10 nm (roughly the diameter of a single hydrogen atom), what is its maximum speed of vibration? (Amazingly, the ear can detect vibrations with amplitudes even smaller than this!)

40. The air pressure variations in a sound wave cause the eardrum to vibrate. (a) For a given vibration amplitude, are the maximum velocity and acceleration of the eardrum greatest for high-frequency sounds or low-frequency sounds? (b) Find the maximum velocity and acceleration of the eardrum for vibrations of amplitude 1.0×10^{-8} m at a frequency of 20.0 Hz. (c) Repeat (b) for the same amplitude but a frequency of 20.0 kHz.

41. A mass oscillates up and down between $y = +A$ and $y = -A$ at the end of a stretched spring. (a) At what point(s) is the kinetic energy maximum? (Give the value(s) of y.) (b) At what point(s) is the gravitational potential energy maximum? (c) At what point(s) is the elastic potential energy maximum? (d) At what point(s) is the total potential energy (gravitational + elastic) *minimum*?

42. A 170 g object on a spring oscillates left to right on a frictionless surface with a frequency of 3.00 Hz and an amplitude of 12.0 cm. (a) What is the spring constant? (b) If the object starts at $x = 12.0$ cm at $t = 0$ and the equilibrium point is at $x = 0$, what equation describes its position as a function of time?

43. Show that, for SHM, the maximum displacement, velocity, and acceleration are related by $v_m^2 = a_m A$.

44. An empty cart, tied between two ideal springs, oscillates with $\omega = 10.0$ rad/s. A load is placed in the cart, making the total mass 4.0 times what it was before. What is the new value of ω?

45. A cart with mass m is attached between two ideal springs, each with the same spring constant k. Assume that the cart can oscillate without friction. (a) When the cart is displaced by a small distance x from its equilibrium position, what force magnitude acts on the cart? (b) What is the angular frequency, in terms of m, x, and k, for this cart? (connect tutorial: cart between springs)

Problems 44 and 45

46. In a playground, a wooden horse is attached to the ground by a stiff spring. When a 24 kg child sits on the horse, the spring compresses by 28 cm. With the child sitting on the horse, the spring oscillates up and down with a frequency of 0.88 Hz. What is the oscillation frequency of the spring when no one is sitting on the horse?

47. A small bird's wings can undergo a maximum displacement amplitude of 5.0 cm (distance from the tip of the wing to the horizontal). If the maximum acceleration of the wings is 12 m/s^2, and we assume the wings are undergoing simple harmonic motion when beating, what is the oscillation frequency of the wing tips?

48. Equipment to be used in airplanes or spacecraft is often subjected to a shake test to be sure it can withstand the vibrations that may be encountered during flight. A radio receiver of mass 5.24 kg is set on a platform that vibrates in SHM at 120 Hz and with a maximum acceleration of 98 m/s^2 (= 10g). Find the radio's (a) maximum displacement, (b) maximum speed, and (c) the maximum net force exerted on it.

49. In an aviation test lab, pilots are subjected to vertical oscillations on a shaking rig to see how well they can recognize objects in times of severe airplane vibration. The frequency can be varied from 0.02 to 40.0 Hz and the amplitude can be set as high as 2 m for low frequencies. What are the maximum velocity and acceleration to which the pilot is subjected if the frequency is set at 25.0 Hz and the amplitude at 1.00 mm?

50. The diaphragm of a speaker has a mass of 50.0 g and responds to a signal of frequency 2.0 kHz by moving back and forth with an amplitude of 1.8×10^{-4} m at that frequency. (a) What is the maximum force acting on the diaphragm? (b) What is the mechanical energy of the diaphragm?

51. An ideal spring has a spring constant $k = 25$ N/m. The spring is suspended vertically. A 1.0 kg object is attached to the unstretched spring and released. (a) What is the magnitude of the acceleration when the extension of the spring is a maximum? (b) What is the maximum extension of the spring?

52. An ideal spring with a spring constant of 15 N/m is suspended vertically. An object of mass 0.60 kg is attached to the unstretched spring and released. (a) What is the extension of the spring when the speed is a maximum? (b) What is the maximum speed?

53. A 0.50 kg object, suspended from an ideal spring of spring constant 25 N/m, is oscillating vertically. How much change of kinetic energy occurs while the object moves from the equilibrium position to a point 5.0 cm lower?

54. A small rowboat has a mass of 47 kg. When a 92 kg person gets into the boat, the boat floats 8.0 cm lower in the water. If the boat is then pushed slightly deeper in the water, it will bob up and down with simple harmonic motion (neglecting any friction). What will the period of oscillation be for the boat as it bobs around its equilibrium position?

55. A baby jumper consists of a cloth seat suspended by an elastic cord from the lintel of an open doorway. The unstretched length of the cord is 1.2 m, and the cord stretches by 0.20 m when a baby of mass 6.8 kg is placed into the seat. The mother then pulls the seat down by 8.0 cm and releases it. (a) What is the period of the motion? (b) What is the maximum speed of the baby?

10.7 Graphical Analysis of SHM

56. The displacement of an object in SHM is given by $y(t) = (8.0 \text{ cm}) \sin [(1.57 \text{ rad/s})t]$. What is the frequency of the oscillations?

57. An object is attached to an ideal spring of spring constant 2.5 N/m. The spring is initially in its relaxed position. The object is then released and oscillates about its equilibrium position. The motion is described by

$$y = (4.0 \text{ cm}) \sin [(0.70 \text{ rad/s})t]$$

What is the maximum kinetic energy?

58. An object of mass 306 g is attached to the base of a spring, with spring constant 25 N/m, that is hanging from the ceiling. A pen is attached to the back of the object, so that it can write on a paper placed behind the mass-spring system. Ignore friction. (a) Describe the pattern traced on the paper if the object is held at the point where the spring is relaxed and then released at $t = 0$. (b) The experiment is repeated, but now the paper moves to the left at constant speed as the pen writes on it. Sketch the pattern traced on the paper. Imagine that the paper is long enough that it doesn't run out for several oscillations.

59. (a) Sketch a graph of $x(t) = A \sin \omega t$ (the position of an object in SHM that is at the equilibrium point at $t = 0$). (b) By analyzing the slope of the graph of $x(t)$, sketch a graph of $v_x(t)$. Is $v_x(t)$ a sine or cosine function? (c) By analyzing the slope of the graph of $v_x(t)$, sketch $a_x(t)$. (d) Verify that $v_x(t)$ is $\frac{1}{4}$ cycle ahead of $x(t)$ and that $a_x(t)$ is $\frac{1}{4}$ cycle ahead of $v_x(t)$. (connect tutorial: sinusoids)

60. A ball is dropped from a height h onto the floor and keeps bouncing. No energy is dissipated, so the ball regains the original height h after each bounce. Sketch the graph for $y(t)$ and list several features of the graph that indicate that this motion is *not* SHM.

61. A 230.0 g object on a spring oscillates on a frictionless horizontal surface with frequency 2.00 Hz and amplitude 8.00 cm. Its position as a function of time is given by $x = A \sin \omega t$. (a) Sketch a graph of the elastic potential energy as a function of time. (b) Graph the system's kinetic energy as a function of time. (c) Graph the sum of the kinetic energy and the potential energy as a function of time. (d) Describe qualitatively how your answers would change if the surface weren't frictionless.

62. ✦ (a) Given that the position of an object is $x(t) = A \cos \omega t$, show that $v_x(t) = -\omega A \sin \omega t$. [*Hint*: Draw the velocity vector for point P in Fig. 10.18b and then find its x-component.] (b) Verify that the expressions for $x(t)$ and $v_x(t)$ are consistent with energy conservation. [*Hint*: Use the trigonometric identity $\sin^2 \omega t + \cos^2 \omega t = 1$.]

10.8 The Pendulum

63. What is the period of a pendulum consisting of a 6.0 kg mass hanging from a 4.0 m long string?

64. A pendulum of length 75 cm and mass 2.5 kg swings with a mechanical energy of 0.015 J. What is the amplitude?

65. A 0.50 kg mass is suspended from a string, forming a pendulum. The period of this pendulum is 1.5 s when the amplitude is 1.0 cm. The mass of the pendulum is now reduced to 0.25 kg. What is the period of oscillation now, when the amplitude is 2.0 cm? (connect tutorial: change in period)

66. A bob of mass m is suspended from a string of length L, forming a pendulum. The period of this pendulum is 2.0 s. If the pendulum bob is replaced with one of mass $\frac{1}{3}m$ and the length of the pendulum is increased to $2L$, what is the period of oscillation?

67. Each of five pendulums has a bob of mass m suspended from a string of length L. Rank them in order of their frequency for small-amplitude oscillations, greatest to smallest.

(a) $m = 300$ g, $L = 1.10$ m
(b) $m = 330$ g, $L = 1.10$ m
(c) $m = 330$ g, $L = 1.00$ m
(d) $m = 330$ g, $L = 1.21$ m
(e) $m = 300$ g, $L = 1.21$ m

68. A pendulum (mass m) moves according to $x = A \sin \omega t$. (a) Write the equation for $v_x(t)$ and sketch one cycle of the $v_x(t)$ graph. (b) What is the maximum kinetic energy?

69. A clock has a pendulum that performs one full swing every 1.0 s (back *and* forth). The object at the end of the pendulum weighs 10.0 N. What is the length of the pendulum?

70. A pendulum of length L_1 has a period $T_1 = 0.950$ s. The length of the pendulum is adjusted to a new value L_2 such that $T_2 = 1.00$ s. What is the ratio L_2/L_1?

71. A pendulum clock has a period of 0.650 s on Earth. It is taken to another planet and found to have a period of 0.862 s. The change in the pendulum's length is negligible. (a) Is the gravitational field strength on the other planet greater than or less than that on Earth? Explain. (b) Find the gravitational field strength on the other planet.

72. A grandfather clock is constructed so that it has a simple pendulum that swings from one side to the other, a distance of 20.0 mm, in 1.00 s. What is the maximum speed of the pendulum bob? Use two different methods. First, assume SHM and use the relationship between amplitude and maximum speed. Second, use energy conservation.

73. Max's arm weighs 34 N and is 63 cm long. If Max lets his relaxed arm swing back and forth like a pendulum, what is the period? Model the arm as a uniform rod.

74. The pendulum for a grandfather clock consists of a thin rigid rod (of negligible mass) with two metal disks attached to it. One disk has mass 0.600 kg and is attached to the rod 0.800 m from the pivot; the other has mass 0.750 kg and is attached 1.20 m from the pivot. Treating the disks as point masses, find the period of the pendulum.

75. ✦ Christy has a grandfather clock with a pendulum that is 1.000 m long. (a) If the pendulum is modeled as a simple pendulum, what would be the period? (b) Christy observes the actual period of the clock, and finds that it is 1.00% faster than that for a simple pendulum that is 1.000 m long. If Christy models the pendulum as two objects, a 1.000 m uniform thin rod and a point mass located 1.000 m from the axis of rotation, what percentage of the total mass of the pendulum is in the uniform thin rod?

76. ✦ A pendulum of length 120 cm swings with an amplitude of 2.0 cm. Its mechanical energy is 5.0 mJ. What is the mechanical energy of the same pendulum when it swings with an amplitude of 3.0 cm?

10.9 Damped Oscillations

77. (a) What is the energy of a pendulum ($L = 1.0$ m, $m = 0.50$ kg) oscillating with an amplitude of 5.0 cm? (b) The pendulum's energy loss (due to damping) is replaced in a clock by allowing a 2.0 kg mass to drop 1.0 m in 1 week. What average percentage of the pendulum's energy is lost during one cycle?

78. The amplitude of oscillation of a pendulum decreases by a factor of 20.0 in 120 s. By what factor has its energy decreased in that time?

79. Because of dissipative forces, the amplitude of an oscillator decreases 5.00% in 10 cycles. By what percentage does its *energy* decrease in ten cycles?

Collaborative Problems

80. When a steel cable supports a heavy load of weight W, its length increases by an amount ΔL compared with its length without a load. Suppose the cable is cut into three equal pieces, and the three resulting cables are used side by side to support the same load. How much is each of the three cables stretched?

81. Ⓒ Martin caught a fish and wanted to know how much it weighed, but he didn't have a scale. He did, however, have a stopwatch, a spring, and a 4.90 N weight. He attached the weight to the spring and found that the spring would oscillate 20 times in 65 s. Next he hung the fish on the spring and found that it took 220 s for the spring to oscillate 20 times. (a) Before answering part (b), determine whether the fish weighs more or less than 4.90 N. Explain your reasoning. (b) What is the weight of the fish?

82. Ⓒ The maximum height of a cylindrical column is limited by the compressive strength of the material; if the compressive stress at the bottom were to exceed the compressive strength of the material, the column would be crushed under its own weight. (a) For a cylindrical column of height h and radius r, made of material of density ρ, calculate the compressive stress at the bottom of the column. (b) Since the answer to part (a) is independent of the radius r, there is an absolute limit to the height of a cylindrical column, regardless of how wide it is. For marble, which has a density of 2.7×10^3 kg/m^3 and a compressive strength of 200 MPa, find the maximum height of a cylindrical column. (c) Is this limit a practical concern in the construction of marble columns?

83. Ⓒ A bungee jumper leaps from a bridge and undergoes a series of oscillations. Assume $g = 9.78$ m/s^2. (a) If a 60.0 kg jumper uses a bungee cord that has an unstretched length of 33.0 m and she jumps from a height of 50.0 m above a river, coming to rest just a few centimeters above the water surface on the first downward descent, what is the period of the oscillations? Assume the bungee cord follows Hooke's law. (b) The next jumper in line has a mass of 80.0 kg. Should he jump using the same cord? Explain.

84. Ⓒ You have a simple pendulum and a mass-spring system in which the mass oscillates vertically. They both oscillate with the same period T. You take them both to the surface of the Moon, where the gravitational field is 1/6 that of Earth. (a) Is the period of the simple pendulum on the Moon greater than, equal to, or less than T? Explain. (b) Find the ratio of the pendulum's period on the Moon to its period on Earth (T_M/T). (c) Is the period of the mass-spring system on the Moon greater than, equal to, or less than T? Explain. (d) Find the ratio of period on the Moon to the period on Earth (T_M/T).

Comprehensive Problems

85. Four people sit in a car. The masses of the people are 45 kg, 52 kg, 67 kg, and 61 kg. The car's mass is 1020 kg. When the car drives over a bump, its springs cause an oscillation with a frequency of 2.00 Hz. What would the frequency be if only the 45 kg person were present?

86. A pendulum passes $x = 0$ with a speed of 0.50 m/s; it swings out to $A = 0.20$ m. What is the period T of the pendulum? (Assume the amplitude is small.)

87. Ⓒ What is the length of a simple pendulum whose horizontal position is described by

$$x = (4.00 \text{ cm}) \cos [(3.14 \text{ rad/s})\ t]?$$

What assumption do you make when answering this question?

88. A naval aviator had to eject from her plane before it crashed at sea. She is rescued from the water by helicopter and dangles from a cable that is 45 m long while being carried back to the aircraft carrier. What is the period of her vibration as she swings back and forth while the helicopter hovers over her ship?

89. An object of mass m is hung from the base of an ideal spring that is suspended from the ceiling. The spring has a spring constant k. The object is pulled down a distance D from equilibrium and released. Later, the same system is set oscillating by pulling the object down a distance $2D$ from equilibrium and then releasing it. (a) How do the period and frequency of oscillation change when the initial displacement is increased from D to $2D$? (b) How does the total energy of oscillation change when the initial displacement is increased from D to $2D$? Give the answer as a numerical ratio. (c) The mass-spring system is set into oscillation a third time. This time the object is pulled down a distance of $2D$ and then given a push downward some more, so that it has an initial speed v_i downward. How do the period and frequency of oscillation compare to those you found in part (a)? (d) How does the total energy compare to when the object was released from rest at a displacement $2D$?

90. A spider's web can undergo SHM when a fly lands on it and displaces the web. For simplicity, assume that a web obeys Hooke's law (which it does not really as it deforms permanently when displaced). If the web is initially horizontal and a fly landing on the web is in equilibrium when it displaces the web by 0.030 mm, what is the frequency of oscillation when the fly lands?

91. ✦ Spider silk has a Young's modulus of 4.0 GPa and can withstand stresses up to 1.4 GPa. A single web strand has a cross-sectional area of 1.0×10^{-11} m^2, and a web is made up of 50 radial strands. A bug lands in the center of a horizontal web so that the web stretches downward. (a) If the maximum stress is exerted on each strand, what angle θ does the web make with the horizontal? (b) What does the mass of a bug have to be in order to exert this maximum stress on the web? (c) If the web is 0.10 m in radius, how far down does the web extend?

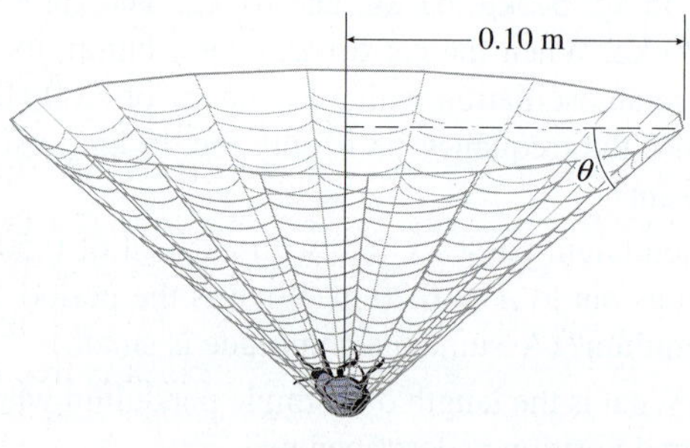

92. A mass-spring system oscillates so that the position of the mass is described by $x = (-10 \text{ cm}) \cos [(1.57 \text{ rad/s})t]$. Make a plot that has a dot for the position of the mass at $t = 0$, $t = 0.2$ s, $t = 0.4$ s, . . . , $t = 4$ s. The time interval between each dot should be 0.2 s. From your plot, tell where the mass is moving fastest and where it is moving slowest. How do you know?

93. A hedge trimmer has a blade that moves back and forth with a frequency of 28 Hz. The blade motion is converted from the rotation provided by the electric motor to an oscillatory motion by means of a Scotch yoke (see Conceptual Question 7). The blade moves 2.4 cm during each stroke. Assuming that the blade moves with SHM, what are the maximum speed and maximum acceleration of the blade?

94. The simple pendulum can be thought of as a special case of the physical pendulum where all of the mass is at a distance L from the rotation axis. For a simple pendulum of mass m and length L, show that the expression for the period of a physical pendulum [Eq. (10-27)] reduces to the expression for the period of a simple pendulum [Eq. (10-26b)].

95. Luke is trying to catch a pesky animal that keeps eating vegetables from his garden. He is building a trap and needs to use a spring to close the door to his trap. He has a spring in his garage, and he wants to determine the spring constant of the spring. To do this, he hangs the spring from the ceiling and measures that it is 20.0 cm long. Then he hangs a 1.10 kg brick on the end of the spring, and it stretches to 31.0 cm. (a) What is the spring constant of the spring? (b) Luke now pulls the brick 5.00 cm from the equilibrium position to watch it oscillate. What is the maximum speed of the brick? (c) When the displacement is 2.50 cm from the equilibrium position, what is the speed of the brick? (d) How long will it take for the brick to oscillate five times?

96. A 4.0 N object is attached to the bottom of an ideal spring of spring constant 250 N/m. The spring is initially in its relaxed position. Write an equation to describe the motion of the object if it is released at $t = 0$. [*Hint*: Let $y = 0$ at the equilibrium point and take $+y =$ up.]

97. A mass-spring system (mass m, spring constant k) oscillates with amplitude A. Show, using dimensional analysis alone, that the frequency f is independent of the amplitude A and is proportional to $\sqrt{k/m}$, assuming that m, k, and A are the only relevant quantities.

98. A horizontal spring with spring constant of 9.82 N/m is attached to a block with a mass of 1.24 kg that sits on a frictionless surface. When the block is 0.345 m from its equilibrium position, it has a speed of 0.543 m/s. (a) What is the maximum displacement of the block from the equilibrium position? (b) What is the maximum speed of the block? (c) When the block is 0.200 m from the equilibrium position, what is its speed?

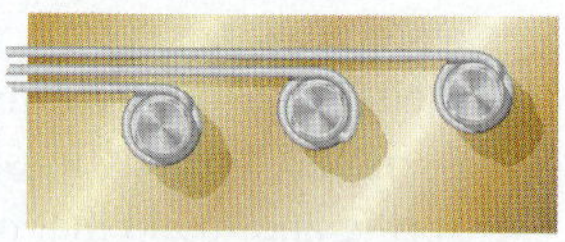

Problems 99 and 100

99. A steel piano wire ($Y = 200$ GPa) has a diameter of 0.80 mm. At one end it is wrapped around a tuning pin of diameter 8.0 mm. The length of the wire (not including the wire wrapped around the tuning pin) is 66 cm. Initially, the tension in the wire is 381 N. To tune the wire, the tension must be increased to 402 N. Through what angle must the tuning pin be turned?

100. When the tension is 402 N, what is the tensile stress in the piano wire in Problem 99? How does that compare with the elastic limit of steel piano wire (826 MPa)?

101. A tightrope walker who weighs 640 N walks along a steel cable. When he is halfway across, the cable makes an angle of 0.040 rad below the horizontal. (a) What is the strain in the cable? Assume the cable is horizontal with a tension of 80 N before he steps onto it. Ignore the weight of the cable itself. (b) What is the tension in the cable when the tightrope walker is standing at the midpoint? (c) What is the cross-sectional area of the cable? (d) Has the cable been stretched beyond its elastic limit (250 MPa)?

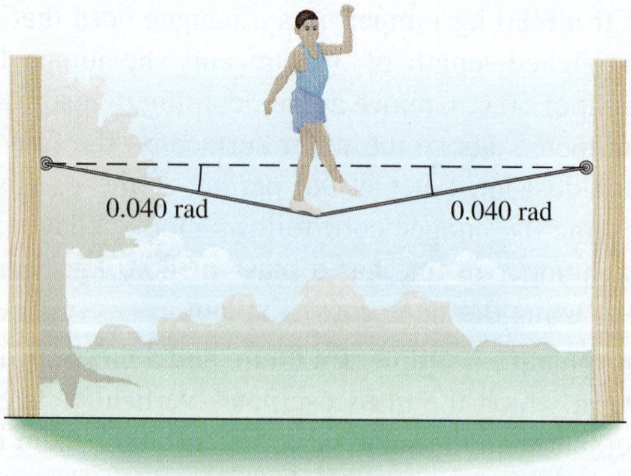

Problem 101 (The 0.040 rad angles are greatly exaggerated.)

102. A gibbon, hanging onto a horizontal tree branch with one arm, swings with a small amplitude. The gibbon's CM is 0.40 m from the branch, and its rotational inertia divided by its mass is $I/m = 0.25\ \text{m}^2$. Estimate the frequency of oscillation.

103. ✦ What is the period of a pendulum formed by placing a horizontal axis (a) through the end of a meterstick (100 cm mark)? (b) through the 75 cm mark? (c) through the 60 cm mark?

104. ✦ The gravitational potential energy of a pendulum is $U = mgy$. (a) Taking $y = 0$ at the lowest point, show that $y = L(1 - \cos\theta)$, where θ is the angle the string makes with the vertical. (b) If θ is small, $(1 - \cos\theta) \approx \frac{1}{2}\theta^2$ and $\theta \approx x/L$ (Appendix A.7). Show that the potential energy can be written $U \approx \frac{1}{2}kx^2$ and find the value of k (the equivalent of the spring constant for the pendulum).

105. ✦ A pendulum is made from a uniform rod of mass m_1 and a small block of mass m_2 attached at the lower end. (a) If the length of the pendulum is L and the oscillations are small, find the period of the oscillations in terms of m_1, m_2, L, and g. (b) Check your answer to part (a) in the two special cases $m_1 \gg m_2$ and $m_1 \ll m_2$.

106. ✦ A thin circular hoop is suspended from a knife edge. Its rotational inertia about the rotation axis (along the knife) is $I = 2mr^2$. Show that it oscillates with the same frequency as a simple pendulum of length equal to the diameter of the hoop.

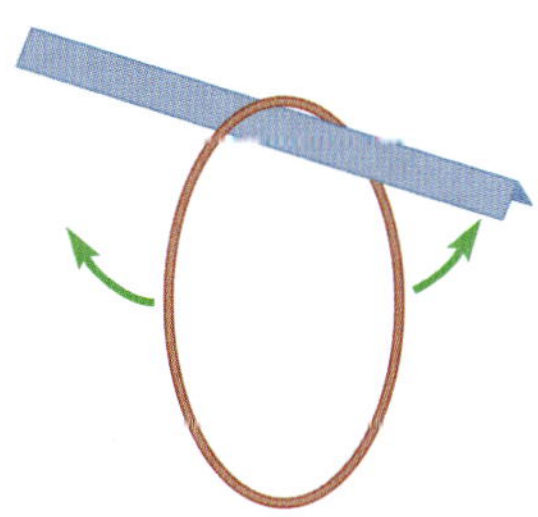

Review and Synthesis

107. By what percentage does the density of water increase at a depth of 1.0 km below the surface?

108. A mass-and-spring system oscillates with amplitude A and angular frequency ω. (a) What is the *average* speed during one complete cycle of oscillation? (b) What is the maximum speed? (c) Find the ratio of the average speed to the maximum speed. (d) Sketch a graph of $v_x(t)$ and refer to it to explain why this ratio is greater than $\frac{1}{2}$.

109. ✦ Ⓒ The motion of a simple pendulum is approximately SHM only if the amplitude is small. Consider a simple pendulum that is released from a horizontal position ($\theta_i = 90°$ in Fig. 10.23). (a) Using conservation of energy, find the speed of the pendulum bob at the bottom of its swing. Express your answer in terms of the mass m and the length L of the pendulum. Do *not* assume SHM. (b) Assuming (incorrectly, for such a large amplitude) that the motion *is* SHM, determine the maximum speed of the pendulum. Based on your answers, is the period of a pendulum for large amplitudes larger or smaller than that given by Eq. (10-26b)?

110. Ⓒ To escape a burning building, Arnold has to jump from a third-story window that is about 10 m above the ground. Arnold is worried about breaking his leg. The largest bone in Arnold's leg is the femur, which has a minimum cross-sectional area of about 5×10^{-4} m and a maximum ultimate strength for compression of about $1.70 \times 10^8\ \text{N/m}^2$. Arnold has a mass of 82 kg. (a) If Arnold lands on the ground with his legs stiff, then his femur can compress only about 5 mm. What will happen to Arnold's femur? (b) Suppose instead of landing on the ground, Arnold lands in deep snow so his legs can move about 30 cm between the time they first hit the snow and the time he comes to a complete stop. What will happen to Arnold's femur in this case?

111. A 5.0 kg block of wood is attached to a spring with a spring constant of 150 N/m. The block is free to slide on a horizontal frictionless surface once the spring is stretched and released. A 1.0 kg block of wood rests on top of the first block. The coefficient of static friction between the two blocks of wood is 0.45. What is the maximum speed that this set of blocks can have as it oscillates if the top block of wood is not to slip?

112. At a grocery store, a spring scale (spring constant = 450 N/m) hangs near the produce section. The spring hangs vertically with a 0.250 kg pan suspended from its lower end. Jenna drops a 2.20 kg bag of oranges from a height of 30.0 cm above the pan. The pan and oranges start oscillating vertically in SHM. (a) What is the velocity of the pan immediately after the oranges land on the pan? Assume a perfectly inelastic collision. (b) How far is the new equilibrium point of the pan (with oranges) below its position before the oranges were dropped on it? (c) What is the amplitude of the oscillations? (d) What is the frequency of the oscillations?

113. A spherical balloon with a radius of 12.0 cm is filled with helium. The bottom of the balloon is attached to a 2.30 m length of ribbon that is anchored to the ground. The balloon alone has a mass of 2.80×10^{-3} kg. Ignore the mass of the ribbon. (a) What is the tension in the ribbon? (b) After the balloon is displaced slightly to the side from its equilibrium position, it oscillates back and forth like an inverted pendulum. What is the period of oscillation? Ignore friction and air resistance.

Answers to Practice Problems

10.1 $2k$ (When the original spring is stretched an amount L, each of the half-springs stretches only $\frac{1}{2}L$. Each of the newly formed springs stretches half as far as the original spring for a given applied force.)

10.2 1.4×10^{-5}

10.3 0.18 mm

10.4 130 MPa

10.5 $-\frac{\Delta P}{B} = -\frac{1.0 \times 10^7 \text{ Pa}}{2.3 \times 10^9 \text{ Pa}} = -0.0043 = \frac{\Delta V}{V}$

and $\Delta V = -0.43\% \times V$

10.6 110 m/s^2 at $x = \pm A$

10.7 $K = 0$, $U_e = 2(mg)^2/k$, $U_g = -(mg)^2/k$, $E = (mg)^2/k$

10.8

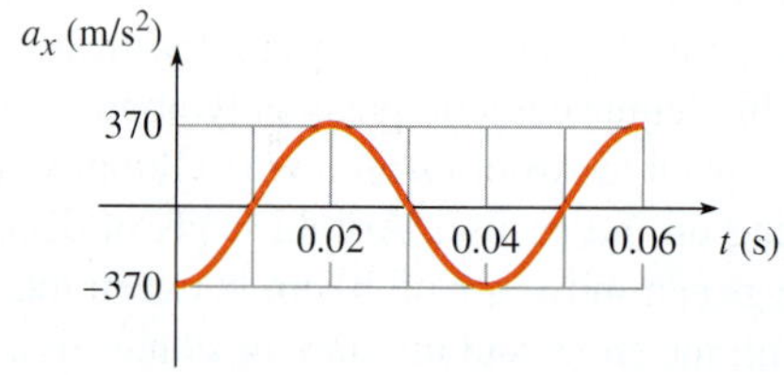

$a_x(t) = -a_m \cos \omega t$, where $\omega = 160$ rad/s and $a_m = 370$ m/s^2.

10.9 1.6 m/s^2 (about 1/6 that of the Earth)

10.10 0.82 m/s or 1.8 mi/h

Answers to Checkpoints

10.2 The two wires are under the same stress (same tensile force and same cross-sectional area). Young's modulus for steel is about $\frac{5}{3}$ times that for copper, so the *strain* for the steel wire is $\frac{3}{5}$ the strain of the copper wire. However, the strain is the *fractional* length change. The steel wire is twice as long, so its length change is 2 × (3/5) times the length change of the copper wire. The steel wire stretches more.

10.3 (a) Young's modulus is the constant of proportionality between stress and strain: stress = Y × strain. Therefore, Y is the slope of the linear part of the stress versus strain graph. Material A has the larger slope so its Young's modulus is larger. (b) The ultimate strength is the largest stress the material can withstand. The graph for material B reaches a larger stress, so it has the higher ultimate strength.

10.5 When the kinetic and potential energies are equal, each is half of the total energy. When $U = \frac{1}{2}kx^2 = \frac{1}{2}E_{\text{total}} = \frac{1}{2}(\frac{1}{2}kA^2)$, $x = \pm A/\sqrt{2}$.

10.6 0.50 Hz

10.7 (a) When the displacement is zero, the potential energy has its minimum value. From conservation of energy, the kinetic energy then has its maximum value. Therefore, the speed has its maximum magnitude ($v = \pm v_m$), as shown in Fig. 10.20. (b) When the speed is zero, the kinetic energy is minimum and the potential energy is maximum. Therefore, the displacement has its maximum magnitude ($x = \pm A$).

10.8 Yes. For a simple pendulum of mass m at the end of a string of length L, the rotational inertia about the pivot is $I = mL^2$. Then

$$T = 2\pi\sqrt{\frac{I}{mgd}} = 2\pi\sqrt{\frac{mL^2}{mgL}} = 2\pi\sqrt{\frac{L}{g}}$$

which agrees with Eq. (10-26b).

CHAPTER

11

Waves

During the 1995 Hanshin earthquake in Japan, sections of an elevated highway collapsed while nearby buildings survived with little damage. What made the highway collapse, and how could it be modified to prevent a collapse in a future earthquake? (See p. 421 for the answer.)

BIOMEDICAL APPLICATIONS

- Sensitivity of the human ear and eye (Sec. 11.1, Prob. 1)
- Seismic waves used by animals to communicate and to monitor their environment (Sec. 11.2)
- Ultrasonography (Ex. 11.5, Prob. 44)

Concepts & Skills to Review

- period, frequency, angular frequency (Section 10.6)
- position, velocity, acceleration, and energy in SHM (Section 10.5)
- resonance (Section 10.10)
- graphical analysis of SHM (Section 10.7)

11.1 WAVES AND ENERGY TRANSPORT

Basic Models: Particles and Waves Physicists use only a few basic models to describe the physical world. One such model is the particle: a pointlike object with no inner structure and with certain characteristics such as mass and electric charge. Another basic model is the **wave**. Water waves are familiar examples. When a pebble is dropped into a pond, it disturbs the surface of the water. Ripples on the surface of the pond travel away from the spot where the pebble landed.

Examples of Waves Any wave is characterized as some sort of "disturbance" that travels away from its source. In Chapters 11 and 12, we concentrate on mechanical waves traveling through a material medium, such as water waves, sound waves, and the seismic waves caused by earthquakes. Particles in the medium are disturbed from their equilibrium positions as the wave passes, returning to their equilibrium positions after the wave has passed. In Chapter 22, we discuss electromagnetic waves such as radio waves and light waves, in which the disturbance consists of oscillating electromagnetic fields. Two of our five human senses are wave detectors: the ear is sensitive to the tiny fluctuations in air pressure caused by compressional waves in air (sound), and the eye is sensitive to electromagnetic waves in a certain frequency range (light).

Energy Transport by a Wave

CONNECTION:

In wave motion, energy is transferred from one oscillating particle to another. Energy is conserved overall, but the energy of any one oscillating particle can change. Mechanical waves carry the same kinds of energy as a simple harmonic oscillator: kinetic energy and potential energy.

Suppose we drop a pebble into a still pond. The kinetic energy of the pebble just before it hits the pond is partly converted into the energy carried off by the water wave. That waves carry energy is clear to anyone who has been surfing or swimming in the ocean. Speaking of surfing, information on the Internet is carried by waves of various sorts: electrical waves in wires, microwaves between Earth and communications satellites, light waves in optical fibers. Microwaves in ovens carry energy from their source to the food; the electromagnetic energy of the microwaves is absorbed by water molecules in the food and appears as thermal energy. Electromagnetic waves from the Sun bring the energy that fuels the growth of green plants. Seismic waves and tsunamis carry energy released by an earthquake far from the point of origin, sometimes with devastating results.

A wave can transmit energy from one point to another *without* transporting any matter between the two points (Fig. 11.1). The sound of thunder travels for kilometers in all directions, but none of the molecules in the air where lightning struck travels more than

Figure 11.1 Two different ways to transfer energy. (a) When a baseball pitcher throws a ball to the catcher, the ball carries energy with it. The pitcher gives the ball kinetic energy; the catcher receives the energy when the ball hits his hand and his hand recoils. (b) Suppose instead that they hold a rope stretched between them. If the pitcher suddenly moves his hand up and down quickly, a wave pulse travels along the rope until it reaches the catcher's hand. Once again, the pitcher sends the energy and the catcher receives it when the rope makes his hand recoil. However, in this case the pitcher is still holding his end of the rope; it never leaves his hand. Energy is transferred *without any matter moving from the pitcher to the catcher.*

a meter or so during the short time it takes the sound to reach our ears. Similarly, seismic waves and tsunamis can wreak havoc hundreds or even thousands of kilometers away from an earthquake without carrying any of the soil or water from their point of origin.

EVERYDAY PHYSICS DEMO

Stretch a heavy rope or belt between yourself and a friend and test out the transfer of energy from one to the other by sending wave pulses down the rope. Can you feel the energy transfer when the pulse arrives?

EVERYDAY PHYSICS DEMO

Observe carefully what happens when you snap your fingers. You start by pressing your thumb against your fingers and then sideways, the thumb in one direction and the fingers in the opposite direction. Initially friction keeps them from moving sideways, but suddenly they slip, releasing the built-up energy.

Similarly, the rocks on two sides of a fault line are pressed together and sideways. Friction keeps them from moving sideways as elastic (or strain) energy builds up. Then suddenly they slip, releasing a tremendous amount of energy largely in the form of seismic waves that carry vibrations far from the focus of the earthquake.

The bow of a stringed instrument such as a violin also uses a stick-slip mechanism to drive the string. The bow carries the string with it until the string suddenly slips, snapping back until the bow catches it again. The player has to carefully control bow speed and downward force on the string to get this to happen.

Intensity

For a wave that travels in a three-dimensional medium (such as sound waves or seismic waves traveling through the Earth), the **intensity** (symbol I, SI unit W/m^2) is a measure of the *average power per unit area* carried by the wave past a surface perpendicular to the wave's direction of propagation.

Intensity

$$I = P/A \tag{11-1}$$

Application: Sensitivity of the Human Ear If a sound wave's intensity is a fairly loud $I = 10^{-5}$ W/m^2 when it reaches the eardrum and the area of the eardrum is $A = 10^{-4}$ m^2, then the power delivered to the eardrum is $P = IA = 10^{-9}$ W (assuming that all the energy incident on the eardrum is absorbed). The energy absorbed by the eardrum at this rate in *one hour* would be

$$10^{-9}\ \text{W} \times 3600\ \text{s} \approx 4\ \mu\text{J}$$

The human ear is a very sensitive detector indeed.

Intensity and Distance from the Wave Source

For most waves, the intensity decreases as the distance from the source increases. Some of the energy can be absorbed (dissipated) by the wave medium. The amount of energy absorbed depends on the medium. Air absorbs relatively little sound energy, which is why we can hear sounds generated far away.

Another reason intensity decreases with distance is that, as the wave spreads out, the energy gets spread over a larger and larger area. Consider a point source emitting a wave uniformly in all directions—an *isotropic* source (Fig. 11.2). The average power

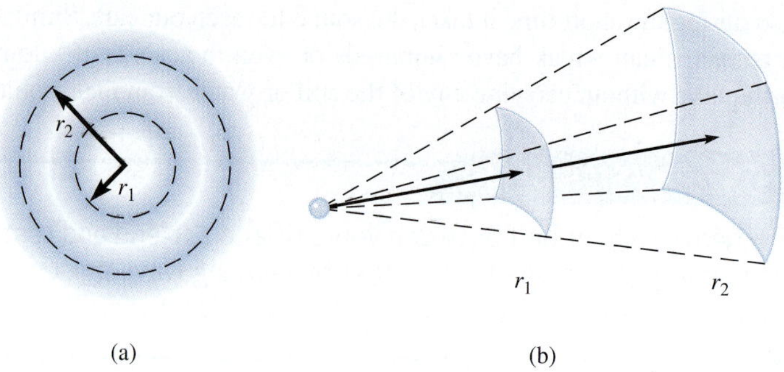

Figure 11.2 (a) A point source of sound radiating energy uniformly in all directions. (b) The intensity at a distance r_2 is smaller than the intensity at a distance r_1 since the same power is spread out over a greater area.

(energy per unit time) emitted is constant. Imagine a sphere surrounding the source; the rate at which energy passes through the surface of the sphere is the same no matter what the radius. The surface area of a sphere is $4\pi r^2$, so as the wave moves farther from the source, the energy spreads out over a larger and larger area. Thus, the power per unit area (intensity) decreases with distance. Assuming that no energy is absorbed by the medium and there are no obstacles to reflect or absorb sound,

Intensity for an isotropic source

$$I = \frac{P}{A} = \frac{P}{4\pi r^2} \quad (11\text{-}2)$$

(assuming no reflection or absorption)

Therefore, if energy absorption by the medium can be ignored, the intensity of the sound is inversely proportional to the square of the distance from the source. This "inverse square law" is the result of a conserved quantity (here, energy) radiating uniformly from a point source in three-dimensional space.

CHECKPOINT 11.1

A siren in a fire tower 20 m high generates a sound wave with intensity 0.090 W/m^2 at a point on the ground below the tower. What is the intensity of the sound wave 2.0 km from the tower? Assume the siren is an isotropic source.

11.2 TRANSVERSE AND LONGITUDINAL WAVES

A Slinky toy can be used to demonstrate two different kinds of waves. In a **transverse** wave, the motion of particles in the medium is perpendicular to the direction of propagation of the wave. To send a transverse wave down a Slinky, wiggle the end of the Slinky back and forth in a direction perpendicular to the length of the Slinky (Fig. 11.3a). In a **longitudinal** wave, the motion of particles in the medium is along the same line as the direction of propagation of the wave. To send a longitudinal wave down the Slinky, jiggle the end in and out along its length to alternately stretch and compress the coils (Fig. 11.3b). A red dot painted on one coil of the Slinky helps illustrate the difference. In a transverse wave, the dot moves back and forth about a fixed position with

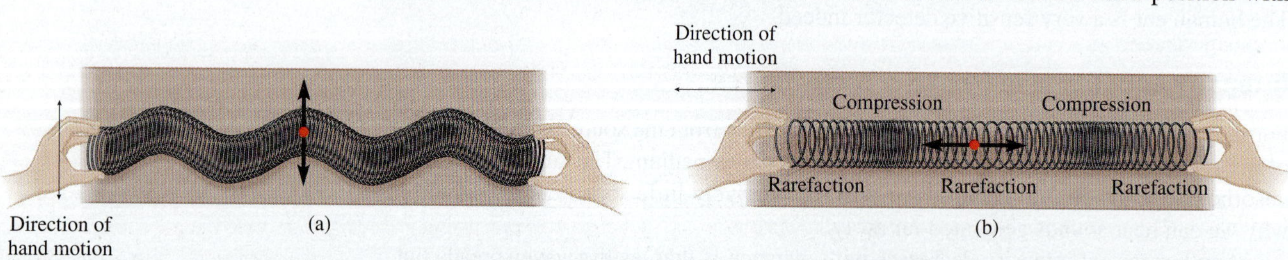

Figure 11.3 (a) Transverse and (b) longitudinal waves on a Slinky.

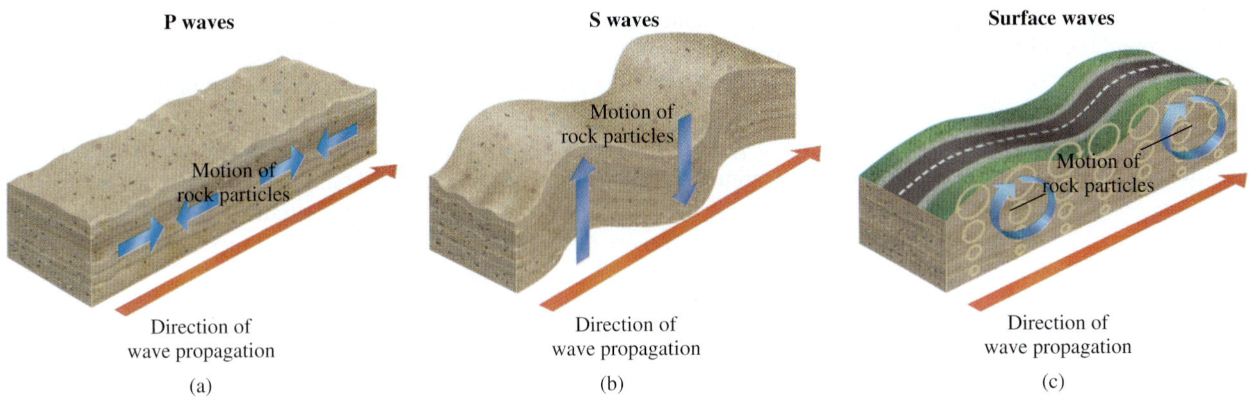

Figure 11.4 Three types of seismic waves. (a) Longitudinal body waves (*P waves*) are the fastest seismic waves (typically 4–8 km/s). They are similar to sound waves in air: particles in the Earth's interior are pushed together and pulled apart in the same direction that the wave propagates. (b) Transverse body waves (*S waves*) travel more slowly—typically 2–5 km/s. In an S wave, particles in the Earth's interior vibrate at right angles to the direction that the wave travels. By measuring the time between the first arrivals of these two types of waves at different detection stations, geologists are able to determine the point of origin of the earthquake. (c) In a surface wave, the motion of the ground combines longitudinal and transverse components.

its motion perpendicular to the direction of propagation of the wave; in a longitudinal wave, the dot also moves back and forth about a fixed position but along the direction of propagation of the wave. In both cases, the wave itself moves from one end of the Slinky to the other while the dot is moving about its fixed position.

The Slinky—or any long spring—is a better approximation to solid materials than the stretched rope. In solids both types of waves can exist; a transverse wave results from a shear disturbance and a longitudinal wave from a compressional disturbance. Therefore, seismic waves can be either longitudinal or transverse (Fig. 11.4).

Fluids can be compressed, but, because they flow, they do not sustain shear stresses. Therefore, longitudinal waves travel through fluids but transverse waves do not. However, gravity or surface tension can provide the transverse restoring force that allows a transverse wave to travel *along the surface* of a liquid.

A sound wave is longitudinal; each small volume of air vibrates back and forth along the direction of travel of the wave. Molecules are compressed together in some places and more thinly spaced (*rarefied*) in others; the air has regions of higher and lower density called **compressions** and **rarefactions** (see Fig. 11.3b).

CHECKPOINT 11.2

When an earthquake occurs, the S waves (transverse waves) are not detected on the opposite side of the Earth, but the P waves (longitudinal waves) are. How does this provide evidence that the Earth's solid core is surrounded by liquid?

Waves That Combine Transverse and Longitudinal Motion

Not all seismic waves are purely transverse or purely longitudinal. In a surface wave, the ground near the surface rolls approximately in a circle. Thus, the motion of the ground has components both parallel and perpendicular to the direction of propagation. The transverse component can either be up and down (as shown in Fig. 11.4c) or side to side. The motion of the ground is greatest at the surface.

Ocean waves are similar to the surface seismic wave shown in Fig. 11.4c. Deep underwater, the wave is mostly longitudinal (Fig. 11.5); as the wave passes, water moves back and forth along the direction of propagation of the wave. Higher up, the wave has both transverse and longitudinal components; water moves in an oval as the wave passes. Water near the surface moves approximately in a circle. The air above the surface

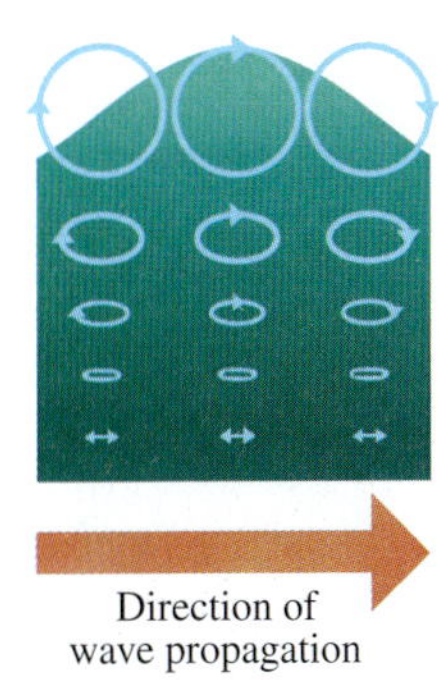

Figure 11.5 The motion of water in an ocean wave combines transverse and longitudinal motion.

presents little resistance, so water swells upward more easily there and then is pulled back downward by gravity (or, for small amplitudes, by surface tension). When the wave gets close to shore, the crest often collapses or *breaks*; the motion of the water particles is then much more complex.

When a guitar string is plucked *gently*, the wave on the string is almost purely transverse; stretching of the string is negligible. When it is plucked more forcefully, the resulting wave is a combination of transverse and longitudinal waves. At any instant, the string is stretched more in some places than in others; a point on the string has longitudinal motion as well as transverse motion.

The Puerto-Rican white-lipped frog (*Leptodactylus albilabris*) communicates using seismic waves.

Application: Animal Communication Using Seismic Waves Animals monitor their surroundings, locate prey, and recognize nearby predators by detecting small-amplitude seismic waves traveling through soil, plant stems, or leaves. Species known to be sensitive to seismic vibrations include snakes, frogs, toads, spiders, birds, elephants, kangaroo rats, worms, and a wide variety of insects. Of terrestrial vertebrates, particularly acute sensitivity to seismic waves has been found in frogs. This sensitivity is due to a special organ in the inner ear (the sacculus) and a set of muscles and bones that connect the inner ear to the pectoral girdle. Many insects have specialized organs in their legs to detect vibration. Some mammals, such as elephants and cats, have fat pads on their feet that are believed to help transmit vibrations to the brain.

Many species of animals generate seismic waves to communicate with members of the same species. Various techniques are used to produce vibrations, including drumming (rhythmically tapping or thumping the substrate with a body part), tremulation (whole-body vibration), and stridulation (rubbing together body parts). The resulting seismic waves are often species-specific and can be used to identify and court potential mates, to warn that a predator is near, to claim territory, or to coordinate activities of a social group. Evidence suggests that elephants may use seismic waves to communicate with other elephants over distances as great as 16 km.

11.3 SPEED OF TRANSVERSE WAVES ON A STRING

The speed of a mechanical wave depends on properties of the wave medium. What properties of a string determine the speed of a transverse wave moving along it? Suppose that a string of length L and mass m is under tension F. In Problem 96, you can show that $\sqrt{FL/m}$ is the only combination of those three quantities with the correct units for speed. There could be a dimensionless constant multiplier, but a derivation using more advanced mathematics shows that the constant is 1; the speed of a transverse wave on a string is

$$v = \sqrt{\frac{FL}{m}}$$

For a given string composition and diameter (say, a yellow brass string of 0.030 in. diameter), the mass of the string is proportional to its length. By defining the **linear mass density** (mass per unit length) of the string to be

$$\mu = \frac{m}{L} \tag{11-3}$$

the speed of a transverse wave on a string can be written

Speed of a transverse wave on a string

$$v = \sqrt{\frac{F}{\mu}} \tag{11-4}$$

Equation (11-4) shows that the wave speed depends on *local* properties of the medium; it does not depend on how much of the medium there is. The wave speed in the vicinity of some point P, for instance, does not depend on how long the string is; only properties of the string in the immediate vicinity of point P can determine how fast the wave travels past that point.

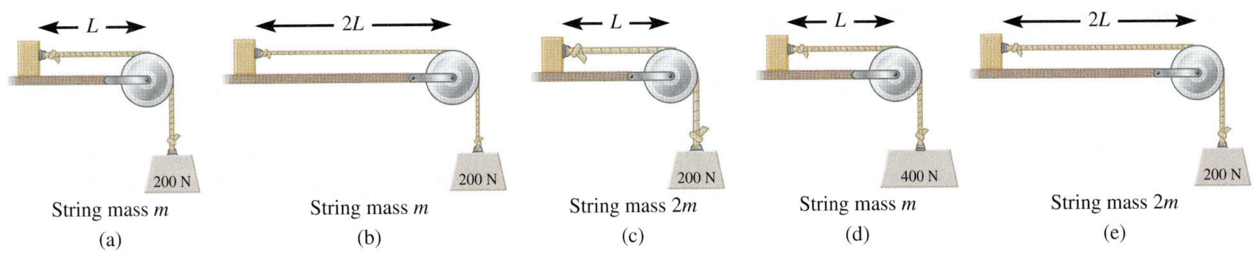

Figure 11.6 Five stretched strings.

Note that as tension increases, wave speed increases; as mass density increases, wave speed decreases. A somewhat more general way to think about it, applicable to other waves as well, is:

More restoring force makes faster waves; more inertia makes slower waves.

CHECKPOINT 11.3

Transverse waves travel on five stretched strings (Fig. 11.6). Rank the strings according to the speed of transverse waves, from largest to smallest.

The speed at which a wave propagates is not the same as the speed at which a particle in the medium moves. Suppose a horizontal string is stretched along the x-axis and a transverse pulse in the y-direction is sent down the string. The speed of propagation of the wave v is the speed at which the *pattern* or disturbance moves along the string (in the x-direction); for a uniform string, the wave speed is constant. A point on the string vibrates up and down in the $\pm y$-direction with a *different* speed that is *not* constant.

Example 11.1

A Piñata

A string of length 2.0 m has a mass of 125 mg. The string is attached to the ceiling and a piñata of mass 4.0 kg hangs from the other end. A child whacks the piñata sideways with a stick; as a result, a transverse pulse travels up the string toward the ceiling. At what speed does the pulse travel?

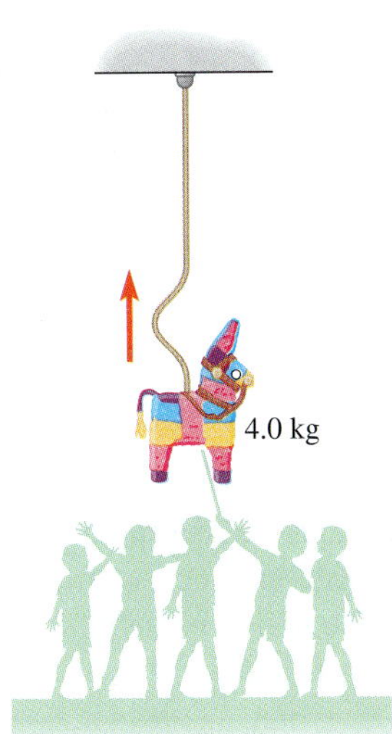

Strategy We start with a diagram of the situation (see the figure). The piñata puts the string under tension. The tension in the string is equal to the weight of the piñata because the weight of the string itself is negligible in comparison. The mass and length of the string are given, so the linear mass density can be found. Then we can find the wave speed.

Solution The speed of a transverse wave on a string is given by Eq. (11-4):

$$v = \sqrt{\frac{F}{\mu}}$$

where F is the tension in the string and μ is the linear mass density of the string. The tension is equal to the weight hanging on the string:

$$F = Mg$$

The linear mass density of the string is mass per unit length ($\mu = m/L$). Substituting the tension and mass density, we have

$$v = \sqrt{\frac{F}{m/L}} = \sqrt{\frac{(Mg)L}{m}}$$

$$= \sqrt{\frac{4.0 \text{ kg} \times 9.8 \text{ m/s}^2 \times 2.0 \text{ m}}{125 \times 10^{-6} \text{ kg}}} = 790 \text{ m/s}$$

continued on next page

Example 11.1 continued

Discussion The *weight* of the string (mg) is negligible in comparison with the weight hanging from the end of the string (Mg). That is not always the case, as can be seen in Practice Problem 11.1.

Practice Problem 11.1 Initial Velocity of Another Wave Pulse Traveling on a String

A string of length 10.0 m has a linear mass density of 25 g/m. The string is fixed at the top and has an object of mass 0.200 kg hanging from the bottom. (a) What is the *initial* wave speed of a pulse sent up the string from the bottom? (b) What is the speed of the pulse as it approaches the top of the string? [*Hint*: Does the weight of the string itself affect the tension in either case?]

11.4 PERIODIC WAVES

A **periodic** wave repeats the same pattern over and over, each repeating section transporting the energy that was used to generate it. A periodic water wave can be produced by steadily dropping a series of pebbles into the water; a wave on a cord can be produced by taking one end of the cord and moving it up and down, over and over, in a repeating pattern. As the wave propagates along the cord, every point on the cord oscillates with the same up and down pattern, though with a time delay that depends on the wave speed. Whereas musical sounds are often periodic waves, noise is *aperiodic*. The human voice makes a periodic sound wave when a vowel is sung at a steady pitch (constant frequency); most of the consonant sounds are aperiodic (Fig. 11.7).

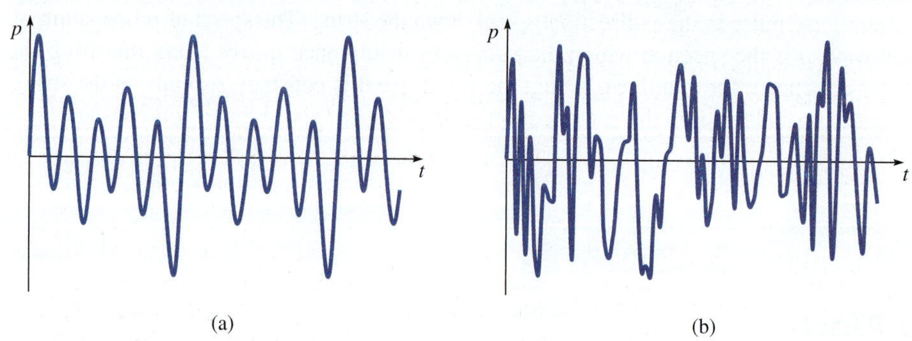

Figure 11.7 (a) Periodic sound wave pattern produced by singing the vowel /e/. (b) Aperiodic sound wave pattern produced by hissing the consonant /s/. (A microphone generates an electrical signal proportional to the pressure variations of the sound wave. The graphs show the pressure as a function of time.)

CONNECTION:

The terminology for periodic waves is similar to that used for uniform circular motion (Chapter 5) and for simple harmonic motion (Chapter 10).

Period, Frequency, Wavelength, and Amplitude At any given point in space, a periodic wave repeats itself after a time T called the **period**. The inverse of the period is the **frequency** f.

$$f = \frac{1}{T} \quad \text{(SI unit Hz = s}^{-1}\text{)} \tag{5-8}$$

The frequency tells how often the pattern of motion repeats itself at any single point. For instance, if the frequency is 20 Hz, then there are 20 repetitions, or cycles, per second. Each cycle takes a time $T = 1/f = 0.05$ s. The angular frequency is $\omega = 2\pi f$ and is measured in rad/s.

During one period T, a periodic wave traveling at speed v moves a distance vT. In Fig. 11.8, note that, at any instant, points separated by a distance vT along the direction of propagation of a wave move "in sync" with each other. Thus, vT is the *repetition distance* of the wave, just as the period is the *repetition time*. This distance is called the **wavelength** (symbol λ, the Greek letter lambda).

Wavelength

$$\lambda = vT \tag{11-5}$$

Combining this relation and the expression for frequency, we obtain

Frequency and wavelength

$$v = \frac{\lambda}{T} = f\lambda \qquad (11\text{-}6)$$

Equations (11-5) and (11-6) are true for all periodic waves, no matter how the wave is produced or what the shape of the wave.

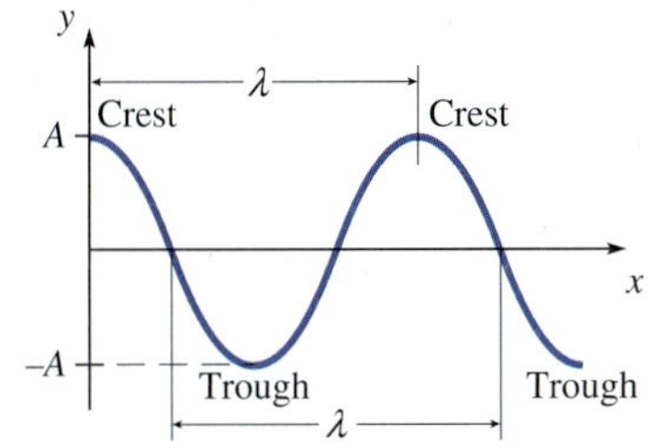

Figure 11.8 Snapshot graph of a sinusoidal wave moving with speed v in the x-direction. The graph shows the displacement y of particles in the wave medium as a function of x, their position along the direction of wave propagation, at one particular time t. The amplitude A and the wavelength λ are shown.

CHECKPOINT 11.4

A seismic wave travels at 4.0 km/s and has a wavelength of 20 km. How long does it take a rock particle to complete one cycle of oscillation?

The maximum displacement of any particle from its equilibrium position is the **amplitude** A of the wave. For a sinusoidal wave traveling along a stretched string in the x-direction, the amplitude A is the maximum displacement of a particle in the positive or negative y-direction. For surface water waves, the amplitude is the height of a crest (a high point) above or the depth of a trough (a low point) below the undisturbed water level.

11.5 MATHEMATICAL DESCRIPTION OF A WAVE

A wave is represented mathematically by a variation in some quantity (e.g., pressure or displacement) that is described as a function of both position and time. For a transverse wave on a guitar string, the function specifies the displacement of each point on the string from its equilibrium position. If the string is oriented along the x-axis and the displacement of any point on the string is in the $\pm y$-direction, then the wave is described by a function of two variables: $y(x, t)$. The notation $y(x, t)$ means that y is a *function* of x and t: the value of y depends on the values of x and t in such a way that only one value of y (the dependent variable) corresponds to a particular choice of x and t (the independent variables).

Traveling Waves

Consider a long stretched string along the x-axis. One end of the string (at $x = 0$) is moved by an external agent according to some function $y = h(t)$; as a result, a transverse wave is produced that travels in the $+x$-direction with wave speed v. A **traveling wave** retains the same shape as it moves in a single direction. For a traveling wave in the $+x$-direction, the motion of any point x on the string copies the motion of the left end after a time delay x/v (the time it takes the wave to travel a distance x at speed v—see Fig. 11.9). Thus, $y(x, t) = h(t - x/v)$. Even though the function that describes the wave has two variables (x and t), these variables must occur in the particular combination $(t - x/v)$ in order to describe a wave that retains its form as it propagates in the $+x$-direction. For a wave moving in the $-x$-direction, the variables would occur in the

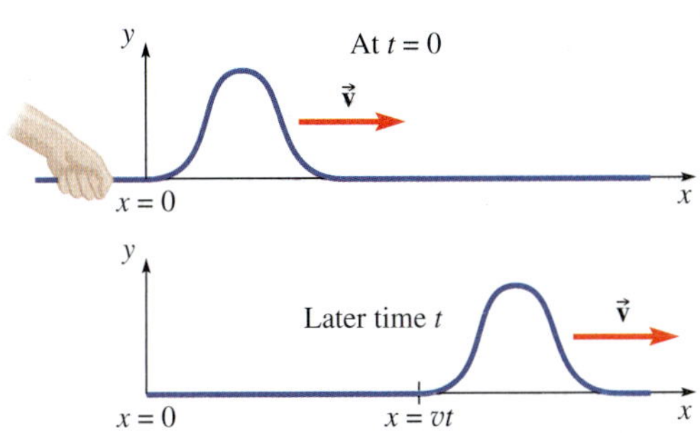

Figure 11.9 A wave pulse, with the same shape, at successive times. The motion of the point x repeats the motion of the point $x = 0$ with a time delay $\Delta t = x/v$.

combination $(t + x/v)$. Note that if the terms containing x and t have opposite signs, the wave moves in the $+x$-direction; if the terms containing x and t have the same signs, the wave moves in the $-x$-direction.

Harmonic Traveling Waves

Harmonic waves are a special kind of periodic wave in which the disturbance is sinusoidal (either a sine or cosine function). In a harmonic traveling wave on a string, every point on the string moves in SHM with the same amplitude and angular frequency, although different points reach their maximum displacements at different times. The maximum speed and maximum acceleration of a point on the string depend on both the angular frequency and the amplitude of the wave:

$$v_m = \omega A \quad (10\text{-}21)$$

$$a_m = \omega^2 A \quad (10\text{-}22)$$

In Eq. (10-21), v_m is the maximum speed at which a point on the string moves in the transverse ($\pm y$) direction. v_m is not the same as v, the speed of wave propagation in the $\pm x$-direction (see Section 11.3).

Intensity and Amplitude Since the total energy of an object moving in SHM is proportional to the amplitude squared (Section 10.5), the total energy of a harmonic wave is proportional to the square of its amplitude. Intensity is the rate at which a wave transports energy per unit area perpendicular to the direction of propagation. The intensity of a harmonic wave is proportional to its total energy and, therefore, is proportional to the square of the amplitude. That turns out to be a general result not limited to harmonic waves:

Intensity and Amplitude

The intensity of a wave is proportional to the square of its amplitude.

Equation of a Harmonic Traveling Wave Suppose the motion of the left end of the string moves in simple harmonic motion: $y = A \cos \omega t$. By substituting $(t - x/v)$ for t, we obtain the function that describes the motion of *any* point $x > 0$:

$$y(x, t) = A \cos [\omega(t - x/v)]$$

To simplify the writing, we introduce a constant called the **wavenumber** (symbol k, SI unit rad/m):

$$k = \frac{\omega}{v} = \frac{2\pi f}{v} = \frac{2\pi}{\lambda} \quad (11\text{-}7)$$

Then the equation for the harmonic wave can be written

$$y(x, t) = A \cos (\omega t - kx) \quad (11\text{-}8a)$$

The symbol k can stand for two different quantities: the wavenumber (SI unit rad/m) and the spring constant (SI unit N/m). Be careful not to mix them up.

CONNECTION:

Note the analogy between ω and k. $\omega = 2\pi/T$, where T is the repeat *time*; $k = 2\pi/\lambda$, where λ is the repeat *distance* ω is measured in radians per *second*; k is measured in radians per *meter*.
(connect tutorial: sine wave)

Equation (11-8a) is not the only way to write a harmonic wave. Simple harmonic motion can be written as a sine function instead of cosine; the only difference is the initial conditions (see Section 10.7). A wave can travel in either direction ($+x$ or $-x$). Equations (11-8b) through (11-8d) also describe harmonic waves.

$$y(x, t) = A \sin (\omega t - kx) \quad (11\text{-}8b)$$

$$y(x, t) = A \cos (\omega t + kx) \quad (11\text{-}8c)$$

$$y(x, t) = A \cos (\omega t + kx) \quad (11\text{-}8d)$$

CHECKPOINT 11.5

Which of Equations (11-8b) through (11-8d) describe harmonic waves moving in the $-x$-direction?

Example 11.2

A Traveling Wave on a String

A wave on a string is described by $y(x, t) = a \sin (bt + cx)$, where a, b, and c are positive constants. (a) Does this wave retain its shape as it travels? (b) In what direction does the wave travel? (c) What is the wave speed?

Strategy We try to manipulate the function to see if it can be written as a function of either $(t - x/v)$ or $(t + x/v)$ as in the general harmonic wave equation $y(x, t) = A \cos \omega (t - x/v)$. The wave speed v does not appear explicitly in the function as written, but it may be some combination of the other constants in the function.

Solution The coefficient of t in our equation should be the constant that represents ω. Factoring out that constant, we have

$$y(x, t) = a \sin b\left(t + \frac{cx}{b}\right) = a \sin b\left(t + \frac{x}{b/c}\right)$$

Now we see that $y(x, t)$ is a function of $t + x/v$, where $v = b/c$:

$$y(x, t) = a \sin b\left(t + \frac{x}{v}\right), \text{ where } v = \frac{b}{c}$$

Therefore: (a) yes, the wave retains its shape since it is a function of $(t + x/v)$; (b) it travels in the $-x$-direction since the t and x/v terms have the same sign; and (c) the wave speed is b/c.

Discussion Before being completely satisfied with this solution, it is a good idea to check that b/c has the right units for wave speed. The two terms bt and cx that are added together must have the same units. In SI, the argument of a sine function is measured in radians, b is measured in rad/s, and c is measured in rad/m. Then the units of b/c are (rad/s)/(rad/m) = m/s, which is correct for wave speed.

Practice Problem 11.2 Another Traveling Wave on a String

A wave on a string is described by

$$y(x, t) = (0.0050 \text{ m}) \sin [(4.0 \text{ rad/s})t - (0.50 \text{ rad/m})x]$$

(a) Does this wave retain its shape as it travels? (b) In what direction does the wave travel? (c) What is the wave speed?

11.6 GRAPHING WAVES

To graph a one-dimensional wave $y(x, t)$, only one of the two independent variables (x, t) can be plotted. The other must be "frozen"; it is treated as a constant. If x is held constant, then one particular point (determined by the value of x) is singled out; the graph shows the motion of *that point* as a function of time (Fig. 11.10a). If instead t is held constant and y is plotted as a function of x, then the graph is like a snapshot—an instantaneous picture of what the wave looks like *at that particular instant* (Fig 11.10b).

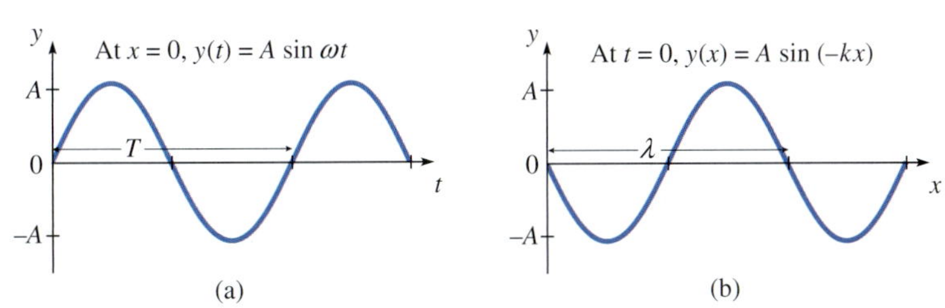

Figure 11.10 Two graphs of a harmonic wave on a string described by the equation $y(x, t) = A \sin (\omega t - kx)$. (a) The vertical displacement y of a particular point on the string ($x = 0$) as a function of time t. (b) The vertical displacement y as a function of horizontal position x at a single instant of time ($t = 0$).

Example 11.3

A Transverse Harmonic Wave

A transverse harmonic wave travels in the $+x$-direction on a string at a speed of 5.0 m/s. Figure 11.11 shows a graph of $y(t)$ for the point $x = 0$. (a) What is the period of the wave? (b) What is the wavelength? (c) What is the amplitude? (d) Write the function $y(x, t)$ that describes the wave. (e) Sketch a graph of $y(x)$ at $t = 0$.

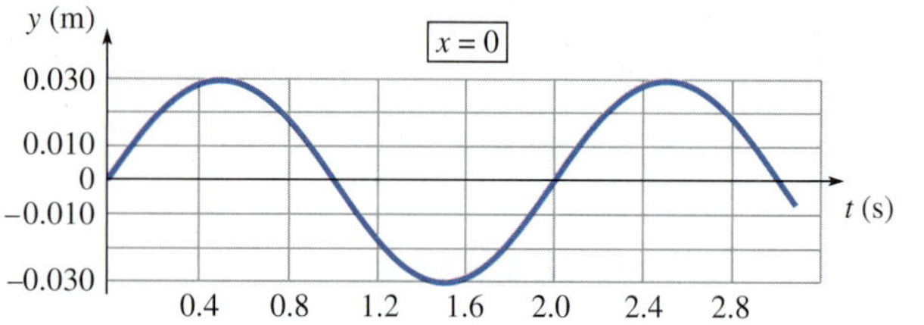

Figure 11.11

Graph of a transverse harmonic wave.

Strategy Since the graph uses time as the independent variable, the period can be read from the graph as the time for one cycle. The wavelength is the distance traveled by the wave during one period. The amplitude can be read from the graph as the maximum displacement. These are all the constants needed to write the function $y(x, t)$. We do have to think about the direction of travel and whether to write sine or cosine.

Solution (a) The period T is the time for one cycle. From the graph, $T = 2.0$ s.

(b) The wavelength λ is the distance traveled by the wave at speed $v = 5.0$ m/s during one period:

$$\lambda = vT = 5.0 \text{ m/s} \times 2.0 \text{ s} = 10 \text{ m}$$

(c) The amplitude A is the maximum displacement from equilibrium. From the graph, $A = 0.030$ m.

(d) Figure 11.11 is a sine function. The motion of the point $x = 0$ is

$$y(t) = A \sin\left(2\pi\frac{t}{T}\right)$$

Since the wave moves in the $+x$-direction, a point at $x > 0$ duplicates the motion of $x = 0$ with a time delay of $\Delta t = x/v$ (the time for the wave to travel a distance x). Then

$$y(x, t) = A \sin\left(2\pi\frac{t - x/v}{T}\right)$$

where $v = 5.0$ m/s and $T = 2.0$ s.

(e) Substituting $t = 0$,

$$y(x) = A \sin\left(-2\pi\frac{x}{vT}\right)$$

Substituting $vT = \lambda$ and using the identity

$$\sin(-\theta) = -\sin\theta$$

(Appendix A.7), we have

$$y(x, t = 0) = -A \sin\left(2\pi\frac{x}{\lambda}\right)$$

A graph of this function is an inverted sine function with amplitude $A = 0.030$ m and wavelength

$$\lambda = vT = 5.0 \text{ m/s} \times 2.0 \text{ s} = 10 \text{ m}$$

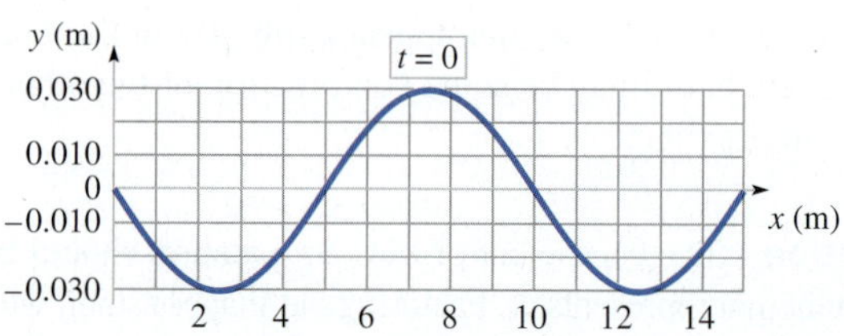

Discussion Figure 11.11 shows that the point $x = 0$ is initially at $y = 0$ and then moves up (in the $+y$-direction) until it reaches the crest (maximum y) at $t = 0.50$ s. Imagine the graph in (e) to represent the first frame (at $t = 0$) of a movie of the wave. Since the wave moves to the right, the sinusoidal pattern shifts a little to the right in each successive frame. The point $x = 0$ moves up until it reaches the crest when the wave has traveled 2.5 m to the right. Since the wave speed is 5.0 m/s, the point $x = 0$ reaches the crest at $t = (2.5 \text{ m})/(5.0 \text{ m/s}) = 0.50$ s.

Practice Problem 11.3 Another Harmonic Transverse Wave

A wave is described by

$$y(x, t) = (1.2 \text{ cm}) \sin[(10.0\pi \text{ rad/s})\, t + (2.5\pi \text{ rad/m})x]$$

(a) Sketch a graph of $y(t)$ at $x = 0$. (b) Sketch a graph of $y(x)$ at $t = 0$. (c) What is the period of the wave? (d) What is the wavelength? (e) What is the amplitude? (f) What is the speed of the wave? (g) In what direction does the wave move?

11.7 PRINCIPLE OF SUPERPOSITION

Suppose two waves of the same type pass through the same region of space. Do the waves affect each other? If the amplitudes of the waves are large enough, then particles in the medium are displaced far enough from their equilibrium positions that Hooke's law (restoring force ∝ displacement) no longer holds; in that case, the waves *do* affect each other. However, for small amplitudes, the waves can pass through each other and emerge *unchanged*. More generally, when the amplitudes are not too large, the principle of superposition applies:

Principle of Superposition

When two or more waves overlap, the net disturbance at any point is the sum of the individual disturbances due to each wave.

Figure 11.12 illustrates the superposition principle for two wave pulses traveling toward each other on a string. The wave pulses pass right through each other without affecting each other; once they have separated, their shapes and heights are the same as before the overlap (Fig. 11.12a). The principle of superposition enables us to distinguish two voices speaking in the same room at the same time; the sound waves pass through each other unaffected.

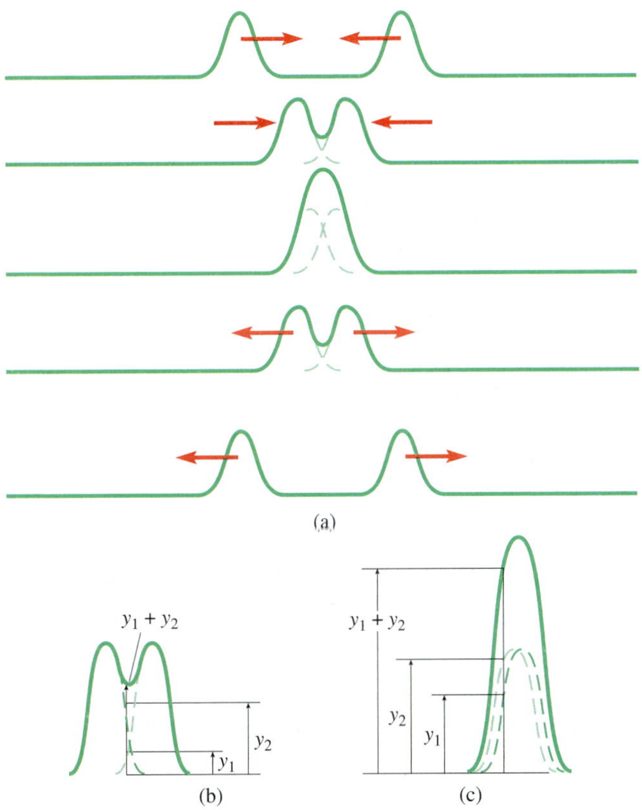

Figure 11.12 (a) Two identical wave pulses traveling toward and through each other. (b), (c) Applying the superposition principle at two different times; in each case, the dashed lines are the separate wave pulses and the solid line is the sum. If one of the pulses (acting alone) would produce a displacement y_1 at a certain point and the other would produce a displacement y_2 at the same point, the result when the two overlap is a displacement of $y_1 + y_2$.

Example 11.4

Superposition of Two Wave Pulses

Two identical wave pulses travel at 0.5 m/s toward each other on a long cord (Fig. 11.13). Sketch the shape of the cord at $t = 1.0$, 1.5, and 2.0 s.

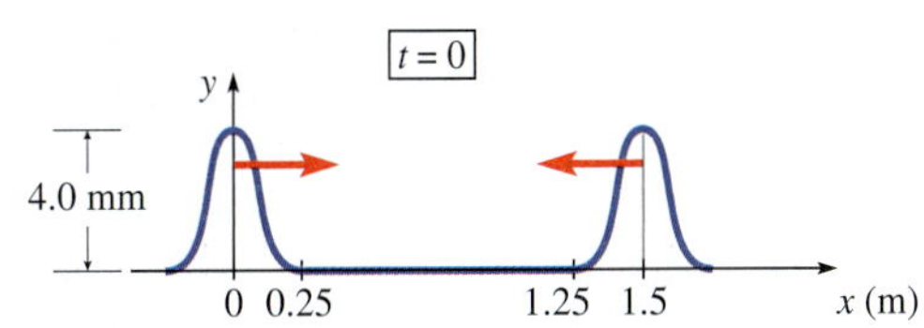

Figure 11.13
Two wave pulses at $t = 0$.

Strategy We start by sketching the two pulses in their new positions at each time given. Wherever they overlap,

continued on next page

Example 11.4 continued

we apply superposition by adding the individual displacements at each point to find the net displacement of the cord at that point.

Solution Using graph paper, we draw the wave pulses at $t = 0$ (Fig. 11.14a). At $t = 1.0$ s, each pulse has moved 0.5 m toward the other. The leading edges of the pulses are just starting to overlap (Fig. 11.14b). At $t = 1.5$ s, each pulse has moved another 0.25 m; the crests overlap exactly. By adding the displacements point by point, we see that the string has the shape of a single pulse twice as high as either of the individual pulses (Fig. 11.14c). At $t = 2.0$ s, the pulses have each moved another 0.25 m (Fig. 11.14d).

Discussion When the two pulses exactly overlap, the displacement of points on the string is larger than for corresponding points on a single pulse because we add displacements *in the same direction* ($y > 0$ for both). However, superposition does not *always* produce larger displacements (see Practice Problem 11.4).

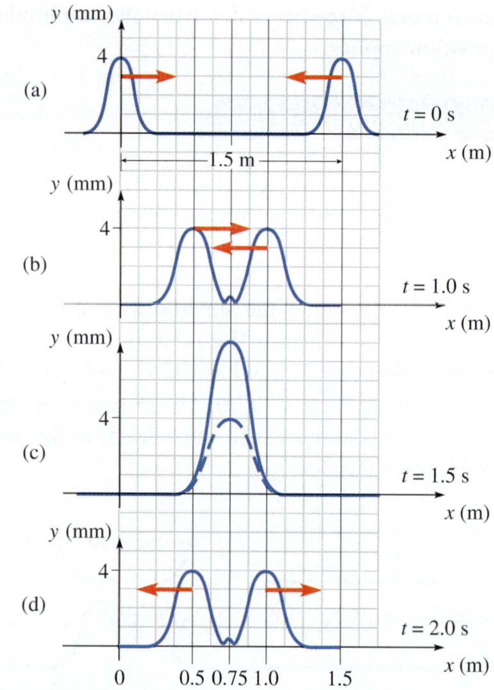

Figure 11.14
Wave positions at times $t = 0$, 1.0, 1.5, and 2.0 s.

Practice Problem 11.4 Superposition of Two Opposite Wave Pulses

Repeat Example 11.4, except now let the pulse on the right be inverted (Fig. 11.15). [*Hint*: Points on the string below the x-axis have negative displacements ($y < 0$).]

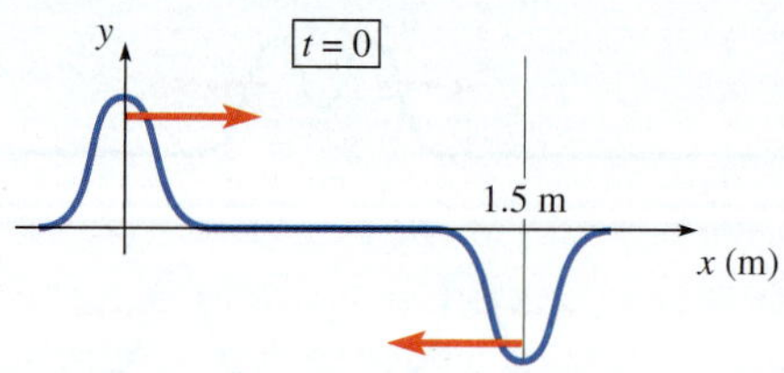

Figure 11.15 Wave pulses for Practice Problem 11.4.

11.8 REFLECTION AND REFRACTION

Reflection

At an abrupt boundary between one medium and another, **reflection** occurs; a reflected wave carrying some of the energy of the incident wave travels backward from the boundary. A sound wave in air, for instance, reflects when it reaches a wall.

A reflected wave can be inverted. Let's look at an extreme example: a string tied to a wall. If you send a wave pulse down the string, the reflected pulse is inverted (Fig. 11.16). By the principle of superposition, the shape of the string *at any point* is the sum of the incident and reflected waves, even at the fixed point at the end. The only way the end can stay in place is if the reflected wave is an upside down version of the incident wave. Another way to understand the inversion is by considering the force exerted on the string by the wall. When an upward pulse reaches the fixed end, the force exerted by the string on the wall has an upward component. By Newton's third law, the wall exerts a force on the string with a downward component. This downward force produces a downward reflected pulse.

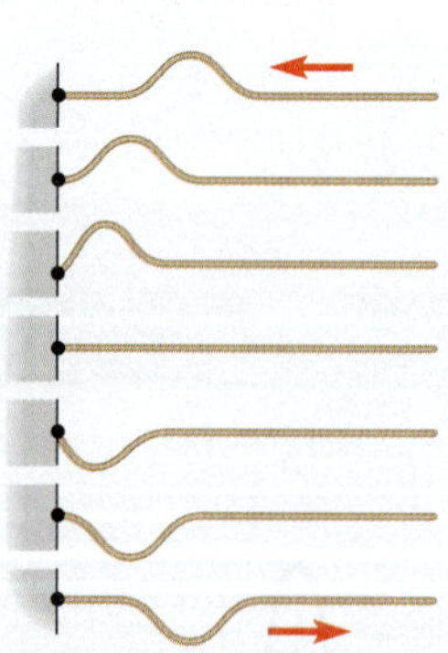

Figure 11.16 Snapshots of the reflection of a wave pulse from a fixed end. The reflected pulse is upside down.

Now, instead of tying the string to the wall, tie it to another string with an enormous linear mass density—so large that its motion is too small to measure. The original string doesn't know the difference; it just knows that one end is fixed in place. The second string with the huge density has a much slower wave speed than the first string. Now make the mass density of the second string not huge, but still greater than the first string. The greater inertia inhibits the motion of the boundary point and causes the reflected wave to be inverted. In general, when a transverse wave on a string reflects from a boundary with a region of slower wave speed, the reflected wave is inverted. On the other hand, when such a wave reflects from a boundary with a region of *faster* wave speed, the reflected wave is *not* inverted.

Change in Wavelength at a Boundary

When there is an abrupt change in wave medium, an incident wave splits up at the boundary; part is reflected and part is transmitted past the boundary into the other medium. The frequencies of both the reflected and transmitted waves are the same as the frequency of the incident wave. To understand why, think of a wave incident on the knot between two different strings. Both the reflected and the transmitted waves are generated by the up-and-down motion of the knot; the knot vibrates at the frequency dictated by the incident wave. However, if the wave speed changes at the boundary, *the wavelength of the transmitted wave is not the same* as the wavelength of the incident and reflected waves. Since $v = \lambda f$ and the frequencies are the same,

$$f = \frac{v_1}{\lambda_1} = \frac{v_2}{\lambda_2} \tag{11-9}$$

Equation (11-9) applies to any kind of wave and is of particular importance in the study of optics.

Example 11.5

Wavelength in Ultrasonography

Ultrasonic imaging is used to detect the presence of gallstones in the gallbladder. A transducer generates ultrasound at a frequency of 6.00 MHz. The speed of sound in the gallstone is 2180 m/s; the speed in the surrounding bile is 1520 m/s. (a) What is the wavelength of the sound wave in the bile? (b) What is the wavelength of the sound wave in the gallstone?

Strategy The *frequency* of the sound wave in water is the same in the two materials. The wavelengths depend on both the frequency and the speed of sound in the medium.

Solution (a) The wavelength in bile is related to the speed of sound in bile and the frequency of the wave:

$$\lambda_b = v_b T = \frac{v_b}{f}$$

Substituting numerical values,

$$\lambda_b = \frac{1520\ \text{m/s}}{6.00 \times 10^6\ \text{Hz}} = 0.253\ \text{mm}$$

(b) The wave in the stone has the *same frequency*, but the speed of sound is different:

$$\lambda_s = \frac{v_s}{f} = \frac{2180\ \text{m/s}}{6.00 \times 10^6\ \text{Hz}} = 0.363\ \text{mm}$$

Discussion As a quick check, the ratio of the wavelengths should be equal to the ratio of the wave speeds:

$$\frac{0.253\ \text{mm}}{0.363\ \text{mm}} = 0.697; \qquad \frac{1520\ \text{m/s}}{2180\ \text{m/s}} = 0.697$$

Practice Problem 11.5 Working on the Railroad

A railroad worker, driving in spikes, misses the spike and hits the iron rail; a sound wave travels through the air and through the rail. (Ignore the *transverse* wave that also travels in the rail.) The wavelength of the sound in air is 0.548 m. The speed of sound in air is 340 m/s; the speed of sound in iron is 5300 m/s. (a) What is the frequency of the wave? (b) What is the wavelength of the sound wave in the rail?

Refraction

A transmitted wave not only has a different wavelength than the incident wave, it also travels in a different direction unless the incident wave's direction of propagation is along the *normal* (the direction perpendicular to the boundary). This change in propagation direction is called **refraction**. (See the text website for a detailed discussion of refraction.)

connect

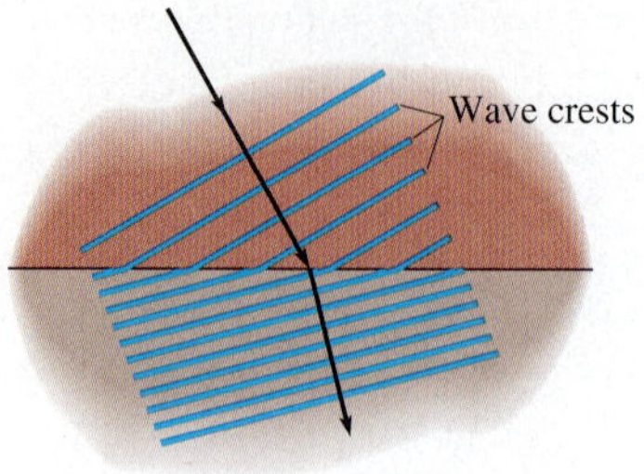

Figure 11.17 Wave crests for a seismic wave incident on a boundary between two different kinds of rock. Not only does the wavelength (distance between wave crests) change at the boundary, the wave also refracts (changes its direction of propagation). The reflected wave is omitted for clarity.

Application: Why Ocean Waves Approach Shore Nearly Head-on If the change in wave speed is gradual, then the change in direction is gradual as well. The speed of ocean waves depends on the depth of the water; the waves are slower in shallower water. As waves approach the shore, they gradually slow down; as a result, they gradually bend until they reach shore nearly head-on.

Application: Seismology A *sudden* change in wave speed, such as when a seismic wave is incident on a boundary between different kinds of rock, causes a sudden refraction (Fig. 11.17). Understanding the propagation of seismic waves, including reflection and refraction due to boundaries between geological features, is an essential part of the effort to reduce damage from future earthquakes. Scientists create small seismic waves with a large vibrator, then use seismographs to record ground vibrations at various locations. The goal is to produce a seismic hazard map so that preventative measures can be targeted to areas with the highest risk of earthquake damage.

11.9 INTERFERENCE AND DIFFRACTION

Interference

The principle of superposition can lead to dramatic effects. Suppose waves with the same frequency f but different amplitudes A_1 and A_2 pass through the same point in space. If the waves are **in phase** at that point, the two waves consistently reach their maxima at exactly the same time (Fig. 11.18a). The superposition of the waves that are in phase with each other is called **constructive interference**; the amplitude of the combined waves is the sum of the amplitudes of the two individual waves ($A_1 + A_2$).

If two waves with the same frequency are **180° out of phase** at a given point, one reaches its maximum when the other reaches its minimum (Fig. 11.18b). The superposition of waves that are 180° out of phase is called **destructive interference**—the amplitude of the combined waves is the *difference* of the amplitudes of the two individual waves ($|A_1 - A_2|$). Constructive interference yields the maximum possible amplitude ($A_1 + A_2$) and destructive interference yields the minimum ($|A_1 - A_2|$).

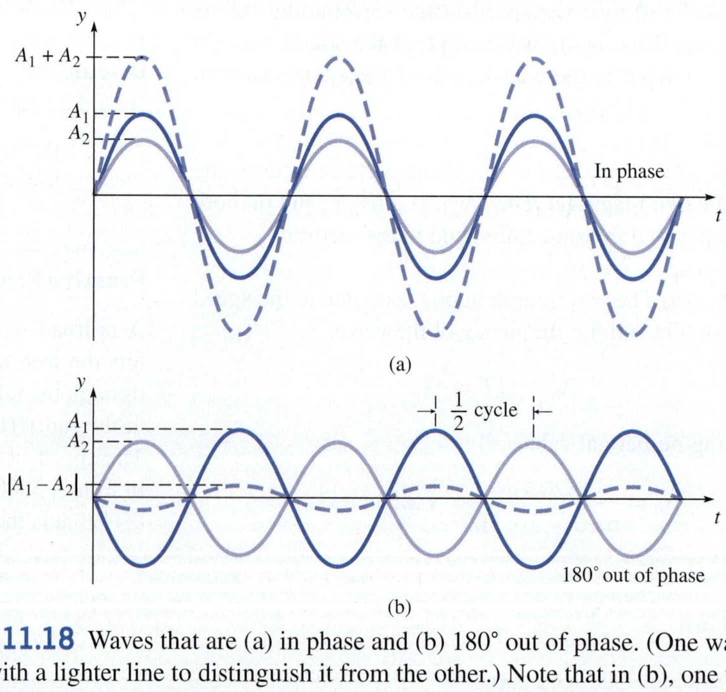

Figure 11.18 Waves that are (a) in phase and (b) 180° out of phase. (One wave is drawn with a lighter line to distinguish it from the other.) Note that in (b), one wave reaches its maximum a half cycle before the other. In both (a) and (b), the dashed curve is the superposition of the two waves. For constructive interference (a), the amplitude is $A_1 + A_2$. For destructive interference (b), the amplitude is $|A_1 - A_2|$.

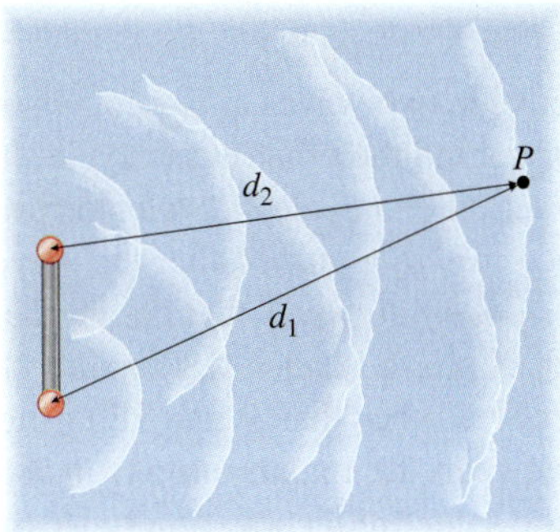

Figure 11.19 Overhead snapshot of two coherent surface water waves. The two waves travel different distances d_1 and d_2 to reach a point *P*. If $d_1 - d_2$ is an integral number of wavelengths, interference is constructive at *P*. If $d_1 - d_2$ is an odd number of half wavelengths, interference is destructive.

In Fig. 11.19, two rods vibrate up and down in step with each other to generate circular waves on the surface of the water. If the two waves travel the same distance to reach a point on the water surface, they arrive in phase and interfere constructively. At points where the distances are not equal, interference can be constructive, destructive, or something in between.

Interference due to path difference

If the path difference $d_1 - d_2$ is an integral number of wavelengths, then constructive interference occurs. If the path difference is $\frac{1}{2}\lambda$, $\frac{3}{2}\lambda$, $\frac{5}{2}\lambda$, . . . , then destructive interference occurs.

Intensity Effects for Interfering Waves When waves interfere, the *amplitudes* add (for constructive interference) or subtract (for destructive interference). However, since intensity is proportional to amplitude *squared*, we cannot simply add or subtract the *intensities* of waves when they interfere.

Example 11.6

Intensity of Interfering Waves

Two waves interfere. The intensity of one of them (alone) is 9.0 times the intensity of the other. What is the ratio of the maximum possible intensity to the minimum possible intensity of the resulting wave?

Strategy The intensity is *not* the sum or difference of the individual intensities. The principle of superposition tells us that the maximum and minimum *amplitudes* of the interfering waves are the sum and difference of the *individual amplitudes*. Intensity is proportional to amplitude squared, so we find the ratio of the amplitudes and then add or subtract them.

Solution The intensities of the two individual waves are related by $I_1 = 9.0I_2$ or $I_1/I_2 = 9.0$. Since intensity is proportional to amplitude squared,

$$\frac{A_1}{A_2} = \sqrt{\frac{I_1}{I_2}} = 3.0$$

Thus, $A_1 = 3.0A_2$. The maximum possible amplitude for the superposition occurs if the waves are in phase:

$$A_{max} = A_1 + A_2 = 4.0A_2$$

The minimum possible amplitude for the superposition occurs if the waves are 180° out of phase:

$$A_{min} = |A_1 - A_2| = 2.0A_2$$

The ratio of the maximum to minimum intensity is

$$\frac{I_{max}}{I_{min}} = \left(\frac{A_{max}}{A_{min}}\right)^2 = \left(\frac{4.0}{2.0}\right)^2 = 4.0$$

Discussion Had we added and subtracted the *intensities* instead of the amplitudes, we would have found a ratio of $10/8 = 1.25$ between the maximum and minimum intensities.

Practice Problem 11.6 Two More Coherent Waves

Repeat Example 11.6, but change the ratio of the individual intensities to 4.0 (instead of 9.0).

Coherence

The *phase difference* $\Delta\phi$ between two waves at a point where they overlap is a measure of how much one is ahead or behind the other in the cycle. It is usually given in degrees or radians rather than as a fraction of a cycle: 1 cycle corresponds to 360°, or 2π rad. In our discussion of interference, we've assumed that the waves are **coherent**—that their phase difference is constant. Coherent waves that are in phase ($\Delta\phi = 0$) *stay* in phase, and waves that are 180° out of phase ($\Delta\phi = 180°$) *stay* 180° out of phase. Coherent waves can have phase differences other than 0 and 180°; then the amplitude when they are added is intermediate between the maximum ($A_1 + A_2$) and minimum $|A_1 - A_2|$. Coherent waves must have the same frequency; otherwise there is no way to maintain a consistent phase difference.

One way to produce coherent waves is to get them from the same source. For example, one could send *the same signal* from an audio amplifier to two speakers. Should some fluctuation occur in the amplifier's circuitry, the same fluctuation occurs in the signal to both speakers and they maintain their coherence.

Waves are **incoherent** if the phase relationship between them varies randomly. Waves from independent sources are incoherent. With incoherent waves, interference effects are averaged out due to the varying phase difference and the total *intensity* is the sum of the individual *intensities*. (As defined here, *coherent* and *incoherent* are idealized extremes.)

Why don't we see and hear interference effects all the time? Light from ordinary sources—incandescent bulbs, fluorescent bulbs, or the Sun—is incoherent because it is generated by large numbers of independent atomic sources. Sound waves from independent sources are also incoherent. Even with a single sound source, in most situations many different sound waves reach our ears after traveling different paths due to the reflection of sound from walls, ceilings, chairs, and so forth. These waves arrive with many different phases, so interference effects may not be noticeable. Also sound waves normally contain many different frequencies, so a point of constructive interference for one frequency is not a point of constructive interference for other frequencies. Nevertheless, sound engineers and acousticians who design classrooms and concert halls must take interference effects into account.

CHECKPOINT 11.9

Two waves have intensities of I_2 and $I_1 = 9.0I_2$ by themselves, as in Example 11.6. (a) What is the intensity of the superposition of the two if they are incoherent? (b) What are the maximum and minimum possible intensities if they are coherent?

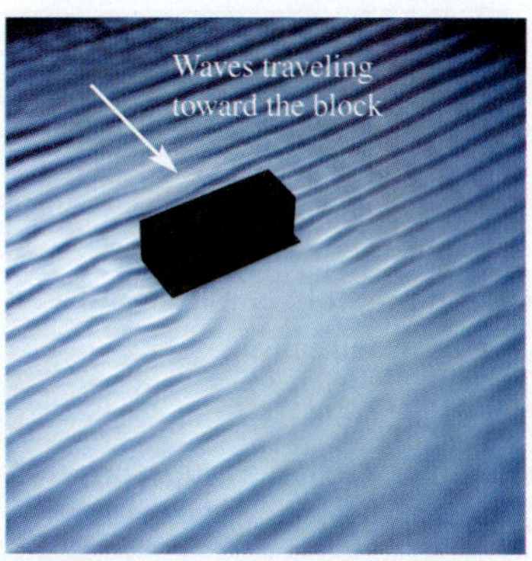

Figure 11.20 Demonstration of diffraction in a ripple tank. Ripples in the water are generated by a bar (not shown) that oscillates up and down. The ripples are visible as shadows when the tank is illuminated. Here, an obstacle is placed in the path of the waves and the waves bend around the obstacle. If the waves traveled in straight-line paths, we would see a shadow with sharp edges behind the obstacle instead.

Diffraction

Diffraction is the spreading of a wave around an obstacle in its path (Fig. 11.20). The amount of diffraction depends on the relative size of the obstacle and the wavelength of the waves. Diffraction enables you to hear around a corner but not to see around a corner. Sound waves, with typical wavelengths in air of around 1 m, diffract around the corner much more than do light waves with much smaller wavelengths (less than 1 μm). We will study interference and diffraction of electromagnetic waves (including light) in detail in Chapter 25.

11.10 STANDING WAVES

Standing waves occur when a wave is reflected at a boundary, and the reflected wave interferes with the incident wave so that the wave appears not to propagate. Suppose that a harmonic wave on a string, coming from the right, hits a boundary where the string is fixed. The equation of the incident wave is

$$y(x, t) = A \sin (\omega t + kx)$$

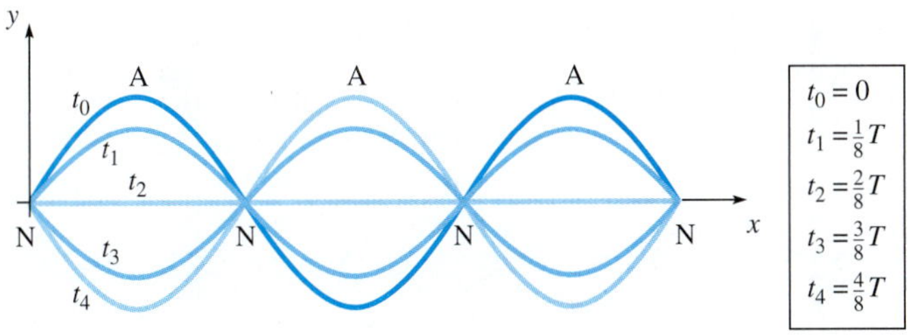

Figure 11.21 A standing wave at various times: $t = 0, \frac{1}{8}T, \frac{2}{8}T, \frac{3}{8}T$, and $\frac{4}{8}T$, where T is the period. The labels "A" and "N" indicate the locations of the antinodes and nodes, respectively. An antinode is a point that vibrates with maximum amplitude; a node is a point that doesn't move (amplitude of zero). The distance between a node and a neighboring antinode is $\lambda/4$; the distance between two adjacent nodes is $\lambda/2$.

The + sign is chosen because the wave travels to the left.

The reflected wave travels to the right, so $+kx$ is replaced with $-kx$; and the reflected wave is inverted, so $+A$ is replaced with $-A$. Then the reflected wave is described by

$$y(x, t) = -A \sin (\omega t - kx)$$

Applying the principle of superposition, the motion of the string is described by

$$y(x, t) = A [\sin (\omega t + kx) - \sin (\omega t - kx)]$$

This can be rewritten in a form that shows the motion of the string more clearly. Using the trigonometric identity (Appendix A.7),

$$\sin \alpha - \sin \beta = 2 \cos [\tfrac{1}{2}(\alpha + \beta)] \sin [\tfrac{1}{2}(\alpha - \beta)]$$

where

$$\alpha = \omega t + kx \qquad \text{and} \qquad \beta = \omega t - kx$$

the motion of the string is described by

$$y(x, t) = 2A \cos \omega t \sin kx \tag{11-10}$$

Notice that t and x are separated. Every point moves in SHM with the same frequency. However, in contrast to a *traveling* harmonic wave, every point reaches its maximum distance from equilibrium *simultaneously*. In addition, different points move with different amplitudes; the amplitude at any point x is $2A \sin kx$.

Figure 11.21 shows the string at time intervals of $\frac{1}{8}T$ where T is the period. What you actually see when looking at a standing wave is a blur of moving string, with points that never move (**nodes**, labeled "N") halfway between points of maximum amplitude (**antinodes**, labeled "A"). The nodes are the points where $\sin kx = 0$. Since $\sin n\pi = 0$ ($n = 0, 1, 2, \ldots$), the nodes are located at $x = n\pi/k = n\lambda/2$. Thus, the distance between two adjacent nodes is $\frac{1}{2}\lambda$. The antinodes occur where $\sin kx = \pm 1$, which is exactly halfway between a pair of nodes. So the nodes and antinodes alternate, with one quarter of a wavelength between a node and the neighboring antinode.

So far we have ignored what happens at the other end of the string. If the other end is fixed, then it is a node. The string thus has two or more nodes, with one at each end. The distance between each pair of nodes is $\frac{1}{2}\lambda$, so

$$n(\lambda/2) = L$$

where L is the length of the string and $n = 1, 2, 3, \ldots$. The possible wavelengths and frequencies for standing waves on a string are

$$\lambda_n = \frac{2L}{n} \quad (n = 1, 2, 3, \ldots) \tag{11-11}$$

$$f_n = \frac{v}{\lambda_n} = \frac{nv}{2L} \quad (n = 1, 2, 3, \ldots) \tag{11-12}$$

There is no need to memorize Eqs. (11-11) and (11-12). Start with a sketch like Fig. 11.22, find the wavelengths, and then use $v = f\lambda$ to find the frequencies.

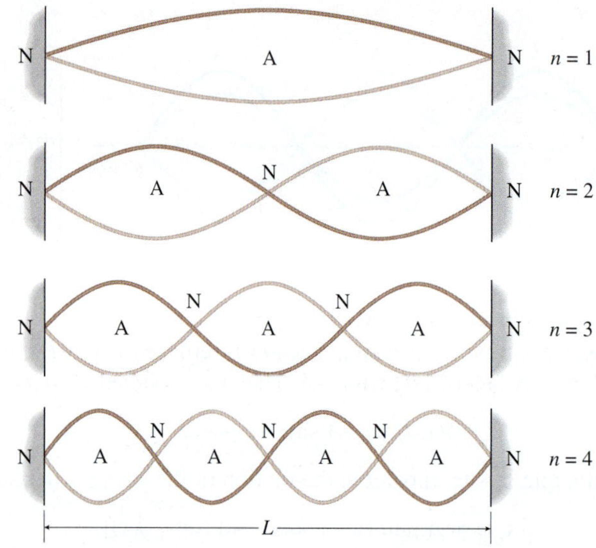

Figure 11.22 Four standing wave patterns for a string fixed at both ends. "N" marks the locations of the nodes and "A" marks the locations of the antinodes. In each case, the node-to-node distance is $\frac{1}{2}\lambda$ and n such "loops" fit into the length L of the string, so $n(\lambda/2) = L$.

CONNECTION:

An ideal mass-spring system has a single resonant frequency (Section 10.10), but extended objects generally have many different resonant frequencies.

The lowest frequency standing wave ($n = 1$) is called the **fundamental**. Notice that the higher frequency standing waves are all integral multiples of the fundamental; the set of standing wave frequencies makes an evenly spaced set:

$$f_1, 2f_1, 3f_1, 4f_1, \ldots, nf_1, \ldots$$

These frequencies are called the *natural frequencies* or *resonant frequencies* of the string. *Resonance* occurs when a system is driven at one of its natural frequencies; the resulting vibrations are large in amplitude compared to when the driving frequency is not close to any of the natural frequencies.

CHECKPOINT 11.10

A standing wave on a string 1.0 m long has four nodes, not including the nodes at the two fixed ends. What is the wavelength?

Figure 11.22 shows the first four standing wave patterns on a string. The two ends are always nodes since they are fixed in place. Notice that each successive pattern has one more node and one more antinode than the previous one. The fundamental has the fewest possible number of nodes (2) and antinodes (1).

Example 11.7

Wavelength of a Standing Wave

A string is attached to a vibrator driven at 1.20×10^2 Hz. A weight hangs from the other end of the string; the weight is adjusted until a standing wave is formed (Fig. 11.23). What is the wavelength of the standing wave on the string?

Strategy The measured distance of 42 cm encompasses six "loops"—that is, six segments of string between one node and the next. Each of the loops represents a length of $\frac{1}{2}\lambda$.

Solution The length of one loop is

$$42 \text{ cm} \times \tfrac{1}{6} = 7.0 \text{ cm}$$

Since the length of one loop is $\frac{1}{2}\lambda$, the wavelength is 14 cm.

Discussion This string is *not* fixed at both ends. The left end is connected to a moving vibrator, so it is not a node. The right end wraps around a pulley; it may not be

continued on next page

Example 11.7 continued

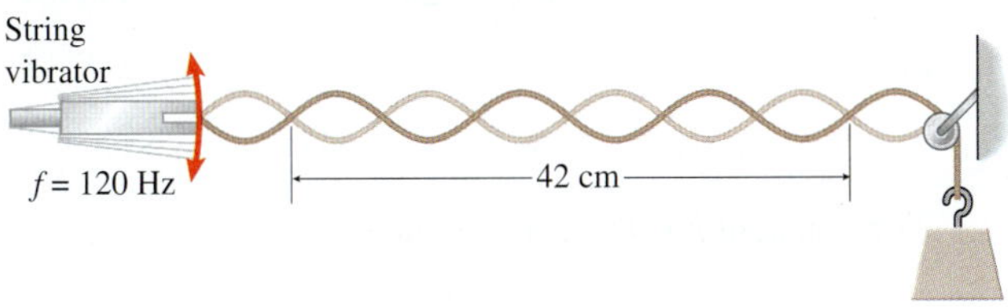

Figure 11.23

Measuring distance between nodes for a standing wave.

easy to determine precisely where the "end" is. For this case, it is more accurate to measure the distance between two actual nodes rather than to assume that the ends are nodes.

Practice Problem 11.7 Standing Wave with Seven Loops

The vibrator frequency is increased until there are seven loops within the 42 cm length. What is the new standing wave frequency for this string (assuming the same tension)?

Application of Resonance: Damage Caused by Earthquakes Resonance is responsible for much of the structural damage caused by seismic waves. If the frequency at which the ground vibrates is close to a resonant frequency of a structure, the vibration of the structure builds up to a large amplitude. Thus, to construct a building that can survive an earthquake, it is not enough to make it stronger. Either the building must be designed so it is isolated from ground vibrations, or a damping mechanism—something like a shock absorber—must be incorporated to dissipate energy and reduce the amplitude of the vibrations. Damping is becoming increasingly common in large buildings since it is just as effective and much less expensive than isolation.

Why did the expressway collapse while nearby buildings were not seriously damaged?

In the 1995 Hanshin earthquake, large sections of the Hanshin expressway collapsed even though nearby buildings and roads suffered little damage. The frequency of vibration of the ground during the earthquake matched closely one of the resonant frequencies of the elevated roadway. The roadway twisted back and forth with increasing amplitude until it collapsed. After the earthquake, rubber base isolators were installed to replace steel bearings connecting the roadway to the concrete piers. Part of their function is to act like shock absorbers to reduce the roadway's vibration amplitude during a future earthquake.

Master the Concepts

- An isotropic source radiates sound uniformly in all directions. Assuming that no energy is absorbed by the medium and there are no obstacles to reflect or absorb sound, the intensity I at a distance r from an isotropic source is

$$I = \frac{P}{A} = \frac{P}{4\pi r^2} \qquad \text{(11-2)}$$

- In a transverse wave, the motion of particles in the medium is perpendicular to the direction of propagation of the wave. In a longitudinal wave, the motion of particles in the medium is along the same line as the direction of propagation of the wave.

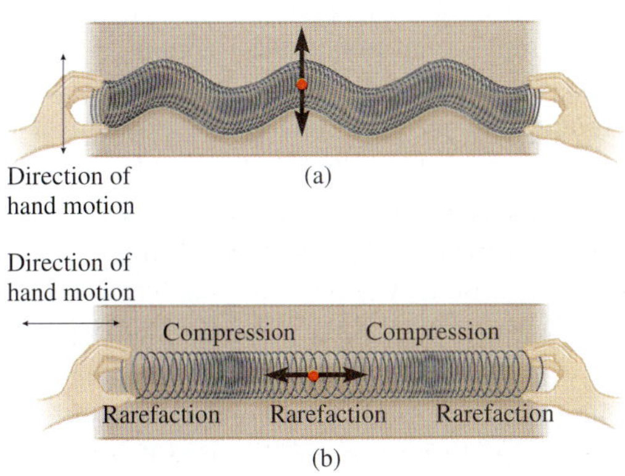

continued on next page

Master the Concepts continued

- The speed of a mechanical wave depends on properties of the wave medium. More restoring force makes faster waves; more inertia makes slower waves.
- The speed of a transverse wave on a string is

$$v = \sqrt{\frac{F}{\mu}} \quad (11\text{-}4)$$

where

$$\mu = m/L \quad (11\text{-}3)$$

- A periodic wave repeats the same pattern over and over. Harmonic waves are a special kind of periodic wave characterized by a sinusoidal function (either a sine or cosine function).
- If a periodic wave has period T and travels at speed v, the repetition distance of the wave is the wavelength:

$$\lambda = vT \quad (11\text{-}5)$$

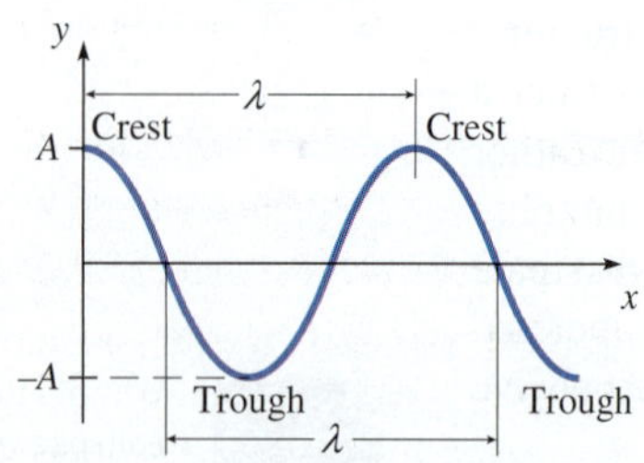

- The principle of superposition: When two or more waves overlap, the net disturbance at any point is the sum of the individual disturbances due to each wave.
- A harmonic traveling wave can be described by

$$y(x, t) = A \cos(\omega t \pm kx) \quad (11\text{-}8)$$

or

$$y(x, t) = A \sin(\omega t \pm kx)$$

The constant k is the wave number

$$k = \frac{\omega}{v} = \frac{2\pi f}{v} = \frac{2\pi}{\lambda} \quad (11\text{-}7)$$

- Reflection occurs at a boundary between different wave media. Some energy may be transmitted into the new medium, and the rest is reflected. The wave transmitted past the boundary is refracted (propagates in a different direction).
- Waves that are in phase with one another interfere constructively; those that are 180° out of phase interfere destructively.
- Diffraction occurs when a wave bends around an obstacle in its path.
- In a standing wave on a string, every point moves in SHM with the same frequency. Nodes are points of zero amplitude; antinodes are points of maximum amplitude. The distance between two adjacent nodes is $\frac{1}{2}\lambda$. The distance between a node and a neighboring antinode is $\frac{1}{4}\lambda$.

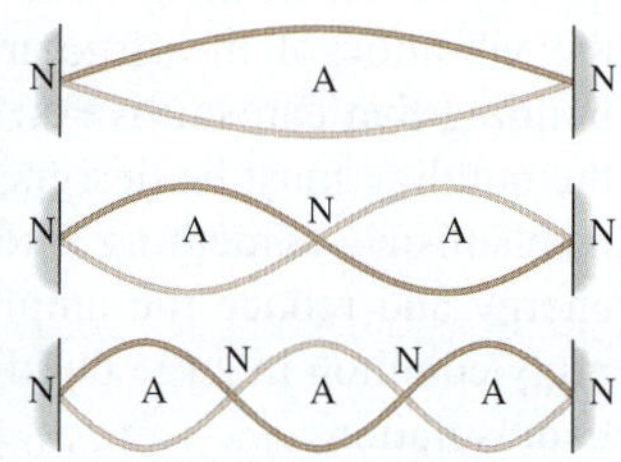

Conceptual Questions

1. Is the vibration of a string in a piano, guitar, or violin a *sound* wave? Explain.
2. The spectators at a sports stadium are "doing the wave": they stand and raise their arms simultaneously with those in front of them and slightly after their neighbors on one side. This gives the appearance of a wave pulse propagation around the stadium. Is "the wave" analogous to a transverse wave or a longitudinal wave? Explain your answer. How would a group of people have to move to simulate the other kind of wave?
3. The piano strings that vibrate with the lowest frequencies consist of a steel wire around which a thick coil of copper wire is wrapped. Only the inner steel wire is under tension. What is the purpose of the copper coil?
4. The wavelength of the fundamental standing wave on a cello string depends on which of these quantities: length of the string, mass per unit length of the string, or tension? The wavelength of the *sound wave* resulting from the string's vibration depends on which of the same three quantities?
5. If the length of a guitar string is decreased while the tension remains constant, what happens to each of these quantities? (a) the wavelength of the fundamental, (b) the frequency of the fundamental, (c) the time for a pulse to travel the length of the string, (d) the maximum velocity of a point on the string (assuming the amplitude is the same both times), (e) the maximum acceleration of a point on the string (assuming the amplitude is the same both times).
6. Why is it possible to understand the words spoken by two people at the same time?
7. A cello player can change the frequency of the sound produced by her instrument by (a) increasing the tension in the string, (b) pressing her finger on the string at different places along the fingerboard, or (c) bowing a different string. Explain how each of these methods affects the frequency.
8. Why is a transverse wave sometimes called a shear wave?

9. A longitudinal wave has a wavelength of 10 cm and an amplitude of 5.0 cm and travels in the y-direction. The wave speed in this medium is 80 cm/s. (a) Describe the motion of a particle in the medium as the wave travels through the medium. (b) How would your answer differ if the wave were transverse instead?

10. Simple ear-protection devices use materials that reflect or absorb sound before it reaches the ears. A newer technology, sometimes called *noise cancellation*, uses a microphone to produce an electrical signal that mimics the noise. The signal is modified electronically, then fed to the speakers in a pair of headphones. The speakers emit sound waves that *cancel* the noise. On what principle is this technology based? What kind of modification is made to the electrical signal?

11. Audio speakers must be connected with the correct polarity so that, if the same electrical signal is sent, they move in the same direction. If the wires going to one speaker are reversed, the listener hears a noticeably weaker bass (low frequencies). Explain what causes this and why low frequencies are affected more than high frequencies.

Multiple-Choice Questions

Student Response System Questions

1. A transverse wave travels on a string of mass m, length L, and tension F. Which statement is correct?
   (a) The energy of the wave is proportional to the square root of the wave amplitude.
   (b) Every point on the string moves with the same speed.
   (c) The wave speed can be calculated from the values of m, L, and F.
   (d) The wave must be periodic.

2. Standing waves on a string are produced by the superposition of two waves with
   (a) the same amplitude, frequency, and direction of propagation.
   (b) the same amplitude and frequency, and opposite propagation directions.
   (c) the same amplitude and direction of propagation, but different frequencies.
   (d) the same amplitude, different frequencies, and opposite directions of propagation.

3. A transverse wave on a string is described by $y(x, t) = A\cos(\omega t + kx)$. It arrives at the point $x = 0$ where the string is fixed in place. Which function describes the reflected wave?
   (a) $A\cos(\omega t + kx)$ (b) $A\cos(\omega t - kx)$
   (c) $-A\sin(\omega t + kx)$ (d) $-A\cos(\omega t - kx)$
   (e) $A\sin(\omega t + kx)$

4. A violin string of length L is fixed at both ends. Which one of these is *not* a wavelength of a standing wave on the string?
   (a) L (b) $2L$ (c) $L/2$ (d) $L/3$ (e) $2L/3$ (f) $3L/2$

5. When a wave passes from one medium into another, which of these quantities *must* stay the same?
   (a) wavelength (b) wave speed
   (c) frequency (d) direction of propagation

6. In a standing wave, what is the distance between two neighboring nodes?
   (a) λ (b) 2λ (c) $\lambda/2$
   (d) $\lambda/4$ (e) 4λ

7. In a transverse wave, the motion of individual particles of the medium is
   (a) circular. (b) elliptical.
   (c) parallel to the direction of the wave's travel.
   (d) perpendicular to the direction of the wave's travel.

8. Which is the only one of these properties of a wave that could be changed without changing any of the others?
   (a) amplitude (b) wavelength
   (c) speed (d) frequency

9. The intensity of an isotropic sound wave is
   (a) directly proportional to the distance from the source.
   (b) inversely proportional to the distance from the source.
   (c) directly proportional to the square of the distance from the source.
   (d) inversely proportional to the square of the distance from the source.
   (e) none of the above.

10. Two successive transverse pulses, one caused by a brief displacement to the right and the other by a brief displacement to the left, are sent down a Slinky that is fastened at the far end. At the point where the first reflected pulse meets the second advancing pulse, the deflection (compared with that of a single pulse) is
    (a) quadrupled. (b) doubled.
    (c) canceled. (d) halved.

11. The drawing shows a complex wave moving to the right along a cord. At the instant shown, which points on the cord are moving downward?
    (a) A (b) B
    (c) C (d) A and C
    (e) A, B, and C

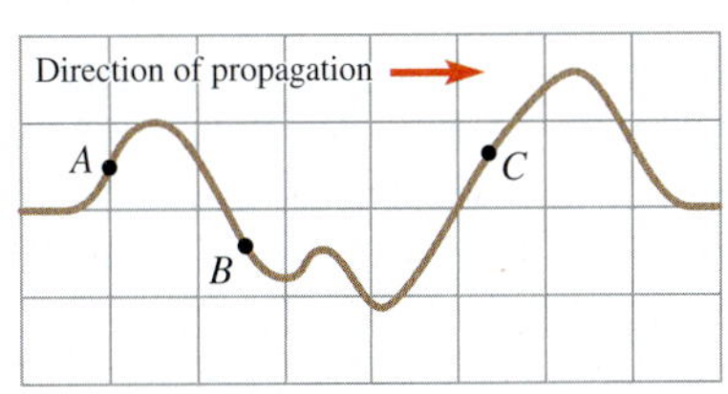

Problems

- C Combination conceptual/quantitative problem
- Biomedical application
- ✦ Challenging
- Blue # Detailed solution in the Student Solutions Manual
- [1, 2] Problems paired by concept
- connect Text website interactive or tutorial

11.1 Waves and Energy Transport

1. The intensity of sunlight that reaches Earth's atmosphere is 1400 W/m^2. What is the intensity of the sunlight that reaches Jupiter? Jupiter is 5.2 times as far from the Sun as Earth. [*Hint*: Treat the Sun as an isotropic source of light waves.]

2. Under favorable conditions, the human eye can detect light waves with intensities as low as 2.5×10^{-12} W/m^2. (a) At this intensity, what is the average power incident on a pupil of diameter 9.0 mm? (b) If this light is produced by an isotropic source 10.0 m away, what is the average power emitted by the source?

3. Michelle is enjoying a picnic across the valley from a cliff. She claps her hands and the echo takes 1.5 s to return. Given that the speed of sound in air is 343 m/s on that day, how far away is the cliff?

4. The intensity of the sound wave from a jet airplane as it is taking off is 100 W/m^2 at a distance of 5.0 m. What is the intensity of the sound wave that reaches the ears of a person standing at a distance of 120 m from the runway? Assume that the sound wave radiates from the airplane equally in all directions.

5. At what rate does the jet airplane in Problem 4 radiate energy in the form of sound waves?

6. The Sun emits electromagnetic waves (including light) equally in all directions. The intensity of the waves at Earth's upper atmosphere is 1.4 kW/m^2. At what rate does the Sun emit electromagnetic waves? (In other words, what is the power output?)

7. Six sources emit sound equally in all directions with average power P. A microphone is placed at a distance d from each source. Rank the situations in order of the intensity at the location of the microphone, smallest to largest.
(a) $P = 10$ W, $d = 2$ m (b) $P = 5$ W, $d = 1$ m
(c) $P = 20$ W, $d = 4$ m (d) $P = 20$ W, $d = 8$ m
(e) $P = 5$ W, $d = 2$ m

11.3 Speed of Transverse Waves on a String

8. Transverse waves travel on five stretched strings with the following properties. Rank the strings according to the time it takes a transverse wave pulse to travel from one end to the other, from largest to smallest.
(a) length L, total mass m, tension F
(b) length $2L$, total mass m, tension F
(c) length L, total mass $2m$, tension F
(d) length L, total mass m, tension $2F$
(e) length $2L$, total mass $2m$, tension F

9. (a) What is the position of the peak of the pulse shown in the figure at $t = 3.00$ s? (b) When does the peak of the pulse arrive at $x = 4.00$ m?

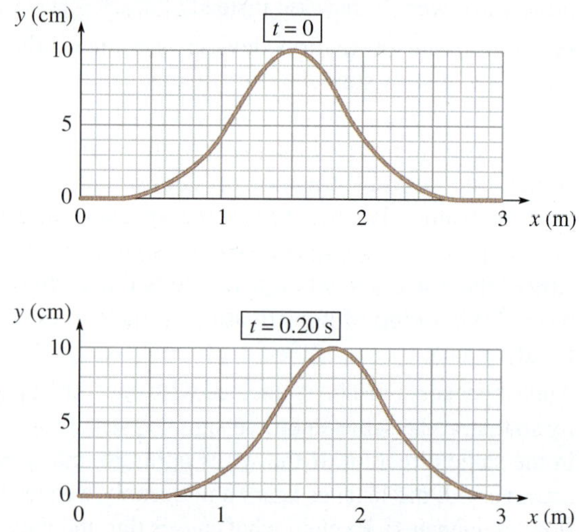

Problems 9, 46–49, and 95

10. When the tension in a cord is 75 N, the wave speed is 140 m/s. What is the linear mass density of the cord?

11. A metal guitar string has a linear mass density of $\mu = 3.20$ g/m. What is the speed of transverse waves on this string when its tension is 90.0 N?

12. Two children are playing with a tin-can telephone. The children are 12 m apart, the string connecting their tin cans has a linear mass density of 1.3 g/m, and it is stretched with a tension of 8.0 N. One child decides to pluck the string. How long does it take for the wave pulse to travel from one child to the other?

13. Two strings, each 15.0 m long, are stretched side by side. One string has a mass of 78.0 g and a tension of 180.0 N. The second string has a mass of 58.0 g and a tension of 160.0 N. A pulse is generated at one end of each string simultaneously. (a) On which string will the pulse move faster? (b) Once the faster pulse reaches the far end of its string, how much additional time will the slower pulse require to reach the end of its string?

14. A uniform string of length 10.0 m and weight 0.25 N is attached to the ceiling. A weight of 1.00 kN hangs from its lower end. The lower end of the string is suddenly displaced horizontally. How long does it take the resulting wave pulse to travel to the upper end? [*Hint*: Is the weight of the string negligible in comparison with that of the hanging mass?]

11.4 Periodic Waves

15. What is the speed of a wave whose frequency and wavelength are 500.0 Hz and 0.500 m, respectively?

16. What is the wavelength of a wave whose speed and period are 75.0 m/s and 5.00 ms, respectively?

17. What is the frequency of a wave whose speed and wavelength are 120 m/s and 30.0 cm, respectively?
18. The speed of sound in air at room temperature is 340 m/s. (a) What is the frequency of a sound wave in air with wavelength 1.0 m? (b) What is the frequency of a radio wave with the same wavelength? (Radio waves are electromagnetic waves that travel at 3.0×10^8 m/s in air or in vacuum.)
19. What is the wavelength of the microwaves transmitted by a cell phone at 900 MHz? (Microwaves travel at 3.0×10^8 m/s.)
20. Light visible to humans consists of electromagnetic waves with wavelengths (in air) in the range 400–700 nm (4.0×10^{-7} m to 7.0×10^{-7} m). The speed of light in air is 3.0×10^8 m/s. What are the frequencies of electromagnetic waves that are visible?
21. A fisherman notices a buoy bobbing up and down in the water in ripples produced by waves from a passing speedboat. These waves travel at 2.5 m/s and have a wavelength of 7.5 m. At what frequency does the buoy bob up and down?

11.5 Mathematical Description of a Wave

22. You are swimming in the ocean as water waves with wavelength 9.6 m pass by. What is the closest distance that another swimmer could be so that his motion is exactly opposite yours (he goes up when you go down)?
23. A transverse wave on a string is described by

$$y(x, t) = A \cos(\omega t + kx)$$

where $A = 0.350$ mm, $\omega = 50.0$ rad/s, and $k = 6.00$ rad/m. Find the (a) wavelength, (b) period, and (c) wave speed. (d) In what direction does the wave travel? (e) What is the maximum transverse speed of a point on the string?

24. A transverse wave on a string is described by

$$y(x, t) = A \cos(\omega t - kx)$$

where $A = 4.0$ mm, $\omega = 740$ rad/s, and $k = 2.8$ rad/m. Find the (a) wavelength, (b) period, and (c) wave speed. (d) In what direction does the wave travel? (e) What is the maximum acceleration of a point on the string?

25. A transverse wave on a string is described by

$$y(x, t) = (0.35\ \text{mm}) \sin\left\{\left(\frac{\pi}{3.0}\frac{\text{rad}}{\text{m}}\right)\left[x - \left(66\ \frac{\text{m}}{\text{s}}\right)t\right]\right\}$$

Find the (a) amplitude, (b) wavelength, and (c) frequency of this wave. (d) In what direction does the wave travel? (e) What is the maximum transverse speed of a point on the string?

26. A transverse wave on a string is described by

$$y(x, t) = (2.20\ \text{cm}) \sin\left[(130\ \text{rad/s})t + (15\ \text{rad/m})x\right].$$

Find the (a) amplitude, (b) wavelength, (c) frequency, and (d) wave speed of this wave. (e) In what direction does the wave travel? (f) What is the maximum acceleration of a point on the string?

27. (a) Write an equation for a harmonic wave with amplitude 1.20 mm, wavelength 30.0 cm, and wave speed 6.40 m/s traveling in the $-x$-direction. At $t = 0$, the point $x = 0$ is moving in the $+y$-direction at its maximum transverse speed. (b) What is the value of the maximum transverse speed?
28. (a) Write an equation for a harmonic wave with amplitude 0.750 mm, frequency 36.0 Hz, and wave speed 144 m/s traveling in the $+x$-direction. At $t = 0$, the point $x = 0$ is at its maximum displacement in the $+y$-direction. (b) What is the maximum acceleration of a point on the string?
29. ✦ Write the equation for a harmonic wave with amplitude 2.50 cm and angular frequency 2.90 rad/s that is moving in the $+x$-direction with a wave speed that is 5.00 times as fast as the maximum transverse speed of a point on the string. At $t = 0$, the point $x = 0$ is at $y = 0$ and is moving in the $-y$-direction.

11.6 Graphing Waves

Problems 30–32. The graphs show displacement y as a function of time t for five transverse waves at a fixed location x. The displacement and time axes use the same scale in each graph.

30. Rank the waves in order of frequency, largest to smallest.
31. Rank the waves in order of amplitude, largest to smallest.
32. Rank the waves in order of maximum transverse speed, largest to smallest.

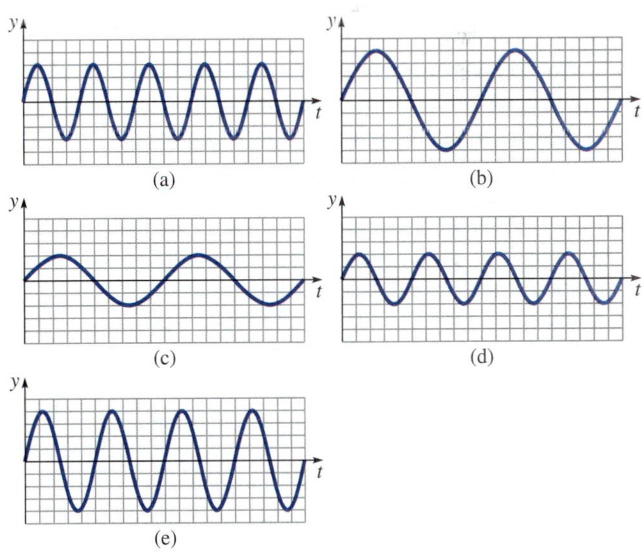

Problems 30–32

33. A sine wave is traveling to the right on a cord. The lighter line in the figure represents the shape of the cord at time $t = 0$; the darker line represents the shape of the cord at time $t = 0.10$ s. (Note that the horizontal and vertical scales are different.) What are (a) the amplitude and (b) the wavelength of the wave?

(c) What is the speed of the wave? What are (d) the frequency and (e) the period of the wave?

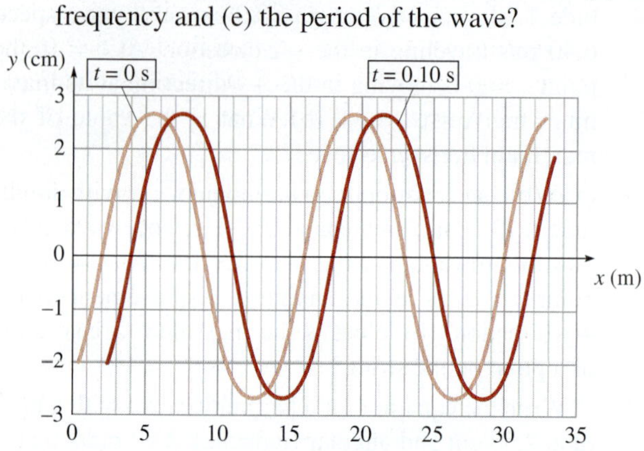

34. (a) Plot a graph for

$$y(x, t) = (4.0 \text{ cm}) \sin [(378 \text{ rad/s})t - (314 \text{ rad/cm})x]$$

versus x at $t = 0$ and at $t = \frac{1}{480}$ s. From the plots determine the amplitude, wavelength, and speed of the wave. (b) For the same function, plot a graph of $y(x, t)$ versus t at $x = 0$ and find the period of the vibration. Show that $\lambda = vT$.

35. For a transverse wave on a string described by

$$y(x, t) = (0.0050 \text{ m}) \cos [(4.0\pi \text{ rad/s})t - (1.0\pi \text{ rad/m})x]$$

find the maximum speed and the maximum acceleration of a point on the string. Plot graphs for one cycle of displacement y versus t, velocity v_y versus t, and acceleration a_y versus t at the point $x = 0$.

36. A transverse wave on a string is described by

$$y(x, t) = (1.2 \text{ mm}) \sin [(2.0\pi \text{ rad/s})t - (0.50\pi \text{ rad/m})x]$$

Plot the displacement y and the velocity v_y versus t for one complete cycle of the point $x = 0$ on the string.

37. Sketch a graph of y versus x for the function

$$y(x, t) = (0.80 \text{ mm}) \sin (kx + \omega t)$$

for the times $t = 0$ and 0.96 s. Make the graphs on the same axes, using a solid line for the first and a dashed line for the second. Use the values $k = (\pi/5.0)$ rad/cm and $\omega = (\pi/6.0)$ rad/s. Is the wave traveling in the $-x$-direction or in the $+x$-direction?

38. ✦ The drawing shows a snapshot of a transverse wave traveling along a string at 10.0 m/s. The equation for the wave is $y(x, t) = A \cos (\omega t + kx)$. (a) Is the wave moving to the right or to the left? (b) What are the numerical values of A, ω, and k? (c) At what times could this snapshot have been taken? (Give the three smallest nonnegative possibilities.)

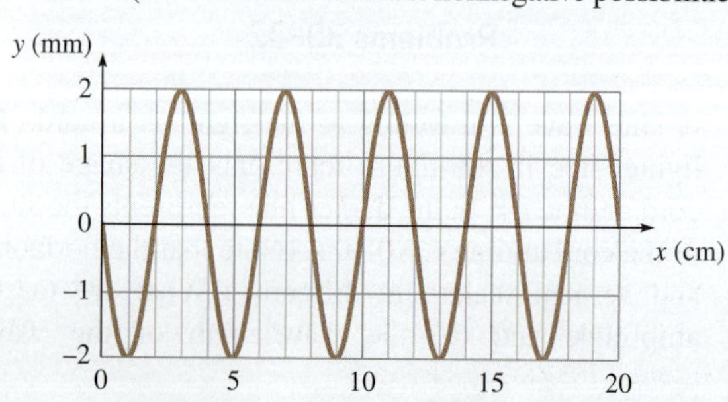

11.7 Principle of Superposition

39. Two pulses on a cord at time $t = 0$ are moving toward each other; the speed of each pulse is 40 cm/s. Sketch the shape of the cord at 0.15, 0.25, and 0.30 s.

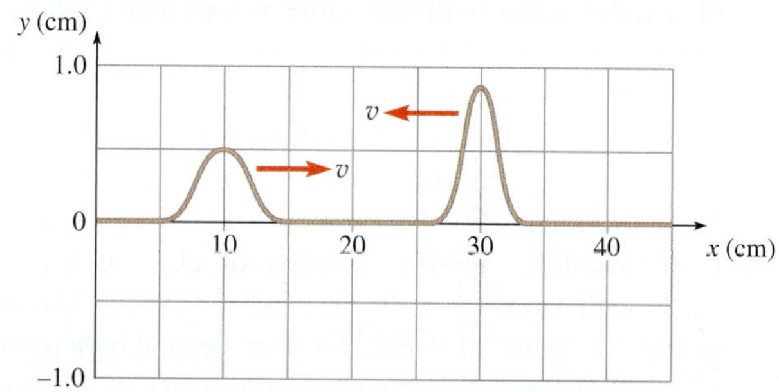

40. Two pulses on a cord at time $t = 0$ are moving toward each other; the speed of each pulse is 2.5 m/s. Sketch the shape of the cord at 0.60, 0.80, and 0.90 s.

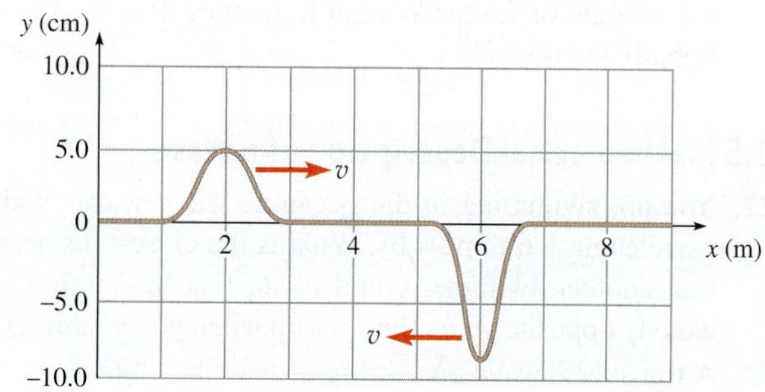

41. Two identical sine waves are described by $y_1 = A \sin (\omega t + kx)$ and $y_2 = A \sin (\omega t + kx + \pi/3)$. Plot graphs of y_1 versus t and y_2 versus t on the same axes for the point $x = 0$. Plot y versus t for the superposition of the two waves at $x = 0$ and estimate its amplitude.

42. ✦ A traveling sine wave is the result of the superposition of two sine waves with equal amplitudes, wavelengths, and frequencies: $y_1 = A \sin (\omega t + kx)$ and $y_2 = A \sin (\omega t + kx - \phi)$. The two component waves each have amplitude $A = 5.00$ cm. If the superposition wave has amplitude 6.69 cm, what is the *phase difference* $\Delta\phi$ between the component waves? [*Hint*: Use the trigonometric identity (Appendix A.7) for $\sin \alpha + \sin \beta$ to find $y = y_1 + y_2$, and identify the new amplitude in terms of the original amplitude.]

11.8 Reflection and Refraction

43. Light of wavelength 0.500 μm (in air) enters the water in a swimming pool. The speed of light in water is 0.750 times the speed in air. What is the wavelength of the light in water?

44. The speed of ultrasound in fat is 1450 m/s, and in muscle it is 1585 m/s. By what percentages do the frequency and wavelength of ultrasound change when passing from fat into muscle?

45. At $t = 0$, the wave pulses shown are moving toward each other on a string. The wave speed is 20 m/s. Use the principle of superposition to sketch the shape of the string at $t = 2.0$ ms.

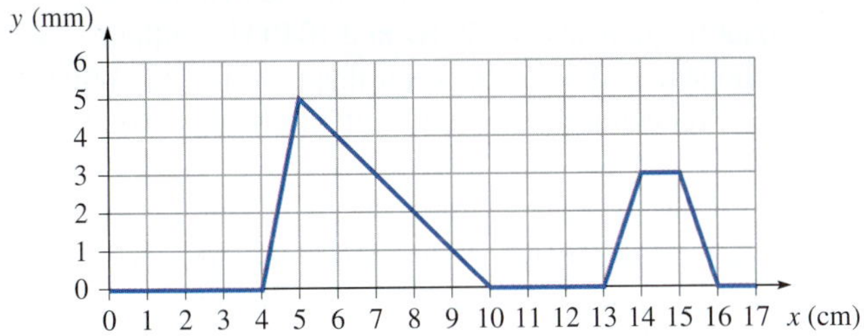

Problems 46–49. The pulse of Problem 9 travels to the right on a string whose ends at $x = 0$ and $x = 4.0$ m are both fixed in place. Imagine a reflected pulse that begins to move onto the string at an endpoint at the same time the incident pulse reaches that endpoint. The superposition of the incident and reflected pulses gives the shape of the string.

46. ✦ When does the string first look completely flat for $t > 0$?

47. When is the first time for $t > 0$ that the string looks exactly as it does at $t = 0$?

48. ✦ Sketch the shape of the string at $t = 2.2$ s.

49. ✦ Sketch the shape of the string at $t = 1.6$ s.

11.9 Interference and Diffraction

50. A sound wave with a frequency of 400.0 Hz is incident upon a set of stairs. The reflected waves from the vertical surfaces of adjacent steps interfere destructively. What is the minimum tread depth of a step for this to occur?

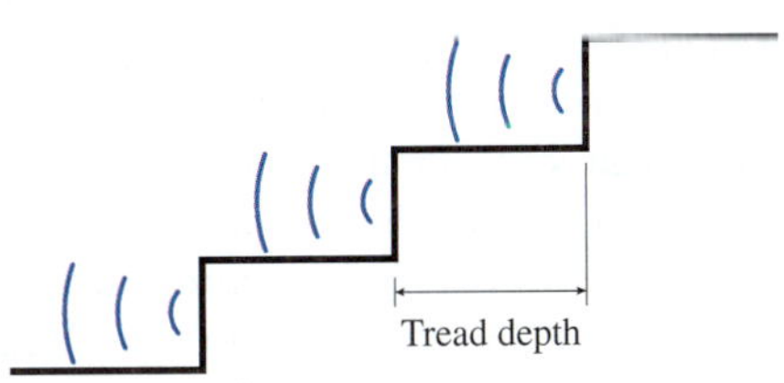

51. Two speakers vibrate in phase with one another at 523 Hz. At certain points in the room, the sound waves from the two speakers interfere constructively. One such point is 2.28 m from speaker #1 and is between 2 m and 4 m from speaker #2. How far is this point from speaker #2? Find all possible distances between 2 m and 4 m. The speed of sound in air is 343 m/s.

52. Two speakers vibrate in phase with each other at 523 Hz. At certain points in the room, the sound waves from the two speakers interfere destructively. One such point is 2.28 m from speaker #1 and is between 2 m and 4 m from speaker #2. How far is this point from speaker #2? Find all possible distances between 2 m and 4 m. The speed of sound in air is 343 m/s.

53. Two waves with identical frequency but different amplitudes $A_1 = 5.0$ cm and $A_2 = 3.0$ cm, occupy the same region of space (i.e., are superimposed). (a) What is the amplitude of the resulting wave if they interfere constructively? (b) What is its amplitude if they interfere destructively? (c) By what factor is the amplitude for constructive interference larger than the amplitude for destructive interference?

54. Two waves with identical frequency but different amplitudes $A_1 = 6.0$ cm and $A_2 = 3.0$ cm, occupy the same region of space (i.e., are superimposed). (a) What is the amplitude of the resulting wave if they interfere constructively? (b) What is its amplitude if they interfere destructively? (c) By what factor is the intensity larger for constructive interference than the intensity for destructive interference?

55. A sound wave with intensity 25 mW/m^2 interferes constructively with a sound wave that has an intensity of 15 mW/m^2. What is the intensity of the superposition of the two? (connect tutorial: superposition)

56. A sound wave with intensity 25 mW/m^2 interferes destructively with a sound wave that has an intensity of 28 mW/m^2. What is the intensity of the superposition of the two?

57. Two coherent sound waves have intensities of 0.040 W/m^2 and 0.090 W/m^2 where you are listening. (a) If the waves interfere constructively, what is the intensity that you hear? (b) What if they interfere destructively? (c) If they were incoherent, what would be the intensity? [*Hint*: If your answers are correct, then (c) is the average of (a) and (b).]

58. ✦ While testing speakers for a concert, Tomás sets up two speakers to produce sound waves at the same frequency, which is between 100 Hz and 150 Hz. The two speakers vibrate in phase with each other. He notices that when he listens at certain locations, the sound is very soft (a minimum intensity compared to nearby points). One such point is 25.8 m from one speaker and 37.1 m from the other. What are the possible frequencies of the sound waves coming from the speakers? (The speed of sound in air is 343 m/s.)

11.10 Standing Waves

59. Five stretched strings have the following properties. Rank the strings according to their fundamental frequencies (for transverse standing waves), from greatest to least.
(a) length L, total mass m, tension F
(b) length $2L$, total mass m, tension F
(c) length L, total mass $2m$, tension F
(d) length L, total mass m, tension $2F$
(e) length $2L$, total mass $2m$, tension F

60. A guitar string has a fundamental frequency f. The tension in the string is increased by 1.0%. Ignoring the very small stretch of the string, how does the fundamental frequency change?

61. A guitar string has a fundamental frequency *f*. The player presses on a fret, reducing the vibrating part of the string to 5/6 of its original length. Ignoring the very small change in tension, by what factor does the fundamental frequency change?

62. A standing wave has wavenumber 200 rad/m. What is the distance between two adjacent nodes?

63. A string 2.0 m long is held fixed at both ends. If a sharp blow is applied to the string at its center, it takes 0.050 s for the pulse to travel to the ends of the string and return to the middle. What are the lowest three standing wave frequencies for this string?

64. The tension in a guitar string is increased by 15%. What happens to the fundamental frequency of the string?

65. In order to decrease the fundamental frequency of a guitar string by 4.0%, by what percentage should you reduce the tension?

66. A harpsichord string of length 1.50 m and linear mass density 25.0 mg/m vibrates at a (fundamental) frequency of 450.0 Hz. (a) What is the speed of the transverse string waves? (b) What is the tension? (c) What are the wavelength and frequency of the sound wave in air produced by vibration of the string? (The speed of sound in air at room temperature is 340 m/s.)

67. A cord of length 1.5 m is fixed at both ends. Its mass per unit length is 1.2 g/m and the tension is 12 N. (a) What is the frequency of the fundamental oscillation? (b) What tension is required to make the $n = 3$ mode have a frequency of 0.50 kHz?

68. Tension is maintained in a string by attaching one end to a wall and by hanging a 2.20 kg object from the other end of the string after it passes over a pulley that is 2.00 m from the wall. The string has a mass per unit length of 3.55 mg/m. What is the fundamental frequency of this string?

69. A guitar's E-string has length 65 cm and is stretched to a tension of 82 N. It vibrates at a fundamental frequency of 329.63 Hz. Determine the mass per unit length of the string.

70. A 1.6 m long string fixed at both ends vibrates at resonant frequencies of 780 Hz and 1040 Hz, with no other resonant frequency between these values. (a) What is the fundamental frequency of this string? (b) If the tension in the string is 1200 N, what is the total mass of the string?

71. In a lab experiment, a string has a mass per unit length of 0.120 g/m. It is attached to a vibrating device and weight similar to that shown in Figure 11.23. The vibrator oscillates at a constant frequency of 110 Hz. How heavy should the weight be in order to produce standing waves in a string of length 42 cm? Give the three largest possibilities.

72. ✦ The longest "string" (a thick metal wire) on a particular piano is 2.0 m long and has a tension of 300.0 N. It vibrates with a fundamental frequency of 27.5 Hz. What is the total mass of the wire?

Collaborative Problems

73. When the string of a guitar is pressed against a fret, the shortened string vibrates at a fundamental frequency 5.95% higher than when the previous fret is pressed. If the length of the part of the string that is free to vibrate is 64.8 cm, how far from one end of the string are the first three frets located?

74. A guitar string has a fundamental frequency of 300.0 Hz. (a) What are the next three lowest standing wave frequencies? (b) If you press a finger *lightly* against the string at its midpoint so that both sides of the string can still vibrate, you create a node at the midpoint. What are the lowest four standing wave frequencies now? (c) If you press *hard* at the same point, only one side of the string can vibrate. What are the lowest four standing wave frequencies?

75. Ⓒ The formula for the speed of transverse waves on a spring is the same as for a string. (a) A spring is stretched to a length much greater than its relaxed length. Explain why the tension in the spring is approximately proportional to the length. (b) A wave takes 4.00 s to travel from one end of such a spring to the other. Then the length is increased 10.0%. Now how long does a wave take to travel the length of the spring? [*Hint*: Is the mass per unit length constant?]

76. Ⓒ The drawing shows a snapshot of a transverse wave moving to the left on a string.

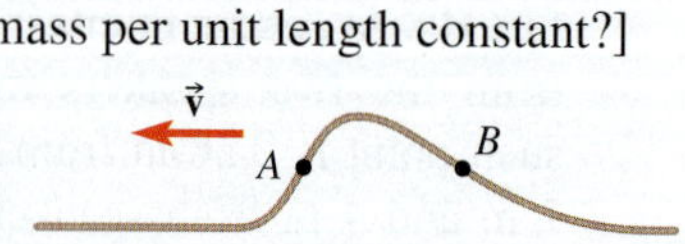

Problems 76 and 94

The wave speed is 10.0 m/s. At the instant the snapshot is taken, (a) in what direction is point *A* moving? (b) In what direction is point B moving? (c) At which of these points is the speed of the string segment (not the wave speed) larger? Explain. (d) How do your answers change if the wave moves to the right instead?

77. ✦ Two speakers spaced a distance 1.5 m apart emit coherent sound waves at a frequency of 680 Hz in all directions. The waves start out in phase with each other. A listener walks in a circle of radius greater than 1 m centered on the midpoint of the two speakers. At how many points does the listener observe destructive interference? The listener and the speakers are all in the same horizontal plane and the speed of sound is 340 m/s. [*Hint*: Start with a diagram; then determine the *maximum* path difference between the two waves at points on the circle.] Experiments like this must be done in a special room so that reflections are negligible.

Comprehensive Problems

78. The speed of waves on a lake depends on frequency. For waves of frequency 1.0 Hz, the wave speed is 1.56 m/s; for 2.0 Hz waves, the speed is 0.78 m/s. The 2.0 Hz waves from a speedboat's wake reach you 120 s after the 1.0 Hz waves generated by the same boat. How far away is the boat?

79. A transverse wave on a string is described by

$$y(x, t) = (1.2 \text{ cm}) \sin [(0.50\pi \text{ rad/s})t - (1.00\pi \text{ rad/m})x]$$

Find the maximum velocity and the maximum acceleration of a point on the string. Plot graphs for displacement y versus t, velocity v_y versus t, and acceleration a_y versus t at $x = 0$.

80. An underground explosion sends out both transverse (S waves) and longitudinal (P waves) mechanical wave pulses (seismic waves) through the crust of the Earth. Suppose the speed of transverse waves is 8.0 km/s and that of longitudinal waves is 10.0 km/s. On one occasion, both waves follow the same path from a source to a detector (a seismograph); the longitudinal pulse arrives 2.0 s before the transverse pulse. What is the distance between the source and the detector?

81. A sign is hanging from a single metal wire, as shown in part (a) of the drawing. The shop owner notices that the wire vibrates at a fundamental resonance frequency of 660 Hz, which irritates his customers. In an attempt to fix the problem, the shop owner cuts the wire in half and hangs the sign from the two halves, as shown in part (b). Assuming the tension in the two wires to be the same, what is the new fundamental frequency of each wire?

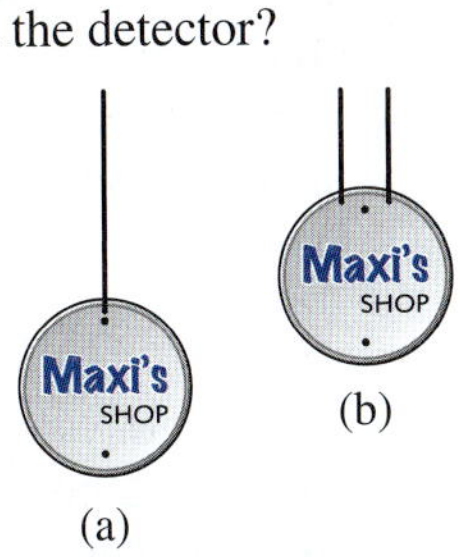

82. (a) Write an equation for a surface seismic wave moving along the $-x$-axis with amplitude 2.0 cm, period 4.0 s, and wavelength 4.0 km. Assume the wave is harmonic, x is measured in meters, and t is measured in seconds. (b) What is the maximum speed of the ground as the wave moves by? (c) What is the wave speed?

83. A seismic wave is described by the equation

$$y(x, t) = (7.00 \text{ cm}) \cos [(6.00\pi \text{ rad/cm})x + (20.0\pi \text{ rad/s})t]$$

The wave travels through a uniform medium in the x-direction. (a) Is this wave moving right ($+x$-direction) or left ($-x$-direction)? (b) How far from their equilibrium positions do the particles in the medium move? (c) What is the frequency of this wave? (d) What is the wavelength of this wave? (e) What is the wave speed? (f) Describe the motion of a particle that is at $y = 7.00$ cm and $x = 0$ when $t = 0$. (g) Is this wave transverse or longitudinal?

84. A stretched string has a fundamental frequency of 847 Hz. What is the fundamental frequency if the tension is increased by a factor of 3.0?

85. A sound wave of frequency 1231 Hz travels through air directly toward a wall, then through the wall out into air again. If the initial speed of the sound wave is 341 m/s and its speed in the wall is 620 m/s, what are (a) the initial wavelength of the sound, (b) the wavelength of the sound in the wall, and (c) the wavelength of the sound when it exits the wall on the other side?

86. When a standing wave is produced in a string fixed at both ends, the string oscillates so fast that it looks like a blur. You want to photograph the string when it is at positions A, B, and C shown in the figure. The tension in the string is 2.00 N and its mass per unit length is 0.200 g/m. The string's length is 0.720 m. Assume that you take your first picture when the string is in position A and let that be time $t = 0$. What are the first two times after $t = 0$ at which you can photograph the string in each of the positions A, B, and C?

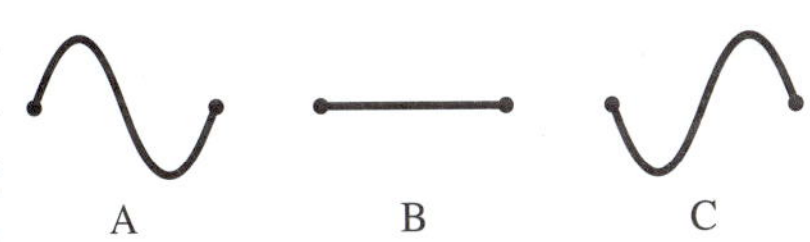

87. Consider the following equations for traveling waves on two different strings:

I. $y(x, t) = (1.50 \text{ cm}) \sin [(4.00 \text{ cm}^{-1})x + (6.00 \text{ s}^{-1})t]$

II. $y(x, t) = (4.50 \text{ cm}) \sin [(3.00 \text{ cm}^{-1})x - (3.00 \text{ s}^{-1})t]$

(a) Which wave has the faster wave speed? What is that speed? (b) Which wave has the longer wavelength? What is that wavelength? (c) Which wave has the faster maximum speed of a point in the medium? What is that speed? (d) Which wave is moving in the positive x-direction?

88. The lowest frequency string on a guitar is 65.5 cm long and is tuned to 82 Hz. (a) If the string has a mass of 3.31 g, what is the tension in the string? (b) By fingering the guitar at the fifth fret, you shorten the vibrating length of the string, thereby changing the

fundamental frequency of this string to match that of the next-highest-frequency string on the guitar, 110 Hz. How long is the lowest frequency string when it is fingered at the fifth fret?

89. ✦ (a) Use a graphing calculator or computer graphing program to plot y versus x for the function

$$y(x, t) = (5.0\text{ cm})\,[\sin(kx - \omega t) + \sin(kx + \omega t)]$$

for the times $t = 0$, 1.0 s, and 2.0 s. Use the values $k = (\pi/5.0)$ rad/cm and $\omega = (\pi/6.0)$ rad/s. (b) Is this a traveling wave? If not, what kind of wave is it?

90. ✦ Show that the amplitudes of the graphs you made in Problem 89 satisfy the equation $A' = 2A\cos(\omega t)$, where A' is the amplitude of the wave you plotted and A is 5.0 cm, the amplitude of the waves that were added together.

91. ✦ The graph shows ground vibrations recorded by a seismograph 180 km from the focus of a small earthquake. It took the waves 30.0 s to travel from their source to the seismograph. Estimate the wavelength.

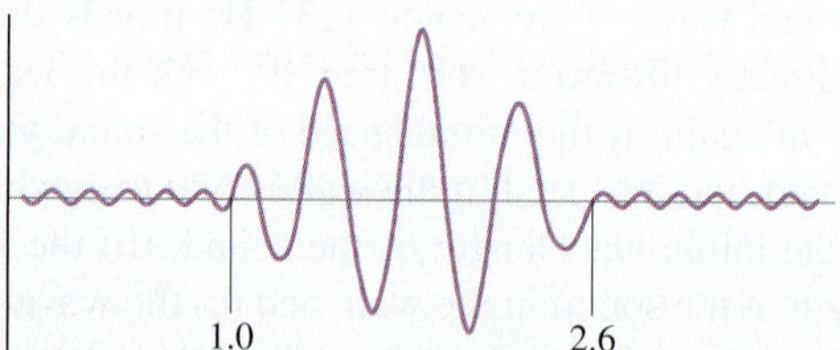

92. Deep-water waves are *dispersive* (their wave speed depends on the wavelength). The restoring force is provided by gravity. Using dimensional analysis, find out how the speed of deep-water waves depends on wavelength λ, assuming that λ and g are the only relevant quantities. (Mass density does not enter into the expression because the restoring force, arising from the weight of the water, is itself proportional to the mass density.)

93. In contrast to deep-water waves, shallow ripples on the surface of a pond are due to surface tension. The surface tension γ of water characterizes the restoring force; the mass density ρ of water characterizes the water's inertia. Use dimensional analysis to determine whether the surface waves are *dispersive* (the wave speed depends on the wavelength) or *nondispersive* (their wave speed is independent of wavelength). [*Hint*: Start by assuming that the wave speed is determined by γ, ρ, and the wavelength λ.]

94. ✦ Consider a point on the flat segment to the left of point A in the drawing with Problem 76. Plot the position of that point and the velocity of that point as a function of time as the wave passes the point.

Review and Synthesis

95. Refer to the pulse in Problem 9. (a) What is the speed of propagation of the pulse? (b) At what average speed does the point at $x = 2.0$ m move during this time interval?

96. ✦ Suppose that a string of length L and mass m is under tension F. (a) Show that $\sqrt{FL/m}$ has units of speed. (b) Show that there is no other combination of L, m, and F with units of speed. [*Hint*: Of the dimensions of the three quantities L, m, and F, only F includes time.] Thus, the speed of transverse waves on the string can only be some dimensionless constant times $\sqrt{FL/m}$.

97. A tetherball set has a ball with mass 0.411 kg and a nylon string with diameter 2.50 mm, Young's modulus 4.00 GPa, and density 1150 kg/m^3. The nylon string has a length of 2.200 m when the ball is at rest (hanging straight down). While playing tetherball, Monty hits the ball around the pole so it moves in a horizontal circle with the string at an angle of 65.0° to the pole. (a) How much does the string stretch compared with when the ball is at rest? (b) What is the ball's kinetic energy? (c) How long would it take a transverse wave pulse to travel the length of the string from the ball to the top of the pole?

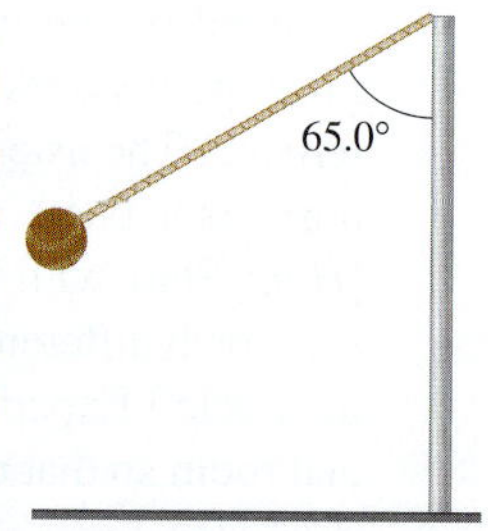

98. A Foucault pendulum has an object with a mass of 15.0 kg hung by a thin 14.0 m wire. (a) What is the oscillation frequency of this pendulum? (b) If the pendulum has a maximum oscillation angle of 6.10°, what is the maximum speed of this pendulum? (c) What is the maximum tension in the wire? (d) If the wire has a mass of 10.0 g, what is the fundamental frequency of a standing wave on the wire when it is at maximum tension?

99. A transverse wave on a string is described by

$$y(x, t) = A\cos(\omega t + kx)$$

where $A = 0.40$ mm, $\omega = 39.27$ rad/s, and $k = 6.0$ rad/m. Draw a motion diagram for the point $x = 0$ for the times $t = 0$, 0.010 s, 0, 0.020 s, 0.030 s, and 0.040 s.

100. A transverse wave on a string has amplitude 4.0 mm, angular frequency 600 rad/s, and wavenumber 6.0 rad/m. (a) What is the maximum transverse speed of a point on the string? (b) What is the *average* transverse speed of a point on the string? [*Hint*: How much time does it take a point to move from $y = 0$ to $y = +A$?] (c) Is the average speed one-half of the maximum? If not, explain why it doesn't have to be.

101. A transverse wave on a string is described by

$$y(x, t) = (2.00\text{ mm})\sin[(157\text{ rad/s})t + (7.85\text{ rad/m})x].$$

Sketch graphs of the transverse velocity and acceleration of the point $x = 0$ as functions of time, showing one complete cycle.

102. ✦ A harpsichord string is made of yellow brass (Young's modulus 90 GPa, tensile strength 0.63 GPa, mass density 8500 kg/m^3). When tuned correctly, the tension in the string is 59.4 N, which is 93% of the maximum tension that the string can endure without breaking. The length of the string that is free to vibrate is 9.4 cm. What is the fundamental frequency?

Answers to Practice Problems

11.1 (a) 8.9 m/s; (b) 13 m/s

11.2 (a) Yes, the traveling wave retains its shape; (b) it travels in the $+x$-direction because the t and x/v terms have *opposite* signs; (c) the wave speed is 8.0 m/s.

11.3

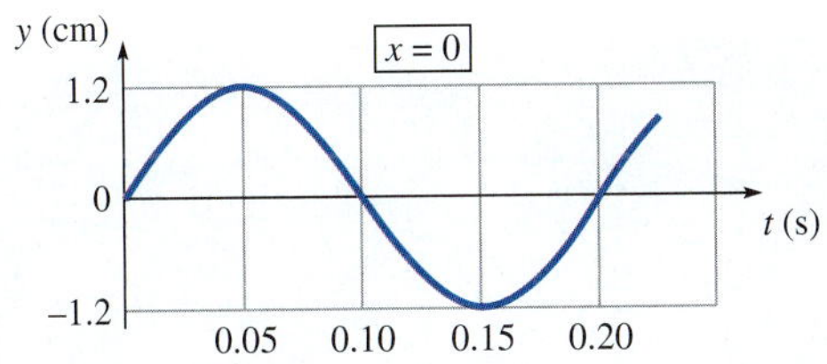

(a)

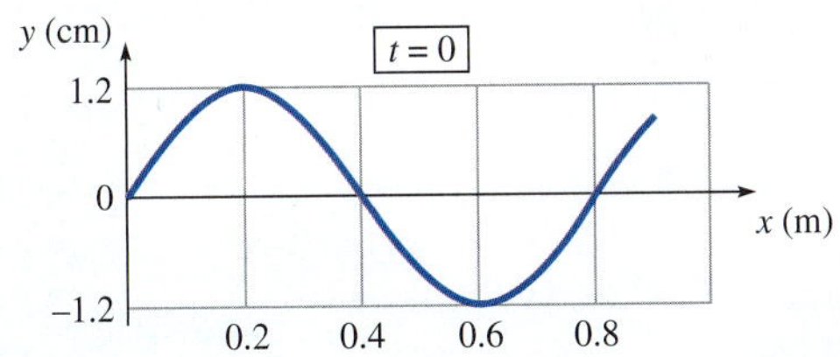

(b)

(c) $T = 0.200$ s; (d) $\lambda = 0.80$ m; (e) $A = 1.2$ cm; (f) $v = 4.0$ m/s; (g) the wave travels in the $-x$-direction because the signs of the terms containing x and t are the same.

11.4

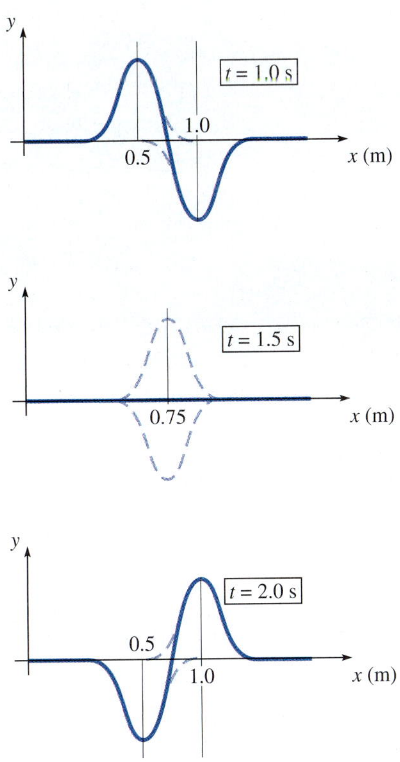

11.5 (a) 620 Hz; (b) 8.5 m

11.6 9.0

11.7 140 Hz

Answers to Checkpoints

11.1 For an isotropic source, $I \propto 1/r^2$. At a distance 10^2 times as far from the tower, the intensity is $10^{-4} \times 0.090\ \text{W/m}^2 = 9.0\ \mu\text{W/m}^2$.

11.2 Since transverse waves do not travel through the core but longitudinal waves do, some part of the core is a liquid that cannot support the transmission of a transverse wave. A longitudinal wave can create compressions and rarefactions in the liquid and travel through the core.

11.3 (b) = (d), (a) = (e), (c)

11.4 The period T is the time for one cycle. During one period, the wave travels 20 km at a speed of 4.0 km/s. Then the period is (20 km)/(4.0 km/s) = 5.0 s.

11.5 Equations (11-8c) and (11-8d). The terms ωt and kx have the same sign.

11.9 (a) The intensity is the sum of the intensities of the two waves: $10.0I_2$. (b) As in Example 11.6, $A_{max} = 4.0A_2$. Intensity is proportional to amplitude squared, so $I_{max}/I_2 = (A_{max}/A_2)^2 = 16.0$ and $I_{min}/I_2 = (A_{min}/A_2)^2 = 4.0$. The maximum and minimum possible intensities are $16.0I_2$ and $4.0I_2$, respectively.

11.10 The nodes are evenly spaced, so the nodes are at $x = 0$, 20 cm, 40 cm, 60 cm, 80 cm, and 100 cm. The distance between nodes is half the wavelength, so the wavelength is 40 cm.

CHAPTER

12 Sound

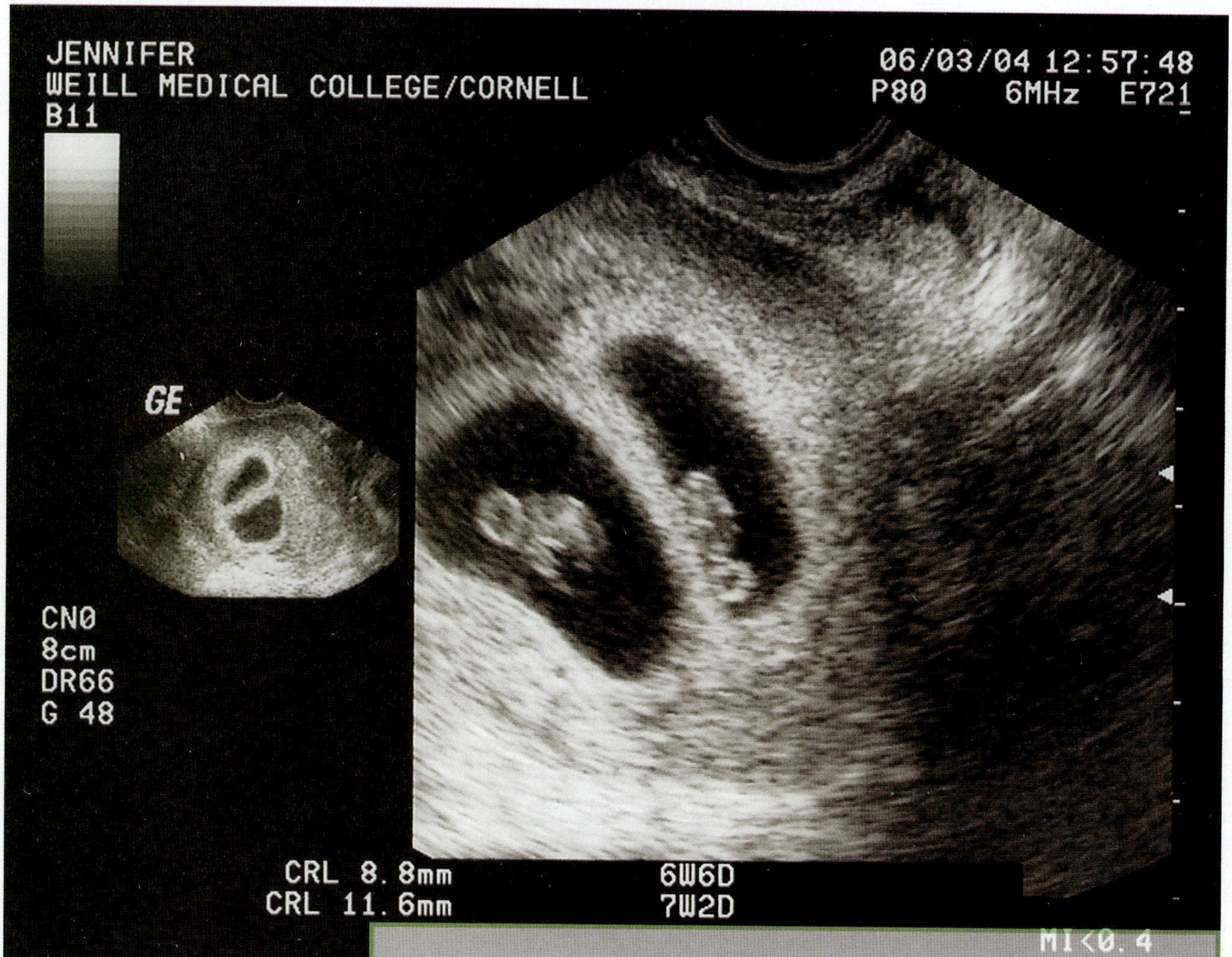

Congratulations! You're expecting twins

Ultrasonic imaging of the fetus is an important part of prenatal care. Could an image of the fetus be produced just as well using sound in the audible range rather than ultrasound? Why is ultrasound used rather than some other imaging technology, such as x-rays? Are there other medical applications of ultrasound? (See pp. 456–457 for the answers.)

BIOMEDICAL APPLICATIONS

- Frequency ranges of animal hearing (Sec. 12.1; Probs. 3 and 26)
- The human ear (Secs. 12.3, 12.6; CQs 4 and 5; Probs. 3, 14–18, 63, 67–69)
- Echolocation (Sec. 12.9; Probs. 4, 5, 55, 56, 70–72)
- Medical applications of ultrasound (Sec. 12.9; CQ 8)
- Angiodynography (Probs. 49, 57, 58)

Concepts & Skills to Review

- gauge pressure (Section 9.5)
- bulk modulus (Section 10.4)
- relation between energy and amplitude in SHM (Section 10.5)
- period and frequency in SHM (Section 10.6)
- longitudinal waves, intensity, standing waves, superposition principle (Chapter 11)
- logarithms (Appendix A.3)

12.1 SOUND WAVES

When a guitar string is plucked, a transverse wave travels along the string. The wave on the string is not what we hear, of course, since the string has no direct connection to our eardrums. The vibration of the string is transmitted through the bridge to the body of the guitar, which in turn transmits the vibration to the air—a sound wave. A transverse wave on a guitar string is not a sound wave, though it does *cause* a sound wave.

In the absence of a sound wave, molecules in the air dart around in random directions. On average, they are uniformly distributed and the pressure is the same everywhere (ignoring the insignificant variation of pressure due to small changes in altitude). In a sound wave, the uniform distribution of molecules is disturbed. A loudspeaker produces pressure fluctuations that travel through the air in all directions (Fig. 12.1). In some regions (*compressions*), the molecules are bunched together and the pressure is higher than the average pressure. In other regions (*rarefactions*), the molecules are spread out and the pressure is lower than average. The sound wave can be described mathematically by the gauge pressure p (the difference between the pressure at a given point and the average pressure in the surroundings) as a function of position and time (Fig. 12.2a).

The speaker cone produces these pressure variations by displacing molecules in the air from their uniform distribution (Fig. 12.2b). When the cone moves to the left of its equilibrium position, air spreads into a region of lower pressure (rarefaction). When the cone moves to the right, air is squeezed together into a region of higher pressure (compression).

Thus, the regions of higher and lower pressure are formed when molecules are displaced from a uniform distribution. A sound wave can be described equally well by

Figure 12.1 The vibrating speaker cones in this boombox create alternating regions of high and low pressure in the air. Air nearby is affected by a net force due to the nonuniform air pressure; as a result, variations in pressure travel in all directions away from the speakers. This traveling disturbance is a sound wave.

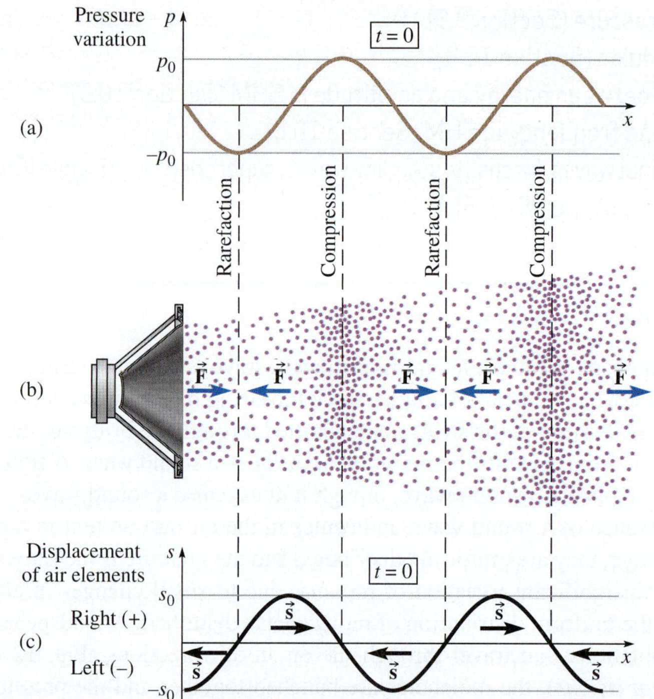

Figure 12.2 A sound wave generated by a loudspeaker. (a) Graph of the pressure variation p of the air as a function of position x. Pressure is high where air is squeezed together and low where it is more spread out. (b) Elements of the air are displaced from their equilibrium positions. Since the pressure is not uniform, air elements experience a net force due to air pressure; the force arrows indicate the direction of this net force. The force is always directed away from a compression (higher pressure) and toward a rarefaction. (c) Graph of the displacement s of an air element from its equilibrium position x as a function of x; the arrows indicate the directions of the displacements in each region. Air elements are displaced leftward or rightward toward compressions and away from rarefactions. Elements at the center of each compression or rarefaction are at their equilibrium positions ($s = 0$).

the displacement s of an *element* of the air—a region of air that can be considered to move together as a unit (Fig. 12.2c). An element is much smaller than the wavelength of the wave, but still large enough to contain many molecules. For a sinusoidal wave, elements at points of maximum or minimum pressure have zero displacement, while the neighboring elements move in toward them (a compression) or away from them (a rarefaction). Conversely, where the gauge pressure is zero, the displacement of an element has its maximum magnitude.

If the pressure is higher on one side than on the other, the net force pushes air toward the side with lower pressure. The uneven distribution of pressure results in air molecules being pushed toward rarefactions and away from compressions, as shown by the force arrows in Fig. 12.2b. Note that the directions of these force arrows, pointing opposite to the displacement arrows in a corresponding region, are such that where there is a compression at a given instant, there will later be a rarefaction, and vice versa; the pressure at a given point fluctuates above and below the average pressure.

Frequency Ranges of Animal Hearing

The human ear responds to sound waves within a limited range of frequencies. We generally consider the **audible range** to extend from 20 Hz to 20 kHz. (The terms **infrasound** and **ultrasound** refer to sound waves with frequencies below 20 Hz and above 20 kHz, respectively.) Very few people can actually hear sounds over that entire range. Even for a person with excellent hearing, the sensitivity of the human ear declines rapidly below 100 Hz and above 10 kHz. Age-related hearing loss is common and affects primarily the high frequencies, which makes it difficult for some older people to understand speech. Repeated or prolonged exposure to loud sounds can also cause hearing loss.

The audible ranges for animals can be quite different. Most mammals can hear frequencies much higher than we can. Dogs can hear frequencies as high as 50 kHz, which is why we can make a dog whistle that is inaudible to humans. Mice can produce and hear sounds with frequencies up to about 90 kHz, higher than what their predators can hear. Bats and bottlenose dolphins can hear frequencies above 100 kHz. Dolphins rely on hearing more than sight for navigation; studies have shown that many dolphins that beach themselves suffer from hearing loss.

Some animals can hear frequencies lower than humans can. Elephants and rhinoceri can hear frequencies down to about 14 Hz and 10 Hz, respectively. Some studies suggest that pigeons and monarch butterflies use infrasound to navigate.

Attenuation of Sound Waves

Attenuation is the decrease in intensity of a sound wave (or any kind of wave) as it propagates. Multiple effects contribute to the attenuation of a sound wave. One is geometric: as the wave propagates away from its source, it spreads out. The intensity decreases because the energy transported by the wave is spread over a larger and larger area. For example, in Section 11.1, we found that the intensity of a wave propagating isotropically away from a point source is proportional to $1/r^2$, where r is the distance from the source. Another cause of attenuation is the absorption of energy by the wave medium. A small fraction of the energy transported by the wave is dissipated in the medium, causing a slight increase in temperature. Thus, the energy transported by the wave decreases as it propagates through the medium. See the text website for more information about attenuation.

connect

12.2 THE SPEED OF SOUND WAVES

For string waves, the restoring force is characterized by the tension in the string F, and the inertia is characterized by the linear mass density μ (mass per unit length). The speed of transverse waves on a string is

$$v = \sqrt{\frac{F}{\mu}} \qquad (11\text{-}4)$$

For sound waves in a fluid, the restoring force is characterized by the bulk modulus B, defined in Section 10.4 as the constant of proportionality between an increase in pressure and the fractional volume change:

$$\Delta P = -B\frac{\Delta V}{V} \qquad (10\text{-}10)$$

The inertia of the fluid is characterized by its mass density ρ. Following our dictum "more restoring force makes faster waves; more inertia makes slower waves," we expect the speed of sound to be faster in a medium with a larger bulk modulus (harder to compress means more restoring force) and slower in a medium with a larger density. By analogy with Eq. (11-4), we might *guess* that

$$v = \sqrt{\frac{\text{a measure of the restoring force}}{\text{a measure of the inertia}}} = \sqrt{\frac{B}{\rho}} \quad \text{(in fluids)} \qquad (12\text{-}1)$$

This guess turns out to be correct; Eq. (12-1) is the correct expression for the speed of sound in fluids.

CONNECTION:

Just as for transverse waves on a string, the speed of sound waves is determined by a balance between two characteristics of the wave medium: the restoring force and the inertia.

Temperature Dependence of the Speed of Sound in a Gas The bulk modulus B of an ideal gas is directly proportional to the density ρ and to T, the *absolute temperature* ($B \propto \rho T$). As a result, the speed of sound in an ideal gas is proportional to

Table 12.1 Speed of Sound in Various Materials (at 0°C and 1 atm unless otherwise noted)

Medium	Speed (m/s)	Medium	Speed (m/s)
Carbon dioxide	259	Blood (37°C)	1570
Air (dry)	331	Muscle (37°C)	1580
Nitrogen	334	Lead	1322
Air (dry, 20°C)	343	Concrete	3100
Helium	972	Copper	3560
Hydrogen	1284	Bone (37°C)	4000
Mercury (25°C)	1450	Pyrex glass	5640
Fat (37°C)	1450	Aluminum	5100
Water (25°C)	1493	Steel	5790
Seawater (25°C)	1533	Granite	6500

the square root of the absolute temperature, but is independent of pressure and density (at a fixed temperature):

$$v = \sqrt{\frac{B}{\rho}} \propto \sqrt{\frac{\rho T}{\rho}} \propto \sqrt{T} \qquad \text{(ideal gas)}$$

The SI unit of absolute temperature is the kelvin (symbol K). To find absolute temperature in kelvins, add 273.15 to the temperature in degrees Celsius:

$$T\ (\text{in K}) = T_C\ (\text{in °C}) + 273.15 \qquad (12\text{-}2)$$

Since $v \propto \sqrt{T}$, the speed of sound in an ideal gas at any absolute temperature T can be found if it is known at one temperature:

Temperature dependence of the speed of sound in a gas

$$v = v_0\sqrt{\frac{T}{T_0}} \qquad (12\text{-}3)$$

where the speed of sound is v_0 at absolute temperature T_0. For example, the speed of sound in dry air (0% humidity) at 0°C (273.15 K) is 331.3 m/s. At 20°C (293.15 K), the speed of sound in dry air is

$$v = 331.3\ \text{m/s} \times \sqrt{\frac{293.15\ \text{K}}{273.15\ \text{K}}} = 343.2\ \text{m/s}$$

An approximate formula that can be used for the speed of sound in dry air is

$$v = 331.3\ \text{m/s} + \left(0.6\,\frac{\text{m/s}}{\text{°C}}\right)T_C \qquad (12\text{-}4)$$

where T_C is air temperature *in degrees Celsius* (see Problem 11). Equation (12-4) gives speeds accurate to better than 1% all the way from −66°C to +89°C.

The speed of sound in air increases slightly with the concentration of water vapor. At 37°C and 100% relative humidity, the speed of sound is about 1% larger than in dry air at the same temperature. Please assume that problems involve dry air unless otherwise stated.

Speed of Sound in a Solid The speed of sound in a *solid* depends on the Young's modulus Y and the shear modulus S. For sound waves traveling along the length of a thin solid rod, the speed is approximately

$$v = \sqrt{\frac{Y}{\rho}} \qquad \text{(thin solid rod)} \qquad (12\text{-}5)$$

Table 12.1 gives the speed of sound in various materials.

Conceptual Example 12.1

Speed of Sound in Hydrogen and Mercury

From Table 12.1, the speed of sound in hydrogen gas at 0°C is almost as large as the speed of sound in mercury, even though the density of mercury is 150 000 times larger than the density of hydrogen. How is that possible? Shouldn't the speed in mercury be much smaller, since it has so much more inertia?

Solution and Discussion The speed of sound depends on *two* characteristics of the medium: the restoring force (measured by the bulk modulus) and the inertia (measured by the density). The bulk modulus of mercury is much larger than the bulk modulus of hydrogen. The bulk modulus is a measure of how hard it is to compress a material. Liquids (e.g., mercury) are much more difficult to compress than are gases. Thus, the restoring forces in mercury are much larger than those in hydrogen; this allows sound to travel a bit faster in mercury than it does in hydrogen gas.

Conceptual Practice Problem 12.1 Speed of Sound in Solids Versus Liquids

Why does sound generally travel faster in a solid than in a liquid?

12.3 AMPLITUDE AND INTENSITY OF SOUND WAVES

Because we can describe a sound wave in two ways—pressure or displacement—the amplitude of a sound wave can take one of two forms: the pressure amplitude p_0 or the displacement amplitude s_0. The pressure amplitude p_0 is the maximum pressure fluctuation above or below the equilibrium pressure; the displacement amplitude s_0 is the maximum displacement of an element of the medium from its equilibrium position. The pressure amplitude is proportional to the displacement amplitude. For a harmonic sound wave at angular frequency ω, an advanced analysis shows that

$$p_0 = \omega v \rho s_0 \qquad (12\text{-}6)$$

where v is the speed of sound and ρ is the mass density of the medium.

Is a larger amplitude sound wave perceived as *louder*? Yes, all other things being equal. However, the relationship between our perception of loudness and the amplitude of a sound wave is complex. Loudness is a subjective aspect of how sound is perceived; it has to do with how the ear responds to sound and how the brain interprets signals from the ear. Perceived loudness turns out to be *roughly* proportional to the logarithm of the amplitude. If the amplitude of a sound wave doubles repeatedly, the perceived loudness does not double; it increases by a series of roughly equal steps.

Discussions of loudness are more often phrased in terms of intensity rather than amplitude since we are interested in how much energy the sound wave carries. The intensity (average power per unit area) of a sinusoidal sound wave is

Intensity and pressure amplitude

$$I = \frac{p_0^2}{2\rho v} \qquad (12\text{-}7)$$

where ρ is the mass density of the medium and v is the speed of sound in that medium. The most important thing to remember is that

Intensity is proportional to amplitude squared.

This is also true for waves other than sound. It is closely related to the fact that energy in SHM is proportional to amplitude squared [see Eq. (10-13)].

Example 12.2

The Brown Creeper

The song of the Brown Creeper (*Certhia americana*) is high in frequency—about 8 kHz. Many people who have lost some of their high-frequency hearing can't hear it at all. Suppose that you are out in the woods and hear the song. If the intensity of the song at your position is 1.4×10^{-8} W/m^2 and the frequency is 6.0 kHz, what are the pressure and displacement amplitudes? (Assume the temperature is 20°C.)

Strategy The displacement and pressure amplitudes are related through Eq. (12-6); the pressure amplitude is related to the intensity through Eq. (12-7). These relationships can be used to solve for both pressure amplitude, p_0, and displacement amplitude, s_0. The density of air at 20°C is $\rho = 1.20$ kg/m^3 (see Table 9.1). The speed of sound in air at 20°C is $v = 343$ m/s. We need to multiply the frequency by 2π to get the angular frequency ω.

Solution Intensity and pressure amplitude are related by

$$I = \frac{p_0^2}{2\rho v} \qquad (12\text{-}7)$$

Solving for p_0,

$$p_0 = \sqrt{2I\rho v}$$
$$= \sqrt{2 \times 1.4 \times 10^{-8}\ \text{W/m}^2 \times 1.20\ \text{kg/m}^3 \times 343\ \text{m/s}}$$
$$= 3.4 \times 10^{-3}\ \text{Pa}$$

The pressure and displacement amplitudes are related by

$$p_0 = \omega v \rho s_0 \qquad (12\text{-}6)$$

Substituting in Eq. (12-7) yields

$$I = \frac{(\omega v \rho s_0)^2}{2\rho v}$$

Solving for s_0,

$$s_0 = \sqrt{\frac{2I}{\rho\omega^2 v}} = \sqrt{\frac{2 \times 1.4 \times 10^{-8}\ \text{W/m}^2}{1.20\ \text{kg/m}^3 \times (2\pi \times 6000\ \text{Hz})^2 \times 343\ \text{m/s}}}$$
$$= 2.2 \times 10^{-10}\ \text{m}$$

Discussion This problem illustrates how sensitive the human ear is. The pressure amplitude is a fluctuation of one part in 30 million in the air pressure. Since the pressure amplitude is 3.4×10^{-3} Pa, the maximum force on the eardrum would be about

$$F_{max} = 3.4 \times 10^{-3}\ \text{N/m}^2 \times 10^{-4}\ \text{m}^2 \approx 3 \times 10^{-7}\ \text{N}$$

which is about the weight of a large amoeba. The displacement amplitude is about the size of an atom.

Practice Problem 12.2 Pressure and Intensity at an Outdoor Concert

At a distance of 5.0 m from the stage at an outdoor rock concert, the sound intensity is 1.0×10^{-4} W/m^2. Estimate the intensity and pressure amplitude at a distance of 25 m if there were no speakers other than those on stage. Explain the assumptions you make.

Sound Intensity Level

The perception of loudness by the human ear is roughly proportional to the *logarithm* of the intensity, making us capable of hearing sound over a wide range of intensities (Table 12.2). By convention, we compare intensities to a reference level $I_0 = 10^{-12}$ W/m^2 known as the **threshold of hearing** because it is roughly the lowest intensity sound wave that can be heard under ideal conditions by a person with excellent hearing.

A sound intensity I is compared with the reference level I_0 by taking the ratio of the two intensities. Suppose a sound has an intensity of 10^{-5} W/m^2; the ratio is

$$\frac{I}{I_0} = \frac{10^{-5}\ \text{W/m}^2}{10^{-12}\ \text{W/m}^2} = 10^7$$

so the intensity is 10^7 times that of the hearing threshold. The power to which 10 is raised is the **sound intensity level** β in units of bels (after Alexander Graham Bell).

Table 12.2 Pressure Amplitudes, Intensities, and Intensity Levels of a Wide Range of Sounds in Air at 20°C (room temperature)

Sound	Pressure Amplitude (atm)	Pressure Amplitude (Pa)	Intensity (W/m^2)	Intensity Level (dB)
Threshold of hearing	3×10^{-10}	3×10^{-5}	10^{-12}	0
Leaves rustling	1×10^{-9}	1×10^{-4}	10^{-11}	10
Whisper (1 m away)	3×10^{-9}	3×10^{-4}	10^{-10}	20
Library background noise	1×10^{-8}	0.001	10^{-9}	30
Living room background noise	3×10^{-8}	0.003	10^{-8}	40
Office or classroom	1×10^{-7}	0.01	10^{-7}	50
Normal conversation at 1 m	3×10^{-7}	0.03	10^{-6}	60
Inside a moving car, light traffic	1×10^{-6}	0.1	10^{-5}	70
City street (heavy traffic)	3×10^{-6}	0.3	10^{-4}	80
Shout (at 1 m); or inside a subway train; risk of hearing damage if exposure lasts several hours	1×10^{-5}	1	10^{-3}	90
Car without muffler at 1 m	3×10^{-5}	3	10^{-2}	100
Construction site	1×10^{-4}	10	10^{-1}	110
Indoor rock concert; threshold of pain; hearing damage occurs rapidly	3×10^{-4}	30	1	120
Jet engine at 30 m	1×10^{-3}	100	10	130

A ratio of 10^7 indicates a sound intensity of 7 bels or, as it is more commonly stated, 70 decibels (dB). Since $\log_{10}(10^x) = x$, the sound intensity level β in decibels is

Sound intensity level

$$\beta = (10 \text{ dB}) \log_{10} \frac{I}{I_0} \qquad (12\text{-}8)$$

(The notation $\log_{10}$ stands for the base-10 logarithm. See Appendix A.3 for a review of the properties of logarithms.) An intensity level of 0 dB corresponds to the threshold of hearing ($I = 10^{-12}$ W/m^2). Although the intensity level is really a pure number, the "units" (dB) remind us what the number means.

Table 12.2 gives the pressure amplitudes, intensities, and intensity levels for a wide range of sounds. Notice that, even for sounds that are quite loud, the pressure fluctuations due to sound waves are small compared to the "background" atmospheric pressure.

CHECKPOINT 12.3

Why doesn't Table 12.2 include a column listing the *displacement* amplitudes of the sound waves?

Example 12.3

Roaring Lion

The sound intensity 0.250 m from a roaring lion is 0.250 W/m^2. What is the sound intensity level in decibels? (Use the usual reference level of $I_0 = 1.00 \times 10^{-12}$ W/m^2.)

Strategy We are given the intensity in W/m^2 and asked for the intensity level in dB. First we find the ratio of the given intensity to the reference level. Then we take the

continued on next page

Example 12.3 continued

logarithm of the result (to get the level in bels) and multiply by 10 (to convert from bels to dB).

Solution The ratio of the intensity to the reference level is

$$\frac{I}{I_0} = \frac{0.250 \text{ W/m}^2}{1.00 \times 10^{-12} \text{ W/m}^2} = 2.50 \times 10^{11}$$

The intensity level in bels is

$$\log_{10} \frac{I}{I_0} = \log_{10} 2.50 \times 10^{11} = 11.4 \text{ bels}$$

The intensity level in decibels is

$$\beta = 11.4 \text{ bels} \times (10 \text{ dB/bel}) = 114 \text{ dB}$$

Discussion As a quick check, 110 dB corresponds to $I = 0.1$ W/m^2 and 120 dB corresponds to $I = 1$ W/m^2; since the intensity is between 0.1 W/m^2 and 1 W/m^2, the intensity level must be between 110 dB and 120 dB.

Practice Problem 12.3 Consequences of a Hole in the Muffler

When rust creates a hole in the muffler of a car, the sound intensity level inside the car is 26 dB higher than when the muffler was intact. By what factor does the intensity increase?

As we saw in Section 11.9, when two sounds are coming from different sources, the waves are incoherent. If we know the intensity of each wave alone at a certain point, then the intensity due to the two waves together at that point is the sum of the two intensities:

$$I = I_1 + I_2 \quad \text{(incoherent waves)}$$

This is *not* true for two coherent waves, where the total intensity depends on the phase difference between the waves. Since there is no fixed phase difference between two incoherent waves, on average there is neither constructive nor destructive interference. The total power per unit area is the sum of the power per unit area of each wave.

Example 12.4

The Sound Intensity Level of Two Lathes

A metal lathe in a workshop produces a 90.0 dB sound intensity level at a distance of 1 m. What is the intensity level when a second identical lathe starts operating? Assume the listener is at the same distance from both lathes.

Strategy The noise is coming from two different machines and, thus, they are incoherent sources. We *cannot* add 90.0 dB to 90.0 dB to get 180.0 dB, which would be a senseless result—two lathes are not going to drown out a jet engine at close range (see Table 12.2). Instead, what doubles is the *intensity*. We must work in terms of intensity rather than intensity level.

Solution First find the intensity due to one lathe:

$$\beta = 90.0 \text{ dB} = (10 \text{ dB}) \log_{10} \frac{I}{I_0}$$

$$\log_{10} \frac{I}{I_0} = 9.00, \qquad \text{so} \qquad \frac{I}{I_0} = 1.00 \times 10^9$$

We could solve for I numerically but it is not necessary. With two machines operating, the intensity doubles, so

$$\frac{I'}{I_0} = 2.00 \times 10^9$$

and the new intensity level is

$$\beta' = (10 \text{ dB}) \log_{10} \frac{I'}{I_0}$$

$$= (10 \text{ dB}) \log_{10} (2.00 \times 10^9) = 93.0 \text{ dB}$$

Discussion The new intensity level is just 3 dB higher than the original one, even though the intensity is twice as big. This turns out to be a general result: a 3 dB increase represents a doubling of the intensity.

Practice Problem 12.4 Intensity Change for an Increment of 5 dB

The maximum recommended exposure time to a sound level of 90 dB is 8 h. For every increase of 5.0 dB in sound level up to 120 dB, the exposure time should be reduced by a factor of 2. (At 120 dB, damage occurs almost immediately; there is no safe exposure time.) By what factor does intensity increase when the intensity level rises 5.0 dB?

Sound intensity level is useful because it roughly approximates the way we perceive loudness. Equal increments in intensity level roughly correspond to equal increases in loudness. Two useful rules of thumb: every time the intensity increases by a *factor* of 10, the intensity level *adds* 10 dB; since $\log_{10} 2 = 0.30$, adding 3.0 dB to the intensity level *doubles* the intensity (see Problem 25). In Example 12.4, when both lathes are running at the same time, the intensity is twice as big as for one lathe, but the two do not sound twice as loud as one. Intensity *level* is a better guide to loudness; two lathes produce a level 3 dB higher than one lathe.

Decibels can also be used in a relative sense; instead of comparing an intensity to I_0, we can compare two intensities directly. Suppose we have two intensities I_1 and I_2 and two corresponding intensity levels β_1 and β_2. Then

$$\beta_2 - \beta_1 = 10 \text{ dB} \left(\log_{10} \frac{I_2}{I_0} - \log_{10} \frac{I_1}{I_0} \right)$$

Since $\log x - \log y = \log \frac{x}{y}$ [see Appendix A.3, Eq. (A-21)],

$$\beta_2 - \beta_1 = (10 \text{ dB}) \log_{10} \frac{I_2/I_0}{I_1/I_0} = (10 \text{ dB}) \log_{10} \frac{I_2}{I_1} \tag{12-9}$$

Example 12.5

Variation of Intensity Level with Distance

At a distance of 30 m from a jet engine, the sound intensity level is 130 dB. Serious, permanent hearing damage occurs rapidly at intensity levels this high, which is why you see airport personnel using hearing protection out on the runway. Assume the engine is an isotropic source of sound and ignore reflections and absorption. At what distance is the intensity level 110 dB—still quite loud but below the threshold of pain?

Strategy The intensity level drops 20 dB. According to the rule of thumb, each 10 dB change represents a factor of 10 in intensity. Therefore, we must find the distance at which the intensity is 2 factors of 10 smaller—that is, $\frac{1}{100}$ the original intensity. The intensity is proportional to $1/r^2$ since we assume an isotropic source [see Eq. (11-2)].

Solution We set up a ratio between the intensities and the inverse square of the distances:

$$\frac{I_1}{I_2} = \left(\frac{r_2}{r_1}\right)^2$$

From the rule of thumb, we know that $I_2 = \frac{1}{100} I_1$. Then

$$\frac{r_2}{r_1} = \sqrt{\frac{I_1}{I_2}} = \sqrt{100} = 10$$

$$r_2 = 10 r_1 = 300 \text{ m}$$

Discussion It is not necessary to use the rule of thumb. Let $\beta_1 = 130$ dB and $\beta_2 = 110$ dB. Then

$$\beta_2 - \beta_1 = -20 \text{ dB} = (10 \text{ dB}) \log_{10} \frac{I_2}{I_1}$$

From this, we find that

$$\log_{10} \frac{I_2}{I_1} = -2 \quad \text{or} \quad \frac{I_2}{I_1} = \frac{1}{100}$$

We can only consider 300 m an estimate. The jet engine may not radiate sound equally in all directions; it might be louder in front than on the side. Sound is partly absorbed and partly reflected by the runway, by the plane, and by any nearby objects. The air itself absorbs some of the sound energy—that is, some of the energy of the wave is dissipated.

Practice Problem 12.5 A Plane as Quiet as a Library

At what distance from the jet engine would the intensity level be comparable to the background noise level of a library (30 dB)? Is your answer realistic?

12.4 STANDING SOUND WAVES

Pipe Open at Both Ends

Recall (Section 11.8) that a transverse wave on a string is reflected from a fixed end. A string fixed at both ends reflects the wave at each end. A standing wave on a string is caused by the superposition of two waves traveling in opposite directions. Standing *sound* waves are also caused by reflections at boundaries. Standing wave patterns for sound waves can be more complex, since sound is a three-dimensional wave. However, the air inside a pipe open at both ends gives rise to standing waves closely analogous to those on a string, as long as the pipe's diameter is small compared with its length. Such a pipe is an excellent model of some organ pipes and flutes.

If the pipe is open at both ends, then the pipe has the same boundary condition at each end. At each open end, the column of air inside the pipe communicates with the outside air, so the pressure at the ends can't deviate much from atmospheric pressure. The open ends are therefore *pressure nodes* (Fig. 12.3). They are also *displacement antinodes*—elements of air vibrate back and forth with maximum amplitude at the ends. Since nodes and antinodes alternate with equal spacing ($\lambda/4$), the wavelengths of standing sound waves in a pipe open at both ends are the same as for a string fixed at both ends (compare Fig. 12.3 with Fig. 11.22), regardless of whether you consider the pressure or the displacement description.

CONNECTION:

The same sketch used to find wavelengths of standing waves for a string fixed at both ends can be used to find the wavelengths for a pipe open at both ends. (The wave speeds are different, however, so a string and pipe of the same length do not have the same standing wave *frequencies*.)

Standing sound waves (thin pipe open at both ends)

$$\lambda_n = \frac{2L}{n} \qquad (11\text{-}11)$$

$$f_n = \frac{v}{\lambda_n} = n\frac{v}{2L} = nf_1 \qquad (11\text{-}12)$$

where $n = 1, 2, 3, \ldots$

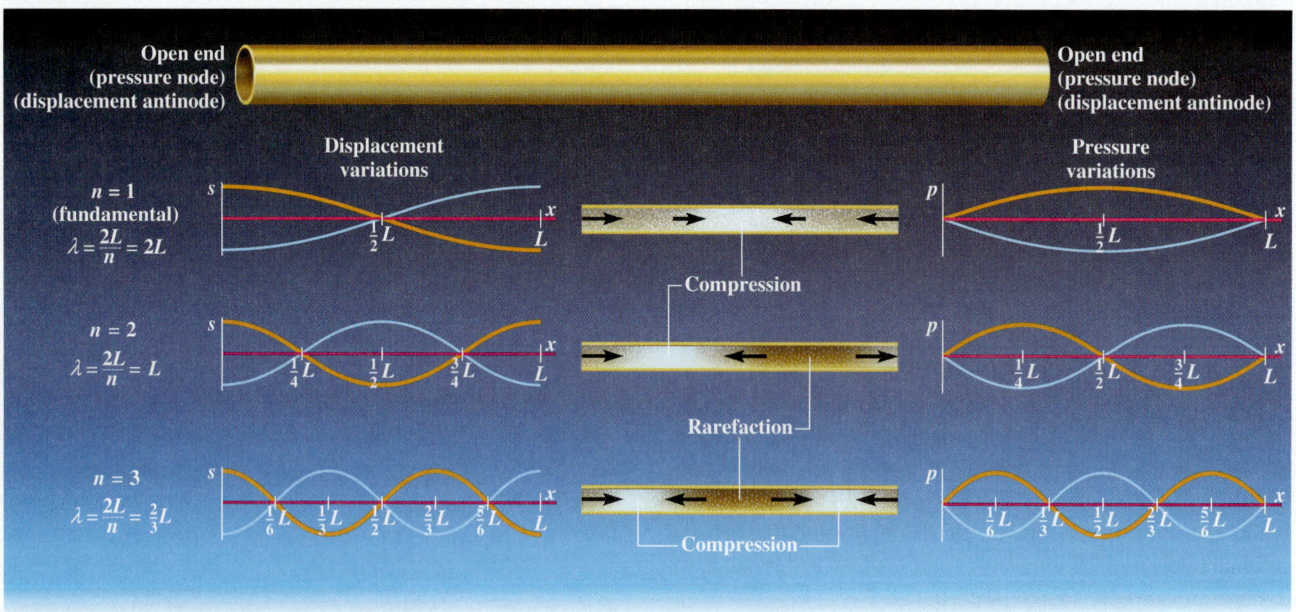

Figure 12.3 The first three standing sound wave patterns for a pipe open at both ends. The sketches in the center show the compressions (maximum pressure) and rarefactions (minimum pressure) at a single instant in time. The black arrows show the air displacement at the same instant. These sketches correspond to the orange graphs of displacement (left) and pressure (right). The red and light blue graphs are one-quarter and one-half period later, respectively. Although the displacement graphs show air displacement s on the vertical axis and x on the horizontal, remember that the displacements are in the $\pm x$-direction, as illustrated by the black displacement arrows.

Pipe Closed at One End

Some organ pipes are *closed at one end* and open at the other (Fig. 12.4). The closed end is a pressure *antinode*; the air at the closed end meets a rigid surface, so there is no restriction on how far the pressure can deviate from atmospheric pressure. The closed end is also a *displacement node* since the air near it cannot move beyond that rigid surface. Some wind instruments are effectively pipes closed at one end. The reed of a clarinet admits only brief puffs of air into the instrument; the rest of the time the reed closes off that end of the pipe. The pressure at the reed end fluctuates above and below atmospheric pressure. The reed end is a pressure antinode and a displacement node.

The wavelengths and frequencies of the standing waves can be found using either the pressure or displacement descriptions of the wave. Using displacement, the fundamental has a node at the closed end, an antinode at the open end, and no other nodes or antinodes (Fig. 12.5). The distance from a node to the nearest antinode is always $\frac{1}{4}\lambda$, so for the fundamental

$$L = \tfrac{1}{4}\lambda \qquad \text{or} \qquad \lambda = 4L$$

Figure 12.4 Some organ pipes are open at the top; others are closed. A pipe closed at one end has a fundamental wavelength twice as large and therefore a fundamental frequency half as large as a pipe of the same length that is open at both ends, assuming the pipes are thin. (For musicians: the pitch of the pipe closed at one end sounds an octave lower than the other, since the interval of an octave corresponds to a factor of 2 in frequency.)

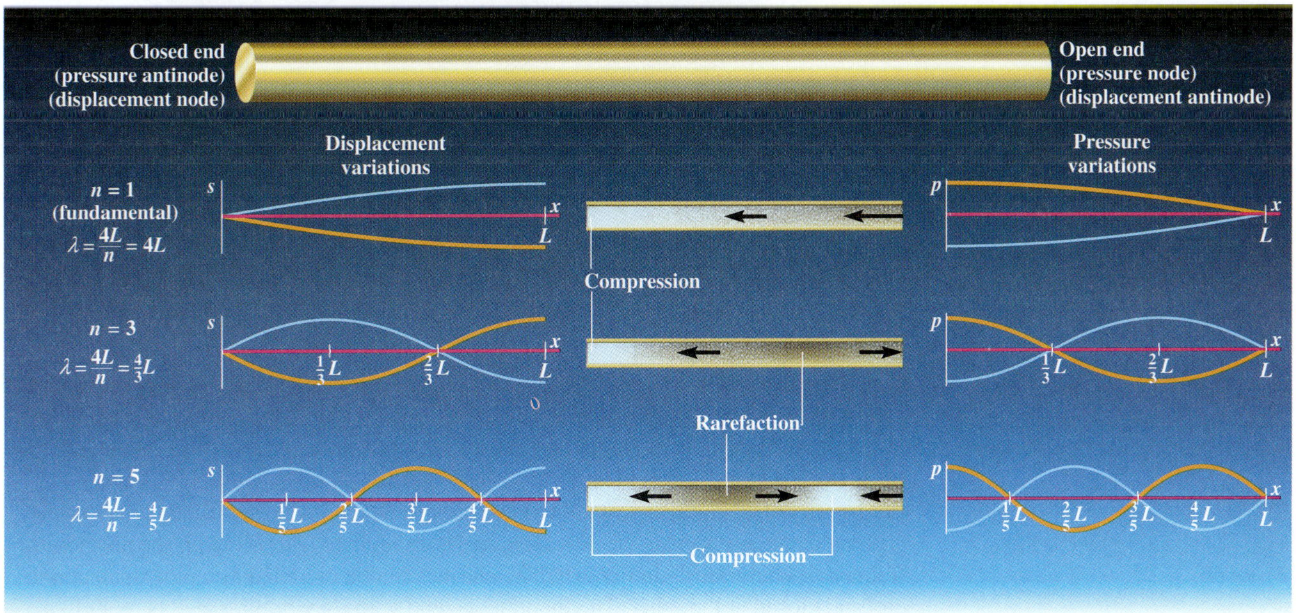

Figure 12.5 The first three standing sound wave patterns for a pipe closed at one end. The sketches in the center show the compressions (maximum pressure) and rarefactions (minimum pressure) at a single instant in time. The black arrows show the air displacement at the same instant. These sketches correspond to the orange graphs of displacement (left) and pressure (right). The red and light blue graphs are one-quarter and one-half period later, respectively. Although the displacement graphs show air displacement s on the vertical axis and x on the horizontal, remember that the displacements are in the $\pm x$-direction, as illustrated by the black displacement arrows.

which is twice as large as the wavelength ($2L$) of the fundamental in a pipe of the same length open at both ends. Two thin organ pipes of the same length, one open at both ends and one closed at one end, do not have the same fundamental wavelength (see Fig. 12.4).

What are the other standing wave frequencies? The next standing wave mode is found by adding one node and one antinode. Then the length of the pipe is 3 quarter-cycles: $L = \frac{3}{4}\lambda$ or $\lambda = \frac{4}{3}L$. This is $\frac{1}{3}$ the wavelength of the fundamental, and the frequency is 3 times that of the fundamental. Adding one more node and one more antinode, the wavelength is $\frac{4}{5}L$. Continuing the pattern, we find that the wavelengths and frequencies for standing waves are

Standing sound waves (thin pipe closed at one end)

$$\lambda_n = \frac{4L}{n} \tag{12-10a}$$

$$f_n = \frac{v}{\lambda_n} = n\frac{v}{4L} = nf_1 \tag{12-10b}$$

where $n = 1, 3, 5, 7, \ldots$

Note that the standing wave frequencies for a pipe closed at one end are only *odd* multiples of the fundamental. The "missing" standing wave patterns for even values of n require a clarinet to have many more keys and levers than a flute (Fig. 12.6). What the keys do is effectively shorten the length of the pipe, making the standing wave frequencies higher.

CHECKPOINT 12.4

Why can't a pipe of length L closed at one end support a standing wave with wavelength $2L$?

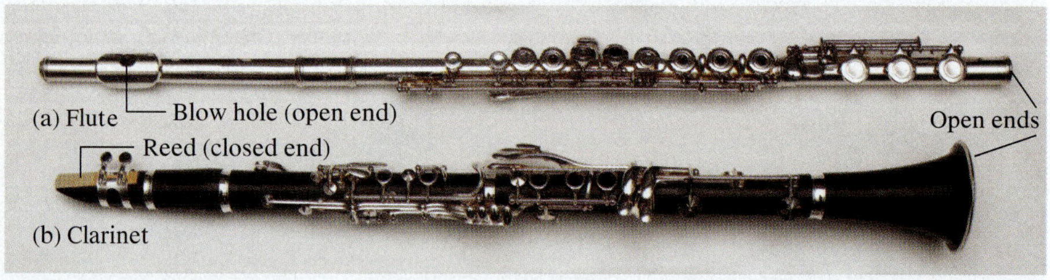

Figure 12.6 A flute can be modeled as a pipe open at both ends, whereas a clarinet can be modeled as a pipe closed at one end. Although the instruments are similar in length, the clarinet can play tones nearly an octave lower than is possible on the flute. (a) The flute's open blow hole serves as one of its open "ends." If a flute's fundamental frequency is f_1 with no keys pressed, the next higher frequency possible without using any keys is $2f_1$. The flute needs enough keys to fill in all the notes with frequencies between f_1 and $2f_1$. (b) The clarinet can be modeled as a pipe open at one end and closed at the other. The mouthpiece end with its vibrating reed is more like a closed end (pressure antinode) than an open end (pressure node). For a clarinet, if the fundamental frequency is f_1 with no keys pressed, the next highest frequency possible without using any keys is $3f_1$. The clarinet must have more keys because it has to accommodate all the notes with frequencies between f_1 and $3f_1$.

Example 12.6

A Demonstration of Resonance

A thin hollow tube of length 1.00 m is inserted vertically into a tall container of water (Fig. 12.7). A tuning fork (f = 520.0 Hz) is struck and held near the top of the tube as the tube is slowly pulled up and out of the water. At certain distances (L) between the top of the tube and the water surface, the otherwise faint sound of the tuning fork is greatly amplified. At what values of L does this occur? The temperature of the air in the tube is 18°C.

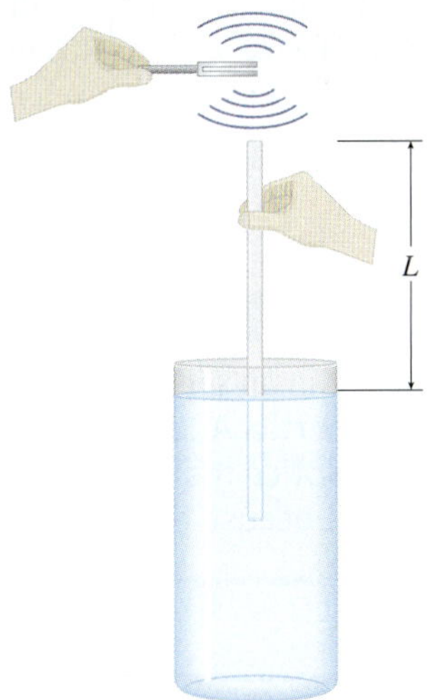

Figure 12.7 Experimental setup for Example 12.6.

Strategy Sound waves in the air inside the tube reflect from the water surface. Thus, we have an air column of variable length L, closed at one end by the water surface and open at the other end. The sound is amplified due to resonance; when the frequency of the tuning fork matches one of the natural frequencies of the air column, a large-amplitude standing wave builds up in the column. For standing waves in a column of air, the wavelength and frequency are related by the speed of sound in air. We start by finding the speed of sound in air from the temperature given. Then we can find the wavelength of the sound waves emanating from the tuning fork. Last, we find the column lengths that support standing waves of that wavelength.

Solution From Eq. (12-4), the speed of sound in air at 18°C is

$$v = 331.3\ \text{m/s} + \left(0.6\frac{\text{m/s}}{^\circ\text{C}}\right)(18^\circ\text{C}) = 342\ \text{m/s}$$

With the speed of sound and the frequency known, we can find the wavelength. The wavelength is the distance traveled by a wave during one period:

$$\lambda = vT = \frac{v}{f}$$

$$\lambda = \frac{342\ \text{m/s}}{520.0\ \text{Hz}} = 0.6577\ \text{m} = 65.77\ \text{cm}$$

The first possible resonance for a tube closed at one end occurs when there is a pressure node at the open end, a pressure antinode at the closed end, and no other pressure nodes or antinodes. Therefore,

$$L_1 = \tfrac{1}{4}\lambda = \tfrac{1}{4} \times 65.77\ \text{cm} = 16.4\ \text{cm}$$

To reach other resonances, the tube must be pulled out to accommodate additional pressure nodes and antinodes. To add one node and one antinode, the additional distance is $\frac{1}{2}\lambda$ = 32.9 cm. The resonances occur at intervals of 32.9 cm:

$$L_2 = 16.4\ \text{cm} + 32.9\ \text{cm} = 49.3\ \text{cm}$$

$$L_3 = 49.3\ \text{cm} + 32.9\ \text{cm} = 82.2\ \text{cm}$$

The next one requires a tube longer than 1.00 m, so there are three values of L that produce resonance in this tube.

Discussion As a check, we can sketch the standing wave pattern for the third resonance (Figs. 12.8a,b). There are 5 quarter-wavelengths in the length of the column, so

$$L_3 = \tfrac{5}{4}\lambda = \tfrac{5}{4} \times 65.77\ \text{cm} = 82.2\ \text{cm}$$

At the open end of the tube, the node for pressure and the antinode for maximum displacement is actually a little *above* the opening. For this reason it is best to measure the distance between two successive resonances to find an accurate value for a half-wavelength rather than measuring the distance for the first possible resonance, the shortest distance between the opening and the water surface, and setting it equal to a quarter-wavelength.

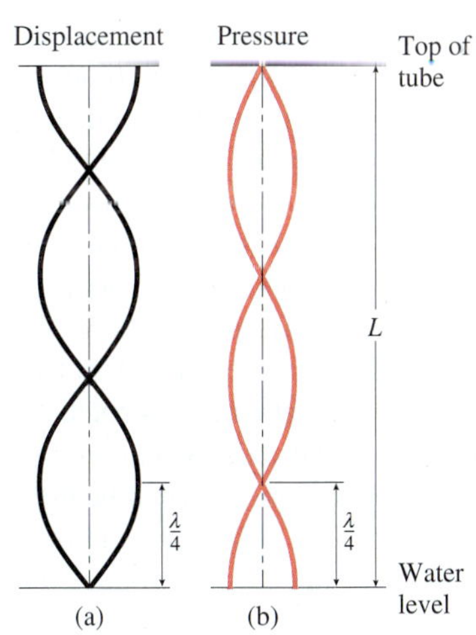

Figure 12.8 (a) Standing wave pattern, showing *displacement* nodes and antinodes, for the third resonance. (b) Standing wave pattern, showing *pressure* nodes and antinodes, for the third resonance.

Practice Problem 12.6 A Roundabout Way to Measure Temperature

A tuning fork of frequency 440.0 Hz is held above the hollow tube in Example 12.6. If the distance ΔL that the tube is moved between resonances is 39.3 cm, what is the temperature of the air inside the tube?

Problem-Solving Strategy for Standing Waves

There is no need to memorize equations for standing wave frequencies and wavelengths. Just sketch the standing wave patterns as in Figs. 12.3 and 12.5. Make sure that nodes and antinodes alternate and that the boundary conditions at the ends are correct. Then determine the wavelengths by setting the distance between a node and antinode equal to $\frac{1}{4}\lambda$. Once the wavelengths are known, the frequencies are found from $v = f\lambda$.

EVERYDAY PHYSICS DEMO

You can set up a resonance in an empty water bottle by blowing horizontally across the top of the bottle. Add varying amounts of water and listen for how the pitch changes. Notice that the shorter the air column within the bottle, the higher the pitch (because the fundamental frequency is higher).

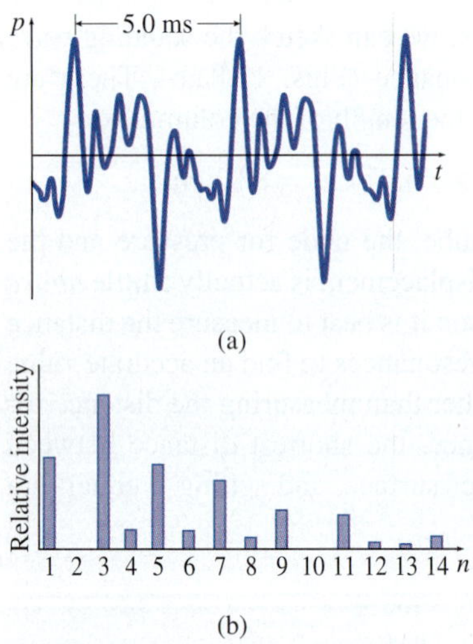

Figure 12.9 (a) A graph of the sound wave produced by a clarinet. (b) A bar graph showing the relative intensities of the harmonics, often called the *spectrum*. The frequency of each harmonic is nf_1, where $f_1 = 200$ Hz. Notice that *odd* multiples of the fundamental dominate the spectrum. A simple pipe closed at one end would show *only* odd multiples in its spectrum. (Data courtesy of P. D. Krasicky, Cornell University.)

12.5 TIMBRE

The sound produced by the vibration of a tuning fork is nearly a pure sinusoid at a single frequency. In contrast, most musical instruments produce complex sounds that are the superposition of many different frequencies. The standing wave on a string or in a column of air is almost always the superposition of many standing wave patterns at different frequencies. The lowest frequency in a complex sound wave is called the fundamental; the rest of the frequencies are called **overtones**. All the overtones of a periodic sound wave have frequencies that are integral multiples of the fundamental; the fundamental and the overtones are called **harmonics**.

Middle C played on an oboe does not sound the same as middle C played on a trumpet, even though the fundamental frequency is the same, largely because the two instruments produce overtones with different relative amplitudes. What is different about the two sounds is the **tone quality**, or **timbre** (pronounced "tamber").

Any periodic wave, no matter how complicated, can be decomposed into a set of harmonics, each of which is a simple sinusoid. The characteristic wave form for a note played on a clarinet, for example, can be decomposed into its harmonic series (Fig. 12.9). This process is called harmonic analysis, or Fourier analysis, in honor of the French mathematician, Jean Baptiste Joseph Fourier (1768–1830), who developed mathematical methods for analyzing periodic functions. Although the spectrum of a periodic wave consists only of members of a harmonic sequence, not all members of the sequence need be present, not even the fundamental (Fig. 12.10).

The opposite of harmonic analysis is harmonic synthesis: combining various harmonics to produce a complex wave. Electronic synthesizers can mimic the sounds of various instruments. Realistic-sounding synthesizers must also allow the adjustment of other parameters such as the attack and decay of the sound.

12.6 THE HUMAN EAR

Figure 12.11 shows the structure of the human ear. The human ear has an external part, or *pinna*, that acts something like a funnel, collecting sound waves and concentrating them at the opening of the auditory canal. The pinna is better at collecting sound coming from in front than from behind, which helps with localization. Resonance in the *auditory canal* (see Problem 67) boosts the ear's sensitivity in the 2 to 5 kHz frequency range—a crucial range for understanding speech.

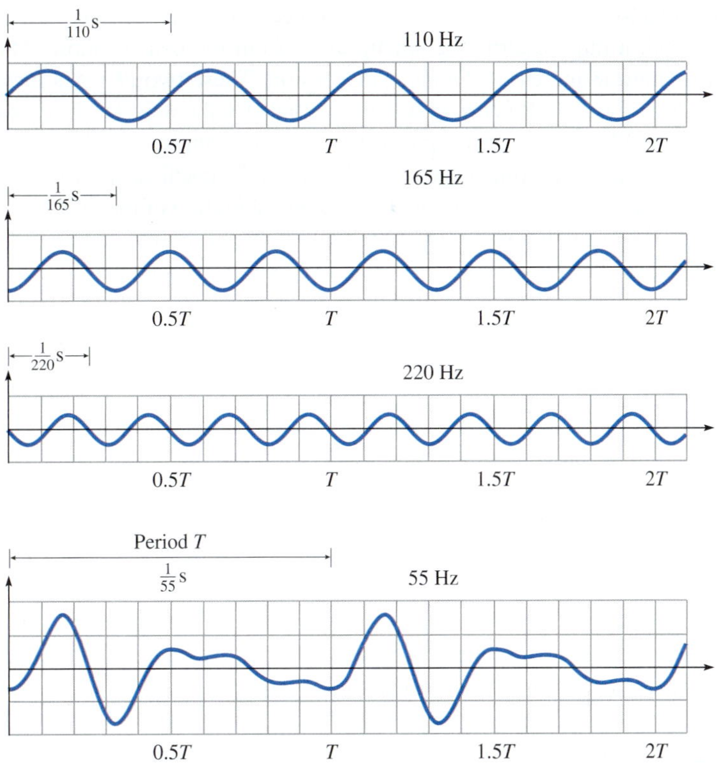

Figure 12.10 Complex wave form (bottom wave) composed by superposition of three sinusoidal waves (three upper waves). A wave with three harmonic components having frequencies of 110, 165, and 220 Hz repeats at a frequency of 55 Hz because each of these three frequencies is an integral multiple of 55 Hz. Even though the fundamental is missing—there is no harmonic component at 55 Hz—the ear is clever enough to "reconstruct" a 55 Hz tone. That's why you can listen to and recognize music on an inexpensive radio whose speaker may reproduce only a small range of frequencies.

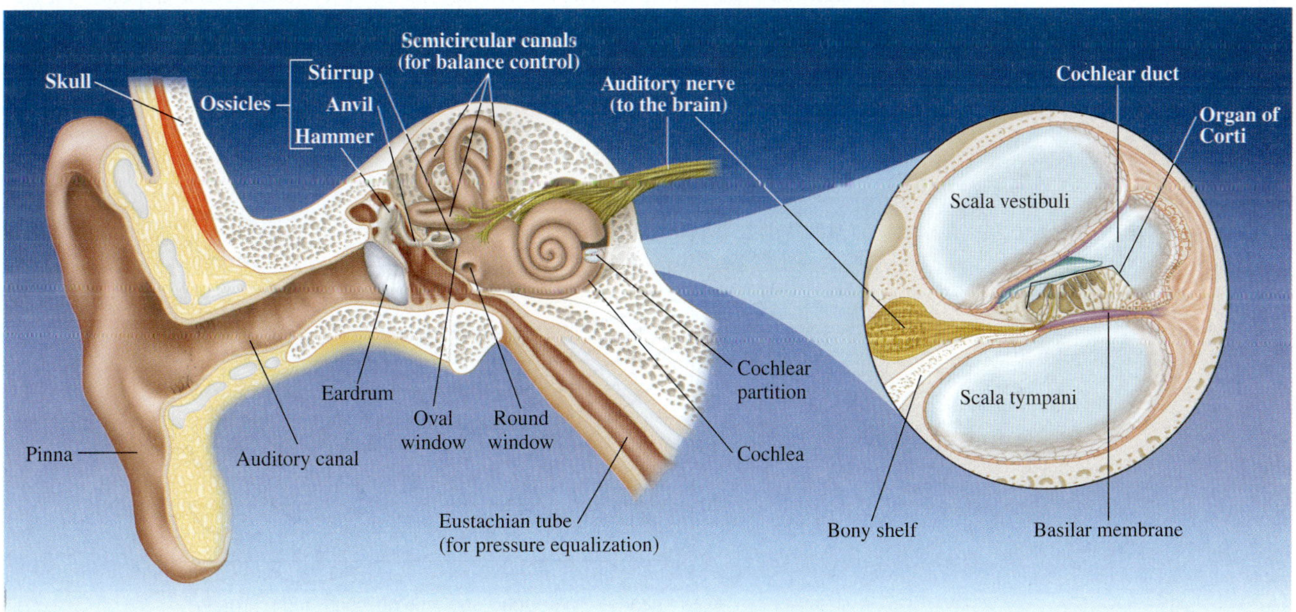

Figure 12.11 Structure of the human ear with a cross section of the cochlea.

At the end of the auditory canal, the eardrum (*tympanum*) vibrates in response to the incident sound wave. The region just beyond the eardrum is called the middle ear. The vibrations of the eardrum are transmitted through three tiny bones of the middle ear (the *auditory ossicles*) to the *oval window* of the *cochlea*, a tapered spiral-shaped organ filled with fluid. The oval window is a membrane that is in contact with the fluid in the cochlea. The ossicles act as levers; the force exerted by the "stirrup" on the oval window is 1.5 to 2.0 times the force the eardrum exerts on the "hammer." The area of the oval window is one-twentieth that of the eardrum, so there is an overall amplification in pressure by a factor of 30 to 40. The ossicles protect the ear from damage: in response to a loud sound, a muscle pulls the stirrup away from the oval window. At the same time, another muscle increases the eardrum tension. These two changes make the ear temporarily less sensitive. It takes a few milliseconds for the muscles to respond in this way, so they provide no protection against *sudden* loud sounds.

The *cochlear partition* runs most of the length of the cochlea, separating it into two chambers (the *scala vestibuli* and the *scala tympani*). Vibration of the oval window sends a compressional wave down the fluid in the scala vestibuli, around the end of the partition, and back up the scala tympani to the *round window*. This wave sets the *basilar membrane*, located on the cochlear partition, into vibration. The basilar membrane is thinnest and under greatest tension near the oval and round windows; it gradually increases in thickness and decreases in tension toward its other end. High-frequency waves cause the membrane to vibrate with maximum amplitude near its thin, high-tension end; low-frequency waves cause maximum amplitude vibrations near its thicker, lower-tension end. The location of the maximum amplitude vibrations is one way the ear determines frequency; for low-frequency sounds (up to about 1 kHz), the ear sends periodic nerve signals to the brain at the frequency of the sound wave. For complex sounds, which consist of the superposition of many different frequencies (see Section 12.5), the ear performs a spectral analysis—it decomposes the complex sound into its constituent frequencies.

Located on the basilar membrane is the sensory organ (the *organ of Corti*). Rows of hair cells on the basilar membrane excite neurons when they bend in response to vibration. These neurons send electrical signals to the brain.

Loudness

Although loudness is most closely correlated to intensity level, it also depends on frequency (as well as other factors). In other words, the sensitivity of the ear is frequency-dependent. Figure 12.12 shows a set of *curves of equal loudness* for a typical person. Each curve shows the intensity levels at which sounds of different frequencies are perceived to be equally loud.

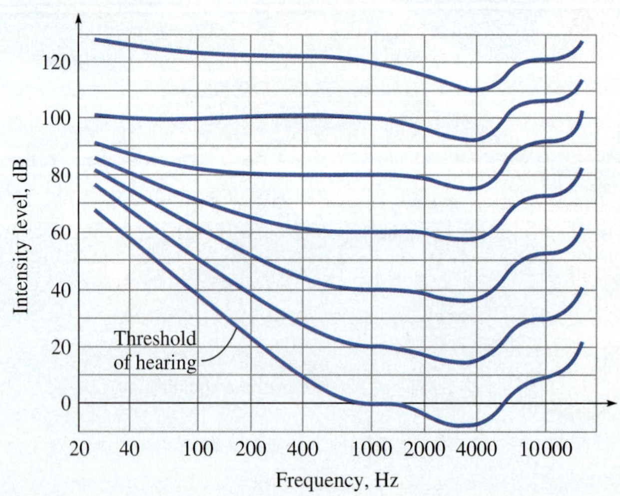

Figure 12.12 Curves of equal loudness. The curves show that the ear is most sensitive to frequencies between 3 kHz and 4 kHz, partly due to resonance in the auditory canal. The ear's sensitivity falls off rapidly below 800 Hz and above 10 kHz. At any given frequency between 800 Hz and 10 kHz, the curves are approximately evenly spaced: equal steps in intensity level produce equal steps in loudness, which is why intensity level is often used as an approximate measure of loudness. In this frequency range, 1 dB is about the smallest change in intensity level that is perceptible as a change in loudness. The threshold of hearing is shown by the lowest curve in the set; a person with excellent hearing cannot hear sounds with intensity levels below this curve. The threshold of hearing is at an intensity level of 0 dB or lower only in the frequency range of about 1–6 kHz.

Pitch

Pitch is the perception of frequency. If you sing or play up and down a scale, it is the pitch that is rising and falling. Although pitch is the aspect of sound perception most closely tied to a single physical quantity, frequency, our sense of pitch is affected to a small extent by other factors such as intensity and timbre (Section 12.5).

Our sense of pitch is a *logarithmic* function of frequency, just as loudness is approximately a logarithmic function of intensity. If you start at the lowest note on the piano (which has a fundamental frequency of 27.5 Hz) and play a chromatic scale—every white and black key in turn—all the way to the highest note (4190 Hz), you hear a series of equal steps in pitch. The frequencies do *not* increase in equal steps; the fundamental frequency of each note is 5.95% higher than the previous note. Under ideal conditions, most people can sense frequency changes as small as 0.3%. A trained musician can sense a frequency change of 0.1% or so.

Localization

How can you tell where a sound comes from? The ear has several different tools it uses to localize sounds:

- The principal method for high-frequency sounds (>4 kHz) is the difference in intensity sensed by the two ears. The head casts a "sound shadow," so a sound coming from the right has a larger intensity at the right ear than at the left ear.
- The shape of the pinna makes it slightly preferential to sounds coming from the front. This helps with front-back localization for high-frequency sounds.
- For lower-frequency sounds, both the difference in arrival time and the phase difference between the waves arriving at the two ears are used for localization.

12.7 BEATS

When two sound waves are close in frequency (within about 15 Hz of each other), the superposition of the two produces a pulsation that we call **beats**. Beats can be produced by any kind of wave; they are a general result of the principle of superposition when applied to two waves of nearly the same frequency.

Beats are caused by the slow change in the phase difference between the two waves. Suppose that at one instant (t = 0 in Fig. 12.13), the two waves are in phase with each other and interfere constructively. The amplitude of the superposition is the sum of the amplitudes of the two waves shown in Fig. 12.13a. However, since the frequencies are different, the waves do not *stay* in phase. The higher-frequency wave has

CONNECTION:

When two waves with different frequencies are superimposed, constructive interference alternates with destructive interference, causing *beats*.

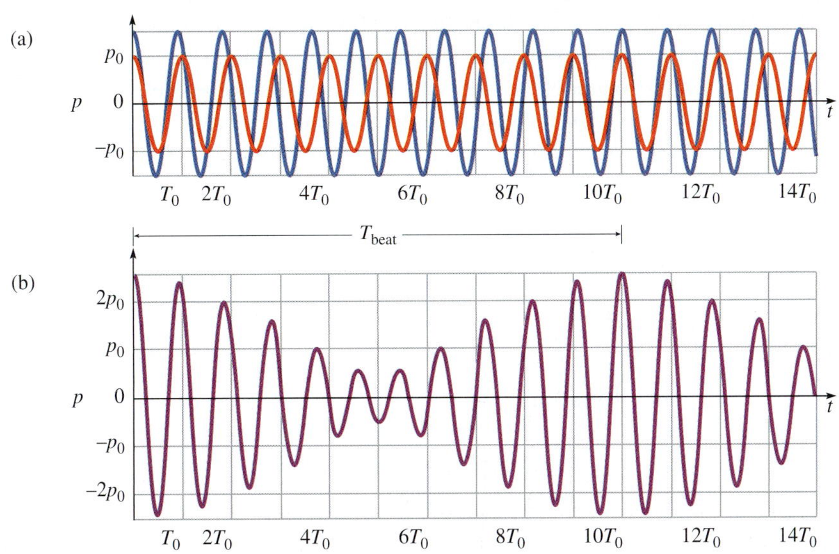

Figure 12.13 (a) Graph (red) of a sound wave with frequency $f_1 = 1/T_0$ and amplitude p_0. Graph (blue) of a second sound wave with frequency $f_2 = 1.1f_1$ and amplitude $1.5p_0$. (b) The superposition (purple) of the two has maximum amplitude $2.5p_0$ and minimum amplitude $0.5p_0$.

a shorter cycle, so it gets ahead of the other one. The phase difference between the two steadily increases; as it does, the amplitude of the superposition decreases. At a later time ($t = 5T_0$), the phase difference reaches 180°; now the waves are half a cycle out of phase and interfere destructively (Fig. 12.13b). Now the amplitude of the superposition is minimum; it is equal to the difference between the amplitudes of the two waves. As the phase difference continues to increase, the amplitude increases until constructive interference occurs again ($t = 10T_0$). The ear perceives the amplitude (and intensity) cycling from large to small to large to small as a pulsation—a repeating alternation of increasing and decreasing loudness.

At what frequency do the beats occur? It depends on how far apart the frequencies of the two waves are. We can measure the time between beats T_{beat} as the time to go from constructive interference to the next occurrence of constructive interference. During that time, each wave must go through a whole number of cycles, with one of them going through one more cycle than the other. Since frequency (f) is the number of cycles per second, the number of cycles a wave goes through during a time T_{beat} is fT_{beat}. (To illustrate: in Fig. 12.13, $T_{beat} = 10T_0$. During that time, wave 1 goes through $f_1T_{beat} = 10$ cycles, while wave 2 goes through $f_2T_{beat} = 1.1/T_0 \times 10T_0 = 11$ cycles.) If $f_2 > f_1$, then wave 2 goes through one cycle more than wave 1:

$$f_2T_{beat} - f_1T_{beat} = (\Delta f)T_{beat} = 1$$

The beat frequency f_{beat} is $1/T_{beat}$:

Beat frequency

$$f_{beat} = 1/T_{beat} = \Delta f \qquad (12\text{-}11)$$

Thus, we obtain the remarkably simple result that the beat frequency is the difference between the frequencies of the two waves. If the difference in frequencies exceeds roughly 15 Hz, then the ear no longer perceives the beats; instead, we hear two tones at different pitches.

CHECKPOINT 12.7

(a) At what time(s) in Fig. 12.13 do the two waves interfere constructively?
(b) At what time(s) do they interfere destructively?

Application: Tuning a Piano Piano tuners listen for beats as they tune. The tuner sounds two strings and listens for the beats. The beat frequency indicates whether the interval is correct or not. If the two strings are played by the same key, they are tuned to the same fundamental frequency, so the beat frequency should be (nearly) zero. If the two strings belong to two different notes, the beat frequency is nonzero. In this case the tuner listens to beats between two *overtones* that are close in frequency.

Example 12.7

The Piano Tuner

A piano tuner strikes his tuning fork ($f = 523.3$ Hz) and strikes a key on the piano at the same time. The two have nearly the same frequency; he hears 3.0 beats per second. As he tightens the piano string, he hears the beat frequency gradually decrease to 2.0 beats per second when the two sound together. (a) What was the frequency of the piano string before it was tightened? (b) By what percentage did the tension increase?

continued on next page

Example 12.7 continued

Strategy The beat frequency is the difference between the two frequencies; we only have to determine which is higher. The wavelength of the string is determined by its length, which does not change. The increase in tension increases the speed of waves on the string, which in turn increases the frequency.

Solution (a) Since the piano tuner heard 3.0 beats per second, the difference in the two frequencies was 3.0 Hz:

$$\Delta f = 3.0 \text{ Hz}$$

Is the piano string's frequency 3.0 Hz higher or 3.0 Hz lower than the tuning fork's frequency? As the tension increases gradually, the beat frequency decreases, which means that the frequency of the piano string is getting *closer* to the frequency of the tuning fork. Therefore, the string frequency must be 3.0 Hz *lower* than the tuning fork frequency:

$$f_{\text{string}} = 523.3 \text{ Hz} - 3.0 \text{ Hz} = 520.3 \text{ Hz}$$

(b) The tension (F) is related to the speed of the wave on the string (v) and the mass per unit length (μ) by

$$v = \sqrt{\frac{F}{\mu}} \qquad (11\text{-}4)$$

The mass per unit length does not change, so $v \propto \sqrt{F}$. The speed of the wave on the string is related to its wavelength and frequency by

$$v = \lambda f$$

The wavelength λ in this expression is the wavelength of the transverse wave on the string, *not* the wavelength of the sound wave in air. Since λ does not change, $v \propto f$. Therefore, $f \propto \sqrt{F}$ or

$$F \propto f^2$$

This means that the ratio of the tension F to the original tension F_0 is equal to the ratio of the frequencies squared:

$$\frac{F}{F_0} = \left(\frac{f}{f_0}\right)^2 = \left(\frac{521.3 \text{ Hz}}{520.3 \text{ Hz}}\right)^2 = 1.004$$

The tension was increased 0.4%.

Discussion We needed to find whether the original frequency was too high or too low. As the beat frequency decreases, the frequency of the string is getting closer to the frequency of the tuning fork. Tightening the string makes the string's frequency increase; since increasing the string's frequency brings it closer to the tuning fork's frequency, we know that the original frequency of the string was lower than the frequency of the tuning fork. Had an increase in tension *increased* the beat frequency instead, we would know that the original frequency was already too high; the tension would have to be relaxed to tune the string.

Practice Problem 12.7 Tuning a Violin

A tuning fork with a frequency of 440.0 Hz produces 4.0 beats per second when sounded together with a violin string of nearly the same frequency. What is the frequency of the string if a slight increase in tension increases the beat frequency?

12.8 THE DOPPLER EFFECT

A police car races by, its sirens screaming. As it passes, we hear the pitch change from higher to lower. The frequency change is called the **Doppler effect**, after the Austrian physicist Johann Christian Andreas Doppler (1803–1853). The observed frequency is different from the frequency transmitted by the source when the source or the observer is in motion relative to the wave medium.

We consider only the motion of the source and observer directly toward or away from each other in the reference frame in which the wave medium is at rest. Velocities of the source and observer with respect to the wave medium are expressed as components along the direction of propagation of the sound wave (from source to observer). A positive component means the velocity is in the direction of propagation of the wave, but a negative component means the velocity is opposite the direction of propagation.

Moving Source

Suppose a source emits a sound wave at frequency f_s. The source is moving with velocity v_s toward an observer on the right (Fig. 12.14). The wave moves with a velocity $v - v_s$ with respect to the source, where v is the velocity of the wave with respect to the medium (here, the air). In the reference frame of the source, the wave moves

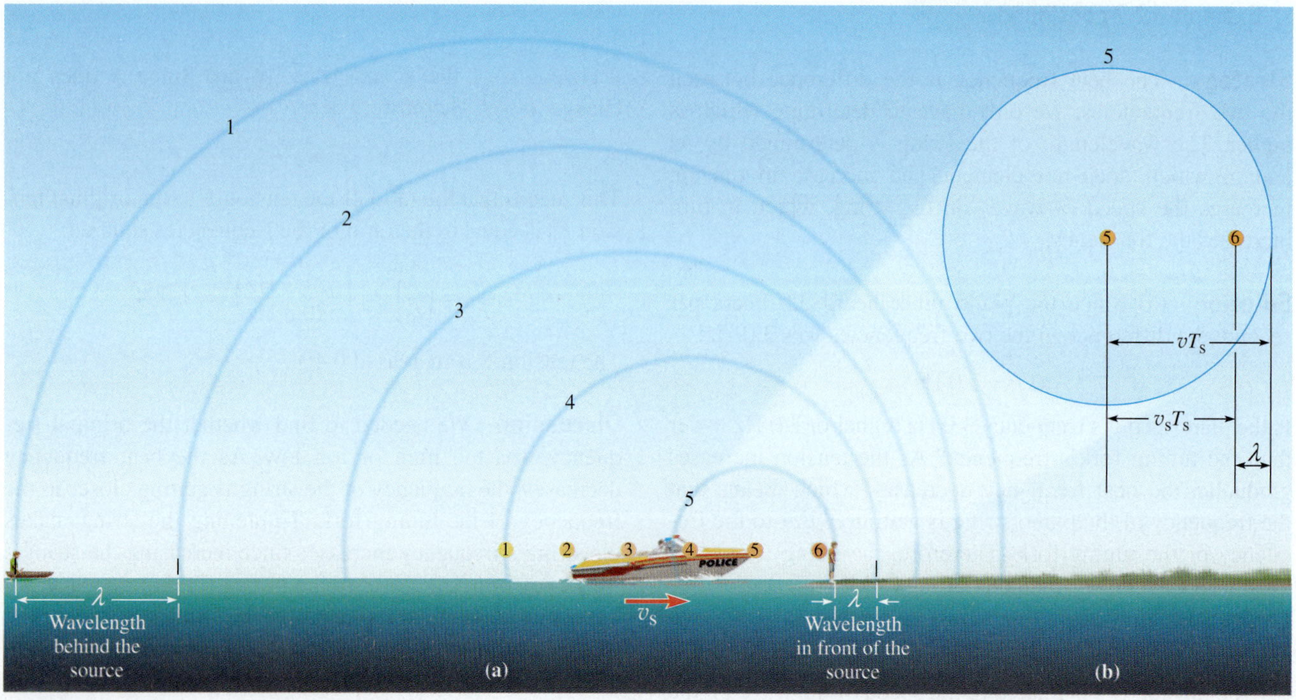

Figure 12.14 (a) A police speedboat is moving to the right at speed v_s (exaggerated for clarity) while it blows its siren. The siren emits wave crests at positions 1, 2, 3, 4, 5, and 6; each wave crest moves outward in all directions, from the point at which it was emitted, at speed v. (b) The wavelength λ is the distance between wave crests. In the reference frame of the wave medium (here, the air), a wave crest moves a distance vT_s and the boat moves a distance v_sT_s during a time interval T_s, so $\lambda = vT_s - v_sT_s$. Equivalently, in the reference frame of the source, the wave moves a distance λ at speed $v - v_s$ in time T_s, so $\lambda = (v - v_s)T_s = (v - v_s)/f_s$. The wavelength is shorter than vT_s in front of the source and longer than vT_s behind the source.

a distance λ in time $T_s = 1/f_s$ at velocity $v - v_s$. The wavelength-frequency relationship is then

$$v - v_s = f_s\lambda \qquad (12\text{-}12)$$

The wavelength is smaller than v/f_s when the source moves toward the observer ($v_s > 0$) and larger than v/f_s when the source moves away from the observer ($v_s < 0$).

Moving Observer

For an observer moving away from the source with velocity v_o, the velocity of the wave relative to the observer is $v - v_o$. In the reference frame of the observer, the wave moves a distance λ in time $T_o = 1/f_o$ at velocity $v - v_o$ (Fig. 12.15). The observed frequency f_o is related to this velocity and the wavelength by

$$v - v_o = f_o\lambda \qquad (12\text{-}13)$$

The observed frequency f_o is smaller than v/λ when the observer moves away from the source ($v_o > 0$) and larger than v/λ when the observer moves toward the source ($v_o < 0$).

Eliminating λ and solving Eqs. (12-12) and (12-13) for f_o yields

Doppler shift (moving source and/or observer)

$$f_o = \frac{v - v_o}{v - v_s} f_s \qquad (12\text{-}14)$$

v_o and v_s are positive in the direction of propagation of the wave (from source to observer) and negative in the opposite direction.

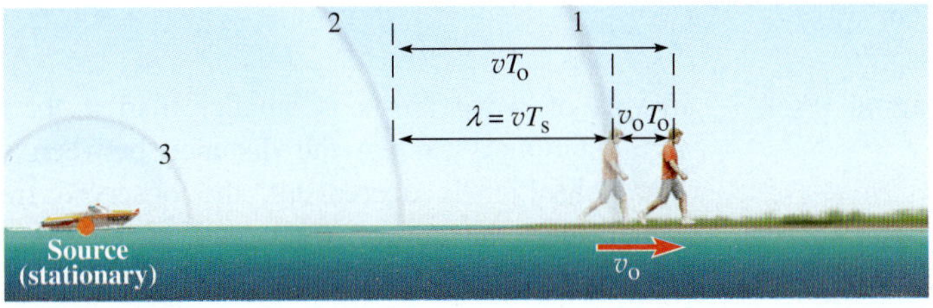

Figure 12.15 An observer moving at speed v_o (exaggerated for clarity) away from a stationary sound source. The wavelength λ is the distance between wave crests. In the reference frame of the wave medium (here, the air), a wave crest moves a distance vT_o and the observer moves a distance v_oT_o during a time interval T_o, so $\lambda = vT_o - v_oT_o$. Equivalently, in the reference frame of the observer, the wave moves a distance λ at speed $v - v_o$ (relative to the observer) in time T_o, so $\lambda = (v - v_o)T_o = (v - v_s)/f_o$.

Problem-Solving Strategy: Doppler Effect

- Determine v_s and v_o, the velocities of the source and observer with respect to the wave medium. v_s and v_o are positive for motion in the direction the wave travels (from source to observer) and negative in the opposite direction (from observer to source).
- No need to memorize Eq. (12-14); you can quickly derive it. Just remember that the wavelength is the same for source and observer and use the frequency-wavelength relationship (wave speed = frequency times wavelength):

$$\lambda = \frac{v - v_s}{f_s} = \frac{v - v_o}{f_o}$$

- Some Doppler effect problems involve reflected waves. One way to handle a reflected wave is to think of the reflecting surface as first observing the wave and then re-emitting it at the same frequency.

CHECKPOINT 12.8

(a) Does the motion of the source of a sound wave affect the wavelength?
(b) Does the motion of the observer affect the wavelength?

Example 12.8

Train Whistle and Doppler Shift

A monorail train approaches a platform at a speed of 10.0 m/s while it blows its whistle. A musician with perfect pitch standing on the platform hears the whistle as "middle C," a frequency of 261 Hz. There is no wind and the temperature is a chilly 0°C. What is the observed frequency of the whistle when the train is at rest?

Strategy In this case, the source—the whistle—is moving and the observer is stationary. The source is moving *toward* the observer, so v_s is *positive*. With the source approaching the observer, the observed frequency is higher than the source frequency. When the train is at rest, there is no Doppler shift; the observed frequency then is equal to the source frequency.

Solution For a moving source, the source (f_s) and observed (f_o) frequencies are related by

$$f_o = \frac{v - v_o}{v - v_s} f_s$$

continued on next page

Example 12.8 continued

where v = 331 m/s (the speed of sound in air at 0°C), $v_s = +10.0$ m/s, and $f_o = 261$ Hz. Solving for f_s,

$$f_s = \frac{v - v_s}{v} f_o$$

$$= \frac{331 \text{ m/s} - 10.0 \text{ m/s}}{331 \text{ m/s}} \times 261 \text{ Hz} = 253 \text{ Hz}$$

The source frequency is less than the observed frequency, as expected. The observed frequency when the train is at rest is equal to the source frequency: 253 Hz.

Discussion When the train is moving toward the platform, the distance between source and observer is decreasing. Wave crests emitted later take *less time* to reach the observer than if the train were at rest, so the time between arrivals of wave crests is smaller than if the train were stationary. When the distance between source and observer is decreasing, the observed frequency is higher than the source frequency; when the distance is increasing, the observed frequency is lower than the source frequency.

Practice Problem 12.8 A Sports Car Racing By

Justine is gardening in her front yard when a Mazda Miata races by at 32.0 m/s (71.6 mi/h). If she hears the sound of the Miata's engine at 220.0 Hz as it approaches her, what frequency does she hear after it passes? Assume the temperature is 20°C and there is no wind.

Example 12.9

Determining Speed from Horn Frequency

Two cars, with equal ground speeds, are moving in opposite directions away from each other on a straight highway. One driver blows a horn with a frequency of 111 Hz; the other measures the frequency as 105 Hz. If the speed of sound is 338 m/s and there is no wind, what is the ground speed of each car?

Strategy The sound wave travels from source to observer. The source moves opposite the direction of the wave, so v_s is negative. The observer moves in the direction of the wave, so v_o is positive. The speeds are the same, so $v_s = -v_o$.

Solution With both the source and observer moving, the frequencies are related by

$$f_o = \frac{v - v_o}{v - v_s} f_s = \left(\frac{1 - v_o/v}{1 - v_s/v}\right) f_s$$

To simplify the algebra, we let $\alpha = v_o/v = -v_s/v$. Then

$$f_o = \left(\frac{1-\alpha}{1+\alpha}\right) f_s$$

Now we solve for α:

$$(1+\alpha)\frac{f_o}{f_s} = 1 - \alpha$$

$$\alpha = \frac{1 - f_o/f_s}{1 + f_o/f_s} = \frac{1 - (105 \text{ Hz})/(111 \text{ Hz})}{1 + (105 \text{ Hz})/(111 \text{ Hz})} = 0.02778$$

Now we can find v_o:

$$v_o = \alpha v = 0.02778 \times 338 \text{ m/s} = 9.4 \text{ m/s}$$

The speed of each car is 9.4 m/s.

Discussion Quick check on the algebra: substituting $v = 338$ m/s, $f_s = 111$ Hz, $v_o = 9.4$ m/s, and $v_s = -9.4$ m/s directly into Eq. (12-14),

$$f_o = \frac{1 - (9.4 \text{ m/s})/(338 \text{ m/s})}{1 - (-9.4 \text{ m/s})/(338 \text{ m/s})} \times 111 \text{ Hz} = 105 \text{ Hz}$$

Practice Problem 12.9 Finding Speed from the Doppler Shift

A car is driving due west at 15 m/s and sounds its horn with a frequency of 260.0 Hz. A passenger in a car heading east away from the first car hears the horn at a frequency of 230.0 Hz. How fast is the second car traveling? The speed of sound is 350 m/s.

Shock Waves

Let's examine two interesting special cases of the Doppler formula [Eq. (12-14)]. First, what if the observer moves away from the source at the speed of sound ($v_o = v$)? The Doppler-shifted frequency would be zero according to Eq. (12-14). What does that

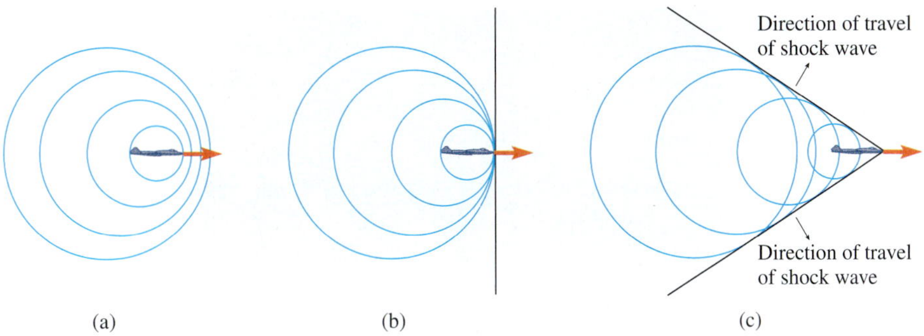

Figure 12.16 (a) Wave crests for a plane moving slower than sound. (b) A plane moving at the speed of sound; the wave crests pile up on one another since the plane moves to the right as fast as the wave crests. (c) Shock wave for a supersonic plane. The wave crests pile up along the cone indicated by the black lines.

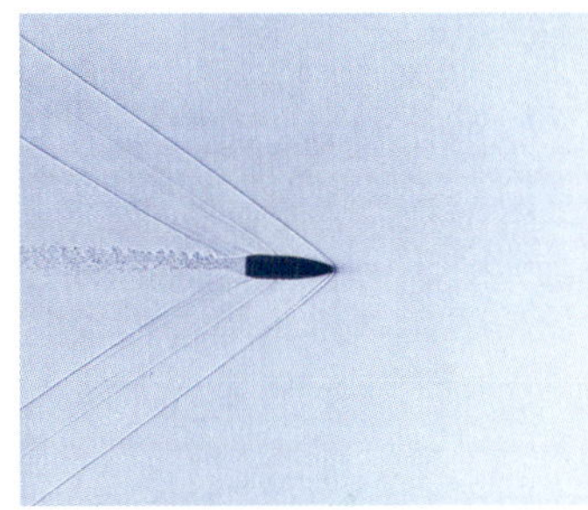

Figure 12.17 A bullet moving through air faster than sound. Notice the two principal shock waves starting at either end of the bullet.

mean? If the observer moves away from the source with a speed equal to (or greater than) the wave speed, the wave crests *never reach the observer.*

Second, what if the source moves toward the observer at a speed approaching the speed of sound ($v_s \to v$)? Then Eq. (12-14) gives an observed frequency that increases without bound ($f_o \to \infty$). Figure 12.16 helps us understand what that means. For a plane moving slower than sound, the wave crests in front of it are closer together due to the plane's motion (Fig. 12.16a). An observer to the right would measure a frequency higher than the source frequency. As the plane's speed increases, the wave crests in front of it get closer and closer together and the observed frequency increases. For a plane moving at the speed of sound (Fig. 12.16b), the wave crests pile up on top of one another; they move to the right at the same speed as the plane, so they can't get ahead of it. An observer to the right would measure a wavelength of zero—zero distance between wave crests—and therefore an infinite frequency.

What happens if the source moves at a speed *greater than* the speed of sound? Figure 12.16c shows that the wave crests pile up on top of one another to form cone-shaped *shock waves*, which travel outward in the direction indicated. There are two principal shock waves formed, one starting at the nose of the plane and one at the tail (Fig. 12.17). The sound of a shock wave is referred to as a *sonic boom*. (See the text website for more information about supersonic flight.)

connect

EVERYDAY PHYSICS DEMO

You can make a visible shock wave by trailing your finger along the surface of the water in a sink or tub. If your finger pushes the water faster than water waves travel, water piles up in front of your finger and forms a V-shaped shock wave. See if you can approximate the case of a plane moving at the speed of sound with rounded waves moving outward from your finger (Fig. 12.16b) instead of a V-shaped wave. The next time you are in a motor boat, or watching one, notice the V-shaped bow wave that extends from the prow of the boat when it moves faster than the speed of water waves.

12.9 ECHOLOCATION AND MEDICAL IMAGING

Application: Animal Echolocation Bats, dolphins, whales, and some birds use *echolocation* to locate prey and to "see" their environment. To find their way around in the darkness of caves, oilbirds of northern South America and cave swiftlets of Borneo and East Asia emit sound waves and listen for the echoes. The time it takes for the echoes to return tells them how far they are from an obstacle or cave wall. Differences between the echoes that reach the two sides of the head provide information on the direction from which the echo comes.

The sounds used by oilbirds and cave swiftlets for echolocation are audible to humans, but dolphins, whales, and most bats use ultrasound (20 to 200 kHz) instead. Bats and dolphins can also determine an object's velocity by sensing the Doppler shift between the emitted and reflected waves—a clear advantage in locating prey that are darting around to avoid being eaten. Some horseshoe bats can detect frequency differences as small as 0.1 Hz.

Prey are not completely helpless. Moths, lacewings, and praying mantises have primitive ears containing a few nerve cells to detect the ultrasound emitted by a nearby bat. A group of moths fluttering about at some distance from a cave may, for no apparent reason, fold their wings and drop suddenly to the ground. Folding their wings both reduces the amount of reflected sound and helps them drop quickly to the ground to evade the swooping bat. The moths' bodies are furry rather than smooth to help absorb some of the sound waves and thus reduce the intensity of reflected sound.

When the tiger moth detects the ultrasound from a bat, it emits its own ultrasound by flexing a part of its exoskeleton. The extra sounds mixed in with the echoes tend to confuse the bat, perhaps encouraging it to hunt elsewhere.

Application: Sonar and Radar Echolocation is a useful navigational tool for seafarers. To find the depth of water below a boat, a *sonar* (**so**und **na**vigation and **r**anging) device sends out ultrasonic pulses (Fig. 12.18). The time delay Δt between an emitted ultrasonic pulse and the return of its reflection is used to determine the distance to the seafloor. Seismic P waves—sound waves traveling through the Earth—generated by explosions or air guns are used to study the interior structure of Earth and to find oil beneath the surface.

Radar is a form of echolocation that uses electromagnetic waves instead of sound waves, but otherwise the concept is similar. Weather forecasting relies on *Doppler radar* to show not only the location of a storm, but also the wind velocity.

Medical Applications of Ultrasound

Millions of expectant parents see their unborn child for the first time when the mother has an ultrasonic examination. Ultrasonic imaging uses a pulse-echo technique similar to that used by bats and in sonar. Pulses of ultrasound are reflected at boundaries between different types of tissue.

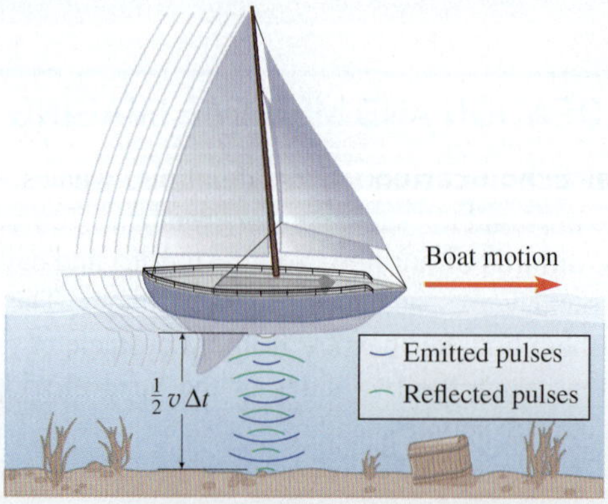

Figure 12.18 A boat with a sonar device to locate the depth of the seafloor; an ultrasound pulse, sent out from the boat by a transmitter, is reflected from the seafloor and detected by a receiver on the boat.

In the early stages of pregnancy (tenth to fourteenth weeks), the scan is used to verify that the fetus is alive and to check for twins. The length of the fetus is measured to help determine the due date more accurately. Some abnormalities can be discovered even at this early stage. For example, some chromosomal abnormalities can be detected by measuring the thickness of the skin at the back of the neck. After the eighteenth week, the fetus can be examined in even more detail. The major organs are examined to be sure they are developing normally. After the thirtieth week, the flow of blood in the umbilical cord is checked to ensure that oxygen and nutrients reach the fetus. The position of the placenta is also checked.

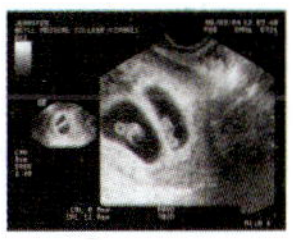

Why is ultrasound used to image the fetus?

Why are sound waves used rather than, say, electromagnetic waves such as x-rays? X-ray radiation is damaging to tissue—especially to rapidly growing fetal tissue. After decades of use, ultrasound has no known adverse effects. In addition, ultrasound images are captured in real time, so they are available immediately and can show movement. A third reason is that regular x-rays detect the amount of radiation that passes through tissue, but cannot resolve details at different depths, and so cannot produce an image of a "slice" of the abdomen; a more complicated and expensive diagnostic tool such as a CT scan (computed tomography) would be required to resolve details at different depths. Fourth, some kinds of tissue are not detected well by x-rays but are clearly resolved in ultrasound.

Why is ultrasound used rather than sound waves of audible frequencies? Sound waves with high frequencies have small wavelengths. Waves with small wavelengths diffract less around the same obstacle than do waves with larger wavelengths (see Section 11.9). Too much diffraction would obscure details in the image. As a rough guideline, the wavelength is a lower limit on the smallest detail that can be resolved. The frequencies used in imaging are typically in the range 1 to 15 MHz, which means that the wavelengths in human tissue are in the range 0.1 to 1.5 mm. As a comparison, if sound waves at 15 kHz were used, the wavelength inside the body would be 10 cm. Higher frequencies give better resolution but at the expense of less penetration; sound waves are absorbed within a distance of about 500λ in tissue.

The medical applications of ultrasonic imaging are not limited to prenatal care. Ultrasound is also used to examine organs such as the heart, liver, gallbladder, kidneys, bladder, breasts, and eyes, and to locate tumors. It can be used to diagnose various heart conditions and to assess damage after a heart attack (Fig. 12.19). Ultrasound can show movement, so it is used to assess heart valve function and to monitor blood flow in large blood vessels. Because ultrasound provides real-time images, it is sometimes used to guide procedures such as biopsies, in which a needle is used to take a sample from an organ or tumor for testing.

Doppler ultrasound is a technique that is used to examine blood flow. It can help reveal blockages to blood flow, show the formation of plaque in arteries, and provide detailed information on the heartbeat of the fetus during labor and delivery. The Doppler-shifted reflections interfere with the emitted ultrasound, producing beats. The beat frequency is proportional to the speed of the reflecting object (see Problem 58).

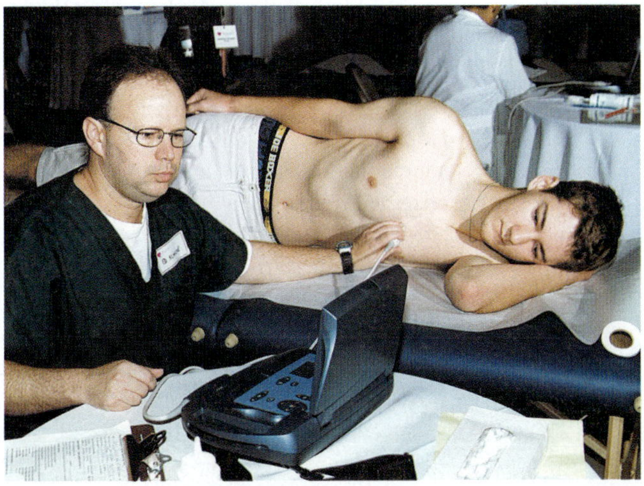

Figure 12.19 Ultrasonic imaging is used to diagnose heart disease.

Master the Concepts

- A sound wave can be described either by the gauge pressure p, which measures the pressure fluctuations above and below the ambient atmospheric pressure, or by the displacement s of each point in the medium from its undisturbed position.

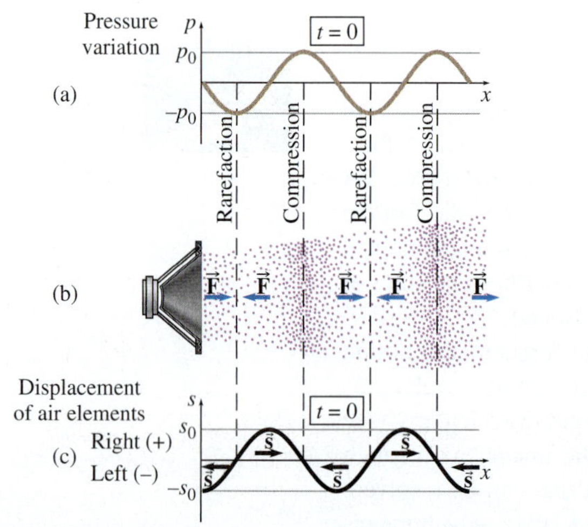

- Humans with excellent hearing can hear frequencies from 20 Hz to 20 kHz. The terms infrasound and ultrasound are used to describe sound waves with frequencies below 20 Hz and above 20 kHz, respectively.
- The speed of sound in a fluid is

$$v = \sqrt{\frac{B}{\rho}} \quad \text{(12-1)}$$

- The speed of sound in an ideal gas at any absolute temperature T can be found if it is known at one temperature:

$$v = v_0\sqrt{\frac{T}{T_0}} \quad \text{(12-3)}$$

where the speed of sound at absolute temperature T_0 is v_0.

- The speed of sound in dry air at 0°C is 331 m/s.
- For sound waves traveling along the length of a thin solid rod, the speed is approximately

$$v = \sqrt{\frac{Y}{\rho}} \quad \text{(thin solid rod)} \quad \text{(12-5)}$$

- The pressure amplitude of a sound wave is proportional to the displacement amplitude. For a harmonic sound wave at angular frequency ω,

$$p_0 = \omega v \rho s_0 \quad \text{(12-6)}$$

where v is the speed of sound and ρ is the mass density of the medium.

- The intensity of a sound wave is related to the pressure amplitude as follows:

$$I = \frac{p_0^2}{2\rho v} \quad \text{(12-7)}$$

where ρ is the mass density of the medium and v is the speed of sound in that medium. The most important thing to remember is that *intensity is proportional to amplitude squared*, which is true for all waves, not just sound.

- Sound intensity level in decibels is

$$\beta = (10\text{ dB}) \log_{10}\frac{I}{I_0} \quad \text{(12-8)}$$

where $I_0 = 10^{-12}$ W/m^2. Sound intensity level is useful since it roughly corresponds to the way we perceive loudness. Equal increments in intensity level roughly correspond to equal increases in loudness.

- In a standing sound wave in a thin pipe, an open end is a pressure node and a displacement antinode; a closed end is a pressure antinode and a displacement node. For a pipe open at both ends,

$$\lambda_n = \frac{2L}{n} \quad \text{(11-11)}$$

$$f_n = n\frac{v}{2L} = nf_1 \quad \text{(11-12)}$$

where $n = 1, 2, 3, \ldots$.
For a pipe closed at one end,

$$\lambda_n = \frac{4L}{n} \quad \text{(12-10a)}$$

$$f_n = n\frac{v}{4L} = nf_1 \quad \text{(12-10b)}$$

where $n = 1, 3, 5, 7, \ldots$.

- When two sound waves are close in frequency, the superposition of the two produces a pulsation called *beats*.

$$f_{beat} = \Delta f \quad \text{(12-11)}$$

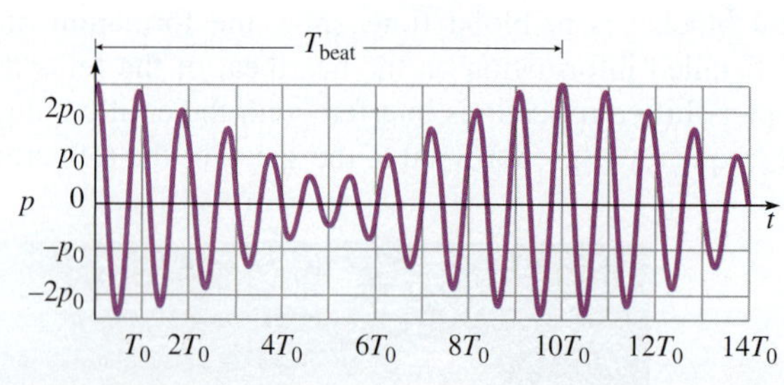

- Doppler effect: if v_s and v_o are the velocities of the source and observer and v is the wave speed, the observed frequency is

$$f_o = \left(\frac{v - v_o}{v - v_s}\right) f_s \quad \text{(12-14)}$$

where v_s and v_o are positive in the direction of propagation of the wave and are measured with respect to the wave medium.

Conceptual Questions

1. Explain why the pitch of a bassoon is more sensitive to a change in air temperature than the pitch of a cello. (That's why wind players keep blowing air through the instrument to keep it in tune.)
2. On a warm day, a piano is tuned to match an organ in an auditorium. Will the piano still be in tune with the organ the next morning, when the room is cold? If not, will the organ be higher or lower in pitch than the piano? (Assume that the piano's tuning doesn't change. Why is that a reasonable assumption?)
3. Many real estate agents have an ultrasonic rangefinder that enables them to quickly and easily measure the dimensions of a room. The device is held to one wall and reads the distance to the opposite wall. How does it work?
4. For high-frequency sounds, the ear's principal method of localization is the difference in intensity sensed by the two ears. Why can't the ear reliably use this method for low-frequency sounds? Doesn't the head cast a "sound shadow" regardless of the frequency? Explain. [*Hint*: Consider diffraction of sound waves around the head.]
5. For low-frequency sounds, the ear uses the phase difference between the sound waves arriving at the two ears to determine direction. Why can't the ear reliably use phase difference for high-frequency sounds? Explain.
6. A sign along the road in Tompkins County reads, "State Law: Noise Limit, 90 decibels." If you were subjected to such a noise level for an extended period of time, would you need to worry about your hearing being affected?
7. Why is it that your own voice sounds strange to you when you hear it played back on a tape recorder, but your friends all agree that it is just what your voice sounds like? [*Hint*: Consider the medium through which the sound wave travels when you usually hear your own voice.]
8. What is the purpose of the gel that is spread over the skin before an ultrasonic imaging procedure? [*Hint*: The speed of sound in the gel is similar to the speed in the body, while the speed in air is much slower. What happens to a wave at an abrupt change in wave speed?]
9. A stereo system whose amplifier can produce 60 W per channel is replaced by one rated 120 W per channel. Would you expect the new stereo to be able to play twice as loudly as the old one? Explain.
10. A moving source emits a sound wave that is heard by a moving observer. Imagine a thin wall at rest between the source and observer. The wall completely absorbs the sound and instantaneously emits an *identical* sound wave. Use this scenario to explain why we can combine the Doppler shifts due to motion of the source and observer as in Eq. (12-14). [*Hint*: What is the net effect of this imaginary wall?]
11. Explain why the displacement of air elements at condensations and rarefactions is zero.
12. Why is the speed of sound in solids generally much faster than the speed of sound in air?
13. If the pressure amplitude of a sound wave is doubled, what happens to the displacement amplitude, the intensity, and the intensity level?
14. The source and observer of a sound wave are both at rest with respect to the ground. The wind blows in the direction from source to observer. Is the observed frequency Doppler-shifted? Explain.
15. Many brass instruments have valves that increase the total length of the pipe from mouthpiece to bell. When a valve is depressed, is the fundamental frequency raised or lowered? What happens to the pitch?

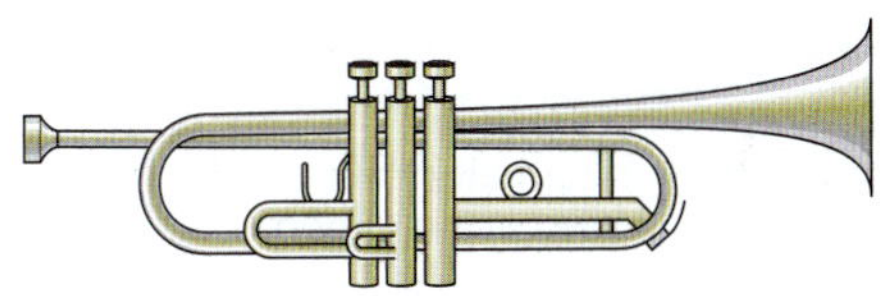

16. When the viola section of an orchestra with six members plays together, is the sound 6 times as loud as when a single viola plays? Explain. Is the intensity 6 times what it would be for a single viola? [*Hint*: The six sound waves are not coherent.]
17. The fundamental frequency of the highest note on the piano is 4.186 kHz. Most musical instruments do not go that high; only a few singers can produce sounds with fundamental frequencies higher than around 1 kHz. Yet a good-quality stereo system must reproduce frequencies up to at least 16 to 18 kHz. Explain.

Multiple-Choice Questions

Student Response System Questions

1. An organ pipe is closed at one end. Several standing wave patterns are sketched in the drawing. Which one is *not* a possible standing wave pattern for this pipe?
2. Of the standing wave patterns sketched in the drawing, which shows the lowest frequency standing wave for an organ pipe closed at one end? (connect tutorial: standing waves)

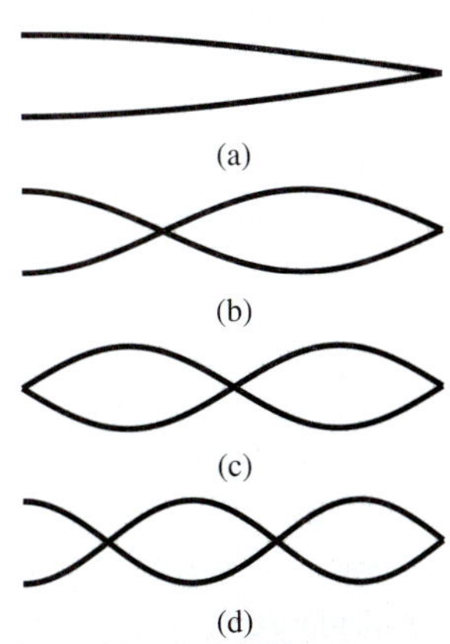

Multiple-Choice Questions 1 and 2

3. The intensity of a sound wave is directly proportional to
   (a) the frequency.
   (b) the amplitude.
   (c) the square of the amplitude.
   (d) the square of the speed of sound.
   (e) none of the above.
4. The speed of sound in water is 4.3 times the speed of sound in air. A whistle on land produces a sound wave with a frequency f_0. When this sound wave enters the water, its frequency becomes
   (a) $4.3f_0$
   (b) f_0
   (c) $\dfrac{f_0}{4.3}$
   (d) not enough information given
5. A source of sound with frequency 620 Hz is placed on a moving platform that approaches a physics student at speed v; the student hears sound with a frequency f_1. Then the source of sound is held stationary while the student approaches it at the same speed v; the student hears sound with a frequency f_2. Choose the correct statement.
   (a) $f_1 = f_2$; both are greater than 620 Hz.
   (b) $f_1 = f_2$; both are less than 620 Hz.
   (c) $f_1 > f_2 > 620$ Hz.
   (d) $f_2 > f_1 > 620$ Hz.
6. A moving van and a small car are traveling in the same direction on a two-lane road. The van is moving at twice the speed of the car and overtakes the car. The driver of the car sounds his horn, frequency = 440 Hz, to signal the van that it is safe to return to the lane. Which is the correct statement?
   (a) The car driver and van driver both hear the horn frequency as 440 Hz.
   (b) The car driver hears 440 Hz, but the van driver hears a lower frequency.
   (c) The car driver hears 440 Hz, but the van driver hears a higher frequency.
   (d) Both drivers hear the same frequency, and it is lower than 440 Hz.
7. A trombone and a bassoon play notes of equal loudness with the same fundamental frequency. The two sounds differ primarily in
   (a) pitch.
   (b) intensity level.
   (c) amplitude.
   (d) timbre.
   (e) wavelength.
8. The fundamental frequency of a pipe closed at one end is f_1. How many nodes are present in a standing wave of frequency $9f_1$?
   (a) 4 (b) 5 (c) 6 (d) 8 (e) 9 (f) 10
9. The length of a pipe closed at one end is L. In the standing wave whose frequency is 7 times the fundamental frequency, what is the closest distance between nodes?
   (a) $\frac{1}{14}L$ (b) $\frac{1}{7}L$ (c) $\frac{2}{7}L$ (d) $\frac{4}{7}L$ (e) $\frac{8}{7}L$
   (f) none of the above
10. The three lowest resonant frequencies of a system are 50 Hz, 150 Hz, and 250 Hz. The system could be
    (a) a tube of air closed at both ends.
    (b) a tube of air open at one end.
    (c) a tube of air open at both ends.
    (d) a vibrating string with fixed ends.
11. The speed of sound in water is _____ than the speed of sound in air because _____.
    (a) faster; water is much harder to compress
    (b) faster; water is much more dense
    (c) slower; water is much easier to compress
    (d) slower; water is much less dense
    (e) equal; the two fluids are at the same pressure

Problems

- Combination conceptual/quantitative problem
- Biomedical application
- Challenging
- Blue # Detailed solution in the Student Solutions Manual
- [1, 2] Problems paired by concept
- connect Text website interactive or tutorial

Note: Assume a temperature of 20.0°C in all problems unless otherwise indicated.

12.2 The Speed of Sound Waves

1. In Death Valley, the highest recorded outdoor temperature so far is 56.7°C (in the shade!). What is the speed of sound in air at that temperature?
2. What is the speed of sound in helium at body temperature (37°C)?
3. What are the wavelengths of sound waves at the lower and upper limits of human hearing (10 Hz and 20 kHz, respectively)?
4. Bats emit ultrasonic waves with a frequency as high as 1.0×10^5 Hz. What is the wavelength of such a wave in air of temperature 15°C?
5. Dolphins emit ultrasonic waves with a frequency as high as 2.5×10^5 Hz. What is the wavelength of such a wave in seawater at 25°C?
6. At a baseball game, a spectator is 60.0 m away from the batter. How long does it take the sound of the bat connecting with the ball to travel to the spectator's ears? The air temperature is 27.0°C.
7. A lightning flash is seen in the sky, and 8.2 s later the boom of the thunder is heard. The temperature of the air is 12°C. (a) What is the speed of sound at that temperature? [*Hint*: Light travels at a speed of 3.00×10^8 m/s.] (b) How far away is the lightning strike?
8. During a thunderstorm, you can easily estimate your distance from a lightning strike. Count the number of seconds that elapse from when you see the flash of lightning to when you hear the thunder. The rule of thumb is

that 5 s elapse for each mile of distance. Verify that this rule of thumb is (approximately) correct. (One mile is 1.6 km and light travels at a speed of 3×10^8 m/s.)

9. A copper alloy has a Young's modulus of 1.1×10^{11} Pa and a density of 8.92×10^3 kg/m^3. What is the speed of sound in a thin rod made from this alloy? Compare your result with the value for copper given in Table 12.1.

10. Find the speed of sound in mercury, which has a bulk modulus of 2.8×10^{10} Pa and a density of 1.36×10^4 kg/m^3.

11. Derive Eq. (12-4): (a) Starting with Eq. (12-3), substitute $T = T_C + 273.15$. (b) Apply the binomial approximation to the square root (see Appendix A.5) and simplify.

12. ✦ Stan and Ollie are standing next to a train track. Stan puts his ear to the steel track to hear the train coming. He hears the sound of the train whistle through the track 2.1 s before Ollie hears it through the air. How far away is the train?

12.3 Amplitude and Intensity of Sound Waves

13. Six sound waves have pressure amplitudes p_0 and frequencies f as given. Rank them in order of the displacement amplitude, largest to smallest.
(a) $p_0 = 0.05$ Pa, $f = 400$ Hz
(b) $p_0 = 0.01$ Pa, $f = 400$ Hz
(c) $p_0 = 0.01$ Pa, $f = 2000$ Hz
(d) $p_0 = 0.05$ Pa, $f = 80$ Hz
(e) $p_0 = 0.05$ Pa, $f = 16$ Hz
(f) $p_0 = 0.25$ Pa, $f = 400$ Hz

14. A sound wave with an intensity level of 80.0 dB is incident on an eardrum of area 0.600×10^{-4} m^2. How much energy is incident on the eardrum in 3.00 min?

15. (a) What is the pressure amplitude of a sound wave with an intensity level of 120.0 dB in air? (b) What force does this exert on an eardrum of area 0.550×10^{-4} m^2?

16. A 40 Hz sound wave is barely audible at a sound intensity level of 60 dB. What is the displacement amplitude of this sound wave? Compare it with the average distance between molecules in air at room temperature, about 3 nm.

17. Chronic exposure to loud noises can be damaging to one's hearing. This can be a problem in occupations in which heavy machinery is used. If a machine produces sound with an intensity level of 100.0 dB, what would its intensity level have to be to reduce the intensity by a factor of 2.0?

18. Table 12.2 lists 120 dB as the intensity level at the threshold of pain for humans. (a) Show that the corresponding pressure amplitude is 29 Pa. (b) What is this pressure amplitude as a fraction of atmospheric pressure?

19. A sound wave in room-temperature air has an intensity level of 65.0 dB and a frequency of 131 Hz. (a) What is the pressure amplitude? (b) What is the displacement amplitude?

20. The sound level 25 m from a loudspeaker is 71 dB. What is the rate at which sound energy is produced by the loudspeaker, assuming it to be an isotropic source?

21. In a factory, three machines produce noise with intensity levels of 85 dB, 90 dB, and 93 dB. When all three are running, what is the intensity level? How does this compare to running just the loudest machine?

22. At the race track, one race car starts its engine with a resulting intensity level of 98.0 dB at point P. Then seven more cars start their engines. If the other seven cars each produce the same intensity level at point P as the first car, what is the new intensity level with all eight cars running?

23. An intensity level change of +1.00 dB corresponds to what percentage change in intensity?

24. (a) Show that if $I_2 = 10.0I_1$, then $\beta_2 = \beta_1 + 10.0$ dB. (A factor of 10 increase in intensity corresponds to a 10.0 dB increase in intensity level.) (b) Show that if $I_2 = 2.0I_1$, then $\beta_2 = \beta_1 + 3.0$ dB. (A factor of 2 increase in intensity corresponds to a 3.0 dB increase in intensity level. connect tutorial: decibels)

25. At a rock concert, the engineer decides that the music isn't loud enough. He turns up the amplifiers so that the amplitude of the sound, where you're sitting, increases by 50.0%. (a) By what percentage does the intensity increase? (b) How does the intensity level (in dB) change?

12.4 Standing Sound Waves

26. Humans can hear sounds with frequencies up to about 20.0 kHz, but dogs can hear frequencies up to about 40.0 kHz. Dog whistles are made to emit sounds that dogs can hear but humans cannot. If the part of a dog whistle that actually produces the high frequency is made of a tube open at both ends, what is the longest possible length for the tube?

27. The figure shows standing wave patterns in five pipes of equal length. Pipes (c) and (e) are open at both ends; the others are closed at one end. Rank the standing waves in order of the frequency, largest to smallest.

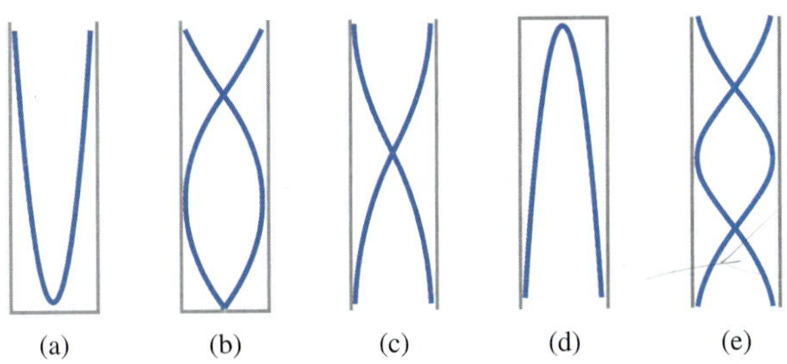

28. A glass tube is closed at one end and has a diaphragm covering the other end. The tube is filled with gas and some sawdust has been scattered along inside the tube. When the diaphragm is driven at a frequency of 1457 Hz, the sawdust forms small piles 20 cm apart.

(a) What is the speed of the sound in the gas? (b) Do the piles of sawdust represent displacement nodes or antinodes in the sound wave? Explain.

29. (a) What should be the length of an organ pipe, closed at one end, if the fundamental frequency is to be 261.5 Hz? (b) What is the fundamental frequency of the organ pipe of part (a) if the temperature drops to 0.0°C?

30. Repeat Problem 29 for an organ pipe that is open at both ends.

31. An organ pipe that is open at both ends has a fundamental frequency of 382 Hz at 0.0°C. What is the fundamental frequency for this pipe at 20.0°C?

32. What is the length of the organ pipe in Problem 31?

33. In an experiment to measure the speed of sound in air, standing waves are set up in a narrow pipe open at both ends using a speaker driven at 702 Hz. The length of the pipe is 2.0 m. What is the air temperature inside the pipe (assumed reasonably near room temperature, 20°C to 35°C)? [*Hint*: The standing wave is not necessarily the fundamental.]

34. When a tuning fork is held over the open end of a very thin tube, as in Fig. 12.7, the smallest value of L that produces resonance is found to be 30.0 cm. (a) What is the wavelength of the sound? [*Hint*: Assume that the displacement antinode is at the open end of the tube.] (b) What is the next larger value of L that will produce resonance with the same tuning fork? (c) If the frequency of the tuning fork is 282 Hz, what is the speed of sound in the tube?

35. Two tuning forks, A and B, excite the next-to-lowest resonant frequency in two air columns of the same length, but A's column is closed at one end and B's column is open at both ends. What is the ratio of A's frequency to B's frequency?

36. How long a pipe is needed to make a tuba whose lowest note is low C (frequency 130.8 Hz)? Assume that a tuba is a long straight pipe open at both ends.

12.7 Beats

37. Ⓒ A violin is tuned by adjusting the tension in the strings. Brian's A string is tuned to a slightly lower frequency than Jennifer's, which is correctly tuned to 440.0 Hz. (a) What is the frequency of Brian's string if beats of 2.0 Hz are heard when the two bow the strings together? (b) Does Brian need to tighten or loosen his A string to get in tune with Jennifer? Explain.

38. A piano tuner sounds two strings simultaneously. One has been previously tuned to vibrate at 293.0 Hz. The tuner hears 3.0 beats per second. The tuner increases the tension on the as-yet untuned string, and now when they are played together the beat frequency is 1.0 s^{-1}. (a) What was the original frequency of the untuned string? (b) By what percentage did the tuner increase the tension on that string?

39. An auditorium has organ pipes at the front and at the rear of the hall. Two identical pipes, one at the front and one at the back, have fundamental frequencies of 264.0 Hz at 20.0°C. During a performance, the organ pipes at the back of the hall are at 25.0°C, while those at the front are still at 20.0°C. What is the beat frequency when the two pipes sound simultaneously?

40. A musician plays a string on a guitar that has a fundamental frequency of 330.0 Hz. The string is 65.5 cm long and has a mass of 0.300 g. (a) What is the tension in the string? (b) At what speed do the waves travel on the string? (c) While the guitar string is still being plucked, another musician plays a slide whistle that is closed at one end and open at the other. He starts at a very high frequency and slowly lowers the frequency until beats, with a frequency of 5 Hz, are heard with the guitar. What is the fundamental frequency of the slide whistle with the slide in this position? (d) How long is the open tube in the slide whistle for this frequency?

41. ✦ A cello string has a fundamental frequency of 65.40 Hz. What beat frequency is heard when this cello string is bowed at the same time as a violin string with frequency of 196.0 Hz? [*Hint*: The beats occur between the third harmonic of the cello string and the fundamental of the violin.]

12.8 The Doppler Effect

42. An ambulance traveling at 44 m/s approaches a car heading in the same direction at a speed of 28 m/s. The ambulance driver has a siren sounding at 550 Hz. At what frequency does the driver of the car hear the siren?

43. At a factory, a noon whistle is sounding with a frequency of 500 Hz. As a car traveling at 85 km/h approaches the factory, the driver hears the whistle at frequency f_i. After driving past the factory, the driver hears frequency f_f. What is the change in frequency $f_f - f_i$ heard by the driver?

44. In parts of the midwestern United States, sirens sound when a severe storm that may produce a tornado is approaching. Mandy is walking at a speed of 1.56 m/s directly toward one siren and directly away from another siren when they both begin to sound with a frequency of 698 Hz. What beat frequency does Mandy hear? (connect tutorial: Doppler effect)

45. A source of sound waves of frequency 1.0 kHz is traveling through the air at 0.50 times the speed of sound. (a) Find the frequency of the sound received by a stationary observer if the source moves toward her. (b) Repeat if the source moves away from her instead.

46. A source of sound waves of frequency 1.0 kHz is stationary. An observer is traveling at 0.50 times the speed of sound. (a) What is the observed frequency if the observer moves toward the source? (b) Repeat if the observer moves away from the source instead.

47. A speedboat is traveling at 20.1 m/s toward another boat moving in the opposite direction with a speed of 15.6 m/s. The speedboat pilot sounds his horn, which has a frequency of 312 Hz. What is the frequency heard by a passenger in the oncoming boat?

48. A source and an observer are *each* traveling at 0.50 times the speed of sound. The source emits sound waves at 1.0 kHz. Find the observed frequency if (a) the source and observer are moving *toward* each other; (b) the source and observer are moving *away* from each other; (c) the source and observer are moving in the same direction.

49. Blood flow rates can be found by measuring the Doppler shift in frequency of ultrasound reflected by red blood cells (known as *angiodynography*). If the speed of the red blood cells is v, the speed of sound in blood is u, the ultrasound source emits waves of frequency f, and we assume that the blood cells are moving directly toward the ultrasound source, show that the frequency f_r of reflected waves detected by the apparatus is given by

$$f_r = f\frac{u+v}{u-v}$$

[*Hint*: There are *two* Doppler shifts. A red blood cell first acts as a moving observer; then it acts as a moving source when it reradiates the reflected sound at the same frequency that it received.]

50. ✦ The pitch of the sound from a race car engine drops the musical interval of a fourth when it passes the spectators. This means the frequency of the sound after passing is 0.75 times what it was before. How fast is the race car moving?

12.9 Echolocation and Medical Imaging

51. A ship is lost in a dense fog in a Norwegian fjord that is 1.80 km wide. The air temperature is 5.0°C. The captain fires a pistol and hears the first echo after 4.0 s. (a) How far from one side of the fjord is the ship? (b) How long after the first echo does the captain hear the second echo?

52. A ship mapping the depth of the ocean emits a sound of 38 kHz. The sound travels to the ocean floor and returns 0.68 s later. (a) How deep is the water at that location? (b) What is the wavelength of the wave in water? (c) What is the wavelength of the reflected wave as it travels into the air, where the speed of sound is 350 m/s?

53. A boat is using sonar to detect the bottom of a freshwater lake. If the echo from a sonar signal is heard 0.540 s after it is emitted, how deep is the lake? Assume the temperature of the lake is uniform and at 25°C.

54. A geological survey ship mapping the floor of the ocean sends sound pulses down from the surface and measures the time taken for the echo to return. How deep is the ocean at a point where the echo time (down and back) is 7.07 s? The temperature of the seawater is 25°C.

55. ✦ A bat emits chirping sounds of frequency 82.0 kHz while hunting for moths to eat. If the bat is flying toward the moth at a speed of 4.40 m/s and the moth is flying away from the bat at 1.20 m/s, what is the frequency of the sound wave reflected from the moth as observed by the bat? Assume it is a cool night with a temperature of 10.0°C. [*Hint*: There are two Doppler shifts. Think of the moth as a receiver, which then becomes a source as it "retransmits" the reflected wave.]

56. The bat of Problem 55 emits a chirp that lasts for 2.0 ms and then is silent while it listens for the echo. If the beginning of the echo returns just after the outgoing chirp is finished, how close to the moth is the bat? [*Hint*: Is the change in distance between the two significant during a 2.0 ms time interval?]

57. ✦ Doppler ultrasound is used to measure the speed of blood flow (see Problem 49). The reflected sound interferes with the emitted sound, producing beats. If the speed of red blood cells is 0.10 m/s, the ultrasound frequency used is 5.0 MHz, and the speed of sound in blood is 1570 m/s, what is the beat frequency?

58. ✦ (a) In Problem 49, find the beat frequency between the outgoing and reflected sound waves. (b) Show that the beat frequency is proportional to the speed of the blood cell if $v \ll u$. [*Hint*: Use the binomial approximation from Appendix A.5.]

Collaborative Problems

59. A certain pipe has resonant frequencies of 234 Hz, 390 Hz, and 546 Hz, with no other resonant frequencies between these values. (a) Is this a pipe open at both ends or closed at one end? (b) What is the fundamental frequency of this pipe? (c) How long is this pipe?

60. ✦ Ⓒ An aluminum rod, 1.0 m long, is held lightly in the middle. One end is struck head-on with a rubber mallet so that a longitudinal pulse—a sound wave—travels down the rod. The fundamental frequency of the longitudinal vibration is 2.55 kHz. (a) Describe the locations of the displacement and pressure nodes and antinodes for the fundamental mode of vibration. (b) Calculate the speed of sound in aluminum from the information given in the problem. (c) The vibration of the rod produces a sound wave in air that can be heard. What is the wavelength of the sound wave in the air? Take the speed of sound in air to be 334 m/s. (d) Do the two ends of the rod vibrate longitudinally in phase or out of phase with each other? That is, at any given instant, do they move in the same direction or in opposite directions?

61. ✦ One cold and windy winter day, Zach notices a humming sound coming from his chimney, which is open at the top and closed at the bottom. He opens the chimney at the bottom and notices that the sound changes. He goes over to the piano to try to match the note that the chimney is producing with the bottom open. He finds that the "C" three octaves below middle "C" matches the chimney's fundamental frequency. Zach knows that the frequency of middle "C" is 261.6 Hz, and each lower octave is $\frac{1}{2}$ of the frequency of the octave above. From this information, Zach finds the height of the chimney and the fundamental frequency of the note that was produced when the chimney was *closed* at the bottom. Assuming that the speed of sound in the cold air is 330 m/s, reproduce Zach's calculations to find (a) the height of the chimney and (b) the fundamental frequency of the chimney when it is *closed* at the bottom.

62. ✦ Your friend needs advice on her newest "acoustic sculpture." She attaches one end of a steel wire, of diameter 4.00 mm and density 7860 kg/m^3, to a wall. After passing over a pulley, located 1.00 m from the wall, the other end of the wire is attached to a hanging weight. Below the horizontal length of wire she places a 1.50 m long hollow tube, open at one end and closed at the other. Once the sculpture is in place, air will blow through the tube, creating a sound. Your friend wants this sound to cause the steel wire to vibrate transversely at the same resonant frequency as the tube. What weight (in newtons) should she hang from the wire if the temperature is 18.0°C?

63. ✦ Ⓒ In this problem, you will estimate the smallest kinetic energy of vibration that the human ear can detect. Suppose that a harmonic sound wave at the threshold of hearing ($I = 1.0 \times 10^{-12}$ W/m^2) is incident on the eardrum. The speed of sound is 340 m/s, and the density of air is 1.3 kg/m^3. (a) What is the maximum speed of an element of air in the sound wave? [*Hint*: See Eq. (10-21).] (b) Assume the eardrum vibrates with displacement s_0 at angular frequency ω; its maximum speed is then equal to the maximum speed of an air element. The mass of the eardrum is approximately 0.1 g. What is the *average* kinetic energy of the eardrum? (c) The average kinetic energy of the eardrum due to collisions with air molecules *in the absence of a sound wave* is about 10^{-20} J. Compare your answer with (b) and discuss.

64. Akiko rides her bike toward a brick wall with a speed of 7.00 m/s while blowing a whistle that is emitting sound with a frequency of 512.0 Hz. (a) What is the frequency of the sound that is reflected from the wall as heard by Haruki, who is standing still? (b) Junichi is walking away from the wall at a speed of 2.00 m/s. What is the frequency of the sound reflected from the wall that Junichi hears?

Comprehensive Problems

65. Kyle is climbing a sailboat mast and is 5.00 m above the surface of the ocean, while his friend Rob is scuba diving below the boat. Kyle shouts to someone on another boat and Rob hears him shout 0.0210 s later. The ocean temperature is 25°C and the air is at 20°C. How deep is Rob below the boat?

66. What are the four lowest standing wave frequencies for an organ pipe that is 4.80 m long and closed at one end?

67. The length of the auditory canal in humans averages about 2.5 cm. What are the lowest three standing wave frequencies for a pipe of this length open at one end? What effect might resonance have on the sensitivity of the ear at various frequencies? (Refer to Fig. 12.12. Note that frequencies critical to speech recognition are in the range 2 to 5 kHz.)

68. A sound wave arriving at your ear is transferred to the fluid in the cochlea. If the intensity in the fluid is 0.80 times that in air and the frequency is the same as for the wave in air, what will be the ratio of the pressure amplitude of the wave in air to that in the fluid? Approximate the fluid as having the same values of density and speed of sound as water.

69. At what frequency f does a sound wave in air have a wavelength of 15 cm, about half the diameter of the human head? Some methods of localization work well only for frequencies below f, whereas others work well only above f. (See Conceptual Questions 4 and 5.)

70. Some bats determine their distance to an object by detecting the difference in intensity between echoes. (a) If intensity falls off at a rate that is inversely proportional to the distance squared, show that the echo intensity is inversely proportional to the fourth power of distance. (b) The bat was originally 0.60 m from one object and 1.10 m from another. After flying closer, it is now 0.50 m from the first and at 1.00 m from the second object. What is the percentage increase in the intensity of the echo from each object?

71. The *Vespertilionidae* family of bats detect the distance to an object by timing how long it takes for an emitted signal to reflect off the object and return. Typically they emit sound pulses 3 ms long and 70 ms apart while cruising. (a) If an echo is heard 60 ms later (v_{sound} = 331 m/s), how far away is the object? (b) When an object is only 30 cm away, how long will it be before the echo is heard? (c) Will the bat be able to detect this echo?

72. Horseshoe bats use the Doppler effect to determine their location. A Horseshoe bat flies toward a wall at a speed of 15 m/s while emitting a sound of frequency 35 kHz. What is the beat frequency between the emission frequency and the echo?

73. According to a treasure map, a treasure lies at a depth of 40.0 fathoms on the ocean floor due east from the lighthouse. The treasure hunters use sonar to find where the depth is 40.0 fathoms as they head east from the lighthouse. What is the elapsed time between an emitted pulse and the return of its echo at the correct depth if the water temperature is 25°C? [*Hint*: One fathom is 1.83 m.]

74. ✦ When playing *fortissimo* (very loudly), a trumpet emits sound energy at a rate of 0.800 W out of a bell (opening) of diameter 12.7 cm. (a) What is the sound intensity level right in front of the trumpet? (b) If the trumpet radiates sound waves uniformly in all directions, what is the sound intensity level at a distance of 10.0 m?

75. ✦ A periodic wave is composed of the superposition of three sine waves whose frequencies are 36, 60, and 84 Hz. The speed of the wave is 180 m/s. What is the wavelength of the wave? [*Hint*: The 36 Hz is not necessarily the fundamental frequency.]

76. ✦ Analysis of the periodic sound wave produced by a violin's G string includes three frequencies: 392, 588, and 980 Hz. What is the fundamental frequency? [*Hint*: The wave on the string is the superposition of several different standing wave patterns.]

77. ✦ During a rehearsal, all eight members of the first violin section of an orchestra play a very soft passage. The sound intensity level at a certain point in the concert hall is 38.0 dB. What is the sound intensity level at the same point if only one of the violinists plays the same passage? [*Hint*: When playing together, the violins are *incoherent* sources of sound.]

Review and Synthesis

78. (a) Show that since the bulk modulus has SI units N/m^2 and mass density has SI units kg/m^3, Eq. (12-1) gives the speed of sound in m/s. Thus, the equation is dimensionally consistent. (b) Show that no other combination of B and ρ can give dimensions of speed. Thus, Eq. (12-1) *must* be correct except for the possibility of a dimensionless constant.

79. A child swinging on a swing set hears the sound of a whistle that is being blown directly in front of her. At the bottom of her swing when she is moving toward the whistle, she hears a higher pitch, and at the bottom of her swing when she is moving away from the swing she hears a lower pitch. The higher pitch has a frequency that is 5.0% higher than the lower pitch. What is the speed of the child at the bottom of the swing?

80. A 30.0 cm long string has a mass of 0.230 g and is vibrating at its third-lowest natural frequency f_3. The tension in the string is 7.00 N. (a) What is f_3? (b) What are the frequency and wavelength of the sound in the surrounding air if the speed of sound is 350 m/s?

81. The A string on a guitar has length 64.0 cm and fundamental frequency 110.0 Hz. The string's tension is 133 N. It is vibrating in its fundamental standing wave mode with a maximum displacement from equilibrium of 2.30 mm. The air temperature is 20.0°C. (a) What is the wavelength of the fundamental mode of vibration? (b) What is the wave speed on the string? (c) What is the linear mass density of the string? (d) What is the maximum speed of any point on the oscillating string? (e) The string transmits vibrations through the bridge to the body of the instrument and then to the air. What is the frequency of the sound wave in air? (f) What is the wavelength of the sound wave in air?

82. A piano "string" is steel wire with radius 0.50 mm and length 1.2 m. It is under 800 N of tension. (a) What is the speed of transverse waves on the string? (b) What is the fundamental frequency for transverse waves? (c) What is the speed of longitudinal waves (i.e., sound) in the wire? Consider the wire to be a thin solid steel rod. (d) Assuming nodes at the ends, what is the fundamental frequency for longitudinal waves? (Longitudinal waves in the wire do contribute to the characteristic sound of the piano.)

Answers to Practice Problems

12.1 Although solids usually have somewhat higher densities than liquids, they have *much* higher bulk moduli—they are much stiffer. The greater restoring forces in solids cause sound waves to travel faster.

12.2 Assumptions: Treat the stage as a point source; ignore reflection and absorption of waves. 4.0×10^{-6} W/m^2, 0.057 Pa.

12.3 400

12.4 a factor of 3.2

12.5 3000 km. No, it is not realistic to ignore absorption and reflection over such a great distance.

12.6 24°C

12.7 444.0 Hz

12.8 182.5 Hz

12.9 27 m/s

Answers to Checkpoints

12.3 The relationship between pressure and displacement amplitudes depends on the frequency and, therefore, does not have a unique value for a given pressure amplitude and intensity.

12.4 A pipe of length L closed at one end has a node at one end and an antinode at the other. The wavelength can be $2L$ only if both ends are nodes (or both are antinodes), because the distance between two successive nodes (or two successive antinodes) is $\frac{1}{2}\lambda$.

12.7 (a) Constructive interference means the two waves are *in phase*, which occurs at $t = 0$ and $t = 10T_0$. At those times, the superposition of the waves has its maximum amplitude. (b) Destructive interference means the two waves are *out of phase*, which occurs at $t = 5T_0$. At this time, the superposition has its minimum amplitude. Destructive interference would next occur at $t = 15T_0$ (not shown on the graph).

12.8 (a) The motion of the source does affect the wavelength: λ is shorter in front of the source and longer behind it (see Fig. 12.14). (b) The motion of the observer does not affect the wavelength, which is the instantaneous distance between two wave crests (see Fig. 12.15).

CHAPTER

13

Temperature and the Ideal Gas

A crocodile basks on a rock in Lake Baringo (Kenya) to get warm.

In homeothermic ("warm-blooded") animals, body temperature is carefully regulated. The hypothalamus, located in the brain, acts as the master thermostat to keep body temperature constant to within a fraction of a degree Celsius in a healthy animal. If the body temperature starts to deviate much from the desired constant level, the hypothalamus causes changes in blood flow and initiates other processes, such as shivering or perspiration, to bring the temperature back to normal. What evolutionary advantage does a constant body temperature give the homeotherms (e.g., birds and mammals) over the poikilotherms (e.g., reptiles and insects), whose body temperatures are not kept constant? What are the disadvantages? (See p. 487 for the answer.)

BIOMEDICAL APPLICATIONS

- Regulation of body temperature (Ex. 13.1; Sec. 13.7; Prob. 106)
- Temperature dependence of biological processes (Sec. 13.7; Probs. 73 and 74)
- Diffusion of O_2, water, platelets (Sec. 13.8; Ex. 13.9; Probs. 80 and 81)
- Breathing of divers, emphysema patients (Ex. 13.6; Probs. 45, 84, 115, and 116)

Concepts & Skills to Review

- energy conservation (Chapter 6)
- momentum conservation (Section 7.4)
- collisions (Sections 7.7 and 7.8)

13.1 TEMPERATURE AND THERMAL EQUILIBRIUM

The measurement of **temperature** is part of everyday life. We measure the temperature of the air outdoors to decide how to dress when going outside; a thermostat measures the air temperature indoors to control heating and cooling systems to keep our homes and offices comfortable. Regulation of oven temperature is important in baking. When we feel ill, we measure our body temperature to see if we have a fever. Despite how matter-of-fact it may seem, temperature is a subtle concept. Although our subjective sensations of hot and cold are related to temperature, they can easily mislead.

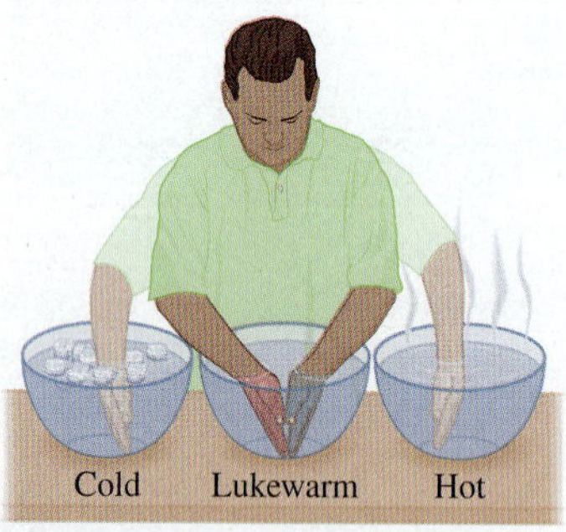

Figure 13.1 It is easy to trick our sense of temperature.

EVERYDAY PHYSICS DEMO

Try an experiment described by the English philosopher John Locke in 1690. Fill one container with water that is hot (but not too hot to touch); fill a second container with lukewarm water; and fill a third container with cold water. Put one hand in the hot water and one in the cold water (Fig. 13.1) for about 10 to 20 s. Then plunge both hands into the container of lukewarm water. Although both hands are now immersed in water that is at a single temperature, the hand that had been in the hot water feels cool but the hand that had been in the cold water feels warm. This demonstration shows that we cannot trust our subjective senses to measure temperature.

The definition of temperature is based on the concept of **thermal equilibrium**. Suppose two objects or systems are allowed to exchange energy. The net flow of energy is always from the object at the higher temperature to the object at the lower temperature. As energy flows, the temperatures of the two objects approach each other. When the temperatures are the same, there is no longer any net flow of energy; the objects are now said to be in thermal equilibrium. Thus, *temperature is a quantity that determines when objects are in thermal equilibrium.* (The objects do *not* necessarily have the same *energy* when in thermal equilibrium.) The energy that flows between two objects or systems due to a temperature difference between them is called **heat**. In Chapter 14 we discuss heat in detail. If heat can flow between two objects or systems, the objects or systems are said to be in **thermal contact**.

To measure the temperature of an object, we put a thermometer into thermal contact with the object. Temperature measurement relies on the **zeroth law of thermodynamics**.

Zeroth Law of Thermodynamics

If two objects are each in thermal equilibrium with a third object, then the two are in thermal equilibrium with each other.

Without the zeroth law, it would be impossible to define temperature, since different thermometers could give different results. The rather odd name *zeroth* law of thermodynamics came about because this law was formulated historically *after* the first, second, and third laws of thermodynamics and yet it is so fundamental that it should come *before* the others. **Thermodynamics**, the subject of Chapters 13 to 15, concerns temperature, heat flow, and the internal energy of systems.

13.2 TEMPERATURE SCALES

Thermometers measure temperature by exploiting some property of matter that is temperature-dependent. The familiar liquid-in-glass thermometer relies on thermal expansion: the mercury or alcohol expands as its temperature rises (or contracts as its temperature drops) and we read the temperature on a calibrated scale. Since some materials expand more than others, these thermometers must be calibrated on a scale using some easily reproducible phenomenon, such as the melting point of ice or the boiling point of water. The assignment of temperatures to these phenomena is arbitrary.

The most commonly used temperature scale in the world is the Celsius scale. On the Celsius scale, 0°C is the freezing temperature of water at $P = 1$ atm (the *ice point*) and 100°C is the boiling temperature of water at $P = 1$ atm (the *steam point*). The Celsius scale is named for Swedish astronomer Anders Celsius (1701–1744), who used a temperature scale that was the reverse of what we use today (water's freezing point at 100° and water's boiling point at 0°).

In the United States, the Fahrenheit scale, named after physicist Daniel Gabriel Fahrenheit (1686–1736), is still commonly used (Fig. 13.2). At 1 atm, the ice point is 32°F and the steam point is 212°F, so the difference between the steam and ice points is 180°F. The size of the Fahrenheit degree interval is therefore smaller than the Celsius degree interval: a temperature difference of 1°C is equivalent to a difference of 1.8°F:

$$\Delta T_F = \Delta T_C \times 1.8 \frac{°F}{°C} \quad (13\text{-}1)$$

Since the two scales also have an offset (0°C is not the same temperature as 0°F), conversion between the two is:

$$T_F = (1.8°F/°C)\, T_C + 32°F \quad (13\text{-}2a)$$

$$T_C = \frac{T_F - 32°F}{1.8°F/°C} \quad (13\text{-}2b)$$

The SI unit of temperature is the **kelvin** (symbol K, *without* a degree sign), named after British physicist William Thomson (Lord Kelvin) (1824–1907). The kelvin has the same degree size as the Celsius scale; that is, a temperature *difference* of 1°C is the same as a difference of 1 K. However, 0 K represents *absolute zero*—there are no temperatures below 0 K. The ice point is 273.15 K, so temperature in °C (T_C) and temperature in kelvins (T) are related.

$$T_C = T - 273.15 \quad (13\text{-}3)$$

Equation (13-3) is the definition of the Celsius scale in terms of the kelvin. Table 13.1 shows some temperatures in kelvins, °C, and °F.

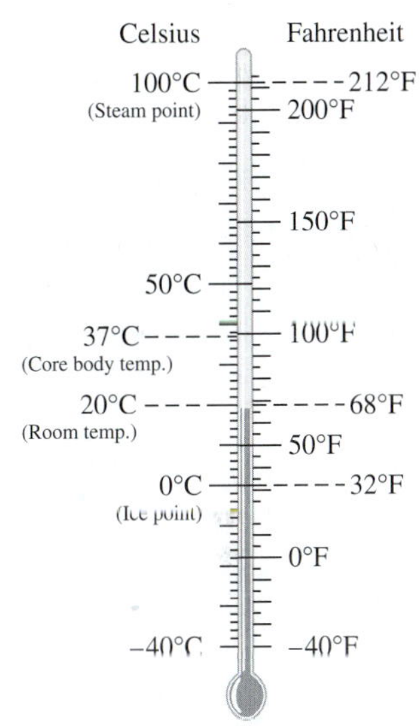

Figure 13.2 The Fahrenheit and Celsius temperature scales.

Table 13.1 Some Reference Temperatures in K, °C, and °F

	K	°C	°F		K	°C	°F
Absolute zero	0	−273.15	−459.67	Water boils	373.15	100.00	212.0
Lowest transient temperature achieved (laser cooling)	10^{-9}			Campfire	1000	700	1300
				Gold melts	1337	1064	1947
Intergalactic space	3	−270	−454	Lightbulb filament	3000	2700	4900
Helium boils	4.2	−269	−452	Surface of Sun; iron welding arc	6300	6000	11000
Nitrogen boils	77	−196	−321				
Carbon dioxide freezes ("dry ice")	195	−78	−108	Center of Earth	16000	15700	28300
				Lightning channel	30000	30000	50000
Mercury freezes	234	−39	−38	Center of Sun	10^7	10^7	10^7
Ice melts/water freezes	273.15	0	32.0	Interior of neutron star	10^9	10^9	10^9
Human body temperature	310	37	98.6				

Example 13.1

A Sick Friend

A friend suffering from the flu feels like she has a fever; her body temperature is 38.6°C. What is her temperature in (a) K and (b) °F?

Strategy (a) Kelvins and °C differ only by a shift of the zero point. Converting from °C to K requires only the addition of 273.15 K since 0°C (the ice point) corresponds to 273.15 K. (b) The °F is a different size than the °C, as well as having a different zero. In the Celsius scale, the zero is at the ice point. First multiply by 1.8°F/°C to find how many °F above the ice point. Then add 32°F (the Fahrenheit temperature of the ice point).

Solution (a) The temperature is 38.6 K *above* the ice point of 273.15 K. Therefore, the kelvin temperature is

$$T = 38.6\text{ K} + 273.15\text{ K} = 311.8\text{ K}$$

(b) First find how many °F above the ice point:

$$\Delta T_F = 38.6°\text{C} \times (1.8°\text{F}/°\text{C}) = 69.5°\text{F}$$

The ice point is 32°F, so

$$T_F = 32.0°\text{F} + 69.5°\text{F} = 101.5°\text{F}$$

Discussion The answer is reasonable since 98.6°F is normal body temperature.

Practice Problem 13.1 Normal Body Temperatures with Two Scales

Convert the normal human body temperature (98.6°F) to degrees Celsius and kelvins.

13.3 THERMAL EXPANSION OF SOLIDS AND LIQUIDS

Most objects expand as their temperature increases. Long before the cause of thermal expansion was understood, the phenomenon was put to practical use. For example, the cooper (barrel maker) heated iron hoops red hot to make them expand before fitting them around the wooden staves of a barrel. The iron hoops contracted as they cooled, pulling the staves tightly together to make a leak-tight barrel.

Linear Expansion

If the length of a wire, rod, or pipe is L_0 at temperature T_0 (Fig. 13.3), then

$$\frac{\Delta L}{L_0} = \alpha \Delta T \quad (13\text{-}4)$$

where $\Delta L = L - L_0$ and $\Delta T = T - T_0$. The length at temperature T is

$$L = L_0 + \Delta L = (1 + \alpha\,\Delta T)\,L_0 \quad (13\text{-}5)$$

The constant of proportionality α is called the **coefficient of linear expansion** of the substance. It plays a role in thermal expansion similar to that of the elastic modulus in tensile stress. If T is measured in kelvins or in degrees Celsius, then α has units of K^{-1} or $°C^{-1}$. Since only the *change* in temperature is involved in Eq. (13-4), either Celsius or Kelvin temperatures can be used to find ΔT; a temperature change of 1K is the same as a temperature change of 1°C.

As is true for the elastic modulus, the coefficient of linear expansion has different values for different solids and also depends to some extent on the starting temperature of the object. Table 13.2 lists the coefficients for various solids.

CONNECTION:

Recall that the fractional length change (strain) caused by a tensile or compressive stress is proportional to the stress that caused it [Hooke's law, Eq. (10-4)]. Similarly, the fractional length change caused by a temperature change is proportional to the temperature change, as long as the temperature change is not too great.

CHECKPOINT 13.3

A steel tower is 150.00 m tall at 40°C. How much shorter is it at −10°C?

Table 13.2 Coefficients of Linear Expansion α for Solids (at $T = 20°C$ unless otherwise indicated)

Material	α (10^{-6} K^{-1})
Glass (Vycor)	0.75
Brick	1.0
Glass (Pyrex)	3.25
Granite	8
Glass, most types	9.4
Cement or concrete	12
Iron or steel	12
Copper	16
Silver	18
Brass	19
Aluminum	23
Lead	29
Ice (at 0°C)	51

Figure 13.4 is a graph of the relative length of a steel girder as a function of temperature over a *wide* range of temperatures. The curvature of this graph shows that the thermal expansion of the girder is in general *not* proportional to the temperature change. However, over a *limited* temperature range, the curve can be approximated by a straight line; the slope of the tangent line is the coefficient α at the temperature T_0. For small temperature changes near T_0, the change in length of the girder can be treated as being proportional to the temperature change with only a small error.

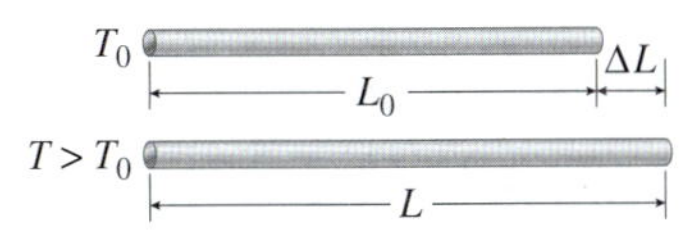

Figure 13.3 Expansion of a solid rod with increasing temperature.

Applications of Thermal Expansion: Expansion Joints in Bridges and Buildings

Allowances must be made in building sidewalks, roads, bridges, and buildings to leave space for expansion in hot weather. Old subway tracks have small spaces left between rail sections to prevent the rails from pushing into each other and causing the track to bow. A train riding on such tracks is subject to a noticeable amount of "clickety-clack" as it goes over these small expansion breaks in the tracks. Expansion joints are easily observed in bridges (Fig. 13.5). Concrete roads and sidewalks have joints between sections. Homeowners sometimes build their own sidewalks without realizing the necessity for such joints; these sidewalks begin to crack almost immediately!

Allowances must also be made for contraction in cold weather. If an object is not free to expand or contract, then as the temperature changes it is subjected to *thermal stress* as its environment exerts forces on it to prevent the thermal expansion or contraction that would otherwise occur.

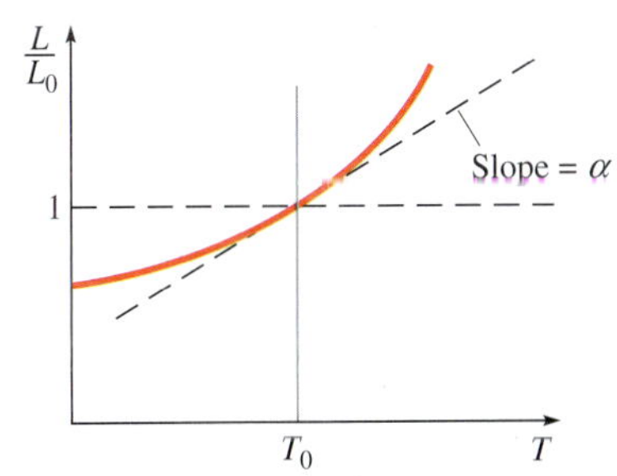

Figure 13.4 The relative length of a steel girder as a function of temperature. The dashed tangent line shows what Eq. (13-4) predicts for small temperature changes in the vicinity of T_0. The slope of this tangent line is the value of α at $T = T_0$.

Figure 13.5 Expansion joints permit the roadbed of a bridge to expand and contract as the temperature changes.

Example 13.2

Expanding Rods

Two metal rods, one aluminum and one brass, are each clamped at one end (Fig. 13.6). At 0.0°C, the rods are each 50.0 cm long and are separated by 0.024 cm at their unfastened ends. At what temperature will the rods just come into contact? (Assume that the base to which the rods are clamped undergoes a negligibly small thermal expansion.)

Strategy Two rods of different materials expand by different amounts. The sum of the two expansions ($\Delta L_{br} + \Delta L_{Al}$) must equal the space between the rods. After finding ΔT, we add it to $T_0 = 0.0°C$ to obtain the temperature at which the two rods touch.

Known: $L_0 = 50.0$ cm, $T_0 = 0.0°C$ for both
Look up: $\alpha_{br} = 19 \times 10^{-6}\ K^{-1}$; $\alpha_{Al} = 23 \times 10^{-6}\ K^{-1}$
Requirement: $\Delta L_{br} + \Delta L_{Al} = 0.024$ cm
Find: $T_f = T_0 + \Delta T$

Solution The brass rod expands by

$$\Delta L_{br} = (\alpha_{br}\,\Delta T)L_0$$

and the aluminum rod by

$$\Delta L_{Al} = (\alpha_{Al}\,\Delta T)L_0$$

The sum of the two expansions is known:

$$\Delta L_{br} + \Delta L_{Al} = 0.024 \text{ cm}$$

Since both the initial lengths and the temperature changes are the same,

$$(\alpha_{br} + \alpha_{Al})\,\Delta T \times L_0 = 0.024 \text{ cm}$$

Solving for ΔT,

$$\Delta T = \frac{0.024 \text{ cm}}{(\alpha_{br} + \alpha_{Al})L_0}$$
$$= \frac{0.024 \text{ cm}}{(19 \times 10^{-6}\ K^{-1} + 23 \times 10^{-6}\ K^{-1}) \times 50.0 \text{ cm}}$$
$$= 11.4°C$$

The temperature at which the two touch is

$$T_f = T_0 + \Delta T = 0.0°C + 11.4°C \rightarrow 11°C$$

Discussion As a check on the solution, we can find how much each individual rod expands and then add the two amounts:

$$\Delta L_{Al} = (\alpha_{Al}\,\Delta T)\,L_0$$
$$= 23 \times 10^{-6}\ K^{-1} \times 11.4 \text{ K} \times 50.0 \text{ cm} = 0.013 \text{ cm}$$
$$\Delta L_{br} = (\alpha_{br}\,\Delta T)\,L_0$$
$$= 19 \times 10^{-6}\ K^{-1} \times 11.4 \text{ K} \times 50.0 \text{ cm} = 0.011 \text{ cm}$$
$$\text{total expansion} = 0.013 \text{ cm} + 0.011 \text{ cm} = 0.024 \text{ cm}$$

which is correct.

Practice Problem 13.2 Expansion of a Wall

The outer wall of a building is constructed from concrete blocks. If the wall is 5.00 m long at 20.0°C, how much longer is the wall on a hot day (30.0°C)? How much shorter is it on a cold day (−5.0°C)?

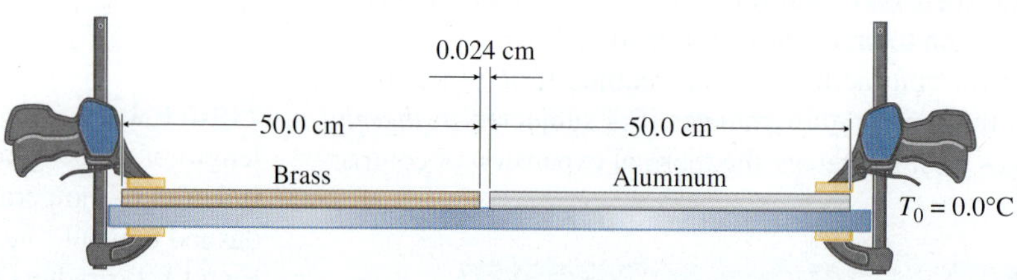

Figure 13.6 Two clamped rods.

Differential Expansion

When two strips made of different metals are joined together and then heated, one expands more than the other (unless they have the same coefficient of expansion). This differential expansion can be put to practical use: the joined strips bend into a curve, allowing one strip to expand more than the other.

Application: Bimetallic Strips The bimetallic strip (Fig. 13.7) is made by joining a material with a lower coefficient of expansion, such as steel, and one of a higher coefficient of expansion, such as brass. Unequal expansions or contractions of the two materials force the bimetallic strip to bend. In Fig. 13.7, the brass expands more than the steel when the bimetallic strip is heated. As the strip is cooled, the brass contracts more than the steel.

The bimetallic strip is used in many wall thermostats. The bending of the bimetallic strip closes or opens an electrical switch in the thermostat that turns the furnace or air conditioner on or off. Inexpensive oven thermometers also use a bimetallic strip wound into a spiral coil; the coil winds tighter or unwinds as the temperature changes.

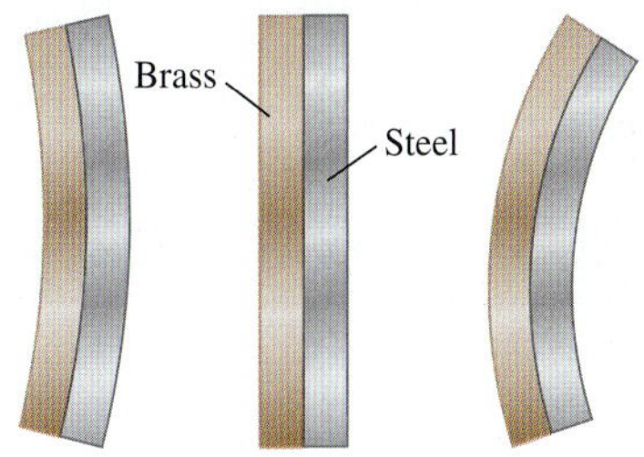

Figure 13.7 A bimetallic strip bends when its temperature changes; brass expands and contracts more than steel for the same temperature change.

Area Expansion

As you might suspect, *each dimension* of an object expands when the object's temperature increases. For instance, a pipe expands not only in length, but also in radius. An isotropic substance expands uniformly in all directions, causing changes in area and volume that leave the *shape* of the object unchanged. In Problem 25, you can show that, for small temperature changes, the area of any flat surface of a solid changes in proportion to the temperature change:

$$\frac{\Delta A}{A_0} = 2\alpha\,\Delta T \qquad (13\text{-}6)$$

The factor of two in Eq. (13-6) arises because the surface expands in two perpendicular directions.

Volume Expansion

The fractional change in volume of a solid or liquid is also proportional to the temperature change as long as the temperature change is not too large:

$$\frac{\Delta V}{V_0} = \beta\,\Delta T \qquad (13\text{-}7)$$

CONNECTION:

Compare Eq. (10-10). There, the fractional volume change is proportional to the *pressure* change; here it is proportional to the *temperature* change.

The coefficient of volume expansion, β, is the fractional change in volume per unit temperature change. For solids, the coefficient of volume expansion is three times the coefficient of linear expansion (as shown in Problem 26):

$$\beta = 3\alpha \qquad (13\text{-}8)$$

The factor of three in Eq. (13-8) arises because the object expands in three-dimensional space. For liquids, the volume expansion coefficient is the only one given in tables. Since liquids do not necessarily retain the same shape as they expand, the quantity that is uniquely defined is the change in volume. Table 13.3 provides values of β for some common liquids and gases.

Expansion of a Cavity When a solid expands, every point in the solid gets farther away from every other point. Therefore, a hollow cavity in a solid expands exactly as if it were filled. For example, the interior of a steel gasoline container expands when its

Table 13.3 Coefficients of Volume Expansion β for Liquids and Gases (at $T = 20°C$ unless otherwise indicated)

Material	β (10^{-6} K^{-1})
Liquids	
Water (at 0°C)*	−68
Mercury	182
Water (at 20°C)	207
Gasoline	950
Ethyl alcohol	1120
Benzene	1240
Gases	
Air (and most other gases) at 1 atm	3340

*Below 3.98°C, water *contracts* with increasing temperature.

temperature increases just as if it were a solid steel block. The steel wall of the can does *not* expand inward to make the cavity smaller. See Conceptual Question 4 for further discussion of cavity expansion.

Application: Thermometers In an ordinary alcohol-in-glass or mercury-in-glass thermometer, it is not just the liquid that expands as temperature rises. The reading of the thermometer is determined by the difference in the volume expansion of the liquid and that of the interior of the glass. The calibration of an accurate thermometer must account for the expansion of the glass. Comparison of Tables 13.2 and 13.3 shows that, as is usually the case, the liquid expands much more than the glass for a given temperature change.

Example 13.3

Hollow Cylinder Full of Water

A hollow copper cylinder is filled to the brim with water at 20.0°C. If the water and the container are heated to a temperature of 91°C, what percentage of the water spills over the top of the container?

Strategy The volume expansion coefficient for water is greater than that for copper, so the water expands more than the interior of the cylinder. The cavity expands just as if it were solid copper. Since the problem does not specify the initial volume, we call it V_0. We need to find out how much a volume V_0 of water expands and how much a volume V_0 of copper expands; the difference is the water volume that spills over the top of the container.

Known: Initial copper cylinder interior volume = initial water volume = V_0

Initial temperature = $T_0 = 20.0°C$

Final temperature = 91°C; $\Delta T = 71°C$

Look up: $\alpha_{Cu} = 16 \times 10^{-6}\ °C^{-1}$; $\beta_{H_2O} = 207 \times 10^{-6}\ °C^{-1}$

Find: $\Delta V_{H_2O} - \Delta V_{Cu}$ as a percentage of V_0

Solution The volume expansion of the interior of the copper cylinder is

$$\Delta V_{Cu} = (\beta_{Cu}\,\Delta T)V_0$$

where $\beta_{Cu} = 3\alpha_{Cu}$. The volume expansion of the water is

$$\Delta V_{H_2O} = (\beta_{H_2O}\,\Delta T)V_0$$

The amount of water that spills is

$$\begin{aligned}\Delta V_{H_2O} - \Delta V_{Cu} &= (\beta_{H_2O}\,\Delta T)V_0 - (\beta_{Cu}\,\Delta T)V_0\\ &= [(\beta_{H_2O} - \beta_{Cu})\Delta T]V_0\\ &= (207 \times 10^{-6}\ °C^{-1} - 3 \times 16 \times 10^{-6}\ °C^{-1})\\ &\quad \times 71°C \times V_0\\ &= 0.011V_0\end{aligned}$$

The percentage of water that spills is therefore 1.1%.

Discussion As a check, we can find the change in volume of the copper container and of the water and find the difference.

$$\begin{aligned}\Delta V_{Cu} &= (\beta_{Cu}\,\Delta T)V_0 = 3 \times 16 \times 10^{-6}\ °C^{-1} \times 71°C \times V_0\\ &= 0.0034V_0\end{aligned}$$

$$\begin{aligned}\Delta V_{H_2O} &= (\beta_{H_2O}\,\Delta T)V_0 = 207 \times 10^{-6}\ °C^{-1} \times 71°C \times V_0\\ &= 0.0147V_0\end{aligned}$$

$$\text{volume of water that spills} = 0.0147V_0 - 0.0034V_0 = 0.0113V_0$$

which again shows that 1.1% spills.

Practice Problem 13.3 Overflowing Gas Can

A driver fills an 18.9 L steel gasoline can with gasoline at 15.0°C right up to the top. He forgets to replace the cap and leaves the can in the back of his truck. The temperature climbs to 30.0°C by 1 P.M. How much gasoline spills out of the can?

13.4 MOLECULAR PICTURE OF A GAS

Number Density As we saw in Chapter 9, the densities of liquids are generally not much less than the densities of solids. Gases are *much* less dense than liquids and solids because the molecules are, on average, much farther apart. The mass density—mass per unit volume—of a substance depends on the mass *m* of a single molecule and the number of molecules *N* packed into a given volume *V* of space (Fig. 13.8). The number of molecules per unit volume, *N/V*, is called the **number density** to distinguish it from mass density. In SI units, number density is written as the number of molecules per

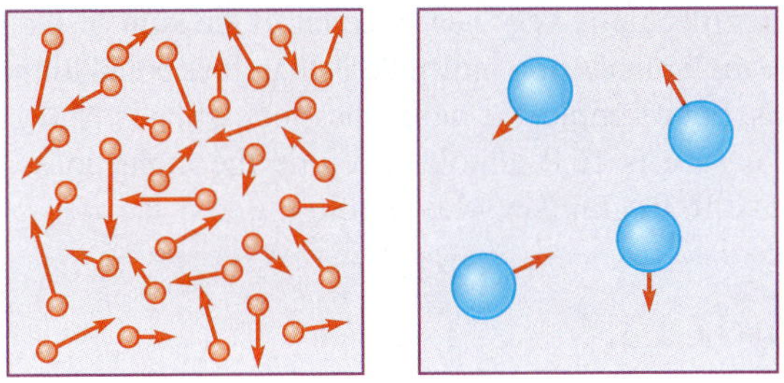

(a) (b)

Figure 13.8 These two gases have the same mass per unit volume but different number densities. The red arrows represent the molecular velocities. In (a), there are a larger number of molecules in a given volume, but the mass of each molecule in (b) is greater.

cubic meter, usually written simply as m^{-3} (read "per cubic meter"). If a gas has a total mass M, occupies a volume V, and each molecule has a mass m, then the number of gas molecules is

$$N = \frac{M}{m} \tag{13-9}$$

and the average number density is

$$\frac{N}{V} = \frac{M}{mV} = \frac{\rho}{m} \tag{13-10}$$

where $\rho = M/V$ is the mass density.

Moles It is common to express the amount of a substance in units of **moles** (abbreviated mol). The mole is an SI base unit and is defined as follows: one mole of anything contains the same number of units as there are atoms in 12 *grams* (not kilograms) of carbon-12. This number is called **Avogadro's number** and has the value

Avogadro's number

$$N_A = 6.022 \times 10^{23}\ \text{mol}^{-1}$$

Avogadro's number is written with units, mol^{-1}, to show that this is the number *per mole*. The number of moles, n, is therefore given by

$$\text{number of moles} = \frac{\text{total number}}{\text{number per mole}}$$

$$n = \frac{N}{N_A} \tag{13-11}$$

Molecular Mass and Molar Mass The mass of a molecule is often expressed in units other than kg. The most common is the **atomic mass unit** (symbol u). By definition, one atom of carbon-12 has a mass of 12 u (exactly). Using Avogadro's number, the relationship between atomic mass units and kilograms can be calculated (see Problem 27):

$$1\ \text{u} = 1.66 \times 10^{-27}\ \text{kg} \tag{13-12}$$

The proton, neutron, and hydrogen atom all have masses within 1% of 1 u—which is why the atomic mass unit is so convenient. More precise values are 1.007 u for the proton, 1.009 u for the neutron, and 1.008 u for the hydrogen atom. The mass of an atom is *approximately* equal to the number of nucleons (neutrons plus protons)—the *atomic mass number*—times 1 u.

Instead of the mass of one molecule, tables commonly list the **molar mass**—the mass of the substance *per mole*. For an element with several isotopes (such as carbon-12, carbon-13, and carbon-14, which all have the same atomic number but different mass numbers), the molar mass is averaged according to the naturally occurring abundance of each isotope. The atomic mass unit is chosen so that the mass of a molecule in "u" is numerically the same as the molar mass in g/mol. For example, the molar mass of O_2 is 32.0 g/mol and the mass of one molecule is 32.0 u.

The mass of a molecule is very nearly equal to the sum of the masses of its constituent atoms. The molar mass of a molecule is therefore equal to the sum of the molar masses of the atoms. For example, the molar mass of carbon is 12.0 g/mol and the molar mass of (atomic) oxygen is 16.0 g/mol; therefore, the molar mass of carbon dioxide (CO_2) is $(12.0 + 2 \times 16.0)$ g/mol = 44.0 g/mol.

CHECKPOINT 13.4

(a) What is the mass (in u) of a CO_2 molecule? (b) What is the mass (in g) of 3.00 mol of CO_2?

Example 13.4

A Helium Balloon

A helium balloon of volume 0.010 m^3 contains 0.40 mol of He gas. (a) Find the number of atoms, the number density, and the mass density. (b) Estimate the average distance between He atoms.

Strategy The number of moles tells us the number of atoms as a fraction of Avogadro's number. Once we have the number of atoms, N, the next quantity we are asked to find is N/V. To find the mass density, we can look up the atomic mass of helium in the periodic table. The mass per atom times the number density (atoms per m^3) equals the mass density (mass per m^3). To find the average distance between atoms, imagine a simplified picture in which each atom is at the center of a spherical volume equal to the total volume of the gas divided by the number of atoms. In this approximation, the average distance between atoms is equal to the diameter of each sphere.

Solution (a) The number of atoms is

$$N = nN_A$$
$$= 0.40 \text{ mol} \times 6.022 \times 10^{23} \text{ atoms/mol}$$
$$= 2.4 \times 10^{23} \text{ atoms}$$

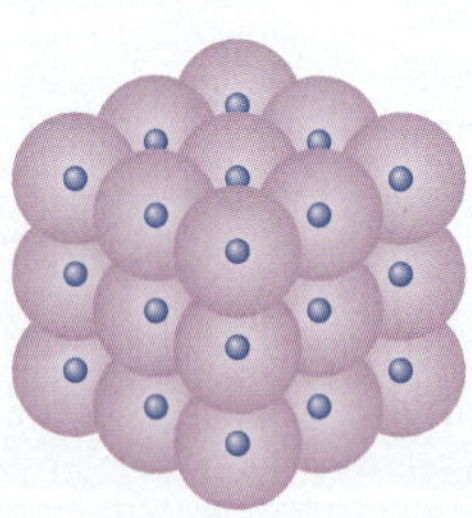

Figure 13.9
Simplified model in which equally spaced helium atoms sit at the centers of spherical volumes of space.

The number density is

$$\frac{N}{V} = \frac{2.4 \times 10^{23} \text{ atoms}}{0.010 \text{ m}^3} = 2.4 \times 10^{25} \text{ atoms/m}^3$$

The mass of a helium atom is 4.00 u. Then the mass in kg of a helium atom is

$$m = 4.00 \text{ u} \times 1.66 \times 10^{-27} \text{ kg/u} = 6.64 \times 10^{-27} \text{ kg}$$

and the mass density of the gas is

$$\rho = \frac{M}{V} = m \times \frac{N}{V}$$
$$= 6.64 \times 10^{-27} \text{ kg} \times 2.4 \times 10^{25} \text{ m}^{-3} = 0.16 \text{ kg/m}^3$$

(b) We assume that each atom is at the center of a sphere of radius r (Fig. 13.9). The volume of the sphere is

$$\frac{V}{N} = \frac{1}{N/V} = \frac{1}{2.4 \times 10^{25} \text{ atoms/m}^3} = 4.2 \times 10^{-26} \text{ m}^3 \text{ per atom}$$

Then

$$\frac{V}{N} = \frac{4}{3}\pi r^3 \approx 4r^3 \quad (\text{since } \pi \approx 3)$$

Solving for r,

$$r \approx \left(\frac{V}{4N}\right)^{1/3} = 2.2 \times 10^{-9} \text{ m} = 2.2 \text{ nm}$$

The average distance between atoms is $d = 2r \approx 4$ nm (since this is an estimate).

Discussion For comparison, in *liquid* helium the average distance between atoms is about 0.4 nm, so in the gas the average separation is about ten times larger.

Practice Problem 13.4 Number Density for Water

The mass density of liquid water is 1000.0 kg/m^3. Find the number density.

13.5 ABSOLUTE TEMPERATURE AND THE IDEAL GAS LAW

We have examined the thermal expansion of solids and liquids. What about gases? Is the volume expansion of a gas proportional to the temperature change? We must be careful; since gases are easily compressed, we must also specify what happens to the pressure. The French scientist Jacques Charles (1746–1823) found experimentally that, if the pressure of a gas is held constant, the change in temperature is indeed proportional to the change in volume (Fig. 13.10a).

$$\text{Charles's law: } \Delta V \propto \Delta T \quad (\text{for constant } P)$$

According to Charles's law, a graph of V versus T for a gas held at constant pressure is a straight line, but the line does not necessarily pass through the origin (Fig. 13.10b).

However, if we graph V versus T (at constant P) for various gases, something interesting happens. If we extrapolate the straight line to where it reaches $V = 0$, the temperature at that point is the *same* regardless of what gas we use, how many moles of gas are present, or what the pressure of the gas is (Fig. 13.10c). (One reason we have to extrapolate is that all gases become liquids or solids before they reach $V = 0$.) This temperature, $-273.15°\text{C}$ or $-459.67°\text{F}$, is called **absolute zero**—the lower limit of attainable temperatures. In kelvins—an *absolute* temperature scale—absolute zero is defined as 0 K (Fig. 13.10d). As long as it is understood that an absolute temperature scale is to be used, then Charles's law can be written

$$V \propto T \quad (\text{for constant } P)$$

EVERYDAY PHYSICS DEMO

Take an empty plastic soda bottle, cap it tightly, and put it in the freezer. Check it an hour later; what has happened? Estimate the percentage change in the volume of the air inside and compare with the percentage change in absolute temperature (if you don't have a thermometer handy, a typical freezer temperature is about −10°C).

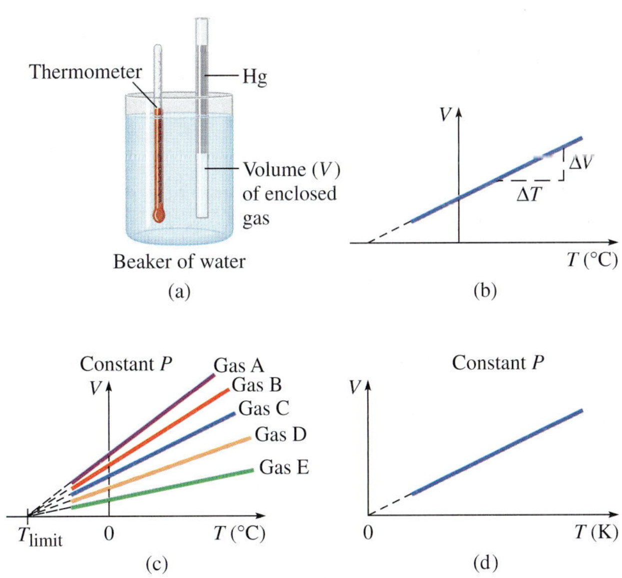

Figure 13.10 (a) Apparatus to verify Charles's law. The pressure of the enclosed gas is held constant by the fixed quantity of mercury resting on top of it and atmospheric pressure pushing down on the mercury. If the temperature of the gas is changed, it expands or contracts, moving the mercury column above it. (b) Charles's law: for a gas held at constant pressure, changes in temperature are proportional to changes in volume. (c) Volume versus temperature graphs for various gas samples, each at a constant pressure, are extrapolated to $V = 0$. The graphs intersect the temperature axis at the same temperature, T_{limit}, even though the gases may differ in composition and mass. (d) An absolute temperature scale sets $T_{limit} = 0$.

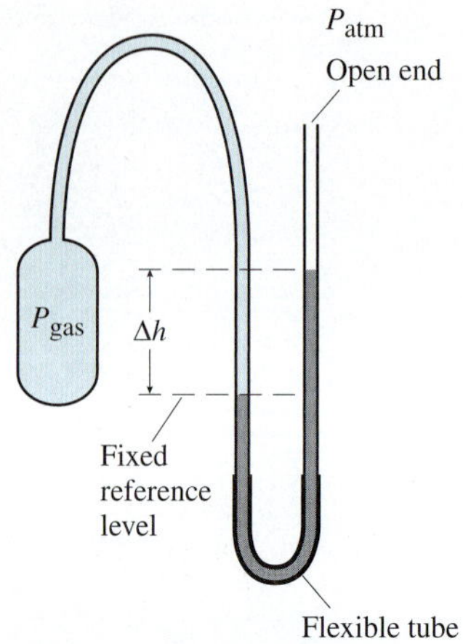

Figure 13.11 A constant volume gas thermometer. A dilute gas is contained in the vessel on the left, which is connected to a mercury manometer. The right side can be moved up or down to keep the mercury level on the left at a fixed level, so the volume of gas is kept constant. Then the manometer is used to measure the pressure of the gas: $P_{gas} = P_{atm} + \rho g\,\Delta h$.

Thermal expansion of a gas can be used to measure temperature. Gas thermometers are universal: it does not matter what gas is used or how many moles of gas are present, as long as the number density is sufficiently low. Gas thermometers give absolute temperature in a natural way and they are extremely accurate and reproducible. The main disadvantage of gas thermometers is that they are much less convenient to use than most other thermometers, so they are mainly used to calibrate other thermometers.

A thermometer based on Charles's law would be called a *constant pressure gas thermometer*. More common is the *constant volume gas thermometer* (Fig. 13.11), which is based on Gay-Lussac's law:

$$P \propto T \quad \text{(for constant } V\text{)}$$

Here we keep the volume of the gas constant, measure the pressure and use that to indicate the temperature. (It is much easier to keep the volume constant and measure the pressure than to do the opposite.)

Both Charles's law and Gay-Lussac's law are valid only for a **dilute** gas—a gas where the number density is low enough (and, therefore, the average distance between gas molecules is large enough) that interactions between the molecules are negligible except when they collide. Two other experimentally discovered laws that apply to dilute gases are Boyle's law and Avogadro's law. Boyle's law states that the pressure of a gas is inversely proportional to its volume at constant temperature:

$$P \propto \frac{1}{V} \quad \text{(for constant } T\text{)}$$

Avogadro's law states that the volume occupied by a gas at a given temperature and pressure is proportional to the number of gas molecules N:

$$V \propto N \quad \text{(constant } P, T\text{)}$$

(A constant number of gas molecules was assumed in the statements of Boyle's, Gay-Lussac's, and Charles's laws.)

One equation combines all four of these gas laws—the **ideal gas law**:

Ideal gas law (microscopic form)

$$PV = NkT \quad (N = \text{number of molecules}) \qquad (13\text{-}13)$$

In the ideal gas law, T stands for *absolute* temperature (in K) and P stands for *absolute* (not gauge) pressure. The constant of proportionality is a universal quantity known as **Boltzmann's constant** (symbol k); its value is

$$k = 1.38 \times 10^{-23}\ \text{J/K} \qquad (13\text{-}14)$$

The macroscopic form of the ideal gas law is written in terms of n, the number of *moles* of the gas, in place of N, the number of molecules. Substituting

$$N = nN_A$$

into the microscopic form yields

$$PV = nN_AkT$$

The product of N_A and k is called the **universal gas constant**:

$$R = N_Ak = 8.31\ \frac{\text{J/K}}{\text{mol}} \qquad (13\text{-}15)$$

Then the ideal gas law in macroscopic form is written

Ideal gas law (macroscopic form)

$$PV = nRT \quad (n = \text{number of moles}) \qquad (13\text{-}16)$$

Problem-Solving Tips for the Ideal Gas Law

- In most problems, some change occurs; decide which of the four quantities (P, V, N or n, and T) remain constant during the change.
- Use the microscopic form if the problem deals with the number of molecules and the macroscopic form if the problem deals with the number of moles.
- Use subscripts (i and f) to distinguish initial and final values.
- Work in terms of ratios so that constant factors cancel out.
- Write out the units when doing calculations.
- Remember that P stands for *absolute* pressure (not gauge pressure) and T stands for *absolute* temperature (in kelvins, not °C or °F).

CHECKPOINT 13.5

Two containers with the same volume are filled with two different gases. The pressure of the two gases is the same. (a) Must their temperatures be the same? Explain. (b) If their temperatures are the same, must they have the same number density? The same mass density?

Example 13.5

Temperature of the Air in a Tire

Before starting out on a long drive, you check the air in your tires to make sure they are properly inflated. The pressure gauge reads 31.0 lb/in.^{2} (214 kPa), and the temperature is 15°C. After a few hours of highway driving, you stop and check the pressure again. Now the gauge reads 35.0 lb/in.^{2} (241 kPa). What is the temperature of the air in the tires now?

Strategy We treat the air in the tire as an ideal gas. We must work with absolute temperatures and absolute pressures when using the ideal gas law. The pressure gauge reads *gauge* pressure; to get absolute pressure we add 1 atm = 101 kPa. We don't know the number of molecules inside the tire or the volume, but we can reasonably assume that neither changes. The number is constant as long as the tire does not leak. The volume may actually change a bit as the tire warms up and expands, but this change is small. Since N and V are constant, we can rewrite the ideal gas law as a proportionality between P and T.

Solution First convert the initial and final gauge pressures to absolute pressures:

$$P_i = 214\text{ kPa} + 101\text{ kPa} = 315\text{ kPa}$$

$$P_f = 241\text{ kPa} + 101\text{ kPa} = 342\text{ kPa}$$

Now convert the initial temperature to an absolute temperature:

$$T_i = 15°\text{C} + 273\text{ K} = 288\text{ K}$$

According to the ideal gas law, pressure is proportional to temperature, so

$$\frac{T_f}{T_i} = \frac{P_f}{P_i} = \frac{342\text{ kPa}}{315\text{ kPa}}$$

Then

$$T_f = \frac{P_f}{P_i} T_i = \frac{342}{315} \times 288\text{ K} = 313\text{ K}$$

Now convert back to °C:

$$313\text{ K} - 273\text{ K} = 40°\text{C}$$

Discussion The final answer of 40°C seems reasonable since, after a long drive, the tires are noticeably warm, but not hot enough to burn your hand.

It is often most convenient to work with the ideal gas law by setting up a proportion. In this problem, we did not know the volume or the number of molecules, so we had no choice. In essence, what we used was Gay-Lussac's law. Starting with the ideal gas law, we can "rederive" Gay-Lussac's law or Charles's law or any other proportionality inherent in the ideal gas law.

continued on next page

Example 13.5 continued

Practice Problem 13.5 Air Pressure in the Tire After the Temperature Decreases

Suppose you now (unwisely) decide to bleed air from the tires, since the manufacturer suggests keeping the pressure between 28 lb/in.^{2} and 32 lb/in.^{2} (The manufacturer's specification refers to when the tires are "cold.") If you let out enough air so that the pressure returns to 31 lb/in.^{2}, what percentage of the air molecules did you let out of the tires? What is the gauge pressure after the tires cool back down to 15°C?

EVERYDAY PHYSICS DEMO

The next time you take a car trip, check the tire pressure with a gauge just before the trip and then again after an hour or more of highway driving. Calculate the temperature of the air in the tires from the two pressure readings and the initial temperature. Feel the tire with your hand to see if your calculation is reasonable.

Example 13.6

Scuba Diver

A scuba diver needs air delivered at a pressure equal to the pressure of the surrounding water—the pressure in the lungs must match the water pressure on the diver's body to prevent the lungs from collapsing. Since the pressure in the air tank is much higher, a regulator delivers air to the diver at the appropriate pressure. The compressed air in a diver's tank lasts 80 min at the water's surface. About how long does the same tank last at a depth of 30 m under water? (Assume that the volume of air breathed per minute does not change and ignore the small quantity of air left in the tank when it is "empty.")

Strategy The compressed air in the *tank* is at a pressure much higher than the pressure at which the diver breathes, whether at the surface or at 30 m depth. The constant quantity is N, the number of gas molecules in the tank. We also assume that the temperature of the gas remains the same; it may change slightly, but much less than the pressure or volume.

Solution Since N and T are constant,

$$PV = \text{constant}$$

or

$$P \propto 1/V$$

The pressure at the surface is (approximately) 1 atm, while the pressure at 30 m under water is

$$P = 1 \text{ atm} + \rho g h$$

$$\rho g h = 1000 \text{ kg/m}^3 \times 9.8 \text{ m/s}^2 \times 30 \text{ m} = 294 \text{ kPa} \approx 3 \text{ atm}$$

Therefore, at a depth of 30 m,

$$P \approx 4 \text{ atm}$$

To match the pressure of the surrounding water, the pressure of the compressed air is four times larger at a depth of 30 m; then the volume of air is one fourth what it was at the surface. The diver breathes the same volume per minute, so the tank will last one fourth as long: 20 min.

Discussion To do the same thing a bit more formally, we could write:

$$P_i V_i = P_f V_f$$

After setting $P_i = 1$ atm and $P_f = 4$ atm, we find that $V_f/V_i = \frac{1}{4}$.

continued on next page

Example 13.6 continued

In this problem, the only numerical values given (indirectly) were the initial and final pressures. Assuming that N and T remain constant, we then can find the ratio of the final and initial volumes. Whenever there *seems* to be insufficient numerical information given in a problem, think in terms of ratios and look for constants that cancel out.

Practice Problem 13.6 Pressure in the Air Tank After the Temperature Increases

A tank of compressed air is at an absolute pressure of 580 kPa at a temperature of 300.0 K. The temperature increases to 330.0 K. What is the pressure in the tank now?

13.6 KINETIC THEORY OF THE IDEAL GAS

In a gas, the interaction between two molecules weakens rapidly as the distance between the molecules increases. In a dilute gas, the average distance between gas molecules is large enough that we can ignore interactions between the molecules except when they collide. In addition, the volume of space occupied by the molecules themselves is a small fraction of the total volume of the gas—the gas is mostly "empty space." The **ideal gas** is a simplified model of a dilute gas in which we think of the molecules as pointlike particles that move *independently* in free space with no interactions except for elastic collisions.

This simplified model is a good approximation for many gases under ordinary conditions. Many properties of gases can be understood from this model; the microscopic theory based on it is called the **kinetic theory** of the ideal gas.

Microscopic Basis of Pressure

The force that a gas exerts on a surface is due to collisions that the gas molecules make with that surface. For instance, think of the air inside an automobile tire. Whenever an air molecule collides with the inner tire surface, the tire exerts an inward force to turn the air molecule around and return it to the bulk of the gas. By Newton's third law, the gas molecule exerts an outward force on the tire surface. The net force per unit area on the inside of the tire due to all the collisions of the many air molecules is equal to the air pressure in the tire. The pressure depends on three things: how many molecules there are, how often each one collides with the wall, and the momentum transfer due to each collision.

We want to find out how the pressure of an ideal gas is determined by the motions of the gas molecules. To simplify the discussion, consider a gas contained in a box of length L and side area A (Fig. 13.12a)—the result does not depend on the shape of the container. Figure 13.12b shows a gas molecule about to collide with the rightmost wall of the container. For simplicity, we assume that the collision is elastic; a more advanced analysis shows that the result is correct even though not all collisions are elastic.

For an elastic collision, the x-component of the molecule's momentum is reversed in direction since the wall is much more massive than the molecule. Since the gas exerts only an outward force on the wall (a static fluid exerts no tangential force on a boundary), the y- and z-components of the molecule's momentum are unchanged. Thus, the molecule's momentum change is $\Delta p_x = 2m|v_x|$.

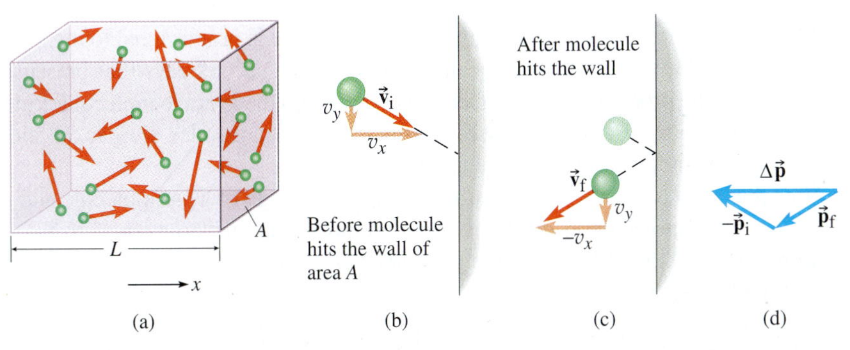

Figure 13.12 (a) Gas molecules confined to a container of length L and area A. (b) A molecule is about to collide with the wall of area A. (c) After an elastic collision, v_x has changed sign, while v_y and v_z are unchanged. (d) The change in momentum due to the collision has magnitude $2|p_x|$ and is perpendicular to the wall.

When does this molecule next collide with the same wall? Ignoring for now collisions with other molecules, its x-component of velocity never changes magnitude—only the sign of v_x changes when it reverses direction (Fig. 13.12*c*). The time it takes the molecule to travel the length L of the container and hit the other wall is $L/|v_x|$. Then the round-trip time is

$$\Delta t = 2\frac{L}{|v_x|}$$

CONNECTION:

We are using the principle that force is the rate of change of momentum (Newton's second law) to draw a conclusion about pressure in a gas.

The *average* force exerted by the molecule on the wall is the change in momentum (Fig. 13.12d) divided by the time for one complete round-trip:

$$F_{\text{av},x} = \frac{\Delta p_x}{\Delta t} = \frac{2m|v_x|}{2L/|v_x|} = \frac{m|v_x|^2}{L} = \frac{mv_x^{\ 2}}{L}$$

The total force on the wall is the sum of the forces due to each molecule in the gas. If there are N molecules in the gas, we can simply multiply N by the *average* force due to one molecule to get the total force on the wall. To represent such an average, we use angle brackets $\langle\ \rangle$; the quantity inside the brackets is averaged over all the molecules in the gas.

$$F = N\langle F_{\text{av}}\rangle = \frac{Nm}{L}\langle v_x^2\rangle$$

The pressure is then

$$P = \frac{F}{A} = \frac{Nm}{AL}\langle v_x^2\rangle = \frac{Nm}{V}\langle v_x^2\rangle \qquad \text{(13-17)}$$

where $V = AL$ is the volume of the box. Eq. (13-17) is correct regardless of the shape of the container enclosing the gas. Since we end up averaging over all the molecules in the gas, the simplifying assumption about no collisions with other molecules does not affect the result.

The product $m\langle v_x^2\rangle$ suggests kinetic energy. It certainly makes sense that if the average kinetic energy of the gas molecules is larger, the pressure is higher. The average translational kinetic energy of a molecule in the gas is $\langle K_{\text{tr}}\rangle = \frac{1}{2}m\langle v^2\rangle$. For any gas molecule, $v^2 = v_x^2 + v_y^2 + v_z^2$, since velocity is a vector quantity. The gas as a whole is at rest, so there is no preferred direction of motion. Then the average value of v_x^2 must be the same as the averages of v_y^2 and v_z^2, so

$$\langle v_x^2\rangle = \tfrac{1}{3}\langle v^2\rangle$$

Therefore,

$$m\langle v_x^2\rangle = \tfrac{1}{3}m\langle v^2\rangle = \tfrac{2}{3}\langle K_{\text{tr}}\rangle$$

Substituting this into Eq. (13-17), the pressure is

$$P = \frac{2}{3}\frac{N\langle K_{\text{tr}}\rangle}{V} = \frac{2}{3}\frac{N}{V}\langle K_{\text{tr}}\rangle \qquad \text{(13-18)}$$

Equation (13-18) is written with the variables grouped in two different ways to give two different insights. The first grouping says that pressure is proportional to the kinetic energy density (the kinetic energy per unit volume). The second says that pressure is proportional to the product of the number density N/V and the average molecular kinetic energy. The pressure of a gas increases if either the gas molecules are packed closer together or if the molecules have more kinetic energy.

Note that $\langle K_{\text{tr}}\rangle$ is the average *translational* kinetic energy of a gas molecule and v is the CM speed of a molecule. A gas molecule with more than one atom (such as N_2), has vibrational and rotational kinetic energy *in addition to* its translational kinetic energy K_{tr}, but Eq. (13-18) still holds.

What about the assumption that the gas molecules never collide with each other? It certainly is *not* true that the same molecule returns to collide with the same wall at a fixed time interval and has the same v_x each time it returns! However, the derivation really only relies on average quantities. In a gas at equilibrium, an average quantity like $\langle v_x^2 \rangle$ remains unchanged even though any one particular molecule changes its velocity components as a result of each collision.

Temperature and Translational Kinetic Energy

The temperature of an ideal gas has a direct physical interpretation that we can now bring to light. We found that in an ideal gas, the pressure, volume, and number of molecules are related to the average translational kinetic energy of the gas molecules:

$$P = \frac{2}{3}\frac{N}{V}\langle K_{tr} \rangle \qquad (13\text{-}18)$$

Solving for the average kinetic energy,

$$\langle K_{tr} \rangle = \frac{3}{2}\frac{PV}{N} \qquad (13\text{-}19)$$

By rearranging the ideal gas law [Eq. (13-13)], we find that P, V, and N occur in the same combination as in Eq. (13-19):

$$\frac{PV}{N} = kT$$

Then by substituting kT for $(PV)/N$ in Eq. (13-19), we find that

$$\langle K_{tr} \rangle = \tfrac{3}{2}kT \qquad (13\text{-}20)$$

Therefore, *the absolute temperature of an ideal gas is proportional to the average translational kinetic energy of the gas molecules.* Temperature then is a way to describe the average translational kinetic energy of the gas molecules. At higher temperatures, the gas molecules have (on average) greater kinetic energy.

CHECKPOINT 13.6

At what temperature in °C do molecules of O_2 have twice the average translational kinetic energy that molecules of H_2 have at 20°C?

RMS Speed The speed of a gas molecule that has the average kinetic energy is called the **rms** (root mean square) **speed**. The rms speed is *not* the same as the average speed. Instead, the rms speed is the square *root* of the *mean* (average) of the speed *squared.* Since

$$\langle K_{tr} \rangle = \tfrac{1}{2}m\langle v^2 \rangle = \tfrac{1}{2}mv_{rms}^2 \qquad (13\text{-}21)$$

the rms speed is

$$v_{rms} = \sqrt{\langle v^2 \rangle}$$

Squaring before averaging emphasizes the effect of the faster-moving molecules, so the rms speed is a bit higher than the average speed—about 9% higher as it turns out.

Since the average kinetic energy of molecules in an ideal gas depends only on temperature, Eq. (13-21) implies that more massive molecules move more slowly on average than lighter ones at the same temperature. If two different gases are placed in a single chamber so that they reach equilibrium and are at the same temperature, their molecules must have the same average translational kinetic energies. If one gas has molecules of larger mass, its molecules must move with a slower average velocity than those of the gas with the lighter mass molecules. In Problem 71, you can show that

$$v_{rms} = \sqrt{\frac{3kT}{m}} \tag{13-22}$$

where k is Boltzmann's constant and m is the mass of a molecule. Therefore, at a given temperature, the rms speed is inversely proportional to the square root of the mass of the molecule.

Example 13.7

O_2 Molecules at Room Temperature

Find the average translational kinetic energy and the rms speed of the O_2 molecules in air at room temperature (20°C).

Strategy The average translational kinetic energy depends only on temperature. We must remember to use absolute temperature. The rms speed is the speed of a molecule that has the average kinetic energy.

Solution The absolute temperature is

$$20°\text{C} + 273\ \text{K} = 293\ \text{K}$$

Therefore, the average translational kinetic energy is

$$\begin{aligned}\langle K_{tr}\rangle &= \tfrac{3}{2}kT \\ &= 1.50 \times 1.38 \times 10^{-23}\ \text{J/K} \times 293\ \text{K} \\ &= 6.07 \times 10^{-21}\ \text{J}\end{aligned}$$

From the periodic table, we find the atomic mass of oxygen to be 16.0 u; the molecular mass of O_2 is twice that (32.0 u). First we convert that to kg:

$$32.0\ \text{u} \times 1.66 \times 10^{-27}\ \text{kg/u} = 5.31 \times 10^{-26}\ \text{kg}$$

The rms speed is the speed of a molecule with the average kinetic energy:

$$\langle K_{tr}\rangle = \tfrac{1}{2}mv_{rms}^2$$

$$v_{rms} = \sqrt{\frac{2\langle K_{tr}\rangle}{m}} = \sqrt{\frac{2 \times 6.07 \times 10^{-21}\ \text{J}}{5.31 \times 10^{-26}\ \text{kg}}} = 478\ \text{m/s}$$

Discussion How can we decide if the result is reasonable, since we have no first-hand experience watching molecules bounce around? Recall from Chapter 12 that the speed of sound in air at room temperature is 343 m/s. Since sound waves in air propagate by the collisions that occur between air molecules, the speed of sound must be of the same order of magnitude as the average speeds of the molecules.

Practice Problem 13.7 CO_2 Molecules at Room Temperature

Find the average translational kinetic energy and the rms speed of the CO_2 molecules in air at room temperature (20°C).

Maxwell-Boltzmann Distribution

So far we have considered only the *average* kinetic energy and *rms* speed of a molecule. Sometimes we may want to know more: how many molecules have speeds in a certain range? The distribution of speeds is called the **Maxwell-Boltzmann distribution**. The distribution for oxygen at two different temperatures is shown in Fig. 13.13. The interpretation of the graphs is that the number of gas molecules having speeds between any two values v_1 and v_2 is proportional to the area under the curve between v_1 and v_2. In Fig. 13.13, the shaded areas represent the number of oxygen molecules having speeds above 800 m/s at the two selected temperatures. A relatively small temperature change has a significant effect on the number of gas molecules with high speeds.

Any given molecule changes its kinetic energy often—at each collision, which means billions of times per second. However, the total number of gas molecules in a given kinetic energy range in the gas stays the same, as long as the temperature is constant. In fact, it is the frequent collisions that maintain the stability of the Maxwell-Boltzmann distribution. The collisions keep the kinetic energy distributed among the gas molecules *in the most disordered way possible*, which is the Maxwell-Boltzmann distribution.

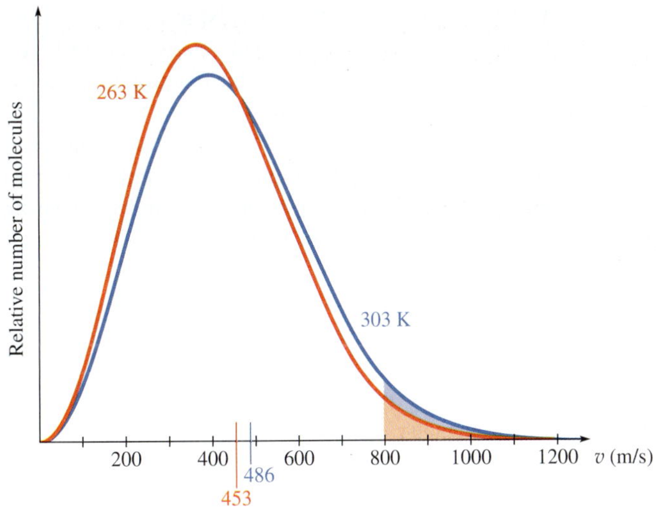

Figure 13.13 The probability distribution of kinetic energies in oxygen at two temperatures: −10°C (263 K) and +30°C (303 K). The area under either curve for any range of speeds is proportional to the number of molecules whose speeds lie in that range. Despite the relatively small difference in rms speeds (453 m/s at 263 K and 486 m/s at 303 K), the fraction of molecules in the high-speed tail is quite different.

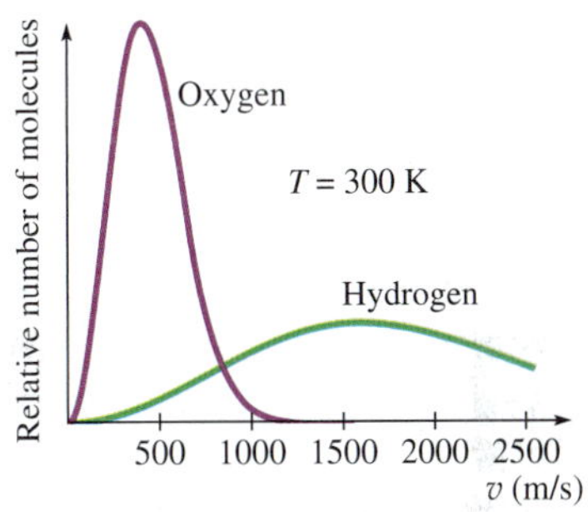

Figure 13.14 Maxwell-Boltzmann distributions for oxygen and hydrogen at $T = 300$ K. Escape speed from Earth is 11 200 m/s (not shown on the graph).

Application of the Maxwell-Boltzmann Distribution: Composition of Planetary Atmospheres The Maxwell-Boltzmann distribution helps us understand planetary atmospheres. Why does Earth's atmosphere contain nitrogen, oxygen, and water vapor, among other gases, but not hydrogen or helium, which are by far the most common elements in the universe? Molecules in the upper atmosphere that are moving faster than the *escape speed* (see Example 6.8) have enough kinetic energy to escape from the planetary atmosphere to outer space. Those that are heading away from the planet's surface will escape if they avoid colliding with another molecule. The high-energy tail of the Maxwell-Boltzmann distribution does not get depleted by molecules that escape. Other molecules will get boosted to those high kinetic energies as a result of collisions; these replacements will in turn also escape. Thus, the atmosphere gradually leaks away.

How fast the atmosphere leaks away depends on how far the rms speed is from the escape speed. If the rms speed is too small compared with the escape speed, the time for all the gas molecules to escape is so long that the gas is present in the atmosphere indefinitely. This is the case for nitrogen, oxygen, and water vapor in Earth's atmosphere. On the other hand, since hydrogen and helium are much less massive, their rms speeds are higher. Though only a tiny fraction of the molecules are above the escape speed, the fraction is sufficient for these gases to escape quickly from Earth's atmosphere (Fig. 13.14). The Moon is often said to lack an atmosphere. The Moon's low escape speed (2400 m/s) allows most gases to escape, but it does have an atmosphere about 1 cm tall composed of krypton (a gas with molecular mass 83.8 u, about 2.6 times that of oxygen).

CONNECTION:

The basic principle behind escape speed is conservation of energy (see Chapter 6). At the escape speed, an atom or molecule has just enough kinetic energy to escape the planet's gravitational pull.

13.7 TEMPERATURE AND REACTION RATES

What we have learned about the distribution of kinetic energies and its relationship to temperature has a great relevance to the dependence of chemical reaction rates on temperature. Imagine a mixture of two gases, N_2 and O_2, which can react to form nitric oxide (NO):

$$N_2 + O_2 \rightarrow 2NO$$

In order for the reaction to occur, a molecule of nitrogen must collide with a molecule of oxygen. But the reaction does not occur every time such a collision takes place. The reactant molecules must possess enough kinetic energy to initiate the reaction, because the reaction involves the rearrangement of chemical bonds between atoms. Some chemical bonds must be broken before new ones form; the energy to break these bonds must come from the energy of the reactants. (Note in this reaction that energy must be *supplied* to break a bond. *Forming* a bond *releases* energy.) The minimum kinetic energy of the reactant molecules that allows the reaction to proceed is called the **activation energy** (E_a).

If a molecule of N_2 collides with one of O_2, but their total kinetic energy is less than the activation energy, then the two just bounce off each other. Some energy may be transferred from one molecule to the other, or converted between translational, rotational, and vibrational energy, but we are still left with one molecule of N_2 and one of O_2.

Now we begin to see why, with few exceptions, rates of reaction increase with temperature. At higher temperatures, the average kinetic energy of the reactants is higher and therefore a greater fraction of the collisions have total kinetic energies exceeding the activation energy. If the activation energy is much greater than the average translational kinetic energy of the reactants,

$$E_a \gg \tfrac{3}{2}kT \qquad (13\text{-}23)$$

then the only candidates for reaction are molecules far off in the exponentially decaying, high-energy tail of the Maxwell-Boltzmann distribution. In this situation, a small increase in temperature can have a dramatic effect on the reaction rate: the reaction rate depends *exponentially* on temperature.

$$\text{reaction rate} \propto e^{-E_a/(kT)} \qquad (13\text{-}24)$$

Although we have discussed reactions in terms of gases, the same general principles apply to reactions in liquid solutions. The temperature determines what fraction of the collisions have enough energy to react, so reaction rates are temperature-dependent whether the reaction occurs in a gas mixture or a liquid solution.

Example 13.8

Increase in Reaction Rate with Temperature Increase

The activation energy for the reaction $N_2O \rightarrow N_2 + O$ is 4.0×10^{-19} J. By what percentage does the reaction rate increase if the temperature is increased from 700.0 K to 707.0 K (a 1% increase in absolute temperature)?

Strategy We should first check that $E_a \gg \frac{3}{2}kT$; otherwise, Eq. (13-24) does not apply. Assuming that checks out, we can set up a ratio of the reaction rates at the two temperatures.

Solution Start by calculating $E_a/(kT_1)$, where $T_1 = 700.0$ K:

$$\frac{E_a}{kT_1} = \frac{4.0 \times 10^{-19}\ \text{J}}{1.38 \times 10^{-23}\ \text{J/K} \times 700.0\ \text{K}} = 41.41$$

So E_a is about 41 times kT, or about 28 times $\frac{3}{2}kT$. The activation energy is much greater than the average kinetic energy; thus, only a small fraction of the collisions might cause a reaction to occur.

At $T_2 = 707.0$ K,

$$\frac{E_a}{kT_2} = \frac{4.0 \times 10^{-19}\ \text{J}}{1.38 \times 10^{-23}\ \text{J/K} \times 707.0\ \text{K}} = \frac{41.41}{1.01} = 41.00$$

The ratio of the reaction rates is

$$\frac{\text{new rate}}{\text{old rate}} = \frac{e^{-41.00}}{e^{-41.41}} = e^{-(41.00-41.41)} = e^{0.41} = 1.5$$

The reaction rate at 707.0 K is 1.5 times the rate at 700.0 K—a 50% increase in reaction rate for a 1% increase in temperature!

Discussion Normally we might suspect an error when a 1% change in one quantity causes a 50% change in another! However, this problem illustrates the dramatic effect of an *exponential* dependence. Reaction rates can be *extremely* sensitive to small temperature changes.

Practice Problem 13.8 Decrease in Reaction Rate for Lower Temperature

What is the percentage decrease in the rate of the same reaction if the temperature is lowered from 700.0 K to 699.0 K?

(a) (b) (c)

Figure 13.15 Warm-blooded animals use different strategies to maintain a constant body temperature. (a) The fur of an Arctic fox serves as a layer of insulation to help it stay warm. (b) Dogs pant and (c) people sweat when their bodies are in danger of overheating. In cases (b) and (c), the evaporation of water has a cooling effect on the body.

Application: Regulation of Body Temperature At the beginning of this chapter, we asked about the necessity for temperature regulation in homeotherms (Fig. 13.15). The temperature dependence of chemical reaction rates has a profound effect on biological functions. If our internal temperatures varied, we would have a varying metabolic rate, becoming sluggish in cold weather.

What are the evolutionary advantages of warm-blooded versus cold-blooded animals?

By maintaining a constant body temperature higher than that of the environment, homeotherms are able to tolerate a wider range of environmental temperatures than poikilotherms (e.g., reptiles and insects). Temperature fluctuation in the environment is much more severe on land than in water; thus, land animals are more likely to be homeothermic than aquatic animals. Keeping muscles at their optimal temperatures contributes to the much larger effort required to move around on land or in the air as opposed to moving through water. Keeping the muscles and vital organs warm allows the high level of aerobic metabolism needed to sustain intense physical activity.

Poikilotherms depend mostly on the environment for temperature regulation. As a crocodile's blood temperature goes down in cold weather, the crocodile becomes inactive and lethargic. Thus, we see a crocodile lying on a rock heated by the Sun in an attempt to keep warm.

However, if environmental conditions become too extreme, it may be difficult for homeotherms to maintain ideal body temperature. Hypothermia occurs when the central core of the body becomes too cold; bodily processes slow and eventually cease. People caught outside in blizzards are urged to stay awake and to keep moving; the metabolic rate during strenuous exercise may be up to 20 times that of the resting body, which can compensate for heat loss in extreme cold.

Homeotherms must consume much more food than poikilotherms of a similar size; metabolic processes in homeotherms act like a furnace to keep the body warm. A human must consume about 6 MJ of food energy per day just to keep warm when resting at 20°C; an alligator of similar body weight needs only about 0.3 MJ/day at rest at 20°C.

Figure 13.16 Successive straight-line paths traveled by a molecule between collisions.

13.8 DIFFUSION

Mean Free Path How far does a gas molecule move, on average, between collisions? The mean (average) length of the path traveled by a gas molecule as a free particle (no interactions with other particles) is called the **mean free path** (Λ, the Greek capital lambda). The mean free path depends on two things: how large the molecules are and how many of them occupy a given volume. A detailed calculation yields

Mean free path

$$\Lambda = \frac{1}{\sqrt{2}\,\pi d^2 (N/V)} \qquad \text{(13-25)}$$

Typically the mean free path is much larger than the average distance between neighboring molecules. Nitrogen molecules in air at room temperature have mean free paths of about 0.1 μm, which is about 25 times the average distance between molecules. Each molecule collides an average of 5×10^9 times per second. (For more information on mean free path, see the text website.)

Diffusion

A gas molecule moves in a straight line between collisions—the effect of gravity on the velocity of the molecule is negligible during a time interval of only 0.2 ns. At each collision, both the speed and direction of the molecule's motion change. The mean free path tells us the *average* length of the molecule's straight line paths between collisions. The result is that a given molecule follows a *random walk* trajectory (Fig. 13.16).

After an elapsed time t, how far on average has a molecule moved from its initial position? The answer to this question is relevant when we consider **diffusion**. Someone across the room opens a bottle of perfume: how long until the scent reaches you? As gas molecules diffuse into the air, the frequent collisions are what determine how long it takes the scent to travel across the room (assuming, as we do here, that there are no air currents). When there is a difference in concentrations between different points in a gas, the random thermal motion of the molecules tends to even out the concentrations (other things being equal). The net flow from regions of high concentration (near the perfume bottle) to regions of lower concentration (across the room) is diffusion.

Consider a molecule of perfume in the air. It has a mean free path Λ. After a large number of collisions N, it has traveled a total *distance* $N\Lambda$. However, its displacement from its original position is much less than that, since at each collision it changes direction. It can be shown using statistical analysis of the random walk that the rms magnitude of its displacement after N collisions is proportional to $\sqrt{N}$. Since the number of collisions is proportional to the elapsed time, the rms displacement is proportional to $\sqrt{t}$.

The root mean squared displacement in one direction is

$$x_{\text{rms}} = \sqrt{2Dt} \qquad \text{(13-26)}$$

where D is a diffusion constant such as those given in Table 13.4. The diffusion constant D depends on the molecule or atom that is diffusing and the medium through which it is moving.

Application: Diffusion of Oxygen Through Cell Membranes Diffusion is crucial in biological processes such as the transport of oxygen. Oxygen molecules diffuse from the air in the lungs through the walls of the alveoli and then through the walls of the capillaries to oxygenate the blood. The oxygen is then carried by hemoglobin in the blood to various parts of the body, where it again diffuses through capillary walls into intercellular fluids and then through cell membranes into cells.

Table 13.4 Diffusion Constants at 1 atm and 20°C

Diffusing Molecule	Medium	D (m^2/s)
DNA	Water	1.3×10^{-12}
Oxygen	Tissue (cell membrane)	1.8×10^{-11}
Hemoglobin	Water	6.9×10^{-11}
Sucrose ($C_{12}H_{22}O_{11}$)	Water	5.0×10^{-10}
Glucose ($C_6H_{12}O_6$)	Water	6.7×10^{-10}
Oxygen	Water	1.0×10^{-9}
Oxygen	Air	1.8×10^{-5}
Hydrogen	Air	6.4×10^{-5}

Diffusion is a slow process over long distances but can be quite effective over short distances—which is why cell membranes must be thin and capillaries must have small diameters. Evolution has seen to it that the capillaries of animals of widely different sizes are all about the same size—as small as possible while still allowing blood cells to flow through them.

Example 13.9

Diffusion Time for Oxygen into Capillaries

How long on average does it take an oxygen molecule in an alveolus to diffuse into the blood? Assume for simplicity that the diffusion constant for oxygen passing through the two membranes (alveolus and capillary walls) is the same: 1.8×10^{-11} m^2/s. The total thickness of the two membranes is 1.2×10^{-8} m.

Strategy Take the x-direction to be through the membranes. Then we want to know how much time elapses until $x_{rms} = 1.2 \times 10^{-8}$ m.

Solution Solving Eq. (13-26) for t yields

$$t = \frac{x_{rms}^2}{2D}$$

Now substitute $x_{rms} = 1.2 \times 10^{-8}$ m and $D = 1.8 \times 10^{-11}$ m^2/s:

$$t = \frac{(1.2 \times 10^{-8}\ \text{m})^2}{2 \times 1.8 \times 10^{-11}\ \text{m}^2/\text{s}} = 4.0 \times 10^{-6}\ \text{s}$$

Discussion The time is proportional to the *square* of the membrane thickness. It would take four times as long for an oxygen molecule to diffuse through a membrane twice as thick. The rapid increase of diffusion time with distance is a principal reason why evolution has favored thin membranes over thicker ones.

Practice Problem 13.9 Time for Oxygen to Get Halfway Through the Membrane

How long on average does it take an oxygen molecule to get *halfway* through the alveolus and capillary wall?

Master the Concepts

- Temperature is a quantity that determines when objects are in thermal equilibrium. The flow of energy that occurs between two objects or systems due to a temperature difference between them is called heat flow. If heat can flow between two objects or systems, the objects or systems are said to be in thermal contact. When two systems in thermal contact have the same temperature, there is no net flow of heat between them; the objects are said to be in thermal equilibrium.

continued on next page

Master the Concepts continued

- Zeroth law of thermodynamics: if two objects are each in thermal equilibrium with a third object, then the two are in thermal equilibrium with each other.
- The SI unit of temperature is the kelvin (symbol K, *without* a degree sign). The kelvin scale is an absolute temperature scale, which means that $T = 0$ is set to absolute zero.
- Temperature in °C (T_C) and temperature in kelvins (T) are related by

$$T_C = T - 273.15 \quad (13\text{-}3)$$

- As long as the temperature change is not too great, the fractional length change of a solid is proportional to the temperature change:

$$\frac{\Delta L}{L_0} = \alpha\,\Delta T \quad (13\text{-}4)$$

The constant of proportionality, α, is called the coefficient of linear expansion of the substance.

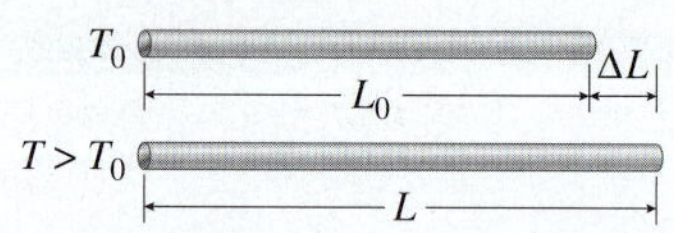

- The fractional change in volume of a solid or liquid is also proportional to the temperature change as long as the temperature change is not too large:

$$\frac{\Delta V}{V_0} = \beta\,\Delta T \quad (13\text{-}7)$$

For solids, the coefficient of volume expansion is three times the coefficient of linear expansion: $\beta = 3\alpha$.

- The mole is an SI base unit and is defined as: one mole of anything contains the same number of units as there are atoms in 12 *grams* (not kilograms) of carbon-12. This number is called Avogadro's number and has the value

$$N_A = 6.022 \times 10^{23}\ \text{mol}^{-1}$$

- The mass of an atom or molecule is often expressed in the atomic mass unit (symbol u). By definition, one atom of carbon-12 has a mass of 12 u (exactly).

$$1\ \text{u} = 1.66 \times 10^{-27}\ \text{kg} \quad (13\text{-}12)$$

The atomic mass unit is chosen so that the mass of an atom or molecule in "u" is numerically the same as the molar mass in g/mol.

- In an ideal gas, the molecules move independently in free space with no interactions except when two molecules collide. The ideal gas is a useful model for many real gases, provided that the gas is sufficiently dilute. The ideal gas law:

$$\text{microscopic form: } PV = NkT \quad (13\text{-}13)$$

$$\text{macroscopic form: } PV = nRT \quad (13\text{-}16)$$

where Boltzmann's constant and the universal gas constant are

$$k = 1.38 \times 10^{-23}\ \text{J/K} \quad (13\text{-}14)$$

$$R = N_A k = 8.31\ \frac{\text{J/K}}{\text{mol}} \quad (13\text{-}15)$$

- The pressure of an ideal gas is proportional to the average translational kinetic energy of the molecules:

$$P = \frac{2}{3}\frac{N}{V}\langle K_{tr}\rangle \quad (13\text{-}18)$$

- The average translational kinetic energy of the molecules is proportional to the absolute temperature:

$$\langle K_{tr}\rangle = \tfrac{3}{2}kT \quad (13\text{-}20)$$

- The speed of a gas molecule that has the average kinetic energy is called the rms speed:

$$\langle K_{tr}\rangle = \tfrac{1}{2}mv_{rms}^2 \quad (13\text{-}21)$$

- The distribution of molecular speeds in an ideal gas is called the Maxwell-Boltzmann distribution.

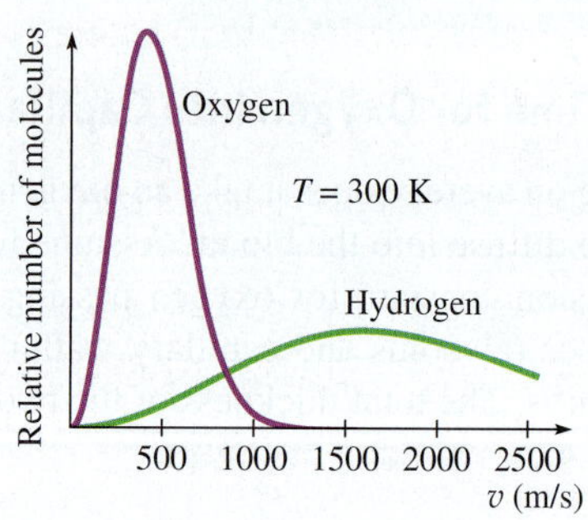

- If the activation energy for a chemical reaction is much greater than the average kinetic energy of the reactants, the reaction rate depends *exponentially* on temperature:

$$\text{reaction rate} \propto e^{-E_a/(kT)} \quad (13\text{-}24)$$

- The mean free path (Λ) is the average length of the path traveled by a gas molecule as a free particle (no interactions with other particles) between collisions:

$$\Lambda = \frac{1}{\sqrt{2}\,\pi d^2 (N/V)} \quad (13\text{-}25)$$

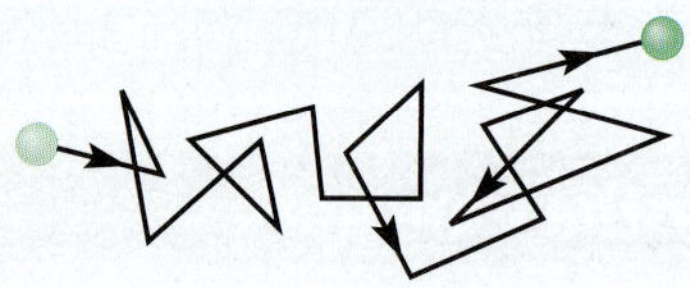

- The root mean square displacement of a diffusing molecule along the x-axis is

$$x_{rms} = \sqrt{2Dt} \quad (13\text{-}26)$$

where D is a diffusion constant.

Conceptual Questions

1. Explain why it would be impossible to uniquely define the temperature of an object if the zeroth law of thermodynamics were violated?
2. Why do we call the temperature 0 K "absolute zero"? How is 0 K fundamentally different from 0°C or 0°F?
3. Under what special circumstances can kelvins or Celsius degrees be used interchangeably?
4. (a) Imagine drawing a circle on the surface of a metal plate. When the temperature increases, what happens to the size of the circle? (b) Instead of drawing a circle, suppose you cut out the circle and then put it back inside the hole in the plate. What would happen to the two pieces when the temperature increases? Does the hole get larger or smaller? Explain.
5. Why would silver and brass probably not be a good choice of metals for a bimetallic strip (leaving aside the question of the cost of silver)? (See Table 13.2.)
6. One way to loosen the lid on a glass jar is to run it under hot water. How does that work?
7. Why must we use absolute temperature (temperature in kelvins) in the ideal gas law ($PV = NkT$)? Explain how using the Celsius scale would give nonsensical results.
8. Natural gas is sold by volume. In the United States, the price charged is usually per cubic foot. Given the price per cubic foot, what other information would you need in order to calculate the price per mole?
9. What are the SI units of mass density and number density? If two different gases have the same number density, do they have the same mass density?
10. Suppose we have two tanks, one containing helium gas and the other nitrogen gas. The two gases are at the same temperature and pressure. Which has the higher number density (or are they equal)? Which has the higher mass density (or are they equal)?
11. The mass of an aluminum atom is 27.0 u. What is the mass of *one mole* of aluminum atoms? (No calculation required!)
12. A ping-pong ball that has been dented during hard play can often be restored by placing it in hot water. Explain why this works.
13. Why does a helium weather balloon expand as it rises into the air? Assume the temperature remains constant.
14. Explain why there is almost no hydrogen (H_2) or helium (He) in Earth's atmosphere, yet both are present in Jupiter's atmosphere. [*Hint*: Escape velocity from Earth is 11.2 km/s and escape velocity from Jupiter is 60 km/s.]
15. Explain how it is possible that more than half of the molecules in an ideal gas have kinetic energies less than the average kinetic energy. Shouldn't half have less and half have more?
16. In air under ordinary conditions (room temperature and atmospheric pressure), the average intermolecular distance is about 4 nm and the mean free path is about 0.1 μm. The diameter of a nitrogen molecule is about 0.3 nm. Explain how the mean free path can be so much larger than the average distance between molecules.
17. In air under ordinary conditions (room temperature and atmospheric pressure), the average intermolecular distance is about 4 nm and the mean free path is about 0.1 μm. The diameter of a nitrogen molecule is about 0.3 nm. Which two distances should we compare to decide that air is dilute and can be treated as an ideal gas? Explain.
18. In air under ordinary conditions (room temperature and atmospheric pressure), the average intermolecular distance is about 4 nm and the mean free path is about 0.1 μm. The diameter of a nitrogen molecule is about 0.3 nm. What would it mean if the intermolecular distance and the molecular diameter were about the same? In that case, would it make sense to speak of a mean free path? Explain.
19. Explain how an automobile airbag protects the passenger from injury. Why would the airbag be ineffective if the gas pressure inside is too low when the passenger comes into contact with it? What about if it is too high?
20. It takes longer to hard-boil an egg in Mexico City (2200 m above sea level) than it does in Amsterdam (parts of which are below sea level). Why? [*Hint*: At higher altitudes, water boils at less than 100°C.]

Multiple-Choice Questions

Student Response System Questions

1. In a mixed gas such as air, the rms speeds of different molecules are
   (a) independent of molecular mass.
   (b) proportional to molecular mass.
   (c) inversely proportional to molecular mass.
   (d) proportional to $\sqrt{\text{molecular mass}}$
   (e) inversely proportional to $\sqrt{\text{molecular mass}}$
2. The average kinetic energy of the molecules in an ideal gas increases with the volume remaining constant. Which of these statements *must* be true?
   (a) The pressure increases and the temperature stays the same.
   (b) The number density decreases.
   (c) The temperature increases and the pressure stays the same.
   (d) Both the pressure and the temperature increase.
3. The absolute temperature of an ideal gas is directly proportional to
   (a) the number of molecules in the sample.
   (b) the average momentum of a molecule of the gas.
   (c) the average translational kinetic energy of the gas.
   (d) the diffusion constant of the gas.

4. Which of these increases the average kinetic energy of the molecules in an ideal gas?
   (a) reducing the volume, keeping P and N constant
   (b) increasing the volume, keeping P and N constant
   (c) reducing the volume, keeping T and N constant
   (d) increasing the pressure, keeping T and V constant
   (e) increasing N, keeping V and T constant
5. The rms speed is the
   (a) speed at which all the gas molecules move.
   (b) speed of a molecule with the average kinetic energy.
   (c) average speed of the gas molecules.
   (d) maximum speed of the gas molecules.
6. An ideal gas has the volume V_0. If the temperature and the pressure are each tripled during a process, the new volume is
   (a) V_0.
   (b) $9V_0$.
   (c) $3V_0$.
   (d) $0.33V_0$.
7. What are the most favorable conditions for real gases to approach ideal behavior?
   (a) high temperature and high pressure
   (b) low temperature and high pressure
   (c) low temperature and low pressure
   (d) high temperature and low pressure
8. If the temperature of an ideal gas is doubled and the pressure is held constant, the rms speed of the molecules
   (a) remains unchanged.
   (b) is 2 times the original speed.
   (c) is $\sqrt{2}$ times the original speed.
   (d) is 4 times the original speed.
9. The average kinetic energy of a gas molecule can be found from which of these quantities?
   (a) pressure only
   (b) number of molecules only
   (c) temperature only
   (d) pressure and temperature are both required
10. A metal box is heated until each of its sides has expanded by 0.1%. By what percent has the *volume* of the box changed?
   (a) −0.3% (b) −0.2% (c) +0.1%
   (d) +0.2% (e) +0.3%

Problems

- Combination conceptual/quantitative problem
- Biomedical application
- ✦ Challenging
- Blue # Detailed solution in the Student Solutions Manual
- [1, 2] Problems paired by concept
- connect Text website interactive or tutorial

13.2 Temperature Scales

1. On a warm summer day, the air temperature is 84°F. Express this temperature in (a) °C and (b) kelvins. (connect tutorial: sun's temperature)
2. The temperature at which liquid nitrogen boils (at atmospheric pressure) is 77 K. Express this temperature in (a) °C and (b) °F.
3. (a) At what temperature (if any) does the numerical value of Celsius degrees equal the numerical value of Fahrenheit degrees? (b) At what temperature (if any) does the numerical value of kelvins equal the numerical value of Fahrenheit degrees?
4. A room air conditioner causes a temperature change of −6.0°C. (a) What is the temperature change in kelvins? (b) What is the temperature change in °F?
5. Aliens from the planet Jeenkah have based their temperature scale on the boiling and freezing temperatures of ethyl alcohol. These temperatures are 78°C and −114°C, respectively. The people of Jeenkah have six digits on each hand, so they use a base-12 number system and have decided to have 144°J between the freezing and boiling temperatures of ethyl alcohol. They set the freezing point to 0°J. How would you convert from °J to °C?

13.3 Thermal Expansion of Solids and Liquids

6. Five slabs with temperature coefficients of expansion α have lengths L at $T_i = 20°C$. Their temperatures then rise to T_f. Rank them in order of how much their lengths increase, greatest to smallest.
   (a) $L = 90$ cm, $T_f = 40°C$, $\alpha = 8 \times 10^{-6}\ K^{-1}$ (granite)
   (b) $L = 90$ cm, $T_f = 50°C$, $\alpha = 8 \times 10^{-6}\ K^{-1}$ (granite)
   (c) $L = 60$ cm, $T_f = 40°C$, $\alpha = 8 \times 10^{-6}\ K^{-1}$ (granite)
   (d) $L = 90$ cm, $T_f = 40°C$, $\alpha = 12 \times 10^{-6}\ K^{-1}$ (concrete)
   (e) $L = 60$ cm, $T_f = 50°C$, $\alpha = 12 \times 10^{-6}\ K^{-1}$ (concrete)
7. A 2.4 m length of copper pipe extends directly from a water heater in a basement to a faucet on the first floor of a house. If the faucet isn't fixed in place, how much will it rise when the pipe is heated from 20.0°C to 90.0°C? Ignore any increase in the size of the faucet itself or of the water heater.
8. Two 35.0 cm metal rods, one made of copper and one made of aluminum, are placed end to end, touching each other. One end is fixed, so that it cannot move. The rods are heated from 0.0°C to 150°C. How far does the other end of the system of rods move?
9. Steel railroad tracks of length 18.30 m are laid at 10.0°C. How much space should be left between the track sections if they are to just touch when the temperature is 50.0°C?

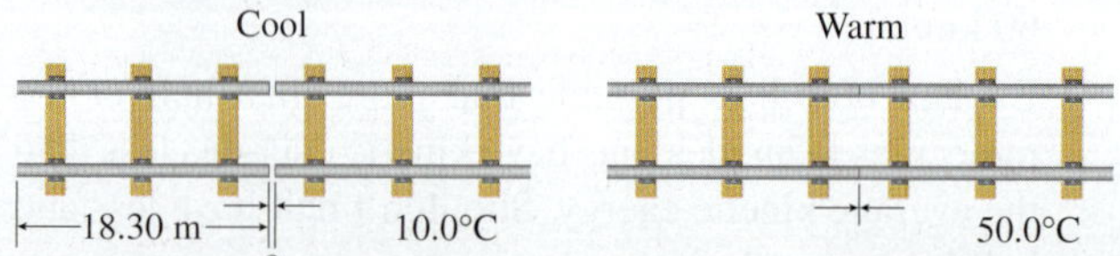

10. A highway is made of concrete slabs that are 15 m long at 20.0°C. (a) If the temperature range at the location of the highway is from −20.0°C to +40.0°C, what size expansion gap should be left (at 20.0°C) to prevent buckling of the highway? (b) How large are the gaps at −20.0°C?

11. A lead rod and a common glass rod both have the same length when at 20.0°C. The lead rod is heated to 50.0°C. To what temperature must the glass rod be heated so that they are again at the same length?

12. The coefficient of linear expansion of brass is $1.9 \times 10^{-5}\ °C^{-1}$. At 20.0°C, a hole in a sheet of brass has an area of 1.00 mm^2. How much larger is the area of the hole at 30.0°C? (connect tutorial: loop around the equator)

13. Aluminum rivets used in airplane construction are made slightly too large for the rivet holes to be sure of a tight fit. The rivets are cooled with dry ice (−78.5°C) before they are driven into the holes. If the holes have a diameter of 0.6350 cm at 20.5°C, what should be the diameter of the rivets at 20.5°C if they are to just fit when cooled to the temperature of dry ice?

14. The George Washington Bridge crosses the Hudson River between New York and New Jersey. The span of the steel bridge is about 1.6 km. If the temperature can vary from a low of −15°F in winter to a high of 105°F in summer, by how much might the length of the span change over an entire year?

15. The fuselage of an Airbus A340 has a circumference of 17.72 m on the ground. The circumference increases by 26 cm when it is in flight. Part of this increase is due to the pressure difference between the inside and outside of the plane and part is due to the increase in the temperature due to air drag while it is flying along at 950 km/h. Suppose we wanted to heat a full-size model of the airbus made of aluminum to cause the same increase in circumference without changing the pressure. What would be the increase in temperature needed?

16. Suppose you have a filling in one of your teeth, and, while eating some ice cream, you suddenly realize that the filling came out. One of the reasons the filling may have become detached from your tooth is the differential contraction of the filling relative to the rest of the tooth due to the temperature change. (a) Find the change in volume for a metallic dental filling due to the difference between body temperature (37°C) and the temperature of the ice cream you ate (−5°C). The initial volume of the filling is 30 mm^3, and its temperature coefficient is $\alpha = 42 \times 10^{-6}\ K^{-1}$. (b) Find the change in volume of the cavity. The temperature coefficient of the tooth is $\alpha = 17 \times 10^{-6}\ K^{-1}$.

17. A cylindrical brass container with a base of 75.0 cm^2 and height of 20.0 cm is filled to the brim with water when the system is at 25.0°C. How much water overflows when the temperature of the water and the container is raised to 95.0°C?

18. An ordinary drinking glass is filled to the brim with water (268.4 mL) at 2.0°C and placed on the sunny pool deck for a swimmer to enjoy. If the temperature of the water rises to 32.0°C before the swimmer reaches for the glass, how much water will have spilled over the top of the glass? Assume the glass does not expand.

19. Consider the situation described in Problem 18. (a) Take into account the expansion of the glass and calculate how much water will spill out of the glass. Compare your answer with the case where the expansion of the glass was not considered. (b) By what percentage has the answer changed when the expansion of the glass is considered?

20. A steel sphere with radius 1.0010 cm at 22.0°C must slip through a brass ring that has an internal radius of 1.0000 cm at the same temperature. To what temperature must the brass ring be heated so that the sphere, still at 22.0°C, can just slip through?

21. A square brass plate, 8.00 cm on a side, has a hole cut into its center of area 4.908 74 cm^2 (at 20.0°C). The hole in the plate is to slide over a cylindrical steel shaft of cross-sectional area 4.910 00 cm^2 (also at 20.0°C). To what temperature must the brass plate be heated so that it can just slide over the steel cylinder (which remains at 20.0°C)? [*Hint*: The steel cylinder is not heated so it does not expand; only the brass plate is heated.]

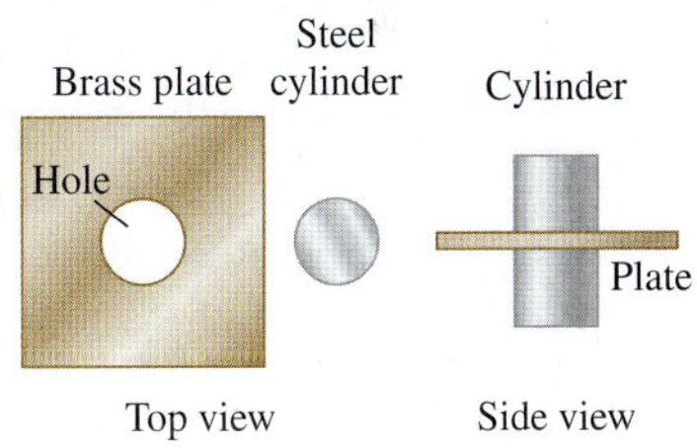

22. A copper washer is to be fit in place over a steel bolt. Both pieces of metal are at 20.0°C. If the diameter of the bolt is 1.0000 cm and the inner diameter of the washer is 0.9980 cm, to what temperature must the washer be raised so it will fit over the bolt? Only the copper washer is heated.

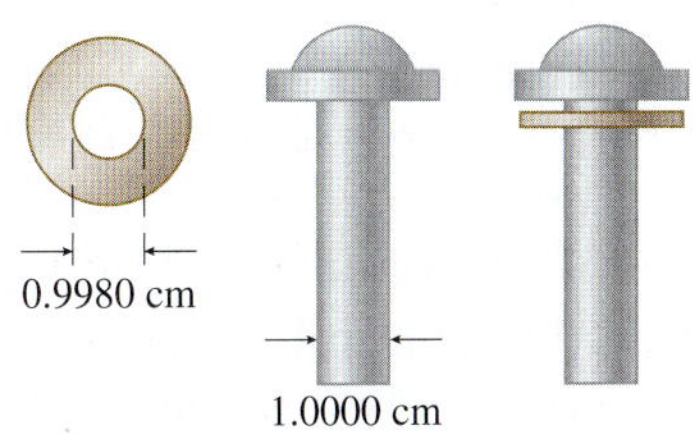

Problems 22 and 23

23. Repeat Problem 22, but now the copper washer and the steel bolt are both raised to the same temperature. At what temperature will the washer fit on the bolt?

24. ✦ A steel rule is calibrated for measuring lengths at 20.00°C. The rule is used to measure the length of a Vycor glass brick; when both are at 20.00°C, the brick is found to be 25.00 cm long. If the rule and the brick are both at 80.00°C, what would be the length of the brick as measured by the rule?

25. ✦ A flat square of side s_0 at temperature T_0 expands by Δs in both length and width when the temperature increases by ΔT. The original area is $s_0^2 = A_0$ and the final area is $(s_0 + \Delta s)^2 = A$. Show that if $\Delta s \ll s_0$,

$$\frac{\Delta A}{A_0} = 2\alpha\,\Delta T \qquad (13\text{-}6)$$

(Although we derive this relation for a square plate, it applies to a flat area of any shape.)

26. ✦ The volume of a solid cube with side s_0 at temperature T_0 is $V_0 = s_0^3$. Show that if $\Delta s \ll s_0$, the change in volume ΔV due to a change in temperature ΔT is given by

$$\frac{\Delta V}{V_0} = 3\alpha\,\Delta T \qquad (13\text{-}7, 13\text{-}8)$$

and therefore that $\beta = 3\alpha$. (Although we derive this relation for a cube, it applies to a solid of any shape.)

13.4 Molecular Picture of a Gas

27. Use the definition that 1 mol of ^{12}C (carbon-12) atoms has a mass of exactly 12 g, along with Avogadro's number, to derive the conversion between atomic mass units and kg.
28. Find the molar mass of ammonia (NH_3).
29. Find the mass (in kg) of one molecule of CO_2.
30. The mass of 1 mol of ^{13}C (carbon-13) is 13.003 g. (a) What is the mass in u of one ^{13}C atom? (b) What is the mass in kilograms of one ^{13}C atom?
31. Estimate the number of H_2O molecules in a human body of mass 80.2 kg. Assume that, on average, water makes up about 62% of the mass of a human body.
32. The mass density of diamond (a crystalline form of carbon) is 3500 kg/m^3. How many carbon atoms per cm^3 are there?
33. How many hydrogen atoms are present in 684.6 g of sucrose ($C_{12}H_{22}O_{11}$)?
34. How many moles of He are in 13 g of He?
35. The principal component of natural gas is methane (CH_4). How many moles of CH_4 are present in 144.36 g of methane?
36. What is the mass of one gold atom in kilograms?
37. Air at room temperature and atmospheric pressure has a mass density of 1.2 kg/m^3. The average molecular mass of air is 29.0 u. How many molecules are in 1.0 cm^3 of air?
38. At 0.0°C and 1.00 atm, 1.00 mol of a gas occupies a volume of 0.0224 m^3. (a) What is the number density? (b) Estimate the average distance between the molecules. (c) If the gas is nitrogen (N_2), the principal component of air, what is the total mass and mass density?
39. Sand is composed of SiO_2. Find the order of magnitude of the number of silicon (Si) atoms in a grain of sand. Approximate the sand grain as a sphere of diameter 0.5 mm and an SiO_2 molecule as a sphere of diameter 0.5 nm.

13.5 Absolute Temperature and the Ideal Gas Law

40. A flight attendant wants to change the temperature of the air in the cabin from 18.0°C to 24.0°C without changing the number of moles of air per cubic meter. What fractional change in pressure would be required?
41. A cylinder in a car engine takes $V_i = 4.50 \times 10^{-2}$ m^3 of air into the chamber at 30°C and at atmospheric pressure. The piston then compresses the air to one-ninth of the original volume ($0.111\,V_i$) and to 20.0 times the original pressure ($20.0\,P_i$). What is the new temperature of the air?
42. A tire with an inner volume of 0.0250 m^3 is filled with air at a gauge pressure of 36.0 psi. If the tire valve is opened to the atmosphere, what volume *outside of the tire* does the escaping air occupy? Some air remains within the tire occupying the original volume, but now that remaining air is at atmospheric pressure. Assume the temperature of the air does not change.
43. Six cylinders contain ideal gases (not necessarily the same gas) with the properties given (P = pressure, V = volume, N = number of molecules). Rank them in order of temperature, highest to lowest.
(a) $P = 100$ kPa, $V = 4$ L, $N = 6 \times 10^{23}$
(b) $P = 200$ kPa, $V = 4$ L, $N = 6 \times 10^{23}$
(c) $P = 50$ kPa, $V = 8$ L, $N = 6 \times 10^{23}$
(d) $P = 100$ kPa, $V = 4$ L, $N = 3 \times 10^{23}$
(e) $P = 100$ kPa, $V = 2$ L, $N = 3 \times 10^{23}$
(f) $P = 50$ kPa, $V = 4$ L, $N = 3 \times 10^{23}$
44. What fraction of the air molecules in a house must be pushed outside while the furnace raises the inside temperature from 16.0°C to 20.0°C? The pressure does not change since the house is not airtight.
45. An emphysema patient is breathing pure O_2 through a face mask. The cylinder of O_2 contains 0.0170 m^3 of O_2 gas at a pressure of 15.2 MPa. (a) What volume would the oxygen occupy at atmospheric pressure (and the same temperature)? (b) If the patient takes in 8.0 L/min of O_2 at atmospheric pressure, how long will the cylinder last?
46. Incandescent lightbulbs are filled with an inert gas to lengthen the filament life. With the current off (at $T = 20.0$°C), the gas inside a lightbulb has a pressure of 115 kPa. When the bulb is burning, the temperature rises to 70.0°C. What is the pressure at the higher temperature?
47. What is the mass density of air at $P = 1.0$ atm and $T =$ (a) −10°C and (b) 30°C? The average molecular mass of air is approximately 29 u.
48. A constant volume gas thermometer containing helium is immersed in boiling ammonia (−33°C), and the pressure is read once equilibrium is reached. The

thermometer is then moved to a bath of boiling water (100.0°C). After the manometer was adjusted to keep the volume of helium constant, by what factor was the pressure multiplied?

49. A hydrogen balloon at Earth's surface has a volume of 5.0 m^3 on a day when the temperature is 27°C and the pressure is 1.00×10^5 N/m^2. The balloon rises and expands as the pressure drops. What would the volume of the same number of moles of hydrogen be at an altitude of 40 km where the pressure is 0.33×10^3 N/m^2 and the temperature is -13°C?

50. An ideal gas that occupies 1.2 m^3 at a pressure of 1.0×10^5 Pa and a temperature of 27°C is compressed to a volume of 0.60 m^3 and heated to a temperature of 227°C. What is the new pressure?

51. In intergalactic space, there is an average of about one hydrogen atom per cm^3 and the temperature is 3 K. What is the absolute pressure?

52. A tank of compressed air of volume 1.0 m^3 is pressurized to 20.0 atm at $T = 273$ K. A valve is opened, and air is released until the pressure in the tank is 15.0 atm. How many molecules were released?

53. A mass of 0.532 kg of molecular oxygen is contained in a cylinder at a pressure of 1.0×10^5 Pa and a temperature of 0.0°C. What volume does the gas occupy?

54. Verify, using the ideal gas law, the assertion in Problem 38 that 1.00 mol of a gas at 0.0°C and 1.00 atm occupies a volume of 0.0224 m^3.

55. ✦ A bubble rises from the bottom of a lake of depth 80.0 m, where the temperature is 4°C. The water temperature at the surface is 18°C. If the bubble's initial diameter is 1.00 mm, what is its diameter when it reaches the surface? (Ignore the surface tension of water. Assume the bubble warms as it rises to the same temperature as the water and retains a spherical shape. Assume $P_{atm} = 1.0$ atm.)

56. ✦ Consider the expansion of an ideal gas at constant pressure. The initial temperature is T_0 and the initial volume is V_0. (a) Show that $\Delta V/V_0 = \beta\, \Delta T$, where $\beta = 1/T_0$. (b) Compare the coefficient of volume expansion β for an ideal gas at 20°C to the values for liquids and gases listed in Table 13.3.

13.6 Kinetic Theory of the Ideal Gas

57. What is the temperature of an ideal gas whose molecules have an average translational kinetic energy of 3.20×10^{-20} J?

58. What is the total translational kinetic energy of the gas molecules of 0.420 mol of air at atmospheric pressure that occupies a volume of 1.00 L (0.00100 m^3)?

59. What is the kinetic energy per unit volume in an ideal gas at (a) $P = 1.00$ atm and (b) $P = 300.0$ atm?

60. Show that, for an ideal gas,

$$P = \tfrac{1}{3}\rho v_{rms}^2$$

where P is the pressure, ρ is the mass density, and v_{rms} is the rms speed of the gas molecules.

61. Rank the six gases of Problem 43 in order of the *total* translational kinetic energy, greatest to least.

62. What is the total internal kinetic energy of 1.0 mol of an ideal gas at 0.0°C and 1.00 atm?

63. If 2.0 mol of nitrogen gas (N_2) are placed in a cubic box, 25 cm on each side, at 1.6 atm of pressure, what is the rms speed of the nitrogen molecules?

64. There are two identical containers of gas at the same temperature and pressure, one containing argon and the other neon. What is the ratio of the rms speed of the argon atoms to that of the neon atoms? The atomic mass of argon is twice that of neon. (connect tutorial: RMS speed)

65. A smoke particle has a mass of 1.38×10^{-17} kg, and it is randomly moving about in thermal equilibrium with room temperature air at 27°C. What is the rms speed of the particle?

66. Find the rms speed in air at 0.0°C and 1.00 atm of (a) the N_2 molecules, (b) the O_2 molecules, and (c) the CO_2 molecules.

67. What are the rms speeds of helium atoms, and nitrogen, hydrogen, and oxygen molecules at 25°C?

68. If the upper atmosphere of Jupiter has a temperature of 160 K and the escape speed is 60 km/s, would an astronaut expect to find much hydrogen there?

69. What is the temperature of an ideal gas whose molecules in random motion have an average translational kinetic energy of 4.60×10^{-20} J?

70. On a cold day, you take a breath, inhaling 0.50 L of air whose initial temperature is -10°C. In your lungs, its temperature is raised to 37°C. Assume that the pressure is 101 kPa and that the air may be treated as an ideal gas. What is the total change in translational kinetic energy of the air you inhaled?

71. ✦ Show that the rms speed of a molecule in an ideal gas at absolute temperature T is given by

$$v_{rms} = \sqrt{\frac{3kT}{m}} \qquad (13\text{-}22)$$

where k is Boltzmann's constant and m is the mass of a molecule.

72. ✦ Show that the rms speed of a molecule in an ideal gas at absolute temperature T is given by

$$v_{rms} = \sqrt{\frac{3RT}{M}}$$

where M is the *molar mass*—the mass of the gas per mole.

13.7 Temperature and Reaction Rates

73. ✦ The reaction rate for the prepupal development of male *Drosophila* is temperature-dependent.

Assuming that the reaction rate is exponential as in Eq. (13-24), the activation energy for this development is then 2.81×10^{-19} J. A *Drosophila* is originally at 10.00°C, and its temperature is increasing. If the rate of development has increased 3.5%, how much has its temperature increased?

74. ✦ The reaction rate for the hydrolysis of benzoyl-L-arginine amide by trypsin at 10.0°C is 1.878 times faster than that at 5.0°C. Assuming that the reaction rate is exponential as in Eq. (13-24), what is the activation energy?

75. ✦ At high altitudes, water boils at a temperature lower than 100.0°C due to the lower air pressure. A rule of thumb states that the time to hard-boil an egg doubles for every 10.0°C drop in temperature. What activation energy does this rule imply for the chemical reactions that occur when the egg is cooked?

13.8 Diffusion

76. Estimate the mean free path of a N_2 molecule in air at (a) sea level ($P \approx 100$ kPa and $T \approx 290$ K), (b) the top of Mt. Everest (altitude = 8.8 km, $P \approx 50$ kPa, and $T \approx 230$ K), and (c) an altitude of 30 km ($P \approx 1$ kPa and $T \approx 230$ K). For simplicity, assume that air is pure nitrogen gas. The diameter of a N_2 molecule is approximately 0.3 nm.

77. About how long will it take a perfume molecule to diffuse a distance of 5.00 m in one direction in a room if the diffusion constant is 1.00×10^{-5} m^2/s? Assume that the air is perfectly still—there are no air currents.

78. Estimate the time it takes a sucrose molecule to move 5.00 mm in one direction by diffusion in water. Assume there are no currents in the water.

79. Your friend is 3.0 m away from you in a room. There are no significant air currents. She opens a bottle of perfume, and you first smell it 20 s later. How long would it have taken for you to smell it if she had been 6.0 m away instead? (connect tutorial: diffusion)

80. Platelet cells in blood play an essential role in the formation of clots and exist in normal human blood at the level of about 200000 per cubic millimeter. In order to illustrate that diffusion alone is not responsible for transporting platelets, consider the following situation. The diffusion constant for platelets in blood is approximately 5×10^{-10} m^2/s. About how long would it take a platelet to diffuse from the center of an artery (diameter 8.0 mm) to a clot forming on one wall of the artery?

81. In plants, water diffuses out through small openings known as stomatal pores. If $D = 2.4 \times 10^{-5}$ m^2/s for water vapor in air, and the length of the pores is 2.5×10^{-5} m, how long does it take for a water molecule to diffuse out through the pore?

Collaborative Problems

82. Agnes Pockels (1862–1935) was able to determine Avogadro's number using only a few household chemicals, in particular oleic acid, whose formula is $C_{18}H_{34}O_2$. (a) What is the molar mass of this acid? (b) The mass of one drop of oleic acid is 2.3×10^{-5} g and the volume is 2.6×10^{-5} cm^3. How many moles of oleic acid are there in one drop? (c) Now all Pockels needed was to find the number of molecules of oleic acid. Luckily, when oleic acid is spread out on water, it lines up in a layer one molecule thick. If the base of the molecule of oleic acid is a square of side d, the height of the molecule is known to be $7d$. Pockels spread out one drop of oleic acid on some water, and measured the area to be 70.0 cm^2. Using the volume and the area of oleic acid, what is d? (d) If we assume that this film is one molecule thick, how many molecules of oleic acid are there in the drop? (e) What value does this give you for Avogadro's number?

83. As a Boeing 747 gains altitude, the passenger cabin is pressurized. However, the cabin is not pressurized fully to atmospheric (1.01×10^5 Pa), as it would be at sea level, but rather pressurized to 7.62×10^4 Pa. Suppose a 747 takes off from sea level when the temperature in the airplane is 25.0°C and the pressure is 1.01×10^5 Pa. (a) If the cabin temperature remains at 25.0°C, what is the percentage change in the number of moles of air in the cabin? (b) If instead, the number of moles of air in the cabin does not change, what would the temperature be?

84. For divers going to great depths, the composition of the air in the tank must be modified. The ideal composition is to have approximately the same number of O_2 molecules per unit volume as in surface air (to avoid oxygen poisoning), and to use helium instead of nitrogen for the remainder of the gas (to avoid nitrogen narcosis, which results from nitrogen dissolving in the bloodstream). Of the molecules in dry surface air, 78% are N_2, 21% are O_2, and 1% are Ar. (a) How many O_2 molecules per cubic meter are there in surface air at 20.0°C and 1.00 atm? (b) For a diver going to a depth of 100.0 m, what percentage of the gas molecules in the tank should be O_2? (Assume that the density of seawater is 1025 kg/m^3 and the temperature is 20.0°C.)

85. If you wanted to make a scale model of air at 0.0°C and 1.00 atm, using ping-pong balls (diameter, 3.75 cm) to represent the N_2 molecules (diameter, 0.30 nm), (a) how far apart on average should the ping-pong balls be at any instant? (b) How far would a ping-pong ball travel on average before colliding with another?

Comprehensive Problems

86. A Pyrex container is filled to the very top with 4.00 L of water. Both the container and the water are at a temperature of 90.0°C. When the temperature has cooled to 20.0°C, how much additional water can be added to the container?

87. A hot air balloon with a volume of 12.0 m^3 is initially filled with air at a pressure of 1.00 atm and a temperature of 19.0°C. When the balloon air is heated, the volume and the pressure of the balloon remain constant because the balloon is open to the atmosphere at the bottom. How many moles are in the balloon when the air is heated to 40.0°C?

88. In a certain bimetallic strip (see Fig. 13.7) the brass strip is 0.100% longer than the steel strip at a temperature of 275°C. At what temperature do the two strips have the same length?

89. The driver from Practice Problem 13.3 fills his 18.9 L steel gasoline can in the morning when the temperature of the can and the gasoline is 15.0°C and the pressure is 1.0 atm, but this time he remembers to replace the tightly fitting cap after filling the can. Assume that the can is completely full of gasoline (no air space) and that the cap does not leak. The temperature climbs to 30.0°C. Ignoring the expansion of the steel can, what would be the pressure of the gasoline? The bulk modulus for gasoline is 1.00×10^9 N/m^2.

90. An iron bridge girder ($Y = 2.0 \times 10^{11}$ N/m^2) is constrained between two rock faces whose spacing doesn't change. At 20.0°C the girder is relaxed. How large a stress develops in the iron if the sun heats the girder to 40.0°C?

91. Consider the sphere and ring of Problem 20. What must the final temperature be if both the ring and the sphere are heated to the same final temperature?

92. Suppose due to a bad break of your femur, you require the insertion of a titanium rod to help the fracture heal. The coefficient of linear expansion for titanium is $\alpha = 8.6 \times 10^{-6}$ K^{-1}, and the length of the rod when it is in equilibrium with the leg bone and muscle at 37°C is 5.00 cm. How much shorter was the rod at room temperature (20°C)?

93. A certain acid has a molecular mass of 63 u. By mass, it consists of 1.6% hydrogen, 22.2% nitrogen, and 76.2% oxygen. What is the chemical formula for this acid?

94. The data in the following table are from a constant-volume gas thermometer experiment. The volume of the gas was kept constant, while the temperature was changed. The resulting pressure was measured. Plot the data on a pressure versus temperature diagram. Based on these data, estimate the value of absolute zero in Celsius.

T (°C)	*P* (atm)
0	1.00
20	1.07
100	1.37
−33	0.88
−196	0.28

95. At a normal body temperature of 37.0°C, (a) what is the average kinetic energy of the gas molecules in the lungs? (b) If a fever increases the temperature to 37.8°C, by what percentage does the average kinetic energy of the molecules increase?

96. The volume of air taken in by a warm-blooded vertebrate in the Andes Mountains is 210 L/day at standard temperature and pressure (i.e., 0°C and 1 atm). If the air in the lungs is at 39°C, under a pressure of 450 mm Hg, and we assume that the vertebrate takes in an average volume of 100 cm^3 per breath at the temperature and pressure of its lungs, how many breaths does this vertebrate take per day?

97. An iron cannonball of radius 0.08 m has a cavity of radius 0.05 m that is to be filled with gunpowder. If the measurements were made at a temperature of 22°C, how much extra volume of gunpowder, if any, will be required to fill 500 cannonballs when the temperature is 30°C?

98. Ten students take a test and get the following scores: 83, 62, 81, 77, 68, 92, 88, 83, 72, and 75. What are the average value, the rms value, and the most probable value, respectively, of these test scores?

99. A hand pump is being used to inflate a bicycle tire that has a gauge pressure of 40.0 psi. If the pump is a cylinder of length 18.0 in. with a cross-sectional area of 3.00 in.^{2}, how far down must the piston be pushed before air will flow into the tire?

100. An ideal gas in a constant-volume gas thermometer (Fig. 13.11) is held at a volume of 0.500 L. As the temperature of the gas is increased by 20.0°C, the mercury level on the right side of the manometer must rise by 8.00 mm in order to keep the gas volume constant. (a) What is the slope of a graph of *P* versus *T* for this gas (in mm Hg/°C)? (b) How many moles of gas are present?

101. A cylinder with an interior cross-sectional area of 70.0 cm^2 has a moveable piston of mass 5.40 kg at the top that can move up and down without friction. The cylinder contains 2.25×10^{-3} mol of an ideal gas at 23.0°C. (a) What is the volume of the gas when the piston is in equilibrium? Assume the air pressure outside the cylinder is 1.00 atm. (b) By what factor does the volume change if the gas temperature is raised to 223.0°C and the piston moves until it is again in equilibrium?

102. Estimate the average distance between molecules in air at 0.0°C and 1.00 atm.

103. Show that, in two gases at the same temperature, the rms speeds are inversely proportional to the square root of the molecular masses:

$$\frac{(v_{\text{rms}})_1}{(v_{\text{rms}})_2} = \sqrt{\frac{m_2}{m_1}}$$

104. The alveoli (see Section 13.8) have an average radius of 0.125 mm and are approximately spherical. If the pressure in the sacs is 1.00×10^5 Pa, and the temperature is 310 K (average body temperature), how many air molecules are in an alveolus?

105. A 10.0 L vessel contains 12 g of N_2 gas at 20°C. (a) Estimate the nearest-neighbor distance. (b) Can the gas be considered to be dilute? [*Hint*: Compare the nearest-neighbor distance to the diameter of an N_2 molecule, about 0.3 nm.]

106. During hibernation, an animal's metabolism slows down, and its body temperature lowers. For example, a California ground squirrel's body temperature lowers from 40.0°C to 10.0°C during hibernation. If we assume that the air in the squirrel's lungs is 75.0% N_2 and 25.0% O_2, by how much will the rms speed of the air molecules in the lungs have decreased during hibernation?

107. ✦ A steel ring of inner diameter 7.000 00 cm at 20.0°C is to be heated and placed over a brass shaft of outer diameter 7.002 00 cm at 20.0°C. (a) To what temperature must the ring be heated to fit over the shaft? The shaft remains at 20.0°C. (b) Once the ring is on the shaft and has cooled to 20.0°C, to what temperature must the ring plus shaft combination be cooled to allow the ring to slide off the shaft again?

108. ✦ The inner tube of a Pyrex glass mercury thermometer has a diameter of 0.120 mm. The bulb at the bottom of the thermometer contains 0.200 cm^3 of mercury. How far will the thread of mercury move for a change of 1.00°C? Remember to take into account the expansion of the glass. (connect tutorial: thermometer)

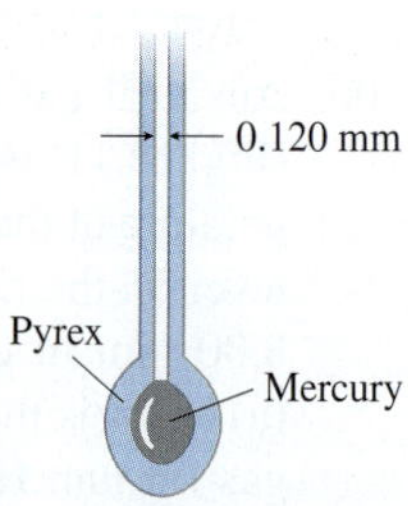

109. ✦ A wine barrel has a diameter at its widest point of 134.460 cm at a temperature of 20.0°C. A circular iron band, of diameter 134.448 cm, is to be placed around the barrel at the widest spot. The iron band is 5.00 cm wide and 0.500 cm thick. (a) To what temperature must the band be heated to be able to fit it over the barrel? (b) Once the band is in place and cools to 20.0°C, what will be the tension in the band?

110. ✦ A bimetallic strip is made from metals with expansion coefficients α_1 and α_2 (with $\alpha_2 > \alpha_1$). The thickness of each layer is s. At some temperature T_0, the bimetallic strip is relaxed and straight. (a) Show that, at temperature $T_0 + \Delta T$, the radius of curvature of the strip is

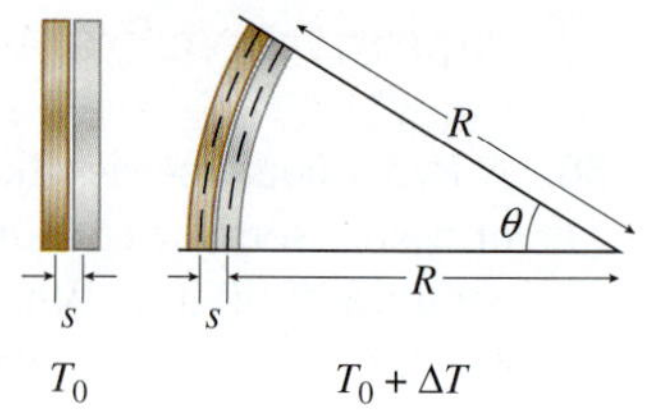

$$R \approx \frac{s}{(\alpha_2 - \alpha_1)\Delta T}$$

[*Hint*: At T_0, the lengths of the two layers are the same. At temperature $T_0 + \Delta T$, the layers form circular arcs of radii R and $R + s$, which subtend the same angle θ. Assume a small ΔT so that $\alpha\ \Delta T \ll 1$ (for either value of α).] (b) If the layers are made of iron and brass, with $s = 0.1$ mm, what is R for $\Delta T = 20.0$°C?

Review and Synthesis

111. Michael has set the gauge pressure of the tires on his car to 36.0 psi (lb/in.^{2}). He draws chalk lines around the edges of the tires where they touch the driveway surface to measure the area of contact between the tires and the ground. Each front tire has a contact area of 24.0 in.^{2} and each rear tire has a contact area of 20.0 in.^{2} (a) What is the weight (in lb) of the car? (b) The center-to-center distance between front and rear tires is 7.00 ft. Taking the straight line between the centers of the tires on the left side (driver's side) to be the y-axis with the origin at the center of the front left tire (positive direction pointing forward), what is the y-coordinate of the car's CM?

112. (a) Calculate Earth's escape speed—the minimum speed needed for an object near the surface to escape Earth's gravitational pull. [*Hint*: Use conservation of energy and ignore air resistance.] (b) Calculate the average speed of a hydrogen molecule (H_2) at 0°C. (c) Calculate the average speed of an oxygen molecule (O_2) at 0°C. (d) Use your answers from parts (a) through (c) along with what you know about the distribution of molecular speeds to explain why Earth's atmosphere contains plenty of oxygen but almost no hydrogen.

113. A long, narrow steel rod of length 2.5000 m at 25°C is oscillating as a pendulum about a horizontal axis through one end. If the temperature changes to 0°C, what will be the fractional change in its period?

114. A temperature change ΔT causes a volume change ΔV but has no effect on the mass of an object. (a) Show that the change in density $\Delta\rho$ is given by $\Delta\rho = -\beta\rho\,\Delta T$. (b) Find the fractional change in density $(\Delta\rho/\rho)$ of a brass sphere when the temperature changes from 32°C to −10.0°C.

115. A diver rises quickly to the surface from a 5.0 m depth. If she did not exhale the gas from her lungs before rising, by what factor would her lungs expand?

Assume the temperature to be constant and the pressure in the lungs to match the pressure outside the diver's body. The density of seawater is 1.03×10^3 kg/m^3.

116. A scuba diver has an air tank with a volume of 0.010 m^3. The air in the tank is initially at a pressure of 1.0×10^7 Pa. Assuming that the diver breathes 0.500 L/s of air, find how long the tank will last at depths of (a) 2.0 m and (b) 20.0 m. (Make the same assumptions as in Example 13.6.)

117. A sealed cylinder contains a sample of ideal gas at a pressure of 2.0 atm. The rms speed of the molecules is v_0. (a) If the rms speed is then reduced to $0.90v_0$, what is the pressure of the gas? (b) By what percentage does the speed of sound in the gas change?

118. ✦ Estimate the percentage of the O_2 molecules in air at 30°C that are moving faster than the speed of sound in air at that temperature (see Fig. 13.13).

119. The diameter of an oxygen (O_2) molecule is approximately 0.3 nm. For an oxygen molecule in still air at 20°C, estimate the average magnitudes of these quantities during a 1.0 s time interval: (a) the distance traveled between collisions with other molecules; (b) the number of collisions; (c) the total distance traveled; (d) the displacement.

120. ✦ A 12.0 cm cylindrical chamber has an 8.00 cm diameter piston attached to one end. The piston is connected to an ideal spring as shown. Initially, the gas inside the chamber is at atmospheric pressure and 20.0°C and the spring is not compressed. When a total of 6.50×10^{-2} mol of gas is added to the chamber at 20.0°C, the spring compresses a distance of $\Delta x = 5.40$ cm. What is the spring constant of the spring?

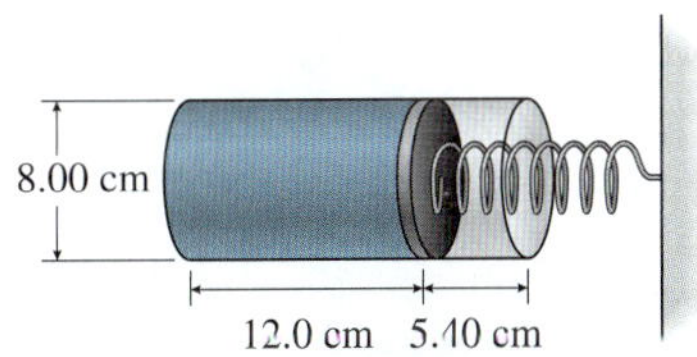

Answers to Practice Problems

13.1 37.0°C; 310.2 K

13.2 0.60 mm longer; 1.5 mm shorter

13.3 0.26 L

13.4 3.34×10^{28} molecules/m^3

13.5 7.9% of the air molecules; 189 kPa (27 lb/in.^{2})

13.6 640 kPa

13.7 $\langle K_{tr} \rangle = 6.07 \times 10^{-21}$ J (same as O_2) and $v_{rms} = 408$ m/s (lower than that of O_2 since the CO_2 molecule is more massive)

13.8 6% decrease

13.9 1.0×10^{-6} s

Answers to Checkpoints

13.3 From Table 13.2, $\alpha = 12 \times 10^{-6}$ K^{-1}. The temperature change is -50°C $= -50$ K and the fractional length change is $\Delta L/L_0 = \alpha\,\Delta T = -6.0 \times 10^{-4}$. Then $\Delta L = -6.0 \times 10^{-4} \times 150.00$ m $= -0.090$ m. The tower is 9.0 cm shorter.

13.4 (a) The molar mass is 44.0 g/mol, so one CO_2 molecule has a mass of 44.0 u. (b) 3.00 mol of CO_2 have a mass of (3.00 mol) × (44.0 g/mol) = 132 g.

13.5 (a) The temperatures do not have to be the same, because they could have different numbers of molecules (N) or moles (n). (b) If the temperatures are the same, then they have the same number of molecules, so they have the same number density N/V. They would have the same mass density only if their molar masses are the same.

13.6 The average translational kinetic energy of an ideal gas depends only on *absolute* temperature. The H_2 is at 20°C = 293 K, so to have twice the translational kinetic energy, the O_2 must be at 2 × 293 K = 586 K = 313°C.

CHAPTER

14 Heat

The weather forecast predicts a late spring hard freeze one night; the temperature is to fall several degrees below 0°C and the apple crop is in danger of being ruined. To protect the tender buds, farmers rush out and spray the trees with water. How does that protect the buds? (See pp. 511-512 for the answer.)

BIOMEDICAL APPLICATIONS

- Why ponds freeze from the top down (Sec. 14.5)
- Forced convection in the human body (Sec. 14.7)
- Convection and radiation in global climate change (Secs. 14.7, 14.8)
- Thermography (Sec. 14.8)
- Heat loss and gain by plants and animals (Exs. 14.12 and 14.14; PPs 14.13 and 14.14; Probs. 30, 31, 36, 46, 47, 51, 78–85, 91–93, 98, 99, 103, 104)
- Metabolism (Probs. 17, 22, 23, 91, 93, 103, 104)
- Insulation value of fur, blubber, down, wool clothing, skin (Probs. 63–67)

Concepts & Skills to Review

- energy conservation (Chapter 6)
- thermal equilibrium (Section 13.1)
- absolute temperature and the ideal gas law (Section 13.5)
- kinetic theory of the ideal gas (Section 13.6)

14.1 INTERNAL ENERGY

From Section 13.6, the average translational kinetic energy $\langle K_{tr} \rangle$ of the molecules of an ideal gas is proportional to the absolute temperature of the gas:

$$\langle K_{tr} \rangle = \tfrac{3}{2}kT \qquad (13\text{-}20)$$

The molecules move about in random directions even though, on a macroscopic scale, the gas is neither moving nor rotating. Equation (13-20) also gives the average translational kinetic energy of the random motion of molecules in liquids, solids, and nonideal gases except at very low temperatures. This random microscopic kinetic energy is *part* of what we call the **internal energy** of the system:

Definition of Internal Energy

The internal energy of a system is the total energy of all of the molecules in the system *except* for the macroscopic kinetic energy (kinetic energy associated with macroscopic translation or rotation) and the external potential energy (energy due to external interactions).

A **system** is whatever we define it to be: one object or a group of objects. Everything that is not part of the system is considered to be external to the system, or in other words, in the surroundings of the system.

CONNECTION:

We've used the idea of a *system* before, for instance when finding the net external force on a system or the momentum change of a system.

Internal energy includes

- Translational and rotational kinetic energy of molecules *due to their individual random motions.*
- Vibrational energy—both kinetic and potential—of molecules and of atoms within molecules due to random vibrations about their equilibrium points.
- Potential energy due to interactions between the atoms and molecules of the system.
- Chemical and nuclear energy—the kinetic and potential energy associated with the binding of atoms to form molecules, the binding of electrons to nuclei to form atoms, and the binding of protons and neutrons to form nuclei.

Internal energy does *not* include

- The kinetic energy of the molecules due to translation, rotation, or vibration of the whole system or of a macroscopic part of the system.
- Potential energy due to interactions of the molecules of the system with something outside of the system (such as a gravitational field due to something outside of the system).

CONNECTION:

Revisit Table 6.1 for an overview of the forms of energy discussed in this book.

Example 14.1

Dissipation of Energy by Friction

A block of mass 10.0 kg starts at point A at a height of 2.0 m above the horizontal and slides down a frictionless incline (Fig. 14.1). It then continues sliding along the horizontal surface of a table that has friction. The block comes to rest at point C, a distance of 1.0 m along the table surface. How much has the internal energy of the system (block + table) increased?

continued on next page

Example 14.1 continued

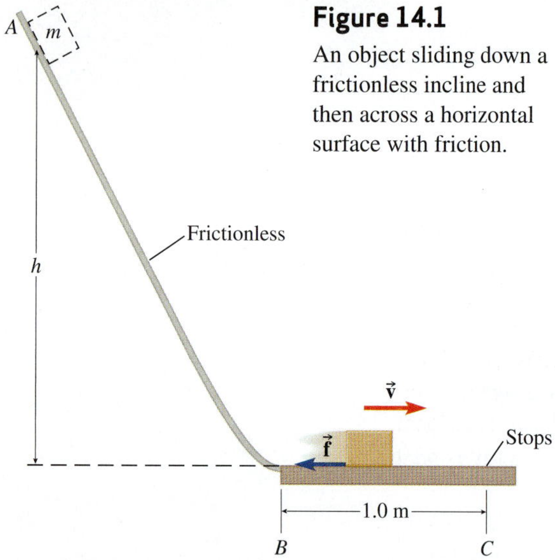

Figure 14.1
An object sliding down a frictionless incline and then across a horizontal surface with friction.

Strategy Gravitational potential energy is converted to macroscopic translational kinetic energy as the block's speed increases. Friction then converts this macroscopic kinetic energy into internal energy—some of it in the block and some in the table. Since total energy is conserved, the increase in internal energy is equal to the decrease in gravitational potential energy:

decrease in PE from A to B = increase in KE from A to B
= decrease in KE from B to C
= increase in internal energy from B to C

Solution The initial potential energy (taking $U_g = 0$ at the horizontal surface) is

$$U_g = mgh = 10.0 \text{ kg} \times 9.8 \text{ m/s}^2 \times 2.0 \text{ m} = 200 \text{ J}$$

The final potential energy is zero. The initial and final translational kinetic energies of the block are both zero. Ignoring the small transfer of energy to the air, the increase in the internal energy of the block and table is 200 J.

Discussion We do not know how much of this internal energy increase appears in the object and how much in the table; we can only find the total. We call friction a *nonconservative* force, but that only means that *macroscopic mechanical* energy is not conserved; total energy is always conserved. Friction merely converts some macroscopic mechanical energy into internal energy of the block and the table. This internal energy increase manifests itself as a slight temperature increase. We often say that mechanical energy is *dissipated* by friction or other nonconservative forces; in other words, energy in an ordered form (translational motion of the block) has been changed into disordered energy (random motion of molecules within the block and table).

Practice Problem 14.1 On the Rebound

If a rubber ball of mass 1.0 kg is dropped from a height of 2.0 m and rebounds on the first bounce to 0.75 of the height from which it was dropped, how much energy is dissipated during the collision with the floor?

A change in the internal energy of a system does not always cause a temperature change. As we explore further in Section 14.5, the internal energy of a system can change while the temperature of the system remains constant—for instance, when ice melts.

Conceptual Example 14.2

Internal Energy of a Bowling Ball

A bowling ball at rest has a temperature of 18°C. The ball is then rolled down a bowling alley. Ignoring the dissipation of energy by friction and drag forces, is the internal energy of the ball higher, lower, or the same as when the ball was at rest? Is the temperature of the ball higher, lower, or the same as when the ball was at rest?

Strategy, Solution, and Discussion The only change is that the ball is now rolling—the ball has macroscopic translational and rotational kinetic energy. However, the definition of *internal energy* does *not* include the kinetic energy of the molecules due to translation, rotation, or vibration *of the system as a whole*. Therefore, the *internal* energy of the ball is the same. Temperature is associated with the average translational kinetic energy due to the *individual random* motions of molecules; the temperature is still 18°C.

Conceptual Practice Problem 14.2 Total Translational KE

Is the *total* translational kinetic energy of the molecules in the ball higher, lower, or the same as when the ball was at rest?

14.2 HEAT

We defined heat in Section 13.1:

Definition of Heat

Heat is energy in transit between two objects or systems due to a temperature difference between them.

Many eighteenth-century scientists thought that heat was a fluid, which they called "caloric." The flow of heat into an object was thought to cause the object to expand in volume in order to accommodate the additional fluid; why no mass increase occurred was a mystery. Now we know that heat is not a substance but is a flow of energy. One experiment that led to this conclusion was carried out by Count Rumford (Benjamin Thompson, 1753–1814). While supervising the boring of cannon barrels, he noted that the drill doing the boring became quite hot. At the time it was thought that the grinding up of the cannon metal into little pieces caused caloric to be released because the tiny bits of metal could not hold as much caloric as the large piece from which they came. But Rumford noticed that the drill got hot even when it became so dull that metal was no longer being bored out of the cannon and that he could create a limitless amount of what we now call internal energy. He decided that "heat" must be a form of microscopic motion instead of a material substance.

It was not until later experiments were done by James Prescott Joule (1818–1889) that Rumford's ideas were finally accepted. In his most famous experiment (Fig. 14.2), Joule showed that a temperature increase can be caused by mechanical means. In a series of such experiments, Joule determined the "mechanical equivalent of heat," or the amount of mechanical work required to produce the same effect on a system as a given amount of heat. In those days heat was measured in calories, where one calorie was defined as the heat required to change the temperature of 1 g of water by 1°C (specifically from 14.5 to 15.5°C). Joule's experimental results were within 1% of the currently accepted value, which is

$$1 \text{ cal} = 4.186 \text{ J} \tag{14-1}$$

The Calorie (with an uppercase letter C) used by dietitians and nutritionists is actually a kilocalorie:

$$1 \text{ Calorie} = 1 \text{ kcal} = 10^3 \text{ cal} = 4186 \text{ J}$$

Although the calorie is still used, the SI unit for internal energy and for heat is the same as that used for all forms of energy and all forms of energy transfer: the joule.

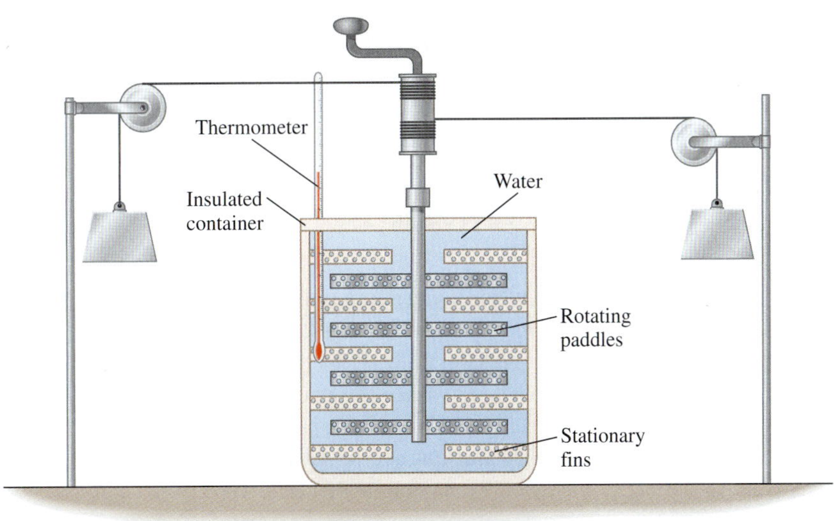

Figure 14.2 Joule's experiment. As the two masses fall, they cause paddles to rotate within an insulated container (not to scale) filled with water. The paddles agitate the water and cause its temperature to rise. By measuring the distance through which the masses fell and the temperature change of the known quantity of water, Joule determined the mechanical work done and the internal energy increase of the water.

CONNECTION:

Heat, like work, is a kind of energy transfer.

Heat and Work Heat and work are similar in that both describe a particular kind of energy *transfer*. Work is an energy transfer due to a force acting through a displacement. Heat is a microscopic form of energy transfer involving large numbers of particles; the exchange of energy occurs due to the individual interactions of the particles. No macroscopic displacement occurs when heat flows, and no macroscopic force is exerted by one object on the other.

It does not make sense to say that a system *has* 15 kJ of heat, just as it does not make sense to say that a system *has* 15 kJ of work. Similarly, we cannot say that the heat of a system has changed (nor that the work of a system has changed). A system can possess *energy* in various forms (including internal energy), but it cannot possess heat or work. Heat and work are two ways of *transferring* energy from one system to another. Joule's experiments showed that a quantity of work done on a system or the same quantity of heat flowing into the system causes the same increase in the system's internal energy. If the internal energy increase comes from mechanical work, as from Joule's paddle wheel, no heat flow occurs.

CHECKPOINT 14.2

Take a rubber band and stretch it rapidly several times. Then hold it against your wrist or your lip. In everyday language, you might say the rubber band "heats up." Is the temperature increase caused by heat flow into the rubber band? If not, what has happened?

Direction of Heat Flow *Heat flows spontaneously from a system at higher temperature to one at lower temperature*. Temperature is associated with the microscopic translational kinetic energy of the molecules; thus, the flow of heat tends to equalize the average microscopic translational kinetic energy of the molecules. When two systems are in thermal contact and no net heat flow occurs, the systems are in thermal equilibrium and have the same temperature.

Example 14.3

A Joule Experiment

In an experiment similar to that done by Joule, an object of mass 12.0 kg descends a distance of 1.25 m at constant speed while causing the rotation of a paddle wheel in an insulated container of water. If the descent is repeated 20.0 times, what is the internal energy increase of the water in joules?

Strategy Each time the object descends, it converts gravitational potential energy into kinetic energy of the paddle wheel, which in turn agitates the water and converts kinetic energy into internal energy.

Solution The change in gravitational potential energy during 20.0 downward trips is

$$\Delta U_g = mg\,\Delta h$$
$$= 12.0 \text{ kg} \times 9.80 \text{ N/kg} \times 20.0 \text{ descents} \times 1.25 \text{ m/descent}$$
$$= 2.94 \text{ kJ}$$

If all of this energy goes into the water, the internal energy increase of the water is 2.94 kJ.

Discussion To perform an experiment like Joule's, we can vary the amount of energy delivered to the water. One way is to change the number of times the object is allowed to descend. Other possibilities include varying the mass of the descending object or raising the apparatus so that the object can descend a greater distance. All of these variations allow a change in the amount of gravitational potential energy converted into internal energy without requiring any changes in the complicated mechanism involving the paddle wheel.

Practice Problem 14.3 Temperature Change of the Water

If the water temperature in the insulated container is found to have increased 2.0°C after 20.0 descents of the falling object, what mass of water is in the container? Assume all of the internal energy increase appears in the water (ignore any internal energy change of the paddle wheel itself). [*Hint*: Recall that the calorie was originally defined as the heat required to change the temperature of 1 g of water by 1°C.]

The Cause of Thermal Expansion

If not to accommodate additional "caloric," then why do objects generally expand when their temperatures increase? (See Section 13.3.) An object expands when the *average* distance between the atoms (or molecules) increases. The atoms are not at rest; even in a solid, where each atom has a fixed equilibrium position, they *vibrate* to and fro about their equilibrium positions. The energy of vibration is part of the internal energy of the object. When heat flows into the object, raising its temperature, the internal energy increases. Some of the increase goes into vibration, so the average vibrational energy of an atom increases with increasing temperature.

The average distance between atoms usually increases with increasing vibrational energy because the forces between atoms are highly asymmetrical. Two atoms separated by *less* than their equilibrium distance repel each other *strongly*, whereas two atoms separated by *more* than their equilibrium distance attract each other much less strongly. Therefore, as vibrational energy increases, the maximum distance between the atoms increases more than the minimum distance decreases; the *average* distance between the atoms increases.

The coefficient of expansion varies from material to material because the strength of the interatomic (or intermolecular) bonds varies. As a general rule, the stronger the atomic bond, the smaller the coefficient of expansion. Liquids have much greater coefficients of volume expansion than do solids because the molecules are more loosely bound in a liquid than in a solid.

CONNECTION:

Section 13.3 discussed thermal expansion. Now we discuss *why* the expansion occurs.

14.3 HEAT CAPACITY AND SPECIFIC HEAT

Heat Capacity

Suppose we have a system on which no mechanical work is done, but we allow heat to flow into the system by placing it in thermal contact with another system at higher temperature. (In Chapter 15, we consider cases in which both work and heat change the internal energy of a system.) As the internal energy of the system increases, its temperature increases (provided that no part of the system undergoes a change of phase, such as from solid to liquid). If heat flows *out* of the system rather than into it, the internal energy of the system decreases. We account for that possibility by making Q negative if heat flows out of the system; since Q is defined as the heat *into* the system, a negative heat represents heat flow *out of* the system.

For a large number of substances, under normal conditions, the temperature change ΔT is approximately proportional to the heat Q. The constant of proportionality is called the **heat capacity** (symbol C):

Definition of heat capacity

$$Q = C\Delta T \qquad (14\text{-}2)$$

The heat capacity depends both on the substance and on how much of it is present: 1 cal of heat into 1 g of water causes a temperature increase of 1°C, but 1 cal of heat into 2 g of water causes a temperature increase of 0.5°C. The SI unit of heat capacity is J/K. We can write J/K or J/°C interchangeably since only temperature *changes* are involved; a temperature change of 1 K is equivalent to a temperature change of 1°C.

The term *heat capacity* is unfortunate since it has nothing to do with a capacity to *hold* heat, or a limited ability to absorb heat, as the name seems to imply. Instead, it relates the heat into a system to the temperature increase. Think of heat capacity as a measure of how much heat must flow into or out of the system to produce a given temperature change.

Table 14.1 Specific Heats of Common Substances at 1 atm and 20°C

Substance	Specific Heat $\left(\frac{\text{kJ}}{\text{kg}\cdot\text{K}}\right)$	Substance	Specific Heat $\left(\frac{\text{kJ}}{\text{kg}\cdot\text{K}}\right)$
Gold	0.128	Pyrex glass	0.75
Lead	0.13	Granite	0.80
Mercury	0.139	Marble	0.86
Silver	0.235	Aluminum	0.900
Brass	0.384	Air (50°C)	1.05
Copper	0.385	Wood (average)	1.68
Iron	0.44	Steam (110°C)	2.01
Steel	0.45	Ice (0°C)	2.1
Flint glass	0.50	Alcohol (ethyl)	2.4
Crown glass	0.67	Human tissue (average)	3.5
Vycor	0.74	Water (15°C)	4.186

Specific Heat

The heat capacity of the water in a drinking glass is much smaller than the heat capacity of the water in Lake Superior. Since the heat capacity of a system is proportional to the *mass* of the system, the **specific heat capacity** (symbol c) of a substance is defined as the heat capacity per unit mass:

Definition of specific heat capacity

$$c = \frac{C}{m} = \frac{Q}{m\,\Delta T} \quad (14\text{-}3)$$

Specific heat capacity is often abbreviated to *specific heat*. The SI units of specific heat are J/(kg·K). In SI units, the specific heat is the number of joules of heat required to produce a 1 K temperature change in 1 kg of the substance. Again, since only temperature changes are involved, we can equivalently write J/(kg·°C).

Table 14.1 lists specific heats for some common substances at 1 atm and 20°C (unless otherwise specified). For the range of temperatures in our examples and problems, assume these specific heat values to be valid. Note that water has a relatively large specific heat compared with most other substances. The relatively large specific heat of water causes the oceans to warm slowly in the spring and to cool slowly as winter approaches, moderating the temperature along the coast.

Rearrangement of Eq. (14-3) leads to an expression for the heat required to produce a known temperature change in a system:

$$Q = mc\,\Delta T \quad (14\text{-}4)$$

Note that in Eqs. (14-3) and (14-4), the sign convention for Q is consistent: a temperature increase ($\Delta T > 0$) is caused by heat flowing *into* the system ($Q > 0$), whereas a temperature decrease ($\Delta T < 0$) is caused by heat flowing *out* of the system ($Q < 0$).

Equations (14-2) through (14-4) apply when no phase change occurs. The value of the specific heat is different for different phases of the same substance. That's why Table 14.1 lists different values for ice, liquid water, and steam.

CHECKPOINT 14.3

A 10 g brass washer at 80°C is dropped into a container of 100 g of water initially at 20°C. Ignoring heat flow to or from the surroundings, will the equilibrium temperature be less than, equal to, or greater than 50°C? Explain.

Example 14.4

Heating Water in a Saucepan

A saucepan containing 5.00 kg of water initially at 20.0°C is heated over a gas burner for 10.0 min. The final temperature of the water is 30.0°C. (a) What is the internal energy increase of the water? (b) What is the expected final temperature if the water were heated for an additional 5.0 min? (c) Is it possible to estimate the flow of heat from the burner during the first 10.0 min?

Strategy We are interested in the internal energy and the temperature of the *water*, so we define a system that consists of the water in the saucepan. Although the pan is also heated, it is not part of this system. The pan, the burner, and the room are all outside the system.

Since no work is done on the water, the internal energy increase is equal to the heat flowing into the water. The heat can be found from the mass of the water, the specific heat of water, and the temperature change. As long as the burner delivers heat at a constant rate, we can find the additional heat delivered in the additional time. Since the temperature change is proportional to the heat delivered, the temperature changes at a constant rate (a constant number of °C per minute). So, in half the time, half as much energy is delivered and the temperature change is half as much.

Solution (a) First find the temperature change:

$$\Delta T = T_f - T_i = 30.0°\text{C} - 20.0°\text{C} = 10.0 \text{ K}$$

(A change of 10.0°C is equivalent to a change of 10.0 K.) The increase in the internal energy of the water is

$$\Delta U = Q = mc\,\Delta T$$
$$= 5.00 \text{ kg} \times 4.186 \text{ kJ/(kg·K)} \times 10.0 \text{ K} = 209 \text{ kJ}$$

(b) We assume that the heat delivered is proportional to the elapsed time. The temperature change is proportional to the energy delivered, so if the temperature changes 10.0°C in 10.0 min, it changes an additional 5.0°C in an additional 5.0 min. The final temperature is

$$T = 20.0°\text{C} + 15.0°\text{C} = 35.0°\text{C}$$

(c) Not all of the heat flows into the water. Heat also flows from the burner into the saucepan and into the room. All we can say is that *more than* 209 kJ of heat flows from the burner during the 10.0 min.

Discussion As a check, the heat capacity of the water is 5.00 kg × 4.186 kJ/(kg·K) = 20.9 kJ/K; 20.9 kJ of heat must flow for each 1.0 K change in temperature. Since the temperature change is 10.0 K, the heat required is

$$20.9 \text{ kJ/K} \times 10.0 \text{ K} = 209 \text{ kJ}$$

Practice Problem 14.4 Price of a Bubble Bath

If the cost of electricity is $0.080 per kW·h, what does it cost to heat 160 L of water for a bubble bath from 10.0°C (the temperature of the well water entering the house) to 70.0°C? [*Hint*: 1 L of water has a mass of 1 kg. 1 kW·h = 1000 J/s × 3600 s.]

Heat Flow with More Than Two Objects Suppose some water is heated in a large iron pot by dropping a hot piece of copper into the pot. We can define the system to be the water, the copper, and the iron pot; the environment is the room containing the system. Heat continues to flow among the three substances (iron pot, water, copper) until thermal equilibrium is reached—that is, until all three substances are at the same temperature. If losses to the environment are negligible, all the heat that flows out of the copper flows into either the iron or the water:

$$Q_{\text{Cu}} + Q_{\text{Fe}} + Q_{\text{H}_2\text{O}} = 0$$

In this case, Q_{Cu} is negative since heat flows *out* of the copper; Q_{Fe} and $Q_{\text{H}_2\text{O}}$ are positive since heat flows into both the iron and the water.

CONNECTION:

Here we apply the principle of energy conservation.

Calorimetry

A calorimeter is an insulated container that enables the careful measurement of heat (Fig. 14.3). The calorimeter is designed to minimize the heat flow to or from the surroundings. A typical constant volume calorimeter, called a *bomb calorimeter*, consists of a hollow aluminum cylinder of known mass containing a known quantity of water;

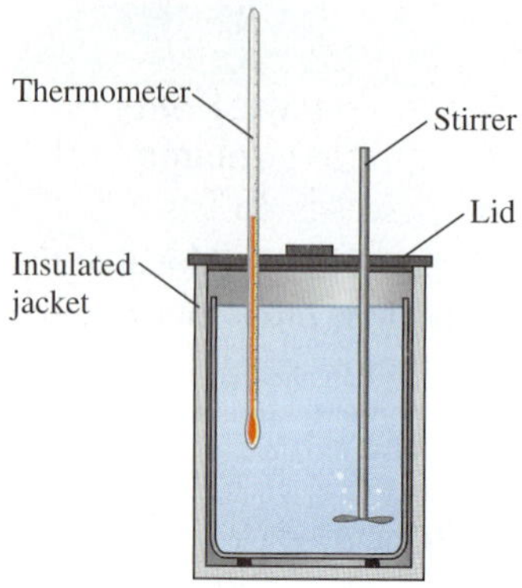

Figure 14.3 A calorimeter.

the cylinder is inside a larger aluminum cylinder with insulated walls. An evacuated space separates the two cylinders. An insulated lid fits over the opening of the cylinders; often there are two small holes in the lid, one for a thermometer to be inserted into the contents of the inner cylinder and one for a stirring device to help the contents reach equilibrium faster.

Suppose an object at one temperature is placed in a calorimeter with the water and aluminum cylinder at another temperature. By conservation of energy, all the heat that flows out of one substance ($Q < 0$) flows into some other substance ($Q > 0$). If no heat flows to or from the environment, the total heat into the object, water, and aluminum must equal zero:

$$Q_o + Q_w + Q_a = 0$$

Example 14.5 illustrates the use of a calorimeter to measure the specific heat of an unknown substance. The measured specific heat can be compared with a table of known values to help identify the substance.

Example 14.5

Specific Heat of an Unknown Metal

A sample of unknown metal of mass 0.550 kg is heated in a pan of hot water until it is in equilibrium with the water at a temperature of 75.0°C. The metal is then carefully removed from the heat bath and placed into the inner cylinder of an aluminum calorimeter that contains 0.500 kg of water at 15.5°C. The mass of the inner cylinder is 0.100 kg. When the contents of the calorimeter reach equilibrium, the temperature inside is 18.8°C. Find the specific heat of the metal sample and determine whether it could be any of the metals listed in Table 14.1.

Strategy Heat flows from the sample to the water and to the aluminum until thermal equilibrium is reached, at which time all three have the same temperature. We use subscripts to keep track of the three heat flows and three temperature changes. Let T_f be the final temperature of all three. Initially, the water and aluminum are both at 15.5°C and the sample is at 75.0°C. When thermal equilibrium is reached, all three are at 18.8°C. We assume negligible heat flow to the environment—in other words, that no heat flows into or out of the system consisting of aluminum + water + sample.

Solution Heat flows out of the sample ($Q_s < 0$) and into the water and aluminum cylinder ($Q_w > 0$ and $Q_a > 0$). Assuming no heat into or out of the surroundings,

$$Q_s + Q_w + Q_a = 0$$

For each substance, the heat is related to the temperature change. Substituting $Q = mc\,\Delta T$ for each gives

$$m_s c_s\,\Delta T_s + m_w c_w\,\Delta T_w + m_a c_a\,\Delta T_a = 0 \qquad (1)$$

A table helps organize the given information:

	Sample	H_2O	Al
Mass (m)	0.550 kg	0.500 kg	0.100 kg
Specific Heat (c)	c_s (unknown)	4.186 kJ/(kg·°C)	0.900 kJ/(kg·°C)
Heat Capacity (mc)	0.550 kg × c_s	2.093 kJ/°C	0.0900 kJ/°C
T_i	75.0°C	15.5°C	15.5°C
T_f	18.8°C	18.8°C	18.8°C
ΔT	−56.2°C	3.3°C	3.3°C

Solving Eq. (1) for c_s,

$$c_s = -\frac{m_w c_w\,\Delta T_w + m_a c_a\,\Delta T_a}{m_s\,\Delta T_s}$$

$$= -\frac{(2.093\text{ kJ/°C})(3.3\text{°C}) + (0.0900\text{ kJ/°C})(3.3\text{°C})}{(0.550\text{ kg})(-56.2\text{°C})}$$

$$= 0.23\,\frac{\text{kJ}}{\text{kg·°C}}$$

By comparing this result with the values in Table 14.1, it appears that the unknown sample could be silver.

Discussion As a quick check, the heat capacity of the sample is approximately $\frac{1}{17}$ that of the water since its temperature change is 56.2°C/3.3°C ≈ 17 times as much—ignoring the small heat capacity of the aluminum. Since the masses of the water and sample are about equal, the specific heat of the sample is roughly $\frac{1}{17}$ that of the water:

$$\frac{1}{17} \times 4.186\,\frac{\text{kJ}}{\text{kg·°C}} = 0.25\,\frac{\text{kJ}}{\text{kg·°C}}$$

That is quite close to our answer.

Practice Problem 14.5 Final Temperature

If 0.25 kg of water at 90.0°C is added to 0.35 kg of water at 20.0°C in an aluminum calorimeter with an inner cylinder of mass 0.100 kg, find the final temperature of the mixture.

14.4 SPECIFIC HEAT OF IDEAL GASES

Since the average translational kinetic energy of a molecule in an ideal gas is

$$\langle K_{tr} \rangle = \tfrac{3}{2}kT \qquad (13\text{-}20)$$

the *total* translational kinetic energy of a gas containing N molecules (n moles) is

$$K_{tr} = \tfrac{3}{2}NkT = \tfrac{3}{2}nRT$$

Suppose we allow heat to flow into a *monatomic* ideal gas—one in which the gas molecules consist of single atoms—while keeping the volume of the gas constant. Since the volume is constant, no work is done on the gas, so the change in the internal energy is equal to the heat. If we think of the atoms as point particles, the only way for the internal energy to change when heat flows into the gas is for the translational kinetic energy of the atoms to change. The rest of the internal energy is "locked up" in the atoms and does not change unless something else happens, such as a change of phase (for example, from gas to liquid) or a chemical reaction—neither of which can happen in an ideal gas. Then

$$Q = \Delta K_{tr} = \tfrac{3}{2}nR\,\Delta T \qquad (14\text{-}5)$$

From Eq. (14-5), we can find the specific heat of the monatomic ideal gas. However, with gases it is more convenient to define the **molar specific heat** at constant volume (C_V) as

$$C_V = \frac{Q}{n\,\Delta T} \qquad (14\text{-}6)$$

The subscript "V" is a reminder that the volume of the gas is held constant during the heat flow. The molar specific heat is the heat capacity *per mole* rather than *per unit mass*. In one case, we measure the amount of substance by the number of moles; in the other case, by the mass.

From Eqs. (14-5) and (14-6), we can find the molar specific heat of a monatomic ideal gas:

$$Q = \tfrac{3}{2}nR\,\Delta T = nC_V\,\Delta T$$

$$C_V = \tfrac{3}{2}R = 12.5\,\frac{\text{J/K}}{\text{mol}} \quad (\textit{monatomic}\text{ ideal gas}) \qquad (14\text{-}7)$$

A glance at Table 14.2 shows that this calculation is remarkably accurate at room temperature for monatomic gases.

Diatomic gases have larger molar specific heats than monatomic gases. Why? We cannot model the diatomic molecule as a point mass; the two atoms in the molecule are separated, giving the molecule a much larger rotational inertia about two perpendicular axes (Fig. 14.4). The molar specific heat is *larger* because not all of the internal energy increase goes into the translational kinetic energy of the molecules; some goes into rotational kinetic energy.

Table 14.2
Molar Specific Heats at Constant Volume of Gases at 25°C

	Gas	$C_V\left(\frac{\text{J/K}}{\text{mol}}\right)$
Monatomic	He	12.5
	Ne	12.7
	Ar	12.5
Diatomic	H_2	20.4
	N_2	20.8
	O_2	21.0
Polyatomic	CO_2	28.2
	N_2O	28.4

CONNECTION:

Specific heat and molar specific heat can be thought of as the same quantity—heat capacity per amount of substance—expressed in different units.

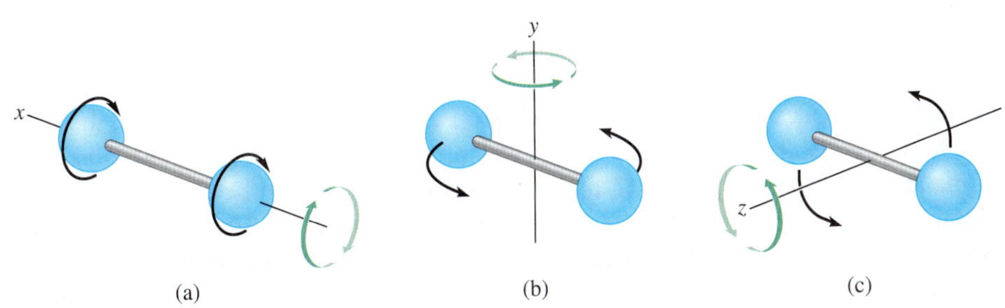

Figure 14.4 Rotation of a model diatomic molecule about three perpendicular axes. The rotational inertia about the x-axis (a) is negligible, so we can ignore rotation about this axis. The rotational inertias about the y- and z-axes (b) and (c) are much larger than for a single atom of the same mass because of the larger distance between the atoms and the axis of rotation.

As we show in Chapter 15, the molar specific heat of a diatomic ideal gas at room temperature is approximately

$$C_V = \tfrac{5}{2}R = 20.8\ \frac{\text{J/K}}{\text{mol}} \qquad (\textit{diatomic}\ \text{ideal gas at room temperature}) \quad (14\text{-}8)$$

Why $\frac{5}{2}R$ instead of $\frac{3}{2}R$? The diatomic molecule has rotational kinetic energy about two perpendicular axes (Fig. 14.4b and c) in addition to translational kinetic energy associated with motion in three independent directions. Thus, the diatomic molecule has five ways to "store" internal energy whereas the monatomic molecule has only three. The theorem of **equipartition of energy**—which we cannot prove here—says that internal energy is distributed equally among all the possible ways in which it can be stored (as long as the temperature is sufficiently high). Each independent form of energy has an average of $\frac{1}{2}kT$ of energy per molecule and contributes $\frac{1}{2}R$ to the molar specific heat at constant volume.

Example 14.6

Heating Some Xenon Gas

A cylinder contains 250 L of xenon gas (Xe) at 20.0°C and a pressure of 5.0 atm. How much heat is required to raise the temperature of this gas to 50.0°C, holding the volume constant? Treat the xenon as an ideal gas.

Strategy The molar heat capacity is the heat required per degree per mole. The number of moles of xenon (n) can be found from the ideal gas law, $PV = nRT$. Xenon is a monatomic gas, so we expect $C_V = \frac{3}{2}R$.

Solution First we convert the known quantities into SI units.

$$P = 5.0\ \text{atm} = 5 \times 1.01 \times 10^5\ \text{Pa} = 5.05 \times 10^5\ \text{Pa}$$

$$V = 250\ \text{L} = 250 \times 10^{-3}\ \text{m}^3$$

$$T = 20.0°\text{C} = 293.15\ \text{K}$$

From the ideal gas law, we find the number of moles,

$$n = \frac{PV}{RT} = \frac{5.05 \times 10^5\ \text{Pa} \times 250 \times 10^{-3}\ \text{m}^3}{8.31\ \text{J/(mol·K)} \times 293.15\ \text{K}} = 51.8\ \text{mol}$$

We should check the units. Since Pa = N/m^2,

$$\frac{\text{Pa} \times \text{m}^3}{\text{J/(mol·K)} \times \text{K}} = \frac{\text{N/m}^2 \times \text{m}^3}{\text{J/mol}} = \frac{\text{N·m}}{\text{J}} \times \text{mol} = \text{mol}$$

For a monatomic gas at constant volume, the energy all goes into increasing the translational kinetic energy of the gas molecules. The molar specific heat is defined by $Q = nC_V\,\Delta T$, where, for a monatomic gas, $C_V = \frac{3}{2}R$. Then,

$$Q = \tfrac{3}{2}nR\,\Delta T$$

where

$$\Delta T = 50.0°\text{C} - 20.0°\text{C} = 30.0°\text{C}$$

Substituting,

$$Q = \tfrac{3}{2} \times 51.8\ \text{mol} \times 8.31\ \text{J/(mol·°C)} \times 30.0°\text{C} = 19\ \text{kJ}$$

Discussion Constant volume implies that all the heat is used to increase the internal energy of the gas; if the gas were to expand, it could transfer energy by doing work. When we find the number of moles from the ideal gas law, we must remember to convert the Celsius temperature to kelvins. Only when an equation involves a *change* in temperature, can we use kelvin or Celsius temperatures interchangeably.

Practice Problem 14.6 Heating Some Helium Gas

A storage cylinder of 330 L of helium gas is at 21°C and is subjected to a pressure of 10.0 atm. How much energy must be added to raise the temperature of the helium in this container to 75°C?

You may wonder why we can ignore rotation for the monatomic molecule—which in reality is not a point particle—or why we can ignore rotation about one axis for the diatomic molecule. The answer comes from quantum mechanics. Energy cannot be added to a molecule in arbitrarily small amounts; energy can only be added in discrete amounts, or "steps." At room temperature, there is not enough internal energy to excite the rotational modes with small rotational inertias, so they do not participate in the specific heat. We also ignored the possibility of vibration for the diatomic molecule. That is fine at room temperature, but at higher temperatures vibration becomes significant, adding two more energy modes (one kinetic and one potential). Thus, as temperature increases, the molar specific heat of a diatomic gas increases, approaching $\frac{7}{2}R$ at high temperatures.

14.5 PHASE TRANSITIONS

If heat continually flows into the water in a pot, the water eventually begins to boil; liquid water becomes steam. If heat flows into ice cubes, they eventually melt and turn into liquid water. A **phase transition** occurs whenever a material is changed from one phase, such as the solid phase, to another, such as the liquid phase.

When some ice cubes at 0°C are placed into a glass in a room at 20°C, the ice gradually melts. A thermometer in the water that forms as the ice melts reads 0°C until all the ice is melted. At atmospheric pressure, ice and water can only coexist in equilibrium at 0°C. Once all the ice is melted, the water gradually warms up to room temperature. Similarly, water boiling on a stove remains at 100°C until all the water has boiled away. Suppose we change 1.0 kg of ice at −25°C into steam at 125°C. A graph of the temperature versus heat is shown in Fig. 14.5. During the two phase transitions, *heat flow continues, and the internal energy changes, but the temperature of the mixture of two phases does not change*. Table 14.3 shows the heat during each step of the process.

Table 14.3 Heat to Turn 1 kg of Ice at −25°C to Steam at 125°C

Phase Transition or Temperature Change	Q (kJ)
Ice: −25°C to 0°C	52.3
Melting: ice at 0°C to water at 0°C	333.7
Water: 0°C to 100°C	419
Boiling: water at 100°C to steam at 100°C	2256
Steam: 100°C to 125°C	50

Latent Heat The heat required *per unit mass* of substance to produce a phase change is called the **latent heat** (L). The word *latent* is related to the lack of temperature change during a phase transition.

Definition of latent heat

$$|Q| = mL \quad (14\text{-}9)$$

The *sign* of Q in Eq. (14-9) depends on the direction of the phase transition. For melting or boiling, $Q > 0$ (heat flows *into* the system). For freezing or condensation, $Q < 0$ (heat flows *out of* the system).

The heat per unit mass for the solid-liquid phase transition (in either direction) is called the **latent heat of fusion** (L_f). From Table 14.3, it takes 333.7 kJ to change 1 kg of ice to water at 0°C, so for water $L_f = 333.7$ kJ/kg. For the liquid-gas phase transition (in either direction), the heat per unit mass is called the **latent heat of vaporization** (L_V). From Table 14.3, to change 1 kg of water to steam at 100°C requires 2256 kJ, so for water $L_V = 2256$ kJ/kg. If 2256 kJ must be supplied to turn 1 kg of water into steam, then 2256 kJ of heat is *released* from 1 kg of steam when it condenses to form water. Table 14.4 lists latent heats of fusion and vaporization for various materials.

CHECKPOINT 14.5

Why is a burn caused by 100°C steam often much more severe than a burn caused by 100°C water?

How does spraying with water protect the buds?

The large latent heat of fusion of water is partly why spraying fruit trees with water can protect the buds from freezing. Before the buds can freeze, first the water must be cooled to 0°C and then it must freeze. In the process of freezing, the water gives up a

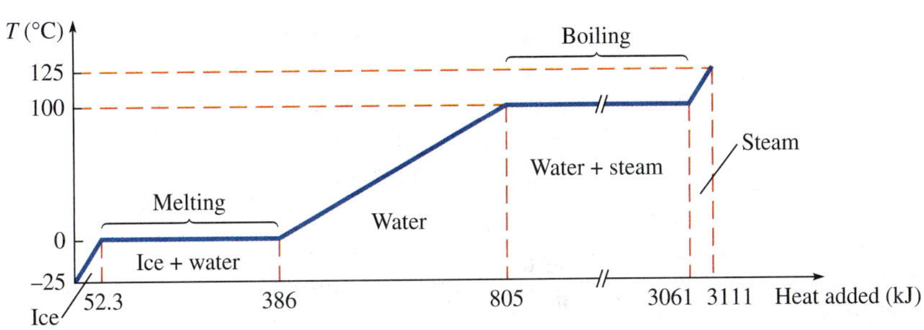

Figure 14.5 Temperature versus heat for 1 kg of ice that starts at a temperature below 0°C. (Horizontal axis not to scale.) During the two phase transitions—melting and boiling—the temperature does not change.

Table 14.4 Latent Heats of Some Common Substances

Substance	Melting Point (°C)	Heat of Fusion (kJ/kg)	Boiling Point (°C)	Heat of Vaporization (kJ/kg)
Alcohol (ethyl)	−114	104	78	854
Aluminum	660	397	2450	11 400
Copper	1083	205	2340	5070
Gold	1063	66.6	2660	1580
Lead	327	22.9	1620	871
Silver	960.8	88.3	1950	2340
Water	0.0	333.7	100	2256

large amount of heat and keeps the temperature of the buds from going below 0°C. Even if the water freezes, then the layer of ice over the buds acts like insulation since ice is not a particularly good conductor of heat.

Microscopic View of a Phase Change To understand what is happening during a phase change, we must consider the substance on the molecular level. When a substance is in solid form, bonds between the atoms or molecules hold them near fixed equilibrium positions. Energy must be supplied to break the bonds and change the solid into a liquid. When the substance is changed from liquid to gas, energy is used to separate the molecules from the loose bonds holding them together and to move the molecules apart. Temperature does not change during these phase transitions because the *kinetic energy* of the molecules is not changing. Instead, the *potential energy* of the molecules changes as work is done against the forces holding them together.

Example 14.7

Making Silver Charms

A jewelry designer plans to make some specially ordered silver charms for a commemorative bracelet. If the melting point of silver is 960.8°C, how much heat must the jeweler add to 0.500 kg of silver at 20.0°C to be able to pour silver into her charm molds?

Strategy The solid silver first needs to be heated to its melting point; then more heat has to be added to melt the silver.

Solution The total heat flow into the silver is the sum of the heat to raise the temperature of the solid and the heat that causes the phase transition:

$$Q = mc\,\Delta T + mL_f$$

The temperature change of the solid is

$$\Delta T = 960.8°\text{C} - 20.0°\text{C} = 940.8°\text{C}$$

We look up the specific heat of solid silver and the latent heat of fusion of silver. Substituting numerical values into the equation for Q yields

$$Q = 0.500\text{ kg} \times 0.235\text{ kJ/(kg·°C)} \times 940.8°\text{C} + 0.500\text{ kg} \times 88.3\text{ kJ/kg} = 110.5\text{ kJ} + 44.15\text{ kJ} = 155\text{ kJ}$$

Discussion An easy mistake to make is to use the wrong latent heat. Here we were dealing with melting, so we needed the latent heat of fusion. Another possible error is to use the specific heat for the wrong phase: here we raised the temperature of *solid* silver, so we needed the specific heat of *solid* silver. With water, we must always be careful to use the specific heat of the correct phase; the specific heats of ice, water, and steam have three different values.

Practice Problem 14.7 Making Gold Medals

Some gold medals are to be made from 750 g of solid gold at 24°C (Fig. 14.6). How much heat is required to melt the gold so that it can be poured into the molds for the medals?

Figure 14.6 A gold medal: the Nobel Prize for physics.

Example 14.8

Making Ice

Ice cube trays are filled with 0.500 kg of water at 20.0°C and placed into the freezer compartment of a refrigerator. How much energy must be removed from the water to turn it into ice cubes at −5.0°C?

Strategy We can think of this process as three consecutive steps. First, the liquid water is cooled to 0°C. Then the phase change occurs at constant temperature. Now the water is frozen; the ice continues to cool to −5.0°C. The energy that must be removed for the whole process is the sum of the energy removed in each of the three steps.

Solution For liquid water going from 20.0°C to 0.0°C,

$$Q_1 = mc_w \Delta T_1$$

where

$$\Delta T_1 = 0.0°C - 20.0°C = -20.0°C$$

Since ΔT_1 is negative, Q_1 is negative: heat must flow *out of* the water in order for its temperature to decrease. Next the water freezes. The heat is found from the latent heat of fusion:

$$Q_2 = -mL_f$$

Again, heat flows *out* so Q_2 is negative. For phase transitions, we supply the correct sign of Q according to the direction of the phase transition (negative sign for freezing, positive sign for melting). Finally, the ice is cooled to −5.0°C:

$$Q_3 = mc_{ice} \Delta T_2$$

where

$$\Delta T_2 = -5.0°C - 0.0°C = -5.0°C$$

We use subscripts on the specific heats to distinguish the specific heat of ice from that of water. The total heat is

$$Q = m\,(c_w \Delta T_1 - L_f + c_{ice} \Delta T_2)$$

Now we look up c_w, L_f, and c_{ice} in Tables 14.1 and 14.4 and substitute numerical values:

$$c_w\,\Delta T_1 = 4.186\,\frac{kJ}{kg{\cdot}K} \times (-20\text{ K}) = -83.72\,\frac{kJ}{kg}$$

$$L_f = 333.7\,\frac{kJ}{kg}$$

$$c_{ice}\,\Delta T_2 = 2.1\,\frac{kJ}{kg{\cdot}K} \times (-5.0\text{ K}) = -10.5\,\frac{kJ}{kg}$$

$$Q = 0.500\text{ kg} \times \left[-83.72\frac{kJ}{kg} - 333.7\frac{kJ}{kg} - 10.5\frac{kJ}{kg}\right] = -214\text{ kJ}$$

So 214 kJ of heat flows out of the water that becomes ice cubes.

Discussion We cannot consider the entire temperature change from +20°C to −5°C in one step. A phase change occurs, so we must include the flow of heat during the phase change. Also, the specific heat of ice is different from the specific heat of liquid water; we must find the heat to cool water 20°C and then the heat to cool ice 5°C.

Practice Problem 14.8 Frozen Popsicles

Nigel pulls a tray of frozen popsicles out of the freezer to share with his friends. If the popsicles are at −4°C and go directly into hungry mouths at 37°C, how much energy is used to bring a popsicle of mass 0.080 kg to body temperature? Assume the frozen popsicles have the same specific heat as ice and the melted popsicle has the specific heat of water.

Example 14.9

Cooling a Drink

Two 50 g ice cubes are placed into 0.200 kg of water in a Styrofoam cup. The water is initially at a temperature of 25.0°C, and the ice is initially at a temperature of −15.0°C. What is the final temperature of the drink? The average specific heat for ice between −15°C and 0°C is 2.05 kJ/(kg·°C).

Strategy Since heat flows out of the water and into ice, $Q_w < 0$ and $Q_{ice} > 0$. Assuming heat flow to or from the environment is negligible, their sum is zero:

$$Q_w + Q_{ice} = 0$$

Each of the quantities Q_w and Q_{ice} includes the heat to change the temperature as well as the latent heat for any phase transitions that occur. To find the final temperature, we have to determine whether the final state is all ice, ice and water in equilibrium, or all water.

Given: $m_{ice} = 0.100$ kg at −15.0°C; $m_w = 0.200$ kg at 25.0°C; $c_{ice} = 2.05$ kJ/(kg·°C)

Look up: L_f for water = 333.7 kJ/kg; $c_w = 4.186$ kJ/(kg·°C)

Find: T_f

continued on next page

Example 14.9 continued

Solution The heat flow out of the water *if it cools all the way to 0°C* is

$$Q = m_w c_w (T_f - T_i)$$
$$= 0.200 \text{ kg} \times 4.186 \text{ kJ/(kg}\cdot\text{°C)} \times (-25.0\text{°C})$$
$$= -20.93 \text{ kJ}$$

Is this enough to bring the ice to 0°C? The heat flow into the ice *if it warms to 0°C* is

$$Q = m_{ice} c_{ice} \, \Delta T$$
$$= 0.100 \text{ kg} \times 2.05 \text{ kJ/(kg}\cdot\text{°C)} \times (+15.0\text{°C})$$
$$= +3.075 \text{ kJ}$$

Heat flow out of the water is more than enough to bring the ice to 0°C, since 20.93 kJ > 3.075 kJ. Therefore, the ice reaches 0°C and starts to melt.

Does *all* of the ice melt? The heat flow required to do that, starting from −15.0°C, is

$$Q = +3.075 \text{ kJ} + m_{ice} L_f$$
$$= +3.075 \text{ kJ} + 0.100 \text{ kg} \times 333.7 \text{ kJ/kg}$$
$$= +36.445 \text{ kJ}$$

Since 20.93 kJ < 36.445 kJ, not all of the ice melts. The final state is ice and water in equilibrium at 0°C.

Discussion The water initially at 25.0°C ends up at 0°C, so $Q_w = -20.9$ kJ. All of this heat flows into the ice, so $Q_{ice} = +20.9$ kJ.

If we had assumed that all the ice melts and the final state is all water, we would have found a final temperature of −12.4°C. That result would not make sense (the water would not be liquid at −12.4°C) so we would know that the assumption was not correct.

Practice Problem 14.9 Melting Ice

How much of the ice of Example 14.9 melts?

Evaporation

If you leave a cup of water out at room temperature, the water gradually evaporates. Recall that the temperature of the water reflects the average kinetic energy of the water molecules; some have higher than average energies and some have lower. The most energetic molecules have enough energy to break loose from the molecular bonds at the surface of the water. As these highest-energy molecules leave the water, the average energy of those left behind decreases—which is why evaporation is a cooling process. Approximately the same latent heat of vaporization applies to evaporation as to boiling, since the same molecular bonds are being broken. Perspiring basketball players cover up while sitting on the bench for a short time during a game to prevent getting a chill even though the air in the stadium may be warm.

When the humidity is high—meaning there is already a lot of water vapor in the air—evaporation proceeds more slowly. Water molecules in the air can also condense into water; the net evaporation rate is the difference in the rates of evaporation and condensation. A hot, humid day is uncomfortable because our bodies have trouble staying cool when perspiration evaporates slowly.

EVERYDAY PHYSICS DEMO

The effects of evaporation can easily be felt. Rub some water on the inside of your forearm and then blow on your arm. The motion of the air over your arm removes the newly evaporated molecules from the vicinity of your arm and allows other molecules to evaporate more quickly. You can feel the cooling effect. If you have some rubbing alcohol, repeat the experiment. Since the alcohol evaporates faster, the cooling effect is noticeably greater.

Phase Diagrams

A useful tool in the study of phase transitions is the **phase diagram**—a diagram on which pressure is plotted on the vertical axis and temperature on the horizontal axis. Figure 14.7a is a phase diagram for water. A point on the phase diagram represents

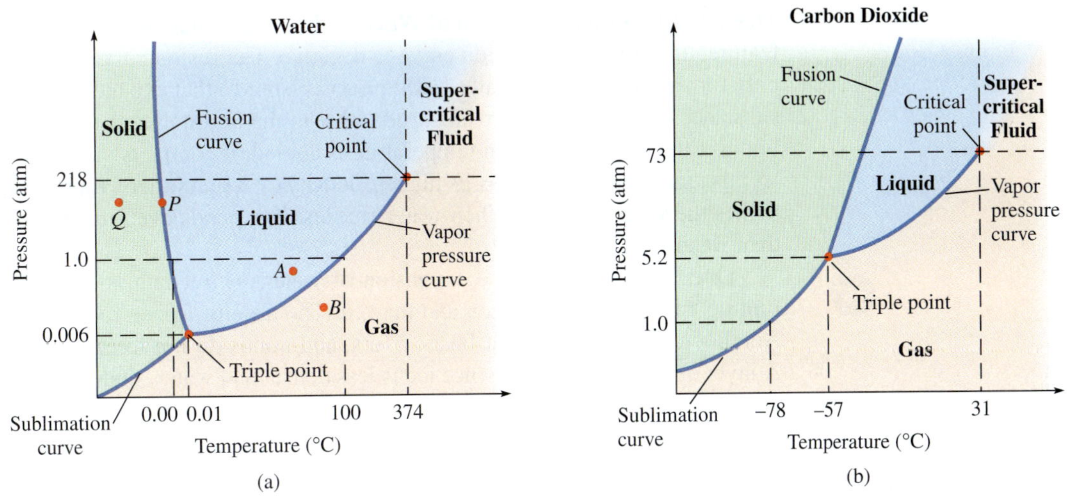

Figure 14.7 Phase diagrams for (a) water and (b) carbon dioxide. (Note that the axes do not use a linear scale.) At a point on the blue curves, two phases of the substance can exist in equilibrium. All three phases (solid, liquid, and gas) can exist in equilibrium only at the triple point. Crossing one of the blue curves represents a phase transition; the latent heat for that phase transition is either absorbed or released (depending on direction).

water in a state determined by the pressure and the temperature at that point. The curves on the phase diagram are the demarcations between the solid, liquid, and gas phases. For most temperatures, there is one pressure at which two particular phases can coexist in equilibrium. Since point P lies on the fusion curve, water can exist as liquid, or as solid, or as a mixture of the two at that temperature and pressure. At point Q, water can only be a solid. Similarly, at A, water is a liquid; at B, it is a gas.

The one exception is at the **triple point**, where all three phases (solid, liquid, and gas) can coexist in equilibrium. Triple points are used in precise calibrations of thermometers. The triple point of water is precisely 0.01°C at 0.006 atm.

From the vapor pressure curve, we see that as the pressure is lowered, the temperature at which water boils decreases. (The term **vapor** refers to a gas below its critical temperature.) It takes longer to cook a hard-boiled egg at high elevations because the temperature of the boiling water is less than 100°C; the chemical reactions proceed more slowly at a lower temperature. It might take as long as half an hour to hard-boil an egg on Pike's Peak, where the average pressure is 0.6 atm.

If either the temperature or the pressure or both are changed, the point representing the state of the water moves along some path on the phase diagram. If the path crosses one of the curves, a phase transition occurs and the latent heat for that phase transition is either absorbed or released (depending on direction). Crossing the fusion curve represents freezing or melting; crossing the vapor pressure curve represents condensation or vaporization.

Notice that the vapor pressure curve ends at the **critical point**. At temperatures above the critical temperature and pressures above the critical pressure, it is impossible to make a clear distinction between the liquid and gas phases; we then refer to the substance as a **supercritical fluid**. If the path for changing a liquid to a gas goes around the critical point without crossing the vapor pressure curve, a *continuous* phase transition occurs with no associated latent heat.

Sublimation occurs when a solid becomes gas (or vice versa) without passing through the liquid phase. An example occurs when ice on a car windshield becomes water vapor on a cold dry day. Mothballs and dry ice (solid carbon dioxide) also pass directly from solid to gas. At atmospheric pressure, only the solid and gas phases of CO_2 exist (Fig. 14.7b). The liquid phase is not stable below 5.2 atm of pressure, so carbon dioxide does not melt at atmospheric pressure. Instead it sublimates; it goes from solid directly to gas. Solid CO_2 is called *dry ice* because it is cold and looks like ice, but does not melt. Sublimation has its own latent heat; the latent heat for sublimation is not the sum of the latent heats for fusion and vaporization.

The Unusual Phase Diagram of Water The phase diagram of water has an unusual feature: the slope of the fusion curve is negative. The fusion curve has a negative slope only for substances (e.g., water, gallium, and bismuth) that expand on freezing. In these substances the molecules are *closer together* in the liquid than they are in the solid! As liquid water starting at room temperature is cooled, it contracts until it reaches 3.98°C. At this temperature water has its highest density (at a pressure of 1 atm); further cooling makes the water *expand*. When water freezes, it expands even more; ice is less dense than water.

One consequence of the expansion of water on freezing is that cell walls might rupture when foods are frozen and thawed. The taste of frozen food suffers as a result. Another consequence is that lakes, rivers, and ponds do not freeze solid in the winter. A layer of ice forms on *top* since ice is less dense than water; underneath the ice, liquid water remains, which permits fish, turtles, and other aquatic life to survive until spring (Fig. 14.8).

Figure 14.8 A Nunavut villager fishing for Arctic char.

14.6 THERMAL CONDUCTION

Until now we have considered the *effects* of heat flow, but not the mechanism of how heat flow occurs. We now turn our attention to three types of heat flow—*conduction*, *convection*, and *radiation*.

The **conduction** of heat can take place within solids, liquids, and gases. Conduction is due to collisions between atoms (or molecules) in which energy is exchanged. If the average energy is the same everywhere, there is no net flow of heat. If, on the other hand, the temperature is not uniform, then on average the atoms with more energy transfer some energy to those with less. The net result is that heat flows from the higher-temperature region to the lower-temperature region.

Conduction also occurs between objects that are in contact. A teakettle on an electric burner receives heat by conduction since the heating coil of the burner is in contact with the bottom of the kettle. The atoms that are vibrating in the object at higher temperature (the coil) collide with atoms in the object at lower temperature (the bottom of the kettle), resulting in a net transfer of energy to the colder object. If conduction is allowed to proceed, with no heat flow to or from the surroundings, then the objects in contact eventually reach thermal equilibrium when the average translational kinetic energies of the atoms are equal.

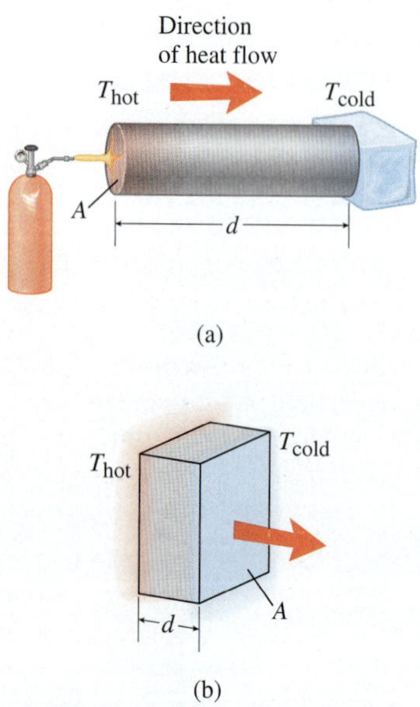

Figure 14.9 (a) Heat conduction along a cylindrical bar of length *d*. (b) Heat conduction through a slab of material of thickness *d*.

Fourier's Law of Heat Conduction Suppose we consider a simple geometry such as an object with uniform cross section in which heat flows in a single direction. Examples are a plate of glass, with different temperatures on the inside and outside surfaces, or a cylindrical bar, with its ends at different temperatures (Fig. 14.9). The rate of heat conduction depends on the temperature difference $\Delta T = T_{hot} - T_{cold}$, the length (or thickness) *d*, the cross-sectional area *A* through which heat flows, and the nature of the material itself. The greater the temperature difference, the greater the heat flow. The thicker the material, the longer it takes for the heat to travel through—since the energy transfer has to be passed along a longer "chain" of atomic collisions—making the rate of heat flow smaller. A larger cross-sectional area allows more heat to flow.

The nature of the material is the final thing that affects the rate of energy transfer. In metals the electrons associated with the atom are free to move about and they carry the heat. When a material has free electrons, the transfer rate is faster; if the electrons are all tightly bound, as in nonmetallic solids, the transfer is slower. Liquids, in turn, conduct heat less readily than solids, because the forces between atoms are weaker. Gases are even less efficient as conductors of heat than solids or liquids since the atoms of a gas are so much farther apart and have to travel a greater distance before collisions occur. The **thermal conductivity** (symbol κ, the Greek letter kappa) of a substance is directly proportional to the rate at which energy is transferred through the substance. Higher values of κ are associated with good conductors of heat, smaller values with *thermal insulators* that tend to prevent the flow of heat. Table 14.5 lists the thermal conductivities for several common substances.

Let $\mathcal{P} = Q/\Delta t$ represent the rate of heat flow (or *power*). (The script $\mathcal{P}$ is used to avoid confusing power with pressure.) The dependence of the rate of heat flow through a substance on all the factors mentioned is given by

Fourier's law of heat conduction

$$\mathcal{P} = \kappa A \frac{\Delta T}{d} \quad (14\text{-}10)$$

where κ is the thermal conductivity of the material, A is the cross-sectional area, d is the thickness (or length) of the material, and ΔT is the temperature difference between one side and the other. The quantity $\Delta T/d$ is called the temperature gradient; it tells how many °C or K the temperature changes per unit of distance moved along the path of heat flow. Inspection of Eq. (14-10) shows that the SI units of κ are W/(m·K).

In Fig. 14.9b, a slab of material is shown that conducts heat because of a temperature difference between the two sides. By rearranging Eq. (14-10) and solving for ΔT,

$$\Delta T = \mathcal{P}\frac{d}{\kappa A} = \mathcal{P}R \quad (14\text{-}11)$$

The quantity $d/(\kappa A)$ is called the **thermal resistance** R.

$$R = \frac{d}{\kappa A} \quad (14\text{-}12)$$

Thermal resistance has SI units of K/W (kelvins per watt). Notice that the thermal resistance depends on the nature of the material (through the thermal conductivity κ) and the geometry of the object (d/A). Equation (14-11) is useful for solving problems when heat flows through one material after another.

Conduction Through Two or More Materials in Series Suppose we have two layers of material between two temperature extremes as in Fig. 14.10. These layers are in series because the heat flows through one and then through the other. Looking at one layer at a time,

$$T_1 - T_2 = \mathcal{P}R_1 \quad \text{and} \quad T_2 - T_3 = \mathcal{P}R_2$$

Then, adding the two together

$$(T_1 - T_2) + (T_2 - T_3) = \mathcal{P}R_1 + \mathcal{P}R_2$$
$$\Delta T = T_1 - T_3 = \mathcal{P}(R_1 + R_2)$$

The rate of heat flow through the first layer is the same as the rate through the second layer because otherwise the temperatures would be changing. For n layers,

$$\Delta T = \mathcal{P}\sum R_n \qquad n = 1, 2, 3, \ldots \quad (14\text{-}13)$$

Equation (14-13) shows that the effective thermal resistance for layers in series is the sum of each layer's thermal resistance.

CHECKPOINT 14.6

In Fig. 14.10, which of the two materials has the larger thermal conductivity?

Table 14.5 Thermal Conductivities at 20°C

Material	$\kappa\left(\frac{\text{W}}{\text{m·k}}\right)$
Air	0.023
Rigid panel polyurethane insulation	0.023–0.026
Fiberglass insulation	0.029–0.072
Rock wool insulation	0.038
Cork	0.046
Wood	0.13
Soil (dry)	0.14
Asbestos	0.17
Snow	0.25
Sand	0.39
Water	0.6
Window glass (typical)	0.63
Pyrex glass	1.13
Vycor	1.34
Concrete	1.7
Ice	1.7
Stainless steel	14
Lead	35
Steel	46
Nickel	60
Tin	66.8
Platinum	71.6
Iron	80.2
Brass	122
Zinc	116
Tungsten	173
Aluminum	237
Gold	318
Copper	401
Silver	429

CONNECTION:

Fourier's law says that the rate of heat flow is proportional to the temperature gradient. Closely analogous is Poiseuille's law for viscous fluid flow [Eq. (9-15)] in which the volume flow rate is proportional to the pressure gradient.

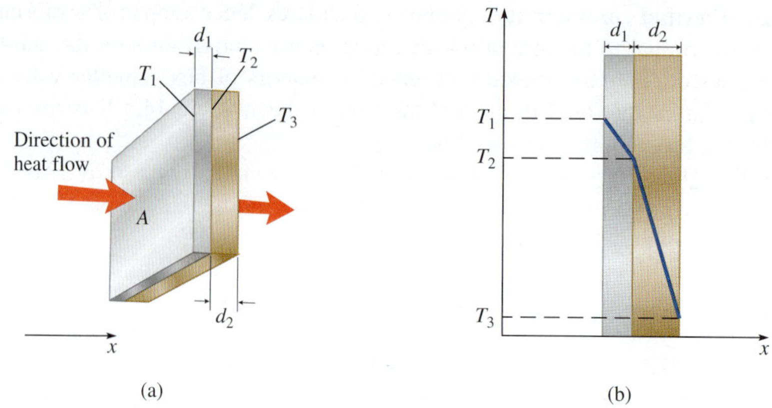

Figure 14.10 (a) Conduction of heat through two different layers ($T_1 > T_2 > T_3$). (b) Graph of temperature T as a function of position x. The slope of the graph in either material is the temperature gradient $\Delta T/d$ in that material. The temperature gradients are not the same because the materials have different thermal conductivities.

Example 14.10

The Rate of Heat Flow Through Window Glass

A windowpane that measures 20.0 cm by 15.0 cm is set into the front door of a house. The glass is 0.32 cm thick. The temperature outdoors is −15°C and inside is 22°C. At what rate does heat leave the house through that one small window?

Strategy We assume one side of the glass to be at the temperature of the air inside the house and the other to be at the outdoor temperature.

Given: $\Delta T = 22°\text{C} - (-15°\text{C}) = 37°\text{C}$; thickness of windowpane $d = 0.32 \times 10^{-2}$ m; area of windowpane $A = 0.200 \text{ m} \times 0.150 \text{ m} = 0.0300 \text{ m}^2$

Look up: thermal conductivity for glass = 0.63 W/(m·K)

Find: rate of heat flow, $\mathcal{P}$

Solution The temperature gradient is

$$\frac{\Delta T}{d} = \frac{37°\text{C}}{0.32 \times 10^{-2} \text{ m}} = 1.16 \times 10^4 \text{ K/m}$$

Now we have all the information we need to find the rate of conductive heat flow:

$$\mathcal{P} = \kappa A \frac{\Delta T}{d}$$

$$= 0.63 \text{ W/(m·K)} \times 0.0300 \text{ m}^2 \times 1.16 \times 10^4 \text{ K/m}$$

$$= 220 \text{ W}$$

Discussion A loss of 220 W through one small window is significant. However, our assumption about the temperatures of the two glass surfaces exaggerates the temperature difference across the glass. In reality, the inside surface of the glass is colder than the air inside the house, and the outside surface is warmer than the air outside.

Practice Problem 14.10 An Igloo

A group of children build an igloo in their backyard. The snow walls are 0.30 m thick. If the inside of the igloo is at 10.0°C and the outside is at −10.0°C, what is the rate of heat flow through the snow walls of area 14.0 m^2?

Thermal Conductivity of Air Air has a low thermal conductivity; it is an excellent thermal insulator *when it is still.* An accurate calculation of the energy loss through a single-paned window *must* take into account the thin layer of stagnant air, due to viscosity, on each side of the glass. If the temperature is measured near a window, the temperature of the air just beside the window is intermediate in value between the temperatures of the room air and the outside air (Fig. 14.11). Thus, the temperature gradient

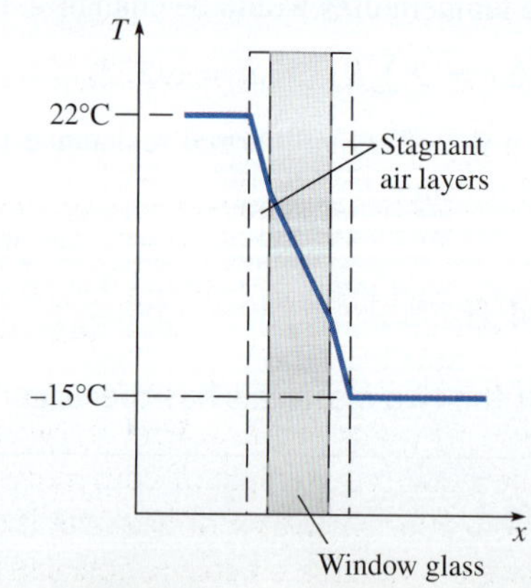

Figure 14.11 Temperature variation on either side of a windowpane. A plot of temperature versus position is superimposed on a cross section of the window glass and the air layers on either side.

across the glass is considerably smaller than the difference between indoor and outdoor temperatures. In fact, much of the thermal resistance of a window is due to the stagnant air layers rather than to the glass.

Example 14.11

Heat Loss Through a Double-Paned Window

The single-paned window of Example 14.10 is replaced by a double-paned window with an air gap of 0.50 cm between the two panes. The inner surface of the inner pane is at 22°C and the outer surface of the outer pane is at −15°C. What is the new rate of heat loss through the double-paned window?

Strategy Now there are three layers to consider: two layers of glass and one layer of air. We find the thermal resistance of each layer and then add them together to find the total thermal resistance. Then we find the temperature difference between the inside of the house and the air outdoors and divide by the total thermal resistance to find the rate at which heat is lost through the replacement window.

Solution For the first layer of glass,

$$R_1 = \frac{d}{\kappa A} = \frac{0.32 \times 10^{-2}\ \text{m}}{0.63\ \text{W/(m·K)} \times 0.0300\ \text{m}^2} = 0.169\ \text{K/W}$$

For the air gap,

$$R_1 = \frac{d}{\kappa A} = \frac{0.50 \times 10^{-2}\ \text{m}}{0.023\ \text{W/(m·K)} \times 0.0300\ \text{m}^2} = 7.246\ \text{K/W}$$

The second layer of glass has the same thermal resistance as the first:

$$R_3 = R_1$$

The total thermal resistance is

$$\sum R_n = 0.169 + 7.246 + 0.169 = 7.584\ \text{K/W}$$

and the rate of conductive heat flow is

$$\mathcal{P} = \frac{Q}{\Delta t} = \frac{\Delta T}{\sum R_n} = \frac{37\ \text{K}}{7.584\ \text{K/W}} = 4.9\ \text{W}$$

Discussion The reduction in the rate of heat loss by replacing a single-paned window with a double pane is significant. This example, however, overestimates the reduction since we assume that heat can only be conducted through the air layer. In reality, heat can also flow through air by convection and radiation. A more accurate calculation would have to account for the other methods of heat flow.

Practice Problem 14.11 Two Panes of Glass Without the Air Gap

Repeat Example 14.11 if the two panes of glass are touching one another, without the intervening layer of air.

R-Factors The U.S. building industry rates materials used in construction with *R-factors*. The R-factor is not quite the same as the thermal resistance; thermal resistance cannot be specified without knowing the cross-sectional area. The R-factor is the thickness divided by the thermal conductivity:

$$\text{R-factor} = \frac{d}{\kappa} = RA$$

$$\frac{\mathcal{P}}{A} = \frac{\Delta T}{\text{R-factor}}$$

Unfortunately, SI units are not used. The R-factors quoted in the United States are in units of °F·ft^2/(Btu/h)! R-factors are added, just as thermal resistances are, when heat flows through several different layers.

14.7 THERMAL CONVECTION

Convection involves *fluid currents* that carry heat from one place to another. In conduction, energy flows through a material but the material itself does not move. In convection, *the material itself moves* from one place to another. Thus, convection can occur only in fluids, not in solids. When a wood stove is burning, convection currents in the air carry heat upward to the ceiling. The heated air is less dense than cooler air, so the buoyant force causes it to rise, carrying heat with it. Meanwhile, cooler air that is more dense sinks toward the floor. An example of convection currents at the seashore is shown in Fig. 14.12. Air is a poor *conductor* of heat, but it can easily flow and carry heat by *convection*. (For more information on convection, see the text website.)

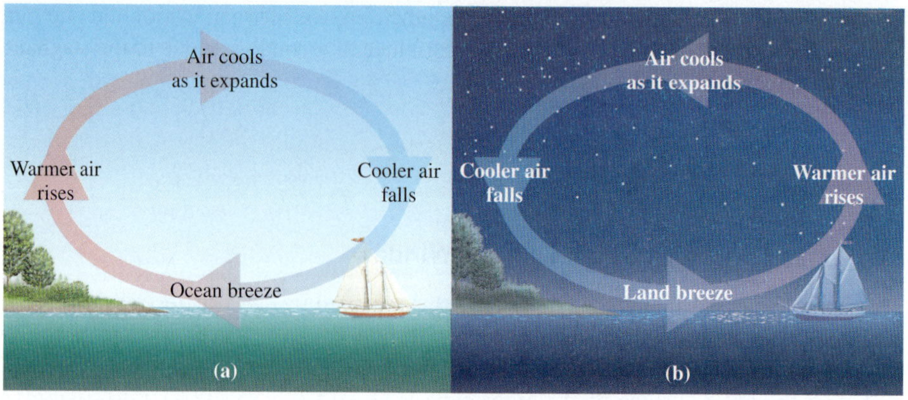

Figure 14.12 (a) During the day, air coming off the ocean is heated as it passes over the warm ground on shore; the heated air rises and expands. The expansion cools the air; it becomes more dense and falls back down. This cycle sets up a convection current that brings cool breezes from the sea to the shore. (b) The reverse circulation occurs at night when the land is cold and the sea is warmer, retaining heat absorbed during the day.

The use of sealed, double-paned windows replaces the large air gap of about 6 or 7 cm between a storm window and regular window with a much smaller gap. The smaller air gap minimizes circulating convection currents between the two panes. Down jackets and quilts are good insulators because air is trapped in many little spaces among the feathers, minimizing heat flow due to convection. Materials such as rock wool, glass wool, or fiberglass are used to insulate walls; much of their insulating value is due to the air trapped around and between the fibers.

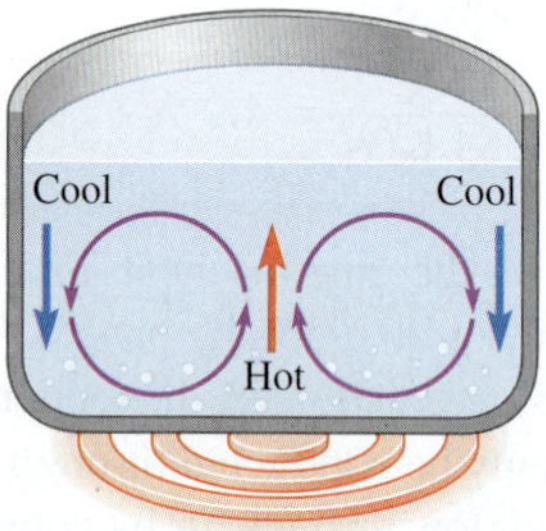

Figure 14.13 Convection currents in heated water. Heat flows through the bottom surface of the pot by conduction and then heats the layer of water in contact with the pot bottom. The heated water is less dense, so buoyant forces make it rise, setting up convection currents.

Natural and Forced Convection In *natural convection*, the currents are due to gravity. Fluid with a higher density sinks because the buoyant force is smaller than the weight; less dense fluid rises because the buoyant force exceeds the weight (Figs. 14.13 and 14.14). In *forced convection*, fluid is pushed around by mechanical means such as a fan or pump. In forced-hot-air heating, warm air is blown into rooms by a fan (Fig. 14.15); in hot water baseboard heating, hot water is pumped through baseboard radiators.

Application: Forced Convection in the Human Body Another example of forced convection is blood circulation in the body. The heart pumps blood around the body. When our body temperature is too high, the blood vessels near the skin dilate so that more blood can be pushed into them by the heart. The blood carries heat from the interior of the body to the skin; heat then flows from the skin into the cooler surroundings. If the surroundings are *hotter* than the skin, such as in a hot tub, this strategy backfires and can lead to dangerous overheating of the body. The hot water delivers heat to the dilated blood vessels; the blood carries the heat back to the central core of the body, *raising* the core temperature.

Figure 14.14 Birds (and people flying sailplanes) take advantage of thermal updrafts.

Application of Convection: Global Climate Change

One worry for scientists studying global climate change is that northern Europe might be plunged into a deep freeze—a seeming contradiction that results from an interruption

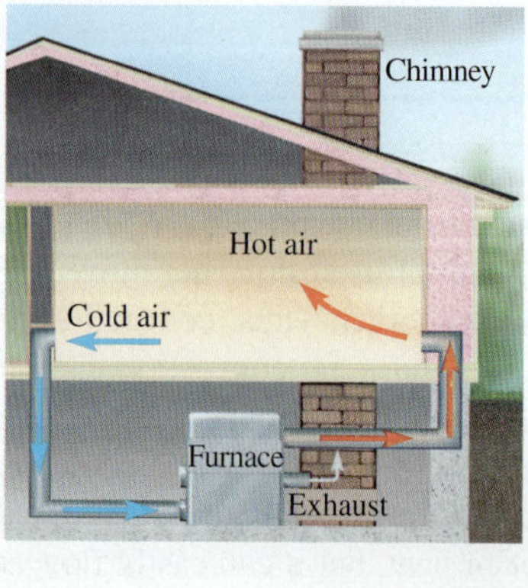

Figure 14.15 Household heating systems rely on forced convection.

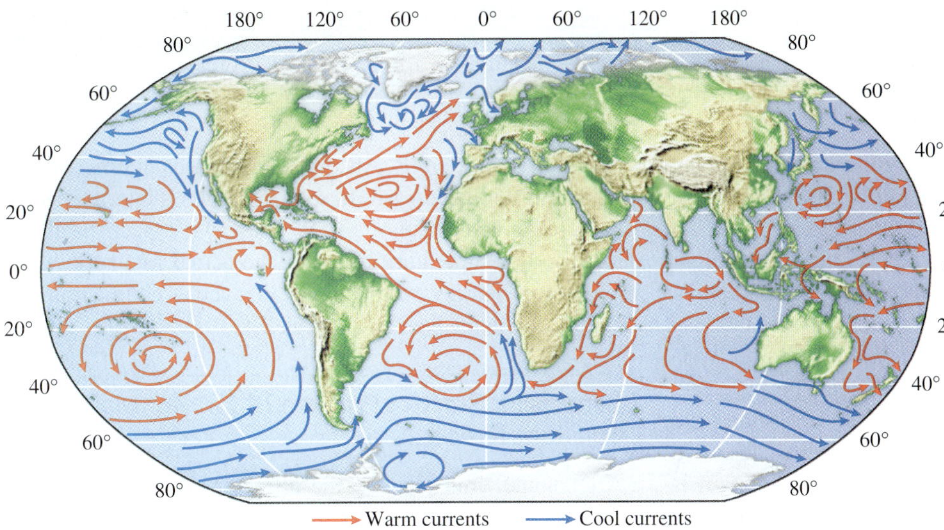

Figure 14.16 Surface convection currents in the oceans. The Gulf Stream is a current of warm water flowing across the Atlantic.

of the natural convection cycle. Earth's climate is influenced by convection currents caused by temperature differences between the poles and the tropics (Fig. 14.16). Massive sea currents travel through the Pacific and Atlantic oceans, carrying about half of the heat from the tropics to the poles, where it is dissipated. Storms moving north from the tropics carry much of the rest of the heat. If the polar regions warm at a faster rate than the tropics, the smaller temperature difference between them changes the patterns of the prevailing winds, the tracks followed by storms, the speed of ocean currents, and the amount of precipitation.

For example, the melting of the ice shelves combined with increased precipitation could lead to a layer of freshwater lying on top of the more dense saltwater in the North Atlantic. Normally, the cold ocean water at the surface sinks and starts the process of convection. With the buoyancy of the less dense freshwater layer keeping it from sinking, the convection currents slow down or are stopped. Without the pull of the convection current, the usual northward movement of water from the warm Gulf Stream would slow or cease, causing *colder* temperatures in northern Europe.

Such an effect on climate is not without precedent. At the end of the last Ice Age, freshwater from melting glaciers flowed out the St. Lawrence River and into the North Atlantic. A freshwater layer, buoyed up by the more dense saltwater, disrupted the usual ocean currents. The Gulf Stream was effectively shut down and Europe experienced a thousand years of deep freeze.

14.8 THERMAL RADIATION

All objects emit energy through electromagnetic radiation due to the oscillation of electric charges in the atoms. Thermal radiation consists of electromagnetic waves that travel at the speed of light. Unlike conduction and convection, radiation does not require a material medium; the Sun radiates heat to Earth through the near vacuum of space.

An object emits thermal radiation while absorbing some of the thermal radiation emitted by other objects. The rate of absorption may be less than, equal to, or greater than the rate of emission. When solar radiation reaches Earth, it is partially absorbed and partially reflected. Earth also emits radiation at nearly the same average rate that it absorbs energy from the Sun. If there were an exact equilibrium between absorption and emission, Earth's average temperature would stay constant. However, increasing concentrations of CO_2 and other "greenhouse gases" in Earth's atmosphere cause energy to be emitted at a slightly lower rate than it is absorbed. As a result, Earth's average temperature is rising. Although the predicted temperature increase may seem small on an absolute scale, it will have dramatic consequences for life on Earth.

Conceptual Example 14.12

An Alligator Lying in the Sun

An alligator crawls out into the Sun to get warm. Solar radiation is incident on the alligator at the rate of 300 W; 70 W of it is reflected. (a) What happens to the other 230 W? (b) If the alligator emits 100 W, does its body temperature rise, fall, or stay the same? Ignore heat flow by conduction and convection.

Solution and Discussion (a) When radiation falls on an object, some can be absorbed, some can be reflected, and—for a transparent or translucent object—some can be transmitted through the object without being absorbed or reflected. Since the alligator is opaque, no radiation is transmitted through it. All the radiation is either absorbed or reflected, so the other 230 W is absorbed. (b) Since 230 W is absorbed while 100 W is emitted, the alligator absorbs more radiation than it emits. Absorption increases internal energy while emission decreases it, so the alligator's internal energy is increasing at a rate of 130 W. Thus, we expect the alligator's body temperature to rise. (The actual rate of increase of internal energy would be smaller since conduction and convection carry heat away as well.)

Conceptual Practice Problem 14.12 Maintaining Constant Temperature

After some time elapses, the alligator's body temperature reaches a constant level. The rate of absorption is still 230 W. If the alligator loses heat by conduction and convection at a rate of 90 W, at what rate does it emit radiation?

Stefan's Radiation Law

An idealized body that absorbs all the radiation incident upon it is called a **blackbody**. A blackbody absorbs not only all visible light, but infrared, ultraviolet, and all other wavelengths of electromagnetic radiation. It turns out (see Conceptual Question 23) that a good *absorber* is also a good *emitter* of radiation. A blackbody emits more radiant power per unit surface area than any real object at the same temperature. The rate at which a blackbody emits radiation per unit surface area is proportional to the fourth power of the *absolute* temperature, as expressed by Stefan's law named for the Slovene physicist Joseph Stefan (1835–1893):

Stefan's law of radiation (ideal blackbody)

$$\mathcal{P} = \sigma A T^4 \tag{14-14}$$

In Eq. (14-14), A is the surface area and T is the surface temperature of the blackbody *in kelvins*. Since Stefan's law involves the absolute temperature and not a temperature difference, °C *cannot* be substituted. The universal constant σ (Greek letter sigma) is called *Stefan's constant*:

$$\sigma = 5.670 \times 10^{-8}\ \text{W/(m}^2\cdot\text{K}^4) \tag{14-15}$$

The fourth-power temperature dependence implies that the power emitted is extremely sensitive to temperature changes. If the absolute temperature of an object doubles, the energy emitted increases by a factor of $2^4 = 16$.

Emissivity Since real bodies are not perfect absorbers and therefore emit less than a blackbody, we define the **emissivity** (e) as the ratio of the emitted power of the body to that of a blackbody at the same temperature. Then Stefan's law becomes

Stefan's law of radiation

$$\mathcal{P} = e\sigma A T^4 \tag{14-16}$$

The emissivity ranges from 0 to 1; $e = 1$ for a perfect radiator and absorber (a blackbody) and $e = 0$ for a perfect reflector. The emissivity for polished aluminum, an excellent reflector, is about 0.05; for soot (carbon black) it is about 0.95. Equation (14-16) is a refinement of Stefan's law, but it is still an approximation because it treats the emissivity as a constant. Emissivity is actually a function of the wavelength of the emitted radiation. Equation (14-16) is useful when the emissivity is approximately constant over the range of wavelengths in which most of the power is radiated.

Human skin, no matter what the pigmentation, has an emissivity of about 0.97 in the infrared part of the spectrum. Many objects have high emissivities in the infrared even though they may reflect much of the visible light incident on them and, therefore, have low emissivities in the visible range.

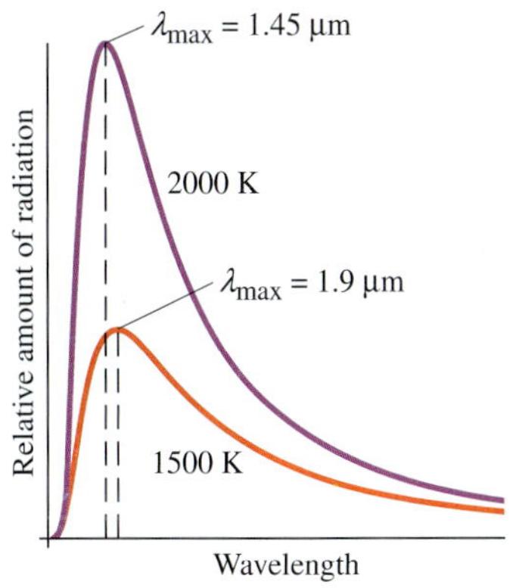

Figure 14.17 Graphs of blackbody radiation as a function of wavelength at two different temperatures. At the higher temperature, the wavelength of maximum radiation is shorter (Wien's law) and the total power radiated, represented by the area under the graph, increases (Stefan's law).

Radiation Spectrum

The electromagnetic radiation we are concerned with falls into three wavelength ranges. Infrared radiation includes wavelengths from about 100 μm down to 0.7 μm. The wavelengths of visible light range from about 0.7 μm to about 0.4 μm. The familiar colors of the visible spectrum from longest to shortest wavelength are red, orange, yellow, green, blue, and violet. Ultraviolet wavelengths are less than 0.4 μm.

The total power radiated is not the only thing that varies with temperature. Figure 14.17 shows the radiation spectrum—a graph of how much radiation occurs as a function of wavelength—for blackbodies at two different temperatures. The wavelength at which the maximum power is emitted decreases as temperature increases. Objects at ordinary temperatures emit primarily in the infrared—around 10 μm in wavelength for 300 K. The Sun, since it is much hotter, radiates primarily at shorter wavelengths. Its radiation peaks in the visible (no surprise there) but includes plenty of infrared and ultraviolet as well. The wavelength of maximum radiation is inversely proportional to the absolute temperature:

Wien's law

$$\lambda_{max} T = 2.898 \times 10^{-3}\ \text{m}\cdot\text{K} \quad (14\text{-}17)$$

where the temperature T is the temperature in kelvins and λ_{max} is the wavelength of maximum radiation in meters. This relationship is named for the German physicist Wilhelm Wien (1864–1928).

As the temperature of a blackbody rises to 1000 K and above, the peak intensity shifts toward shorter wavelengths until a significant fraction of the emitted radiation falls in the visible part of the spectrum. The longest visible wavelengths are for red light, so the blackbody appears dull red. As the temperature of the blackbody continues to increase, the red glow becomes brighter red, then orange, then yellow-white, and eventually blue-white. "Red-hot" is not as hot as "white-hot."

An incandescent light bulb is a common example of thermal radiation. Electric current passes through a thin tungsten filament, which becomes hot enough to emit a significant fraction of its thermal radiation in the visible part of the spectrum.

EVERYDAY PHYSICS DEMO

Locate an incandescent light bulb controlled by a dimmer switch. The dimmer works by reducing the current through the filament; as a result the filament's temperature drops and the power radiated decreases (Stefan's law). Observe that as the light dims, its color also changes, becoming more reddish. At a lower temperature, the spectrum has a reduced fraction of its visible radiation in the shorter wavelengths (blue, violet) and an increased fraction in the longer wavelengths (red, orange).

CHECKPOINT 14.8

A distant star looks reddish in color. How does its surface temperature compare with the Sun's? Can you determine which star emits radiation at a higher rate?

Example 14.13

Temperature of the Sun

The maximum rate of energy emission from the Sun occurs in the middle of the visible range—at about $\lambda = 0.5\ \mu\text{m}$. Estimate the temperature of the Sun's surface.

Strategy We assume the Sun to be a blackbody. Then the wavelength of maximum emission and the surface temperature are related by Wien's law.

Solution Given: $\lambda_{max} = 0.5\ \mu\text{m} = 5 \times 10^{-7}$ m. Then from Wien's law, we know that the product of the wavelength for maximum power emission and the corresponding temperature for the power emission is

$$\lambda_{max} T = 2.898 \times 10^{-3}\ \text{m}\cdot\text{K}$$

We can solve for the temperature since we know λ_{max}:

$$T = \frac{2.898 \times 10^{-3}\ \text{m}\cdot\text{K}}{5 \times 10^{-7}\ \text{m}}$$
$$= 6000\ \text{K}$$

Discussion Quick check: an object at 300 K has $\lambda_{max} \approx 10\ \mu\text{m}$, which is 20 times the λ_{max} in the radiation from the Sun (0.5 μm). Since λ_{max} and T are inversely proportional, the Sun's surface temperature is 20 times 300 K = 6000 K.

Practice Problem 14.13 Wavelengths of Maximum Power Emission for Skin

The temperature of skin varies from 30°C to 35°C depending on the blood flow near the skin surface. What is the range of wavelengths of maximum power emission from skin?

Simultaneous Emission and Absorption of Thermal Radiation

An object simultaneously emitting and absorbing thermal radiation has a *net* rate of heat flow due to thermal radiation given by $\mathcal{P}_{net} = \mathcal{P}_{emitted} - \mathcal{P}_{absorbed}$. Suppose an object with surface area A and temperature T is bathed in thermal radiation coming from its surroundings in all directions that are at a *uniform* temperature T_s. Then the *net* rate of heat flow due to emission and absorption of thermal radiation is

$$\mathcal{P}_{net} = e\sigma AT^4 - e\sigma AT_s^4 = e\sigma A(T^4 - T_s^4) \qquad (14\text{-}18)$$

An object emits energy even if it is at the same temperature as its surroundings; it just emits at the same rate that it absorbs, so $\mathcal{P}_{net} = 0$. If $T > T_s$, the object emits more thermal radiation than it absorbs. If $T < T_s$, the object absorbs more thermal radiation than it emits.

Why is the rate of *absorption* proportional to the *emissivity*? Because a good emitter is also a good absorber. The emissivity e measures not only how much the object emits compared to a blackbody, it also measures how much the object *absorbs* compared with a blackbody. A blackbody at the same temperature as its surroundings would have to absorb radiation at the rate $\mathcal{P}_{absorbed} = \sigma AT_s^4$ to exactly balance the rate of emission. However, emissivity does depend on temperature. Equation (14-18) assumes the emissivity at temperature T is the same as the emissivity at temperature T_s. If T and T_s are very different, we would have to modify Eq. (14-18) to use two different emissivities.

Do not substitute temperature in Celsius degrees into Eq. (14-18). The quantity inside the parentheses might look like a temperature difference, but it is not. The two kelvin temperatures are raised to the fourth power, *then* subtracted—which is not the same as the corresponding two Celsius temperatures subjected to the same mathematical operations. By the same token, do not subtract the temperatures in kelvins and then raise to the fourth power. The difference of the fourth powers is not equal to the difference raised to the fourth power, as can be readily demonstrated:

$$(2^4 - 1^4) = 15 \quad \text{but} \quad (2 - 1)^4 = 1$$

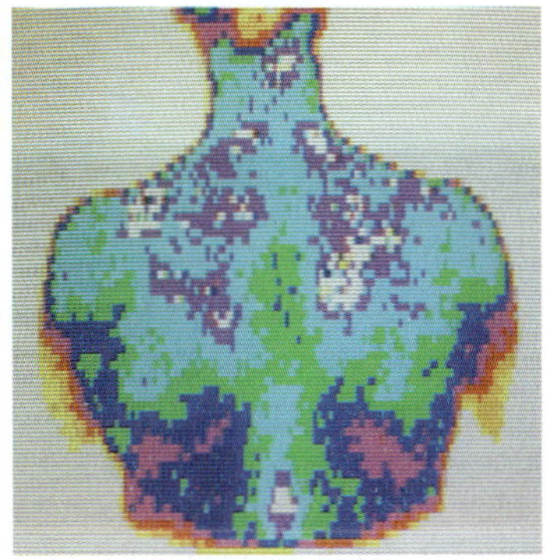

Figure 14.18 Thermography of a backache. The magenta areas are warmer than the surrounding tissue, revealing the location of the source of pain.

Medical Applications of Thermal Radiation

Thermal radiation from the body is used as a diagnostic tool in medicine. "Instant-read" thermometers work by measuring the intensity of thermal radiation in the patient's ear. A thermogram shows whether one area is radiating more heat than it should, indicating a higher temperature due to abnormal cellular activity. For example, when a broken bone is healing, heat can be detected at the location of the break just by placing a hand lightly on the area of skin over the break. Infrared detectors, originally developed for military uses (nightscopes, for example), can be used to detect radiation from the skin. The radiation is absorbed and an electrical signal is produced that is then used to produce a visual display (Fig. 14.18). Thermography has been used to screen travelers at airports for the high fever that accompanies infection with severe acute respiratory syndrome (SARS).

Example 14.14

Thermal Radiation from the Human Body

A person of body surface area 2.0 m^2 is sitting in a doctor's examining room with no clothing on. The temperature of the room is 22°C and the person's average skin temperature is 34°C. Skin emits about 97% as much as a blackbody at the same temperature for wavelengths in the infrared region, where most of the emission occurs. At what *net* rate is energy radiated away from the body?

Strategy Both radiation and absorption occur in the infrared—the absolute temperatures of the skin and the room are not very different. Therefore, we can assume that 97% of the incident radiation from the room is absorbed. Equation (14-18) therefore applies. We must convert the Celsius temperatures to kelvins.

Given: surface area, $A = 2.0$ m^2; $T_{\text{room}} = 22$°C; skin temperature, $T = 34$°C; fraction of energy emitted, $e = 0.97$

To find: net rate of energy transfer, $\mathcal{P}_{\text{net}}$

Solution The temperature of the skin surface is

$$T = 273 + 34 = 307 \text{ K}$$

and of the room is

$$T_s = 273 + 22 = 295 \text{ K}$$

The net rate of energy transfer between the room and the body is

$$\mathcal{P}_{\text{net}} = e\sigma A(T^4 - T_s^4)$$

Substituting,

$$\mathcal{P}_{\text{net}} = 0.97 \times 5.67 \times 10^{-8}\ \text{W/(m}^2\text{·K}^4) \times 2.0\ \text{m}^2 \times (307^4 - 295^4)\ \text{K}^4$$
$$= 140 \text{ W}$$

Discussion 140 W is a significant heat loss because the body also loses about 10 W by convection and conduction. To stay at a constant body temperature, an inactive person must give off heat at a rate of 90 W to account for basal metabolic activity; if the rate of heat loss exceeds that, the body temperature starts to drop. The patient had better wrap a blanket around his body or start running in place.

We need only the fraction of energy emitted and absorbed by the body; the emissivity of the walls of the room is irrelevant. If the walls are poor emitters, then they also absorb poorly, so they reflect radiation. The amount of radiation incident on the body is the same.

Practice Problem 14.14 The Roller Blader Radiates

Find how much energy per unit time a roller blader loses by radiation from her body. Her skin temperature is 35°C and the air temperature is 30°C. Her surface area is 1.2 m^2, of which 75% is exposed to the air. Assume skin has $e = 0.97$.

Example 14.15

Radiative Equilibrium of Earth

Radiant energy from the Sun reaches Earth at a rate of 1.7×10^{17} W. An average of about 30% is reflected, and the rest is absorbed. Energy is also radiated by the atmosphere. Assuming equal rates of absorption and emission, and that the atmosphere emits as a blackbody in the infrared ($e = 1$), calculate the temperature of the atmosphere. (The Sun's radiation peaks in the visible part of the spectrum, but Earth's radiation peaks in the infrared due to its much lower surface temperature.)

Strategy Earth must radiate the same power as it absorbs. We use Stefan's law to find the rate at which energy is radiated as a function of temperature and then equate that to the rate of energy absorption.

Solution Earth absorbs 70% of the incident solar radiation. To have a relatively constant temperature, it must emit radiation at the same rate:

$$\mathcal{P} = 0.70 \times 1.7 \times 10^{17}\text{ W} = 1.2 \times 10^{17}\text{ W}$$

From Stefan's law,

$$\mathcal{P} = e\sigma A T^4$$

where we take $e = 1$ since the atmosphere is assumed to emit as a blackbody. Earth's surface area is

$$A = 4\pi R_E^2$$

Solving Stefan's law for T yields

$$T = \left(\frac{\mathcal{P}}{e\sigma A}\right)^{1/4}$$

Now we substitute numerical values:

$$T = \left(\frac{\mathcal{P}}{e\sigma 4\pi R_E^2}\right)^{1/4}$$

$$= \left[\frac{1.2 \times 10^{17}\text{ W}}{1 \times 5.67 \times 10^{-8}\text{ W/(m}^2\cdot\text{K}^4) \times 4\pi\,(6.4 \times 10^6\text{ m})^2}\right]^{1/4}$$

$$= 253\text{ K} = -20°\text{C}$$

Discussion Remember that −20°C is supposed to be the average temperature of the *atmosphere*, not of Earth's surface. This relatively simple calculation gives impressively accurate results. To find the temperature of Earth's surface, we must take the greenhouse effect into account.

Practice Problem 14.15 Reflecting Less Incident Radiation

If Earth were to reflect 25% of the incident radiation instead of 30%, what would be the average temperature of the atmosphere?

Application of Thermal Radiation: Global Climate Change

Earth receives heat by radiation from the Sun. The atmosphere helps trap some of the radiation, acting rather like the glass in a greenhouse. When sunlight falls on the glass of a greenhouse, most of the visible radiation and short-wavelength infrared (*near-infrared*) travel right on through; the glass is transparent to those wavelengths. The glass absorbs much of the incoming ultraviolet radiation. The radiation that gets through the glass is mostly absorbed inside the greenhouse. Since the inside of the greenhouse is much cooler than the Sun, it emits primarily infrared radiation (IR). The glass is not transparent to this longer-wavelength IR; much of it is absorbed by the glass. The glass itself also emits IR, but in both directions: half of it is emitted back inside the greenhouse. The absorption of IR by the glass keeps the greenhouse warmer than it would otherwise be. (The glass in a greenhouse has a second function not mirrored in Earth's atmosphere—it prevents heat from being carried away by convection.)

Earth is something like a greenhouse, where the atmosphere fulfills the role of the glass. Like glass, the atmosphere is largely transparent to visible and near IR; the ozone layer in the upper atmosphere absorbs some of the ultraviolet. The atmosphere absorbs a great deal of the longer-wavelength IR emitted by Earth's surface. The atmosphere *radiates* IR in two directions: back toward the surface and out toward space (Fig. 14.19). "Greenhouse gases" such as CO_2 and water vapor are particularly good

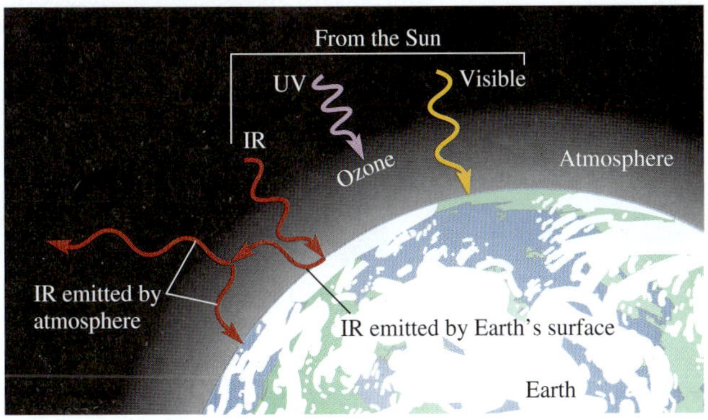

Figure 14.19 The global greenhouse effect. In this *simplified* diagram, all the UV from the Sun is absorbed by the atmosphere, while all the visible and IR from the Sun is transmitted. Earth absorbs the visible and IR and radiates longer-wavelength IR. The longer-wavelength IR is absorbed by the atmosphere, which itself radiates IR both back toward the surface and out toward space.

absorbers of IR. The higher the concentration of greenhouse gases in the atmosphere, the more IR is absorbed and the warmer Earth's surface becomes. Even small changes in the average surface temperature can have dramatic effects on climate.

In applying Stefan's radiation law to Earth, there are some complications. One is the effect of the cloud cover. Clouds are quite reflective, but they are sometimes there and sometimes not. The heating of the lakes and oceans causes water to evaporate and form clouds. The clouds then serve as a screen and reflect sunlight away from Earth, reducing the temperature again.

Master the Concepts

- The internal energy of a system is the total energy of all of the molecules in the system except for the macroscopic kinetic energy (kinetic energy associated with macroscopic translation or rotation) and the external potential energy (energy due to external interactions).
- Heat is a *flow* of energy that occurs due to a temperature difference.
- The joule is the SI unit for all forms of energy, for heat, and for work. An alternative unit sometimes used for heat and internal energy is the calorie:

$$1\ \text{cal} = 4.186\ \text{J} \qquad (14\text{-}1)$$

- The ratio of heat flow into a system to the temperature change of the system is the heat capacity of the system:

$$Q = C\,\Delta T \qquad (14\text{-}2)$$

- The heat capacity per unit mass is the specific heat capacity (or specific heat) of a substance:

$$c = \frac{Q}{m\,\Delta T} \qquad (14\text{-}3)$$

- The *molar specific heat* is the heat capacity per mole:

$$C_V = \frac{Q}{n\,\Delta T} \qquad (14\text{-}6)$$

At room temperature, the molar heat capacity at constant volume for a monatomic ideal gas is approximately $C_V = \frac{3}{2}R$, and for a diatomic ideal gas it is approximately $C_V = \frac{5}{2}R$.

- Phase transitions occur at constant temperature. The heat *per unit mass* that must flow to melt a solid or to freeze a liquid is the latent heat of fusion L_f. The latent heat of vaporization L_v is the heat *per unit mass* that must flow to change the phase from liquid to gas or from gas to liquid.

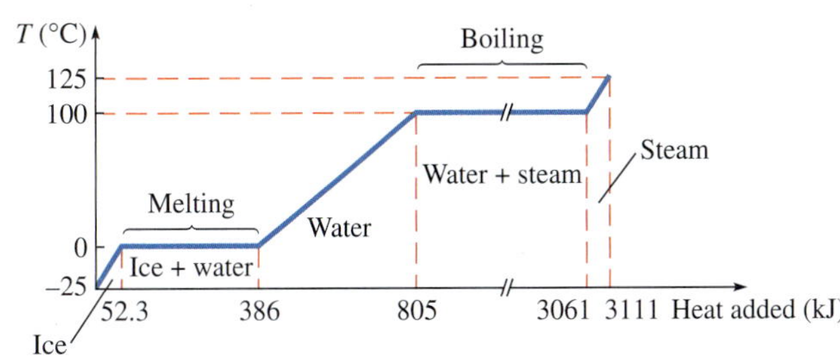

- *Sublimation* occurs when a solid changes directly to a gas without going into a liquid form.
- A phase diagram is a graph of pressure versus temperature that indicates solid, liquid, and gas regions for a

continued on next page

Master the Concepts continued

substance. The sublimation, fusion, and vapor pressure curves separate the three phases. Crossing one of these curves represents a phase transition.

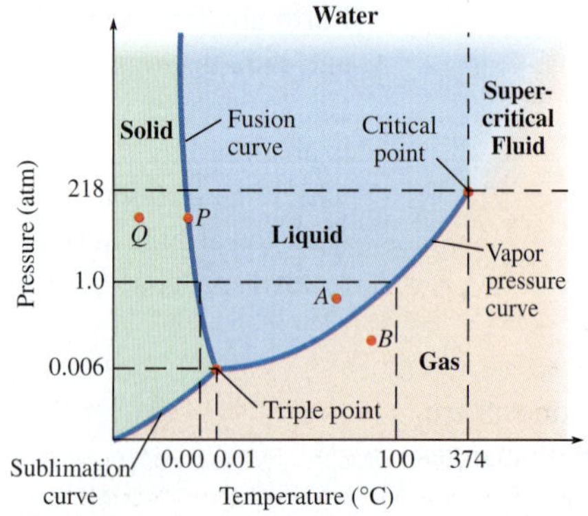

- Heat flows by three processes: conduction, convection, and radiation.
- Conduction is due to atomic (or molecular) collisions within a substance or from one object to another when they are in contact. The rate of heat flow within a substance is:

$$\mathcal{P} = \kappa A \frac{\Delta T}{d} \quad \text{(14-10)}$$

where $\mathcal{P}$ is the rate of heat flow (or power delivered), κ is the thermal conductivity of the material, A is the cross-sectional area, d is the thickness (or length) of the material, and ΔT is the temperature difference between one side and the other.

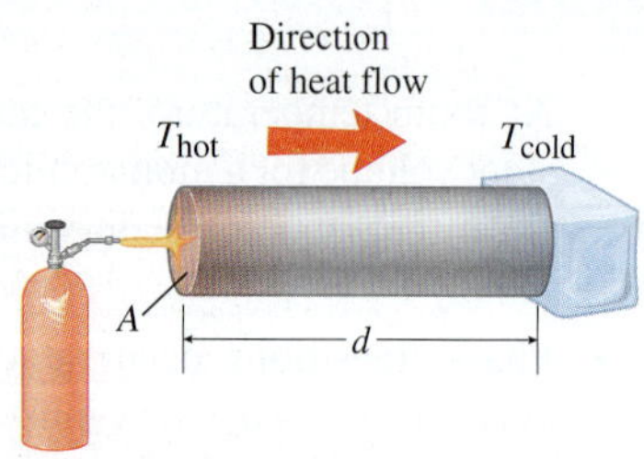

- Convection involves *fluid currents* that carry heat from one place to another. In convection, the material itself moves from one place to another.

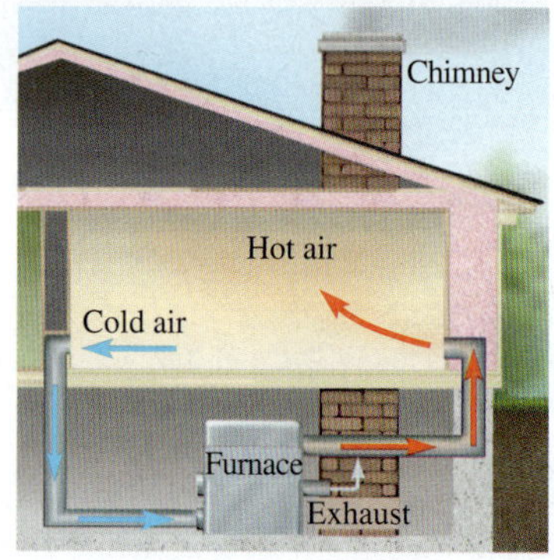

- Thermal radiation does not have to travel through a material medium. The energy is carried by electromagnetic waves that travel at the speed of light. All bodies emit energy through electromagnetic radiation. An idealized object that absorbs all the radiation incident on it is called a blackbody. A blackbody emits more radiant power per unit surface area than any real object at the same temperature. Stefan's law of thermal radiation is

$$\mathcal{P} = e\sigma A T^4 \quad \text{(14-16)}$$

where the emissivity e ranges from 0 to 1, A is the surface area, T is the surface temperature *in kelvins*, and Stefan's constant is $\sigma = 5.670 \times 10^{-8}\ \text{W/(m}^2\cdot\text{K}^4)$. The wavelength of maximum power emission is inversely proportional to the absolute temperature:

$$\lambda_{max} T = 2.898 \times 10^{-3}\ \text{m}\cdot\text{K} \quad \text{(14-17)}$$

The difference between the power emitted by the object and that absorbed by the object from its surroundings is the net power emitted:

$$\mathcal{P}_{net} = e\sigma A(T^4 - T_s^4) \quad \text{(14-18)}$$

Conceptual Questions

1. What determines the direction of heat flow when two objects at different temperatures are placed in thermal contact?
2. When an old movie has a scene of someone ironing, the person is often shown testing the heat of a hot flat iron with a moistened finger. Why is this safe to do?
3. Why do lakes and rivers freeze first at their surfaces?
4. Why is drinking water in a camp located near the equator often kept in porous jars?
5. Why are several layers of clothing warmer than one coat of equal weight?
6. Why are vineyards planted along lakeshores or riverbanks in cold climates?
7. A metal plant stand on a wooden deck feels colder than the wood around it. Is it necessarily colder? Explain.
8. Near a large lake, in what direction does a breeze passing over the land tend to blow at night?
9. What is the purpose of having fins on an automobile or motorcycle radiator?
10. Why do roadside signs warn that bridges ice before roadways? Explain.
11. Why do cooking directions on packages advise different timing to be followed for some locations?
12. Explain the theory behind the pressure cooker. How does it speed up cooking times?
13. When you eat a pizza that has just come from the oven, why is it that you are apt to burn the roof of your mouth with the first bite although the crust of the pizza feels only warm to your hand?
14. Explain why the molar specific heat of a diatomic gas such as O_2 is larger than that of a monatomic gas such as Ne.

15. At very low temperatures, the molar specific heat of hydrogen (H_2) is $C_V = 1.5R$. At room temperature, $C_V = 2.5R$. Explain.

16. When the temperature as measured in °C of a radiating object is doubled (such as a change from 20°C to 40°C), is the radiation rate necessarily increased by a factor of 16?

17. A cup of hot coffee has been poured, but the coffee drinker has a little more work to do at the computer before she picks up the cup. She intends to add some milk to the coffee. To keep the coffee hot as long as possible, should she add the milk at once, or wait until just before she takes her first sip?

18. Would heat loss be reduced or increased by increasing the usual air gap, 1 to 2 cm, between commercially made double-paned windows? Explain your reasoning. [*Hint*: Consider convection.]

19. A study of food preservation in Britain discovered that the temperature of meat that is kept in transparent plastic packages and stored in open and lighted freezers can be as much as 12°C above the temperature of the freezer. Why is this? How could this be prevented?

20. Which possesses more total internal energy, the water within a large, partially ice-covered lake in winter or a 6 cup teapot filled with hot tea? Explain.

21. A room in which the air temperature is held constant may feel warm in the summer but cool in the winter. Explain. [*Hint*: The walls are not necessarily at the same temperature as the air.]

22. Many homes are heated with "radiators," which are hollow metal devices filled with hot water or steam and located in each room of the house. They are sometimes painted with metallic, high gloss silver paint so that they look well polished. Does this make them better radiators of heat? If not, what might be a more efficient finish to use?

23. Two objects with the same surface area are inside an evacuated container. The walls of the container are kept at a constant temperature. Suppose one object absorbs a larger fraction of incident radiation than the other. Explain why that object must emit a correspondingly greater amount of radiation than the other. Thus a good absorber must be a good emitter.

24. Even though heat is not a fluid, Eq. (14-11) has a close analogy in Poiseuille's law, which describes the viscous flow of a fluid through a pipe (see Problem 9.69). (a) Explain the analogy. (b) For two or more thermal conductors in series, the total thermal resistance is just the sum of the thermal resistances [Eq. (14-13)]. Is the total fluid flow resistance for two or more pipes in series equal to the sum of the resistances? Explain.

25. In a conventional exterior wall, a 2 × 6 wooden stud is placed every 16 in. (A *stud* is an upright support.) The stud runs all the way from the exterior siding to the interior wall and the spaces between studs are filled with insulation. In offset stud wall construction, 2 × 4 studs are staggered as shown in the figure. Each stud connects with the exterior siding *or* the interior wall, but not both. Explain why an offset stud wall is much more energy efficient than a conventional wall of the same thickness and with the same insulation material.

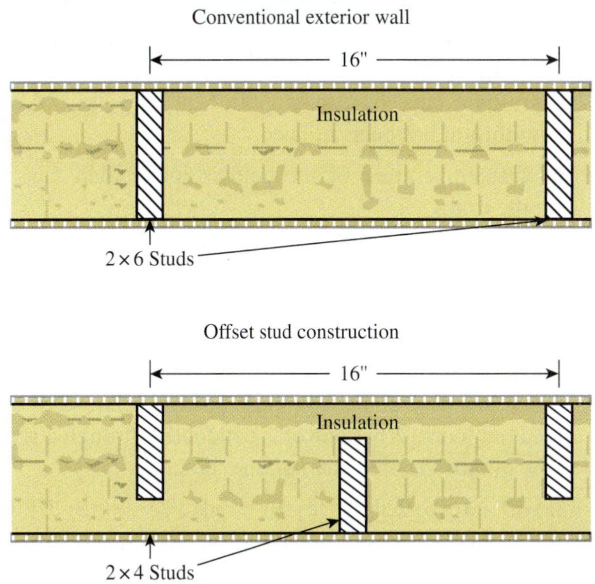

26. (a) Why is the coolant fluid in an automobile kept under high pressure? (b) Why do radiator caps have safety valves, allowing you to reduce the pressure before removing the cap? [*Hint*: See Fig. 14.7a, the phase diagram for water.]

Multiple-Choice Questions

Student Response System Questions

1. The main loss of heat from Earth is by
(a) radiation.
(b) convection.
(c) conduction.
(d) All three processes are significant modes of heat loss from Earth.

2. The average temperature of Earth's atmosphere is 253 K. What would be the eventual average temperature of Earth's atmosphere if the power radiated by the Sun were to decrease by 10%?
(a) 253 K
(b) $(0.90)^{1/4} \times 253\ \text{K} = 246\ \text{K}$
(c) $0.90 \times 253\ \text{K} = 228\ \text{K}$
(d) $(0.90)^4 \times 253\ \text{K} = 166\ \text{K}$

3. Which term best represents the relation between a blackbody and radiant energy? A blackbody is an ideal _____ of radiant energy.
(a) emitter (b) absorber
(c) reflector (d) emitter and absorber

4. If Mars orbits the Sun with an orbital radius that is 1.5 times the orbital radius of Earth about the Sun, what

is the approximate atmospheric temperature of Mars? The atmospheric temperature of Earth is 253 K.

(a) (253 K)/1.5 = 170 K (b) (253 K)/1.5^2 = 112 K
(c) (253 K)/1.5^4 = 50 K (d) (253 K)/1.5$^{1/2}$ = 207 K

5. Iron has a specific heat that is about 3.4 times that of gold. A cube of gold and a cube of iron, both of equal mass and at 20°C, are placed in two different Styrofoam cups, each filled with 100 g of water at 40°C. The Styrofoam cups have negligible heat capacities. After equilibrium has been attained,
   (a) the temperature of the gold is lower than that of the iron.
   (b) the temperature of the gold is higher than that of the iron.
   (c) the temperatures of the water in the two cups are the same.
   (d) Either (a) or (b), depending on the mass of the cubes.
6. A window conducts power $\mathcal{P}$ from a house to the cold outdoors. What power is conducted through a window of *half* the area and *half* the thickness?
   (a) $4\mathcal{P}$ (b) $2\mathcal{P}$ (c) $\mathcal{P}$ (d) $\mathcal{P}/2$ (e) $\mathcal{P}/4$
7. If you place your hand underneath, but not touching, a kettle of hot water, you *mainly* feel the presence of heat from
   (a) conduction.
   (b) convection.
   (c) radiation.
8. Two thin rods are made from the same material and are of lengths L_1 and L_2. The two ends of the rods have the same temperature difference. What should the relation be between their diameters and lengths so that they conduct equal amounts of heat energy in a given time?
   (a) $\frac{L_1}{L_2} = \frac{d_1}{d_2}$ (b) $\frac{L_1}{L_2} = \frac{d_2}{d_1}$
   (c) $\frac{L_1}{L_2} = \frac{d_1^2}{d_2^2}$ (d) $\frac{L_1}{L_2} = \frac{d_2^2}{d_1^2}$
9. Sublimation is involved in which of these phase changes?
   (a) liquid to gas (b) solid to liquid
   (c) solid to gas (d) gas to liquid
10. When a gas condenses to a liquid,
    (a) its internal energy increases.
    (b) its temperature rises.
    (c) its temperature falls.
    (d) it gives off internal energy.
11. When a substance is at its triple point, it
    (a) is in its solid phase.
    (b) is in its liquid phase.
    (c) is in its gas phase.
    (d) may be in any or all of these phases.
12. The phase diagram for water is shown in the figure. If the temperature of a certain amount of ice is increased by following the path represented by the horizontal dashed line from A to B through point P, which of the graphs of temperature as a function of heat added is correct?

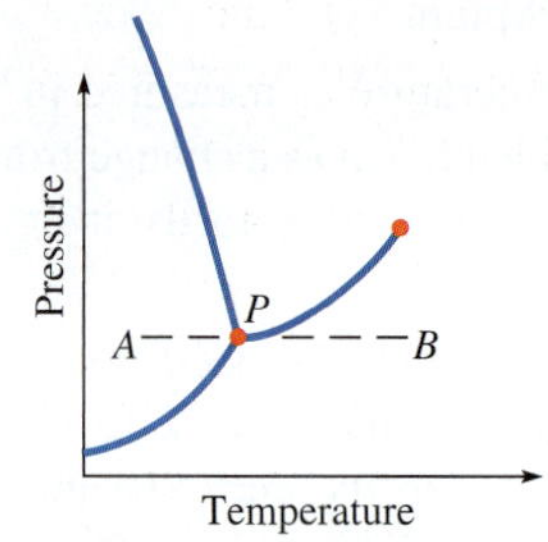

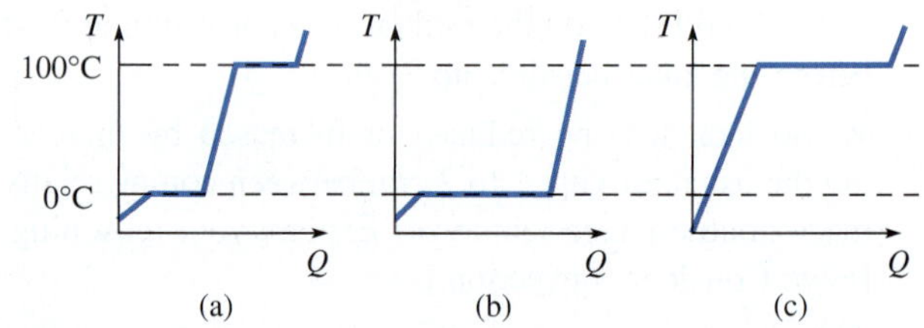

Problems

C Combination conceptual/quantitative problem
Biomedical application
✦ Challenging
Blue # Detailed solution in the Student Solutions Manual
[1, 2] Problems paired by concept
connect Text website interactive or tutorial

14.1 Internal Energy

1. A mass of 1.4 kg of water at 22°C is poured from a height of 2.5 m into a vessel containing 5.0 kg of water at 22°C. (a) How much does the internal energy of the 6.4 kg of water increase? (b) Is it likely that the water temperature increases? Explain.
2. The water passing over Victoria Falls, located along the Zambezi River on the border of Zimbabwe and Zambia, drops about 105 m. How much internal energy is produced per kilogram as a result of the fall?
3. How much internal energy is generated when a 20.0 g lead bullet, traveling at 7.00×10^2 m/s, comes to a stop as it strikes a metal plate?
4. Nolan threw a baseball, of mass 147.5 g, at a speed of 162 km/h to a catcher. How much internal energy was generated when the ball struck the catcher's mitt?
5. A child of mass 15 kg climbs to the top of a slide that is 1.7 m above a horizontal run that extends for 0.50 m at the base of the slide. After sliding down, the child comes to rest just before reaching the very end of the horizontal portion of the slide. (a) How much internal energy was generated during this process? (b) Where did the generated energy go? (To the slide, to the child, to the air, or to all three?)

6. A 64 kg sky diver jumped out of an airplane at an altitude of 0.90 km. She opened her parachute after a while and eventually landed on the ground with a speed of 5.8 m/s. How much energy was dissipated by air resistance during the jump?

7. ✦ During basketball practice Shane made a jump shot, releasing a 0.60 kg basketball from his hands at a height of 2.0 m above the floor with a speed of 7.6 m/s. The ball swooshes through the net at a height of 3.0 m above the floor and with a speed of 4.5 m/s. How much energy was dissipated by air drag from the time the ball left Shane's hands until it went through the net?

14.2 Heat; 14.3 Heat Capacity and Specific Heat

8. What is the heat capacity of 20.0 kg of silver?

9. What is the heat capacity of a gold ring that has a mass of 5.00 g?

10. What is the heat capacity of a 30.0 kg block of ice?

11. What is the heat capacity of 1.00 m^3 of aluminum?

12. Convert 1.00 kJ to kilowatt-hours (kWh).

13. If 125.6 kJ of heat are supplied to 5.00×10^2 g of water at 22°C, what is the final temperature of the water?

14. Rank these six situations in order of the temperature increase, largest to smallest.
    (a) 1 kJ of heat into 400 g of steel with $c = 0.45$ kJ/(kg·K)
    (b) 2 kJ of heat into 400 g of steel
    (c) 2 kJ of heat into 800 g of steel
    (d) 1 kJ of heat into 400 g of aluminum with $c =$ 0.90 kJ/(kg·K)
    (e) 2 kJ of heat into 400 g of aluminum
    (f) 2 kJ of heat into 800 g of aluminum

15. What is the heat capacity of a system consisting of (a) a 0.450 kg brass cup filled with 0.050 kg of water? (b) 7.5 kg of water in a 0.75 kg aluminum bucket?

16. A 0.400 kg aluminum teakettle contains 2.00 kg of water at 15.0°C. How much heat is required to raise the temperature of the water (and kettle) to 100.0°C?

17. How much heat is required to raise the body temperature of a 50.0 kg woman from 37.0°C to 38.4°C?

18. It takes 880 J to raise the temperature of 350 g of lead from 0°C to 20.0°C. What is the specific heat of lead?

19. A mass of 1.00 kg of water at temperature T is poured from a height of 0.100 km into a vessel containing water of the same temperature T, and a temperature change of 0.100°C is measured. What mass of water was in the vessel? Ignore heat flow into the vessel, the thermometer, and so on.

20. An experiment is conducted with a Joule apparatus (see Fig. 14.2), where a mass is allowed to descend by 1.25 m and rotate paddles within an insulated container of water. There are several different sizes of descending masses to choose among. If the investigator wishes to deliver 1.00 kJ to the water within the insulated container after 30.0 descents, what descending mass value should be used?

21. It is a damp, chilly day in a New England seacoast town suffering from a power failure. To warm up the cold, clammy sheets, Jen decides to fill hot water bottles to tuck between the sheets at the foot of the beds. If she wishes to heat 2.0 L of water on the wood stove from 20.0°C to 80.0°C, how much heat must flow into the water?

22. An 83 kg man eats a banana of energy content 418 kJ (100 kcal). If all of the energy from the banana is converted into kinetic energy of the man, how fast is he moving, assuming he starts from rest?

23. A high jumper of mass 60.0 kg consumes a meal of 3.00×10^3 kcal prior to a jump. If 3.3% of the energy from the food could be converted to gravitational potential energy in a single jump, how high could the athlete jump?

24. A thermometer containing 0.10 g of mercury is cooled from 15.0°C to 8.5°C. How much energy left the mercury in this process?

25. A bit of space debris penetrates the hull of a spaceship traversing the asteroid belt and comes to rest in a container of water that was at 20.0°C before being hit. The mass of the debris is 1.0 g and the mass of the water is 1.0 kg. If the space rock traveled at 8.4×10^3 m/s with respect to the spaceship and if all of its kinetic energy is used to heat the water, what is the final temperature of the water?

26. A 7.30 kg steel ball at 15.2°C is dropped from a height of 10.0 m into an insulated container with 4.50 L of water at 10.1°C. If no water splashes, what is the final temperature of the water and steel?

27. A heating coil inside an electric kettle delivers 2.1 kW of electric power to the water in the kettle. How long will it take to raise the temperature of 0.50 kg of water from 20.0°C to 100.0°C? (connect tutorial: heating)

14.4 Specific Heat of Ideal Gases

28. A cylinder contains 250 L of hydrogen gas (H_2) at 0.0°C and a pressure of 10.0 atm. How much energy is required to raise the temperature of this gas to 25.0°C?

29. A container of nitrogen gas (N_2) at 23°C contains 425 L at a pressure of 3.5 atm. If 26.6 kJ of heat are added to the container, what will be the new temperature of the gas?

30. Imagine that 501 people are present in a movie theater of volume 8.00×10^3 m^3 that is sealed shut so no air can escape. Each person gives off heat at an average rate of 110 W. By how much will the temperature of the air have increased during a 2.0 h movie? The initial pressure is 1.01×10^5 Pa and the initial

temperature is 20.0°C. Assume that all the heat output of the people goes into heating the air (a diatomic gas).

31. A resting person takes in about 0.5 L of air in a single breath. In an hour of breathing, approximately 5 m^3 of air will be breathed. (a) If air is taken in at 20°C and raised in temperature to 35°C before being exhaled, how much heat leaves the body in a day due to breathing alone? Ignore the humidification of the air in the lungs, treat air as an ideal diatomic gas, and assume the pressure is 101 kPa. (b) Is this heat loss significant relative to the total heat loss from the body in one day, about 9 MJ? (c) What is the average power (rate of heat loss) due to breathing alone?

32. A chamber with a fixed volume of 1.0 m^3 contains a monatomic gas at 3.00×10^2 K. The chamber is heated to a temperature of 4.00×10^2 K. This operation requires 10.0 J of heat. (Assume all the energy is transferred to the gas.) How many gas molecules are in the chamber?

14.5 Phase Transitions

33. As heat flows into a substance, its temperature changes according to the graph in the diagram. For which sections of the graph is the substance undergoing a phase change? For the sections you identified, what kind of phase change is occurring? (connect tutorial: temperature graph)

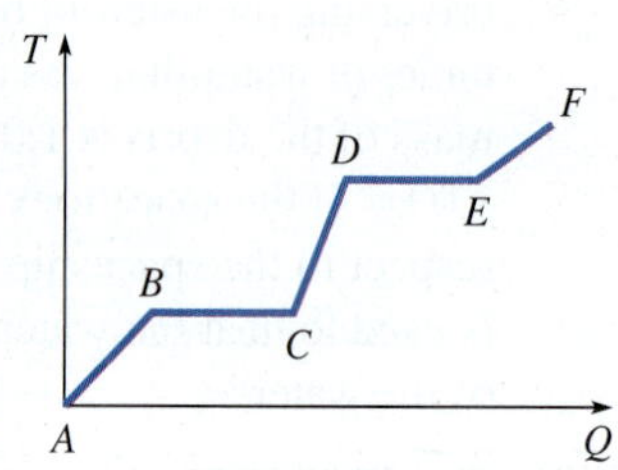

34. Given these data, compute the heat of vaporization of water. The specific heat capacity of water is 4.186 J/(g·K).

Mass of calorimeter = 3.00×10^2 g	Specific heat of calorimeter = 0.380 J/(g·K)
Mass of water = 2.00×10^2 g	Initial temperature of water and calorimeter = 15.0°C
Mass of condensed steam = 18.5 g	Initial temperature of steam = 100.0°C
	Final temperature of calorimeter = 62.0°C

35. Given these data, compute the heat of fusion of water. The specific heat capacity of water is 4.186 J/(g·K).

Mass of calorimeter = 3.00×10^2 g	Specific heat of calorimeter = 0.380 J/(g·K)
Mass of water = 2.00×10^2 g	Initial temperature of water and calorimeter = 20.0°C
Mass of ice = 30.0 g	Initial temperature of ice = 0°C
	Final temperature of calorimeter = 8.5°C

36. In an emergency, it is sometimes the practice of medical professionals to immerse a patient who suffers from heat stroke in an ice bath, a mixture of ice and water in equilibrium at 0°C, in order to reduce her body temperature. (a) If a 75 kg patient whose body temperature is 40.8°C must have her temperature reduced to the normal range, how much heat must be removed? (b) If she is placed in a bath containing 7.5 kg of ice, will there be ice remaining in the bath when her body temperature is 37.0°C? If so, how much? If not, what will the final water temperature be?

37. In a physics lab, a student accidentally drops a 25.0 g brass washer into an open dewar of liquid nitrogen at 77.2 K. How much liquid nitrogen boils away as the washer cools from 293 K to 77.2 K? The latent heat of vaporization for nitrogen is 199.1 kJ/kg.

38. What mass of water at 25.0°C added to a Styrofoam cup containing two 50.0 g ice cubes from a freezer at −15.0°C will result in a final temperature of 5.0°C for the drink?

39. How much heat is required to change 1.0 kg of ice, originally at −20.0°C, into steam at 110.0°C? Assume 1.0 atm of pressure.

40. Ice at 0.0°C is mixed with 5.00×10^2 mL of water at 25.0°C. How much ice must melt to lower the water temperature to 0.0°C?

41. Tina is going to make iced tea by first brewing hot tea, then adding ice until the tea cools. How much ice, at a temperature of −10.0°C, should be added to a 2.00×10^{-4} m^3 glass of tea at 95.0°C to cool the tea to 10.0°C? Ignore the temperature change of the glass. (connect tutorial: iced tea)

42. Repeat Problem 41 without ignoring the temperature change of the glass. The glass has a mass of 350 g and the specific heat of the glass is 0.837 kJ/(kg·K). By what percentage does the answer change from the answer for Problem 41?

43. The graph shows the change in temperature as heat is supplied to a certain mass of ice initially at −80.0°C. What is the mass of the ice?

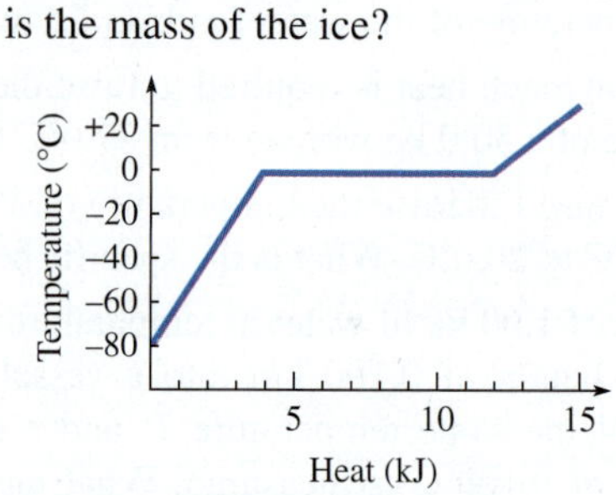

44. How many grams of aluminum at 80.0°C must be dropped into a hole in a block of ice at 0.0°C to melt 10.0 g of ice?

45. Is it possible to heat the aluminum of Problem 44 to a high enough temperature so that it melts an equal mass of ice? If so, what temperature must the aluminum have?

46. If a leaf is to maintain a temperature of 40°C (reasonable for a leaf), it must lose 250 W/m^2 by transpiration (evaporative heat loss). Note that the leaf also loses heat by radiation, but we will ignore this. How much water is lost after 1 h through transpiration only? The area of the leaf is 0.005 m^2.

47. A birch tree loses 618 mg of water per minute through transpiration (evaporation of water through stomatal pores). What is the rate of heat lost through transpiration?

48. (a) How much ice at −10.0°C must be placed in 0.250 kg of water at 25.0°C to cool the water to 0°C and melt all of the ice? (b) If half that amount of ice is placed in the water, what is the final temperature of the water?

49. A 75 g cube of ice at −10.0°C is placed in 0.500 kg of water at 50.0°C in an insulating container so that no heat is lost to the environment. Will the ice melt completely? What will be the final temperature of this system?

50. A 0.360 kg piece of solid lead at 20°C is placed into an insulated container holding 0.980 kg of liquid lead at 420°C. The system comes to an equilibrium temperature with no loss of heat to the environment. Ignore the heat capacity of the container. (a) Is there any solid lead remaining in the system? (b) What is the final temperature of the system?

51. ✦ A dog loses a lot of heat through panting. The air rushing over the upper respiratory tract causes evaporation and thus heat loss. A dog typically pants at a rate of 670 pants per minute. As a rough calculation, assume that one pant causes 0.010 g of water to be evaporated from the respiratory tract. What is the rate of heat loss for the dog through panting?

52. ✦ You are given 250 g of coffee (same specific heat as water) at 80.0°C (too hot to drink). In order to cool this to 60.0°C, how much ice (at 0.0°C) must be added? Ignore heat content of the cup and heat exchanges with the surroundings.

53. ✦ A phase diagram is shown. Starting at point *A*, follow the dashed line to point *E* and consider what happens to the substance represented by this diagram as its pressure and temperature are changed. (a) Explain what happens for each line segment, *AB*, *BC*, *CD*, and *DE*. (b) What is the significance of point *a* and of point *b*?

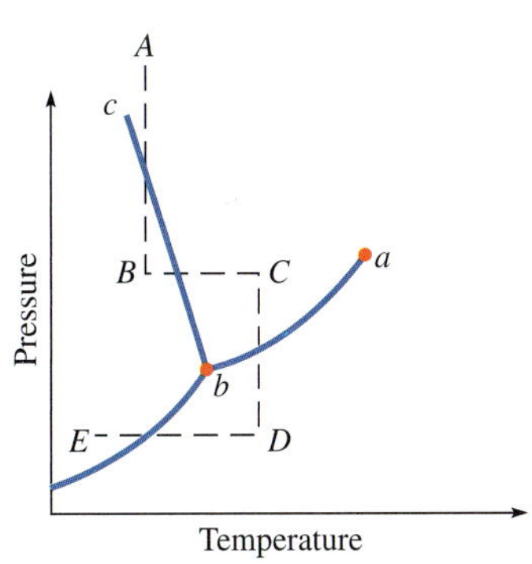

54. ✦ Compute the heat of fusion of a substance from these data: 31.15 kJ will change 0.500 kg of the solid at 21°C to liquid at 327°C, the melting point. The specific heat of the solid is 0.129 kJ/(kg·K).

14.6 Thermal Conduction

55. (a) What thickness of cork would have the same R-factor as a 1.0 cm thick stagnant air pocket? (b) What thickness of tin would be required for the same R-factor?

56. A metal rod with a diameter of 2.30 cm and length of 1.10 m has one end immersed in ice at 32.0°F and the other end in boiling water at 212°F. If the ice melts at a rate of 1.32 g every 175 s, what is the thermal conductivity of this metal? Identify the metal. Assume there is no heat lost to the surrounding air.

57. Given a slab of material with area 1.0 m^2 and thickness 2.0×10^{-2} m, (a) what is the thermal resistance if the material is asbestos? (b) What is the thermal resistance if the material is iron? (c) What is the thermal resistance if the material is copper?

58. A copper rod of length 0.50 m and cross-sectional area 6.0×10^{-2} cm^2 is connected to an iron rod with the same cross section and length 0.25 m. One end of the copper is immersed in boiling water and the other end is at the junction with the iron. If the far end of the iron rod is in an ice bath at 0°C, find the rate of heat transfer passing from the boiling water to the ice bath. Assume there is no heat loss to the surrounding air. (connect tutorial: composite rod)

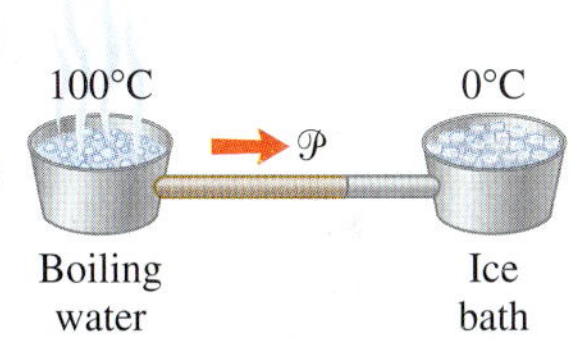

59. A wall that is 2.74 m high and 3.66 m long has a thickness composed of 1.00 cm of wood plus 3.00 cm of insulation (with the thermal conductivity approximately of wool). The inside of the wall is 23.0°C and the outside of the wall is at −5.00°C. (a) What is the rate of heat flow through the wall? (b) If half the area of the wall is replaced with a single pane of glass that is 0.500 cm thick, how much heat flows out of the wall now?

60. Boiling water in an aluminum pan is being converted to steam at a rate of 10.0 g/s. The flat bottom of the pan has an area of 325 cm^2 and the pan's thickness is 3.00 mm. If 27.0% of all heat that is transferred to the pan from the flame beneath it is lost from the sides of the pan and the remaining 73.0% goes into the water, what is the temperature of the base of the pan?

61. Your hot water tank is insulated, but not very well. To reduce heat loss, you wrap some old blankets around it. With the water at 81°C and the room at 21°C, a thermometer inserted between the outside of the original tank and your blanket reads 36°C. By what factor did the blanket reduce the heat loss?

62. A copper rod has one end in ice at a temperature of 0°C, the other in boiling water. The length and diameter of the rod are 1.00 m and 2.00 cm, respectively. At what rate in grams per hour does the ice melt? Assume no heat flows out the sides of the rod.

63. The thermal conductivity of the fur (including the skin) of a male Husky dog is 0.026 W/(m·K). The dog's heat output is measured to be 51 W, its internal temperature is 38°C, its surface area is 1.31 m^2, and the thickness of the fur is 5.0 cm. How cold can the outside temperature be before the dog must increase its heat output?

64. The thermal resistance of a seal's fur and blubber combined is 0.33 K/W. If the seal's internal temperature is 37°C and the temperature of the sea is about 0°C, what must be the heat output of the seal in order for it to maintain its internal temperature?

65. A hiker is wearing wool clothing of 0.50 cm thickness to keep warm. Her skin temperature is 35°C and the outside temperature is 4.0°C. Her body surface area is 1.2 m^2. (a) If the thermal conductivity of wool is 0.040 W/(m·K), what is the rate of heat conduction through her clothing? (b) If the hiker is caught in a rainstorm, the thermal conductivity of the soaked wool increases to 0.60 W/(m·K) (that of water). Now what is the rate of heat conduction?

66. Find the temperature drop across the epidermis (the outer layer of skin) under these conditions: the rate of heat flow via conduction through a 10.0 cm^2 area of the epidermis is 50 mW; the epidermis is 2.00 mm thick and has thermal conductivity 0.45 W/(m·K).

67. ✦ One cross-country skier is wearing a down jacket that is 2.0 cm thick. The thermal conductivity of goose down is 0.025 W/(m·K). Her companion on the ski outing is wearing a wool jacket that is 0.50 cm thick. The thermal conductivity of wool is 0.040 W/(m·K). (a) If both jackets have the same surface area and the skiers both have the same body temperature, which one will stay warmer longer? (b) How much longer can the person with the warmer jacket stay outside for the same amount of heat loss?

68. Five walls of a house have different surface areas, insulation materials, and insulation thicknesses. Rank them in order of the rate of heat flow through the wall, greatest to smallest. Assume the same indoor and outdoor temperatures for each wall.

(a) area = 120 m^2; 10 cm thickness of insulation with thermal conductivity 0.030 W/(m·K)
(b) area = 120 m^2; 15 cm thickness of insulation with thermal conductivity 0.045 W/(m·K)
(c) area = 180 m^2; 10 cm thickness of insulation with thermal conductivity 0.045 W/(m·K)
(d) area = 120 m^2; 10 cm thickness of insulation with thermal conductivity 0.045 W/(m·K)
(e) area = 180 m^2; 15 cm thickness of insulation with thermal conductivity 0.030 W/(m·K)

69. For a temperature difference $\Delta T = 20.0$°C, one slab of material conducts 10.0 W/m^2; another of the same shape conducts 20.0 W/m^2. What is the rate of heat flow per m^2 of surface area when the slabs are placed side by side with $\Delta T_{tot} = 20.0$°C?

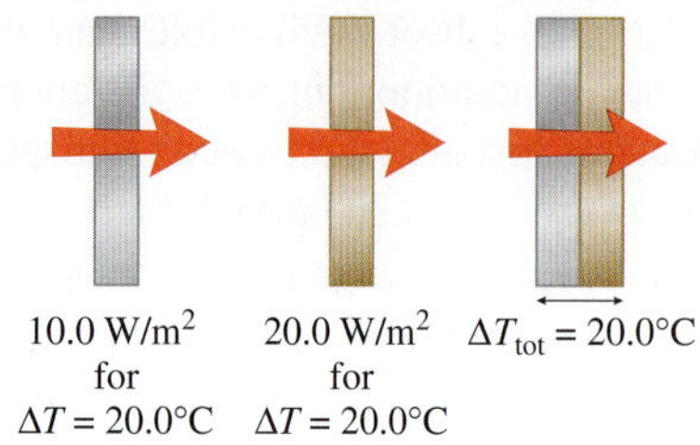

70. A wall consists of a layer of wood and a layer of cork insulation of the same thickness. The temperature inside is 20.0°C, and the temperature outside is 0.0°C. (a) What is the temperature at the interface between the wood and cork if the cork is on the inside and the wood on the outside? (b) What is the temperature at the interface if the wood is inside and the cork is outside? (c) Does it matter whether the cork is placed on the inside or the outside of the wooden wall? Explain.

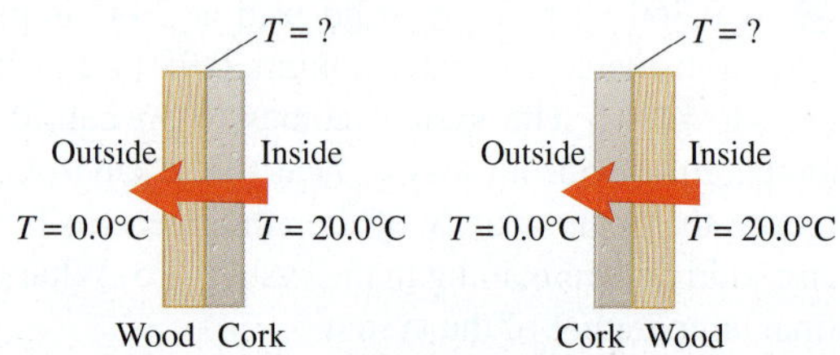

71. A brick wall with thermal conductivity $\kappa = 1.3$ W/(m·K) is covered completely with a sheet of foam of the same thickness as the brick, but with $\kappa = 0.025$ W/(m·K). How is the rate at which heat is conducted through the wall changed by the addition of the foam?

14.8 Thermal Radiation

72. If a blackbody is radiating at $T = 1650$ K, at what wavelength is the maximum intensity?

73. Wien studied the spectral distribution of many radiating bodies to finally discover a simple relation between wavelength and intensity. Use the limited data shown in Fig. 14.17 to find the constant predicted by Wien for the product of wavelength of maximum emission and temperature.

74. Six wood stoves have total surface areas A and surface temperatures T as given. Rank them in order of the power radiated, from greatest to least. Assume they all have the same emissivity.

(a) $A = 1.00\ m^2, T = 227$°C (b) $A = 1.01\ m^2, T = 227$°C
(c) $A = 1.05\ m^2, T = 227$°C (d) $A = 1.00\ m^2, T = 232$°C
(e) $A = 0.99\ m^2, T = 232$°C (f) $A = 0.98\ m^2, T = 232$°C

75. A sphere with a diameter of 80 cm is initially at a temperature of 250°C. If the intensity of the radiation detected at a distance of 2.0 m from the sphere's center is 102 W/m^2, what is the emissivity of the sphere?

76. An incandescent lightbulb has a tungsten filament that is heated to a temperature of 3.00×10^3 K when an electric

current passes through it. If the surface area of the filament is approximately 1.00×10^{-4} m^2 and it has an emissivity of 0.32, what is the power radiated by the bulb?

77. A tungsten filament in a lamp is heated to a temperature of 2.6×10^{-3} K by an electric current. The tungsten has an emissivity of 0.32. What is the surface area of the filament if the lamp delivers 40.0 W of power?

78. A person of surface area 1.80 m^2 is lying out in the sunlight to get a tan. If the intensity of the incident sunlight is 7.00×10^2 W/m^2, at what rate must heat be lost by the person in order to maintain a constant body temperature? (Assume the effective area of skin exposed to the Sun is 42% of the total surface area, 57% of the incident radiation is absorbed, and that internal metabolic processes contribute another 90 W for an inactive person.)

79. A student wants to lose some weight. He knows that rigorous aerobic activity uses about 700 kcal/h (2900 kJ/h) and that it takes about 2000 kcal per day (8400 kJ) just to support necessary biological functions, including keeping the body warm. He decides to burn calories faster simply by sitting naked in a 16°C room and letting his body radiate calories away. His body has a surface area of about 1.7 m^2, and his skin temperature is 35°C. Assuming an emissivity of 1.0, at what rate (in kcal/h) will this student "burn" calories?

80. A student in a lecture hall has 0.25 m^2 of skin (arms, hands, and heads, at 34°C, with emissivity of 97%) exposed. The temperature of the room is 20°C (air, walls, ceiling, and floor all at the same temperature). (a) At what rate does the skin emit thermal radiation? (b) At what rate does the skin absorb thermal radiation? (c) What is the net rate of heat flow from the body due to thermal radiation? Compare this to the *total* rate of heat flow from the body, about 100 W.

81. It is often argued that the head is the most important part of the body to cover when out in cold weather. Estimate the total energy loss by radiation if a person's head is uncovered for 15 min on a very cold, −15°C day, assuming he is bald, his skin temperature is 35°C, and that skin has an emissivity (in the infrared) of 97%.

82. Consider the *net* rate of heat loss by radiation from exposed skin on a cold day. By what factor does the rate for an outdoor temperature of 0°C exceed the rate at 5°C? Assume an initial skin temperature of 35°C.

83. ✦ A lizard of mass 3.0 g is warming itself in the bright sunlight. It casts a shadow of 1.6 cm^2 on a piece of paper held perpendicularly to the Sun's rays. The intensity of sunlight at the Earth is 1.4×10^3 W/m^2, but only half of this energy penetrates the atmosphere and is absorbed by the lizard. (a) If the lizard has a specific heat of 4.2 J/(g·°C), what is the rate of increase of the lizard's temperature? (b) Assuming that there is no heat loss by the lizard (to simplify), how long must the lizard lie in the Sun in order to raise its temperature by 5.0°C?

84. ✦ If the total power per unit area from the Sun incident on a horizontal leaf is 9.00×10^2 W/m^2, and we assume that 70.0% of this energy goes into heating the leaf, what would be the rate of temperature rise of the leaf? The specific heat of the leaf is 3.70 kJ/(kg·°C), the leaf's area is 5.00×10^{-3} m^2, and its mass is 0.500 g.

85. ✦ Consider the leaf of Problem 84. Assume that the top surface of the leaf absorbs 70.0% of 9.00×10^2 W/m^2 of radiant energy, while the bottom surface absorbs all of the radiant energy incident on it due to its surroundings at 25.0°C. (a) If the only method of heat loss for the leaf is thermal radiation, what would be the temperature of the leaf? (Assume that the leaf radiates like a blackbody.) (b) If the leaf is to remain at a temperature of 25.0°C, how much power per unit area must be lost by other methods such as transpiration (evaporative heat loss)?

86. An incandescent light bulb radiates at a rate of 60.0 W when the temperature of its filament is 2820 K. During a brownout (temporary drop in line voltage), the power radiated drops to 58.0 W. What is the temperature of the filament? Ignore changes in the filament's length and cross-sectional area due to the temperature change. (connect tutorial: light bulb)

87. If the maximum intensity of radiation for a blackbody is found at 2.65 μm, what is the temperature of the blackbody?

88. A black wood stove has a surface area of 1.20 m^2 and a surface temperature of 175°C. What is the net rate at which heat is radiated into the room? The room temperature is 20°C.

89. ✦ At a tea party, a coffeepot and a teapot are placed on the serving table. The coffeepot is a shiny silver-plated pot with emissivity of 0.12; the teapot is ceramic and has an emissivity of 0.65. Both pots hold 1.00 L of liquid at 98°C when the party begins. If the room temperature is at 25°C, what is the rate of radiative heat loss from the two pots? [*Hint*: To find the surface area, approximate the pots with cubes of similar volume.]

Collaborative Problems

90. ✦ A copper bar of thermal conductivity 401 W/(m·K) has one end at 104°C and the other end at 24°C. The length of the bar is 0.10 m, and the cross-sectional area is 1.0×10^{-6} m^2. (a) What is the rate of heat conduction, $\mathcal{P}$, along the bar? (b) What is the temperature gradient in the bar? (c) If two such bars were placed in series (end to end) between the same temperature baths, what would $\mathcal{P}$ be? (d) If two such bars were placed in parallel (side by side) with the ends in the same temperature baths, what would $\mathcal{P}$ be? (e) In the series case, what is the temperature at the junction where the bars meet?

91. ✦ Small animals eat much more food per kg of body mass than do larger animals. The basal metabolic rate (BMR) is the minimal energy intake necessary to sustain life in a state of complete inactivity. The table lists the BMR, mass, and surface area for five animals. (a) Calculate the BMR/kg of body mass for each animal. Is it true that smaller animals must consume much more food per kg of body mass? (b) Calculate the BMR/m^2 of surface area. (c) Can you explain why the BMR/m^2 is approximately the same for animals of different sizes? Consider what happens to the food energy metabolized by an animal in a resting state.

Animal	BMR (kcal/day)	Mass (kg)	Surface Area (m^2)
Mouse	3.80	0.018	0.0032
Dog	770	15	0.74
Human	2050	64	2.0
Pig	2400	130	2.3
Horse	4900	440	5.1

92. ✦ Imagine a person standing naked in a room at 23.0°C. The walls are well insulated, so they also are at 23.0°C. The person's surface area is 2.20 m^2, and his basal metabolic rate is 2167 kcal/day. His emissivity is 0.97. (a) If the person's skin temperature were 37.0°C (the same as the internal body temperature), at what net rate would heat be lost through radiation? (Ignore losses by conduction and convection.) (b) Clearly the heat loss in (a) is not sustainable—but skin temperature is less than internal body temperature. Calculate the skin temperature such that the net heat loss due to radiation is equal to the basal metabolic rate. (c) Does wearing clothing slow the loss of heat by radiation, or does it only decrease losses by conduction and convection? Explain.

93. For a cheetah, 70.0% of the energy expended during exertion is internal work done on the cheetah's system and is dissipated within his body; for a dog only 5.00% of the energy expended is dissipated within the dog's body. Assume that both animals expend the same total amount of energy during exertion, both have the same heat capacity, and the cheetah is 2.00 times as heavy as the dog. (a) How much higher is the temperature change of the cheetah compared to the temperature change of the dog? (b) If they both start out at an initial temperature of 35.0°C, and the cheetah has a temperature of 40.0°C after the exertion, what is the final temperature of the dog? Which animal probably has more endurance? Explain.

Comprehensive Problems

94. A hotel room is in thermal equilibrium with the rooms on either side and with the hallway on a third side. The room loses heat primarily through a 1.30 cm thick glass window that has a height of 76.2 cm and a width of 156 cm. If the temperature inside the room is 75°F and the temperature outside is 32°F, what is the approximate rate (in kJ/s) at which heat must be supplied to the room to maintain a constant temperature of 75°F? Ignore the stagnant air layers on either side of the glass.

95. While camping, some students decide to make hot chocolate by heating water with a solar heater that focuses sunlight onto a small area. Sunlight falls on their solar heater, of area 1.5 m^2, with an intensity of 750 W/m^2. How long will it take 1.0 L of water at 15.0°C to rise to a boiling temperature of 100.0°C?

96. Five ice cubes, each with a mass of 22.0 g and at a temperature of −50.0°C, are placed in an insulating container. How much heat will it take to change the ice cubes completely into steam?

97. A 10.0 g iron bullet with a speed of 4.00×10^2 m/s and a temperature of 20.0°C is stopped in a 0.500 kg block of wood, also at 20.0°C, which is fixed in place. (a) At first all of the bullet's kinetic energy goes into the internal energy of the bullet. Calculate the temperature increase of the bullet. (b) After a short time the bullet and the block come to the same temperature T. Calculate T, assuming no heat is lost to the environment.

98. If the temperature surrounding the sunbather in Problem 78 is greater than the normal body temperature of 37°C and the air is still, so that radiation, conduction, and convection play no part in cooling the body, how much water (in liters per hour) from perspiration must be given off to maintain the body temperature? The heat of vaporization of water is 2430 J/g at normal skin temperature.

99. Many species cool themselves by sweating, because as the sweat evaporates, heat is given up to the surroundings. A human exercising strenuously has an evaporative heat loss rate of about 650 W. If a person exercises strenuously for 30.0 min, how much water must he drink to replenish his fluid loss? The heat of vaporization of water is 2430 J/g at normal skin temperature.

100. A wall consists of a layer of wood outside and a layer of insulation inside. The temperatures inside and outside the wall are +22°C and −18°C; the temperature at the wood/insulation boundary is −8.0°C. By what factor would the heat loss through the wall increase if the insulation were not present?

101. If 4.0 g of steam at 100.0°C condenses to water on a burn victim's skin and cools to 45.0°C, (a) how much heat is given up by the steam? (b) If the skin was originally at 37.0°C, how much tissue mass was involved in cooling the steam to water? See Table 14.1 for the specific heat of human tissue.

102. If 4.0 g of boiling water at 100.0°C was splashed onto a burn victim's skin and if it cooled to 45.0°C on

the 37.0°C skin, (a) how much heat is given up by the water? (b) How much tissue mass, originally at 37.0°C, was involved in cooling the water? See Table 14.1. Compare the result with that found in Problem 101.

103. The amount of heat generated during the contraction of muscle in an amphibian's leg is given by

$$Q = 0.544 \text{ mJ} + (1.46 \text{ mJ/cm})\ \Delta x$$

where Δx is the length shortened. If a muscle of length 3.0 cm and mass 0.10 g is shortened by 1.5 cm during a contraction, what is the temperature rise? Assume that the specific heat of muscle is 4.186 J/(g·°C).

104. ✦ The power expended by a cheetah is 160 kW while running at 110 km/h, but its body temperature cannot exceed 41.0°C. If 70.0% of the energy expended is dissipated within its body, how far can it run before it overheats? Assume that the initial temperature of the cheetah is 38.0°C, its specific heat is 3.5 kJ/(kg·°C), and its mass is 50.0 kg.

105. Two 62 g ice cubes are dropped into 186 g of water in a glass. If the water is initially at a temperature of 24°C and the ice is at −15°C, what is the final temperature of the drink?

106. A 0.500 kg slab of granite is heated so that its temperature increases by 7.40°C. The amount of heat supplied to the granite is 2.93 kJ. Based on this information, what is the specific heat of granite?

107. A spring of force constant $k = 8.4 \times 10^3$ N/m is compressed by 0.10 m. It is placed into a vessel containing 1.0 kg of water and then released. Assuming all the energy from the spring goes into heating the water, find the change in temperature of the water.

108. One end of a cylindrical iron rod of length 1.00 m and of radius 1.30 cm is placed in the blacksmith's fire and reaches a temperature of 327°C. If the other end of the rod is being held in your hand (37°C), what is the rate of heat flow along the rod? The thermal conductivity of iron varies with temperature, but an average value between the two temperatures is 67.5 W/(m·K). (connect tutorial: conduction)

109. A blacksmith heats a 0.38 kg piece of iron to 498°C in his forge. After shaping it into a decorative design, he places it into a bucket of water to cool. If the available water is at 20.0°C, what minimum amount of water must be in the bucket to cool the iron to 23.0°C? The water in the bucket should remain in the liquid phase.

110. The student from Problem 79 realizes that standing naked in a cold room will not give him the desired weight loss results since it is much less efficient than simply exercising. So he decides to "burn" calories through conduction. He fills the bathtub with 16°C water and gets in. The water right next to his skin warms up to the same temperature as his skin, 35°C, but the water only 3.0 mm away remains at 16°C. At what rate (in kcal/h) would he "burn" calories? The thermal conductivity of water at this temperature is 0.58 W/(m·K). [*Warning*: Do not try this. Sitting in water this cold can lead to hypothermia and even death.]

111. A 2.0 kg block of copper at 100.0°C is placed into 1.0 kg of water in a 2.0 kg iron pot. The water and the iron pot are at 25.0°C just before the copper block is placed into the pot. What is the final temperature of the water, assuming negligible heat flow to the environment?

112. A piece of gold of mass 0.250 kg and at a temperature of 75.0°C is placed into a 1.500 kg copper pot containing 0.500 L of water. The pot and water are at 22.0°C before the gold is added. What is the final temperature of the water?

113. A 20.0 g lead bullet leaves a rifle at a temperature of 87.0°C and hits a steel plate. If the bullet melts, what is the minimum speed it must have?

114. The inner vessel of a calorimeter contains 2.50×10^2 g of tetrachloromethane, CCl_4, at 40.00°C. The vessel is surrounded by 2.00 kg of water at 18.00°C. After a time, the CCl_4 and the water reach the equilibrium temperature of 18.54°C. What is the specific heat of CCl_4?

115. On a hot summer day, Daphne is off to the park for a picnic. She puts 0.10 kg of ice at 0°C in a thermos and then adds tea initially at 25°C. How much tea will just melt all the ice?

116. It requires 17.10 kJ to melt 1.00×10^2 g of urethane $[CO_2(NH_2)C_2H_5]$ at 48.7°C. What is the latent heat of fusion of urethane in kJ/mol?

117. ✦ A stainless steel saucepan, with a base that is made of 0.350 cm thick steel [$\kappa = 46.0$ W/(m·K)] fused to a 0.150 cm thickness of copper [$\kappa = 401$ W/(m·K)], sits on a ceramic heating element at 104.00°C. The diameter of the pan is 18.0 cm, and it contains boiling water at 100.00°C. (a) If the copper-clad bottom is touching the heat source, what is the temperature at the copper-steel interface? (b) At what rate will the water evaporate from the pan?

118. ✦ Ⓒ A scientist working late at night in her low-temperature physics laboratory decides to have a cup of hot tea, but discovers the lab hot plate is broken. Not to be deterred, she puts about 8 oz of water, at 12°C, from the tap into a lab dewar (essentially a large thermos bottle) and begins shaking it up and down. With each shake the water is thrown up and falls back down a distance of 33.3 cm. If she can complete 30 shakes per minute, how long will it take to heat the water to 87°C? Would this really work? If not, why not?

Review and Synthesis

119. A 20.0 g lead bullet leaves a rifle at a temperature of 47.0°C and travels at a speed of 5.00×10^2 m/s until it hits a 6.0 kg block of ice at 0°C that is initially at rest on a frictionless surface. The bullet becomes embedded in

the ice. (a) How fast is the the block of ice moving after the bullet is embedded? (b) How much ice melts?

120. A star's spectrum emits more radiation with a wavelength of 700.0 nm than with any other wavelength. (a) What is the surface temperature of the star? (b) If the star's radius is 7.20×10^8 m, what power does it radiate? (c) If the star is 9.78 ly from Earth, what will an Earth-based observer measure for this star's intensity? Stars are nearly perfect blackbodies. [*Note*: ly stands for light-years.]

121. Ⓒ A 3.0 L container of nitrogen gas (N_2) and a 5.0 L container of oxygen gas (O_2) are both at 20°C and 1.0 atm. (a) Which gas has the larger rms speed? Explain. (b) At what temperature will oxygen gas have the same rms speed as nitrogen when the nitrogen is at 20°C? (c) How much heat must flow into or out of the container of oxygen to change its temperature from 20°C to the temperature you found in part (b)?

122. Ⓒ Two aluminum blocks are in thermal contact. (a) Are the blocks necessarily in physical contact? Explain. (b) If they have the same temperature, do they necessarily have the same internal energy? Explain. (c) If their internal energies are not equal, is there necessarily a net energy transfer between the two blocks? Explain. (d) One block has mass 1.00 kg and temperature 40.0°C. The other has mass 3.00 kg and temperature 20.0°C. Find the final equilibrium temperature and the changes in internal energy of each block.

123. A 60.0 g piece of ice slides 5.00 m down an icy roof inclined at 27.0° to the horizontal. The magnitude of its acceleration is 4.10 m/s^2. All the ice is at 0°C. How much ice melts?

124. ✦ A 75 kg block of ice at 0.0°C breaks off from a glacier, slides along the frictionless ice to the ground from a height of 2.43 m, and then slides along a horizontal surface consisting of gravel and dirt. Find how much of the mass of the ice is melted by the friction with the rough surface, assuming 75% of the internal energy generated is used to heat the ice.

Answers to Practice Problems

14.1 4.9 J

14.2 Higher. The molecules have the same amount of *random* translational kinetic energy plus the additional kinetic energy associated with the ball's translation and rotation.

14.3 350 g

14.4 at least $0.89

14.5 48°C

14.6 92 kJ

14.7 150 kJ

14.8 40 kJ

14.9 53.5 g

14.10 230 W

14.11 110 W

14.12 To maintain constant temperature, the net heat must be zero. The rate at which energy is emitted is 140 W.

14.13 9.4 μm (at 35°C) to 9.6 μm (at 30°C)

14.14 28 W

14.15 −16°C

Answers to Checkpoints

14.2 No, the temperature increase is not caused by heat flow. When you stretch the rubber band, you do work on it. This increases its internal energy and its temperature. (If you now put the rubber band down, heat does flow *out* of the rubber band, decreasing its internal energy and its temperature until it is in thermal equilibrium with its surroundings.)

14.3 The heat capacity of the washer is smaller than the heat capacity of the water, both because brass has a smaller specific heat than water and because the mass of the washer is less than the mass of the water. Therefore, heat flow out of the washer causes more of a temperature change than the same amount of heat flow into the water. The final temperature is less than 50°C.

14.5 The steam releases a large quantity of heat as it condenses into water on the skin. Much more energy is transferred to the skin than would be the case for the same amount of water at 100°C.

14.6 The rate of heat flow through the two materials is the same, so the material with the larger thermal conductivity has the smaller temperature gradient. Figure 14.10b shows that the temperature gradient is smaller in the material on the left, so it has the larger thermal conductivity.

14.8 The red star's surface temperature is lower than the Sun's because the peak wavelength of emitted light is longer. We can't tell which emits radiation at a higher rate because that depends on surface area as well as on surface temperature.

CHAPTER

15

Thermodynamics

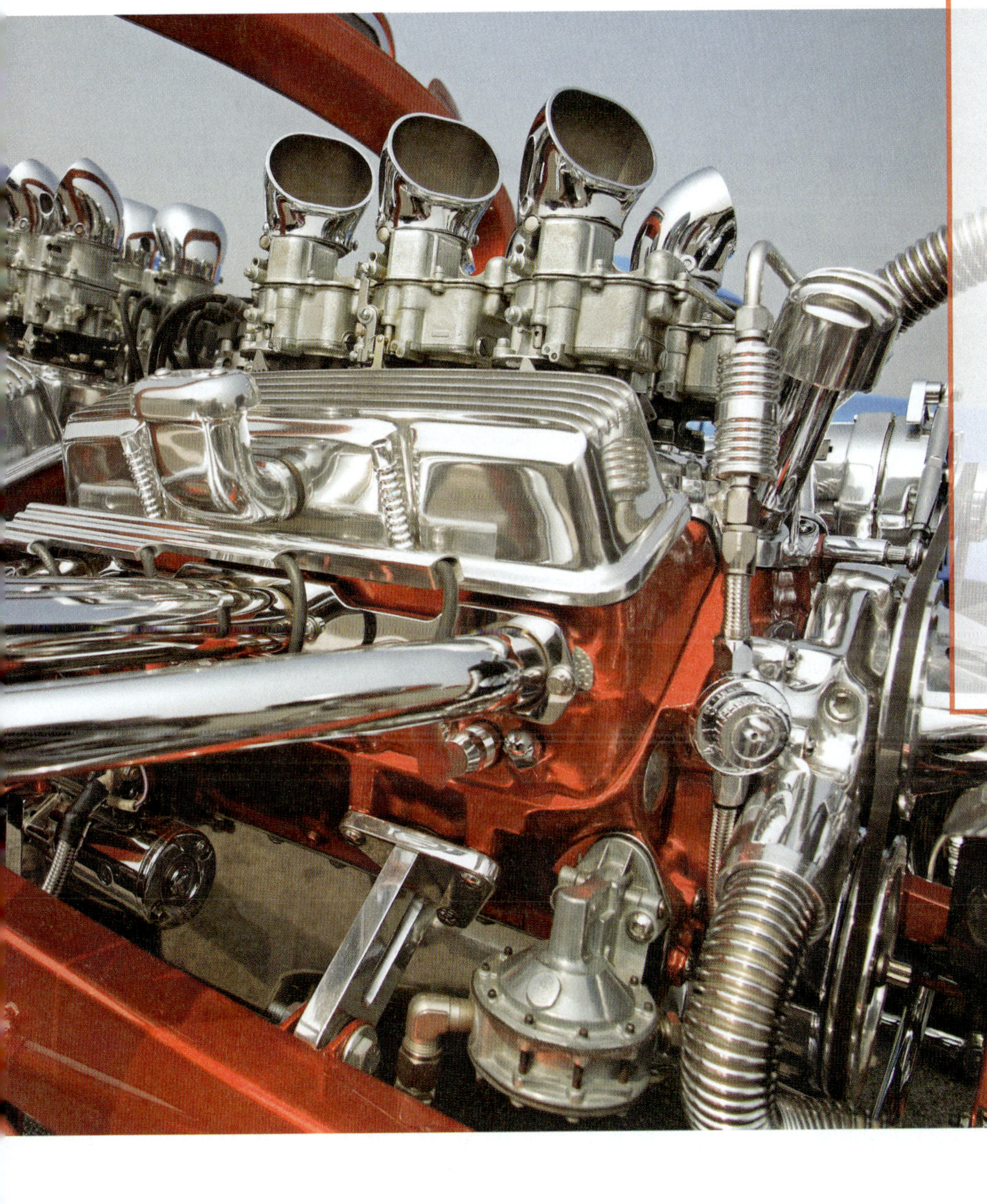

The gasoline engines in cars are terribly inefficient. Of the chemical energy that is released in the burning of gasoline, typically only 20% to 25% is converted into useful mechanical work done on the car to move it forward. Yet scientists and engineers have been working for decades to make a more efficient gasoline engine. Is there some fundamental limit to the efficiency of a gasoline engine? Is it possible to make an engine that converts all—or nearly all—of the chemical energy in the fuel into useful work? (See p. 556 for the answer.)

BIOMEDICAL APPLICATIONS

- Entropy and evolution (Sec. 15.8)
- Changes in internal energy and entropy for biological processes (Ex. 15.1; Probs. 67–70, 78, 85, 96)

Concepts & Skills to Review

- conservation of energy (Section 6.1)
- internal energy and heat (Sections 14.1–14.2)
- zeroth law of thermodynamics (Section 13.1)
- a system and its surroundings (Section 14.1)
- work done is the area under a graph of F_x versus x (Section 6.6)
- heat capacity (Section 14.3)
- the ideal gas law (Section 13.5)
- specific heat of ideal gases at constant volume (Section 14.4)
- natural logarithm (Appendix A.3)

15.1 THE FIRST LAW OF THERMODYNAMICS

Both work and heat can change the internal energy of a system. Work can be done on a rubber ball by squeezing it, stretching it, or slamming it into a wall. Heat will flow into the ball if it is left out in the Sun or put into a hot oven. These two methods of changing the internal energy of a system lead to the **first law of thermodynamics**:

First Law of Thermodynamics

The change in internal energy of a system is equal to the heat flow into the system plus the work done *on* the system.

The choice of a *system* is made in any way convenient for a given problem.

CONNECTION: The first law is not a new principle—just a specialized form of energy conservation.

The first law is a specialized statement of energy conservation applied to a thermodynamic system, such as a gas inside a cylinder that has a movable piston. The gas can exchange energy with its surroundings in two ways. Heat can flow between the gas and its surroundings when they are at different temperatures, and work can be done on the gas when the piston is pushed in.

In equation form, we can write

First law of thermodynamics

$$\Delta U = Q + W \qquad (15\text{-}1)$$

CONNECTION: Our sign conventions for Q and W are consistent with their definitions in previous chapters (Chapter 6 for work and Chapter 14 for heat).

In Eq. (15-1), ΔU is the change in internal energy of the system. (The symbol U, previously used for potential energy, is used exclusively for *internal energy* in this chapter.) The internal energy can increase or decrease, so ΔU can be positive or negative. The signs of Q and W have the same meaning we have used in previous chapters. If heat flows into the system, Q is positive, but if heat flows out of the system, Q is negative. W represents the work done *on* the system, which can be positive or negative, depending on the directions of the applied force and the displacement. Using the example of the gas in a cylinder, if the piston is pushed in, then the force on the gas due to the piston and the displacement of the gas are in the same direction (Fig. 15.1a) and W is positive. If the piston moves out, then the force and the displacement are in opposite directions, because the piston still pushes inward on the gas, and W is negative (Fig. 15.1b). Table 15.1 summarizes the meanings of the signs of ΔU, Q, and W.

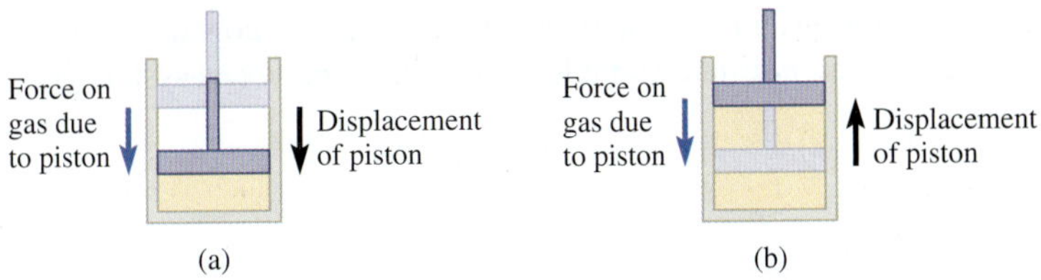

Figure 15.1 (a) When a gas is compressed, the work done *on* the gas is positive. (b) When a gas expands, the work done *on* the gas is *negative*.

Table 15.1 Sign Conventions for the First Law of Thermodynamics

Quantity	Definition	Meaning of + Sign	Meaning of − Sign
Q	Heat flow into the system	Heat flows *into* the system	Heat flows *out of* the system
W	Work done *on* the system	Surroundings do *positive* work on the system	Surroundings do *negative* work on the system (system does positive work on the surroundings)
ΔU	Internal energy change	Internal energy *increases*	Internal energy *decreases*

Example 15.1

Working Out

Katie works out on an elliptical trainer for 30 min. During the workout, she does work (pushing the machine with her feet) at an average rate of 220 W. Heat flows from her body into the surroundings by evaporation, convection, and radiation at an average rate of 910 W. (a) What is the change in her internal energy during the workout? (b) One serving of pasta supplies 1.0 MJ (240 kcal) of internal energy. How many servings of pasta would supply enough internal energy for the workout?

Strategy From conservation of energy, the internal energy changes due to both the work done and the heat flow.

Solution (a) The internal energy is decreasing by 220 J/s due to the work done and by 910 J/s due to the heat flow. Therefore the rate of change of internal energy is

$$-(220\text{ J/s} + 910\text{ J/s}) = -1130\text{ J/s}$$

The rate is negative because the internal energy is *decreasing*. In 30 min, the internal energy change is

$$(-1130\text{ J/s}) \times (30\text{ min}) \times (60\text{ s/min}) = -2.0\text{ MJ}$$

(b) Each serving of pasta supplies 1.0 MJ, so 2.0 servings would supply enough energy for the workout.

Discussion In Eq. (15-1), Q and W stand for the heat flow *into* the system and the work done *on* the system, respectively. In our example heat flows *out* of the system and work is done *by* the system, so both Q and W are negative—they represent energy transfers *out* of the system.

Conceptual Practice Problem 15.1 Changing Internal Energy of a Gas

While 14 kJ of heat flows into the gas in a cylinder with a moveable piston, the internal energy of the gas increases by 42 kJ. Was the piston pulled out or pushed in? Explain. [*Hint*: Determine whether the piston does positive or negative work on the gas.]

15.2 THERMODYNAMIC PROCESSES

A thermodynamic process is the method by which a system is changed from one *state* to another. The state of a system is described by a set of **state variables** such as pressure, temperature, volume, number of moles, and internal energy. State variables describe the state of a system at some instant of time but not how the system got to that state. Heat and work are *not* state variables—they describe *how* a system gets from one state to another.

The *PV* Diagram

If a system is changed so that it is always very near equilibrium, the changes in state can be represented by a curve on a plot of pressure versus volume (called a ***PV* diagram**). Each point on the curve represents an equilibrium state of the system. The *PV* diagram is a useful tool for analyzing thermodynamic processes. One of the chief uses of a *PV* diagram is to find the work done on the system.

Work and Area Under a *PV* Curve Figure 15.2a shows the expansion of a gas, starting with volume V_i and pressure P_i; Fig. 15.2b is the *PV* diagram for the process. In Fig. 15.2, the force exerted by the piston on the gas is downward, and the displacement of the gas is upward, so the piston does negative work on the gas. This work represents a transfer of energy from the gas to its surroundings. (Equivalently, we can say the gas does positive work on the piston.) The piston pushes against the gas with a force of magnitude $F = PA$, where P is the pressure of the gas and A is the cross-sectional area of the piston. This force is not constant since the pressure decreases as the gas expands. As was shown in Section 6.6, the work done by a variable force is the area under a graph of F_x versus x. Can we find the work done on the gas from the area under a graph of P vs. V?

CONNECTION:

In Chapter 6, we saw that work is represented by the area under a graph of force versus displacement. Here we use the same concept; we just modify which variables are being graphed.

Imagine that the piston moves out a *small* distance *d*—small enough that the pressure change is insignificant. The work done on the gas is

$$W = Fd \cos 180° = -PAd$$

The volume change of the gas is

$$\Delta V = Ad$$

So the work done on the gas is $W = -P\,\Delta V$. The same reasoning applies to the work done on *any* system, as long as the pressure is constant.

Work done on a system (constant pressure)

$$W = -P\,\Delta V \qquad (15\text{-}2)$$

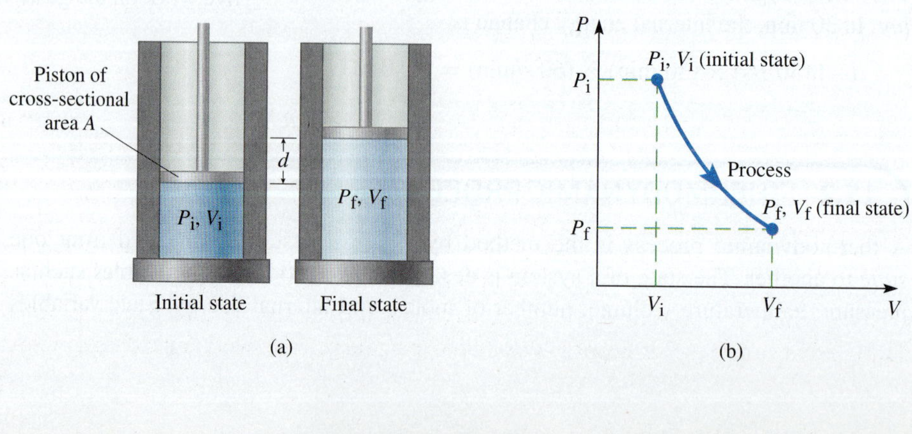

Figure 15.2 (a) Expansion of a gas from initial pressure P_i and volume V_i to final pressure P_f and volume V_f. During the expansion, *negative* work is done on the gas by the moving piston because the force exerted on the gas and the displacement are in opposite directions. (b) A *PV* diagram for the expansion shows the pressure and volume of the gas starting at the initial values P_i, V_i, and ending at the final values P_f, V_f.

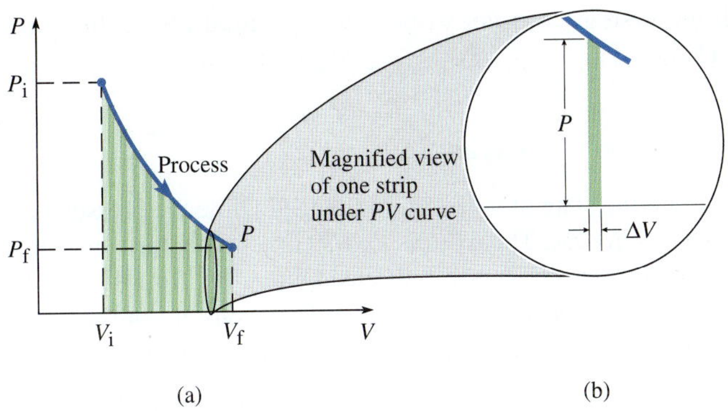

Figure 15.3 (a) The area under the *PV* curve is divided into many narrow strips of width ΔV and of varying heights P. The sum of the areas of the strips is the total area under the *PV* curve, which represents the magnitude of the work done on the gas. (b) An enlarged view of one strip under the curve. If the strip is very narrow, we can ignore the change in P and approximate its area as $P\,\Delta V$.

To find the *total* work done on the gas, we add up the work done during each small volume change. During each small ΔV, the magnitude of the work done is the area of a thin strip of height P and width ΔV under the *PV* curve (Fig. 15.3). Therefore,

The magnitude of the total work done on a system is the area under the *PV* curve.

Volume change	Work done on the system
Increase	Negative
Decrease	Positive
No change	Zero

The magnitude of the work done on a system depends on the *path* taken on the *PV* curve. Figures 15.4a and 15.4b show *PV* diagrams for two other possible paths between the same initial and final states as those of Fig. 15.3. The work done differs from one process to another, even though the initial and final states are the same in each case.

Work Done During a Closed Cycle Because the work done on a system depends on the path on the *PV* diagram, the net work done on a system during a **closed cycle**—a series of processes that leave the system in the same state it started in—can be nonzero. The magnitude of the net work done during a cycle is the area *inside* the cycle on the *PV* diagram because it is the sum of the negative work done during expansion and the positive work done during compression (Fig. 15.4c). A closed cycle

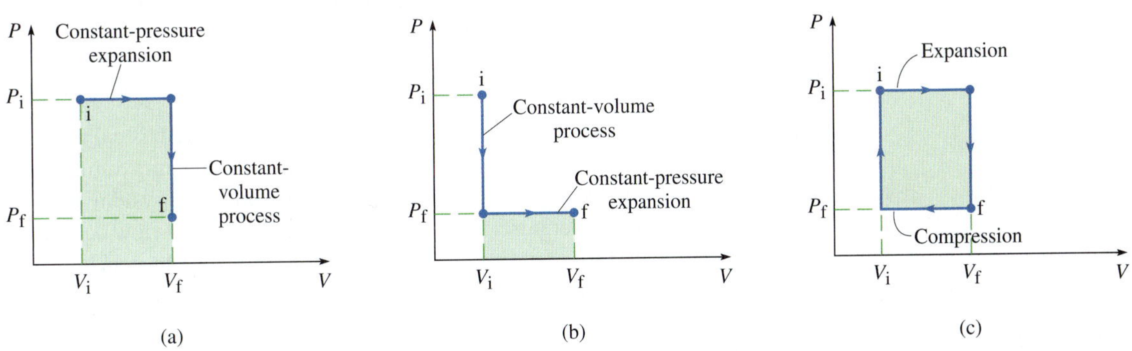

Figure 15.4 (a) and (b) Two paths between the same initial and final states as the process shown in Fig. 15.3. The magnitude of the work done on the gas is equal to the area under the graph, and is negative because the volume increases. For (a), $W = -P_i(V_f - V_i)$, and for (b), $W = -P_f(V_f - V_i)$. Note that the work done depends on the path taken between the initial and final states. (c) A closed cycle. The work done on the gas from i to f is $-P_i(V_f - V_i)$, as in (a). The work done from f back to i is $+P_f(V_f - V_i)$, the same magnitude as in (b) but opposite in sign because we have reversed the process (compression instead of expansion). The net work done during the cycle is the sum of these: $W_{net} = -P_i(V_f - V_i) + P_f(V_f - V_i) = -(P_i - P_f)(V_f - V_i)$. The magnitude of W_{net} is the area of the shaded rectangle, and the sign is negative because the negative work done during expansion (i to f) is larger in magnitude than the positive work done during compression (f to i).

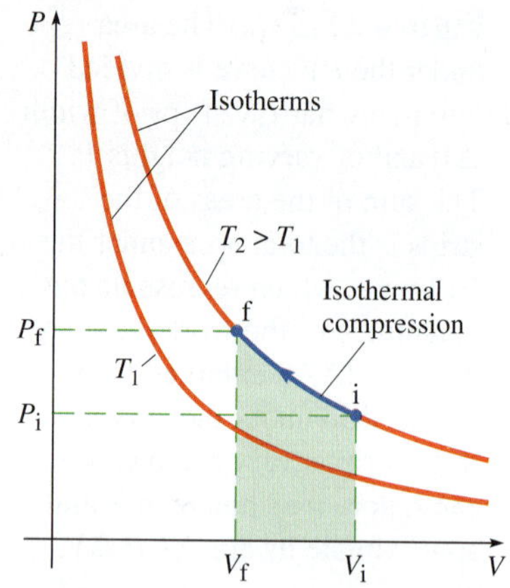

Figure 15.5 Isotherms for an ideal gas at two different temperatures. Each isotherm is a graph of $P = nRT/V$ for a constant temperature. The shaded area represents the work done by the gas during an isothermal compression at temperature T_2, which is positive. (The work done by the gas during an isothermal *expansion* would be *negative*.)

during which the system does net work is the essential idea behind the heat engine (see Section 15.5).

Constant-Pressure Processes

A process by which the state of a system is changed while the pressure is held constant is called an *isobaric* process. The word *isobaric* comes from the same Greek root as the word "barometer." In Fig. 15.4a, the first change of state from V_i to V_f along the line from 1 to 2 occurs at the constant pressure $P = P_i$. A constant-pressure process appears as a horizontal line on the *PV* diagram. The work done on the gas is

$$W = -P(V_f - V_i) = -P\Delta V \quad \text{(constant pressure)} \tag{15-3}$$

Constant-Volume Processes

A process by which the state of a system is changed while the *volume* remains constant is called an *isochoric* process. Such a process is illustrated in Fig. 15.4a when the system moves along the line from 2 to 3 as the pressure changes from P_i to P_f at the constant volume V_f. No work is done during a constant-volume process; without a displacement, work cannot be done. The area under the *PV* curve—a vertical line—is zero:

$$W = 0 \quad \text{(constant volume)} \tag{15-4}$$

If no work is done, then from the first law of thermodynamics, the change in internal energy is equal to the heat flow into the system:

$$\Delta U = Q \quad \text{(constant volume)} \tag{15-5}$$

Constant-Temperature Processes

A process in which the temperature of the system remains constant is called an **isothermal** process. On a *PV* diagram, a path representing a constant-temperature process is called an **isotherm** (Fig. 15.5). All the points on an isotherm represent states of the system with the same temperature.

How can we keep the temperature of the system constant? One way is to put the system in thermal contact with a **heat reservoir**—something with a heat capacity so large that it can exchange heat in either direction without changing its temperature significantly. Then as long as the state of the system does not change too rapidly, the heat flow between the system and the reservoir keeps the system's temperature constant.

Adiabatic Processes

A process in which no heat is transferred into or out of the system is called an **adiabatic** process. An adiabatic process is *not* the same as a constant-temperature (isothermal) process. In an isothermal process, heat flow into or out of a system is necessary to maintain a constant temperature. In an adiabatic process, *no* heat flow occurs, so if work is done, the temperature of the system may change. One way to perform an adiabatic process is to completely insulate the system so that no heat can flow in or out; another way is to perform the process so quickly that there is no time for heat to flow in or out.

For example, the compressions and rarefactions caused by a sound wave occur so fast that heat flow from one place to another is negligible. Hence, the compressions and rarefactions are adiabatic. Isaac Newton made a now-famous error when he assumed that these processes were isothermal and calculated a speed of sound that was about 20% lower than the measured value.

From the first law of thermodynamics, when no heat flows the change in internal energy is equal to the work done on the system:

$$\Delta U = W \quad \text{(adiabatic)} \tag{15-6}$$

Table 15.2 Summary of Thermodynamic Processes

Process	Name	Condition	Consequences
Constant temperature	Isothermal	T = constant	(*For an ideal gas*, $\Delta U = 0$)
Constant pressure	Isobaric	P = constant	$W = -P\,\Delta V$
Constant volume	Isochoric	V = constant	$W = 0$; $\Delta U = Q$
No heat flow	Adiabatic	$Q = 0$	$\Delta U = W$

Table 15.2 summarizes all of the thermodynamic processes discussed. (See also the interactive: thermodynamics.)

connect

CHECKPOINT 15.2

(a) Can an adiabatic process cause a change in temperature? Explain. (b) Can heat flow during an isothermal process? (c) Can the internal energy of a system change during an isothermal process?

15.3 THERMODYNAMIC PROCESSES FOR AN IDEAL GAS

Constant Volume

Figure 15.6 is a *PV* diagram for heat flow into an ideal gas at constant volume. Since the temperature of the gas changes, the initial and final states are shown as points on two different isotherms. (Note that the higher-temperature isotherm is farther from the origin.) The area under the vertical line is zero; no work is done when the volume is constant. With $W = 0$, the heat flow increases the internal energy of the gas, so the temperature increases.

In Section 14.4, we discussed the molar specific heat of an ideal gas at constant volume. The first law of thermodynamics enables us to calculate the internal energy change ΔU. Since no work is done during a constant-volume process, $\Delta U = Q$. For a constant-volume process, $Q = nC_V\,\Delta T$ and, therefore, $\Delta U = nC_V\,\Delta T$.

Internal energy is a state variable—its value depends only on the current state of the system, not on the path the system took to get there. Therefore, as long as the number of moles is constant, *the internal energy of an ideal gas changes only when the temperature changes*. Equation (15-7) therefore gives the internal energy change of an ideal gas for *any* thermodynamic process, not just for constant-volume processes.

$$\Delta U = nC_V\,\Delta T \quad \text{(ideal gas, } any \text{ process)} \qquad (15\text{-}7)$$

CONNECTION:

Section 15.2 described some general aspects of various thermodynamic processes. Now we find out what happens when the system undergoing the process is an ideal gas.

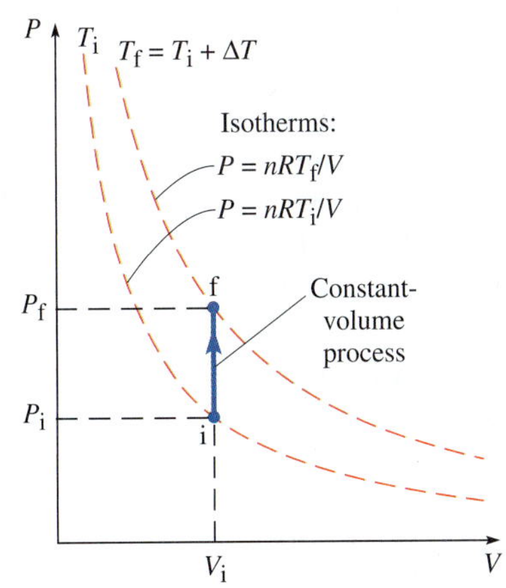

Figure 15.6 A *PV* diagram for a constant-volume process for an ideal gas. Every point on an isotherm (red dashed lines) represents a state of the gas at the same temperature.

Constant Pressure

Another common situation occurs when the *pressure* of the gas is constant. In this case, work is done because the volume changes. The first law of thermodynamics enables us to calculate the molar specific heat at constant pressure (C_P), which is different from the molar specific heat at constant volume (C_V).

Figure 15.7 shows a *PV* diagram for the constant-pressure expansion of an ideal gas starting and ending at the same temperatures as for the constant-volume process of Fig. 15.6. Applying the first law to the constant-pressure process requires that

$$\Delta U = Q + W$$

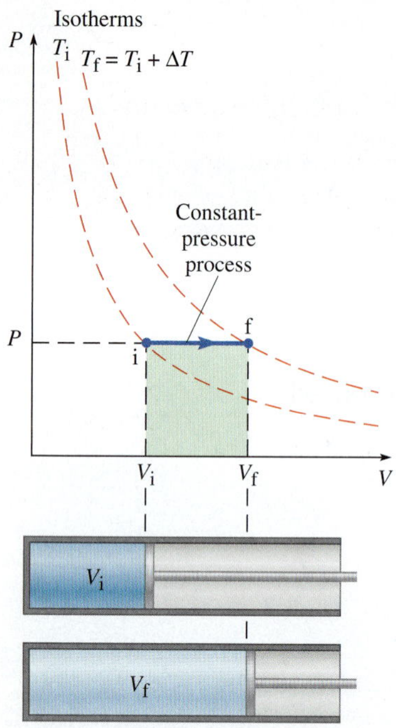

Figure 15.7 A *PV* diagram of a constant-pressure expansion of an ideal gas. Heat flows into the ideal gas ($Q > 0$). The increase in the internal energy ΔU is less than Q because negative work is done on the expanding gas by the piston. The work done by the gas is the negative of the shaded area under the path.

where the work done on the gas is, from the ideal gas law,

$$W = -P\,\Delta V = -nR\,\Delta T$$

The definition of C_P is

$$Q = nC_P\,\Delta T \tag{15-8}$$

Substituting Q and W into the first law, we obtain

$$\Delta U = nC_P\,\Delta T - nR\,\Delta T$$

Since the internal energy of an ideal gas is determined by its temperature, ΔU for this constant-pressure process is the same as ΔU for the constant-volume process between the same two temperatures:

$$\Delta U = nC_V\,\Delta T \tag{15-7}$$

Then

$$nC_V\,\Delta T = nC_P\,\Delta T - nR\,\Delta T$$

Canceling common factors of n and ΔT, this reduces to

$$C_P = C_V + R \qquad \text{(ideal gas)} \tag{15-9}$$

Since R is a positive constant, the molar specific heat of an ideal gas at constant pressure is larger than the molar specific heat at constant volume.

Is this result reasonable? When heat flows into the gas at constant pressure, the gas expands, doing work on the surroundings. Thus, not all of the heat goes into increasing the internal energy of the gas. More heat has to flow into the gas at constant pressure for a given temperature increase than at constant volume.

Example 15.2

Warming a Balloon at Constant Pressure

A weather balloon is filled with helium gas at 20.0°C and 1.0 atm of pressure. The volume of the balloon after filling is measured to be 8.50 m^3. The helium is heated until its temperature is 55.0°C. During this process, the balloon expands at constant pressure (1.0 atm). What is the heat flow into the helium?

Strategy We can find how many moles of gas n are present in the balloon by using the ideal gas law. For this problem, we consider the helium to be a system. Helium is a monatomic gas, so its molar specific heat at constant *volume* is $C_V = \frac{3}{2}R$. The molar specific heat at constant *pressure* is then $C_P = C_V + R = \frac{5}{2}R$. Then the heat flow into the gas during its expansion is $Q = nC_P\Delta T$.

Solution The ideal gas law is

$$PV = nRT$$

Using the initial volume and temperature, we can solve for the number of moles:

$$n = \frac{PV_i}{RT_i}$$

For an ideal gas at constant pressure, the heat required to change the temperature is

$$Q = nC_P\,\Delta T$$

where $C_P = \frac{5}{2}R$. The temperature change is

$$\Delta T = 55.0°\text{C} - 20.0°\text{C} = 35.0\text{ K}$$

The initial pressure, volume, and temperature are P = 1.0 atm = 1.01×10^5 Pa, $V_i = 8.50$ m^3, and T_i = 273 K + 20.0° C = 293 K. Then

$$Q = nC_P\,\Delta T = \left(\frac{PV_i}{\cancel{R}T_i}\right)\left(\frac{5}{2}\cancel{R}\right)\Delta T$$

$$= \frac{5}{2}\left(\frac{1.01 \times 10^5\text{ Pa} \times 8.50\text{ m}^3}{293\text{ K}}\right)(35.0\text{ K}) = 260\text{ kJ}$$

continued on next page

Example 15.2 continued

Discussion We do not have to find the work done on the gas separately and then subtract it from the change in internal energy to find Q. The work done is *already* accounted for by the molar specific heat at constant pressure. This simplifies the problem since we use the same method for constant pressure as we use for constant volume; the only change is the choice of C_V or C_P.

Practice Problem 15.2 Air Instead of Helium

Suppose the balloon were filled with dry air instead of helium. Find Q for the same temperature change. (Dry air is mostly N_2 and O_2, so assume an ideal diatomic gas.)

Constant Temperature

For an ideal gas, we can plot isotherms using the ideal gas law $PV = nRT$ (see Fig. 15.5). Since the change in internal energy of an ideal gas is proportional to the temperature change,

$$\Delta U = 0 \quad \text{(ideal gas, isothermal process)} \qquad (15\text{-}10)$$

From the first law of thermodynamics, $\Delta U = 0$ means that $Q = -W$. Note that Eq. (15-10) is true for an *ideal gas* at constant temperature. Other systems can change internal energy without changing temperature; one example is when the system undergoes a phase change.

It can be shown (using calculus to find the area under the PV curve) that the work done on an ideal gas during a constant-temperature expansion or contraction from volume V_i to volume V_f is

$$W = nRT \ln \frac{V_i}{V_f} \quad \text{(ideal gas, isothermal)} \qquad (15\text{-}11)$$

In Eq. (15-11), "ln" stands for the natural (or base-e) logarithm.

Example 15.3

Constant-Temperature Compression of an Ideal Gas

An ideal gas is kept in thermal contact with a heat reservoir at 7°C (280 K) while it is compressed from a volume of 20.0 L to a volume of 10.0 L (Fig. 15.8). During the compression, an average force of 33.3 kN is used to move the piston a distance of 0.15 m. How much heat is exchanged between the gas and the reservoir? Does the heat flow into or out of the gas?

Strategy We can find the work done on the gas from the average force applied and the distance moved. For isothermal compression of an ideal gas, $\Delta U = 0$. Then $Q = -W$.

Solution The work done on the gas is

$$W = fd = 33.3 \text{ kN} \times 0.15 \text{ m} = 5.0 \text{ kJ}$$

This work adds 5.0 kJ to the internal energy of the gas. Then 5.0 kJ of heat must flow out of the gas if its internal energy

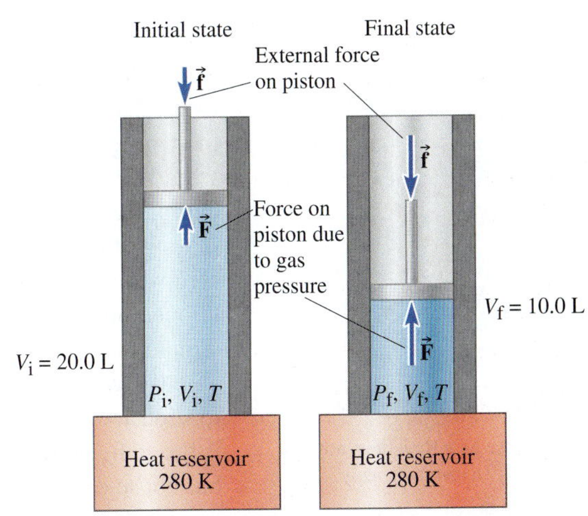

Figure 15.8
Isothermal compression of an ideal gas. Thermal contact with a heat reservoir keeps the gas at a constant temperature.

continued on next page

Example 15.3 continued

does not change. The work done on the gas is positive since the piston is pushed with an inward force as it moves inward.

$$Q = -W = -5.0 \text{ kJ}$$

Since positive Q represents heat flow *into* the gas, the negative sign tells us that heat flows out of the gas into the reservoir.

Discussion Although the temperature remains constant during the process, it does not mean that no heat flows. To maintain a constant temperature when work is done on the gas, some heat must flow out of the gas. If the gas were thermally isolated so no heat could flow, then the work done on the gas would increase the internal energy, resulting in an increase in the temperature of the gas.

Practice Problem 15.3 Work Done During Constant-Temperature Expansion of a Gas

Suppose 2.0 mol of an ideal gas are kept in thermal contact with a heat reservoir at 57°C (330 K) while the gas expands from a volume of 20.0 L to a volume of 40.0 L. Does heat flow into or out of the gas? How much heat flows? [*Hint*: Use $W = nRT \ln (V_i/V_f)$, which applies to an ideal gas at constant temperature.]

15.4 REVERSIBLE AND IRREVERSIBLE PROCESSES

Have you ever wished you could make time go backward? Perhaps you accidentally broke an irreplaceable treasure in a friend's house, or missed a one-time opportunity to meet your favorite movie star, or said something unforgivable to someone close to you. Why can't the clock be turned around?

Imagine a perfectly elastic collision between two billiard balls. If you were to watch a movie of the collision, you would have a hard time telling whether the movie was being played forward or backward. The laws of physics for an elastic collision are valid even if the direction of time is reversed. Since the total momentum and the total kinetic energy are the same before and after the collision, the reversed collision is physically possible.

Figure 15.9 Spontaneous heat flow goes from warm to cool; the reverse does not happen spontaneously.

The perfectly elastic collision is one example of a **reversible** process. A reversible process is one that does not violate any laws of physics if "played in reverse." Most of the laws of physics do not distinguish forward in time from backward in time. A projectile moving in the absence of air resistance (on the Moon, say) is reversible: if we play the movie backward, the total mechanical energy is still conserved and Newton's second law ($\Sigma\vec{\mathbf{F}} = m\vec{\mathbf{a}}$) still holds at every instant in the projectile's trajectory.

Notice the caveats in the examples: "perfectly elastic" and "in the absence of air resistance." If friction or air resistance is present, then the process is **irreversible**. If you played *backward* a movie of a projectile with noticeable air resistance, it would be easy to tell that something is wrong. The force of air resistance on the projectile would act in the wrong direction—in the direction of the velocity, instead of opposite to it. The same would be true for sliding friction. Slide a book across the table; friction slows it down and brings it to rest. The macroscopic kinetic energy of the book—due to the orderly motion of the book in one direction—has been converted into disordered energy associated with the random motion of molecules; the table and book will be at slightly higher temperatures. The reversed process certainly would never occur, even though it does not violate the first law of thermodynamics (energy conservation). We would not expect a slightly warmed book placed on a slightly warmed table surface to spontaneously begin to slide across the table, gaining speed and cooling off as it goes, even if the total energy is the same before and after. It is easy to convert ordered energy into disordered energy, but not so easy to do the reverse. *The presence of energy dissipation (sliding friction, air resistance) always makes a process irreversible.*

As another example of an irreversible process, imagine placing a container of warm lemonade into a cooler with some ice (Fig. 15.9). Some of the ice melts and the lemonade gets cold as heat flows out of the lemonade and into the ice. The reverse would never happen: putting cold lemonade into a cooler with some partially melted ice, we would never find that the lemonade gets warmer as the liquid water freezes. *Spontaneous heat flow from a hotter system to a colder system is always irreversible.*

Conceptual Example 15.4

Irreversibility and Energy Conservation

Suppose heat *did* flow spontaneously from the cold ice to the warm lemonade, making the ice colder and the lemonade warmer. Would conservation of energy be violated by this process?

Solution and Discussion Heat flow from the ice to the lemonade would increase the internal energy of the lemonade by the same amount that the internal energy of the ice would decrease. The total internal energy of the ice and the lemonade would remain unchanged—energy would be conserved. The process would never occur, but *not* because energy conservation would be violated.

Conceptual Practice Problem 15.4 A Campfire

On a camping trip, you gather some twigs and logs and start a fire. Discuss the campfire in terms of irreversible processes.

As we will see later in this chapter, irreversible processes such as the frictional dissipation of energy and the spontaneous heat flow from a hotter to a colder system can be thought of in terms of a change in the amount of order in the system. A system never goes *spontaneously* from a disordered state to a more ordered state. Reversible processes are those that do not change the total amount of disorder in the universe; irreversible processes increase the amount of disorder.

Second Law of Thermodynamics According to the **second law of thermodynamics**, the total amount of disorder in the universe never decreases. Irreversible processes increase the disorder of the universe. Spontaneous heat flow from a colder system to a hotter system does not occur because it would decrease the total disorder in the universe. One statement of the second law, phrased in terms of heat flow, is:

Second Law of Thermodynamics (Clausius Statement)

Heat never flows spontaneously from a colder system to a hotter system.

The second law of thermodynamics determines what we sense as the direction of time—none of the other physical laws we have studied would be violated if the direction of time were reversed.

CHECKPOINT 15.4

A perfectly elastic collision is reversible. What about an inelastic collision? Explain.

15.5 HEAT ENGINES

We said in Section 15.4 that it is far *easier* to convert ordered energy into disordered energy than to do the reverse. Converting ordered into disordered energy occurs spontaneously, but the reverse does not. A **heat engine** is a device designed to convert disordered energy into ordered energy—loosely speaking, to "convert heat into work." Fuel is burned, and the energy released is used to do some useful work (such as generating electricity or propelling an automobile.) We will see that the second law of thermodynamics places a fundamental limitation on how much work can be produced by a heat engine from a given amount of heat.

The development of practical steam engines—heat engines that use steam as the working substance—around the beginning of the eighteenth century was one of the crucial elements in the industrial revolution. These steam engines were the first machines that produced a sustained work output using an energy source other than muscle, wind, or moving water. Steam engines are still used in many electric power plants.

The source of energy in a heat engine is most often the burning of some fuel such as gasoline, coal, oil, natural gas, and the like. A nuclear power plant is a heat engine using energy released by a nuclear reaction instead of a chemical reaction (as in burning). A geothermal engine uses the high temperature found beneath the Earth's crust (which comes to the surface in places such as volcanoes and hot springs).

Cyclical Engines Practical engines operate in cycles. Each cycle consists of several thermodynamic processes that are repeated the same way during each cycle. The *magnitudes* of the energy transfers during each cycle are as follows (Fig. 15.10): the heat input to the engine is Q_H, the net work done *by* the engine is W_{net}, and the heat exhausted from the engine is Q_C. (The subscripts "H" and "C" remind us that heat is taken in from something hot and exhausted to something colder.)

Sign convention for heat engines

Q_H, W_{net}, and Q_C represent *magnitudes* (all three are positive)

W_{net} is the net work done *by* the engine

In order for these processes to repeat the same way, the engine must end each cycle in the same state in which it started. In particular, the internal energy of the engine must be the same at the end of a cycle as it was in the beginning. Then for one complete cycle, $\Delta U = Q_H - W_{net} - Q_C = 0$. Rearranging the terms, we find

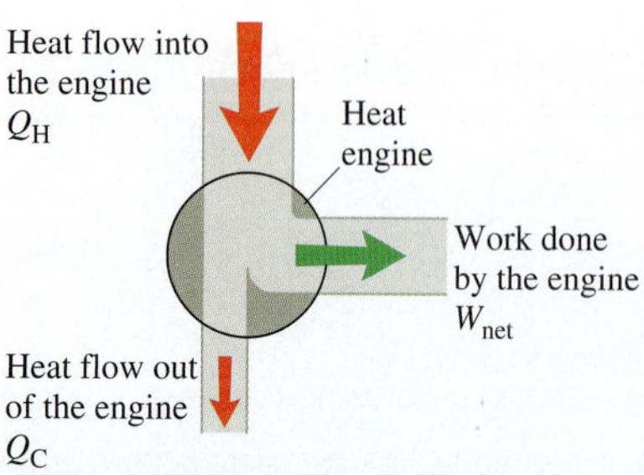

Figure 15.10 Diagram of the energy transfers in a heat engine. The engine is represented by a circle, and the arrows indicate the directions of the energy transfers. The total energy entering the engine during one cycle (Q_H) equals the total energy leaving the engine during the cycle ($W_{net} + Q_C$).

Energy conservation in a cyclical engine

$$Q_H = W_{net} + Q_C \qquad (15\text{-}12)$$

In a steam engine (Fig. 15.11), the heat input Q_H comes from burning fuel or from a nuclear reaction. Pressurized steam does work W_{net} as it pushes against a piston or, more commonly, a turbine. Exhaust heat Q_C is carried off by cooling water or by release of the steam itself. The coal burned, for example, releases heat that is used to make steam; the steam is the working substance of the engine that does work to drive the turbines.

Application: The Internal Combustion Engine One familiar engine is the internal combustion engine found in automobiles. *Internal* combustion refers to the fact that gasoline is burned inside a cylinder; the resulting hot gases push against a piston and do work. (A steam engine is an *external* combustion engine.)

Most automobile engines work in a cyclic thermodynamic process shown in Fig. 15.12. Of the energy released by burning gasoline, only about 20% to 25% is

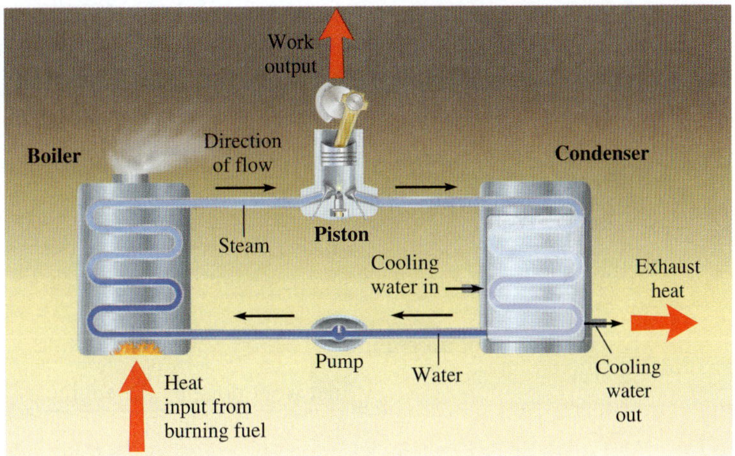

Figure 15.11 Diagram of an idealized condensing steam engine. Burning fuel releases heat that is used to boil water and make high-pressure steam. The steam does work on a piston or turbine as it expands; this is the useful work output of the engine. In the condenser, the steam is turned back into water; cooling water carries away the exhaust heat. (Additional exhaust heat is carried away by the combustion gases in the boiler.) The water is then pumped back into the boiler to start the cycle again. (Not all steam engines have condensers. In some, a water source feeds water into the boiler and the steam is released into the environment after driving the piston or turbine.)

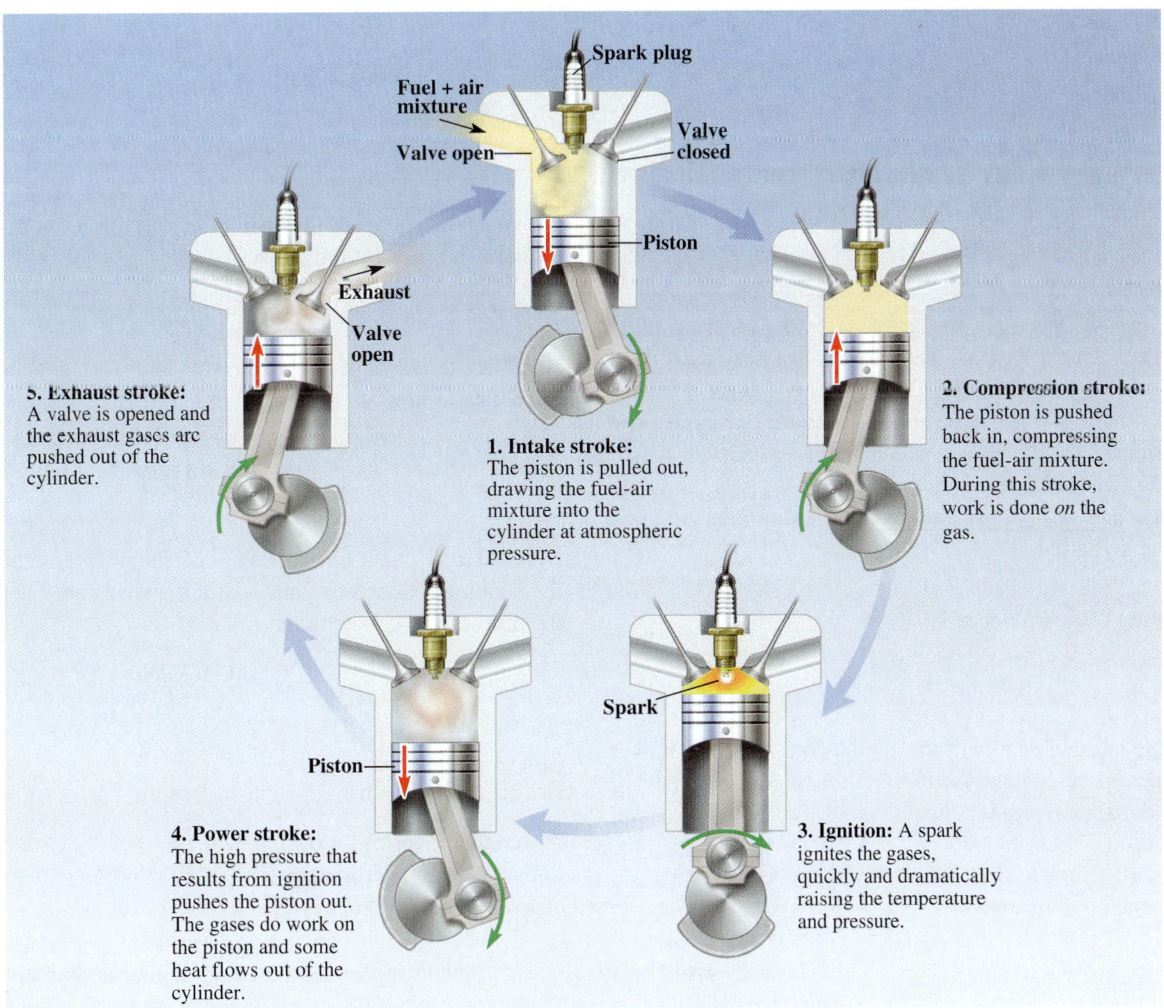

Figure 15.12 The four-stroke automobile engine. Each cycle has four strokes during which the piston moves (steps 1, 2, 4, and 5).

turned into mechanical work used to move the car forward and run other systems. The rest is discarded. The hot exhaust gases carry energy out of the engine, as does the liquid cooling system.

Efficiency of an Engine

To measure how effectively an engine converts heat into mechanical work, we define the engine's **efficiency** e as what you get (net useful work) divided by what you supply (heat input):

Efficiency of an engine

$$e = \frac{\text{net work done by the engine}}{\text{heat input}} = \frac{W_{net}}{Q_H} \tag{15-13}$$

If an engine does work at a constant rate and its efficiency does not change, then it also takes in and exhausts heat at constant rates. The work done, heat input, and heat exhausted during any time interval are all proportional to the elapsed time. Therefore, all the same relationships that are true for the amounts of heat flow and work done apply to the *rates* at which heat flows and work is done. For example, the efficiency is

$$e = \frac{\text{net work done}}{\text{heat input}} = \frac{\text{net } rate \text{ of doing work}}{rate \text{ of taking in heat}} = \frac{W_{net}/\Delta t}{Q_H/\Delta t}$$

Example 15.5

Rate at Which Heat Is Exhausted from an Engine

An engine operating at 25% efficiency produces work at a rate of 0.10 MW. At what rate is heat exhausted into the surroundings?

Strategy The problem gives the *rate* at which work is done, which we can write $W_{net}/\Delta t$, where W_{net} is the net work done per cycle and Δt is the elapsed time for one cycle. The problem asks for the rate at which heat is exhausted, which is $Q_C/\Delta t$. The efficiency is defined as $e = W_{net}/Q_H$. To relate Q_H to Q_C and W_{net}, apply energy conservation (the first law of thermodynamics).

Solution The efficiency is

$$e = \frac{W_{net}}{Q_H}$$

Since the internal energy of the engine does not change over a complete cycle, energy conservation (or the first law of thermodynamics) requires that

$$Q_H = W_{net} + Q_C$$

We now solve for Q_C, using the definition of efficiency to make the substitution $Q_H = W_{net}/e$:

$$Q_C = Q_H - W_{net} = \frac{W_{net}}{e} - W_{net} = W_{net}\left(\frac{1}{e} - 1\right)$$

Then the *rates* of heat exhausted and work done and are related as follows:

$$\frac{Q_C}{\Delta t} = \frac{W_{net}}{\Delta t}\left(\frac{1}{e} - 1\right) = 0.10 \text{ MW} \times \left(\frac{1}{0.25} - 1\right) = 0.30 \text{ MW}$$

Discussion As a check: 25% efficiency means that $\frac{1}{4}$ of the heat input does work and $\frac{3}{4}$ of it is exhausted. Therefore, the ratio of work to exhaust is

$$\frac{1/4}{3/4} = \frac{1}{3} = \frac{0.10 \text{ MW}}{0.30 \text{ MW}}$$

Practice Problem 15.5 Heat Engine Efficiency

An engine "wastes" 4.0 J of heat for every joule of work done. What is its efficiency?

Efficiency and the First Law According to the first law of thermodynamics, the efficiency of a heat engine cannot exceed 100%. An efficiency of 100% would mean that all of the heat input is turned into useful work and no "waste" heat is exhausted. It might seem theoretically possible to make a 100% efficient engine by eliminating all of the imperfections in design such as friction and lack of perfect insulation. However, the second law of thermodynamics requires that the efficiency of even an *ideal* engine be less than 100%, as we see in Section 15.7.

15.6 REFRIGERATORS AND HEAT PUMPS

The second law of thermodynamics says that heat cannot *spontaneously* flow from a colder system to a hotter system; but machines such as refrigerators and heat pumps can make that happen. In a refrigerator, heat is pumped out of the food compartment into the warmer room. That doesn't happen by itself; it requires the *input* of work. The electricity used by a refrigerator turns the compressor motor, which does the work required to make the refrigerator function (Fig. 15.13). An air conditioner is essentially the same thing: it pumps heat out of the house into the hotter outdoors.

The only difference between a refrigerator (or an air conditioner) and a heat pump is which end is performing the useful task. Refrigerators and air conditioners pump heat out of a compartment that they are designed to keep cool. Heat pumps pump heat from the colder outdoors into the warmer house. The idea is not to cool the outdoors; it is to warm the house.

Notice that the energy transfers in a heat pump are reversed in direction from those in a heat engine (Fig. 15.14). In the heat engine, heat flows from hot to cold, with work as the output. In a heat pump, heat flows from cold to hot, with work as the *input*. We use the same symbols Q_H, Q_C, and W_{net} to stand for the magnitudes of the energy transfers, but all the directions are opposite: W_{net} is the work *input*, Q_C is the heat removed from something cold, and Q_H is the heat exhausted to something hotter.

CONNECTION:

A refrigerator or heat pump is like a heat engine with the directions of the energy transfers *reversed*.

Sign convention for engines, refrigerators, and heat pumps

Q_H, Q_C, and W_{net} are all positive.

Exactly as for a heat engine, conservation of energy requires that

$$Q_H = W_{net} + Q_C \tag{15-12}$$

for a refrigerator or heat pump

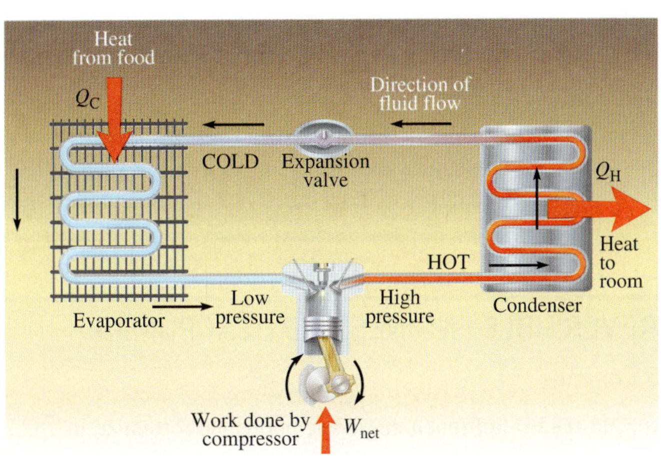

Figure 15.13 In a refrigerator, a fluid is compressed, increasing its temperature. Heat is exhausted as the fluid passes through the condenser. Now the fluid is allowed to expand; its temperature falls. Heat flows from the food compartment into the cold fluid. The fluid returns to the compressor to begin the same cycle again.

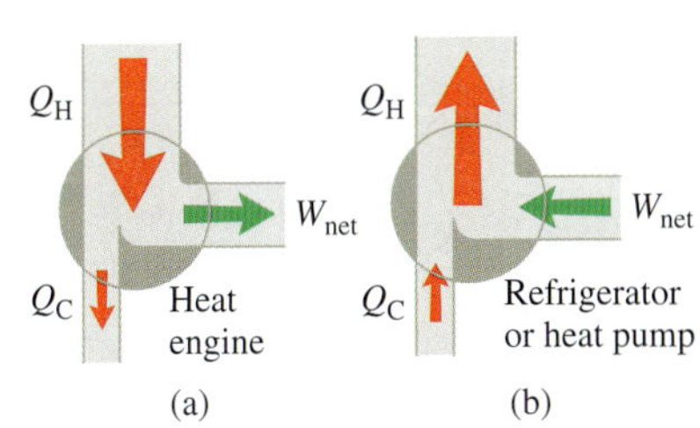

Figure 15.14 Energy transfers during one cycle for (a) a heat engine and (b) a refrigerator or heat pump. The directions of all three energy transfers for the refrigerator are opposite those for the engine. With our definition of Q_H, Q_C, and W_{net} as positive quantities, in either case conservation of energy requires that $Q_H = W_{net} + Q_C$.

Coefficient of Performance To measure the performance of a heat pump or refrigerator, we define a **coefficient of performance** K. Just as for the efficiency of an engine, the coefficients of performance are ratios of what you get divided by what you pay for:

Coefficient of performance for a heat pump:

$$K_p = \frac{\text{heat delivered}}{\text{net work input}} = \frac{Q_H}{W_{net}} \quad (15\text{-}14)$$

Coefficient of performance for a refrigerator or air conditioner:

$$K_r = \frac{\text{heat removed}}{\text{net work input}} = \frac{Q_C}{W_{net}} \quad (15\text{-}15)$$

A higher coefficient of performance means a better heat pump or refrigerator. Unlike the efficiency of an engine, coefficients of performance can be (and usually are) *greater than* 1. The second law says that heat cannot flow spontaneously from cold to hot—we need to do some work to make that happen. That's equivalent to saying that the coefficient of performance can't be infinite.

Example 15.6

A Heat Pump

A heat pump has a performance coefficient of 2.5. (a) How much heat is delivered to the house for every 1.0 J of electrical energy consumed? (b) In an electric heater, for each joule of electric energy consumed, one joule of heat is delivered to the house. Where does the "extra" heat delivered by the heat pump come from?

Strategy First, we identify the energy transfers. The work input is the electricity used to run the heat pump, so $W_{net} = 1.0$ J. The heat delivered to the house is Q_H, which we are to find. Second, we must be careful to use the correct coefficient of performance. For a heat pump, whose object is to deliver heat to the house, the coefficient of performance is the heat delivered (Q_H) per unit of net work done to run the pump.

Solution (a) As a heat pump,

$$K_p = \frac{\text{heat delivered}}{\text{net work input}} = \frac{Q_H}{W_{net}} = 2.5$$

$$Q_H = 2.5W_{net}$$

For every 1.0 J of electric energy, 2.5 J of heat are delivered to the house.

(b) The 2.5 J of heat delivered include the 1.0 J of work input plus 1.5 J of heat pumped in from the outside. The electric heater just transforms the joule of work into a joule of heat.

Discussion One thing that makes a heat pump economical in many situations is that the same machine can function as a heat pump (in winter) and as an air conditioner (in summer). As a heat pump, it delivers heat Q_H to the interior of the house, while as an air conditioner, it removes heat Q_C.

Practice Problem 15.6 Heat Exhausted by Air Conditioner

An air conditioner with a coefficient of performance $K_r = 3.0$ consumes electricity at an average rate of 1.0 kW. During 1.0 h of use, how much heat is exhausted to the outdoors?

15.7 REVERSIBLE ENGINES AND HEAT PUMPS

Reversible Engines

Sadi Carnot (1796–1832), a French engineer, published a treatise in 1824 that greatly expanded the understanding of how heat engines work. His treatment introduced a hypothetical, ideal engine using an ideal gas as working substance. The engine assumes the existence of two reservoirs, a hot reservoir at absolute temperature T_H and a cold reservoir at absolute temperature T_C (where $T_C < T_H$). The engine takes its heat input from the hot reservoir and exhausts heat into the cold reservoir (Fig. 15.15). Recall that a *heat reservoir* is a system with such a large heat capacity that it can exchange heat in either direction with a negligibly small temperature change. Therefore, the cold reservoir stays at temperature T_C, and the hot reservoir stays at temperature T_H.

In the ideal engine, no irreversible processes take place, so we call it a **reversible engine**. We must assume that all friction has somehow been eliminated—otherwise an irreversible dissipation of energy would occur. We also must avoid heat flow across a nonzero temperature difference, which would be irreversible. Whenever the engine takes in or gives off heat, the gas must be at the same temperature as the reservoir with which it exchanges energy.

Using the second law of thermodynamics, Carnot proved that:

- All reversible engines operating between the same two reservoirs have the same efficiency.
- The efficiency of a reversible engine depends only on the absolute temperatures of the two reservoirs.
- The efficiency of any real engine that exchanges heat with two reservoirs cannot be greater than the efficiency of a reversible engine using the same two reservoirs.

(See the text website for details.) Carnot also showed that the efficiency of a reversible engine is given by this remarkably simple expression:

Efficiency of a reversible engine

$$e_r = 1 - \frac{T_C}{T_H} \tag{15-16}$$

Remember that the temperatures in Eq. (15-17) must be *absolute* temperatures. Absolute temperature is also called *thermodynamic temperature* because the efficiency of reversible engines can be used to define a temperature scale. In fact, the definition of the kelvin is based on Eq. (15-17).

The efficiency of a reversible engine is always less than 100%, assuming that the cold reservoir is not at absolute zero. Even an ideal, perfectly reversible engine must exhaust some heat, so the efficiency can never be 100%, *even in principle*. Efficiencies of real engines cannot be greater than those of reversible engines, so the second law of thermodynamics sets a limit on the theoretical maximum efficiency of an engine: $e < 1 - T_C/T_H$.

Using Eq. (15-17), the ratio of the heat exhaust to the heat input for a reversible engine is

$$\frac{Q_C}{Q_H} = \frac{Q_H - W_{net}}{Q_H} = 1 - \frac{W_{net}}{Q_H} = 1 - e_r = \frac{T_C}{T_H} \tag{15-17}$$

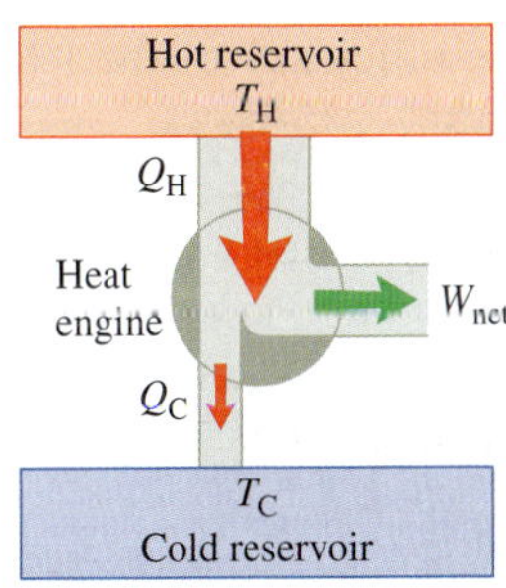

Figure 15.15 Energy transfers in a reversible heat engine. Heat flows into the engine from a reservoir at temperature T_H, and heat flows out of the engine into a reservoir at T_C. The heat transfers are *isothermal*—in other words, the working substance in the engine is at the same temperature as the reservoir when heat is exchanged. Energy transfers in a reversible heat pump or refrigerator are in the opposite directions from those for the engine.

The Carnot Cycle Carnot's reversible engine operates in a four-step cycle. During two of the steps, heat is exchanged between the gas and one of the reservoirs. For the heat exchange to be reversible, the gas must be at the same constant temperature as the reservoir, so these two steps are *isothermal* processes.

How is it possible to get heat to flow without a temperature difference? Imagine putting the gas in good thermal contact with a reservoir at the same temperature. Now *slowly* pull a piston so that the gas expands. As long as the expansion occurs slowly, heat flows into the gas fast enough to keep its temperature nearly constant. The slower the expansion, the closer we get to the behavior of the *idealized* reversible engine.

The other two steps in the cycle must change the gas temperature from T_H to T_C and back to T_H. These processes must be *adiabatic* (no heat flow) since otherwise an irreversible heat flow would occur. (For more detail on the Carnot cycle, see the text website.)

Reversible Refrigerators and Heat Pumps

Equation (15-17) also applies to reversible heat pumps and refrigerators because they are just reversible engines with the directions of the energy transfers reversed.

$$\frac{Q_C}{Q_H} = \frac{T_C}{T_H} \qquad (15\text{-}17)$$

(reversible engine, refrigerator, or heat pump)

Using Eq. (15-17) and the first law, we can find the coefficients of performance for reversible heat pumps and refrigerators (see Problems 56 and 57):

$$K_{p,\,rev} = \frac{1}{1 - T_C/T_H} \quad \text{and} \quad K_{r,\,rev} = \frac{1}{T_H/T_C - 1} \qquad (15\text{-}18)$$

Real heat pumps cannot have coefficients of performance greater than that of a reversible heat pump operating between the same two reservoirs. The same is true for refrigerators:

$$K_p < K_{p,\,rev} \quad \text{and} \quad K_p < K_{p,\,rev}$$

Example 15.7

Efficiency of an Automobile Engine

In an automobile engine, the combustion of the fuel-air mixture can reach temperatures as high as 3000°C and the exhaust gases leave the cylinder at about 1000°C. (a) Find the efficiency of a *reversible* engine operating between reservoirs at those two temperatures. (b) Theoretically, we might be able to have the exhaust gases leave the engine at the temperature of the outside air (20°C). What would be the efficiency of the hypothetical reversible engine in this case?

Strategy First we identify the temperatures of the hot and cold reservoirs in each case. We must convert the reservoir temperatures to kelvins in order to find the efficiency of a reversible engine.

Solution (a) The reservoir temperatures in kelvins are found using

$$T = T_C + 273\text{ K}$$

Therefore,

$$T_H = 3000°\text{C} = 3273\text{ K}$$

$$T_C = 1000°\text{C} = 1273\text{ K}$$

The efficiency of a reversible engine operating between these temperatures is

$$e_r = 1 - \frac{T_C}{T_H} = 1 - \frac{1273\text{ K}}{3273\text{ K}} = 0.61 = 61\%$$

(b) The high-temperature reservoir is still at 3273 K, whereas the low-temperature reservoir is now

$$T_C = 293\text{ K}$$

This gives a higher efficiency:

$$e_r = 1 - \frac{T_C}{T_H} = 1 - \frac{293\text{ K}}{3273\text{ K}} = 0.910 = 91.0\%$$

Can an engine be 100% efficient?

Discussion As mentioned in the chapter opener, real gasoline engines achieve efficiencies of only about 20% to 25%. Although improvement is possible, the second law of thermodynamics limits the theoretical maximum efficiency to that of a reversible engine operating between the same temperatures. The *theoretical* maximum efficiency can only be increased by using a hotter hot reservoir or a colder cold reservoir. However, practical considerations may prevent us from using a hotter hot reservoir or colder cold reservoir. Hotter combustion gases might cause engine parts to wear out too fast, or there may be safety concerns. Letting the gases expand to a greater volume would make the exhaust gases colder, leading to an increase in efficiency, but might reduce the *power* the engine can deliver. (A reversible engine has the theoretical maximum efficiency, but the *rate* at which it does work is vanishingly small because it takes a long time for heat to flow across a small temperature difference.)

Practice Problem 15.7 Temperature of Hot Gases

If the efficiency of a reversible engine is 75% and the temperature of the outdoor world into which the engine sends its exhaust is 27°C, what is the combustion temperature in the engine cylinder? [*Hint*: Think of the combustion temperature as the temperature of the hot reservoir.]

Example 15.8

Coal-Burning Power Plant

A coal-burning electrical power plant burns coal at 706°C. Heat is exhausted into a river near the power plant; the average river temperature is 19°C. What is the minimum possible rate of thermal pollution (heat exhausted into the river) if the station generates 125 MW of electricity?

Strategy The minimum discharge of heat into the river would occur if the engine generating the electricity were *reversible*. As in Example 15.5, we can take all of the rates to be constant. The rate at which electrical energy is generated is $W_{net}/\Delta t = 125$ MW. The question asks for the rate of heat exhausted, which is $Q_C/\Delta t$.

Solution First find the absolute temperatures of the reservoirs:

$$T_H = 706°C = 979 \text{ K}$$

$$T_C = 19°C = 292 \text{ K}$$

The efficiency of a reversible engine operating between these temperatures is

$$e_r = 1 - \frac{T_C}{T_H} = 1 - \frac{292 \text{ K}}{979 \text{ K}} = 0.702$$

We want to find the rate at which heat is exhausted, which is $Q_C/\Delta t$. The efficiency is equal to the ratio of the net work output to the heat input from the hot reservoir:

$$e = \frac{W_{net}}{Q_H}$$

Conservation of energy requires that

$$Q_H = Q_C + W_{net}$$

Solving for Q_C,

$$Q_C = Q_H - W_{net} = \frac{W_{net}}{e} - W_{net} = W_{net}\left(\frac{1}{e} - 1\right)$$

Assuming that all the rates are constant,

$$\frac{Q_C}{\Delta t} = 125 \text{ MW} \times \left(\frac{1}{0.702} - 1\right) = 53 \text{ MW}$$

The rate at which heat enters the river is 53 MW.

Discussion We expect the actual rate of thermal pollution to be higher. A real, irreversible engine would have a lower efficiency, so more heat would be dumped into the river.

Practice Problem 15.8 Generating Electricity from Coal

What is the minimum possible rate of heat *input* (from the burning of coal) needed to generate 125 MW of electricity in this same plant?

15.8 ENTROPY

When two systems of different temperatures are in thermal contact, heat flows out of the hotter system and into the colder system. There is no change in the total energy of the two systems; energy just flows out of one and into the other. Why then does heat flow in one direction but not in the other? As we will see, heat flow *into* a system not only increases the system's internal energy, it also increases the *disorder* of the system. Heat flow *out of* a system decreases not only its internal energy but also its disorder.

The **entropy** of a system (symbol *S*) is a quantitative measure of its disorder. Entropy is a state variable (like *U*, *P*, *V*, and *T*): a system in equilibrium has a unique entropy that does *not* depend on the past history of the system. (Recall that heat and work are *not* state variables. Heat and work describe *how* a system goes from one state to another.) The word *entropy* was coined by Rudolf Clausius (1822–1888) in 1865; its Greek root means *evolution* or *transformation*.

If an amount of heat *Q* flows into a system at constant absolute temperature *T*, the entropy change *of the system* is

Entropy change (constant temperature)

$$\Delta S = \frac{Q}{T} \quad (15\text{-}19)$$

The SI unit for entropy is J/K. Heat flowing into a system increases the system's entropy (both ΔS and Q are positive); heat leaving a system decreases the system's entropy (both ΔS and Q are negative). Equation (15-19) is valid as long as the temperature of the system is constant, which is true if the heat capacity of the system is large (as for a reservoir), so that the heat flow Q causes a negligibly small temperature change in the system.

Note that Eq. (15-19) gives only the *change* in entropy, not the initial and final values of the entropy. As with potential energy, the *change* in entropy is what's important in most situations.

If a small amount of heat Q flows from a hotter system to a colder system ($T_H > T_C$), the *total* entropy change of the systems is

$$\Delta S_{tot} = \Delta S_H + \Delta S_C = \frac{-Q}{T_H} + \frac{Q}{T_C}$$

Since $T_H > T_C$, the increase in the colder system's entropy is larger than the decrease of the hotter system's entropy and the total entropy increases. Every irreversible process increases the total entropy of the universe. A process that would decrease the total entropy of the universe is impossible. A reversible process causes no change in the total entropy of the universe. We can restate the second law of thermodynamics in terms of entropy:

Second Law of Thermodynamics (Entropy Statement)

The entropy of the universe never decreases.

$$\Delta S_{tot} > 0 \quad \text{(irreversible process)}$$

$$\Delta S_{tot} = 0 \quad \text{(reversible process)}$$

For example, a reversible engine removes heat Q_H from a hot reservoir at temperature T_H and exhausts Q_C to a cold reservoir at T_C. The entropy of the engine itself is left unchanged since it operates in a cycle. The entropy of the hot reservoir decreases by an amount Q_H/T_H and that of the cold reservoir increases by Q_C/T_C. Since the entropy of the universe must be unchanged by a reversible engine, it must be true that

$$\Delta S_{tot} = -\frac{Q_H}{T_H} + \frac{Q_C}{T_C} = 0 \qquad (15\text{-}18)$$

as we found in Eq. (15-17).

Entropy is *not* a conserved quantity like energy. The entropy of the universe is always increasing. It is possible to decrease the entropy of a *system*, but only at the expense of increasing the entropy of the surroundings by at least as much (usually more).

CHECKPOINT 15.8

The entropy of a system increases by 10 J/K. Does this mean the process is necessarily irreversible? Explain.

Example 15.9

Entropy Change of a Freely Expanding Gas

Suppose 1.0 mol of an ideal gas is allowed to freely expand into an evacuated container of equal volume so that the volume of the gas doubles (Fig. 15.16). No work is done on the gas as it expands, since there is nothing pushing against it. The containers are insulated so no heat flows into or out of the gas. What is the entropy change of the gas?

continued on next page

Example 15.9 continued

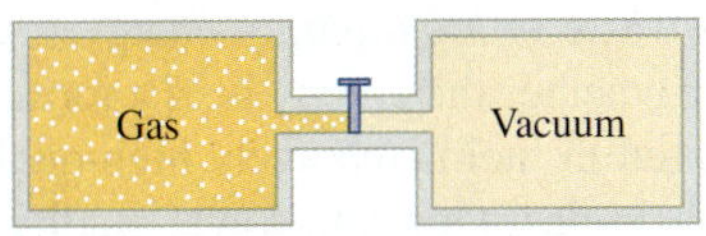

Figure 15.16

Two chambers connected by a valve. One chamber contains a gas, and the other has been evacuated. When the valve is opened, the gas expands until it fills both chambers.

Strategy The only way to calculate entropy changes that we've learned so far is for heat flow at a constant temperature. In free expansion, there is no heat flow—but that does not necessarily mean there is no entropy change. Since entropy is a state variable, ΔS depends only on the initial and final states of the gas, not the intermediate states. We can therefore find the entropy change using *any* thermodynamic process with the same initial and final states. The initial and final temperatures of the gas are identical since the internal energy does not change; therefore we find the entropy change for an *isothermal* expansion.

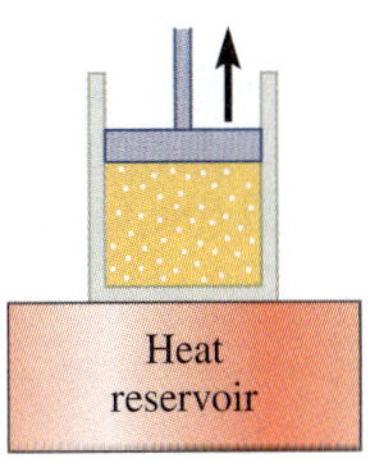

Figure 15.17

As the gas in the cylinder expands, heat flows into it from the reservoir and keeps its temperature constant.

Solution Imagine the gas confined to a cylinder with a moveable piston (Fig. 15.17). In an isothermal expansion, heat flows into the gas from a reservoir at a constant temperature T. As the gas expands, it does work on the piston. If the temperature is to stay constant, the work done must equal the heat flow into the gas:

$$\Delta U = 0 \quad \text{implies} \quad Q + W = 0$$

In Section 15.3, we found the work done by an ideal gas during an isothermal expansion:

$$W = nRT \ln\left(\frac{V_i}{V_f}\right)$$

The volume of the gas doubles, so $V_i/V_f = 0.50$:

$$W = nRT \ln 0.50$$

Since $Q = -W$, the entropy change is

$$\Delta S = \frac{Q}{T} = -nR \ln 0.50$$

$$= -(1.0 \text{ mol}) \times \left(8.31 \frac{\text{J}}{\text{mol·K}}\right) \times (-0.693) = +5.8 \text{ J/K}$$

Discussion The entropy change is positive, as expected. Free expansion is an irreversible process; the gas molecules do not spontaneously collect back in the original container. The reverse process would cause a decrease in entropy, without a larger increase elsewhere, and so violates the second law.

Practice Problem 15.9 Entropy Change of the Universe When a Lump of Clay Is Dropped

A room-temperature lump of clay of mass 400 g is dropped from a height of 2 m and makes a totally inelastic collision with the floor. Approximately what is the entropy change of the universe due to this collision? [*Hint*: The temperature of the clay rises, but only slightly.]

Application of the Second Law to Evolution

Some have argued that evolution cannot have occurred because it would violate the second law of thermodynamics. The argument views evolution as an increase in order: life spontaneously developed from simple life forms to more complex, more highly ordered organisms.

However, the second law says only that the *total entropy of the universe* cannot decrease. It does not say that the entropy of a particular system cannot decrease. When heat flows from a hotter system to a colder system, the entropy of the hotter system decreases, but the increase in the colder system's entropy is greater, so the entropy of the universe increases. A living organism is not a closed system and neither is the Earth. An adult human, for instance, requires roughly 10 MJ of chemical energy from food per day. What happens to this energy? Some is turned into useful work by the muscles, some more is used to repair body tissues, but most of it is dissipated and leaves the body as heat. The human body therefore is constantly increasing the entropy of its environment. As evolution progresses from simpler to more complicated organisms, the increase in order within the organisms must be accompanied by a larger increase in disorder in the environment.

Application of the Second Law to the "Energy Crisis"

When people speak of "conserving energy," they usually mean using fuel and electricity sparingly. In the physics sense of the word *conserve*, energy is *always* conserved.

Burning natural gas to heat your house does not change the amount of energy around; it just changes it from one form to another.

What we need to be careful not to waste is *high-quality* energy. Our concern is not the total amount of energy, but rather whether the energy is in a form that is useful and convenient. The chemical energy stored in fuel is relatively high-quality (ordered) energy. When fuel is burned, the energy is degraded into lower-quality (disordered) energy.

Statistical Interpretation of Entropy

Thermodynamic systems are collections of huge numbers of atoms or molecules. How these atoms or molecules behave statistically determines the disorder in the system. In other words, the second law of thermodynamics is based on the statistics of systems with extremely large numbers of atoms or molecules.

The **microstate** of a thermodynamic system specifies the state of each constituent particle. For instance, in a monatomic ideal gas with N atoms, a microstate is specified by the position and velocity of each of the N atoms. As the atoms move about and collide, the system changes from one microstate to another. The **macrostate** of a thermodynamic system specifies only the values of the macroscopic state variables (e.g., pressure, volume, temperature, and internal energy).

Statistical analysis is the microscopic basis for the second law of thermodynamics. It turns out, remarkably, that the number of microstates corresponding to a given macrostate is related to the entropy of that macrostate in a simple way. Letting Ω (the Greek capital omega) stand for the number of microstates, the relationship is

CONNECTION:

In Chapter 13, we learned about the distribution of speeds in gas molecules (the Maxwell-Boltzmann distribution). Why should the speeds be distributed in this particular way? Because it has the highest entropy. The Maxwell-Boltzmann distribution can be calculated statistically by maximizing the number of microstates for a given macrostate.

Statistical basis of entropy

$$S = k \ln \Omega \quad (15\text{-}20)$$

where k is Boltzmann's constant. Equation (15-20) is inscribed on the tombstone of Ludwig Boltzmann (1844–1906), the Austrian physicist who made the connection between entropy and statistics in the late nineteenth century. The relationship between S and Ω has to be logarithmic because entropy is additive: if system 1 has entropy S_1 and system 2 has entropy S_2, then the total entropy is $S_1 + S_2$. However, the number of microstates is *multiplicative*. Think of dice: if die 1 has 6 microstates and die 2 also has 6, the total number of microstates when rolling two dice is not 12, but $6 \times 6 = 36$. The entropy is additive since $\ln 6 + \ln 6 = \ln 36$. (For more information on entropy and statistics see the text website.)

connect

15.9 THE THIRD LAW OF THERMODYNAMICS

Like the second law, the third law of thermodynamics can be stated in several equivalent ways. We will state just one of them:

Third Law of Thermodynamics

It is impossible to cool a system to absolute zero.

Although it is impossible to *reach* absolute zero, there is no limit on how *close* we can get. Scientists who study low-temperature physics have attained equilibrium temperatures as low as 1 μK and have sustained temperatures of 2 mK; transient temperatures in the nano- and picokelvin range have been observed.

Master the Concepts

- The first law of thermodynamics is a statement of energy conservation:

$$\Delta U = Q + W \qquad (15\text{-}1)$$

where Q is the heat flow *into* the system and W is the work done *on* the system.

- Pressure, temperature, volume, number of moles, internal energy, and entropy are state variables; they describe the state of a system at some instant of time but *not* how the system got to that state. Heat and work are *not* state variables—they describe *how* a system gets from one state to another.
- The work done on a system when the pressure is constant—or for a volume change small enough that the pressure change is insignificant—is

$$W = -P\Delta V \qquad (15\text{-}2)$$

The magnitude of the work done is the total area under the PV curve.

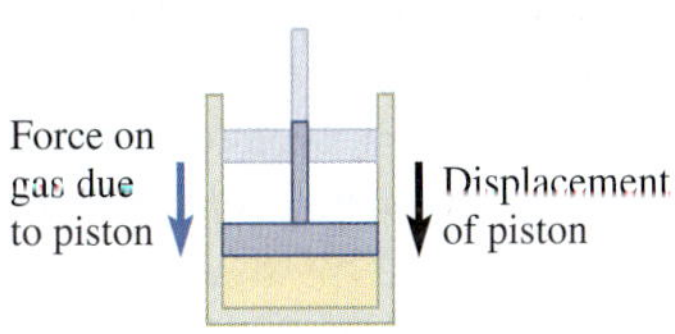

$W > 0$ for compression

- The change in internal energy of an ideal gas is determined solely by the temperature change. Therefore,

$$\Delta U = 0 \quad \text{(ideal gas, isothermal process)} \qquad (15\text{-}10)$$

- A process in which no heat is transferred into or out of the system is called an adiabatic process.
- The molar specific heats of an ideal gas at constant volume and constant pressure are related by

$$C_P = C_V + R \qquad (15\text{-}9)$$

- Spontaneous heat flow from a hotter system to a colder system is always irreversible.

Spontaneous heat flow

- For one cycle of an engine, heat pump, or refrigerator, conservation of energy requires

$$Q_H = W_{net} + Q_C$$

where Q_H, Q_C, and W_{net} are defined as positive quantities.

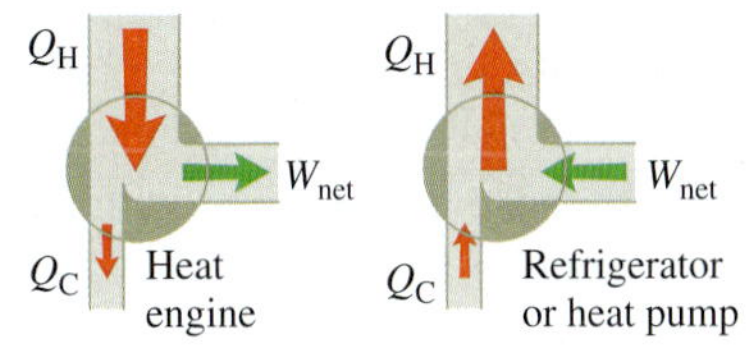

- The efficiency of an engine is defined as

$$e = \frac{W_{net}}{Q_H} \qquad (15\text{-}13)$$

- The coefficient of performance for a heat pump is

$$K_p = \frac{\text{heat delivered}}{\text{net work input}} = \frac{Q_H}{W_{net}} \qquad (15\text{-}14)$$

- The coefficient of performance for a refrigerator or air conditioner is

$$K_r = \frac{\text{heat removed}}{\text{net work input}} = \frac{Q_C}{W_{net}} \qquad (15\text{-}16)$$

- A reservoir is a system with such a large heat capacity that it can exchange heat in either direction with a negligibly small temperature change.

Reverse heat flow does not happen spontaneously

- The second law of thermodynamics can be stated in various equivalent ways. Two of them are: (1) heat never flows spontaneously from a colder system to a hotter system, and (2) the entropy of the universe never decreases.
- The efficiency of a *reversible* engine is determined only by the *absolute* temperatures of the hot and cold reservoirs:

$$e_r = 1 - \frac{T_C}{T_H} \qquad (15\text{-}17)$$

The efficiency of a real engine is less than the efficiency of a reversible engine operating between the same temperature reservoirs.

- The coefficient of performance of a reversible heat pump or refrigerator is determined only by the absolute temperatures of the reservoirs:

$$K_{p,\,rev} = \frac{1}{1 - T_C/T_H} \quad \text{and} \quad K_{r,\,rev} = \frac{1}{T_H/T_C - 1} \qquad (15\text{-}18)$$

continued on next page

Master the Concepts continued

The coefficient of performance of a real heat pump is less than that of a reversible heat pump operating between the same temperature reservoirs. The same is true for a refrigerator.

- If an amount of heat Q flows into a system at constant absolute temperature T, the entropy change of the system is

$$\Delta S = \frac{Q}{T} \tag{15-20}$$

- The third law of thermodynamics: it is impossible to cool a system to absolute zero.

Conceptual Questions

1. Is it possible to make a heat pump with a coefficient of performance equal to 1? Explain.
2. An electric baseboard heater can convert 100% of the electric energy used into heat that flows into the house. Since a gas furnace might be located in a basement and sends exhaust gases up the chimney, the heat flow into the living space is less than 100% of the chemical energy released by burning. Does this mean that electric heating is better? Which heating method consumes less fuel? In your answer, consider how the electricity might have been generated and the efficiency of that process.
3. A whimsical statement of the laws of thermodynamics—probably not one favored by gamblers—goes like this:
   I. You can never win; you can only lose or break even.
   II. You can only break even at absolute zero.
   III. You can never get to absolute zero.
   What do we mean by "win," "lose," and "break even"? [*Hint*: Think about a heat engine.]
4. Why must all reversible engines (operating between the same reservoirs) have the same efficiency? Try an argument by contradiction: imagine that two reversible engines exist with $e_1 > e_2$. Reverse one of them (into a heat pump) and use the work output from the engine to run the heat pump. What happens? (If it seems fine at first, switch the two.)
5. When supplies of fossil fuels such as petroleum and coal dwindle, people might call the situation an "energy crisis." From the standpoint of physics, why is that not an accurate name? Can you think of a better one?
6. If you leave the refrigerator door open and the refrigerator runs continuously, does the kitchen get colder or warmer? Explain.
7. Most heat pumps incorporate an auxiliary electric heater. For relatively mild outdoor temperatures, the electric heater is not used. However, if the outdoor temperature gets very low, the auxiliary heater is used to supplement the heat pump. Why?
8. Why are heat pumps more often used in mild climates than in areas with severely cold winters?
9. Are entropy changes always caused by the flow of heat? If not, give some other examples of processes that increase entropy. (connect tutorial: reversibility)
10. Can a heat engine be made to operate without creating any "thermal pollution," that is, without making its cold reservoir get warmer in the long run? The net work output must be greater than zero.
11. A warm pitcher of lemonade is put into an ice chest. Describe what happens to the entropies of lemonade and ice as heat flows from the lemonade to the ice within the chest.
12. A new dormitory is being built at a college in North Carolina. To save costs, it is proposed to not include air conditioning ducts and vents. A member of the board overseeing the construction says that stand-alone air conditioning units can be supplied to each room later. He has seen advertisements that claim these new units do not need to be vented to the outside. Can the claim be true? Explain.
13. After a day at the beach, a child brings home a bucket containing some saltwater. Eventually the water evaporates, leaving behind a few salt crystals. The molecular order of the salt crystals is greater than the order of the dissolved salt sloshing around in the seawater. Is this a violation of the entropy principle? Explain.
14. Explain why the molar specific heat at constant volume is not the same as the molar specific heat at constant pressure for gases. Why is the distinction between constant volume and constant pressure usually insignificant for the specific heats of liquids and solids?

Multiple-Choice Questions

Student Response System Questions

1. A heat engine runs between reservoirs at temperatures of 300°C and 30°C. What is its maximum possible efficiency?
   (a) 10% (b) 47% (c) 53%
   (d) 90% (e) 100%
2. The PV diagram illustrates several paths to get from an initial to a final state. For which path does the system do the most work?

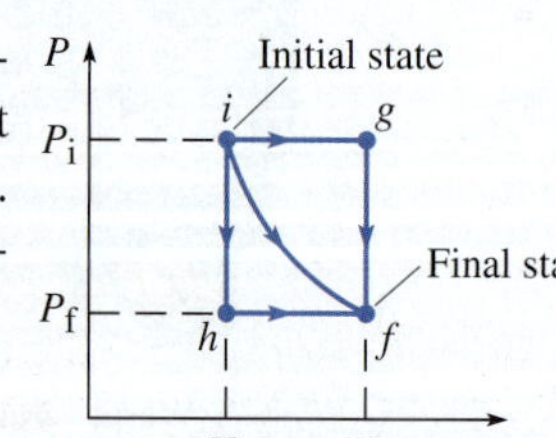

   (a) path *igf*
   (b) path *if*
   (c) path *ihf*
   (d) All paths represent equal work.

3. If two different systems are put in thermal contact so that heat can flow from one to the other, then heat will flow until the systems have the same
(a) energy.
(b) heat capacity.
(c) entropy.
(d) temperature.

4. When the first law of thermodynamics ($\Delta U = Q + W$) is applied to a system S, the variables Q and W stand for
(a) the heat flow *out of* S and the work done *on* S.
(b) the heat flow *out of* S and the work done *by* S.
(c) the heat flow *into* S and the work done *by* S.
(d) the heat flow *into* S and the work done *on* S.

5. As a system undergoes a constant-volume process
(a) the pressure does not change.
(b) the internal energy does not change.
(c) the work done on the system is zero.
(d) the entropy stays the same.
(e) the temperature of the system does not change.

6. As an ideal gas is adiabatically expanding,
(a) the temperature of the gas does not change.
(b) the internal energy of the gas does not change.
(c) work is not done on or by the gas.
(d) no heat is given off or taken in by the gas.
(e) both (a) and (d)
(f) both (a) and (b)

7. An ideal gas is confined to the left chamber of an insulated container. The right chamber is evacuated. A valve is opened between the chambers, allowing gas to flow into the right chamber. After equilibrium is established, the temperature of the gas _____. [*Hint*: What happens to the internal energy?]
(a) is lower than the initial temperature
(b) is higher than the initial temperature
(c) is the same as the initial temperature
(d) could be higher than, the same as, or lower than the initial temperature

8. Which choice correctly identifies the three processes shown in the diagrams?
(a) I = isobaric; II = isochoric; III = adiabatic
(b) I = isothermal; II = isothermal; III = isobaric
(c) I = isochoric; II = adiabatic; III = isobaric
(d) I = isobaric; II = isothermal; III = isochoric

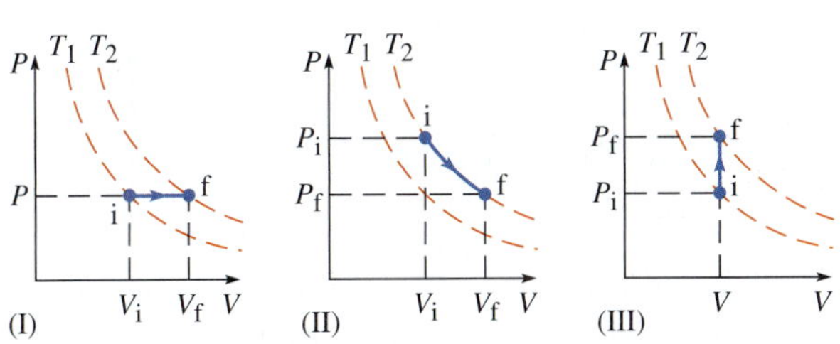

Question 8; Problem 12

9. As an ideal gas is compressed at constant temperature,
(a) heat flows out of the gas.
(b) the internal energy of the gas does not change.
(c) the work done on the gas is zero.
(d) none of the above
(e) both (a) and (b)
(f) both (a) and (c)

10. As moisture from the air condenses on the outside of a cold glass of water, the entropy of the condensing moisture
(a) stays the same.
(b) increases.
(c) decreases.
(d) not enough information

11. Given 1 mole of an ideal gas, in a state characterized by P_A, V_A, a change occurs so that the final pressure and volume are equal to P_B, V_B, where $V_B > V_A$. Which of these is true?
(a) The heat supplied to the gas during the process is completely determined by the values P_A, V_A, P_B, and V_B.
(b) The change in the internal energy of the gas during the process is completely determined by the values P_A, V_A, P_B, and V_B.
(c) The work done by the gas during the process is completely determined by the values P_A, V_A, P_B, and V_B.
(d) All three are true.
(e) None of these is true.

12. On a summer day, you keep the air conditioner in your room running. From the list numbered 1 to 4, choose the hot reservoir and the cold reservoir.
1. the air outside
2. the compartment inside the air conditioner where the air is compressed
3. the freon gas that is the working substance (expands and compresses in each cycle)
4. the air in the room

(a) 1 is the hot reservoir, 2 is the cold reservoir.
(b) 1 is the hot reservoir, 3 is the cold reservoir.
(c) 1 is the hot reservoir, 4 is the cold reservoir.
(d) 2 is the hot reservoir, 3 is the cold reservoir.
(e) 2 is the hot reservoir, 4 is the cold reservoir.
(f) 3 is the hot reservoir, 4 is the cold reservoir.

13. Which of these statements are implied by the *second* law of thermodynamics?
(a) The entropy of an engine (including its fuel and/or heat reservoirs) operating in a cycle never decreases.
(b) The increase in internal energy of a system in any process is the sum of heat absorbed plus work done on the system.
(c) A heat engine, operating in a cycle, that exhausts no heat to the low-temperature reservoir is impossible.
(d) Both (a) and (c).
(e) All three [(a), (b), and (c)].

Problems

Ⓒ Combination conceptual/quantitative problem
Biomedical application
✦ Challenging
Blue # Detailed solution in the Student Solutions Manual
[1, 2] Problems paired by concept
connect Text website interactive or tutorial

15.1 The First Law of Thermodynamics; 15.2 Thermodynamic Processes; 15.3 Thermodynamic Processes for an Ideal Gas

1. On a cold day, Ming rubs her hands together to warm them up. She presses her hands together with a force of 5.0 N. Each time she rubs them back and forth, they move a distance of 16 cm with a coefficient of kinetic friction of 0.45. Assuming no heat flow to the surroundings, after she has rubbed her hands back and forth eight times, by how much has the internal energy of her hands increased?

2. A system takes in 550 J of heat while performing 840 J of work. What is the change in internal energy of the system?

3. The internal energy of a system increases by 400 J while 500 J of work are performed on it. What was the heat flow into or out of the system?

4. Verify the units in Eq. (15-2)—that is, show that the SI units of pressure times volume is equals to the SI unit of work.

5. How much does the internal energy change for 1.00 m^3 of water after it has fallen from the top of a waterfall and landed in the river 11.0 m below? Assume no heat flow from the water to the air.

6. A pot containing 2.00 kg of water is sitting on a hot stove, and the water is stirred violently by a mixer that does 6.0 kJ of mechanical work on the water. The temperature of the water rises by 4.00°C. What quantity of heat flowed into the water from the stove during the process?

7. A contractor uses a paddle stirrer to mix a can of paint. The paddle turns at 28.0 rad/s and exerts a torque of 16.0 N·m on the paint, doing work at a rate

$$\text{power} = \tau\omega = 16.0\ \text{N·m} \times 28.0\ \text{rad/s} = 448\ \text{W}$$

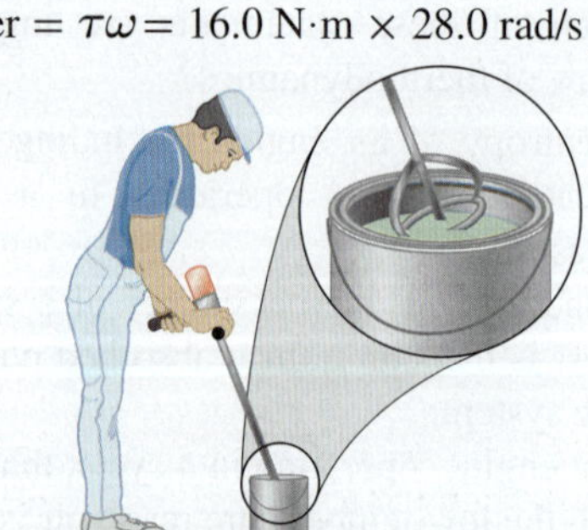

An internal energy increase of 12.5 kJ causes the temperature of the paint to increase by 1.00 K. (a) If there were no heat flow between the paint and the surroundings, what would be the temperature change of the paint as it is stirred for 5.00 min? (b) If the actual temperature change was 6.3 K, how much heat flowed from the paint to the surroundings?

8. The figure shows *PV* diagrams for five systems as they undergo various thermodynamic processes. Rank them in order of the work done on the system, from greatest to least. Rank positive work higher than negative work.

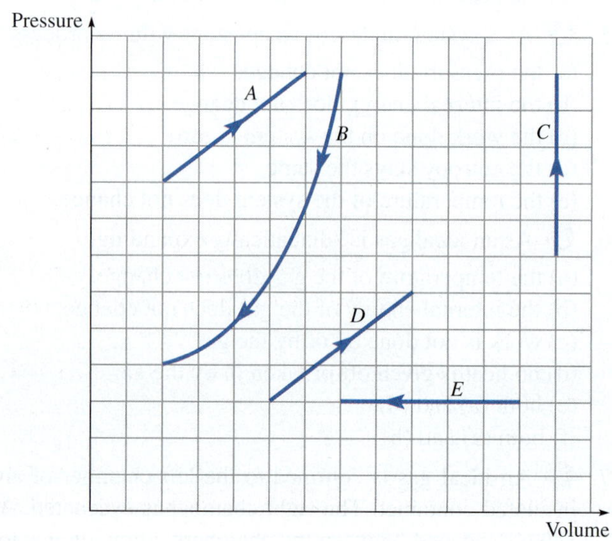

9. A monatomic ideal gas at 27°C undergoes a constant-pressure process from *A* to *B* and a constant-volume process from *B* to *C*. Find the total work done during these two processes.

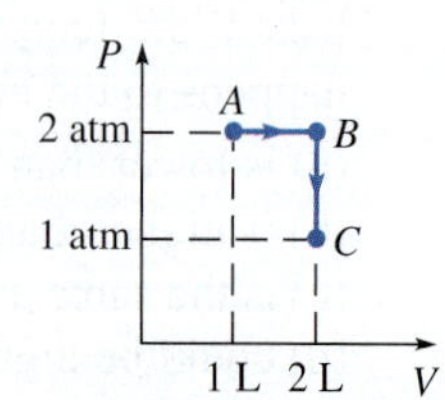

10. A monatomic ideal gas at 27°C undergoes a constant-volume process from *A* to *B* and a constant-pressure process from *B* to *C*. Find the total work done during these two processes.

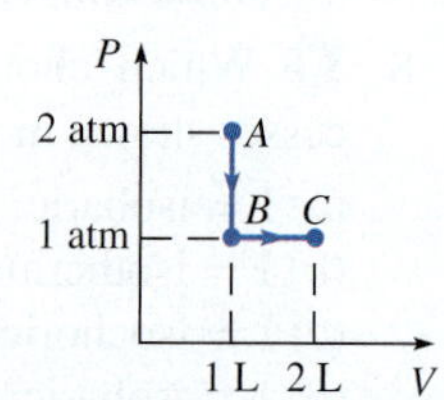

11. An ideal monatomic gas is taken through the cycle in the *PV* diagram. (a) If there are 0.0200 mol of this gas, what are the temperature and pressure at point *C*? (b) What is the change in internal energy of the gas as it is taken from *A* to *B*? (c) How much work is done by this gas per cycle?

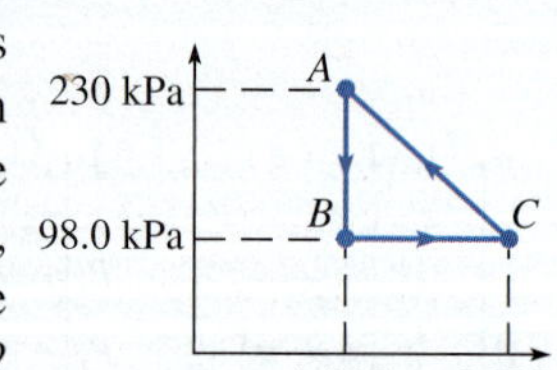

(d) What is the total change in internal energy of this gas in one cycle?

12. The three processes shown with Multiple Choice Question 8 involve a diatomic ideal gas. Rank them in order of the change in internal energy, from greatest to smallest.

13. In a refrigerator, 2.00 mol of an ideal monatomic gas are taken through the cycle shown in the figure. The temperature at point A is 800.0 K. (a) What are the temperature and pressure at point D? (b) What is the net work done on the gas as it is taken through four cycles? (c) What is the internal energy of the gas when it is at point A? (d) What is the total change in internal energy of this gas during four complete cycles?

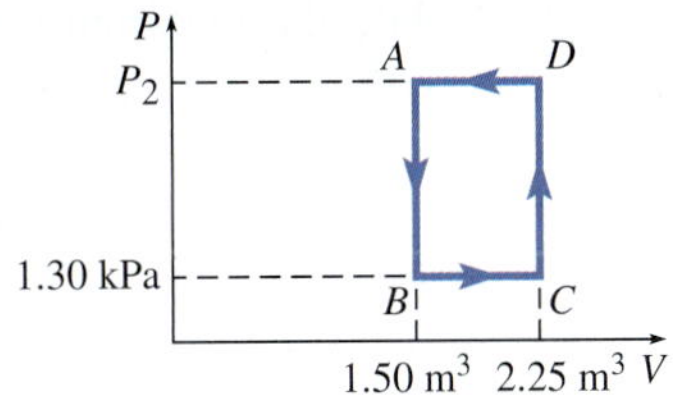

14. A balloon contains 200.0 L of nitrogen gas at 20.0°C and at atmospheric pressure. How much energy must be added to raise the temperature of the nitrogen to 40.0°C while allowing the balloon to expand at atmospheric pressure?

15. An ideal gas is heated at a constant pressure of 2.0×10^5 Pa from a temperature of −73° C to a temperature of +27° C. The initial volume of the gas is 0.10 m^3. The heat energy supplied to the gas in this process is 25 kJ. What is the increase in internal energy of the gas?

16. If the pressure on a fish increases from 1.1 atm to 1.2 atm, its swim bladder decreases in volume from 8.16 mL to 7.48 mL while the temperature of the air inside remains constant. How much work is done on the air in the bladder?

17. An ideal gas is in contact with a heat reservoir so that it remains at a constant temperature of 300.0 K. The gas is compressed from a volume of 24.0 L to a volume of 14.0 L. During the process, the mechanical device pushing the piston to compress the gas is found to expend 5.00 kJ of energy. How much heat flows between the heat reservoir and the gas, and in what direction does the heat flow occur?

18. Suppose 1.00 mol of oxygen is heated at constant pressure of 1.00 atm from 10.0°C to 25.0°C. (a) How much heat is absorbed by the gas? (b) Using the ideal gas law, calculate the change of volume of the gas in this process. (c) What is the work done by the gas during this expansion? (d) From the first law, calculate the change of internal energy of the gas in this process.

19. ✦ Suppose a monatomic ideal gas is changed from state A to state D by one of the processes shown on the PV diagram. (a) Find the total work done on the gas if it follows the constant-volume path AB followed by the constant-pressure path BCD. (b) Calculate the total change in internal energy of the gas during the entire process and the total heat flow into the gas.

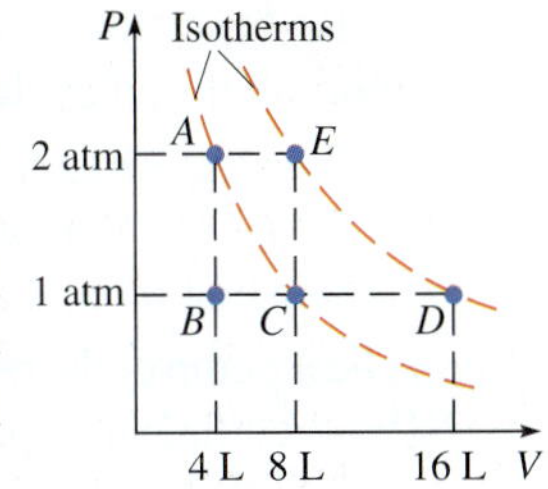

Problems 19–21

20. ✦ Repeat Problem 19 for the case when the gas follows the constant-temperature path AC followed by the constant-pressure path CD.

21. ✦ Repeat Problem 19 for the case when the gas follows the constant-pressure path AE followed by the constant-temperature path ED.

15.5 Heat Engines; 15.6 Refrigerators and Heat Pumps

Problems 22–23. The figure shows PV diagrams for five cyclical processes. (Cycle D is a simplified model of a steam engine cycle.)

22. Rank the processes in order of the net work done *by* the system per cycle, from greatest to least. Rank positive work done by the system (as in an engine) higher than negative work done by the system (as in a heat pump).

23. (a) Which might describe a heat engine? (b) Which might describe a heat pump? (c) Which might describe a refrigerator? Explain.

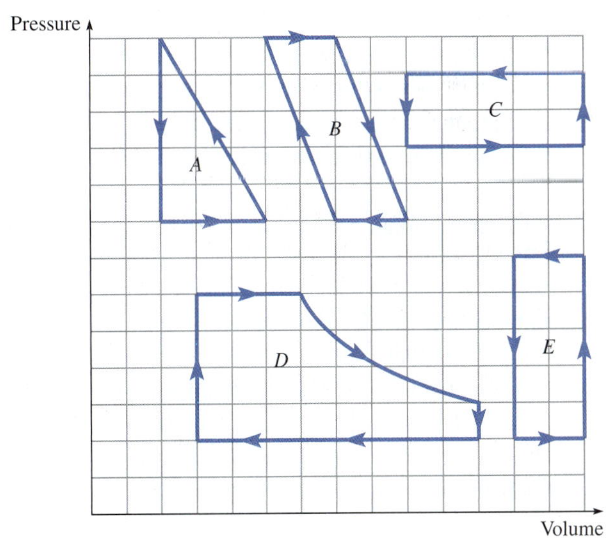

Problems 22–23

24. A heat engine follows the cycle shown in the figure. (a) How much net work is done by the engine in one cycle? (b) What

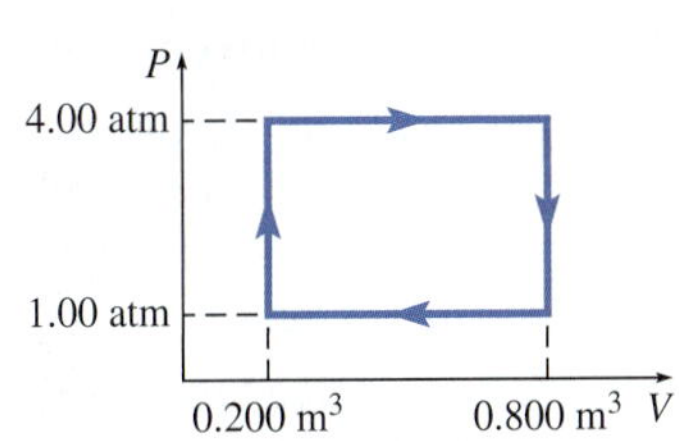

is the heat flow *into* the engine per cycle? (connect tutorial: closed cycle)

25. What is the efficiency of an electric generator that produces 1.17 kW·h per kg of coal burned? The heat of combustion of coal is 6.71×10^6 J/kg.

26. A heat pump delivers heat at a rate of 7.81 kW for 10.0 h. If its coefficient of performance is 6.85, how much heat is taken from the cold reservoir during that time?

27. (a) How much heat does an engine with an efficiency of 33.3% absorb in order to deliver 1.00 kJ of work? (b) How much heat is exhausted by the engine?

28. The efficiency of an engine is 0.21. For every 1.00 kJ of heat absorbed by the engine, how much (a) net work is done by it and (b) heat is released by it?

29. The United States generates about 5.0×10^{16} J of electric energy a day. This energy is equivalent to work, since it can be converted into work with almost 100% efficiency by an electric motor. (a) If this energy is generated by power plants with an average efficiency of 0.30, how much heat is dumped into the environment each day? (b) How much water would be required to absorb this heat if the water temperature is not to increase more than 2.0°C?

30. The intensity (power per unit area) of the sunlight incident on Earth's surface, averaged over a 24 h period, is about 0.20 kW/m^2. If a solar power plant is to be built with an output capacity of 1.0×10^9 W, how big must the area of the solar energy collectors be for photocells operating at 20.0% efficiency?

31. An engine releases 0.450 kJ of heat for every 0.100 kJ of work it does. What is the efficiency of the engine?

32. How much heat does a heat pump with a coefficient of performance of 3.0 deliver when supplied with 1.00 kJ of electricity?

33. An air conditioner whose coefficient of performance is 2.00 removes 1.73×10^8 J of heat from a room per day. How much does it cost to run the air conditioning unit per day if electricity costs \$0.10 per kilowatt-hour? (Note that 1 kilowatt-hour = 3.6×10^6 J.)

34. A steam engine has a piston with a diameter of 15.0 cm and a stroke (the displacement of the piston) of 20.0 cm. The average pressure applied to this piston is 1.3×10^5 Pa. What operating frequency in cycles per second (Hz) would yield an average power output of 27.6 kW?

15.7 Reversible Engines and Heat Pumps

35. An ideal engine has an efficiency of 0.725 and uses gas from a hot reservoir at a temperature of 622 K. What is the temperature of the cold reservoir to which it exhausts heat?

36. A heat engine takes in 125 kJ of heat from a reservoir at 815 K and exhausts 82 kJ to a reservoir at 293 K. (a) What is the efficiency of the engine? (b) What is the efficiency of an ideal engine operating between the same two reservoirs?

37. In a certain steam engine, the boiler temperature is 127°C and the cold reservoir temperature is 27°C. While this engine does 8.34 kJ of work, what minimum amount of heat must be discharged into the cold reservoir?

38. Calculate the maximum possible efficiency of a heat engine that uses surface lake water at 18.0°C as a source of heat and rejects waste heat to the water 0.100 km below the surface where the temperature is 4.0°C.

39. An ideal refrigerator removes heat at a rate of 0.10 kW from its interior (+2.0° C) and exhausts heat at 40.0°C. How much electrical power is used?

40. A heat pump is used to heat a house with an interior temperature of 20.0°C. On a chilly day with an outdoor temperature of −10.0° C, what is the minimum work that the pump requires in order to deliver 1.0 kJ of heat to the house? (connect tutorial: heat pump)

41. An ideal refrigerator keeps its contents at 0.0°C and exhausts heat into the kitchen at 40.0°C. For every 1.0 kJ of work done, (a) how much heat is exhausted? (b) How much heat is removed from the contents?

42. The outdoor temperature on a winter's day is −4° C. If you use 1.0 kJ of electric energy to run a heat pump, how much heat does that put into your house at 21°C? Assume that the heat pump is ideal.

43. The motor that drives a refrigerator produces 148 W of useful power. The hot and cold temperatures of the heat reservoirs are 20.0°C and −5.0° C. What is the maximum possible amount of ice it can produce in 2.0 h from water that is initially at 8.0°C?

44. A new organic semiconductor device is able to generate electricity (which can be used to charge a battery or light an LED) using the warmth of human skin. If your skin temperature is maintained by your body at 35°C and the temperature of the surroundings is 20°C, what is the maximum possible efficiency for this device?

45. The human body could potentially serve as a very good thermal reservoir, as its internal temperature remains quite constant at around 37°C and is stabilized by continual intake of food. Suppose an inventor designed microscopic engines that could be implanted under the skin in order to charge batteries or power other equipment (e.g., pacemakers, or other necessary medical devices) by using the temperature difference between the interior of the body and the outside temperature. If such an engine were capable of half the Carnot efficiency and were able to store 5.0 nJ of energy in a battery in the course of a day, at what rate would the body be supplying energy if the temperature of the surroundings were 20°C?

46. Two engines operate between the same two temperatures of 750 K and 350 K, and have the same rate of heat input. One of the engines is a reversible engine with a power output of 23 kW. The second engine has

an efficiency of 42%. What is the power output of the second engine?

47. (a) Calculate the efficiency of a reversible engine that operates between the temperatures 600.0°C and 300.0°C. (b) If the engine absorbs 420.0 kJ of heat from the hot reservoir, how much does it exhaust to the cold reservoir?

48. A reversible engine with an efficiency of 30.0% has $T_C = 310.0$ K. (a) What is T_H? (b) How much heat is exhausted for every 0.100 kJ of work done?

49. An electric power station generates steam at 500.0°C and condenses it with river water at 27°C. By how much would its theoretical maximum efficiency decrease if it had to switch to cooling towers that condense the steam at 47°C?

50. An oil-burning electric power plant uses steam at 773 K to drive a turbine, after which the steam is expelled at 373 K. The engine has an efficiency of 0.40. What is the theoretical maximum efficiency possible at those temperatures?

51. An inventor proposes a heat engine to propel a ship, using the temperature difference between the water at the surface and the water 10 m below the surface as the two reservoirs. If these temperatures are 15.0°C and 10.0°C, respectively, what is the maximum possible efficiency of the engine?

52. A heat engine uses the warm air at the ground as the hot reservoir and the cooler air at an altitude of several thousand meters as the cold reservoir. If the warm air is at 37°C and the cold air is at 25°C, what is the maximum possible efficiency for the engine?

53. A reversible refrigerator has a coefficient of performance of 3.0. How much work must be done to freeze 1.0 kg of liquid water initially at 0°C?

54. ✦ An engine operates between temperatures of 650 K and 350 K at 65.0% of its maximum possible efficiency. (a) What is the efficiency of this engine? (b) If 6.3×10^3 J is exhausted to the low temperature reservoir, how much work does the engine do?

55. Ⓒ (a) For a reversible engine, will you obtain a better efficiency by increasing the high-temperature reservoir by an amount ΔT or decreasing the low-temperature reservoir by the same amount ΔT? Explain. (b) To illustrate your answer to this question, calculate the efficiencies of a reversible engine that initially uses reservoirs at 373 K and 923 K for $\Delta T = 50$ K.

56. Show that the coefficient of performance for a reversible heat pump is $1/(1 - T_C/T_H)$.

57. Show that the coefficient of performance for a reversible refrigerator is $1/[(T_H/T_C) - 1]$.

15.8 Entropy

58. Rank these in order of increasing entropy: (a) 0.5 kg of ice and 0.5 kg of (liquid) water at 0°C; (b) 1 kg of ice at 0°C; (c) 1 kg of (liquid) water at 0°C; (d) 1 kg of water at 20°C.

59. Rank these in order of increasing entropy: (a) 1 mol of water at 20°C and 1 mol of ethanol at 20°C in separate containers; (b) a mixture of 1 mol of water at 20°C and 1 mol of ethanol at 20°C; (c) 0.5 mol of water at 20°C and 0.5 mol of ethanol at 20°C in separate containers; (d) a mixture of 1 mol of water at 30°C and 1 mol of ethanol at 30°C.

60. An ice cube at 0.0°C is slowly melting. What is the change in the ice cube's entropy for each 1.00 g of ice that melts?

61. From Table 14.4, we know that approximately 2256 kJ is needed to transform 1.00 kg of water at 100°C to steam at 100°C. What is the change in entropy of 1.00 kg of water evaporating at 100.0°C? (Specify whether the change in entropy is an increase, +, or a decrease, −.)

62. What is the change in entropy of 10 g of steam at 100°C as it condenses to water at 100°C? By how much does the entropy of the universe increase in this process?

63. A large block of copper initially at 20.0°C is placed in a vat of hot water (80.0°C). For the first 1.0 J of heat that flows from the water into the block, find (a) the entropy change of the block, (b) the entropy change of the water, and (c) the entropy change of the universe. Note that the temperatures of the block and water are essentially unchanged by the flow of only 1.0 J of heat.

64. A large, cold (0.0°C) block of iron is immersed in a tub of hot (100.0°C) water. In the first 10.0 s, 41.86 kJ of heat is transferred, although the temperatures of the water and the iron do not change much in this time. Ignoring heat flow between the system (iron + water) and its surroundings, calculate the change in entropy of the system (iron + water) during this time.

65. On a cold winter day, the outside temperature is −15.0° C. Inside the house the temperature is +20.0° C. Heat flows out of the house through a window at a rate of 220.0 W. At what rate is the entropy of the universe changing due to this heat conduction through the window?

66. Within an insulated system, 418.6 kJ of heat is conducted through a copper rod from a hot reservoir at +200.0° C to a cold reservoir at +100.0° C. (The reservoirs are so big that this heat exchange does not change their temperatures appreciably.) What is the net change in entropy of the system, in kJ/K?

67. A student eats 2000 kcal per day. (a) Assuming that all of the food energy is released as heat, what is the rate of heat released (in watts)? (b) What is the rate of change of entropy of the surroundings if all of the heat is released into air at room temperature (20°C)?

68. Humans cool off by perspiring; the evaporating sweat removes heat from the body. If the skin temperature is 35.0°C and the air temperature is 28.0°C, what

is the entropy change of the universe due to the evaporation of 150 mL of sweat? Take the latent heat of vaporization of sweat to be the same as that for water.

69. Polypeptides are transformed from a random coil into their characteristic three-dimensional structures by a process called *protein folding*. In the folded state, the proteins are highly ordered (low in entropy). Through the application of heat, proteins can become *denatured*—a state in which the structure unfolds and approaches a random structure with higher entropy. This denaturation often is observable macroscopically, as in the case of the difference between uncooked and cooked egg albumin. Suppose a particular protein has a molar mass of 33 kg/mol. (a) If, at a constant temperature of 72°C, a sample of 45 mg of this protein is denatured such that the sample's entropy change is 2.1 mJ/K, how much heat did this process require? (b) How much did the energy of each molecule increase?

70. Suppose you have a 35.0 g sample of a protein for which denaturation required an input of 2.20 J at a constant temperature of 60.0°C. If the molar mass of the protein is 29.5 kg/mol, what is the entropy change per protein molecule?

Collaborative Problems

71. A coal-fired electrical generating station can use a higher T_H than a nuclear plant; for safety reasons the core of a nuclear reactor is not allowed to get as hot as burning coal. Suppose that $T_H = 727°$ C for a coal station but $T_H = 527°$ C for a nuclear station. Both power plants exhaust waste heat into a lake at $T_C = 27°$ C. How much waste heat does each plant exhaust into the lake per day to produce 1.00×10^{14} J of electricity per day? Assume both operate as reversible engines. (connect tutorial: power stations)

72. ✦ A town is planning on using the water flowing through a river at a rate of 5.0×10^6 kg/s to carry away the heat from a new power plant. Environmental studies indicate that the temperature of the river should only increase by 0.50°C. The maximum design efficiency for this plant is 30.0%. What is the maximum possible power this plant can produce?

73. Ⓒ A 0.500 kg block of iron at 60.0°C is placed in contact with a 0.500 kg block of iron at 20.0°C. (a) The blocks soon come to a common temperature of 40.0°C. *Estimate* the entropy change of the universe when this occurs. [*Hint*: Assume that all the heat flow occurs at an average temperature for each block.] (b) Estimate the entropy change of the universe if, instead, the temperature of the hotter block increased to 80.0°C while the temperature of the colder block decreased to 0.0°C. Explain how your answer indicates that the process is impossible.

74. ✦ A town is considering using its lake as a source of power. The average temperature difference from the top to the bottom is 15°C, and the average surface temperature is 22°C. (a) Assuming that the town can set up a reversible engine using the surface and bottom of the lake as heat reservoirs, what would be its efficiency? (b) If the town needs about 1.0×10^8 W of power to be supplied by the lake, how many cubic meters of water does the heat engine use per second? (c) The surface area of the lake is 8.0×10^7 m^2 and the average incident intensity (over 24 h) of the sunlight is 200 W/m^2. Can the lake supply enough heat to meet the town's energy needs with this method?

Comprehensive Problems

75. On a day when the temperature is 19°C, a 0.15 kg baseball is dropped from the top of a 24 m tower. After the ball hits the ground, bounces a few times, and comes to rest, approximately how much has the entropy of the universe increased?

76. Suppose you mix 4.0 mol of a monatomic gas at 20.0°C and 3.0 mol of another monatomic gas at 30.0°C. If the mixture is allowed to reach equilibrium, what is the final temperature of the mixture? [*Hint*: Use energy conservation.]

77. A balloon contains 160 L of nitrogen gas at 25°C and 1.0 atm. How much energy must be added to raise the temperature of the nitrogen to 45°C while allowing the balloon to expand at atmospheric pressure?

78. Suppose you inhale 0.50 L of air initially at 20°C and 100 kPa pressure. While holding your breath, this air is warmed at constant pressure to 37°C. Treating the air as an ideal diatomic gas, how much heat flows from the body into the air?

79. Ⓒ On a hot day, you are in a sealed, insulated room. The room contains a refrigerator, operated by an electric motor. The motor does work at the rate of 250 W when it is running. The refrigerator removes heat from the food storage space at a rate of 450 W when the motor is running. In an effort to cool the room, you open the refrigerator door and let the motor run continuously. At what *net* rate is heat added to (+) or subtracted from (−) the room and all of its contents?

80. (a) What is the entropy change of 1.00 mol of H_2O when it changes from ice to water at 0.0°C? (b) If the ice is in contact with an environment at a temperature of 10.0°C, what is the entropy change of the universe when the ice melts?

81. Estimate the entropy change of 850 g of water when it is heated from 20.0°C to 50.0°C. [*Hint*: Assume that the heat flows into the water at an average temperature of 35.0°C.]

82. For a more realistic estimate of the maximum coefficient of performance of a heat pump, assume that a

heat pump takes in heat from the outdoors at *10°C below* the ambient outdoor temperature, to account for the temperature difference across its heat exchanger. Similarly, assume that the output must be *10°C hotter* than the house (which is kept at 20°C) to make the heat flow into the house. Make a graph of the coefficient of performance of a reversible heat pump under these conditions as a function of outdoor temperature (from −15° C to +15° C in 5°C increments).

83. A container holding 1.20 kg of water at 20.0°C is placed in a freezer that is kept at −20.0° C. The water freezes and comes to thermal equilibrium with the interior of the freezer. What is the minimum amount of electrical energy required by the freezer to do this if it operates between reservoirs at temperatures of 20.0°C and −20.0° C?

84. A reversible heat engine has an efficiency of 33.3%, removing heat from a hot reservoir and rejecting heat to a cold reservoir at 0°C. If the engine now operates in reverse, how long would it take to freeze 1.0 kg of water at 0°C, if it operates on a power of 186 W?

85. ✦ A fish at a pressure of 1.1 atm has its swim bladder inflated to an initial volume of 8.16 mL. If the fish starts swimming horizontally, its temperature increases from 20.0°C to 22.0°C as a result of the exertion. (a) Since the fish is still at the same pressure, how much work is done by the air in the swim bladder? [*Hint*: First find the new volume from the temperature change.] (b) How much heat is gained by the air in the swim bladder? Assume air to be a diatomic ideal gas. (c) If this quantity of heat is lost by the fish, by how much will its temperature decrease? The fish has a mass of 5.00 g, and its specific heat is about 3.5 J/(g·°C).

Review and Synthesis

Problems 86–88. The *PV* diagram shown is for a heat engine that uses 1.000 mol of a diatomic ideal gas as its working substance. In the constant-temperature processes *A* and *C*, the gas is in contact with reservoirs at temperatures 373 K and 273 K, respectively. In constant-volume process *B*, the gas temperature decreases as heat flows into the cold reservoir. In constant-volume process *D*, the gas temperature increases as heat flows from the hot reservoir.

86. (a) Find the work done by the engine during each of the four steps and the net work done for the cycle. (b) If the heat input per cycle is 2770 J, what is the efficiency of the engine? (c) Compare the efficiency to that of an ideal engine using the same reservoirs.

87. Find the change in internal energy of the gas during each of the four steps.

88. (a) Find the heat flow into or out of the gas during each of the four steps. (b) What is the net heat flow into the gas per cycle? (c) Calculate the change in entropy *of the cold reservoir* (not of the gas) in steps *B* and *C* and the change in entropy *of the hot reservoir* in steps *A* and *D*. (d) What is the total entropy change of the universe per cycle?

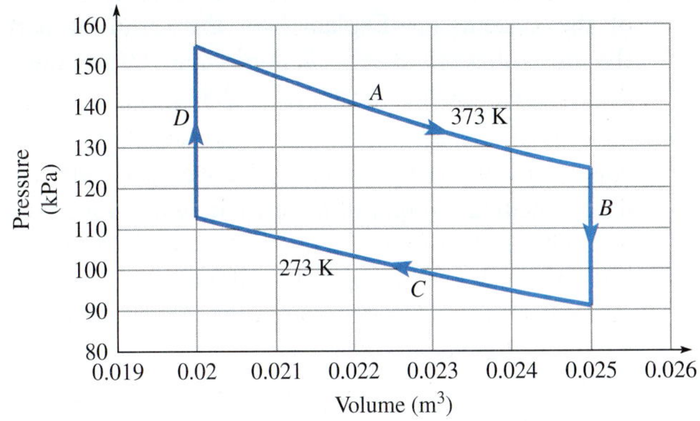

Problems 86–88

Problems 89–91. In a heat engine, 3.00 mol of a monatomic ideal gas, initially at 4.00 atm of pressure, undergoes an isothermal expansion, increasing its volume by a factor of 9.50 at a constant temperature of 650.0 K. The gas is then compressed at a constant pressure to its original volume. Finally, the pressure is increased at constant volume back to the original pressure.

89. Draw a *PV* diagram to illustrate the cycle for this engine. Label the axes with numerical values.

90. (a) Calculate the work done by the engine during each step and the net work done per cycle. (b) If the heat input per cycle is 58.3 kJ, what is the efficiency?

91. Ⓒ (a) Find the heat flow into or out of the gas during each step. (b) Find the entropy change of the gas during the isothermal step. (c) What is the entropy change of the gas for a complete cycle? Is it equal in magnitude to the entropy change of the environment per cycle? Explain.

92. A model steam engine of 1.00 kg mass pulls eight cars of 1.00 kg mass each. The cars start at rest and reach a velocity of 3.00 m/s in a time of 3.00 s while moving a distance of 4.50 m. During that time, the engine takes in 135 J of heat. What is the change in the internal energy of the engine?

93. A certain engine can propel a 1800 kg car from rest to a speed of 27 m/s in 9.5 s with an efficiency of 27%. What are the rate of heat flow into the engine at the high temperature and the rate of heat flow out of the engine at the low temperature?

94. An engine works at 30.0% efficiency. The engine raises a 5.00 kg crate from rest to a vertical height of 10.0 m, at which point the crate has a speed of 4.00 m/s. How much heat input is required for this engine?

95. Ⓒ A 0.50 kg block of iron [$c = 0.44$ kJ/(kg·K)] at 20.0°C is in contact with a 0.50 kg block of aluminum [$c = 0.900$ kJ/(kg·K)] at a temperature of 20.0°C. The

system is completely isolated from the rest of the universe. Suppose heat flows from the iron into the aluminum until the temperature of the aluminum is 22.0°C. (a) From the first law, calculate the final temperature of the iron. (b) Estimate the entropy change of the system. (c) Explain how the result of part (b) shows that this process is impossible. [*Hint*: Since the system is isolated, $\Delta S_{\text{System}} = \Delta S_{\text{Universe}}$.]

96. The efficiency of a muscle during weight lifting is equal to the work done in lifting the weight divided by the total energy output of the muscle (work done plus internal energy dissipated in the muscle). Determine the efficiency of a muscle that lifts a 161 N weight through a vertical displacement of 0.577 m and dissipates 139 J in the process.

97. A power plant burns coal to produce pressurized steam at 535 K. The steam then condenses back into water at a temperature of 323 K. (a) What is the maximum possible efficiency of this plant? (b) If the plant operates at 50.0% of its maximum efficiency and its power output is 1.23×10^8 W, at what rate must heat be removed by means of a cooling tower? (c) During periods of low demand, the steam engine is used to pump water 380 m uphill from one reservoir to another. (Then during periods of high demand, the water is released to drive turbines and generate electricity.) What is the maximum possible rate (in m^3/s) at which water can be pumped uphill?

Problems 98–100. The figure shows a *PV* diagram for an engine that uses a monatomic ideal gas as the working substance. The temperature at point *A* is 470.0 K.

98. (a) How much net work does this engine do per cycle? (b) Assuming that the efficiency of the engine is 0.444, what is the heat *input* into the gas per cycle? (c) How much heat is exhausted per cycle? (d) It takes 3.0 s for the engine to go through each cycle. The engine is used to drive a turbine that spins at 3000 rev/min. What is the average torque exerted on the turbine?

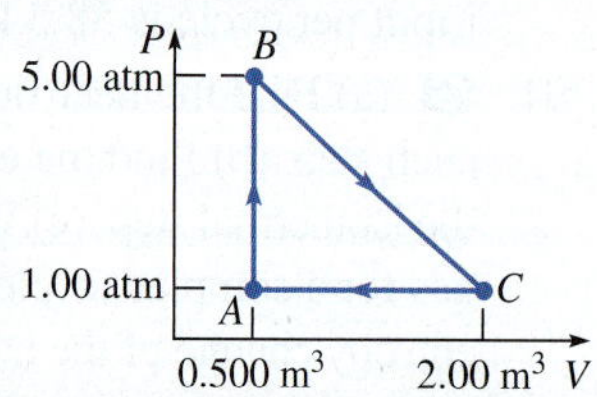

Problems 98–100

99. (a) What is the maximum temperature of the gas? (b) What would be the efficiency of an *ideal* engine with reservoirs at the maximum and minimum temperatures of this engine? Compare this to the actual efficiency, 0.444.

100. (a) How many moles of gas are used in this engine? (b) Calculate the heat flow into or out of the gas in steps *AB* and *CA*. (c) Calculate the work done by the gas during each step.

Answers to Practice Problems

15.1 The internal energy increase is greater than the heat flow into the gas, so positive work was done on the gas. Positive work is done by the piston when it moves inward.

15.2 360 kJ

15.3 Heat flows into the gas; $Q = 3.8$ kJ.

15.4 The fire is irreversible: smoke, carbon dioxide, and ash will not come together spontaneously to form logs and twigs.

15.5 20%

15.6 4.0 kW·h = 14 MJ

15.7 1200 K

15.8 178 MW

15.9 0.03 J/K

Answers to Checkpoints

15.2 (a) Yes. The heat flow during an adiabatic process is zero ($Q = 0$), but work can be done. The work done on the system changes its internal energy, which can cause a temperature change. (b) Yes. If a system is in thermal contact with a heat reservoir, heat flows between the reservoir and the system to keep the temperature constant. (c) Yes. During a phase transition such as freezing or melting, the internal energy of the system changes but the temperature does not.

15.4 An inelastic collision involves the conversion of kinetic energy into internal energy, an irreversible process.

15.8 No, in an irreversible process the *total* entropy of the universe increases. If the entropy of one system increases by 10 J/K while the entropy of its surroundings decreases by 10 J/K, the process would be reversible ($\Delta S_{\text{tot}} = 0$).

Appendix A

Mathematics Review

A.1 ALGEBRA

There are two basic kinds of algebraic manipulations.

- The same operation can always be performed on both sides of an equation.
- Substitution is always permissible (if $a = b$, then any occurrence of a in any equation can be replaced with b).

Products distribute over sums

$$a(b + c) = ab + ac \tag{A-1}$$

The reverse—replacing $ab + ac$ with $a(b + c)$—is called *factoring*. Since dividing by c is the same as multiplying by $1/c$,

$$\frac{a + b}{c} = \frac{a}{c} + \frac{b}{c} \tag{A-2}$$

Equation (A-2) is the basis of the procedure for adding fractions. To add fractions, they must be expressed with a *common denominator*.

$$\frac{a}{b} + \frac{c}{d} = \frac{a}{b} \times \frac{d}{d} + \frac{c}{d} \times \frac{b}{b} = \frac{ad}{bd} + \frac{bc}{bd}$$

Now applying Eq. (A-2), we end up with

$$\frac{a}{b} + \frac{c}{d} = \frac{ad + bc}{bd} \tag{A-3}$$

To divide fractions, remember that dividing by c/d is the same as multiplying by d/c:

$$\frac{a}{b} \div \frac{c}{d} = \frac{\frac{a}{b}}{\frac{c}{d}} = \frac{a}{b} \times \frac{d}{c} = \frac{ad}{bc}$$

A product in a square root can be separated:

$$\sqrt{ab} = \sqrt{a} \times \sqrt{b} \tag{A-4}$$

Pitfalls to Avoid

These are some of the most common *incorrect* algebraic substitutions. Don't fall into any of these traps!

$$\sqrt{a + b} \neq \sqrt{a} + \sqrt{b}$$

$$\frac{a}{b + c} \neq \frac{a}{b} + \frac{a}{c}$$

$$\frac{a}{b} + \frac{c}{d} \neq \frac{a + c}{b + d}$$

$$(a + b)^2 \neq a^2 + b^2$$

In the last one, the cross term is missing: $(a + b)^2 = a^2 + 2ab + b^2$.

Graphs of Linear Functions

If the graph of y as a function of x is a straight line, then y is a *linear function* of x. The relationship can be written in the standard form

$$y = mx + b \tag{A-5}$$

where m is the *slope* and b is the *y-intercept*. The slope measures how steep the line is. It tells how much y changes for a given change in x:

$$m = \frac{\Delta y}{\Delta x} = \frac{y_2 - y_1}{x_2 - x_1} \quad \text{(A-6)}$$

The y-intercept is the value of y when $x = 0$. On the graph, the line crosses the y-axis at $y = b$. See Section 1.9 for more information on graphs.

Example A.1

What is the equation of the line graphed in Fig. A.1?

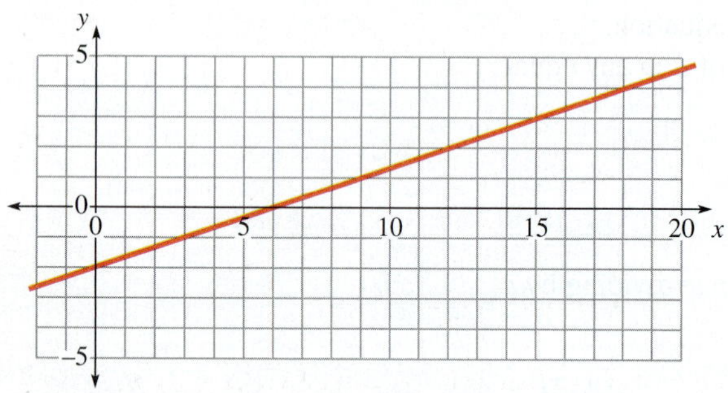

Figure A.1

Solution The y-intercept is –2. To find the slope, we choose two points on the line and then divide the "rise" (Δy) by the "run" (Δx). Using the points (0, –2) and (18, 4),

$$m = \frac{\text{rise}}{\text{run}} = \frac{y_2 - y_1}{x_2 - x_1} = \frac{4 - (-2)}{18 - 0} = \frac{1}{3}$$

Then $y = mx + b = \frac{1}{3}x - 2$.

A.2 SOLVING EQUATIONS

Solving an equation means using algebraic operations to isolate one variable. Many students tend to substitute numerical values into an equation as soon as possible. In many cases, that's a mistake. Although at first it may seem easier to manipulate numerical quantities than to manipulate algebraic symbols, there are several advantages to working with symbols:

- Symbolic algebra is much easier to follow than a series of numerical calculations. Plugging in numbers tends to obscure the logic behind your solution. If you need to trace back through your work (to find an error or review for an exam), it'll be much clearer if you have worked through the problem symbolically. It will also help your instructor when grading your homework papers or exams. When your work is clear, your instructor is better able to help you understand your mistakes. You may also get more partial credit on exams!
- Symbolic algebra lets you draw conclusions about how one quantity depends on another. For instance, working symbolically you might see that the horizontal range of a projectile is proportional to the *square* of the initial speed. If you had substituted the numerical value of the initial speed, you wouldn't notice that. In particular, when an algebraic symbol cancels out of the equation, you know that the answer doesn't depend on that quantity.
- On the most practical level, it's easy to make arithmetic or calculation errors. The later on in your solution that numbers are substituted, the fewer number of steps you have to check for such errors.

When solving equations that contain square roots, be careful not to assume that a square root is positive. The equation $x^2 = a$ has *two* solutions for x, $x = \pm\sqrt{a}$. (The symbol $\pm$ means *either* + *or* –.)

Solving Quadratic Equations

An equation is quadratic in x if it contains terms with no powers of x other than a squared term (x^2), a linear term (x^1), and a constant (x^0). Any quadratic equation can be put into the standard form:

$$ax^2 + bx + c = 0 \quad \text{(A-7)}$$

The quadratic formula gives the solutions to any quadratic equation written in standard form:

$$x = \frac{-b \pm \sqrt{b^2 - 4ac}}{2a} \qquad \text{(A-8)}$$

The symbol "±" (read "plus or minus") indicates that in general there are two solutions to a quadratic equation; that is, two values of x will satisfy the equation. One solution is found by taking the + sign and the other by taking the − sign in the quadratic formula. If $b^2 - 4ac = 0$, then there is only one solution (or, technically, the two solutions happen to be the same). If $b^2 - 4ac < 0$, then there is no solution to the equation (for x among the real numbers).

The quadratic formula still works if $b = 0$ or $c = 0$, although in such cases the equation can be solved without recourse to the quadratic formula.

Example A.2

Solve the equation $5x(3 - x) = 6$.

Solution First put the equation in standard quadratic form:

$$15x - 5x^2 = 6$$

$$-5x^2 + 15x - 6 = 0$$

We identify $a = -5$, $b = 15$, $c = -6$. Then

$$x = \frac{-b \pm \sqrt{b^2 - 4ac}}{2a}$$

$$= \frac{-15 \pm \sqrt{15^2 - 4 \times (-5) \times (-6)}}{-10}$$

$$\approx \frac{-15 \pm 10.25}{-10} = 0.475 \text{ or } 2.525$$

Solving Simultaneous Equations

Simultaneous equations are a set of N equations with N unknown quantities. We wish to solve these equations *simultaneously* to find the values of all of the unknowns. We *must* have at least as many equations as unknowns. It pays to keep track of the number of unknown quantities and the number of equations in solving more challenging problems. If there are more unknowns than equations, then look for some other relationship between the quantities—perhaps some information given in the problem that has not been used.

One way to solve simultaneous equations is by *successive substitution*. Solve one of the equations for one unknown (in terms of the other unknowns). Substitute this expression into each of the other equations. That leaves $N - 1$ equations and $N - 1$ unknowns. Repeat until there is only one equation left with one unknown. Find the value of that unknown quantity, and then work backward to find all the others.

Example A.3

Solve the equations $2x - 4y = 3$ and $x + 3y = -5$ for x and y.

Solution First solve the second equation for x in terms of y:

$$x = -5 - 3y$$

Substitute $-5 - 3y$ for x in the first equation:

$$2 \times (-5 - 3y) - 4y = 3$$

This can be solved for y:

$$-10 - 10y = 3$$

$$-10y = 13$$

$$y = \frac{13}{-10} = -1.3$$

Now that y is known, use it to find x:

$$x = -5 - 3y = -5 - 3 \times (-1.3) = -1.1$$

It's a good idea to check the results by substituting into the original equations.

A.3 EXPONENTS AND LOGARITHMS

These identities show how to manipulate exponents.

$$a^{-x} = \frac{1}{a^x} \tag{A-9}$$

$$(a^x) \times (a^y) = a^{x+y} \tag{A-10}$$

$$\frac{a^x}{a^y} = (a^x) \times (a^{-y}) = a^{x-y} \tag{A-11}$$

$$(a^x) \times (b^x) = (ab)^x \tag{A-12}$$

$$(a^x)^y = a^{xy} \tag{A-13}$$

$$a^{1/n} = \sqrt[n]{a} \tag{A-14}$$

$$a^0 = 1 \quad (\text{for any } a \neq 0) \tag{A-15}$$

$$0^a = 0 \quad (\text{for any } a \neq 0) \tag{A-16}$$

A common mistake to avoid: $(a^x) \times (a^y) \neq a^{xy}$ [see Eq. (A-10)].

Logarithms

Taking a logarithm is the inverse of exponentiation:

$$x = \log_b y \quad \text{means that} \quad y = b^x \tag{A-17}$$

Thus, one undoes the other:

$$\log_b b^x = x \tag{A-18}$$

$$b^{\log_b x} = x \tag{A-19}$$

The two commonly used bases b are 10 (the *common* logarithm) and $e = 2.71828\ldots$ (the *natural* logarithm). The common logarithm is written "$\log_{10}$," or sometimes just "log" if base 10 is understood. The natural logarithm is usually written "ln" rather than "$\log_e$."

These identities are true for any base logarithm.

$$\log xy = \log x + \log y \tag{A-20}$$

$$\log \frac{x}{y} = \log x - \log y \tag{A-21}$$

$$\log x^a = a \log x \tag{A-22}$$

Here are some common mistakes to avoid:

$$\log (x + y) \neq \log x + \log y$$

$$\log (x + y) \neq \log x \times \log y$$

$$\log xy \neq \log x \times \log y$$

$$\log x^a \neq (\log x)^a$$

Semilog Graphs

A semilog graph uses a logarithmic scale on the vertical axis and a linear scale on the horizontal axis. Semilog graphs are useful when the data plotted is thought to be an exponential function. Note that $y = ax^3$ is *not* an exponential function of x. An exponential function of x must have the x in the exponent, as in this example:

$$y = y_0 e^{ax}$$

Taking the natural logarithm of both sides of this exponential function yields

$$\ln y = ax + \ln y_0$$

Therefore, a graph of $\ln y$ versus x will be a straight line with slope a and y-intercept $\ln y_0$.

Rather than calculating $\ln y$ for each data point and plotting on regular graph paper, it is convenient to use special semilog paper. The vertical axis is marked so that the values

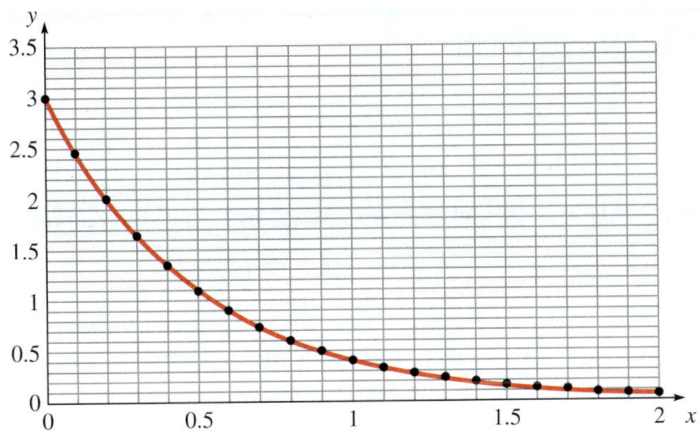

Figure A.2 Graph of the exponential function $y = 3e^{-2x}$ on linear graph paper.

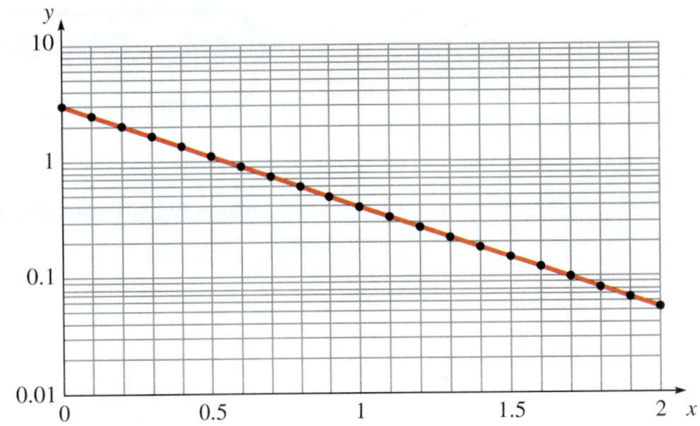

Figure A.3 Graph of the exponential function $y = 3e^{-2x}$ on semilog graph paper.

of y can be plotted directly, but the markings are spaced proportional to the log of y. (If you are using a plotting calculator or a computer to make the graph, log scale should be chosen for one axis from the menu of options.) The slope a on a semilog graph is *not* $\Delta y/\Delta x$ since the logarithm is actually being plotted. The correct way to find the slope is

$$a = \frac{\Delta(\ln y)}{\Delta x} = \frac{\ln y_2 - \ln y_1}{x_2 - x_1}$$

Note that there cannot be a zero on a logarithmic scale.

Figures A.2 and A.3 are graphs of the function $y = 3e^{-2x}$ on linear and semilog axes, respectively.

Log-Log Graphs

A log-log graph uses logarithmic scales for both axes. Log-log graphs are useful when the data plotted is thought to be a power function

$$y = Ax^n$$

For such a function,

$$\log y = n \log x + \log A$$

so a graph of $\log y$ versus $\log x$ will be a straight line with slope n and y-intercept $\log A$. The slope (n) on a log-log graph is found as

$$n = \frac{\Delta(\log y)}{\Delta(\log x)} = \frac{\log y_2 - \log y_1}{\log x_2 - \log x_1}$$

Figures A.4 and A.5 are graphs of the function $y = 130x^{3/2}$ on linear and log-log axes, respectively.

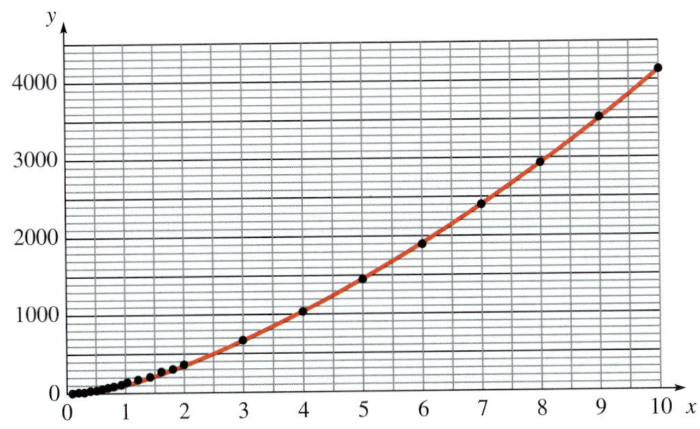

Figure A.4 Graph of the power function $y = 130x^{3/2}$ on linear graph paper.

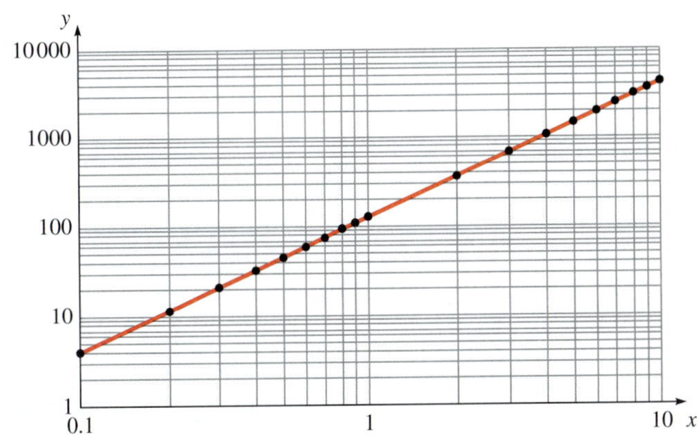

Figure A.5 Graph of the power function $y = 130x^{3/2}$ on log-log graph paper.

A.4 PROPORTIONS AND RATIOS

The notation

$$y \propto x$$

means that y is directly proportional to x. A proportionality can be written as an equation

$$y = kx$$

if the constant of proportionality k is written explicitly. Be careful: an equation can *look like* a proportionality without being one. For example, $V = IR$ means that $V \propto I$ *if and only if* R is constant. If R depends on I, then V is not proportional to I.

The notation

$$y \propto \frac{1}{x}$$

means that y is inversely proportional to x. The notation

$$y \propto x^n$$

means that y is proportional to the nth power of x.

Writing out proportions as ratios usually simplifies solutions when some common items in an equation are unknown but we do know the values of all but one of the proportional quantities (connect tutorial: ball toss). For example if $y \propto x^n$, we can write

$$\frac{y_1}{y_2} = \left(\frac{x_1}{x_2}\right)^n$$

Percentages

Percentages require careful attention. Look at these four examples:

"B is 30% of A" means $B = 0.30A$

"B is 30% larger than A" means $B = (1 + 0.30)A = 1.30A$

"B is 30% smaller than A" means $B = (1 - 0.30)A = 0.70A$

"A increases by 30%" means $\Delta A = +0.30A$, so $A_{\text{final}} = 1.30A_{\text{initial}}$

Example A.4

If $P \propto T^4$, and T increases by 10.0%, by what percentage does P increase?

Solution

$$\Delta T = +0.100T_{\text{i}}$$

$$T_{\text{f}} = T_{\text{i}} + \Delta T = 1.100T_{\text{i}}$$

$$\frac{P_{\text{f}}}{P_{\text{i}}} = \left(\frac{T_{\text{f}}}{T_{\text{i}}}\right)^4 = 1.100^4 \approx 1.464$$

Therefore, P increases by about 46.4%.

A.5 APPROXIMATIONS

Binomial Approximations

A binomial is the sum of two terms. The general rule for the nth power of an algebraic sum is given by the binomial expansion:

$$(a + b)^n = a^n + na^{n-1}b + \frac{n(n-1)}{1 \times 2}a^{n-2}b^2 + \frac{n(n-1)(n-2)}{1 \times 2 \times 3}a^{n-3}b^3 + \cdots$$

The binomial approximations are used when a binomial in which one term is much smaller than the other is raised to a power n. Only the first two terms of the binomial expansion are of significant value; the other terms are dropped. A common case for physics problems is that in which $a = 1$, or can be made equal to one by factoring. The basic approximation forms are then given by

$$(1 + x)^n \approx 1 + nx \quad \text{when } |x| \ll 1 \tag{A-23}$$

$$(1 - x)^n \approx 1 - nx \quad \text{when } |x| \ll 1 \tag{A-24}$$

The power n can be any real number, including negative as well as positive numbers. It does not have to be an integer. (connect tutorial: small percentage changes) An *estimate* of the error—the difference between the approximation and the exact expression—is given by the third term in the expansion:

$$\text{error} \approx \tfrac{1}{2}n(n-1)x^2 \tag{A-25}$$

Of course, the larger term in a binomial is not necessarily 1, but the larger term can be factored out and then Eq. (A-23) or Eq. (A-24) applied. For instance, if $A \gg b$, then

$$(A + b)^n = \left[A \times \left(1 + \frac{b}{A}\right)\right]^n = A^n\left(1 + \frac{b}{A}\right)^n$$

Another common expansion is

$$e^x = 1 + x + \frac{x^2}{2!} + \frac{x^3}{3!} + \cdots$$

where, for any integer n, $n! = n \times (n-1) \times (n-2) \times \ldots \times [n - (n-1)]$; for example, $3! = 3 \times 2 \times 1 = 6$. From this expansion, we have an approximation for e^x:

$$e^x \approx 1 + x \qquad \text{when } |x| \ll 1$$

Small-Angle Approximations

Approximations for small angles appear in Section A.7 on trigonometry.

A.6 GEOMETRY

Geometric Shapes

Table A.1 lists the geometric shapes that commonly appear in physics problems. It is often necessary to determine the area or volume of one of these simple shapes to complete the solution of a problem. The formulae for the properties associated with each geometric form are listed in the column to the right.

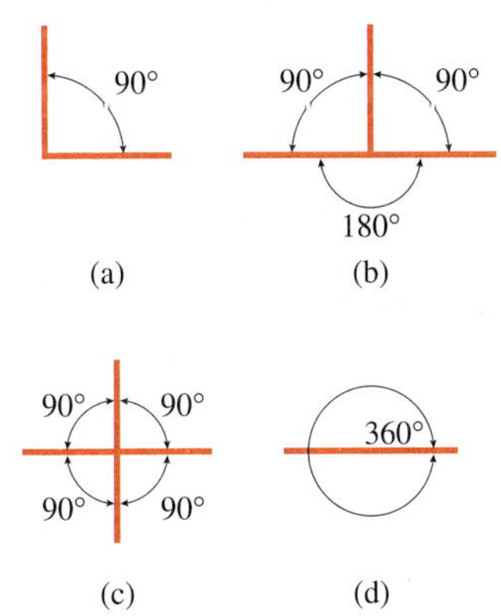

Figure A.6 (a) A right angle; (b) two adjacent right angles, or a straight line; (c) four adjacent right angles; (d) a full circle.

Angular Measure

An angle between two straight lines that meet at a single point can be specified in degrees. If the two lines are perpendicular to each other, as shown in Fig. A.6a, the angular separation is said to be a right angle or 90°. In Fig. A.6b two such 90° angles are placed side by side; they add to 180°, so a straight line represents an angular separation of 180°. When four right angles are grouped as shown in Fig. A.6c, the angles add to 360° and a full circle contains 360° as shown in Fig. A.6d.

When two lines meet, as shown in Fig. A.7, there are two possible angles that might be specified. When two angles placed adjacent to each other form a straight line, they are called supplementary angles; angles α and β in Fig. A.7 are supplementary angles. Supplementary angles always add to 180°. When two angles placed adjacent to each other form a right angle, they are called complementary angles. Complementary angles add to 90°.

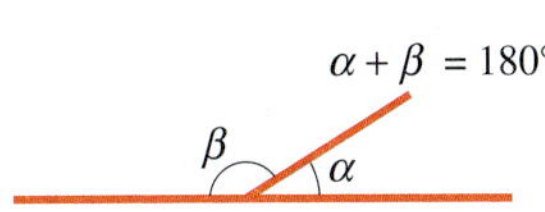

Figure A.7 Supplementary angles add to 180°.

Various triangles are shown in Fig. A.8. The sum of the interior angles of any triangle is 180°. An isosceles triangle has two sides of equal length; the angles opposite to the equal sides are equal angles. An equilateral triangle has all three sides of equal

Table A.1 Properties of Common Geometric Shapes

Geometric Shape	Properties	Geometric Shape	Properties
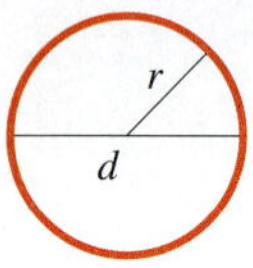 Circle	Diameter $d = 2r$ Circumference $C = \pi d = 2\pi r$ Area $A = \pi r^2$	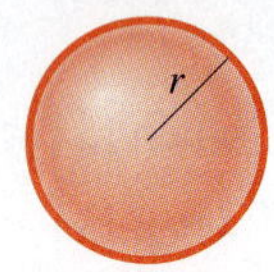 Sphere	Surface area $A = 4\pi r^2$ Volume $V = \frac{4}{3}\pi r^3$
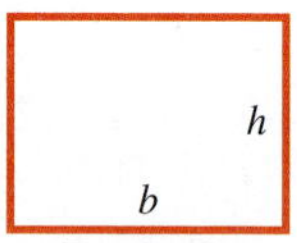 Rectangle	Perimeter $P = 2b + 2h$ Area $A = bh$	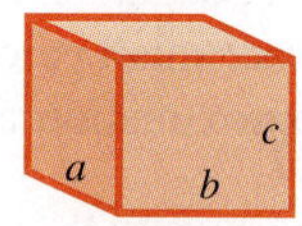 Parallelepiped	Surface area $A = 2(ab + bc + ac)$ Volume $V = abc$
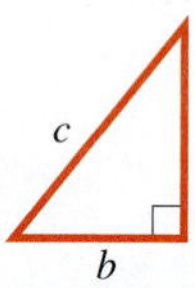 Right triangle	Perimeter $P = a + b + c$ Area $A = \frac{1}{2}\text{base} \times \text{height} = \frac{1}{2}ba$ Pythagorean theorem $c^2 = a^2 + b^2$ Hypotenuse $c = \sqrt{a^2 + b^2}$	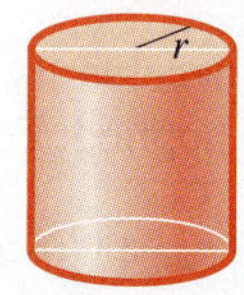 Right circular cylinder	Surface area $A = 2\pi r^2 + 2\pi rh$ $= 2\pi r(r + h)$ Volume $V = \pi r^2 h$
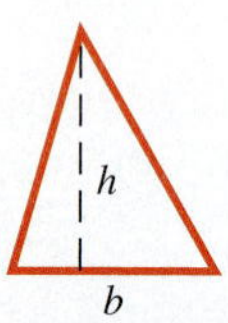 Triangle	Area $A = \frac{1}{2}bh$		

length; it is also equiangular. Right triangles have one right angle, 90°, and the sum of the other two angles is 90°, so those angles are acute angles.

Triangles are similar when all three angles of one are equal to the three angles of the other. If two angles of one triangle are equal to two angles of the other, the third angles are necessarily equal and the triangles are similar. The ratio of corresponding sides of similar triangles are equal, as shown in Fig. A.9. Similar triangles of the same size are called congruent triangles.

Figure A.10 shows other useful relations among angles between intersecting lines.

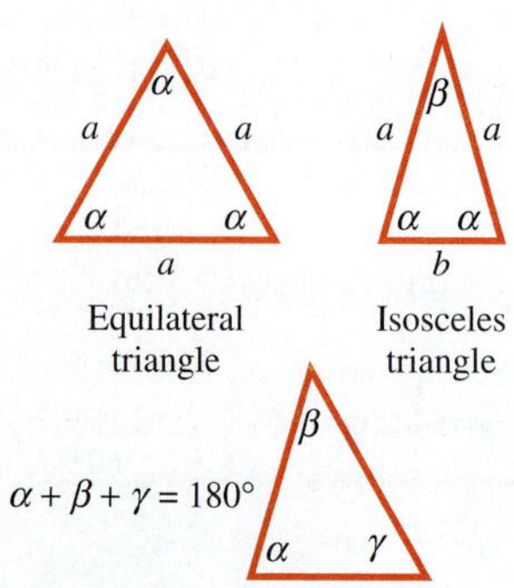

Figure A.8 Triangles.

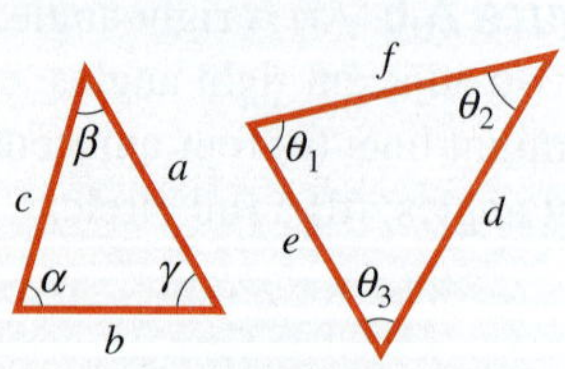

If $\alpha = \theta_1$ and $\beta = \theta_2$, then $\gamma = \theta_3$

and

$$\frac{a}{d} = \frac{b}{e} = \frac{c}{f}$$

Figure A.9 Similar triangles.

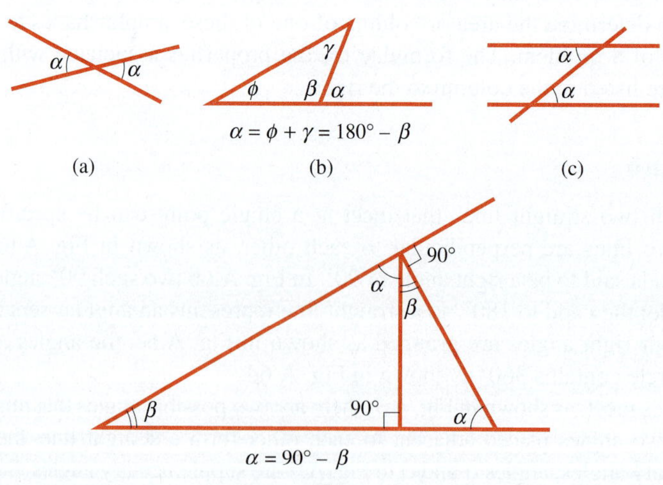

Figure A.10 Angles formed by intersecting lines. (a) Vertically opposite angles are equal. (b) The external angle α is supplementary to β and equal to the sum of the other two interior angles of the triangle. (c) Alternate interior angles are equal. (d) Because α and β are complementary, these three triangles are similar.

For many physics problems it is convenient to use angles measured in radians rather than in degrees; the abbreviation for radians is rad. The arc length s measured along a circle is proportional to the angle between the two radii that define the arc, as shown in Fig. A.11. One radian is defined as the angle subtended when the arc length is equal to the length of the radius.

For θ measured in radians,

$$s = r\theta$$

When the angle subtended is 360°, the arc length is equal to the circumference of the circle, $2\pi r$. Therefore,

$$360° = 2\pi \text{ rad}$$

$$1 \text{ rad} = \frac{360°}{2\pi} \approx 57.3° \qquad \text{and} \qquad 1° = \frac{2\pi}{360°} \approx 0.01745 \text{ rad}$$

Note that the radian has no physical dimensions; it is a ratio of two lengths so it is a pure number. We use the term *rad* to remind us of the angular units being used.

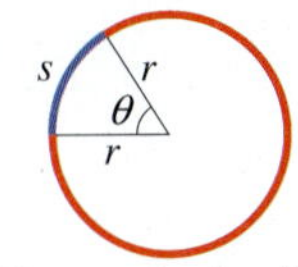

If $s = r$, $\theta = 1$ radian

Figure A.11 Radian measure.

A.7 TRIGONOMETRY

The basic trigonometric functions used in physics are shown in Fig. A.12. Note that in determining the function values, the units of length cancel, so the sine, cosine, and tangent functions are dimensionless.

The side opposite and the side adjacent to either of the acute angles in the right triangle are of lesser length than the hypotenuse, according to the Pythagorean theorem. Therefore, the absolute values of the sine and cosine cannot exceed 1. The absolute value of the tangent can exceed 1.

Figure A.13 shows the signs (positive or negative) associated with the trigonometric functions for an angle θ located in each of the four quadrants. The hypotenuse r is positive, so the sign for the sine or cosine is determined by the signs of x or y as measured along the positive or negative x- and y-axis. The sign of the tangent then depends on the signs of the sine and cosine. The angle θ is measured in a counterclockwise direction starting from the positive x-axis, which represents 0°. Angles measured from the x-axis going in a clockwise direction (below the x-axis) are negative angles; an angle of −60°,

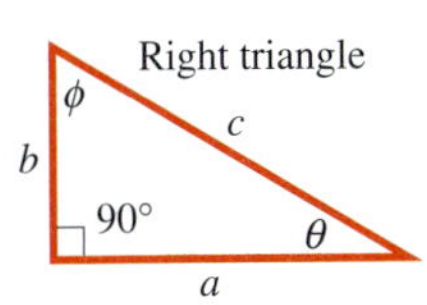

$\phi = 90° - \theta$

$$\sin\theta = \frac{\text{side opposite } \theta}{\text{hypotenuse}} = \frac{b}{c} = \cos\phi$$

$$\cos\theta = \frac{\text{side adjacent } \theta}{\text{hypotenuse}} = \frac{a}{c} = \sin\phi$$

$$\tan\theta = \frac{\text{side opposite } \theta}{\text{side adjacent } \theta} = \frac{b}{a} = \frac{\sin\theta}{\cos\theta} = \frac{1}{\tan\phi}$$

Figure A.12 Trigonometric functions used in physics problems; angles θ and ϕ are complementary angles.

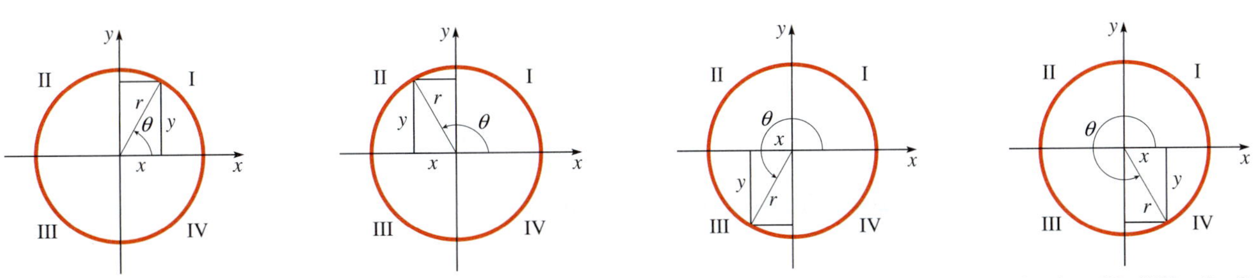

Quadrant I: $0 < \theta < 90°$
$\sin\theta = y/r$ is positive
$\cos\theta = x/r$ is positive
$\tan\theta = y/x$ is positive

Quadrant II: $90° < \theta < 180°$
$\sin\theta = y/r$ is positive
$\cos\theta = x/r$ is negative
$\tan\theta = y/x$ is negative

Quadrant III: $180° < \theta < 270°$
$\sin\theta = y/r$ is negative
$\cos\theta = x/r$ is negative
$\tan\theta = y/x$ is positive

Quadrant IV: $270° < \theta < 360°$
$\sin\theta = y/r$ is negative
$\cos\theta = x/r$ is positive
$\tan\theta = y/x$ is negative

Figure A.13 Signs of trigonometric functions in various quadrants.

Table A.2 Useful Trigonometric Identities

$\sin^2\theta + \cos^2\theta = 1$	$\tan 2\theta = \dfrac{2\tan\theta}{1-\tan^2\theta}$
$\sin(-\theta) = -\sin\theta$	
$\cos(-\theta) = \cos\theta$	$\sin(\alpha \pm \beta) = \sin\alpha\,\cos\beta \pm \cos\alpha\,\sin\beta$
$\tan(-\theta) = -\tan\theta$	$\cos(\alpha \pm \beta) = \cos\alpha\,\cos\beta \mp \sin\alpha\,\sin\beta$
$\sin(180° \pm \theta) = \mp\sin\theta$	$\tan(\alpha \pm \beta) = \dfrac{\tan\alpha \pm \tan\beta}{1 \mp \tan\alpha\tan\beta}$
$\cos(180° \pm \theta) = -\cos\theta$	
$\tan(180° \pm \theta) = \pm\tan\theta$	$\sin\alpha + \sin\beta = 2\sin\left[\frac{1}{2}(\alpha+\beta)\right]\cos\left[\frac{1}{2}(\alpha-\beta)\right]$
$\sin(90° \pm \beta) = \cos\beta$	$\sin\alpha - \sin\beta = 2\cos\left[\frac{1}{2}(\alpha+\beta)\right]\sin\left[\frac{1}{2}(\alpha-\beta)\right]$
$\cos(90° \pm \beta) = \mp\sin\beta$	$\cos\alpha + \cos\beta = 2\cos\left[\frac{1}{2}(\alpha+\beta)\right]\cos\left[\frac{1}{2}(\alpha-\beta)\right]$
$\sin 2\theta = 2\sin\theta\,\cos\theta$	$\cos\alpha - \cos\beta = -2\sin\left[\frac{1}{2}(\alpha+\beta)\right]\sin\left[\frac{1}{2}(\alpha-\beta)\right]$
$\cos 2\theta = \cos^2\theta - \sin^2\theta = 2\cos^2\theta - 1 = 1 - 2\sin^2\theta$	

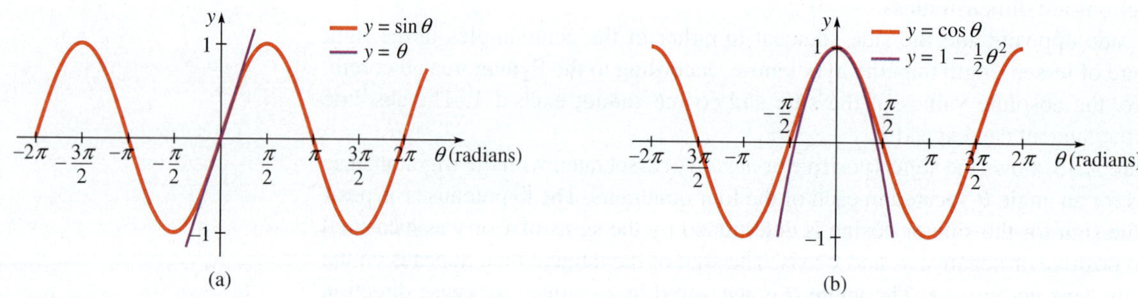

Figure A.14 (a) Graphs of $y = \sin\theta$ and $y = \theta$. Note that $\sin\theta \approx \theta$ for small θ. (b) Graphs of $y = \cos\theta$ and $y = 1 - \frac{1}{2}\theta^2$. Note that $\cos\theta \approx 1 - \frac{1}{2}\theta^2$ for small θ.

which is located in the fourth quadrant, is the same as an angle of +300°. Figure A.14 shows graphs of $y = \sin\theta$ and $y = \cos\theta$ as functions of θ in radians. Also graphed are two functions that are useful approximations for the sine and cosine functions when $|\theta|$ is sufficiently small.

Table A.2 lists some of the most useful trigonometric identities.

Inverse Trigonometric Functions The inverse trigonometric functions can be written in either of two ways. To use the inverse cosine as an example: $\cos^{-1} x$ or arccos x. Both of these expressions mean *an angle whose cosine is equal to x*. A calculator returns only the *principal value* of an inverse trigonometric function (Table A.3), which may or may not be the correct solution in a given problem.

Law of Sines and Law of Cosines These two laws apply to any triangle labeled as shown in Fig. A.15:

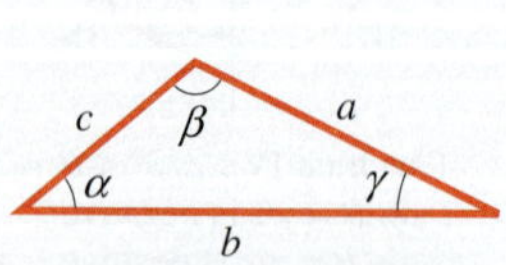

Figure A.15 A general triangle.

$$\text{Law of sines:}\quad \frac{\sin\alpha}{a} = \frac{\sin\beta}{b} = \frac{\sin\gamma}{c}$$

Law of cosines: $c^2 = a^2 + b^2 - 2ab\cos\gamma$ (where γ is the interior angle formed by the intersection of sides a and b)

Table A.3 Inverse Trigonometric Functions

Function	Principal Value Range	(Quadrants)	To Find Value in a Different Quadrant
$\sin^{-1}$	$-\frac{\pi}{2}$ to $\frac{\pi}{2}$	(I and IV)	Subtract principal value from π
$\cos^{-1}$	0 to π	(I and II)	Subtract principal value from 2π
$\tan^{-1}$	$-\frac{\pi}{2}$ to $\frac{\pi}{2}$	(I and IV)	Add principal value to π

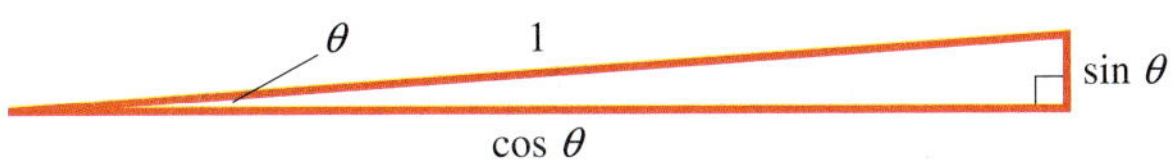

Figure A.16 Illustration of the small angle approximations $\sin\theta \approx \theta$ and $\cos\theta \approx 1 - \frac{1}{2}\theta^2$ (for θ in radians) using a right triangle with $\theta \ll 1$ rad.

Small-Angle Approximations

These approximations are written for θ *in radians* and are valid when $\theta \ll 1$ rad.

$$\sin\theta \approx \theta \tag{A-26}$$

$$\cos\theta \approx 1 - \tfrac{1}{2}\theta^2 \tag{A-27}$$

$$\tan\theta \approx \theta \tag{A-28}$$

The sizes of the errors involved in using these approximations are roughly $\frac{1}{6}\theta^3$, $\frac{1}{24}\theta^4$, and $\frac{2}{3}\theta^3$, respectively. In *some* circumstances it may be all right to ignore the $\frac{1}{2}\theta^2$ term and write

$$\cos\theta \approx 1 \tag{A-29}$$

The origin of these approximations can be illustrated using a right triangle of hypotenuse 1 with one very small angle θ (Fig. A.16). If θ is very small, then the adjacent side ($\cos\theta$) will be nearly the same length as the hypotenuse (1). Then we can think of those two sides as radii of a circle that subtend an angle θ. The relationship between the arc length s and the angle subtended is

$$s = \theta r$$

Since $\sin\theta \approx s$ and $r = 1$, we have $\sin\theta \approx \theta$. To find an approximate form for $\cos\theta$ (but one more accurate than $\cos\theta \approx 1$), we can use the Pythagorean theorem:

$$\sin^2\theta + \cos^2\theta = 1$$

$$\cos\theta = \sqrt{1 - \sin^2\theta} \approx \sqrt{1 - \theta^2}$$

Now, using a binomial approximation,

$$\cos\theta \approx (1 - \theta^2)^{1/2} \approx 1 - \tfrac{1}{2}\theta^2$$

A.8 VECTORS

The distinction between vectors and scalars is discussed in Section 3.1. Scalars have magnitude, whereas vectors have magnitude and direction. A vector is represented graphically by an arrow of length proportional to the magnitude of the vector and aligned in a direction that corresponds to the vector direction.

In print, the symbol for a vector quantity is sometimes written in bold font, or in roman font with an arrow over it, or in bold font with an arrow over it (as done in this

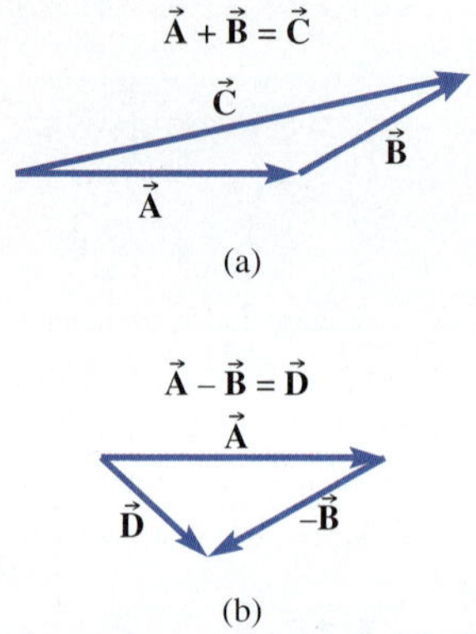

Figure A.17 Graphical (a) addition and (b) subtraction of two vectors.

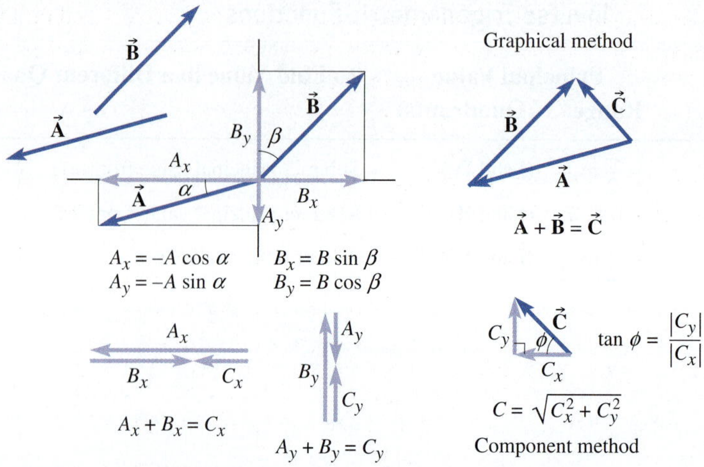

Figure A.18 Adding two arbitrary vectors by two different methods.

Figure A.19 Multiplication of a vector by a scalar.

book). When writing by hand, a vector is designated by drawing an arrow over the symbol: $\vec{A}$. When we write just plain A, that stands for the *magnitude* of the vector. We also use absolute value bars to stand for the magnitude of a vector, so $A = |\vec{\mathbf{A}}|$.

Addition and Subtraction of Vectors

When vectors are added or subtracted, the magnitudes and directions must be taken into account. Details on vector addition and subtraction are found in Sections 3.1 and 3.2. Here we provide a brief summary.

The graphical method for adding vectors involves placing the vectors tip to tail and then drawing from the tail of the first to the tip of the second, as shown in Fig. A.17. To subtract a vector, add its opposite. In Fig. A.17, $-\vec{\mathbf{B}}$ has the same magnitude as $\vec{\mathbf{B}}$ but is opposite in direction. Then $\vec{\mathbf{A}} + \vec{\mathbf{B}} = \vec{\mathbf{A}} - (-\vec{\mathbf{B}})$.

Figure A.18 shows both the graphical and component methods of vector addition.

Product of a Vector and a Scalar

When a vector is multiplied by a scalar, the magnitude of the vector is multiplied by the absolute value of the scalar, as shown in Fig. A.19. The direction of the vector does not change unless the scalar factor is negative, in which case the direction is reversed.

Scalar Product of Two Vectors

One type of product of two vectors is the *scalar product* (also called the *dot product*). The notation for it is

$$C = \vec{\mathbf{A}} \cdot \vec{\mathbf{B}}$$

As its name implies, the scalar product of two vectors is a scalar quantity; it can be positive, negative, or zero but has no direction.

The scalar product depends on the magnitudes of the two vectors and on the angle θ between them. To find the angle, draw the two vectors starting *at the same point* (Fig. A.20). Then the scalar product is defined by

$$\vec{\mathbf{A}} \cdot \vec{\mathbf{B}} = AB \cos \theta$$

Reversing the order of the two vectors does not change the scalar product: $\vec{\mathbf{B}} \cdot \vec{\mathbf{A}} = \vec{\mathbf{A}} \cdot \vec{\mathbf{B}}$. The scalar product can be written in terms of the components of the two vectors

$$\vec{\mathbf{A}} \cdot \vec{\mathbf{B}} = A_x B_x + A_y B_y + A_z B_z$$

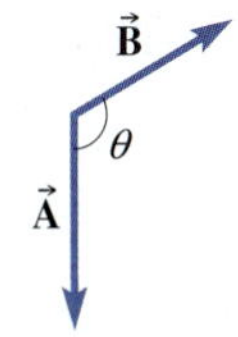

Figure A.20 Two vectors are drawn starting at the same point. The angle θ between the vectors is used to find the scalar product and the cross product of the vectors.

Cross Product of Two Vectors

Another type of product of two vectors is the *cross product* (also called the *vector product*), which is introduced in Chapter 19. It is denoted by

$$\vec{\mathbf{A}} \times \vec{\mathbf{B}} = \vec{\mathbf{C}}$$

The cross product is a *vector* quantity; it has magnitude and direction. $\vec{\mathbf{A}} \times \vec{\mathbf{B}}$ is read as "$\vec{\mathbf{A}}$ cross $\vec{\mathbf{B}}$."

For two vectors, $\vec{\mathbf{A}}$ and $\vec{\mathbf{B}}$, separated by an angle θ (with θ chosen to be the *smaller* angle between the two as in Fig. A.20), the magnitude of the cross product $\vec{\mathbf{C}}$ is

$$|\vec{\mathbf{C}}| = |\vec{\mathbf{A}} \times \vec{\mathbf{B}}| = AB \sin \theta$$

The direction of the cross product $\vec{\mathbf{C}}$ is one of the two directions perpendicular to both $\vec{\mathbf{A}}$ and $\vec{\mathbf{B}}$. To choose the correct direction, use the right-hand rule explained in Section 19.2.

The cross product depends on the order of the multiplication.

$$\vec{\mathbf{A}} \times \vec{\mathbf{B}} = -\vec{\mathbf{B}} \times \vec{\mathbf{A}}$$

The magnitude is $AB \sin \theta$ in both cases, but the direction of one cross product is opposite to the direction of the other.

A.9 SELECTED MATHEMATICAL SYMBOLS

$\times$ or $\cdot$	multiplication
Δ	change in, small increment, or uncertainty in
$\approx$	is approximately equal to
$\neq$	is not equal to
$\leq$	is less than or equal to
$\geq$	is greater than or equal to
$\ll$	is much less than
$\gg$	is much greater than
$\propto$	is proportional to
$\lvert Q \rvert$	absolute value of Q
$\lvert \vec{\mathbf{a}} \rvert$	magnitude of vector $\vec{\mathbf{a}}$
$\perp$	perpendicular
$\parallel$	parallel
∞	infinity
$'$	prime (used to distinguish different values of the same variable)
Q_{av}, $\overline{Q}$, or $\langle Q \rangle$	average of Q
Σ	sum
Π	product
$\log_b x$	the logarithm (base b) of x

$\ln x$	the natural (base e) logarithm of x
$\pm$	plus or minus
$\mp$	minus or plus
$\ldots$	ellipsis (indicates continuation of a series or list)
$\angle$	angle
$\Rightarrow$	implies
$\therefore$	therefore
$\lim_{\Delta t \to 0} Q$	the limiting value of the quantity Q as the time interval Δt approaches zero
$\bullet$ or $\odot$	represents a vector arrow pointing out of the page
$\times$ or $\otimes$	represents a vector arrow pointing into the page

Table of Selected Nuclides

Atomic Number Z	Element	Symbol	Mass Number A	Mass of *neutral atom* (u)	Percentage Abundance (or Principal Decay Mode)	Half-life (if unstable)
0	(Neutron)	n	1	1.008 664 9	β^-	10.17 min
1	Hydrogen	H	1	1.007 825 0	99.9885	
	(Deuterium)	(D)	2	2.014 101 8	0.0115	
	(Tritium)	(T)	3	3.016 049 3	β^-	12.32 yr
2	Helium	He	3	3.016 029 3	0.000 137	
			4	4.002 603 3	99.999 863	
3	Lithium	Li	6	6.015 122 9	7.6	
			7	7.016 003 4	92.4	
4	Beryllium	Be	7	7.016 929 7	EC	53.24 d
			8	8.005 305 1	2α	6.7×10^{-17}s
			9	9.012 183 1	100	
5	Boron	B	10	10.012 936 9	19.9	
			11	11.009 305 4	80.1	
6	Carbon	C	11	11.011 433 6	β^+	20.334 min
			12	12.000 000 0	98.93	
			13	13.003 354 8	1.07	
			14	14.003 242 0	β^-	5730 yr
			15	15.010 599 3	β^-	2.449 s
7	Nitrogen	N	12	12.018 613 2	β^+	11.00 ms
			13	13.005 738 6	β^+	9.965 min
			14	14.003 074 0	99.634	
			15	15.000 108 9	0.366	
8	Oxygen	O	15	15.003 065 4	EC	122.24 s
			16	15.994 914 6	99.757	
			17	16.999 131 8	0.038	
			18	17.999 159 6	0.205	
			19	19.003 578 0	β^-	26.88 s
9	Fluorine	F	19	18.998 403 2	100	
10	Neon	Ne	20	19.992 440 2	90.48	
			22	21.991 385 1	9.25	
11	Sodium	Na	22	21.994 437 4	β^+	2.6027 yr
			23	22.989 769 2	100	
			24	23.990 963 0	β^-	14.9590 h
12	Magnesium	Mg	24	23.985 041 7	78.99	
13	Aluminum	Al	27	26.981 538 5	100	
14	Silicon	Si	28	27.976 926 5	92.230	
15	Phosphorus	P	31	30.973 762 0	100	
			32	31.973 907 6	β^-	14.997 d
16	Sulfur	S	32	31.972 071 2	95.02	
17	Chlorine	Cl	35	34.968 852 7	75.77	
18	Argon	Ar	40	39.962 383 1	99.6003	
19	Potassium	K	39	38.963 706 5	93.2581	
			40	39.963 998 2	0.0117; β^-	1.248×10^9 yr
20	Calcium	Ca	40	39.962 590 9	96.94	
24	Chromium	Cr	52	51.940 506 2	83.789	
25	Manganese	Mn	54	53.940 357 6	EC	312.12 d
			55	54.938 043 9	100	
26	Iron	Fe	56	55.934 936 3	91.754	

(continued)

Atomic Number Z	Element	Symbol	Mass Number *A*	Mass of *neutral atom* (u)	Percentage Abundance (or Principal Decay Mode)	Half-life (if unstable)
27	Cobalt	Co	59	58.933 194 3	100	
			60	59.933 816 3	β^-	5.271 yr
28	Nickel	Ni	58	57.935 342 4	68.077	
			60	59.930 785 9	26.223	
29	Copper	Cu	63	62.929 597 7	69.17	
30	Zinc	Zn	64	63.929 142 0	48.63	
36	Krypton	Kr	84	83.911 497 7	57.0	
			86	85.910 610 6	17.3	
			92	91.926 173 1	β^-	1.840 s
37	Rubidium	Rb	85	84.911 789 7	72.17	
			93	92.922 039	β^-	5.84 s
38	Strontium	Sr	88	87.905 612 5	82.58	
			90	89.907 730 0	β^-	28.90 yr
39	Yttrium	Y	89	88.905 840 3	100	
			90	89.907 143 9	β^-	64.053 h
47	Silver	Ag	107	106.905 091 6	51.839	
50	Tin	Sn	120	119.902 201 6	32.58	
53	Iodine	I	131	130.906 126 3	β^-	8.0252 d
55	Cesium	Cs	133	132.905 452 0	100	
			141	140.920 045	β^-	24.84 s
56	Barium	Ba	138	137.905 247 0	71.698	
			141	140.914 403	β^-	18.27 min
60	Neodymium	Nd	143	142.909 820 0	12.2	
62	Samarium	Sm	147	146.914 829 2	14.99; α	1.06×10^{11} yr
79	Gold	Au	197	196.966 568 8	100	
82	Lead	Pb	204	203.973 044 0	1.4; α	$\geq 1.4 \times 10^{17}$ yr
			206	205.974 465 7	24.1	
			207	206.975 897 3	22.1	
			208	207.976 652 5	52.4	
			210	209.984 188 9	β^-	22.20 yr
			211	210.988 737 1	β^-	36.1 min
			212	211.991 897 7	β^-	10.64 h
			214	213.999 805 9	β^-	26.8 min
83	Bismuth	Bi	209	208.980 399 1	100	
			211	210.987 270	α	2.14 min
			214	213.998 712	β^-	19.9 min
84	Polonium	Po	210	209.982 874 1	α	138.376 d
			214	213.995 201 7	α	164.3 μs
			218	218.008 973 5	α	3.098 min
86	Radon	Rn	222	222.017 578 2	α	3.8235 d
88	Radium	Ra	226	226.025 410 3	α	1600 yr
			228	228.031 070 7	β^-	5.75 yr
90	Thorium	Th	228	228.028 741 3	α	1.91 yr
			232	232.038 055 8	100; α	1.40×10^{10} yr
			234	234.043 601	β^-	24.10 d
92	Uranium	U	235	235.043 930 1	0.7204; α	7.04×10^{8} yr
			236	236.045 568 2	α	2.342×10^{7} yr
			238	238.050 788 4	99.2742; α	4.468×10^{9} yr
			239	239.054 293 5	β^-	23.45 min
93	Neptunium	Np	237	237.048 173 6	α	2.144×10^{6} yr
94	Plutonium	Pu	239	239.052 163 6	α	2.411×10^{4} yr
			242	242.058 742 8	α	3.75×10^{5} yr
			244	244.064 205	α	8.00×10^{7} yr

EC = electron capture; β^+ = both positron emission and electron capture are possible.

Answers to Selected Questions and Problems

CHAPTER 1

Multiple-Choice Questions

1. (b) **3.** (a) **5.** (d) **7.** (b) **9.** (b)

Problems

1. 2.5 m **3.** 7.7% **5.** 7.2 **7.** 10^{-8} **9.** 11.8 yr **11.** 36.0% **13.** 4.0 **15.** 0.704 **17.** (a) 1.29×10^8 kg (b) 1.3×10^8 m/s **19.** (a) 3.63×10^7 g (b) 1.273×10^2 m **21.** 1.7×10^{-10} m^3 **23.** 459 m/s **25.** (a) 3; 5.74×10^{-3} kg (b) 1; 2 m (c) 3; 4.50×10^{-3} m (d) 3; 4.50×10^1 kg (e) 4; 1.009×10^5 s (f) 4; 9.500×10^3 mL **27.** (b), (a), (e), (d), (c) **29.** 2.8×10^{-7} in. **31.** (a) 4.8361×10^2 m (b) 1.834×10^3 m **33.** (a) 180 mi/h (b) 8.0 cm/ms **35.** 0.12 or 12% **37.** 4.9 L/min **39.** 1.7×10^{-10} km^3 **41.** (a) 2.7×10^{-3} ft/s (b) 1.9×10^{-3} mi/h **43.** kg·m^2·s^{-2}

45. $[T]^2 = \dfrac{[L]^3}{\dfrac{[L]^3}{[M][T]^2} \times [M]} = \dfrac{[L]^3}{[M]} \times \dfrac{[M][T]^2}{[L]^3} = [T]^2$ **47.** (a) $[L^3]$ (b) volume **49.** 30 to 40 cm **51.** 10 kg **53.** 100 m

55.

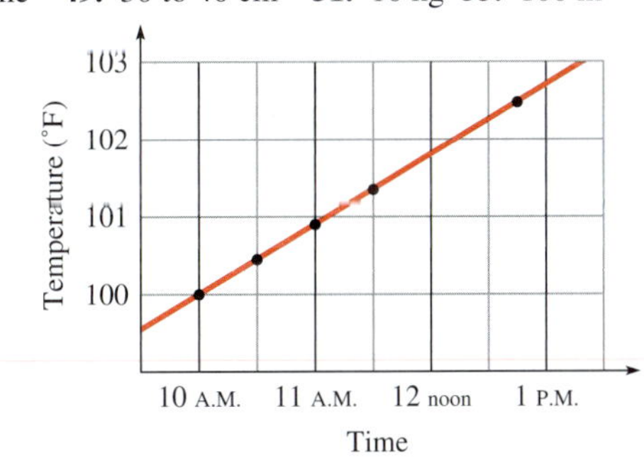

(a) 101.8°F (b) 0.90°F/h (c) No; the patient would die before the temperature reached 113°F.

57. (a) a (b) $+v_0$ **59.** (a) 1.6 km/h; 3.0 km (b) speed; starting position **61.** $x = (25\ \text{m/s}^4)t^4 + 3$ m

63. (a)

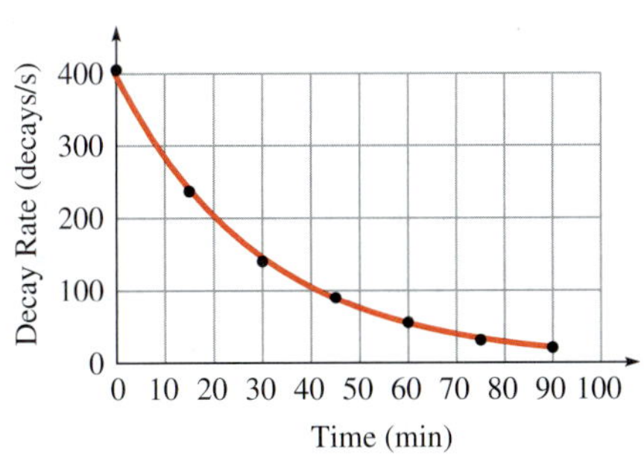

(b)

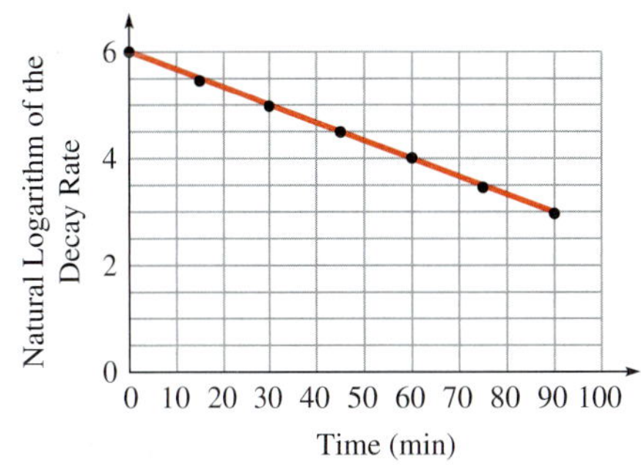

The presentation is useful because the graph is linear, showing that the decay rate is an exponetial function of time.

65. $v = \omega r$

67. (a)

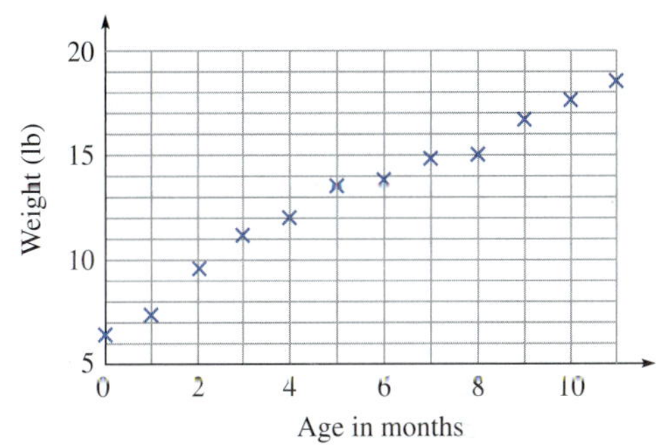

(b) 1.4 lb/mon (c) 0.78 lb/mon (d) 210 lb **69.** 10^5 hairs **71.** 10^4 viruses **73.** (a) 33.5 m (b) 4.2 bus lengths **75.** 5 L **77.** 434 m/s **79.** The conversion factor is (2.24 mi/h)/(1 m/s). For a quick, approximate conversion, multiply by 2.

81. $\dfrac{\text{kg·m}}{\text{s}^2}$ **83.** \$59 000 000 000 **85.** (a) 2.4×10^5 km/h (b) 10 min **87.** 2.6 N **89.** 10^{11} gal/yr **91.** (a) $\sqrt{\dfrac{hG}{c^5}}$ (b) 1.3×10^{-43} s **93.** 0.46 s^{-1} **95.** By a factor of 4.9

97. (a)

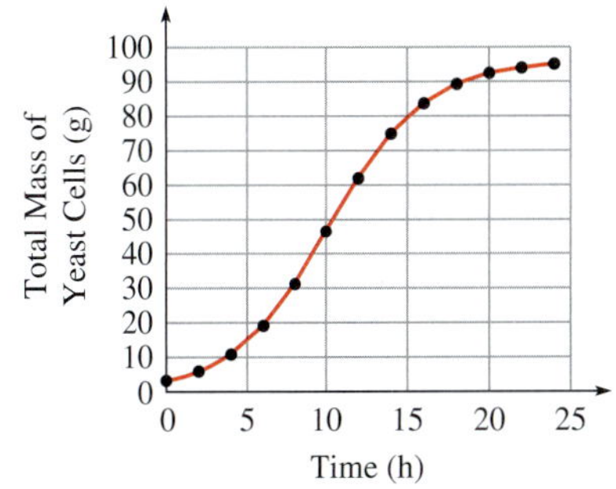

(b) about 100 g

(c)

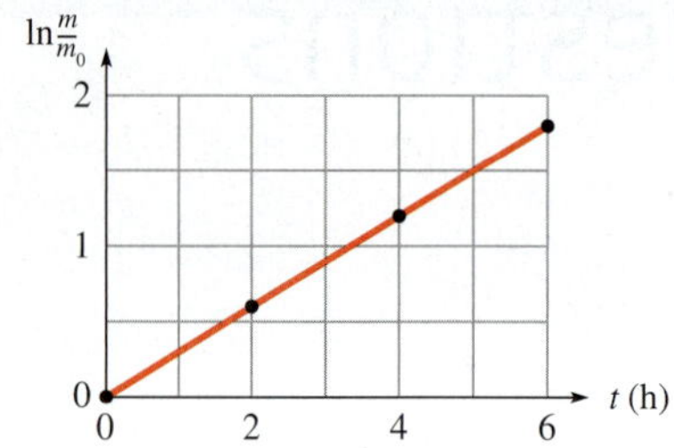

$r \approx 0.30\ \text{h}^{-1}$

CHAPTER 2

Multiple-Choice Questions

1. (c) **3.** (a) **5.** (c) **7.** (b) **9.** (a) **11.** (a) **13.** (d) **15.** (d) **17.** (b) **19.** (d) **21.** + **23.** $-x$ **25.** $-$ **27.** $-$ **29.** $-x$

Problems

1. 16 cm, east **3.** (a) -80 m, or 80 m west (b) -20 m, or 20 m west (c) $+80$ m, or 80 m east (d) 240 m **5.** 30 km/h east **7.** 98 m/s or 220 mi/h due north **9.** 32 s **11.** (a) 8 m (b) $t = 10$ s to $t = 14$ s

(d)

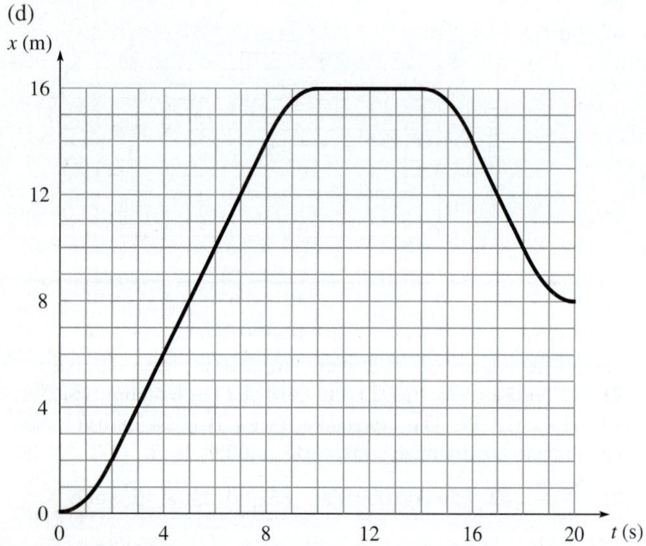

13. 2–3 s; 0–1 s, 1–2 s, and 3–4 s are equal; 5–6 s; 4–5 s **15.** 20 m **17.** 27 m/s west **19.** 16.5 m **21.** 1.0 m/s **23.** (a) 170 cm to the left (b) 28 cm/s (c) 9.4 cm/s to the left **25.** 1.05 m/s to the north **27.** 13 s **29.** $a(5.5\ \text{s})$, $a(0.5\ \text{s})$, $a(1.5\ \text{s}) = a(2.5\ \text{s})$, $a(3.5\ \text{s}) = a(4.5\ \text{s})$ **31.** (a) 4.4 m/s² forward **33.**

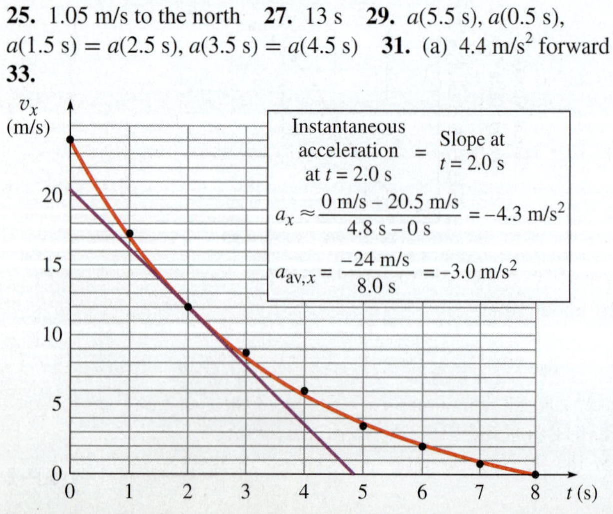

(a) $a_{\text{av},x} = -3.0\ \text{m/s}^2$ where the negative sign means the average acceleration is directed to the northwest
(b) $a_x \approx -4.3\ \text{m/s}^2$ (northwest)
35. (a) to the right with increasing speed (b) to the right with decreasing speed (c) to the left with constant speed (d) to the right with decreasing speed, turns around at point 4, and then to the left with increasing speed **37.** (a) 2 m/s² (b) 9.0 m/s (c) 9.8 m/s (d) 2 m/s (e) 69 m **39.** (a) -10 m/s² (b) 0

(c)

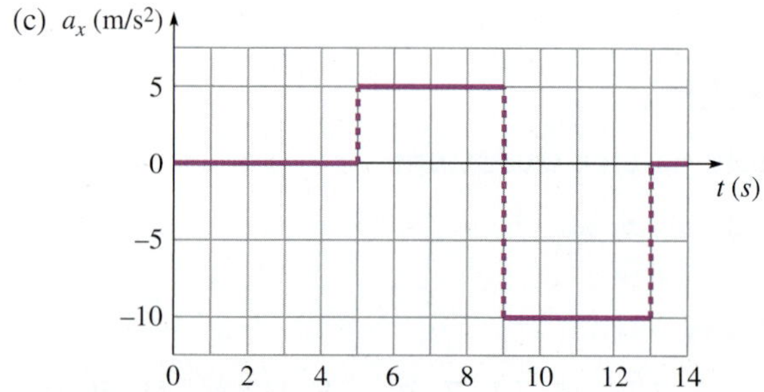

(d) 5.0 m
41. (a), (d), (b) = (c) **43.** (a) 5.7 s (b) 2.6 km
45. 1.5 m/s² northeast

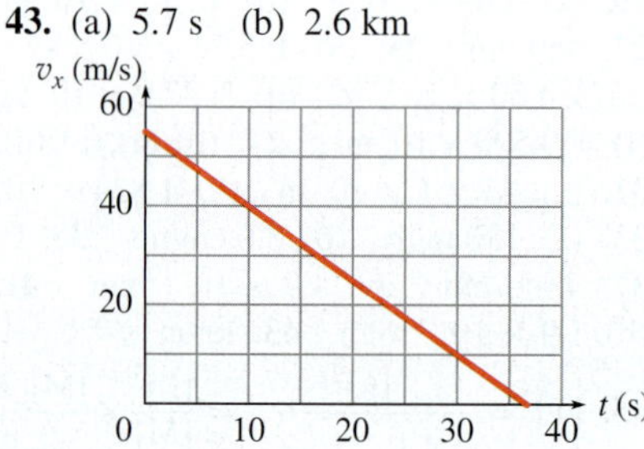

47. (a) 9.20 s (b) 212 m
49. You will not hit the tractor. 52.1 m, 11.5 m **51.** No; it takes 236 m for the train to stop. **53.** 80 m
55. (a)

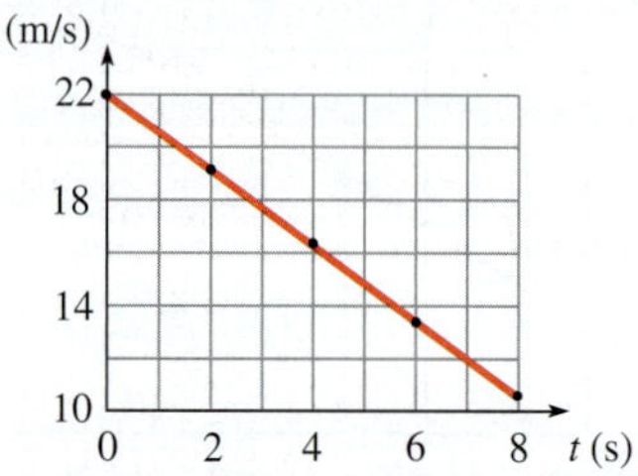

(b) 11 m/s (c) 130 m
(d)

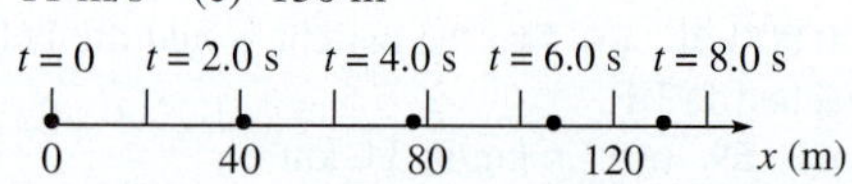

57. 85.0 m/s down **59.** 5.0 m/s **61.** 1.22 s **63.** (a) 44 m (b) 7.0 m/s (c) 29 m/s (d) 17.1 m below the top of the tower **65.** (a) 120 m/s² toward Lois (b) 170 m/s **67.** 46 m **69.** $2v$ **71.** No, it either fell from 133 m high or it was thrown downward. **73.** (a) 1.7 m/s (b) The swimmer pushes off from each end of the pool and he goes faster during the push-off than when swimming. **75.** 40 m/s² in the direction of motion **77.** (a) 330 m/s up (b) 16 m/s² up **79.** (a) 17.6 m/s downward (b) 97.0 m **81.** (a) 181 s (b) 2.76 m/s **83.** (a) 0.30 s (b) 0.05 s (c) 0.45 m (d) 10 m/s² down (e) 120 m/s² up **85.** (a) t_3 and t_4 (b) t_0, t_2, t_5, and t_7 (c) t_1 and t_6 (d) t_0, t_3, and t_7 (e) t_6
87. (a) 1.0 mm/s (b) 20.2 ms (c) 99 m/s

CHAPTER 3

Multiple-Choice Questions

1. (d) **3.** (b) **5.** (a) **7.** (e) **9.** (a) **11.** (d) **13.** (b) **15.** (c) **17.** (d)

Problems

1. (a) 10.00 km west (b) 4.88 km east (c) 4.88 km west
3. (a) same direction (b) perpendicular (c) opposite directions; 1.0
5.

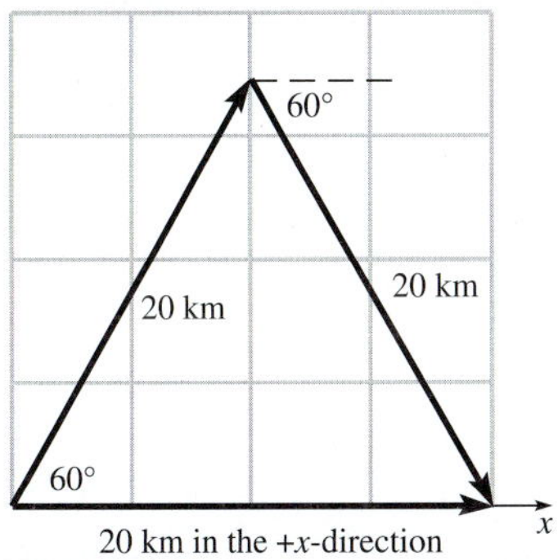

7. $B = C, A$ **9.** 14 N to the east **11.** (a) about 1.4 cm (b) about 7.9 cm **13.** 2.0 km at 20° east of south **15.** B_x, C_x, A_x **17.** $B_x + C_x$, $A_x + B_x$, $A_x + C_x$ **19.** 8.7 units **21.** (a) 5.0 m/s^2 (b) 37° CCW from the $+y$-axis **23.** 7.9 cm **25.** They both double, without changing sign. **27.** (a) 31.0 m/s (b) 58.1° with the $+x$-axis and 31.9° with the $-y$-axis **29.** $A_x = -14.3$ cm and $A_y = 17.0$ cm **31.** 0.283 mi at 45.0° north of west **33.** 2.0 km at 20° east of south **35.** 4.4 mi at 58° north of east **37.** (a) 6.7 m/s (b) 0 **39.** (a) 70 mi/h (b) 59 mi/h at 14° south of west **41.** (a) 110 km/h (b) 97 km/h at 35° north of east **43.** (a) 21 m/s (b) 16 m/s **45.** 0.70 m/s^2 south **47.** (a) 3.33 m/s at 45° north of east (b) 1.84 m/s^2 at 45° south of east (c) Changing the direction of the velocity requires an acceleration.
49. (a)

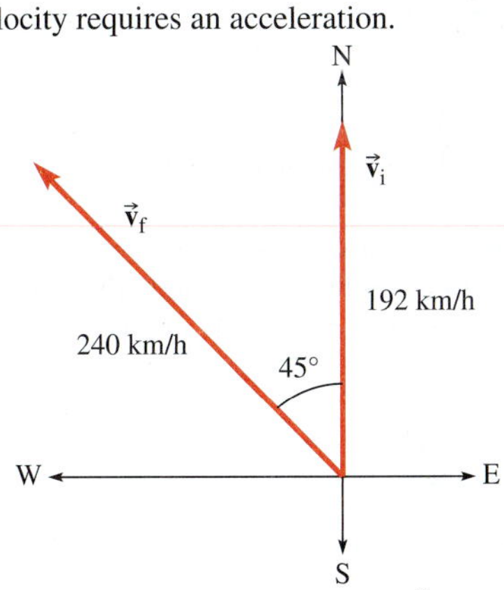

(b) 170 km/h at 7° south of west (c) 57 km/h^2 at 7° south of west **51.** 44.7 m/s at 26.6° south of east **53.** 15°, 30°, 45°, 60°, 75° **55.** It is on the ground after 1.32 s, so the horizontal distance along the ground is 26.3 m. **57.** (a) 5.9 m (b) 17.0 m/s **59.** 12.5 m/s **61.** (a) 37 m (b) 170 m (c) 32 m/s; −27 m/s
63. 21 m/s **65.** (a) $\dfrac{2v_i \sin\theta}{g}$
67. (a) $v_x = v_i \cos\theta$ and $v_y = 0$ (b) $(v_i \sin\theta)/g$ **69.** (a) 3.49 s (b) 2.01 s (c) 80.6 m **73.** 254 s **75.** 0.42 km/h
77. (a) 39.0 m/s (b) 7.4° south of west **79.** 27° upstream
81. (a) 1.80 mi/h (b) 48.0 min (c) 0.800 mi upstream (d) 32.2° upstream **83.** (a) 1.00 m/s (b) 1.12 m/s
85. (a)

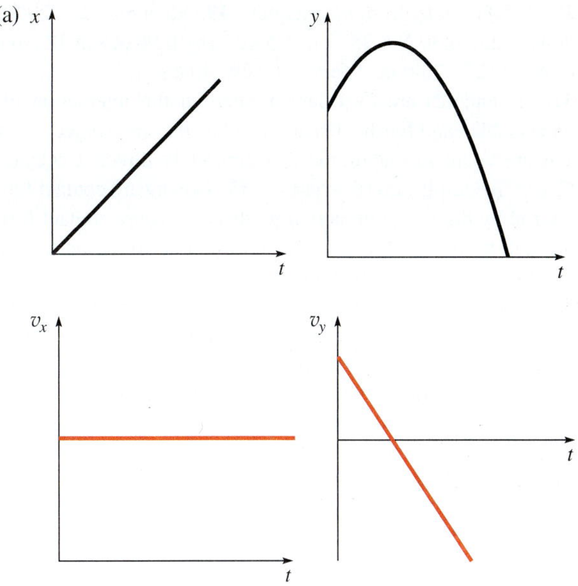

(b) 27.6 m/s at 25.0° above the horizontal (c) 37.5 m (d) 44.4 m above the ground **87.** 23 m
89. (a)

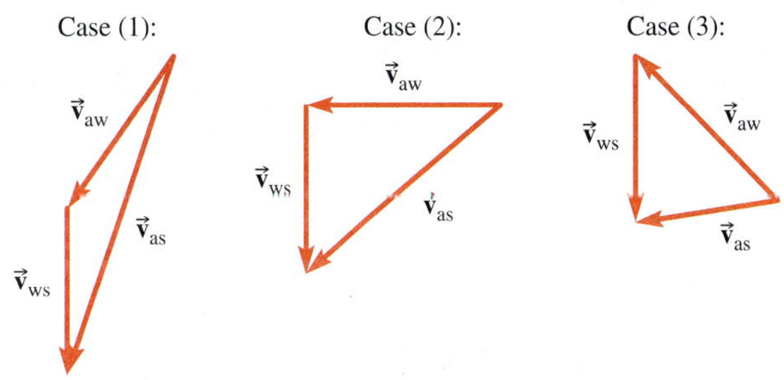

(b) 1 and 2 (c) all three cases **91.** (a) 6.33 km at 29.6° north of east (b) 22.1 min **93.** Your favorite team will win. **95.** (a) 68.5 km/h at 12.5° north of east (b) 68.5 km/h at 12.5° south of west **97.** (a) 4.5 s (b) 81 m **99.** 200 km
101. 12 m east and 40 m north **103.** (a) 28.6 cm (b) smaller (c) larger (d) $H = 21.3$ cm; $R = 85.1$ cm **105.** (a) 7.67 m (b) 13.9 m/s (c) 12.5 m/s **107.** 0.13 ms, 3.0 × 10^6 m/s^2
109. 32.0° **111.** (a) 2.02 s (b) 2.02 s (c) 1.5 s
113. (a) 1.3 m (b) 0.2 m (c) 2 m/s (d) 0.6 m/s

CHAPTER 4

Multiple-Choice Questions

1. (b) **3.** (a) **5.** (b) **7.** (b) **9.** (b) **11.** (e) **13.** (a) **15.** (c) **17.** (a) **19.** (a) **21.** (b) **23.** (a) **25.** (b)

Problems

1. 4.45 lb **3.** 20 N in the positive x-direction **5.** yes; $F_S > F_B$

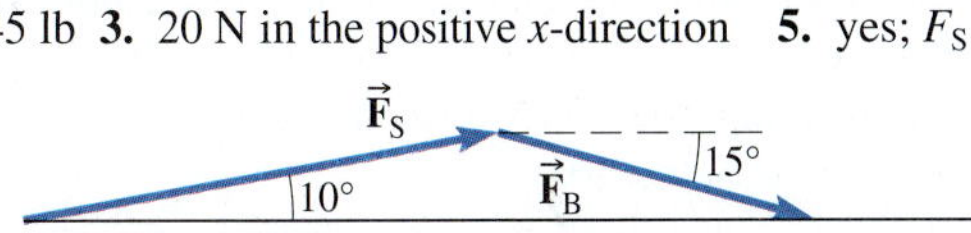

7. 157 N, 57° below the $+x$-axis **9.** 2 N to the east
11. 120 N north **13.** 13° from the vertical
15. (a) 30 N to the right (b) 0 (c) 18 N downward

17. 1550 N away from the racquet **19.** 80 N up **21.** 2980 N down **23.** 0.30 N **25.** (a) 1.3 kg (b) 0.90 m/s at 18° south of west **27.** 2.40 m/s^2 forward **29.** 40 kg
31. (1) and (2) are third-law partners in the interaction of two objects, bike and Earth. (1) and (3) act on the same object, the bike, and are equal in magnitude and oppositely directed because the bike is in equilibrium (first law). **33.** Downward contact force on the rod by the line (interaction partner = upward contact force on the line by the rod); downward gravitational force on the rod by the Earth (interaction partner = upward gravitational force on the Earth by the rod); upward contact force on the rod by the fisherman's hands (interaction partner = downward contact force on the fisherman's hands by the rod).
35. (a) Gravitational force exerted on the skydiver by the Earth; drag exerted on the skydiver by the air; tension exerted on the skydiver by the parachute
(b)

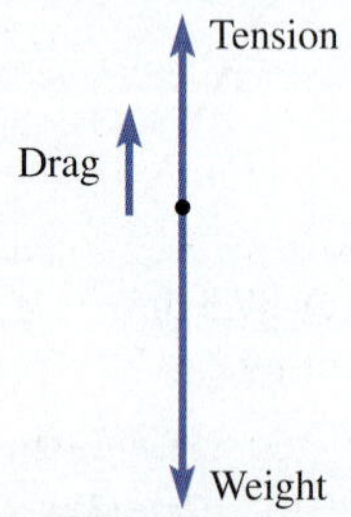

(c) 30 N (d) Gravitational force exerted on the Earth by the skydiver, 650 N upward; drag exerted on the air by the skydiver, 30 N downward; tension exerted on the parachute by the skydiver, 620 N downward

37.

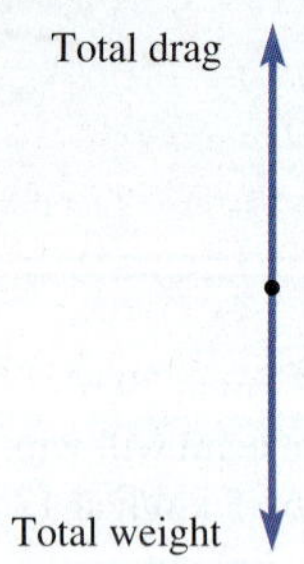

39. (a) 670 N (b) 2.5 N (c) a stick of butter **41.** 82 kg
43. (a)

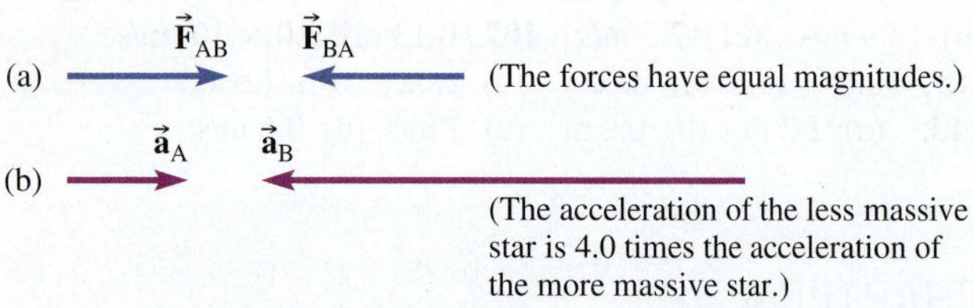

45. 1.1 **47.** 640 N on Earth (a) 240 N (b) 580 N (c) 100 N His weight on Earth is 2.6, 1.1, and 6.1 times his weight on Mars, Venus, and the Moon, respectively. **49.** 3.6 km
51. (a) 1.98 × 10^{20} N (b) the same **53.** 32.1 kg **55.** 22.7 kN upward **57.** 2.27 m/s **59.** 20 N **61.** (a) 6.29 × 10^{20} N (b) 2.37 × 10^{20} N **65.** The normal force is perpendicular to and away from the ramp in all three cases. The frictional force is upward along the ramp for (a) and (c) and downward along the ramp for (b). **67.** (a) 160 N up the slope (b) 0.19
69. 88 N up the ramp **71.** 61 N down the ramp
73. (a) yes (b) unnecessary

75. (a) zero (b) $T/(mg)$ **77.** (a) Static friction; the apples are not moving relative to the belt. (b) $\mu_s \geq 0.40$. No conclusion about μ_k. **79.** (a) $g(\sin\theta_2 - \cos\theta_2 \tan\theta_1)$ (b) 3.8 m/s^2
81. 850 N, due west **83.** 400 N **85.** Both scales read 120 N.
87. Scales A and B both read 120 N. **89.** (a) 34 N (b) 39 N
91. $T_{15} = 30$ N; $T_{25} = 18$ N **93.** 19 N toward the back of the mouth
95. $\dfrac{m_1}{m_1 + m_2}$ **97.** 3.5 N **99.** (a) m_1: 2.5 m/s^2 up; m_2: 2.5 m/s^2 down (b) 37 N **101.** 2.1 m/s^2 in the direction of motion **103.** $T_1 = 3.2$ kN, $T_2 = 1.1$ kN **105.** 642 N
107. 960 N upward **109.** (a) 1.4 m/s^2 downward (b) no
111. 620 N **113.** the weak force **115.** electromagnetic and gravitational forces **117.** the strong force
119. (a)

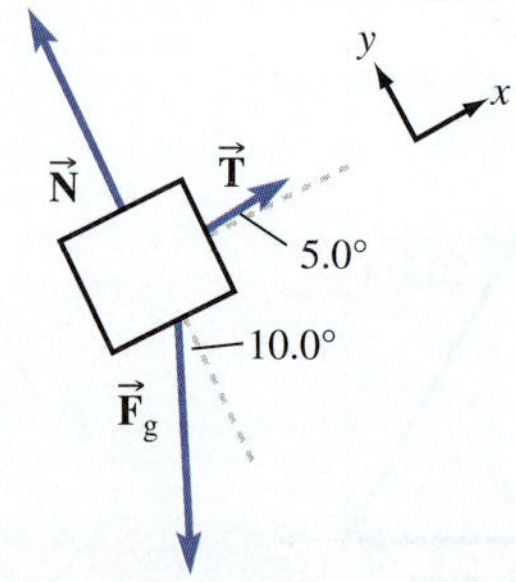

(b) Choose the coordinate system with the x-axis in the direction of the slope and the y-axis in the direction of the normal force. Then the problem can be solved by just summing the forces in the x-direction. We don't need to find the normal force so there is no need to sum the forces in the y-direction. (c) 2100 N
121.

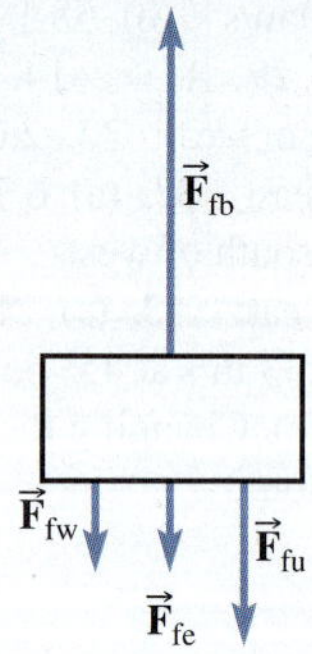

123. (a) 16 N (b) the block will accelerate (c) 1.3 m/s^2
125. (a) 15 N (b) 8.8 N
127. (a) $\vec{a}_1$ = 3.9 m/s^2 to the right; $\vec{a}_2$ = 3.9 m/s^2 downward (b) 4.7 m/s to the right (c) 2.8 m to the right (d) block 1: 0.31 m to the right; block 2: 0.31 m down
129.

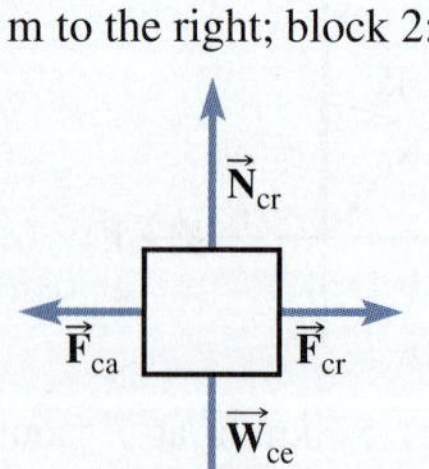

131. (a) zero (b) 2.6 × 10^4 N
133. (a)

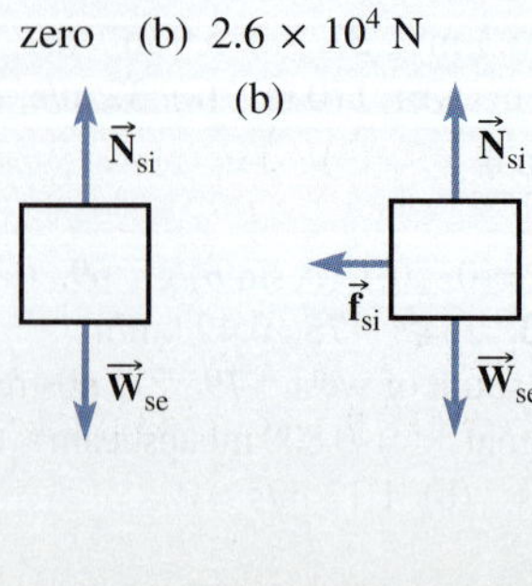

(c)

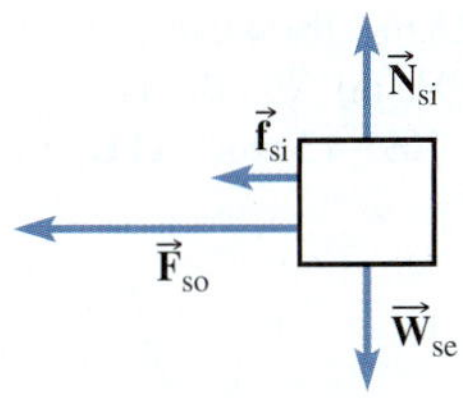

135. 0.49% **137.** $mg/(\cos\theta)$ **139.** $F/3$ to the right **141.** (a) 22° with respect to the horizontal (b) 0.9 m/s^2 down the incline **143.** 120 N **145.** 90.0% of the Earth-Moon distance **147.** (a) 110.0 N (b) $T_A = 115.0\text{ N} = T_C$ and $T_B = 110.0\text{ N} = T_D$. **149.** (a) 1) the gravitational forces between the magnet and the Earth 2) the contact forces, normal and frictional, between the magnet and the photo 3) the magnetic forces between the magnet and the refrigerator

(b)

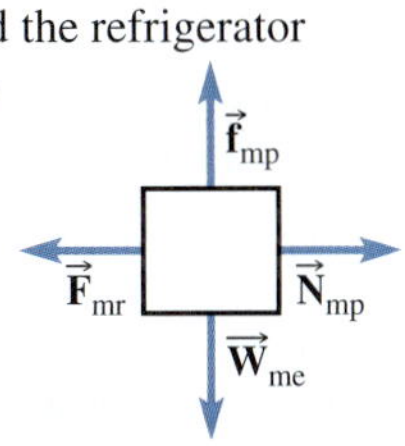

(c) The long-range forces are gravity and magnetism. The contact forces are friction and the normal force. (d) $W_{me} = 0.14$ N, $F_{mr} = 2.10$ N, $f_{mp} = W_{me} = 0.14$ N, and $N_{mp} = F_{mr} = 2.10$ N **151.** (a) A = 137 N; B = 39 N (b) A = 147 N; B = 39 N **153.** (a) 15.1 N (b) 34.3 N **155.** 480 N **157** (a) $mg\tan\theta$ (b) $mg\tan\theta$ (c) $mg\tan\theta + \dfrac{ma}{\cos\theta}$ **159.** (a) 1.10mg (b) 1.10mg **161.** (a) 2.60×10^8 m from Earth (b) away from **163.** (a) 5.0° from the vertical (b) 0.52 m/s^2 (c) 0.02 m/s^2 **165.** 4.22 km **167.** (a) 0.34 m/s^2, where the watermelon moves up and to the left (b) 1.5 cm (c) 6.8 cm/s **169.** (a) 80 N directed down the incline (b) 8.0 m/s^2 directed down the incline (c) 1.2 s

(d)

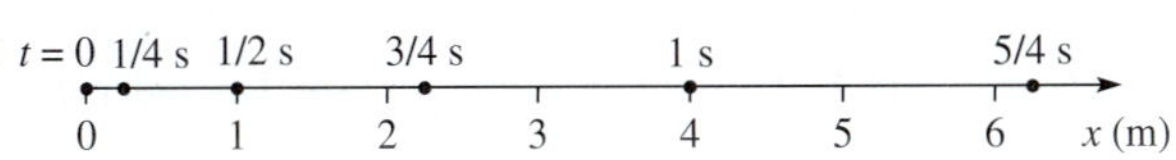

(e)

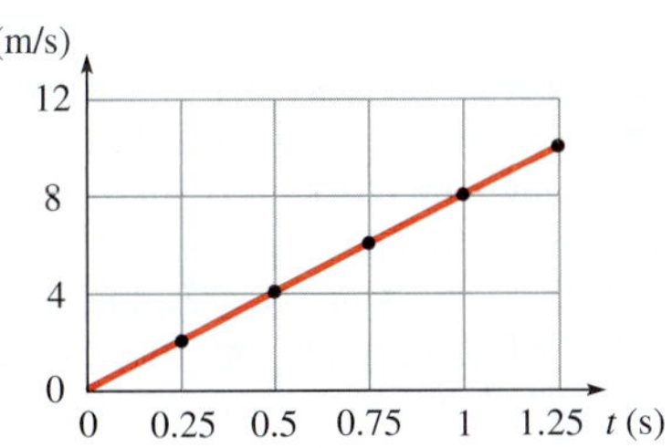

171. (a) 1.8 m/s^2 down (b) 8.7 m/s **173.** (a) 19 m (b) 3.6 m/s **175.** 36 m **177.** 2.02 s, 1.65 m to the left of B's initial position **179.** 700 kN

CHAPTER 5

Multiple-Choice Questions

1. (f) **3.** (b) **5.** (b) **7.** (b) **9.** (b) **11.** (c)

Problems

1. 17 m **3.** 0.105 rad/s; 0.52 rad **5.** (a) 160 rad (b) 4700 cm/s (c) 25 Hz **7.** 26 rad/s **9.** 3800 ft **11.** Linear speed at the rim is proportional to radius/period, so arrange these quotients in descending order: (c), (a) = (d), (b), (e). **13.** 3.37 cm/s^2 **15.** (a) 890 m (b) 490 m

17.
$$v\omega: \frac{[\mathrm{L}]}{[\mathrm{T}]}\cdot\frac{1}{[\mathrm{T}]} = [\mathrm{L}]/[\mathrm{T}]^2$$
$$\frac{v^2}{r}: \left(\frac{[\mathrm{L}]}{[\mathrm{T}]}\right)^2\cdot\frac{1}{[\mathrm{L}]} = \frac{[\mathrm{L}]^2}{[\mathrm{T}]^2}\cdot\frac{1}{[\mathrm{L}]} = [\mathrm{L}]/[\mathrm{T}]^2$$
$$\omega^2 r: \left(\frac{1}{[\mathrm{T}]}\right)^2\cdot[\mathrm{L}] = \frac{1}{[\mathrm{T}]^2}\cdot[\mathrm{L}] = [\mathrm{L}]/[\mathrm{T}]^2$$

19. (a)

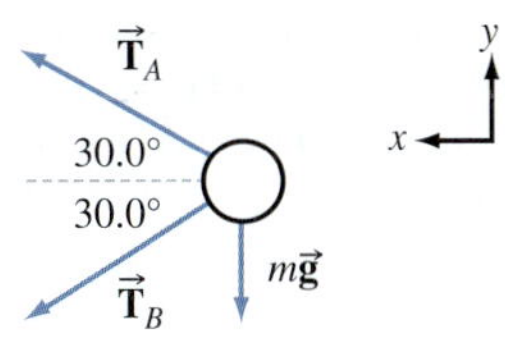

(b) $T_A = 3.64$ N, $T_B = 1.68$ N

21. (a) $\dfrac{mg}{\cos\phi}$ (b) $2\pi\sqrt{\dfrac{L\cos\phi}{g}}$

23. (a) 5.08° **25.** 3900 N inward at 53° above the horizontal **27.** (a) 3500 N (b) No. $\mu_s N$ would be the frictional force if the car were traveling at the maximum speed without skidding. **29.** 11.5 m/s **31.** $\tan^{-1}\dfrac{v^2}{rg}$ **33.** 2.99×10^4 m/s **35.** 130 h **37.** $r_{Io} = 420\,000$ km; $r_{Europa} = 670\,000$ km **39.** 17.0 m/s **41.** 438 N **43.** 0.39 rad/s^2 **45.** 190 rad/s^2 **47.** (a) 73 rad/s^2 (b) 23 rev **49.** (a) 17.7 m/s (b) 6.28 m/s^2 (c) 6.59 m/s^2 at an angle of 17.7° east of south **51.** (a) 170 rad/s^2 (b) 2.2 m/s^2 **53.** (a) 1.3×10^6 s (b) 5.0×10^{10} rev **55.** $a_t = 2.54$ m/s^2; $a_r = 2.45$ m/s^2; 11.9 N **57.** 16g **59.** 7.0 rad/s **61.** (a) $m(g - \omega^2 R)$ (b) $m(g + \omega^2 R)$ **63.** 2.04×10^7 m **65.** (a) 7.3×10^{-5} rad/s (b) about 0.02 rad (c) about 5 min **67.** (a) 14 m/s^2 (b) 1.4g (c) 33 m/s **69.** 464 m/s **71.** 1.80×10^6 degrees **73.** (a) 0.48 m/s (b) 12 rad/s clockwise **75.** 200 km/s **77.** 3.8 N; 2.0 N **79.** 110 μm/s **81.** 8.1 cm **83.** 1.04 rad/s **85.** (a) 3.6×10^7 m (b) 55 N; above the car **87.** 1.4 rev/s **89.** 0.40ω_0 **91.** (a) 0.034 m/s^2 (b) less (c) 0.34% smaller (d) at the poles **93.** 42 200 km **95.** (a) 17.3 N (b) 10.0 N (c) tension 10.0 N, radial accel 0, tangential accel 8.49 m/s^2

CHAPTER 6

Multiple-Choice Questions

1. (c) **3.** (b) **5.** (c) **7.** (c) **9.** (b) **11.** (f)

Problems

1. 75 J **3.** No work is done. **5.** 210 kJ **7.** (a) 0 (b) 8.8 J **9.** 153 J **11.** +224 J **13.** +52.6 J **15.** 6.09 kJ; 0 J **17.** 720 kJ **19.** Murphy: 27.2 kJ; Howard: 163 kJ **21.** 0.12 N **23.** −550 J **25.** (a) −50 MJ (b) 600 kN opposite the plane's direction of motion **27.** 54.8 kJ **29.** $E, C, B = D, A$ **31.** $A = B = C = D = E$ **33.** 11 h **35.** (a) 0 (b) −2.9 J **37.** (a) 1.88 kJ (b) 1.88 kJ (c) 8.00 m **39.** (a) 14.3 m/s (b) Yes, the cart will reach position 4. **41.** 8.42 m/s **43.** −52 kJ **45.** (a) $\sqrt{v^2 + 2gh}$ (b) The final speed is independent of the angle. **47.** 2.6 m/s **49.** 60.0 km/s **51.** 22.4 km/s **53.** 11.2 km/s **55.** 55 km/s **57.** 10 500 m/s **59.** 1.6 J **61.** 5.2 J

63. (a) 4.9 cm (b) 1.4 N/cm (c) 88 mJ **65.** (a) 1.9 N/cm (b) 0.49 J (c) 2.4 kg **67.** zero **69.** (a) 1.1 kN (b) 4.2 J **71.** $4h$ **73.** 8.7 cm **75.** (a) 2.2 m/s (b) 0.21 m (c) 0.50 m **77.** 13.0 s **79.** 4.08 min **81.** 150 W **83.** 5.2 W **85.** (a) 510 W (b) No, the body would have to be 100% efficient. **87.** (a) 8.8 kW (b) 6.4 kW **89.** (a) −500 J (b) 3 GW (c) 300 000 households **91.** No **93.** 27 N **95.** (a) 10 kW (b) 5.8° **97.** (a) 500 m^3 (b) 600 kg (c) 30 kJ (d) 12 kW (e) The power output is $\frac{1}{8}$ as large. Wind turbines are best located where steady winds prevail, since modest changes in wind speed produce large changes in power output. **99.** (a) 0.15 N/cm (b) 12 cm **101.** 53 kJ **103.** 2.3 m **105.** 4.8 m/s **107.** 16 m/s **109.** 6.0 m/s **111.** (a) 3.45 kJ (b) 4.96 kJ (c) −1.52 kJ (d) 187 N **113.** (a) 124 J (b) 10 300 fastballs **115.** 5.8 m/s **117.** 1.3 cm, 32 J **119.** The kinetic energy cannot be negative so, it must remain in the $x < 3$ cm region. **123.** 1.6 m/s **125.** (a) 2.62 kW (b) 7.85 kW **127.** (a) $k = k_1 + k_2$ (b) 0.16 J **131.** (a) 2.0 m/s (b) 2.0 m/s The energy method is easier. Using Newton's laws, we first have to find the acceleration and then apply constant-acceleration kinematics to find the final speed. **133.** (a) 1.7 m/s (b) 0.76 m/s^2, 1.7 m/s **135.** (a) A = B = C (b) Energy is easier since we are concerned only with the final speeds. By applying conservation of mechanical energy, we find that the final speeds are the same in a single step. We can reach the same conclusion using constant-acceleration kinematics and a few steps of algebra, but this approach doesn't provide the same insight into why the speeds are equal. **137.** $\frac{2}{3}R$

CHAPTER 7

Multiple-Choice Questions

1. (d) **3.** (c) **5.** (b) **7.** (f) **9.** (d) **11.** (d)

Problems

1. 0 **5.** 3 kg·m/s north **7.** (b), (a) = (e), (c), (d), (f) **9.** 20 kg·m/s in the $-x$-direction **11.** 1.0×10^2 kg·m/s downward **13.** (d), (b) = (c), (a), (e) **15.** 320 s **17.** 6.0×10^3 N opposite the car's direction of motion **19.** (a) 750 kg·m/s upward (b) 990 N·s downward (c) 2500 N downward **21.** 0.12 m/s **23.** 1.8 m/s **25.** 0.010 m/s **27.** 100 m/s (224 mi/h). Dash will not succeed. **29.** 0.10 m/s **31.** (8.0 cm, 20 cm) **33.** 1.9 cm **35.** 4.0 cm in the positive x-direction **37.** (0.900 m, −2.15 m) **39.** 98.0 m/s downward **41.** 5.0 m/s west **45.** −0.15 m/s **47.** 3.0 m/s east **49.** 350 m/s **51.** 0.066 m/s **53.** The 300 g ball moves at 2.50 m/s in the $+x$-direction, and the 100 g ball moves at 2.50 m/s in the $-x$-direction. **55.** 4.8 m/s **57.** 0.49 m **59.** 5.0 m/s **61.** 170 m/s **63.** (a) $\Delta p_{1x} = -1.00 m_1 v_i$; $\Delta p_{1y} = 0.751 m_1 v_i$ (b) $\Delta p_{2x} = m_1 v_i$; $\Delta p_{2y} = -0.751 m_1 v_i$; the momentum changes for each mass are equal in magnitude and opposite in direction. **65.** $1.73 v_{1f}$ **67.** 8.7 kg·m/s **69.** 6.0 m/s at 21° south of east **71.** 1.7 m/s at 30° below the x-axis **73.** 0.64 m/s at 73° above the $+x$-axis **75.** (a) 2.4 m/s (b) 3.7 m/s (c) 19° **77.** 29 m/s **79.** (a) 0.30 m/s to the left (b) The final kinetic energy is greater than the initial kinetic energy. The extra kinetic energy comes from the elastic potential energy stored in the spring. **81.** 0.83 m/s **83.** 37 m/s in the $+x$-direction **85.** (a) 11 kg·m/s (b) 11 kg·m/s (c) 3.8 kN **87.** (5.00 cm, 6.67 cm) **89.** 410 N in the direction of water flow **91.** (a) 0.01 kg·km/h opposite the car's motion (b) 0.01 kg·km/h along the car's velocity (c) 10^5 flies **93.** 2.8 m/s **95.** (a) 2.5 m (b) 4.0 m **97.** 10^{-18} N **99.** $\frac{1}{9}h$ **101.** 10 m/s **103.** (a) $\frac{111}{2}$ (b) 1 (c) $\frac{111}{2}$ **105.** 1.0 m **107.** 1.27 m **109.** 13 m/s **111.** 0.73 m

CHAPTER 8

Multiple-Choice Questions

1. (d) **3.** (a) **5.** (e) **7.** (a) **9.** (c)

Problems

3. (a) reduced by a factor of 8 (b) reduced by a factor of 32 **5.** (b), (a) = (c) **7.** 0.0512 J **9.** 0.019 **11.** 570 J **13.** 4.5 N·m **15.** 0.30 N·m **17.** (e), (a) = (b) = (d), (c) **19.** (a) 0 (b) 790 N·m **21.** (a) 58.5 N·m (b) 39.9 N·m (c) 0 **23.** 5.83 m **25.** (0.42s, 0.58s) **27.** (a) 3.14 m (b) 15.7 J (c) 2.50 N·m (d) 6.28 rad (e) $\tau\Delta\theta$ = (2.50 N·m)(6.28 rad) = 15.7 J = W **29.** (a) 53.0 kJ (b) 1.51 MN·m **31.** 200 N **33.** (a) 540 N (b) 390 N **35.** Left support: 2.2 kN downward; right support: 3.4 kN upward **37.** (a) 730 N (b) 330 N at 19° above the horizontal

39. $\dfrac{mg/2 + W}{\tan\theta}$; for $\theta = 0$, $T \to \infty$, and for $\theta = 90°$, $T \to 0$.

41. 1.3 m **43.** 640 N **45.** 7.0 kN **47.** (a) 330 N (b) 670 N **51.** 0.0012 N·m **53.** (a) 13 rad/s (b) 16 N·m (c) 15 m to the same height, plus about another meter if released 1 m above the ground.

55. 1.5 N·m **57.** (a) 48 N·m (b) 19 N

59. (a) $a = R\alpha$ (b) $(T_1 - T_2)R$, CCW (c) If $m_1 \neq m_2$, the blocks accelerate, so the pulley has an angular acceleration. Since a nonzero net torque is required for the pulley to accelerate, $T_1 - T_2 \neq 0$, thus $T_1 \neq T_2$.

(d) $a = \dfrac{(m_1 - m_2)g}{\frac{1}{2}M + m_1 + m_2}$ $\quad T_1 = m_1(g - a) \quad T_2 = m_2(g + a)$

61. 4.0 m/s^2 **63.** solid sphere: $K = \frac{7}{10}mv^2$; solid cylinder: $K = \frac{3}{4}mv^2$; hollow cylinder: $K = mv^2$ **65.** (a) 3.0 m/s (b) 8.4 N (c) 5.6 m/s^2 down **67.** (a) $\frac{5}{2}r$ (b) $\frac{27}{10}r$ **69.** h will decrease. The smaller the rotational inertia, the less gravitational energy will go into rotational energy and the more will go into translational energy. Problem 68 had a minimum of $h = 3r$. With a solid sphere, the minimum is $h = 2.7r$, which is a little less than $3r$. **71.** 0.0864 kg·m^2/s **73.** 1.4×10^7 kg·m^2/s **75.** 1.60 s **77.** 1.5 rev/s **79.** 70.3 J **81.** 0.125 rad/s **83.** (a) 3.0 (b) 1.6 **85.** 2.10×10^6 N·m **87.** 3.0 kN; about 4.7 times larger **89.** (a) 2.6×10^{29} J (b) The length of the day would increase by 7 minutes. (c) 2.6 million years **91.** (a) 9.6 m/s (b) 3.1 m/s (c) 21 m/s **93.** 0.44 N·m **95.** $\sqrt{3gL}$ **97.** T_1 = 67 kN; T_2 = 250 kN; $\vec{\mathbf{F}}_p$ = 380 kN at 51° with the horizontal **99.** (a) 6.53 m/s^2 down (b) 4.2 N **101.** 1.79 m/s **103.** The vertical force exerted by the door on each hinge is 27.4 N down. The upper and lower horizontal forces are 14.2 N toward the door and 14.2 N away from the door, respectively.

105. (a) $\dfrac{\omega_i}{1 + \dfrac{mr^2}{MR^2}}$

(b) The total angular momentum does not change, since no external torques act on the system. (c) Yes; the kinetic energy changes. It decreases because of because of frictional work done by the contact forces between the disks, changing mechanical energy into internal energy.

107. (a) 0.96 m from the RH edge (b) 0.58 m from the LH edge **111.** 110 N **113.** (a) 1.35×10^{-5} kg·m^2 (b) 524 N **115.** (a) 6.28 rad/s (b) 0.955 kg·m^2/s (c) friction (d) 0.300 N **117.** (b) $L = I\omega = mr^2\omega$ (c) $A = \frac{1}{2}r^2\omega\Delta t$ (d) $A/\Delta t = L/(2m) =$ constant

119. (a) $\left(\frac{I_i}{I_i + mR^2}\right)\omega_i$ (b) Before, $K_i = \frac{1}{2}I_i\omega_i^2$ and $L_i = I_i\omega_i$. After, $K_f = \frac{1}{2}\left(\frac{I_i^2}{I_i + mR^2}\right)\omega_i^2$ and $L_f = L_i = I_i\omega_i$

121. 23 N **123.** 1.3 rev/s **125.** (a) $\left(\mu_s \frac{M + m}{M}\tan\theta - \frac{m}{2M}\right)L$ (b) Since tan θ increases as θ increases on the interval $0 \leq \theta < 90°$, and since d increases as tan θ increases [see the answer to (a)], placing the ladder at a larger angle θ increases d, the distance the person can climb up the ladder. (c) 63° **127.** 0.792 m **129.** (a) 62 N (b) 4.8 rad/s^2 (c) 0.76 N·m (d) 35 N **131.** (a) 55 rad/s^2 (b) 83 N·m (c) 7.3 rad (d) 0.52 m/s **133.** (a) 52 m/s (b) The stone hits the window.

CHAPTER 9

Multiple-Choice Questions

1. (d) **3.** (b) **5.** (a) **7.** (a) **9.** (d)

Problems

1. 49 atm **3.** (a) 1.0×10^5 N (b) 2.2×10^4 lb (c) The pressure of the air under the desktop pushes upward, counteracting the downward force. **5.** (a) 420 N (b) No force is needed. **7.** 88.0 kPa **9.** 31 m **11.** 0.126 **13.** (b) = (e), (a) = (c), (d) **15.** 35 kPa **17.** (a) 539 kN (b) Parts of this container push *down* on the fluid (the top part of the larger cylinder). The bottom of the cylinder must push up with a force greater than the fluid's weight to compensate for this downward force. **19.** 10 km **21.** 114.0 cm Hg **23.** (a) 5.6 cm (b) 0.37 cm **25.** 211 mm Hg **27.** (a) 2.2×10^5 Pa (b) 1700 torr (c) 2.2 atm **29.** (a) aluminum since it is less dense it occupies more volume (b) wood since it displaces more water than the steel (c) Lead: 0.87 N, Aluminum: 3.6 N Steel: 1.2 N, Wood: 9.8 N **31.** 1.5 m **33.** (a) 140 kg/m^3 (b) 18% **35.** 0.74 g/cm^3 **37.** (a) 0.910 (b) 1.28 cm (c) 0.13 cm **39.** (a) 89% (b) In (2), ignoring a very small buoyant force due to air, the buoyant force due to water is equal to the weight W of the plastic. In (3), the *total* buoyant force due to water *and oil* is equal to W. The buoyant force due to water alone is then less than W, so a smaller volume of plastic is submerged in water. (c) 35% **41.** 1.2×10^{-3} m^3 **43.** yes **45.** (a) 9.8 m/s^2 upward (b) 3.3 m/s^2 upward (c) 68.6 m/s^2 upward **47.** 50 m/s **49.** (a) 39.1 cm/s (b) 78.5 cm^3/s (c) 78.5 g/s **51.** 1.12×10^5 Pa **53.** 1.9×10^5 N **55.** 310 kPa **57.** 8.6 m **61.** (a) 6850 Pa (b) 0.685 N **63.** 12 m/s **65.** 17 atm **67.** (a) 50 Pa (b) 1100 Pa (c) approximately 13 kPa **69.** $R = 8\eta L/(\pi r^4)$ **71.** 0.4 Pa·s **73.** 1.5 cm/s **75.** (a) 120 m/s (b) Too fast to be reasonable **77.** 5 Pa **81.** 230 kg **83.** (a) 26 m/s (b) 2.6 m/s **85.** (a) 220% (b) 0.68 **87.** (a) 81.1 g (b) 55.6 g and 485.5 g **89.** For the pine, the scale reading doesn't change. For the steel, the scale reading will increase by 0.538 N. **91.** 19 m/s **93.** 23.0 m **95.** 110 m **97.** 27 kPa **99.** (a) 2.2 m/s up (b) 22 kPa/s **101.** (a) 0.600W (b) 0.64W **103.** 8.7 kg **105.** (a) 10.0 cm (b) 0.814 (c) 0.545 **107.** 0.83 g/cm^3 **109.** (a) The only vertical forces on such a column are the upward force $P_{atm}A$ due to atmospheric pressure at the bottom and the downward force W due to gravity. These forces must be equal in magnitude, so $P_{atm} = W/A$. (b) 8.0 km (c) lower limit **111.** 1.0 m **113.** (a) 1.3×10^{-10} N (b) 2.6×10^{-14} W **115.** (a) a straight line with negative slope (b) The acceleration (slope of the graph) is not constant. (c) The value of $|a_y|$ when $v_y = 0$ is only about 2.0 m/s^2. (d) $F_B \approx 0.8mg$ (e) about 2.4 s **117.** (a) 41.7 cm/s, 118 kPa (b) 5.98 cm

CHAPTER 10

Multiple-Choice Questions

1. (c) **3.** (b) **5.** (a) **7.** (c) **9.** (c) **11.** (f) **13.** (e) **15.** (f) **17.** (c) **19.** (j)

Problems

1. 0.097 mm **3.** (a) 0.0046 cm (b) -4.6×10^{-5} **5.** 76.9 GPa **7.** 0.80 mm **9.** (a) 5.1 N (b) 7.7×10^{-2} J **11.** 5.0 mm **13.** tension: 15 GPa; compression: 9.0 GPa **15.** (a) tendon: 7.3×10^4 Pa·m^3/kg; steel: 6.5×10^4 Pa·m^3/kg; tendon is stronger than steel. (b) bone: 1.0×10^5 Pa·m^3/kg; concrete: 1.5×10^5 Pa·m^3/kg; concrete is stronger than bone. **17.** human: 3 cm^2; horse: 7.1 cm^2 **19.** 630 N **21.** 4.0×10^8 Pa **23.** (a) 1.3 mm; 84 MPa (b) 570 N **25.** decrease by 57×10^{-6} cm^3 **27.** 3.3×10^{-5} **29.** -6.71 cm^3 **31.** (a) 2.8 mm (b) 20 kPa **33.** 7.0 cm/s **35.** 0.63 m/s **37.** (c), (a) = (b), (d) = (e) **39.** 2.5 μm/s **41.** (a) $y = 0$ (b) $y = +A$ (c) $y = -A$ (d) $y = 0$ **45.** (a) $2kx$ (b) $\sqrt{2k/m}$ **47.** 2.5 Hz **49.** $v_m = 0.157$ m/s, $a_m = 24.7$ m/s^2 **51.** (a) g (b) 0.78 m **53.** -0.031 J **55.** (a) 0.90 s (b) 0.56 m/s **57.** 2.0 mJ

59. (a)

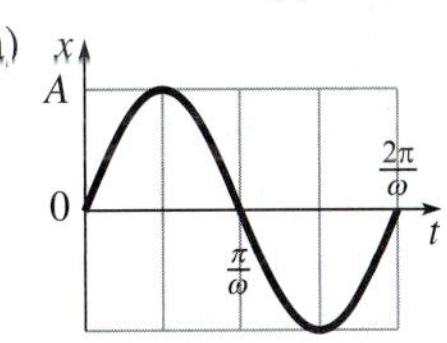

(b)

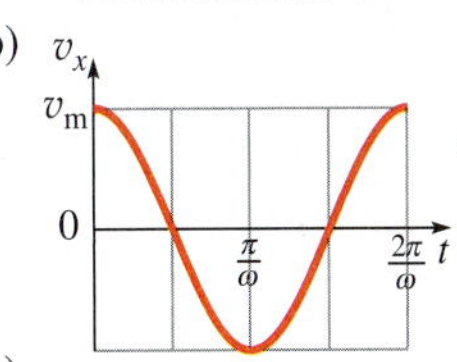

a cosine function

(c)

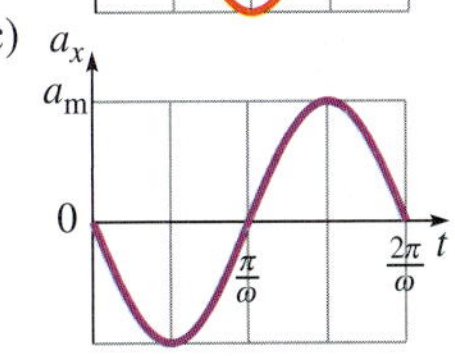

61. (a)

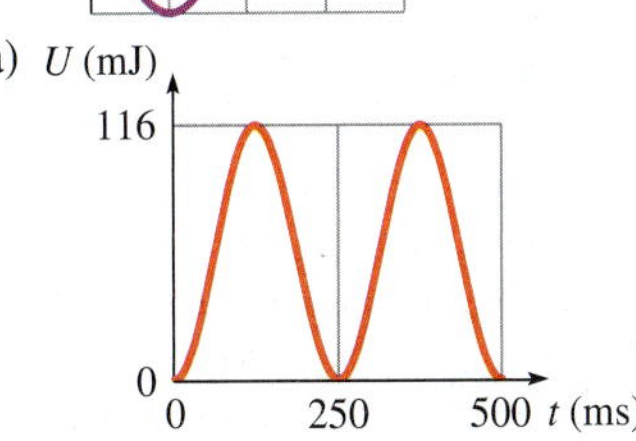

(b)

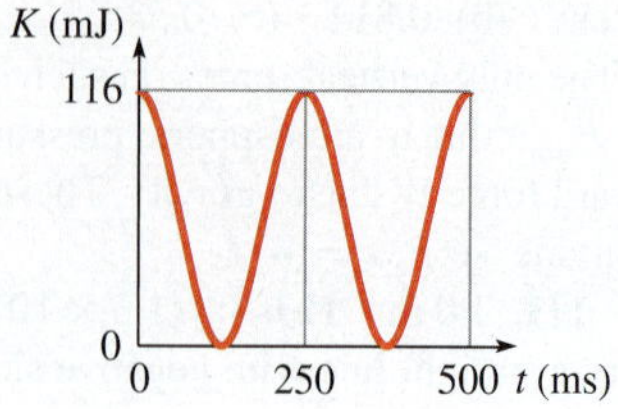

(c)

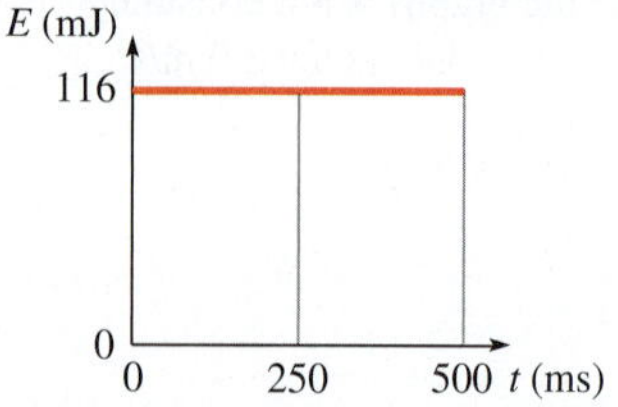

(d) U, K, and E would gradually be reduced to zero. **63.** 4.0 s **65.** 1.5 s **67.** (c), (a) = (b), (d) = (e) **69.** 0.25 m **71.** (a) less (b) 5.57 m/s^2 **73.** 1.3 s **75.** (a) 2.01 s (b) 11.3% **77.** (a) 6.1 mJ (b) 1.1% **79.** −9.75% **81.** (a) more, because now the period is longer (b) 56 N **83.** (a) 3.42 s (b) No, the greater initial potential energy means the cord would stretch more. **85.** 2.16 Hz **87.** 0.994 m, assuming $g = 9.80$ m/s^2 **89.** (a) No change. The frequency and period don't depend on amplitude. (b) The total energy for an amplitude of $2D$ is four times that for an amplitude of D. (c) The frequency and period are still the same. (d) The energy is greater when given an initial push, since it starts with the same initial potential energy but also some initial kinetic energy. The energy is greater by $\frac{1}{2}mv_i^2$ **91.** (a) 42.2° (b) 48 g (c) 9.1 cm **93.** 2.1 m/s; 370 m/s^2 **95.** (a) 98.0 N/m (b) 0.472 m/s (c) 0.409 m/s (d) 3.33 s **99.** 2.0° **101.** (a) 8×10^{-4} (b) 8.0 kN (c) 5×10^{-5} m^2 (d) No **103.** (a) 1.64 s (b) 1.53 s (c) 1.94 s

105. (a) $2\pi\sqrt{\dfrac{2L(m_1 + 3m_2)}{3g(m_1 + 2m_2)}}$

(b) For $m_1 \gg m_2$, $T = 2\pi\sqrt{\dfrac{2L}{3g}}$, and for $m_1 \ll m_2$, $T = 2\pi\sqrt{\dfrac{L}{g}}$.

107. 0.45% **109.** (a) $\sqrt{2gL}$ (b) $\frac{\pi}{2}\sqrt{gL}$; larger

111. 0.88 m/s **113.** (a) 5.13×10^{-2} N (b) 2.69 s

CHAPTER 11

Multiple-Choice Questions

1. (c) **3.** (d) **5.** (c) **7.** (d) **9.** (d) **11.** (d)

Problems

1. 52 W/m^2 **3.** 260 m **5.** 31 kW **7.** (d), (c) = (e), (a), (b) **9.** (a) 6.0 m (b) 1.7 s **11.** 168 m/s **13.** (a) The pulse moves faster on the second string. (b) 6.9 ms **15.** 250 m/s **17.** 400 Hz **19.** 33 cm **21.** 0.33 Hz **23.** (a) 1.05 m (b) 0.126 s (c) 8.33 m/s (d) $-x$ (e) 1.75 cm/s **25.** (a) 0.35 mm (b) 6.0 m (c) 11 Hz (d) $+x$ (e) 2.4 cm/s **27.** (a) $y(x, t) =$ (1.20 mm) sin [(134 rad/s)t + (20.9 rad/m)x] (b) 16.1 cm/s **29.** $y(x, t) =$ (2.50 cm) sin[(8.00 rad/m) x − (2.90 rad/s)t] **31.** (b) = (e), (a), (c) = (d) **33.** (a) 2.6 cm (b) 14 m (c) 20 m/s (d) 1.4 Hz (e) 0.70 s

35. $v_m = 0.063$ m/s; $a_m = 0.79$ m/s^2

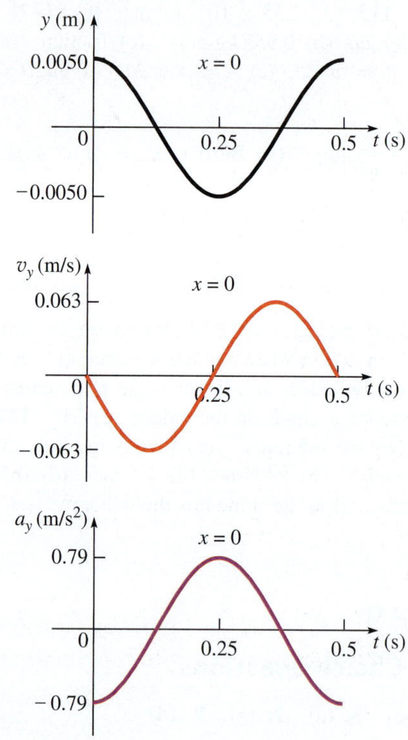

37.

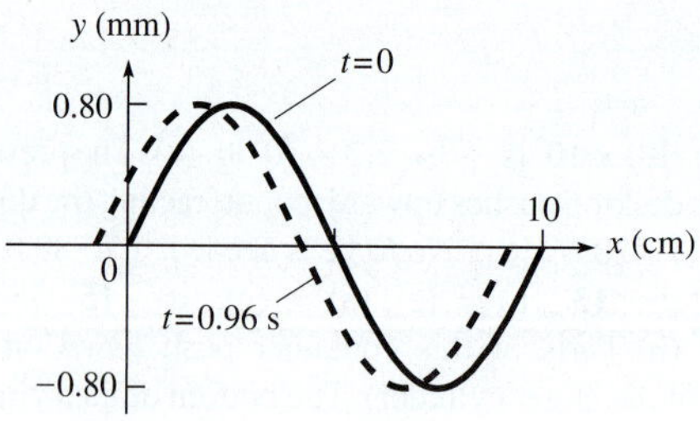

The wave travels in the $-x$-direction.

39.

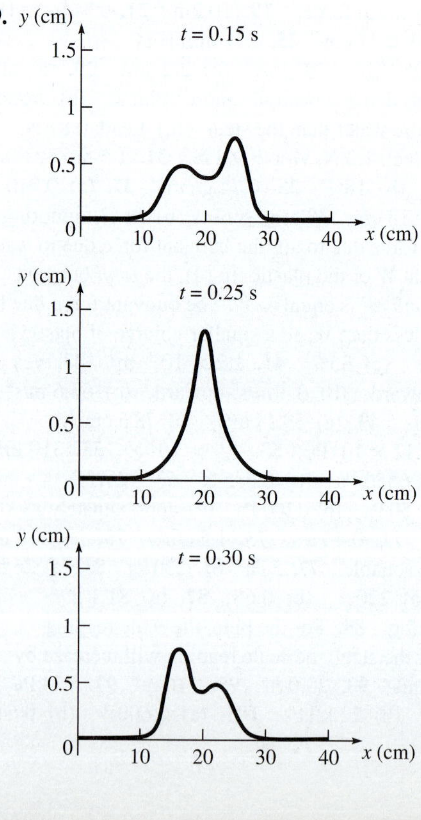

41. The amplitude of the superposition is about $1.7A$.

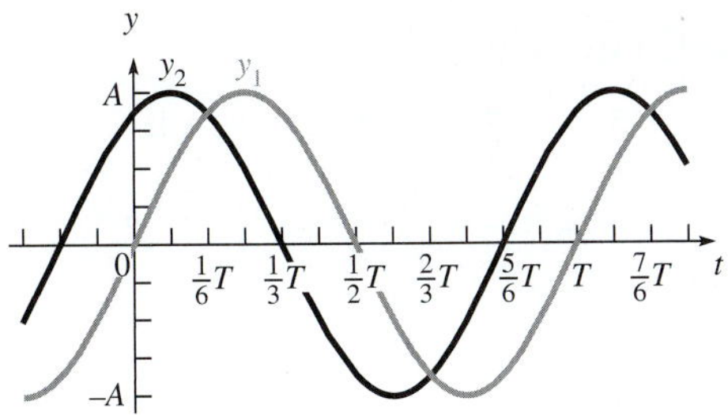

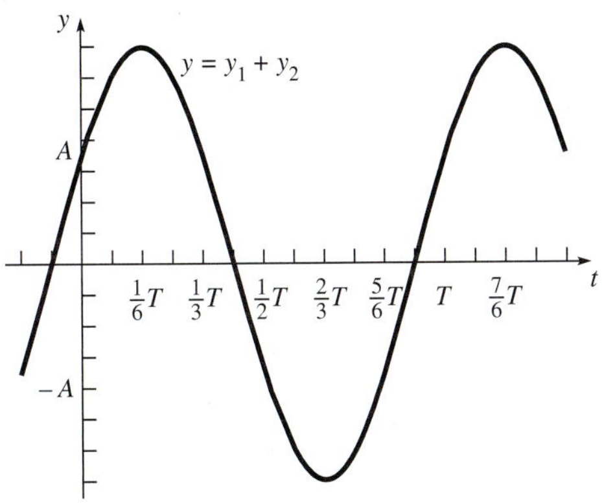

43. 375 nm

45.

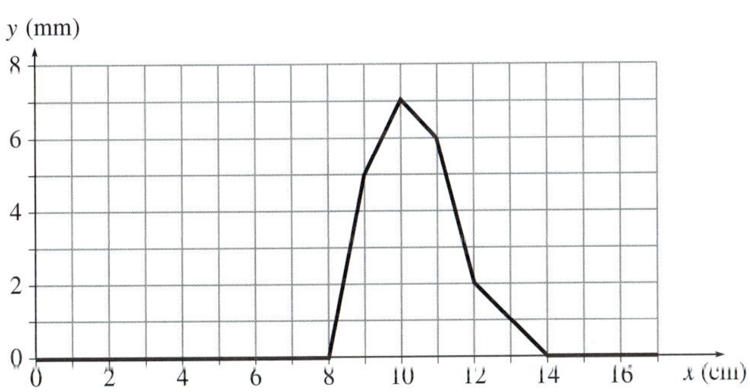

47. 5.3 s

49.

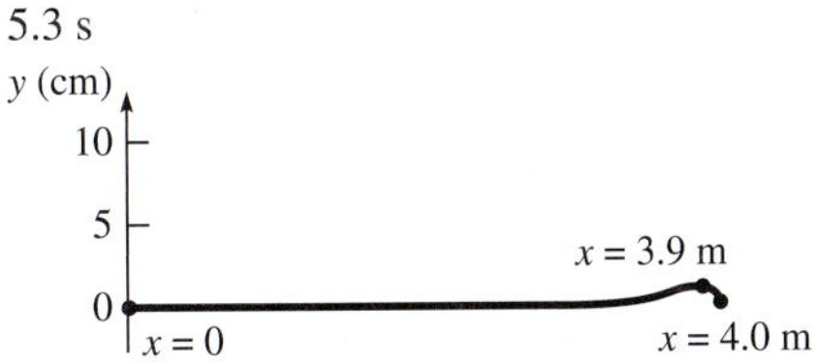

51. 2.28 m, 2.94 m, and 3.59 m **53.** (a) 8.0 cm (b) 2.0 cm (c) 4.0 **55.** 79 mW/m^2 **57.** (a) 0.25 W/m^2 (b) 0.010 W/m^2 (c) 0.130 W/m^2 **59.** (d), (a), (b) = (c), (e) **61.** $f_2 = (6/5)f_1$ **63.** $f_1 = 10$ Hz, $f_2 = 20$ Hz, $f_3 = 30$ Hz **65.** 7.8% **67.** (a) 33 Hz (b) 300 N **69.** 4.5×10^{-4} kg/m **71.** 1.0 N, 0.26 N, and 0.11 N **73.** 3.64 cm, 7.07 cm, 10.32 cm **75.** (a) Hooke's law: $T = k(x - x_0) \approx kx$ for $x \gg x_0$. (b) 4.00 s **77.** 12 **79.** $v_m = 1.9$ cm/s; $a_m = 3.0$ cm/s^2

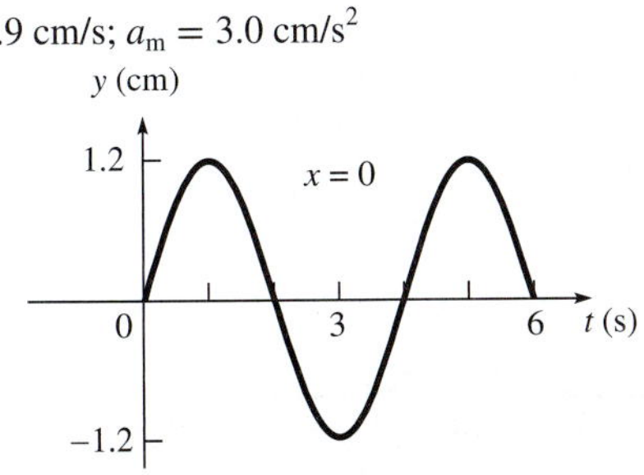

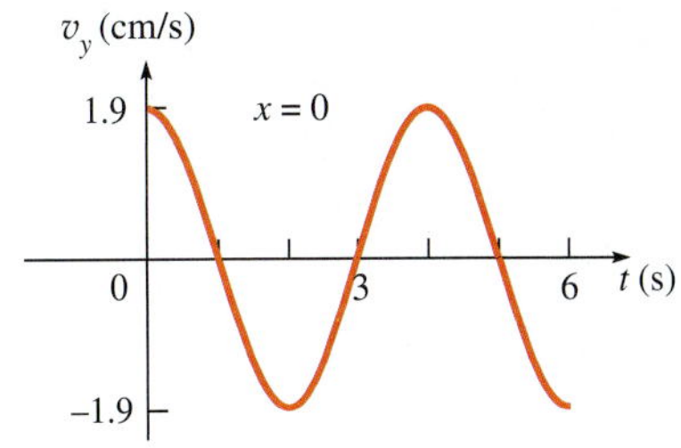

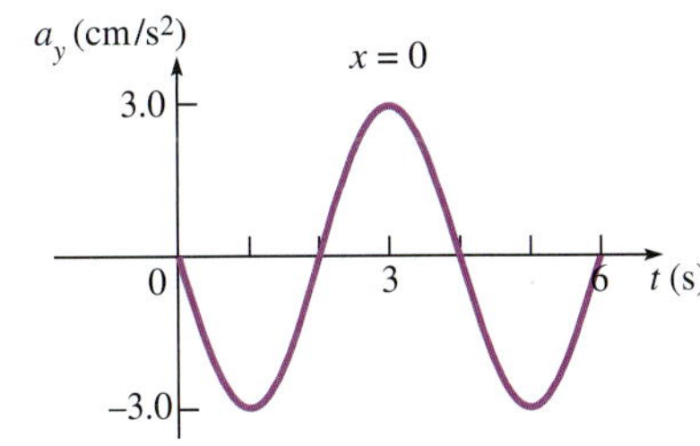

81. 930 Hz **83.** (a) left (b) 7.00 cm (c) 10.0 Hz (d) 0.333 cm (e) 3.33 cm/s (f) oscillates sinusoidally along the y-axis about $y = 0$ with an amplitude of 7.00 cm (g) transverse **85.** (a) 27.7 cm (b) 50 cm (c) 27.7 cm **87.** (a) Eq. I; 1.50 cm/s (b) Eq. II; 2.09 cm (c) Eq. II; 13.5 cm/s (d) Eq. II **89.** (a)

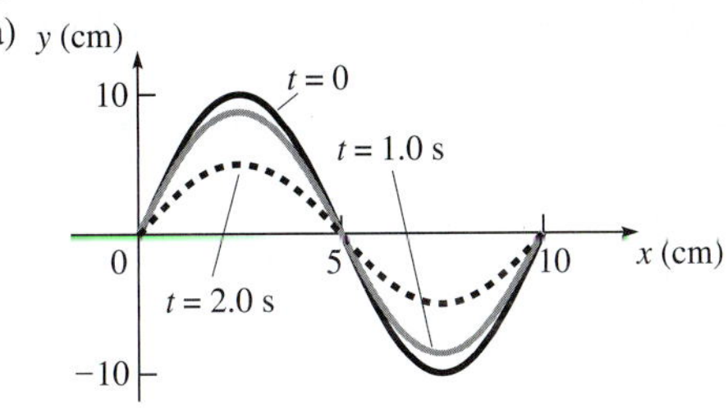

(b) standing wave

91. 2.4 km **93.** $v \propto \sqrt{\frac{\gamma}{\lambda\rho}}$, dispersive **95.** (a) 1.5 m/s (b) 21 cm/s **97.** (a) 6.17×10^{-4} m (b) 8.61 J (c) 0.0536 s **99.**

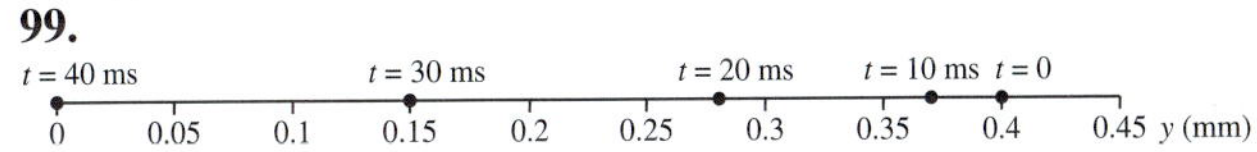

101.

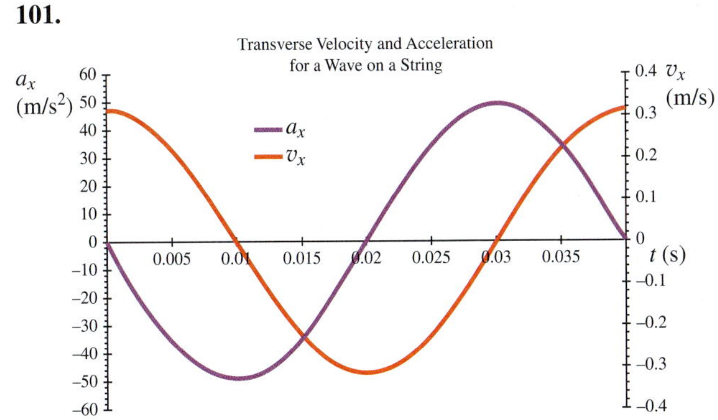

CHAPTER 12

Multiple-Choice Questions

1. (c) **3.** (c) **5.** (c) **7.** (d) **9.** (c) **11.** (a)

Problems

1. 364.1 m/s **3.** for 10 Hz, 34 m; for 20 kHz, 1.7 cm **5.** 6.1 mm **7.** (a) 338 m/s (b) 2.8 km **9.** 3.5 km/s **13.** (e), (d) = (f), (a), (b), (c) **15.** (a) 28.7 N/m^2 (b) 1.58 mN **17.** 97.0 dB **19.** (a) 0.0510 Pa (b) 0.151 μm **21.** 95 dB, not much different than with only one machine running **23.** 26.0% **25.** (a) 125.0% (b) 3.522 dB **27.** (e), (b), (c), (a) = (d) **29.** (a) 32.8 cm (b) 252.4 Hz **31.** 396 Hz **33.** 33°C **35.** 3/4 **37.** (a) 438.0 Hz (b) tighten **39.** 2 Hz **41.** 0.2 Hz **43.** −69 Hz **45.** (a) 2.0 kHz (b) 670 Hz **47.** 346 Hz **51.** (a) 670 m (b) 2.8 s **53.** 403 m **55.** 83.6 kHz **57.** 640 Hz **59.** (a) closed at one end (b) 78.0 Hz (c) 1.10 m **61.** (a) 5.05 m (b) 16.35 Hz **63.** (a) 6.7×10^{-8} m/s (b) 1×10^{-19} J (c) The ear is about as sensitive as possible. **65.** 9.8 m **67.** f_1 = 3.4 kHz, f_3 = 10 kHz, and f_5 = 17 kHz; 3.4 kHz enhances the sensitivity of the ear **69.** 2.3 kHz **71.** (a) 9.9 m (b) 1.8 ms (c) No **73.** 0.0955 s **75.** 15 m **77.** 29.0 dB **79.** 8.4 m/s **81.** (a) 1.28 m (b) 141 m/s (c) 6.71 g/m (d) 1.59 m/s (e) 110.0 Hz (f) 3.120 m

CHAPTER 13

Multiple-Choice Questions

1. (e) **3.** (c) **5.** (b) **7.** (d) **9.** (c)

Problems

1. (a) 29°C (b) 302 K **3.** (a) −40 (b) 575 **5.** $T_J = (0.750°J/°C)\, T_C + 85.5°J$ **7.** 2.7 mm **9.** 8.8 mm **11.** 113°C **13.** 0.6364 cm **15.** 650 K **17.** 15.8 cm^3 **19.** (a) 1.44 mL (b) 14% less water is predicted to be spilled when the expansion of the glass is ignored. **21.** 26.8°C **23.** 520°C **27.** 1 u = $(1 \times 10^{-3}$ kg$)/(6.022 \times 10^{23})$ = 1.66×10^{-27} kg **29.** 7.31×10^{-26} kg **31.** 1.7×10^{27} **33.** 2.650×10^{25} atoms **35.** 8.9985 mol **37.** 2.5×10^{19} molecules **39.** 10^{18} atoms **41.** 400°C **43.** (b) = (d), (a) = (c) = (e) = (f) **45.** (a) 2.55 m^3 (b) 5.3 h **47.** (a) 1.3 kg/m^3 (b) 1.2 kg/m^3 **49.** 1.3×10^3 m^3 **51.** 4×10^{-17} Pa **53.** 0.38 m^3 **55.** 2.1 mm **57.** 1550 K **59.** (a) 1.52×10^5 J/m^3 (b) 4.559×10^7 J/m^3 **61.** (b), (a) = (c) = (d), (e) = (f) **63.** 370 m/s **65.** 3.00 cm/s **67.** He: 1360 m/s; N_2: 515 m/s; H_2: 1920 m/s; O_2: 482 m/s **69.** 2220 K **73.** 0.14°C **75.** 1.3×10^{-19} J **77.** 1.25×10^6 s **79.** 80 s **81.** 1.3×10^{-5} s **83.** (a) The number of moles decreases by 25%. (b) −48° C **85.** (a) 52 cm (b) 12 m **87.** 467 mol **89.** 140 atm **91.** 165°C **93.** HNO_3 **95.** (a) 6.42×10^{-21} J (b) 0.3% **97.** 80 cm^3 **99.** 13.2 inches **101.** (a) 50.9 cm^3 (b) increases by a factor of 1.675 **105.** (a) 4.2 nm (b) the gas is dilute **107.** (a) 44°C (b) −21°C **109.** (a) 27.4°C (b) 4.5 kN **111.** (a) 3170 lb (b) −3.18 ft **113.** -1.5×10^{-4} **115.** 1.50 **117.** (a) 1.6 atm (b) −10% **119.** (a) 0.1 μm (b) 5×10^9 (c) 500 m (d) 1 cm

CHAPTER 14

Multiple-Choice Questions

1. (a) **3.** (d) **5.** (b) **7.** (c) **9.** (c) **11.** (d)

Problems

1. (a) 34 J (b) Yes; the increase in internal energy increases the average kinetic energy of the water molecules, so the temperature is slightly increased. **3.** 4.90 kJ **5.** (a) 250 J (b) all three **7.** 5.4 J **9.** 6.40×10^{-4} kJ/K **11.** 2430 kJ/K **13.** 82°C **15.** (a) 0.38 kJ/K (b) 32 kJ/K **17.** 250 kJ **19.** 1.34 kg **21.** 0.50 MJ **23.** 700 m **25.** 28.4°C **27.** 80 s **29.** 44°C **31.** (a) 2 MJ (b) 20%, which is significant (c) 30 W **33.** B to C, solid to liquid; D to E, liquid to gas **35.** 330 J/g **37.** 10.4 g **39.** 3100 kJ **41.** 179 g **43.** 24 g **45.** 371°C, yes **47.** 23.2 W **49.** The ice will melt completely; 32°C **51.** 250 W **53.** (a)

Segment	Pressure	Temperature	Phase Change
AB	Decreases	Constant	Fluid to solid
BC	Constant	Increases	Solid to liquid
CD	Decreases	Constant	Liquid to Gas
DE	Constant	Decreases	Gas to solid

(b) *a* is the critial point; *b* is the triple point **55.** (a) 2.0 cm (b) 29 m **57.** (a) 0.12 K/W (b) 2.5×10^{-4} K/W (c) 5.0×10^{-5} K/W **59.** (a) 320 W (b) 18 kW **61.** reduced to 75% of the original **63.** −37° C **65.** (a) 300 W (b) 4500 W **67.** (a) the skier with the down jacket (b) The person with the down jacket can stay outside 6.4 times longer. **69.** 6.67 W/m^2 **71.** reduced by a factor of 0.019 **73.** 2.9×10^{-3} m·K **75.** 0.60 **77.** 4.8×10^{-5} m^2 **79.** 170 kcal/h **81.** 28 kJ **83.** (a) 0.0089 °C/s (b) 9.4 min **85.** (a) 39°C (b) 182 W/m^2 **87.** 1090 K **89.** Coffeepot: 4.5 W; teapot: 24 W **91.** (a) true (c) Rate of heat flow from the body is proportional to surface area.

Animal	(a) BMR/kg	(b) BMR/m^2
Mouse	210	1200
Dog	51	1000
Human	32	1000
Pig	18	1000
Horse	11	960

93. (a) 7.00 times higher (b) 35.7°C; the dog is a much better regulator of temperature and, as a result, has more endurance. **95.** 320 s **97.** (a) 180°C (b) 20.9°C **99.** 480 g **101.** (a) 9.9 kJ (b) 360 g **103.** 0.0065°C **105.** 0°C **107.** 0.010°C **109.** 6.3 kg **111.** 35°C **113.** 330 m/s **115.** 0.32 kg **117.** (a) 103.81°C (b) 0.565 g/s **119.** (a) 1.7 m/s (b) 7.86 g **121.** (a) Nitrogen; at the same temperature they have the same average KE, so the less massive molecule has a higher rms speed. (b) 61.9°C (c) 180 J **123.** 3.1×10^{-4} g

CHAPTER 15

Multiple-Choice Questions

1. (b) **3.** (d) **5.** (c) **7.** (c) **9.** (e) **11.** (b) **13.** (d)

Problems

1. 2.9 J **3.** 100 J of heat flows out of the system. **5.** 108 kJ **7.** (a) 10.8 K (b) 56 kJ **9.** 202.6 J **11.** (a) 98.0 kPa; 1180 K (b) −200 J (c) 66 J (d) $\Delta U = 0$ because $\Delta T = 0$ in a cycle. **13.** (a) 8.87 kPa; 1200 K (b) 23 kJ (c) 20.0 kJ (d) 0 **15.** 15 kJ **17.** −5.00 kJ; out of the gas and into the reservoir **19.** (a) −1216 J (b) $\Delta U = 1216$ J; $Q = 2431$ J **21.** (a) −1934 J (b) $\Delta U = 1216$ J; $Q = 3149$ J **23.** (a) *D*, *B*; cycle moves clockwise. (b) *A*, *C*, *E*; cycle moves counterclockwise. (c) *A*, *C*, *E*; heat pumps and refrigerators work the same way. **25.** 0.628 **27.** (a) 3.00 kJ (b) 2.00 kJ **29.** (a) 1.2×10^{17} J (b) 1.4×10^{13} kg **31.** 0.182 **33.** \$2.4 **35.** 171 K **37.** 25.0 kJ **39.** 14 W **41.** (a) 7.8 kJ (b) 6.8 kJ **43.** 31 kg **45.** 2.11 pW **47.** (a) 0.3436 (b) 275.7 kJ **49.** 4.2% **51.** 0.0174 **53.** 110 kJ **55.** (a) Decreasing T_C improves the efficiency more. In $e_r = (T_H - T_C)/T_H$ it increases the numerator without changing the denominator. Increasing T_H produces the same increase in the numerator while also increasing the denominator, for a smaller overall change in e_r. (b) The initial efficiency is 0.596. Lowering T_C gives 0.650, for an increase of 0.054. Raising T_H gives 0.617, an improvement only of 0.021. **59.** (c), (a), (b), (d) **61.** +6.05 kJ/K **63.** (a) 3.4×10^{-3} J/K (b) -2.8×10^{-3} J/K (c) 6×10^{-4} J/K **65.** 0.102 J/(K·s) **67.** (a) 97 W (b) 0.33 W/K **69.** (a) 0.72 J (b) 8.8×10^{-19} J/molecule **71.** Coal: 4.3×10^{13} J; nuclear: 6.0×10^{13} J **73.** (a) 0.90 J/K (b) −2.7 J/K. Since the entropy of the universe decreases, the process is impossible. **75.** 0.12 J/K **77.** 3.8 kJ **79.** 250 W **81.** 350 J/K **83.** 87.1 kJ **85.** (a) 6.2 mJ (b) 22 mJ (c) 1.2 mK **87.** (a) Step A: 0; Step B: −2080 J; Step C: 0; Step D: +2080 J **89.**

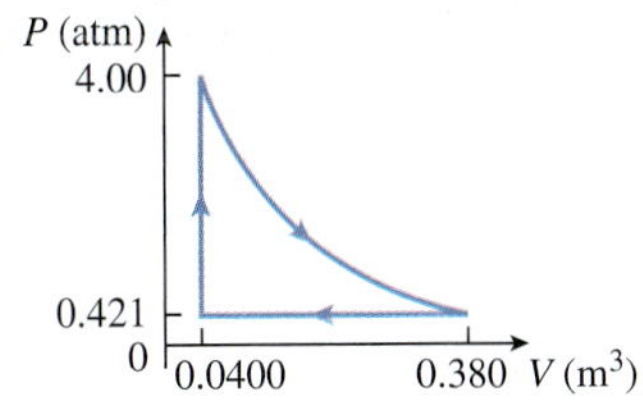

91. (a) Step 1: 36.5 kJ; Step 2: −36.3 kJ; Step 3: 21.8 kJ (b) 56.2 J/K (c) Entropy is a state variable so $\Delta S_{gas} = 0$ for a complete cycle. No: not a reversible engine, so $\Delta S_{environment} > 0$. **93.** high temperature: 2.6×10^5 W; low temperature: 1.9×10^5 W **95.** (a) 15.9°C (b) −0.03 J/K (c) the entropy of the universe cannot decrease **97.** (a) 39.6% (b) 498 MW (c) 33.0 m³/s **99.** (a) 2350 K (b) $e_r = 0.800 = 1.80e$

Credits

About the Authors

Photo courtesy of Phil Krasicky, Department of Physics, Cornell University.

Chapter 1

Opener: NASA/JPL-Caltech/MSSS; p. 2: © Royalty-Free/Corbis; 1.1a: © Andrew Dunn/Alamy RF; 1.1b: Photo by Jennifer Merlis; 1.1c: NASA Earth Observatory image by Robert Simmon with data courtesy of the NASA/NOAA GOES Project Science team; 1.1d: © Photo Courtesy of NASA/Corbis RF; 1.1e: © Hans Gelderblom/Getty Images; 1.1f: © Jonathan Blair/Corbis; 1.1g: © Photo Courtesy of NASA/Corbis RF; p. 9: NASA/JPL-Caltech/MSSS; p. 12: © Ryan McVay/Getty Images RF; 1.3: © Dennis Kunkel Microscopy, Inc./ Visuals Unlimited/Corbis.

Chapter 2

Opener & p. 35: © J & A Scott/NHPA/ Photoshot.

Chapter 3

Opener: © Peter Johnson/Corbis; p. 60: © Oliver Benn/Getty Images; p. 68: © Stockbyte/Getty Images; 3.19: © Loren Winters/Visuals Unlimited/ Corbis; 3.20: © Peter Johnson/Corbis; p. 77: © Universal History Archive/ Getty Images; p. 79: © Peter Johnson/ Corbis; p. 90: © Kevin Schafer/ Alamy.

Chapter 4

Opener: Courtesy Durango Soaring Club, Inc.; 4.1: © Franck Seguin/ TempSport/Corbis; p. 99(top): © Karl Weatherly/Getty Images; 4.9: NASA; p. 104: © Felicia Martinez/PhotoEdit; p. 126: Courtesy Durango Soaring Club, Inc.; p. 148: © Yusuke Nakanishi/ AFLO/Newscom.

Chapter 5

Opener: © AFP/Getty Images; p. 158: © Russell Illig/Getty Images RF; 5.9: NASA; p. 163: © AFP/Getty Images; 5.15: © Chris Sattlberger/Getty Images; p. 177: © Corbis Premium/ Alamy RF; p. 178: © Angelo Hornak/ Corbis; 5.25: © Photos 12/Alamy; p. 184(top): © James L. Amos/Corbis; p. 184(bottom): © Matthew Stockman/ Allsport/Getty Images.

Chapter 6

Opener: © Tier Und Naturfotografie J und C Sohns/Getty Images; 6.1: © Owaki-Kulla/Corbis; 6.2: © Koji Sasahara/AP Photo; 6.25: © David Cousins; 6.33a-c: © Dwight Kuhn; p. 215(bottom): © Tier Und Naturfotografie J und C Sohns/Getty Images; p. 220: © Brand X Pictures/Getty Images RF; p. 222: Courtesy Pileco, Inc.; p. 223: U.S. Navy Photo; p. 227: © Free Agents Limited/Corbis.

Chapter 7

Opener: © Laramie Boomerang, Andy Carpenean/AP Photo; 7.3: © Scott Leva - Precision Stunts; 7.8: © Photo Courtesy of NASA/Corbis RF; 7.10: © David Fleetham/Alamy; 7.11: © Michael Steele/Allsport/Getty Images; 7.17: © Visuals Unlimited/ Corbis; p. 253: © Laramie Boomerang, Andy Carpenean/AP Photo; p. 257: © Rubberball/SuperStock RF; p. 261: © Stephen Mallon/Getty Images.

Chapter 8

Opener: © Rick Rycroft/AP Photo; 8.3: © AP Photo; 8.20: © View Pictures/ Edmund Sumner/VIEW/Newscom; 8.27: © Mike Powell/Allsport/Getty Images; p. 290: © Rick Rycroft/AP Photo; 8.34: © Michael Newman/ PhotoEdit; p. 294: © Kevin Fleming/ Corbis; 8.38a-b: © Photography by Leah; p. 306: © Doug Pensinger/ Allsport/Getty Images; p. 315a: © Jonathan Daniel/Allsport/Getty Images; p. 315b: © Tony Duffy/ Allsport/Getty Images; p. 316: © Gilbert Lundt, Jean-Yves Ruszniewski/ TempSport/Corbis.

Chapter 9

Opener: © Peter Johnson/Corbis RF; 9.15: James Gathany/CDC; p. 336(top): © Peter Johnson/Corbis RF; p. 336(bottom): © Carl & Ann Purcell/Corbis RF; 9.18: © Getty Images RF; 9.19: © culture-images GmbH/Alamy; 9.20: © Gary S. Settles/Science Source; 9.23: © Michael Bodmann/E+/Getty Images RF; 9.31: © Visuals Unlimited/Corbis.

Chapter 10

Opener: © Hubert Stadler/Corbis; 10.1: © Gonzalo Fuentes/Reuters/Corbis; 10.5: © Prefabricats Pujol; p. 370(top): © Greater Houston Convention and Visitors Bureau; p. 370(bottom): © John Springer Collection/Corbis; 10.10a: © Cameron Spencer/Getty Images; 10.10b: © Dr. P. Marazzi/Science Source; 10.17: © Don Farrall/Getty Images; 10.29a: © Everett Collection/ Newscom; 10.29b: © Topham/The Image Works; p. 388: © Hubert Stadler/ Corbis; p. 390(top): © Ray Malace Photo; p. 390(bottom): © Adam Eastland/Alamy.

Chapter 11

Opener: © Chiaki Tsukumo/AP Photo; p. 402: © Patrick Clark/Getty Images RF; p. 406: © Alberto L. Lopez-Torres; 11.20: © Richard Megna/Fundamental Photographs, NYC; p. 421: © Chiaki Tsukumo/AP Photo; p. 428: © Dorling Kindersley/Getty Images.

Chapter 12

Opener: © Betty Richardson; p. 438: © Brian E. Small/www.briansmallphoto.com; p. 441: U.S. Navy photo by Ensign

John Gay; 12.4: © billkelleyphoto.com/Bill Kelley; 12.6: © Jill Braaten; 12.17: © Science Source; p. 456: © Stephen Dalton/Science Source; p. 457: © Betty Richardson; 12.19: © AP Photo.

Chapter 13

Opener: © Stone/Getty Images; 13.5: © Tom Uhlman/Alamy RF; p. 480: © Geoff Tompkinson/Science Source; p. 487: © Stone/Getty Images; 13.15a: © Royalty-Free/Corbis; 13.15b: © Dex Image/Getty Images RF; 13.15c: © Brand X Pictures/PunchStock RF; p. 487: © Stone/Getty Images..

Chapter 14

Opener & p. 511: © ARCO/W Layer/agefotostock; 14.6: Photo by author R.C. Richardson; 14.8: © Staffan Widstrand/Corbis; 14.18: © 2008 David Montrose, M.D./Custom Medical Stock Photo.

Chapter 15

Opener & p. 556: © Photographer's Choice/Getty Images RF.

Index

A

B

C

D

E

F

H

I

J

K

L

M

N

O

Q

R

S

T

U

X

Y

Z

MAIN-GROUP ELEMENTS · TRANSITION ELEMENTS · MAIN-GROUP ELEMENTS

- Metals (main-group)
- Metals (transition)
- Metals (inner transition)
- Metalloids
- Nonmetals

1 **H** 1.00794	
	Atomic number
	Symbol
	Atomic mass

For elements without stable isotopes, the number shown in parentheses is the mass number of the longest-lived known isotope.

Period	1A (1)	2A (2)	3B (3)	4B (4)	5B (5)	6B (6)	7B (7)	8B (8)	8B (9)	8B (10)	1B (11)	2B (12)	3A (13)	4A (14)	5A (15)	6A (16)	7A (17)	8A (18)
1	1 **H** 1.00794																	2 **He** 4.00260
2	3 **Li** 6.941	4 **Be** 9.01218											5 **B** 10.811	6 **C** 12.011	7 **N** 14.0067	8 **O** 15.9994	9 **F** 18.99840	10 **Ne** 20.1797
3	11 **Na** 22.98977	12 **Mg** 24.3050											13 **Al** 26.98154	14 **Si** 28.0855	15 **P** 30.97376	16 **S** 32.065	17 **Cl** 35.453	18 **Ar** 39.948
4	19 **K** 39.0983	20 **Ca** 40.078	21 **Sc** 44.95591	22 **Ti** 47.867	23 **V** 50.9415	24 **Cr** 51.9961	25 **Mn** 54.93805	26 **Fe** 55.845	27 **Co** 58.93320	28 **Ni** 58.6934	29 **Cu** 63.546	30 **Zn** 65.409	31 **Ga** 69.723	32 **Ge** 72.64	33 **As** 74.92160	34 **Se** 78.96	35 **Br** 79.904	36 **Kr** 83.798
5	37 **Rb** 85.4678	38 **Sr** 87.62	39 **Y** 88.90585	40 **Zr** 91.224	41 **Nb** 92.90638	42 **Mo** 95.94	43 **Tc** (98)	44 **Ru** 101.07	45 **Rh** 102.90550	46 **Pd** 106.42	47 **Ag** 107.8682	48 **Cd** 112.411	49 **In** 114.818	50 **Sn** 118.710	51 **Sb** 121.760	52 **Te** 127.60	53 **I** 126.90447	54 **Xe** 131.29
6	55 **Cs** 132.90545	56 **Ba** 137.327	57 **La** 138.9055	72 **Hf** 178.49	73 **Ta** 180.9479	74 **W** 183.84	75 **Re** 186.207	76 **Os** 190.23	77 **Ir** 192.217	78 **Pt** 195.084	79 **Au** 196.96657	80 **Hg** 200.59	81 **Tl** 204.3833	82 **Pb** 207.2	83 **Bi** 208.98040	84 **Po** (209)	85 **At** (210)	86 **Rn** (222)
7	87 **Fr** (223)	88 **Ra** (226)	89 **Ac** 227.0278	104 **Rf** (261)	105 **Db** (262)	106 **Sg** (266)	107 **Bh** (264)	108 **Hs** (277)	109 **Mt** (268)	110 **Ds** (271)	111 **Rg** (272)	112 **Cn** (285)		114 **Fl** (289)		116 **Lv** (293)		

INNER TRANSITION ELEMENTS

6	Rare Earths (Lanthanides)	58 **Ce** 140.116	59 **Pr** 140.90765	60 **Nd** 144.24	61 **Pm** (145)	62 **Sm** 150.36	63 **Eu** 151.964	64 **Gd** 157.25	65 **Tb** 158.92535	66 **Dy** 162.500	67 **Ho** 164.93032	68 **Er** 167.259	69 **Tm** 168.93421	70 **Yb** 173.04	71 **Lu** 174.967
7	Actinides	90 **Th** 232.0381	91 **Pa** 231.036	92 **U** 238.0289	93 **Np** (237)	94 **Pu** (244)	95 **Am** (243)	96 **Cm** (247)	97 **Bk** (247)	98 **Cf** (251)	99 **Es** (252)	100 **Fm** (257)	101 **Md** (258)	102 **No** (259)	103 **Lr** (262)

Claims for the discovery of elements 113, 115, and 117–120 are under consideration by the International Union of Pure and Applied Chemistry.